For the Love of Science

Roderick W. Home, Isabel M. Malaquias &
Manuel F. Thomaz (eds.)

For the Love of Science

The Correspondence of J. H. de Magellan (1722–1790)

Vol. 1

PETER LANG
Berlin · Bern · Bruxelles · New York · Oxford

Bibliographic information published by die Deutsche Nationalbibliothek
Die Deutsche Nationalbibliothek lists this publication in the Deutsche Nationalbibliografie; detailed bibliographic data is available on the Internet at ‹http://dnb.d-nb.de›.

British Library Cataloguing-in-Publication Data: A catalogue record for this book is available from The British Library, Great Britain

Library of Congress Cataloging-in-Publication Data: 2016960803

Cover illustration: Magellan's diagram of the double sextant he invented, from his *Description des octants et sextants anglois* (1775), Plate 2.

ISBN 978-3-0343-3716-8 hb.
E-ISBN 978-3-0343-3723-6 MOBI
DOI 10.3726/b14851
E-ISBN 978-3-0343-3721-2 eBook
E-ISBN 978-3-0343-3722-9 EPUB

Wabernstrasse 40, CH-3007 Bern, Switzerland
info@peterlang.com, www.peterlang.com

Printed in Germany

Frontispiece. Baro-Thermometer or Αεροσταθμιον invented by J. H. de Magellan and patented by him and Henry Pyefinch in 1765; see Magellan to the Earl of Morton, 6 June 1765, and Magellan to James Ferguson, 8 November 1765 (© Royal Society of London).

Table of Contents

Volume I

Volume 2

Introduction

The surviving correspondence of the *abbé* J.H. de Magellan (1722-1790) that is brought together for the first time in this edition is one of the most interesting science-related correspondences of the late eighteenth century.[1] Magellan was not himself a front-rank contributor to scientific thought. He was, however, intelligent, ingenious and interested in everything going on around him, and he was also very well connected, being acquainted with most of the leading men of science of his day and the leading instrument-makers, as well as members of the upper classes, including even the aristocracy and higher nobility, in several European countries. Moreover, he made it his business to keep himself well informed about the latest scientific and technical developments, and to pass on the information to his correspondents all over Europe. For, as we shall see, his letter-writing served a more than social purpose, it also brought him an income.

Magellan and his correspondence in fact occupy an important moment in the history of science. As the scientific world expanded during the second half of the eighteenth century, it became increasingly dependent on the efficient interchange of techniques and information between the growing numbers of people involved in scientific work. Magellan represents an intermediate state between an earlier and more leisurely period in science where information travelled relatively slowly between the small number of individuals involved, and the new, enlarged community that began to look for a much more rapid dissemination of news. Eventually the needs of that community were met by the development of a new kind of scientific journal, epitomized by the one founded by François Rozier in Paris in 1772, that appeared much more frequently than the old-style volumes of academic proceedings and carried letters and short reports rather than full-scale memoirs. For a brief moment, until those journals took over, there was a gap that Magellan's incessant letter-writing helped to plug. His correspondence is thus of much more than biographical interest, it offers us a window on to a wide range of late-eighteenth-century scientific practice and technological innovation.

---

1 I. Malaquias and M.F. Thomaz, "Scientific Communication in the 18th Century: the Case of Jean Hyacinthe de Magellan", *Physis*, 31 (1994), 817-834.

Magellan himself was especially interested in the development of scientific instruments, and most of his own publications, and also many of his letters, were devoted to describing improvements he had devised to one or another measuring device. Some of these improvements were taken up by the English instrument-makers whose workshops he frequented, who led the world at the time. The exciting developments taking place in chemistry in those years – especially in relation to the chemistry of gases – also figure large in the correspondence.[2] As we shall see, Magellan played a significant part in these developments by serving as one of the principal conduits by which information about the latest discoveries flowed between the leading French and British chemists of the day. He also took a close interest in the dramatic changes taking place in British agricultural practice in these years and numbered leading reformers such as Arthur Young and John Arbuthnot among his friends; and he went out of his way to provoke discussion of these matters among his acquaintances on the Continent. Letters that he wrote in his capacity as Continental agent for the pioneer Birmingham industrialists Matthew Boulton and James Watt shed new light on the spread of steam-engine technology, and the industrial revolution that this underpinned, from England to the rest of Europe. However, these are but some of the highlights of the correspondence; Magellan's letters range much more widely in their subject-matter and, in the process, offer us a fascinating new view of the intellectual life of his day.

Generally speaking, Magellan wrote in French, the *lingua franca* of the day; but when he wrote to his Portuguese contacts, he naturally wrote in Portuguese, and to British correspondents and to those in what had been Britain's colonies in North America, he of course wrote in English. In this remarkable correspondence, the metaphorical Republic of Letters becomes real. Yet the correspondence that has survived, rich though it is, is far from complete. In the letters that we have, many others are referred to that are now missing; moreover, we know that many letters perished, along with other papers, in a fire in Magellan's lodgings in December 1772.

2 I. Malaquias, "Aspects of the Scientific Network and Communication of John Hyacinth de Magellan in Britain, Flanders and France", *Ambix*, 55 (2008), 255-273.

Figure 1. View of Aveiro in the eighteenth century. Engraving printed in the review *O Panorama*, 2nd series, vol. 2 (1843). The sixteenth-century walls and the aqueduct, demolished in the 1850s, are still seen in this engraving.

Early years

Magellan was born on 4 November 1722 in the small town of Aveiro in northern Portugal, a country without a solid tradition in science, especially in the experimental sciences. His family background is not disclosed or even referred to in his correspondence. However, in a couple of his publications he claimed a relationship with the famous sixteenth-century navigator Ferdinand de Magellan and, in confirmation of this, from national and parochial archives in Portugal his ancestry has been traced to the explorer's younger brother.[3] The fact that he never refers to his parents is comprehensible because he never knew them: his mother died five days after he was born and his father died less than a year later. He was their only

3 Magellan, *Description des octants et sextants anglais* (London, 1775), p. vii; idem, *Notice des instrumens d'astronomie, de géodesie, de physique, &c., .. pour la Cour d'Espagne* (London, 1780), p. 193; M. F. Thomaz, "João Jacinto de Magalhães e a sua ascendência", *Revista lusófona de genealogia e heráldica*, 2 (2007), 165-80. A full family tree is given on a fold-out sheet that accompanies this article.

child.[4] During his earliest years he was probably looked after by relatives. Though of a noble line, his family was not wealthy. However, he inherited several properties and the income from these enabled him to be enrolled at 11 years of age in the Colégio da Sapiência, a college of the Augustinian Monastery of the Holy Cross in Coimbra famous throughout Portugal for its rigorous teaching and its modern and well-stocked library where even the more modern scientific and philosophical books could be read. On 23 April 1743, at the age of 21, Magellan took his vows of poverty and chastity and joined the Augustinian Order, taking the name Frei João de Nossa Senhora do Desterro (*Fr. Joannes a Domina Nostra de Exilio*).[5] On 7 April 1744 he became a regular canon, which entitled him to use the title "Dom" instead of "Friar". Then in 1751, at the age of 29, he was ordained a priest in the Augustinian Order. For at least two years he was enrolled as a student in the Faculty of Canon Law at the neighbouring University of Coimbra, Portugal's principal university at the time.[6] However, he is not recorded as having sat the examinations and in any event his studies there would not have contributed in any way to his knowledge of current science. Not only was science not a part of the curriculum in canon law, the university as a whole remained mired in the ancient tradition: no modern science was taught in any part of it until after the reform of the institution in 1772, long after Magellan had left Coimbra.

Somehow, nevertheless, Magellan managed during the period of more than twenty years that he spent with the Augustinians in Coimbra to develop a taste for science and a good acquaintance with it. His preference for science may be inferred from his being the person designated to assist the French astronomer Gabriel de Bory when he visited Coimbra in 1753. Bory went to Portugal to observe the total solar eclipse that was calculated to be visible at Aveiro in October of that year. Once the eclipse had passed, Bory travelled to Coimbra, where he visited the Augustinians. It was Magellan, we learn from the report that Bory later presented to his colleagues in Paris, who showed him around "that magnificent house".[7] The monastery, Bory reported, was a closed community and the monks adhered strictly to this; it was their doing so, he thought, that had led them to cultivate the sciences.

---

4 This was his father's second marriage; no record has been found of any children of the first marriage.

5 Brother John of Our Lady of the Exile.

6 University of Coimbra Archives, Student Processes.

7 *Histoire de l'Académie Royale des Sciences*, 1772 (part 2), Hist. pp. 112-126; p. 115.

Not long after Bory's visit and perhaps partly inspired by it, Magellan apparently sought permission to leave the congregation under the terms of a general dispensation issued by Pope Benedict XIV as part of a wider reform of the religious orders in Portugal. We have not found the documentation relating to Magellan's leaving the Order but it would presumably have reflected a lack of vocation on his part.[8] Under the terms of a decree issued some years earlier by the Reformer General appointed to oversee the Augustinians of Santa Cruz, canons who opted to leave were entitled to receive a pension of Rs. 30,000[9] per annum (later increased to Rs. 36,000 per annum) from the congregation.[10] They were still regarded as priests – Magellan was generally referred to as *abbé* – and were required to make themselves available to the highest church official (the bishop or his representative) in the town in which they took up residence. For some reason, for over twenty years Magellan did not receive the pension to which he was entitled. When he did eventually receive what was owed to him, he presented it to the recently founded Lisbon Academy of Sciences, of which he had been appointed a member, to endow a prize for agricultural improvements.[11]

Finding a new career

Until recently, little was known of Magellan's whereabouts during his first years outside the monastery. Neither did we know how he supported himself, but presumed that he must have been receiving sufficient income from his estates or elsewhere to maintain himself in a respectable lifestyle. He later stated on a couple of occasions that he was in Lisbon on 1 November 1755 and witnessed the destruction caused by the great earthquake and subsequent tsunami that struck the city that day.[12] We also knew that he

8 The published report of Bory's visit to Portugal (*ibid.*) stated that Magellan was released from the Order in 1744, which is obviously wrong. Several authors have suggested that the date was a misprint and that 1754 was intended.

9 Approximately £9 in English currency.

10 Portuguese National Archives of Torre do Tombo: Sta Cruz de Coimbra, Livº 67, ff. 138v-141.

11 Magellan to the Portuguese Secretary of State for Foreign Affairs, Ayres de Sá e Mello, 17 February 1784 and 5 April 1785.

12 *Journal étranger*, April 1760, 196-234; pp. 198, 208; Le Begue du Presle, *Relation ou notice des derniers jours de Mons. Jean-Jacques Rousseau … avec une addition … par J.H. de Magellan …* (London, 1778), p. 33.

afterwards left Portugal, supposedly on a "tour philosophique" with the stated aim of becoming better acquainted with the scientific and technical advances being made in other parts of Europe. Throughout his life, these were the things that interested him most. By contrast, music and the fine arts figure scarcely at all in his surviving letters – although, astonishingly, we learn from his will, drawn up in 1788, that he owned a piano, a newly developed instrument at the time and a rather expensive item then as now.[13]

Recently, a fortunate reading of London's *Morning Herald* newspaper, founded in 1780, brought to our attention a previously unremarked column entitled "Memoirs of Eminent Living Persons" that appeared from time to time in this journal in its earliest days. One such memoir, seemingly one of the first written specifically for the journal, was devoted to "Hyacinth de Magellan"; and even though this includes some inaccuracies, it also includes enough pertinent detail that there can be no doubt that it was compiled by whoever wrote it from information supplied by Magellan himself.[14] The inaccuracies, which relate chiefly to chronology, are presumably a result of the author's running together incorrectly things that Magellan had told him. However, their presence means that we must be cautious about other things that are said in the article, that may also incorporate misunderstandings on the author's part.

Subject to that qualification, we learn from the article that Magellan fled Portugal when a person with whom he had been associating was arrested:

> Magellan as he walked the street was told of his friend's fate, and knowing that it was an universal rule with the minister [the Marquis de Pombal] to imprison every man who had ever been in correspondence with any person he seized, and the unhappy innocent having letters of his in his possession, immediately went home, put up what money and valuables he had by him, and without the delay of an instant, set off for Spain.

Magellan's hasty departure reduced him, we are told, to poverty, and he was forced to travel on foot, walking all the way to Madrid. Here, fortunately, he met a man who had known him in earlier life as a respectable person, who provided him with introductions that enabled him to live decently. He stayed there for about a year and a half, long enough to learn Spanish, before moving on, still on foot because he remained very poor but carrying letters of recommendation from his Spanish friends. His route

13 Magellan, last will and testament, 9 August 1788.

14 *Morning Herald*, 26 October 1781.

took him across Spain and southern France to Marseilles "where he staid long enough upon his recommendations, and his unremitting industry in teaching Latin and Portuguese, to write to Paris". The reply he received was not encouraging, however, so he continued instead to Italy, first to Turin and then through Lombardy to Rome "where his letters procured him some attention". He spent two years in Rome, teaching natural philosophy and mechanics as well as Latin and Portuguese, while learning Italian ("which language he has ever since spoken better than any other") and "studying those arts and sciences, by which he hoped to support himself in future". Eventually he moved on again, travelling to Paris via Geneva where he stayed "long enough to make himself a master of all they knew in watch making" and where his mechanical abilities created such a good impression that one of the town's master watchmakers offered him a place at his house and table for as long as he liked to stay. Paris, however, proved a disappointment, and so after staying "long enough for acquiring the French tongue" and to meet the famous philosopher-mathematician Jean d'Alembert "who shewed him many civilities on finding that he was no inconsiderable mathematician", he moved on to London. There is no mention of his ever returning to Portugal, even though we know he did so, in June 1761, before ever he settled in London.[15]

Magellan certainly visited Rome at some stage during his travels[16] but otherwise there is no supporting evidence for the story just told until 1759, when we know he was in Paris for a while. No documents of any kind in his hand from earlier in his life have been found. The Portuguese Minister in Paris, Mgr. Pedro da Costa de Almeida Salema, reported that he succoured Magellan, who had reached Paris in dire need, but always kept a wary eye on him "because he touched and wanted to know our business more than it was lawful to ask".[17] His earliest surviving letters date from this time, when he was already 37 years old and had been out of the monastery in Coimbra for several years. From one of these letters we learn that he was living at the time at the very respectable Hôtel des Prouvaires,

15 The Portuguese Minister in Paris, Mgr. Pedro da Costa de Almeida Salema, reported to D. Luiz da Cunha, the Secretary of State in Lisbon, on 18 June 1761 that Magellan "left last week for Lisbon through Le Havre" (Portuguese National Archives of Torre do Tombo, MNE, Cx. 566); quoted in Maximiano de Lemos, "Portuguezes illustres em França: Soares de Barros, João Jacinto de Magalhães e Ribeiro Sanches", *Boletim de Segunda Classe da Academia das Sciencias de Lisboa*, 3 (1910), 431-63. Internal evidence shows that Magellan's letter to Sanches, 20 April 1762, was written from Portugal.

16 Magellan to Sanches, 19 May 1778.

17 Salema to D. Luiz da Cunha, 18 June 1761 (see note 15).

in the St Eustache quarter of the city.[18] In addition, we learn from one of Magellan's publications from later in his life that he visited the workshop of the well-known Parisian instrument-maker Claude-Siméon Passement at this period.[19] There is, however, no mention in any other source we have found of his meeting d'Alembert.

The correspondence that has survived from Magellan's time in Paris – fragments in each case rather than full letters – reveal that he had established contact with some of the leading astronomers in the French capital and was making himself useful to them by translating into French reports of astronomical observations made in Portugal. Almost certainly he was introduced into Parisian astronomical circles by his old acquaintance, Gabriel de Bory. Now, Magellan also established links with Joseph-Nicolas Delisle, who was perhaps the leading French astronomer of the day, and Charles Messier with whom he was subsequently to correspond for many years. The observations he passed on were due to the Oratorian priest Jean Chevalier, an enthusiast for astronomy who later moved to Brussels and became Keeper of the Royal Library there. Chevalier had been a corresponding member of the Académie Royale des Sciences since 1753, and Delisle was his designated correspondent there. The fragments that have survived all report observations of comets. Despite Chevalier's failure to provide precise information about the position and motion of the comets in question, such reports would have been particularly welcomed by Messier, who had already embarked on what was to become a life-long career as a comet-hunter.

Magellan while in Paris also met and became close friends with the distinguished Portuguese physician and man of letters, António Nunes Ribeiro Sanches (1699-1783). Sanches had studied medicine at Leiden under Herman Boerhaave and been recommended by him to a position in Russia, where he rose to become in due course personal physician to the Empress Elisabeth. In 1747, the Empress had allowed him to retire and leave Russia, and he had settled in Paris where he was later supported in part by a pension from the Portuguese Government that after he died was transferred to Magellan.[20] His patronage was crucial to the subsequent development

18 Magellan to Messier, 26 May 1759.

19 Magellan, *Description des baromètres nouveaux...* (London, 1779), p. 144.

20 Magellan to Sanches, 17 December 1782. Initially, Sanches also received a pension from Russia, but this was annulled by the Empress when she was persuaded – falsely – that he was not a Christian but a practising Jew. The pension was restored fifteen years later by a new Empress, Catherine the Great, out of gratitude to Sanches whom she credited with saving her life when she fell dangerously ill, shortly after she first arrived in Russia. See David Willemse, *António Nunes Ribeiro Sanches – élève de Boerhaave – et son importance pour la Russie* (Leiden, 1966).

of Magellan's career, for he was able to introduce Magellan into both the topmost levels of Parisian society and scientific and philosophical circles in the capital.

Magellan was probably still in Paris in April 1760, when the *Journal étranger* published his lengthy summary for French readers of an account published in Portugal of the measures taken by the authorities there to restore order after the Lisbon earthquake, as well as to establish the plans and to provide the means to rebuild the shattered city. The account, supposedly written by Francisco José Freire, was actually the work of Portugal's reforming but increasingly despotic Prime Minister, Sebastião José de Carvalho e Melo, Count of Oeiras (later Marquis de Pombal). Not surprisingly, it was lavish in its praise for the government's response to the crisis, and so, too, was Magellan's summary.

Other publications by Magellan during this period, anti-Jesuit in flavour, suggest that he would have been very much in sympathy with Pombal's campaign to break the power of the Jesuits, longstanding rivals of the Augustinians, over civil and political life in Portugal. He may even have been a pawn in that campaign. One publication was a translation into Portuguese, published initially in Paris and later in Lisbon, of the Greek grammar used by the Jesuits' chief rivals in France, the Jansenists, in their famous Parisian college of Port-Royal.[21] Another, also published in Paris, was a new edition of a seventeenth-century biography of a famous pre-Jesuit Portuguese theologian.[22] A third, much more potent from a political point of view, was a Portuguese translation, published in Lisbon and dedicated to Pombal, of a virulent anti-Jesuit work by a French Capuchin monk generally known as the *abbé* Platel, whom the Prime Minister had invited to Portugal a couple of years earlier.[23]

Magellan's preface to the Greek grammar sheds an interesting light on his manner of working. In order to learn more about the pronunciation of the Greek language, he said, he had profited whenever he could, in different parts of Europe, from any contacts he could make with Greek nationals or people recommended to him who had travelled in Greece.[24]

Whether or not Magellan was a pawn in Pombal's campaign against the Jesuits, it seems that once he returned to Portugal, he soon became disillu-

21 *Novo epitome da grammatica Grega de Porto-real…* (Paris/Lisbon, 1760).

22 Frei Luiz de Souza, *Vida do veneravel D. Frei Bartolomeu dos Mártyres da Ordem dos Pregadores…* (new edition, 2 vols, Paris, 1760).

23 Abbé Platel [i. e. Pierre Parisot], *A fé dos catholicos: obra dirigida a instruir e confirmar na sua crença de catholicos…* (Lisbon, 1763).

24 *Novo epitome da grammatica Grega de Porto-real*, Prologue, p. vii.

sioned with Pombal's despotic behaviour and by early 1762 he was telling Sanches that he had decided to leave again and settle in a country where he could live more freely – in Paris, perhaps, or England or Italy, particularly Venice. Before leaving, however, he spent considerable time and energy in trying, apparently at some risk to himself, to resolve problems that Sanches was having with the Portuguese authorities.[25] It was probably on this account that Sanches afterwards paid him an annual pension that Magellan continued to receive until Sanches' death, more than twenty years later.

On leaving Portugal, Magellan once more made his way to Paris. He later claimed to have contemplated settling in America.[26] Instead, however, with Sanches' advice and assistance he moved soon afterwards to London where he was to make his home for the rest of his life, interspersed by regular visits to Paris and the Low Countries.

Settling in London

In London, Magellan gradually worked his way into the heart of the British scientific community centred on the Royal Society of London. As we learn from the Society's *Journal Book*, he took to attending the Society's meetings regularly as a guest of one or other of the Fellows – he first attended a meeting on 10 November 1763 – and in due course he began presenting reports to the meetings, many of these being extracts from letters he had received from his astronomer friends in Paris with details of their latest observations. Most notably, most of the French observations of the transit of Venus across the face of the Sun in 1768 reached the Royal Society through Magellan. The Society's meetings also became a principal source of the scientific news that he took to spreading to his ever-widening circle of correspondents on the Continent.

One of Magellan's earliest contacts in the Royal Society was Emanuel Mendes da Costa (1717-1791), the English-born son of a Sephardic Jew of Portuguese descent who had settled in England in the 1690s. Da Costa was an enthusiastic naturalist and an expert on fossils who had been elected a Fellow of the Royal Society in 1747 and in February 1763 was appointed its clerk, librarian and housekeeper. (He was later, however, dismissed and

25 Magellan to Sanches, 20 April 1762; "Notes on Ribeiro Sanches" (84.02.28), final note.
26 Magellan to Sanches, 28 April 1778.

sent to jail for defrauding the Society.) Magellan's earliest surviving letters after he arrived in London show him facilitating da Costa's contacts, on both the Society's behalf and his own, with naturalists in Portugal.

London was at this period the world's leading centre for the manufacture of scientific instruments, and Magellan quickly became a familiar figure in the workshops of the leading makers. As we shall discuss further below, this not only enabled him to indulge his passion for invention, the connections that he established became a means by which he could earn significant sums of money. In some cases, those business connections developed into personal friendships sufficiently close that when, for example, near the end of his life, Magellan drew up his will, the witnesses to his signature were the renowned instrument makers John and Edward Troughton and the well-known optician Thomas Hennings.

One of those with whom Magellan became associated at an early stage was Henry Pyefinch, a maker of optical and mathematical instruments with whom Magellan appears to have lived for a time, and with whom he collaborated in developing an improved "barothermometer" or "aerostathmion" that in due course they patented. Magellan presented a description of this device to the Royal Society in mid-1765 after reading a paper to the Society earlier that year in which he discussed the deficiencies of existing barometers. His reading of his new paper was not, however, a success, perhaps partly because of Magellan's still imperfect English (which he would no doubt also have spoken with a thick accent at the time) but chiefly because of the minute and tedious detail into which he entered in explaining the operation of the instrument. His reading extended over two meetings of the Society, on 13 and 20 June 1765, yet still he had not reached the end. No further time was allocated to him, however, until after he submitted a much-abbreviated version of the paper in the following November, and the paper was never published in the Society's *Philosophical Transactions*. In response to the protests of Magellan and his friends about the way he had been treated, the Society introduced new procedures to ensure that henceforth papers were read in the order in which they were received by the Secretaries; but the Secretaries were also instructed to peruse all papers before they were read, thus introducing a degree of quality control that if in place earlier might have led to Magellan's being required to edit his paper before reading it.

Some years later, Magellan again ran into difficulties when he presented a paper to the Royal Society. On this occasion the paper, which described improvements Magellan had devised for angle-measuring instruments such as octants and sextants, was subjected to heavy criticism by the

Astronomer Royal, Nevil Maskelyne, who favoured alternative improvements devised by the optical instrument-maker Peter Dollond. Dollond had patented an articulated arm to support the side mirrors of the instruments, whereas Magellan's proposal made use of a standard toothed circle to direct the mirrors. Maskelyne's criticism, which Magellan regarded as self-interested and completely unjustified, was sufficient to ensure that Magellan's paper did not appear in the *Philosophical Transactions*, whereupon he arranged for it to be printed separately instead, in London but in French for the benefit of the international market in scientific instruments.[27] Other works of his were later published in the same way as slim books, in French, thirty or forty pages long;[28] indeed, Magellan seems at this time to have abandoned all idea of having his work published by the Royal Society. Most of these publications were concerned with improvements to various scientific instruments, several of which were taken up by the leading instrument-makers of the day. One tract described Magellan's improved method of aerating artificial soda water, which was widely adopted. Perhaps the most influential work from a scientific point of view was his account, published in 1780 and discussed further below, of recent British advances in the study of heat, that was instrumental in bringing these to the attention of Lavoisier and the other leading French chemists with epoch-making consequences.

Despite these occasional setbacks, Magellan continued attending meetings of the Royal Society very regularly and on 21 April 1774 he was elected a Fellow on the home list, that is, as a "domestic" rather than a foreign member. The signatories on his nomination certificate included such luminaries as Benjamin Franklin, Joseph Priestley, Joseph Banks and Daniel Solander and, as Magellan later reported to the Genevan astronomer Jacques-André Mallet, an overwhelming majority of votes were cast in his favour despite the efforts of a few members who, Magellan said, caballed against him – eight, he reported, as against 62 who voted in his favour.[29]

---

27 Magellan, *Description des octants et sextants anglois* (London, 1775). Magellan read his paper to the Royal Society on 17 December 1772 and Maskelyne his criticism on 21 January 1773; see the Society's *Journal Book Copy*, vol. XXVII, pp. 300, 333-349. See also Magellan to N. Pigott, 23 March 1773.

28 German translations of several of these works also appeared in due course, probably independently of Magellan.

29 Magellan to J.-A. Mallet, 3 June 1774. Others who signed the nomination certificate were Daines Barrington, Samuel Harper, William Hunter, William Jones, Edward King, John Letch, Matthew Maty, William Watson and Peter Woulfe.

What led a group of Fellows, albeit a small one, to oppose Magellan's election is not known. His being a Catholic priest is a possible explanation,[30] given the staunchly anti-Catholic views of many Englishmen at the time, although other priests – the Jesuit Roger Boscovich, for example – had been elected before him. Magellan, however, had "form" in this regard, because in 1768 he had been indicted, with four others, by an informer before the Court of King's Bench at Westminster, accused of being a priest and having said mass – this being a felony in England at the time.[31] Though the case collapsed on a legal technicality and he was acquitted, mud may have stuck, at least in the eyes of those holding more extreme anti-Catholic views.

Joseph Banks's signing Magellan's nomination form is but one indication that the two men were on good terms at the time. Earlier, when it appeared that Banks, having accompanied James Cook on his first great voyage of discovery, 1768-70, would also accompany him on his next voyage, Magellan applied to join Banks's party.[32] Some years afterwards, following the death of the French botanist Jean Aublet, he facilitated Banks's purchasing of Aublet's notable herbarium, and he diligently passed on the congratulations of several of his foreign correspondents when Banks was elected President of the Royal Society in 1778.[33] Later, however, the relationship between the two men soured as Magellan became disenchanted, as did a number of other Fellows of the Society, with Banks's leadership. When matters came to a head in early 1784 with an open challenge to Banks's authority by a group of "mathematician" Fellows, Magellan was very much on the side of the dissidents, broadcasting their complaints about "this arrogant (*orgueilleux*) President" to his correspondents abroad.[34] Even after the initial challenge failed, he continued to look forward – in vain, as it turned out – to a new President's being elected at the Society's next anniversary meeting, when the election of officebearers was traditionally held.[35] He seems never to have reconciled himself thereafter

30 According to the article in the *Morning Herald*, Maskelyne was one of those who opposed his election, on the grounds that he was a Jesuit.

31 James Barnard, *The Life of the Venerable and Right Reverend Richard Challoner...* (London, 1784), pp. 149-50.

32 Magellan to Banks, 14 December 1771.

33 Magellan to Banks, 28 December 1778.

34 Magellan to Johann III Bernoulli, 19 December 1783. On the upheavals within the Royal Society at this time, see J.L. Heilbron, "A Mathematicians' Mutiny, with Morals", pp. 81-129 in P. Horwich, ed., *World Changes: Thomas Kuhn and the Nature of Science* (Cambridge, Mass., 1993).

35 Magellan to A. Volta, 22 June 1784; Magellan to J.-A. Euler, 20 July 1784.

to Banks's continuing in office but was reduced to presenting the Society in a very jaundiced light in his letters, while sniping ineffectually at the President and his supporters.[36]

In addition to becoming involved with the Royal Society soon after he settled in London, Magellan became a devotee of the Society for the Encouragement of Arts, Manufactures and Commerce. Formed only a few years earlier, in 1754, this body (now the Royal Society of Arts) provided a forum at which inventors could bring their inventions before the public – indeed, it encouraged them to do so by offering premiums for those deemed worthy of support. This was grist to the mill for a man such as Magellan, with his lively interest in novel technology. He attended the Society's meetings and passed on details of inventions that particularly caught his eye to his correspondents abroad. He was elected a member of the Society on 2 May 1770 and a week later presented an invention of his own, an improved form of crane.[37] He continued to be actively involved in the Society thereafter, for example proposing new members from time to time[38] and serving as a referee on matters submitted to the Society.[39]

In the 1780s, Magellan was one of the three founding members of the less formal Coffee House Philosophical Society that took to meeting fortnightly except during the summer, usually at the Chapter coffee house near St Paul's Cathedral in London and later at the Baptist's Head coffee house in Chancery Lane, to discuss questions of natural philosophy. The moving spirit of the Society was the chemist Richard Kirwan, who usually presided at the meetings, and chemistry dominated its discussions; and the Society did not long survive Kirwan's moving to Dublin in 1787. (The third founding member was Magellan's friend Dr Adair Crawford.) Until the Society ceased functioning, Magellan attended meetings very regularly except when he was away from London and presided many times in Kirwan's absence. Thanks to his international network of correspondents, he was well placed to bring forward items of scientific news that he had received from abroad. The growing challenge from Lavoisier and his colleagues to chemical theories based on the concept of phlogiston was a frequent topic for discussion. Kirwan was at this period one of the foremost critics of the new ideas emanating from Paris and Magellan, equally

---

36 See for example Magellan to A. Young, 19 July 1786, and Magellan to M. van Marum, 17 April 1787.

37 Magellan to Samuel More, 9 May 1770.

38 Magellan to More, 22 February 1775 and 8 October 1787.

39 Magellan to More, 1 March 1780.

unpersuaded by Lavoisier's ideas, diligently disseminated Kirwan's arguments hither and yon.[40]

During his early years in London, Magellan struck up a close friendship with Dr Gowin Knight, Britain's leading authority on magnetism at the time. Knight was famous for his discovery of a method of making artificial magnets much superior to even the strongest of natural magnets, and of magnetizing not only soft iron but hardened steel bars in which, if they were stored correctly, the magnetism induced would persist almost indefinitely. The use of small steel bars magnetized in this way in place of the traditional soft-iron needles revolutionized the construction of the mariner's compass and other magnetic instruments. Knight's technique made use of an enormously powerful "magnetical machine" – a large assembly of artificial magnets strapped together – that he kept during his lifetime in the British Museum, where he lived and worked as the Museum's principal librarian. Conscious of the commercial possibilities of his discovery, he refused to divulge details of the process but instead negotiated a contract to supply mariner's compasses, made exclusively to his design by the instrument-maker George Adams, to the Royal Navy. Before he died in June 1772, Knight passed on his secrets to Magellan who kept the contract with the Royal Navy going, supplying magnetized needles in the way that Knight had done and certifying compasses made by Adams, and after Adams' death in October 1772 by his son George Adams Jr. A compass inscribed "I hereby certify that I have examined this compass and that it appears to be rightly constructed. No. 636. J.H. de Magalhaens" still survives in the Mitchell Library in Sydney. Magellan had one section of Knight's "magnetical machine" moved to his house so that he could continue producing the needles, and presented the other section to the Royal Society. Unfortunately, the section Magellan had was destroyed, along with most of his belongings, in a catastrophic fire at his house in December 1772, while he was away in Paris. In testimony, however, to his mastery of Knight's technique, he was able to build a replacement that approached Knight's original in strength.

Throughout his years in London, Magellan lived close to the centre of things and to the things that interested him most. We can trace his movements from the addresses that he gave on his letters. He was apparently already in London and lodging with Henry Pyefinch in Cornhill

40 T. Levere and G. L'E. Turner, eds, *Discussing Chemistry and Steam: the Minutes of a Coffee House Philosophical Society, 1780-1787* (Oxford, 2002). See also Magellan to J.-A. Euler, 9 September 1785; Magellan to Volta, 11 October 1785; Magellan to van Marum, 24 May 1787; Magellan to Fabbroni, 17 April 1787.

when the French astronomer Jérôme Lalande visited London in the spring and early summer of 1763, because Lalande's list of contacts included "M. J.H. de Magalhaens, ami du docteur Sanchez en cornhill, chés un mathematical instrument maker". [41] However, Pyefinch's house was one of those destroyed in a fire in Cornhill on 7 November 1765, and Magellan probably lost many of his belongings. In 1766, he gave Leadenhall Street – the eastwards continuation of Cornhill – as his address, while in 1769 he was writing from a few hundred metres north and west of there at 98 Wood Street. From 1770 until mid-1772 he gave his address as 188 Fleet Street, close to the rooms that the Royal Society occupied in Crane Court until it moved along the street in 1780 to new premises in Somerset House. By August 1772 Magellan had moved around the corner out of Fleet Street to 25 Fetter Lane, still very close to Crane Court; but it was here, in December of that year, that most of his belongings were destroyed in another fire. Forced to move again, Magellan took up temporary quarters a little further away, in Old North Street, Red Lion Square; but by the end of 1773 he was back in Fetter Lane, at 12 Nevil's Court. This remained his residence for the rest of his days, conveniently close not just to the Royal Society but to the Society of Arts and to the shops of the booksellers, instrument-makers and other suppliers that he frequented. For a number of years he even employed an instrument-maker of his own to help produce some of the innovative apparatus that he developed, "my Workman Clockmaker William Hutton" whom he specified as one of the beneficiaries of his will.[42]

Establishing an international correspondence

Once Magellan was securely established in London, he began making visits to the Netherlands – as a companion, according to his obituary in the *Gentleman's Magazine*,[43] to "young foreigners" undertaking a finishing tour. We have no detailed information about these journeys but some, at least, evidently also took him to Paris and enabled him to consolidate his scientific and social contacts in the French capital while making himself useful to them. We have seen already that Magellan became the conduit

41 Lalande, *Journal d'un voyage en Angleterre, 1763* (Oxford, 1980), p. 107.
42 Magellan, last will and testament, 9 August 1788.
43 *Gentleman's Magazine*, 60 (part 1) (1790), 184.

by which the leading French astronomers communicated many of their observations to their British counterparts. Traffic also went in the other direction. For example, the Berlin astronomer Johann III Bernoulli, who met Magellan while visiting London in 1768-9 and subsequently became a regular correspondent of his, reported that Magellan had carried one of John Dollond's telescopes to Paris for the wealthy astronomical enthusiast Bochart de Saron (1730-1794), president of the Parlement de Paris, and had conceived several improvements to its fitting-out.[44] Magellan and Bochart de Saron became lifelong friends.

During these same years, Magellan also formed a connection with a senior Minister in the French Government, the Intendant des Finances, Trudaine de Montigny (1733-1777). The correspondence published here includes no fewer than 43 letters that he wrote to Trudaine's secretary, Louis-Henri Duchesne, between late 1769 and early 1777, as well as a letter and associated memoir addressed to Trudaine himself. The French authorities were evidently well aware of the technological advances being made in Britain at the time and, anxious to know more about them, had established an extensive programme of industrial espionage run through Trudaine's department.[45] Magellan became a welcome new source of information, which he supplied so assiduously that some historians have been inclined to see him as an industrial spy working on behalf of the French Government.[46] However, even though Trudaine does appear to have arranged for him to receive a gratification at one point,[47] there is not the slightest hint in the letters that Magellan expected to be paid by the French for his services – only reimbursed for the expenses he incurred in sending things to France – or that he had any other motive but his oft-expressed belief that the benefits flowing from human ingenuity should be made available to all mankind.

Magellan's letters to Duchesne record an astonishing range of things being sent to Paris, sometimes on Magellan's own initiative, at other times in response to a request. Some were new inventions: a new kind of "chemical paper" on which one could write without using pencil or ink, a new stocking frame for knitting socks, a new kind of jack, a brush for separat-

---

44 Johann III Bernoulli, *Lettres astronomiques* (Berlin, 1771), p. 88-90.

45 J.R. Harris, *Industrial Espionage and Technology Transfer: Britain and France in the Eighteenth Century* (Aldershot, 1997).

46 Henry Guerlac, *Lavoisier – the Crucial Year: the Background and Origin of His First Experiments on Combustion in 1772* (Ithaca, N.Y., 1961), pp. 41-2; Eric Robinson and Douglas McKie, eds, *Partners in Science: Letters of James Watt and Joseph Black* (London, 1970), p. 77.

47 Magellan to Duchesne, 17 January 1772.

ing flour, an improved form of gimlet, a loom of a new design, a desalination apparatus of Magellan's own devising for rendering salt water sweet. Others reflected recent progress in British agriculture: new varieties of potato and turnip, seeds of a promising new species of barley from Siberia that Magellan said was being widely adopted in Britain. Early on, he sent a suite of specimens of different timbers used in shipbuilding that he had assembled at the request of the engineer Antoine-Gabrielle Jars, who had died before he could carry out the tests he had designed to see whether the chemical treatment to which they had been subjected had the preservative effect claimed for it. Sometimes Magellan supplied raw materials that were unobtainable from French sources, for example, flint glass – essential for the manufacture of achromatic lenses – for the Parisian optical manufacturer Étienne George, cylinders of sheet metal for the engineer Vaucanson, best-quality tea for Trudaine himself as well as exotic newly introduced foodstuffs such as Chinese vermicelli and soy sauce. Sometimes he sent information rather than objects: news of a new kind of suspension for carriages or a new kind of portable stove; a catalogue of new inventions published by the Society of Arts; a flyer for a soon-to-be-published edition of voyages to the South Pacific; copies of the latest issues of the Royal Society's *Philosophical Transactions* and of Joseph Priestley's famous first paper on different kinds of air; plans of the new Plymouth hospital, built to a radically new design. There were many items, some mere curiosities, sent personally for either Trudaine or Duchesne: a supply of English mustard, printed diaries for the forthcoming year, a globular glass fish tank and lorgnettes for Mme Trudaine, copies requested by Duchesne of some of the great English novels of the eighteenth century, ormolu ware from Matthew Boulton's Soho manufactory and decorated glassware from other manufacturers, a sample of the cloth used by the natives of Tahiti, brought back to Europe by Joseph Banks. At Trudaine's request, Magellan sent him a hydrometer of the kind used by English customs officials for determining the alcohol content of fluids so that they could be taxed at the appropriate rate; but at the same time he offered to explain to Trudaine, during his forthcoming visit to Paris, how this might be done more efficiently. And he also sent a great variety of other scientific instruments and apparatus supplied by London makers – telescopes, pendulums, barometers, thermometers, proportional compasses, "pocket laboratories" for assaying mineral ores on the spot after the Swedish chemist Cronstedt's design, an air pump, navigational instruments (quadrants, sextants, compasses, etc.), even the first of the celebrated circular dividing engines constructed by the famous instrument-maker Jesse Ramsden.

Many of the instruments and scientific books that Magellan sent to Duchesne were intended for others – Sanches, Rozier, the astronomer Bourriot, various "Messrs. de l'Académie R^{le}. des Sciences" including Bory, Messier, Bochart de Saron, the Marquis de Courtanvaux who was also an enthusiastic astronomer, the chemists Lavoisier and Macquer, the Chevalier de Borda who was an authority on navigational science – and were sent through Trudaine's office, with his blessing, in order to travel free of customs and postal charges. Traditionally, such charges were levied on the recipient rather than the sender, and Magellan became adept at saving his correspondents money by sending things to them via official channels. His letters are also filled with exhortations to his correspondents as to how, if they were sending things to him, to do so in a way that would minimize what he would have to pay. In mid-1779, he was surprised to receive a letter from Paris, from Sanches, with "the novelty of the pre-payment of the post",[48] an early experiment by the French postal authorities with the system that was eventually adopted universally in the 1840s and afterwards. In England, however, in Magellan's day, the system continued as before.

The value to French science of Magellan's activities was recognised by those who benefited from them and was soon given concrete expression when Trudaine, who had already served a term as President of the Académie Royale des Sciences and was soon to serve again, arranged for him to be elected, on 4 September 1771, as a corresponding member of that body, with Bory as his designated correspondent (see 71.09.04 and Figure 2). Thereafter Magellan sent reports to Bory from time to time for communicating to the Academy, and he also made a point of attending meetings of the Academy whenever he was in Paris. He presented to the Academy an improvement of Hadley's quadrant and a memoir on a double sextant that received a favourable report by an Academy commission comprising Borda, Bory and the engineer Étienne Bézout. His desalination apparatus attracted considerable interest from the French Navy and was investigated by an Academy commission headed by Lavoisier.

On at least one occasion, Magellan's enthusiastic shipping of things to France very nearly brought him undone. In July 1771, he tried to send to Paris a complete, fully mounted loom of a new design, notwithstanding that, in order to protect British woollen manufacturers during a period of dramatic technical development, the British Government had imposed an embargo on exporting any kind of wool-handling machinery. Unfortunately, Magellan reported to Duchesne, the customs officials at Dover had

48 Magellan to Sanches, 6 August 1779.

reccognised the machine for what it was and seized it, and were threatening to impose a large fine.[49] Magellan was alarmed at first and mobilized "quelques amis" to intervene on his behalf, but soon the crisis had passed and Magellan was able to tell Duchesne that the officials, who had carefully not reported their discovery, had been persuaded to take "quelque petit present" (that is, a bribe), and that the crate with the loom inside had been returned to him intact. There was no further talk of a fine. The now chirpy Magellan had a new plan: instead of sending the loom fully mounted, which was clearly far too dangerous a proceeding, he would take it apart, number the parts so that they could be reassembled easily, and ship them in a number of separate boxes, hidden beneath books and other things. If, when the boxes arrived, people couldn't see how the parts went together, these could be simply set aside until his next visit to Paris, when he himself would reassemble them.

François Rozier's founding in 1772 of his journal, *Observations sur la physique, sur l'histoire naturelle et sur les arts*, provided Magellan with new opportunities to disseminate information. In contrast to the weighty volumes filled with formal scientific memoirs that were issued annually or semi-annually by institutions such as the Royal Society of London or the various Continental academies of science, Rozier's journal appeared monthly and typically published not lengthy memoirs but letters or short reports on recent developments in science. Magellan became, in effect, the London correspondent for Rozier's publication. His letters to his various friends in Paris, many directed to Rozier himself and clearly written in the expectation that they would be published, were manna from heaven for Rozier and his successor as editor, Jean-André Mongez, as they sought suitable material to fill each month's issue. They published a round dozen of them over the course of a decade or so, as well as several papers (as distinct from letters) that Magellan also sent them. Some of these items were subsequently re-published in Italian translation in Milan, in a journal founded in imitation of Rozier's, *Opuscoli scelti sulle scienze e sulle arti*.

Henry Guerlac long ago drew attention to Magellan's having played a pivotal role in bringing the work on "airs" of British investigators – especially Joseph Black, Henry Cavendish and Joseph Priestley – to the attention of French chemists.[50] He noted, in particular, a sequence of letters in 1771 and early 1772 from Magellan to perhaps the leading French chemist of the day, Pierre Macquer, and, with much greater effect, Magellan's send-

49 Magellan to Duchesne, 2 August 1771.
50 Guerlac, *op. cit.*

ing a copy of Priestley's booklet on the making of artificial soda-water to Trudaine de Montigny in July 1772. Along with the booklet, Magellan sent a brief but incisive summary of the work that lay behind Priestley's invention[51] that Lavoisier – who had himself recently begun thinking about the role of air in chemical processes and was shortly to make his first, famous experiments on combustion – promptly read to the Académie Royale des Sciences.

Magellan sent copies of Priestley's booklet to many of his other correspondents as well, thereby bringing Priestley's work to the attention of chemists across Europe. Later Magellan published a booklet of his own in which he described various improvements – including several that he himself had devised – to the method Priestley had described for making aerated water.[52] He also included in this booklet an account of some newly developed eudiometers "for ascertaining the wholesomeness of respirable air", using the nitrous-air test that Priestley had described.

A few years after Magellan sent news of Priestley's work on gases to Paris, he played an equally pivotal role in informing Lavoisier and his colleagues about the work of Black and others on latent and specific heats. When his friend Adair Crawford's book, *Experiments and Observations on Animal Heat*, which included an exposition of Black's ideas, was published in 1779, he was assiduous in singing its praises to his correspondents on the Continent. Then, when Crawford's book quickly sold out, he published, in French for the benefit of his Continental readers, his own synopsis, *Essai sur la nouvelle théorie du feu élémentaire, & de la chaleur des corps* (London, 1780). This was immediately re-published in Rozier's journal and became the principal vehicle by which Black's ideas became known in France.[53] Magellan in this tract actually invented the term "specific heat" (*chaleur specifique*).

In the Netherlands as in Paris, Magellan would no doubt have sought out leading Dutch and Belgian practitioners of science. We have no information, however, about any such contacts that he might have made there in the 1760s, the earliest surviving correspondence being letters he wrote early in the following decade. There is, for example, a sequence of letters commencing in 1773 to the professor of experimental physics at Louvain, Joannes Franciscus Thijsbaert. By then, Magellan clearly knew Thijsbaert

51 Magellan to Trudaine de Montigny, 5 July and 7 July 1772.

52 Magellan, *Description of a Glass Apparatus for making Mineral Waters like those of Pyrmont, Spa, Seltzer, etc....* (London, 1777).

53 Henry Guerlac, "Chemistry as a Branch of Physics: Laplace's Collaboration with Lavoisier", *Historical Studies in the Physical Sciences*, 7 (1976), 193-276, especially pp. 231-4.

well, addressing him as "mon cher ami" before reporting progress on a suite of scientific apparatus that he had ordered in London at Thijsbaert's request. Magellan was also by this time corresponding with the astronomer Nathaniel Pigott, an English Catholic who on account of his religion lived for a number of years in the Austrian Netherlands. In this case, the earliest letter that we have dates from 1772, when Magellan was helping Pigott acquire the instruments he needed to carry out his planned determination of the precise geographical locations of the principal towns in the region. The letter reveals in passing that Magellan was also by this time in friendly contact with the English Catholic priest and notable experimental philosopher John Turberville Needham, who was also based in Belgium. In 1773, Needham became director of the Académie Impériale des Sciences et Belles-Lettres when this was established in Brussels. There was a suggestion that Magellan be appointed a member of the new Academy at its foundation;[54] and although this came to nothing at the time – the limited number of places available for foreign members had apparently already been spoken for – he did become a member in 1785, an appointment that carried a substantial pension with it and also entitled him, at least within the bounds of the Austrian Empire, to wear a sword in public.

Magellan's Belgian links were not confined to men of science. Though we have no information as to how it came about, by the mid-1770s he was corresponding in the friendliest manner with one of the country's most distinguished citizens, Charles-Marie-Raymond, 5th Duke of Arenberg. Magellan had clearly become a friend of the family – the contact continued after the Duke died and was succeeded by his son Louis-Engelbert – and whenever he visited the Netherlands he stayed, as a matter of course, at the Arenberg palace in Brussels or at the family's principal country seat, at Enghien. During one such visit he seems deliberately to have provoked a discussion of the merits of enclosure, the consolidating of small rural landholdings into large estates that was transforming the English countryside at the time; and this led in due course to vigorous debates on the subject in the Académie des Sciences et Belles-Lettres.[55] Magellan's contacts in the Austrian Netherlands extended even to royalty. He undertook commissions for Prince Charles of Lorraine, who for many years governed the territory on behalf of the Empire,[56] and he met and conversed with the

54 Magellan to N. Pigott, 1 June 1773.

55 Claude Bruneel, *L'hostilité à l'égard des grands fermes: un aspect du populationnisme dans les Pays-Bas Autrichiens* (Louvain-la-Neuve, 1990). See also Magellan to the Duke of Arenberg, 22 December 1776, and Magellan to Sanches, 28 January 1777.

56 Magellan to N. Pigott, 12 October 1773.

Austrian Emperor Joseph II himself when the latter made a tour of inspection of his Belgian domains in 1781.[57]

Magellan's Catholicism and his clerical status no doubt helped open doors for him in Catholic Belgium. He also regularly visited the Protestant part of the Netherlands, the United Provinces, and established friendships there, though these are not so well represented in the surviving correspondence. His Dutch contacts included a number of university professors at Leiden and Utrecht in addition to Jan Henrik van Swinden at Franeker and later Amsterdam, with whom an extensive correspondence survives. He also counted among his friends the celebrated director of Teyler's Museum in Haarlem, Martinus van Marum, the naval officer Count Frederik Sigismund van Bylandt who sought Magellan's help when a watch by the London watchmaker Josiah Emery failed to live up to expectations,[58] and the long-serving Russian Ambassador to The Hague, Prince Dimitrii Gallitzin, who was a great enthusiast for science. For Gallitzin, with other science-minded friends in attendance, Magellan repeated some of Priestley's latest experiments, having sent the necessary apparatus ahead from England.[59] To van Swinden and van Marum, and no doubt to other Dutch friends as well, he sent a steady stream of scientific news, books, instruments and apparatus.

It seems that Magellan never visited Germany but he met the Berlin astronomer Johann III Bernoulli when the latter visited England in 1768-9 as part of an extensive tour of European observatories. Eighteen months later, Bernoulli wrote suggesting that they should correspond. Magellan's reply, in which he agreed at once to Bernoulli's proposal, epitomised his approach to his correspondence more generally. For someone like himself, he wrote, "un homme, qui se fait un devoir de supleer au defaut de la science, par des services qu'il est bien aise de faire à tous ceux qui la possedent", there could be nothing more flattering than what Bernoulli had suggested: "Ainsi Vous n'avez qu'à me donner vos ordres sur quelques objects que Vous souhaiterez; & comptez toujours que je n'oublierai rien de ma part pour accomplir vos souhaits".[60] The two men corresponded for many years after that and Magellan was as good as his word. Indeed, as well as keeping up a steady flow of scientific news, and procuring books and instruments for Bernoulli from the London suppliers, he placed copies of various of

57 Magellan to Sanches, 17 July 1781.
58 Jonathan Betts, "Josiah Emery, Watchmaker of Charing Cross. Part 2: Precision Pioneer", *Antiquarian Horology*, 22 (1995-6), 510-523; p. 523.
59 Magellan to Sanches, 31 October 1776; Magellan to Priestley, 30 November 1776.
60 Magellan to Bernoulli, 25 January 1771.

Bernoulli's publications with London booksellers and monitored the sales of these. Later, he helped Bernoulli in negotiations with Britain's Board of Longitude that led to the Board's publishing Bernoulli's *Sexcentenary Table* (1779) and rewarding Bernoulli with a gratuity for his work.[61]

Magellan also from time to time offered advice to the sometimes indiscreet Berlin academician. When Britain's Astronomer Royal, Nevil Maskelyne, complained about the unfavourable comparison Bernoulli had drawn between the tables he had published in the *Nautical Almanac* and those presented by Lalande in the Paris Observatory's *Connoissance des tems*, Magellan served as an intermediary between the two men. Bernoulli, it emerged, had misunderstood the basis upon which Maskelyne's tables had been calculated. Magellan passed on to Bernoulli the terms of the apology that Maskelyne demanded, which Bernoulli promptly published.[62] Bernoulli also offended the optical instrument maker Peter Dollond by declaring that his achromatic lenses had been developed by mere trial and error, without any theoretical insight. Magellan pointed out a number of errors, including this one, in what Bernoulli had written about Dollond, leading the latter to acknowledge, in print, that several of his comments had been mistaken. Magellan also gave the gossipy Bernoulli some wise general advice: "en verité il faut etre très delicat, en tout ce qu'on publie sur les qualités d'un autre: & plutot ne dire pas ce qu'on sait de disavantageux, même lorsqu'on en pourroit donner toutes les preuves possibles."[63]

It was probably through Bernoulli that Magellan began what was evidently a frequent correspondence with other Berlin academicians – the mathematician Johann Castillon, the chemist Franz Achard and later the astronomer Johann Elert Bode – although only a fragment of the correspondence has been found in each case. Bode, who became the editor of the Berlin Academy's *Astronomisches Jahrbuch*, included in the annual volumes of this work extracts in German translation from several of Magellan's letters to him. These provided, among other things, news of some of the remarkable astronomical discoveries being made by William Herschel at this period – discoveries that featured large in Magellan's correspondence more generally in these years. One letter published by Bode included a vivid account of Herschel's observing practice in conjunction with his sister Caroline, as Magellan had seen it during a night he spent at Herschel's observatory near Windsor.[64]

61 Magellan to Bernoulli, 30 January 1778.

62 Magellan to Bernoulli, 29 October 1771, 21 April 1772.

63 Magellan to Bernoulli, 7 April 1772.

64 Magellan to Bode, 21 January 1785.

Elsewhere in Germany, Magellan is known to have corresponded with the Göttingen professor Abraham Kästner and with Kästner's colleague Georg Christoph Lichtenberg whom he probably met when Lichtenberg visited London in 1770;[65] with successive directors of the observatory at Mannheim, Christian Mayer and Karl König;[66] and with Georg Forster who after accompanying James Cook on his second voyage around the world settled in Cassel.[67] No doubt there were others with whom he also exchanged letters, no trace of which survives. Beyond Germany, he corresponded with Nathanael Wolf in Danzig and the Swedish-born astronomer Jöns Mattias Ljungberg in Kiel (this being in what was then a part of Denmark), both of whom he had met in London;[68] and with Torbern Bergman, Gustav von Engestrom and Carl Wilhelm Scheele in Sweden. Another of his contacts was the Lithuanian Prince-Bishop Ignacy Massalski, whom he provided with introductions to the leading English agricultural reformers when the Prince visited England in 1772.[69] He also had several correspondents in Strasbourg, and others in Geneva; many of the letters he exchanged with the enthusiastic Genevan astronomer Jacques-André Mallet have survived.

If Magellan's correspondence with Berlin began with an approach from Bernoulli, that with the Academy of Sciences in St Petersburg was initiated by Magellan himself when, in September 1776, he wrote offering the St Petersburg academicians "tout ce qui peut etre de leur service dans ce pais". As he went on to explain, "ma connexion avec la plupart de nos Savants, & l'employ que je fais continuellem^t^. des artistes les plus habiles en toute sorte des instruments de Physique & d'Astronomie, m'ont affranchi la grande correspondance qui j'entretiens avec beaucoup de gens de lettres dans la plupart de l'Europe, aux quels je ne laisse pas d'etre souvent serviable".[70] He would be delighted, he said, if the St Petersburg academicians made use of him in the same way. To provide a taste of the services he could provide, he announced that he had recently despatched a box to St Petersburg that included several

---

65 Kästner to Magellan, 29 October 1770. No correspondence with Lichtenberg has been found but Lichtenberg referred in other correspondence to letters received from and sent to Magellan in the early 1780s; see G. C. Lichtenberg, *Briefwechsel*, ed. U. Joost and A. Schöne (Munich, 1983-92), vol. 2, pp. 494, 503, 505, 523.

66 König to Magellan, 21 August 1784.

67 Magellan to Forster, 13 April 1780.

68 König to Magellan, 21 August 1784.

69 Magellan to Arthur Young, 19 August 1772.

70 Magellan to J.-A. Euler, 17 September 1776. See also R.W. Home, "Scientific Links between Britain and Russia in the Second Half of the Eighteenth Century", pp. 212-224 in A.G. Cross, ed., *Great Britain and Russia in the Eighteenth Century: Contacts and Comparisons* (Newtonville, Mass, 1979).

recent publications, including one of his own, that were intended as presents for the Academy. The box was addressed to General Baur and also contained an astronomical pendulum of Magellan's own improved and well-received design that Baur had apparently ordered. Magellan also took up in his letter the question of how to make the work being done in St Petersburg better known in England, "parce qu'on ignore ici pour la plupart les publications qu'on fait chez vous". One of London's leading booksellers, John Nourse, wanted, he reported, to stock the Academy's journal, its *Commentarii*, and also Leonhard Euler's *Lettres philosophiques*, published by the Academy, provided they could be supplied at the normal trade discount. Finally, in a postscript, Magellan added news of an exciting new discovery that had been made in London only a few days earlier – news that in the ordinary course of events would have taken months to reach St Petersburg. It was a brilliant letter, perfectly calculated to capture the interest of those to whom it was addressed, isolated as they were in the Russian capital from the scientific communities of western Europe: it revealed Magellan as a serious contributor to science in his own right but one who was also prepared to go out of his way to procure books and instruments for St Petersburg and, more generally, to facilitate the flow of scientific information in both directions.

While the reply Magellan received from Russia has not been found, it is clear from the next letter he wrote to St Petersburg that his offer had been accepted enthusiastically. Now he knew to whom to direct his letters, namely the secretary of the Academic Conference, Johann Albrecht Euler, son of the famous mathematician and pillar of the St Petersburg Academy, Leonhard Euler, and he was able to report good progress on the commissions he had been given and to suggest workable arrangements for the transmission of books and apparatus. He also included news of an improved air pump that had been developed by the London instrument-maker Edward Nairne, a specimen of which he could obtain for St Petersburg if they wished, and of the publication of the latest volume of Joseph Priestley's researches on airs.[71] A long series of letters to the younger Euler, equally informative and helpful and extending almost to the time of Magellan's death, has survived in St Petersburg and is included in this edition. Magellan's services were valued sufficiently highly that he was appointed a corresponding member of the St Petersburg Academy in 1778 and a few years later was awarded a pension of 200 roubles per year as an additional inducement to him to keep up his letter-writing.[72] Those services extended

71 Magellan to Euler, 14 March 1777.
72 Euler to Magellan, 15 July 1783.

beyond the Academy itself: for example, when a party of young Russians was sent to England to learn about the new agricultural practices being employed there, he helped arrange for them to spend time on the estates of his friends John Arbuthnot and Arthur Young.[73]

Supplier of services

Magellan's letter-writing thus brought him both high-level scientific recognition and financial reward. The financial reward was, however, on a considerably larger scale than he obtained from the various pensions he gradually accumulated: as the journalist J.P. Brissot de Warville noted in his journal after meeting Magellan in London, "en se faisant le correspondant de tous les savants, il était parvenu a son indépendance chérie, à une indépendance honorable".[74] In short, Magellan had hit on a novel way of making a living, one that fitted perfectly with his interest in and enthusiasm for science, that was fully compatible with his status as a gentleman, and that took advantage of the fact that educated people all over Europe who were technically aware were keen to gain access to the interesting things they knew were happening in British science and technology at the time. In his letters, Magellan brought to the attention of those to whom he was writing the latest developments in British science – the latest publications, the latest doings at the Royal Society, the latest developments in the scientific instrument market. Then he served as an agent, supplying to his many correspondents the books and instruments they requested. And, as Brissot hints at, in doing so he received – most discreetly, of course, so as not to infringe upon his status as a gentleman – commissions from those with whom he placed the orders. There is no doubt that he earned his money because, once he had placed an order for instruments, he took great pains to ensure that these were supplied to the highest quality, constantly visiting the workshops and keeping a close eye on the progress of the work.

The second half of the eighteenth century witnessed a blossoming of astronomical observatories in many parts of Europe, some sponsored by the state but others established by private entities and individuals. Much of the equipment installed was of British manufacture, this being generally

73 Magellan to Sanches, 18 February 1777.

74 J.P. Brissot, *Mémoires, 1754-1793*, ed. C. Perroud (Paris, 1911), *I*, 363.

seen as the best available, and from Magellan's correspondence we discover that he often served as intermediary in the purchasing of it. However, his interventions were by no means confined to astronomical instruments, but on the contrary ranged wisely across the sciences.

Some of the tasks he took on were very large, as no doubt were the corresponding commissions. For example, when the governments of Spain and Portugal finally agreed, after years of bickering, to have the boundary between their respective empires in South America re-surveyed, Magellan was commissioned to supply the survey teams from the two sides with similar sets of astronomical and surveying equipment. The resulting expeditions, in addition to fixing the boundaries of Brazil more securely, added substantially to knowledge of the topography and meteorology of that vast territory.[75] Again, when the Austrian Empress, Maria Theresa, decided to reinvigorate the university that served her Italian domains, the University of Pavia, and the university's new professor of experimental physics, Alessandro Volta, was given an effectively unlimited budget to acquire the instruments and apparatus that he needed, Volta asked Magellan to purchase most of them for him in London. Another substantial commission came from the Portuguese government in 1782, when he was asked to assemble a large suite of astronomical instruments and apparatus, to be taken to China by the newly appointed Bishop of Peking, D. Alexandre de Gouveia.[76]

Sometimes, however, Magellan over-reached himself. In April 1783 he wrote to Euler *fils* in St Petersburg, with a personal letter sent separately to the Academy's newly appointed Director, Princess Dashkova, whom he had met in Paris a couple of years earlier, suggesting that the St Petersburg Academy might like to mark Dashkova's elevation by acquiring the great burning lens recently manufactured by the London optical instrument-maker William Parker, the asking price for which was over £700. He had, he said, persuaded Parker to let the lens go to Russia, so that all that was required was for the Empress to agree to the purchase. The academicians were not impressed, advising Dashkova that not only was the price ex-

75 M.F. Thomaz and I. Malaquias, "Aspectos científicos das expedições de demarcação de limites na América Meridional – Scientific aspects of the demarcation expeditions in South America", *Proceedings of the First Luso-Brazilian Congress of History of Science and Technology, Évora and Aveiro, October 2000*, pp. 201-13; I. Malaquias, "Instruments in Transit: The Santo Ildefonso Treaty and the Brazilian Border Demarcations", in M. Granato & M.C. Lourenço (eds), *Scientific Instruments in the History of Science: Studies in Transfer, Use and Preservation* (Rio de Janeiro, 2014), pp. 99-113.

76 Magellan to Luiz Pinto de Sousa, 24 September 1782.

orbitant, the famous burning mirror that the Academy already owned, that had once belonged to Ehrenfried von Tschirnhaus, was capable of doing everything that Parker's lens could do.[77]

Magellan's supplying instruments and apparatus to Alessandro Volta is documented in the extensive surviving correspondence between the two men, commencing in late 1778.[78] Some of the things supplied, including some very expensive items, were intended to be used in up-to-date demonstration experiments accompanying Volta's lectures; they included an orrery to demonstrate the motions of the planets, microscopes, an apparatus to demonstrate the refraction of light, burning lenses, a "pneumatic machine" (that is, an air pump) by Nairne, one of the latest frictional electrical machines by Adams, a telescope by Ramsden, and the machine recently invented by the Cambridge mathematician George Atwood to demonstrate the laws of motion, a device about which Magellan waxed lyrical both in his letters and in a booklet that he published about it.[79] Other items were clearly intended to support Volta's research, including barometers, standard weights and measures, special glassware for experiments on gases, and parts for the eudiometer that Volta had invented to test the "goodness" of different airs. Magellan took great pains over the supply of all this equipment, including constant supervision of the makers and a detailed inspection of the final product before it was shipped to Genoa. He also sent Volta books and scientific news.

From Volta's point of view, his correspondence with Magellan was very rewarding. He clearly felt very isolated in Italy, far from the leading scientific centres of the day in northern Europe, and he came to rely very heavily on his contact with Magellan, which gave him for the first time a dependable link with English science. The relationship between the two blossomed still further during a year-long tour of the scientific centres of northern Europe that Volta undertook in 1781-2. Magellan was in Brussels when Volta first arrived there and the two men toured Holland together, with Magellan introducing Volta to several of his Dutch scientific acquaintances before Volta settled in Paris for several months. Then, when Volta was ready to move on to England, the two met up again in Brussels and travelled together to London. There Magellan introduced Volta to his

77 Euler to Magellan, 15 July 1783.

78 Magellan to Volta, 27 October 1778 and subsequently. See also R.W. Home, "Volta's English Connections", pp. 115-132 in F. Bevilacqua and L. Fregonese, eds, *Nuova Voltiana: Studies on Volta and His Times*, vol. 1 (Pavia, 2000).

79 Magellan, *Description d'une machine nouvelle de dynamique, inventée par Mr. G. Atwood...* (London, 1780).

scientific friends and to Joseph Banks as a necessary preliminary to Volta's attending meetings of the Royal Society as Magellan's guest. Magellan and Volta seem to have got on extremely well – so well, indeed, that they then undertook an extensive tour together of the new manufacturing centres in the English Midlands. There is a new warmth to their letters after this time that obviously reflects the personal friendship they had established. The last letter we have in this fascinating sequence is dated October 1785. Doubtless there were others afterwards but probably not very many because Magellan suffered persistent ill-health in the later 1780s that greatly interfered with his letter-writing, and for the last year of his life – he died in February 1790 – he was a hopeless invalid.

Volta was not Magellan's only Italian correspondent, although many more of the letters exchanged have survived in Volta's case than in others. For example, while only two of Magellan's letters to the professor of experimental physics at the Brera college in Milan, Marsilio Landriani, have survived – and these only in Italian translations published by Landriani in the Milanese journal *Opuscoli scelti sulle scienze e sulle arti* – and only three letters from Landriani to Magellan, it is clear from what has survived that the two men were on the friendliest of terms and that many more letters passed between them over the years. Landriani was, like Volta, a beneficiary of the Empress Maria Theresa's commitment to upgrading educational institutions in Lombardy, and Magellan spent large sums on his behalf on instruments and apparatus supplied by London makers. A single surviving letter to Giuseppe Morozzo, vice-president of the Royal Academy in Turin, suggests a more extensive correspondence with Turin.[80] Magellan also had several correspondents in Florence. He became friends with the *abbé* Felice Fontana[81] and the young Giovanni Fabbroni, co-founders of the Royal Museum of Physics and Natural History in that city, while they were in England for almost a year in 1778-9 – he may, indeed, have met them in Paris before that – and a number of his subsequent letters to Fabbroni have survived, chiefly concerned with the preparation of Magellan's new edition of Axel Cronstedt's *Essay towards a System of Mineralogy*, for which Fabbroni furnished extensive notes. Magellan is also known to have been corresponding with Alessandro Bicchierai, clinical professor at the hospital of Santa Maria Nuova,[82] although none of the letters they exchanged have survived; with the English aristocrat Earl Cowper

80 Magellan to Morozzo, 11 May 1784.
81 See Magellan to Turgot, 13 November 1778.
82 Magellan to Fabbroni, 4 February 1780, note 12.

who lived in Florence for a number of years; and with the Liceo delle Scuole Pie where Gaetano Del Ricco, who taught mathematics and physics, sought his help in obtaining "a good astronomical quadrant".[83]

Reference was made earlier to Magellan's being commissioned to supply surveying instruments to the governments of Spain and Portugal for the re-surveying of the boundary between their respective empires in South America. Magellan published detailed accounts of the equipment supplied, the most important components of which were portable quadrants for determining latitude and longitude, such as he had already supplied to several correspondents elsewhere in Europe.[84] The correspondence published here provides further details about the purchase of the equipment for Portugal[85] but does not add any new information concerning that supplied to Spain. However, Magellan's supplying of further large collections of instruments and other items to the Spanish Court some years later is thoroughly documented in his correspondence with the then Spanish Ambassador in London, the Marquis Bernardo del Campo. In this case, the equipment was destined for the astronomical observatories in Madrid and Cadiz and a new one being established at Santa Fé de Bogotá in South America. Other letters document Magellan's filling new orders for the Portuguese Navy – including a specimen of a new and improved iron stove recently approved for use on ships of the British Navy, together with detailed instructions for installing it[86] – and supplying instruments and apparatus for technical and higher educational institutions in Portugal, namely the University of Coimbra,[87] the Royal College at Mafra,[88] and the Marine Guards and the newly established Academy of Sciences in Lisbon.[89] He was also commissioned by the Portuguese Government to purchase items

83 C. A. Guadagni to Magellan, 21 April 1789; G. Del Ricco to Magellan, 6 November 1789.

84 Magellan, *Description et usages des instrumens d'astronomie et de physique faits à Londres, par ordre de la Cour de Portugal ...* (London, 1779); idem, *Notice des instrumens d'astronomie, de geodesie, de physique, etc., faits dernierement par ordre de la Cour d'Espagne ...* (London, 1780).

85 Magellan to Luiz Pinto de Sousa Coutinho, 14 July 1778 and 2 July 1779.

86 Magellan to Martinho de Mello e Castro, 5 August 1783 and 14 January 1784; Magellan to Ayres de Sá e Mello, 14 January 1784.

87 José Monteiro da Rocha to Magellan, 26 March 1787.

88 Joaquim d'Assumpção Velho to Magellan, undated (1784?).

89 Magellan to Dom Garrellon, 22 July 1788. See also Isabel Malaquias, "International Networks of Production and Distribution of Scientific Instruments in Eighteenth Century Europe", *Portuguese Studies of Social Science* (in press), and Isabel Malaquias, "J. H. de Magellan and the Royal/Real Dissemination of Instruments all over Europe", in M. C. Lourenço (ed.), *Royal Cabinets of Physics in Portugal, 18th-19th centuries* (Lisbon, in press).

from the philosophical cabinet of Prince Charles of Lorraine, which was auctioned off after the Prince's death, to be used in the education of the heir to the Portuguese throne, the Prince of Brazil.[90]

A much more exotic correspondent of Magellan's was the Hungarian adventurer Count Maurice de Benyowski. It is not clear how the two men became acquainted but in due course Magellan was persuaded to invest a considerable sum in an expedition that Benyowski was planning to Madagascar.[91] He also engaged in negotiations with the Portuguese Government on Benyowski's behalf.[92] In due course Benyowski proclaimed himself King of Madagascar and appointed Magellan and an English MP, William Morton Pitt, his official representatives in London.[93] Soon afterwards, however, Benyowski was killed and his "kingdom" was no more. In the meantime, Magellan tried unsuccessfully to arrange the publication of Benyowski's memoirs; these eventually appeared in print after being edited by Magellan's friend William Nicholson.

Magellan's earliest contacts with the pioneering Birmingham industrialist Matthew Boulton were a result of his taking over Gowin Knight's contract to supply mariner's compasses to the Royal Navy. After the "magnetical machine" that he used to magnetize the compass needles according to Knight's method was damaged in the disastrous fire in his lodgings in December 1772, he engaged Boulton to re-harden and polish the steel bars of which the machine chiefly consisted.[94]

A much more substantial connection developed some years later between Magellan, Boulton, and James Watt who in the meantime had gone into partnership with Boulton to exploit the improved steam engine he had invented. Magellan had in his usual way been spreading news of Watt's engine to his acquaintances abroad. The first hint in the surviving correspondence of a response is a letter that Magellan wrote to Watt in December 1777, asking whether Watt would be free to meet one of Magellan's foreign correspondents who was proposing to visit Birmingham, accompanied by Magellan, to talk to Watt about the engine.[95] It was perhaps this that first opened the eyes of Boulton and Watt both to the possibility of finding an expanded market for the new engines on the Continent, and to

---

90 Magellan to Ayres de Sá e Mello, 17 April 1782.

91 W. Nicholson, ed., *Memoirs and Travels of Mauritius Augustus Count de Benyowsky...* (London, 1790), pp. xi-xii.

92 Magellan to Martinho de Mello e Castro, 14 October and 30 December 1783.

93 Benyowski to Magellan and Pitt, 15 December 1785.

94 Magellan to Boulton, undated 1773 and 9 September 1773.

95 Magellan to Watt, 8 December 1777.

Magellan's potential value as an intermediary on account of his network of influential contacts in several different countries. Be that as it may, only a few weeks later we find Magellan writing to both Boulton and Watt about negotiations that by then were already well in train with the authorities in France to obtain an exclusive privilege there (the equivalent of patent protection) for Watt engines. In early March 1778 he was also able to forward to Birmingham a formal proposal from the director of mines in Prussia to buy four engines to pump water from mines there,[96] and he later introduced would-be purchasers from other parts of the Continent to Boulton and Watt.

The initial French negotiations were with the elderly Comte d'Hérouville, who wished to instal a Watt engine to drain the marshes on his land at Les Moeres near Dunkirk and who, in order to do so, promptly obtained for Boulton and Watt an *arrêt* of the kind they demanded to protect their patent rights. There were complications, however, because the *arrêt* specified that it would not come into effect until the advantages of the new engine had been demonstrated by running it alongside a traditional engine at either Les Moeres or Paris, and Boulton and Watt proved reluctant to travel to either site to carry out the demonstration, especially after France joined the war being fought between Britain and her American colonies. Later, d'Hérouville generously helped persuade the French authorities to broaden the terms of the *arrêt* to include the possibility of conducting the required demonstration in Brittany, at a coal mine owned by Joseph Jary, who had in the meantime visited England and been greatly impressed by what he saw of Watt's engines after being given a letter of introduction by Magellan. Eventually, three engines went to France, one to Jary and two to Paris, to the company responsible for that city's water supply, where, as Boulton had suspected would happen, J.-C. Périer promptly breached the Boulton and Watt patent by building machines modelled on theirs.[97] Magellan's correspondence with Boulton, Watt and d'Hérouville that is published here reveals how he worked tirelessly to keep the negotiations going, passing information from one side to the other, smoothing ruffled feathers when necessary – he was on the friendliest of terms with d'Hérouville and his family – and advising Boulton and Watt at every step on how to proceed.

Magellan also became involved in the marketing to his friends abroad of another of Watt's inventions, a machine for making copies of letters,

96 Magellan to Boulton and Watt, 3 March 1778.

97 J.R. Harris, *Industrial Espionage and Technology Transfer: Britain and France in the Eighteenth Century* (Aldershot, 1998), chap.13.

using a special ink and specially prepared paper.[98] He himself used one of the earliest of these machines, finding it very useful in keeping track of his correspondence, and did not hesitate to sing its praises, even as he discussed with Watt ways of improving its performance. Later, however, he had to report problems with the rollers on some of the machines he had sold.[99] Unfortunately, Watt's response has not been found.

Magellan's correspondence with Watt includes a fascinating exchange of views about the attributing of credit for discoveries in science. The exchange was initiated by Magellan in early 1780 when, at the time he was preparing his *Essai sur la nouvelle théorie du feu élémentaire*, he sought detailed information from Watt about Joseph Black's "discoveries and system ... relative to his Latent heat". In particular, he wanted to know when Black made his discovery and whether he had ever published anything about it. He was asking, he said, "because I see by a late work of Bergman, the famous Swedish Chymist, that Professor Wilcke of Stockholm did publish nearly the same facts long while ago in the Memoirs of Stockholm Acad^{y}."[100] In reply, Watt told Magellan that knew nothing of any Stockholm Academy publication on this subject and that, while Black had published nothing himself on the matter, "I know positively that [he] was the inventor of these Experiments and Theorys & that, prior to the year 1763, and that he has publickly taught them ever since, and I believe before that Æra."[101]

Notwithstanding Watt's assurance, Magellan remained troubled about Black's claim to the discovery. "I am deeply concerned", he told Watt, "for the carelessness of D^{r}. Black's having made so long ago his discoveries without publishing them to y^{e}. world. Professor Wilcke at Stockholm has published the same discovery: and as far as I can learn, independent of any knowledge of those made by D^{r}. Black. So that if this last pretends to the priority, the Public will never thank him for his coveteousness, in making a real monopole of them amongst his Scholars". As a result of Black's failure to publish, he continued, in the essay he himself was about to publish, "I am oblidged to quote the discovery, as made by Prof^{or}. Wilcke, who published it; but I mention D^{or}. Black as far as his scholars have quoted him".[102] Watt, alarmed by this news, immediately set about collecting further information, including from Black himself, to support the latter's

98 Magellan to Watt, 16 September and 6 October 1780.
99 Magellan to Watt, 28 August 1781.
100 Magellan to Watt, 22 February 1780.
101 Watt to Magellan, 1 March 1780.
102 Magellan to Watt, 7 March 1780.

claim to priority, but Magellan remained firm in his view of the matter and published accordingly.[103] For this he has been condemned by later British writers on the subject.[104] Watt, however, while annoyed, recognised that Magellan's intentions were honourable and acknowledged his right to his opinion: "I find that we are not yet exactly agreed in our sentiments on the Devoir of publication however don't lett that make us differ in other respects".[105] Nor was this empty rhetoric on Watt's part, for the two continued to correspond in the friendliest manner. Magellan's stance on the matter was in fact entirely consistent with his wider commitment to promoting the dissemination of scientific information. Two years later, he made exactly the same criticism of James Price, FRS, for not publishing the details of his supposed discovery of a method of turning mercury into gold – concluding, however, in this case that Price's failure to publish indicated that he was a charlatan.[106] The withholding of information by failing to make one's discoveries public was anathema to everything Magellan stood for.

An incident occurred in December 1785 of which no trace appears in Magellan's correspondence but that would have sorely tested Magellan's friendship with both Boulton and Watt. It related to an attempt by the Swiss physicist Ami Argand (1750-1803) to protect the British patent he had been granted for the improved lamp he had invented, now generally known as the Argand lamp or Argand burner. Argand had gone into partnership with Boulton and the optical instrument maker William Parker to manufacture and sell lamps built according to his design. On 13 December 1785 the partners brought a suit for breach of Argand's patent against competitors making very similar lamps. Magellan was called as a witness by the defendants and testified that in November 1783 – that is, some months before Argand was granted his patent – he had received a letter from Paris describing Argand's lamp and had had one made for himself in London. The judge thereupon instructed the jury to declare a non-suit, on the ground that Magellan's testimony showed that Argand's invention was

---

103 Watt to Magellan, 9 March, 20 March and 29 March 1780; Magellan, *Essai sur la nouvelle théorie du feu élémentaire, & de la chaleur des corps* (London, 1780), pp. 165-6. See also *The Correspondence of Joseph Black*, ed. Robert G. W. Anderson and Jean Jones (Farnham, Surrey, 2012), pp. 412-5, 420-1 (Watt to Black, 9 March 1780; Black to Watt, 15 March 1780; Watt to Black, 17 May 1780).

104 See especially Douglas McKie and Niels H. de V. Heathcote, *The Discovery of Specific and Latent Heats* (London, 1935).

105 Watt to Magellan, 29 March 1780.

106 Magellan to Watt, 20 September 1782.

already in the public domain at the time he was granted his patent. However, evidence subsequently assembled by Argand and his friends with a view to launching an appeal cast doubt on the date when Magellan's lamp was constructed and hence on the testimony he had given, leading one recent author in discussing the case to describe Magellan's evidence without qualification as perjury.[107] For Magellan to have deliberately perjured himself in a situation like this would, however, have been completely out of character with the person who emerges from his correspondence. Moreover, it is extremely difficult to conceive of any reason why he would have done so.[108]

Boulton and Watt were members of the so-called Lunar Society of Birmingham, a group of enterprising technologists and enthusiasts for science who met monthly at the time of the full moon. Magellan was acquainted with several other members of the group including Joseph Priestley, the chemist James Keir and the physician William Withering, and he procured foreign books for the Society as well as for individual members of it.[109] In London, long before the formation of the Coffee House Philosophical Society, he had become friendly with several members of the liberal-minded group who styled themselves the Club of Honest Whigs, including the effective leader of the group, the schoolmaster and authority on electricity John Canton; Joseph Priestley; the renowned electrical investigator and statesman Benjamin Franklin, who resided in London for many years as agent for Britain's North American colonies; and the dissenting preacher and writer on the theory of probability Richard Price. Magellan shared the liberal political outlook of the Honest Whigs, often declaring that it was on account of the prevailing political freedom that he had settled in "my England, the only place where I breathe quietly & without political fear"[110]; and when the crisis came and war broke out between Britain and her erstwhile colonies, his sympathies, like those of other Honest Whigs, lay with the Americans. He kept in touch with Franklin after the latter became Plenipotentiary of the American colonies in Paris, and when he

107 John J. Wolfe, *Brandy, Balloons & Lamps: Ami Argand, 1750-1803* (Carbondale, Illinois, 1999), p. 70; see also Michael Schrøder, *The Argand Burner: Its Origin and Development in France and England, 1780-1800* (Odense, 1968), pp. 150ff.

108 Magellan refers to what is evidently an Argand's lamp in a letter to the Academy of Sciences in Lisbon that we have dated to some time in 1784.

109 Magellan to Watt, 22 December 1783. See also Robert E. Schofield, *The Lunar Society of Birmingham: a Social History on Provincial Science and Industry in Eighteenth-Century England* (Oxford, 1963).

110 Magellan to Sanches, 31 October 1776.

himself was in Paris he spent much time in Franklin's company. In England, he worried about the restrictions on political freedom being imposed as a result of the war: "Here things are going to hell", he told Sanches in June 1777, "A separated America will be the sole asylum for philosophical spirits, as soon as peace is achieved".[111] Even the war, however, could not permanently dampen his spirits. Some months later, when an unidentified critic denounced him to Sanches as a heretic, he responded: "Tell him that I live here in this country of freedom because I am a man & not a slave, like all the Gensfoutres [jackasses] like him and his fellow creatures".[112]

Franklin had no doubt fostered Magellan's interest in the scientific society that he had helped found in Philadelphia, the American Philosophical Society – an interest that was only enhanced so far as Magellan was concerned by the political struggles of the Americans. During a visit to Paris in the early 1780s, Magellan gave Franklin some books as a present for the Society and expressed an interest in becoming a member, and when – no doubt due to wartime disruptions to communications – this came to nothing, he wrote directly once peace was declared, sending more copies of the books and seeking information about the working of the Society. He also hinted at the possibility of a legacy. "Happily", he wrote, "I have outlived the noble strugles of America, and behold with overjoyed heart the magnificent prospect of the great scene, which is there open'd for the progress of Science, and of whatever is great in human affairs!"[113] Three years later, after some intervening correspondence, he formally announced the donation of a sum of 200 guineas to endow an annual prize of 10 guineas to be awarded by the Society to "the Author of the best discovery, or most usefull improvement relating to Navigation, or to Natural Philosophy".[114] The prize thus established, the Magellanic Premium, was announced in the second volume of the Society's *Transactions*, published in mid-1786, and has been awarded regularly ever since. Magellan's election as a member of the Society was announced in the same volume.

111 Magellan to Sanches, 27 June 1777.

112 Magellan to Sanches, 28 April 1778.

113 Magellan to the American Philosophical Society, 17 June 1783.

114 Magellan to the American Philosophical Society, 27 September 1785 and 7 July 1786. In the latter letter, Magellan asked that "Astronomy" be inserted after "Navigation".

Improvements to scientific instruments

Magellan was not a fundamental contributor to new devices or instruments but he had instrumentation in his blood and engaged with some of the important problems relating to instruments in his day.[115] He kept closely in touch with the latest developments and some of the improvements that he claimed as his own were in fact adaptations of the ideas of others, modifications that he conceived to facilitate the performance of the instrument concerned. He commissioned large numbers of instruments and other apparatus for his network of correspondents, and he often had the makers incorporate his ideas into the things he ordered from them. Some of these objects that have survived carry the signature, "J.H. de Magellan Lusitanus Invenit atque Fieri Curavit Londini" ("J.H. de Magellan Portuguese devised and had made at London") or just "J.H. de Magellan Fieri Curavit Londoni".[116] In his enthusiasm for instrumentation, he often provided in his own publications a fascinating quasi-historical record of novelties previously introduced by others, but he was also prone to giving long-winded explanations of how his own improvements functioned in practice, with the idea of helping users far from the London workshops to master the equipment and overcome any difficulties that arose.

In his modifications to barometers, Magellan was dealing with questions relating to their portability as well as to their scales and the most precise ways to get good readings from them. His earliest presentation on the subject was a critical analysis of the performance of existing barometers that he read to the Royal Society in early 1765, several months before he began reading his paper on the *Aerostathmion*, and many of the advantages claimed for the latter depended simply on paying careful attention to detail so as to avoid the problems he had identified. A sliding sighting vane ensured that all readings were taken exactly horizontally. Making allowance for the variation with temperature of the density of mercury and hence in the height of the mercury column for a given atmospheric pressure was a particular focus. A novel feature favourably remarked on by later historians was to graduate the two arms of the barometer in op-

115 I. Malaquias, "A obra de João Jacinto de Magalhães no contexto da ciência de séc. XVIII" (Aveiro, 1994).

116 Inscriptions on the pendulum of the clock in the Atwood machine in the Museu da Ciência, Coimbra, and on the pendulum bob of a clock in the Musée d'Histoire des Sciences in Geneva; Jonathan Betts, "John Hyacinth de Magellan (1722-1790), Part 1: Horological and Scientific Agent", *Antiquarian Horology*, September 2003, pp. 509-17.

posite directions.[117] Later developments involved the variable volume of the mercury container of a portable barometer, for which Magellan had to ask the instrument-maker Edward Nairne to construct a screw specially adapted to it, and also a tripod support that coud be used either to support the barometer in the field or as its container when travelling, or to render it more stable when used at sea.[118]

His work with barometers led Magellan to conceive a curious device that could register different meteorological parameters such as pressure (using a static or balance barometer), time, temperature, height of tides, wind speed, humidity, rainfall and the amount of evaporation over an extended period. He called it a constant meteorograph, and saw as a great advantage the fact that it needed to be adjusted only after eight days, or even a month, when the pendulum was rewound.[119] While an interest in recording instruments was awakening in the 1780s, a multi-function register instrument like the one conceived by Magellan, which presaged those proposed by Secchi in the mid-nineteenth century, was not known. Secchi, at least, appears to have been unaware of Magellan's proposal, though a subsequent paper by Radau referred to Magellan's work on barometers and his proposed registering machine.[120]

As noted above, Magellan's improvements to the sextant and octant were severely criticized by Britain's Astronomer Royal, and as a result his paper describing them was not published in the *Philosophical Transactions*. In addition to arranging instead for his paper to be published separately, Magellan sent a full description of his improved versions of both instruments to the Académie Royale des Sciences in Paris, with details of how they could be used to obtain one's latitude and longitude. He also sent an account of a new double sextant that he had invented. The Academy, as was its custom, established an expert commission to evaluate Magellan's proposals. In this case, the commission comprised the mathematician Étienne Bézout and two members of the Academy who were also senior serving naval officers, Charles de Borda and Magellan's official correspondent at the Academy, Gabriel de Bory. The commission reported that

117 W. E. K. Middleton, *The History of the Barometer* (Baltimore, 1964), p. 142.

118 Magellan, *Description et usages des nouveaux barometres…* (London, 1779).

119 Ibid.

120 Isabel Malaquias and M. F. Thomaz, "Contributos de João Jacinto de Magalhães para o desenvolvimento da Meteorologia", *3rd Symposium of Meteorology and Geophysics of the APMG and 4th Luso-Spanish Meeting of Meteorology* (Lisbon, 2004), pp. 19-24; R. Radau, "Sur un météorographe ancien et sur la théorie du baromètre statique", *Comptes rendus hebdomadaires de l'Académie Royale des Sciences*, 65 (1867), 360-4.

the subject's importance and its presentation deserved the Academy's approval and compliments, as well as the honour of being printed under its privilege. Regarding the double sextant, the commission noted that "it has some relation with an Instrument presented by one of us [i.e. Borda] at the Academy on February 5, 1774. But it seems that M. Magellan arrived at the same ideas separately and without communication."

Astronomical clocks were another class of instrument that attracted Magellan's attention. In particular, he pursued the idea of a pendulum the length of which would not be subject to temperature variations, claiming that with those he used, made of wood with straight, parallel fibres, cut and prepared in a special form, any thermal expansion or contraction was lateral rather than lengthwise and thus without effect on the period of vibration. He also insisted on using pallets of "hard stones" such as chalcedony or ruby. By careful attention to such matters, he wrote, the large number of clocks he had supplied to different parts of Europe "ont prouvé excellentment bien".[121] These clocks had rope for eight days rather than the original thirty hours, thanks to Magellan's "coupling the driving pulley to the hour arbour via two different gears of a 2×1 ratio".[122] Jonathan Betts in a careful study of surviving clocks signed by Magellan has concluded that with the earliest of them, Magellan "was to some extent feeling his way with his designs, taking advice where he could get it, but apparently managing without a sound knowledge of what a good regulator was really like"; but that the commissions he obtained from the Portuguese and Spanish Governments in 1778 resulted in "an altogether more professional movement being produced". The clocks supplied still, however, "boasted a number of Magellan's 'special features'", especially the specially-shaped wooden pendulum. Then in the 1780s, Magellan became a leading advocate of the newly developed detached escapement, before becoming involved in the successful development and dissemination of the lever escapement that made the construction of portable precision watches possible.[123]

A similar concern for precision underpinned Magellan's work on the beam balance, where he was concerned with the deformation of the body

---

121 Magellan to Turgot, 17 July 1778.

122 A. A. Miguel, "Some Modifications and Perfectings introduced by J. J. M. in Pendulum Clocks", pp. 227-8 in L. Alte da Veiga et al., eds, *João Jacinto de Magalhães (John Hyacinth de Magellan): Conference on Physical Sciences in the XVIII Century, 7-10 November 1990* (Coimbra, 1994).

123 Jonathan Betts, "John Hyacinth de Magellan (1722-1790), Part 2: The Early Clocks", *Antiquarian Horology*, June 2004, 17383; "Part 3: The Later Clocks and Watches", *ibid.*, March 2007, 25-44; "Part 4: The Precision Pioneers", *ibid.*, September 2007, 365-75.

of the balance due to temperature variations, successfully maintaining the centre of gravity unchanged and avoiding other, collateral mechanical deformations. Other features were the use of hard stone – agate – in the suspension system, and a mechanism to stabilize the balance when not in use. Magellan also proposed a method of comparing masses that avoided problems arising from a possible inequality in the balance arms. With the study object in one pan, copper shavings were added to the other pan until balance was achieved; the study object was then removed from its dish and known weights added to this until balance was once again achieved.

In his enlarged and rearranged edition of Cronstedt's *Mineralogy*,[124] Magellan not only updated many entries, he also included a description of a portable chemical laboratory that could be used in the field to undertake both dry and wet tests and operations with the blowpipe and lamp-furnace. One of the problems that arose when using the blowpipe was condensation of vapour in the tube. Magellan made some modifications to Cronstedt's blowpipe, using a hollow ball of two hemispheres, the lower one of which received the moisture, and some small caps with little holes of different sizes that would enable different flames. In the same publication, Magellan presented the hydrometer recently developed by William Nicholson for measuring specific gravities of liquids (1785) and his own method of producing precise sets of weights, a subject to which he had previously made reference in a publication in *Observations sur la Physique*.[125]

Personality and character

The anonymous obituary notice of Magellan that was published in the *Gentleman's Magazine* describes him as follows: "He was a studious, ingenious, and learned man, particularly distinguished among the Literati in this and other enlightened countries for his intimate acquaintance with most branches of Natural Philosophy, and no less ingenious in his experiments therein, particularly in mechanics.... All the Literati in Europe knew something of his merit, and the most noted of them were desirous to know more; it was really great."[126] The notice also tells us what he looked like:

124 A. Cronstedt, *An Essay towards a System of Mineralogy*, 2nd ed., greatly enlarged and improved by J.H. de Magellan (London, 1788).

125 Magellan to F. Rozier, January 1781.

126 *Gentleman's Magazine*, 60 (pt. 1) (1790), 184.

"His height was about six feet one or two inches; a bony and rather bulky man; plain in his dress, unaffectedly mild and decent in his whole demeanour." As to his religious outlook, after recording that Magellan had at one time been an Augustinian monk, the unknown obituarist states: "He was a very mild Christian, having many years renounced the Popish faith".

The claim that Magellan renounced Catholicism has since been repeated by several other writers.[127] It is, however, incorrect, for Magellan remained a Catholic and indeed a priest to the end of his days. Yet the obituarist's assertion suggests, at least, that Magellan acted in a way that gave rise to the idea that he had abandoned the faith in which he had been brought up. Magellan was, indeed, highly critical of certains aspects of Catholic religious practice and in at least some respects very lax in his own observance, so that the obituarist's mistake can be readily excused.

Magellan's attitude emerges most clearly from his letters to Sanches, to whom he was obviously very close and to whom he wrote in the frankest of terms. One letter makes it clear that he no longer attended church regularly: "I'm going to write this while you may be attending Holy Mass, since it's Sunday morning, a time when I'm idle and alone in my room".[128] On the other hand, his account to Sanches of his journey across Flanders in November 1776 makes it clear that he still regarded himself as a Catholic, even as it displays his disregard for the Church's rule about eating fish on Fridays: "I took the stage-coach at 4 in the morning & arrived at Breda before the gates were open. I immediately got into the coach to Antwerp where I stayed that night. It was Friday and I couldn't find anything for dinner except salted fish. Damned Austrian hypocrisy! Fortunately the innkeeper in Antwerp took me for what I am not (Protestant) and gave me a piece of cold veal for dinner in my room."[129] Similarly, when some years later he solicited James Watt's opinion on ideas he had set out in a paper that he had sent him, he assured Watt that "altho' I am a Papist (not very strict indead! as our good friend D^r^. Priestley may testify), nevertheless I dont pretend to be a Pope, viz infailible".[130] Magellan's eventually receiving the benefit specified at the time he left the monastery in Coimbra indicates that, so far as the Church authorities were concerned, he was

127 "S. P. O.", in *Dictionary of National Biography* (London, 1885-1900), vol. XXXV, pp. 317-8; Ricardo Jorge, "Amigos de Ribeiro Sanches: J. H. de Magellan", *Medicina Contemporânea* (Lisbon, 1910); Stuart Pierson, in *Dictionary of Scientific Biography*, vol. 9, pp. 5-6; Weiss in Michaud, *Biographie universelle ancienne et moderne*, vol. 26, p. 119.

128 Magellan to Sanches, 16 June 1776.

129 Magellan to Sanches, 15 November 1776.

130 Magellan to Watt, 22 December 1783.

still a priest and thus entitled to the money. In further confirmation of his status, in August 1785, Magellan was the successful applicant for a position that had become vacant as a canon in the Royal Chapter of the Church of St Vincent, at Soignies in the Austrian Netherlands. The position was a sinecure – Magellan continued to live in London but became entitled to a share of the income of the congregation – but a canon was a superior grade of priest and being in holy orders was a prerequisite of the appointment.[131]

In other letters to Sanches, Magellan inveighed against what he regarded as mere superstition associated with the Catholic Church: for example, after the very devout Queen Maria I succeeded to the Portuguese throne, he wrote sarcastically to Sanches: "You don't know how much I've laughed at the rosary's being sung by our soldiers. It's said that the Queen never signs any decree or order, without having her hand covered with relics of saints. It's the safest way to have the heavens on her side in everything. Fortunate government …!"[132] He also wrote bitterly about the tyranny of the Inquisition, which seemed to be flourishing anew in Portugal under the new Queen: he had, he said, been told "that there was again great vigour in the Holy Inquisition: and that several persons had been arrested, even military officers, because they read Montesquieu, the works of Rousseau, Voltaire &c. &c.".[133] "May God will", he wrote on another occasion, "that the sanbenito is not set on fire, since the Inquisition & the monks will soon return to their zenith".[134]

In his personal life, Magellan certainly did not live as a Catholic priest should, because his letters to Sanches are filled with accounts of his sexual exploits with what he called "my Bettys, Nellys, & Fannys",[135] some of whom were clearly regular partners while others he seems to have recruited from the street on the spur of the moment. This was, he conceded, a "weakness",[136] but there is no suggestion that he sought to overcome it. On the contrary, he took great delight in such activity – for some time he had actually living with him a young woman named Sally "who gives me more pleasure than any other thing"[137] – and he constantly boasted about

131 Ame Demeuldre, *Le Chapitre de Saint-Vincent à Soignies-ses dignitaires et ses chanoines* (Soignies, 1902).
132 Magellan to Sanches, 15 August 1777.
133 Magellan to Sanches, 13 March 1778.
134 Magellan to Sanches, 15 August 1777. The *sanbenito* was the garment worn by heretics condemned by the Inquisition to be burnt at the stake.
135 Magellan to Sanches, 12 July 1776.
136 Magellan to Sanches, 28 April 1778.
137 Magellan to Sanches, 16 May 1780.

his prowess, teasing Sanches, who was considerably older than him, for no longer being sexually active.

Thus while Magellan considered himself a Catholic and was from a formal view still a priest, he was clearly neither strict nor orthodox in his way of thinking but felt free to adopt the principles that seemed to him correct, even if these were drawn from non-Catholic sources. He himself was explicit about this, writing cheerfully to Sanches at one point that "everything is going well and better all the time, because I know how to take the best part from everything, with a sprinkle of the martinian [i.e. Lutheran] philosophy that is my Mother Celestine".[138] Magellan was, in fact, a typical man of the Enlightenment, inquisitive and questioning, confident in the ability of human beings to improve their lot through their own efforts, who accepted Christian doctrine at a foundational level but who rejected much of the institutional structure and formalism that had been built around this.

Magellan's correspondence makes it clear that he was also a very sociable person who made friends easily at all levels of society, from members of the aristocracy to the humblest artisan. Whether in London or abroad, he clearly led a very active social life. A remarkable feature of the letters he wrote to people whom he had met in person and into whose homes he had been invited is that he often sent friendly greetings not just to the person himself but to members of the person's household, in a way that suggests that he was interested in them individually and did not see them as mere appendages of the person with whom he was corresponding.

Magellan's letter-writing was in fact an embodiment of his sociability, for despite what has been said above, it would be a mistake to conclude that he conducted his vast correspondence for wholly commercial reasons, he also had higher motives than this. No doubt he enjoyed the extra income that his letters generated – it was presumably this that allowed him, for example, to acquire things like his piano and to survive losing a small fortune in Benyowski's dubious enterprises – but even though it is not entirely clear how he managed to do so, he had been able to live the comfortable leisured life of an eighteenth-century gentleman, long before he began earning money through his letter-writing. Magellan wrote because he saw science as the key to progress and was genuinely and deeply committed to the principle of the free sharing of information for the betterment of

138 Magellan to Sanches, 7 July 1775. Mother Celestine is a character of Spanish and Portuguese literature of the 16th and 17th centuries who does marvellous things by means of witchcraft.

humanity. Johann Bernoulli undoubtedly spoke for many when he wrote of "la singuliere complaisance avec la quelle l'estimable & savant M. de Magellan se prête à tout ce qui peut contribuer aux progrès des Sciences & à la satisfaction de ses amis".[139] The theme is ever-present in his letters. He wanted the French to have the new cotton looms, for example, and also Watt's steam engine, not because he could make money thereby but in order to improve the lot of the French people. He disseminated scientific news both in order to advance the general level of understanding and in the hope of prompting further discoveries. Consistent with this, where scientific malpractice or charlatanry was discovered – as for example he thought it had been in Benjamin Wilson's spectacular public demonstrations at the London Pantheon in 1777 in opposition to the use of pointed lightning rods – he trumpeted it to the skies, to prevent people being misled by the malefactor's claims.[140]

In addition to these serious concerns, however, Magellan also did what he did because he enjoyed it, "having no other Profession but the Study of Sciences and Liberal Arts" as he put it in the will that he drew up near the end of his life.[141] He clearly loved fiddling with scientific instruments and trying to improve them. More generally, he had a wide interest in matters scientific and technical, and he derived much pleasure from keeping up to date with the latest developments and from helping other people to do so, as well as from the standing that he gained thereby. "I had the instruments executed that various friends ask me for," he reported to Sanches, "which amuses me and makes me feel, living this kind of philosophical independence, greatly esteemed by the good ones & far enough away from the wickedness of the bad."[142] Above all, he enjoyed the letter-writing itself: "Communication of ideas with true friends whom we know to be such creates the most lasting pleasure, & gives the highest satisfaction & most agreeable sensation that I know in life."[143] Those to whom he wrote had reason to be grateful that he felt that way.

In his last years, Magellan suffered several ailments that affected him severely, especially pains in the head and eyes that inhibited his reading

139 J. Bernoulli, *Recueil pour les astronomes*, vol. 3 (1776), p. 298.

140 Magellan to J.-A. Euler, 7 October 1777. See also R. W. Home, "Points or Knobs: Lightning Rods and the Basis of Decision Making in Late Eighteenth Century British Science", in P. Heering, O. Hochadel and D. Rhees, eds, *Playing with Fire: Histories of the Lightning Rod* (Philadelphia, 2009), pp. 97-120.

141 Magellan, Last Will and Testament, 9 August 1788.

142 Magellan to Sanches, 28 January 1777.

143 Magellan to Sanches, 31 October 1776.

and writing, sometimes for months at a time. In early 1789, he suffered a seizure of some kind, perhaps a stroke, and seems to have been a helpless invalid thereafter. By December of that year, it was reported that he "avait perdu complétement l'esprit".[144] He died on 7 February 1790 at the age of 67 years. According to his obituarist in the *Gentleman's Magazine*, "He glided gently out of life, resigned and thankful; and, in comparing his exit with others, we may say

> 'Omnibus est eadem Lethi via;
> Non tamen unus est vitæ cunctis
> Exitiique modus!'"[145]

The correspondence

There is no central cache of Magellan's letters and papers. While it is clear from comments he makes in surviving letters that he kept extensive files of the correspondence he received and also copies of the letters he wrote – the latter being produced from the early 1780s onwards using the copying machine he acquired from its inventor, James Watt – very little of this material has survived. Nor have any other papers of his, or the notebooks to which he regularly referred in which he recorded the details of the books, instruments and other items that he shipped abroad and of the financial arrangements involved in his doing so. Any papers that he had relating to his early life, along with such correspondence as had accumulated up to that time, appear to have been destroyed in the fire that consumed his lodgings in December 1772. What happened to the papers and letters that he accumulated after that, however, is much less clear. The only known remnant is a collection of 59 letters written to Magellan that is now housed in the Bodleian Library in Oxford, that reached the library from the collection of the Oxford astronomer and scientific antiquarian Stephen Peter Rigaud (1774-1839). It is not known how Rigaud acquired these. Remarkably, the 59 letters are from no fewer than 49 different correspondents, which suggests that the collection was not separated from the rest of Magellan's

144 Éd. Mailly, *Histoire de l'Académie Impériale et Royale des Sciences et Belles-Lettres de Bruxelles* (Brussels, 1883), p. 562.

145 "Death's road is the same for all; nevertheless there is no single mode for all of living and departing" (Maximianus, Elegy 6, lines 5-6). Though printed on three lines in the obituary, the verse was written as a couplet.

papers as a unit but was deliberately formed by someone – probably not Rigaud, since he was only 16 years old when Magellan died – carefully selecting from among Magellan's papers single letters (or just occasionally more than one) from different files, as if to represent in some sense the range of Magellan's correspondence.

Almost all the other manuscript letters that we have found are letters written by Magellan or drafts of letters to him, filed with the papers of the person or institution with whom or with which he was corresponding. Sometimes, however, such papers, with Magellan's letters among them, are now located at some remove from where the person concerned was based at the time. For example, the correspondence with J.H. van Swinden, who was based at Franeker and later in Amsterdam, is now with van Swinden's papers in Leiden; the letters to Johann III Bernoulli, who was based in Berlin, are now in the Bernoulli Archive in Basel; and the letters to Ribeiro Sanches, who was living in retirement in Paris at the time Magellan was writing to him, are now in Vienna, having been purchased in Paris along with other papers of Sanches's in the early nineteenth century to form part of the newly founded autograph collection of Austria's National Library. Among some papers of France's one-time Controller-General of Finances, Turgot, that somehow came into the possession of the nineteenth-century Portuguese journalist and historian Sousa Viterbo (1845-1910) were ten letters that Magellan wrote to Turgot that are now among Sousa Viterbo's papers at the Museu Nacional de Arte Antiga in Lisbon. A number of letters held by the American Philosophical Society were acquired by the Society when they came up for sale in France some years ago, the Society having a particular interest in Magellan as a result of his endowing its Magellanic Premium.

The present edition is the result of systematic searching in libraries and archives around the world. We here publish the text of all letters that we have found, written either by Magellan or to him, and whether what has survived is an actual letter or a draft retained by the writer. (Among the latter is at least one item probably drafted by Magellan as a *jeu d'esprit* and never sent.) Where the original letter has not been found but a published version has come down to us, this has also been included. So, too, have a number of other documents that we have located, not strictly letters, relating to Magellan's activities, such as his last will and testament, the letters patent granted to him and Henry Pyefinch for the Aerostathmion and the specification they submitted for this, and sundry bills of lading and bank documents. A number of the letters have been published previously: unless otherwise indicated, these have been transcribed afresh from the originals for this edition.

Acknowledgements

Publication of this work has been made possible by a generous grant from the Fundação para o Estudo e Desenvolvimento da Região de Aveiro (FEDRAVE). In Australia, research was supported by the University of Melbourne and the Australian Research Council. In Portugal, visits to local and foreign institutions benefitted from the support of the Portuguese Foundation for Science and Technology, the Calouste Gulbenkian Foundation, and the University of Aveiro Research Centre "Didactics and Technology in Teacher Education".

A number of colleagues at the University of Melbourne – Dino Bressan, Gianluca Caputo, John Hajek, Peter McPhee, Colin Nettelbeck – helped resolve difficulties in transcribing or translating uncertain passages in various languages. Roger Scott, in particular, has been patience personified in identifying and translating passages written in Latin and Greek. Dr Carlos Mora from the University of Aveiro also helped with translations from Latin. Tom Darragh from the Melbourne Museum provided translations of material in German.

Scholars too numerous to acknowledge individually, from many parts of the world, drew our attention to relevant materials – including in some cases letters we had not previously known about – or helped clarify matters discussed in letters. The comprehensiveness of our edition, the accuracy of our transcriptions of unfamiliar terms and the richness of our annotations have benefitted greatly from such assistance. Anita McConnell, Gerard Turner and Robert Anderson were especially helpful in this regard and more generally in their encouragement of our efforts.

Librarians and archivists from many institutions around the world, also too numerous to acknowledge individually, have facilitated access to documents for the purpose of preparing this edition, and have provided photocopies, microfilm or scanned images of the documents and in many cases also advice about difficult readings. We acknowledge the following for permitting publication of documents in their care: the late Dr W.A. Smeaton; Mr and Mrs N. Arnold-Forster; Academia das Ciências de Lisboa; American Philosophical Society Library, Philadelphia; Annenberg Rare Book & Manuscript Library, University of Pennsylvania, Philadelphia; Archief Van Arenberg, Enghien/Edingen; Archives de l'Académie des Sciences, Paris; Archives départementales du Bas-Rhin, Strasbourg; Archives Générales du Royaume, Brussels; Archives Nationales, Paris; Archivio di Stato di Milano, Milan; Archivo General de Simancas; Arquivo Histórico do Museu Bocage, Universidade de Lisboa, Lisbon; Arquivo Histórico Ultramarino, Lisbon; Arquivo Nacional da Torre do Tombo, Lisbon; Biblioteca da Ajuda, Lisbon; Biblioteca do Museu Nacional de Arte Antiga, Lisbon; Biblioteca Pública Municipal do Porto; Bibliothèque de Genève, Geneva; Bibliothèque de la Faculté de Médecine, Paris; Bibliothèque de l'Institut de France, Paris; Bibliothèque Nationale, Paris; Birmingham Central Library; Bodleian Library, Oxford; British Library, London; County Record Office, Hertfordshire; Glasgow University Library; Istituto Lombardo, Milan; Leiden University Library; Mitchell Library, State Library of New South Wales, Sydney; Museo Naval, Madrid; National Archives, London; Niedersächsische Staats- und Universitätsbibliothek Göttingen; Öffentliche Bibliothek der Universität Basel; Österreichische Nationalbibliothek, Vienna; Rijksarchief Noord-Holland, Haarlem; Royal Astronomical Society, London; Royal Library, Copenhagen; Royal Society of Arts, London; Royal Society of London; Royal Society of Medicine, London; Royal Swedish Academy of Sciences, Stockholm; Smithsonian Institution Libraries, Washington; Staatsbibliothek Preussischer Kulturbesitz, Berlin; St Petersburg Branch, Archives Department, Russian Academy of Sciences; Tartu University Library; University Library, Université de Liège; University Library, Uppsala; Wedgwood Museum, Stoke-on-Trent; Wellcome Institute Library, London; Westminster City Archives, London; York City Archives, York.

We are also grateful for information and assistance provided by many other institutions around the world. We especially acknowledge the Arquivo Distrital de Aveiro and the Arquivo Distrital de Leiria, whose help was fundamental to establishing Magellan's ancestry. The Real Academia Medica Matritense provided information about Magellan's affiliation to this academy.

Editorial Conventions

Transcriptions are made up of the following elements:

Item number entry. Each item has a unique number based on the date of the item, e.g. 77.06.27a = second item of date 27 June 1777. Incomplete elements of the date are represented by '00' e.g. 69.06.00 = item of date June 1769. Item numbers are used in cross-references in the footnotes.

Correspondent entry. Provides information on the individual(s) to whom the item is addressed or from whom it comes, e.g. *To Louis-Henri Duchesne*, or gives a descriptive title e.g. *Bill of Exchange.*

Location entry. The archival source of the item, e.g. Öffentliche Bibliothek der Universität Basel, MS L I a 709, f. 72. Where the item is known only in a published form, details of the publication are given in a footnote.

Text. Letters are transcribed into a standardized format consisting of address and date, salutation, letter paragraphs, valediction, addressee and postscripts. Marginal notes are interpolated in the body of the letter following the author's marks or, in the absence of such marks, are treated as postscripts and placed at the end of the letter.

Spelling and punctuation follow the MS. Where letters are ill-formed, the writer is given the benefit of the doubt if that will yield a standard spelling. Corrections such as deletions and overwritings are usually noted only where they have a number of possible readings. Editorial comments on the text of the MS are given in footnotes.

Square brackets are used to perform a variety of editorial functions but mean *uncertain reading* unless otherwise specified in a footnote such as '*editorial addition*' or '*illegible* – obscured by binding tape'.

Items in languages other than English or French are followed by translations into English. The translations are intended to assist readers to gain an understanding of the contents; we have not standardized the translating style of the various editors. Physical information about the MS may be provided in a footnote at the end of the correspondent entry.

Where letters are accompanied by other documents such as invoices or receipts, transcriptions of these follow the text of the letter as appendices, separated from each other and from the text of the letter by ruled lines.

Footnotes. Wherever possible, persons mentioned in the text are identified in footnotes. References to sources used by the editors are also given in footnotes, while full citations are provided in the Bibliography.

Illustrations

Calendar of Correspondence

Volume I

59.05.21 from Jean Chevalier
59.05.26 to Charles Messier
60.01.08 from Jean Chevalier
62.04.20 to Ribeiro Sanches
63.12.10 to Luiz Soares de Lima Brandão
64.03.11 from João Mendes Sachetti Barbosa
65.03.28 Letters Patent
65.06.06 to the Earl of Morton
65.07.24 Patent Specification
65.09.25 to John Canton
65.09.26 to John Canton
65.11.05 to Thomas Birch
65.11.08 to James Ferguson
66.01.09 to Emanuel Mendes da Costa
66.05.14 to Matthew Maty
68.05.16 to John Nourse
69.06.00 from Charles Messier
69.07.17 to Nevil Maskelyne
69.11.02 from Ribeiro Sanches
69.11.21 to Ribeiro Sanches
69.12.26 to Louis-Henri Duchesne
70.00.00 to [Louis-Henri Duchesne]
70.05.09 to Samuel More
70.08.18 to Louis-Henri Duchesne
70.10.02 to Louis-Henri Duchesne
70.10.29 from Abraham Gotthelf Kästner
70.11.30 to Louis-Henri Duchesne
70.12.17 from Raymond-François Bourriot
70.12.28 to [Louis-Henri Duchesne]
71.01.00 from Charles Messier
71.01.12 from Charles Messier
71.01.25 to Johann III Bernoulli
71.02.01 to Louis-Henri Duchesne
71.03.05 to Louis-Henri Duchesne
71.03.15 to Johann III Bernoulli
71.03.22 to Louis-Henri Duchesne
71.04.13 to Matthew Maty
71.04.16 to Johann III Bernoulli
71.04.19 to Louis-Henri Duchesne
71.04 23 to Louis-Henri Duchesne
71.05.04 to Pierre Macquer
71.05.12 from Charles Messier
71.05 30 to Louis-Henri Duchesne
71.06.28 to Louis-Henri Duchesne
71.07.04 from Nathanael Wolf
71.08.02 to Louis-Henri Duchesne
71.08.20 to Johann III Bernoulli
71.08.20a to Pierre Macquer
71.09.04 to Jean-Paul Grandjean de Fouchy
71.10.01 to Louis-Henri Duchesne
71.10.05 to Pierre Macquer
71.10.25 to Pierre Macquer
71.10.29 to Johann III Bernoulli
71.11.14 to Matthew Maty
71.12.14 to Joseph Banks
72.00.00 to Nathaniel Pigott
72.01.17 to Louis-Henri Duchesne
72.01.28 to Johann III Bernoulli
72.03.13 to Louis-Henri Duchesne
72.03.17 to Jacques-André Mallet
72.03.20 to Pierre Macquer
72.03.20a to Louis-Henri Duchesne
72.03.21 to Jesse Ramsden
72.03.27 to Johann III Bernoulli
72.03.30 from Jacques-André Mallet
72.03.31 to Louis-Henri Duchesne
72.04.07 to Johann III Bernoulli
72.04.10 to Louis-Henri Duchesne
72.04.21 to Johann III Bernoulli
72.04.24 to Domenico Vandelli
72.05.04 to Louis-Henri Duchesne
72.05.26 to Louis-Henri Duchesne
72.06.01 to Louis-Henri Duchesne
72.06.09 to Louis-Henri Duchesne
72.07.05 to Trudaine de Montigny
72.07.07 to Trudaine de Montigny
72.07.07a to Louis-Henri Duchesne
72.07.21 to Johann III Bernoulli
72.08.19 to Arthur Young
72.09.04 to the Royal Society of London
72.10.06 to Johann III Bernoulli
72.10 09 to Arthur Young
72.10 16 to Louis-Henri Duchesne
72.10.26 to Johann III Bernoulli
72.10.27 to Louis-Henri Duchesne
72.12.03 to Jacques-André Mallet

72.12.07 to Louis-Henri Duchesne
73.00.00 to Matthew Boulton
73.02.17 from Jacques-André Mallet
73.03.05 to Louis-Henri Duchesne
73.03.23 to Pierre Macquer
73.03.23a to Nathaniel Pigott
73.03.23b to Johann III Bernoulli
73.04.04 to Pierre Macquer
73.04.29 to Johann III Bernoulli
73.04.30 to Nathaniel Pigott
73.05.11 to Louis-Henri Duchesne
73.05.14 to Pierre Macquer
73.06.01 to Nathaniel Pigott
73.06.08 to Louis-Henri Duchesne
73.06.25 to Nathaniel Pigott
73.06.25a to Johann III Bernoulli
73.07.23 to Louis-Henri Duchesne
73.07.27 to Louis-Henri Duchesne
73.08.03 to Joannes Franciscus Thijsbaert
73.08.03a to Louis-Henri Duchesne
73.08.27 to Louis-Henri Duchesne
73.09.09 to Matthew Boulton
73.10.07 to John Fothergill
73.10.12 to Nathaniel Pigott
73.10.29 to Nathaniel Pigott
73.11.02 to Louis-Henri Duchesne
73.11.05 to Joannes Franciscus Thijsbaert
73.11.12 to Louis-Henri Duchesne
73.11.27 to Johann III Bernoulli
73.12.14 to Louis-Henri Duchesne
73.12.17 from Jacques-André Mallet
73.12.22 to Matthew Boulton
73.12.27 from Nathaniel Pigott
73.12.30 to Johann III Bernoulli
74.01.03 to Jacques-André Mallet
74.01.07 to Louis-Henri Duchesne
74.01.18 to Nathaniel Pigott
74.01.18a to Louis-Henri Duchesne
74.01.24 from Jacques-André Mallet
74.01.25 to Nathaniel Pigott
74.01.25a to Joannes Franciscus Thijsbaert
74.01.25b to Gabriel de Bory
74.01.25c to François Rozier
74.01.25d to Louis-Henri Duchesne
74.02.00 from Jacques-André Mallet
74.02.11 to Antoine-Laurent Lavoisier
74.02.22 to Jacques-André Mallet
74.03.00 to Charles Messier
74.03.11 from Jacques-André Mallet
74.03.22 to Nathaniel Pigott
74.04.15 to Jacques-André Mallet
74.04.29 to Johann III Bernoulli
74.05.16 to Johann III Bernoulli
74.05.16a to Antoine-Laurent Lavoisier
74.06.03 to Jacques-André Mallet
74.08.02 to Benjamin Franklin
74.09.18 to Johann Castillon
74.10.08 to Joseph Banks
74.11.04 from Jacques-André Mallet
74.11.24 to Antoine-Laurent Lavoisier
74.11.28 to Jacques-André Mallet
75.00.00 to Joannes FranciscusThijsbaert
75.01.02 to Nathaniel Pigott
75.02.08 from Jacques-André Mallet
75.02.22 to Samuel More
75.02.26 to Nathaniel Pigott
75.03.05 to Jacques-André Mallet
75.03.15 from Jacques-André Mallet
75.03.31 to Jacques-André Mallet
75.03.31a to Nathaniel Pigott
75.04.01 to Pierre Macquer
75.04.07 from Jacques-André Mallet
75.04.08 to Nathaniel Pigott
75.04.10 to Jacques-André Mallet
75.04.17 from Jacques-André Mallet
75.05.24 to Jacques-André Mallet
75.06.23 to the Duke Charles-Marie-Raymond d'Arenberg
75.06.23a to Ribeiro Sanches
75.07.01 to Antoine-Laurent Lavoisier
75.07.07 to Ribeiro Sanches
75.07.14 to Antoine-Laurent Lavoisier
75.07.14a to Antoine-Laurent Lavoisier
75.07.14b to the Duke Charles-Marie-Raymond d'Arenberg
75.07.24 to Ribeiro Sanches
75.08.00 to [François Rozier]
75.08.25 to Ribeiro Sanches
75.09.00 from Jacques-André Mallet
75.09.01 to Jacques-André Mallet
75.09.06 to Benjamin Wilson
75.09.16 to Ribeiro Sanches
75.09.29 to Johann III Bernoulli
75.09.30 to the Duke Charles-Marie-Raymond d'Arenberg
75.10.13 to Jacques-André Mallet
75.10.27 to Ribeiro Sanches
75.11.13 to Antoine-Laurent Lavoisier
75.11.13a from Jacques-André Mallet
75.11.17 to Ribeiro Sanches

75 11.20 to Joseph Priestley
75 11.24 to Pierre Macquer
75 12.00 to Antoine-Laurent Lavoisier
75 12.12 to Ribeiro Sanches
75 12.15 to Marie Lavoisier
75 12.19 to the Duke Charles-Marie-Raymond d'Arenberg
76 00.00 to Joseph Banks
76 01.01 to Ribeiro Sanches
76 01.04 to Charles Messier
76 01.16 to Antoine-Laurent Lavoisier
76 01.18 to the Duke Charles-Marie-Raymond d'Arenberg
76 01.18a to Ribeiro Sanches
76 01.26 to Joannes Franciscus Thijsbaert
76 02.02 to Ribeiro Sanches
76 02.23 to Ribeiro Sanches
76 02.27 to Ribeiro Sanches
76 04.02 to Antoine-Laurent Lavoisier
76 04.07 from Nathanael Wolf
76 04.07a from Nathanael Wolf
76 04.16 to Ribeiro Sanches
76 04.25 to Alessandro Volta
76 05.21 to the Duke Charles-Marie-Raymond d'Arenberg
76 05.29 from Charles Messier
76 06.16 to Ribeiro Sanches
76 06.21 to Louis-Henri Duchesne
76 07.12 to Ribeiro Sanches
76 07.23 to the Duke Charles-Marie-Raymond d'Arenberg
76 07.29 to Jacques-André Mallet
76 08.06 to Johann III Bernoulli
76 08.09 to Joannes Franciscus Thijsbaert
76 08.13 to Horace Bénédict de Saussure
76 09.05 from Jacques-André Mallet
76 09.10 to Antoine-Laurent Lavoisier
76 09.12 to John Walsh
76 09.13 to Johann III Bernoulli
76 09.13a to Ribeiro Sanches
76 09.17 to Johann Albrecht Euler
76 09.27 to the Duke Charles-Marie-Raymond d'Arenberg
76 10.11 to the Duke Charles-Marie-Raymond d'Arenberg
76 10.12 to [Charles Messier]
76 10.12a to Johann III Bernoulli
76 10.15 to [...] Wingate
76 10.31 to Ribeiro Sanches
76 11.15 to Ribeiro Sanches
76.11.27 to the Duke Charles-Marie-Raymond d'Arenberg
76.11.30 to Joseph Priestley
76.12.17 to Ribeiro Sanches
76.12.22 to the Duke Charles-Marie-Raymond d'Arenberg
77.00.00 to Nathaniel Pigott
77.01.28 to Ribeiro Sanches
77.01.31 to [Gabriel de Bory]
77.02.00 to Gabriel de Bory
77.02.03 to Nathaniel Pigott
77.02.04 to the Duke Charles-Marie-Raymond d'Arenberg
77.02.18 to Ribeiro Sanches
77.02.25 to Joannes Franciscus Thijsbaert
77.03.00 to Johann Castillon
77.03.11 to the Marquis de Condorcet
77.03.11a from Johann Castillon
77.03.14 to Johann Albrecht Euler
77.03.14a to Johann III Bernoulli
77.03.21 to Ribeiro Sanches
77.04.00 to Joseph Priestley
77.04.15 to Ribeiro Sanches
77.04.22 to Louis-Henri Duchesne
77.05.20 to Ribeiro Sanches
77.06.20 to Jean-Baptiste Le Roy
77.06.27 to Johann Albrecht Euler
77.06.27a to Ribeiro Sanches
77.07.04 to Ribeiro Sanches
77.07.11 to Ribeiro Sanches
77.07.21 to the Marquis de Condorcet
77.08.05 to Nathaniel Pigott
77.08.08 to Nathaniel Pigott
77.08.15 to Ribeiro Sanches
77.08.28 to Nathaniel Pigott
77.09.11 to Nathaniel Pigott
77.09.21 to Ribeiro Sanches
77.09.30 to Ribeiro Sanches
77.10.07 to Johann Albrecht Euler
77.10.17 to the Marquis de Condorcet
77.11.07 from Jan Henrik van Swinden
77.11.14 to Johann Albrecht Euler
77.11.14a to Ribeiro Sanches
77.12.08 to James Watt
78.01.13 to Ribeiro Sanches
78.01.22 to the President and Council, Royal Society of London
78.01.30 to Johann III Bernoulli
78.01.30a to James Watt
78.02.02 from James Watt

78.02.10 to Ribeiro Sanches
78.02.20 to [...]
78.02.28 to Matthew Boulton
78.03.03 to Matthew Boulton and James Watt
78.03.12 to James Watt
78.03.13 to Johann Albrecht Euler
78.03.13a to Ribeiro Sanches
78.03.21 to James Watt
78.03.23 to the President and Council, Royal Society of London
78.03.23a to Matthew Boulton
78.03.24 from James Watt
78.03.27 to Johann III Bernoulli
78.04.17 to Matthew Boulton
78.04.17a to the Comte d'Hérouville
78.04.22 from Matthew Boulton
78.04.24 to the Comte d'Hérouville
78.04.28 to Ribeiro Sanches
78.05.01 to the Comte d'Hérouville
78.05.02 from Matthew Boulton and James Watt
78.05.05 to Jan Henrik van Swinden
78.05.17 from the Comte d'Hérouville
78.05.19 to Ribeiro Sanches
78.05.22 from the Boulton & Watt company
78.05.22a from Jan Henrik van Swinden
78.05.29 to Matthew Boulton
78.06.00 to Gabriel de Bory
78.06.20 to Benjamin Franklin
78.06.20a to Ribeiro Sanches
78.06.30 from Nathanael Wolf
78.07.14 to Luiz Pinto de Sousa Coutinho
78.07.17 to the Comte d'Hérouville
78.07.17a to the Marquis de Puységur
78.07.17b to Anne-Robert-Jacques Turgot
78.07.21 to Matthew Boulton
78.08.18 to [François Rozier]
78.08.22 from Matthew Boulton
78.08.26 from John Walsh
78.09.01 to the Comte d'Hérouville
78.09.01a to Ribeiro Sanches
78.09.08 to Johann III Bernoulli
78.09.08a to Ribeiro Sanches
78.10.09 to Ribeiro Sanches
78.10.12 from Matthew Boulton
78.10.13 to Luiz Pinto de Sousa Coutinho
78.10.15 to Johann III Bernoulli
78.10.16 to Matthew Boulton
78.10.16a from Matthew Boulton
78.10.23 to Comte Joseph Velasco del Pico
78.10.23a to the Duke Louis-Engelbert d'Arenberg
78.10.27 from Johann Albrecht Euler
78.10.27a Bill of Exchange
78.10.27b to Jan Henrik van Swinden
78.10.27c to Alessandro Volta
78.10.30 to Ribeiro Sanches
78.11.04 to Ribeiro Sanches
78.11.06 to Matthew Boulton
78.11.10 to Comte Joseph Velasco del Pico
78.11.12 to Ribeiro Sanches
78.11.13 to Anne-Robert-Jacques Turgot
78.11.16 to the Comte d'Hérouville
78.11.16a to Anne-Robert-Jacques Turgot
78.11.20 to Anne-Robert-Jacques Turgot
78.11.24 to Comte Joseph Velasco del Pico
78.12.02 from Casimiro Gomez de Ortega
78.12.04 to James Watt and Matthew Boulton
78.12.08 to Ribeiro Sanches
78.12.09 to Johann Albrecht Euler
78.12.09a to the President and Members, St Petersburg Academy of Sciences
78.12.12 to Joseph Banks
78.12.15 to Comte Joseph Velasco del Pico
78.12.28 to Joseph Banks
78.12.29 to the Duke Louis-Engelbert d'Arenberg
79.01.00 to Giovanni Fabbroni
79.01.05 to Anne-Robert-Jacques Turgot
79.01.05a to Ribeiro Sanches
79.01.13 to Matthew Boulton and James Watt
79.01.13a to the Comte d'Hérouville
79.01.15 from Jan Henrik van Swinden
79.01.15a to Comte Joseph Velasco del Pico
79.01.19 to the Comte d'Hérouville
79.01.21 to James Watt
79.01.26 to Anne-Robert-Jacques Turgot
79.01.29 to Nathaniel Pigott
79.01.29a to Joseph Banks
79.02.19 to Anne-Robert-Jacques Turgot
79.02.23 to Anne-Robert-Jacques Turgot
79.02.26 to Ribeiro Sanches
79.02.27 to Joseph Priestley
79.03.02 to the Duke Louis-Engelbert d'Arenberg
79.03.02a to Comte Joseph Velasco del Pico
79.03.06 to Joseph Banks
79.03.09 to Johann III Bernoulli
79.03.11 to Ribeiro Sanches

79.03.23 to Jan Henrik van Swinden
79.04.09 to Alessandro Volta
79.04.13 to Anne-Robert-Jacques Turgot
79.04.13a to Ribeiro Sanches
79.04.29 to James Watt
79.05.02 from Jan Henrik van Swinden

Volume II

79.05.13 to Benjamin Franklin
79.05.21 to Anne-Robert-Jacques Turgot
79.05.21a to Ribeiro Sanches
79.05.24 to Ribeiro Sanches
79.06.15 to Ribeiro Sanches
79.06.22 from the Boulton & Fothergill company
79.06.22a to Benjamin Franklin
79.07.02 to Luiz Pinto de Sousa Coutinho
79.07.02a to Luiz Pinto de Sousa Coutinho
79.08.05 to Alessandro Volta
79.08.06 to Ribeiro Sanches
79.08.17 from the Boulton & Fothergill company
79.09.04 to Ribeiro Sanches
79.09.14 to Alessandro Volta
79.09.14a to Ribeiro Sanches
79.09.28 to Alessandro Volta
79.10.19 to Ribeiro Sanches
79.11.09 to Johann III Bernoulli
79.11.09a to Marsilio Landriani
79.11.26 from Jan Henrik van Swinden
79.12.07 from Marsilio Landriani
79.12.21 from Charles Messier
79.12.28 to Johann III Bernoulli
79.12.28a to Alessandro Volta
80.01.14 to Ribeiro Sanches
80.02.01 from Charles Messier
80.02.04 to Giovanni Fabbroni
80.02.22 to James Watt
80.03.01 from James Watt
80.03.01a to the Society of Arts
80.03.07 to James Watt
80.03.09 from James Watt
80.03.13 to Jan Henrik van Swinden
80.03.17 to James Watt
80.03.20 from James Watt
80.03.22 from the Boulton & Fothergill company
80.03.27 to James Watt
80.03.29 from James Watt
80.04.03 to John Hodges
80.04.13 to Georg Forster
80.04.14 to [François Rozier]
80.04.17 to Matthew Boulton
80.04.18 to Johann III Bernoulli
80.04.22 to James Watt
80.05.03 from James Watt
80.05.05 to Matthew Boulton
80.05.16 to Ribeiro Sanches
80.05.22 from the Count von Brühl
80.06.09 to Ribeiro Sanches
80.07.00 to [François Rozier]
80.07.28 from Jan Henrik van Swinden
80.07.29 to Ribeiro Sanches
80.08.08 to James Watt
80.08.16 to Ribeiro Sanches
80.08.29 to Torbern Bergman
80.08.29a to Martinho de Mello e Castro
80.08.29b to Carl Wilhelm Scheele
80.09.01 from James Watt
80.09.16 to James Watt
80.09.20 from James Watt
80.10.06 to James Watt
80.10.16 from James Watt
80.11.21 to Alessandro Volta
80.12.08 to Benjamin Franklin
80.12.14 from Jérôme de la Lande
80.12.20 to Giovanni Fabbroni
80.12.20a to James Watt
80.12.20b from Jan Henrik van Swinden
80.12.21 from James Watt
80.12.29 to Alessandro Volta
80.12.31 to James Watt
81.01.00 to François Rozier
81.01.02 to Alessandro Volta
81.01.13 to James Watt
81.02.12 from Marsilio Landriani
81.02.26 from Félix Vicq d'Azir
81.03.00 to the Editor of the *Critical Review*
81.03.13 from James Watt
81.03.29 Declaration
81.04.00 to [Jean-André Mongez]
81.04.26 from Jean-Cyrille Rigaud
81.04.30 to Jozé Correa da Serra
81.05.00 to Jean-André Mongez
81.06.09 to [the Royal Academy of Sciences of Lisbon]
81.07.08 from [...] Hoffmann
81.07.17 to Ribeiro Sanches

81.08.17 to Johann III Bernoulli
81.08.27 to Ayres de Sá e Mello
81.08.28 to James Watt
81.10.15 from Benjamin Vulliamy
81.10.20 from Edme-Sébastien Jeaurat
81.11.13 to Johann Albrecht Euler
81.11.17 to Jan Henrik van Swinden
81.11.23 to Johann III Bernoulli
81.11.26 from Jan Henrik van Swinden
81.11.30 to Ribeiro Sanches
82.00.00 to Thomas Henry
82.00.00a from Pierre Levêque
82.03.05 from Edward Nairne
82.03.28 to Gabriel de Bory
82.04.13 to Benjamin Franklin
82.04.17 to Ayres de Sá e Mello
82.04.20 to Ayres de Sá e Mello
82.05.00 from Thomas Henry
82.06.00 from Franz Carl Achard
82.07.16 to Johann III Bernoulli
82.07.23 to Ribeiro Sanches
82.08.09 from Jerome Allen
82.08.16 to Ribeiro Sanches
82.08.27 to William Herschel
82.09.20 to James Watt
82.09.24 to Luiz Pinto de Sousa Coutinho
82.10.00 from James Watt
82.10.01 to Luiz Pinto de Sousa Coutinho
82.10.05 to James Watt
82.10.11 to Ribeiro Sanches
82.10.12 to James Watt
82.10.15 to Johann III Bernoulli
82.10.18 to Alessandro Volta
82.11.00 to [...]
82.11.05 to Ribeiro Sanches
82.11.06 to Ribeiro Sanches
82.11.16 to Johann III Bernoulli
82.11.26 to Torbern Bergman
82.12.17 to Ribeiro Sanches
82.12.17a to Martinho de Mello e Castro
82.12.31 to Alessandro Volta
82.12.31a to Ayres de Sá e Mello
83.01.03 to Johann III Bernoulli
83.01.14 to Ayres de Sá e Mello
83.01.15 from [Adair Crawford]
83.01.24 from Gustav von Engestrom
83.02.07 to [...] Grand
83.02.21 to Ribeiro Sanches
83.03.00 to [Jean-André Mongez]
83.03.08 to James Watt
83.03.10 to Johann Albrecht Euler
83.03.11 to Matthew Boulton and James Watt
83.03.11a to Matthew Boulton
83.04.04 to Johann Albrecht Euler
83.04.08 from the Duc de la Rochefoucauld
83.04.18 to Ribeiro Sanches
83.04.26 to Johann Albrecht Euler
83.05.00 to Marsilio Landriani
83.05.02 to Arthur Young
83.05.06 to Gabriel de Bory
83.05.18 to Johann III Bernoulli
83.05.18a to Johann Elert Bode
83.05.22 from James Watt
83.05.30 to Johann Albrecht Euler
83.05.30a to Ribeiro Sanches
83.06.17 to the American Philosophical Society
83.06.17a to Ribeiro Sanches
83.06.24 to Ribeiro Sanches
83.06.26 from Étienne-Alexandre-Jacques Anisson-Duperron
83.07.08 to Ribeiro Sanches
83.07.15 from Johann Albrecht Euler
83.07.22 to Ribeiro Sanches
83.07.23 to William Herschel
83.07.29 from Marsilio Landriani
83.08.05 to Martinho de Mello e Castro
83.08.08 to Alessandro Volta
83.08.12 to Ribeiro Sanches
83.08.13 to Arthur Young
83.08.15 to Johann Albrecht Euler
83.08.19 to Johann III Bernoulli
83.08.29 to Anders Lexell
83.08.31 to Martinho de Mello e Castro
83.09.01 to Martinus van Marum
83.09.15 from Johann Albrecht Euler
83.09.25 to Joannes Franciscus Thijsbaert
83.09.26 to Johann III Bernoulli
83.09.30 from Anders Lexell
83.10.03 from Charles Messier
83.10.04 from Count Maurice de Benyowski
83.10.05 to Martinus van Marum
83.10.14 to Martinho de Mello e Castro
83.10.18 from Thomas Bond
83.10.21 to Jacques Reinbold Spielmann
83.10.24 to Johann Albrecht Euler
83.10.28 from Alessandro Volta
83.11.11 to Joseph Banks
83.11.17 to Armand Baudouin de Guémadeuc
83.11.22 from Edward Pigott

83.11.24 to Johann Albrecht Euler
83.12.01 to William Withering
83.12.02 to William Withering
83.12.18 from James Watt
83.12.19 to Johann III Bernoulli
83.12.22 to James Watt
83.12.28 from Edme-Sébastien Jeaurat
83.12.30 to Martinho de Mello e Castro
84.00.00 to [the Royal Acadmy of Sciences of Lisbon]
84.00.00a from Joaquim d'Assumpção Velho
84.01.01 from James Watt
84.01.12 from Charles Messier
84.01.14 to Ayres de Sá e Mello
84.01.14a to Martinho de Mello e Castro
84.01.30 to [...] Grand
84.02.06 to Johann Albrecht Euler
84.02.17 to Ayres de Sá e Mello
84.02.24 to James Watt
84.02.28 Notes on Ribeiro Sanches
84.03.21 to Jozé Correa da Serra
84.03.23 to Johann III Bernoulli
84.03.23a to James Watt
84.03.25 to Arthur Young
84.03.25a from Jacques Pierre Brissot de Warville
84.05.11 to Giuseppe Morozzo
84.05.18 from William Herschel
84.05.24 from Martinus van Marum
84.05.28 to Johann Albrecht Euler
84.06.00 to Johann Elert Bode
84.06.01 from William Nicholson
84.06.15 to Matthew Boulton and James Watt
84.06.18 to Johann Albrecht Euler
84.06.18a to Johann III Bernoulli
84.06.22 to Alessandro Volta
84.06.27 from James Watt
84.06.27a from James Watt
84.06.29 to Johann Albrecht Euler
84.07.08 to Martinus van Marum
84.07.08a to Christianus van der Aa
84.07.17 to Martinus van Marum
84.07.20 to Johann Albrecht Euler
84.08.00 to Bernardo del Campo
84.08.03 to Johann Albrecht Euler
84.08.06 from Franz Carl Achard
84.08.10 to [Armand Baudouin de Guémadeuc]
84.08.14 to William Herschel
84.08.20 from Jean-Baptiste Le Roy
84.08.21 from Karl König
84.09.03 to James Watt
84.09.05 to Tomas Perez
84.09.06 to Bernardo del Campo
84.09.23 to Dom L. Garrellon
84.09.24 to Johann Albrecht Euler
84.09.25 to Bernardo del Campo
84.10.15 to Johann Albrecht Euler
84.10.22 to Johann Albrecht Euler
84.10.22a from Jan Henrik van Swinden
84.11.08 to William Temple Franklin
84.11.30 from Étienne-Alexandre-Jacques Anisson-Duperron
84.12.10 to Dom L. Garrellon
85.00.00 from Jean-Étienne Montucla
85.00.00a from Jean-Baptiste-Gaspard Bochart de Saron
85.01.00 from Charles Messier and Jean-Baptiste-Gaspard Bochart de Saron
85.01.01 from Armand Baudouin de Guémadeuc
85.01.03 to [...]
85.01.08 from Charles Messier
85.01.21 to Johann Elert Bode
85.01.27 to Bernardo del Campo
85.02.08 to Bernardo del Campo
85.02.08a to Bernardo del Campo
85.02.09 to [Joseph Banks]
85.02.17 from Peter & John Dollond
85.02.20 from Anne-Robert-Jacques Turgot
85.02.22 to Alessandro Volta
85.04.05 to Ayres de Sá e Mello
85.04.30 from Antoine-Laurent Lavoisier
85.05.25 to Manuel Figueiredo
85.05.31 to Johann Albrecht Euler
85.06.07 to Emanuel de Benyowski
85.06.28 to Bernardo del Campo
85.06.29 to Louis Bernard Guyton de Morveau
85.06.30 from Martinus van Marum
85.07.15 to Johann III Bernoulli
85.07.29 to Martinus van Marum
85.07.30 to Bernardo del Campo
85.08.09 to Bernardo del Campo
85.08.11 to William Withering
85.08.13 to Bernardo del Campo
85.09.09 to Johann Albrecht Euler
85.09.17 to the American Philosophical Society
85.09.20 to Armand Baudouin de Guémadeuc and Dom L. Garrellon

85.09.26 from Johann Albrecht Euler
85.09.29 to Luiz Pinto de Sousa Coutinho
85.10.07 to Martinus van Marum
85.10.11 to Alessandro Volta
85.10.22 from Johann III Bernoulli
85.10.25 to Martinus van Marum
85.11.14 to Martinho de Mello e Castro
85.11.18 to Johann Albrecht Euler
85.11.23 to Bernardo del Campo
85.11.26 to Arthur Young
85.12.15 from Count Maurice de Benyowski
85.12.20 to Thomas Bugge
85.12.24 from [François Barthélemy]
85.12.26 to George Bew
86.01.22 to Bernardo del Campo
86.01.24 from Benjamin Franklin
86.01.26 to Bernardo del Campo
86.03.09 to Martinus van Marum
86.03.12 to Bernardo del Campo
86.03.14 from Bernard Pieter van Lelyveld
86.03.31 from Bernardo del Campo
86.04.01 to William Withering
86.04.04 to Luiz Pinto de Sousa Coutinho
86.04.06 from Bernardo del Campo
86.04.12 to Tomas Perez
86.04.13 to Bernardo del Campo
86.04.13a from Bernardo del Campo
86.04.15 to Johann Albrecht Euler
86.04.30 from Roger Joseph Boscovich
86.06.02 from J. F. da Fonseca
86.06.08 from Bernard Pieter van Lelyveld
86.07.05 to Arthur Young
86.07.05a to Johann III Bernoulli
86.07.07 to the American Philosophical Society
86.07.07a to Bernardo del Campo
86.07.13 to Bernardo del Campo
86.07.19 to Arthur Young
86.07.19a to William Withering
86.07.25 from Alexandre-Guy Pingré
86.07.27 to Bernardo del Campo
86.07.29 to Josiah Wedgwood
86.07.30 to Bernardo del Campo
86.08.03 to Nevil Maskelyne
86.08.07 to Thomas Byerley
86.08.08 to Johann III Bernoulli
86.08.11 to Emanuel de Benyowski
86.08.14 to Thomas Byerley
86.09.14 from Pierre-François-André Méchain
86.09.16 to Bernardo del Campo
86.09.27 to Bernardo del Campo
86.10.03 to Luiz Pinto de Sousa Coutinho
86.10.03a to Bernardo del Campo
86.11.06 to Jan Henrik van Swinden
86.11.17 from Stepan Rumovskii
86.11.18 to Bernardo del Campo
86.11.18a to Thomas Byerley
86.11.21 to Johann III Bernoulli
86.11.21a from George Gilpin
86.11.26 to Luiz Pinto de Sousa Coutinho
86.11.30 from Jan Henrik van Swinden
86.12.06 to Bernardo del Campo
86.12.12 to Jan Henrik van Swinden
86.12.12a to Martinus van Marum
86.12.19 from Ernst Franz von Platen, Count of Hallermund
87.01.01 to Luiz Pinto de Sousa Coutinho
87.01.19 to Bernardo del Campo
87.02.02 to Thomas Byerley
87.02.05 to Thomas Byerley
87.02.14 to Thomas Byerley
87.02.18 to Bernardo del Campo
87.02.20 from Bernardo del Campo
87.02.21 to Bernardo del Campo
87.02.22 Bill of Lading
87.02.22a from Bernardo del Campo
87.03.01 to Bernardo del Campo
87.03.03 to Bernardo del Campo
87.03.08 to Bernardo del Campo
87.03.11 to Bernardo del Campo
87.03.12 to Bernardo del Campo
87.03.20 to Bernardo del Campo
87.03.20a to Thomas Byerley
87.03.26 from José Monteiro da Rocha
87.03.26a to Luiz Pinto de Sousa Coutinho
87.03.26b to Bernardo del Campo
87.04.02 from Jan Ingenhousz
87.04.08 from Bernardo del Campo
87.04.09 to Bernardo del Campo
87.04.10 from Martinus van Marum
87.04.17 to Giovanni Fabbroni
87.04.17a to Martinus van Marum
87.04.26 to Bernardo del Campo
87.04.27 from Jean-Baptiste Le Roy
87.05.24 to Martinus van Marum
87.06.18 to George, 3rd Earl Cowper
87.07.07 from the Chevalier de Chateaubourg
87.07.13 to George, 3rd Earl Cowper
87.07.14 from Bernardo del Campo
87.07.18 to Bernardo del Campo

87.07.23 from José de Magalhães de Castel-Branco
87.09.30 to Luiz Pinto de Sousa Coutinho
87.10.08 to Samuel More
87.10.09 Bill of Lading
87.10.16 to José de Magalhães de Castel-Branco
87.11.08 from Count François de Mercy-Argenteau
88.01.01 to Luiz Pinto de Sousa Coutinho
88.02.02 from Charles Messier
88.02.04 from José Espinosa y Tello
88.02.06 to Johann Elert Bode
88.03.07 to Johann III Bernoulli
88.03.31 to Luiz Pinto de Sousa Coutinho
88.04.00 from Armand Baudouin de Guémadeuc
88.04.10 from Dom L. Garrellon
88.04.18 to Bernardo del Campo
88.04.23 from Bernardo del Campo
88.04.26 to Bernardo del Campo
88.04.28 to Bernardo del Campo
88.05.13 to Bernardo del Campo
88.05.19 to Bernardo del Campo
88.05.22 from Bernardo del Campo
88.06.06 to Johann III Bernoulli
88.06.21 from Bernardo del Campo
88.06.30 to Luiz Pinto de Sousa Coutinho
88.07.13 from Jean-Dominique de Cassini
88.07.17 to Bernardo del Campo
88.07.22 to Dom L. Garrellon
88.07.22a to Johann Albrecht Euler
88.08.09 Last Will and Testament
88.08.26 to Dom L. Garrellon
88.10.01 to Cypriano Ribeiro Freire
88.11.09 from Martinus van Marum
88.12.30 to Martinus van Marum
89.01.04 to Cypriano Ribeiro Freire
89.01.27 from Martinus van Marum
89.02.03 from Josef Fidel Escolá
89.02.03a to Josef Fidel Escolá
89.02.19 from John Vaughan
89.04.21 from Carlo Alfonso Guadagni
89.06.19 to Martinus van Marum
89.11.06 from Gaetano Del Ricco
89.12.28 from Antoine-Laurent Lavoisier

Undated letters

00.00.00 from Sir George Shuckburgh
00.06.18 from [...] Herrenschneider

Correspondence

From Jean Chevalier[1] 59.05.21

Archives Nationales, Paris, MSS Marine 2 JJ 66, f. 112

Au 28. Avril on a decouvert par la deuxieme fois la Comete qu'on avoit perdû de vuë sur notre Orizont ici à Lisbone le 22 de ce mois. je l'ai observe le 29. & les jours suivants que le temps me l'eût permis, excepté le comencement de ce mois de Mai, oû le tems pluvieux & les brouillards m'ont empechè. elle a suivi presque la route que vous venez d'annoncer dans votre lettre de 30 Avril. hier [20. Mai] on la voioit encore à simple vuë sans lunete, & avoit aussi une quëue sensible: Je crois de la pouvoir observer à l'aide des Instruments assez long temps, au moins pendant grande partie du Mois de Juin. quand elle disparoitra tout à fait, j'aurai l'honneur de vous communiquer le resultat de mes observations. &ca.

1 This extract from Chevalier's letter is preserved among the correspondence of J.N. Delisle, with the heading: "Extrait d'une lettre du P. Chevalier ècrite du 21. Mai à l'abbé Magaillhens communiquée le 18 Juin 1759". Chevalier was at this period an Oratorian priest in the congregation of S. Filipe Nery in Lisbon, and was an enthusiastic astronomer. He later became Keeper of the Royal Library in Brussels.

To Charles Messier[1] 59.05.26

Archives Nationales, Paris, MSS Marine 2 JJ 66, f. 98

Monsieur

En consequence de ce que m'a ordoné l'autre jour Mr. de l'Isle,[2] j'ai l'honneur de vous Rendre en françois le mieux que je peux le sens de

1 Annotated: "reçu le 26 may 1759"; published in Rómulo de Carvalho, "Quatro cartas inéditas de João Jacinto de Magalhães", pp. 185-98 in L. Alte da Veiga et al., eds, *João Jacinto de Magalhães (John Hyacinth de Magellan): Conference on Physical Sciences in the XVIII Century, 7-10 November 1990* (Coimbra, 1994).

2 The astronomer Joseph-Nicolas Delisle (1688-1768).

59.05.26 l'article de la Lettre de Mr. le P. Chevalier du 24 Avril par Raport à ce qui concerne la Comete. Voici donc ses propres termes.

> "Il y a 15. jours que j'ai ècrit à Monsr. de Barros[3] pour lui annoncer que j'avois decouvert la comete dont vous me parlez. Mais j'aprends par la votre qu'il y a deja long tems qu'on l'observoit à Paris. Sur la fin du mois dernier [Mars] j'entendis dire à quelques gens de mer que l'on avoit vû avant le lever du Soleil une grande etoile avec de l'eclat. D'abord je fis des Recherches; mais je n'y pûs reussir à cause du tems brouillé, que le 5me. de ce mois ici, oû je l'apercûs par la 1re. fois. Les jours suivant je continuai les observations autant que le tems me l'eût permis: à l'exception de quelques uns, oû la fievre m'a pris, mais cependant je le fis observer par un de mes amis. Le 22 du courant mois fut la derniere fois qu'on l'a pû voir fort proche de l'Orisont & avec une grande latitud Meridionale. Ainsi on la pourra voir plus long tems dans les terres Meridionales. Je souhaiterois bien sçavoir les observations qu'on a fait dans ces pais la sur cette Comete & aussi le jugement que les Astronomes font la dessus sur l'identite de ce comette avec celle qu'on atendoit[4] &a."

C'est tout ce que j'ai à vous dire sur cette article. Si vous juger à propoz de lui rendre quelque Reponse la dessus: Vous aurez la bonté de l'avoir faite pour Lundi: oû j'aurai l'honneur d'aller chez vous. en atendant je vous souhaite une bone & parfaite santé: & en vous priant de presenter mes respects à Monsr. de l'Isle, je suis

Monsieur
Votre tres humble & obeissante Servr.
J.H. de Magalhaëns

Hotel des Prouvaires
aujourd'hui le Samedi

A Monsieur
Monsieur Messier
dans Le College Royal[5]

---

3 José Joaquim Soares de Barros e Vasconcellos (1721-1793), Portuguese mathematician and astronomer who studied and worked with Delisle. In 1757 he was elected a Corresponding Member of the Académie Royale des Sciences in Paris, Delisle being his designated correspondent within the Academy.

4 Edmond Halley had predicted that the comet that now bears his name would return in 1758 or 1759.

5 There follows on the sheet a draft of Delisle's reply to Chevalier.

From Jean Chevalier[1] 60.01.08

Archives Nationales, Paris, MSS Marine 2 JJ 66, f. 165

Le 7me. de ce mois (Janvier) au soir la nuit etant assez claire apres d'autres nuits prècedentes de ciel couvert par le mauvais temps, on a decouvert à Lisbone une comète que l'on voyoit assez bien sans Lunette. elle n'avoit point de queüe: mais seulement une Cheveleure assez distincte autour d'un Noyau qui étoit plus clair. Sur le 9. heures du Soir elle etoit assez elevèe sur l'horizon et située à la fin de la Constellation du Navire Argos, dirigeant sa course vers le grand Chien: on l'a pû observer pendant plusieurs heures; ce qui a fait connoitre que son mouvement etoit retrograde & assez prompt. dans la poste suivante je vous dirai ce que j'aurai observé la nuit prochaine &^{a}. je vous prie de communiquer cet avis à M^{r}. De L'Isle.

---

1 This extract from Chevalier's letter is preserved among the correspondence of J. N. Delisle, with the heading: "Article d'une lettere du P. Chevalier ècrite de Lisbone le 8 Janvier 1760 à M^{r} l'abbé Magaillans". It is annotated: "communiquée le 2 fevrier 1760".

To Ribeiro Sanches 62.04.20

Österreichische Nationalbibliothek, Vienna
A. N. Ribeiro Sanches Korrespondenz, codex 12714, ff. 523-523v

20. de Abril 1762

S^{or}. D^{or}. Ant^{o}. Rib^{o}. S[...]

Am^{o}. e S^{or}. do C. Recebi com summo gosto a de V. M. de 29 do passado, sem q a cuberta e lacre o diminuisse, nem aumentasse o preço do porte. Isto mostra a regularidde. do correio, q generozam^{te}. quis desprezar estas minúcias. Alegro me m^{to}. q V. M. escrevesse o q me diz ao nosso Ministro; porem quizera, q lhe falasse mais claro e em outro tom, ponderando-lhe a contradição de obrar a resp^{o}. de V. M., comparando o prez^{te}. ao passado; dizendo-lhe q a perseguição feita pelo Oliv^{a}. a V. M., de sua caza nacia; descobrindo-lhe nominatim os conciabulos do seo Secret^{o}. com Oliv^{ra}. e as suas machinaçoens; e q ninguem era tolo p^{a}. intender q Oliv^{ra}. se atrevia a atacar a V. M., sem hũ consentim^{to}. e aprovação (tacita ao menos) de Sua Illma. & este devia ser o tom da carta: mas estaremos a ver o exito desta q V. M. escrever.

62.04.20 pª. o correio espero noticia da recepção da justificação q a V.M. mandei; e q aviza e vá avizando do êxito desta sua contenda com esse casquilho.

a 14 do corrente fui falar a D. Luiz da Cunha, pª. me dar hũ passaporte pª. sahir daqui; porem se tive o desgosto de não o alcançar, dizendo-me q não podia ser na ocazião prez^{te}.; tive a satisfação de o ver afeiçoado a V.M.: elle me disse q brevem^{te}. havia de escrever a Mons^{or}. Salema pª. q V.M. fosse pago da sua pensão em sua própria caza, e concluindo = pois q pode fazer esse Olivª.? não lhe pode fazer nada. = foi bom e óptimo q o S^{r}. Alcaçova lhe tivesse falado e o tivesse disposto nesta forma. V.M. lhe deve dar os agradecim^{tos} em hũa brevissima carta, dizendo-lhe o q contra V.M. se tem armado e o perigo a q está exposto, e fale-lhe claram^{te}. nas maquinaçoens do Secretario de Salema, coligado com o Olivª. com Alvares & Cª.; pois ainda q não fosse verd^{e}., sempre se devia carregar o criado e não o amo: e toque-lhe em mim, dizendo como tãobem eu sou o alvo das furias destes patifes.

Com q meo amigo, aqui estaremos athe q os affaires publiques me dem lugar pª. sahir pª. lá, ou pª. Inglaterra, ou pª. italia, particularm^{te}. Veneza. a seo tempo avizarei.

Philoctetes não deixa de dar pena a todos os amigos, se he q os há como nós-ambos. eu não lhe posso dar a V.M. ideias mais claras do q as q tenho. o q se vio, foi: q elle sahio dessa sem ordem; q chegou aqui, onde não obteve particular audiencia do Ministro, nem foi escutado: andou á prática, athe q dezinganado retirou-se a sua caza, onde vive com a sua costumada taciturnidde, em hũa inteira solidão, dentro do seo quarto, lendo por algum livro. todos assentão q obrou como hũ doido, e q perdeo a cabeça neste negocio: q não conheceo q coiza era servir no Ministerio: e q totalm^{te}. se inganou, cuidando q havia de ser escutada a razão de hũ dezatino. disserão-me q em Pariz fora falar a Choiseul, sem estar ainda caracterizado e aprezentado á Corte, sobre as suas dezavenças com Salema: e q já quando chegou a Hespanha fora mal recebido pelo Emb^{or}. Peçanha. eu q ignorava tudo, cuidei ao principio q tudo hia bem e elle pela constancia com q se portava fazia semblante de não ter nada de funesto e obrou como philosofo, viverá como tal: e se obrou como honrado, viverá como pobre; q já hoje he ordinº. premio da virtude a pobreza. isto he só o q delle sei. elle não escreve a ninguém e ainda q escrevesse, não havia de dizer mais do q costume q^{do}. fala. eu delle nunca pude saber o fundo, e nem ainda a superficie do seo negocio. Elle he inabordavel e incomunicavel em similhantes materias e eu me acho tanto em jejum neste ponto, como se nunca lhe tivesse falado. prezentem^{te}. não sei onde se acha. eu lhe escrevi à Mafra e a Setubal, avizando-o de ter posto a carta de V.M. na mão de Ciera; mas nem tive reposta nem fez buscar a dita carta de V.M.

será bom q V.M. escreva ao d°. D^{or}. Ciera; por q suposto seja homem de 62.04.20
coração sem caroço, he sojeito q reina. Verd^{a}.m^{te}. he bom homem, incapaz de fazer mal a ninguem, mas prim°. a si próprio: digo q não se molestará por amigos. he hum bom italiano e basta mas a sua correspondencia [é] boa. ofereça-lhe os seos prestimos e q lhe dê a regra q quer se observa na correspondencia literaria [...] hontem estive com o bom Sachetti: elle me dêo esse titulo de outro milhor palito, he obra do S^{r}. seo primo de V.M.: tem coizas galantissimas. a dedicatoria he hum chef d'oeuvre: aqui sim q terá V.M. m^{to}. q notar. Vá-se preparando q há de ver lindas coizas. eu ahi lhe mando huma autorid^{e}. ou passage do tal livro, mas tem outras m^{to}. mais galantes esquipaçoens e asneiradas. avize V.M. o Doutor Sachetti, porq se o autor não mandar a V.M. hum exemplar, elle quer-lhe mandar.

torno a advertir q quando falar na sua carta das Maquinaçoens do Olivr^{a}. e do Conego não se esqueça do Alvares e do Bahia e do Xavier. he necess°. falar claro e nomear toda a canalhada, q intenta destruir os homens de bem p^{a}. prevenir os seos golpes. hũ sojeito me disse q o Alvares tinha escrito ao cirurgião-mór: porem q elle ainda não tinha aberto a sua carta, segundo elle mesmo disse.

a Madame Le Barthz morreo-lhe seo marido, e torna p^{a}. esta terra, segd°. me diz agora numa carta: e chegará em Junho ou Julho a esta. Com o q perdemos esta boa gente no Havre p^{a}. a correspondencia dos paquettes grossos. eu neste prim°. navio hei de mandar a V.M. algum ou alguns exemplares da oração inaugural do cirurgião Gouveia, q cá fez hum curso de [operaçoens][1] com aplauzos e vivas: queira Deos q não fique só com isso, pois he pobre, e merece pelo seo zelo algũa pensão da Corte. [...] imprimir os principios de M^{r}. de la Faye p^{a}. a cirurgia traduzidos porem em portuguez. eu mando a V.M. esta lista de nomes franceses p^{a}. q V. [M.] concorra p^{a}. esta obra de utilid^{e}. nacional, mandando-me a tradução genuina em Portuguez, ou ao menos em latim a Ds

Vale & iterum vale

Recordaçoens e saudades ás Mlles. e q Ivan va à maravilha &^{a}. Saud^{es}. tãobem aos amigos &^{a}.

A Monsieur
Monsr. Le Docteur Sanchez
Rue de Michel le Comte
à Paris

1 Conjectural reading.

62.04.20 Illustração Medica, ethico-politica, historico-sistematica, sceptico-eclectica, physico-analytica e theorico-practica, ou, Reflexão critica ás consideraçoens medicas sobre o methodo de conhecer, curar e preservar as epidemias, ou febres malignas podres, pestilenciaes, contagiosas, &c; dividida em dois tomos oferecido o primeiro ao Sereniss°. S[or]. Infante D. Manoel por Duarte Rebello de Saldanha, medico nesta Corte. Lisboa. 1761. in 8°. de 620 pag.

Com duas cartas censorias ao principio e aprovaçoens, q não entrão na conta, e hũa excelente dedicatoria, e hũ antelóquio da mesma libré, q só elle ocupa [18] paginas.[2] isto sim q he hũ admiravel Palito Moderno. eizaqui hũ retalhinho. na Dialexe 3ª. Essencia e diversa sorte de febres pestilenciaes. no n°. 152 = He a peste aquella infermid[e]. q com a sua furia nos estragos q cauza, imita as do Cocîto: nem estas trez celebradas filhas de Acheronte e da Noute produzem mais estragos do q a peste em favor da Lebitine: he Alecto na entranhavel voracid[e]. com q destróe: he Thisiphone na ligeireza com q precipita: he Megera na implacavel arrogancia com q sepulta: he de mais Hydra na venenoza natureza com q mata: e he Herculea na ampla jurisdição com q trancende os mais izentos e reconditos azylos. ense recidendum est ne pars sincera trahatur q lá a perder de vista o celebrado Protheu, quando a peste muda em hum instante mais de mil formas á figura, e nunca jamais apareceu de hũa mesma reprezentação: sendo tão variavel como isto a sua natureza, só conhecida entre os doutos pelo vario, e indifinido modo com q se manifesta.

lista de nomes Francezes, pª. obter a tradução e em portuguez ou pelo menos em latim.

Dartres (mal cutaneo)? 2. gruau (certo grão)? 3. framboises (certo fruto) 4. estas hervas oseille , pour[pier][3], cresson, estragon, persil? 5. trufes (são tuberes da terra?)? 6. ecrouelles (são alporcas?) 7. virus[...]rico[4]? 8. lochies? 9. errhines (rem°. pª. espirrar)? 10. estes anodynos, coçoins de tripes, pulpe de casse, frai de [gren]ouille[5]? 11. estes narcóticos la jusquiame, bella-dona, morelle, pomme-épineuse? 12. repercussivos simples la [terre] simolée, lentille déau, la joubarde, les limeçons, (são lesmas ou caracois?) l'Argentine? 13. l'eau de [Plan]tin[6]? 14. estes emolientes simples, le bouillon

2 Obliterated margin. From the book itself, we learn that the prologue occupied 18 pages.
3 Conjectural reading.
4 Illegible.
5 Obliterated margin. Conjectural reading.
6 Obliterated margin. Conjectural reading.

blanc, la poirée, la mercuriale, séneçon, Belle-dame, bran[che ursine][7], la Bette, le bon-Henri, la violette (he o bonefe? ou violas) le thim, oignon de lys, le peuplier, graine [de] lin[8] (he a linhaça?) le son (he o farello?) le bouillon de tripes? 15. rezolutivos simples; fleurs de melicot, de camomille, semence de Dancus, d'aneth, cumin, anis, fenoüil, marrube, sureau, Hyeble, la mente, ou menthe, le calament? 16. estas plantas aromaticas: l'orignan, Pouliot, souge, serpolet, lavande, bardane, scrophulaire, herbe de S^t^. Etienne, millespertuis, sceau de Salomon, Persicaire, marc du vin, (são as borras?) lessive de cendre de sarment? 17. rezolutivos, gomme ammoniac, odelium, la moëlle des animaux (são os tutanos ou o cerebro?) farinha d'orobe? 18. rezolutivos compostos unguento Martiatum, de Strirax, de la Mere, huil de vers (he óleo de minhocas?) de camomille, aspic, Romarin, pétrole, emplastre de Bétoine, melilot, And[ré][9] de la Croix, de Diachylum, de Vigo, triapharmacum de Massue, du cimbre? 19. detersivos. mille folium, aigremoine, la Renoncule, savoniere, la liere, les Ronces, la mirrhe? 20. detersivos comp^{os}. l'eau phagedenique, huil de gayac, collyre de Lanfranc, unguent des Apotres, mondificatif d'ache, le beaume de Madame Feuille d'aiguille, huil d'Hypericum? 21.sarcoticos beaume de copü, blanc de Tolus, du comendeur? 22. corrozivos, poudre de sabine, l'ocre, le precipité rouge & blanc, sublimé corrosif, l'eau mercuriale, la pierre à cautere, 23. cicatrizantes Plomb brulé, litharge, la ceruse? onguent blanc de Rhasis, de Pompholix, [eau][10]-vulneraire, trochisques blancs de Rhasis? 24. astringentes la vesse de loup, la craye, l'amidon, l'eau de Rabel? 25. ophtalmiques moelle de casse, l'eau de guimauve, les mucilages de Plyllium, le Macis, l'oliban, la pierre admirable, la ceruse? 26. l'oxicrat, du vieux-oing, bayes de genievre, huil d'hypericum, castor, mastic en larmes, concombre (he o pepino?) verd de gri[...][11] bugle, sanicle, mouron à fleurs rouges, éclair, Pigeonneau vivant & plumé (he hum pombinho vivo com pennas e tudo?) poudre de semelles de vieilles savates brulees (he pós de solas velhas queimadas?)? feuilles d'amolie, de sauge frises, & de menthe? graine de psylium, sirop de mûres (he enxarope ou he anobe de amoras?) sommités de rues (são olhos de arruda?) de l'orgeat? Raifort sauvage, bardane, fumeterre, cresson, becabunga, quando fala das sanbixugas diz q a boca he hum demi-cercle godronne: como traduzirei isto? Como traduzirei o banho de injeção a q chamão douche? enchifrenement he o 62.04.20

7 Conjectural reading.
8 Obliterated margin.
9 Conjectural reading.
10 Obliterated margin. Conjectural reading.
11 Obliterated margin.

62.04.20 estar entupido dos narizes? Quando diz q a saliva he hum humor muito delayé, hei de dizer q he m^{to}. liquido ou q he diluente como na verdade parece q devia dizer? de todas estas coizas, quizera q V.M. me fizesse hum indece com a sua tradução portugueza, ou ao menos com os nomes latinos, p^{a}. cá ver se os posso achar nas farmacopeias em Portuguez. tãobem se V.M. tivesse algũa noticia bem fundada dos banhos e agoas thermais q há em Portugal, e das suas qualid^{es}. e enfermid^{es}. p^{a}. q servem estimaria esta noticia p^{a}. a pôr em lugar dos banhos e agoas thermais q há em França e q o Autor poem no seo dito tratado. eizaqui tem V.M. no q agora me ocupo, p^{a}. hir comendo á custa do meo trabalho: mas como elle há de sahir em nome de outro, fique isto entre nós. V.M. mandará a reposta do q aqui lhe pergunto pelo Havre de grace; pois como há de ser voluminoza, custaria m^{to}. pela posta ordinária exceto se V.M. a fosse mandando aos poucos; em as cartas q me for escrevendo, fazendo em cada huma hum §. desta matéria. no Havre de grace há Monsr. la Priere, q se receber franco o cartaz, não deixará de o remeter por via dalgum navio dos S^{res}. Valles e Peres, em cuja logea eu terei cuid^{o}. de o hir buscar: e se [acazo ainda][12] não tiver partido o yate S. Francisco Xavier, ainda lá achará Madame le Barthz q he m^{to}. minha affeiçoada e q me trará o d^{o}. cartaz. Vale. &^{a}.

20 April 1762

D^{r}. Ant^{o}. Rib^{o}. S[anches]

My dear friend, I have received with enormous pleasure yours of the 29th of last month,[13] the cover and sealing wax not having decreased or increased the price of the postage. This shows the regularity of the post, which generously neglected these details. I am happy about what you say you wrote to our Minister but I wish you had spoken even more clearly and in another tone, making him ponder the contradiction of the procedure in what concerns you by comparing the present with the past; telling him that the persecution made by Oliveira[14] against you originated in his own home; showing him nominatim[15] the collusions of his secretary with Oliveira and his intrigues; and that nobody was foolish enough to think that Oliveira would dare to attack you without the consent and approval (at least tacit) of His Excellency. & this should be the tone of the letter: but let us wait to see the effect of what you wrote.

12 Obliterated margin. Conjectural reading.
13 Letter not found.
14 Not identified.
15 by giving the names.

In the next post I hope for news of the reception of the justification that I sent you;[16] and that you will report the success of this fight with that fop. 62.04.20

On the 14th of this month I went to meet with D. Luiz da Cunha[17] to obtain a passport to get out of here;[18] but though I had the annoyance of not getting it, being told that it could not be issued at present, I had the satisfaction of seeing his affection for you: he told me that soon he would write to Monsignor Salema[19] in order that you be paid your pension at home, summing up: "what can Oliveira do then? He can do nothing". It was good, and excellent that Mr. Alcaçova[20] had spoken to him and prepared him in this way. You should thank him in a short letter, telling him what had been framed against you and the danger to which you are exposed, and explain to him clearly the intrigues of Salema's secretary, in partnership with Oliveira and Alvares[21] &c.; because even if it should not be always true, the servant must be charged and not the master: and mention to him briefly about me, saying how I am also the target of these villains.

With this, my friend, here we are until the affaires publiques give me an opportunity to go out there, or to England, or to Italy, particularly to Venice. In the proper time I will warn you.

Philotectes[22] cannot but bring sorrow to all his friends, provided there are some as both of us are. I cannot give you clearer ideas than the ones I have. What we saw was that he escaped from that disorder; that he arrived here, where he did not obtain a private audience with the Minister, nor was heard: he kept in the same way until, feeling disillusioned, he retired to his home where he lives with his usual taciturnity in absolute solitude, inside his room, reading some book. Everyone agrees that he acted like a fool and lost his head in this matter: that

---

16 Letter not found.

17 D. Luiz da Cunha Manuel (?-1775), nephew of the famous Ambassador in London, D. Luiz da Cunha, served in important political positions under Pombal's government. At the time of this letter he was in charge of Foreign Affairs and War.

18 Magellan was trying to obtain a passport to leave Portugal after travelling there from Paris in June 1761.

19 Monsignor Pedro da Costa de Almeida Salema, Portuguese Envoy to the French Court from 1757 to 1762.

20 Gonçalo Xavier de Alcaçova (1712-1785), Portuguese physician who corresponded with Ribeiro Sanches and tried to have Sanches' valuable library transferred to Portugal.

21 In spite of the bad opinion of Magellan, Sanches considered Dr. J. Alvares (fl. 1760-70) as his friend.

22 Greek hero abandoned by the Greek chiefs and later called to help in the Trojan war. Magellan uses this epithet to represent the Portuguese mathematician and astronomer Soares de Barros, who in a way was also abandoned by his countrymen and lived in solitude in spite of his value as a scientist (see also Magellan to C. Messier, 26 May 1759, note 3). He was disgraced because, while serving as Portuguese Chargé d'Affaires in Paris, he had received the Abbé Chevalier at his home after Chevalier fled from Portugal after having been imprisoned. In addition, Barros had left the Embassy in Paris and returned to Lisbon without permission.

62.04.20 he did not know how to serve in the ministry and was totally mistaken, thinking that the reason for his folly would be listened to. I was told that in Paris he went to speak with Choiseul,[23] without having been empowered and presented at court, about his dissentions with Salema: and when he arrived in Spain he was badly received by the Ambassador Peçanha.[24] As I was ignorant of everything, I thought that all was going well, and by the constancy of his behaviour it seemed that he was not at all dismal and that he acted as a philosopher and will live as such: and if he acted as an honest man he will live poor; since today poverty is the common reward of virtue. This is all I know of him. He does not write to anyone and even if he wrote he would say no more than what he used to say when he spoke. From him I could never know the essence and not even the surface of what occupied him. He is inaccessible and incommunicative in such matters and I remained as ignorant at this point as if I had never spoken to him. At present I don't know where he is. I wrote to him to Mafra[25] and to Setubal,[26], giving him notice that I had put your letter in Ciera's[27] hands; but he neither answered, nor collected the said letter of yours.

It will be kind if you write to Dr. Ciera since he is supposedly a man whose heart is good and not hardened, he is a considerate person. Truly he is a good man, unable to do harm to anyone, but above all to himself: I say that he will not annoy himself for his friends. He is a good Italian and that is enough but [is] his correspondence good Offer him your services and ask him the rule that one ought to observe in literary correspondence [...] Yesterday I met the good Sachetti:[28] he gave me the title of another, better palito;[29] it is the work of your cousin.[30] It has the most graceful things. The dedication is a chef d'oeuvre: yes, here you will have much to note. Prepare yourself that you will see beautiful things. I am sending you herewith a passage from that book[31] but it has other, much more handsome extravagancies and absurdities. Warn Dr. Sachetti because if the author does not send you a copy, he wishes to send one.

---

23 Etienne François, duc de Choiseul (1719-1785) was France's Foreign Minister between 1758 and 1761.

24 D. José da Silva Peçanha (or Pessanha) (1717-1774/1775), Portuguese Envoy to the United Provinces, Plenipotentiary in Naples and Ambassador in Spain.

25 Town 30 km north of Lisbon.

26 Town 40 km south of Lisbon where Soares de Barros was born. Letters not found.

27 Miguel Antonio Ciera (fl. 1770), an Italian engineer from Piedmont, went to Portugal to collaborate in the demarcation of the boundaries of the South American colony of Brazil. Later he was invited to teach at the College of Noblemen in Lisbon and to help in the 1772 reform of the University of Coimbra.

28 João Mendes Sachetti Barbosa (1714-1774), Portuguese physician who corresponded with Ribeiro Sanches and Emanuel Mendes da Costa. He wrote a book on epidemic fevers as well as some articles in the *Philosophical Transactions* of the Royal Society.

29 An old Portuguese familiar expression meaning that something is a cause of mockery.

30 Duarte Rebello de Saldanha, physician to Queen Maria I of Portugal.

31 See attachment below.

I keep warning you that when you mention in your letters the intrigues of Oliveira and the Canon[32] don't forget Alvares and Bahia[33] and Xavier.[34] It is necessary to speak clearly and name that entire rabble, which aims to destroy men of good will to prevent their blows. A chap told me that Alvares had written to the head surgeon: but that he had not yet opened the letter, according to his own words. 62.04.20

Madame Le Barthz'[35] husband died and she is coming back to this country, according to what she says in a letter to me:[36] and she will arrive here in June or July. And we lose these good people in the Havre for the correspondence of large packets. By the next ship I will send you one or more copies of the inaugural speech of surgeon Gouveia,[37] who gave a course here on operations[38] with applause and cheers: may God grant that he does not get only this since he is poor and for his zeal deserves some pension from the Court. [...] print the principles of Mr. de la Faye[39] for surgery but translated into Portuguese. I send you this list of French names[40] in order that you contribute to this work of national utility, sending me the genuine translation into Portuguese or at least into Latin.

vale & iterum vale[41]

Remembrances and compliments to the young ladies and that Ivan[42] is going well &ª. Also compliments to the friends &ª.

A Monsieur
Monsr. Le Docteur Sanchez
Rue de Michel le Comte
à Paris

---

32 Probably Magellan refers to Mons. Salema.

33 Felis (or Felix) António Bahia was a not very honest Portuguese citizen living in Paris.

34 Not identified.

35 Not identified.

36 Letter not found.

37 Filippe José de Gouveia, Portuguese army surgeon who studied in Spain and in France where he worked with La Faye.

38 Most probably Magellan refers to *Oração inaugural do primeiro curso de operações cirúrgicas, que publicamente fez no Hospital real de Lisboa, no ultimo de Janeiro de 1762* (Lisbon, 1762).

39 Georges de la Faye, *Principes de chirurgie ...* (6th ed., Paris, 1761). There was a Spanish translation of this work in 1761, but the first Portuguese translation dates from 1787. Therefore Magellan's attempt to promote a Portuguese translation, possibly edited by Gouveia, came to nothing. In fact Magellan about one year later was already in London and Gouveia died early.

40 See below.

41 Farewell and again farewell.

42 Probably Jean Pernelet, a relative of Sanches' housekeeper Marie Jeanne Pernelet and her cousin Antoinette who was also part of Sanches' household; see Maximiano Lemos, *Estudos de História da Medicina Peninsular* (Porto, 1916), pp. 187-190.

62.04.20 *Illustração Medica, ethico-politica, historico-sistematica, sceptico-electica, physico-analytica e theorico-practica, ou, Reflexão critica ás consideraçoens medicas sobre o methodo de conhecer, curar e preservar as epidemias, ou febres malignas podres, pestilenciaes, contagiosas, &c;* dividida em dois tomos oferecido o primeiro ao Sereniss°. S^{or}. Infante D. Manoel por Duarte Rebello de Saldanha, medico nesta Corte. Lisboa. 1761. in 8°. de 620 pag.

With two censorial letters in the beginning and approbations that don't count, and an excellent dedication and a prologue of the same attire which by itself occupies [18] pages. Yes this is an admirable Modern Palito.[43] Here is a small extract. In the 3rd Dialexe,[44] Essence and diverse sorts of pestilential fevers, in no. 152 = The plague is that infirmity that with its fury in the damage it causes, imitates those of the Cocito:[45] not even these three celebrated daughters of Acheron and Nox[46] produce more damage than the plague as compared with the Lebitine:[47] it is Alecto in the deep voraciousness with which it destroys: it is Thisiphone[48] in the swiftness that it precipitates: it is Megera in the implacable arrogance with which it buries: it is also Hydra[49] in the venomous nature to kill: it is Hercules in the large jurisdiction to transcend the most immune and concealed refuges. ense recidendum est ne pars sincera trahatur.[50] That the celebrated Proteus[51] stays far away, when the plague suddenly changes its aspect into more than a thousand forms, and never appears in the same aspect: its nature being as variable as this, it is only known among the learned through the variable and indefinite way in which it manifests itself.

List of French names to obtain the translation into Portuguese or at least into Latin.[52]

1. dartres (disease of the skin)? 2. gruau (a certain grain)? 3. framboises (a certain fruit) 4. these herbs oseille, pourpier, cresson, estragon, persil? 5. trufes (are they tubers from the earth?)? 6. ecrouelles (is it scofula?) 7. virus[...] 8. lochies? 9. errhines (medicine to sneeze)? 10. these anodynes, coçoins de tripes,[53] pulpe

---

43 See note 29.

44 Lecture?

45 Cocytus, the "river of lamentations" in Greek mythology, one of the rivers of Hades.

46 In Greek mythology, the Erinyes (Furies) were the three daughters of Acheron and Nox (Night, or Noute, in Portuguese): Alecto, Megaira and Tisiphone.

47 Vipera Lebitina (or Levantine) – a species of very venomous viper.

48 Tisiphone, one of the three Furies.

49 Lernean Hydra – mythological beast with numerous heads and poisonous breath.

50 "[The incurable part] must be cut out so that the healthy part is not infected" – Ovid, *Metamorphoses*, *1*, 191.

51 Proteus – an old sea god who was able to change his shape at will.

52 The terms that follow correspond to old pharmaceutical terms that have not been translated since they were found in eighteenth-century dictionaries and pharmacopoeias. Only terms that are wrongly written are mentioned with the correct spelling.

53 Coçoin (probably coction) de tripes – boiled tripes.

de casse, frai de [...]ouille?[54] 11. these narcotics la jusquiame, bella-dona, morelle, pomme-épineuse? 12. Simple repercussives la [terre] simolée,[55] lentille d'eau, la joubarde, les limeçons,[56] (are these snails or snail-shells?) l'Argentine? 13. l'eau de [...]tin?[57] 14. these simple emollients, le bouillon blanc, la poirée, la mercuriale, séneçon, Belle-dame, bran[...],[58] la Bette, le bon-Henri, la violette (is it the bonefre?[59] or violas[60]), le thim,[61] oignon de lys, le peuplier, graine [de] lin[62] (is it linseed?), le son (is it bran?), le bouillon de tripes? 15. simple resolutives; fleurs de melicot, de camomille, semence de Daucus, d'aneth, cumin, anis, fenoüil, marrube, sureau, Hyeble,[63] la mente, ou menthe, le calament? 16. these aromatic plants: l'orignan,[64] Pouliot, sauge, serpolet, lavande, bardane, scrophulaire, herbe de S^t^. Etienne, millespertuis, sceau de Salomon, Persicaire, marc du vin (is it the lees?), lessive de cendre de sarment? 17. resolutives, gomme ammoniac, ocelium, la moëlle des animaux (is it the marrow or the brain?), flour of orobe? 18. compound resolutive unguents Martiatum, de Stirax, de la Mere, huil de vers (is it oil of earthworms?), de camomille, aspic, Romarin, pétrole, plaster of Bétoine, melilot, And[...] de la Croix,[65] of Diachylum, of Vigo, triapharmacum de Massue, of cinabre? 19. detersives. millefolium, l'aigremoine, la renoncule, savoniere, la liere,[66] les ronces, la mirrhe? 20. compound detersives l'eau phagedenique, huil de gayac, collyre de Lanfranc, unguent des Apotres, mondificatif d'ache, le beaume de Madame Feuille,[67] d'aiguille, huil d'hypericum? 21. sarcotics beaume de copü,[68] blanc de Tolus,[69] du comendeur?[70] 22. corrosives, poudre de sabine, l'ocre, le precipité rouge & blanc, sublimé corrosif, l'eau mercuriale, la pierre à cautere, 23. cicatrisants Plomb brulé, litharge, la ceruse? onguent blanc de Rhasis, de Pompholix, [...]-vulneraire,[71] trochisques blancs de Rhasis?[72] 24. astringents la vesse de loup, la craye, l'amidon, l'eau de Rabel? 25. ophthalmiques moelle 62.04.20

54 frai de grenouille– sperm of frog?
55 terre simolée – *terra sigilata*?
56 limaçons?
57 eau de Plantin?
58 branche ursine?
59 A scrophulariaceous plant.
60 Viola – Latin name of violet.
61 Thym.
62 linseed? Obliterated margin. Conjectural reading.
63 Hieble.
64 Origan?
65 André de la Croix?
66 Lierre?
67 Baume de Madame Feuillet?
68 copahu?
69 Tolu?
70 Commandeur?
71 Eau vulnéraire?
72 Trochisques de blanc Rhasis.

62.04.20 de casse, l'eau de guimauve, les mucilages de Psyllium, le Macis, l'oliban, la pierre admirable,[73] la ceruse? 26. l'oxicrat, du vieux-oing, bayes de genievre, huil d'hypericum, castor, mastic en larmes, concombre (is it cucumber?) verd de gri[...][74] bugle, sanicle, mouron à fleurs rouges, éclair, Pigeonneau vivant & plumé (is it a little dove alive with feathers and all?) poudre de semelles de vieilles savates brulees (is it powdered burned old soles?)? feuilles d'amolie,[75] de sauge frises, & de menthe? graine de psylium, sirop de mûres (is it syrup or juice of mulberry?) sommités de rues (is it the extremities of rues?) de l'orgeat? raifort sauvage, bardane, fumeterre, cresson, becabunga. When speaking of leeches it is said that the mouth is a demi-cercle godrone:[76] how can I translate this? How can I translate the injection bath that they call douche? Enchifrenement is it having a blocked nose? When it is said that saliva is a humour very delayé, should I say that it is very fluid or that it is diluent as truly it seems should be said? Of all these things I would like you make an index or table of contents with their Portuguese translation or at least the Latin names to see if here I can find them in the Portuguese pharmacopoeias. Also if you have any well-founded news about the baths and thermal waters that exist in Portugal, as well as their qualities and the infirmities they are used for, I would appreciate this news to put in instead of the baths and thermal waters in France that the author includes in his treatise. Here is as you see what now occupies my time to get something to eat, at the cost of my work: but since it will be published in the name of someone else, let this stay between us. You will send the answer to what I here ask you through the Havre de grace;[77] because as it will be voluminous it would cost a lot by the common post unless you were to send one §. of this matter in each one. At the Havre de grace there is Mr. La Priere[78] who, if he receives the volume exempt of taxes, cannot but remit it through some ship of Messrs. Valles and Peres,[79] from whose shop I shall take care to fetch it: and if the yacht S. Francisco de Xavier[80] has not departed yet, Madame le Barthz who is very affectionate to me will still be there and she will bring the said volume to me. Vale &ª.

73 Pierre admirable (or pierre divine).
74 Vert-de-gris?
75 ancolie?
76 Probably referring to the lamprey eel, whose mouth is like a milling cutter of round shape.
77 Old name of Le Havre.
78 Not identified.
79 The Portuguese firm Valle & Peres was in the shipping business to foreign countries as well as to the Portuguese colonies.
80 This yacht was a corvette or sloop of war that was also used to carry slaves between the west coast of Africa and Brazil.

British Library, Add. Ms. 28,539, E. M. da Costa Correspondence, Vol. VI, ff. 128-128v

Post Scriptum da Carta q escrevi [o Abbade Magalhaens][1] a o Autor da Gazeta Literaria Luiz Suares de Lima Brandão em 10 de Dez^{bro}. 1763.

P.S. Depois de ter feito esta p^{a} partir pelo prim^{o}. navio, Recebo hum bilhete do mesmo S^{r}. M^{el}. Mendes da Costa em data de 12 do Corrente, com huma ideia geral do Seo empenho. Consiste este em alcançar todo o genero de produçoens naturaes q forem próprias e particulares desse nosso paiz Portuguez. 1° alguns peixes, aves, animaes, ou insectos raros eg. borboletas especiaes, bezoiros, escaravelhos, bichos, lagartixas, lagartos & outros reptis extraordinarios nestes paizes do Norte &^{a}.

2°. todo o genero de vegetais raros: arvores, plantas, arbustos, musgos, fungoens, cogumelos, frutos, flores, sementes, &^{a}.

O mesmo douto Naturalista me aponta neste Seo bilhete a Serra de Cintra e Suas vizinhanças, como hum dos sitios mais abundantes de raridades nesta Classe e na seg^{te}.: e o mesmo conceito me lembro q formava destes sitios hum nosso portuguez m^{to}. capaz e instruido, verdadeiram^{te}. digno de mais venturozos sucessos, q conheci e tratei com particular amizade em Pariz, chamado Joze Joaquim de Barros, o qual me assegurou ter achado por aquellas p^{tes}. muitas e muitas especies raras e preciozas, com q talvez se singularizão, por falta de conhecim^{to}. entre nós, algumas Provincias m^{to}. Remotas.

3° todo o genero de terras, barros, bólos, grédas, marmores, pederneiras, petrificaçoens; pedras configuradas, minerais, metais, &^{a}.

Esta he a sustancia do q aquele erudito Naturalista me diz no Seo bilhete: concluindo q a mesma e ainda maior diligencia estimará q se faça a Resp^{o}. das produçoens do nosso Brazil e conquistas da Africa e Asia, por meio de bons correspondentes e amigos, assistentes naquellas p^{tes}. ou q fazem a navegação de Portugal p^{a}. ellas &^{a}.

Em fim adverte q p^{a}. as coisas secas, pouco cuid^{o}. basta p^{a}. serem transportadas em bom estado, embrulhando-as em papeis, ou metendo-as em algodão qd^{o}. a Sua delicadeza o Requerer. Porem p^{a}. a maior p^{te}. das especies pertencentes á 1^{a}. e 2^{a}. clase, he percizo q sejão metidas em frascos, garrafinhas, ou mangas de vidro, cheias de algum licor espirituozo, como o espirito de vinho, agoa ardente fina de cana, de milho &^{a}. a fim de se

1 The text between brackets is in a different handwriting.

63.12.10 conservarem e serem transportadas com a sua natural frescura &ª. Notando geralm^te^. q em cada papel ou vidro deve pôr-se o nome da coiza contida, do sitio e lugar em q costuma achar-se, e todas as mais particularid^es^. q lhe tocarem &ª.

Postscript to the letter that I [Abbot Magalhaens] wrote to the author of the Gazeta Literaria, Luiz Suares de Lima Brandão, 10 December 1763.[2]

P.S. After having written this to send by the first ship, I received a note from the abovementioned Mr. Manuel Mendes da Costa dated the 12th of the current [month], giving a general idea of his interest. This amounts to obtaining every kind of natural product that is proper and specific to that land of Portugal.

1°. some fishes, birds, animals, or rare insects, e.g. special butterflies, beetles, scarabs, worms, small lizards, lizards, & other reptiles unknown in these northern countries &ª.

2°. every kind of rare vegetable: trees, plants, shrubs, mosses, fungi, mushrooms, fruits, flowers, seeds, &ª.

The same learned naturalist indicates in this note the Cintra Mountain ridge and its neighbourhood as one of the places abounding in this class of rarities and of the next: and the same idea, I remember, was held by a very able and learned Portuguese, truly worthy of more fortunate successes, whom I met and developed a particular friendship with in Paris, named Joze Joaquim de Barros, who assured me he had found in those places a great many rare and precious species that perhaps are peculiar to some very remote provinces, of which we are lacking knowledge.

3°. all kinds of earths, clays, bole, argil, marble, flint, petrifications, figured stones, minerals, metals, &ª.

This is the contents of what that learned naturalist tells me in his note: concluding that he wishes that the same and even greater diligence be taken concerning the products of our Brazil and our conquests in Africa and Asia, by means of good correspondents and friends, assistants in those parts, or those who find their way from Portugal to those places &ª.

Finally he warns that for dry things, a little care is enough for them to be transported in good condition, wrapping them in paper, or putting them in cotton

2 Luiz Soares de Lima Brandão (1731-?) was a priest and canon at the Oporto cathedral with a Canon's degree taken in Coimbra. It was his brother Francisco Bernardo de Lima who was the editor of the short-lived (1761-2) periodical *Gazeta Literária*. Magellan seems to have anticipated that his letter would be published in the *Gazeta Literária* and that this postscript would be added to it prior to publication. However, there were no further issues of the *Gazeta Literaria* published after 1762, so that neither Magellan's letter nor the postscript ever appeared in print. The letter itself has not been found; the copy of the postscript that we have used appears to be one that Magellan gave to Mendes da Costa.

when their fragility thus requires. However for most of the species belonging to the 1st and 2nd classes, it is necessary to put them in flasks, little bottles or glass tubes, filled with some spirituous liquor like spirit of wine, fine brandy from sugar cane or wheat, &^{a}. in order to be conserved and transported in their natural freshness &^{a}. Noticing generally that on each paper or glass there must be written the name of the thing inside, the place and locale where it is usually found, and all other particulars concerning it &^{a}. 63.12.10

From João Mendes Sachetti Barbosa[1] 64.03.11

British Library, Add. Ms. 28,534, E. M. da Costa Correspondence, f. 120

11 de March de 1764

... Assegure os meos Resp^{os}. ao Bibl^{o}. da Real Socied^{e}. o S^{r}. M^{el}. Mendes da Costa e lhe agradeça da minha p^{te}. o fazer-me na Lista da mesma Socied^{e}.; na qual me pode condecorar com os titulos de Cavalheiro da Ordem de Christo, e Academico da Real Academia Medica de Madrid, e Socied^{e}. das Ciencias de Sevilha, q tãobem gozo.

... Pergunte-lhe se a R. Socied^{e}. estimará q lhe communique hum manuscrito unico e original sobre a Agricultura das Palmeiras nos nossos Estados da Índia, composto por hũ Jezuita, q lá viveo m^{tos}. annos ?

... Q lhe mandarei alguns Ramos da espécie de cicuta q aqui tem obrado maravilhas, deq tenho observaçoens verídicas, cuja espécie he m^{to}. parecida com aq estampou Stork.

... Fiz a Analyse das agoas minerais das Alcaçarias desta Corte, alguãs das quaes achei tão quentes no seo manancial q chegão ao calor de 80 gráos de Farenheit.

... Todas estas agoas, lançando-lhe a Solução de prata, precipitão ao fundo flocos branquissimos: porem na agoa do Chafariz D'ElRey, edo Chafariz dos páos, e no da Praya, tanto q passão 3 minutos, começão a

---

1 Letter not found. MS is an extract in Magellan's handwriting, headed "Artigo da Carta do Dr. Sachetti Barboza de 11 de March de 1764" [Extract from the letter of Dr. Sachetti Barboza of 11th March 1764]. In a different hand, most probably that of E.M. da Costa, are added the missing names of Sachetti, "João Mendes" and, after Sachetti, "Barboza", and also the annotation: "in his Letter to the Abbot Magalhaens".

64.03.11 tomar huã cor roxa agradável. ... Depois de alguãs horas, decantando a agoa, e enxuto o d°. Sedim^{to}. ou flocos brancos, fica esta da cor de ferrugem de ferro, e dá a mesma côr a q^{to}. se esfrega com ella. os ditos flocos ou Sedim^{to}. são reputados por uma materia Salina, neutra ou marina. Porem

... Pergunto agora 1°. se esta procede de haver algum espirito sulfureo na dita agoa, q produza esta côr?

Pergunto 2do. Se a prata q vai dissolvida na d^{a}. tintura será q^{m}. recebe a dita côr? ou a materia Salina? ou huã e outra?

... Give my respects to the librarian of the Royal Society, Mr. Manuel Mendes da Costa, and thank him on my part to put me in the list of the same Society,[2] where he may distinguish me with the titles of Knight of the Order of Christ, and Academician of the Royal Medical Academy of Madrid and of the Society of Sciences of Seville, to which I also belong.

... Ask him if the Society would like me to communicate to it a unique and original manuscript about the palm trees of our Indian States, composed by a Jesuit who lived there for many years?

... That I will send some branches of the species of cicuta that has worked marvellously here. Of which I have true observations, the species being very likely the one published by Stork.[3]

... I made the analysis of the mineral waters of the springs of this Kingdom, some of which I found so hot at their source that they reach 80 degrees Fahrenheit.

... All these waters, on adding silver solution, precipitate very white flakes at the bottom: but with the water of the Kingdom's fountain, and of the sticks fountain, and that of the beach, as soon as 3 minutes have passed, they start getting a nice purple colour ... After a few hours, after decanting the water and drying the said sediment or white flakes, it takes on the colour of rusty iron and gives the same colour to anything that is rubbed with it. The said flakes are considered as a salt matter, neutral or marine. However

... I ask now 1°. if this proceeds from a sulphurous spirit in the said water, that produces this colour? I ask 2°. Is it the silver that is dissolved in the said tincture that receives the said colour? Or a saline matter? Or both?[4]

2 João Mendes Sachetti Barboza (1714-1774), physician at Portugal's Royal Hospital in Lisbon, had been elected FRS in 1750. He appears in the Royal Society's records as Joahnnes Mendez Saquet Barboza.

3 Antonius Storck, *Libellus de cicuta* (Vienna, 1761).

4 There is a further annotation in da Costa's hand: "respondido em m^{a}. carta a o D^{r}. Sachetti dattada do 29 Mayo de 1764" [answered in my letter to Dr Sachetti dated 29 May 1764].

Letters Patent 65.03.28

National Archives, London, C66/3700, no. 11

George the Third by the Grace of God etc

To all to whom these presents shall come Greeting whereas Henry Pyefinch of Cornhill in our City of London Optician[1] and John Hyacinth de Magalheans of the same place Gentleman have by their petition humbly represented unto us that they have after long study application and great Expense found out and brought to perfection a new constructed Instrument by which is shown the true Effect of the weight of the atmosphere with the variation which is caused by the Heat and Cold and likewise the Quantity of that variation so as to become of general utility and benefit to our subjects of this our Kingdom Wherefore and in regard the petitioners are the first and true Inventors thereof They most humbly prayed us to grant unto them their Executors Adm˜ors[2] and Assigns our Royal Letters patent for the sole use and benefit of their said Invention within that part of our Kingdom of Great Britain called England our Dominion of Wales and Town of Berwick upon Tweed and also within our plantations and colonies abroad for fourteen years according to the Statute in such case made and provided We being willing to give Encouragement to all Arts and Inventions which may be for the publick good are graciously pleased to condescend to their request Know ye therefore that We of our Especial Grace certain knowledge and [...] Motion have Given and granted and by these presents for Us our Heirs and Successors Do Give and Grant unto the said Henry Pyefinch and John Hyacinth de Magalheans their Executors Adm˜ors and Assigns our Especial Licence full power sole priviledge and Authority that they the said Henry Pyefinch and John Hyacinth de Magalheans their Executors Adm˜ors and Assigns and every of them by themselves or by their Deputy or Deputys Servants or Agents or such others as they the said Henry Pyefinch and John Hyacinth de Magalheans their Executors Adm˜ors or Assigns shall at any time agree with no others from time to time and at all times hereafter during the Terms of Years herein expressed shall and lawfully may make use Exercise and vend their

1 After being apprenticed to Francis Watkins in 1753, Pyefinch became a freeman of the Spectaclemakers' Guild in 1763 and well known as an optical and mathematical instrument maker. He died in 1790.

2 Administrators.

65.03.28 said Invention within that part of our Kingdom of Great Britain called England our Dominion of Wales and Town of Berwick upon Tweed and also in our plantations and colonies abroad in such manner as to them the said Henry Pyefinch and John Hyacinth de Magalheans their Executors Adm~ors and Assigns or any of them shall in their Discretions seem meet And that they the said Henry Pyefinch and John Hyacinth de Magalheans their Executors Adm~ors and Assigns shall and lawfully may have and enjoy the whole profit benefit commodity and advantage from time to time coming growing accrewing and arising by reason of the said Invention for and during the Term of Years herein mentioned To Have hold Exercise and Enjoy the said Licence powers priviledges and Advantages hereinbefore granted or mentioned to be granted unto the said Henry Pyefinch and John Hyacinth de Magalheans their Executors Adm~ors or Assigns for and during and unto the full End and Term of fourteen years from the date of these presents next and immediately ensuing and fully to be compleat and ended according to the Statute in such case made and provided and to the End that they the said Henry Pyefinch and John Hyacinth de Magalheans their Executors Adm~ors or Assigns and every of them may have and enjoy the full benefit and the sole use and Exercise of the said Invention according to our Gracious Intention hereinbefore declared We do by these presents for Us our Heirs and Successors require and strictly com̄and all and every person and persons Bodies politick and Corporate and all other our subjects whatsoever of what Estate Quality Degree Name or condition soever they be within that said part of our Kingdom of Great Britain called England our Dominion of Wales Town of Berwick upon Tweed and also in our plantations and colonies abroad aforesaid that neither they nor any of them at any time during the continuance of the said Term of fourteen years hereby granted either directly or indirectly do make use or put in practice the said Invention or any part of the same so attained unto by the said Henry Pyefinch and John Hyacinth de Magalheans as aforesaid nor in any wise counterfeit imitate or resemble the same nor shall make or cause to be made any addition thereunto or subtraction from the same whereby to pretend himself or themselves the Inventor or Inventors Devisor or Devisors thereof without the Licence Consent or Agreement of the said Henry Pyefinch and John Hyacinth de Magalheans their Executors Adm~ors or Assigns in Writing under their Hands and Seals first had and obtained in that behalf upon such pains and penalties as can or may be justly inflicted on such offenders for their contempt of this our Royal Command and further to be answerable to the said Henry Pyefinch and John Hyacinth de Magalheans their Executors Adm~ors or Assigns

according to Law for their Damages thereby occasioned And Moreover 65.03.28
We do by these presents for Us our Heirs and Successors Will and Command all and singular the Justices of the Peace Mayors Sheriffs Bailiffs Constables Headboroughs and all other Officers and Ministers whatsoever of Us our Heirs and Successors for the time being that they or any of them do not nor shall at any time hereafter during the said Term hereby granted in any wise molest Trouble or hinder the said Henry Pyefinch and John Hyacinth de Magalheans their Executors Adm˜ors or Assigns or any of them or their Deputys Servants or Agents in or about the due and lawfull use or Exercise of the aforesaid Invention or any thing relating thereto provided always and these our Letters patent are and shall be upon this condition that if at any time during the said Term hereby granted it shall be made appear to Us our Heirs and Successors or any six or more of our or their Privy Council that this our Grant is contrary to Law or prejudicial or inconvenient to our Subjects in General or that the said Invention is not a new Invention as to the publick use and Exercise thereof in that said part of our Kingdom of Great Britain called England our Dominion of Wales Town of Berwick upon Tweed and also in our plantations and colonies abroad aforesaid or not invented and found out by the said Henry Pyefinch and John Hyacinth de Magalheans as aforesaid Then upon signification or Declaration thereof to be made by Us our Heirs or Successors under our or their Signet or Privy Seal or by the Lords and others of our or their Privy Council or any six or more of them under their Hands these our Letters patent shall forthwith cease determine and be utterly void to all intents and purposes any thing hereinbefore contained to the contrary thereof in any wise notwithstanding provided also that these our Letters patent or any thing herein contained shall not extend or be construed to extend to give priviledge unto the said Henry Pyefinch and John Hyacinth de Magalheans their Executors Adm˜ors or Assigns or any of them to use or imitate any Invention or work whatsoever which hath heretofore been found out or invented by any other of our subjects whatsoever and publicly used or Exercised in that said part of our Kingdom of Great Britain called England our Dominion of Wales or Town of Berwick upon Tweed or in our plantations and Colonies abroad aforesaid unto whom like Letters patent or priviledges have been already granted for the sole use Exercise and benefit thereof It being our will and pleasure that the said Henry Pyefinch and John Hyacinth de Magalheans their Executors Adm˜ors or Assigns and all and every other person and persons to whom like Letters patent or priviledges have been already granted as aforesaid shall distinctly use and practice their several Inventions by them invented and found out

65.03.28 according to the true intent and meaning of the same respective Letters patent and of these presents provided likewise nevertheless and these our Letters patent are upon this express Condition that if the said Henry Pyefinch and John Hyacinth de Magalheans their Executors or Adm˜ors or any person or persons which shall or may at any time or times hereafter during the Continuance of this Grant have or claim any Right Title or Interest in Law or Equity of in or to the power priviledge and Authority of the sole Use and benefit of the said Invention hereby granted shall make any Transfer or Assignment or any pretended Transfer or Assignment of the said Liberty priviledge or any share or shares of the benefit or profit thereof or shall declare any Trust thereof to or for any number of persons exceeding the Number of five or shall open or cause to be opened any Book or Books for publick Subscription to be made by any Number of persons exceeding the Number of five in order to the raising any sum or sums of Money under pretence of carrying on the said Liberty or priviledge hereby granted or shall by themselves or their Agents or Servants receive any Sum or Sums of Money whatsoever of any number of persons exceeding in the whole the number of five for such or the like intents or purposes or shall presume to act as a Corporate Body or shall Divide the benefit of these our Lrts[3] patent for the Liberty and priviledges hereby by Us granted into any number of Shares exceeding the number of five or shall commit or Do or procure to be committed or Done any Act Matter or thing whatsoever during such time as such person or persons shall have any Right or Title either in Law or Equity in or to the said premisses which will be contrary to the true intent and meaning of a certain Act of parliament made in the sixth year of the Reign of our late Royal Great Grand Father King George the First Intituled an Act for the better securing certain powers and priviledges intended to be granted by his Majesty by two Charters for Assurance of Ships and Merchandizes at Sea and for lending Money upon Bottomry and for restraining several extravagant and Unwarrantable practices therein mentioned or in case the said power priviledge or authority shall at any time hereafter become vested in or in Trust for more than the Number of five persons or their Representatives at any one time (reckoning Executors or Admors as and for the single person whom they represent as to such Interest as they are or shall be intitled to in right of such their Testator or Intestate) That then and in any of the said Cases these our Letters patent and all Liberties and advantages whatsoever

3 Letters.

hereby granted shall utterly cease determine and become void anything hereinbefore contained to the contrary thereof in any wise Notwithstanding provided also that if the said Henry Pyefinch and John Hyacinth de Magalheans or one of them shall not particularly describe and ascertain the Nature of their said Invention and in what manner the same is to be performed by an Instrument in writing under their Hands and Seals or the Hand and Seal of one of them and cause the same to be inrolled in our high Court of Chancery within four Kalendar Months next and immediately after the Date of these our Letters patent That then these our Letters patent and all Liberties and advantages whatsoever hereby granted shall utterly cease determine and become void anything hereinbefore contained to the contrary thereof in any wise Notwithstanding[4] And lastly We do by these presents for us our Heirs and Successors Grant unto the said Henry Pyefinch and John Hyacinth de Magalheans their Executors Adm˜ors or Assigns that these our Letters patent or the Inrollment or Exemplification thereof shall be in and by all things good firm valid sufficient and effectual in the Law according to the true intent and meaning thereof and shall be taken construed and adjudged in the most favourable and beneficial sense for the best advantage of the said Henry Pyefinch and John Hyacinth de Magalheans their Executors Adm˜ors or Assigns as well in all our courts of Record as elsewhere and by all and singular the Officers and Ministers whatsoever of Us our Heirs and Successors in that part of our said Kingdom of Great Britain called England our Dominion of Wales and Town of Berwick upon Tweed and also in our plantations and colonies abroad and amongst all and every the Subjects of Us our Heirs and Successors whatsoever and wheresoever Notwithstanding the not full and certain describing the Nature or Quality of the said Invention or of the Materials thereto conducing and belonging In Witness Witness ourself at Westminster the twenty Eighth Day of March 65.03.28

By writt of Privy Seal

4 To satisfy this proviso, Pyefinch submitted specifications for the invention on his own and Magellan's behalf, a few days before the period of four months allowed for doing so expired; see Specification of Patent, 24 July 1765.

65.06.06 *To the Earl of Morton*

Royal Society of London, Letters & Papers, IV, 288

An Account of the new Baro-Thermometer; invented by John Hyacinth de Magalhaens, addressed to the Right Honble. the Earl of Morton, President of the Royal Society.

My Lord,

In the Memoir I had the honour of presenting to the Royal Society on the 21st. February last, I pointed out the greatest faults, to which the best Barometers hitherto used were subject. I now beg leave to submit to this illustrious Body the means I made use of to bring this Philosophical instrument to perfection, so as justly to merit the name of an exact Αεροσταθμιον or Air-Balance, with y^{e}. addition also of being a true Thermometer, the most proper of all instruments for Medical Observations. But as a minut account of these means may be tedious to the generality of Readers, I shall finish this second Memoir by a recapitulation or Abridgement in which the substance of the whole will be considered.

Abridgement of the two Memoirs.

To correct the Barometer, from those essential faults which hitherto prevented its not answering the proposed end (since I have proved in these two memoirs and will yet shew in the following Notes, that the mistakes thereof may amount sometimes to 2½ inches at least in com̄on observations, and to much more than to 3½ inches in the extraordinary ones) it is requisite that this Philosophical instrument should have the following properties.

1st. Both extremities or ends of the mercurial Column should be as near of the same size as possible, in order that the least change of the mercurial height may be always perceived & easily observed in each of them[(x)].

(x) The errors in not observ^{g}. the rise and fall of Quicksilver in the Cistern, owing to its different heights in the Torricellian tube, may amount to some tenths in com̄on observations. For the proportion between the bores of the tubes of the Barometers, com̄only made in London, & the width of their Cisterns, is as 9 to 200: & consequently if the differences of the mercurial heights are but 3 inches; the rise & fall caused thereby on the Surface of the Mercury in the Cistern, must be 0,135: for as 200 : 9 :: 3 : 0,135. But if the differences of the mercurial Columns were 12. inches, as it will happen at least in experiments like those made by the French Academiciens upon the mountains of Quito in America, & by P. Feuillée upon the Pike of Teneriff (where he found 10 inches & 7 lines

2ly. That the Mercurial Column should be thick enough to overcome any cohesion of the particles to the inside of the tube; 3dly. That all the inequalities of the inside of the tube, must be we[ll] measured & ascertained, to the end that the Dimensions of every part may be exactly settled and marked thereon. For otherwise we never can judge with any certainty the true expansions and contractions of the quicksilver, they being not proportional to the lengths of the mercurial Columns, but to its bulk or Mass. 4thly. That the instrument should be so contrived, that the Observer's sight may be confined at every observation in a true horizontal line to each surface or extremity of the mercurial Column. For otherwise there may happen a mistake of some decimals. 5thly. that it must be contrived in such a manner, that the smallest differences in the mercurial heights, arising from the pressure of the Atmosphere, may be very distinctly seen by the Observer. 6thly. That the expansions & contractions of the mercury owing to the effect of different temperatures of the air, must be very exactly distinguished at every different height of the mercurial Columns(y). 65.06.06

---

less on the Barometer Memoirs of the French Acad. for the year 1733. pag. 45 [J. Cassini, "Réflexions sur l'hauteur du barometre observée sur diverses montagnes", *Histoire de l'Académie Royale des Sciences*, 1733, Mém. pp. 40-48] answering to 11,288 of English measure) compared with Experiments made at the level of the Sea, & at the bottom of Mines or other great depths: then if such experiments have been made (according to the directions of Dr. Desagulliers [J. T. Desaguliers, *A Course of Experimental Philosophy* (London, 1734), vol. 2, p. 256]) with portable Barometers, wherein these differences of the rise & fall of the mercurial surface in the Cistern, are not at all observable: then the mistakes would be more than half an inch. Since 3 : 12 :: 0,135 : 0,540.

(y) The differences of the Bulk of the Quicksilver, when it stands at the mean height of 30 inches in the Barometer, from the heat of a common Summer to the cold of a common Winter in our Climates, may be about 3 lines, according to Monsr. Amontons Memoir of French Acad. for 1704. pag. 2 ["Sur le barometre rectifié", *Histoire de l'Académie Royale des Sciences*, 1704, Hist. pp. 1ff, summarizing two papers by Amontons]: or two tenths of an inch, according to Dr. Halley Philos. Transact. Abridg. by Lowthorp. Vol. 2d. pag. 35 [John Lowthorp, ed., *Philosophical Transactions and Collections, to the End of the Year 1700, abridg'd and disposed under General Heads...*(London, 1716, vol. 2, p. 35]. So that 30 inches observed in the Summer, answer but to 29.8 (= 30 – 0,2) in the Winter, & 30 inches in this cold, answer to 30,2 in the Summer: from whence it appears that there may happen sometimes a mistake of near half an inch (0,4 = 30,2 – 29,8) in the mean heights of the Barometer, observed in the different temperatures of our climate. But if experiments were made in greater heats, such as at Syria, or at the least at Paris in the year 1753, & were to be compared with other equal observations made in greater colds, viz. at Siberia, or at the least at Paris in the year 1709: then these differences would be more than one inch (1,3353) in the first case, & above half an inch (0,566) in the second. For by the Memoirs of the French Academy for the year 1736 (pag. 489) [R.-A. Réaumur, "Observations du thermometre...", *Histoire de l'Académie Royale des Sciences*, 1736, Mém. pp. 469-502] we are told that the heat was at Marra in Syria, about nine o'clock

65.06.06 7thly. That the quantity of quicksilver imployed in each Barometer, being once settled according to the Dimensions of the tube, must always remain

---

in the afternoon, at 35 degrees on Reaumur's Thermometer; which must have been far more than at 45 degrees 6. or 7. hours before. However to avoid all kind of uncertainty, let us take the heat of Senegal, which was but 38¼ degrees upon the same Thermometer, & the cold of Yeniseisk in Siberia, which was 70. degrees below 0 (see the s^d. Memoir for the year 1738. pag. 401 [Réaumur, "Observations du thermometre pendant l'année DCCXXXVIII...", *Histoire de l'Académie Royale des Sciences*, 1738, Mém. pp. 387-403], & for 1749. pag. 11 [J.-N. Delisle, "Observations du thermometre faites pendant les grands froids de la Sibérie", *ibid.*, 1749, Mém. pp. 1-14]) the whole sum will then amount to 108¼ degrees of Reaumur, which answer to 243½ of Fahrenheit's scale: for as 80 : 180 (= 212 – 32) :: 108,225 : 243,50625. Now supposing that the Quicksilver, at 0 of this last Thermometer, contains but 11124 times the bulk of each degree (according to D^r. Boerhaave pag. 174 of his Chemistry Lugd. Batavor. 1732 [i. e. H. Boerhaave, *Elementa chemiae* (Leiden, 1732), p. 174]): it is clear that a Cylinder of 30½ inches of quicksilver, which is not an extraordinary height for the Barometer, will expand or contract itself, from one of these temperatures to the other, $^{66765}/_{100000}$ of an inch: for $^{1}/_{11124}$ × 243,50625 × 30,5 = $^{7426,940625}/_{11124}$ = 0,66765. And consequently if there were two observations both of 30½ inches, one made in the heat of Senegal, & the other in the Siberian cold: I say that the differences or mistakes between these similar observations, would be more than one inch, since the 30½ inches observed at Siberia, were answering to 31,16765 (= 30,5 + 0,66765) in Senegal, & the 30½ inches there, will be equivalent only to 29,8323 (= 30,5 – 0,66765) at Siberia. Then 31,16765 – 29,8323 = 1,3353. The heat in the Summer of 1753 at Paris has arrived at 30½ degrees of Reaumur's Thermometer; and the cold in the Winter of 1709, at 15½ degrees below 0 (see the s^d. French Memoir for the year 1753 pag. 589 [J.-P. Grandjean de Fouchy, "Observations météorologiques faites à l'Observatoire de Paris", *Histoire de l'Académie Royale des Sciences*, 1753, Mém. pp. 589-90]; and for 1740, pag. 548 [Réaumur, "Observations du thermometre faites en MDCCXL...", *ibid.*, 1740, Mém. pp. 539-66]). The differences will be 46. degrees answering to 108,5 of Fahrenheits: then the expansion or contraction from one extremity to the other is $^{288}/_{1000}$: for $^{1}/_{11124}$ × 103,5 × 30,5 = $^{3156,75}/_{11124}$ = 0,283: & consequently the observation of 30½ in the said heat at Paris, answers but to 30,217 inches (= 30,5 – 0,283) in the cold of the said winter: & the same altitude of 30½ inches in this cold, answers to 30,783 (= 30,5 + 0,283) inches in the said heat. Then the differences or mistakes between these two similar observations, in the said different temperatures at Paris, were more than half an inch; for 30,783 – 30,217 = 0,566. Observe nevertheless that applying the same reasoning to the most common & mean differences of heat and cold in our climate, that is from 32 to 78 degrees of Fahrenheit's, we shall find but $^{252}/_{1000}$ of an inch: since $^{1}/_{11124}$ × 46 (= 78 – 32) × 30,5 = $^{1403}/_{11124}$ = 0,126: & 30,626 (= 30,5 + 0,126) – 30,374 (= 30,5 – 0,126) = 0,252. From whence it appears that the proportion of D^r. Boerhaave, made use of in these Calculations, is really too small in regard to the experiments made by D^r. Halley & by Monsr. Amontons: & consequently, that all those Calculations are very far from being urged to extremity.

N.B. Tho' D^r. Martine points out, in his Essays [G. Martine, *Essays Medical and Philosophical* (London, 1740), p. 203], the 80th. degree of Reaumur's Thermometer, made with sprit of Win, to be not answering to the 212 degree of Fahrenheit's: nevertheless the same thermometers also have been made long since with Mercury; and perhaps such were those of the experiments mentioned before, in which case the said proportion 80 : 180 &c ought by no means to be altered.

unaltered: for otherwise it will be almost impossible to ascertain the variation owing to the different temperatures[z]. 65.06.06

8thly. That the proportional Specific gravity of the mercury, imployed in each Barometer must be exactly known, to judge of the true pressure of the Atmosphere: since its gravities differ very considerably, as may be seen by the Tables of Dr. Musschenbroek and others[aa]. 9thly. That the measure of the mercurial heights must begin, either from the top of the lower surface to the top of the upper one, or from the lateral contact of the

---

(z) As true cylindrical tubes (I mean, of an equal size in every part of their Lenght) are not to be found; it follows that the expansions and contractions of the quicksilver in every tube must be different from the other: since their capacities and dimensions may never happen to be exactly the same. And consequently if these variations of the mercurial tube are not ascertained in each tube upon a constant and invariable quantity of quicksilver, it will be impossible to determine them truely. Observe also for the same reasons, that it is impossible to correct the inconveniences attending the common Barometer or Torricellian Tubes, upon this account, by any kind of calculations: since they are supposed to be of a perfect cylindrical form, which in reality they are not. Monsr. J. A. de Luc in his new improvement upon the Barometer, prescribes the Tube to be perfectly cylindrical, at least in that part of it's Lenght, which is to be used: as Monsr. de la Lande mentions in his book of the Mouvemens Celestes for 1765 pag. 204 [i.e. *Connoissance du tems*, 1765, p. 204; De Luc did not publish his own account until 1772]. But by my method all these inconveniences, of the inequalities of the tube, are fully remedied. I was very glad on finding, that this ingenious Author agrees with me in rejecting the use of cisterns, and using Siphons instead of them. But in other respects our improvements are far different one from the other, as it will easily appear by comparing this memoir with the said Account: and I have besides attended to many other requisites which I apprehed wery usefull and yet necessary to be observed in this Instrument, as to rely upon it in Philosophical experiments.

(aa) According to this Author (§ 703 of his Nat. Phil. [Petrus van Musschenbroek, *The Elements of Natural Philosophy*, trans. J. Colson (London, 1744), vol. 1, p. 296]) the differences of the Specific gravities of the quicksilver, even after being purified and distilled many times, may be from 14,110 to 13,550. But quicksilver as it is sold in the Shops, is very often of a less gravity. Then supposing one Barometer A filled with the first kind of mercury, and another Barometer B filled with the second, it is evident, that when B shews a Column of 30 inches, the Barometer A cannot shew more than 28,809: and when A shews the same Column of 30 inches, B will shew 31,239 for as 14,110 : 13,550 :: 30 : 28,809 and 13,550 : 14110 :: 30 : 31,239. So that when the two observers find the same effect from the aerial pressure in their respective instruments, they will have a mistake of almost 2½ inches (2,430 = 31,239 – 28,809). From whence it appears, what great errors must inevitably have been made in barometrical observations only upon this single article; and how well grounded the suspicions of Dr. Wallis were on this head, in the Scholium of the 11 Prop. of his Hydrostatiks. Tom. 1. pag. 1044 [John Wallis, "De Hydrostaticis", in his *Opera mathematica*, vol. 1 (Oxford, 1695), p. 1044; previously published in his *Mechanica* (London, 1670-71)].

65.06.06 lower surface to the like contact of the other. For otherwise there is always a loss of about two tenths of an inch in the convexities, which the mercury naturally has in both surfaces, by the mutual attraction of its particles.

10^{thly}. That the resistance, which the mercury must meet with, passing through the mouth of the tube in the rising and falling (being commonly so low, as to touch the bottom of the Cistern) must be avoided.

11^{thly}. That the heights of the mercurial Columns must be taken in the line truely perpendicular to the horizon.

12^{thly}. That the tubes must be well examined, whether they have any extraordinary influence of attraction or repulsion upon the Quicksilver; as it sometimes happens so as to hinder in some degree the influence of the Atmosphere[bb].

13th. That the Barometers should be so constructed as to avoid their being emptied when removed; the necessary conveniencies for refilling them with due care not always offering; besides the tubes getting sometimes damp or dirt in the inside which must make great variations in the mercurial heights. 14th. And finally that they must be of so simple a construction, that they are less liable to be so soon out of order; and not so difficult in their construction as those of Messrs. Amontons, Hugens, de la Hire, Cassiny, and Bernouilli.[1]

These were the requisites wanting in this Philosophical instrument, which I now apprehend have been fully obtained. The great difficulty I met with in this attempt was to find an able hand, by which my thoughts might be put into practice. M^{r}. Henry Pyefinch a very ingenious Optician and Philosophical instrument-Maker has already executed in different manners some of these instruments, which are now exposed to the Public,

---

(bb) See Holman's account read to the Royal Society in the year 1742. Num. 464 of Phil. Transact. [Samuel Christian Hollmann, "Epistola de differentiis altitudinum barometrorum", *Philosophical Transactions*, 42 (1742), 116-22] and the History of the French Academy for the year 1705 pag. 21 ["Sur un irregularité de quelques barometres", *Histoire de l'Académie Royale des Sciences*, 1705, Hist. pp. 20ff, summarizing four papers by Amontons]. And also the Treatise De l'Aurora Boreale by Monsr. de Mairan pag. 43 [Dortous de Mairan, *Traité physique et historique de l'Aurore Boréale* (Paris, 1733), p. 43] where many considerable differences in the heights of the Barometer, are only imputed to the particular influence of their tubes upon the Quicksilver.

1 i.e. Guillaume Amontons, Christiaan Huygens, Philippe de La Hire, Jean Dominique Cassini, Johann I Bernoulli. These and several other barometers were described and commented on by J. T. Desaguliers in his *A Course of Experimental Philosophy* (London, 1734), vol. 2, pp. 262ff. They are also described in W.E.K. Middleton, *The History of the Barometer* (Baltimore, 1964).

fully answering the aforesaid requisites, as will clearly appear by the following description.[2] 65.06.06

AKBR (fig. 1.) is a bent tube or Siphon hermetically sealed at the top A: the diameter of whose bore is between ¼ & ⅓ of an inch: and its internal dimensions are so carefully measured, that every part of its length is settled, at proper distances, in proportion to the capacity of the Short Leg viz. the space or width C, E, 30, 31, is equal to the space D, Z, 31, 30, and the space contained between R, H, 28, 27, to the other space between G, X, 27, 28; &c: so that the boundaries of these Columns 27, 28, 29, 30 and 31 inches are exactly answering to one another in each Leg. By these circumstances (which are the most difficult to be performed) it appears that the three first necessary qualities are obtained[(cc)].

The Siphon being adjusted, as above described, it is to be carefully filled with mercury well purified by distillation and filtration from all moisture and other heterogeneous particles, and then boiled for that purpose in the same tube, or at least in a stone-vessel; which when thus done, is poured warm into the Siphon in such a manner, as not to let in the smallest bubbles of air.

Fig. 2[d]. represents a frame where the Siphon is fixt into a proper groove with the loops δ and is adjusted in a closet where the temperature of the air is uniformly brought to 55 degrees of Fahrenheit's thermometer, and then as much quicksilver only is suffered to remain in the tube, as to make both surfaces exactly answer to the same number in each Leg.

Fig. 3[d]. is a double Sight-vane, which slides upon the short Leg in the groove LR (fig. 2[d].): the extreams of whose internal angles x, n serve to direct the visual ray horizontally from the lower surface of the mercury to the Observer's eye, by which means the 4th. necessary quality is fully obtained, and if a magnifying Lens should be placed within this vane, all the observations will be made so accurately, that the 5th. necessary quality will be accomplished. And the method to raise or lower this Sight-vane is done by means of the milled-head H, which is furnished with a pinion and Slider LS.

From the vane E goes a watch-chain D, which surrounds the wheel G at the bottom of the hand or index GMO, and is fastened to the spiral spring T: and in order to keep this Chain parallele to the column of mercury in the short Leg, the little pully V is applyed. According to this movement, the

2 See also the specification lodged by Pyefinch and Magellan, 24 July 1765, to support the patent issued to them, 28 March 1765.

(cc) The settling the internal dimensions or capacities of the tube, may be performed by adding equal quantities of Quicksilver, into every part of its length, marking it from time to time. But I am now aiming at a better and easier method, which, if it proves successfull, shall communicate to the public at a proper time.

65.06.06 index or hand GMO will shew, upon the Dial NPQ, the smallest differences which happen in the heigths of the mercurial surface in the short Leg from S to R, by very large spaces: viz. shewing by its point the number of inches and tenths of the mercurial Columns; and by the horse hair MO, streched in the middle of it, the hundreth parts of an inch upon the transversal or diagonal lines thereon. By these means all the smallest differences of the mercurial heights may be very conspicuously shewn (according to the 5th. article) by the equal rising or sinking of the Surface in the short Leg in proportion to that in the long Leg: and all the inconveniences of Dr. Hook's wheel-Barometer,[3] pointed out by Drs. Desaguliers,[4] Muschenbroek[5] and others, are perfectly avoided. But the same end of obtaining the 5th. necessary quality and the third also, will be more fully accomplished upon the Vertical-Aerostathmion which is described hereafter.

The button or handle γ serves likewise to move the upper Sight-vane αν, through which the upper surface of the mercury is seen: and by the chain αβ this movement is communicated, in the same manner as is before described, to the index θλ, so as to make one whole turn upon the dial φω, whilst the sight vane moves throuh the space answering to one inch of the mercurial Column. By these means all the differences, (if there are any) caused by the heat and cold upon the bulk of the quicksilver, will be shewn in this dial: and by the rule of proportion, the quantity, belonging to the small column of the short leg, may be exactly corrected. From thence it appears that the sixth requisite is then fully obtained: for as we have regulated and settled all the measures of the mercurial columns in the mean temperature of 55. degrees of Fahrenheit's scale; we at once clearly know what is the excess or diminution of the mercurial Bulk at every heigth of its columns upon the upper Dial, by comparing the numbers there indicated with the other numbers shewn in the lower Dial: and consequently we may correct them perfectly. But as the proportion of these variations in both legs, is in a direct ratio to their different lengths, and we only reckon the mercurial heighths by the sinking and the rising, caused thereby in the surface of the short leg; we shall (by the calculation as above) com̄only find the exact quantity to be allowed, so small, that it may be rejected without any sensible error in ordinary observations[(dd)].

---

3 See Robert Hooke, *Philosophical Transactions*, 1 (1666), 218-9.

4 Desaguliers, *A Course of Experimental Philosophy*, vol. 2, pp. 263-4.

5 Musschenbroek, *Elements of Natural Philosophy*, vol. 2, p. 171.

(dd) Suppose that by the lower dial we find 30,45 inches, & by the upper one 30,56: Then the bulk of quicksilver is expanded $^{11}/_{100}$ of an inch. Now suppose that by measuring the two

Now since we have a very sure method to measure the real expansions and contractions of the quicksilver, contained in each Aerostathmion, from the mean temperature of 55 degrees of Fahrenheit's to any other degree whatsoever of heat or cold; therefore by marking the true quantity, answering to each degree, above & below the said temperature, upon the external moveable ring μξβ, we shall have a true Thermometer in the same tube, together with the Barometer. 65.06.06

For putting the 0 of this ring against the number, known before by the index of the lower dial: the other index will shew, according to the movement of the upper sight-vane, the degree of heat or cold at that time[ee].

---

whole Columns of mercury, we find 33 inches from n to C; & 2½ inches from C to *x*: then the proportion 33 : 2,5 :: 0,11 : 0,0083 shews that the quantity which ought to be corrected, in the length of the mercurial column, is but $^{83}/_{10000}$ so that instead of 30,45 inches, we shall know the aerial pressure to be at 30,4417 = 30,45 – 0,0083 inches. On the contrary, if the lower surface was at 30,56, and the upper surface at 30,45: by the same proportion we should know, that the true length of the mercurial column, answering to the pressure of the Atmosphere, was 30,5683 = 30,56 + 0,0083 inches. From whence appears the great advantage of not reckoning the mercurial heighths upon the real column of the mercury, as it is done with the Torricellian tubes or common Barometers; but to reckon its different appearances in the short Leg of a Siphon, after it is regulated & ascertained by my method described in this manner. Since if we chuse to make the aforesaid easy calculation, we may have the true height of the mercurial columns with the greatest exactness. But if not regarded, we have only a small & im̄aterial error of less than a centesimal in the greatest warmth or cold of a common year in our climates.

(ee) According to the common ideas we have of heat & cold, this is the best method, by which the Thermometrical degrees may be generally understood: since I call the increasing numbers downwards (after the 0 answering to the mean temperature, or 55 degrees of Fahrenheits scale), degrees of cold, which freezing point remains at the 23 degree, answering to 32 of Fahrenheits: and the numbers upwards, degrees of heat, which in com̄on Sum̄ers arrives at the 23d. degree, answering to 78th. of Fahrenheit's scale, & equally distant from the said mean temperature, as the freezing point. But as these thermometrical differences in com̄on Siphons are very small, they may be enlarged at pleasure by making the Siphons according to some of the forms & shapes represented by fig. 5. 6. 8. 9. 10. or 11 for then the bulk of the quicksilver will be greater in regard to the width of that part of the tube, where these degrees will be reckoned, & consequently the expansions & contractions will be more & more perceivable: N.B. Tho' the temperature of the Caves of the Royal Observatory at Paris is constantly at 9¼ degree of Reaumur, or at 53. of Fahrenheit's (Dr. Martine's Essays pag. 225 [Martine, *Essays Medical and Philosophical*, p. 225, where however the constant temperature is stated to be 10¼ degrees]), it doesn't interfere with my plan: since *we* are not sure (says Dr. Martine pag. 184 [*Ibid.*, p. 184]) that other deep pits would coincide precisely with the temperature of these places. The difference of soils & different depths might occasion a considerable variation. And besides I don't aim to put the 0 of my Thermometer, but at that degree equally distant from whence our bodies feel a sensible heat or cold, which by a com̄on agreement is fixed long since at the 55. degree of Fahrenheit's scale.

65.06.06 Tho' the mercurial expansions and contractions of this Thermometer, are freely performed in the vacuum at the top of the tube; nevertheless the same mercurial fluid is really bearing (through the mouth of the short Leg) the whole pressure of the Atmosphere, in so much as to be equally affected by it, as the vessels of our Bodies are. Therefore the differences of the expansions and contractions of fluids, owing to different degrees of heat and cold under divers pressures of the Atmosphere, being very considerable, as may be found in D^r^. Boerhaave's Chemistry, D^r^. Martyne's Essays, Musschenbroek,[6] and others who wrote on this subject: now it is very plain, that this kind of Thermometer is the best of any hitherto invented, to make medical observations. Since by the same reasons all other Thermometers (which are hermetically sealed) not being equally affected as our Bodies, by the different pressures of the Atmosphere combined with different degrees of heat and cold, they will often be found contradictory in their appearances, shewing many times the same degree, when the true expansions or contractions of our own fluids are very different from those of their mercury.

The 7th necessary quality is easily obtained by taking a due care to keep always the exact quantity of quicksilver, with which every Aerostathmion was at first graduated. The 8th. requisite may be also easily obtained by inscribing upon each Aerostathmion the specifick gravity of the quicksilver. But as metals & water have their specific gravities very variable, according to their mixtures, and to other circumstances, which are not commonly known as appears by the different tables published by many Authors: I believe it would be better to ascertain the said gravity of the quicksilver by comparing it relatively to the specific weight of the common gold coins of Europe, their standards being constantly and carefully kept in the respective mints[(ff)].

---

6 Magellan is presumably referring to Herman Boerhaave, *Elementa chemiae* (Leiden, 1732) or to one of the English translations of this; to George Martine, *Essays Medical and Philosophical* (London, 1740); and to Petrus van Musschenbroek, *The Elements of Natural Philosophy*, trans. John Colson (London, 1744), the edition of Musschenbroek's work that he cited elsewhere in his paper. However, we have been unable to identify the specific passages in these works that Magellan had in mind.

(ff) Besides there must be observed a constant term in regard to the temperature and pressure of the Atmosphere, whilst these hydrostatical tryals are made: since from these two principles arise some differences not inconsiderable (see Philos. Commerc. by D^r^. Lewis tom. 1. pag. 42 & 43) [William Lewis, *Commercium Philosophico-Technicum; or, the Philosophical Commerce of Arts*, vol. 1 (London, 1763), pp. 42-3]. As to myself I fix these terms at 30. inches in the Aerostathmion, which mercury has the specific gravity to that of the gold of a Guinea, as 13,60 to 18,10: and at the 0 of the aforesaid medical-Thermometer, or at the 55. degree of Fahrenheit's scale, which will be the same.

By the form of the Siphons the 9th. requisite is obtained; Since the surfaces of both extreams of the mercurial Column, as well as their contacts with the tube, are confined in its inside, and equally visible in each Leg, in so much as to make the true distance between them easily observeable. 65.06.06

The 10th. requisite will be found in the nature of the Siphons also, there not being such a free passage through the mouths of the other kinds of tubes when near to the bottom of the Cistern, as it is in the bent part of the Siphons, which is always equally oppen and smooth in its inside.

As to the 11th. requisite it will be easily obtained by marking upon the instrument the plumb line ω, σ, G, Z (fig. 2^{d}.), or by adding to it a spirit-Level ω, Z (fig. 14).

And to obtain the 12th. requisite, there must be made a just comparison of each Aerostathmion with a good one, kept as a standard, in some Museum Viz. at the Royal Society of London &c: and if there is found some difference, to mark it upon the instrument with the sign add, if less; or subtract, if more.

The 13th. requisite, that is to convey this instrument from one place to another without emptying of it, is also obtained by the nature and figure of the Siphons. For I find it may be done with safety, if the Siphon is put in a position almost parallele to the horizon, due care being taken in inclining it down, and in getting it upright by a vertical motion upon the large Leg, and in such a manner that the short one be always over the other, that the air does not rush into the long Leg. Add to it, that, if a bit of leather is carefully tied round the mouth of the short Leg; the instrument may be carried more commodiously in an inverse position. But when the distances are so great as not to allow there being carried by hand; in such case let the tube with its contents of quicksilver be carefully and exactly weighed, then by tying a small bladder or a wooden box about one inch below the mouth of the short Leg, as it is seen in fig. 4., the whole is then filled with Quicksilver. In such a condition it may conveniently be packed in a box, and sent to the intended place without danger of any air rushing in it, the mouth of the tube being always immersed to the middle of the Quicksilver in the Wooden box or in the Bladder. At its arrival this apparatus is thrown aside, and as much quicksilver is taken out of the tube, as to reduce it to its former weight. Otherwise the same end, I apprehend, may be well obtained, by pushing into the short Leg a kind of piston, which will keep the quicksilver sufficiently close to prevent its breaking the siphon.

In regard to the last requisite, I believe, it cannot be imagined after this description, but that it is obtained: chiefly considering, that this

65.06.06 instrument may be reduced to a more simple construction as in fig. 12, where the Nonius-divisions answer to the end of the two dials. But if the Observer likes better to see the variations of the mercurial heights more fully by whole inches and their divisions: then a brass rule must be placed there, with the real measure of 31 or more inches, as in fig. 13. And in such a case the lowest boundaries of the quicksilver must be ascertained at B, that the 6th. requisite may be obtained. For inclining laterally the instrument by means of the slide-screw P, the thermometrical differences will be found in the little dial NR, at the top of the brass-rule: and by the rule of proportion may be known the just correction, which must be done in the whole mercurial Column, already measured.

To distinguish these different species of aerostathmions, the first (fig. 2^{d}.) is called the complete: the second (fig. 12.) by the name of simple aerostathmion: and the third (fig. 13.) by the name of Philosophical. The fourth species is represented in the fig. 9. and is called the Medical, it being only destined for the observations upon the heat and cold, combined with the different pressures of the Atmosphere. There is no apparatus for Barometrical experiments and to make the degrees of heat and cold appear more distinct, two pieces of thin tubes must be added, one to the top, and the other immersed in the short Leg, according to what is said before in the Note ee.

Besides these four species, there is another contrivance of this new instrument, distinguished by the name of the vertical Aerostathmion, which I apprehend to be the best. It is made upon the same principles: and by it the 3rd. and the 5th. requisites are more fully obtained. It is represented in fig. 14. First the Siphon and the quantity of the quicksilver is adjusted by the foregoing rules; and fixt in a frame so contrived that C may be the Center of motion. Then the lower surface of the mercury is to be kept in the said Center C, which may be done by means of a fixt sight-vane as that of fig. 3^{d}.; the tube being moved towards one or the other side in a vertical manner, with the handle H.

The long Leg of the Siphon, being furnished with a hand, becomes an index to the great arch SS, which may be made of any dimension, and is divided by the radius, thrown from the Center C and passing by the end of the sines answering to the different altitudes of the mercurial columns; as it may be easily understood by inspecting the figure. The cross OBEG is supported upon a steady pin x in the back-side of the top of the instrument: this movement being communicated by the watch-chain GV, to the double pully W, and to the axle of the index θλ, each tenth of an inch is divided into hundred parts upon the dial ϕω. the two spiral springs T and R serve to keep the cross OB close to the steady pin x̲. And a second Sight-vane is added to the place A, which communicating in the backside with the pin x̲, shews the small expansions and contractions of the quicksilver, upon the said dial ϕλ, to the thousandth part of an inch. 65.06.06

I am
My Lord
Your Lordship's
Most Obedient and most humble Servant
John Hyacinth de Magalhaens

Cornhill, in London,
June the 6th. 1765.[7]

7 The reading of Magellan's paper to the Royal Society was begun at its meeting on 13 June 1765 and continued at the meeting on 20 June, but not completed until 19 December 1765; cf. Magellan to T. Birch, 5 November 1765. As Magellan notes, he had already presented a paper to the Society, read at the meeting on 21 February 1765, in which he had discussed at length the deficiencies of existing barometers. An anonymous four-page pamphlet was subsequently published – presumably once it became clear that Magellan's description of his new instrument was not going to be published in the *Philosophical Transactions* – that gave a summary account of it (*Notice sommaire d'un Baro-Thermometre nouvellement inventé à Londres en 1764, & rendu public l'année suivante sous le nom de AEROSTATHMION, ou vraye Balance de l'Atmosphere* [London, 1765]).

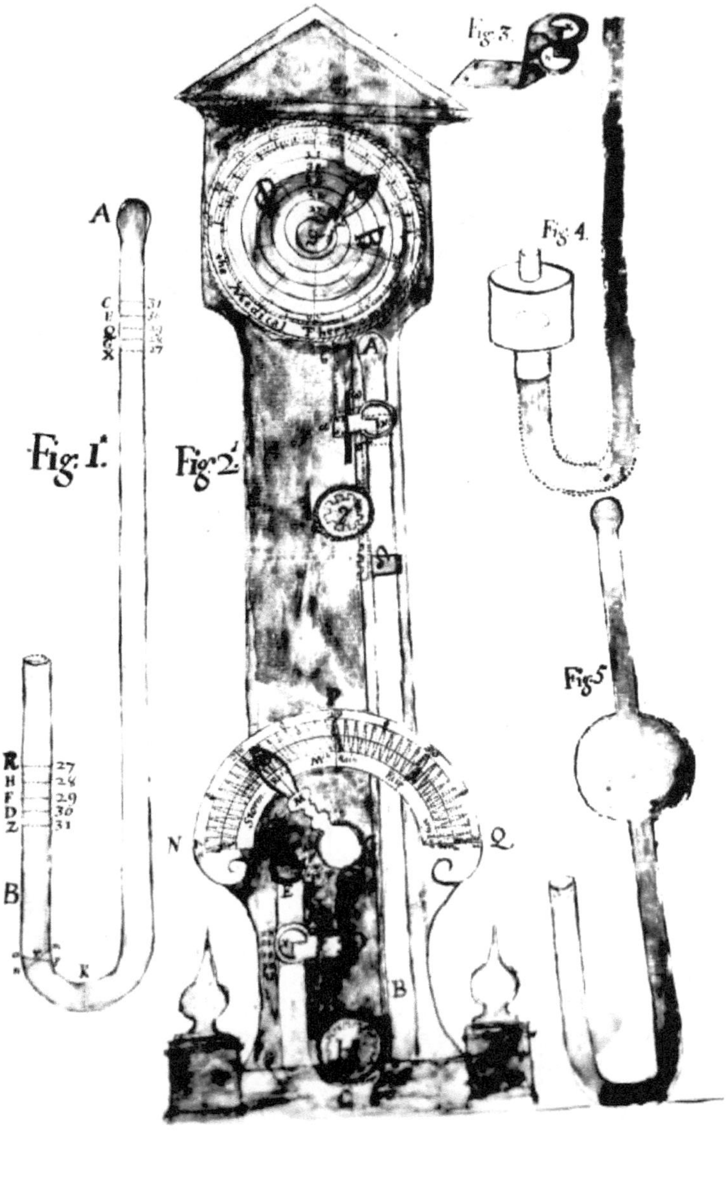
Fig. 1.
Fig. 2.
Fig. 3.
Fig. 4.
Fig. 5
A
B
N
Q

Fig. 6
Fig. 7
Fig. 8.
Fig. 9.
Fig. 10.
A
Æ
C
P
B
D
E

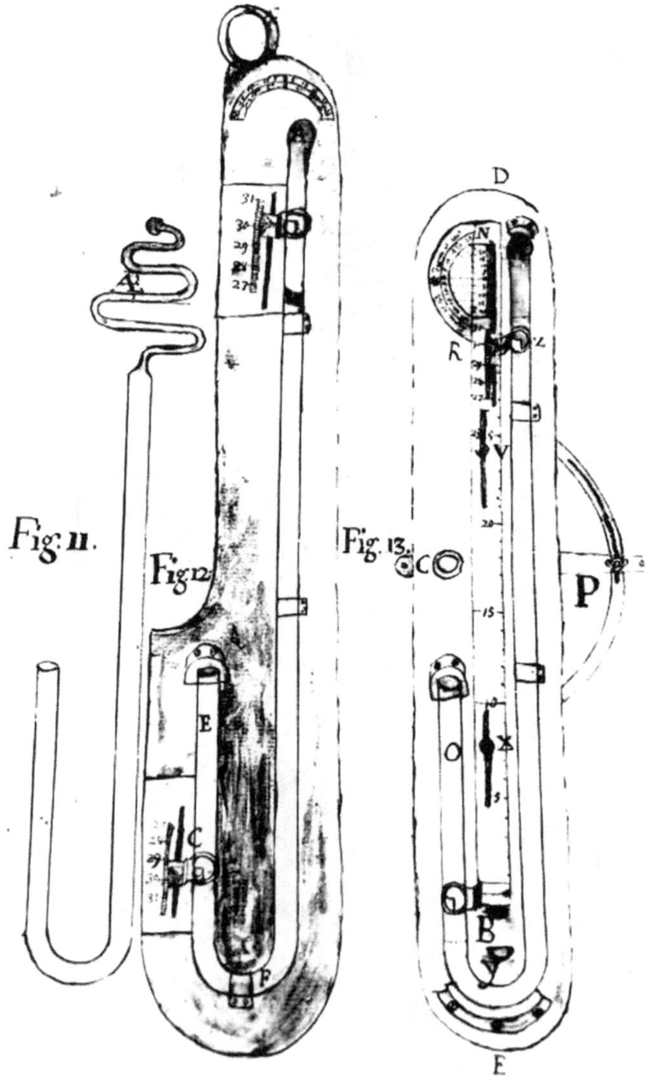
Fig. 11.
Fig. 12
Fig. 13.
A
E
C
F
D
N
R
V
X
O
B
P
E

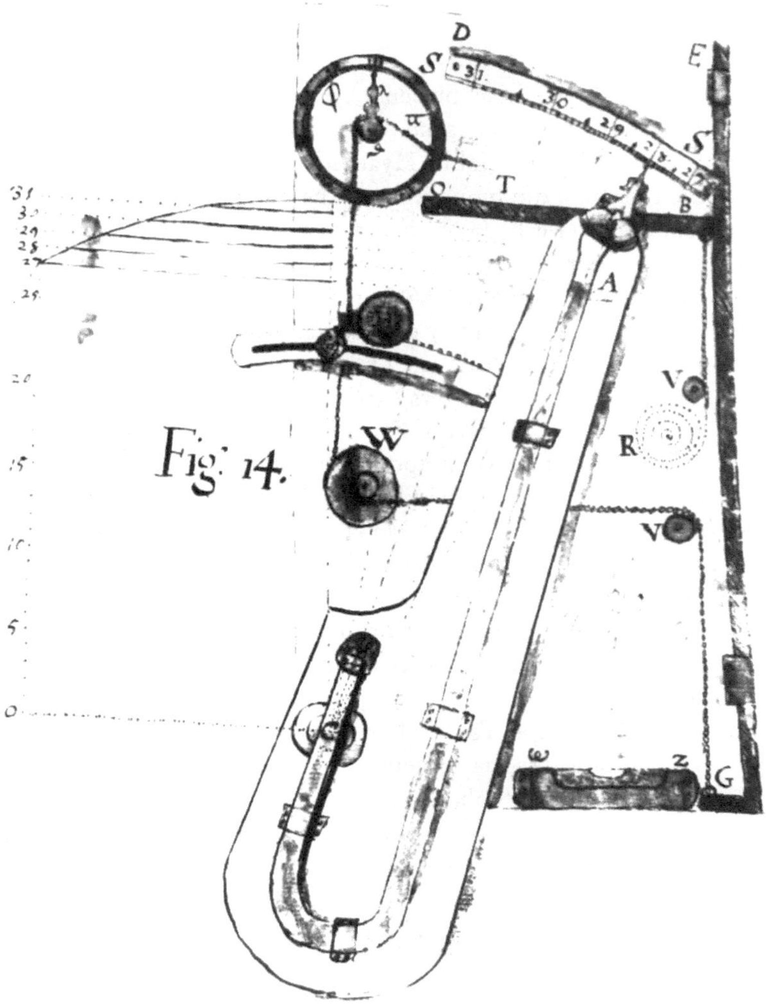

Fig: 14.

65.07.24 *Patent Specification*
National Archives, London, C210/7[1]

To All to whom these presents shall come I Henry Pyefinch of Cornhill London Optician send Greeting Whereas I the said Henry Pyefinch together with John Hyacinth de Magalhaens of Cornhill aforesaid Gentleman did by petition to the Kings most Excellent Majesty humbly shew that we had found out and brought to perfection a new constructed Instrument by which is shewn the true effect of the Weight of the Atmosphere with the Variation which is caused by the Heat and Cold and likewise the Quantity of that Variation and therefore as well as for other reasons contained in the said petition did pray his said Majesty to grant unto us his Royal Letters Patent for the sole use and benefit of the said Invention within that part of Great Britain called England, Dominion of Wales and Town of Berwick upon Tweed and also within his Majesty's Plantations and Colonies abroad for the Term of fourteen Years To the prayer of which petition his said Majesty was most graciously pleased to condescend and Letters patent agreable thereto have accordingly passed the Great Seal[2] Now Know Ye that I the said Henry Pyefinch in compliance and in performance of a proviso or condition contained in the said Letters Patent Do hereby describe and ascertain the said Invention in manner following vizt.

1st, Instead of a simple Torricellian Tube we make use of one bent or in form of a Siphon for this Instrument

2^d^, The Inside of this Siphon is to be exactly measured by any method able to settle the Capacity of its Bore and account for the different Sizes thereof which may be done by pouring therein a certain Quantity of Quicksilver adding as often thereto the same Quantity till the whole be filled making[3] at each time the space occupied thereby

3^d^ To observe in temperate weather vizt at fifty five degrees of Fahrenheits Scale the true Bulk of the Quicksilver which is necessary for the filling the Siphon and settle every different Altitude of the various Mercurial Columns that may happen throughout the year vizt from twenty

1 Abridgement published in *Patents for Inventions: Abridgements of Specifications relating to Optical, Mathematical and Other Philosophical Instruments, 1636-1866* (London, 1875).

2 See Letters Patent, 28 March 1765.

3 marking?

seven or twenty eight Inches to thirty one according to the Tubes Inside which will be known by the second Article 65.07.24

4th, To make very exact Observations with such Siphon in all the different Temperatures of the Weather and to calculate upon them all the different Degrees of the various Heighths of the Mercurial Columns already settled by Article 3, which Numbers of the Degrees must be marked in an inverse order vizt directly down upon the short Leg and inversely right up in the long Leg always keeping the true mean temperate Bulk of the places where are marked the Numbers belonging to the longest Leg

5th, To Examine with a good Hydrostatical ballance what is the true specific weight of cubic Inch of the Quick Silver contained in each Siphon

And I the said Henry Pyefinch do hereby avow and declare that the above is a true full and perfect description of the nature and practice of the said Invention In witness whereof I the said Henry Pyefinch have hereunto set my Hand and Seal the twenty fourth day of July one thousand seven hundred and sixty five

Henry Pyefinch[4]

4 Marked: 'Sealed and delivered being first duly stampt in the presence of Luke Currit | Nat^{l}. Nicholls' and 'And be it Remembred that on the same twenty fourth day of July in the Year abovesaid the aforesaid Henry Pyefinch came before our said Lord the King in his Chancery and acknowledged the specification and all and every thing therein contained & specified in form abovewritten, And also the specification aforesaid was stampt according to the tenor of the Statute made in the sixth year of the Reign of the late King & Queen William & Mary of England & so forth | Inrolled the day and year last abovewritten'. A signature, 'J. Eames', appears in the margin beside this passage.

65.09.25 *To John Canton*

Royal Society of London, Canton Papers, II, 54

Sir

Although I have not the honour of your acquaintance, nevertheless I flater my-self that your kindness will not deny me the favour of one Leisure hour when soever you shall please to appoint, to obtain the satisfaction I wish for, upon some particulars relative to Philosophical experiments, with which I hear You are skillfully delighted.[1] I hope you never will regret the having obliged a foreigner, who possesses the qualities, which make a good friend, tho' otherwise destitute of Learning or other personal merit.

I am
Sir
Your most obed^{t} hble serv^{t}.
John Hyacinth de Magalhaens

At M^{r}. Pyefinch's in
Cornhill. 25 Sept^{ber}. 1765

To M^{r}. John Canton
Spittel Square

1 Canton evidently gave Magellan some time; cf. Magellan to J. Canton, 26 September 1765.

To John Canton 65.09.26

Royal Society of London, Canton Papers, II, 55

Sir

There is but two hours past, that I came from your kind company, and I cannot tranquilize my self without writing to you again – Your experiments, about the compressibility of water,[1] strok me so hardly, that I should not sleep, if I put not or lay down on the paper the thoughts, which are swiming now on my brains in despite of my self. You perhaps will call it a foly, but no mater: I cannot help it . if it is so; I should be not the only one, who obtains but non sense, after very good Leçons. – But to avoid the tediousness of my coarse language, and to spare your precious time, Let me go speedily to my thoughts, which perhaps are alwais the first, You have meet with at every time that You have used the same kindness with other blockheads as me.

How might it be so, dear Sir, that a body in the same act of it's Compression, be able to receive a new addition of mater in it's masse, which was not there before! – You comprime with your hands the sphere of glass, full till it's neck with water: and by this compression, the water which was in the neck, is received or absorved in it! – What? – You diminish the capacity of the ball (or phial for I do not which is the proper term): and it is able yet to receive more water? – my head is out of it's place, and my brains turn over with such experiment. – Good God! it is quite the contrary effect of all others. When I comprime some soft or fat mater or body in my hands, it scapes out between my fingers. – I think that if there was a such neck communicating with the golden ball, which the Florentine Academians tryed in the past time; the water, instead of sweting out between it's pores, would undoubtedly rush out by the neck. – Why the water does not springs out by the neck of your phial? Since there is not any valve, which hindres it. – No, my dear Sir; I fear that there is some

1 Canton was awarded the Royal Society's Copley Medal for 1764 for his experiments demonstrating the compressibility of water, in opposition to what had been generally accepted following the experiments of the Florentine Accademia del Cimento in the early 1660s (cf. *Philosophical Transactions*, 52 [1761-2], 640-3, and 54 [1764], 261-2). Canton's interpretation of his experiments was, however, controversial, and the award of the medal was delayed for a year while the experiments were further investigated by a committee headed by Lord Charles Cavendish that was established by the Society for this purpose. The award was finally confirmed at the Society's Anniversary Meeting on 30 November 1765.

65.09.26 what hide from our (at least certainly from mine) eyes in this experiment. There is some other play of the Laws of the Nature, which I can not guess. – But if it is not so: be so kind, I pray, as to lighten me about it. And if you chuse better, by sake of sparing your time, write me two Lines (for I am pretty sure of you having thought upon these difficulties, a thousand times before) and be so obliging as to send me with it, the three numbers, about which we have spoken, found by your own experiments, of which if you give me leave, I'll use with the due acknowledgement.

1^{rs}. the fraction answering to the increasement or expansion of the mercury on each degree of Fahrenheit's scale: for I fear by forget, that it is not 185/10,000 as it seems me now. 2^{d}. the specific weight of the mercury, and also of the golden guinea, if you have tryed it. An 3^{d}. the precise differences You found in the degrees of the boiling water under different pressures of the atmosphere.

I am
Sir
Your most obed^{t}. h^{ble}. serv^{t}.
John Hyacinth de Magalhaens

P.S.
Perhaps your experiment proves that there is not at all any elasticity in the water. Ther[ef]ore[2] being forced into a narrower space, the water of the neck goes to fill the remainder. But why does return the same to refill again the neck, after it's freedom? – is it not it's own spring or elasticity, which does it?

At the 26 in this morning.

My dear Sir

As soon as I get up, the first I did, was to read over again those Lines: for I find my phantasia not less full (or fool) of your Phial as yesterday night. I have dreamed with it. and I fear that the first trouble I will give you in the next, will be to seek for me a place in Bethlem Hospital. – Devil the phial! – it is not for me as that I have read of long ago in the Ariosto Poems, where the Astolphe found his good sense, or where good

2 Paper damaged.

judgement was shot up.[3] – If I was not so warm about it in my head; I will not sent to you this sheet: for I would be shamed of. add to it the impudence of my beggary about the fruit of your experiments; I mean the 3 numbers, I ask for! But there is all my pretension or claim to Philosophy: Viz. to tell sincerely and freelly all my wishes and thoughts. – that is the case. I will have those numbers, and have not the pacience of concealing it from you. Nevertheless if there is any reason on the contrary, or if it does not please to you: then send not them. and be sure that it shall not change nothing in the worth sentiments, steem, respect, and friendship, I am possessed of in your regard. 65.09.26

I forgot yesterday to tell You, why I called my thermometer on the Aerostathmion, the medical one.[4] You will see in this paper print the motif of it, which I wish you may find in the right
I am &c[a].

at M[r]. Pyefinch in Cornhill
26 Sept[r]. 1765.

To M[r]. John Canton
in Spittal Square[5]

3 Ludovico Ariosto, *Orlando furioso* (1516), Canto 34, 82-87, where Astolpho retrieves his good sense, which he has found bottled up in a jar.

4 Cf. Magellan to the Earl of Morton, 6 June 1765.

5 This letter was published in full by Augustus de Morgan in his account of the Canton Papers in *The Athenaeum*, 17 February 1849, "to teach people that the letters written by foreigners in novels are not impossible". He adds, "We have a great respect for the writer", and afterwards gives the following not altogether accurate biographical notes:

> This Magellaens was a person of some mark. He was a Portuguese and a Jesuit, descended from the great navigator of the same name. He resided in England, – where, we believe, he died in or shortly before 1790. He had very early information on scientific matter from abroad, – and was frequently employed in procuring English instruments for foreigners. In 1768, just after Mr. Malony had been convicted at the suit of a common informer under the statute against saying mass, Magellaens and three others were indicted for the same offence. But Lord Mansfield and all the judges held that it was necessary that the accused should be *proved* to be a priest, as well as to have said mass, – and the second attempt failed. The penalty was nothing less than perpetual imprisonment; but Mr. Malony had received a pardon, it is said, before the decision of the judges could help him.

65.11.05 *To Thomas Birch*

British Library, Add. MS 4444, f. 83

Sir

I here send to you this new Abridgment of my two memoirs upon the Barometer; desiring You'll please to read it, in the place of what is yet remaining from the Last Memoir:[1] having found, on consulting some friends, that the other abridgement, I have delivered to You, was wrote in bad English and not sufficient to give a full idea of this contrivance. Therefore should be glad, when You are at Leisure to have the said two Memoirs with the first Abridgement returned.

When this new abridgement is read before the R[l]. Society, the Notes therein must be omitted; they containing only some calculations and other niceties, commonly disagreable to the hearers: and as to the figures or drawings, they must be taken off from the end of the former abridgement.

I hope the Committee will not judge these sheets unworthy of being printed in the Transactions of this Year, there being some new improvements, I apprehend, not a little usefull to experimental Philosophy.[2] In this case

---

1 The first of Magellan's two memoirs was read to the Royal Society on 21 February and the other on 13 and 20 June 1765. As Magellan here notes, the reading of the second paper (of which Magellan to the Earl of Morton, 6 June 1765, is an abridgement) had not at that time been completed. The new version was read to the Society at its meeting on 19 December 1765 (Royal Society, *Journal Book Copy*, XXV, 691-7). The whole episode evidently gave rise to difficulties in the Society, to judge by two resolutions adopted by the Council of the Society at its meeting on 28 November 1765, viz. (1) "that it be a general Instruction to all future reading Secretaries of the Society, that they peruse the several papers communicated to the Society, before the said papers are read in Public – Excepting in such particular cases as the President or Vice President in the Chair shall occasionally direct to the Contrary" and (2) "that the said Secretary note on each paper the date of its receipt – and read the Papers according to the Order of Time in which they have been received, unless the President or Vice President shall otherwise direct" (Royal Society, *Council Minutes Copy*, V, 143).

2 Magellan's hope was not fulfilled: his papers were not printed in the *Philosophical Transactions*. Extracts were, however, printed separately as an undated four-page pamphlet headed "Notice Sommaire d'un Baro-Thermometre nouvellement inventé à Londres en 1764, & rendu public l'année suivante sous le nom de AEROSTATHMION, ou vray *Balance de l'Atmosphere*", a copy of which is held by the Royal Society. The pamphlet ends with an advertisement for Henry Pyefinch, mathematical instrument maker in Cornhill, who, it was proclaimed, had an exclusive licence to sell Aerostathmions according to Magellan's design.

I shall be very glad to have a review of the cuts answering to my drawings, to see if the engraver has executed them truely. 65.11.05

I am
Sir
Your most humble Servant
John Hyacinth de Magalhaens

Cornhill 5 Nov^{bre}. 1765.

To James Ferguson[1] 65.11.08

Royal Society of London, Letters & Papers, IV, 279

I apprehend, dear Sir, that this contrivance is, in part, the same Improvement I have found long since, to enlarge very considerably the smallest differences of the mercurial heights of my Aerostathmion; tho' I have not yet spoken of it in the memoir I presented to the Royal Society,[2] because I did not choose to give any account of what I was not able to shew reduced into perfect practice.

1 An extract from a letter to Ferguson, to which is prefaced the following letter of explanation addressed to Thomas Birch, Secretary of the Royal Society of London: "Reverend Sir, | At the last meeting of the Royal Society, I observed a Barometer lying on the Table before the Chair, but did not understand the Construction of it. | Mr. Magalhenes, who was then present, observed that I was at a loss how to explain that Barometer to some Gentlemen, after the President had left the Chair. – On the next Day Mr. Magalhenes sent me a Letter containing an Explanation of the said Barometer, with a proposed Improvement of his own; which he desired me to communicate at next meeting of the Royal Society, if I thought proper. | In order therefore to satisfy those now whom I could not at last meeting, I shall here transcribe that part of Mr. Magalhanes's Letter which relates to the Barometer, with his proposed Improvement thereof; and am with the greatest respect, | Reverend Sir, | Your most humble Servant, | James Ferguson. | Little Castle Street | Novr. 14th 1765". The document is annotated: "Read at R. S. 14 Novemb. 1765."

2 Cf. Magellan to the Earl of Morton, 6 June 1665.

65.11.08

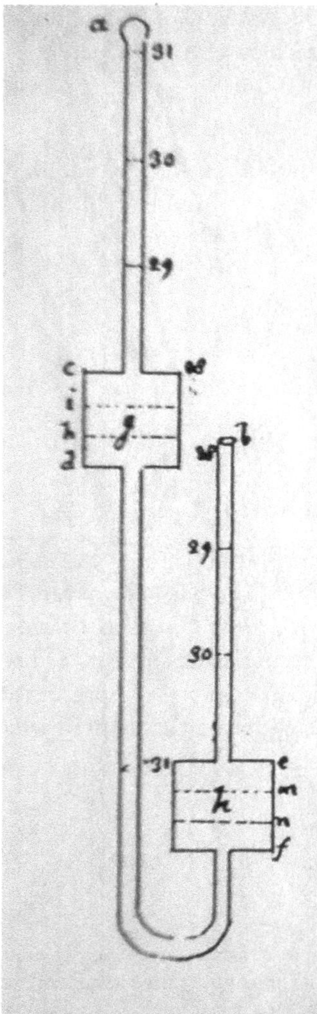

Let acb be a bended tube, like a syphon, with two boxes g and h; and that from c to f be the perpendicular distance of 31 inches. – Suppose the two Boxes to be of the same figure and size, as well as the two Tubes ac and eb; or, at least, that their irregular capacities are proportionally ascertained by any method.

Now suppose the Space contained between c and f to be filled with mercury; and that the tube ac, as well as the space of the cylinder h, are filled with coloured water or by any other liquor not too liable to coagulate or freeze by great colds.

Then it is plain, that when the mercury falls from c to i (equal to half an inch) the upper water will fall from 31 to 30, which space might be of five or more inches, according to the proportion between the tube and of the cylinder; and the under water will make the like rising in the lower tube: and consequently we shall have the smallest differences in the mercurial heights shewn by as large spaces as we please. 65.11.08

This is the contrivance which I have not yet fully reduced to practice, in such a manner as that it might be exposed to the public view; but which I believe I can bring to a sufficient degree of perfection, since I have the means of avoiding the fault arising from the evaporation of the water, or any other liquid, in the lower tube. And I have also the means of making appear in the two tubes all the different degrees of expansion or contraction either of the quicksilver or the other liquid, in such a manner, that the real effect of the pressure of the atmosphere be clearly shewn, distinct from them, as in my mercurial Aerostathmions; and that they may serve to shew the temperature of the air, as true Thermometers.

I think proper to communicate this to you, to satisfy the curiosity you shewed yesterday of knowing the aforesaid contrivance; which, you see, seems to be but half of the Improvement I have here described; that is, the said curved tube without the lower box h and its small tube eb. I am, with great respect,

Sir,
Your very humble Servant,
John Hyacinth de Magalhanes

Mark Lane
Novr. 8th 1765.[3]

3 Magellan gives a different address from his previous address at Cornhill (see, for example, his letter to Thomas Birch written three days earlier) as a result of a major fire that occurred in Cornhill on 7 November 1765.

66.01.09 *To Emanuel Mendes da Costa*[1]

Royal Society of London, Letters & Papers, IV, 292

Dear Sir

This serves to accompany a mahogany case, containing 36 Roman medals, more valuable I think for the particularity of the place where found, and the Remarks on them here subjoined, than for their Rareness. I desire You will do me the honour of presenting them to the Illustrious Royal Society in my name, to deposit in their curious Museum, now so elegantly and well disposed by Your great assiduity, care, and knowledge; by which means it becomes, not only an honour to the Society, but of real utility to the Learned.[2]

These medals I lately received from my Learned contry-man and good friend Dr. Antony Ribeiro Sanches, now living at Paris, and who formerly was Physician to the Emperess of Russia, and fellow of the Academy at Petersburg. He describes them to me in his Letter in the following Manner.

"In the year 1754, when I went to drink the mineral waters at Bourbon L'Ancy (a Town in Burgundy), the States of that Province caused the great well or Bath, built by the Romans, to be cleansed and repaired. This well was like an inverted cone, and built of free stone faced with white marble. They cleansed it to about the depth of 36 or 37 feet, for they never could empty it to the botom. But when they got to the depth of 10 or 12 feet, they met with a cavity in the wall, in which in a hollow stone there placed they found deposited a great number of copper and brass medals, and some of silver of the lower Empire. I collected from the workmen and the Director of the work, these I have the honour to send You. They also found one of very pure gold of Constantine the Great, of about a Ducat weight. This the Director

1 Annotated: "Received Jany. 9. 1765. Read the same day" and "Read Jan: 9. 1765 not to be printed April 7. 1766." The date 1765 is a mistake; 1766 was intended (Royal Society, *Journal Book Copy*, XXV, 709-12).

2 Emanuel Mendes da Costa (1717-1791) was Librarian and Clerk of the Royal Society and, as such, Keeper of its museum. During the summer of 1765 the Society's museum had been cleaned and set in order and an inventory compiled under the direction of a committee of three Fellows (Henry Baker, James Parsons and William Hudson), who in their report (*Journal Book Copy*, XXV, 632-40), had paid fulsome tribute to the diligence with which da Costa had assisted them. In December 1767, however, he was summarily dismissed by the Society after he was found to have misappropriated its funds, and he subsequently spent several years in prison. In 1781 the Society's collections were transferred to the British Museum; however, the medals Magellan describes cannot now be identified in the Museum's collection.

kept to present to the States. As these waters corrode Iron very soon, the stones of the well were crampt or linkt together with copper and Lead, of which I send You also a piece of each. this is an exact account of the affair." 66.01.09

It would be perhaps very needless to remark how much less corroded some of these medals are than others, as well as the piece of copper cramp. But I think the two following remarks, I have made upon these medals, will not be unacceptable to You. Viz that some of the brass ones have very considerable spots or spaces of copper in them, their surfaces appearing nevertheless very even and without any sensible sign of incrustation or of any new matter added. Which can only proceed (I should think) from the waters causing the spelter (mixed with the copper to make brass) to fly off from it.

The other Circumstance is, that on strongly heating three or four of them with a blow-pipe, many very small drops of a white metallic substance issued from their pores; which by it's maleability, abiding aqua Regia, dissolving intimately with aqua fortis, and giving a like solution to that of a piece of silver dissolved in this acid, proves it to be the same metal, which might be originaly in the copper. But what extraordinary I think,[3] is that by so simple a process, these metals should be separated from each other, after having been united by fusion. These metallic particles, I have hamered, and laid in a bit of paper under one of the medals at the corner of the box.

I am Sorry that of all the specimens, I mentioned to You (promised me by my friends at Portugal and the Brasils) none are come to hand but these four pieces of portugal marbles, which I take the Liberty of sending You also for the Said Museum: and beg leave to assure You, as soon as I receive any thing else, worthy to be deposited in your collection, shall think my self highly honor'd by their acceptance. I am

Dear Sir
Your very oblig^{d}. Servt. & friend
John Hyacinth de Magalhaens

Leadenhall Street
9 January 1766

To M^{r}. Emanuel Mendes da Costa
Librarian to the Royal Society
Crane Court[4]

3 Magellan has inserted superscript numbers to indicate that the correct order in which to read these words is "But what I think extraordinary…".

4 Additional annotations near the address: "January 9 1766" and "Not to be printed April 17. 1766".

66.05.14 *To Matthew Maty*[1]

Royal Society of London, Letters & Papers, IV, 329

Monsieur

Je viens de recevoir une Lettre de mon ami M^r. Messier[2] avec un article contenant la theorie & les Ephemerides de la comete dernierement decouverte, calculées sur ses observations pour le mois de May & le suivant oû elle pourra etre encore observée. Je crois, que cette communication vous fera du plaisir: Et je vous prie d'en faire part à la Soc^e. R^le. selon que le meme Astronome le desire. Voici ses propres paroles. La date est du 1er du courant mois.

"Je ne doute pas que vous n'aiez vu dans les nouvelles publiques la decouverte de deux nouvelles cometes; la premiere le 8 Mars; & la seconde un mois après, l'8 Avril. dans le tems j'en ai ecrit à Monsr. le D^r. Matty Secretaire de la Soc^e. R^le. pourqu'il voulut bien en faire part à la Societé, & qu'on puisse les observer – La prem^re. de ces cometes a commencé à paroitre dans la Constelation des Poissons. je l'ai observée le 8, 9, 10, 11, 12, 13, 14, 15, et le 17 de Mars. Le crepuscule m'a empeché de la voir plus long tems. Elle ne reparaitra pas le matin à sa sortie des raions du Soleil, parce qu'elle sera trop éloignée de la Terre. La seconde comete decouverte le 8 Avril au Soir, etoit considerable. on la voioit aisément à la vüe simple. Le 11 elle avoit une queue de 7 à 8 degrés de longueur. Je n'ai pu l'observer que le 8, 9, 10, 11, et le 12 d'Avril. Il n'a pas eté possible de la voir plus long tems à cause du grand crepuscule du soir dans lequel elle se couchoit. De ces cinq jours d'observations, Monsr. Pingré en a déduit les elements de sa theorie, qu'il donne comme très aprochant et que voici

1 This letter is accompanied by an English translation of the first section, down to but not including the actual elements of the comet's orbit as calculated by Pingré. The translation is headed "A Letter from M^r. de Magarlhans to D^r. Maty M.D. Sec.R.S." and is marked "Read May 15 1766". It was not, however, published in the *Philosophical Transactions*; instead, a composite account of Messier's discoveries was published under the title "A Discovery, with Observations, of two new Comets in the Marine Obvatory (*sic*) at Paris" (*Philosophical Transactions*, 56 [1766], 57-65). This did not include any of the text transmitted by Magellan but did give much the same information together with a detailed set of observations; presumably it was taken from another account of his discovery that Messier sent to London. The elements of the comet's orbit as calculated by Pingré and the ephemeris he had computed for May and June 1766 were included in the published report, and these may have been taken from Magellan's letter.

2 Letter not found.

Lieu du noeud ☊ de la seconde comete de 1766 - ♉ 17° 22' 19'' 66.05.14

Inclinaison de l'orbite - 8 18 45

Lieu du Périhelie - ♎ 26 5 13
Passage au Perihelie le 17 Avril à - - - - - - - - - - - - - 0^h 26' 13''
Distance au perihelie - 0,636825
Logarithme de la distance Perihelie - - - - - - - - - - 9,804020
Le mouvement Direct

De ces Elements il a reconnu que La Comete se reverroit le matin à sa sortie des raions du Soleil: en consequence il a calculé des Ephemerides pour le mois de Mai & de Juin; par le moyen des quels il sera aisé de la retrouver. Les voici

Ephemerides de la 2de. Comete de 1766 pour Le mois de May à 16^h. tems moien au Meridien de Greenwich

<u>1766</u>

jours H	Levé de la ☄	Longit. de la ☄	Latit. de la ☄	Ascension dr. de la ☄	Declin. de la ☄	Dist. de la ☄ à la terre	Dist. de la ☄ au ☉
Mai. 1. 16	15 52	♈ 5° 2'	3° 43 A	6° 3'	1° 23 A	0.437	0.709
3	15 45	3 0	4 37	4 37	3 3	0.455	0.729
5	15 39	1 19	5 26	3 24	4 28	0.473	0.750
7	15 32	♓ 29 54	6 10	2 23	5 42	0.492	0.773
9	15 26	28 41	6 51	1 32	6 48	0.510	0.797
11	15 20	27 39	7 28	0 50	7 46	0.528	0.821
13	15 14	26 47	8 1	0 16	8 38	0.546	0.847
15	15 8	26 3	8 32	359 47	9 24	0.564	0.874
17	15 2	25 24	9 1	359 23	10 6	0.581	0.901
19	14 55	24 49	9 28	359 3	10 45	0.597	0.929
21	14 49	24 18	9 54	358 45	11 20	0.613	0.957
23	14 42	23 49	10 18	358 29	11 53	0.628	0.986
25	14 35	23 22	10 40	358 13	12 25	0.642	1.016
27	14 29	22 56	11 2	357 58	12 56	0.656	1.046
29	14 22	22 31	11 23	357 44	13 26	0.669	1.076
31	14 15	22 07	11 42	357 30	13 56	0.682	1.105
Juin 5				356 48	15 03		
10				355 58	16 13		
15				354 57	17 24		
20				353 41	18 37		
25				352 10	19 53		
30				350 22	21 10		

66.05.14 La Comete doit être visible non seulement pendant le mois de Mai, mais aussi pendant celui de Juin; puisqu'à la fin de Juin sa distance à la terre ne sera que de 0,8 tandis que la precedente comete étoit eloignée de la Terre à mon observation de 15 Mars de 1,43 qui sans la lune auroit eté visible plus long tems. Vous pourriez communiquer ces Ephemerides à M^r^. le Secretaire de la Societé Royale pourque les Astronomes en prissent connoissance, & qu'on puisse revoir cette comete le matin. Les observations qu'on en pou[rra][3] faire seront très utiles pour verifier les elements qui ont eté tirés de mes observations qui sont en très petit nombre[.]"[4]

J'ai l'honneur d'etre avec un vray respect & beaucoup d'attachement

Monsieur
Votre tres humble & obeiss^t^. serv^r^.
Jean Hyacinthe de Magalhaens

Leadenhall Street 14 Mai 1766

3 Part of word obscured by binding.

4 Following this communication of Messier's, Magellan wrote out but then crossed out several further lines he had received from Messier concerning the publication of his observation of the comet of 1759, i.e. Halley's comet (see *Philosophical Transactions*, 55 [1765], 294-325).

68.05.16 *To John Nourse*

Wellcome Institute Library, London[1]

Received back from M^r^. Nourse[2] the sum of six pound sterling, the price I poured in his hands for a set of books I took upon condition of 4 days in the Saturday past. Today the 16 of May 1768

J.H. de Magelhaens

1 Letter transcribed in 1977 but now missing.

2 John Nourse, the well-known bookseller and publisher.

I observed the transit of Venus, June 3, 1769, at the College of Louis le Grand at Paris, which is 2'' to the east of the meridian of the Royal Observatory. I had an achromatic telescope, of 12 feet focus, which had an aperture of 3¾ inches, and magnified 180 times, with the view of making my observation more correspondent to that which M. the Abbé Chappe was to make, in California,[2] with a telescope of the same length, the same magnifying power, and equal goodness. The first contact could not be seen, on account of a very thick cloud; there even fell some rain. I waited for the second; the Sun then was pretty clear. But there were some vapours, which caused such great undulations as to hinder me from seeing the disc of the Sun, and that of Venus, well defined. At 7^h 38' 45'', apparent time, or 7^h 38' 3'', reduced to the meridian of the Royal Observatory, the second contact appeared decisively to me. Two seconds after, a very fine thread of light appeared between the limb of Venus and that of the Sun; so that in my observation there is not an uncertainty of two seconds in the moment of the internal contact. After this observation, I viewed the Sun with different glasses, which rendered him alternately red and white. I saw Venus, with this last colour, with a crescent of a blueish colour; and a little inclined towards the limb of the Sun: with the glass which made the image of the Sun red this crescent disappeared; but I saw Venus flattened in the direction of the crescent. I measured the greatest and least diameter; the greatest was 56''½, and the least 53''½. Perhaps this crescent was only visible by the effect of some optic illusion: but I relate only what I saw. At 7^h 52' 8'', apparent time, I measured the interval between the limb of Venus and that of the Sun, which I found 46''½; and at 7^h 58' 4'', the first limb of Venus touched the horizon.[3]

---

1 Letter not found. The original French version of Messier's report was included in Magellan's letter to the Astronomer Royal, Nevil Maskelyne, dated 17 July 1769. The translation given here was published in *Philosophical Transactions*, 59 (1769), 376-7. The letter was presumably written shortly after Messier made the observations he reports.

2 See Magellan to N. Maskelyne, 17 July 1769, n. 4.

3 Immediately following the extract is the following, also drawn from Magellan's letter to Maskelyne:

> By a letter from M. l'Abbé Bourriot to Mr. Magalhaens it appears, that Mess. de Fouchy and Bailly, at the Meute, each made use of reflecting telescopes of 30 inches focus and 4½ inches aperture; that M. Bory made use of an achromatic telescope of 5 feet focus, and 2 inches aperture; and M. l'Abbé Bourriot made use of a very good achromatic telescope of 6 feet long, and 2¼ inches aperture, made by himself, magnifying 120 times.

The letter from Bourriot has not been found.

69.07.17 *To Nevil Maskelyne*

Royal Society of London, Letters & Papers, V, 132

Wood Street n°. 98.
July 17. 1769

Monsieur

Je comptois pouvoir vous parler la derniere assemblée de la Societé Royale lors de mon retour de Paris, pour vous remercier de la part de M^{r}. Messier & de M^{r}. le President de Saron,[1] par les soins que vous eutes La bonté de prendre dans l'examen de la lunete achromatique, que je aportai à Paris, & qui a eté admirée de tout le monde par sa bonté. Mais vous ayant [...][2] dû de vûe, & ne pouvant pas pour le present allez à Greenwich moi-meme, je prends la liberté de vous temoigner le contenu ci dessus, en ajoutant aussi les compliments de M^{r}. de la Lande, dont je vous envoyai d'abord par la petite Poste une Lettre lors de mon arrivée ici, qu'il m'avoit conseigné pour vous.

Je viens de recevoir une copie du Memoire de M^{r}. Messier lû dernierement par Lui meme à L'Academie Royale, sur son observation du passage de Venus; & je vais vous en donner le precis, croyant vous fair plaisir.[3] = J'ai observé (dit il) le passage de Venus du 3 Juin 1769 au College de Louis Le Grand à Paris, qui est 2''. à l'orient de le Meridienne de l'Observatoire Royal... J'avois une Lunete achromatique de <u>12</u> pieds de foyer, qui portoit 39 lignes d'ouverture & qui grossissoit 180 fois le diametre des objets.... dans le dessein de rendre mon observation plus correspondante à celles que M^{r}. L'Abbé Chappe devoit faire dans Le Californie[4] avec une Lunete de meme longueur, egale force, & meme bonté... Le premier contact ne put avoir Lieu à cause d'un nuage fort epais il tomboit meme de la pluie... J'attendis le second.... pour lors le ciel oû se trouvoit le soleil, etoit assez serein; mais il y avoit des vapeurs, & les ondulations excessives empechoient

1 Jean-Baptiste-Gaspard Bochart de Saron (1730-1794), Président of the Parlement de Paris.

2 Paper damaged, word illegible.

3 Maskelyne arranged for an English translation of Messier's report, together with a summary of Bourriot's account, to be published in *Philosophical Transactions*, 59 (1769), 376-7. See above, 69.06.00.

4 Jean-Baptiste Chappe d'Auteroche (1728-1769) went to southern California to observe the transit of Venus in 1769. Having made the observations, he and all but one of his companions died there of an unknown epidemic disease.

de voir le disque du soleil & celui de Venus terminés. à $7^h.38'.45''$. du tems vray, ou $7^h.38'.43''$. reduit au Meridien de l'Observatoire Royal, le second contact se decida. deux seconds après on voioit un filet de lumiere, très deliée entre le bord de Venus & celui du Soleil: de maniere que dans mon observation i[l] ne reste qu'une incertitude de 2''. sur le veritable moment du contact interieur. Après cette observation, Je regardai le soleil avec differents verres, qui me le rendoient alternativement rouge & blanc. Je vis Venus dans cette derniere couleur avec un croissant de couleur bleuatre; et un peu incliné vers le bord du soleil. le verre qui me rendoit l'image du soleil rouge, ce croissant disparoissoit, mais je voiois Venus ap[la]tie dans le sens du croissant. Je mesurai le grand & le petit diametre. Le grand fut trouvé de 56½'': & le petit de 53½''. (Ce croissant n'etoit peutetre apparent, que par l'effet d'une ilusion optique; mais je raporte ce que j'ai observé). À $7^h.52'.8''$. tems vrai, je mesurai l'intervale qu'il avoit entre le bord de Venus & celui du soleil, que je trouvai de 46½'': & à $7^h.58'.4''$. le premier bord de Venus touche l'horizon. = 69.07.17

J'ai reçu aussi une Lettre de M^r. l'Abbé Bourriot sur ce sujet, dont voici les paroles = Vous trouverez cy près celle (observation du passage de Venus) que M^r. de Fouchy Secretaire de l'Academie R^{le}.,[5] M^r. Bailly,[6] M^r. Bory Chef d'Escadre,[7] tous deux aussi Academiciens, & votre serviteur (le même Abbé Bourriot) ont faite au chateau de La meute,[8] 14''. à l'ouest de l'Observatoire Royal, & 16''. egalement à l'ouest de celui, où M^r. Messier a observé.[9] Messrs. de Fouchy & Bailly se sont servis chacun d'un telescope de 30 pouces de foyer, & portant 4½ d'ouverture. M^r. Bory d'une lunete achromatique de 5 pieds, & de 24 lignes d'ouverture. on ignore la force amplificante de ces 3 instruments. Je me servois de celle que J'ai faite, &

5 Jean-Paul Grandjean de Fouchy (1707-1788), astronomer and Secretary of the Académie Royale des Sciences in Paris since 1743.

6 Jean-Sylvain Bailly (1736-1793).

7 The chevalier Gabriel de Bory (1720-1801).

8 "La Muette", the royal pavilion at Passy, where the King's *Cabinet de Physique* was kept.

9 The observations of the transit and of the solar eclipse the following day made by de Fouchy, de Bory and Bailly were published in *Histoire de l'Académie Royale des Sciences*, 1769, Mém. 531-8; the presence at the time of Bourriot and also of Dom Noël, Keeper of the *Cabinet de Physique*, was noted in the report but only their observations of the eclipse, not of the transit, were recorded. Though many other observations of the transit were also published in this volume of the Academy's *Mémoires*, Messier's for some reason were not; they were however referred to by Pingré in his critical survey of all the reported observations, published in the *Mémoires* for the following year (pp. 558-83).

69.07.17 que vous connoissez (c'est un for bon achromatique à double objectif). elle a 6 pieds de long, 27 lignes d'ouverture, & grossit 120 fois. nos quatre observations se reduisent à très peu de chose près à celles cy. 1er. contact à $7^h.21'.6''$. tems vray. 2^d. contact à $7^h.38'.31''$. On a observé aussi l'Eclipse du Soleil dans le jour suivant. Le commencement fut à $6^h.46'.32$. tems vray. & la fin à $8^h.27'.4''$. La part obscurcie fut 5 doigts & 20'. = Cependant pour ce que regarde cette derniere observation de l'Eclipse du soleil, je vous prie de ne compter pas là dessus, car il semble que par quelqu'accident, ou meprise dans celui qui comptoit à la pendule, elle se trouve en defaut. Vous pouvez compter sur la exactitude avec laquelle je transcris les nombres; & en faire l'usage que vous jugerez à propos.

Recevez donc cette communication, que je vous fais de ces observations, comm'une marque de l'estime que je vous professe, & de l'[...]tion[10] avec laquelle je suis

Monsieur
Votre très humble & obeiss^t. serv^r.
J.H. de Magalhaens

To the Rev^d. M^r. Nevil Maskelyne
Astronomer Royal
at Greenwich

10 Word partly illegible.

From Ribeiro Sanches[1] 69.11.02

Glasgow University Library, Ms. Hunter H139

[...][2]

In the Last Transactions of ye Royal Society, you sent me, I read a Paper by D[or]. Hunter, in which he occasionally takes notice of the Bones and Teeth of that animal, the Russians call (I know not if rightly) Mammouth.[3] You know what Mess[rs]. de Buffon,[4] Sloane,[5] Gmelin,[6] and Daubenton[7] have said about them: I'll tell you now some particulars I know relative to this Subject.

During my residence in Russia, about the year 1745, on my coming to the House of M[r]. Tatixof, Chamberlain to the Empress & Lieutenant-General of the Police at Petersbourg,[8] I saw on a Table a monstruous Large Tooth: and knowing him to be a Learned gentleman, conversant in these maters, I expressed my curiosity to him about it. To this he satisfied me, saying that tooth had been found at Novogorod, about 30 Leagues to the East of Petersbourg, at the depth of 30 feet near the River Wolkof[9] in diging a canal of communication with the Wolga:[10] and sent, by the Director of that work, to him as a present.

1 Annotated in Magellan's hand: "Extract of a Letter from D[r]. Sanchez of Paris, formerly first Physician to the Empress of Russia, to his friend M[r]. Magalhaens at London, dated from Paris 2 Nov[ber]. 1769."

2 Unknown amount of text not copied by Magellan, here and at the end of the extract.

3 William Hunter, "Observations on the Bones, commonly supposed to be Elephants Bones, which have been found near the River Ohio in America", *Philosophical Transactions*, 58 (1768), 34-45.

4 Buffon, Georges Louis Leclerc, comte de, *Histoire naturelle, générale et particulière* (Paris, 1749-67), vol. 11, pp. 86ff.

5 Sir Hans Sloane (1660-1753), British physician, naturalist and collector, sent two memoirs to the Paris Academy of Sciences on the subject: "Observations sur une paire de cornes d'une grandeur et figure extraordinaires," *Histoire de l'Académie Royale des Sciences*, 1727, Mém. pp. 206-29, and "Mémoire sur les dents et autres ossemens de l'élephant, trouvés dans la terre," *ibid.*, pp. 305-34.

6 Johann Georg Gmelin, *Reise durch Sibirien, von dem Jahr 1733 bis 1743* (Göttingen, 1751-2); French translation, *Voyage en Sibérie* (2 vols, Paris, 1767).

7 Louis-Jean-Marie Daubenton, "Mémoire sur des os et des dents remarquables par leur grandeur", *Histoire de l'Académie Royale des Sciences*, 1762, Mém. pp. 206-29. On the different ideas concerning the origin of these bones, see Claudine Cohen, *The Fate of the Mammoth: Fossils, Myth, and History* (Chicago, 2002).

8 The courtier Alexei Danilovich Tatishchev (1697-1760), chief of police of St Petersburg, 1745-60.

9 i.e. Volkhov.

10 i.e. Volga River.

69.11.02 This was all he told me: and, as I attended him and his family when sick, and he honour'd me with great friendship, he presented me with this curious specimen. It's weight might be, as near as can recolect, between 20 and 30 pounds. The colour was yelowish: and the substance like to Ivory. had 4 Large roots: and it's form shew'd it was a grinder.

Not long afterwards an Elephant died, the largest of those sent as a present to the Russian Emperor, by Thaman Kolikam.[11] I had seen this Elephant and his companions many times, which when alive was 14 feet high. Abraham Kaau Boerhaave, the same who wrote that book you sent me Lately de perspiratione dicta Hippocrati &c[a],[12] wery well versed in Anatomy, undertook the Dissection of this Elephant, and at my request made a carefull examination of that tooth already mentioned comparing it with those of this Elephant: which having done, he declared as his opinion, there was not any likeness between them: and that this could by no means be an Elephant's tooth. In this opinion I always acquiesced, nor could I ever be persuaded to the contrary by the reasonings of others.

When I left Petersbourg in September 1747, I made a present of the tooth, with some other rarities of Natural History, to the Academy of Sciences there: and fixt on it a tiket, containing the above: from whence, I dare say, may be obtained without much Dificulty, a drawing of this tooth, or a model in plaster, which Last would be still better to Shew it's true Shape.[13]

As far as I can Recolect, this tooth is not intirely like to those drawn in the Plate anexed to D[r]. Hunter's Account in the Transactions; as the points or elevated parts of that grinder were not so remarcably high, as there represented.

If you think this article may please D[r]. Hunter's Curiosity on that subject, you are at Liberty to acquaint him therewith.

---

11 In 1741 the Shah of Persia, Nader – known before his crowning as Tahmasp Qoli Khan – sent a caravan of fourteen elephants to St Petersburg as a gift, to join the one sent some years earlier. Sanches's saying that the gift was to "the Russian Emperor" implies that it was made during the brief reign of the infant Tsar Ivan VI, 1740-1, prior to the palace revolution in 1741 that installed Peter the Great's daughter Elisabeth as Empress in his place.

12 Abraham Kaau Boerhaave, *Perspiratione Hippocrati dicta anatomice illustrata* (Leiden, 1745).

13 The zoological collections of the Kunstkamera, the museum of the St Petersburg Academy of Sciences, were transferred to the Zoological Museum in St Petersburg in 1832. There is, however, no catalogue of the items transferred. The Zoological Museum has a mammoth tooth (inventory number a.21) that is thought to have come from the Kunstkamera, but it is impossible to know if it is the one that once belonged to Sanches. (Information kindly supplied by Nadezhda Slepkova.)

To Ribeiro Sanches[1] 69.11.21

Bibliothèque de la Faculté de Médecine, Paris, Journal, Ms. N° 2015, f. 61v

[Se aprezentou na Sociedade Real ha poucos dias hua rais de Rhuibarbo palmato crecido neste Reyno: he de 6 annos; e não esta perfeito ainda: a raiz pezava 12 arrateis.]

[A root of Rhuibarbo palmato grown in this kingdom was presented at the Royal Society a few days ago: it is 6 years old and is not yet fully mature: the root weighed 12 arrateis.[2]]

1 Letter not found. MS is an entry in Sanches's journal, 26 November 1769, headed "Lettre de M^r^. Magalhaens. Nov^bre^ 21 1769".

2 A Portuguese unit of weight; 1 arratel = 16 ounces.

To Louis-Henri Duchesne[1] 69.12.26

Archives Nationales, Paris, MS T/160 (10-12): lettres de J. H. de Magalhaens

Londres, 26 Dec^bre^. 1769

Mon cher Monsieur

Il n'est pas aisé de vous exprimer Le grand plaisir, que me causa La votre du 5 du courant,[2] oû vous m'aprennez le beau commencement du Retablissement de la Sante de M^r^. Trudaine. J'etois très persuadé que le voyager seroit fort capable de le guerir: & je sens dans ce moment les plus flateuses idées de voir acomplie cette presensation, si je puis l'apeller ainsi, de mon espirit.

Ce que feu notre bon ami[3] se proposoit de faire avec ces bois preparés & non preparés, n'etoit d'autre chose, que seulement Les essaier rigureusement

1 Several documents were enclosed with this letter that are presented here as appendices.

2 Letter not found.

3 The French mining engineer Antoine-Gabriel Jars (1732-1769), who had died the preceding August; see Magellan to Duchesne, 1 February 1771. Jars travelled extensively around Europe (Germany, Sweden, Scotland, England, Norway) to observe mining methods and steel production, and no doubt met Magellan while he was in London, where he was elected a corresponding member of the Society of Arts in 1765.

69.12.26 en les metant à la fois deux le chaque espece dans le meme endroit, pour voir de combien le morceau preparé resistoit plus à la putrefaction. C'est pour celà qu'il me donna l'ordre de les couper en deux, comme vous le verrez dans le recit que j'avois preparé pour le lui envoyer, & que vous trouverez, ci joint:[4] aussi bien c'etoit pourquoi il me chargea de les faire peser & marquer soigneusement, comme vous y trouverez. je ne doute gueres que en chargeant quelque sur intendant, ou officier de ceux qui president dans les chantiers, oû l'on fabrique des vaisseaux du Roy, il pourra faire des epreuves decisives sur La qualité de cette preparation pour preserver Les bois. je crois meme qu'il seroit suffissant de les faire enturer[5] à demi dans quelque marage,[6] ou terre pourrie, en les laissant demi exposés au dehors de la terre: & que au bout d'une année ou deux on pourroit voir decidement si les morceaux du meme bois preparés se conservoient incorrupts & sans pourriture, au contraire des autres non preparés. C'est une chose si aisé à concevoir, & à eprouver, que je crains d'y ajouter des nuages en tachant de plus dire sur ce qui de soi meme est si simple & aisé. pour ce qui Regarde mes deboursements les voicy. par Le Reçu du Charpentier, vous verrez que je lui payai il y a plus de quatre mois £8 :3^{s} :9^{d}.[7] outre celà je demandai les depenses qu'il faut payer au Waterman, qui les garde dans son magazin jusqu'à l'ordre de les embarquer: & lui m'envoya la petite memoire marquée n°. 2, qui vá a 11 shillings;[8] mais il faudra encore ajouter environ 3 shilligs pour le magazinage jusqu'à ce qu'ils partent effetivem^{t}.: je depensai de plus en portes de lettres, & quelques messages, que je n'ai pas pû faire moi meme, 6^{s}:8^{d}. Outre ceci je viens de vous expedier 21 cahiers de papier de la nouvelle invention qu'on appelle chimique,[9] que Selon le detail de la quittance marquée n°. 3, vá a £2:18:2 le tout ensemble fait £12:2^{s}:7^{d} Esterl. qui font en argent de France 284$^{\#}$. 3^{s} Tournoises. pour Les quelles je vous envoye icy La quitance, que vous me demandez. Seulement pour vous aiser l'envoy de cette bagatelle, peut etre il vous sera plus commode de la faire conseigner à notre common ami Le D^{or}. Sanchez dans la Rue des Blancs–Manteaux à

4 See below, Annex 1.
5 entourer?
6 marécage?
7 See below, Annex 2.
8 See below, Annex 3. In the 18th century passengers and goods were carried from the dock to ships moored in the Thames by watermen and lightermen (later organized in the Company of Watermen and Lightermen of the River Thames).
9 See Annex 5. Nathaniel Bermingham (fl.1740-1780), whose advertisement for "prepared paper" Magellan enclosed with his letter, was an Irish artist living in England.

Paris.[10] Pour ce qui Regarde cette rame de papier, vous me disiez dans votre Lettre, de vous envoyer une demi rame comme le papier de la Lettre meme: & l'autre demi comme la feuille jointe. or quand votre Lettre est venue entre mes mains, je ne trouvai pas aucune autre feuille: ainsi il faut conclure, ou que vous la avez oublieé, ou qu'on l'a oté dans La poste. quoiqu'il en soit je ai taché de remedier à cela en vous envoyant du plus aprochant en qualité & de plus d'usage. Voicy les qualités. je vous envoye 8 cahiers du papier appelée fullscap,[11] qui est le plus commun pour des Lettres. autant de ce qu'on appelle thik post,[12] qui est un peu plus grand & plus fort. quelques cahiers sont coupés, & quelques autres point du tout. outre celà je vous envoye deux cahiers du papier Imperial de la plus grande marque, qui sert à faire des desseins &plans dessus: & deux autres cahiers du papier Royall qui sert egalement pour dessigner dessus. enfin je vous envoye un cahier appellé thin-post qui est transparent, & seulement preparé d'un coté. Celui cy sert pour tirer des copies des plans ou desseins, en le mettant dessus, & en marquant les contours des figures qui paroissent à travers de sa pelucidité. Je ne doute que quelque personne curieuse pourra venir à bout d'imiter cette manufacture de papier qui est reelement assez avantageuse. Les petits essais que j'ai fait, me font voir, que tout Le mistere consiste en ajouter quelque poudre tres sutile & apre, comme Vg. de la poudre du pomex,[13] ou cerusa,[14] ou quelqu'autre semblable, de la ajouter je dis, dans la cole qu'on donne au papier dans La fabrique & par La quelle il devient capable de soutenir l'encre sans y tremper. Car vous conceverez aisement, que d'abord qu'il y aura une matiere apre, capable de derober les particles metaliques de la pointe d'argent, ou cuivre qui passent dessus, elles y paroitront d'une couleur obscure, comme s'on y ecrivoit avec du crayon. 69.12.26

10 The file includes, on a separate page, a note by Duchesne, as follows:

Copie de la lettre ecrite a Mr de St Laurent le 20 janvier 1770.

je vous prie M. de vouloir bien payer a Mr de Magalhaens ou au porteur de la quittance une somme de 284.#3. qui lui sont dues pour des avances qu'il a été autorisé de faire Cette somme sera comprise dans l'etat des depenses sur le demi pour cent de la presente année. je suis &c

----

J'ay envoyé cette lettre et ma quittance le 20 janvier a Vailland afin qu'il retire cet argent et le porta au docteur Sanches.

11 Foolscap.

12 See below, Annex 4. Writing paper, like all other paper, was heavily taxed, the sizes being specified as follows (largest first): Imperial, Super Royal, Royal, Medium, Demy paper, Thick post, Thin post, Small post, Foolscap, Pot paper.

13 Pumice.

14 Ceruse = white lead.

69.12.26 La caisse oû je viens de vous envoyer ces 21 cahiers de papier Chymique, est mise à bord du pakeboat de Calais nomé l'Eagle, & je viens d'ecrire à mon correspondant de Calais de l'envoyer tout de suite, d'abord qu'il arriverá, à Paris selon L'adresse qu'y est dessus, c'est-à-dire = à Monsr. Trudaine de Montigny, Intendant des Finances à Paris = ainsi vous n'avez qu'à en avertir ceux qui sont chargès de Recevoir ce qui vá adressé a M^{r}. Trudaine, de vous faire tenir cette caisse dans L'endroit qu'il vous plaira. à Dieu mon cher Monsieur: conservez moi votre amitié & croyez moi a toujours

Votre de coeur

J.H. de Magalhaens

Annex 1[15]

Je vous ecrirai icy La Relation en detail des mesures & moyens qu'on a pris avec ces bois pour les distinguer, & etre sur de luers[16] qualitès après la preparation.

1°. J'ai ordonné deux pieces de chaque espece de bois de l'epaisseur que vous m'avez marquée, c'est à dire d'environ 6 pouces de pied de Roy, & d'une Longueur suffissante pour Les couper en milieu, en les sciant des deux cotés, & en lessant[17] l'intervalle du milieu pour qu'on puisse voir que chaque piece du bois preparé se raporte à celle de la meme marque qui n'est pas preparée: après celà Je Les ai fait peser avec soin une chacune; Les 8 pieces ont eté mises dans une caisse & clouées, & tout de suite envoyées au Magazin du Waterman: & les autres 8, à la Fabrique pour y etre preparées: & après mises dans un autre Caisse & envoyées au meme magazin pour y attendre l'ordre de les embarquer: soit par la voye du havre de grace,[18] ou par Calais. Voicy le poids de chaque piece de bois.

---

15 As Magellan explained to Duchesne, he prepared this statement with a view to sending it to Gabriel Jars who, however, had died before he could do so.

16 leurs?

17 laissant?

18 The former name of Le Havre.

La piece d'oak, marquée O.h. (c'est-à-dire Oak hard) avec une barre au dessous, pesa Lb. d'Angleterre apellées avoir du poids	81.¼	69.12.26
l'autre piece marquée de même (O.h), mais sans aucune barre ou ligne au dessous, pesa	86.¼	
La piece ou morceau marqué O.T. [c'est-à-dire Oak Tough] avec une ligne au dessous	76.¼	
L'autre piece marquée de meme, mais sans aucune ligne au dessous, pesa	78.¼	
Le morceau d'Elm tree marqué E.H., pesa	75¼	
L'autre marqué E.H., pesa	76¼	
L'autre marqué E.T., pesa	70¼	
L'autre marqué E.T., peza	70¼	
Le morceau de Beech tree marqué B.H.	53. ¼	
L'autre marqué B.H.	58.¼	
L'autre marqué B.T.	68.¼	
L'autre marqué B.T.	60.¼	
Le morceau de Fir [je crois que vous L'appelez Sapin] marqué F.H., pesa	40.¼	
L'autre marqué F.H., pesa	46.¼	
L'autre marqué F.T.	51¼	
& l'autre marqué F.T.	48.¼	

N.B. il ne faut pas prendre les ¼ de Livre pesante ci-dessus que seulement, comme La quantité du poids pour tourner La balance decisivement: car dans des grosses Balances on ne peut pas peser aucun gros poids, qu'avec une incertitude d'une fraction plus ou mois grand selon que La Constrution des mêmes balances est plus ou moins parfaite: & meme je ne crois pas qu'il soit possible d'avoir au monde aucune grande balance, qui puisse montrer des poids considerables, avec une parfaite exactitude.

July 16th., 1769

Mr Magaline[19] Dr. to Wm North	£	S	D
For two Pieces of Sound oake, Each 7 ft. ½ Long and 7 Inches [s]quare, cut out from two Beams at Mr Crowell near Stafford --------------------------------	1	16	0
Do of Elms, Bought in wite Chaple[20]	1	8	0
Do of fir, in ye Same Place ----------------------------------	0	18	0
Do of Beach[21] Bought, in Puddell Dock[22] ------------------	0	16	0
for my Trouble & son in feching the above, through allmost all in London and in the Nabrouing Coorntry[23] by which we Bouth Lost our time in Drifrent Days	0	16	0
for Drifrent Carrys of the wood to my Yard	0	10	0
for Cuting in 16 Peices & marking	0	8	0
Carridg to the weight house	0	3	0
for the weight Master	0	2	0
Carridg to the Mannufattur' for Prepairing 8 Pieces of Do	0	2	0
Carridg to the waterman House Mr Little & two Larg Packing Cases	1	4	9
£	8	3	9

Rec.d the Content of this Bill
in full on the 23 of August 1769
Wm North

19 Magellan?
20 Whitechapel?
21 Beech?
22 Puddle Dock, on the north bank of the Thames below Ludgate Hill.
23 Neighbouring Country?

Annex 3 2 69.12.26

Cockett	3 – 6
Searchs	= - 6
Wharfig & portrg[24]	2 – 0
Warehouse	3 - =
Warterig[25]	2

	£ — 11 —

Annex 4 3

Mr. Messuland[26]

London 26 Decr. 1769

Bought of Bloxam & Fourdrinier,
Stationers, &c.
No. 11, the Corner of the Post Office Lombard Street.

Chemical Paper

8. qrs. Fo. Cap	1/6	0. 12. 0
8. qrs. thick post	2/3	0. 18. 0
2. qrs. Imperial	7/	0. 14. 0
2. qrs. Royall	4	0. 8. 0
1. qr. thin post done on one side		0. 2. 0
Box		0 1. 8

	£	2. 15. 8

Recd. ye. same time ye. Contents
For Messrs. Bloxam & Fourdrinier
Pr. Wm. Bloxam

Porteur, bateau & lading bills		2. 6
		----------
total	£	2. 18. 2[27]

24 Wharfage & porterage?
25 Warderage?
26 Magellan?
27 The last item and final total have been added by Magellan.

SOLD HERE
NATH. BERMINGHAM's
ROYAL CHARTA CHYMICA,
OR
PREPARED PAPER,

So peculiarly adapted for

Tracing, Drawing, Pocket-Books, and various other Uſes.

THIS Paper, tho' ſold at a very moderate Rate, may be deemed almoſt invaluable, as by the Help thereof, every Purpoſe of the Pen and Pencil is ſupplied, and both the Expence and Trouble of procuring good black Lead Pencils totally obviated, the Nature of it being ſuch that it may be written or drawn upon with the greateſt Eaſe, by any kind of Metal made into the Form of a Pencil, Iron alone excepted.

To [Louis-Henri Duchesne][1]

70.00.00

Archives Nationales, Paris, MS T/160 (10-12): lettres de J. H. de Magalhaens

Il a eté arreté à Calais une lunette achromatique de poche avec diferents instruments, principalement de chirurgie qui appartenoient à Mr. Righi, cirurgien[2] qui voyageoit par ordre de Son Altesse L'Infant de Parme Lors de son passage d'Angleterre par Paris, pour retourner à Parme. On croit ce paquet sous L'adresse de Mr. Messier.

Monsieur Trudaine est prié vouloir bien donner ses Ordres pour faire Rendre ces effects, qui etant pour l'usage d'un particulier Etranger, ne paroissent pas etre dans Le cas da La Saisie, & on Lui aura La plus grand Obligation de Sa bonne Volonté à son égard.

---

1 The letter is in Magellan's handwriting but is neither addressed nor dated. It was no doubt sent by Magellan to Duchesne, who was Trudaine de Montigny's secretary, so that he could bring the problem to Trudaine's attention. Its position in the file of letters from Magellan to Duchesne at the Archives Nationales in Paris, between letters dated 26 December 1769 and 20 March 1770, in a file in which everything is otherwise in chronological order, suggests that it may have been written in early 1770. The letter has been annotated "Mr. Magalheens" and, in Duchesne's handwriting, "les marchandises ont èté rendues, dù a Mr Magalhens".

2 Jacopo Antonio Righi (1736-1802), professor at the Università di Parma from 1767, who in 1769 was sent by Ferdinando de Bourbon-Parma (1751-1802), Duke of Parma and Piacenza and Infant of Spain, to inspect medical facilities in France and elsewhere. He returned to Italy in 1771, and this letter must have been written before that. It may have been written some time before, however, because Righi may not have travelled directly from Calais to Parma, but may have stayed for a while at Paris or at other places along the way.

70.05.09 *To Samuel More*[1]

Royal Society of Arts, London, RSA/PR/GE/110/27/150[2]

Fleet Street, May 9 1770

Sir,

I send you to be laid before the Society of Arts, the Model of a Crane, with an additional Contrivance of mine, to preserve the Men at Work from any ill accident, if either the Rope shou'd happen to break, or the Wheel be so overpower'd by the Weight, as to be forced back. There is likewise a contrivance to encrease the Power of the Men[3]

Perhaps you may remember that when the other method was applied to this Machine to perform the same by a double Bellows, I then told you, it cou'd be more easily and effectually obtain'd in some other way.

To shew the truth of this Assertion, I caus'd one Sawyer, a Cabinet maker in Butcherhall Lane, near Newgate, to make me this Model which he did about the beginning of December 1767. My intention was to present it, when compleated to this honourable Society for Public Use. However hearing the former improvement had met with great approbation, I declin'd doing it at that time, that no Suspicion of opposition or ill nature might be suppos'd on my part.[4] But after so long a time is elapsed, as nobody else that I know of has turn'd his thoughts on this Subject, I don't see any impropriety in Communicating my idea to this Society, of which I have now the honour of being become a Member.[5]

---

1 More was elected Secretary of the Society of Arts in February 1770.

2 Annotated "with a model of a crane" and, on the reverse, "Maghaellaens | on a crane | to, Comttee. of Mechanics | May. 9. 1770 | Read Novr. 21".

3 Magellan's letter and model were considered at a meeting of the Society's Committee of Mechanics on 22 November 1770 and again a week later at a meeting that Magellan attended by invitation, when it was recommended that the Society accept Magellan's model "as the Committee apprehend some usefull Hints may be taken from it" (Royal Society of Arts, Minutes of Committee of Mechanics). The Society displayed models such as this in its Repository until the mid-nineteenth century, at which time all that survived were transferred to the Science Museum, South Kensington.

4 Magellan is probably referring here to the Society's consideration in May 1767 of a contrivance for stopping cranes that had been invented by a Mr Pinchbeck and was at that time "in common use in a crane at Dice's Key". At its meeting on 16 May 1767 the Society's Committee of Mechanics resolved concerning Pinchbeck's device "that it appears to the Committee that it will fully Answer under all Circumstances, and is to be constructed at an easy Expence" (ibid.).

5 Magellan had been elected a member of the Society of Arts a week earlier, on 2 May 1770, on the proposal of Henry Baker, seconded by A. Kellet and T. Stewart.

The advantages of this Contrivance are as follows: 70.05.09

1st The immediate Stopping of the Crane, in case the rope shou'd break with the weight of the Load, depends not on any Machinery, capable of being altered by the weather, or wanting any new adjustment to each degree of Velocity in the Wheel.

2d This effect is almost instantaneously produced. The Wheel of the Model is able to go no farther in such case than the 31th part of a whole Turn but the Stopers, or Teeth may be put so close in a great Wheel, that it will be stopp'd before the 80th or 100th part of its Circumference be gone round.

3th There is also another Catch to stop the Wheel with the burthern rised to any height, and prevent its turning back, should it be overpowered by the Weight.

4th The Wheel may be free'd at pleasure from this Catch, by the Men within the same pulling a little rope or String. So the man may go out of the Wheel to turn the moveable arm, or do any other necessary service; and when he comes back may free the wheel, and let the Weight descend into its proper place.

5th When there is no Weight acting in Connection with the Wheel, the above two Catches are to be freed from touching it, by another small rope or String: So that it may turn in both ways. But as soon as the burthern begins to stretch the rope, Then both are put in action, one to hinder its going on, in Case the Rope breaks. And the other Continually falling upon the Stopers, to hinder the Wheel returning back, in Case it be overpower'd by the Weight of the burthen. therefore when the Crane is loaded only can go on, but cannot return back, and if the rope breaks, then can neither go on, nor return back.

6th The Additional pieces to perform the above on the Crane, are so simple that any Common Carpenter can make them in full perfection.

7th This Apparatus may be added to any old Crane, without altering in the least its former Construction.

8th The additional Cost can be but very trifling. I am told six or seven Shillings will be more than sufficient to pay the whole expence for Cheap pieces of Common Deal will sufficiently answer the purpose.

9th And finally if the Wheel is so made, as to have one side without any Cross beams, as hinted in this Model; then one man alone may rise more than the double of the usual Weight, by adding to his own gravity the action of pulling with his hands the Cross beam of a small post fixed on the floor for that purpose.

70.05.09 NB The pieces belonging to this improvement may be made much slender than represented in the Model without any prejudice to their effect. And the Lead added to the end of each arm near the Rope will be unnecessary when the Crane is made of a full size; for then the pieces will have weight enough to keep the rope bent, whenever there is no burthen to stretch it.

If the Society shall deem this Contrivance usefull for the purpose already mentioned, I desire the acceptance of the Model to be kept hereafter with the others in their Repository for Public Use.

But if I am mistaken in my intentions, I then desire to have it return'd back; and in either Case shall ever be with great regard & every good wish

Dear Sir
Your most humble Servt.
John Hyacinth de Magalhaens

M^{r}. Samuel More
Secretary of the Society of Arts &c
Strand

70.08.18 *To Louis-Henri Duchesne* [1]

Archives Nationales, Paris, MS T/160 (10-12): lettres de J. H. de Magalhaens

Blancs-manteaux[2]
18 Aoust 1770

Mon cher Monsr.

Je vous envoye avec celle cy un de ces cuirs à Rasoir, dont Je vous ai parlé; & que J'espere vous trouverez excellent, comme tous ceux qui en ont fait l'experience. Le plus singulier est La qualité de rendre Les rasoirs, même ceux d'un accier fort commun, presqu'aussi bons à bien raser, que ceux de la meilleure espece; comme Je l'ai experimenté differentes fois.

1 Annotated by Duchesne: "Rep. le 2. 8e." Letter not found.
2 The address in Paris of Ribeiro Sanches.

Le porteur de celle ci est Le Sieur Sylva, Portugais;[3] dont Je vous ai parlé mecredi dernier: et dont Les qualités de probité, honneur, exactitude & fidelité sont à toute epreuve. Il a eté malhereusement La victime de son attachement pour ceux, qui avoient agi en vrais amis pour Lui; & celà avec une constance & desinterêt, qui me l'ont rendû cher pour La vie. C'est comme tel que J'ose vous le recommender: car Je sçais bien, que Le vrai honnet homme est un tresor assez rare, & souvent fort precieux pour qui Le trouve. Il est capable, pour toute Sorte, de petites commissions: & outre celà Il entend fort bien ce qui regarde les chevaux, La tapisserie, & La dorure: Si par sa bonne fortune, quelque occasion se presentera pour l'employer, Je me flate, que vous serez aussitôt convaincû de la verité de ce que Je viens d'avancer en sa faveur. Et pour ce qui me regarde croyez aussi que c'est avec Le plus sincere attachement & vraye estime que Je me suscris 70.08.18

Mon cher Monsieur

Votre très humble & très obeiss^{t}. Serv^{r}.

J.H. de Magalhaens

P.S. Je partirai dimanche matin sans faute pour Londres: oû je me attends vos ordres, & ne me oublierai pas de celles que vous m'avez deja données.

A Monsieur
Monsieur Duchenne Secret^{re}.
De M^{r}. Trudaine de Montigny
Rue des quatre-filles vis-à-vis
La porte des Capucins

3 Polycarpo da Sylva; but see also Magellan to Duchesne, 23 April 1771 and 14 December 1773.

70.10.02 *To Louis-Henri Duchesne*[1]

Archives Nationales, Paris, MS T/160 (10-12): lettres de J.H. de Magalhaens

Londres 8bre. 2 – 1770

Monsieur & Cher Ami

De jour en jour vous voyez mon retard à vous acuser mon arrivée ici. mais ce fut ce même de jour en jour qui m'a trompé dans mes atteintes, croiant à chaque poste pouvoir vous ecrire, en vous annonçant l'envoy de quelque chose, ensemble avec Le verre que vous m'avez [or]donée pour M^r. George.[2] Le commis de la verrerie me dit chaque semaine d'attendre pour La suivante à fin de l'avoir bon, & c'est sur ce pied que J'ai eté mis d'un jour à l'autre. Cependant si je ne puis pas en avoir du bon jusqu'à Samedi je ne manquerai pas de vous envoyer quelque chose toute seule par Calais. par exemple je pense vous envoyer deux petits vases d'alabastre violet de la fabrique d'Angleterre, qui est une fort belle pierre: & je serois tenté de vous envoyer aussi d'une fabrique en terre noire, & a terre vitrée dont on fait des pieces dans Le gout des antiques. Mais craignant que celà Soit trop du commun, j'attendrai que vous me le disiez; car il vous sera aisé de passer chez Granchez[3] au petit Dunquerque, quai de Conti, presque vis a vis le Pont neuf, & d'y voir La qualité: Seulement la diference peut etre par Le prix: car je ne doute guere que ces Marchands ne demandent bien au dela de la moitie d[u] premier prix. Je vous envoyerai aussi des Macaroni transparents[4] que je vous ai fait voir; une bouteille de Sauie[5] qui est une espece de sauce qui vient des Indes, & qui done un gout fort agreable (du

1 Annotated by Duchesne: "Rep le 25 8e." Letter not found.

2 The well-known French optician Étienne François George, who had a shop on the Quai de Conti in Paris. At this period, French opticians were dependent on England for the flint glass they needed to make achromatic lenses. See also Magellan to Duchesne, 30 November 1770 and 30 May 1771.

3 The merchant Granchez was established at the Quai de Conti au Petit Dunquerque at the end of the Pont Neuf in Paris. He opened a renowned and luxurious boutique in 1767 named *Perle de l'Orient* that sold gold and silver objects, mainly imported from England, and that had "all the curiosities from India and China".

4 See Magellan to Duchesne, 30 November 1770, where Magellan refers to "des macarroni ou soit vermicelli de l'Indie". Magellan is probably referring to the variety of noodles made from mung beans, later often known as "Chinese vermicelli", that becomes translucent when cooked.

5 Pronounced saw-ee? i.e. soy; see Magellan to Duchesne, 30 November 1770. Magellan had evidently heard the name but had not yet seen it in writing.

moins pour plusieurs personnes) lorsqu'on Le mele avec du poisson à la table: & une pair de pomes ananás, qui croissent icy en perfection chez nos Jardiniers à Serres.[6] tout ceci n'est que pour essayer si cela vous fera du plaisir: & peut etre que Mad^{me}. le goute aussi. alors, avec votre avis, vous en aurez autant qu'il vous plaira. aussitôt que les Kalendriers des agenda paroitront, je vous en envoyerai aussi de ceux de la espece[7] la plus propre: on en fait a present en Anglois & françois qui seront fort commodes. au reste, mon Cher ami, ne doutez pas que Je veille à vous rendre Service & à vous faire plaisir: et que je serai bien faché s'il m'echapera aucun object que je crois propre à celà. Obligez moi, Je vous prie, de m'en donner L'idee, & Celà me charmera on ne peut plus. 70.10.02

en attendant croyez moi sans fard, & tres sincerement pour La vie[8]

A Monsieur
[Monsieur Duchen]ne Secretaire
[de M^{r}. Trudaine de] Montigny
Paris

6 Jardiniers à serres = greenhouse gardeners. Tropical fruits such as pineapples were cultivated in greenhouses.

7 Magellan goes from one line to the next between "la" and "espece".

8 The letter is unsigned but is in Magellan's handwriting.

70.10.29 *From Abraham Gotthelf Kästner*[1]
Bodleian Library, Oxford, MS Rigaud, 38, no. 6[2]

Sir

I think it to be my duty to offer to You those mathematical dissertations Mr. Lichtenberg, who hath the honour to be known to You,[3] is going next Year to Osnabrug,[4] to determine the latitude and longitude of this place I wish't we had a quadrant proper to such an expedition. We have yet no moveable quadrant which may be made use of. There is one to be made here under my direction, which I'm sure will be very good, but as he was originally design'd for our observatory, he will be great and heavy, and will suffer I fear by being transported.

Besides that, he would be very necessary here, and it is a very great disappointment to be oblig'd to give away for more than half a Year an instrument, i have some years wisht for, in the very moment I get it finish't.

If besides the moveable quadrant, i have spoken of we had another of some less size, and made in a manner to be commodiously transported on astronomical expeditions, such an instrument would be of great use in ascertaining geographical situations, there being, excepted Gottingen, no place in His Majestys German dominions, of which the longitude and the latitude be known.[5]

I'm with the sincerest esteem

Sir
Your most humble and
most obedient servant
Kaestner

Gottingen
Oct. 29. 1770.[6]

1 Professor of mathematics and physics at Göttingen since 1756 and also, following the death of Tobias Mayer in 1764, director of the Göttingen Observatory.

2 Published by Joaquim de Carvalho, "Correspondência científica dirigida a João Jacinto de Magalhães", *Revista da Faculdade de Ciências da Universidade de Coimbra*, 20 (1951), 93-283.

3 Georg Christoph Lichtenberg (1742-1799) had visited England earlier in 1770, shortly before being appointed professor of philosophy at Göttingen, and doubtless met Magellan while he was there.

4 i. e. Osnabrück. In the event, Lichtenburg's journey to Hannover and Osnabrück in order to determine their geographical positions was delayed until 1772.

5 The position of Göttingen had been determined some years earlier by Tobias Mayer.

6 No addressee is indicated, but the letter has been preserved in a collection of letters written to Magellan.

To Louis-Henri Duchesne 70.11.30

Archives Nationales, Paris, MS T/160 (10-12): lettres de J.H. de Magalhaens

Londres 30 Nov^{bre}. 1770

Mon cher Monsieur

Voilà un fort grand interval depuis le 12 Oct^{bre}. que Je recus votre obligeante reponse,[1] jusqu'à present, que je vous ecris celle ci pour vous annoncer Le depart d'une caisse contenant le flint glass que vous m'avez ordonné pour M^{r}. George[2] & quelques autres bagatelles mais je me flate que vous ne me refuserez pas la justice de croir, que ce n'a pas eté à cause de ma nonchalance, ou peu de soin. Par Le Reçu ci joint du capit^{n}., vous voyez que Je l'envoye par Rouen à votre consignation & adressée sous Le nom de M^{r}. Trudaine. vous n'avez qu'à l'envoyer à M^{r}. Chateaugiron[3] ou quelqu'autre de vos amis à Rouen, pour la debarquer du vaisseau, en lui recommendant de la faire partir par eau toujours jusqu'à Paris, oû je vous prie de la faire manier (cette caisse) avec soin, pour ne pas casser les verres & d'autres choses qu'y sont contenues: dont voicy Le detail. Cette caisse contient dedans, une autre marquee n°. 1. qui contient cinq vases de pierre violete, qu'on appelle racine d'amethiste.[4] trois sont de la meilleure sorte, & couterent £.3:15:0 Esterl. sans etre garnis en or moulû: car s'ils l'etoyent, couteroient plus du double. & les autres deux d'une qualité inferieur, couterent 26 shillings. Outre ces 5. vases, vous trouverez une bouteille de soy du Japan,[5] qui est une Liqueur qu'on apporte des Indes Orientales, & que les Heliogabales Anglois[6] trouvent fort delicieuse, en La mellant dans La sauce pour manger

---

1 Letter not found.

2 See Magellan to Duchesne, 2 October 1770.

3 The *Manuel de Bibliographie Normande* refers to a Capt. Chateaugiron, born in the middle of the 18th century, who was “fils du directeur de la douane de Rouen”. The latter is presumably the person to whom Magellan is referring, but we have found no biographical data about him.

4 The fluorspar that was abundant in Derbyshire was used to make vases and other ornamental objects, turning the stone in a lathe. Its resemblance in colour and structure to some varieties of amethyst was so great that the vases were often sold under the name, “root of amethyst”.

5 See Magellan to Duchesne, 2 October 1770. Soy sauce was first imported into Europe from Japan in the 18th century. Diderot defined it in his *Encyclopédie* as “a kind of Japanese sauce that enhances the taste of meat dishes”.

6 From the name of the Roman emperor Heliogabalus (reigned 218-222 AD), who was renowned for hosting orgies at which guests were served all sorts of exotic delicacies. Magellan probably means to refer to contemporary English lovers of exotic foods.

70.11.30 du poisson. Je crois que ceci n'est pas fort commun à Paris, & que poura faire du plaisir le faire gouter à La Table de M^{dme}. Trudaine.

une autre caisse marquée n°. 2. contient 60 Livres pesantes de Flint glass pour M^{r}. George, dont Les depenses du premier achat dans La verrerie, & le Rest y compris celles Respectives pour La metre a bord du vaisseau &c. font la some de £7:5:0 C'est à dire 165.$^{\#}$13^{s}. de France. outre ces deux petites caisses vous y trouverez une autre mince & plus Longue, marquée n°. 3. Rue du Bacq qui contient un barometre pour M^{r}. Le President de Saron.[7] et je vous prie de la faire manier avec le plus de Soin posible, & tres doucement, pour ne pas le casser, etant un instrument fort delicat, & aisé à etre derangé. une autre petite marquée n°. 4. Rue du Bacq, un pacquet marqué aussi n°. 6 Rue du Bacq & une petite caisse ou boete marquée egalement Rue du Bacq n°. 7. toutes celles ci contient differents articles de mecanique & de curiosité, que je vous prie avoir soin de faire tenir au meme Mons^{r}. Le Presid^{t}. de Saron, en chargeant par la part Respective des depenses du port: ou autrement vous n'avez qu'à me les charger, en me les rebatant des 165$^{\#}$.13^{s}. de M^{r}. George. Ce sont tous Les articles contenus dans cette caisse par Rouen. Mais considerant qu'elle tardera peut etre plus de 2 mois à arriver à Paris, Je vous ai deja empaquetés quelques autres petits articles, dans une autre petite boete qui partira par Calais, & que outre n'etre pas bien pesants, ne pourroient demeurer si long tems commodement jusqu'à Paris. Cette petite caisse ou boete par Calais, est egalement adressée à M^{r}. Trudaine de Montigny, marquée n°. 5.: elle partira par le premier vaiseau, dans La semaine prochaine, & M^{r}. Morelle du Faux[8] vous la envoyera selon mon avis. en voici le contenû dans cette boete. un pacquet marqué n°. 1, qui contient un Journal de poche relie en Marroquin avec des agrapes[9] d'argent, pour L'annee suivant 1771, & pour l'usage de M^{r}. Trudaine; cela etant fort commode pour y marquer toute sorte de rendezvous, agenda &^{a}. etant en françois & Anglois: et de plus 3 autres des memes Journaux reliés en cuire rouge, pour votre propre usage & de quelqu'autre de vos amis. un autre paquet, marqué n°. 2 qui contient de la moutarde fine renfermée dans une vessie pour la conserver mieux. J'ai crû que cela pourra vous faire plaisir, à cause que la moutarde en France ne vaut rien, ni la fabriquent comm'il faut. au contraire celle ci d'Angleterre, on l'envoye & on La cherche de

7 Jean-Baptiste-Gaspard Bochart de Saron (1730-1794), President of the Parliament of Paris, was a noted patron of science with a passion for astronomy and an excellently equipped private observatory.

8 Benjamin Morel Dufaux and his widowed mother were partners in the mercantile firm of Veuve Dominique Morel et Fils in Dunkirk.

9 agrafes?

tous les endroits. un autre pacquet marqué n°. 3 contient aussi de la même moustarde, & est destiné pour Le President de Saron. plus deux autres marqués avec les n°s. 5 & 6. contient des macarroni ou soit vermicelli de l'Indie,[10] dont vous garderez une pour vous, & donnerez l'autre pour le meme Presid[t]. de Saron, aussi bien que le petit pacquet marqué n°. 7, lequel contient quelques compas de Reduction.[11] enfin il y a aussi un autre dernier pacquet, marqué n°. 8 qui contient deux Livres & quelques journaux de poche, & je vous prie de le faire tenir à notre bon ami le D[r]. Sanchez dans La Rue des Blancs–manteaux. 70.11.30

Vous disposerez de tous les articles que je vous adresse, selon qu'il vous plaira & dans La maniere, que vous semblera mieux. & pour l'argent de la part de M[r]. George, vous n'aurez qu'à Le consigner entre les mains de Messrs. Banquet & Page, Banquiers,[12] qui demeurent dans le cul de sac de l'Etoile, rue Tevenot,[13] par celle de S[t]. Martin, en le leur donnant pour le compte de M[r]. Alexandre Aubert[14] de Londres, sans rien de plus, & seulement qu'on vous donne un Reçu, le quel vous m'envoyerez; & c'est suffissant pour que je le Reçoit ici sans autre trouble de Lettres de change &[a].

Je vous prie de presenter mes très humbles Respects à M[r]. Trudaine, auquel Je serois charmé pouvoir exprimer ma gratitude par les dispositions dont il se sent pour me favouriser: & que je souhaite les savoir meriter par des services bien reels & d'une avantage fort eclatante. Je pense que si l'on faisoit executer ici une des nouvelles machines à faire des bas,[15] de la nouvelle construction, & dont je parlai à M[r]. Trudaine, pour la faire introduire chez Les manufacturiers de France, que celà seroit fort avantageux: car elles coutent moins, & travaillent plus aisement. on les a experimentées ici, & on en est fort convincû. Comme je suis membre de la Societé des Arts, oû l'original de cette machine se trouve, je pourois

10 Probably the translucent noodles made from mung beans, also known as Chinese vermicelli. At a meeting of the Society of Arts in April 1770, reference was made to "a memorial from Messrs. Levi and Co. on vermicelli".

11 A mathematical instrument used to reduce measurements by a designated factor.

12 Pierre Banquet (1713-1798) and Jean-Pierre-Louis Pache (1736-?), partners in the Parisian banking firm of Banquet et Pache, were both of Genevan descent. Following Banquet's retirement in 1775, the firm became known as Pache frères & Compagnie.

13 i.e. Thévenot.

14 Alexander Aubert, FRS (1730-1805) was a London banker as well as one of the leading English astronomers of the day.

15 At a meeting of the Society of Arts on 25 April 1770, a prize of twelve guineas was awarded to a Mr Whyman, a land surveyor from Aston, near Nottingham, for his improved stocking frame. Whyman had already received a premium of £100 from the Society in 1766 for his new design of stocking frame.

70.11.30 la faire faire aisement. je crois qu'elles coutent 12 guinees icy, lorsque les autres vont au delà du double. du moins ne seroient les protecteurs de M^{r}. le Fabre, l'honnete criminel,[16] tentés de lui faire un si avantageux present?

J'ai vû dernierement les carrosses de M^{r}. O'Keeffe[17] icy, dont la construction est telle, que si les roues du devant rencontrent un obstacle, meme plus haut que l'ixieu,[18] peuvent le passer dessus sans renverser la carrosse: & l'arbre est tellement arrangé, qu'on tourne de coté soudain dans la moitié de l'espace requis pour Les carroces ordinaires. il avoit deja finie dans son attelier une de campagne pour Monsg^{r}. le Prince de (je crois Bourbon) pour envoyer à Paris. elles sont très fortes & fermes, & fort avantageuses: & leur prix n'augmente au delà de l'ordinaire, que de 5 guinees.

On vient d'inventer aussi une espece de poele mouvable, ou rechaud, qu'on pretend fort avantageux dans l'hiver. Car on peut la metre à l'endroit qu'on veut, & porter sucessivement pour echaufer differentes chambres. J'ira[19] un de ces jours pour en voir l'experience; & si l'incovenient de la fumée, & du vapeur du charbon allumé, y est remedie, assurement ce sera une tres belle invention.

Voilà, mon cher Monsr., une Lettre bien Longue, avec tres peu de matiere importante: Mais souvenez vous que quand Le Coeur parle, ce n'est pas toujours arreglé aux principes de l'etiquete, mais bien si à ceux de l'amitie, & confidence. à Dieu encore une fois: conservez vous, & ne souhaitez plus de bonheur que vous souhaite celui qui est pour La vie

Tout à vous
J. H. de Magalhaens

A Monsieur
Monsr. Duchesne Secretaire
De Monsr. Trudaine de Montigny
à Paris

16 The sad story of Jean Fabre (1717-1797), from Bézouce in Nîmes, was the basis of a drama written by Charles-George Fenouillot de Falbaire de Quingey (1727-1800), *L'Honnête criminel, ou l'Amour filial* (1767). Fabre offered himself to be arrested in place of his old father for being a protestant and was condemned to the galleys. His imprisonment lasted for seven years, until he was pardoned by the Duc de Choiseul. He was then compelled, however, to go into exile in the country. There he established a small workshop manufacturing silk stockings, which is what Magellan is alluding to.

17 The coach-maker Owen O'Keefe was awarded a prize by the Society of Arts for his improvements to carriage wheels. He was elected a member of the society on 12 February 1772 and was last listed as a member in 1775.

18 essieu?

19 i.e. J'irai.

From Raymond-François Bourriot[1] 70.12.17

Royal Society of London, Letters & Papers, V, 204[2]

The 7th December instant the Journals and Manuscripts of the late Abbé Chappe[3] were deposited at the R. Observatory, with Mr. Cassini de Thury, by the Sieur Pauli, one of the Kings Engineers and Geographers who had accompanied the Abbé in his Voyage to California. Mr. Pauli[4] relates that M. Chappe chose to stay at St. Joseph, a small Village ten Leagues from St. Lucar, notwithstanding the contagious Disease prevailing there, and relying on his own good Constitution, because he had no more than eight days to prepare for his Observation.

Eight days after the Transit, he sickened, yet continued his Observations to the 18th of July, and a little before his Death left his Materials in writing, put into a Box, with M Pauli, to be delivered to the Royal Academy. He died the 1st. of August, as did about the same time the Clock-maker, the Interpreter, one of the two Spanish Officers, besides 12 Soldiers, and four Officers sent from Mexico, and about 50 Indians.

The first internal Contact was at 0h. 17m. 27s. The second Contact at 5h. 54m. 50s 3/10. The Duration 5h. 37m. 23s 3/10. The Latitude of the Place

---

1 Letter not found. The text is from a document headed: "Extract of a Letter, dated Paris Decr. 17th. 1770 to Mr Magalhaens; from M. Buriot, by which the small Differences between the Abbé Chappe's real observation of the late Transit of Venus, and communicated to the R. S. by the Spanish Ambassador, may be accounted for … Translated by Dr. Bevis F.R.S." Annotated on verso: "To Dr Morton or Dr Maty" and "Jan 10 | Messier on Chappe's Observation".

Raymond-François Bourriot (1734-1796) came from a wealthy family from Bazas, in Gironde. He spent twenty years in Paris, where he became a canon and acquired a reputation as an astronomer, before returning to the town of his birth. In the 1760s he succeeded in making achromatic lenses, using flint glass imported from England; he also developed the first spherometer that could measure the curvature of lens surfaces with precision. For these details, see Arthur Birembaut, "Sur les lettres du physicien Magellan conservées aux Archives Nationales", *Revue d'histoire des sciences*, 9 (1956), 150-161; pp. 158-9. See also Ducotray de Blainville, *Observations sur la physique*, January 1781, p. 48, and Magellan to Gabriel de Bory, February 1777. Johann III Bernoulli in his *Liste des astronomes connus actuellement vivants* (1776) lists Bourriot as an astronomer living in Paris.

2 Published in *Philosophical Transactions*, 60 (1770), 551-2, under the same heading as given in n. 1 (with the spelling of Bourriot's name corrected).

3 The French astronomer Jean-Baptiste Chappe d'Auteroche (1728-1769) died, together with all but one of his companions, during an expedition to southern California to observe the transit of Venus across the face of the Sun in 1769.

4 Pauly, Chappe's engineer-geographer, was the only survivor of the expedition.

70.12.17 23°. 03' 37''. Lastly (according to M. de la Lande) the Parallax of Venus 8'' 1/2, and her Distance about 35000000 Leagues of 2283 Toises each, upon a Mean of Comparisons with Observations made in the North of Europe at Cajaneburg[5] and Wardhus.[6]

5 Now Kajaani, Finland?
6 The province in Lapland that included the North Cape.

70.12.28 *To [Louis-Henri Duchesne]*[1]

Archives Nationales, Paris, MS T/160 (10-12): lettres de J. H. de Magalhaens

de Londres 28. X^bre^. 1770

On viens de présenter a la Societé des arts une espece de jack[2] c'est la machine assès communément employée pour lever a quelque petite hauteur des grands poids par exemple des gros blocs de marbre &c^a^. dont la manivelle reste fixe dans l'endroit qu'on veut d'elle meme par le moyen d'une vis sans fin qui meut la tige laquelle est cranellée en coche,[3] lui sauve beaucoup de mauvais accidents.

On y presenta aussi un frolen[4] ou machine pour separer differentes sortes de farines a la fois avec beaucoup d'expedition. 21 livres de farines furent separées en cinq differentes qualités dans le son (furfur)[5] en 4 minutes de tems. C'est fort commode pour des familles particulieres et des Boulangers. Cette machine ne coute que cinq guinèes.

1 This document, which is in Duchesne's handwriting and is located in his file of correspondence received from Magellan, would appear from both its location and its contents to be information he received from Magellan.
2 The new jack, invented by Abraham Staghold, was discussed at several meetings of the Society of Arts during 1770.
3 cranellée en coche = toothed, notched.
4 brush?
5 "Furfur" is the Latin word for bran.

From Charles Messier

Royal Society of London, Letters & Papers, V, 209.[1]

Without what I had done the 10th of January for determining the Position and forming any tolerable conjecture of the Direction of the Motion of the Comet,[2] it would have been impossible for me to have found it afterward; for from the 10th to the 16th the Sky was every night quite clouded, during which the light of the Comet had extremely abated, and its motion decreased, insomuch that between the 16th and 17th of Jan, the Sky clearing, I sought for it two hours, without finding it; but tho' with little hopes of seeing it, I sought for it in Perseus and Andromeda. After a world of useless pains and look-outs, at length I saw it again, between the Horns of the Bull, of a very feeble light: and it was necessary to be acquainted with the Heavens as well as I was, to find where it was, nor have any of our astronomers, as far as I can yet learn, succeeded in their attempts. I observed it Jany 10th, 16th, 17th and 20th on which last day it was extremely close to the Planet Mars less than a minute of time, both in R^{t}. Ascension and Decl^{n}. and its light so languid as to be in a manner extinguished by that of Mars. Without that excellent Telescope which you procured for M. le President de Saron,[3] I should not have perceived it with any of my Instruments. Here follow my Observations, which may possibly turn out the only good ones, and which I intreat you to lay before the R. Society.

---

1 Letter not found. The extract given here is a translation prepared by John Bevis that was read to the Royal Society on 1 February 1771. It was subsequently published, together with an extract of Messier's earlier letter announcing the discovery of the comet (71.01.12), in *Philosophical Transactions*, 61 (1771), 104-6, under the heading: "Extract of two Letters from M. Messier, of the Royal Academy of Sciences, and F. R. S. to M. de Magalhaens, on a new Comet: Translated by Dr. Bevis, F. R. S." There is an annotation: "Read Jan. 24, and Feb. 1, 1771."

2 See Messier to Magellan, 12 January 1771.

3 Jean-Baptiste-Gaspard Bochart de Saron (1730-1794), president of the Parlement of Paris and an enthusiastic amateur astronomer. Bernoulli in his *Lettres astronomiques*, I, 87-9, discusses the telescope that Magellan purchased from Dollond for Saron and compares it favourably to the similar instrument acquired by Greenwich Observatory.

71.01.00

1771	True Time	R^{t}. Ascension	Nor. Declinas^{n}.	
	h ′ ″	° ′ ″	° ′ ″	
Jan. 10	8 44 24	122 27 46	5 4 37	Estimate position to δ Hydrae.
	10 16 45	121 47 16	5 21 15	Determined the position by a * of the 7th mag. then unknown, as also by δ Hyd.
	10 46 31	121 35 31	5 36 51	By δ Hydrae
	12 48 58	120 49 16	6 1 47	By the same * of Hydr, repeated.
	13 19 5	120 24 31	6 14 46	By Procyon
16	7 16 4	84 3 17	22 39 21	By ζ the Southn. Horn of the Bull.
	7 41 39	84 1 32	22 40 51	By the same.
17	6 27 35	80 41 58	23 45 25	By Flamsteed's 121 of Taurus and y^{e}. 163 of La Caille
	7 2 37	80 37 13	23 46 51	By the same
	7 21 23	80 35 28		By the same *.
20	8 18 30	72 52 2	25 55 25	By a * of the 8 mag. observ'd on the Merid.
	8 18 30	72 52 25	25 55 28	By the Planet Mars.
	8 56 49	72 47 47	25 56 47	By the above * of the 8th mag.
	8 56 49	72 48 17	25 56 46	By the Planet Mars
	9 19 24	72 47 17	27 57 3	By the above * of the 8th mag.
	9 19 24	72 47 32	25 57 4	By Mars.

Such are my Observations, from whence M. Pingré[4] has deduced the following Elements of its Orbit

	s ° ′ ″
Ascending ☊	3. 18 42 10
Inclination of the Orbit	31 25 55
Place of the Perihelion	8. 28 22 44
Log. of the Perihel. Dist.	9,722833

Passed the Perihel. the 22d Novemb. 1770, at 22^{h} 5' 48" M. T. at the Royal Observatory, motion Retrograde. He adds, "that this comet resembles none of those whose Elements are determined. On comparing its motion with the Places of its Perihel and ☊, tis easy to see that it was impossible to discover it at Paris before the year 1771, and it may even be added, that it must frequently have passed in the Sun's neighbourhood, imperceptible to the Northern parts of the Earth."

4 The astronomer Alexandre-Guy Pingré (1711-1796).

From Charles Messier 71.01.12

Royal Society of London, Letters & Papers, V, 209.[1]

... C'est pour vous prier, mon bon ami, d'annoncer à la Societé Royale la découverte d'une nouvelle cométe que J'ai commencée à Voir le 10 de ce mois vers les 8 heures du soir: elle paroissoit entre la téte de l'Hydre & le petit Chien, sur le Parallele de Procyon. J'ai determine sa position en comparant son noyau à cette Etoile, et à l'Etoile δ de l'Hydre. Voicy ces Observations.

1re. observ.
Le 10 Janvier 1771 à 10^h. 16'. 45''. tems vray
Ascension droite de la Comete 121°. 47'. 16''.
Declinaison Boreale ------------- 5°. 21'. 15''.

2^d. Observation
La meme nuit à ------------------ 13^h. 19'. 5''.
Ascension droite de la Comete 120°. 24'. 31''.
Declinaison Boreale ------------- 6°. 4'. 46''.

On voit par ces observations qu'en 3^h. 2'. 20''. de tems, le mouvement de cette comete en ascension droite a eté de 1°. 22'. 45''.: et 43'. 31''. en declinaison. On voit cette Comete à la vuë simple à la Lunete, son noyau est brillant, de coleur blanchatre, sans etre terminé, et environné d'une atmosphere qui a plusieurs minutes d'etenduë avec une queuë d'une Lumiere foible de 5 à 6 dégrés de longueur; son mouvement apparent parmi les Etoiles fixes se fait contre l'Ordre des signes en s'ecartant de l'Equateur vers le Pole Boreal.

Cette Comete, Cher ami, fait la douzieme que J'ai decouverte, et que J'observe depuis treize ans.

---

1 The text is in Magellan's handwriting and is headed: "Article d'une lettre de M^r. Messier Astronome F.R.S. & membre de l'Academie R^{le}. des Sciences de Paris adressée à J. H. de Magalhaens, pour etre communiquée à la Societé R^{le}., en date du 12 du courant Janvier 1771." A translation of Messier's report by John Bevis accompanies Magellan's MS; this was read to the Royal Society at its meeting on 24 January 1771 and subsequently published in the *Philosophical Transactions*, together with an extract of a second letter (see Messier to Magellan, January 1771) that was read to the Society on 1 February 1771; see *Philosophical Transactions*, 61 (1771), 104-6.

71.01.25 *To Johann III Bernoulli*

Öffentliche Bibliothek der Universität Basel, MS L I a 709, f. 11-12

Londres 25. Janvier 1771
at n°. 188 Fleet Street.

Monsieur

C'est avec un extreme plaisir que Je viens de reçevoir La Votre du 12me. du courant,[1] aiant depuis Long tems au coeur un grand desir de savoir de vos nouvelles.[2] La correspondence que Vous me proposez a tant d'apas du coté de vos aimables qualités, qui vous siedent comme hereditaires avec votre nom, que vous ne pouviez pas trouver d'autre compliment aussi flateux, que celui ci l'est pour un homme, qui se fait un devoir de supleer au defaut de la science, par des services qu'il est bien aise de faire à tous ceux qui La possedent. Ainsi Vous n'avez qu'à me donner vos ordres sur quelques objects que Vous souhaiterez; & comptez toujours que je n'oublierai rien de ma part pour accomplir vos souhaits, avec d'autant plus de plaisir, que Vos ordres seront les plus precis, & les plus clairs. Comme Je suis resolu de passer le reste de mes jours dans Ce pays Philosophique, & dans le meme tems me trouve en Correspondence avec diferents amis, repandus par La plupart des meilleurs endroits de l'Europe, peut etre me trouverai je à meme de vous servir mieux que quelques autres: quoique Le principal que Je dois ofrir, & sur que vous pouvez compter, est tout ce que depend proprement de l'activité de mon personnel.

Le Recueil Astronomique me semble un ouvrage très interessant.[3] aussitôt que Vous pouvez me faire passer ici quelques copies, Je les metterai Chez un Libraire des plus fameux, & le plus à propos, pour Les vendre pour votre compte, & selon les ordres que vous donnerez. Vous savez bien qu'il faudra Lui donner un tant pour Cent pour son trouble: & je tacherai de le reduire au moins possible. Je crois aussi que Les Lettres sur l'Etat actuel de l'Astronomie[4]

1 Letter not found.

2 Magellan and Bernoulli met during the latter's visit to England in 1769.

3 The first volume of Bernoulli's *Recueil pour les astronomes* was published in early 1771, Bernoulli expressing the hope in the preface that, if the work were well received, subsequent volumes would appear every six months. However, the second volume did not appear for eighteen months, its preface being dated 8 May 1772, while a third volume was not published until 1776 (the preface being dated 26 April of that year). In late 1779, Bernoulli brought together, as a volume entitled *Supplément au recueil pour les astronomes*, the sixth and last number of his *Nouvelles littéraires de divers pays* and the five numbers of this publication previously issued, 1776-9.

4 Bernoulli, *Lettres astronomiques où l'on donne une idée de l'état actuel de l'astronomie pratique dans plusieurs villes de l'Europe* (Berlin, 1771).

auront egalem[t]. un grand debit. Mais pour la souscription de Vos Memoires, Je pense un peu differem[t]. Car J'observe que Le public est si surchargé de scuscriptions, dont quelques unes n'ont pas manqué de le prejudiquer contre cette methode de publiquer des ouvrages, que Je doute beaucoup du succès correspondant à votre expectation: d'autant plus que Je vois de ce coté de l'eau, les esprits si detournès des objets especulatifs, pour les objets materiaux de l'Histoire Naturelle, qu'il semble on va ensevelir dans le cahos de l'oubli, tout ce qui employe autre faculté, que La memoire de l'homme. neanmoins on ne perd rien à distribuer les Prospectus, & à ouvrir la souscription. pour cette derniere Je me charge de la mettre entre de bonnes mains, c'est à dire ou du meme Libraire ci dessus, ou de quelqu'autre dans La meme voye: & en envoyant un certain nombre de reçus signés icy il n'y peut avoir dans La suite aucune difficulté. Car on retournera après la meme quantité numerique soit en reçus, soit en argent. L'affaire est de me les faite parvenir au plutot, & avec La moindre depense possible.[5] 71.01.25

Je parlai hier au D[r]. Matty[6] & à M[r]. Robertson.[7] on a trouvé effectivem[t]. que l'Acad. de Berlin n'a pas encore eu les Transactions de 1768 ni de l'annee derniere. il parlera au Conseil sur ce point; & ne manquera pas de me Les delivra, pour vous Les envoyer. Le meme D[r]. Matty m'a dit avoir deja reçu le dernier volume de Votre Academie, qu'il avoit encore chez soi, & l'apporteroit bientôt à La Societe R[le]. Ce Matin Je parlai aussi avec un des plus fameux de nos Noursery & Seeds Men: qui me dit avoir reçu à toute l'heure une bonne quantité de Semences de l' Amerique: & me promit de garder une partie à mes ordres.[8] J'ai ici avec moi Son Catalogue imprimé, que Je souhait bien pouvoir vous envoyer. mais contenant environ 60 pages d'impression in 8[o]., il n'est pas à propos pour l'envoyer par la poste à un particulier. neanmoins si Vous voulez prevenir Le Ministre, ou m'indiquer quelque voye commode, je vous l'envoyerai au plutôt.

---

5 Magellan's doubts about Bernoulli's proposed publication seem to have been shared by others, since the work never appeared in print. There is nothing in Magellan's later letters to Bernoulli to suggest that the latter even took up his suggestion of testing the market by distributing a prospectus.

6 Matthew Maty (1718-1776), physician and librarian at the British Museum, and Secretary of the Royal Society of London from 1765 until his death.

7 John Robertson (1712-1776) was Clerk, Housekeeper and Librarian of the Royal Society of London from 1768, being succeeded at his death by his eldest son, also named John.

8 From subsequent letters we learn that the nurseryman was James Gordon (*ca.* 1708-1780). The printed catalogue to which Magellan goes on to refer, *A Catalogue of Seeds Plants, Fruit-trees, and Flower-roots, sold by James Gordon, Seedsman…* (London, *ca.* 1769), is said to have been "probably the first nurseryman's list in botanical form" (Ray Desmond, *Dictionary of British and Irish Botanists and Horticulturalists* [London, 1977], p. 258).

71.01.25 Je crois que notre commun ami Messier,[9] vous aura communiqué la decouverte de sa 12me. Comete depuis les derniers 13 années. Ce fut hier aussi que Je l'ai communique de Sa part à La Soc. Royale.[10] Ce fut le 10 du courant à 10h.16'.45'' tems vray – Ascens. droite 121°.47'.16'' declinaison boreale 5°.21'.15''. La meme nuit à 13h.19'.5'' Asc. dr. 120°.2[4'.31'']¹¹ [de]clinaison boreale 6°.4'.46''. de façon qu'en 3h.2'.20'' elle parcourû 1°.22'.[45'']¹² & 43'.31'' en declinaison. on peut La voir à la vue simple. par le Telescope elle paroissoit d'une couleur blanchatre, avec un noyau mal terminé, de la Chevelure, & une queue entre 5 & 6 degres. elle avoit eté vue la meme nuit a Canterbury & fut annoncée dans La gazete.[13] mais Je ne sai pas que personne l'aye vue ici à Londres. Mr. Maskelyne[14] n'est pas venu hier à La Societé, ainsi je ne sai encore s'il l'a trouvée, ou non.

Il est tems de finir celle ci, en Vous reiterant les souhaits sinceres que J'ai pour que Vous jouissiez une parfaite Santé & beaucoup de bonheur, pour poursuivre la glorieuse Carriere que vous allez ouvrir pour Le bien de la Republique Savante, & benefice du Public, et en attendant, que vous me croyez avec les plus sinceres sentimts. & vraye stime

Monsieur
Votre tres humble & obeisst. Servr.
Jean Hyacinth de Magalhaens.

P.S. Je m'oubliois la remarque du botaniste. C'est qu'on perd bien souvent du tems & des depenses à faire elever des plantes par semence. le plus courte (mais beaucoup plus chere à la fois) est d'avoir les plantes les plus dificiles, deja un peu robustes. J'en ai envoyees d'ici une bonne quantité pour le Jardin botanique d'Estrasbourg, à mon ami Le Dr. Spielman;[15] qui y ont tres bien reussi, arrivant en fort bon etat de vegetation &c. reflechissez donc sur ce point, & prennez Votre resolution là dessus

A Monsieur
Monsr. Bernoulli de l'Acade. Rle. des Sciences
à Berlin

---

9 Charles Joseph Messier, the well known French astronomer.

10 See 71.01.12. Bernoulli published Messier's initial set of observations in his *Recueil pour les astronomes*, *I*, 230-1, including Magellan's report that the comet had also been seen at Canterbury.

11 Paper torn away from seal; missing numbers taken from printed reports.

12 Paper torn away; missing number taken from printed report.

13 No report has been found in either the *London Gazette* or the *Public Advertiser.*

14 Nevil Maskelyne, Astronomer Royal.

15 Jacques Reinbold Spielmann (1722-1783), professor of medicine and director of the botanic garden at Strasbourg from 1759.

To Louis-Henri Duchesne[1] 71.02.01

Archives Nationales, Paris, MS T/160 (10-12): lettres de J.H. de Magalhaens

Londres 1 Fevrier 1771

Je viens d'apprendre mon Cher Monsieur, La facheuse nouvelle de l'exil de M^{r}. Le President de Saron, quoiqu'on a pas pû me dire Le nom de l'endroit de son exil.[2] à cet cause il faut vous prevenir sur ce qu'on doit faire des boetes ou petites caisses que Je lui adressois au dedans La caisse, qui est partie par Rouen, & que peutetre n'est pas encore arrivée chez vous.

Vous vous souvenez bien que les caisses & pacquets marquées avec les nombres 3, 4, 6, 7. sont destinées pour Lui: mais pour Le present, ne sachant pas comme il faudra les lui envoyer à son exil, & de plus y aiant quelques autres petites bagatelles pour d'autres amis, dont je l'avois prevenû de les leur envoyer, Je crois necessaire de vous prier par celle-ci de Les conseigner toutes quatre (celles marquées comme cidessus avec 3, 4, 6 & 7 dessus) de les delivrer je dis, à M^{r}. Messier Astronome au College Royal de France, qui est un de nos communs amis, & que Je charge dans une je vais lui ecrire, avec toutes les minucies du detail ; ne voulant pas vous troubler avec de pareiles bagateles. Seulem^{t}. Je vous prie d'avoir soin comme on manie La caisse marquée n°. 3 qui contient un Barometre assez delicat, & fort aissé à se casser. aussitôt que La grande caisse sera arrivée à Paris, vous n'aurez qu'à faire aviz par La petite poste à M^{r}. Messier & lui aura soin de venir chez vous pour s'en charger.

Vous aiant ecrit au Large dans ma derniere sur les articles de La votre, Je n'ai rien à ajouter, que seulem^{t}. un mot sur Les bois preparés,[3] dont vous me demandiez quelque information au Large pour Montrer à M^{r}. de Montigny de l'Ac. R^{le}. en voicy La substance de ce que regarde les bois preparés en Angleterre. il y a 3 ou 4 ans ou meme plus qu'un chymiste de Londres trouva après plusieurs experiences, une certaine matiere qu'il garde en secret, qui etant dissoue dans l'eau, oû l'on fera bouillir des planches de bois, leur donne La qualité de Les faire beaucoup plus durables, & une

1 Annotated by Duchesne: "Rep. le 16. et accusé la reception de la caisse venue par Rouen en bon etat". Letter not found.

2 On 21 January 1771, following a long-running dispute between the French Crown and the judiciary, the Chancellor, Maupeou, in a show of strength on the part of the Crown, issued lettres de cachet to every member of the Parlement de Paris, declaring their posts confiscated and exiling them to different parts of France.

3 See Magellan to Duchesne, 26 December 1769.

71.02.01 espece incorruptibilité.[4] Les experiences ont reussi jusqu'au present si bien, que plusieurs particuliers en font usage du bois qu'il prepare, & on a fait construire, ou batir, un vaisseau, dont la principale part de son bois a etée preparée par ce chymiste; & outre celà on batit un autre avec du bois entremelée avec celui ci, pour mieux en faire La comparaison de sa durabilité. Reellem^{t}. ce chimiste à fait batir de chaudrons oblongs, d'une telle longueur, que 45 & meme plus personnes y peuvent se mettre à Table, & effetivem^{t}. il y donna un dinner à un tel nombre. outre ces enormes depenses, qui certainem^{t}. ne seroient pas entreprises sans quelque bonne fundation & esperance du succès, on peut bien voir, en coupant le bois ainsi preparée, que La matiere y a penetrée jusqu'au interieur des plus grosses planches, car en y regardant avec une loupe on y voit (à la lumier du soleil) des particules luisantes, dont tous les interstices des pores du bois sont parsemées. Or l'idee de feu notre bon ami Jars,[5] etoit de mettre des morceaux de ce bois ainsi preparés, à l'epreuve, aiant d'autres morceaux de la meme qualité, mais non preparés, demi ensevelis, comme les autres dans le meme endroit, pour examiner la difference qu'il y avoit dans la corruption des uns à l'egard des autres: & outre celà il pensoit aussi se mettre à travailler chymiquement, pour trouver, ou la meme matiere, ou quelque autre semblable, qui pût produire le meme effet. mais la mort nous a privé de ses recherches. les bois vous les avez mis comme vous me le marquez dans une Lettre, à demi sous terre: & au bout de quelque tems on peut voir comm'ils preuvent de leurs qualités. N.B le meme Chymiste commence depuis peu de tems à preparer aussi du bois à froid, avec un succès reel selon qu'il le dit

a Dieu.

Je suis
Mon cher Monsr.
Tout à vous
Magalhaens

4 The chemist to whom Magellan refers was Humphry Jackson (1717-1801), who "enrolled a patent in 1768; his method being to boil the wood, for several hours, in a very strong solution of calcareous earth, in water, or acid of vitriol" (*Minutes of Proceedings of the Institution of Civil Engineers*, Session 1852-53, p. 216). Jackson was elected FRS in 1772. In A. Hayward, ed., *Autobiography, Letters and Literary Remains of Mrs. Piozzi (Thrale)* (London, 1861), II, 257, Mrs. Piozzi, widow of Henry Thrale, MP, describes Jackson as "a vulgar fellow who persuaded her husband to build a copper somewhere in East Smithfield, the very metal of which cost £2000, whereas this Jackson was to make experiments and conjure some curious stuff, which would preserve ship's bottom from the worm; gaining from Government money to defray these mad expenses."

5 Antoine-Gabriel Jars (1732-1769); see Magellan to Duchesne, 26 December 1769.

To Louis-Henri Duchesne[1] 71.03.05

Archives Nationales, Paris, MS T/160 (10-12): lettres de J.H. de Magalhaens

Londres Mars 5 1771

La Votre derniere, Mon cher Monsr., n'est pas arrivée que la poste de vendredy[2] trop tard pour vous Repondre dans la même; & par l'aticle, qui suit, vous verrez que Je n'en suis pas faché pour profiter l'ocasion de vous parler plus foncierem^{t}. du jugem^{t}. que fut fait hier dans La Societé des Arts sur l'Orge de Siberie[3] dont Je vous envoye quelques grains ci dedans, pour que vous les communiquez a M^{r}. Trudaine, & consultiez Là-dessus des personnes capables de savoir, si cette espece de grains est, ou non cultivée en France. Car elle fait un des plus interessants objects pour le bien du Public. premierem^{t}. on La seme en Avril, & elle est tout à fait meurit au milieu d'Aoust. 2^{d}. Cet orge produit fort considerablem^{t}. plus que L'ordinaire. 3me. il produit assez bien dans des terres maigres, & qui ne sont bien bonnes pour produire du bled. 5me.[4] en le reduisant à farine, il donne plus de farine & beaucoup moins de son que pareille quantité de bled. 6me. le pain qu'on en fait, est presque aussi bon que celui du froment pour La qualité de la blancheur, & beaucoup mieux pour conserver la fraicheur 3 & 4 fois plus long tems, que le pain ordinaire du froment ou bled; & par consequent il est plus agreable à manger. 7me. en fin cet orge etant generalem^{t}. cultivé, sera d'un grand secours pour le peuple, sur tout lorsque il y aura peu de bled dans quelques annees: & sert egalem^{t}. & avantageusem^{t}. pour faire la biere, & pour tous les autres usages de l'orge ordinaire. Je souhaite qu'on s'informe par example de M^{r}. Duhamel,[5] si ce grain est cultivé en France: car s'il ne l'est pas, ce sera d'une grande utilité publique que de l'encourager: et je pourrai vous envoyer peut etre un boisseau pour etre semé cette année. Voilà une communication bien interessante! Je vous envoye aussi quelques grains de l'avoine nude,[6] qu'on

1 Annotated by Duchesne: "Rep. et envoyé 180 livres reçu de Messrs. Pache et Banquet le 13 Mars." Letter not found.
2 Letter not found.
3 See Magellan to Duchesne, 19 April 1771, 20 March 1772 and 31 March 1772.
4 *sic.*
5 Henri-Louis Duhamel du Monceau (1700-1782), renowned French botanist and agriculturist.
6 A hull-less variety of oat originating in Central Asia, which had been grown in England since the 16th century.

71.03.05 cultive en quelques provinces d'Angleterre: & qui est beaucoup plus avantageuse que la ordinaire.

Les Livres de Tom Jones, Pamela, Clarissa, Grandisson & Andrew,[7] sont deja achetés, & empaquetés ici dans ma chambre pour partir cette semaine par le premier vaisseau. Seulem^{t}. les deux Laboratoires, que vous m'avez demandés, ne sont pas encore arrivés de chez L'ouvrier: mais il me promit de les donner finis vers La fin de cette semaine.

Je viens de donner des ordres pour Le metier à bas.[8] Celui pour des bas ordinaires de laine, est le plus bon marché & le plus simple. l'ouvrier y travaille de la meme maniere que dans les ordinaires, mais avec plus d'aisance, & moins de mechanisme. Je ne manquerai pas d'y ajouter les instructions necessaires, de meme que pour le nouveau metier à des toiles ou doublures de laine: mais touts deux sont si evidents dans leur mechanique, que une seule vüe suffit pour les comprendre.

Je suis très charmé que vous soyez à la veille de Recevoir la caisse que Je vous envoyai par Rouen: & Je souhaite que Le tout soit si bien manié, que L'instrum^{t}. pour M^{r}. Le Presid^{t}. de Saron (qui est un barometre) ne soufre du derangem^{t}. Je vous prie de m'envoyer aussitôt possible quelque remise d'argent: car payant icy, comme Je le fais tout en argent comptant; même ce peu de ces sommes, me rend difficile d'attendre long tems pour le reboursem^{t}., n'aiant pas de fond pecuniaire de mon propre cre[dit.] Voyez: les Livres Anglois ci dessus, sont en tous 25: & voulant les envoier tous reliés netem^{t}. & d'une maniere propre & uniforme Je les payai cent Shellings, comme vous voyez par le Reçu ci dedans.[9] les 4 petites caisses ou boetes des laboratoires de poche couteront 6 guinées: & ceci avec les parcelles deja expediés par Rouen vont bien au dela de vint[10] & tant de louis d'or. un autre, que vous, riroit à ce detail. mais un ami intime comme vous

7 Henry Fielding, *The History of Tom Jones, a Foundling* (1749); Samuel Richardson, *Pamela, or Virtue Rewarded* (1740); Richardson, *Clarissa, or the History of a Young Lady* (1748); Richardson, *The History of Sir Charles Grandison* (1753); and Fielding, *The History of the Adventures of Joseph Andrews, and His Friend Mr. Abraham Adams* (1742).

8 Stocking frame. See Magellan to Duchesne, 30 November 1770.

9 The receipt from the London bookseller J. Johnson, dated 6 February 1770, for having the 25 volumes involved (*Grandison* 7 vols, *Clarissa* 6 vols, *Pamela* 6 vols, *Tom Jones* 4 vols, *Joseph Andrews* 2 vols) "elegantly bound" in calf at a cost of £5 is annexed to the present letter.

10 vingt?

ne doit pas ignorer les particuliers d'un autre aussi sincere & devoûé pour La vie, comme 71.03.05

Votre vray ami & Serv^r^.
J.H. de Magalhaens

N.B. mon adress est = at n°. 188 Fleet Street London =.

a Monsieur
Monsr. Duchesne Secretaire
de Monsr. Trudaine de Montigny
à Paris

To Johann III Bernoulli 71.03.15

Öffentliche Bibliothek der Universität Basel, MS L I a 709, f. 13-14

Londres 15 Mars 1771

Monsieur

J'ai reçu La votre datee du 2 du courant Jeudy dernier; & Je me trouve en perte (at a Lost) sur L'article des semences de l'Amerique, que Vous souhaitez: car ne sachant pas celles que vous eutes l'année derniere, il seroit une sotise de vous envoyer Les mêmes de nouveau. ainsi il faut que Vous me donniez une Liste de celles que vous avez eû, & qui vous ont reussi; ce que vous pouvez très aisem^t^. faire dans une Lettre en petit caractere: & celà doit etre au plutôt, car La saison avance, & J'aurai dans un mois d'ici, un Vaisseau qui va en droiture à Stetin. en 2^d^. Lieu il faut vous dire, que cette année on ne peut pas avoir des demicaisses: & il faut on prendre une entiere, ou seulem^t^. une quantité à choisir de celles qu'on souhaite; & J'estime plus cette derniere methode; car en prennant des caisses on court Le risque d'avoir plusieurs inutiles &c. ainsi ayez La bonté de me donner la resolution du Monsr. Directeur de votre Jardin Botanique: en me marquant quelle des deux lui plait Le plus: & meme pour combien d'argent il veut employer à peut près.[1] Je serai aussi charmé de savoir combien a couté

1 The director the Berlin Academy's botanic garden at this period was Johann Gottlieb Gleditsch (1714-1786).

71.03.15 l'autre demicaisse, si celà est la premiere quantité que Vous avez acheté: car J'aime à prendre toutes Les mesures possibles pour servir mes amis, sans impositions de la part des Shop-keepers ou d'autre quelconque.

Pour ce que regarde mes reboursem^ts^. il faudra sans doute, que Vous cherchiez quelque maniere de me les faire tenir aussitôt ici; car n'aiant pas des aisances extraordinaires, & au delà du precis pour ma subsistance, il ne m'est pas possible de les faire: & meme quand vos Livres seroient arrivés ici, Je n'en pourrois tirer aucun parti, sans vous couter des pertes considerables dans leur vente; n'ayant point de libraire, qui ofre aucun prix raisonable, par des livres deja imprimés, dans Les quels lui ne puisse pas gagner peutetre cent pour cent. ainsi si vous faites que quelque correspondant de M^r^. Ernst (dont vous m'avez parlé dans votre premiere Lettre)[2] ou quelqu'autre March^t^. ici, n'accepte mes Lettres sur Lui, ou sur Vous, jusqu'à une certaine some vg. 20 ou plus Livres Esterlines, alors le tout sera aisem^t^. arrangé. Celle ci est La methode de la plupart de mes Correspond^ts^., excepté ceux qui envoyent d'avance quelques sommes d'argent, & qui restent à mes ordres chez un bon Negociant jusqu'à ce que ces commissions sont remplies. Ce Negociant est M^r^. Aubert (Alexandre) à Austin friars: oû je garde une memoire detailée des personnes à qui les sommes, qu'il a entre ses mains, appartiennent.[3]

En attendant Votre Reponse sur ces deux articles ci dessus, Je tacherai d'avoir les livres & autres commissions pretes pour le premier vaisseau, qui comme Je vous l'ai dit partira d'ici à un mois pour Stetin.

J'ai reçu le detail des observations de cette derniere comete par M^r^. Messier; & je les communiquai à La Societé Royale le 21 de Fevrier, avec les elem^ts^. de Son Orbite par M^r^. Pingré:[4] si vous les voulez, Je vous en envoyerai une Copie.

---

2 Letter not found. Ernst was apparently a merchant in London, probably one of the partners in the firm of Smith & Ernst, merchants in Great St Helen's.

3 The banker Alexander Aubert (1730-1805), governor of the London Assurance Company, was also an enthusiastic astronomer. He built up an exceptionally well equipped observatory at Loampit Hill, Deptford, which he subsequently in 1788 transferred to Highbury House, Islington. He was elected a Fellow of the Royal Society in 1772 and was active in the Society's affairs, being seen by some as a possible successor to Sir John Pringle as President when Pringle resigned in 1778.

4 The details were published, together with those reported by Magellan to the Royal Society on Messier's behalf on 24 January 1771, in the Society's *Philosophical Transactions*, 61 (1771), 104-6. Alexandre-Guy Pingré (1711-1796) was one of the leading French astronomers of this period, being particularly well known for his computations of cometary orbits.

Mr. Landen vient de publier une brochure sur la distance du Soleil par Stewart:[5] que Je compte de Vous envoyer. Vous savez qu'il a publié un traité en 1764 On the Residual Analysis[6] par oû il pretend autant presque comme par le Calcul differential: & encore un autre intitulé Mathem[l]. Lucubrations dans l'annee 1755.[7] 71.03.15

Dans l'année derniere paroit un petit commentaire assez joli, par Mr. Emerson sur le Principia de Newton. in 8°. à la fin est une espece de vindiciœ oû il repond à quelques objections de Messrs. Votre Pere & Oncle contre Newton.[8]

les oeuvres Mathemes. du meme Emerson vient d'etre completées en 11 volumes reliés in 10, in 8°. & coutent £3:12:0.[9] mais il y en a en outre sa Mechanique in 4°.,[10] & le Methode des Incremts.[11] la premiere coute 14s:0d & la seconde 7s.6d.

Mr. Ludlam vient de publier aussi deux brochures algebraiques,[12] selon qu'on m'a dit, car je ne les ai pas encore vues. Monsr. Maskelyne a publiés ou taché de faire publier des Tables fort utiles pour Les Calculations Astronomiques,[13] dont je vous donnerai quelque detail dans ma prochaine. Vous m'obligerez de me dire si vous souhaitez que Je joigne quelques uns de ces Livres dont je viens de vous parler à ceux, que vous me marquez dans votre Lettre.

---

5 John Landen, *Animadversions on Dr Stewart's Computation of the Sun's Distance from the Earth* (London, 1771), criticizing Matthew Stewart's *The Distance of the Sun from the Earth determined by the Theory of Gravity* (Edinburgh, 1763).

6 Landen, *The Residual Analysis: Book I* (London, 1764).

7 Landen, *Mathematical Lucubrations; containing New Improvements in Various Branches of the Mathematics* (London, 1755).

8 William Emerson, *A Short Comment on Sir I. Newton's Principia, containing Notes upon Some Difficult Places of That Excellent Book* (London, 1770). Emerson sought to rebut objections due not to Bernoulli's father, Johann II Bernoulli, but to his grandfather Johann I Bernoulli and his uncle Daniel Bernoulli.

9 William Emerson, *Cyclomathesis: or an Easy Introduction to the Several Branches of the Mathematicks...* (10 vols., London, 1770). Individual volumes were also sold separately, without the series title.

10 Emerson, *The Principles of Mechanics; explaining and demonstrating the General Laws of Motion, the Laws of Gravity, Motion of Descending Bodies, etc.* (London, 1754; 2nd ed. 1758).

11 Emerson, *The Method of Increments, wherein the Principles are demonstrated, and the Practice thereof shewn in the Solution of Problems* (London, 1763).

12 William Ludlam, *Two Mathematical Essays; the first, on Ultimate Ratios; the second, on the Power of the Wedge* (Cambridge, 1770).

13 Presumably the work eventually published as Maskelyne's *Tables for computing the Apparent Places of the Fixt Stars, and reducing Observations of the Planets...published by the President and Council of the Royal Society, at the Public Expence* (London, 1774).

71.03.15 Je ne me souviens à present d'aucune nouvelle interessante, si ce n'est pas qu'un tel M^{r}. Kenrick pretend avoir trouvé le mouvem^{t}. perpetuel: il a deja fait deux lectures pour en demontrer la possibilité, qui ont plu à plusieurs.[14]

on commence à cultiver en Angleterre une espece de bled, qu'on appelle Orge de Siberie, qui se produit en trois mois après semé, & qui semble preferable au bled ordinaire en certaines circumstances.[15] à dieu. le papier est plein: & je suis

Votre t. h. serv^{r}.
Magalhaens.

A Monsieur
Monsr. Bernoulli de l'Acad^{e}. R^{le}. des Sciences
à Berlin

---

14 William Kenrick, *A Lecture on the Perpetual Motion*, 2 parts (London, 1771).

15 Moses Bartram, *Some Account of the Naked or Siberian Barley* (1771). There is a copy of this anonymously published broadside at the American Philosophical Society Library, which gives the author's name and date of publication in its catalogue. For a digital image taken from another copy, see <http://www.nal.usda.gov/speccoll/collect/history/mars17pg.htm>.

71.03.22 *To Louis-Henri Duchesne*

Archives Nationales, Paris, MS T/160-(10-12): lettres de J. H. de Magalhaens

Londres 22 Mars 1771

Mon Cher Monsieur

Je viens de Recevoir cet après diner La votre datée du 13 du courant,[1] avec La Lettre d'echange, ou Reçu de Messrs. Banquet & Pache[2] pour 480$^{\#}$ Tournoises, par Lesquelles Je vous fais bien de Remerciem^{ts}., car J'en avois bien besoin. voilà ce que vaut avoir un ami sensible comme vous, qui ne s'oublie de ceux qui l'oeuvrent Leur Coeur avec sinceretė. Croyez de ma part que Je vous tiens dans une place bien distinguée à l'egard de mon estime: & que dans quelqu'occasion que ce soit, vous me verez toujours agir à Votre egard en consequence de ces sentiments.

---

1 Letter not found.

2 See Magellan to Duchesne, 30 November 1770, n. 12.

Le nouveau bled, ou soit orge de Siberie,[3] a fait un grand bruit parmi nos cultivateurs. on en est charmé: & on croit avoir un grain très precieux pour le bien du public & particulier de la Nation. J'ai ordonné d'abord quelque quantité pour distribuer à mes amis: & je commence à m'atrister pour n'avoir pas Reçu encore aucune Reponse. Le cas est qu'il a encore fort peu de ce grain en Angleterre: & que plusieurs l'ont demandé de tous les cotés pour le multiplier. La premiere semence qu'on a eu, ne fut pas tant que l'echantillon que je vous ai envoyés. ainsi Je vous prie de les faire semer par votre jardinier soigneusem^{t}., au bout ou meme dans le milieu du mois prochain, en donnant environ 6 pouces de distance entre chaque grain: car on a vû quelques uns produire 12, 15, & meme plus de plantes. Mais soyez assuré, que si je puis en obtenir en propre tems je ne manquerai pas de vous en envoyer, autant que Je pourai: au pis aller Je vous envoyerai quelques boisseaux en August ou Sept^{bre}. prochain, pour qu'on puisse les distribuer en differents endroits de France. 71.03.22

Les Livres que vous m'avez ordonnez sont il y a presqu'une semaine, chez le Water-man pour les embarquer dans le premier vaisseau pour Calais: & sans doute vous en aurez bientôt l'avis de M^{r}. Morelle du Faux.[4] ils sont empaquetés dans une boete adressée a M^{r}. Trudaine: vous y trouverez les deux assortim^{ts}. des Laboratoires de poche[5] selon la methode de Cronstedt que vous m'avez ordonnés; & de plus un pair de bas de Soye, faites dans une des nouvelles machines ou metiers à Bas,[6] pour que M^{r}. Trudaine en voye la qualité. ceux ci sont de la machine La plus chere: car l'autre ne sert, comme je vous l'ai deja marque, que pour y faire des bas ordinaires, plutot grossiers, de laine: mais en revanche cette machine grossiere a l'avantage de pouvoir etre travaillée par eau, ou par mouvem^{t}. de cheval, à ce qu'on m'a assurée. lorsqu'elle sera faite Je vous en parlerai plus en detail.

les deboursem^{ts}. pour Le present sont pour les 25. vol. de ces Livres[7] ci-dessus £5 :0^{s} :0^{d}: les deux sets de Laboratoires: c'est à dire une pour les experiences à sec & l'autre avec des acids, à 3 guinees chaque, font £6 :6^{s} :0^{d}: Les bas de soye 15^{s} :0^{d}. caisse 2^{s}: porteur 1^{s}. depeches & depenses à la douane

3 See Magellan to Duchesne, 5 March 1771.
4 See Magellan to Duchesne, 30 November 1770, n. 8.
5 These "pocket laboratories", portable sets of apparatus for testing minerals *in situ*, had been devised by the Swedish chemist Axel Cronstedt and were described in an appendix added by Gustav von Engestrom to his English translation of Cronstedt's *Essay towards a System of Mineralogy* (London, 1770).
6 i. e. stocking frames.
7 See Magellan to Duchesne, 5 March 1771.

71.03.22 jusqu'à bord du vaisseau 4^{s} :8^{d}. Mais pour le total Laissez moi Recevoir Les metiers ci dessus & je vous en rendrai [mon] compte du tout.

On vient de comuniquer à La Societé des Arts une nouvell[e] espece de giblet[8] ou pierçoir,[9] dont tous les artistes sont ravis; cet instrum^{t}. fait Les trous dans Les planches des vaisseaux, ou des batiments, meme les plus epais, sans avoir besoin de l'oter, pour decharger le bois vermolû, ce qui arrive dans tous ceux qu'on connoit jusqu'à Present: car c'est une lame d'acier regulierem^{t}. tourtillée, qui laisse echaper à chaque tour le bois, quelle coupe. C'est un pauvre Charpentier qui en est l'inventeur. aussitot qu'on en fera pour l'usage du public, j'envoyerai un de ces instrum^{ts}. à M^{r}. Trudaine, pour Le faire communiquer aux batisseurs de la marine &^{a}.

Vous aurez la bonté de souvenir à M^{r}. Trudaine que Je ne laisse pas echaper aucune nouvelle utile, dont je puisse penser de lui faire plaisir en La lui communiquant.

Au rest conservez moi la place dont vous m'honorer dans votre estime: & croyez moi pour La vie

Mon cher Monsieur
Votre tres humble et obeiss^{t}. serv^{r}.
J.H. de Magalhaens

N.B. Selon L'avis que M^{r}. Le Chev^{r}. de Bory vient de me faire, je lui adresse dessous La couverte pour M^{r}. Trudaine une Lettre;[10] & je ferai autant pour M^{r}. Messier qui me dit avoir obtenu aussi cette permission du meme Monsieur Trudaine.

A Monsieur
Monsieur Duchesne Secretaire
de M^{r}. Trudaine de Montigny
à Paris

8 gimlet?

9 i.e. perçoir.

10 Letter not found.

To Matthew Maty[1]

71.04.13

Royal Society of London, Letters & Papers, V, 230

J'ai reçu hier soir, mon cher Doct[r]., l'article suivant dans une Lettre de Monsr. Messier, que Je vous communique cidessous, dans l'idée, que Vous serez bien aise peut etre de le faire savoir à quelques Astronomes en plutôt possible, & avant L'assemblee prochaine de la Societé Royale, pour qu'ils puissent profiter des avantages du tems dans leur Observations:[2] et ce fut par cette meme raison, que Je l'ai deja communiqué hier soir par La petite poste à M[r]. Maskelyne.

Il me reste à Vous remercier, de tout mon mieux possible, pour La maniere obligeante avec laquelle vous avez parlé à mon egard dans la derniere assemblée de la Soc[te]. R[le], à l'occasion de mon papier scelé.[3] Soyez bien assuré que Je ne m'oublierai jamais de cette faveur aussi bien que de toutes les precedentes, dont Vous m'avez honoré: et que Je serai très charmé de trouver quelques occasions pour pouvoir Vous marquer en tout tems, combien Je suis sincerem[t]. & avec toute la consideration possible

Monsieur
Votre très humble & obeiss[t]. serv[r].
Jean Hyacinth Magellan

Fleet Stree[t] 13 Avril 1771

To Doct[r]. Matty Secretary to y[e]. R[l]. Society
at y[e]. British Museum[4]

1 Secretary of the Royal Society of London. Attached to the letter is an unidentified clipping of a printed English translation of the report from Messier that Magellan appended to his letter.

2 Messier's communication was reported at the Royal Society's meeting on 18 April 1771; see Royal Society of London, *Journal Book Copy*, XXVII, 67. It was not, however, published in the *Philosophical Transactions*.

3 Magellan had sought and been granted permission to deposit a sealed envelope with the Royal Society. Some years later, he applied for it to be returned to him, and this was done (Royal Society, *Council Minutes*, 11 March 1779).

4 Annotated: "April 18 Messier on new comet".

71.04.13 Article d'une Lettre de Monsr. Messier, Astronome F.R.S. & Membre de l'Acad^{e}. Roy^{le}. des Sciences de Paris & plusieurs autres adressée à Jean Hyacinth Magellan, à Londres, en date de Paris du 6me du courant Avril 1771.

"Le sujet de ma Lettre, est de vous prier de faire part à la Soc^{e}. Roy^{le}. de la decouverte d'une nouvelle comete, que J'ai commencé à voir le premier de ce mois, vers les 8 heures du soir. Elle paroissoit à l'occident, à 12 ou 15 degr. au dessus de l'horizont: au dessous & vers la droite des Pleyades entre les etoiles nu & epsilon de la constellation du Belier. On la voioit difficilem^{t}. à la vüe simple: à la Lunette le noyau etoit brillant sans etre terminé, environné de nebulosité, avec une queüe de 2 degr. ½ environ de longeur, dirigée vers les Pleyades. Depuis le premier Avril jusqu'aujourd'hui Je l'ai observée chaque jour. Il suffira de vous envoyer les deux observations extremes pour jouger de sa route, ou du chemin qu'elle tient parmi les etoles fixes. Le 1er. Avril à 8^{h}.46'.52'' du soir tems vrai, l'ascension droite du noyau de la comete etoit de 38°.47'.47'', & sa declinaison Boreale de 20. deg. 18'.17''. Le 6 de ce meme mois, son Asc. droite etoit de 44°.17'.57''. et sa decl^{n}. boreale de 22°.25'.50''. Cette derniere determination fut faite à 8^{h}.17'.6''. du soir tems vrai. Cette comete suit l'ordre des signes: & son mouvem^{t}. paroit uniforme. en 24 heures elle parcourt un degré 10 min. en Ascens. droite; et environ 26 min. en declinaison en s'elevant vers le Nord. Depuis le premier de ce mois jusqu'aujourd'hui sa Lumiere m'a paruë constament La meme: et il n'a pas eté possible de m'assurer par son mouvem^{t}. observé, si elle s'approche de la Terre, ou si elle s'en eloigne. Je soupçonerois cependant qu'elle s'en approche."

To Johann III Bernoulli

Öffentliche Bibliothek der Universität Basel, MS L I a 709, f. 15-16

Londres 16 Avril 1771

Monsieur

La Votre du 2 du Courant m'est parvenue le 12 au soir:[1] tout de suite J'ai fait chercher les ouvrages de Landen & Ludlam[2] que vous me marquiez, & changer celui d'Waring,[3] que J'avois deja, car J'avois tout le rest prêt pour ne perdre point de tems. Le Samedy matin Je me portai chez le Seeds-man, M^{r}. Gordon,[4] qui est le plus fameux dans cette Capital; & Je lui ai fait ouvrir une caisse de l'Amerique, entiere, car on n'en trouve à present des demicaisses ni quart de caisses: & la trouvant en bonne condition, c'est à dire sans apparence des semences etre gatées, J'ai aprete le tout en deux caisses: & hier matin Je les ai fait metre à bord du premier Vaisseau, qui partira (la semaine prochaine) d'ici pour Stetin, oû Vous donnerez vos ordres pour les tirer du vaisseau & vous les envoyer, en vertu du Lading bill ou Reçu du Capitain que Je Vous envoye ici. Les deux caisses sont marquées I.G. N°. 1 & n°. 2. La premiere contient les semences Americaines: La 2de. contient les Livres suivants. Les Transact. Philos. pour 1768 & 1769. / Ionian Antiquities[5] que Je payay £1:16^{s}:0^{d}. / Search's Light of Nature 5 vol.[6] – £1:0^{s}:0^{d} / Reid on human mind[7] – 5^{s}.0^{d}. / Elem^{ts}. of Criticism. 2v.[8] – 12^{s}.0^{d} / Nautical alman. for 1772 – 3^{s}. 6^{d}. / Moivre on Chances, last edit.[9] – £1:1^{s}.0^{d}. / Franklin on Electricity last Ed.[10] – 10^{s}.0^{d}. / les 2 pamplets de Ludlam[11] –

1 Letter not found.
2 See 71.03.15, notes 5 and 12 respectively.
3 Possibly Edward Waring, *Meditationes algebraicae* (Cambridge, 1770).
4 James Gordon (*ca.* 1708-1780) had a nursery at Mile End, London, from 1742, and from the mid-1760s also had a seed shop in Fenchurch Street. He introduced many exotics to England, including the camellia.
5 *Ionian Antiquities, published with permission of the Society of Dilettanti* (London, 1769).
6 Edward Search, *The Light of Nature Pursued* (London, 1768).
7 Thomas Reid, *An Inquiry into the Human Mind, on the Principles of Common Sense*, 3rd ed. (Edinburgh/London, 1769).
8 Henry Home, Lord Kames, *Elements of Criticism*, 4th ed., 2 vols. (Edinburgh, 1769).
9 Probably the 1756 ed. of Abraham de Moivre, *The Doctrine of Chances: or, a Method of Calculating the Probability of Events in Play.*
10 Benjamin Franklin, *Experiments and Observations on Electricity, made at Philadelphia in America*, 4th ed. (London, 1769).
11 See Magellan to Bernoulli, 15 March 1771, n. 12.

71.04.16 2^{s}.6^{d} / Sterling's lineæ 3^{ii} ordinis[12] qui est un petit livre tres rare à present, & celui ci est l'unique qu'on trouve à Londres dans aucune boutique par 3^{s}.6^{d}. / Landen's Residuals[13] – 6^{s}.0^{d}. / d^{o}. Lucubrations[14] – 6^{s}.0^{d} / d^{o}. on Stewart's sun's distance[15] – 1^{s}.0^{d}. / pour la caisse 2^{s}.10^{d} / diferents messages & porteurs 2^{s}.6^{d}. / de porter les deux caisses à la douane 1^{s}.8^{d}. / registre, searchers, wharphage (depenses à la douane) & bateau à bord du vaisseau 5^{s}.6^{d}. / Lading bills – 9^{d} / Cette Lettre payera 1^{s}.0^{d}. à la poste. payai celles du 12 Janvier & reponse, du 2 mars & reponse, & du 2 Avril 5^{s}.0^{d}. payai pour La Caisse des semences de l'Amerique £5:5^{s}:0^{d}, come vous montrerez par le Reçu, car sachant que celà n'est pas pour Vous, je vous l'envoye dans un petit paquet, avec un petit livre de poche pour cette année, dont Je vous fais present, etant fort commode pour faire des memorandums & marquer les rendezvous &c. Je crois le donner au capitain, qui le delivrera à qui le lui demandera de Votre part: &c. La some totale des articles ci dessus est £12:11^{s}:3^{d}.

avant de fermer celle ci Je parlerai à M^{r}. Ernst, dont Je recevrai cet argent en cas, qu'il aye reçu les ordres de Mess^{r}. Splittgerb & Daum.[16] autrem^{t}. Je tirerai sur vous pour cette some

Je ne sai pas quelle espece de demicaisse on vous a envoyé d'ici auparavant avec des semences d'Amerique par 2 guinées. Car M^{r}. Gordon, assurem^{t}. le plus fameux Seeds-man de Londres, m'assure que l'année derniere il n'en a jamais vendu à moins de 6 & 7 guinees les caisses entieres! Je serois charmé de savoir au juste le prix de celle là & le nom du vendeur s'il etoit possible car Je me mefie toujours des affirmations de ces Shop-keepers.

Je n'ai pas reçu aucun billet de vous, outre vos lettres par la Poste. M^{r}. le Marquis Caraccioli[17] aparem^{t}. s'en a oublie, ou son domestique.

M^{r}. Maskelyne n'a pas encore publié ses Tables, seulem^{t}. on a fait part à un de ses amis intimes (M^{r}. Aubert) qui en est de moi aussi. dans la premiere occasion Je Vous en ferai quelque detail.[18]

---

12 James Stirling, *Lineæ tertii ordinis Neutonianæ, sive illustratio tractatus D. Neutoni de enumeratione linearum tertii ordinis...* (Oxford, 1717).

13 See Magellan to Bernoulli, 15 March 1771, n. 6.

14 Ibid., n. 7.

15 Ibid., n. 5.

16 Splitgerber und Daum was a well known banking and industrial firm in Berlin, founded in 1712.

17 The marquis Domenico Caraccioli (1715-1789), Neapolitan ambassador to London, 1763-71.

18 Maskelyne's tables were not published until 1774; see 71.03.15, n. 13.

On vend ici les Tables de Mayer[19] par 10s.0d / & sa Theorie de la Lune[20] par 2s.6d. 71.04.16

Il ne faut pas vous decourager pour la vente de vos ouvrages ici. J'aime à donner toujours des idees au pis aller, pour eviter du repentis à mes amis, en cas de mauvais succès. D'abord que vous m'en envoyerez, je prendrai les meilleures voyes pour y reussir. & je parlerai en cas de besoin au Dr. Merdock:[21] quoique je pense que son authorite ou amitié avec les libraires n'aura jamais plus de poids, que leur interet. Outre celà Je connois parfaitemt. bien le caractere de Cadel,[22] au quel J'achete très souvent, aussi bien qu'à plusieurs autres libraires, beaucoup plus generaux & good-natured que lui. Car la grande quantité de Correspon^ds que J'ai en diferents Pays, m'oblige à etre ici avec des commissions continuelles en toute sorte de choses.

le point est que Vous me dissiez votre derniere prix; & alors je arrangerai votre affaire au mieux possible avec les libraires ici.

Vous trouverez dans la caisse n°. 2. le Catalogue imprimé des semences & Plantes de Gordon[23] & avec lui vous trouverez le catalogue manuscript des semences de l'Amerique contenues dans La Caisse n°. 1.

Je viens de parler à Mr. Ernst, lui a reçu ce matin l'ordre de me payer La some que vous m'avez annoncée dans votre derniere: ainsi J'en ai recu les £12:11s:3d. ci dessus, dont je lui donnai une double quitance.

Vous aurez deja apris que Mr. Messier vient de decouvrir une nouvelle Comete, ce qu'il m'a communiqué pour la Societé Royale, aussitôt j'en fis avis à Mr. Maskelyne, & au Dr. Matty,[24] pour ne perdre pas le tems de la pouvoir chercher avant jeudy, sa lettre ayant arrivée icy Vendredi soir dans le meme tems que la votre derniere. Le 1er. de Ce mois elle avoit 38°.47'.47'' d'Asc. droite: & 20°.18'.17'' de declin. Boreal, à 8h.46'.52'' du soir tems vray. [Sa 2e][25]. observation, le meme jour qu'il m'ecrivit, fut faite le 6 du Courant à 8h.17'.6'' du soir tems vray: alors la Comete avoit 44°.17'.57'' d'Asc. droite: & 22°.25'.50'' declin. boreale. Son mouvemt. est fort uniforme d'environ

---

19 Tobias Mayer, *Tabulum motuum solis et lunae novae et correctae...quibus accedit methodus longitudinum promota, eodem auctore*, ed. N. Maskelyne (London, 1770).

20 Tobias Mayer, *Theoria lunae juxta systema Newtonianum* (London, 1767).

21 Probably the clergyman-mathematician Patrick Murdoch, who had earlier lived for some time in Berlin. He would thus have been known to Bernoulli, who has evidently suggested that Magellan should seek his help in placing Bernoulli's books with suitable booksellers.

22 Thomas Cadell, bookseller in the Strand, 1767-93.

23 *A Catalogue of Seeds Plants, Fruit-trees, and Flower-roots, sold by James Gordon, Seedsman...* (London, *ca.* 1769).

24 Matthew Maty, Secretary of the Royal Society.

25 Conjectural reading.

71.04.16 1°.10' par jour en Asc. droite, & environ 16' de declinaison. on la voit mal à la vue simple. mais à La Lunete le noyau est brillant, environé d'une nebulosité, & une queüe d'environ 2 ½ degres, vers Les Pleyades. on ne pouvoit pas encore decidé si elle l'aproche, ou non de la terre.[26]

Pour la Comete preced^{te}. que le meme M^{r}. Messier decouvrit le 10 Janvier dernier, le P. Pingré a calculé les elem^{ts}. de Son Orbite sur les observations du meme Messier:[27] & trouva le Neud ascendant de 3^{s}.18°.42'.10'' l'inclinaison de l'Orbite de [31°]25'.55''.[28] Le lieu du Perihelie 8^{s}.28°.22'.44''. Logarithme de la distance Perihelie 9.722833. Passage au Perihelie le 22 Nov^{bre}. a 22^{h}.5'.48'' tems moyen à l'Observatoire de Paris. Le mouvem^{t}. Retrograde. Cette Comete ne se ressemble point à aucune des Cometes dont les elem^{ts}. sont connus. en comparant son mouvem^{t}. aux lieux de son Perihelie & de son Neud ascendant, elle ne pouvoit pas etre vûe à Paris avant L'entrée de l'année 1771: car elle passa proche du soleil sans pouvant etre aperçûe de la partie Septentrionale de la Terre.

Je saurois le jeudi prochain si M^{r}. Maskelyne ou quelqu'autre de la S. R.[29] a aperçûe cette derniere Comete. Si vous ecrivez à M^{r}. Maskelyne pour avoir son Correspondance Astronomique, J'en aurai soin de lui demander ses decouvertes: mais Je crains que Vous n'y trouverez pas beaucoup: car c'est principalem^{t}. au calcul qu'il s'aplique à ce qu'il semble se contentant des observations journalieres qu'il & son Adjoint font continuellem^{t}. à Greenwich.

Le D^{r}. Bevis[30] ne fait rien, ni est en etat de le faire, se trouvant reclus dans sa chambre & ne pouvant pas se tenir debout après une deslocation ou luxation qu'il eut dans une cuisse: Cet accid^{t}. lui arriva dans sa meme Chambre, etant pour prendre l'hauteur meridien du Sol. Nous somes en peine pour le D^{r}. Solander, M^{r}. Banks, & les Astronomes qui furent aux

---

26 Magellan reported Messier's new comet to Maty on 13 April 1771 (see 71.04.13) and to the Royal Society at its meeting on 18 April 1771 (*Journal Book Copy*, XXVII, f. 67). Unlike Messier's observations of the comet he had discovered in January 1771, this new report was not published in the *Philosophical Transactions*, but an English translation was published in one of the London journals (see 71.04.13, n. 1). The elements of this comet's orbit, as calculated by Pingré on the basis of Messier's observations, were published in *Journal des sçavans*, June 1771, p. 443; an improved set based on additional observations was given in *ibid.*, August 1771, p. 574.

27 Magellan communicated Messier's observations of this comet to the Royal Society on 24 January 1771, with further details, together with Pingré's computations, on 1 February; see 71.01.00 and 71.01.12.

28 The number 31° is now illegible in the original MS, and is taken from the printed report.

29 i.e. Société Royale.

30 John Bevis (1693-1771) was a well known astronomer with his own observatory at Stoke Newington; see 72.01.28 for an account of his life.

Isles de la mer Pacifique pour y observer la derniere passage de Venus. on les croit perdus, n'en ayant pas aucune nouvelle après leur atouchem[t]. au Bresil. C'est une grande perte pour l'Histoire Naturelle, & Astronomie...![31] 71.04.16

il n'est pas possible de trouver le Dilenii Hist[a]. Muscorum.[32] C'est un livre très rare & très recherché: quand il paroit quelqu'exemplaire d'hazard, on le vend à 4 guinés & meme au delà. Mais on trouve ici une edition, qui en contient seulem[t]. Les planches, avec quelques petites explications en Anglois,[33] qu'on vend plus raisonablem[t]. Je crois, que par une guinee. Vous me direz ce que vous souhaitez Là dessus.

De meme on ne peut pas avoir l'essay de Wood on y[e]. orig[l]. genius of Homer, mais si M[r]. Le Secretaire ou Vous voulez m'envoyer une Lettre pour l'Auteur, je tenterai avec cette autorité de voir si lui me donera pour L'Acad[e].[34]

Je suis bien aise, que vous vous passez du Brunker's Algebre[35] car c'est un Livre très rare, dont a peine on connoit ici le titre.

Vous trouverez dans votre caisse n[o]. 2. les figures des brochures d'Estewart,[36] que M[r]. Wilson[37] m'a donnez pour vous.

la 2[d]. edition de la Phil[a]. Britanica de Martin fut en 1759:[38] & ce fut dans cette meme année qu'il imprima ses Elem[ts]. of Optics part 1.[39] ainsi je n'ose pas vous les envoyer. Quant à la 2de. part de cet ouvrage, elle parut il y a

---

31 Daniel Carl Solander and Joseph Banks had sailed with James Cook on board HMS "Endeavour" in August 1768 on the first of his three great voyages of discovery in the South Seas. A major objective was to observe the transit of Venus across the face of the Sun, expected in June 1769, from a site in the southern Pacific Ocean. Despite the fears that Magellan here reports, the expedition was not lost: Cook brought his ship safely back to England in June 1771, having circumnavigated New Zealand and charted the east coast of Australia en route.

32 Johann Jakob Dillenius, *Historia muscorum in qua circiter sexcentae species veteres et novae ad sua genera relatae describuntur et iconobis genuinis illustrantur* (Oxford, 1741).

33 Dillenius, *Historia Muscorum: A General History of Land and Water etc. Mosses and Corals, containing all the known Species, exhibited by about 1000 figures* (London, 1768).

34 The M.P. Robert Wood (?1717-1771) was a respected writer on Greek antiquities. Only seven copies of his *Original Genius of Homer* were printed when it was first published in 1767. An enlarged, anonymous edition was published in 1769.

35 Possibly Johann Heinrich Rann, *An Introduction to Algebra, translated out of the High-Dutch into English by T. Brancker* (London, 1668).

36 Some of the figures probably refer to Matthew Stewart, *The Distance of the Sun from the Earth determined, by the Theory of Gravity* (Edinburgh, 1763).

37 Probably Alexander Wilson, professor of astronomy at the University of Glasgow, acting on behalf of his ailing colleague Stewart, professor of mathematics at Edinburgh.

38 Benjamin Martin, *Philosophia Britannica; or, a New System of the Newtonian Philosophy, Astronomy and Geography* (Reading, 1747; 2nd ed. London, 1759).

39 Benjamin Martin, *New Elements of Optics* (London, 1759).

71.04.16 environ 2 ans.[40] lorsque Vous me direz quelle vous voulez de celles ci, je vous les envoyerai. en verité cet Auteur est un rambler, & n'a pas quelque consideration d'aucun poids ici: une espece de charlatan en physique & Mathematique. neanmoins quelques choses peuvent etre utiles.[41]

Je crois avoir satisfait à tous les articles de vos 3 lettres: du moins je ne me rapelle aucune chose de nouveau. Mr. Aubert, un de mes bons amis, dont vous verez son observation du passage de Venus dans Les Transactions,[42] à pris une certain methode fort aisée pour reduire toutes ses observations, trouver la valeur des arcs de l'azimuth, les refractions &cc. ce qu'il fait par le moien de certaines petites tables qu'il calcula sur cette latitude, par oû il abrege fort le resultat de quelqu'observation. Je pense à l'engager à me les communiquer pour Vous, si vous en serez content. quoiqu'il est si modest, qu'il y aura de la reluctance à Le faire.[43] Mr. [Narne][44] a fait ici dernieremt. des machines Equatoriales, qu'il appelle observatoires mouveables, dont beaucoup du mond est fort content. on soupçone que quelqu'ami l'a aidé dans La theorie, car il n'en a pas beaucoup. pour moi Je soupçone cet Alexdre. Aubert dont je vous parle, qui en est de ses bons amis: mais celà est absolumt. caché entre les deux: & plusieurs n'y soupçonnent rien. Cependant Ramsden, ce fameux Artiste, qui sans doute est à present le meileur de l'Europe, ayant assez de theorie, & une excellente practique en a fait beaucoup meilleurs: & leur a ajouté une methode pour corriger la refraction, en telle sorte qu'on y trouve tout de suite en plein jour les etoiles dans le cham[p] de l'Instrumt.[45]

40 The second part of this work, embracing Parts IV-VI and "containing the theory of Catoptrics and Dioptrics deduced from Physical and Mathematical Principles, and its application to the Construction and Uses of Optical Instruments", was published some years after the first, but without any date of publication being shown. A date of "ca. 1769" is given by John R. Millburn in the bibliography of Martin's publications included in his *Benjamin Martin: Author, Instrument-Maker and 'Country Showman* (Leiden, 1976).

41 Benjamin Martin was an itinerant lecturer on experimental philosophy and a prolific author of works on scientific subjects, but the fact that he never became a Fellow of the Royal Society suggests that Magellan's low opinion of his scientific merits was shared by others. For an account of his life and a bibliography of his numerous publications, see John R. Millburn, op. cit. (n. 40).

42 *Philosophical Transactions*, 59 (1769), 378.

43 Aubert appears never to have published his method.

44 i.e. the instrument maker Edward Nairne; the paper is damaged, but Magellan has clearly mis-spelled the name. See "Description and Use of a new constructed Equatorial Telescope or portable Observatory, made by Mr. Edward Nairne, London", *Philosophical Transactions*, 61 (1771), 107-13.

45 Ramsden likewise published an account of his instrument, initially in French; see Jesse Ramsden, *Description d'un nouvel instrument, appellé Equatorial Universel, fait par M. J. Ramsden à Londres, avec la manière de l'ajuster pour s'en servir.* This work was dated January 1773; a slightly extended English version appeared bearing the date May 1774.

A present je vais vous parler de deux ou 3 articles des plus interessants pour Le Public, & que Je souhaite que vous en puissiez fournir les idees à Votre Ministre. 1[remt]. on vient d'inventer ici une espece de metier pour faire de la toile en laine, ou lin &c. qui est portable. l'ouvrier peut l'aprocher du [feu][46] ou du Jour tout entier. il est fort petit, de façon qu'a peine ocupe la moitié d'un metier ordinaire & a toutes les avantages qu'on peut souhaiter, beaucoup plus parfaitem[t]. que les autres metiers. 2[d]. on vient d'inventer un auger, ou drille, c'est un outil pour faire des trous grands dans la boiserie, particulierem[t]. pour batir les vaisseaux, & meme pour percer les tuyaux des pompes de bois, qui a l'avantage de faire sortir continuellem[t]. en arriere tout le boit vermolû, qu'il va coupant, sans avoir besoin d'etre tiré à reprises, comm'à l'ordinaire pour le decharger du bois qu'il a coupé. on est très flate de cet outil, & on espere qu'on l'employera avec tres grande avantage dans de gros ouvrages. 3me. on vient d'anoncer au public la culture d'une espece nouvelle de bled, qui frutifie beaucoup, donne plus de farine & fait meilleur pain, que l'ordinaire. son grain est façoné comme celui de l'orge, mais il est tout a fait nud: & pour celà on l'apelle the naked barley or Siberian barley. il fait plus de biere lorsqu'on le reduit à malt &c. On le seme à present, & munit avant l'orge ordinaire. Je vous envoye un recit imprimé de ce nouveau bled,[47] avec le pocket-book ci dessus que je consignai aujourd'hui au Capitain George Simpson pour vous. Or tous ces objets etant d'une avantage publique, je pourrai vous les faire passer chez vous, en prennant des mesures à propos pour ne pas exciter la jalousie Angloise, qui en general n'aime point que de pareilles choses nouvelles sortent & soyent communiquées au dehors. Je suis membre de la Societé des Arts & metiers:[48] & c'est dans ces Comitées ou assemblees, que Je me trouve à meme de pouvoir communiquer tout ce qui paroit de nouveau util pour le public.[49] Vous y penserez, & si pouvez, & le juger à propos 71.04.16

A more elaborate English-language account was published under a slightly different title in 1791.

46 Conjectural reading.

47 Probably Moses Bartram, *Some Account of the Naked or Siberian Barley* (1771); see Magellan to Johann III Bernoulli, 15 March 1771, n. 15.

48 Magellan was elected a member of the Society of Arts on 2 May 1770.

49 The loom described by Magellan had been invented by John Almond, who had brought it to London at the request of the Society of Arts, to be examined by the Society's Committee of Mechanics. Magellan was not present at the Committee's meeting on 8 January 1771 when the loom was very favourably received and the Committee recommended that Almond be granted a bounty of 50 guineas by the Society; but he had been at an earlier meeting, on 13 December 1770, at which it was agreed to invite Almond to display his machine. The auger that Magellan discusses had been invented by Phineas Cooke. It was

71.04.16 faites cette ouverture de mes offres à celui, ou à ceux aux quels ces choses peuvent avoir du raport, pour l'interet & avantage de Votre Nation. à Dieu, mon Cher Monsieur; excusez La longeur de cette Lettre: & soyez assuré que personne vous servira de meilleur Coeur, que celui qui a l'honneur d'etre de Coeur

Monsieur
Votre tres humble & obeis^{t}. serv^{r}.
J.H. Magellan

A Monsieur
Monsr. Bernoulli de l'Academie Royale des Sciences
à Berlin
Prusse

considered by the Committee of Mechanics on 11 April 1771, with Magellan present, and a bounty of 30 guineas was recommended to be paid by the Society. Magellan had also been present at a meeting of the Society's Committee of Agriculture on 25 March 1771 at which reports were received of the successful cultivation of Siberian barley by Thomas Widdens of Liverpool and by William Halliday. The Committee recommended that both men should be awarded the Society's Silver Medal.

71.04.19 *To Louis-Henri Duchesne*

Staatsbibliothek Preussischer Kulturbesitz, Berlin, Sammlung Darmstädter G2 1770(1)

Londres 19 Avril 1771

Je profite, Mon Cher Monsieur, de cette occasion pour vous envoyer le recit ci joint qu'on a imprimé sur les qualités & avantages du grain nouveau, apellé l'orge de Siberie, & que par ces belles avantages semble en certaine façon preferable, au froment ordinaire.[1] Je vous prie de le communiquer

1 Probably Moses Bartram, *Some Account of the Naked or Siberian Barley* (1771); see Magellan to Johann III Bernoulli, 15 March 1771, n. 15.

à Monsr. Trudaine,[2] en quelque moment de loisir, car je crois qu'il ne manquera pas de se jouir de cette communication pour le bien publique. On attend ici quelque pluie pour le semer, & meme quelques curieux l'ont deja semé. Je crois que dans la recolte suivante (vers Aoust prochain) je pourai vous envoyer quelques boisseaux. 71.04.19

Pour les autres points en commission, le metier de bass,[3] & l'autre portable, je ne puis encore vous fixer le tems; n'ayant pas encore reçu des reponses bien decidés sur le tems qu'ils seront finis: car ces choses qui regardent les manufactures ici, ont besoin d'etre menagées par troisieme personne pour eviter toute sorte de jalousie: moi etant etranger, aisement les ouvriers aprehendent tout de suite, que c'est pour en envoyer au dehors, &c. ainsi je prends les mesures prudentes pour ne pas alarmer leur confidence. &c.

Je crois qu'à cette heure vous avez reçu bien conditionés les livres que je vous ai envoyés, & les Laboratoires de Poche:[4] & je espere aussi que la grande caisse par Rouen[5] est arrivée deja à Paris, & que tout y etoit bien conditioné.

Il n'a rien de nouveau ici assez interessant pour vous en entretenir. Je vous envoye avec celle ci, une lettre pour M^r^. Messier, qui selon l'avis du Chev^er^. de Bory[6] a permission de Monsr. Trudaine pour les recevoir sous son adresse. aussi bien que le meme Chev^er^. & l'Abbé Bouriot:[7] Mais en cas d'y avoir quelque meprise sur cette permission, vous aurez la bonté de m'en aviser, pour eviter toute sorte d'abus là dessus

Je suis pour la vie
Mon cher Monsieur
Votre
Magelhaens
ou Magellan qui est le meme nom

A Monsieur
Monsr. Duchesne
Secretaire de Monsr. Trudaine de Montigny
à Paris

2 Jean-Charles-Philibert Trudaine de Montigny (1733-1777), Intendant des Finances in Paris and honorary member of the Académie Royale des Sciences, by whom Duchesne was employed as secretary.

3 i. e. metier à bas; see Magellan to Duchesne, 22 March 1771.

4 See Magellan to Duchesne, 22 March 1771.

5 See Magellan to Duchesne, 1 February 1771.

6 Magellan's correspondent at the Académie Royale des Sciences in Paris, the chevalier Gabriel de Bory (1720-1801). See also Magellan to Duchesne, 22 March 1771.

7 All three letters not found.

71.04.23 *To Louis-Henri Duchesne*[1]

Archives Nationales, Paris, MS T/160 (10-12): lettres de J. H. de Magalhaens

Londres 23 Avril 1771

Vous me faites un fort sensible plaisir, mon cher Monsieur, avec L'anonce d'avoir Reçu en bonne condition Les choses envoyées par Rouen: & plus encore en aprouvant Les vases d'Ametiste-root.[2] Je pourrois vous les envoyer d'une grandeur beaucoup plus considerable, & garnis en ormoulû:[3] car nous en avons ici très superbes; mais aussi Le prix en est considerable. Il y en a de 30, 40, & plus Louis d'or La pair: et ce sont des prix que Je n'oserai jamais donner, sans un ordre exprès, & meme celà avec un peu de crainte; car toutes ces choses etant de pure fantaisie, bien souvent manquent dans le point de contenter le gout d'un chacun.

Les idées que vous me donnez par Raport à L'Acad^e^. Roy^le^., marquent bien La bonté de votre Coeur pour moi, & caracterisent très notam^t^. La sinceretè de l'amitié dont me vous m'honorez.[4] Il y a quelque tems, que je m'aperçus de pouvoir reussir, peut etre, dans cette ateinte, ayant egard aux diferents amis, qui semblent etre portés à ma faveur dans L'Academie. Même quelques uns me l'avoient fait entendre: mais ayant fait la reflexion, que celà m'apporteroit un degré de consideration, & d'honeur tout sec, qui pour un Philosophe dans ma situation, & de ma façon de penser, me seroit plus à charge, & non pas d'aucune façon avantageux, Je n'ai jamais eu la moindre envie de l'entreprendre. Pour le present, L'affaire a une apparence tout à faite differente, par Les esperances, que Vous me marquez dans Votre Lettre: & outre celà, c'est assez que M^r^. Trudaine l'approuve, pour que je prendre courage & me jete entierem^t^. entre Les dispositions & avis d'un tel ami comme Vous. Ainsi Vous pouvez compter, que Je ne manquerai d'ebaucher une petite Memoire, sur un objet, qui me semble interess^t^. dans La mechanique, & que Je tacherai de vous envoyer au plutôt:[5] en vous

1 Annotated by Duchesne: " Rep. le 12 May". Letter not found.

2 See Magellan to Duchesne, 30 November 1770.

3 Presumably supplied from Matthew Boulton's Soho manufactory; see N. Goodison, *Ormolu: the Work of Matthew Boulton* (London, 1974).

4 Duchesne had probably indicated to Magellan the possibility that he might be elected a corresponding member of the Académie Royale des Sciences, as indeed he was on 4 September 1771.

5 Duchesne had evidently advised Magellan that, in order to strengthen his case at the Académie Royale, he should present a memoir to the Academy. There is, however, no record in the Academy's *procès-verbaux* of such a paper being presented prior to his election.

priant d'en corriger Le Language, avec quelque ami intime sur lequel vous pouver compter. Je pense pouvoir le finir dans La quinzaine ou 3 semaines: & aussitôt Je vous L'envoyerai. 71.04.23

Je suis etonné, que vous n'ayez pas encore Reçu La petite caisse avec vos Livres.[6] en tout cas, si vous n'en avez aucune nouvelle à l'arrivée de celle-ci, ayez La bonté de me Le marquer, ou plutôt ecrivez à M[r]. Morelle du Faux à Calais, pour savoir si cette caisse a eu quelqu'acident: car il y a Long tems qu'elle partit d'ici, selon que Je vous l'ai marqué dans mes precedentes. N.B. J'ai mis dans la meme caisse, un petit pacquet rond, que Je vous priois de donner au Portugais nomé Policarpo da Sylva, le meme pauvre bon homme que Je vous recommendai, pour vous en servir (en cas en avoir ocasion de l'occuper) dans l'eté dernier.[7]

Je vous prie de faire tenir bientot La Lettre ci jointe, aux Banquiers, selon son adresse Le marque;[8] elle contient un petit affaire qui me presse à me prevaloir de votre manniere obligeante. C'est tems de finire celle-ci, malgre moi, car je ne puis vous exprimer tous Les sentim[ts]. dont je suis penetré, par Le zele, & soin dont veillez pour obliger celui qui est pour La vie

tout à Vous
Magalhaens

A Monsieur
Monsr. Duchesne Secretaire
de M[r]. Trudaine de Montigny
Cons[r]. d'Etat &c. &c.
à Paris

6 See Magellan to Duchesne, 5 March 1771.
7 See Magellan to Duchesne, 18 August 1770.
8 Letter not found.

71.05.04 *To Pierre Macquer*

Bibliothèque Nationale, Paris, MS fr. 12306, f. 11-12

Monsieur

Londres 4 May 1771

Je me souviens fort bien Lors que Je fus à Paris dans L'Eté dernier, de vous avoir promis quelques specimens ou morceaux de la gome elastique, dont M^r^. de Condamine parle dans son voyage d'Amerique par la Riviere des Amazones.[1] Mais il ne me fut pas possible de Les avoir plutôt, que le mois dernier, quoique Je ne manquai pas d'en donner mes ordres pour celà à un de mes amis en Portugal. La grande elasticité de cette gome pouroit peut etre la rendre utile a quelques objects de mechanique & dans Les Arts ou metiers. si l'on connoissoit comment la dissoudre pour la moudeller de nouveau sans rien perdre de son elasticité.

Je me doute guere que Vous en ferez quelques experiences Là dessus avec succès, ou du moins qu'on obtiendra par votre travail des idees plus nettes sur La nature de cette Substance singuliere.[2]

Vous aurez sans doute deja entendu, qu'on fait ici actuellem^{t}. L'edition de Votre excellent dicionaire de Chymie traduit en Anglois, avec des notes &c. ainsi dans cette supposition je ne vous envoye pas ici Le prospecttus de cette entreprise.[3]

Quant à moi Je ne compte pas, que sur L'original: & seulem^{t}. que je souhaite, seroit de voir deja La 2de edition qu'on m'avoit dit à Paris que vous pensiez de donner de cet excellent ouvrage. L'empressem^{t}. avec lequel le Public le reçu, devroit sans doute vous engager de faire cette 2de. edition au plutôt.[4]

Au reste si puis vous etre bon à quelque chose dans ce pais-ci, n'epargnez pas Les bonnes dispositions avec Lesquelles Je suis de Coeur

Monsieur

Votre très humble & obeiss^{t}. Serviteur

Jean Hyacinthe de Magellan

A Monsieur

Monsr. Macquer, Professeur de

Chymie de l'Acad^{e}. R^{le}. des Ciences

à Paris

1 Charles Marie de la Condamine (1701-1774), French mathematician, physicist and explorer, after an astronomical mission to Peru, sailed the Amazon River in 1743-4 and carried caoutchouc, or rubber, back to France.

2 Macquer had been investigating the properties of rubber for some time; see Henry Guerlac, *Lavoisier – the Crucial Year* (Ithaca, NY, 1961), p. 43.

3 The first edition of the translation by James Keir of Macquer's *Dictionnaire de Chimie* was issued in London in 1771.

4 A second edition of Macquer's work was not published until 1778.

From Charles Messier 71.05.12

Royal Society of London, Letters & Papers, V, 249[1]

M. Pingré, from the observations I have already made, has deduced the following Elements of the comet

The ☊ asc. node ♈ 27°. 51'. 00''
Inclinat. of y^{e}. Orbit 11°. 15. 29
Place of the Perihel. ♋ 13. 28. 13
Perihel Dist. 0.90576, Log. 9.957013
Pass'd the Perihel 18 April 1771 at 22^{h}. 14'. 27'' M T[2] Merid of Paris Its Motion Direct.

These Elements will be yet susceptible of some corrections, if a greater number of Observations can be procured. I hope to observe it again till the beginning of August; when it will be too remote from the Earth to be seen much longer. The weak light of it was far from depriving us of insight, since that continually grew less and less to the last.

As to Temporary Stars, I have Observations of several which have disappeared. I have indeed employed a good deal of Time on Changeable Stars; but I still want a supply of more perfect Instruments than any I have, to pursue my Undertaking; but I lament that I am without the Faculties of procuring them. However I have a considerable Work about Nebulous Stars, and on Clusters of Stars I have discovered above the horizon of Paris. I have gone into Divers Particulars, as to their R^{t}. Ascensions, & Declinations, Light &c. In a word, I gave into the Academy a Memoire on this Subject the 16th. of February this year.[3]

It is incontestable that the Wires of a Micrometer with a double Eye-glass appear crooked as the focus is more or less. A single Eye-glass shews them streighter. But I take a more commodious Method when I measure the Diameters of the Sun or Moon, by placing two Lenses one on either side, so that their Centres may answer to the Extremities of the Diameter, and I have no more to do, but to shift my Eye from the one to the other.

---

1 Letter not found. The text is a translated extract headed: "Extract of a Letter from Mr Messier of the R. Academy of Sciences and of the R. Society, to M^{r}. Magalahens. Dated Paris, May 12, 1771. Translated by D^{r}. Bevis." There is an annotation: "Read Jun 13, 1771".

2 Mean Time.

3 Messier's famous catalogue of nebulae, published in *Histoire de l'Académie Royale des Sciences, avec les mémoires*, 1771, pp. 435-461.

71.05.30 *To Louis-Henri Duchesne*

Archives Nationales, Paris, MS T/160 (10-12): lettres de J. H. de Magalhaens

Londres 30 May 1771

Je pensois, mon Cher Monsieur & bon ami pouvoir vous donner cette poste le detail que vous me demandiez dans votre derniere,[1] touchant Les empletes que J'ai faites pour Monsr. Trudaine; mais n'aiant pas Reçu L'avis que J'attendois pour L'envoi des choses que J'ai ordonnées dans la campagne, il ne sera pas possible d'etre aussi exact comme Je le souhaitois. ainsi Je vous dirai à peu près ce que será le montant de mes deboursem^{ts}. aussitot que Le metier du Tisserant arrivera & l'instrument à faire percer les grands trous[2] &c. Car pour celui (le metier) à faire des bas,[3] on ne m'a pas encore donné aucune Reponse positive. Je comprends fort bien le soin que vous avez mon bon ami, de ne me laisser pas exposé à quelqu'ambaras, à cause de votre voyage à Spa avec Monsr. Trudaine: & je vous suis sensiblem^{t}. obligé, on ne peut plus, par cette delicatesse de votre part.

La memoire de ce que Je vous ai deja envoyé, est comme suit. Les 5 vases de racine d'amethiste[4] selon qu'on les apelle ici, couterent comme par les Reçus que Je vous ai envoyés, £5: 1^{s} :0^{d} / le verre ou flint glass pour l'opticien (Jeorge, Je crois c'est son nom[5]) & depenses respectives £7: 5^{s} :8^{d} (ou en Argent de France 164$^{\#}$. 13^{sous}). les ouvrages de Clarissa, Pamella, Tom Jones, Grandisson, & Andrew, selon le bill du Libraire que J'envoyai avec,[6] couterent £5: 0^{s} :0^{d} / les deux jeux ou sets des Laboratoires de Poche[7] £6: 6^{s} :0^{d} / Les bas de Soye 15^{s} :0^{d} / La caisse & porteur jusqu'à la Douane 3^{s} :4^{d} / Registre, revüe, & autres depeches, & bateau à bord du vaisseau 4:9./ total £24: 15^{s} :9: or aiant Reçu en Avril dernier l'envoy de 20 Louis, ce qui fait en notre Argent £21: 0^{s} :0^{d} Esterl. Il me vient encore £3: 15: 9 pour la balance à mon faveur. pour ce que regarde les choses que J'ai ordonnées, mentionées ci dessus, Je crois qu'elles iront à peu pres à 14 ou 15 louis d'or: ainsi si vous m'envoyez 17 ou 18 louis d'or, avant de partir pour Spá, Je crois que Je pourrai me tirer d'affaire: car assurem^{t}. il faut paier tout de

1 Letter not found.
2 See Magellan to Duchesne, 22 March 1771.
3 See Magellan to Duchesne, 30 November 1770.
4 Ibid.
5 The optician's name was George; see Magellan to Duchesne, 2 October 1770.
6 eux *omitted*? See also Magellan to Duchesne, 5 March 1771.
7 See Magellan to Duchesne, 22 March 1771.

suite ici leur prix, aussitôt que Je Recevrai les choses qui sont ordonnées 71.05.30
de la Campagne.

J'espere que vous n'oublierez pas de me dire avant votre depart, sous quelle adresse, & à qui fau-til[8] que J'envoye d'ici ces choses après le depart de M^{r}. Trudaine: ou si vous voulez que Je les garde dans le Magazin, jusqu'à son Retour à Paris.

Je serai charmé si le Memoir, que Je vous ai envoyé pour l'Academie R^{le}. etoit corrigé, & même Lû à L'Academie avant votre depart pour Spá: & si le Monsr. votre ami, & vous aviez acompli de me faire son Correspond^{t}., alors je pourrois lui ecrire en droiture, lors de votre absence en cas qu'il y auroit quelque chose de nouveau: mais pour tout cet affaire, c'est de vous entierem^{t}. que Je depends pour des instructions à fin de m'y conformer.[9]

Je dois envoyer bientôt d'ici quelque Instrument Astronomique pour Monsr. Le Presid^{t}. de Saron: & je pensois de l'envoyer sous l'adresse de M^{r}. Trudaine, selon ce que nous avions agrée en des pareils cas: mais à present que Vous & Monsr. allez partir pour Spá, Je crains que celà ne peut pas avoir Lieu. ainsi je vous prie de me dire, si non obstante votre absence & de M^{r}. Trudaine, il sera egalem^{t}. faisable que de mettre en sureté le dit Instrument à L'adresse de Monsr. Trudaine; ou non?

Enfin, mon cher ami, je finis celle-ci, en vous souhaitant le plaisir d'un voyage hereux à tout egard, & principalem^{t}. pour la Santé precieuse de Monsr. Trudaine, pour Laquelle je ne cesse de faire les plus ardents voeux au Ciel. presentez Lui mes humbles Compliments: & assurez vous du plus sincere attachem^{t}. pour la vie de Celui, qui est de Coeur

Mon cher Monsieur

votre très humble & obeiss^{t} serv^{r}.

Magellan

A Monsieur
Monsr. Duchesne Secretaire
de Monsr. Trudaine de Montigny
Cons^{r}. d'Etat &c.
à Paris

---

8 faut-il?

9 There is no record in the *procès-verbaux* of the Académie Royale des Sciences of a paper by Magellan being read at the Academy at this time. He was, however, elected a corresponding member of the Academy on 4 September 1771. A paper by Magellan was communicated soon after 27 Mars 1772 to Bory, by then his official correspondent at the Academy. Bernoulli also received Magellan's communication and announced it in Vol. 2 of *Recueil pour les Astronomes*, published in Berlin in 1772.

71.06.28 *To Louis-Henri Duchesne*[1]

Archives Nationales, Paris, MS T/160 (10-12): lettres de J. H. de Magalhaens

Londres 28 Juin 1771

Je vous ai ecrites deux Lettres, mon cher ami, sur la fin du mois dernier, la premiere contenant le petit memoire que vous m'aviez conseillé pour L'Academie Royale[2] & l'autre avec l'etat de mes deboursements[3] selon que vous le souhaitiez savoir dans la derniere de vos Lettres.[4] Mais n'aiant pas Reçue aucune Reponse, Je commence à etre fort inquiet à votre egard, dans L'aprehension de quelque mauvais acident. ainsi Je vous prie instament, de me donner vos nouvelles aussitot, que vous Recevrez celle-ci, & de me dire aussi si vous avez Reçû ou non Les deux Lettres ci dessus, & ce que s'est passé là dessus à L'egard du contenû de la premiere. Je vous merite cette complaisance, par L'interet que J'ai à tout ce qui vous Regarde, & disez moi aussi comme se trouve la Santé de M^r^. Trudaine, & s'il est parti en effet pour prendre les eaux de Spá. en attendant Je suis de Coeur & pour La Vie

Mon Cher Monsieur

Tout à Vous

Magellan

P.S.

Je m'obliois de vous dire, que J'ai deja ici chez moi Le nouveau metier, tout monté & avec de Laine comme on la travaille, de façon qu'il n'y a qu'y faire asseoir un tisserant pour y travailler sans d'autre trouble que seulem^t^. couper quelques ficelles qui rafermissent les pieces du metier ensemble. le tout monté, est renfermé dans une caisse d'environ 2 ½ pieds sur 3 ½ ou meme moins de longueur: & c'est de la sorte que Je l'eus par le Carabane ou Chariot de la deligence de chez l'inventeur: & je l'envoyerai de meme par Rouen dans quelque vaisseau.[5] J'ai aussi deja l'auger ou machine pour faire des trous dont je vous ai parlé.[6] J'attends avec impacience vos ordres Là dessus, & vous prie de me dire au plutot si je les puis envoyer tout de suite adressés a Monsieur de Chateau Giron Directeur des fermes à Rouen,[7]

1 Annotated by Duchesne: "Rep: le 15 aoust 1771". Letter not found.
2 Letter not found. Regarding the memoir, see Magellan to Duchesne, 23 April 1771 and 30 May 1771, n. 9.
3 See Magellan to Duchesne, 30 May 1771.
4 Letter not found.
5 On the fate of this machine, see Magellan to Duchesne, 2 August 1771.
6 See Magellan to Duchesne, 22 March 1771.
7 See Magellan to Duchesne, 30 November 1770.

pour avoir soin de les envoyer chez vous à Paris. ayez donc la bonté de me Repondre à cet article au plutôt. 71.06.28

N.B. on est charmé de voir l'effet de ces deux machines, chacune en son genre. quelques amis dont je puis me confier, ont eté ravis de les voir operer. L'ouvrier lui meme est venu à Londres & donna pleine satisfaction aux curieux, sur son Invention.

A Monsieur
Monsieur Duchenne Secretaire
de M^r^. Trudaine de Montigny Conseill^r^. d'Etat &c
à Paris

From Nathanael Wolf 71.07.04

Royal Society of London, MS. 13[1]

J'admire la delicatesse de M^r^. Baker en vous delivrant mon manuscript. Reelement il est plein de defauts & d'imperfections. Son unique merite est un Plan nouveau pour universaliser, & conserver les sciences humaines une fois acquises: & pour en abreger l'etude. Les Chinois ont 300 caracteres radicales, dont ils composent toute leur langue: et J'ai fait voir dans ma preface, que nos idees primitives ne vont pas au delà de 100: & que nous n'en employons que la moitié de ce nombre dans tout ce que nous savons. C'est la Logique de la Nature, que J'ai donné, commune non seulem^t^. aux hom̃es, mais à tous les animaux, à laquelle personne n'avoit pas encore pensé, & qui pourtant est la seule vraye, etant celle des sens. Il n'y a rien d'arbitraire dedans, même l'ordre dans la combinaison des idées radicales, pour en composer le nombre infini contenu dans nôs langues. Cet ordre est la methode naturelle. Les caracteres de ces idées primitives sont arbitraires. Les hommes attachent à chaque idée deux caracteres, dont l'un est le substitut de l'autre. Le premier est la voix articulée: & l'autre

1 MS is in Magellan's hand and headed: "Article d'une Lettre de l'Auteur, adressée à J. H. M., au sujet de sa Botanique caracteristique. Datée le 4 Juillet 1771." MS is bound with Wolf's handwritten 97-page "Tentamen Botanicae Characteristicae sistens Genera Plantarum perfectiorum", 1770.

71.07.04 l'ecriture. Il faut donc, que la même ecriture reponde toujours à la meme articulation de la voix, & que toutes les deux indiquent toujours la même idée. Mais il n'y a pas de langue, qui observe cette Loi. Les sons & les articulations des sons changent com̄e les modes, de village en village, & dans la même ville, de siecle en siecle, de sorte, qu'à la fin, il n'y a plus la moindre ressemblance, ni à la prononciation primitive, ni à l'ecriture, qui pourtant le doit indiquer, ni à l'idée, qu'elle designa originairement. Il est indigne aux Philosophes de se sousmettre à ces inconstances ridicules. Il faut fixer chaque son & le designer toujours avec la meme Lettre, & nous serons delivrés à jamais de cette critique, qui est la plus juste & terrible des sciences humaines.

Quant à mes caracteres, je les ai pris, comme je les ai trouvés dans les langues Europeennes, le Latine, la Gothique, & l'Esclavone: & je les ai exemplifiés par des mots tirés de ces memes langues. Mais ils sont universels pour toutes les langues possibles. Com̄ençons par les voyelles. Il n'y a pas de langue, qui n'employe au moins neuf voyelles. J'ai choisi en Latin a, ae, ei, e, i, o, oi, u, au, toutes très distinctes & simples. Car les dipthongues ae, ei, oi, au, sont des sons aussi simples que les autres, formés par l'ouverture & le prolongement plus ou moins grand du Larinx, & ils ne sont doubles voyelles, que dans la Fantasie des Gram̄ariens.

Chaque Nation devoit donc attacher un caractere ecrit, à chaque de ces 9 differens sons, & ne les jamais confondre ensemble: Loi que le Vulgaire n'observera jamais, mais que les Savants devoient observer. Ordonnez à votre libraire de prononcer ces neuf voyelles sans les meler avec des consones, & vous verrez que ces sons lui seront connus, & tout à fait distincts l'un de l'autre: mais d'abord, qu'il les voudra prononcer avec des consones differentes, il les confondra tellement selon l'usage etablie dans sa langue, qu'un Etranger, en l'écoutant, entendra successivement neuf voyelles, quand votre libraire dira n'avoir prononce qu'une seule meme. Cette inconstance admise dans toute langue, & diferente dans chaque ville, famille, & generation, reste toujours un abus detestable, ne laissent rien de certain & de fixe dans nos connoissances.

J'ai emploié pour les pretendus dipthongues les caracteres de differentes alphabetes, parce qu'il etoit necessaire de les presenter sous la forme de simples voyelles, & il est fort indifferent, si la figure est tirée de l'alphabet Greque, ou Gothique, ou tel autre, qu'on voudra, et les imprimeurs possedent toujours, outre les caracteres latins, encore les greques, & les gothiques: de sorte que cette difficulté d'impression est de peu d'importance.

Quant aux accents ā, â, ah; il faut ne pas admettre la musique, si l'on nie la distinction des Tons dans la let-Re-mi-fa-sol: & il faut nier la Prosodie, si l'on nie qu'une même voyelle peut etre longue ou courte. L'aspiration ah, je l'avoue, est un peu difficile à la plupart des Europeens; mais les Anglois & les Alemands la connoissent, de même que les Grecs & les Turcs en font un continuel usage. 71.07.04

D'ailleurs l'incertitude dans la prononciation de ces accents, n'est pas une objection d'importance. Le premier donne une erreur de dix en dix, le seconde une erreur de plusieurs choses ensemble, jointes ou separées. La troisieme donne l'incertitude sur l'egalité & l'inegalité de ces mêmes choses. L'ecriture, ou une prononciation lente, ôtera cette incertitude. Il faut surement une attention très particuliere, si l'on veut connoitre la chose par ces caracteres. Mais une fois rendue familiere, on la prononce, sans y porter plus d'attention, que dans nos discours familiers.

Venons aux consones. Les Labiées sont de trois sortes. b, p, proprement labiées, qui se prononcent en ouvrant moins ou plus vite nos lèvres: les Saxons sont incapables de distinguer ces deux consones, quoique tout le genre humain les distingue aisément. 2) la nasale m, en fermant les levres, & poussant le son par le nez. 3) les souflantes (Flantes) f, v, w, qui se distinguent par la force du soufle. Il faut surément de l'attention pour distinguer les deux dernieres; mais les Anglois & les Alemands les distinguent bien: & l'usage les rendra bientôt familieres & distinctes, pourvû qu'on ne les confonde pas ensemble. par example les Russes confondent le b avec le v, ce qu'aucune autre Nation ne fait point. Je peux donc pretendre, que chaque Nation prononce six labiées differentes.

Les Linguales d, t, θ, se forment en appuyant la langue contre les dents, & la retirant ensuite avec moins ou plus de repidité: c'est pourquoi on les appelle dentales. Tout le monde distingue aisément deux de ces dentales excepté les Saxons, qui les confondent toutes. Les Anglois, les Alemands, & les Grecs distinguent toutes les trois. Il m'etoit donc permis de les adopter pour telles. Les autres Linguales sont en même tems palatines, L, Ł, n, ń. Les Polonois se moquent des gens, qui ne savent pas prononcer ces 4 lettres. Reellem[t]. toutes les Nations distinguent le double L: mais ils ne l'employent qu'avec certaines voyelles. ń est très distinct chez les Espagnols & chez les Polonois. Les autres Nations l'employent com̄e une double consone, composé de n & de g. Celà n'empeche point, qu'elle ne soit une consone distincte. La tremula v est encore une Linguale, mais j'espere hors de dispute.

Les gutturales j, g, x, k sont formées dans le gosier, en appuyant la racine de la langue contre le Palais, plus ou moins près du Velum palatinum, & poussant le son avec plus ou moins de force. Les Latins confondent le j & le g. Ils disent jam, Jehu, Jota, Judas; mais ils ecrivent gibbus, oû ils devoient ecrire jibbus. Ils disent galea, goliat, gula: mais devant e & i, ils changent

71.07.04 le g en j, com̄e genus, gibbus. Il est intollerable que les Anglois & les François prononcent le g & le j des latins par ź, ou par dź. Ils disent jam; mais au lieu de Jesu, ils lisent źesu ou dźesu. Marquez qu'ils peuvent bien prononcer ces consones, mais que la mode de leur propre langue les fait changer en d'autres, qui leur sont d'un plus frequent usage.

Les Sibilantes z, s & les Superrantes[2] ź, & ś sont très distinctes chez les Polonois, & d'un usage continuel. Les autres Nations les confondent, ou ils les ecrivent par d'autres lettres. La plupart prononcent la S com̄e un S après la voyelle; mais com̄e un z, avant la voyelle, e.c. subtus, salus. L'exemple des Polonois fait voir, qu'on peut prononcer le S scindens avant la voyelle avec la meme force, qu'après la voyelle: & le S serratum après la voyelle avec la meme douceur qu'avant la voyelle. Donc ce sont deux consones distinctes l'une de l'autre. Les ź & ś chicottent avec moins ou plus de force. Les François ecrivent gêne & chene, qu'il falloit prononcer jene & xene. mais ils la prononcent źene & śene: ils ecrivent jadis & challons; & ils prononcent źadi & śallong.

Vous voyez donc, cher ami, que toutes ces consones sont distinctes l'une de l'autre, & pratiquées dans toutes les langues, & que ce n'est que la mode ridicule, qui les a changées de leur signification primordiale. Un Grand, ou une Femme à la mode, ayant un defaut de langue, change les lettres qui lui sont difficiles en d'autres. Je ne saurai donc retrancher aucune de ces consones. Peutetre le θ & le v sont celles auxquelles on pourroit faire l'objection, de ne pas etre assez distinctes de t & de w: & pourtant votre libraire ne fait aucune mention de ces deux lettres, parcequ'elles sont reçues dans la langue Angloise.

Enfin il n'est pas necessaire, de faire fondre de nouveaux caracteres, ni de les faire graver, il y peut substituer tel autre caractere qu'il oudra. S'il manque des lettres Greques, comme α, ψ, [...],[3] il pourra mettre le gothique *α, o, w*; ou mieux encore, il pourra effacer toutes les voyelles, & leur substituer les chiffres 1, 2, 3, 4, 5, 6, 7, 8, 9; & leur donner la prononciation des voyelles. au lieu de *θ*, il pourra mettre le t gothique; au lieu de Ł, il pourra mettre le L gothique: au lieu de S, le ʃ long: au lieu de z, le s petit: au lieu de ź, le z gothique: au lieu de χ, le x latin. Aucune imprimerie ne manque pas du circumflexe ˆ: au lieu de ń, il pourra mettre le w gothique, Les Accents etant de simples traits se trouvent toujours dans l'imprimerie: pour les placer, il ne faut que de laisser un peu plus d'intervalle entre les lignes.

2 Spirantes?

3 Letter over-written, illegible.

Le Libraire a raison de dire, que plusieurs terms de Linneus manquent de caractere avec moi. Ce sont les notions comparatives, coṁe Spatulaformis, Lyraeformis, deltoideo cruciformis &c. Car il nous donne quelque fois d'etranges comparaisons. Je n'ai adopté pour terms de comparaison, que les figures mathematiques determinées; oû il n'entre aucune fantaisie. La comparaison cloche toujours; mais dans ma façon, elle ne cloche que d'un coté, au lieu que ses comparaisons clochent souvent des deux. 71.07.04

Les Libraires aiment les gros volumes, que Je haisse. J'ai dit tant, qu'il faut, pour mettre toute l'histoire naturelle ou en general, toutes nos verités en caracteres. Mais pour le bien faire, il faut posseder toutes les branches dans la plus grande perfection, & j'avoue volontiers, que celà est au dessus de mes forces. Je ne pretends qu'indiquer la route, & je suis très éloigné d'en tirer de la gloire, desirant plutôt de rester inconnû. Mais je suis très persuade que mon Plan ne restera pas là. Quelques genies superieurs le mettront bientôt en execution. Les vieux Botanistes n'ont que faire de ces caracteres. Addonnés à cette seule science, leur Memoire suffit pour contenir tout ce qu'on en sçait. Mais les jeunes botanistes, & en general tous ceux, qui ne se bornent pas à la seule botanique, y trouveront leur compte: & la posterité ne se cassera pas eternellement la tete, pour savoir ce que c'etoit, que le stelebore.

J'ai deja traduite la moitie des Species Plantarum de Linneus avec ma caracteristique. La plupart n'ont que 3 Syllabes: 1) le tronc: 2) les feuilles: 3) la florescence. Celà suffit le plus souvent. Car il s'entend qu'il est impossible de dire en trois syllabes tant, que Linneus dit en trois pages. L'intention n'est pas de donner une description complête. Car dans ce cas il faudroit 40 syllabes; parceque les Botanistes employent jusqu'à 40 parties des plantes dans leurs descriptions. Je ne fais usage que de 4 ou 5 parties pour les genres: & de 3 ou 5 parties pour les especes. Tote la description est donc renfermée en 7 syllabes, rarement en 8 ou 9, que je distribue en 3 mots; dont le premier renferme la classe & la famille: le second, le genre: & le troisieme, l'espece. La prononciation aisée est une Loi, à laquelle doit ceder la distinction: & il vaut mieux de laisser quelque incertitude & obscurité, que de rendre le mot non prononciable. Aussi crois je que les mots, que j'ai donnés, se prononcent plus aisément, que la plupart de Linneus, sans qu'ils laissent une incertitude considerable. Un botaniste plus accompli que moi, les rendra encore bien plus exacts & aisés.

Depuis que J'ai envoyé mon manuscript, J'ai changé de classes entieres, & J'ai encore beaucoup à changer. C'est un fil de Penelope. pour composer de mots exacts, il les faut tirer des plantes vives. Car il faut que la difference soit tirée de 7 ou 9 parties, que j'ai adoptées pour mon Sisteme. Les

71.07.04 botanistes les tirent de 40 parties, tantôt de l'une, tantôt de l'autre; de sorte, qu'on trouve rarement dans leurs livres la description juste de la partie qu'il faut.

Je veux bien que mon Manuscript soit presenté à la Societé Royale. Ce sera assez d'honneur si l'on le jugera digne d'etre deposé dans leurs armoires.[4] Il est glorieux de former & de tenter un si grand Plan: mais je cede volontiers cette gloire à celui, qui lui donnera après moi un tollerable degré de perfection. Il ne faut pas esperer, que des gens bornés à la Botanique seule, lui donneront leur suffrage. Je ne connois, que trois personnes, qui m'honorent de leur amitié, capables d'en juger avec equité. Ce sont M^{r}. le D^{r}. Morton, qui a lui meme travaillé beaucoup dans la recherche des caracteres de la langue universelle,[5] & à qui J'avois com̄uniqué mon Plan il y en a 12 ans: M^{r}. de Jussieux,[6] & M^{r}. Addanson[7] à Paris. Je voudrois que M^{rs}. Morton & Addanson l'examinent: car M^{r}. de Jussieux est dela trop vieux. Je souhaite aussi que la preface soit inserée dans quelque Magazine: & pour obvier aux difficultés de l'impression on y pourroit faire les changements, que j'ai indiqués plus haut, c'est à dire de mettre les chiffres en place des voyelles, & employer des consones gothiques pour les Greques, ou tel autre caractere qu'on trouvera de la meme grandeur. J'avoue que Je suis las de ce travail infini: & je ne sais pas, si je le retoucherai encore, & si je finirai les especes, & la cryptogamie. Reste encore le Regne animal & mineral: pour les quels J'ai bien formé le Clef: mais la chose est encore trop rude, pour en faire mention. Je ne doute pas, que ma preface, si elle verra le jour, incitera quelque Naturaliste habile à employer son tems dans un ouvrage si honorable & si utile.[8]

---

4 While Wolf's manuscript was presented to the Royal Society, there is no record in the Society's *Journal Book* of its being read at a meeting.

5 Charles Morton (1716-1799), head of the Department of Manuscripts at the British Museum and an authority on oriental languages.

6 Bernard de Jussieu (1699-1777), the leading figure at the Jardin du Roi in Paris who is credited with developing the natural system of classification.

7 The French naturalist Michel Adanson (1727-1806), famous for his *Histoire naturelle de Sénégal* (1757), an advocate of natural systems of classification.

8 Magellan has added the following note at the end: "N.B. il y avoit plus de 4 ou 5 ans que l'auteur avoit communiqué son projet de la Botanique caracteristique à plusieurs savants, lors de son voyage à Paris & en Angleterre." See also N. Wolf, *Genera et species plantarum vocabulis characteristicis definita* (Marienwerder, 1781).

To Louis-Henri Duchesne 71.08.02

Archives Nationales, Paris, MS T/160 (10-12): lettres de J. H. de Magalhaens

Londres 2 Aoust 1771
at n°. 188 Fleet street

Je Reçus la poste derniere, la Votre, mon Cher & bon ami, datée du 15 du mois de Juillet;[1] qui en verité m'a livrée de plusieurs aprehensions pour votre Sante & celle de M^r^. Trudaine. aussitôt J'ordonnai La caisse, oû j'avois fait empaqueter le nouveau metier du Tisserant, tout entier & aussi bien monté, qu'il n'y eut La moindre chose à faire, que seulem^t^. couper quelques morceaux de ficelle, & commencer à travailler;[2] Je ordonnai, dis-je, de la faire mettre à bord du premier vaisseau pour Rouen. Mais malhereusem^t^. Les gardes de la douane voulurent l'examiner, & trouvant, que c'etoit une machine des manufactures de laine, qui sont defendues d'envoyer hors d'ici aux Royaumes Etrangers, il s'en sont saisis: & meme pertendent[3] de me faire payer L'amende de 200 Livres Esterlines. Vous pouvez bien concevoir combien J'ai eté alarmé avec cet acident. aussitot Je tachai d'y faire interesser quelques amis & espere que le tout se racommodera, sans grande depense: meme un de mes amis vient de me quiter, en me disant que J'aurai le metier retourné sans aucun domage: car les officiers ne firent point de raport à la Commitée: & le tout Je espere se racomodera sans fracas. Je pensois vous envoyer celle-ci à present; mais Je veux attendre un autre poste, pour pouvoir vous dire ce que sera de cet affaire.

12 Aoust. Les officiers delateurs contre moi ont trouvé mieux d'avoir quelque petit present, que de me faire du mal. ainsi J'ai deja en mon pouvoir Le tout, c'est à dire la caisse avec le metier dedans, que J'ai mis dans un magazin, pour attendre de vos nouvelles: car à present il n'y aura d'autre Remede que le demonter, & l'empaqueter en deux ou 3 caisses avec des Livres & autres choses dessus: & vous les envoyer de la sorte; pour y etre monté comm'auparavant. J'aurai soin d'y coler des numeros & Caracteres de raport pour pouvoir le faire monter aisem^t^. mais au pis aller, il n'y aura qu'à attendre jusqu'à mon premier voyage à Paris, & laisser rester Le tout comme je vous l'envoyerai jusqu'à ce tems Là. Car pour ce que regarde à

1 Letter not found.
2 See Magellan to Duchesne, 28 June 1771.
3 prétendent?

71.08.02 le faire mettre à bord du vaisseau tout entier, comme je voulois Le faire, il semble entierem^{t}. impracticable. en verité Je l'ai echapé bonne: & il ne sera pas prudent de l'attenter de nouveau. & cet affaire est si serieux ici, que Je crois que meme l'Ambassadeur d'une Tete Coronnée, ne sera pas capable d'obtenir du Ministre Anglois un pareil faveur de lui permetre d'envoyer au dehors une pareille Machine, à moins de s'y prendre adroitement sous quelqu'autre titre especieux &c. Ayez donc La bonté de m'ecrire aussitot de votre arrivée à Paris, pour voir Le parti que je dois prendre avec cette machine.

Pour ce que Regarde L'autre du metier à bas[4] J'en ai suspendu l'execution, jusqu'à savoir comment me tirer d'affaire avec celle-ci. Car je crains qu'on ne me la permetra non plus de la embarquer toute montée: & je crains qu'en l'envoyant demontée, on aura de la peine à l'arranger comm'il faut. Outre cette raison, J'ai entendu qu'on avoit fait quelques nouvelles alterations & ameliorations, & qu'on pensoit d'en rendre part à La Societé des Arts. ainsi je crois qu'on gagnera en attendant quelque tems pour voir Le Resultat. pour ce que Regarde mes deboursements, un de mes amis m'a fourni, par emprunt, l'argent dont j'avois besoin jusqu'au mois d'8bre. prochain. ainsi vous aurez le tems de me le Remetre lors de votre Retour: & c'est après la derniere Resolution sur le metier ci-dessus, que le total pourra etre connu. à dieu mon bon & cher ami. portez vous bien & croyez moi pour La vie

tout à vous

Magellan

A Monsieur
Monsieur Duchesne Secretaire
de M^{r}. Trudaine de Montigny
Conseiller d'Etat &c
à Paris

4 See Magellan to Duchesne, 30 November 1770.

To Johann III Bernoulli[1] 71.08.20

Öffentliche Bibliothek der Universität Basel, MS L I a 709, f. 17

Londres 20 Aoust 1771

Monsieur

J'ai reçu en son tems La votre datée du 13 Juillet,[2] avec une dedans pour Monsr. Wood:[3] que je fermai, avec une autre contenant mon adresse, & je la fis porter Chez Lui. en Consequence de celà il est venu hier en ville, se trouvant actuellem^t^. en Campagne, & me Laissa chez moi deux Lignes que voici apres avoir dit qu'il etoit venu chez moi il ajoute = pour le prier de vouloir bien assurer M^r^. Bernoulli de ses Respects, & qu'il ne manquera pas de lui faire tenir un exemplaire de Son Essay aussitot qu'il sera publié. M^r^. Wood profite du loisir dont il jouit à present dans les vacances, pour tacher de le donner au public avant La Seance prochaine du Parlement &c. = ainsi vous voiez par ces paroles que l'edition de cet ouvrage n'est pas faite, & meme je l'ai ainsi entendu dire, que l'auteur seulem^t^. imprime quelques copies en fort petit nombre, pour faire consulter avec ses amis.[4]

Je suis faché de ce que les semences & caisses n'arriverent pas en parfait etat chez Vous: & je prendrai les mesures que vous me marquez pour l'avenir. Le pocket-book (avec le reçu de la caisse des semences si je me souviens bien) Je le donnai de ma main au capitain du vaisseau: qui me promit de le delivrer au Correspond^t^. à Stetin. mais apparem^t^. il s'est egaré: & celà ne vaut pas La peine d'y penser.

J'ai communique votre avis à differents amis: & en consequence vous pouvez envoyer tout de suite, ou aussitôt qu'il sera convenable une douzaine d'exemplaires de Votre ouvrage: que je suis faché n'a pas eté expedié ici aussitôt qu'il est sorti de la presse. Cependant je espere que je pourai en disposer ici.

Vous trouverez ci près quelques lignes de la main du D^r^. Bevis:[5] qui depuis plusieurs mois se trouve circonscript à garder sa chambre à cause

---

1 Annotated "rep. le 21. Sept.". Letter not found.

2 Letter not found.

3 The M.P. Robert Wood (?1717-1771).

4 This information was correct; see 71.04.16, n. 34. It seems that Wood was planning to release a more public edition of his work, but his death in September 1771 put a stop to this. A posthumous edition appeared in 1775, edited by Jacob Bryant.

5 John Bevis (1693-1771) died soon after this letter was written; see 72.01.28 for his obituary.

71.08.20 d'une Chute chez soi par laquelle il a desloqué le femur; mais il se trouve presque retabli. Les tabletes de Aubert[6] [...] sont d'autre chose, que [...] calculs, qu'il s'est fait [...] son usage [...][7] s'epargne beaucoup de trouble pour mettre au net ses observations. il les a ecrits sur des cartes comme celles à jouer, & qu'il porte dans la poche lorsqu'il veut faire quelqu'observation hors de chez soi. je lui en ai demandé une Copie: il me l'a promis avec reluctance: car il ne veut pas croire, que cela vaille La peine: & sa modestie s'oppose à les communiquer. aussitôt que je les aurai, je vous en ferai part. par Là vous voyez que le D^{r}. Bevis ne peut pas savoir que ce que ces tabletes sont. pour celles de Maskelyne, il ne les a pas encore fait publiques. & seulem^{t}. je sai qu'il a donné une Copie à un de ses bons amis:[8] mais Celui-ci ne veut pas les faire voir à personne avant que l'auteur les publie. le but de ces Tables est pour faciliter les calculs Astronomiques.

on vient de publier ici une methode pour deduire la Latitude, seulement de deux observations du ☉ dans quelque tems du jour, sans celle du meridien. mais c'est une brochure tout à fait de practique, sans donner aucune theorie ni aucune demonstration. elle contient diferent Tables, du sines naturels &c.[9] on y dit que cette methode fut trouvée par un tel M^{r}. Cornelis Douwes, Mathematicien d'Amsterdam.[10] On les vend seulem^{t}. 2^{s}. ou demi Courone. si vous le voulez je Vous les envoyerai.

aussi on vient de faire quelque changemen^{t}. dans l'Hadley's Quadrant, qui me paroit avoir quelque avantage. J'ai parlé ce matin avec l'auteur, qui me promit de me donner une description, & je ne manquerai de Vous L'envoyer.[11]

La description de l'observatoire mobile fait par Nairne n'est autre chose qu'un instrument equatorial. Cette description à ce que le meme Nairne me dit, sera inserée dans Les Transactions de l'annee prochaine.[12]

---

6 Alexander Aubert.

7 Paper damaged, text unreadable.

8 In 71.04.16, Magellan identifies Maskelyne's friend as Alexander Aubert. Maskelyne's tables were not published until 1774.

9 N.D. Falck, *Ready Observator* (London, 1771).

10 Cornelis Douwes (1712-1773) was a Dutch mathematician and examiner of sea officers at Amsterdam. According to Taylor (*Mathematical Practitioners of Hanoverian England*, p. 233), what became well known among seamen as the "Douwes method" of obtaining one's latitude had already been published a number of years earlier by Richard Harrison, but without the necessary tables that Falck supplied. No reference is given, however, to Harrison's publication.

11 Cf. 71.10.29, where Magellan reports that the changes proposed have proved to be of little value.

12 The description was published in *Philosophical Transactions*, 61 (1771), 107-13.

Cette semaine sera publié le volume des Transact. de l'année derniere & je prendrai un pour votre Academie, si vous le voulez: & vous l'envoyerai avec ce que vous ordonnerez de plus, aussitôt & par La voye que vous m'indiquerez. 71.08.20

Mr. Shepard me dit la derniere fois que Je le vis, qu'il publiera bientôt Son traité sur les Refractions Astronomiques: mais il n'est pas encore publie à ce que je sache.[13]

Le flint-glass[14] est le verre blanc d'Angleterre qu'on appelle generalmt. Cristal d'Angleterre, & est Le meme dont on fait nos beaux flacons, verres à boire &c.

le tems presse pour porter celle ci à La poste – à Dieu

Je suis de Coeur
Votre
Magellan

N.B. Mr. Dollond a fait aussi un joli instrumt. equatorial, dont il a imprimé la description:[15] [je] vous envoyerai un copie de celle-ci. pour l'information [...] voyez ce que le Dr. Bevis vous dit là dessus[.] [...] me communique une [...][16] pour prendre le contour de tout sorte d'objets, vz. un paisage, ou un tableau: qui reellem. est assez ingenieuse.[17] il n'a pas pû la faire executer en perfection: mais je la crois très aisée d'executer & meme à peu de frais. on n'y fait poin d'usage d'aucun verre. peut etre que je me metrai un jour à le faire executer devant mes yeux par quelque artiste.

13 The treatise by Shepherd, Plumian Professor of Astronomy at Cambridge, was published in the following year as *Tables for correcting the apparent Distance of the Moon and a Star from the Effects of Refraction and Parallax.*

14 Used, together with crown glass, in making achromatic lenses.

15 P. and J. Dollond, *Description and Uses of the New Invented Universal Equatorial Instrument, or, Portable Observatory: with the Divided Object-Glass Micrometer* (London, n.d. [1771]).

16 Paper damaged, text unreadable.

17 Magellan gives further details of this contour-drawing machine, developed by William(?) Hirst, in 71.10.29.

71.08.20a *To Pierre Macquer*

Bibliothèque Nationale, Paris, MS fr. 12306, f. 13-14

Londres 20 Aoust 1771

Monsieur

Je viens de Recevoir L'honneur de la votre du 30 du mois dernier,[1] & Je vous ai deja acheté les cahiers qui sont deja imprimés de La traduction du Dictionnaire de Chymie.[2] ils sont jusqu'à present 25, & forment 792 pages in 4°. Le libraire m'assura qu'il ne manque plus que les 3 derniers cahiers: de façon que le livre fera un volume d'environ 888 pages. mais que je ne manquerai pas de vous envoyer par la premiere occasion ce que se trouve imprimé, le Reste pouvant etre remis à un second envoy. peut etre la semaine prochaine partira d'icy une caisse pour M^r^. Trudaine de Montigny, dans Laquelle je metterai ce paquet pour vous; & son Secretaire M^r^. Duchesne aura soin de vous L'envoyer chez vous. Reellem^t^. c'est necessaire que vous voyez les Notes du Traducteur, avant de publier la 2de. edition:[3] mais ne vous y attendiez pas de fort grandes choses: car à l'exception de quelques notes assez interessantes sur le travail des mines à fer & de l'acier & quelques autres relatives à L'air fixe, il n'y a pas beaucoup à apprendre.

J'ai taché de savoir le nom du Traducteur: mais je n'en ai pas pu obtenir autre information que seulem^t^.: que lui est un gentil-home de Birmingham: & qu'il fait travailler dans les mines de fer pour son compte: & en verité on voit par ses notes qu'il est un metalurgiste:[4] cependant on ne peut pas dire qu'il est ignorant dans La Chymie. Je vais vous copier un des articles de son prospectus, pour voir son but. Après avoir parlé avec tout egard & sinceres eloges de l'original, il ajoute = ... at the same time it must be acknowledged that as the subject perhaps was too extensive for the labors of one man, many articles have been omitted which ought to have been committed, most of which have proceeded from the Author not having been acquainted with some very late discoveries, specialy those important ones concerning fixable air, made by Doctor Black, Professor of Chemistry in the University of Edimbourgh, by Doct.

1 Letter not found.

2 P. J. Macquer, *Dictionnaire de chimie* (Paris, 1766); English translation by James Keir, *A Dictionary of Chemistry* (London, 1771).

3 Macquer did not publish a second edition of his *Dictionnaire* until 1778.

4 James Keir (1735-1820), FRS, an English chemist, geologist and industrialist, was in fact born in Scotland but lived in Birmingham where he was a member of the Lunar Society.

Macbride (of Dublin),[5] and by the Honorable M^r^. Cavendish.= de façon que le traducteur se proposa de supléer à ces petites omissions: mais je laisserai à vous à juger s'il a acompli sa parole entierem^t^. 71.08.20a

Je suis bien faché de n'avoir pas pû reussir à dissoudre la Resine elastique, selon la methode que vous avez annoncé dans Les Memoires (ou Histoire) de l'Acad. Royale[6] & ce fut La raison pourquoi je vous en demandai la methode dans ma Letre precedente. apparement, que ni l'ether, ni l'essence de terebinthine, que j'appliquai, etoient assez depurés pour pouvoir y reussir.

Lors que vous verrez M^r^. Duhamel ayez La bonté de lui dire de ma part, que Je dois lui envoier une copie de la nouvelle edition du Dictionaire Botanique de Philippe Miller,[7] qui me l'a confiée depuis peu, pour La Lui envoyer: & j'aurai soin de le faire partir avec Les Cahiers cidessus de votre dictionaire de Chymie.

J'ai vu hier chez un ami l'excellente Memoire de M^r^. … (je crois Archet)[8] contenant Les experiences sur le resultat des diamants & autres pierres precieuses exposées à un grand & violent feu.[9] reellement ce sont des travaux très precieux & des experiences poursuivies avec beaucoup de jugement & peut etre qu'on la rendra traduit en Anglois. Je n'ai pas pu m'empecher de rire avec le coup qu'il donne au raport fait par Le Doct^r^. Hill sur l'immutabilité du diamant exposé au feu violent.[10]

Celuici est le plus grand scribler[11] que nous avons ici. il ecrit presque sur tous les sujets, beaucoup & fort vivement: C'est un vrai charlatan en histoire Naturelle Chimie & en botanique principalem^t^.; quoique vraiment on ne peut pas Lui refuser beaucoup d'esprit & quelques talents. Je l'ai deja entendu nomer plus d'une fois le Lalande de Londres. il decrit les plantes quelques fois

---

5 David Macbride (1726-1778), surgeon in Dublin, published a collection of five essays on medical and chemical subjects, including fixed air, under the general title *Experimental Essays on the following Subjects* (London, 1764).

6 Macquer, "Memoire sur un Moyen de dissoudre la Résine Caoutchouc, connue présentement sous le nom de Résine élastique de Cayenne et de la faire reparoître avec tous ses qualités", *Histoire de l'Académie Royale des Sciences* (1768), Mémoires, pp. 209-217.

7 Probably the 8th edition of Miller's *The Gardener's Dictionary*, London, 1768.

8 Jean d'Arcet (1724-1801), French physician and chemist.

9 J. d'Arcet, "Mémoire sur l'action d'un feu egal, violent, et continue …", *Histoire de l'Académie Royale des Sciences de Paris* (1766), Hist. pp. 75-9. Lavoisier reviewed earlier work on the combustion of diamonds, including that by d'Arcet and Rouelle, in the first of his two memoirs on the subject, *Histoire de l'Académie Royale des Sciences de Paris* (1772), Mém. pp. 564ff.

10 Theophrastus, *Traité des pierres*, traduit du Grec avec des notes physiques et critiques par [John] Hill (Paris, 1754), p. 66. The original English edition of this work was published in London in 1746.

11 scribbler.

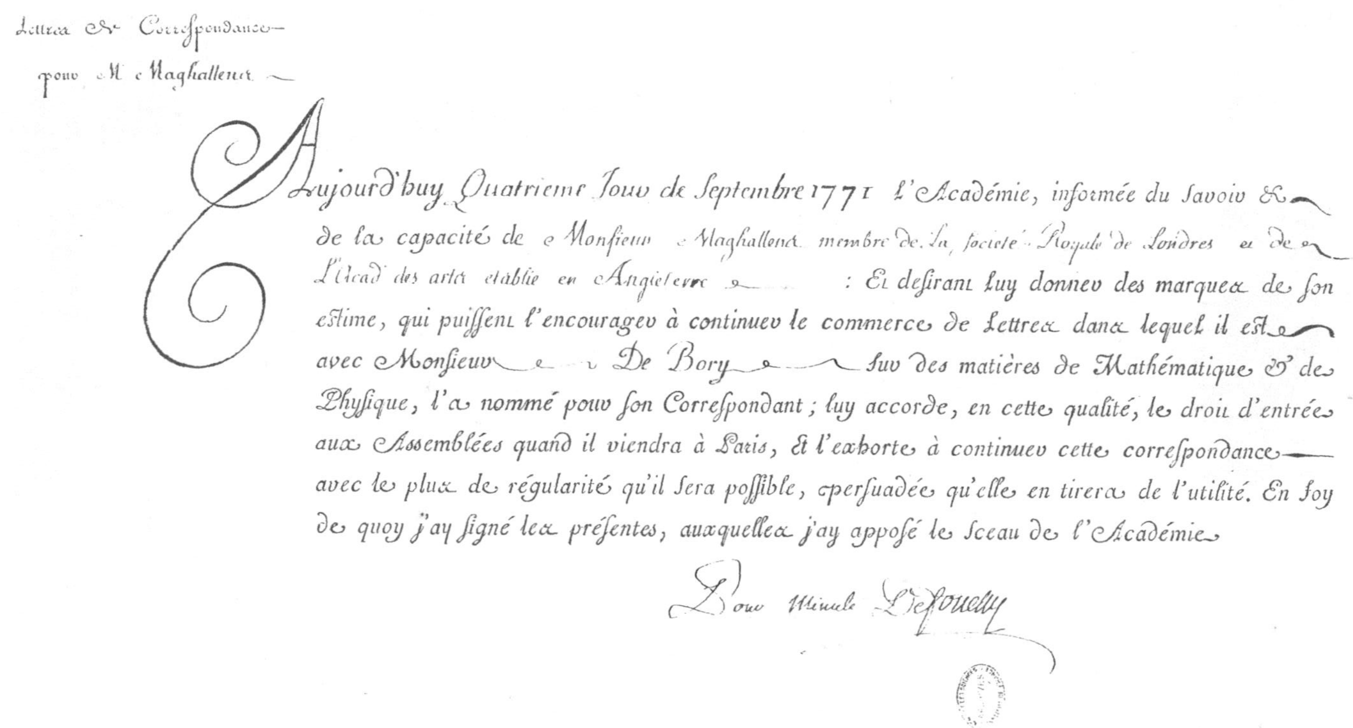

Lettres de Correspondance pour M. Maghallens

Aujourd'huy Quatrieme Jour de Septembre 1771 l'Académie, informée du savoir & de la capacité de Monsieur Maghallens membre de la Societé Royale de Londres et de l'Acad. des arts établie en Angleterre : Et desirant luy donner des marques de son estime, qui puissent l'encourager à continuer le commerce de Lettres dans lequel il est avec Monsieur De Bory sur des matières de Mathématique & de Physique, l'a nommé pour son Correspondant ; luy accorde, en cette qualité, le droit d'entrée aux Assemblées quand il viendra à Paris, et l'exhorte à continuer cette correspondance avec le plus de régularité qu'il sera possible, persuadée qu'elle en tirera de l'utilité. En foy de quoy j'ay signé les présentes, auxquelles j'ay apposé le sceau de l'Académie

Pour Minute Defouchy

Figure 2. Magellan's official notification of his appointment as a corresponding member of the Académie Royale des Sciences, Paris, 4 September 1771. Courtesy Archives de l'Académie des Sciences, Paris.

dont il a revé, & raporte des experiences qu'il n'a jamais vû ni personne non plus. ainsi si par hasard ses ouvrages vous tombent entre les mains, ne prenniez pas beaucoup de peine à concilier ses raports & resultats avec les votres. 71.08.20a

Je vais finir celle-ci en vous priant de me croire sincerement

Monsieur

Votre tres humble & très obeiss[t]. serviteur

Magellan

a Monsieur
Monsr. Macquer Professeur de
Chimie & Membre de l'Academie
Royale des Sciences
Rue Saint Sauveur
à Paris

From Jean-Paul Grandjean de Fouchy 71.09.04

Archives de l'Académie des Sciences, Paris, Dossier Magalhaens[1]

Aujourd'huy Quatrieme Jour de Septembre 1771 l'Académie, informée du savoir & de la capacité de Monsieur Maghallens membre de La Societé Royale de Londres et de L'Acad[e] des arts etablie en Angleterre:[2] Et desirant Luy donner des marques de son estime, qui puissent l'encourager à continuer le commerce de Lettres dans lequel il est avec Monsieur De Bory sur des matières de Mathématique & de Physique, l'a nommé pour son Correspondant;[3] luy accorde, en cette qualité, le droit d'entrée aux Assemblées quand il viendra à Paris, Et l'exhorte à continuer cette correspondance avec le plus de régularité qu'il sera possible, persuadée qu'elle en tirera de l'utilité. En foy de quoy j'ay signé les présentes, auxquelles j'ay apposé le sceau de l'Académie

Pour Minute De Fouchy[4]

---

1 The original certificate has not been found. The text is taken from a file copy, marked "Lettre de Correspondance pour Mr Maghallens", retained by the Académie des Sciences.

2 Magellan was elected a Fellow of the Society of Arts in 1770 but, although he regularly attended meetings of the Royal Society, he was not elected to this society until 1774.

3 Magellan had been proposed for election on 31 July 1771. No discussion of his candidature is recorded in the *procès-verbaux* of the Academy's meetings either on that date or on the date of his election.

4 Sécretaire perpétuel of the Académie Royale des Sciences, 1743-76.

71.10.01 *To Louis-Henri Duchesne* [1]

Archives Nationales, Paris, MS T/160 (10-12): lettres de J. H. de Magalhaens

Londres 1. Oct^{bre}. 1771

Je vous ecris mon cher Monsieur & bon ami, dans L'incertitude de vous etre ou non de ronde de retour à Paris, en vous priant de me donner de vos nouvelles, & de Monsr. Trudaine, qui selon m'a dit Le Chev^{er}. Laurent,[2] par l'avoir vu à Spá, commençoit alors de Recevoir considerable benefit de son voiage. Vous aurez deja sçû que mon election en correspondent de l'Academie est deja faite:[3] & je suis bien faché d'avoir manqué en partie notre bût: qui etoit d'avoir M^{r}. Courniere[4] pour Recevoir ma correspondence. mais un acident, dont je ne veux pas lui imputer la faute, en empecha l'effet: car Le Chevalier Bory en passant chez lui, a vû le petit Memoire, que Je Vous avois envoyé, & tout de suite m'ecrivit une Lettre en se plaignant,

---

1 Annotated by Duchesne: "Rep. Le 8. 8e. 1771".
Annexed to this letter is the following note written by Duchesne, presumably a memorandum to Trudaine:

> "M. Magellan n'a point encore pû envoyer le metier a Bas, le metier a Tisserant, et la machine a forer de nouvelles inventions qu'il avoit annoncé a cause de l'opposition qu'il a trouvé pour leur embarquement; mais il espere les envoyer incessament.
>
> Il luy est dût dix huit Louis pour les avances qu'il a fait a ce sujet; il desireroit en etre payé. Monsieur est prié de vouloir bien donner des ordres pour son remboursement.
>
> Monsieur avoit aussi promis qu'il accorderoit a M^{r}. Magellan quelques gratifications pour les soins qu'il se donne pour faire connoitre les découvertes utiles qui se font en Angleterre; c'est un homme de merite, mais qui n'a point de fortune; son association dans la Societé des arts, ses Liaisons avec les Sçavants de l'angleterre, et des pais etrangers peuvent rendre son zele utile; et les secours quel'on luy accorderoit l'aideroit a suporter les memes dépenses qu'il a a faire.
>
> Il vient d'etre nommé correspondant de l'Académie des Sciences".

2 Pierre Joseph Laurent (1713-1773), French engineer who designed many hydraulic works. He was created Chevalier by Louis XV in 1756.

3 Magellan was elected a corresponding member of the Académie Royale des Sciences on 4 September 1771, the Chevalier de Bory being named as his official correspondent at the Academy.

4 There was no member of the Académie Royale des Sciences with this name. Duchesne had probably suggested Charles Penot de Tournière (?-1772) as Magellan's correspondent, but his writing is not always easy to read and Magellan evidently misread the name. Tournière was a chemist and also treasurer and receiver-general of rents at the Paris City Hall; he had been an *associé libre* of the Académie Royale des Sciences since 1761.

que je lui avois manqué, en ne pas acceptant les ofres qu'il m'avoit faits (comme je vous l'ai insinué dans une de mes Letres) pour me proposer son correspondent. me voiant donc dans cet embaras, il m'a falu de lui repondre avec Le recit de mon ignorance, en disant que Je ne l'avois pas compris, & que je me trouvois bien au dessous de me considerer capable de l'honneur qu'il m'avoit ofert: & que c'etoit dans cette idée que je vous avois envoyé le dit mémoire, me trouvant assez hereux que M[r]. Courniere voulut prendre la peine rebutante de Le corriger & presenter à L'academie sans etre connu de lui &c. mais Je n'ajoutai rien de plus. Cependant Le Chev[er]. arrangea L'affaire comm'il lui a plu, que reelem[t]. a fait voir bien plus d'egard & de bonté, que je ne lui merite point, & par La quelle je Lui suis infinitment obligé; mais dans Le meme tems je ne puis pas m'empecher d'etre mortifié pour ne pas avoir reussi entierement dans ce que vous m'aviez avisé de faire. Ce ne fut point ma faute: & si j'oserois le dire, M[r]. Courniere ne devroit pas montrer au Chev[r]. le mémoire que vous Lui aviez mis entre ses mains. Mais laissons Là, ce qui est passé. Car il faut se soumettre à son Sort. Je espere cependant que vous me continuerez vos bons avis, & de me guider comm'auparavant, etant bien assuré, que je ne sais pas manquer d'egard ni de consideration pour ceux qui comme vous se sont amparés par Leurs bonnes manieres de mon Coeur. 71.10.01

Aussitôt que j'aurai de vos nouvelles de Paris, je ferai empaqueter Le metier en pieces, avec des marques & Lettres de raport: & Je vous L'envoyerai comme si c'etoient des Instruments de mathematique: car comme je vous l'ai dit, il n'est pas possible de l'envoyer, comme je le voulois faire tout entier, & monté.[5]

à Dieu mon cher ami, portez Vous bien & soiez assuré de l'atachement avec lequel je suis pour La Vie.

tout à Vous

Magellan

A Monsieur
M[r]. Duchesne Secretaire de M[r]. Trudaine de Montigny,
Conseilleur d'Etat, & Intendant des Finances
Paris

5 See Magellan to Duchesne, 2 August 1771.

71.10.05 *To Pierre Macquer*

Bibliothèque Nationale, Paris, MS fr. 12306, f. 15

Londres 5 Oct.^{bre} 1771
at n.° 188 Fleet Steet

Monsieur

Celle-ci sert à acompagner Les deux derniers nombres du Dictionaire de Chimie traduit en Anglois, qui vient de sortir de la presse,[1] & completent le total avec Les 26 que Je vous ai deja envoyés. Chaque nombre couta un shilling, & la depense proportionelle de Calais jusqu'à Paris a eté 5^{s}:3^{d}: ce qui fait 33^{s}.3^{d} d'Angleterre, equivalentes en argent de france à 37$^{\#}$:19½^{solds} que vous aurez La bonté de remettre à votre commodité, pour mon compte à M^{r}. L'Abbé Bouriot, qui demeure dans La Communauté des Cudistes Rue des Postes à L'Estrapade, qui me les remettera avec d'autres petites sommes d'autres petits envoys pour differents amis. Dans quelque autre occasion que Je puisse vous etre de quelque service ici, ne doutez pas de me trouver prêt à vous Le rendre de tout mon coeur. On attende ici avec empressement La nouvelle edition que vous preparez de ce Dictionaire: dont la grande etendue de son objet ne pouroit etre mieux aprofondie, que par un homme comme vous. [2] Continuez donc à rendre des services aussi precieux au public & croiez que personne au monde prendra jamais autant de part à votre sucès comme celui qui a L'honeur d'etre

Monsieur

votre très humble & obeiss^{t}. serv^{r}.

Magellan

A Monsieur
Monsr. Macquer Doct^{r}. Reg^{t}.
de Medicine, Professeur de Chimie, de
L'Acad^{e}. Royale des Cienses
Rue S^{t}. Sauveur
à Paris[3]

---

1 P.J. Macquer, *Dictionnaire de chimie* (Paris, 1766); English translation by James Keir, *A Dictionary of Chemistry* (London, 1771). See Magellan to Macquer, 20 August 1771.

2 Macquer did not publish a new edition of his dictionary until 1778.

3 On the address page there is the following note:

> je reconnois avoir Reçu de Monsieur Maquer de L academie des Sciences La somme de trente sept Livres dix neuf Sous Six deniers que M. L'abbe de Magalhaens m'a chargé de Retirer de ses mains, et que je dois faire passer a Londres.
>
> Bourriot
>
> a Paris ce 28 9bre. 1771

Bibliothèque Nationale, Paris, MS fr. 12306, f. 16-17v

Londres 25 8bre. 1771

Monsieur

Lorsque Je Reçus La votre du 27 du mois dernier,[1] j'avois deja envoyés les derniers nombres de la tradution du Dict. Chye.[2] & je crois que vous les avez deja Reçus à cet heure, avec leur depense. C'est pourquoi je ne m'empressais pas de vous repondre à votre obligeante Lettre & offres du precieux present de la nouvelle edition de votre ouvrage, que j'accepterai comm'une marque de votre egard & gracieuse estime pour moi qui je ne puis pas offrir de ma part qu'un Coeur bien disposé à tout que sera de Votre service.

J'ai eu ici une conversation avec un ami, dont peut etre il ne vous sera pas desagreable d'apprendre le sujet, & peut etre y a-til quelque notice, qui n'est pas connu en france de touts Les curieux. lavoicy. Mr. Mayer Apothicaire d'Osnabruck ecrivit un traité sur La chaux,[3] dans Lequel il s'empressa de prouver que la cause des phenomenes qui se montrent dans La chaux vive, sont causés par une matiere appellée par Lui acidum pingue.[4] Le Dr. Black d'Edimbourgh tache d'expliquer les memes phenomenes par son système de l'air fixe ou fixable. Mr. Rousseau à Muncken en Baviere,[5] a fait de la chaux vive, avec le feu solaire d'un miroir ardent: & il trouva qu'il etoit impossible de faire du Lapis infernalis avec cette chaux vive: ce qui semble prouver que Le feu culinaire en chassant des pierres calcaires quelques particules d'eau & d'air fixe, y depose en meme tems quelques unes de ses particules, ou un certain principe phlogistique & inflamable. &c. voilà une experience bien digne d'etre repetée & bien examinée par un homme comme vous.

Mr. Jacquin[6] publia à Vienne 1769 une brochure Latine examen chemicum doctrinae Mayerianæ de acido pingui & Blackianæ de aere fixo, respectu caloris. & Mr. Crantz publia en 1770 une autre examinis chemici doctrinæ Mayerianne &c. rectificatio.[7] Je puis vous envoyer ces deux

1 Letter not found.

2 See Magellan to Macquer, 5 October 1771.

3 Johann Friedrich Meyer, *Chymischen Versuche zur näheren Erkenntniss des ungelöschten Kalchs* ... (Leipzig, 1764). A French translation had been published in Paris in 1765.

4 "fatty acid".

5 Probably Georg Ludwig Claudius Rousseau (1724-1794), German chemist, apothecary and later professor of chemistry and natural philosophy at the Bavarian University of Ingolstadt; see H. Guerlac, *Lavoisier – the Crucial Year* (Ithaca, NY, 1961).

6 Nikolaus Joseph Jacquin (1727-1817), Austrian chemist and botanist.

7 Leipzig, 1770. Heinrich Johann Nepomuk von Crantz (1722-1797) was a professor of medicine at Vienna.

71.10.25 brochures d'ici, en cas que vous ne les ayez pas à Paris. Je souhaite que vous examinez bien cette matiere: & ce qu'on a ecrit sur elle: vous savez que Macbride en a traité[8] & dans les Transact. Philosoph. dernierement (il y a environ 3 ou 4 annees) il y a des bonnes experiences sur celà par M^{r}. Cavendish:[9] & une Memoire de M^{r}. Lane, qui a fait des eaux calybeates par le moien de l'air fixe &c.[10] Daignez Recevoir cette communication quoique peut-etre sans aucune nouveauté ni interet, du moins comme une preuve des sentiments d'estime & regard, que vous porte de coeur

Monsieur

Votre tres humble & obeiss^{t}. Serv^{r}.

Magellan

P.S.

L'histoire de M^{r}. de Buffon sur les oiseaux vient d'etre censurée par Le Critical Review ici.[11] M^{r}. de Buffon y est ataqué un peu durement, & j'ai eté bien faché d'entendre à quelques uns qu'il l'avoit merité.

N.B. pourriez vous m'envoyer L'information suivante, en recommendant quelqu'un de vos eleves de l'informer à la Bibliotheque du Roy. 1°. s'il y a quelque Bible entiere en Copte? 2do. si celle du Manuscript du P. Montfaucon[12] est toute entiere ou non? s'il y a des glosaires coptes vulgaires ou en Dialecte Thebaique i.e. Saïdique;[13] & quelle langue y est pour l'explication des mots coptes? 3°. y a-t-il d'autres manuscripts coptes sur d'autres matieres?

a Monsieur
Monsr. Macquer D^{r}. Regent de la Faculté de Medecine
Profess^{r}. de chimie & de
l'Acad. R^{le}. des Ciences
Rue S^{t}. Sauveur
à Paris

8 David Macbride, *Experimental Essays on the following Subjects. I. On the Fermentation of Alimentary Mixtures. II. On the Nature and Properties of Fixed Air. III. ...IV. On the Scurvy; with a Proposal for trying new Methods to prevent or cure the same, at Sea. V. On the Dissolvent Power of Quick-Lime* (London, 1764).

9 Henry Cavendish, "Three Papers, containing Experiments on Factitious Airs", *Philosophical Transactions*, 56 (1766), 141-84.

10 Timothy Lane, "A Letter ... on the solubility of Iron in simple Water, by the Intervention of fixed Air", *Philosophical Transactions*, 59 (1769), 216-27.

11 *Critical Review*, 37 (1771), 209-15.

12 Bernard de Montfaucon (1655-1741), Benedictine monk and antiquarian scholar.

13 The dialect in which most Coptic texts are written.

To Johann III Bernoulli

Öffentliche Bibliothek der Universität Basel, MS L I a 709, f. 18-19

Londres 29. 8bre. 1771

Monsieur

La votre du 21 de 7bre. m'est parvenue Le 30,[1] & J'ai eté attendant quelque notice des Livres que Vous m'envoyez, pour vous en donner des nouvelles. enfin J'ai reçu une de M^r. Lutyens de Hambourg avec le Lading bill, & ordre de payer les petites depenses avec votre caisse, à M^r. Lutyens ici,[2] ce que J'ai deja acompli. Ces petites depenses les voici. per fret ici à Hambourg – 4.8 + frais jusqu'à bord, & port de letre – 2.14 = 7.6 ℔ 125 g^{ℓ} 5.14 B^o.[3] + Provision 2.– some 7.14 B^o. ce qu'a 32/2 ϐ[4] fait £-:13^s:1^d. Esterling & C'est la some que je payai ici à M^r. Lutyens pour compte de celui d'Hambourg.

Hier J'ai eté tout exprès sur La Riviere pour voir si le Navire La Paix oû cette caisse a etée chargée, etoit deja arrivé: mais un des Commis de la douane m'a dit qu'on l'attendoit tous les jours. ainsi j'ai donné Le Bill of lading à mon batellier pour aller abord aussitot; & voir, si le Capitain vouloit Lui donner cette petite caisse, sans la faire passer par les depeches de la douane, ce qui coutera peutetre plus que Vous ne pensez.

Vous m'avez dit, il est vrai, dans Votre lettre que vous defrayeriez les depenses de cette caisse jusqu'à Hambourg. mais voyant que M^r. Lutyens me les chargeoit, J'ai cru que vous aviez changé de resolution & qu'il ne me siedoit pas bien de les lui refuser, en aceptant Le lading bill. au reste s'il y a quelque meprise on pourra la rectifier en tout tems; car j'ai le reçu des 13^s. 1^d que je payai ici pour ce compte.

L'Observatoire mobile de Dollond, aussi bien qui celui de Nearne,[5] ne sont proprement autre chose que des machines Equatoriales, dont vous avez sans doute vû quelques unes ici de differentes constructions. & le tout consiste à avoir un mouvement vertical pour l'elevation du pole, & un autre circulaire sur un plan qui represente l'Equateur. un autre plan horizontal,

---

1 Letter not found.

2 The Lutyens family, probably of Dutch origin, were merchants in Hamburg. Some members of the family migrated to England and established a similar business there; Nicholas Lutyens was a merchant in Lime Street, London.

3 7.6 Pounds, 125 Groschen, 5.14 Pfennigs Banco. We are grateful to Dr Willem Korthals Altes of Amsterdam for help in deciphering the currency symbols.

4 Schillings.

5 i.e. Edward Nairne.

71.10.29 & un cercle de declinaison come vous savez. toute La dificulté consiste à en simplifier le mecanisme, le faire assez Leger, & bien equilibrer Les pieces. reelement celui de Dollond est fort joli, fort simple, & fort bien fini. Je vous envoyerai la description imprimée, que je vous garde pour Le premier envoy que Je Vous ferai.[6] Mais après tout il reste encore la refraction à surmonter dans ces instruments. M^{r}. Ramsden, qui est le plus ingenieux Mechaniste que nous avons & qui peut etre existe au jour d'hui dans toute L'Europe, joinant beaucoup de genie avec une excellente main d'oeuvre & theorie, a trouvé moiens d'ajouter une petite piece à vis, dans la Lunete de ses Instruments Equatoriaux, par la quelle on corrige tout de suite la diference causée par La Refraction.[7]

Je n'ai pas encore pû avoir Les tabletes de M^{r}. Aubert[8] pour Messier: mais aussitôt, je ne manquerai de Vous envoyer une copie. Cependant il est si modeste, & mefiant de ses propres lumieres, qu'à peine pourai je le faire acomplir sa promesse.

La machine inventée par Hirst[9] pour prendre toute sorte de contours, est tout à fait diferente de celle de Guyot dans le Vol. 3me. pag. 33.[10] elle consistoit à attacher une pointe de lapis ou crayon, qui par un fil communiquoit avec une autre pointe de metal qui servoit de guide à l'oeil, & se mouvoit en parcourant tous les contours qu'on vouloit de l'objet. De façon que La main tenant Le craion, marquoit sur le papier dans un plan horizontal, tous les contours de l'objet, par les quels la guide passoit dans un plan vertical. pour moi je crois que C'etoit une nouvelle invention. Au present il y a un curieux Ecossez qui fait une autre espece de machines, dont l'effet est le meme, & sont moins volumineuses que celle de M^{r}. Hirst, mais elles ne me plaisent au tant.

---

6 P. and J. Dollond, *Description and Uses of the New Invented Universal Equatorial Instrument, or, Portable Observatory: with the Divided Object-Glass Micrometer* (London, n. d. [1771]). See also "Description and Use of a new constructed Equatorial Telescope or portable Observatory, made by Mr. Edward Nairne, London", *Philosophical Transactions*, 61 (1771), 107-13.

7 See 71.04.16, note 45.

8 See Magellan's account of these in 71.08.20.

9 Probably the astronomer William Hirst, who assisted Maskelyne in observing the transit of Venus in 1769 at Greenwich and afterwards drew up an account that was published in *Philosophical Transactions*, 59 (1769), 228-35. Some months later he was lost, together with the rest of the ship's company, when the frigate "Aurora" vanished without trace after leaving the Cape of Good Hope bound for India. On the contour-drawing machine, see also Magellan to Johann III Bernoulli, 20 August 1771.

10 Guyot, *Nouvelles récréations physiques et mathematiques*, vol. 3 (Paris, 1770), p. 33.

Je ne crois pas que les notices de M[r]. Hirst pussent faire aucun article assez interessant dans Votre ouvrage. ainsi il ne vaut pas La peine de les Soliciter plus. 71.10.29

J'ai eté il y a quelque tems avec M[r]. Maskelyne, qui s'a plaint de ce que vous dites dans votre Livre "que Les tables du Nautical almanak ne sont pas si correctes, comme celles de la Connoissance des Tems de M[r]. de la Lande. & que vous y trouviez des erreurs de je ne sai pas combien de seconds (ou meme minuts, car je ne me souviens pas tout à fait) au lieu que celles de de la Lande ne s'eloignoient pas de beaucoup près &c." Comme je n'ai pas encore votre ouvrage, je ne puis pas etre exact; mais vous devez savoir ce que vous y dites à cet egard: & si celà est vrai, vous avez tort d'annoncer au public ces preferences, ou decrier persone sans au même tems en donner Les preuves: & c'est ce que M[r]. Maskelyne demande de Vous.[11] dans des ouvrages pareils il est bon d'annoncer les defauts; mais il faut les marquer & en donner les preuves, autrement on tombe dans l'esprit de parti, & de predilection pour des vues personelles & particulieres. pour moi je ne sais rien de cet âfaire, ni suis je astronome pour m'en meler ou pouvoir doner mon opinion Là dessus. Cependant il faut vous dire, que toute la reste de l'Europe difere de Votre opinion à l'egard de l'exactitude des calculs de M[r]. Lalande. C'est un grand ecrivain par egard à la quantité, mais non pas à l'egard de la qualité, ou pour le dire comme je l'ai entendu plusieurs fois, c'est un vrai bruillon encyclopedique. Je suis faché de Vous repeter tout ce discours que J'ai entendu; mais etant votre sincere ami, je ne voudrois que vous sacrifiasses votre Caractere personnel aux interets de la Reputation de persone, & moins encore de M[r]. La Lande.

Les refractions aiant une si grande Connexion avec les densités & pesanteur de l'air, vous aurez un bon article à faire sur La perfection, à Laquelle on a poussé ici la Construction des Barometres. Je vous en envoyerai une description imprimée,[12] qui vous metera au fait. Ceux de cette construction montrent les hauteurs les plus precises de la colomne du mercure en commençant de la surface dans le Reservoir, jusqu'à la surface superieur dans Le tube, & non pas jusqu'au contact Lateral avec le tube, comme dans Les barometres ordinaires. on y evite toute sorte de

11 Bernoulli had made his criticism in his *Recueil pour les astronomes*, vol. 1, p. 21. It eventually emerged that he had misunderstood the basis upon which the tables in the *Nautical Almanac* had been calculated, and for this he made a public apology in the next volume of his work; see 72.04.21.

12 Probably *Description of the Portable Barometer as made and sold by J. Ramsden* (London, 1772); see Magellan to Johann III Bernoulli, 28 January 1772.

71.10.29 paralaxe provenant de ne pas mettre l'oeil bien horizontalem^{t}. vis à vis de l'objet qu'on veut mesurer avec un index, qui en est à quelque distance (l'epaisseur du verre du tube): on y voit cette colomne mesuré à la fois en pouces, Lignes, & 10mes. de ligne, tant du pied de France, comme de l'anglois: & on y met à coté un petit thermometre à double echelle de Fahrenheit & de Reaumur, avec une 3me. qui marque la propre Correction pour les expansions & contractions du mercure, selon la chaleur ou froid de l'atmosphere. & enfin ces barometres sont egalem^{t}. à cisterne ouverte, & neanmoins sont portables & fort commodes pour les voyages.

La nouvelle comparaison du pied françois & anglois poura fournir un autre article. Par une declaraison du Roy de France, donnée à Versailles le 16 May 1766 la toise Françoise doit etre egale à celle qui a servi sous L'Equateur aux Academiciens qui allerent pour la mesure de la terre.[13] elle se trouve depositée dans L'Acad^{e}. Royale: & lorsque j'y etois l'autre annee (dans l'eté de 1770) j'y ai vu le raport qu'on avoit fait de la comparaison de cette toise, avec l'autre qui servit aux Academiciens qui allerent au Nord. Celle ci etoit un peu plus courte d'environ un vingtieme de Ligne. Or dans les Transactions philosophiques se trouve un Memoire sur Le raport de cette mesure avec celle d'Angleterre.[14] Voulant l'examiner de nouveau, J'ai ordonné une mesure du pied François, pris soigneusem^{t}. sur le meme etalon & par Le meme ouvrier employé par L'academie, qui est M^{r}. Canivet.[15] Celui-ci le fit avec le soin dont il etoit capable, dans la temperature de 13 degrès de Reaumur. Mais de la Lande avec un tour d'adresse qui lui est ordinaire, prit cette mesure disant qu'il me la envoyeroit, & en fit present à M^{r}. Maskelyne, sans lui dire un mot de m'apartenir à moi. M^{r}. Maskelyne donc fit cette comparaison, & par une Lettre datée du 13 Sept^{bre}. de cette annee, il me dit, qu'etant comparée avec l'etalon (standard) de la Societé Royale par M^{r}. Bird,[16] ce pied est egal à 12,785 ⅔ pouces d'Angleterre. & selon la comparaison faite par M^{r}. Sisson[17] ce pied est egal à 12,7845 pouces

13 The French expeditions sent to Peru and Lapland in the mid-1730s to measure the length of an arc of the meridian near the equator and near the pole, and thus to determine the shape of the Earth, had each carried with them a standard measure of length.

14 *Philosophical Transactions*, 42 (1742-43), 185-88.

15 Jacques Canivet (d. 1774), "Ingénieur du Roi et de l'Académie", had made the eighty copies of the "Peru toise" that had been distributed to the various *Parlements* of France in 1766 in connection with the adoption of this measure as the French standard of length.

16 John Bird (1709-1776).

17 Jonathan Sisson (1690-1749) who while working for George Graham divided the Standard Yard that the latter presented to the Royal Society in 1742.

Anglois. Je sai d'aileurs que M[r]. Aubert en fit aussi la comparaison avec la mesure standard pareille à celle qui se trouve dans l'exchiquier à Londres, & il trouva le meme, c'est à dire 12,785.[18] Dans Le meme tems l'Abbé Bouriot de Paris, mon intime ami,[19] m'envoya ici une pareille mesure qu'il avoit fait faire sur le meme etalon, & moi je l'ai comparée avec une mesure du pied Anglois (qui appartenoit jadis à Graham[20]) ensemble avec M[r]. Ramsden: & nous trouvames aussi le meme resultat, savoir 12,785 pouces Anglois. N.B. il y a quelques annees que M[r]. Bird fut ordonné de faire une regle bien divisée pour servir dans L'exchiquier à Londres comme standard. il en fit deux, qu'il presenta comme tout à fait pareilles & exactem[t]. faites.[21] une reste dans l'exchiquier: & l'autre entre Les mains de M[r]. Harris[22] le directeur de la Minte (ou monnoye) à Londres, & par sa mort, est devenue en possession de M[r]. Alexander Aubert, qui la guarde comme un tresor. elle a un excellent atirail de compasses &c. & un Nonius de 10 pouces, par Lequel chaque pouce est divisée en milliemes. C'est un chef d'oeuvre que ces divisions. On voit par Là qu'on peut prendre comme bien assurée la proportion de 12,785 : 12,000 ou 10,655 : 10000 entre le pied françois & celui d'Angleterre. 71.10.29

J'ai parlé à la Personne qui pretendoit avoir fait quelque changement ou alteration avantageuse dans l'Adley's Quadrant.[23] mais il n'ose à m'en donner La Description, craignant qu'on lui rit au nez. C'etoit un certain Ecossois nommé W[m]. Grant.[24] J'ai parlé neanmoins avec l'ouvrier que lui fit ce quadrant: & le tout se reduit à employer un demi cercle au lieu d'un quadrant, mais d'un raion d'environ 7 à 8 pouces. il y met le zero au milieu de ce demicercle, & croit par là pouvoir faire plus avantageusement les observations par derriere. un autre pretendu ameliorateur, met le petit miroir de l'horizont au bout inferieur de l'index, & le grand

---

18 As Magellan goes on to explain, Aubert owned a copy, certified by Bird, of the Standard Yard made for the House of Commons in 1758. A further copy was made from this after Aubert's death and became the base measure for the Great Trigonometrical Survey of India.

19 For information about Bourriot, see his letter to Magellan, 17 December 1770, note 1.

20 George Graham (1673-1751).

21 Bird made the Standard Yard for the House of Commons in 1758, having earlier trained under George Graham at about the time Graham made the Royal Society's Standard Yard in 1742.

22 Joseph Harris (1702-1764) became Master of the Mint in 1748.

23 i.e. Hadley's quadrant.

24 William Grant went to Philadelphia in 1769. A brief notice of his modification of the quadrant was included in a paper by John Ewing published in *Transactions of the American Philosophical Society*, 1 (1769-71), Appendix, 25-27.

71.10.29 miroir objectif il le met fixe dans le centre du Quadrant. mais touts ces changements ne produisent pas meilleur effet que la construction ordinaire de ce quadrant.

Vous savez sans doute que M^{r}. Dollond il y a quelques années avoit adapté un niveau à esprit de vin au Quadrant d'Adley, fort ingenieusement qui par le moien d'une lentille convexe faisoit apercevoir dans le petit miroir horizontal, la boule du niveau, & par Là on avoit un horizont artificiel. mais l'experience a desabusé Le mond, que cette invention ne produit pas un bon effet, et personne au present en fait aucun usage du tout.[25]

J'ai reçu les observations (Seulem^{t}. 6) des satellites de Jupiter faites cette annee à Dantzic (ou à peu de distance) avec deux occultations de la Lune. Si celà vous fera plaisir, Je vous Les envoyerai.[26]

à Dieu mon cher Monsieur. Conservez moi vos amitiés, & croyez moi si[n]cerm^{t}. à Vous

Magellan.

a Monsieur
Monsr. Bernoulli de l'Academie R^{le}. des Ciences
à Berlin

25 To judge from Magellan's description, he may be referring to a bubble level first patented by John Hadley in 1734 and subsequently improved by Thomas Winter, John Dollond, Daniel Scatliffe and Henry Gregory; see E.G.R. Taylor and M.W. Richey, *The Geometrical Seaman: A Book of Early Nautical Instruments* (London, 1962), pp. 82-3. An example survives on an octant by Cole, ca. 1760, at the National Maritime Museum, London (NAV1291).

26 Bernoulli evidently asked Magellan to send him these observations, which he did in 72.04.07. Bernoulli subsequently published them in the second volume of his *Recueil pour les astronomes*, p. 325.

To Matthew Maty[1]

71.11.14

Royal Society of London, Letters & Papers, V, 278

Dear Sir

Here I take the Liberty of sending You the following articles taken from the Letters of some of my friends to me, relating to matters of Natural Philosophy, that You may judge if they are worth of being noticed to the Royal Society. If so, beg the favour of your communicating them to that Respectable & learned Body:[2] otherwise beg pardon for this trouble: and am with great Regard and sincere atachment

Dear Sir
your most Obed^{t}. humble Servt.
John Hyacinth Magellan

Fleet street 14 Nov^{ber}. 1771.

1. Article of a Letter of Monsr. L'Abbé Bourriot at Paris dated y^{e}. 29 Juillet 1771

"Le 17 de ce mois vers les 10½ heures du soir, le tems etant très serein à l'exception de quelques nuages qui bordoient l'horizont du coté du couchant, on vit ici tout d'un coup, au Nordouest dans la moyenne Region de l'air une Lumiere qui croissoit à mesure qu'elle avançoit. elle parut d'abord sous la forme d'un globe, ensuite avec une queue semblable à celle d'une comete. Ce globe aiant traversé avec assez de rapidité une partie du Ciel du Nord-ouest au Sud-ouest, en s'approchant de l'horizont, repandit, comm'en s'ouvrant, une Lumiere si vive & si brillante, que la plupart de ceux qui La virent, du nombre des quels J'étois, ne pouvent en supporter l'eclat. Cette Lumiere rassembloit à celles des Bombes lumineuses d'artifice. Au dernier instant de son apparition ce globe prit la forme de poire, & sa lumiere etoit d'un blanc pareil à celui d'un metal en fusion. Cependant il sembloit que dans quelques endroits , cette lumiere etoit plus rouge, & l'on y voioit des especes de bouillonements avec une matiere fumeuse. La queue etoit d'une couleur plus rougeatre. La durée de ce phenomene n'a gueres eté que de quelques secondes. La grosseur extraordinaire de ce globe & son

1 Secretary of the Royal Society.
2 Magellan's news was read to the Royal Society at its meeting on 20 February 1772 (*Journal Book Copy*, XXVII, 179-81).

71.11.14 extreme hauteur ont fait croire à la plupart de ceux qui l'on vu, qu'il etoit tombé au dessus d'eux. Il a eté vu non seulem^{t}. ici & aux environs, comm'à Versailles, à S^{t}. Germain &c. mais encore à Senlis, à Compiegne, à Rouen, à Argenteu, à Tours, dans La Champagne, dans La Bourgogne, & dans plusieurs autres endroits à plus de 70 Lieues de cette capitale. Deux ou trois minutes àpres sa disparition on entendit un bruit sourd, approchant de celui d'un coup de Tonerre qui eclate au loin. on assure que dans plusieurs endroits les vitres & les meubles ont tremblé. Ce Meteore est un des plus considerables, qu'il ait paru depuis Long tems."

2. another article on the same subject, from a Letter of M^{r}. Hornsby[3] Professor of Astronomy at Oxford dated the 11th. Instant.

"I did not see the Meteor on y^{e}. 17th. of July. I was then on my way to y^{e}. North, having passed thro' London two days before. After much Inquiry I met with a very Intelligent Person, who being then in y^{e}. fields about half a Mile to y^{e}. North of this Place, saw it very distinctly. "it was oblong, according to his description, in y^{e}. direction of it's passage, & in that direction nearly as large as the full-Moon. It gave a considerable light of a colour more reddish than that of y^{e}. Moon: and had Something Like a blaze or tail behind it. It fell obliquely, & disappeared in the S.E. behind some buildings about ¼ after ten in the evening. This Sir is all the account which I am able to procure for you. at this time I myself was in a Chaise travelling with my face to the North, & saw nothing of it."

3. upon another Meteore, from the same Letter.

"But on Friday Last (the 8 Instant) at 29 minutes past five app^{t}. time I saw a Meteor which moving from y^{e}. WSW passed on to y^{e}. N.E. at y^{e}. altitude of about 50°. emitting a considerable number of sparks from it's head, which continued visible at y^{e}. distance of 3 or 4 degrees. It's head was very small, not Larger than Jupiter, & of a reddish colour, as were the sparks. I judged it to be very high in the atmosphere. I have not hear'd that it was seen by many Persons."

4. taken from a Letter of D^{r}. Wolfe[4] dated Danzic the 15 Mars 1771

"Nous avons ici un hiver très rude & très Long. Le thermometre de Reaumur depuis le 12 Janvier a eté constament depuis 4 degr. jusqu'à 20. & le vent de Sud nous apporte ici le plus grand froid. A Varsovie le froid etoit

3 Thomas Hornsby (1733-1810).

4 Dr Nathanael Matthew Wolf (1724-1784).

constament entre 8 degr & 18½. Le Barometre y a varié entre 28 pouces 8½ lignes de Paris, & entre 26 pouc. 5 lign. à Danzic le Barometre etoit de meme fort haut, pendant tout ce tems Là. avec tout celà nous n'avons pas eu beaucoup de jours clairs." 71.11.14

5 of a Letter from Monsr. Nicolaÿ Professor at y^{e}. university of Strasbourg and now Secretary to his Imp^{l}. Highness the Grand Duc of Russie,[5] dated at Petersbourg the 6/17 April 1771.

"J'ai touts prets pour vous envoyer par La premiere occasion deux morceaux de la pierre du piedestal de Pierre le Grand.[6] L'un n°. 1 qui n'est qu'un granite fort compacte & ordinaire, vous montrera quelle est La nature de la masse entiere de cette pierre. L'autre n°. 2 qui n'est qu'un essay d'une veine qui se trouve à un Coin de la pierre & qui sera entierement otée en arrangeant la pierre suivant la plan de M^{r}. Falconet, est d'une nature differente, & a donné lieu à tous ces comptes qu'on a fait de la pierre, appellée dans ce pays Krom, qui veut dire tonerre. Des morceaux de cette Veine on a fait des boulles d'oreilles, des garnitures &c. Elle contient de petits morceaux comme vitrifications, ressemblantes à des grenates, comme j'en ai vu dans Les Laves du Vesuve. Cependant il est bien sur que nous n'avons pas de Vesuve ici, & il est fort singulier, que cette masse se soit trouvée toute isolée dans un pays, oû il n'y a ni montagne, ni Rocher, ni Carriere, ni rien qui Lui ressemble."

N.B. the specimens here mentioned are deliver'd with this paper: and if worth keeping, Magellan desires they may be kept in the R^{l}. Society's Museum among others of y^{e}. kind.

5 Louis-Henri Nicolay (1737-1820), raised in Strasbourg, went to St Petersburg where in 1769 he was placed in charge of the education of the Grand Duke Paul, son of the Empress Catherine and later Tsar.

6 The famous statue by Falconet in St Petersburg of the Tsar mounted on his horse.

71.12.14 *To Joseph Banks*[1]

Mitchell Library, State Library of New South Wales, Sydney, Banks papers, Series 6, 06.169

Dear Sir

Since I heard of your new intended voyage,[2] many times wished for the possessing of those qualifications, which should enable me to engage, as volontier, in so noble an expedition, that I might be of some help towards this [worth][3] attempt, and, if possible, preform as much in the service of England, as the brother of my fifth grandfather did in y^{e}. service of Spain: England being the only country I've chosen amongst others, did see in Europe, for the remaining part of my life, as Spain too had been for my said ancestor. However knowing well the deficiency of my habilities, thought improper to acquaint you till this moment with my above wishes. But hearing now the dificulties, met with to find out some person, willing enough to undertake the Astronomical and Philosophical observations of your next voyage and myself being not intirily unacquainted with these maters, I take the liberty of opening my mind to you: and if approved of, on you pointing one interview, will be very readily waited on by

dear Sir
your most obed^{t}. humble serv^{t}.
John Hyacinth de Magalhaens

N.B. I have no terms to ask, but only the necessaries for y^{e}. voyage, nor advantages to require, but the honour of sharing your troubles in the pursuit of your undertaking. Am 49 years old; but enjoy a good state of health, and pretty strong constitution.

Fleet Street n°. 188
this 14 Dec^{ber}. 1771.

To M^{r}. Banks F.R.S.

---

1 Annotated: 'M^{r}. Magalhaens. Decr. 15. 1771. Answ^{d}. y^{e}. 17th.' Letter not found.

2 Banks, having returned with James Cook to England in July 1771 at the end of Cook's first great voyage of discovery to the South Seas, was assembling an entourage to accompany Cook on his next expedition. Cook, however, refused to allow the extensive additional accommodation on board that Banks demanded for his party. Banks eventually pulled out and was replaced as naturalist on the voyage by Johann Reinhold Forster.

3 worthy?

71.12.14

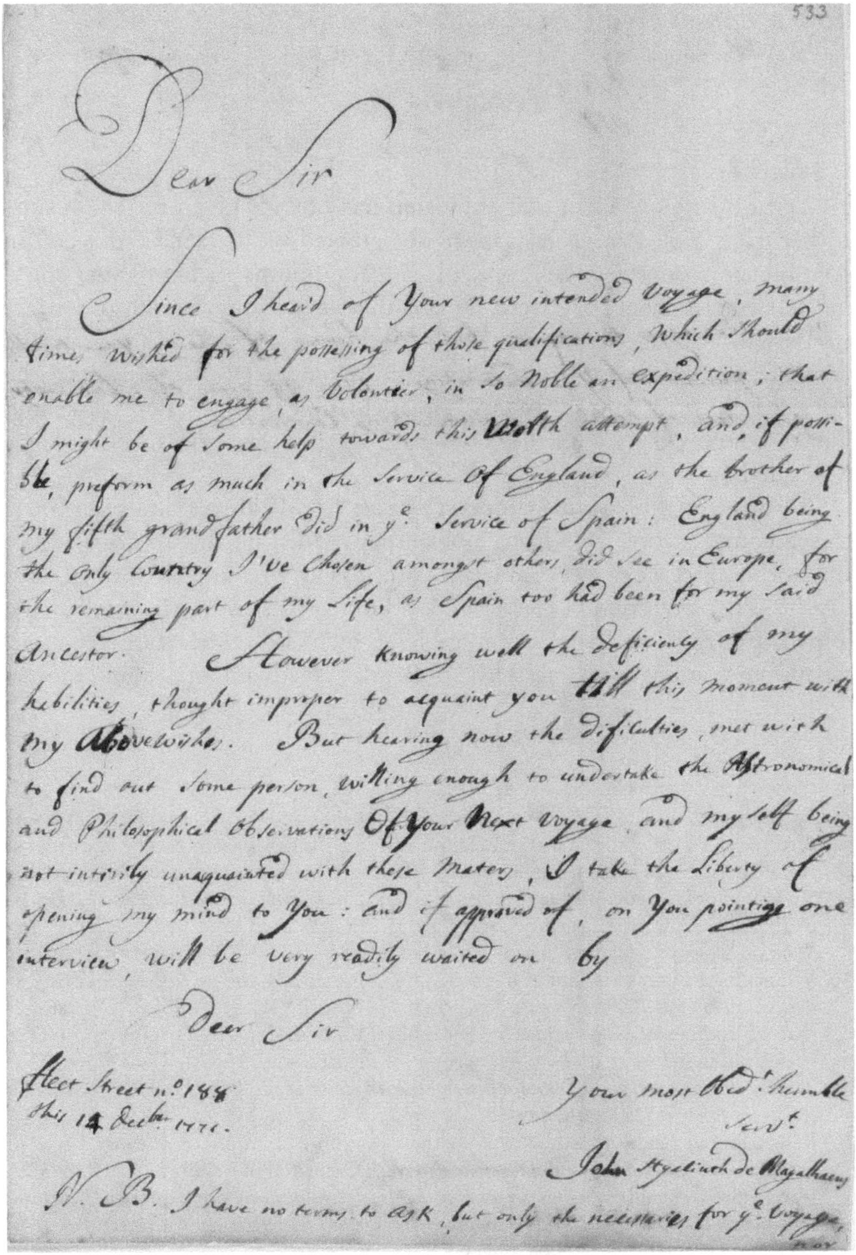

533

Dear Sir

Since I heard of Your new intended Voyage, many times wished for the possessing of those qualifications, which should enable me to engage, as Volontier, in so Noble an expedition; that I might be of some help towards this Worth attempt, and if possible, preform as much in the Service Of England, as the brother of my fifth grandfather did in ye Service of Spain: England being the only Country I've Chosen amongst others, did see in Europe, for the remaining part of my Life, as Spain too had been for my said Ancestor. However knowing well the deficiency of my habilities, thought improper to acquaint you till this moment with my Abovewishes. But hearing now the difficulties met with to find out some person, willing enough to undertake the Astronomical and Philosophical Observations Of Your Next Voyage, and my self being not intirily unacquainted with these maters, I take the Liberty of opening my mind to You: and if approved of, on You pointing one interview, will be very readily waited on by

Dear Sir

Your most Obed.t humble Serv.t

John Hyacinth de Magalhaens

Fleet Street n.o 188
this 14 Decbr 1771.

N. B. I have no terms to ask, but only the necessaries for ye Voyage, nor

Figure 3. Magellan's handwriting: his letter to Joseph Banks, 14 December 1771. The postscript continues on the reverse of the sheet. Reproduced courtesy of the State Library of New South Wales.

72.00.00 *To Nathaniel Pigott*[1]

Royal Astronomical Society, London, MSS Pigott, 45

Dear Sir

You do suppose well, that Ramsden has not yet given me the results of his trials à S^{t}. Pauls.[2] he appear not pleased with them: and when in compliance to my request, took his memorandums in his hands, could not tell exactly what they were. he says that very soon other more acurate tryals will be made in the same spot: and promises to communicate the result thereof. the same is the case with his engraved plate, of which he does not chuse as yet to give any copies. I think that y^{e}. calculations and Tables contained in y^{e}. work of Monsr. de Luc[3] are much more to be depended on.

it was impossible to me till now, to pass by M^{r}. Bird's house, in order to see your Astron^{l}. Quadrant: nor perhaps will I be able to do it before the midle of y^{e}. next week. But am sure it is as well and as perfect, as it can be. Pray my compl^{s}. to M^{dam}. & son; See if there is any part of the joined french letter[4] not intelligible, which I'll endeavour to clear, whenever you acquaint me with. N.B. M^{r}. Needham[5] has some money

1 No addressee or date is shown on this letter, but the content makes it clear that it was directed to Pigott while he was in Belgium awaiting the new barometers he had ordered from Ramsden, and also the quadrant by John Bird that he was to be lent by the Royal Society in order to determine the precise geographical locations of the principal towns of the Low Countries. (The quadrant had previously been used by J. Dymond and William Wales during their expedition to Hudson's Bay in 1768 to observe the transit of Venus.) Pigott in his account of his work (*Mémoires de l'Académie Impériale et Royale des Sciences et Belles-Lettres de Bruxelles*, 1 [1777], I-VIII and 1-25; p. V) indicates that he commenced operations in September 1772, so Magellan's letter must have been written before that. A briefer account of Pigott's work appeared in *Philosophical Transactions*, 66 (1776), 182-95; for biographical details, see A. Brech and A. McConnell, *Recusant History*, 25 (2001), 449-60.

2 Ramsden had made measurements on the variation in barometer readings between ground level and the top of the dome of St Paul's Cathedral in London.

3 Jean André Deluc, *Recherches sur les modifications de l'atmosphere, contenant l'histoire critique du barometre et du thermometre, un traité sur le construction de ces instrumens, des experiences relatives à leurs usages...* (Geneva, 1772).

4 Letter not found.

5 John Turberville Needham (1713-1781), director of the Imperial Academy of Sciences in Brussels.

to send to me: perhaps it may be more convenient to send the whole together. 72.00.00

I sincerily am
Your most obd^{t}. h^{ble}. serv^{t}.
Magalhaens.

Pray my complim^{ts}. to Miss Cardoza, M^{r}. Forcca,[6] the Bishop of Ypres and his brothers,[7] whenever you hepon[8] to meet them, or any other of those who honour me with their memory & favours.

---

6 Cardoza and Forcca not identified.

7 Félix-Joseph-Hubert de Wavrans was Bishop of Ypres, 1762-1784. His brother Louis-François-Ghislain de Wavrans (1715-1785) was president of the Chambre des Comptes in Brussels from 1759; another brother, Henri-Jacques-Hyacinthe (d. 1776) was a Privy Councillor; while the youngest brother, Chrétien, was also in holy orders.

8 happen?

To Louis-Henri Duchesne 72.01.17

Archives Nationales, Paris, MS T/160 (10-12): lettres de J.H. Magalhaens

Londres 17 Janvier 1772
at n°. 188 Fleet Street

Je viens de Recevoir, mon bon ami, La votre du 25 du mois dernier[1] avec La lettre de change pour Les mille Livres, dont M^{r}. Trudaine se digne par votre moien de me favourir.[2] C'est à vous que Je dois ce secour, & ce sera un souvenir que durera autant que ma vie, pour La sensation de ma Reconnoissance, sans rien ôter de celle que je dois au Maitre genereux qui nous protege. Le malheur arrivé au metier en question, n'empechera pas d'en introduire L'usage chez vous: car J'ai etudié ce mechanisme assez, pour etre à meme d'en faire de nouveaux: outre celà il se trouvera très distinctem^{t}. decrit dans Le Livre, qui va paroitre des Machines de la Societé des Arts,[3]

---

1 Letter not found.

2 i.e. Livres Tournois. This payment appears to be a gratification, not just a reimbursement of expenses.

3 William Bailey, *The Advancement of Arts, Manufactures and Commerce, or Description of the Useful Machines and Models contained in the Repository of the Society for the Encouragement of Arts, Manufactures and Commerce* (London, 1772).

72.01.17 dans Lequel j'ai fait en sorte, qu'on d'y ajoutat, quoiqu'il soit posterieur au prospectus de cet ouvrage. Je crois vous avoir dit que Le Reste de ce metier est entre mes mains: & j'ai deja ici deux caisses pareilles, pour mettre dans une un Barometre que Monsr. Le Marquiz de Courtenvaux[4] m'a ordonné, & dans l'autre cette partie du metier. Car j'espere, qu'en faisant examiner celle du Barometre à la douane les officiers laisseront passer l'autre tout a fait pareille sans L'ouvrir. Mais au pis aller je suis en etat de le faire executer tout entier chez vous: aussi bien que les autres machines de la manufacture du Linge, dont je vous ai parlé dans mes dernieres. Je compte partir d'ici vers La fin de Mars, pour Paris: & alors nous arrangerons cet affaire comme vous le jugerez plus à propos. Cet argent que vous venez de m'envoyer est venû on ne peut plus à propos: car je me trouvois fort endeté sur ces depenses du metier: & sur celle que J'ai faite pour secourir La famille du manufacturier dont je vous parlai dans ma derniere:[5] car j'ai crû, & le crois fermement, que c'est un affaire d'importance, & qui peut etre fini avec peu de depense. mais ce sera à la vue, que je pourrai vous endonner tout Le detail.

Je vous ai dit deja dans ma derniere, que Je vous allois expedier le jour suivant, comm'en effet je le fis, une petite boete de Livres pour Le D^r^. Sanchez: en l'ouvrant vous trouverez tout au dessus les deux brochures du voyage autour du Monde,[6] que M^r^. Trudaine souhaite voir. peut etre qu'à L'arrivée de celle ci vous les aurez deja, ou du moins vous en aurez Reçu l'avis, car dans ces jours de fete, les capitains des vaisseaux, aiment plutôt les passer avec leurs familles. & ordinairem^t^. il y a quelque retard; comme en effet je trouvai qu'il avoit avec L'envoy du Perçoir ou Auger:[7] car lorsque j'allai chez le batelier, je le trouvai encore chez lui & voulant Le gronder là dessus, il m'assura, que c'etoit à cause du retarde du vaisseau, qui ne mettroit à la voile que la semaine derniere. ainsi j'ai mis tout ensemble (la boete des Livres, & celle du perçoir ou Auger) & j'ai ecrit à M^r^. Carnier[8] d'avoir soin de toutes les deux, pour eviter de doubler la depense, en l'adressant à Mr Caffieri[9] celle des Livres toute seule.

J'ai trouvé par l'information de mon batelier, que les officiers qui firent La prise de la boiserie du metier avoient deja cessé de la tache qu'ils avoient

4 François-César Le Tellier, marquis de Courtanvaux (1718-1781) was elected an honorary member of the Académie Royale des Sciences in 1765.

5 The letter to which Magellan here refers has not been found.

6 Presumably seeking subscriptions to J. Hawkesworth, ed., *An Account of the Voyages undertaken by the Order of His Present Majesty, for making Discoveries in the Southern Hemisphere, and Successively Performed by Commodore Byron, Captain Wallis, Captain Carteret, and Captain Cook …* (London, 1773).

7 See Magellan to Duchesne, 22 March 1771.

8 Not identified.

9 Charles Augustin Caffieri (c.1739-?) was Directeur de Postes at Calais.

prise pour decouvrir, qui en etoit la personne qui l'envoyoit au dehors. en effet ils s'adresserent à lui (mon batelier) pour faire cette decouverte; mais il leur assura fort adroitment que La personne etoit deja partie pour La campagne, & alloit s'embarquer pour les Indes. Vous voyez combien il est utile de gagner l'affection de ces gens, qui souvent (comme dans mon cas) vous rendent un fort grand service avec peu de paroles...! 72.01.17

Je suis bien aise que vous ayez ecrit pour l'information du capitain du vaisseau á bord duquel les searchers firent La prise de la caisse qui contenoit la boiserie du mettier: mais ce capitain avec son Navire l'amicable society (vous avez le nom des deux, c'est-à-dire du Capitain & du vaisseau) doit etre arrivé depuis long tems à Rouen: & c'est là que vous devez ecrire, pour en avoir quelque reponse.

J'aurai soin desormais de ne pas hazarder dans les demarches que je ferai contre les ordres ou lois de ce pais jaloux: & la caisse ci dessus, dans la quelle je meterai le reste du metier, ne sera pas entrée dans les depeches de la douane que sous un nom supposé.

à Dieu Mon cher & bon ami. portez vous bien, & croiez moi pour Le reste de la vie

tout à vous
Magellan

N.B. ayez La bonté de me marquer le tems que M^r^. Trudaine compte sortir de Paris, dans le printems prochaine à fin de mesurer mon voiage de façon à l'y trouver avec Vous. [10]

A Monsieur
Monsr. Duchenne Secretaire
de Monsieur Trudaine de Montigny
à Paris

---

10 Filed with this letter at the Archives Nationales is an undated and apparently unrelated note to Duchesne, as follows:

> Melle de Chenoise fait mils compliments a m du chesne elle lui envoie cette note pour m^me^ de bellegarde qu'elle contoit lui porter elle meme aujourdui qu'elle devroit etre a paris pour faire ces pacquets pour chambery, mais au moment de partir elle reçoit une lettre de m^me^ de [...] qui elle devoit partir lundy prochain qui lui mande par depart retardé par la maladie de la personne de chambre, m^elle^ de chenoise ignore donc quand elle partira, mais ce ne sera pas [moment] pour prendre les commissions de m du chesne et lui renouveller les assurances de mon attachement | ce samedy matin

[Mlle Chenoise was probably Marie Magdeleine Charlotte de Castille Chenoise (1722-?), the friend of Mme Helvétius. Mme de Bellegarde was either the famous courtisan and writer, Elisabeth Françoise Sophie de Lalive de Bellegarde, Comtesse d'Houdetot (1730-1813), or a relation of hers. The Bellegarde and Chenoise families were connected.]

72.01.28 *To Johann III Bernoulli*

Öffentliche Bibliothek der Universität Basel, MS L I a 709, f. 20-22

Londres 28 Janvier 1772.

Monsieur

Il y a environ 3 mois que Je n'ai pas aucune nouvelle de vous, quoique Je m'en attendois à la occasion de la complainte de M^{r}. Maskelyne,[1] qui peut etre n'est pas fondeé que sur quelque expression mal entendue, ou tout au plus negligement annoncée. Quoiqu'il en soit, J'espere que vous n'etes point faché contre moi, pour vous repeter ce que J'avois entendu & pensé Là dessus: tout au contraire Je suis persuadé que ma franchise devroit vous convaincre de mon amitie, & egard pour Vous, & de mon zele pour Votre reputation.

Le vaisseau qui apporte vos Livres d'Hambourgh, est enfin arrivé ici après un très long voyage. meme on le soupconoit perdu. J'envoyai un batelier à bord pour avoir La caisse, mais Le capitain n'a pas voulû La delivrer sans Les formalités de la douane, ce qui coutera plus d'argent. le Commis à qui Je l'ai recomendée, me permit de me l'envoyer dans deux jours. aussitôt j'en disposerai selon vos ordres.

Voyant par Vos annonces que le 2^{d}. volume de votre Recueil paroitra dans peu de mois Je vais vous donner quelques articles, qui peut etre vous y voudrez faire entrer.

1. Je vous envoye ici la Liste des errata des Tables Logarith. de la belle edition de Gardiner,[2] qui sans doute seront bien agreables à tous les Calculateurs, qui auront ces Tables sans les corrections. & si vous demandez aux amateurs de vous communiquer toutes les autres qu'ils auront trouvées de plus pour Les faire publiques, là sera d'un grande avantage, & servira d'aiguiser encore un peu la curiosite pour votre receuil.[3] L'edition que le P. Pezenas vient de faire à Avignon de ces Tables, a eté fort mal reçue en

1 See 71.10.29.

2 William Gardiner, *Tables of Logarithms, for all Numbers, from 1 to 102,100, and for the Sines and Tangents to every Ten Seconds of each Degree in the Quadrant...* (London, 1742).

3 Bernoulli published Magellan's list of corrections in the second volume of his *Recueil pour les astronomes* (1772), pp. 314-7, and expressed the hope that those who found further corrections that needed to be made should likewise wish to publish them. He did not however propose, as Magellan here suggests, that such corrections should be sent to him.

Angleterre, à cause du grande nombre de fautes & errata de cette edition.[4] 72.01.28
Je voudrois bien les avoir pour vous les communiquer: mais quoique J'ai parlé à plusieurs, pas un me les montra. L'affaire est qu'on ne peut pas faire usage de cette edition, tandis qu'on peut avoir la premiere. Mais bientôt on sera forcé l'employer: & pour Lors il sera très utile d'avoir les errata qu'on y trouvera; ce sera une invitation, que vous pouvez faire au public dans votre Receuil, & qui lui sera très utile. Voici les errata de l'edition de Gardiner, que Vous examinerez sur Votre exemplaire. Les figures souslignées avec une barre dessous, sont les corrections.

Numbers.	Logarith.	Numbers	Logar.
1308	5427	71492	2574
6719	8202	74096	7948
4728	6798	73983	1319
5707	7534	74294	9537
4200	4223	76000	8136
4087	0929	76037	9907
4740	3048	76041	0478
5640	1686	75561	2977
5630	1024	82958	8583
7329	2022	86880	.. .. ..
		9[ ]370[5]	.. .. ..

4 Gardiner, *Tables de logarithmes contenant les logarithmes des nombres depuis 1 jusqu'à 102,100...*, nouvelle édition par les PP. Pezenas, Dumas et Blanchard (Avignon, 1770). Magellan is evidently responding to Bernoulli's favourable notice of this edition in his *Recueil pour les astronomes*, *I*, 181. Esprit Pezenas (1692-1776), a Jesuit, was director of the observatory at Marseilles. He published French translations of a number of English scientific works during the middle decades of the eighteenth century.

5 Second numeral illegible. Bernoulli published this number as 91570.

		logar.
88°.20'.30"	tang.	11. 538
86.39.40	tan.	11.23
73.18.0	tan.	10.5228
68.19.20	tan.	10.400663
44.12.20	tan.	9.987
43.56.30	tan.	9.9839
37.27.10	cosine diff.	162
37.26.50	sin. diff.	275
37.26.20	sin. diff.	275
34.16.50	cosin.	9.91713
32.3.50	sin.	9.724
24°.55'.30"	cos. diff.	{98
		{97
14.42.40	sin.	9.4047405
		dif. 802
14°.6'.20"	cosin diff.	52
13.27.30	sin.	9.36
13.21.30	tan.	9.37
12.37.40	cos.	9.9893
12.58.20	cot.	10.63759
11.24.10	sin.	9.2960174
10.13.50	tan.	9.25
8.36.20	tan.	9.17993
7.59.20	cos.	9.995
5.52.50	cot.	10.98717
3°.21'.0	tang.	8.76
3.11.10	sin. diff.	3784
2.40.40	cot.	11.3300
2.4.0	sin.	8.5570536
1.24.0	sin.	8.3879622
1.7.40	sin.	8.29

Mr. Ornsby Professeur d'Astr. à Oxford[6] me dona dans une Lettre les 4 errata suivants, que Je n'ai pu examiner. Savoir. Dans la pag. 3. la differ 432 opposite to 2, must be 87. / pag. 5. dif. 385. opp. to 9 must be 346. / the number 8688 is wrongly printed. / tangent of 43°.56'.30 must be 9.9839523. / Le même Mr. Orsnby me pria de Vous presenter ses très humbles Respects.

6 Thomas Hornsby (1733-1810), Savilian Professor of Astronomy at Oxford from 1763. Bernoulli did not include Hornsby's corrections in the list that he published.

N.B. peut-etre que vous ne savez pas que M^r^. Gardiner, après que l'edition de ses Tables in 4°. fut finie, il prit le trouble de l'examiner chaque nombre soigneusem^t^. & y trouva la plupart des errata ci dessus. pour lors il fit un avertissement par lequel il s'engagea de les corriger à la main dans tous les examplaires qui lui seroient envoyés, par je crois une demie guinee de plus, & dans le meme tems il promit de payer 50 livres Esterlines pour chaque errata qu'on decouvriroit de plus. Cependant on y a decouvert après sa mort 4 ou 5, qui sont marqués dans La Liste ci dessus. On demande si le P. Pezenas osera faire un ofre pareil pour son edition. du moins si vous adoptez le plan que Je vous ofre ci dessus, on obtiendra la meme fin moiennant la publication de Votre Recueil. 72.01.28

N.B. Les corrections, que M^r^. Gardiner fit de sa main dans plusieurs examplaires de ses Tables, sont si parfaitement executées, qu'il faut avoir de bons yeux, & meme employer une bonne loupe pour distinguer les nombres corrigés.

Vous savez que le D^r^. Bevis vient de mourir depuis 2 mois. Voici les notices que j'ai pu avoir de sa vie, communiquées par M^r^. Horsefall F.R.S. son intime ami, & Testamentaire.[7] Je vous les copierai par ses propres paroles.[8] = The Late D^r^. Bevis M.D. F.R.S. was born on y^e^. last day of October 1695. old stile near y^e^. city of Old Sarum in y^e^. county of Wilts. His father distinguished himself remarkably at y^e^. time of y^e^. Revolution, by rising & maintaining, at his own expense a Company of Foot to assist William, to y^e^. amount of near £2000 Est., which was never repaid by Government. After having given his Son a proper Scholastic education to fit him for y^e^. University; he was entered one of X^t^. Church Oxford, where he applied himself closely, not only to y^e^. study of Physic (for which he was intended) but of every Liberal Science. He was there particularly fond of Sir Isaak Newton's Optics, in so much that he always had it in his pocket: and when we consider y^e^. importance of that science for y^e^. perfection of his favorite study Astronomy, we need not wonder. Not content with making himself Master of y^e^. Theory, he applied himself (whilst at College) to y^e^. Practical part, and made a great proficiency in y^e^. actual grinding of optick glasses with his own hands. Having taken his several degrees to that of M.D. in that University, he then made y^e^. Tour of France & Italy; & upon his return, began to profess Medicine, & had abundant practice

7 James Horsfall (d. 1785) was librarian and of the Middle Temple. Bevis had been one of his sponsors at the time of his election to the Royal Society in 1768.

8 Bernoulli translated this obituary of Bevis into French and included it in the second volume of his *Recueil pour les astronomes*, pp. 331-6. Magellan's inserted comments were published as footnotes to the text of the obituary.

72.01.28 in that Profession, as I have been informed by one, who knew him about y^{e}. years 1730, 1731, 32 &c. But this was not his natural element, for in y^{e}. year 1738, I find by papers now in my hands, he had furnished himself with an excellent Apparatus of Astronomical Instruments to be placed in a New Observatory built under his direction at Stoke Newington near London; where he was a most indefatigable Observer of hevenly Bodies, having from 6th. March in that year to y^{e}. 6th. March in the year following, filled 3 folio volumes with his own observations; from whence he selected in his own manuscript some of y^{e}. best of 'em into one Volume, consisting of 196 pages in Large paper. In this Selected number, sometimes in the course of one night, you may count the transit of above 160 (one hundred & sixty) Stars &c. which he observed. From this time to y^{e}. Year 1745 he continued an assiduous observer, having then acquired materials sufficient, he undertook a most laborious Task, intitled Uranographia Britanica, or An exact Survey of y^{e}. Heavens, on 52 large coper Plates; on which are delineated all y^{e}. constellations & fixed stars to that time observed by all Astronomers, besides a Considerable Number of new ones first observed by himself: and two hemispheres, representing y^{e}. constellations according to y^{e}. Antient Astronomers. To each plate he wrote a particular Explanation, with remarks thereon, & subjoined a Complete Catalogue of all y^{e}. Stars, with their Magnitudes & Places to that time &c. Although these Plates (which would have been an honor to his Country) have been engraved near 20 years, the Public have hitherto been deprived of that valuable work. For the Person concerned (son nom etoit Jean Neale[9]) to make the money advances, who had received several hundred Pounds Est. of y^{e}. subscription-money became Bankrupt, whereby that work fell into y^{e}. hands of strangers (les assinises ou procureurs de cette banqueroute, & des interessés dans les affaires de ce M^{r}. Neale, qui est mort depuis long tems) and was at last lock'd up in Court of Equity: so that the Author never reaped the benefit of his Labour."[10] &c. qui est pire, peut etre les souscripteurs de cet excellent ouvrage aiant perdu leur agent, peut etre ont soupçoné le pauvre D^{r}. Bevis de quelque mauvaise conduite Là dessus. Sans doute cet affaire l'a rendu fort chagrin pour le rest de sa vie. il ne parloit jamais de ce malhereux affaire sans sortir de son humeur. Cependant le grand amour que M^{r}. Horsefall ha pour L'Astronomie, qu'il cultive hereusement,

9 John Neale was a watchmaker with a shop at the King's Arms and Dial, Fleet Street. His bankruptcy was announced in *Gentleman's Magazine*, October 1750, p. 477.

10 Though Bevis's star catalogue was never published, a number of impressions were printed from the plates before they disappeared; see following note.

l'induira sans doute à achever ce procès avec toute l'expedition possible. lui 72.01.28
me dit ce matin d'avoir deja ecrit aux commissaires de cette Banqueroute & procureurs de ce procès de la Chancelerie qu'il etoit prêt comme legataire & executeur du Testam^{t}. du D^{r}. Bevis, à faire de son coté [t]oute l'aisance possible pour le finir au plutôt, & meme ceder autant qu'il sera possible à son interet pour le finir à l'amiable & pour que cet ouvrage soit publié: puisqu'il n'y manque que très peu de chose à finir pour La presse, toutes les planches etant achevées & parfaitem^{t}. finies.[11] le pauvre D^{r}. Bevis y avoit pris des soins extraordinaires: etant un tres grand nombre des jours entiers à coté du graveur, & y faisant Luimeme les corrections necessaires, pour l'exactitude de la position de chaque etoile. "the late learned D^{r}. Hadley, with whom our deceased friend was intimately acquainted, having left his Astronomical tables without precepts for using them – in the year 1749 he (D^{r}. Bevis) undertook the drawing 'em up; and also added [a] few necessary Tables, such as are generally inserted in works of that kind.[12] The only thing that appears under his name is his Sattelite Sliding rule for finding y^{e}. Imers. & Emersions of Jup^{r}. 4 Satellites;[13] yet he has been y^{e}. author of divers works, which have been well received by y^{e}. Public: but as his modesty chose to conceal the real Author, it would be unwarentable in me [to] divulge it. on y^{e}. 11th. of June 1750, his Diploma from y^{e}. R^{l}. Acad^{y}. of Scienc. &c. of Prussia is dated: which was accompanied with a very polite letter from M^{r}. Maupertuis then Presid^{t}.,[14] where he takes notice of y^{e}. work above mentioned, & stiles it his inimitable Atlas. on y^{e}. 12 of

11 Despite Horsfall's hopes, the project was never completed. Several sets survive of the plates that had been printed before Neale went bankrupt; see William B. Ashworth, Jr., "John Bevis and his *Uranographia* (ca. 1750)", *Proceedings of the American Philosophical Society*, 125 (1981), 52-73, and Ruth Wallis, "John Bevis, M.D., F.R.S. (1695-1771), Astronomer Loyal", *Notes and Records of the Royal Society of London*, 36 (1981-2), 211-25.

12 The tables in question were in fact prepared by the late Astronomer Royal, Edmond Halley (1656-1742), whose name Bernoulli followed Magellan in mis-spelling; see *Edmundi Halleii...Tabulae astronomicae, accedunt de usu tabularum praecepta* (London, 1749). An English version of Bevis's edition of Halley's tables was also published, in 1752, and a French edition, with additions, came out in two parts in 1754 and 1759.

13 No copy of this work has been located, nor was it listed in the sale catalogue of Horsfall's library, into which Bevis's library was absorbed after his death. Wallis suggests that it was perhaps a pasteboard instrument with accompanying instructions; see Ruth Wallis, "John Bevis, M.D., F.R.S. (1695-1771), Astronomer Loyal", *Notes and Records of the Royal Society*, 36 (1981), 211-25, and her typescript "Bibliography of John Bevis...", held by the Royal Society.

14 Pierre-Louis Moreau de Maupertuis (1698-1759), appointed President of the Académie Royale des Sciences et Belles-Lettres in Berlin by Frederick the Great in 1746 as part of his reform of the Academy.

72.01.28 July 1768 he was elected a corresponding member of y^{e}. Royal Academy of Sciences at Paris. He was of a cheerfull, benevolent disposition, could not endure to see a Person in distress, of what Country or Religion soever, without Lending his assistance. His death was brought on by a fall, which happen'd as he was turning suddenly to his clock, when making an Astronomical observation (pour prendre le hauteur meridiene du Soleil, chez soi) whereby he dislocated his hip, & was never afterwards able to stand. He died perfectly resigned & sensible with y^{e}. fortitude of a Philosopher & Christian on 6 Nov^{ber}. 1771 (s'il auroit vecu plus jusqu'au 11 de ce mois, il auroit 76 ans, aiant né le dernier 8bre. de 1695 vieux stile) and was interred [in][15] y^{e}. Temple church yard close by his learned friend & intimate Acquaintance D^{r}. Gregory Sharpe, Late Master of y^{e}. Temple London.[16] C'est aujourd'hui le 29 du courant. Je viens de recevoir Votre caisse de la dou[ane][17] oû je trouve 30 copies de vos Lettres Astronomiques, & 22 de votre Recueil en feuilles.[18] & de plus 7 de Vos Lettres brochées pour differents de Vos amis, que je delivrerai, & vous remercie pour celle que Vous me destinez: celle pour feu D^{r}. Bevis Je la donnerai à M^{r}. Horsefall son Testamentaire qui me fournit les notices ci dessus. tout de suite je passai chez M^{r}. White Libraire,[19] qui prendra une demidouzaine de chacun. pour le rest je parlai ce soir à M^{r}. Elmsley, successor de Vaillant,[20] qui Les debitera pour votre compte. Lorsque Je lui parlai sur celà, lui me repondit I am sorry: for the book is hurted. par ces mots J'ai compris que la complainte de M^{r}. Maskelyne a eté divulgee ou communiquée à quelques autres personnes qu'à moi: car je n'en ai parlé ici à personne sur ce point. Je viens de lire votre article sur le Nautical Almanack, & sur les Tables de Mayer, & je n'y trouve pas le moindre mot contre Lui. mais Je pense de lire attentivement le Recueil & les Lettres d'un bout à l'autre pour chercher quelle est la passage qui le chagrine. en attendant si vous en soupçonnez quelque mot, Je serois bien aise que Vous Lui ecrivisses & vous expliquasses avec Lui.

Selon mon calcul chaque copie de votre receuil devient à environ 4 shillings, & les Lettres à 18^{d}.: mais J'ai crû qu'il est mieux de fixer les

---

15 Ink smudged, illegible.

16 Gregory Sharpe (1713-1771), chaplain to Frederick, Prince of Wales, and to George III, Master of the Temple from 1763. Elected FRS 1754.

17 Paper missing.

18 Respectively Bernoulli's *Lettres astronomiques* and the first volume of his *Recueil pour les astronomes.*

19 Benjamin White, bookseller in Fleet Street, c. 1749-1792.

20 Paul Vaillant, bookseller in the Strand, 1750-67, was succeeded by Peter Elmsley who continued in the business until *ca.* 1797. Elmsley became noted as an importer of foreign books.

Recueils à 3s.6d: & les lettres à 2s:0d: & ce sera Le prix que je fixerai à Mr. Elmsley pour le public. Quant à la douzaine que Mr. White a pris, je lui montrerai votre annonce & la depense dès Hambourg jusqu'ici, & lui les payera sur ce pied. 72.01.28

C'est le 31 Janvier. Je vais finir celle-[ci] pour la mettre à La poste. J'ai lû hier soir dans votre recueil: oû je trouvai l'affaire de la complainte de Maskelyne. Je crois la racomodation ou conciliation très aisé. marquez lui un assez bon nombre des lieux du ⊙ à midy vrai pour Greenwich, oû vous avez trouvez jusqu'a 13'' d'erreur: & autant pour le meridien de Paris, oû il n'y a d'erreur, ou zero selon la Connoissance des Mouvemts. Celestes, & la controverse sera finie.[21]

Je crois vous avoir deja parlé de l'improvement du micrometre objective de Dollond, qui les fait à present avec un foyer negatif: & par là au lieu de racourcir le foier composé du telescope ou Lunete, il l'elongue: & lui donne, à cet objectif negative, une grande ouverture; de façon que dans le plus grand ecartement des deux moities, toute l'ouverture de la lunete est egalemt. pleine de lumiere, au lieu qu'auparavant à peine il y avoit pour lors Le 3 ou 4me. partie de la Lumiere. par consequent ce micrometre, aiant muni d'une croix filaire la lunete au foyer des oculaires, devient un excellent micrometre tant pour les Asc. droites que pour les declinaisons &c.[22]

J'ai vû ici the Transactions of ye. American Society of Philadelphie. vol. 1. from 1769 to 1771 oû se trouvent differents articles relatives à l'Astronomie.[23] Voici un. à pag. 2.[24] Mr. John Ewing a fait quelques alterations [au] octant de Hadley qu'il pretend fort avantageuses. le limbe contient 120 degrés entiers. le zero est au milieu de l'arc, & de là on marque chaque demi degré avec des nombres entiers (comm'à l'ordinaire) vers chaque bout de l'arc de façon, que les 120°. se trouvent à chaque bout. il met 2 miroirs sur l'index inclinés de 60°. l'un à l'autre. lorsque le soleil est within 30°. from ye. Zenith, ye. double sextant will give his altitude,

21 Bernoulli in his *Recueil*, vol. I, p. 21 reported having calculated a large number of positions of the Sun at noon: "Dans l'Almanac nautique, les longitudes du Soleil m'ont paru fort exactes pour l'usage auquel cet ouvrage est destiné; les erreurs ne se sont jamais trouvés nulles, mais aussi la plus grande n'a pas été que de 13". J'ai remarqué deux erreurs plus fortes dans la connoissance des Tems, mais en revanche l'erreur assés souvent étoit 0''. Magellan is thus asking him to produce the evidence to back up this claim.

22 N. Maskelyne, "Description of a Method of Measuring Differences of Right Ascension and Declination, with Dollond's Micrometer, together with Other New Applications of the Same", *Philosophical Transactions*, 61 (1771), 536-46.

23 *Transactions of the American Philosophical Society*, vol. 1, 1771.

24 Ewing's paper is actually in the Appendix to Section I of the volume, pp. 21-27.

72.01.28 either above the southern or northern horizon, in y^{e}. same manner as in y^{e}. fore-observation and half y^{e}. difference between their Sum & 180°. being added to y^{e}. lesser altitude, when the Sum is less than 180°., or subtracted, if greater, will give y^{e}. true altitude more acurately than either separately. Avec ce quadrant on observe le meme astre de 4. façons, dont les unes corrigent les autres. Savoir 1. on observe comme par devant. 2. on pousse [l']index au delà du zero & on observe avec l'autre miroir. 3. on toure le limbe ou Cercle du Quadrant en haut, & [le] centre en bas, & on fait la 3me. & la Quatrieme avec l'instrument dans cette position. il faut avoir quelques p[...][25] à coulisse sur le Limbe pour marquer les endroits des trois premieres observations, à fin de ne perdre du tems entre les observations: & dans La suite les coulisses montrent la place ou degré de chacune. Cet improvem^{t}. est different de celui de M^{r}. Grant,[26] which had y^{e}. horizon-glass & place of y^{e}. eye fixed on y^{e}. Arch, without y^{e}. 2^{d}. speculum in the index (ou alidade) & par consequent ne peut fournir que seulem^{t}. deux observations du meme astre, au lieu que [celui] de M^{r}. Ewing donne 4 observations.[27]

Feu D^{r}. Bevis avoit inventé une nouvelle espece de micrometre, qui consiste en different Cercles concentriques peints à email sur un verre concave. lui communiqua par ma voye cette construction à M^{r}. Messier: mais à vous dire le vrai, Je n'ai pas pû bien concevoir les avantages de ce micrometre.

je vous envoye la description du Barometre de Ramsden,[28] qui est excellent. Cet ingenieux artiste a fait une machine à diviser des quadrants Astronomiques très differente de celle du Duque de Chaulnes.[29] on y fait

25 Text bound in, unreadable.

26 Cf. Magellan to Bernoulli, 29 October 1771, note 23. Ewing's paper included a brief account of Grant's invention.

27 Bernoulli, who clearly had not yet seen the volume in which Ewing's paper appeared, alluded to Magellan's description of Ewing's instrument in his *Recueil pour les astronomes*, *II*, 308, as follows: "Pour ce qui regarde le nouveau Quartier, qui est de M. Ewing, on me dit dans une lettre particulière que c'est un arc de 120°. avec lequel on observe un astre de 4 façons, de sorte que ces 4 observations se corrigent les unes les autres; on me donne aussi une idée de la manière de s'en servir, mais on me laisse indécis s'il mérite réellement que les Navigateurs y fassent attention".

28 *Description of the Portable Barometer as made and sold by J. Ramsden* (London, 1772).

29 Michel-Ferdinand d'Albert d'Ailly, duc de Chaulnes (1714-1769) was an Honorary Member of the Académie Royale des Sciences in Paris and well known for his improvements to various scientific instruments and apparatus. On his dividing engine here mentioned, see his *Nouvelle méthode pour diviser les instruments de mathématique et d'astronomie* (Paris, 1768).

toute sorte de divisions, sans regarder les points ou divisions de la plaque original, car elle de soi meme s'arrete à chaque pas sur le point precis, qu'on a choisi au commencem^{t}. i.e. soit degrés, ou demi degres, ou 4^{s} de degrès. Ses garçons, & meme sa femme y divida des Instruments aussi bien que le meilleur Artiste. C'est avec cette plateforme qu'il divisa le petit quadrant de poche, que j'envoyai à M^{r}. le President de Saron[30] cette année, d'environ 3½ pouces de raion: avec lequel on distingue & observe jusqu'à des minuts. Il vient de faire diviser devant moi par son aprentif un cercle de 15°. en 60 divisions, ou 4mes. de degré, & un Nonius à coté de 61 divisions, ce que fait 15'' pour chaque, le radius etant seulem^{t}. [3][31] pouces d'Angleterre: & par consequent chaque division excede celle des premieres par une quantité d'environ une 4000me. partie de pouce; puisque dans l'arc de 60°. il y a 3600': & 14400 fois 5''; ces divisions sont si petites qu'il faut les soumetre à une loupe très convexe, ou sous un microscope, pour voir l'augmentation des premieres divisions. Je pense aller ce printems à Paris, & y aporterai cette curieuse piece de division. Le petit Quadrant de poche que J'envoyai à M^{r}. Le Presid^{t}. de Saron etoit acompagné d'un horizont artificiel, qui consistoit dans un petit miroir dont Les surfaces etoient très paralleles: & une nouvelle espece de niveau rond, de l'invention du meme Ramsden. il avoit environ une pouce & ¾ de diametre (comme une tabatiere) avec un verre concave desus excellentment travaillé, avec 3 ou 4 points marqués à angles droits: de façon qu'etant rempli avec de l'esprit de vin, la boule d'air qui etoit fort grande & applatie au dessous du verre concave se tenoit entre ces 4 points, & par là on arrangeoit le petit miroir parfaitem^{t}. parallel à l'horizont.[32] M^{r}. le President m'a deja ecrit qu'il a essayé attentivem^{t}. ce quadrant & horizont artificiel, & par son moien il peut s'assurer de l'hauteur du pole 72.01.28

---

30 Jean-Baptiste-Gaspard Bochart de Saron (1730-1794), President of the Parlement of Paris, was a noted patron of science with a passion for astronomy and an excellently equipped private observatory.

31 Text bound in, reading uncertain.

32 A description of what appears to be the same device was published soon afterwards by Dom François Bedos de Celles, *La gnomonique pratique, ou l'art de tracer les cadrans solaires* (2nd edition, Paris, 1774), p. 356, when he wrote of "une autre espèce de niveau qui paroit nouvellement inventé. C'est une boïte ronde en cristal, recouverte par une glace un peu concave en-dessous. On y met de l'esprit de vin en dedans. L'on fait une boîte de cuivre dans laquelle on enferme la boîte de verre qui n'est découvert qu'en dessus. Cette invention est encore trop récente pour être assez perfectionnée & pour être d'un prix moderé." This passage does not occur in the first (1760) edition of the book. In the diagram that Bedos gives, two circles are drawn on the glass rather than the four points that Magellan mentions. (We are grateful to Anthony Turner for this reference.)

72.01.28 à une minute près; ce qui est etonant vue La petitesse de cet instrument. Ramsden est le plus habile homme qui vive à present dans ce genre. C'est d'après ces qualités, que tous Les autres artistes le haissent: meme son beau frere Dollond. Lui parle ni le voir de plus plusieurs années. J'ai vû cette semaine chez Lui une lunete achromatique de 16 pouces de foyer à triple objectif, avec 2½ pouces d'ouverture, qui egaloit, ou meme surpassoit le celebre Tellescope de la nouvelle & derniere façon de feu M^{r}. Short,[33] avec lequel je l'ai comparé! quelques Artistes de Londres voudroient bien avoir une pareille plateforme à La Sienne, pour diviser les Adley's[34] Quadrant ou Sextant mais il n'a jamais voulû par aucun argent leur faire ce Service. Celle qu'il a à present ne peut diviser que des instrum^{ts}. d'environ 10 pouces de raion: mais il a commencé deja une plus grande qui servira pour des instrum^{ts}. de 2 pieds & meme 42 pouces de raion, qui lui coutera plus de 300 Livres Esterlines.[35] Je crois qu'il se soumeteroit à faire une pareille pour quelque Academie Etrangere, & meme à deceller la maniere de la faire à quelque Etranger (mais non pas ici aux ouvriers Anglois, qui l'ont taché de Le ruiner & qui par jalousie ne cessent de le decrier) moienant une somme par example 500 Livres Esterlines. M^{r}. Maskelyne, qui se doutoit de l'operation de cette Machine, fut convaincû en La voiant operer: Car Ramsden y adapte une loupe, & à mesure que le plateforme s'arretoit à chaque pas, on y voyoit precisem^{t}. le point de la division desirée tomber tout a fait sous la croix filaire de la Loupe. Il vient d'inventer une nouvelle espece de raporteur, qui est très

---

33 James Short (1710-1768). See G. L'E. Turner, "James Short, F.R.S., and His Contribution to the Construction of Reflecting Telescopes", *Notes and Records of the Royal Society*, 24 (1969), 91-108.

34 John Hadley (1682-1744).

35 Ramsden made two dividing engines, the first of which was completed in 1767. Despite what Magellan says about its performance, Ramsden was not completely satisfied. The second machine, to Ramsden's plans for which Magellan here alludes, was not only larger but was much superior in every way. It seems to have become operational in about 1775 (Allan Chapman, *Dividing the Circle: The Development of Critical Angular Measurement in Astronomy, 1500-1850* [Chichester, 1995], p. 130). Ramsden was afterwards persuaded to publish his method by the award of a prize of 600 guineas from the Board of Longitude; see his *Description of an Engine for Dividing Mathematical Instruments* (London, 1777). Under the terms of the award, he assigned ownership of the dividing engine to the Commissioners; agreed to give instruction to up to ten instrument makers nominated by the Commissioners, within a two-year period commencing in October 1775, in the making and using of it, "sufficient to enable any intelligent workman to construct and use other Engines of the same kind"; and undertook, so long as the Commissioners left him in possession of the engine, for a specified fee to divide with it all octants and sextants brought to him by other makers.

commode:[36] & J'espere pouvoir avoir une de cette espece, lorsque J'irai en France. Mais Le desavantage de cet ouvrier est qu'on ne put pas aracher de l'ouvrage de ses mains. J'ai toujours quelques instruments ordonnés chez lui: J'y vais presque tous Les jours, & cependant il faut des mois entiers après le tems qu'il fixe. il y a plus de 2 mois que Je devrois avoir un ces [nouveaux] Barometres pour Mr. Le Marquiz de Courtanvaux[37] & un Sextant de Adley: mais a peine aurai je le premier La semaine prochaine, & le second au bout de ce mois suivant. 72.01.28

à Dieu Mon Cher Monsieur. excusez la longueur de celle ci, que Je crois pouvoir contenir quelques notices pour Votre 2de. volume du Recueil.

J'ai l'honneur d'etre
Votre de Coeur
Magalhaens.

Mr. Banks & Dr. Solander entreprendrent leur second Voyage, allant par les Indes Orientales & revenant par Cap d'Horn, au commencemt. d'Avril prochain.[38] C'est Mr. Bailly[39] & Mr. Wale[40] qui vont avec eux en qualité d'Astronomes.

A Monsieur
Monsr. Bernoulli
Astronome Royal & de l'Acade. Rle. de Prusse
à Berlin

36 Not identified.

37 François-César Le Tellier, marquis de Courtanvaux (1718-1781), was an Honorary Member of the Académie Royale des Sciences in Paris. He erected a private observatory at Colombes, published several scientific papers on chemical and astronomical subjects, and encouraged the development of French scientific instrument-making.

38 In the event, Banks and Solander did not go with James Cook on his second great voyage of exploration, Cook refusing to allow Banks to endanger the stability of his ship by building the large additional cabin accommodation that he demanded. Johann Reinhold Forster went as naturalist in their place.

39 William Bayly (1727-1810).

40 William Wales (1734-1798).

72.03.13 *To Louis-Henri Duchesne*[1]
Archives Nationales, Paris, MS T/160 (10-12)

Londres 13 Mars 1772

J'ai Reçu mon Cher & bon ami La votre du 1er. de ce mois[2] avec un extreme plaisir. le nouveau perçoir[3] a entré en vogue ici, & on en a fait plusieurs cents, qu'on vend à present assez à bon marché. le pauvre inventeur[4] en a eu très peu de profit Là dessus. lui m'ècrit dernieremt. qu'il avoit imaginé quelques alterations ou changements avantageux, pour le faire en grand pour percer des corps de pompes &c. mais qu'il ne les fera publies qu'àprès en avoir une grande quantité de faits: âfin qu'on ne les contrefasse: ainsi J'attendrai ce tems Là pour en parler precisemt. Quant à celui que vous avez Reçu, il faut en juger en travaillant avec, c'est à dire en coupant un trou sur une grosse planche ou perche bien epaisse: cars alors vous verrez que le bois sort hors du trou de soimeme à mesure que l'instrumt. le coupe. & je vous prie d'en avertir Mr. le Chevr. de Bory de celà. La lettre ci jointe[5] vous aurez La bonté de Ia Lire à votre Loisir: si vous le croiez à propoz, montrez la à Mr. Trudaine. dans la suite fermez la & envoyez la au dit Mr. Bory.

Je viens de faire mettre à Bord du premier vaisseau une caisse, que vous será envoyée de Calais marquée **R M** n°.1. oû vous trouverez deux douzaines de gans des couleurs & mesure que vous m'avez envoyez, & de la meilleure sorte qu'on fait ici. je souhaite qu'ils soient du gout desiré. après que vous les oterez de la caisse, Je vous prie de donner Le reste à Mr. l'Abbé Bourriot: car il y a differents articles pour differents amis, & j'ai cru que vos occupations ne vous permetroient pas de les distribuer. Lui vous paiera les frais qu'on vous aura chargé: & Je lui ecris cette poste sur celà. C'est par La meme Raison que Je ne l'envoye pas au Dr. Sanchez, quoique la plupart des livres qu'y sont contenus, lui appartiennent. il faut menager les affaires, & la santé de ceux que nous aimons le plus: & employer ceux qui sont plus à La portée de s'en charger.

---

1 Annotated by Duchesne: "Rep le 28 Mars". Letter not found.
2 Letter not found.
3 See Magellan to Duchesne, 22 March 1771 and 17 January 1772.
4 The inventor was Phineas Cook, who was awarded a bounty of £30 by the Society of Arts in return for making his invention freely available to the public; see minutes of the meeting of the Society's Committee of Mechanics, 11 April 1771.
5 Letter not found.

La Collection des Machines de la Societé des Arts[6] ne sortira qu'au bout de ce mois: & je vous envoyerai au plutot votre copie. Je me suis informé de nouveau sur le metier à bas dans la douane, & un officier de ma connoissance & fort honnete home[7] m'assura qu'on ne peut pas en envoyer au dehors sans danger du moins de le perdre sans Resource: à moins de l'envoyer avec les equipages de quelque Ministre Etranger. ainsi J'attendrai que vous voyez la description de ces nouveaux metiers dans La Collection ci dessus: & pour lors on pensera à ce qui faudra faire. Car il est bien triste de risquer de l'argent sans etre sur de venir à son but. au reste je ferai comme vous me l'aviserez. 72.03.13

J'attends Recevoir bientôt quelque boisseau du bled de Siberie,[8] que Je ne manquerai pas d'envoier aussitôt à M^r^. Trudaine. C'est une excellente aquisition en Agriculture, & d'une grande Resource pour le monde dans des années de mauvaise racolte; car celle ci ne manque jamais de Reussir. nous avons eu du pain fait avec cet grain dans La Societé des Arts: & trouvames qu'il etoit superieur de beaucoup à celui fait avec de l'orge commun. on assure & est très probable, que ce pain est beaucoup plus nutritive que celui du bled ordinaire.

Je verrai si je puis trouver des potatoes ou pommes de terre dont Je vous parlai. ils sont beaucoup plus avantageux que Les ordinaires.

Je crains que mon voiage à Paris se retarde de plus en plus, & Je ne pourrai partir d'ici que vers le milieu ou meme la fin de May: aiant tant de choses à arranger que Je me vois chaque jour plus embarassé pour mes amis. Je suis charmé des bonnes nouvelles de la Santé de M^r^. Trudaine: & vous La souhaite parfaite à tous deux, etant de coeur

Mon cher ami

Votre pour La vie Magellan

P.S. Le globe de verre pour Les poissons &c n'est qu'une simple bagatelle amusante. elle couta ici environ 4 guinées: & reellm^t^. il y a de le hazard à etre cassée dans Le transport. neanmois si vous Le jugez à propoz je vous en envoyerai une.

---

6 William Bailey, *The Advancement of Arts, Manufactures and Commerce, or Description of the Useful Machines and Models contained in the Repository of the Society for the Encouragement of Arts, Manufactures and Commerce* (London, 1772). See Magellan to Duchesne, 17 January 1772.

7 homme?

8 See Magellan to Johann III Bernoulli, 15 March 1771, and Magellan to Duchesne, 19 April 1771.

72.03.13 La Semence des Choux Navetes,[9] qui sont à ce qu'on m'assure très differents de ceux qu'on connoissoit auparavant, ne manquera pas de partir ensenble avec l'orge de Siberie.

iterum vale.[10]

9 Literally "cabbage-turnip", probably the species of turnip introduced into England at about this time that became known as Swedish turnip.

10 again farewell.

72.03.17 *To Jacques-André Mallet*[1]

Bibliothèque de Genève, Geneva, Ms suppl. 1654, fol. 19

Londres 17 Mars 1772
at N°. 188 Fleet Street

Monsieur

Vous aurez deja appris la facheuse nouvelle de la mort de notre commun ami feu D^{r}. Bevis, arrivée le 6 Nov^{bre}. dernier dans le 76me. de son age presque complete.[2] Monsr. Horsefall,[3] son executeur testamentaire, reçut dans la suite la votre du 9 dec^{bre}. suivant: & je me chargeai très volontiers de vous faire reponse là dessus; car outre votre liaison avec cet ami commun, Je me souviens avec plaisir de vous avoir vû & parlé differents fois à la Soc^{e}. R^{le}. lors de votre sejour à Londres;[4] & ce qui plus est, votre caractere & talents vous rendent generalem^{t}. recommendable à tous ceux qui aiment la vertu & le genie. Ainsi Je vous ofre mes services pour l'objet de votre lettre, en vous assurant de faire tout ce que je pourra pour y reussir: même pour surmonter la dificulté de faire signer votre certificat chez vous, vous n'avez qu'à l'envoyer de ma part à M^{r}. Messier à Paris, pour qu'il le signe, & le fasse signer par deux autres des Membres Etrangers

1 Annotated: "de Londres M^{r}. Magalhaens Reçu le 28 mars 1772 Répondu". See Mallet to Magellan, 30 March 1772.

2 For an obituary of Bevis, see Magellan to Johann III Bernoulli, 28 January 1772.

3 James Horsfall (died 1785).

4 Mallet visited France and England in 1765. In England, he became particularly friendly with Bevis.

de la Societé R^{le}. de Londres, dont il y a plusieurs à Paris, qui vous connoissent.[5] $Autrem^{t}$. Je prendrai ce soin moimeme, car je pense d'y aller faire un tour cet Eté.

Et pour tout autre Objet literaire, ou commission, soit $d'instrum^{ts}$. d'Astronomie & $Mathemat^{e}$., ou des livres, ou toute autre chose, soyez persuadé de me trouver toujours disposé à vous rendre touts les services dont je serais capable, avec autant de plaisir & d'exactitude, que vous n'ayez pas à regreter la faute de feu notre ami, moi etant ambitieux de pouvoir le remplacer par ces deux qualités, au defaut de toutes les autres qu'il possedoit.

En attendant J'ai l'honneur d'etre avec toute consideration possible

Monsieur
Votre très humble & très $obeiss^{t}$. $serv^{r}$.
J.H. Magalhaens

A Monsieur
$Mons^{r}$. Mallet $Prof^{r}$. d'Astronomie
Cour de S^{t}. Pierre
à Geneve
en Suisse

5 Mallet had evidently inquired, in his letter to Bevis, about being proposed for election to the Royal Society. He was never elected or, apparently, nominated.

72.03.20 *To Pierre Macquer*

Bibliothèque Nationale, Paris, MS fr. 12306, f. 18

Londres 20 Mars 1772

Monsieur

Il y a une meprise dans ma Lettre precedente,[1] dont je ne suis pas Le Coupable. en voici La correction. Les fluores dont M^r^. Scheel (non pas Sheel) Suedois,[2] a extrait l'acide, qui etant uni avec les vapeurs de l'eau pure bouillante, produisent des vrais quartz,[3] sont Les fluores phosphoriques & non pas les vitrescentes.

on continua hier à Lire l'excellent Memoire sur l'air fixe & le Meuphitique, par M^{r}. Priestely dans La Soc^{e}. Royale,[4] & on y a lu aussi la communication d'un Medecin ou Cirurgien, qui cura d'une terrible fievre putride, un malade en Lui appliquant par L'anus l'air fixe qui se degageoit de la mixture de l'acide vitriolique avec le chalk ou pierre à chaux moienant un tube, qui etoit adapté à un verre oû la dite misture ou solution se faisoit.[5]

Je n'ai pas plus de loisir que me dire

Mon [...] Monsieur

votre de Coeur

Magellan

a Monsieur
Monsr. Macquer Reg^{t}. de Medecine
& Professeur de Chymie
Rue S^{t}. Sauveur
à Paris

---

1 The previous letter to Macquer in our list (Magellan to Macquer, 25 October 1771) does not deal with the question referred to here. Therefore there is a letter missing that was written between October 1771 and March 1772. Macquer read the letter that is now missing to the Académie Royale des Sciences on 1 April 1772 (*Procès-verbaux*, 91, 1 April 1772), and an extract from it was published by Rozier in the April 1772 number of his journal, *Observations sur la physique, sur l'histoire naturelle et sur les arts*. This was republished in D. McKie, "The '*Observations*' of the Abbé François Rozier (1734-93) – I", *Annals of Science*, 13 (1957), 73-89, at pp. 85-6, and in H. Guerlac, *Lavoisier – the Crucial Year* (Ithaca, NY, 1961), p. 54.

2 Carl Wilhelm Scheele (1742-1786).

3 See Partington, *History of Chemistry*, vol. 3, p. 214.

4 J. Priestley, "Observations on Different Kinds of Air", *Philosophical Transactions*, 62 (1772), 147-264. The reading of Priestley's paper occupied four successive sessions at the Royal Society, on 5, 12, 19 and 20 March 1772.

5 William Hey, FRS (1736-1819), surgeon at Leeds, wrote to Priestley on 15 February 1772, describing this treatment. His letter was published as an appendix (pp. 256-64) to Priestley's paper.

To Louis-Henri Duchesne 72.03.20a

Archives Nationales, Paris, MS T/160 (10-12): lettres de J. H. de Magalhaens

Londres 20 Mars 177[2][1]

Celle ci mon cher ami, est seulem^{t}. à cause de celle pour Le Chev^{r}. Bory,[2] que Je vous prie cacheter, après l'avoir Lue s'il vous plait, & la lui faire tenir.

Le retard de vos gans[3] fut causé par le deuil par la mort de la Princesse de Galles,[4] qui occupe tous Les ouvriers long tems ici pour servir à ce pressant objet.

J'atends tous les moments le bled de Siberie:[5] & aussitot Je l'envoyerai.

adieu mon cher, escusez, car je n'ai plus de tems à present. mais toujours

Votre pour La vie
Magellan

---

1 Magellan appears to have written "1770", but the date of Princess Augusta's death makes it clear that the letter was written in 1772.

2 Letter not found.

3 See Magellan to Duchesne, 13 March 1772.

4 Augusta, dowager Princess of Wales, died on 8 February 1772.

5 i.e. Siberian barley; see Magellan to Duchesne, 5 March 1771, 19 April 1771, 13 March 1772 and 31 March 1772.

72.03.21 *To Jesse Ramsden*[1]

Westminster City Archives, London, Box 74, N°. 8

Sir,

Since you have given leave to direct to you, for further information, relatively to an Advertisement in the Newspapers, I think it necessary to make you Master of the Particulars thereof, which the modesty of the Advertiser could not allow him to mention to the Public, though easily averred in the most satisfactory manner. The pleasure I feel, in giving this testimony to real merit and virtue, and the prospect of procuring by this means, some suitable protection and employment, to a gentlemen well qualified for rendering considerable services to those, who often are at a loss to find out a proper person, is an anticipated Reward I receive for my solicitude; and I desire you will be so good as to impart, to any inquirer, the following information, Grounded on the recommendations I have received from persons of very distinguished character abroad, besides many others, who I know, have strongly recommended this Gentlemen to several friends in London, and amongst others, to Dr Franklin,[2] and Dr Matty,[3] who are generally known by their learned professions, and moreover to George Aufrere Esq.,[4] John Marsh Esq.,[5] William Baker Esq.,[6] – Legros Esq.,[7] Moore jun. Esq.[8], etc.

Mr Dazeux is a gentlemen of genteel liberal education; who has always lived amongst people of fashion, and possesses many good qualities, which makes him fit for such employments, where personal Accomplishments

1 A printed sheet headed "Copy of a letter directed to Mr Ramsden, Mathematical Instrument Maker and Philosophical Instrument Maker, near the Little Theatre in the Haymarket, on the subject of a public advertisement."

2 Benjamin Franklin (1706-1790). Dazeux had carried a letter from Jean-Baptiste Le Roy in Paris to Franklin, then resident in London; see Le Roy to Franklin, 5 March 1772, and Franklin to Le Roy, 20 April 1772, in *The Papers of Benjamin Franklin* (New Haven, 1959-), vol. 19, pp. 84, 111.

3 Matthew Maty (1718-1776), Secretary of the Royal Society.

4 The wealthy merchant, member of Parliament and member of the Society of Arts, George René Aufrère (1715-1801).

5 John March, Esq. of Waresly, Huntingdonshire, a member of the Society of Arts?

6 Possibly the printer and essayist William Baker (1742-1785).

7 John Le Gros, merchant of Bishopgate Street, London and member of the Society of Arts. Magellan evidently did not know his given name.

8 Not identified.

and Moral Qualifications are required. He is master of the Latin language, 72.03.21
speaks Italian and English tolerably, besides his own (the French) which he writes with elegance. Is well acquainted with the polite literature, as well as tactics and other branches of military knowledge. He is of mild decent behaviour and sedate temper. Has a tolerable taste for drawing and thorough knowledge of music, especially the Italian, and plays on the violin with delicacy. Is very sober, and as I am informed, has always been remarkable for his good morals. But, different Accidents and adversities have reduced him to the necessity of availing himself of his personal abilities; and a Philosophical turn of mind, has easily persuaded him to resort to this country, to pass the remainder of his life, where the rights of man are best maintained by the laws, and where he hopes to find some person to employ him as a secretary, or as a tutor to youth. He would even be ready to go abroad in that capacity.

This is what at present occurs to say, by way of information about this Gentlemen, whose age is a little above thirty; he enjoys a good state of health: his religion is the Protestant: and true philosophy has so properly disposed his mind, that his present expectations go no further, than to be received into a decent family, where he may make himself useful and agreeable.

Any person who desires to see this Gentlemen may direct to his lodgings at Mrs Clarke's Lace-shop, the corner of Great-Turnstile and Lincoln's Inn Fields, or if more convenient, may call on him.

I am Sir, with due regard and great Esteem,
your most obedient and humble servant
J H Magellan

Fleet Street, March 21st, 1772.

72.03.27 *To Johann III Bernoulli*

Öffentliche Bibliothek der Universität Basel, MS L I a 709, f. 23-24

Londres 27 Mars 1772

Je reçois, Monsieur la Votre du 17 du courant dans cet instant,[1] & je m'empresse de vous ecrire celle-ci pour Vous donner les deux articles suivants pour Votre Recueil. mais je suis très faché d'aprendre le mauvais etat de votre santé: & voudrois vous voir plutot en querelle avec M^rs^. Maskelyne,[2] Dollond & tout le rest du monde que non pas avec vous meme & votre bien etre. assurem^t^. vous ne vous menagez point ce don le plus precieux & le moins connu que les hommes possedent, & qu'il est hors de leur pouvoir de supleer par toute autre consideration de bonheur. Je communiquerai à M^r^. Maskeylne (avec lequel je m'avois rencontré chez Dollond cet après midy) & sans doute il ne manquera pas d'etre satisfait. Dollond aussi s'a plaint vivement de vos Lettres Astronomiques.[3] mais tout ces articles & ceux de votre Lettre resteront pour une autre fois. allons aux deux articles.

1. Vous savez qu'on n'a pas toujours sur mer un horizon assez net au dessous de l'astre qu'on veut observer. pour lors on tourne le dos à l'astre, & on observe le Complement de sa hauteur, dès l'astre jusqu'à la partie opposée de l'horizon. C'est ce que nous appelons <u>back observation</u>: & pour la faire on fait usage d'un second petit miroir horizontal <u>a</u>, qui fait un angle droit avec le miroir <u>b</u> de l'index <u>on</u>. Mais pour rectifier cet angle & que le miroir n'a pas alteré de position, il n'y a d'autre moien que de le rectifier lorsque le horizon de devant est aussi decouvert que celui de derriere. C'est à dire, on ne peut pas s'en assurer de sa position, lorsqu'on en a le plus grand besoin. C'est à ce defaut du <u>Hadley's Quadrant</u> que M^r^. Dollond remedia, en ajoutant la queue <u>Z</u> au miroir <u>a</u>. on la met sur

1 Letter not found.

2 On Maskelyne's complaint against Bernoulli, see 71.10.29.

3 Bernoulli's *Lettres astronomiques* (Berlin, 1771) includes a chapter (pp. 65-76) on "Des Opticiens de Londres", several pages of which are devoted to not very flattering comments on Dollond. A second chapter (pp. 126-33) is devoted to English instrument makers. For the specifics of Dollond's complaint about Bernoulli's treatment of him, see 72.04.07.

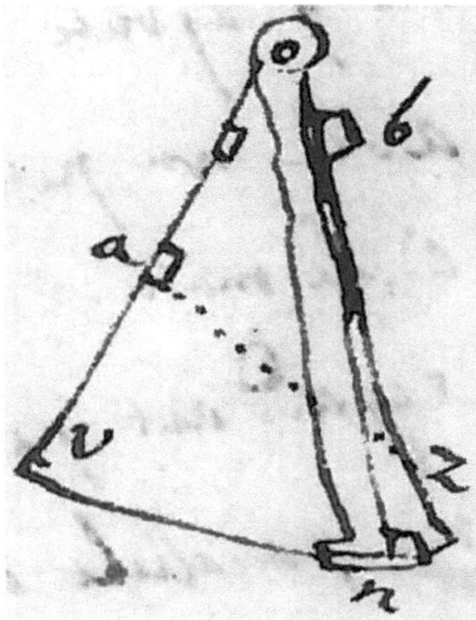

une marque dans l'endroit v. alors on regarde par une pinule qui est proche de Z, & on met ce miroir a parallele à celui b de l'index on. en faisant coincider l'image reflechie du soleil avec celle vue par les raions directs. Celà fait, on porte La queue Z à un point qu'y est marque proche de Z, dans Le quel l'artiste trouve, que l'angle de 90°. aboutit, & par là le miroir a devient à angle droit vis à vis du miroir b. Ce qu'il faut avoir pour le back observation celle ci est la principale invention ou amelioration (improvement) de M^r^. Dollond: avoir moien de s'assurer, sans le secours des deux horizons, que le miroir a est à angle droit avec le miroir b lorsque l'Index est à zero. Pour les autres petits improvem^ts^. ils ne sont pas aussi eclatants que celui ci.

2. Cependant ni moi, ni Ramsden, nous n'avons jamais été bien contents de cet improvement. il est purem^t^. mechanique, & depend de l'exactitude de l'ouvrier, sans que l'observateur s'en puisse bien assurer dans tous les cas. outre celà il est impossible de s'assurer à une minute près de mettre la Ligne de la queue, dont le raion est assez petit, sur la marque que l'artiste y a faite: on ne sait non plus si ce mouvement est ou non parallele au plan de l'instrument, i.e. si le pivot dans le quel tourne le miroir a, est ou non perpendiculaire au plan de l'instrument. ainsi Je me suis mis à penser là dessus: & enfin Je viens de trouver il y a deux jours, un moyen, qui selon l'opinion de connoisseurs à qui Je l'ai communiqué, est le plus excellent & notable improvem^t^. que cet Instrument a jamais eu depuis son invention; car outre l'avantage de fixer ce miroir à un angle droit precis & demonstrable, J'obtiens dans le meme tems l'avantage de pouvoir racourcir considerablem^t^. le limbe du Quadrant, & neanmoins prendre avec lui les

72.03.27 angles les plus grands. il n'y a que 2 jours d'ecoulés depuis ma premiere idee de cette methode. ce fut Le 25 de ce mois au matin: & peut etre n'ai je pas encore developé tous les avantages que cette methode me fournit. la voici.[4]

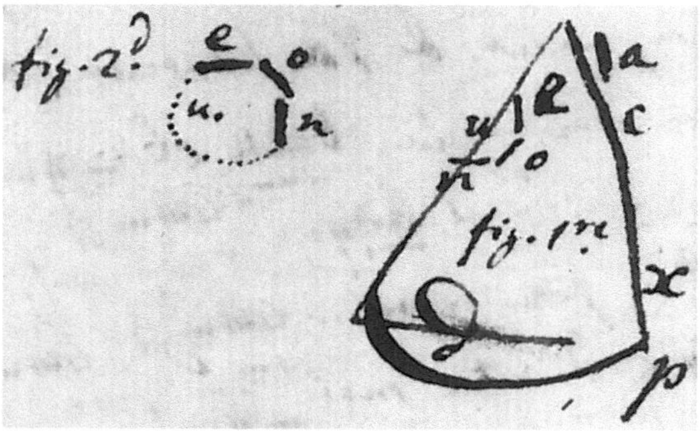

Qu'au lieu d'un miroir horizontal, pour la back observation on y ajoute une platine, qui puisse tourner dans le centre u: avec un petit miroir e fixe, & avec deux autres o.n. mobiles à la main. Celà suppose metez l'index ap du quadrant au zero & regardant par une pinule C, tournez la platine de façon que le miroir e devienne parallele au miroir a, moienant la coincidence des deux images. dans la suite mettez l'index sur le 45°. du limbe. regardez alors par une autre pinule X sur le miroir o, & sans toucher l'autre miroir e, tournez à la main ce miroir o jusqu'à ce que la coincidence des deux images vous montre qu'il est parallele au miroir a. remetez de nouveau l'index au zero, voiez si e conserve le meme parallelisme, & en tournant toute la platine sur son centre u, faitez que le miroir o soit parallele à a: & pour lors vous ne pouvez pas douter que L'autre miroir e a fait un tour exact de 45°.: dans la suite mettez encore l'index au 45°. du limbe: tournez uniquement le miroir n pour qu'il soit parallele a a: remetez l'index a zero, observez si o garde encore sa position parallele: tournez la platine toute sur son centre

4 Bernoulli published a brief announcement of Magellan's invention in the second volume of his *Recueil pour les astronomes*, published later in 1772 (p. 309), and promised to give a full description in the next volume. The publication of this was, however, long delayed. Magellan also sent an account of the invention to the Chevalier de Bory, in Paris, and subsequently an example of the instrument itself for presentation to the Académie Royale des Sciences, where it was reported on very favourably in January 1773 by de Bory and Le Monnier, acting as commissioners. Two years later, Magellan himself published a description of his invention in his *Description des octants et sextants anglois…* (London, 1775).

jusqu'à ce que n devienne parallele à a: & pour Lors vous ne pouvez point 72.03.27
douter que le miroir e ne soit pas exactement à 90°. d'inclinaison à l'egard du miroir a de l'index lorsqu'il est à zero, ce qu'il faloit avoir exactement pour faire l'observation par derriere sans aucune dependence des deux horizonts. Outre cet avantage, on a encore celui de pouvoir prendre des angles les plus grands avec un instrument dont le limbe ne soit pas au delà de 45°. & encore moins: mais pour lors il faudra avoir 4 miroirs horizontaux, au lieu de 3, pour le Back observation. Par exemple suposons un instrument avec un limbe pas plus grand que 30°. (qui par consequent peut avoir un raion bien plus long que les ordinaires) & qu'on veut prendre avec lui un angle de 78°. par une observation par devant. pour celà metez le miroir e parallele à a, lorsque l'index est à zero: menez l'index sur le 20°. & en regardant par la pinule X, metez l'autre miroir o parallele, sans pour tant toucher e, au miroir a de l'index. remettez l'index à zero: & pour lors vous etes sur que le premier e a fait un tour de 20°. pour celà si vous observez un angle de 58° sur le limbe (car les angles y sont doubles par la double reflexion) vous y ajoutez les 20°. & vous aurez reellem^{t}. observé un angle de 78°. N.B. il est evident que les miroirs o & n (ou du moins o) doivent etre otés lorsqu'on observe par derriere, avant de faire l'observation, aiant dejà ajusté le miroir e comm'il faut: car autrem^{t}. ils empecheroient de voir l'horizont. N.B. 2do il est clair que en observant la meme image sur un miroir qui tourne sur un axe, & celà de deux diferents endroits, ce miroir absolument a tourné sur un axe parallele au plan qui passe par ces deux endroits ou pinules: & par consequent on est assuré de la vraie perpendicularité des miroirs, ce que ne paroit pas dans l'invention de Dollond. Lui avoit tirée une Patente pour la sienne: & J'ai deja eté avec Lui, pour lui ofrire la mienne qui sans doute l'effacera. Je crois qu'il tirera une autre pour elle. à Dieu que la poste va se fermer.

Je suis de coeur
votre &c.
Magalhaens.

a Monsr.
M^{r}. Bernoulli Astronome
Royal & de l'Acad^{e} R^{le}. de Prusse
à Berlin

72.03.30 *From Jacques-André Mallet*

Bibliothèque de Genève, Geneva, Ms suppl. 1654, fol. 46[1]

A M^{r}. Magalhaens at
N°. 188 Fleet Street
London

Geneve le 30 Mars 1772

M^{r}.

J'ai appris la mort de notre ami le Dr Bevis c'est par la lettre que vous m'avez fait l'honneur de m'écrire,[2] que j'en ai eté vraim^{t}. affligé & je regrette bien sincer^{t}. ce brave & dig. homme; son dernier accid^{t}. à un age aussi avancé m'a bien fait craindre qu'il ne put y résister.

Je suis ex^{ment} sensible M^{r}. de l'attention que vous avez eu de me communiq. cette nouvelle, & je suis extrement sensible à toutes les choses b'c.[3] trop obligeantes que vous me faites l'honn. de me dire dans votre lettre. J'accepte avec reconn. l'offre que vous avez la bonté de me faire de vos recommandat. & de vos soins, p^{r}. m'aider à reussir dans le désir que j'ai d'etre aggregé à votre savante Societé. Je procurerai d'ici au mois de 9bre, le certificat necess. & je prend. la liberté de vous l'addresser, je pourrois meme vous l'envoyer à Paris, si vous y faites ce voy. cet été, car je pense que quoiqu'on n'exige que la signat de 3 membres etrang. seulem^{t}. celle d'un plus grand n^{bre} ne peut que faire un bon effet.

Je vous remercie aussi très sincerem^{t}. M^{r}. de l'offre oblig^{te}. que vous me faites de vous charger des petits comm. que je percois etre dans le cas de faire faire à Londres, je connois votre zèle p^{r}. obliger vos amis, & je serai bien flatté que vous veuilliez me mettre du n^{bre}. Mr Bernoulli de Berlin, avec qui je suis etroitment lié, m'a souvent parlé de vous dans ses lettres, la rareté & la brieveté de celles du pauvre D^{r} me faisoit souhaite depuis longtems d'avoir en Angleterre un correspond^{t}. moins occupé, qui de t. en t.[4] put me comuniq. les nouvelles littéraires de ce pays là, et surtout ce qui a rapport à l'astronomie, dont je fais à prés^{t}. presque mon unique occupation; Je me suis procuré une collection d'Instr. d'astron. suffis^{te}. p^{r}. meubler un observatoire assez [passable] p^{r}. un particulier, j'ai eté depuis mon retour de Russie[5] occupé à solliciter auprès de mon Etat un batim^{t}. propre à y

1 Letter not found. The text is from a draft kept by Mallet.
2 See Magellan to Mallet, 17 March 1772.
3 beaucoup?
4 de temps en temps.
5 Mallet visited Russia in 1768-9 to observe the transit of Venus across the face of the Sun.

placer ces Instrumens, le succes n'a point repondu à ce que j'espérois d'une 72.03.30
ville ou l'on se pique d'aimer les sciences, c'est ce qui m'a determiné enfin à faire moimeme les fraix de ce petit batiment au quel je vais faire travailler incessam^{t}. vous aurez peutetre vu le plan que j'ai imaginé, je l'avois envoyé au D^{r}. Bevis, il me paroit reunir toutes les commodites possibles si vous en etre curieux, je pourrois vous en envoyer une esquisse.

La situation incommode & peu solide du galetas oû sont actuellem^{t}. mes Instrum. ne m'a pas permis jusqu'a prés^{t}. de faire les observations que j'aurois desire, ce n'est gueres que des eclipses de Satellites qui me serviront pour^{t}.[6] à [construer] la longit. de notre ville que je crois mal déterminée. Vous me feriez bien plaisir de m'en procurer de correspond^{tes}., j'en ai deja qlq'unes que M^{r}. Messier a eu la bonté de me communiquer, voici les Tables des miennes.

1768	Fevr.	15 mat.	Immers. du	1^{r}
	Mars	2 soir	"	2^{d}
	"	10 mat.	"	2^{d}
	"	10 mat.	"	3^{e}
	"	17 mat.	"	2^{d}
	"	18 mat.	–	1^{r}
	–	25 mat.	–	1^{r}
	–	26 soir	–	1^{r}
	–	27 soir	–	2^{d}.
1771	May.	22. mat.	Immers. du	3e.
	–	23 mat.	du	1^{r}.
	Juin	3 mat	- - - - - - - - -	2^{d}.
	- - - -	7 soir	- - - - - -	1^{r}.
	- - - -	15 mat.	- - - - - -	1^{r}.
	Aout	17 soir	Emers. du	1^{r}.
	- - - -	30 soir	Em. - - - -	2^{d}.

C'est p^{r}. abreger que je ne vous envoyerois les observat. elles mêmes mais si vous les souhaitez, elles sont bien a votre service.

Je souhaite beaucoup M^{r}. que l'occasion se présentera oû je puis vous etre de qlq. uti[lité][7] soit ici soit ailleurs, Je vous prie de disposer de moi, à votre volonté. Je serois charmé de pouvoir vous prouver plus particul. l'estime & la consid^{n} avec lesquelles J'ai l'honneur d'etre[8]

---

6 pourtant?

7 Cut off at margin.

8 The draft is unsigned.

72.03.31 *To Louis-Henri Duchesne*

Archives Nationales, Paris, MS T/160 (10-12)

Londres le 31 Mars 1772

C'est pour vous donner avis, mon cher ami, de l'envoy que Je viens de faire à M^{r}. Trudaine d'une petite quantite environ un demi bushel ou boisseau du bled de Siberie, qui est L'unique que J'ai pu obtenir malgre toutes Les promesses & diligences que J'ai faites.[1] Je vous prie de donner tout de suite des ordres pour preparer La terre, & pour Le semer aussitot qu'il arrivera Car La saison avance: mais il arrivera encore en bon tems. tout le mois d'avril est assez bon pour Le semer. on n'a pas besoin d'autres attentions que celles qu'on donne a l'orge ordinaire. vous verrez que ce grain est tres beau: & en gardant La semence avec soin on pourra en peu d'annees avoir assez pour en reprendre La culture de ce precieux grain par tout Le Royaume. Je suis bien faché de ne pas pouvoir envoyer une plus grande quantité: mais le fournier qui me l'avoit promis & à un de mes amis intimes, nous a assuré jusquà ce moment, & profita de l'occasion de vendre ce qu'il nous avoit promis à des autres qui Lui donnerent à raison d'une guinee par boisseau. Je suis egalem^{t}. en defaut pour les potatoes, dont il ne m'a pas eté possible d'avoir, malgré toutes Les promesses qu'on m'avoit faites de differents parts.

Le morceau de toile (ou plutot papier) ci joint est de la espece dont les habitants de Otahitee[2] font leur habillement. peut etre M^{r}. de Bougainville[3] en a rependu par curiosité chez vous. pour celui ci ce fut M^{r}. Banks qui m'en donna mercredy dernier allant chez Lui à Lui faire mes àdieux: Lui & Solander vont à partir pour son voyage nouveau dans 15 jours vers Les terres Australes.[4]

---

1 See Magellan to Duchesne, 20 March 1772.

2 Tahiti.

3 The French explorer Louis Antoine de Bougainville (1729-1811) visited Tahiti in 1768.

4 Joseph Banks and Daniel Solander, who had accompanied James Cook on his first round-the-world voyage in 1768-70, were expecting also to accompany him on his second expedition, which was due to depart. Cook, however, rejected the elaborate alterations to his ship that Banks had had implemented and in the end they did not go. Johann Reinhold Forster took their place as naturalist on the expedition, assisted by his 17-year-old son Georg. By the time Banks' alterations were reversed, it was July before *Resolution* and its companion ship, *Adventure*, set sail from Plymouth.

Je viens de faire quelques ameliorations notables sur Le Quadrant marin, dont j'ai donné compte La poste derniere a L'Acad^{e}.,[5] & j'en ferai autant jeudy prochain à La Sociéte Royalle.[6] J'aurai bientot quelques instrum^{ts}. faits de La Sorte. & le plus qu'on y pense, le plus on trouve bonne L'amelioration. 72.03.31

Je n'ai plus de tems que à me dire pour La vie

Mon cher ami

Votre de coeur

Magalhaens

N.B. par Le billet ci joint,[7] que Je viens de Recevoir dans La minute, vous verez que La personne qui devoit me fournir Le bled de Siberie, a trouvé mieux de le vendre a une guinée par bushel; & combien mon ami Arbuthnot[8] en est fache: & moi egalement.

A Monsieur
Monsr. Duchesne Secretaire
de M^{r}. Trudaine de Montigny
a Paris

5 Magellan sent an account of his invention to the Chevalier de Bory, in Paris, and subsequently an example of the instrument itself for presentation to the Académie Royale des Sciences, where it was reported on very favourably in January 1773 by de Bory and Le Monnier, acting as commissioners. See also Magellan to J. Bernoulli, 27 March 1772.

6 In the event, Magellan's presentation to the Royal Society was much delayed, his paper describing his invention not being read until 17 December 1772; see Magellan to the Royal Society of London, 4 September 1772.

7 Notice not found.

8 John Arbuthnot, FRS.

72.04.07 *To Johann III Bernoulli*

Öffentliche Bibliothek der Universität Basel, MS L I a 709, f. 25-26

Londres 7 Avril 1772

Je vais Vous ecrire, mon cher Monsr., moins à la date que la poste preced^t^., mais pas assez pour vous faire une reponse complete à la Votre du 17 Mars,[1] d'autant plus que Je n'ai pas eu encore aucune occasion de m'informer sur quelques uns des points dont vous me parler. J'ai deja dit à M^r^. Maskelyne, que J'avois reçu une lettre de vous sur l'affaire en question, mais je n'ai pas pû la lui montrer, etant dans la Societé R^le^. & ne l'aiant point dans ma poche. mais je le verrai jeudy prochain.

J'ai presenté Vos compl^ts^. à M^r^. Valtravers.[2]

les changem^ts^. du micrometre de Dollond, sont qu'il disperse les raions comm'un concave. par là le foyer de la lunete est plus eloigné, & ce qui plus est il peut travailler la Lentille (composée, car elle est achromatique, avec un verre de flint glass, & l'autre de crown glass) d'un grand diametre, ce qui donne un grande avantage.[3] Car dans la fig.^e^. 1re. Vous voyez que dans

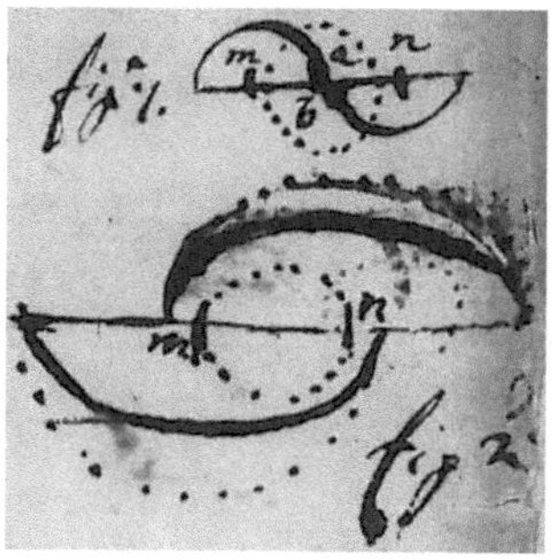

le plus grand écart du micrometre, la lumiere qui entre dans la lunete n'est à peine la moitié: puisque les deux secteurs a & b restent couverts avec la platine de cuivre, & par consequent les deux images sont tres foibles. mais si l'on employe des verres dont Les foyers soient negatifs, il est possible de

1 Letter not found.

2 Rodolph Valltravers, FRS (1723-c1815), on whom see G. R. de Beer, "Rodolph Valltravers, F.R.S.", *Notes and Records of the Royal Society of London*, 4 (1946), 216-26 and addendum, *Ibid.*, 8 (1950-51), 116-9.

3 Cf. Maskelyne, "Description of a Method of measuring Differences of Right Ascension and Declination, with Dollond's Micrometer, together with other new Applications of the same", *Philosophical Transactions*, 61 (1771), 536-46.

les travailler sous des dimensions aussi grands qu'on veut, voyez la fig. 2^{d}. 72.04.07
& observez que la même ecartment, ou separation du foyers mn (ou le même foier divisé en deux) etant aussi ecarté comme celui de la figure 1re., n'empeche pas que l'ouverture de la lunete ne soit pas entierement pleine avec les deux branches ou portions de lumiere, refringée par les deux motiés de l'objectif, qui composent le micrometre. Par consequent l'objet doit etre beaucoup plus lumineux, malgré la portion de lumiere qui se perd par la divergence des raions. mais ce qui vaut encore plus, C'est que les objectifs, dont le foier est racourci par le micrometre positif, soufrent dans la clarté, & reproduissent quelques fois des couleurs, qui nuisent la distincte vision &c. particulierement lorsque les lunetes ont un foyer assez court, comme celles à triple objectif achromatiques.

Je ne sai pas comment mon improvem^{t}. dans le Hadley's Quadrant sera reçu du public. meme je n'attends que les partisans de M^{r}. Dollond le decrieront, pour ne pas faire manquer le soin.[4] mais j'attends paisiblem^{t}. le jugement de ceux qui jugeront sans passion. Depuis ma derniere, je trouvai une plus belle position des 3 miroirs horizontaux par la quelle on n'a pas besoin d'oter aucun d'eux pour faire quelqu'observation. Voici la figure, qui par la seule inspection vous convaincra de cette verité. Soit un

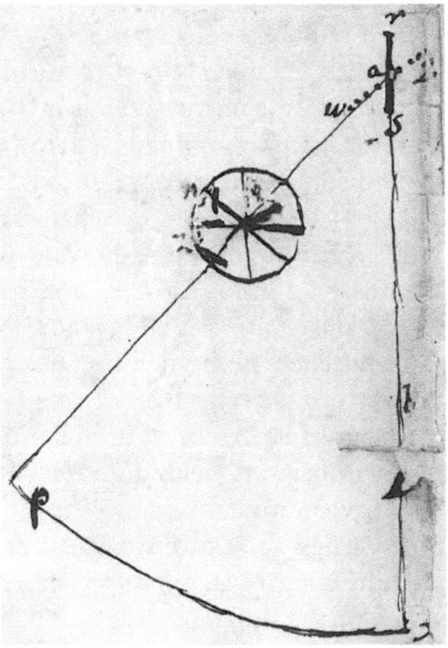

4 Cf. Peter Dollond, "Letter…describing some Additions and Alterations made to Hadley's Quadrant, to render it more serviceable at Sea", *Philosophical Transactions*, 62 (1772), 95-8.

72.04.07 octant (de 45°.) p. o. x. dont l'index porte le miroir r.s dans le centre de l'arc px. ajoutez sur le raion a p, un petit cercle n. o. z. m, qui puisse tourner dans un centre e. fixez dans de ce petit cercle les 3 petits miroirs horizontaux n.o.z. dans La situation qu'on y voit, les uns à l'egard des autres. C'est à dire le miroir z faisant angle droit avec le miroir o: & le miroir n faisant un angle de 45°. avec le miroir z. pour Lors, si vous metez l'index r.x à zero: vous trouverez le miroir o parallele au miroir rs. Ce que vous prouverez en regardant par la pinule b vers le soleil, de façon que l'image reflexie tombe parfaitement sur l'objet vu par vue directe. Ce qui est un probleme qui a eté mile fois demontré dans la Catoptrique. Car d'abord que l'objet vu par des raions directs, coincide avec l'image vue par la reflection de deux miroirs, il est evident que ces miroirs sont entre euxmemes parfaitement paralleles. aiant donc le petit miroir o parallele au miroir rs : portez l'index au degre 45°., C'est à dire à p, & en regardant de quelque part proche de p, sur le miroir n jusqu'à ce que les deux images du soleil vu par raions directs, & par reflection, coincident exactem[t]., Ce miroir n sera parallele au miroir de l'index qui pour lors se[ra] dans la position w.q, & par consequent fera un angle de 45° avec le miroir o. ainsi si vous remettez l'index a zero en x & faites mouvoir le petit circle sur son axe, jusqu'à ce que en x & faites mouvoir le petit cercle sur son axe, jusqu''a ce que en regardant par la pinule mise à l'endroit (vg. q) du raion ax, vous vous assurez qu'il est parallele à ce raion; pour lors vous etes sur que le premier miroir o a reculé de 45°. Faitez La meme operation avec Le miroir z, en le mettant parallele à celui de l'index mis à p: & retournant avec l'index à zero (à x) tourner le petit cercle jusqu'à ce que z lui soit parallele: & pour lors vous etes sur que le miroir o a reculé de 90°. vis à vis du miroir rs & par consequent est à angle droit avec lui. Or il est bien aisé de concevoir que si les 3 petits miroirs sont dans cette position les uns vis à vis des autres, le miroir z restera dans sa propre place pour faire les observations par devant sans aucun obstacle: & dans le meme tems il n'empechera point que Les raions reflechis du miroir r.s tombent sur Le miroir o & se reflechissent vers la pinule qui sera au dehors (à peu pres en o) pour faire L'observation par derriere, sans aucun besoin d'observer les deux horizons (celui du devant & celuy de derriere) pour rectifier & preparer avec sureté son improvement.

En fin pour ce qui regarde l'autre proprieté de mon improvement, qui est celle de pouvoir prendre avec un octant de 45°. des angles aussi grandes & meme plus que ceux qu'on peut prendre avec un sextant de 60°. ou meme avec un quadrant, il est bien aisé à concevoir, que si, tandis que vous avez le miroir z parallele à celui rs de l'index à zero, vous tournez le miroir o ou n de façon qu'il devienne parallele à celui de l'index (après l'avoir mis vg. à 40°.) et que pour lors vous remetez l'index à zero, & tournez le petit cercle sur son axe, jusqu'à ce que ce meme petit miroir o ou n, lui devienne parallele, moienant La coincidence des deux images vues par des raions directs, & par des raions reflechis: pour lors vous serez sur que le miroir z a reculé de 40°.: & par consequence tous les angles que vous observerez avec votre instrument, auront 40°. de plus & au delà de ceux qui y sont marqués sur Le limbe. par exemple un angle ainsi observé avec L'index sur 94°. sera reelement un angle de 134°. 72.04.07

Voici les petites observations faites à Danzic par mon bon ami le Dr. Wolfe.[5]

Observationes factæ Dirsaviæ, quod oppidum fere sub meridiano Gedanensi, à quo urbe 3 milliar. Germ. distat versus Austrum. tubo Shortiano 200 ties augente.

1771	tempus verum	Iovis
Maj.	21d.15h.16'.44''	Im̄. Sat. III fasciæ distinctæ, non obstante crepusculo
	22.14.37.5.	Im̄. Sat. 1. fasciæ mediocriter videatur
Jun.	7.12.49.51.	Im̄. Sat. 1. fasciæ distinctæ satis.
	14.14.41.48.	Im̄. Sat. 1. fasciæ distinctæ.
	26.11.1.35.	Im̄. Sat. III. – minus distinctæ.
Jul.	4.12.37.59.	Im̄. Sat. II. valde dubia ob coelum obscurum & inter curr… tes nubes.
	4.13.44.57.	Im̄. δ ♓ in ☾
	4.14.52.40.	Em̄. δ ♓ ex parte obscura disci Lunæ, dubia ad aliquod secunda.

Stella circiter 1'.50''. borealior centro Lunæ, nam Linea transitus inter Cardanum & Aristarchem, atque inter Copernicum & Erasthostenem apparebat.

Le Dr. Long est mort.[6] mais Je n'ai pas encore pu avoir aucune notice de sa vie pour vous envoyer.

il est vrai que vous vous avez trompé sur quelques points dans vos Lettres Astr. relatifs à Dollond & à Ramsden. Je suis bien faché que vous

5 Bernoulli published Wolf's observations in his *Recueil pour les astronomes*, 2, 325.
6 Roger Long, D.D. (1680-1770), Master of Pembroke Hall, Cambridge, and from 1750 Lowndean Professor of Astronomy in the University of Cambridge.

72.04.07 n'aiez pas faite une intime connoissance avec ce dernier: & aussi que vous ayez touché si fortment sur [l']ignorance theoretique du premier. Lui (& moi aussi) a eté bien faché, mais pour moi je connois que vous ne l'avez pas fait à dessein de le degrader. en verité il faut etre très delicat, en tout ce qu'on publie sur les qualités d'un autre: & plutot ne dire pas ce qu'on sait de disavantageux, même lorsqu'on en pourroit donner toutes les preuves possibles. mais nous parlerons peutetre plus au Large sur cet affaire dans une autre Lettre. Neanmoins voici quelques mots, qui l'ont offensé. a pag. 65. ...il est etonant (en parlant de Dollond) que Le tatonemant l'ait pû conduire si loin.... Dollond ne part que des regles de theorie la plus commune = donc, mon ami, s'il a de la theorie, meme La plus commune, pourquoi lui dire que c'est le tatonement qui l'a mené si loin? Ceci semble une contradiction digne d'un coeur malevolent, & non pas du Votre. Ce fut un Lapsus pennae d'autant plus que La raison qu'il vous donna sur ce que les oculaires (à pag. 67) ne peut rien faire pour corriger les abberations, à cause de la petitesse de leurs foyers, n'est pas tout à fait une raison de tatonement. un peu plus bas vous dites que cet admirable artiste vous satisfit sur le peu de Succès dans les achromatiques, entrepris par ceux qui prendent la theorie pour leur guide unique. or voilà un artiste qui connoit jusqu'à quel point la theorie peut guider, & consequement n'est pas un simple tatonem. pag. 68 C'est trop fort de dire que les ouvrages chez lui sont ceux faits à La douzaine. Car premierem^{t}. il n'a pas chez lui d'ouvriers pour des instruments de mathematique. peut etre que vous y avez vû quelque petit ou petis ouvriers pour des [...][7] [ac]commodages: mais tous les bons instruments qu'il vend sont toujours faits au dehors. Celui ci est Le Cas du fameux Nairne[8] & de tous les autres excepté Ramsden. car celui ci est par profession faiseur d'instruments de Mathematique &c. & dans le meme tems il s'a melé avec un extraordinaire Sucès aussi de l'Optique, & de faire des achromatiques, malgrè & contre La Volonté de Dollond. mais ni celui ci ni Nairne ni Pyefinch,[9] ne sont que simples opticiens. il est vrai que Ramsden a travaillé longtems pour Dollond: mais il est faux qu'il fasse depuis deux ou 3 ans, meme je crois du tems de Votre Sejour ici, aucune chose pour Dollond. ainsi à pag. 69 vous ne dev[ez] dire que Ramsden vend des tellescopes qu'on dit bons. peut etre les deux beaux freres s'entre aident

7 Illegible word; paper damaged.
8 Edward Nairne (1726-1806).
9 Henry Pyefinch, optician and instrument maker.

=. Car Ramsden est egalement capable de faire aussi bons tellescopes, & meme peutetre meilleurs que Dollond. Lui meme les a poussés aussi loin que Short:[10] car il en a fait comme les deux derniers de ce fameux Artiste, qui avoient une ouverture très grande dont les miroirs avoient une vraie curvature parabolique; car il ne seroit pas poss[ible] avec une portion de pure esphere Leur donner une pareille ouverture. Je crois que de ces deux derniers telescopes à Reflexion, fut acheté par le Earl Bute.[11] 72.04.07

à pag. 71 vous ne parlez pas en connoisseur des achromatiques. Car vous ne faites aucune difference ni mention des achromatiques à Large ouverture, dont l'effet est très superieur à ceux d'une ouverture commune, quoique du meme foier.

pag. 75. Dollond s'appelle Pierre, & non pas Jean. Celuici est un cadet, qui est en compagnie avec l'ainé Pierre: & qu'on croit n'avoir pas aucun grand genie, comme son frere.[12] pag. 154. Vous vous trompez disant que la pendule de Julien Le Roi[13] à l'observatoire de M^r. Messier, aie un joli compteur. Je connois parfaitem^t. bien cet observatoire & cette pendule. le compteur est une machine tout à fait diferente, qui est mise dans un coin & n'a pas la moindre Liaison avec cette pendule.[14]

pag. 167 le joli Equatorial de M^r. [le][15] Presid^t. de Saron,[16] fut fait sous ses yeux & selon les modelles qu'il coupa lui meme en carton, dans la

---

10 James Short, FRS (1710-1768).

11 The Earl of Bute's large collection of scientific instruments was auctioned after his death, and the sale catalogue included two reflecting telescopes of 2-ft focal length, one identified as being by James Short. However, these would not have had a large enough aperture to have been the ones described by Magellan. Cf. G. L'E. Turner, "The Auction Sales of the Earl of Bute's Instruments, 1793", *Annals of Science*, 23 (1967), 213-42.

12 Bernoulli corrected Dollond's name in the second volume of his *Recueil pour les astronomes*, p. 361, and also acknowledged, though without going into details, that some of his comments on Dollond and Ramsden had been mistaken.

13 Julien Le Roi (1686-1759) was a famous Parisian clockmaker and watchmaker who gave birth to a family of clockmakers.

14 Bernoulli also corrected this statement in the second volume of his *Recueil pour les astronomes*, p. 361, and then quoted Magellan's reports (which he attributed to "un de mes amis") about the telescopes of Saron and Noël. In his *Lettres astronomiques*, he had said he did not know who had made Saron's "joli instrument", and had discussed Noël's project of building a 23-foot Gregorian telescope in sceptical tones, and as if it had barely begun.

15 Paper damaged.

16 Jean-Baptiste-Gaspard Bochart de Saron (1730-1794), Président of the Parlement de Paris, an honorary member of the Académie Royale des Sciences and an enthusiastic astronomer.

72.04.07 conva[le]scence d'une maladie: & je crois que lui y mit aussi La main: car il est très ingenieux: & je sçais qu'il travaille de sa main le miroir du beau Tellescope de 30 pouces dont vous parlez à pag. 155, & dont il a fait dans la suite un present à Mr. Messier. pag. 173 le grand tellescope du P. Noel[17] etoit deja presque fini la derniere fois que Je fus à Paris (en 1769) car J'ai Lû un papier de lettre assez menue à une considerable distance, avec ce Telescope. Le P. Noel etoit alors après la preparation & ajustement des oculaires. mais je ne sai pas s'il y a deja mis la derniere main, ou s'il trouve son compte à le finir entierement. Voilà tout ce que J'ai remarqué en Lisant Vos Lettres Astronomiques. Vous prendrez mes remarques, comme de la part d'un a[mi] & non pas d'un censeur.

J'ai entendu que Mr. Bailly Acad. de Paris propose aux Astronomes une methode pour regler les observations & les reduire à un etalon, ce qu'il dit de faire, en ajoutant autant de cercles à l'embouchure des instruments, jusqu'à ce que la Lumiere du Satelite qu'on veut observer soit egale à celle d'une etoile d'une certaine magnitude.[18] &c. mais Je ne puis pas vous donner aucune idée fixe de ce projet. peut etre Vous en avez eu deja quelque notice.

Je viens de voir une petite lunete achrom. chez Ramsden, d'environ <u>17</u> pouces de foier, à triple objectif, & avec une ouverture d'environ 2½ pouces Angloises, qui fait un effet surprenant. elle me paroit beaucoup plus distincte & nette qu'un excelent tellescope de deux pieds de foyer. Lui garde cette lunete pour un equatorial fort ingenieux qu'il fait pour Mr. Mackenzie,[19] frere de Earl Bute, favori du Roy. Ce diable de Ramsden excelle en touts les genres: mais [en][20] revanche il est un si grand menteur, qu'on ne peut venir à bout de lui arracher l'ouvrage d'entre ses mains. il y a plus de 4 mois que Je suis après lui pour avoir [un] des ses beaux Barometres pour Mr. le Marquiz de Courtanvaux de Paris,[21] & ce ne fut que ce matin que J'ai pu le lui arracher. Je lui ai ordonné <u>6</u> autres, &

---

17 The Benedictine monk Dom Noël was director of the royal physics cabinet and instrument-making workshop at the Château de La Muette, at Passy. He had for many years been building a Gregorian telescope of unprecedented size, but it was still not in operation when Noël was made to retire in 1774.

18 Jean-Sylvain Bailly, "Mémoire sur les inégalités de la lumière des satellites de Jupiter, sur la mesure de leurs diamètres, & sur un moyen, aussi simple que commode, de rendre les observations comparables, en remédiant à la différence des vues & des lunettes", *Histoire de l'Académie Royale des Sciences*, 1771, Mém. pp. 580-667.

19 Hon. James Stuart Mackenzie (?1719-1800), MP and astronomical enthusiast.

20 Paper damaged.

21 François-César Le Tellier, marquis de Courtanvaux (1718-1781).

plusieurs autres instruments dès long tems: & que Je ne puis pas encore lui faire achever & finir. 72.04.07

à Dieu mon Cher. portez Vous bien, tachez de recouvrir de la Santé.

Croyez moi
pour la vie
Votre de Coeur
Magalhaens

N.B. Je ne sais pas encore quel nombre de votre Receuil se trouve vendu chez le libraire. Mais je crois que Vous pouvez bien envoyer autant comme du premier. C'est à dire 18 ou 20 exemplaires. mais il faudra les envoyer aussitôt qu'ils sortiront de la presse. Je serois charme que mon improvement du Quad^{t}. de Hadly y trouvat de la place dans ce volume.[22]

A Monsieur
Monsr. Bernoulli Astronome R^{l}. & de l'Acad. des Sciences de
Berlin
en Prusse.

22 Despite this plea, Bernoulli merely announced Magellan's improvement to the quadrant but did not publish an account of it, apparently because he regarded the subject as too controversial. Magellan therefore arranged to have an account published in Rozier's journal (*Observations sur la physique*, 4 [1774], 112-43), and separately a year later under the title *Description des octants et sextants anglois...précédée d'un mémoire sur une nouvelle construction de ces instruments...* (Paris/London, 1775), remarking in the preface to this latter work that he was doing so "parce que je n'étois pas satisfait de celle qui en avoit déja paru dans un Ouvrage périodique".

72.04.10 *To Louis-Henri Duchesne*[1]

Archives Nationales, Paris, MS T/160 (10-12)

Londres 10 Avril 1772

Je viens d'envoyer abord du vaisseau Betsey[2] Cp^{t}. Cuthbert Dunn qui va partir incessament pour Rouen, une Caisse, oû vous trouverez La Collection imprimée (que vient de paroitre) des Machines de La Societé des Arts,[3] pour Laquelle J'ai Souscrit 2 guinées, pour Monsr. Trudaine de Montigny. vous y trouverez aussi Les deux brochures de Bird[4] pour la division des instruments de Mathematique, que Je payai 2^{s} :6^{d}. Cette caisse est marquée MR ·Φ· N n.° 1.: & elle contient dedans un Barometre dans une seconde caisse bien empaqueté, qui est pour M^{r}. Le Marquiz de Courtenvaux,[5] qui m'ecrivit depuis quelque tems d'avoir deja parlé sur celà à M^{r}. Trudaine & comme cette sorte d'instruments est si delicate, & aisée à casser, J'ai choisi de l'envoyer par Rouen à fin qu'elle aille toujours par eau (& sans les cahos & violentes secousses des carriages). Je n'ai pas pu avoir ce soir le reçu du Capitain pour vous L'envoyer ici: mais je l'aurai pour La poste suivante, & àlors J'ecrirai à M^{r}. Le Marquiz de Courtanvaux & Le lui envoierai dans Sa lettre: & lui aura soin de vous le remettre & arranger cet affaire &c. Vous trouverez de plus dans La meme caisse une autre jeu de la Collection des memes machines de la Societé des Arts, qui est pour Le meme Marquiz: & un rouleau de papier pour M^{r}. Messier, outre un petit pour notre bon ami Le D^{r}. Sanchez, que Je vous prie de Lui faire remettre.

pour ce qui regarde Les Lettres de Franklin,[6] je me suis apparem^{t}. oublié de vous dire qu'on attendoit bientot une nouvelle edition plus ample: & que c'etoit à cette cause que Je n'avois pas acheté La vieille edition. Cependant si Monsr. Trudaine n'aime pas L'attendre La nouvelle edition, je Lui envoierai avec votre avis, celle qu'on trouve encore a present. Je ne m'oublierai pas de vous

---

1 Annotated by Duchesne: "Rep. le [7] May". Letter not found.

2 The merchant ship *Betsey* was the setting for a murder case in 1770 when Captain Ferguson was accused of murdering his 13-year-old cabin boy and was in due course convicted. Captain Cuthbert Dunn must have succeeded to Ferguson's command.

3 William Bailey, *The Advancement of Arts, Manufactures and Commerce, or Description of the Useful Machines and Models contained in the Repository of the Society for the Encouragement of Arts, Manufactures and Commerce* (London, 1772).

4 John Bird, *The Method of Dividing Astronomical Instruments* ... (London, 1767).

5 François César Le Tellier, Marquis d Courtanvaux (1718-1781).

6 Benjamin Franklin, *Experiments and Observations on Electricity made at Philadelphia in America* ... (4th ed., London, 1769). Another edition was issued in 1774.

envoier les deux petits perçoirs dont vous me parlez. Je n'ai pas encore pu savoir si on en faisoit deja plus petits. pour ce qui regarde le globe à poissons, je crois qu'en l'envoyaiant par eau jusqu'à Paris comme cette derniere caisse, il n'y aura pas du grand risque à courrir. Mon voyage devient chaque jour plus obstruit. J'ai tant de petites choses à arranger pour mes amis & pour mes petits affaires ici, que je ne me crois pas en etat de partir avant la fin de Mai. 72.04.10

à Dieu mon cher & tendre ami. Recevez La Reconnoissance correspondente de mon coeur pour vos amitiés & croiez moi pour La vie

Votre

Magalhaens

N.B. La carte à jouer, que vous avez mise dans votre lettre derniere,[7] fut assez pour qu'on en chargeat ici le postage double. ainsi le mieux est d'ecrire tout dans un papier entier, & non pas en morceaux, qui meme etant tres petits, on charge chacun comm'une Lettre entiere. vous trouverez ceci bien ridicule mais tel est L'usage de cet office ici.

A Monsieur
Monsr. Duchesne Secretaire de
M^r^. Trudaine de Montigny
à Paris

7 Letter not found.

To Johann III Bernoulli 72.04.21

Öffentliche Bibliothek der Universität Basel, MS L I a 709, f. 27-28

Londres 21 Avril 1772

Il faut vous ecrire encore, mon cher Monsieur cette lettre, aussitot pour L'affaire à l'egard de M[r]. Maskelyne. Je l'ai visité dernierem[t]. à Greenwich; & le pressai de me donner par ecrit ce qu'il souhaiteroit que vous sçusses, pour etre inseré dans Votre Recueil. mais Lui me dit = qu'il suffissoit de Vous faire observer que Votre meprise à L'egard du Nautical Almanak rouloit sur ce que vous aviez pensé que ces tables avoient eté calculées sur celles de l'Abbé de la Caille: au lieu qu'elles ne l'etoient que sur celles de Mayer, desquelles

72.04.21 les Tables en question ne different de beaucoup près. = dans Le jour suivant Je passai chez M^{r}. Wales,[1] qui s'avoit chargé de les revoir: & voici ce qu'il me donna après un examen très escrupuleux. en voici la copie.

		days	errors	
1767	January	the 11	1"-	
		16	1+	
	February	1	1-	
		22	1+	
	March	8	2-	
	June	16	1+	
	July	11	1-	
		27	1+	
	Nov^{ber}.	1	1+	
	Dec^{ber}.	11	1+	
		19	1+	
		24	9,8-	recomputed to tenths of seconds
1768	February	8	1-	
		15	1+	
		19	1+	
	March	3	1+	
	April	2	1-	
		12	1-	
		21	2+	
	July	28	2-	
1769	April	1	1-	
		22	1-	
	May	4	40	recomputed to tenths of seconds
		11	1-	
	Nov^{ber}.	1	1-	
1770	March	28	1+	
	May	31	1+	
	Sept^{ber}.	12	1-	
1771	March	4	1-	
		24	1+	
	Nov^{ber}.	21	2-	
	Dec^{ber}.	24	1-	

1 William Wales (1734-1798), astronomer on James Cook's second voyage of exploration.

There are no errors that exceed half a second in the first five years of the Nautical Almanac, except the above. I have called all those one second, which exceed half one, and are less than 1 ½''. And all that exceed this quantity, are called 2'': Although I believe none rise to quite so much. None of these therefore can properly be called errors, as it is very easy in the application of so many corrections to err 2''.; as every person who is conversant in those matters will allow. There can therefore only be said to have happened two errors and those very obvious ones in five years. 72.04.21

Wm. Wales.

N.B. the computations of ☉ & ☾'s Longitude, in the Nautical Almanac, are made from the Tables of Mr. Mayer, sent over to the Commissioners of Longitude by his widow, after Professor Mayer's death.[2]

Je garde auprès moi L'original du detail de Mr. Wales cidessus, signé de Sa propre main: & Je compte aussi fort sur son exactitude, que je n'ai pas La moindre doute de son assertion; c'est à dire qu'il n'y a dans Les Tables en question plus d'erreurs que les deux qu'il marque, c'est à dire celui de 9'' en moins dans Le 24 Decbre. 1767: & celui de 40''. dans Le 4 Mai 1769. Car Les autres ne meritent pas le nom d'erreurs.

Mr. Ramsden vient d'inventer une piece fort commode & avantageuse pour observer Les astres vers le Zenith, car au Lieu du miroir metalique qu'on employe ordinairement dans ce cas au dedans du tuyau, faisant un angle de 45°. pour jeter L'image de coté, il y applique une nouvelle espece d'oculaire qui fait le meme effet. Voici comm' il l'y a pris, après avoir trouvé l'embarras de pareils miroirs, qui absorbent beaucoup de Lumiere, qui se ternissent avec le tems, & qui ne peuvent pas etre aisemt. usés avec des oculaires d'un foyer fort petit. Vous savez que La lumiere qui tombe obliquement sur un Verre au delà d'un certain degré, est reflechie entierement Sans passer cette surface: comm' il arrive dans Les prismes rectangles, dans Les quels on observe qu'en les tournant, ils reflechissent La Lumiere sur La surface inferieure &c. C'est d'après ce principe que Ramsden a formé son nouveau oculaire, en prennant une lentille comme celle de la fig. 1. fort epaisse, comm'on le voit dans La fig. 2d., de laquelle

2 Bernoulli published a correction in the terms demanded by Maskelyne in the second volume of his *Recueil pour les astronomes*, pp. 350-2, including the full text of Wales's report.

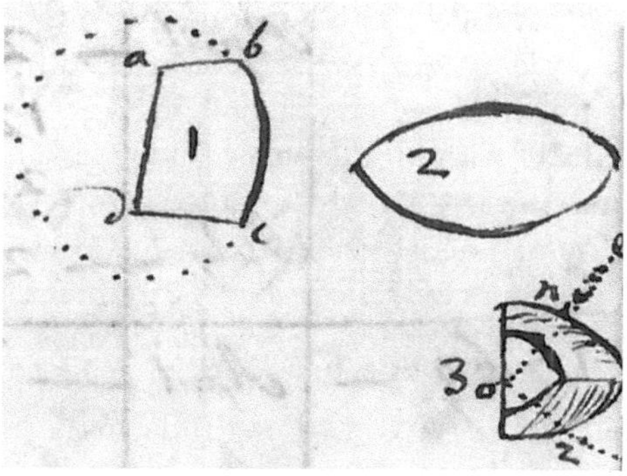

il n'a pris que La portion a.b.c.d, qui etant vue de coté, est representée dans La fig. 3me. ainsi si l'epaisseur de cette Lentille est assez considerable, on trouvera que Le raion e tombant sur la surface n se refrange, & tombe dans Le point o; mais y trouvant une surface plane sous un angle trop aigû, au lieu de passer en outre, il se reflechit vers le point z, & de là sort vers le point f. faisant un angle droit, à peu près avec le raion incident en. N.B. il ne sera pas difficile de calculer quelle doit etre La courbure des deux surfaces de cette nouvelle espece d'oculaires, pour que L'image soit reflechie à angles droits avec la Ligne de Colimation.

Ramsden appliqua fort heureusemt. ce nouveau oculaire à l'admirable Lunete achromatique qu'il vient de faire, qui a environ 15 pouces ou 16 de foier, à triple objectif, avec 2 ½ pouces d'ouverture, & qui est adaptée à un instrument equatorial pour Mr. Mackenzie, frere de l'Earl Bute, qui est un grand amateur de l'astronomie.[3] J'ai comparé hier soir cette lunete avec un excellent telescope de feu Mr. Short, qui a 18 pouces de foier avec 4 ¾ d'ouverture: & je l'ai trouvée beaucoup superieur au Tellescope: car je pouvois voir un des Satelites de Saturne constament avec La Lunete, qu'à peine je pouvois entrevoir à reprises, & très faint, avec Le tellescope.

Dans Le changement ou improvement que J'ai fait au Hadley's Octant, il y a encore une autre avantage très considerable, dont je ne vous ai pas parlé dans ma derniere.[4] C'est que par cette construction on peut rectifier avec toute precision possible, les erreurs de la division de l'instrument. Car il est certain que bien souvent les Quadrants faits par les meilleurs artistes mesurent des angles plus grandes, que les nombres qu'ils marquent. ce qui arrive de ce que le pointe du compas à verge, qui sert de centre tandis que l'autre pointe decrit la peripherie, n'est pas dans le meme plan mathematiquement parlant,

3 Hon. James Stuart Mackenzie (?1719-1800), MP and astronomical enthusiast.
4 See Magellan to J. Bernoulli, 7 April 1772.

dans Le quel l'arc est decrit; mais plutot un peu au dessous. ainsi lors qu'on prend avec ce raion, comme chorde de 60°., un arc de ce cercle ainsi decrit, on prend un arc reellement plus grand que <u>60°</u>. Or il est bien aisé de voir, que si l'on ajoute quelques petits miroirs horizontaux de plus, ou qu'on fait des trous pour Les y mettre (ce que revient au meme) après l'operation precedente selon ma methode: on peut par cette methode faire tourner le premier miroir <u>360°</u>.; & par consequent on verra s'il est precisem^t^. parallele pour Lors à celui de l'index sur zero. Car si par exemple on trouve qu'il faut mettre l'index plus au delà de zero vg. 40'., alors on sera convaincû que l'arc de l'octant ou <u>45°</u>. est plus grande de 5'.: & par là on obtiendra d'avoir (moienant une table faite soigneusem^t^. pour chaque instrument) des observations les plus exactes avec des instruments, meme très mal divisés. Car il est bien evident que si l'on a La patience de repeter cette operation pour chaque degre ou demi-degré, on sera à meme de connoitre touts Les erreurs de la division de chaque demidegré, & en former une table, qui servira de corriger & dresser les observations qu'on aura faites avec l'instrument. Je ne sai pas si vous saisissez mon idée par defaut d'expression. Je vais l'enlargir encore. Suposons qu'on veut examiner chaque <u>5°</u>. de cet octant, après qu'on a connu sa vraie dimention, ou autrement, quel est l'endroit du Limbe, oû le degre <u>45°</u>. tombe precisement. pour cet effet prenner 9 fois la distance de <u>5°</u>. sur le limbe de l'instrument: & voiez dans la derniere operation si le premier miroir horizontal reste parallele à celui de l'index mis à 45°. s'il y a quelques minutes de plus, en prennant la 9me. partie moins vous rajusterez àlors l'endroit de chaque 5°. avec precision: & ainsi de suite jusqu'à chaque demi-degré. Or Vous concevez aisement que cette methode est generale, & peut etre appliquée à tous Les Quadrants Astronomiques, & meme aux quadrants muraux; & par Là vous serez sensible de Son avantage. Je confesse qu'elle est un peu annuyeuse: mais je crois que tout observateur curieux ne doit pas se ref[user][5] à ce travail, quoiqu'insipide, lorsqu'il s'agit d'avoir toute La precision possible dans ses instruments. N.B. 1. pour mettre le 1er. miroir horizontal à angles droits avec celui de l'index lorsqu'il est à zero, vous n'avez selon ma methode qu'à mener une fois uniquement l'alidade ou index au degré <u>45°</u>. de l'arc: car si par exemple vous avez ajusté ce premier miroir parallele à celui de l'index à zero, & dans la suite vous menez l'index au 45°. pour lui mettre parallele le 2de. petit miroir; il est certain que si pour lors vous tournez le petit cercle de ces miroirs horizontaux jusqu'à ce que le 3me. miroir devienne aussi parallele à celui de l'index, & retournez à zero; pour lors, si vous y voiez que Le 2^d^. miroir lui est parallele, vous concluez

5 Conjectural reading; paper damaged.

72.04.21 qu'il a reculé 45°. & par consequence, en tournant ce petit cercle jusqu'à ce que le 3me. petit miroir soit parallele au meme de l'index (qui est à zero), pour lors le premier miroir à recule encore 45°. de plus & par consequent est à rectangle avec celui de l'index. N.B. 2^{d}. il est fort aisé d'eviter que l'index ne soit pas empeché par le petit cercle des miroirs horizontaux, pour etre mis au 45°. Car 1°. on peut le faire courbé dans l'endroit qui corresponde au petit cercle. 2^{d}. on peut faire en sorte que l'index passe au dessous de ce petit cercle. & 3°. encore mieux on peut arranger ce petit cercle à une distance convenable, pour ne pas empecher l'index de parcourir jusqu'au bout du 45°.; ce qu'on obtiendra encore mieux en le faisant tourner sur un centre assez proche du premier petit miroir horizontal, qui sert pour Les observations avec le dos tourné au Soleil. N.B. 3°. en fin on voit par ce que Je viens de dire, & qui me semble demontré, que cette meme nouvelle methode porte avec ellememe une nouvelle & plus sure maniere de diviser toute sorte d'instruments astronomiques avec toute La precision possible.

Je suis à près les ouvriers pour avoir quelques uns de ces instruments finis selon mon idée: mais il est très dificile de les faire changer de routine.

Je vous ai recomendé par une Lettre du 19 de ce mois, M^{r}. Wolfe, Chymiste,[6] qui va passer quelque tems chez vous. C'est un digne homme fort ingenieux dans sa branche, & de mes bons amis. il n'aura pas besoin que de votre avis pour trouver une famille honnette, dans Laquelle lui puisse etre, en paiant honnettement par sa depense, le tems qu'il compte de s'y entretenir.

Je vous souhaite une bonne sante, & tout Le bonheur possible & suis de Coeur

Mon cher Monsieur
Votre t. h. & t. o. serv^{r}.
Magellan

N.B. on me demanda à la Societé Royale (ce fut Le D^{r}. Matty[7] qui me le demanda) si vous n'y aviez fait present de votre Recueil & lettres[8] pour la meme Societé. Je lui repondit qu'apparement vous me l'ordonnerez dans La premiere occasion.

iterum vale[9]

A Monsieur
Monsr. Bernoulli, Astronome
Royal & de l'Acad^{e}. des Sciences de Prusse
à Berlin

---

6 Peter Woulfe, FRS (?1727-1803).

7 Matthew Maty, Secretary of the Royal Society.

8 i.e. Bernoulli's publications, *Recueil pour les astronomes* and *Lettres astronomiques*.

9 Farewell again.

To Domenico Vandelli 72.04.24

Arquivo Histórico do Museu Bocage, Universidade de Lisboa, CE/M-40

Londres 24 Avril 1772
at n°. 188 fleet-street

Monsieur

Il y a Long tems que J'attends de vos nouvelles, en repondant aux deux dernieres que Je vous ai ecrites L'année derniere.[1] mais dans L'apprehension que mes Lettres ne vous vient point parvenues, je prends cette voye de La poste, qui est La plus sure pour eviter de pareils acidents; quoique elle soit La plus dispendieuse. en voici Le precis de ces deux Lettres. dans La premiere Je vous annonçois La facheuse nouvelle de la mort du Respectable viellard M^r^. Miller:[2] & vous asserois dans Le meme tems que, malgré cet accident, vous ne manqueriez pas d'avoir toute La Correspondence que Vous souhaiterois de ce pais; car plusieurs de mes amis qui aiment La Botanique, seront fort aisés de vous rendre tout Le service que vous souhaitez, sans d'autre avantages que seulem^t^. l'exchange des semences de ce pais & de l'Amerique Angloise, pour celles du Portugal & Bresil & que vous Leur envoyeriez. & entre eux je vous nommoit quelques noms des principaux savants & curieux botanistes de ma connoissance comme le D^r^. Fothergill,[3] le D^r^. Pitcairn,[4] M^r^. Hudson,[5] Le Duc de Northumberland,[6] le hon^ble^. M^r^. Beauclerck,[7] le Baron de Poederlé de Bruxelles,[8] Le Duc d'Arenberg du meme endroit, & M^r^. Richard Jardinier du Roi à Paris,[9] le D^r^. Solander, qui va partir incessament avec M^r^. Banks pour les Terres Australes &^a^.[10] Mais dans le meme tems je vous priois de

1 Letters not found.

2 The well-known gardener Philip Miller, FRS (1691-1771), chief gardener at London's Chelsea Physick Garden, died on 18 December 1771.

3 Dr John Fothergill, FRS (1712-1780) had a celebrated garden at Upton, West Ham.

4 Dr William Pitcairn (1711-1791) established a botanic garden at Islington.

5 William Hudson, FRS (1734-1793), lecturer at the Chelsea Physic Garden, 1765-71.

6 Hugh Percy (1714-1786), first Duke of Northumberland, developed the gardens at Syon House by the Thames opposite Kew Gardens. The grounds contained more than 200 species of rare trees.

7 Hon. Topham Beauclerk, FRS (1739-1780), a wealthy patron of science and literature.

8 See Magellan to J. Banks, 28 December 1778, note 9.

9 Claude Richard (1705-1784) and his son Antoine Richard (1734/5-1807) were the officially designated King's Gardeners in Paris.

10 Joseph Banks and Daniel Carl Solander accompanied James Cook on his first voyage of circumnavigation of the globe, 1768-71. It was planned that they would also accompany Cook on his next voyage but, in the end, they did not.

72.04.24 vouloir bien prendre des mesures propres, pour que Vos envoys arrivent ici dans un tems propre pour faire usage des sciences &c[a].

Dans La Seconde Lettre Je vous marquois que M[r]. Gordon[11] avoit prets pour vous envoyer Les 3 plantes suivantes, qui sont rares & curieuses, savoir Kalmia Latifolia, Kalmia angustifolia & l'Azalia viscosa. elles sont mises separement dans des baskets & peuvent passer chez vous sans Le moindre risque de perir. Mais il faut savoir si vous les avez ou non: car dans ce dernier cas ce seroit vous envoier inutilement des choses que vous ne souhaitez pas. Ainsi aiez La bonté de me faire part par La poste pour agir en consequence. Je vous prie de mes Complim[ts]. au D[r]. Ciera[12] & à M[r]. de la Bella:[13] & donnerez moi des nouvelles du D[r]. Franzini,[14] si vous en avez, car il y a plusieurs mois que Je n'en ai pas aucune.

Au reste si je puis vous etre bon à quelque chose dans ce paiz, vous n'avez qu'à compter sur Les bonnes dispositions de mon Coeur & vous trouverez toujours que Je suis sincerem[t].

Monsieur
Votre très humble & ob. serv[r].
J.H. Magalhaens

P.S. dites-moi si l'histoire Naturele du Portugal & ses productions &c[a]. que vous annoncez dans votre dissertation,[15] est deja publié. pour ce que regarde les brochures que vous m'avez envoyées, elles ne sont pas encore toutes vendues, ni peut etre le seront sans l'autre ouvrage.

Monsieur
Monsr. Le D[r]. Vandelli Professeur Royal de Botanique
au Jardin du Roy
à Lisbone
Portugal.

11 The London nurseryman James Gordon (*ca.* 1708-1780).

12 See Magellan to Ribeiro Sanches, 20 April 1762, note 27.

13 Giovanni Antonio dalla Bella (1730-c.1823), physicist at Padua, was contracted by the Marquis de Pombal to teach experimental physics in the new programs of the Royal College of Noblemen in Lisbon and later at the University of Coimbra.

14 See Magellan to Ribeiro Sanches, 13 January 1778, note 14.

15 Presumably Domenico Vandelli, *Dissertatio de studio Historia Naturalis necessario in medicina, oeconomia, agricultura, artibus, et commercio*, published together with his *Dissertatio de arbore Draconis seu Dracoena* (Lisbon, 1768).

To Louis-Henri Duchesne[1] 72.05.04

Archives Nationales, Paris, MS T/160 (10-12)

Londres 4 May 1772

C'est, mon cher ami, pour vous annoncer que J'ai Reussi dans Les experiences pour faire une grande quantité d'eau douce, avec un apparatus fort aisé, de fort peu Volume, & peu de depense, à bord des vaisseaux.[2] J'avois annoncé cet objet à Monsr. Trudaine,[3] mais Je n'ai pas osé me flater de son effet jusqu'à ce que mes experiences m'en convainquirent. J'ai eté plusieurs fois à bord de differents vaisseaux, & particulierement des deux qui sont destinés pour cette expedition de Messrs. Banks & Solander,[4] pour y observer tout ce qui pouroit etre utile & particulierement pour ce qui regardoit cet objet: & toutes les observations que J'ai pu faire, jointes à mes tentatives & experiences, m'ont conduit à deceler cette methode precieuse: dont je vous prie de rendre part à M^r^. Trudaine pour L'assurer du bon usage que Je sais faire du donatif qu'il a eu La bonté de me faire,[5] en l'emploiant à des objets avantageux au service du public. Je pense aller toujours à Paris aussitôt que Je pourai me degager de mes petits engagements & arranger quelques petits affaires; particulierement faire executer en perfection l'avantageuse alteration que Je viens de faire au Quadrant Marin, comme je vous l'ai deja dit & dont J'ai donné une ample description à L'Acad. R^le^.[6]

J'ai observé à bord d'un de ces vaisseaux une machine très simple pour rendre bonne L'eau pourie dans les voyages trop Longs. tout ceci sont des objets utiles, que je crois ne manqueront pas de faire plaisir à M^r^. Trudaine dont Le Coeur est aussi porté pour Les avantages du Public. Lorsque nous nous verrons, nous prendrons les mesures plus propres pour practiquer ce

---

1 Annotated by Duchesne: "joindre".

2 An example of Magellan's desalination apparatus was displayed in Paris in 1773 and the Académie Royale des Sciences appointed a commission, of which Lavoisier was a leading member, to investigate the question. The commission's report, dated 1775 but unsigned, survives among Lavoisier's papers; it was published in *Observations sur la physique, sur l'histoire naturelle et sur les arts*, 18 (1781), 3-19, and republished in Lavoisier's *Oeuvres*, vol. 4, pp. 717-36.

3 Letter not found.

4 Joseph Banks and Daniel Charles Solander had travelled with James Cook on his first voyage around the world, 1768-1770, and were expecting to travel with him on his next expedition, due to depart as soon as preparations were complete. In the event, however, they did not go.

5 See Magellan to Duchesne, 17 January 1772.

6 See Magellan to Duchesne, 31 March 1772.

72.05.04 qu'il y peut avoir de plus interessant dans Les acquisitions & decouvertes ou observations que J'ai faites. en attendant assistez moi avec vos avis & directions pour mieux conduire ces affaires: ne me refusez point la satisfaction d'entendre votre opinion franchement en tout ce que je vous communique: & croiez que personne sait mieux reconnoitre dans le monde tout Le prix d'un ami si attaché comme vous & si obligeant: & que c'est pour La vie que je me flate de vous meriter Le titre d'etre

Votre entierem^{t}.

Magalhaens

N.B. ne vous obliez pas de me dire si M^{r}. Trudaine n'aime point attendre La nouvelle edition des Lettres de M^{r}. Franklin:[7] ce qui peut etre tardera encore plusieurs mois. Car si celà est, je vous envoyerai au plutôt Les deux exemplaires dont vous m'avez parlé.

A Monsieur
Monsr. Duchenne Secretaire
de M^{r}. Trudaine de Montigny
à Paris

7 See Magellan to Duchesne, 10 April 1772.

72.05.26 *To Louis-Henri Duchesne*[1]

Archives Nationales, Paris, MS T/160 (10-12)

Londres 26 May 1772

Je viens de Recevoir, mon cher ami, une Lettre de M^{r}. L'Abbé Bourriot,[2] dans Laquelle il me demande pour qui Sont Les deux douzaines de gants qu'il a trouvées dans La caisse derniere avec Les autres articles. Je suis on ne peut plus mortifié que vous n'aiez pas cherché cet article dans La dite

1 Annotated by Duchesne: "Rep. le 15 juin." Letter not found.
2 Letter not found.

caisse, comme je vous l'avisai dans ma Lettre Lors de son depart;[3] ce fut dans cette idée que je ne le marquai point dans La Liste des autres choses pour differents amis, dont Le dit commun ami Bourriot s'en chargeat. ainsi aiez La bonté de le lui faire savoir, car il est bien afligeante pour moi, qu'outre le delai que L'ouvrier a mis ici, à cause de la mort de la Mere du Roy (de son deuil)[4] il arrive encore de plus celui Là, sans que J'y aie pensé. 72.05.26

La petite Lettre ci jointe pour M^r^. Le Chev^er^. de Bory[5] contient le Reçu du Capitain pour La caisse avec La tole Cylindrée que M^r^. Vaucanson[6] m'a ordonnée ici: & que Selon l'avis & le consentement de M^r^. Trudaine, je l'ai envoyée dans une grande Caisse par Rouen.

Il ne m'est pas encore possible de fixer mon voyage à Paris; du moins je ne pense encore de le manquer cet eté: Continuez moi vos amitiés & soiez sur que personne en sçait connoitre mieux tout Le prix, ni plus sincerement correspondre avec un attachement que ne cessera qu'avec La vie.

de votre

Magalhaens

P.S. M^r^. Banks ni M^r^. Solander, ni le D^r^. Lind, n'iront pas à faire le voyage au tour du mond qu'il etoient determinés de faire.[7] le vaisseau est trop haut de bord. on va lui oter le pont qu'on lui avoit ajouté pour l'accomodation de ces Messieurs: on ne veut pas Leur donner un autre vaisseau plus commode: ainsi ils ne veuillent pas aller sans leur propre accomodation &c. Voila comme cet objet est evanoui. Cependant les deux vaisseaux iront toujours: mais sans Banks ni Solander, ni le D^r^. Lind, &c.

3 See Magellan to Duchesne, 13 March 1772.

4 Princess Augusta of Saxe-Gotha, mother of King George III, died on 8 February 1772.

5 Letter not found.

6 The engineer and member of the Académie Royale des Sciences Jacques de Vaucanson (1709-1782).

7 Joseph Banks and Daniel Carl Solander had been planning to travel as naturalists with James Cook on his second round-the-world voyage of exploration, and had invited the young Scottish doctor and amateur astronomer.James Lind (1736-1812) to go with them. In the event, however, they did not go. As Magellan indicates, the additional accommodation that Banks had insisted on having constructed for his party had rendered the ship top-heavy and unsafe, and Cook had had it demolished.

72.06.01 *To Louis-Henri Duchesne*[1]
Archives Nationales, Paris, MS T/160 (10-12)

Londres 1 Juin 1772

Ayez la bonté mon cher ami, de lire les deux dernieres pages de cette lettre pour Mr. de Montigny:[2] après celà vous la fermerez &la lui donnerez de ma part. il s'agit d'arranger les droits sur les eaux de vie: Je n'entends sur ces matieres & puis dire de les Connoitre à fond. si Mr. Trudaine le souhaite, je n'aurai la moindre dificulté d'aller à Paris, oû je monterai la methode la plus aisée & certaine de construire des instruments pour cet objet beaucoup meilleurs que ceux qu'on use en Angleterre dans les douanes: & dont j'envoie un à Mr. de Montigny qu'il me demanda: apparament c'est pour la compte de Mr. Trudaine. la caisse arrivera dans peu de jours chez vous. Je n'ai pas a present du tems pour vous donner le detail: car je me trouve très fatigué & il est trop tard. Je vous ecrirai la poste suivante

à Dieu mon cher

votre

Magalhaens

A Monsieur
Monsr. Duchenne
Secretaire de Mr. Trudaine de Montigny
à Paris

1 Annotated by Duchesne: "Rep. le 15 juin." Letter not found.
2 Letter not found.

To Louis-Henri Duchesne[1] 72.06.09

Archives Nationales, Paris, MS T/160 (10-12)

Londres 9 Juin 1772

Je suis tres sensible, mon cher ami, à La perte que vous venez de faire du pauvre M^r. de Courmiere.[2] qu'il seroit doux de pouvoir aller ensemble, au lieu de nous arracher une partie du coeur par la liaison de ceux qui le partagent en sincere amitié! mais il faut se soumettre tranquilement à La Loi, dont il n'est possible de nous soustraire, & se premunir du confort de la saine Philosophie contre ces coups irreparables du Sort, qui malgré nos meilleurs Desirs & souhaits, ne se flechit jamais contre nos cris & continue toujours à exercer son pouvoir sans aucun control sur nous autres pauvres jouets de la Nature.

Je ne sais pas si je vous ai deja dit, que dans La caisse de la Tole cylindrée, envoyée par Rouen pour M^r. de Vaucanson,[3] il y a un auger ou perçoir pour vous, selon que vous m'avez dit dans une de vos lettres que vous desiriez de l'avoir.[4] ainsi vous aurez soin de vous le faire donner lorsque La caisse arrivera à Paris. Ce perçoir est le plus mince qu'on fait de cette espece.

Je suis bien aise que vous ayez Reçu le pacquet ou sac avec l'orge de Siberie:[5] & J'espere que vous l'aurez fait deja semer tout de suite selon les instructions que Je vous communiquai L'année derniere. autrement Je crains qu'il seroit trop tard à present: & il vaudroit mieux le Laisser pour L'année suivante car son propre tems est de le semer jus[qu']au bout d'avril ou milieu de May. L'année prochaine je puis vous envoyer une plus grande quantité. reelement c'est un grain très precieux & d'un grand avantage pour Le public: on ne sauroit trop encourager La culture. on en peut faire du pain presqu'aussi bon qu'avec du bled. il donne plus de biere que l'orge ordinaire; & à tous egards est preferable à lui. Car il produit beaucoup, & ne manque presque jamais de donner une bonne racolte.

---

1 Annotated by Duchesne: " Rep. le 15 juin." Letter not found. At the bottom left of the first page, Magellan has written "M^r. Duchenne."

2 i.e. Charles Penot de Tournière, who died on 21 May 1772. See Magellan to Duchesne, 1 October 1771.

3 See Magellan to Duchesne, 26 May 1772.

4 Letter not found, but see Magellan to Duchesne, 13 March 1772.

5 See Magellan to Duchesne, 31 March 1772.

assurement Je n'irai pas à Paris que vers 7bre. ou 8bre. prochain: car je ne voudrois m'y trouver sans vous voir, & causer avec vous sur Les objets de mes communications utiles. ainsi vous aurez La bonté de m'avertir precisement Le tems de votre Retour, pour prendre mes mesures.

en attendant marquez moi si je puis vous envoier sous votre adresse (Je veux dire de M^{r}. Trudaine) quelques choses que je dois faire expedier: ou s'il vaudra mieux attendre votre Retour. Car par example Le globe à Poissons pour M^{dme}. Trudaine ou pour qui vous jugerez à propos, est presque prêt pour etre empaqueté: & pour eviter quelqu'acident je pense le faire partir par Rouen àfin de aller par eau jusqu'à Paris. marquez moi là dessus votre opinion.

La lorgnete dont vous me parlez pour L'espectacle, ne devroit pas etre achromatique: car les simples font meilleur effet dans la nuit aux espectacles: mais pour les autres usages les achromatiques sont Les meilleures. vous me direz encore un mot sur ce que vous choisissez dans ce cas.

Je vous prie de faire parvenir cette Lettre au chev^{er}. de Bory,[6] aiant soin de la cacheter. car je l'envoye ouverte pour que vous puissiez La lire, en cas que vous aiez assez de patience & de tems: il y a je crois quelques choses bien utiles: mais le language est si degoutante, que je Rougis de ma plume.

à Dieu mon Cher & bon ami: portez vous bien. allez vous amuser un peu avec vos amis & parents dans votre pais: & si vous y voiez M^{r}. Mallet à Geneve Astronome,[7] donnez Lui de mes complim^{ts}. & que je ne manquerai pas de lui ecrire aussitôt que J'aurai les observations qu'il souhaite pour constater La Longitude de Son observatoire.

à Dieu encore mon Cher ami, aimez moi, & sachiez que personne surpasse en amitie pour vous

Le Coeur de votre
Magalhaens

6 Letter not found.
7 See J. A. Mallet to Magellan, 30 March 1772.

To Trudaine de Montigny 72.07.05

Archives de l'Académie des Sciences, Paris, Fond Lavoisier, 1642[1]

Monsieur

Vous savez que l'Air-fixe a eté depuis quelques années l'objet des Recherches de nos Philosophes, & que leurs resultats sont devenus interessants. Le Chev^{r}. Pringle aiant observé avant aucun autre qu'on sache, que le progrès de la putrefaction etoit arreté par la fermentation;[2] le D^{r}. Macbride de Dublin decouvrit, que cet effet venoit de l'air-fixe qui se degage dans cette operation:[3] & ce fut d'après ce principe, qu'il recommanda le mout de bière (je crois que c'est le propre signification du mot Anglois wort) comm'un antiseptique, pour preserver & curer l'escorbut des equipages en long voyages sur mer; car ce fluide est capable de produire une grande quantité d'Air-fixe dans l'estomac par la fermentation, comm'il arrive en mangeant des vegetaux, qui, comme tout le monde le sçait, sont d'un grand secours dans cette maladie. En effet les expériences, qu'on a faites à bord des vaisseaux Anglois avec le Wort confirmerent abondament cette verité.

Dans la suite le D^{or}. Black trouva, que les pierres à chaux, comme le craion blanc & toutes les substances calcaires, contiennent une grande quantité de cet Air-fixe, dont la privation les rende caustiques, au lieu que la presence

---

1 Corrected version of the text published in *Oeuvres de Lavoisier: Correspondance*, ed. René Fric, fascicule II, 1770-1775 (Paris, 1957), pp. 356-66. The document is in Magellan's hand, unaddressed, with the heading (also in Magellan's hand):

> Copie d'une Lettre de J. H. de M. Resident à Londres, pour un de ses amis de Paris, datée le 5 Juillet 1772: contenant un detail de la methode pour faire aisement des Eaux artificieles, comme celles de Pyrmont, Spa &c par le moien de l'Air-fixe; dont Les qualités antiseptiques les rendent egalement propres contre l'Escorbut en long voyages sur mer, & pour d'autres maladies: avec un precis de la brochure du D^{r}. Priestley sur ce Sujet.

It is evidently the paper to which Magellan refers in his letter to Trudaine de Montigny of 7 July 1772, that Trudaine forwarded to Lavoisier together with a copy of Priestley's brochure on 14 July 1775, asking him to repeat the experiments for himself; cf. *ibid.*, p. 368. It is also evidently the "lettre de M. Magellens au sujet de L'air fixe" that Lavoisier read to the Académie Royale des Sciences on 18 July 1772 (*Procès-verbaux*, vol. 91, 1772, fol. 254^{v}; cited by Guerlac, *Lavoisier – the Crucial Year* [Ithaca, NY, 1961], p. 59).

2 Sir John Pringle, *Observations on the Diseases of the* Army, 4th ed. (London, 1764), Appendix, paper IV, pp. xl-li.

3 David Macbride, *Experimental Essays on the following Subjects: I. On the Fermentation of Alimentary Mixtures. II. On the Nature and Properties of Fixed Air. III. On the Respective Powers, and Manner of Acting, of the different Kinds of Antiseptics. IV. On the Scurvy; with a Proposal for trying new Methods to prevent or cure the same at Sea. V. On the Dissolvent Power of Quick-Lime* (London, 1764), especially pp. 147ff.

72.07.05 de cet air les adoucit & les rende innocentes.[4] Le D^{r}. Brownrigg[5] decouvrit, que les eaux minérales de Pyrmont & autres de ce genre, qui ont un gout aciduleux, contiennent une grande quantité de cet air fixe; & que c'est de lui, que leurs vertus principalement dependent.[6] Monsr. Lane de Londres trouva que le même Air-fixe etoit le vrai dissolvant de la substance ferrugineuse, qui compose les eaux minerales connus sous le nom de Calybeates, dont on fait usage avec sucès en plusieurs infirmités.[7] Enfin le D^{r}. Priestley communiqua dernierement à la Societé R^{le}. de Londres un gran nombre d'experiences fort interessantes sur l'air-fixe qui se degage des fermentations; l'air-mephitique surchargé des vapeurs & exhalations des matieres corruptes; l'air brulé par le feu; et l'air inflamable, qui resulte des solutions, dans lesquelles il y a quelque dissipation considerable du Phlogistique.[8] Dans le cours de ces experiences il trouva le moien d'impregner l'eau commune avec l'air-fixe, d'une maniere fort aisée: & en suivant les principes du D^{r}. Macbride & du D^{r}. Brownrigg, il a appliqué cette eau, ainsi preparée, aux memes usages du Wort ci dessus & des eaux minerales. Ce fut dans ces vues qu'il proposa à l'Amirauté d'Angleterre la methode qu'il avoit trouvée, laquelle fut examinée & approuvée par le College des Medecins de Londres, & publiée à Leeds par l'Auteur, avec une lettre adressée au Comte de Sandwich, premier Lord de l'Amirauté, en date du 4 de Juin dernier.[9] Mais ce ne fut que 20 jours àprès, que J'ai pû voir une de ces brochures, envoiée par l'Auteur à M^{r}. Forster, mon particulier ami dans la veille de son depart pour le voyage aux Terres Australes & Isles de la Mer Pacifique, oû il est chargé d'aller par le Roy, pour y faire des recherches de Histoire Naturelle & de Physique, entre lesquelles

4 Joseph Black, *Dissertatio medica inauguralis, de humore acido a cibis orto, et magnesia alba* (Edinburgh, 1754); the results to which Magellan refers were presented in an appendix to the thesis. They were subsequently published in English as "Experiments upon Magnesia Alba, Quicklime, and Some Other Alcaline Substances", *Essays and Observations, Physical and Literary, read before a Society in Edinburgh* (Edinburgh, 1756), II, 157-225.

5 William Brownrigg, MD, FRS (1711-1800).

6 See Brownrigg, "An Experimental Enquiry into the Mineral Elastic Spirit, or Air, contained in Spa Water… ", *Philosophical Transactions*, 55 (1765), 218-43; also published separately, London, 1666.

7 Timothy Lane, "A Letter … on the Solubility of Iron in Simple Water, by the Intervention of Fixed Air", *Philosophical Transactions*, 59 (1769), 216-27.

8 Joseph Priestley, "Observations on Different Kinds of Air", *Philosophical Transactions*, 62 (1772), 147-264. Priestley's paper was read to the Royal Society during the course of four successive meetings in March 1772.

9 Despite what Magellan says, Priestley's 22-page brochure was published not in Leeds but in London; Joseph Priestley, *Directions for Impregnating Water with Fixed Air…* (London, 1772).

on lui a recommandé de faire des experiences sur cette eau-artificiele.[10] 72.07.05
En parcourant cette brochure avec attention, J'ai crû qu'on pouvoit rendre encore cette methode un peu plus aisée & commode dans la practique, tant à bord des vaisseaux, que sur terre. en effet je l'ai repetée plusieurs fois à differents amis, qui la trouvent simple & aussi practiquable que l'autre, quoiqu'elle n'en difere aucunement pour l'essentiel.[11] Je vais vous en faire ici le detail, ensemble avec le precis du contenû dans la brochure du Dr. Priestley, qui vient de me faire gratieusement le present d'une, par son libraire d'ici.

Prennez du craion blanc, que vous casserez en petits morceaux, sans le cuire au feu ni aucune autre preparation; avec les quels vous remplirez une bouteille ordinaire de verre, jusqu'à une pouce au dessous du Col: & jettez y de l'eau commune, autant qu'il faut pour couvrir à peine le craion. Prennez un siphon de verre; ou, si vous n'en avez point, prennez deux petits batons, de la grosseur du petit doigt, l'un de 3 pouces & l'autre de 6 pouces de longueur, que vous percerez tout au long avec un foret, ou avec une verge de fer rougi au feu; & coupant en biais ou chanfrant[12] deux de leurs bouts, liez-les ensemble de façon qu'ils forment un angle aigû, en les entourant, dans la jointe commune, avec un morceau ou petite bandelete de vessie, & du fil ciré, ensorte que l'air ne puisse point echaper par cet endroit; & celà vous rendra un si bon service comme le meilleur siphon de verre. Attachez, à la branche plus longue, une vessie aussi grande que vous pouvez trouver; & une autre plus petite, que vous couperez à la distance de 2 pouces, ou seulemt. un morceau circulaire de vessie, avec un trou dans le milieu, dans la branche la plus courte du siphon, en l'introduisant par le trou, & l'y liant avec du fil ciré. Jettez un peu d'acide vitriolique, sur le craion dans la bouteille: et ayant chassé tout l'air de la vessie, mettez l'autre bout du siphon dans la bouche de la bouteille, en arrangeant autour de son col, le morceau de vessie, qui y est en forme de manchete, & que vous y lierez, aussi avec du fil ciré. Aussitôt la vessie commencera à se gonfler, avec l'air fixe, qui se degage du craion, par l'action de l'acide. En attendant, vous remplirez, d'eau commune de fontaine ou rivierre, mais qui soit pure, une petite cuve, ou un seau, dans lequel vous plongerez une autre bouteille ordinaire, que vous renverserez toute pleine d'eau dans le seau, avec la bouche en bas. Lorsque la vessie sera presque pleine d'air-fixe, prennez-la par le col, que vous serrerez entre les doigts pour que l'air ne s'echape point: & en detachant l'autre bout qui etoit

10 Johann Reinhold Forster travelled as naturalist with James Cook on his second voyage around the world, 1772-5.

11 Magellan published his improved apparatus in *Description of a Glass Apparatus for Making Mineral Waters...* (London, 1777).

12 chanfrein?

72.07.05 mis dans la bouteille de craion, vous y laisserez sa manchette de vessie, et vous metterez ce meme bout du siphon, par dessous l'eau du seau, dans la bouche de l'autre bouteille, qui est renversée toute pleine d'eau aiant soin de ne pas hausser la bouche de cette bouteille au dessus de l'eau du seau; car alors l'air exterieur y entreroit, & elle se vuideroit. Aussitôt que le bout du siphon est dans l'embouchure de cette bouteille, pressez la vessie, qui est pleine d'air-fixe, entre vos mains, & l'air sortira par le bout du siphon, qui est dans la bouteille renversée: il y montera en haut jusqu'à son fond; & en fera descendre l'eau à mesure qu'il y entre. Lorsque la bouteille sera la moitié vuide, otez en le bout du siphon. Liez le col de la vessie pour y garder le reste d'air-fixe: & prennant le fond de la bouteille renversée avec la main, secouez vivement l'eau qui y est, tenant toujours l'embouchure de la bouteille sumergée dans l'eau du seau: & au bout de 3 ou 4 minutes, vous trouverez que l'eau, au dedans de la bouteille, a monté jusqu'en haut, & a absorbé presque tout l'air fixe, que vous y aviez jetté: car moienant le mouvement & secousses de l'eau, ses differentes particules se sont touchées avec celles de l'air fixe, & se sont atirées mutuellement. Remetez droite la bouteille dans le seau pour que l'air immiscible, qui y est puisse sortir (car il y a toujours dans la vessie quelqu'air commun, melé avec l'air fixe): aussitot remplissez cet espace avec de l'eau: renversez encore la bouteille une autre fois dans le seau: jettez y autant de l'air-fixe, que vous avez gardé dans la vessie: secouez-la bien dans la meme position & avec les memes precautions qu'auparavant: et vous verrez que cette eau absorbe encore une grande quantité d'air fixe: de façon que le total absorbé par l'eau, sera autant & meme plus que son volume. Celà fait, voiant que l'eau n'absorbe plus d'air, ce qui arrivera au bout de 4 ou 5 minutes de secouement, vous oterez la bouteille, en y mettant le doigt sur la bouche par dessous l'eau du seau, pour la remettre droite sans verser ce qu'elle a dedans: & vous en ferez usage tout de suite. Car si vous la laissez exposée à l'air commun, l'air fixe qui y est melé, s'envolera en peu de tems. Mais si vous voudrez garder cette eau pour quelque tems, ou pour l'envoyer aileurs, pour lors il faut preparer par la meme methode une autre, ou plusieurs autres bouteilles, en tenant la premiere, ou les premieres avec leurs bouches tournées en bas, dans l'eau (c'est à dire couvertes ou sumergées dans l'eau) tandis que vous preparez & impregnez d'air-fixe les suivantes, par la meme methode ci dessus. Quand l'operation est finie, vous prennez une bouteille à la fois, la redressant avec le doigt dans la bouche, comm'on l'a deja dit: vous la remplissez tout à fait avec une autre de la meme eau, & vous la bouchez avec un bon bouchon de liege, sans y laisser dedans aucun espace vuide avec de l'air: & vous les gardez tournées avec la bouche en bas: meme il est à propos de couvrir de resine ou avec du poix leurs bouchons &c. Cette eau acquiert par l'impregnation de cet air-fixe, un gout acciduleux, & une espece d'esprit petillant, comme les eaux minerales de Pyrmont. Mais si vous voudrez lui donner les qualités des eaux Calybées, vous n'avez qu'a y meler en chaque bouteille, une petite pincée de limaille de fer ou d'acier, avant de la boucher: & pour lors vous trouverez que cette eau se charge très fort de la sustance metalique en peu de tems (en 8 ou 12 heures & meme bien moins de celà) de façon qu'en la melant avec de l'infusion de noix de gales, ou meme du thé ordinaire, elle devient noire comme de l'encre; ce qui est la preuve demonstrative de la solution du fer. On aura donc fort aisement & à très peu de frais, de ces eaux artificielles, meme à bord des vaisseaux en voyages longs de mer, dont on pourra tirer les mêmes avantages, qu'on peut attendre des vraies eaux acidules & Calybées: & meme preserver & curer l'escorbut, selon les principes etablis ci-dessus. 72.07.05

N.B. 1. il est avantageux de secouer la bouteille, qui contient le craion, pour augmenter l'effervescence avec l'acide, & aider à degager l'air fixe. 2. après chaqu'operation, il est nécessaire, pour exciter une autre effervescence, renouveller l'eau qui est avec le craion, en jettant celle qui a deja servi, en y mettant une autre nouvelle eau, avec une autre pareille quantité d'acide vitriolique: & meme il faudra après un certain tems, changer & renouveller le craion. 3. Les eaux qui sont impregnées avec de l'acier, en vertu de leur union avec l'air fixe, & non pas par un acide etranger, sont beaucoup plus salutaires, & agreables à l'estomac, que les autres. 4. par la methode ci dessus on peut aisement impregner avec l'air-fixe tout autre liqueur, comme le vin, la biere, &c. Ce sera avec le tems & à force d'en repeter les experiences, qu'on viendra à connoitre tous les avantages qu'on en peut tirer: du moins il est presque demontré, que toutes les maladies provenantes d'un principe de putrefaction, comme l'escorbut, &c. reçoivent un grand secours par l'application de l'air fixe. 5. Le Chevalier Pringle dit, qu'en ajoutant 8 ou 10 goutes de tinctura Martis cum spiritu salis,[13] à chaque pinte de cette eau artificielle, elle devient fort semblable à celle de Pyrmont. Mais il paroit certain, que les vertus de ces eaux ne dependent principalement que de l'air-fixe, qui y est melé avec elles. Cependant on les fera aussi Calybées lorsqu'on voudra, selon ce qu'on a deja dit ci dessus. 6. M^r^. Hay cirurgien de reputation à Leeds, appliqua avec sucès, l'air-fixe en forme de clystere, à un malade d'une grande fievre putride dont les evacuations etoient deja noires & très fetides.[14] Si un seul example n'est pas suffisant pour decider,

13 tincture of iron in spirit of salt.

14 Joseph Priestley, *Experiments and Observations on Different Kinds of Air*, vol. 1 (London, 1774), appendix 2, pp. 292-9, "A Letter from Mr. Hey to Dr. Priestley, concerning the Effects of Fixed Air applied by Way of Clyster". William Hey (1736-1819)

72.07.05 on peut du moins conclure que cette application de l'air-fixe n'est point du tout nuisible. 7. on a essaié de faire respirer cet air fixe à des personnes, qui ont des ulceres dans les poumons, en les mettant avec la tête panchée sur des vaisseaux avec des liqueurs en fermentation. Le D^r^. Percival a temoigné le bon sucès de ce remede en trois cas. un des malades guerit parfaitement: & les autres deux eurent du soulagement considerable.[15] on faisoit couvrir, avec un grand entonnoir renversé, un vaisseau ou terrine, oû il y avoit du craion en effervescence avec de l'huile de vitriole, delaié avec de l'eau: & le malade respiroit les fumes qui sortoient par le trou de l'entonnoir. 8. Le meme D^r^. Percival dit aussi, que la sanie d'un cancer a eté fort adoucie par l'application de l'air-fixe, que les douleurs en etoient mitigués, & la digestion produite de façon qu'on attendoit une cure parfaite. 9. on a appliqué l'air fixe, qui sortoit d'une mixture de jus de limons ou citrons aigres avec le sel d'absinthium (wormwood en anglois) dans les ulceres de la gorge, avec un avantage decidé. 10. Le D^r^. Macbride observa qu'en exposant à l'air fixe un morceau de viande demi-pourrie, elle devenoit ferme, & la pourriture disparoissoit. ne pourroit-on pas attendre quelque bon sucès, en exposant à cette espece d'air, les plaies, ou membres attaqués de la pourriture, meme lorsque tout le sisteme y tendroit? Car il n'est gueres dificile d'exposer à cet air fixe un membre, ou meme tout le corps entier, en conservant la tête dehors &c. Peut etre le principe vital, ou vis vitae, aidé par cette assistance externe de l'air fixe, sera pour lors en etat de surmonter les attaques de la disolution putride. On dit de quelques Indiens, qu'ils enterrent jusqu'au cou, ceux qui sont attaqués par des maladies putrides, dans une fosse faite en terre fraiche, laquelle, nous savons d'aileurs, a la qualité d'oter la puanteur de la viande qui commence à etre gatée. Ne-seroit ce pas l'air-fixe, absorbé par les pores de la peau, la vraie cause du sucès de cette operation Indienne. On sçait par tradition, que nos vieux gents conseilloient ceux qui etoient attaqués dans la poitrine, ou de Phtisie, d'aller dans les champs qu'on labouroit, et suivre la charrue, & d'aller demeurer dans les endroits oû il y avoit des fourneaux, qui bruloient de la chaux. Bien souvent on s'apperçoit de l'avantage de pareilles practiques dont on ne sçait point donner aucune raison: & ce n'est pas qu'après une longue suite du tems, qu'on vient à en decouvrir la cause.

---Fin---

---

was surgeon and from 1773 senior surgeon at the Leeds Infirmary; in 1768 he founded the Leeds Medical Society. He was a close friend of Priestley's.

15 Thomas Percival, *Essays Medical and Experimental* (London, 1772), II, 71-80.. A somewhat elaborated account of Percival's experiments was published in Priestley's *Experiments and Observations on Different Kinds of Air*, vol. 1 (London, 1774), as appendix 3, pp. 300-14.

To Trudaine de Montigny

Archives de l'Académie des Sciences, Paris, Fond Lavoisier, 1642[1]

Monsieur

Le papier ci-joint renferme La methode pour faire des eaux acidules & Calybées, que J'ai abregé en françois & generalisée à la portée de tout le monde, d'après celle du Dr Priestley,[2] dont je eus l'honneur de vous envoyer des copies imprimées par la poste derniere. Je pense avoir detaillée assez nettement ma particuliere methode, un peu diferente de celle de l'Inventeur, pour ne pas avoir besoin d'aucune planche pour que tout le monde la comprende. Si vous pensez si avantageusement de cette decouverte, comme plusieurs le pensent icy (& je suis du nombre aussi) Je souhaiterois qu'on La fisse imprimer tout de suite chez Vous, car ce ne fera pas au delà d'une feuille d'impression, & qu'on la repandit gratis par toute la France, pour que tous soient à meme de s'en utiliser.[3] mais il faudra pour lors en faire corriger le francois, sans en alterer le sens: car je sais bien que mon mauvais language ne peut pas manquer d'etre degoutant à la delicatesse Francoise. Je vous demande mil escuses pour vous troubler en deux postes consecutives avec Le même objêt. Mais Lorsque J'ai quelque chose que je crois pouvoir etre utile au public, je ne puis pas me tranquiliser sans faire tous mes eforts Là dessus. C'est par cette meme raison que J'ai envoiée une autre copie de ce papier à un autre de mes amis de Paris,[4] & que je ne manquerois de multiplier, s'il m'etoit possible dans si peu de tems de le repandre par toute L'Europe.

---

1 Corrected version of the text published in *Oeuvres de Lavoisier: Correspondance*, ed. René Fric, fascicule II, 1770-1775 (Paris, 1957), pp. 366-8. Since the letter is now among Lavoisier's papers, Trudaine evidently forwarded it to him, as he did Magellan's letter of 5 July, the paper to which Magellan here refers.

2 Joseph Priestley, *Directions for Impregnating Water with Fixed Air...* (London, 1772).

3 Trudaine arranged instead for a French translation of Priestley's brochure to be published in the August 1772 issue of Rozier's newly founded monthly journal of scientific news; see *Introduction aux observations sur la physique, sur l'histoire naturelle et sur les arts* (1771-2), vol. 2, pp. 323-31. A footnote briefly summarized Magellan's improvement of Priestley's procedure.

4 Not identified.

72.07.07 J'ai l'honneur d'etre avec tous Les sentimens de Respect & de gratitude pour les bontés dont vous vous daignez m'honnorer

Monsieur
Votre très humble & tres obeiss^t. serviteur
Magalhaens

Londres 7 Juillet 1772

A Monsieur
Monsr. Trudaine de Montigny
Conseiller d'Etat & Intendant des Finances
à Paris

72.07.07a *To Louis-Henri Duchesne*[1]

Archives Nationales, Paris, MS T/160 (10-12)

Londres 7 Juillet 1772

Je vous envoye ici, mon cher & bon ami, une nouveauté qui me semble fort utile au public. Je crois que vous ne sera pas faché d'en avoir L'idée pour La communiquer à vos amis dans votre pais, oû je crois que cette Lettre vous trouvera deja. dans La poste derniere j'envoiai deux brochures angloises avec ce detail à M^r. Trudaine & une pour L'Academie Royale.[2] dans Le samedi je me suis mis à ecrire quelques copies comme celle ci pour envoyer par La poste à differentes amis, mais j'ai encore rendu hier cette operation un peu plus simple que celle ci dans cette Lettre. J'en envois par cette meme poste une copie manuscrite de ma derniere methode à Monsr. Trudaine[3], & je souhaiterois qu'elle fut imprimée, après que le françois en seroit corrigé, & qu'on La Repandit tout partout. mais peut etre on ne fera Rien à moins que par hazard vous vous trouviez encore à Paris, ce que Je

1 Annotated by Duchesne: "Rep. le 24 aoust". Letter not found.
2 Joseph Priestley, *Directions for Impregnating Water with Fixed Air...* (London, 1772).
3 Magellan to Trudaine de Montigny, 7 July 1772.

doute beaucoup selon ce que vous me disiez dans vos Lettres dernieres.[4] 72.07.07a
Quoiqu'il en soit, je espere que vous me donnerez des nouvelles Là dessus; & me dissiez lorque vous penserez de revenir à Paris, dans quel tems vous le ferez pour voir si je m'arrange en consequance pour vous y embrasser car pour Le present je me trouve de plus en plus embarassé ave[c] diferents objets qu'il faut poursuivre. Le D^{r}. Knight un de mes intimes amis vient de mourir, & il faut mettre en ordre quelques choses qui le regardent, & particulierement La superintendence des Bussoles des Vaisseaux du Roy, qui peut etre me sera confié aiant eté instruit à fond sur cet objet par le dit mon ami, qui etoit Le seul qui La possedoit.[5] mais tout celà est encore incertain. mon nouveau Quadrant est encore un autre objet qu'il me faut mettre en ordre avec Le Public &c. ainsi j'aurai bien de la besogne jusqu'au autumne qui vient c'est à dire jusqu'à 7 ou 8bre. &c.

à Dieu mon cher, portez vous bien & Croyez moi pour La vie

Votre de Coeur

Magalhaens

P.S.

J'envoie aussi par cette poste une copie, comme Celle de M^{r}. Trudaine, à notre bon ami le D^{r}. Sanchez. Si par hazard cette Lettre vous trouve encore à Pariz Lui vous La communiquera bien volontiers. Car elle est bien diferente, & beaucoup meileur que celle-ci, que je vous envoie seulem^{t}. dans l'idée de vous arriver en Savoie: au lieu que pour etre imprimée, c'est la derniere que devroit l'etre.

A Monsieur
Monsr. Duchenne Secretaire
de M^{r}. Trudaine de Montigny
a Paris

4 Letters not found.

5 Dr Gowin Knight, FRS died on 8 June 1772. Having devised a method of making artificial magnets much more powerful than naturally occurring ones, he had designed a much improved mariner's compass using steel needles magnetized by his method, and had been given a contract by the Royal Navy to certify the compasses made for the Navy according to his method by the instrument maker George Adams. After Knight's death, the contract for certifying Adams' compasses was indeed transferred to Magellan as he here indicates might happen.

72.07.21 *To Johann III Bernoulli*

Öffentliche Bibliothek der Universität Basel, MS L I a 709, f. 30-31

Londres 21 Juillet 1772

J'ai reçu, Mon cher Monsieur, en propre tems La Votre du 19 de Mai dernier:[1] & je vous aurois repondu tout de suite si ce n'etoit pour attendre les Tables pour corriger la paralaxe & la refraction, qui furent imprimées sous les yeux du D^{r}. Shepherd par ordre du Board of Longitude:[2] & que Lui me promit d'envoyer pour Vous. mais jusqu'à Ce moment je ne les ai encore reçues: & aussitôt qu'elles arriveront je les ferai partir avec Les autres Livres que J'ai achetés deja. Cependant le retard fera que je pourai vous envoier les transact. Philos. de l'année derniere, qui paroitront dans peu de jours. M^{r}. le D^{r}. Franklin[3] doit m'envoyer aussi le 1er. Vol. des Trans. de la Societé de Philadelphie,[4] que cette Societé presente à L'Academ. R^{l}. de Berlin. Je viens de recevoir l'avis de la caisse avec les copies de votre 2^{d} vol. du Recueil pour Les Astronomes. mais le Navire n'est pas encore arrivé. en attendant Je voudrois savoir au plutôt quel est Le prix que je dois fixer au Libraire pour l'avoir net. Car Ceux de l'année derniere furent fixés arbitrairement, & je n'aime pas à disposer, que par principes certains de ces choses. vous savez que selon votre annonce les recueils de l'année derniere revenoient ils à environ 3^{s}: 6^{d}. or le Libraire qui les vend ne peut pas Le faire sans avoir un profit raisonable que je ne puis pas arbitrer de moi meme. ainsi il faudra que vous Lui donniez le dernier du premier cost, de façon qu'en y ajoutant Les depenses du carriage & transport &c. le libraire ici sache à quoi se tenir, & à quel prix il faut qu'il les vende pour avoir un profit raisonable.

Je ne crois pas que Je fasse aucun traité sur L'amelioration que Je fis sur l'Octant de Hadley. Je suis content que vous l'aiez annoncé: & ne manquerai pas de vous donner un precis detaillé de ses avantages, comme Je l'ai deja communiqué à l'Acad. R^{l}. des sciences de Paris.[5]

1 Letter not found.

2 *Tables for correcting the Apparent Distance of the Moon and a Star, from the Effects of Refraction and Parallax, published by order of the Commissioners of Longitude* (Cambridge, 1772). Anthony Shepherd (1721-1796) was Plumian professor of astronomy at Cambridge from 1760.

3 Benjamin Franklin (1706-1790).

4 *Transactions of the American Philosophical Society*, vol. 1, 1771.

5 Disappointed, once he saw it, by the brief notice Bernoulli accorded his invention, Magellan changed his mind and published an account of his own, initially in Rozier's *Observations sur la physique*, 4 (1774), 112-43, and then separately as *Description des octants et sextants anglois...* (London, 1775).

l'ouvrage de M^{r}. Wood sur le Genie d'Homere ne paroitra qu'à la fin de l'année. lui est mort depuis quelques mois: & peut etre que c'est la veuve celle qui fait achever l'edition.[6]

Ramsden n'a pas publié rien sur sa methode de diviser les instruments; ni je crois qu'il publiera jamais aucune chose sur ses inventions. il n'est pas pour s'annoncer. on lui prend tous les jours de ses inventions; & lui se laisse aller sans revendiquer ce que lui appartient.

Je ne vous dis rien plus des tabletes de M^{r}. Aubert:[7] parceque sa modestie l'empeche d'en faire aucun cas; & meme pour Les communiquer à ses amis. je l'ai pressé bien pour celà: & lui insiste qu'elles ne valent La peine. il est à present ocupé à batir un petit observatoire pour s'amuser en Campagne: du moins on aura de lui des observations bonnes & exactes.

il n'a pas de Journaux ici qui aient l'Astronomie pour objet principal. quelques fois il y a dans les magazins des petits chantillons, mais rien de bien interessant. Nous n'avons non plus aucune journal Necrologue comme Les François. Ceux ci sont Les gens qui savent faire des Journaux ephemeres des ciences. mais pour les Anglois, ils dedaignent de ce meler de ces taches journalieres. a peine avons nous le Critical & le Monthly Review, qui parlent des Livres, selon que les libraires entrepreneurs le veuillent. mais on n'en fait pas beaucoup de cas de ces Reviseurs ici.

Je comptois aller dans ce tems ci à Paris. mais differents affaires m'ont fait remetre mon Voyage jusqu'au mois de 8bre. prochain. alors je pense porter avec moi une toise de fer, que Je comparerai moi meme avec celle de l'Acad. Royale, qui servit dans les Observat. au Peru:[8] & je ne manquerai pas de vous donner Le Resultat.

Ce M^{r}. Wolfe dont Je Vous parlai,[9] est un Chymiste fort entendu, mais different de l'autre D^{r}. Wolfe de Danzic.[10] lui est à present à Paris, & se trouve incommodé de façon que je crains qu'il ne puisse aller passer l'eté comm'il le vouloit à Berlin.

J'ai presenté à La S. R^{le}. vos 2 brochures, qui vous ont retourné leurs remerciem^{ts}. & Je lui presenterai le 2^{d}. Vol. de Votre Recueil,[11] si vous l'ordonnerez dans Nov^{bre}. prochain. Car La S. R^{le}. est deja en vacances.

---

6 The work was in fact not published until 1775; Robert Wood, *An Essay on the Original Genius and Writings of Homer* (London, 1775).

7 See Magellan's account of these in 71.08.20.

8 The "Peru toise", which in 1766 had been declared the standard French measure of length.

9 See 72.04.21.

10 i. e. Nathanael Wolf.

11 Bernoulli, *Recueil pour les astronomes*, vol. 2.

72.07.21 J'ai parlé pour le change des transactions de votre ami contre celles qui lui manquent. mais on n'en a ici aucun Libraire, qui puisse l'accepter. Car les nombres qu'il a ne sont point ceux qui sont les plus rares, & ceux qui lui manquent, sont ceux dont on ne veut point se defaire pour ne pas gater les jeux entiers des transactions, dont on n'a que fort peu de complets. J'ai rendu Votre Letter à M^r^. Valtravers[12] comme Vous souhaitiez tout de suite: & peut etre en avez vous deja la reponse. M^r^. Horsely a publié the Sieve of Eratosthenes, being an account of his Method of finding the prime Numbers.[13] on a publié aussi Apollonius on Tangencies 2^d^. edition, to which is added Fermat on Spherical Tangencies.[14] likewise Apollonius on Determinate Section. by Snellius. the same also restored by W. Wales; bu[t] an intire new work.[15] = Voilà toutes nouvelles qu'il y a pour le present des livr[es] de Mathematiques nouveaux. Le tems & le papier finissent ai[nsi] je conclus cette Lettre en vous souhaitant perfecte santé & beaucoup de Bonheur & suis de Coeur

Votre très humble & [...][16] serv^r^.
Magellan

A Monsieur
Monsr. Bernoulli Astronome Royal
& de l'Acad^ie^. Royale des sciences de Prusse
à Berlin

12 Rodolph Valltravers, FRS.

13 Samuel Horsley, "ΚΟΣΚΙΝΟΝ ΕΡΑΤΟΣΘΕΝΟΥΣ; or, The Sieve of Eratosthenes: being an account of his Method of finding all the Prime Numbers", *Philosophical Transactions*, 62 (1772), 327-47.

14 *The Two Books of Apollonius Pergaeus, concerning Tangencies...[translated] with a supplement by John Lawson. – 2nd ed., to which is now added... Fermat's Treatise on Spherical Tangencies* (London, 1771).

15 *The Two Books of Apollonius Pergaeus, concerning Determinate Section, as they have been restored by Willberordus Snellius. By John Lawson ... To which are added, the same Two Books, by William Wales, being an Entire New Work* (London, 1772).

16 Word illegible.

To Arthur Young

72.08.19

British Library, Add. MS 35126, f. 144

Sir

Although I have not the pleasure of a particular acquaintance with you, hope however you'll look on the contents of this as a mark of my great regard and steem, by the desire I have of making you acquainted with Prince Masalsky, who came to this country, and wishes to see and speak with you on the favourit objects of agriculture.[1] the Prince is to make the tour of England, and should be glad to be a eye witness of some parts where Agriculture flourishes. I have recomended your works[2] for his directory, which he has already bought for that purpose: and thinks proper to begin by paying a visit to you in your farme. he intends to set out within two days, and that hinders me from taking a letter of Recom̄endation from our common worthy friend M^r^. Arbuthnot,[3] with who I have passed the sunday last, but knew not yet at that time of the intentions of the Prince: I intend to company him to Northmimes,[4] in order to introduce him to you: then will take your advice for the tour of the Prince, and hope your favour of recommending to him the more advantageous places for his curiosity on agriculture. Now advance this notice in order to know if we may find you at home by the end of this week. therefore see if you can contrive to be there, and be sure of not being displeased with this new acquaintance.

---

1 Ignace Massalski, Prince-Bishop of Vilna, in Poland, had been living for the previous few months in self-imposed exile in Paris following the Russian invasion of his country that he had helped to stir up. An enthusiastic advocate of the physiocratic doctrines propounded by Quesnay and others, he visited England in August-September 1772 before returning to Poland following the first of the two eighteenth-century partitions of the country (Ambroise Jobert, *Magnats polonais et physiocrates français (1767-1774)* [Paris, 1941]).

2 Magellan doubtless here refers to Young's *Course of Experimental Agriculture* (2 vols. in 4 parts, London, 1770), and probably also to his *Farmer's Letters to the People of England* (London, 1767; 3rd ed. 1768).

3 John Arbuthnot (died 1830?), elected FRS on 22 November 1770 on the basis of having made "several considerable improvements in husbandry", was subsequently author of *An Inquiry into the Connection between the Present Price of Provisions, and the Size of Farms* (London, 1773).

4 i.e. North Mimms, Hertfordshire, where Young farmed for nine years, 1768-77.

72.08.19 I am with due regard and full steem

Sir

your most obed[t]. humble Serv[t].

Magalhaens

August 19 – 1772

at n°. 25. Fetter Lane

To M[r]. A. Young

at Northmimes Near

Barnet in Hartfordshire[5]

---

5 Note added: "through Barn[et] [t]ake the right hand till the 16 milles po[t]... then to y[e] left".

72.09.04 *To the Royal Society of London*[1]

Royal Society of London, Letters & Papers, V, 323

Account of a new construction of the Sea-Quadrant, invented by John Hyacinth de Magalhaens

Communicated by . . . .[2]

Although the principles of Hadley's Quadrant, and the disposition of it's essential parts, have been Several times thorougly considered by many ingenious Persons: yet we may assert, that no very material improvements

---

1 Annotated: "Read Dec 17 1772". A cover sheet carries the request, "The Author desires that this paper may be read to y[e]. R[l]. Society before the end of this year 1772". The paper was read at the Society's meeting on 17 December and then referred to Nevil Maskelyne, the Astronomer Royal, who reported most unfavourably on 21 January 1773, to the effect that Magellan's instrument was distinctly inferior to the one already developed by Peter Dollond (see *Journal Book*, XXVIII, 10 and 37-51). In the light of Maskelyne's report, the paper was never published in the *Philosophical Transactions*. Magellan believed that Maskelyne's action reflected his personal commitment to protecting Dollond's invention rather than the actual merits of the case; see Magellan to Johann III Bernoulli, 23 March 1773.

2 A space is left at this point for a name to be added.

have been made till now on this most usefull Instrument. The Little 72.09.04
success however of y^{e}. preceeding attempts, did not discourage me from undertaking the task: and in fact on the 27th of Last March a new method for constructing this Instrument occurred to me, that appears to be the most advantageous. For such is the superiority of this construction, that the Observers will have in their power to make right Observations, even if the instrument was wrongly divided, or injured by any acident: as they will be enabled thereby to know the real value of the measured angles, without depending, as before, upon the accuracy of y^{e}. instrument-maker, or upon y^{e}. unalterable position of every part thereof, which, tho' always supposed, many times does not exist. This is a new quality, which I apprehend to be of the utmost importance in an Instrument, so generally imployed at Sea, where there is no Conveniency either for ascertaining it's accidental faults, or for mending them. A Quadrant of this sort, tho' indifferently made, may be so serviceable, as a better one (and indeed there a[re][3] but very few to be had well executed, even at a great price): and those already made in the usual manner, may easily be so altered, as to answer the same purposes.

Besides these qualities, there are still five or six more, arising from this construction, which afford very material advantages. Greater angles may be measured with an octant of this sort, than with a common Sextant. The adjustment for the back-observation, is therein strictly set by one single objet, without looking at y^{e}. opposite parts of the horizon, nor depending on the accuracy of y^{e}. maker, or being Liable to any accidental alteration, which may escape the eyes of the Observer: and a single adjustment may be set at once upon Sextants so constructed, both for the fore and back observation. By the same method the real value of any arch in Astronomical Quadrants, may be effectually ascertained, tho' there should be faults in the work it self, or any occasioned by accident. And finally this Invention furnishes us with a new method for dividing any arch, or whole Circle, with great accuracy.

The construction required for obtaining these advantages, consists in fixing the frame of y^{e}. fore-horizon glass C (see fig. 1st. & 2^{d}.)[4] well tight, by means of a single center-screw, to the end of a steady small arm CA, which is moveable on the axis of y^{e}. circle ZP parallel to y^{e}. plane of y^{e}. Instrument. The frame of y^{e}. back-horizon glass A is steadily screwed to the axis of the same circle ZP, which is raised enough to let the index pass under it's part P, when brought to y^{e}. end of the arch. There is another

3 Text obscured in binding.

4 See below.

72.09.04 small arm BA like the first, which carries a spare-horizon glass B. But this may be omitted in Sextants, & even in Octants, if the observer be contented with less advantages. the whole must be so contrived, that each arm may be stoped and fixed at pleasure on any point of the circle ZP, by means of a clamp or otherwise: and in the same manner the circle ZP must be so disposed, as to be fixed at pleasure in any point of it's revolution. The slow movement of each part may be contrived by the known methods of endless and adjusting screws, as well as by Levers, or pinions. These last were applied with success to the first instrument of this kind, almost finished by Mr. Ramsden, Mathematical instrument-maker, well known by his abilities. But I apprehend, that the endless screws, if well executed with a discharging spring to each, will do better, as having at the same time the properties of clamps, which may then be spared, and render the machine more simple. Figure the first represents the plan of an Octant, furnished with this new apparatus. Figure the second represents the plan of ye. same apparatus on a larger scale and the other two figures represent the plans of two Sextants, made upon the same principles, which are much more simple in their construction, and contain the Chief properties of the first.

The advantages of this construction, at Least the principal ones, will fully appear by the following Problems, never yet resolved by any other instrument we know of. these demonstrations depend on those generally known to be the fundation of Hadley's Quadrant: and may be seen in Smith's Opticks,[5] Book 3 Chap. 12 viz = that any ray of Light, after a double reflexion in two parallel planes, proceeds in a direction parallel to ye. former: but, if the planes are inclined to each other, then ye. angle of it's direction, will be double of ye. angle of their inclination.

Problem 1st. To adjust the new Octant for taking an angle above 120°. by the fore observation:

Solution. Make the glass C (fig. 1st.) parallel to N, when the index NM is at 90°., by the known method of bringing the reflected image of the Sun, Moon, or a Star, to coincide with the same object, seen by direct vision. Fix C in this position, by means of it's screw or clamp. Bring the index to 0°.: make the glass B parallel to N, and fix it likewise. Then loosen this clamp X: turn the circle ZP, till C becomes parallel to N: and the instrument will be ready for measuring any angle above 90°. For C will be then at T, and B at V, where it would be parallel to N, was the index NM to have retroceeded 45°. beyond 0°., viz to W at right angle with the side NR of the

5 Robert Smith, *A Compleat System of Opticks...* (London, 1738).

instrument. Upon tryal it is found, that an angle of 160°. and even more, may be well observed by this fore-observation. 72.09.04

N.B. 1st. you may begin the adjustment by puting B parallel to N, with the index at 0°.: then carry it to 90°., and make C parallel to N there. afterwards bring it back to 0°., examine if B keeps it's parallelism: and turn the whole circle ZP, till C becomes parallel to N &c.

2 If instead of an Octant, use is made of a Sextant, furnished with a single arm: then this Problem is solved by making A parallel to N, the index being at 60°.: afterwards B parallel to the same with the index at 0°.: and by turning the circle ZP till A becomes parallel to N at 0°., the glass B will be then at V, where it would be parallel to N, was the index NM at 30°. beyond the 0°. of the arch: and consequently the Sextant will be then fit for measuring any angle from 60°. to 160°. or more, by the fore-observation.

3. Proper marks are made upon the Limb of ye. Instrument, as EFGHL, where the moveable sight, or a Telescope, is to be fixed at once, for finding the reflected image, in each position of the horizon glasses.

4. These different positions of ye. horizon-glasses prove at the same time, that the index-glass, as well as the others, are perfectly perpendicular to the plane of the instrument: and that this plane is a perfect one, which are two very material circumstances, absolutely necessary in the Sea Quadrant, as well as in any other, for making good observations. And from this it may be assured, that this new instrument carries with it self the demonstration of it's being right: a circumstance which till now was only supposed: there being but very few observers who examine their Quadrants by means of the additional apparatus, used for that purpose by the instrument makers, but almost unknown to ye. generality of observers.

Problem 2 To adjust the horizon glass A for the back-observation, by a single object, such as the Sun, Moon, or a Star.

Solution. Make A parallel to N, the index being at 0°.: then repeat once more the preceeding Problem: and the glass A will be exactly twice 45°. viz 90°. from N, when the index is at 0°.: which is the required position for making the Back observation.

Problem 3. To know, either at Land or at Sea, if the arch of the Octant measures, or not, exactly 90°.: and by what quantity it differs, if any?

Solution. Repeat four times the operation of Problem ye. Second: by which means the glass A will run the whole 360°.: and if there be any difference to it's perfect parallelism with N at 0°. in the last operation, this difference being divided by 8, will shew how much the arch of ye. Octant differs from 90°.

72.09.04 N.B. If instead of an Octant, we have a Sextant to rectify, then the operation may be shorter, as we may take every time 60°., instead of 45°., and compleat the whole in 6 operations, instead of 8.

Problem 4 To know the real value of every 30°. 15°. 10°. 5°. or any other division, marked in the arch of the instrument.

Solution. Bring the arm CA almost behind the glass A: fix it there; and turn the frame of glass C in it's own axix or center screw, till the Coincidence of the two images, proves it to be parallel to N, the index being at 0°. This done, bring the index to 30°., 15°., or to any other compleate division of the arch; make the glass B parallel to N: and turn the circle ZP till you see the coincidence of the two images on the glass C. If you move the index further, till you find the same coincidence in the glass B; this space run by the index, will shew an angle equal to the first: and, if you repeat this operation, untill the index comes to the end of the arch, and mark down the number of degrees & minutes, shewn by y^e. index at each operation, you will have a table of equal parts, viz of every 30°. 15°. 10°. 5°. &c of your arch. But if the index in the Last operation, falls short, or goes farther than the 90°., already determined & known by the Last Problem: then each interval must be corrected by the proportional part of the whole deficiency or surplus, found at the end of the operations. This Table being compleated for every degree or 20'., any angle may be measured thereby with the greatest certainty, tho' in fact the arch of the instrument was wrongly divided at first or afterwards hurt by some acident.

N.B. The two Problems, y^e. first and y^e. Second, depend intirely upon a previous and exact performance of these two Last Problems; which for this Reason must preceed them in the order of trials.

Problem 5. To find out the errors of Astronomical Quadrants &c.

Solution. Fix a glass like N to the end of the index, or of the moveable telescope of the Quadrant: and an Apparatus, like that represented in fig. second, to it's Side. Then by the two Last Problems (the 3 & the 4th.) the errors, if any, of each division will clearly be discovered, and a Table of Correction effectually made for the use of y^e. Instrument.

Problem 6. To ascertain the Length of an arch of 60°., and to divide it into degrees &c.

Solution. Apply a glass like N to an index in the center of the given arch: and an Apparatus, like that represented in fig. y^e. second, to the Side or radius of it. Then take the length of y^e. radius, as a chord, upon the given arch: and examine, by Problem the 3^d. if this arch contains exactly 60°. If not, you may find it out, by taking a little more or less on the arch, and repeating the trials from end to end, untill the true point (where the images

coincide in the horizon-glass A, after the 6th. operation) is ascertained. 72.09.04
this being done, you may find by the same method every 20°. of this arch, making use of Problem the 4th.: afterwards every 10°. 5°. 1°. 20'. &c.

N.B. 1. one Sextant of a whole circle being once ascertained and divided, each division may be easily transfered to the other 5 Sextants, by the method of Problem the 4th. and the whole circle will be then thoroughly divided.

2. When the given arch only contains about 30°., you may ascertain and divide it, by taking the half radius, and drawing an arch with it from the extremity of the given one. Then let a tangent to the small arch, fall from the center of the great arch: and the part cut off by this tangent of the small (which is a radius of the other) will shew nearly the point of 30°. I say nearly, for I know the dificulty attending the practice in drawing these Lines by the true points. this arch you must ascertain and divide, as directed by these two last Problems: but the whole performance will be longer, as there are 12 operations to be made in each tryal, for ascertaining the real Length of 30°.

Problem 7. To construct a Sextant, upon the same Principles above mentioned, in such a manner that the fore and the back observations be adjusted by one operation, without any dependence from the horizon.

Solution. Let the horizon-glasses B & C (see fig. 3) be connected by a small arm or bar, so that C be the center of ye. movement of B: and both at an angle of 60°. between them-self. This being so constructed, adjust the parallelism between B & N, by turning it's proper screw. then the whole operation will be, to turn only the arm, untill C becomes parallel to N, for the fore-observation. For in this case B will be removed from A, where it was at 60°. with C, to the place B, where it must make also an angle of 60°. with N at 0°., which is parallel to C: and consequently, if you bring the index to ye. 60°. of the arch (which really are but 30°., according to the nature of these instruments) the Sextant will be fit for making the back-observation without any further adjustment: Since B will make then an angle of 90°. (60°. + 30°.) with the glass N of the index at 60°. Any angle therefore above 120°. may be taken by the back-observation in this Sextant, as far as the head of the observer stands not in the way. Angles of 200°. and more degrees could easily be taken with a rough model I had of this instrument, which is now in hand at the instrument-maker.

N.B. 1. When B is made parallel to N for the adjustmt., it receives the reflected rays passing freely near the uper side of the glass C, which is then very oblique, as it is seen by the pointed shape in figure the 3d.: and after

72.09.04 the second reflection the same rays pass below the under side of that glass C, to the observer's eye in S.

2 The glass N of the index is set at an angle of 6 or 7 degrees from the radius drawn by the 0°. of the arch, in order to have the glass C near enough to N, for the sake of getting a good field.

3. If the movement of the two connected glasses C & B, be made to turn on a center G (see fig. 4.) about the middle line of the instrument: then the reflected image from B, when parallel to N at 0°. for the adjustment, will fall on T: which cannot be deemed inconvenient, as there may be eassily fitted a sight, or vane, occasionally for that purpose.

4 The angles of 120°. are taken in this instrument upon the same division of the arch, both by the fore- and back-observation; which will infallibly ascertain the true Length of the whole arch. and may be examined many times at Sea, with no more trouble, than turning the instrument, whenever the distance of two heavenly objects or the Sun's altitude is 120°.: and observing it both ways on the two horizon-glasses.

5. Every division of the arch may be equally examined and ascertained in this Sextant, as directed by Problem the 4th. by turning the adjusting screw of y^{e}. glass B, and putting it at any angle with C.

6 It will be convenient to have two rows of numbers engraved on the arch of this Sextant. The uper one shewing the degrees from 0°. to 120°., as usual for the fore-observation: and the other under it, shewing the degrees from 120°. to 180°., which will fall under the 60°. of the first row, and so on to the begining, as it is called, of the arch. these numbers will shew the real degrees, not their suplements, taken by the back observation. There should be likewise another set of numbers on the Nonius of this Instrument in a contrary order. However it is no matter how the numbers are marked, provided the Observer knows well, how to make a proper use of his instrument.

the End

Fetter Lane the 4 September 1772.

72.09.04

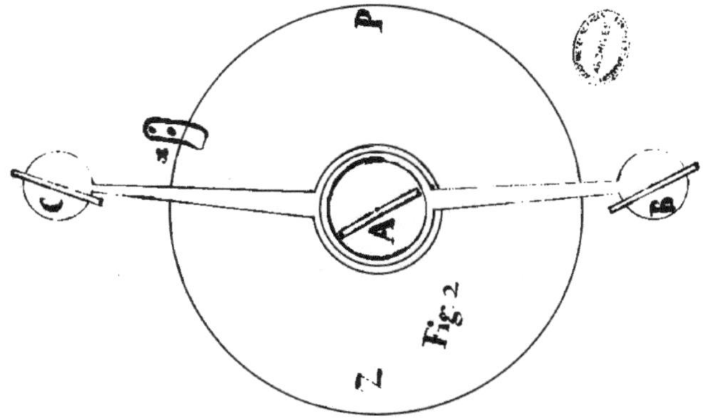

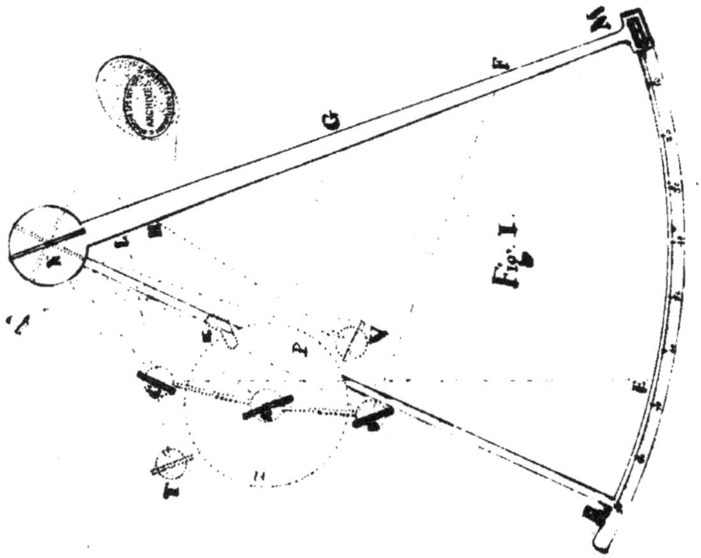

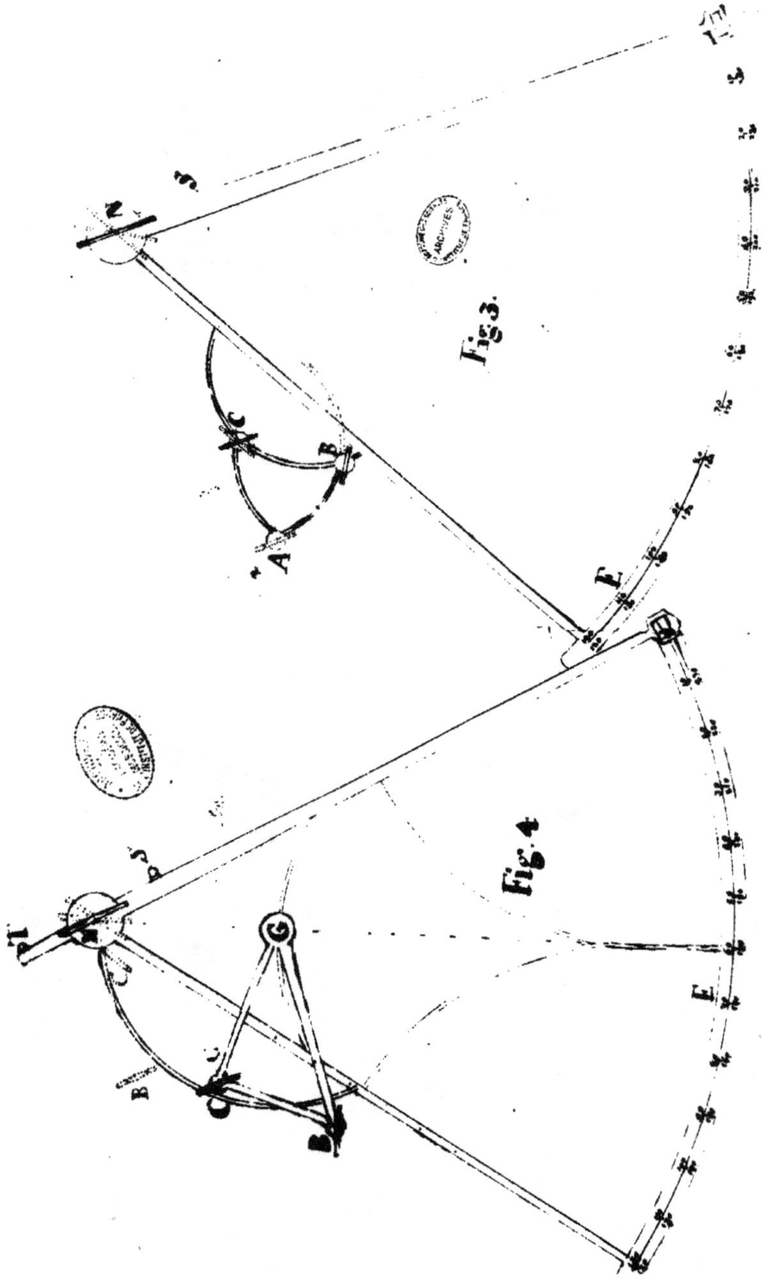
Fig. 3.
Fig. 4

To Johann III Bernoulli

Öffentliche Bibliothek der Universität Basel, MS L I a 709, f. 33-34

Londres 6. Oct^{bre}. 1772
at n°. 25. Fetter Lane

Mon cher Monsieur

Le billet (ou Reçu) ci-joint[1] regarde une caisse de Livres que je vous ai expediée le mois dernier. mais ne pouvant pas avoir aussitôt Le reçu avant d'aller en Campagne, j'ai differé de vous ecrire jusqu'à ce moment, etant aussi à La veille de partir pour Bruxelles & de là pour Paris; oû je ne resterai que jusqu'à la derniere Semaine de Novembre, à cause de quelques petits affaires qui demandent mon Retour avant La seconde Semaine de Decembre. en cas que vous aiez à m'ecrire tandis que Je serai à Paris, mon adresse sera = Chez Le Chev^{er}. de Bory Chef d'Esquadre, de la Acad. R^{le}. des Sciences, rue vivienne = sur cette adresse il faudra mettre une seconde enveloppe, adressée à M^{r}. Trudaine de Montigny, Conseiller d'Etat & Intend^{t}. des Finances à Paris = A present je reprendrai Le Contenû dans votre caisse, qui est comme suit. Bird's Method of Dividing.[2] 2. pamphlets. 2^{s}:6^{d}. / Simpson's Annuities & Apendix[3] 3^{s}.6. Nautical Almanak pour 1773. 3^{s}.6^{d} / vous trouverez aussi celui pour 1774, dont M^{r}. Maskelyne vous fait un present. peutetre merite-il que vous Lui en fassiez du 2^{d}. vol. de votre Recueil. / N.B. on a eté generalem^{t}. fort content de la satisfaction que Vous donnez au public sur les articles du premier volume & de Vos Lettres qui regardoient le Nautical Almanak & M^{r}. Dollond.[4] / Harrison's Time Keeper[5] – 5^{s}:0^{d} / Account of going[6] d^{o}. – 2.6. / Mayer Theoria Lunæ[7] 2^{s}.6^{d}. / Dillenius Muscorum[8] – £1:11^{s}:6^{d}. / Priestley on fixed air[9] – 1^{s}.0^{d}. N.B. cette

---

1 Receipt not found.

2 John Bird, *The Method of Dividing Astronomical Instruments...* (London, 1767).

3 Thomas Simpson, *The Doctrine of Annuities and Reversions* (London, 1742), and *An Appendix, containing some Remarks on a Late Book on the Same Subject...* (London, 1743).

4 See 72.04.07.

5 John Harrison, *The Principles of Mr. Harrison's Timekeeper* (London, 1767).

6 Nevil Maskelyne, *An Account of the Going of Mr. John Harrison's Watch...* (London, 1767).

7 Tobias Mayer, *Theoria lunae juxta systema Newtonianum* (London, 1767).

8 Johann Jakob Dillenius, *Historia muscorum: in qua circiter sexcentae species veteres et novae ad sua genera relatae describuntur et iconibus genuinis illustrantur* (Oxford, 1741).

9 Joseph Priestley, *Directions for Impregnating Water with Fixed Air...* (London, 1772).

72.10.06 petite brochure contient La methode pour impregner L'eau Commune avec l'air fixe: de façon qu'elle devient comme celle de Pyrmont & Spa &c. on espere meme qu'elle sera un remede contre L'escorbut. mais ce sera après une suite d'observations bien constatée qu'on pourra prononcer sur Le merit & avantages de cette decouverte. Cependent elle est très curieuse en elle meme, & vaut Les peines du Philosophe Observateur. / Vous y trouverez les Transactions Phil. de la Soc^{e}. R^{le}. en 2 vol. qui sont pour Votre Acad. R^{le}. & de meme Les Transact. de la Societe Philosophique de Philadelphie,[10] dont elle fait present à votre Academie. Ce fut Le D^{r}. Franklin qui me chargea de vous envoier Cette premier des travaux philosophiques de nos Americains. Vous y trouverez en fin deux petits morceaux d'une gome Elastique, avec La quelle on ôte toutes Les marques du Craion lorsqu'on en a fait ou ecrit sur Le papier. il n'y a qu'à froter dessus avec cette matiere, dont les proprietés sont fort peu communes. Vos Livres aiant arrivé ici, je les ai envoyés tous Chez M^{r}. Elmsley, pour Les vendre pour votre compte. Je vous avois demandé quel etoit le prix que vous aviez fixé pour cet ouvrage. mais observant que vous y repetez la même annonce, je conclus, que celui-ci etoit Le prix que le Libraire devoit vous donner; & qu'il devoit les vendre en y ajoutant le surplus necessaire pour paier Les depenses de l'envoy & la commission de son travail. or comme j'ai paie pour votre compte les depenses des deux caisses que vous m'avez envoyées, je trouve que les recueils doivent vous reporter 3^{s}:4½^{d} pour chaque copie & les Lettres à raison de 1^{s}:8¾^{d}: ainsi celle-ci fut l'ordre que je lui donnai, c'est à dire de les vendre en sorte que vous aiez ces prix pour un chacun respectivement. ce qui sera executé avec ceux ci & avec ceux qui restent encore de l'envoy de l'année derniere. mais ceux qui l'avoient vendû auparavant, ils ne vous reporteront autant, car si vous vous souvenez j'avois crû que les Lettres à 2^{s}: & les recueils à 3^{s}.6^{d} etoient raisonables: & vous avez aprouvé ces prix, dans les quels le Libraire doit avoir quelque chose pour sa commission.

Les copies reçues dans cette caisse furent 24. copies du 2^{d}. vol. du Recueil. 3 copies du premier volume: & 3 copies de vos lettres. J'ai fait un etat courant de mes petits deboursements jusqu'à present; & je les ai mis en compte à M^{r}. Elmsley, pour les charger sur le votre. le tout monta à £4.12^{s}.5^{d}. dont voici le detail comme je le trouve dans mon Livre. Lettre double du 13 Juillet 1771 – 1^{s}.0^{d} / reponse du 20 Aoust. – 1^{s}.0^{d} / lettre du 21 Sept^{bre}. & sa reponse – 2^{s}.0^{d} / lettre de M^{r}. Luytens avec le premier envoie de vos livres

10 *Transactions of the American Philosophical Society*, vol. 1 (1771).

– 1^{s}.0^{d} / je payai à son correspond^{t}. ici pour ses deboursements – 13^{s}:1^{d}. / lettre double avec quelques notices pour votre usage du 22 Janv. 1772 – 2^{s}.0^{d} / depenses à la Douane, fret, porterage &c. avec le premier caisse de vos livres – 12^{s}.8^{d} / lettre du 14 May avec une dedans pour M^{r}. Valtravers – 2^{s}.0^{d} / lettre que J'envoiai du 21 Juillet – 1^{s}.0^{d} / lettre de M^{r}. Luytens sur l'envoy de cette 2de. caisse – 1^{s}.0^{d} / paiai à son correspond^{t}. ici pour les deboursem^{ts}. avec cette meme caisse jusqu'à Hambourg – 10^{s}.9^{d}. / depenses à la douane pour cette meme 2^{d}. caisse, fret du vaisseau, porterage &c. 14^{s}.6^{d}. / Caisse & porterage pour l'envoy dernier de les Livres ci dessus – 3^{s}:4^{d}. / depeches à la douane pour l'embarquer 4^{s}.0^{d} / je paierai pour cette Lettre, & lading bills – 1^{s}.9^{d}. Some totale avec les livres ci dessus 6.3^{s}.1^{d}. or aiant vendu une demi douzaine du premier vol. de Votre receuil à M^{r}. White, Libraire, par £1:0^{s}.3½ & une autre demi-douzaine de Vos Lettres par 10^{s}:4^{d}. le total est £1:10:7½ qui otés de 6:3^{s}:1^{d}: je etois crediteur des £4:12^{s}:5^{d}½, que M^{r}. Elmsley chargea sur votre compte, en me donnant credit pour la meme some dans Le mien. 72.10.06

Le Bibliothecaire de la Soc. Royale me dit de savoir si les Memoires de votre Academie sont complete dès 45 jusqu'à *66*, car celles-ci sont les seules qu'il a dans La Bibiotheque. peut etre vous avez publiés d'autres vol. qu'il n'a point ici?

M^{r}. le D^{r}. Shepard m'a dit depuis fort Long tems qu'il m'envoyeroit Les Tables qu'on a imprimées pour la correction des Paralaxes & refractions des observations celestes,[11] pour vous les envoyer. mais jusqu'à ce moment je ne Les ai pas encore reçues. reelem^{t}. elles ne sont pas encore publiées. mais on m'a dit qu'elles le seront dans peu de jours. Je crois vous avoir dit qu'un M^{r}. Ferguson, qui tient une Academie dans le Wapping, a inventé une machine pour faire cette operation sans aucun calcul. il lui donne le nom de Rotula:[12] & si elle est si exacte comm'on Le dit, elle epargnera la peine & depense de ces tables. Lui m'a promis une pour L'Acad. R^{le}. de Paris. sans doute elle y sera examinée: & je ne manquerai pas de vous donner Le jugement qu'on en fera.

11 *Tables for correcting the Apparent Distance of the Moon and a Star from the Effects of Refraction and Parallax*, ed. Anthony Shepherd (London, 1772).

12 The list of subscribers to T.H. Croker et al., *The Complete Dictionary of Arts and Sciences...* (London, 1764-6) includes "James Ferguson, Master of the Academy in Newhermitage St., Wapping". This, however, was a different person from James Ferguson, FRS (1710-1776), who was not a schoolmaster (cf. Ebenezer Henderson, *Life of James Ferguson, F.R.S.* [Edinburgh, 1870]). See also James Ferguson, *The Description and Use of the Astronomical Rotula...* (London, 1775).

72.10.06 il n'a pas à present rien de nouveau qui vaille La peine dans la Literature. Je suis en attendant mon Quadrant marin, amelioré selon ma methode pour L'aporter à L'Acad. Le menteur Ramsden m'a trompé tout ce tems sans le finir. Cependant je l'ai vû hier fort avancé. en attendant j'ai fait faire par un autre ouvrier ordinaire la meme ajustement sur un de ces quadrants qu'on vend à la douzaine: & je trouva avec satisfaction que toutes Les proprietés dont je vous ai donné compte, s'y trouvent à merveille. J'espere que vous ne manquerez pas d'en donner une idee complete dans Le 3me. vol. de votre Recueil. Car pour moi je ne pense d'en faire imprimer aucun traité là dessus.[13] Lorsque Les Commissaires de l'Academie Royale de Paris auront fait Leur raport (ce que j'ai prié de souspendre jusqu'à voir l'instrument tout fini & complet, que je vais Leur apporter) je ne m'oublierai point de vous en faire part.[14] en attendant donnez moi des nouvelles de votre sante, que Je vous souhaite retablie de tout mon Coeur.

& suis toujours
tout a vous
Magalhaens.

A Monsieur
Monsr. Bernoulli Astronome Royal
à Berlin
Prusse

---

13 The publication of the third volume of Bernoulli's *Recueil* was long delayed, and so Magellan was driven to publish his own account of his invention in his *Description des octants et sextants anglois...* (London, 1775). And just as well that he did, for when Bernoulli's third volume did eventually appear, in 1776, it still did not include an account of Magellan's device!

14 The commissioners appointed by the Academy to investigate Magellan's invention, de Bory and Le Monnier, reported favourably on it in January 1773.

To Arthur Young 72.10.09

British Library, Add. 35126, f. 146

London 9 Oct^{ber}. 1772
at n°. 25 Fetter Lane

Dear Sir

It is a Long while I had the pleasure of Receiving your's, with the account relative to y^{e}. culture of Cabages.[1] I did communicate it's contents soon to the Prince Massalski,[2] who expressed the greatest obligations for your kind regard and care. But his affairs calling him, sooner than he thought, to France, charged me with the return of his thanks, which I have retarded till now in hopes of Receiving some further news from him, and at the same time some Queries, to which he told me his desire of getting proper answers to, they being chiefly related to Husbandery. I made the Tour of Norfolk, and that of y^{e}. Iland of Thanet[3] with the Prince: companied him to Calais, and went back to London, where I have not yet received any news of him. the late disposal and sharing of the different provinces of Poland amongst Russia, Austria, and Prussia has brought the utmost confusion into this Prince affairs, whose states being scatered in different parts of Poland, make him a subject to four different Powers. any person in such situation as his, must be much distracted to keep any regular correspondence about maters of different kind. Now I am setting out to Flandres in two days hence, and from Bruxells to Paris, where I intend to stay till the 25 of November, so as to be bak in London by the begining of December next. If you have any kind of commands towards those parts, please to honour me therewith, as no body will more heartly fullfill them. you may direct your Letters to me till the end of this month at Bruxelles at his Highness the Duke of Aramberg.[4] this Nobleman is very fond of encouraging husbandery in his country: and just now I carry with me some bushels of Siberian Barley[5] to be distributed amongst his farmers. but in the month of November, then my Letters must be directed to Paris, under a second

1 Letter not found.
2 See Magellan to A. Young, 19 August 1772, n. 1.
3 At the eastern extremity of the county of Kent.
4 i.e. Arenberg.
5 See Magellan to J. Bernoulli, 15 March 1771.

72.10.09 cover, with the following direction in the outside = A Monsr. Trudaine de Montigny, Conseiller d'Etat & Intendant des Finances à Paris =

Pray present my best Respects to M^{rs}. Young, and wishing you both and family true happiness, am

dear Sir
Your most obed^{t}. humble serv^{t}.
Magalhaens

To Arthur Young Esqr.
Northmimes near Barnet
Hertfordshire

72.10.16 *To Louis-Henri Duchesne*[1]

Archives Nationales, Paris, MS T/160 (10-12)

Londres 16 Octobre 1772
at n°. 25 Fetter Lane

Monsieur & Cher ami

Selon L'avis de votre derniere[2] vous devez vous trouver deja de Retour à Paris dans ce tems-ci. ainsi je m'adresse à vous pour dire que je pense envoier bientôt une caisse avec diferents instruments & d'autres choses pour L'Acad. Royale des Sciences, & aussi pour quelque ami, comme par exemple Le Chev^{er}. de Bory, qui m'ordonna quelques aunes[3] de tricoté de laine pour un habit. Cette caisse será adressée à M^{r}. Trudaine de Montigny, & arrivera à Paris bien auparavant que moi. car je dois aller par Bruxelles, & je ne puis pas partir encore que vers La fin de ce mois. Or comm'il y aura dans cette caisse, quelques instruments delicats, & entre autres mon nouveau Quadrant marin,[4] Je souhaite que personne n'y touche, & que vous La conserviez fermée jusqu'à ce que J'arrive. Si celà est faisable vous m'obligeriez beaucoup de faire en sorte, que celà soit ainsi.

---

1 Annotated by Duchesne: "Rep le 24. 8e". Letter not found.
2 Letter not found.
3 Aune (or aulne) – a measure of length of 3 feet, 7 inches and 3 lines (of Paris); an ell.
4 See Magellan to Duchesne, 31 March 1772.

Voyez en attendant si vous voulez quelque chose d'ici, & ne me Refusez Le plaisir de vous montrer en quelque façon combien je suis 72.10.16

Mon cher Monsieur

Votre pour La vie

Magalhaens

A Monsieur
Monsr. Duchesne Secretaire de
M[r]. trudaine de Montigny
à Paris

To Johann III Bernoulli 72.10.26

Öffentliche Bibliothek der Universität Basel, MS L I a 709, f. 35-36

Londres ce 26 Oct[bre]. 1772

Monsieur & très cher ami

La votre du 29 du mois dernier[1] n'arriva entre mes mains qu'une Semaine après que je vous ecrivis ma derniere.[2] La nouvelle de vos indispositions & mauvaise santé, me fache beaucoup, contre vous, qui ne vou[le]z point vous epargner ce precieux tresor de la vie, pour votre famille & pour vos bons amis, dont je presume [etre][3] un des plus sinceres. M[r]. Elmsley est parti pour Paris depuis peu, ainsi il n'est pas possible de vous donner aucun notice [...][4] exacte de la vente de vos Livres. par ma derniere vous avez vû ce que J'avois arrangé sur cet article. Je suis bien [a]ise de voir que les prix que J'y avois fixés, n'etoient pas au dessous de ceux que vous me marquez. le 2[d]. vol. du Recueil sera vendu par le meme prix du premier. Je ne crois pas pouvoir vendre le reste tout d'un coup ici, comme vous le souhaitez. lorsque M[r]. Elmsley sera de retour, Je lui parlerai sur celà, & vous marquerai sa reponse. au reste vous ferez bien de vous arranger avec quelque Libraire pour le continuation de votre ouvrage. Le D[r]. Shepherd n'a pas encore envoyé les Tables qu'il m'a promis pour vous. lorsque Je les aurai

1 Letter not found.
2 See 72.10.06.
3 Conjectural reading.
4 Word indecipherable.

72.10.26 reçues, j'y ajouterai le Sieve d'Erathostnes.[5] Le 2^{d}. vol. de votre Receuil est deja entre les mains du Biblioth^{e}. de la S. R^{le}. Je suis très charmé que vous aiez M^{r}. Bode pour vous aider dans les calculs &c.[6] J'espere que Vous ne manquerez pas d'inserer la description de mon nouveau Quadrant marin dans le 3me. Vol. de votre Receuil.[7] mais de crainte que la derniere relation que Je vous en ai faite, ne soit pas claire, je vous ajoute ici une plus correcte & exacte avec l'idee d'un nouveau Sextant beaucoup plus simple, construit sur les mêmes principes. Je la copierai de celle que J'ai faite en Anglois[8] = ... By this instrument it will be always in the power of y^{e}. Observer to make right observations, even when the instrum^{t}. is wrongly divided, since he will be enabled hereafter, either at Land or at Sea, to know the real value of y^{e}. measured angles, without depending, as before, upon the accuracy of y^{e}. Instrum^{t}. maker, or upon the unalterable position of every part thereof, which tho' always supposed, many times did not exist. Such an instrum^{t}. therefore, tho' indifferently executed may be as serviceable, as a better one: and if already made in the usual manner, may easily be altered with no great expense, so as to answer the same purposes. Greater angles may be measured with one octant of this sort, than with a common Sextant. the adjustm^{t} . for y^{e}. back observation is therein strictly settled, without looking at y^{e}. horizon: nor does it depend at all upon the maker's accuracy, nor is it liable to any acidental alteration. by this improvem^{t}. the real value of any arch in the Astronomical Quadrants, may be easily ascertained, whether the faults arise from y^{e}. bad work it self, or from some acidental injury: & finally this improvem^{t}. furnishes a new method for dividing any arch or circle, with very great accuracy. this construction consists in fixing the horizon-glass A, used for y^{e}. back observation, upon a small circle zp. (fig. 1^{rst}. & 2^{d}.) parallel to y^{e}. plane of y^{e}. instrument, & moveable round it's own axis. it is rised so that it's part p passes over the index, when brought to 90°. the small arm CA carries the horizon-glass used in the fore observation at it's end C: and is moveable round the axis A. Likewise the arm BA carries another glass B: and is equally moveable round the axis A. Each arm has a little clamp and a pinion, to fix or to move it to any point of the circle zp. and x represents another pinion & clamp,

5 See 72.07.21, notes 2 and 13.

6 The young Johann Elert Bode (1747-1826) had recently been engaged by the Berlin Academy as a calculator, with a view to improving the quality of the annual almanacs sold by the Academy, on which much of its income depended but which were held in low repute because of their many inaccuracies.

7 Despite Magellan's hopes, Bernoulli did not publish a description of his invention.

8 Cf. 72.09.04.

to move or fixe the circle in any point of its revolution. The Second arm BA & y^e^. circle zp. may be omitted in sextants, as will be seen hereafter. the following Problems will demonstrate it's advantages. figure the 1rst. represents an Octant with the necessary apparatus just now described. fig. ye. second represents the same apparatus on a larger scale. and fig. the third represents the new sextant. 72.10.26

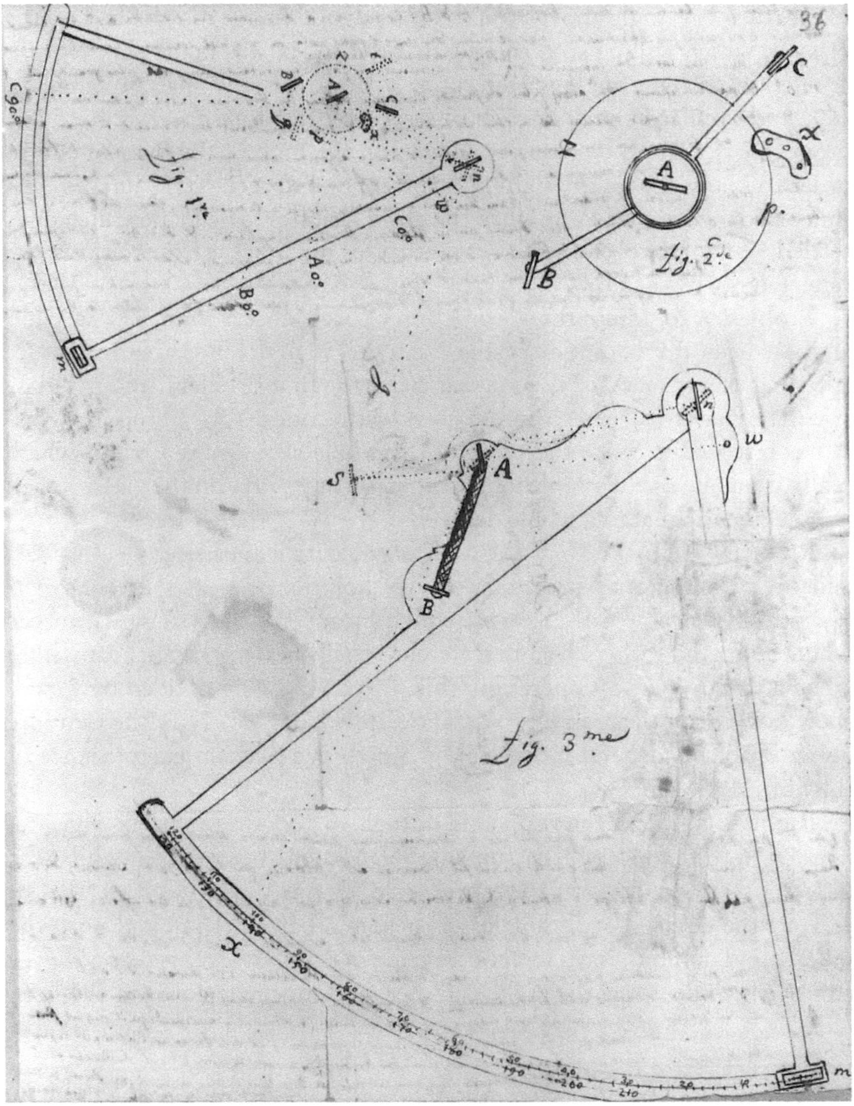

72.10.26 this new Octant. Solution. Make c parallel to n, when the index is at 90°., by the known method of bringing the reflected image of y^{e}. Sun, Moon, or Star, to coincide with that seen by direct vision, applying the eye at the place marked C90°: fix the glass A in this position, by means of it's screw or clamp. bring the index to 0°. apply the eye at B0°. and make the glass B parallel to [...][9]: fix it there likewise. then applying the eye at C0°., turn the circle zp, till C becomes parallel to n. and the instrument will be ready to measure any angle above 90°., as far as a good image may be had on the glass B, which will be so disposed as if the Index was made parallel to it at n.g: viz. at right angle with n. upon tryal it is found that any angle of 160° and even 165°. may be well observed by this fore observation. N.B. proper marks are to be made at once upon the limb of the instrument, in order to know, where the sight must be applied to receive the images each time in adjusting, as well as when the observation is to be made which once settled, remain for ever. the place w is that for making the observations, the reflected ray being n.q.w.

Probl. 2^{d}. To adjust the glass A for ye back observation. viz. at right angles with n at 0°. Solution. make A parallel to n, the index being at 0°. & applying the eye at A0°.: then repeat twice the first Problem, and the glass A will be exactly turned 45°. from n, when the index is at 0°. Since the glass A has retroceded twice the value of the whole arch of the Octant which really is but 45°.; tho' by the double reflexion of y^{e}. rays on the glasses, each half degree is equal to a whole degree.

Problem 3^{d}. To know if the arch of y^{e}. octant measures or not 90°., and by what quantity it differs, if any? Solution. repeat four times the operation of Problem the 2^{d}., by which means the glass A will run the whole 360°: and if there be any difference to it's perfect parallelism with n when at 0°. in the Last operation: this difference being divided by 8, will shew how much the arch of y^{e}. octant differs from 45°. N.B. the two fore going Problems depend intirely upon a previous & exact performance of this third Problem.

Probl. 4. to know the real value of each 10°. 5°. or any other division, marked in the arch of y^{e}. Instrument. Sol. Make C parallel to n whilst the index is at 0°.: bring it to 10°., or other division, and make B parallel to n. Turn the circle zp. till the glass C becomes parallel to n, and push the index till you find the two images coincide in the glass C. this second space run by the index over the arch, will measure one angle equal to the first.

9 Text obliterated.

if you repeat the same operation untill the index comes to 90°., and mark down the numbers answering to each operation upon the arch, you will thereby a table of equal parts, viz. of every 10°. 5°. or 2°. &c. but if in the Last operation the index falls short, or goes further than the point of 90°. already known by the Last Problem; then each interval must be corrected by the proportional part of y^e^. whole deficiency or excess, found at the end of the operations. this Table being compleated for every degree or half degree, any angles may be measured thereby, with y^e^. greatest accuracy, tho' in fact the divisions of y^e^. instrument be at first set down intirely wrong. 72.10.26

Problem 5. To find out the errors of Astronomical Quadrants. Sol. Fix a glass Like n to y^e^. end of ye index, or Telescope of y^e^. Quadrant: and an Apparatus like that represented in fig. 2^d^. to it's side. Then by the two Last Problems, the errors, if any, of each division will clearly appear: and a Table of correction effectually made for it.

Problem 6. To ascertain the Lenght of an arch of 60°, and every division thereof. Solution. Apply a glass n to an index moveable in the centre of y^e^. arch: and an apparatus like that of fig. 2^d^. to the side or radius of it. then taking the Lenght of y^e^. radius in the arch, examine by Problem 3^d^. if this arch is of 60°. exactly. if not, you may find out by repeating the tryals from end to end, the true point, where the images coincide in the horizon glass after 6 operations. this being done, you may find by the same method, every 10°. of this arch, as shewn in the Problem. 4: afterwards every 5°. 2°. 1°. and every 20'. &c. N.B. every arch of a whole circle, subtended by its radius, may be ascertained and divided in the same manner; and so on till the whole circle is finished. but a second index must be then provided, in order to carry the apparatus of figure the second.

La fig. premiere bien considerée, ne peut pas manquer de faire voir la demonstration de mes principes. car on voit que le petit miroir C etant parallele au miroir n de l'index à 90°.: & le miroir B parallele au meme miroir n à 0°. ils font un angle de 45°. entre eux memes. ainsi en tournant le circle zp oû ils sont attachés, jusqu'à ce que C devient parallele à n dans la meme position, pour lors C se trouvera à t & B se trouvera a q. de façon qu'il seroit parallele au miroir n en cas qu'on poussat l'index jusqu'à g. C'est à dire au bout d'un arc de 90°. &c.

mon nouveau sextant que j'ai rendu plus simple, ne contient que seulement le miroir horizontal A (fig. 3me.) qui sert pour observer par devant avec un petit bras AB, qui porte le miroir B pour observer avec le dos tourné à l'objet. les degrés y sont marqués à l'ordinaire dans la ligne d'en haut; jusqu'au 120° comm'à l'ordinaire. mais dans la rangée d'en bas

72.10.26 ils vont depuis 120°. jusqu'à 180°. & meme 200°. en voici l'usage. menez l'index au degré 120°., tournez le bras avec les deux petits miroirs, jusqu'à ce que le miroir A est parallele à n. fixez-le par le moien de son ecrou dans cette position. menez l'index à zero. 0°.: faitez le miroir B (qui pour lors est à S) parallele à n. tournez le bras, avec les miroirs ainsi disposés, jusqu'à ce que A devient parallel à n: & vous aurez ce meme miroir pret pour observer par devant jusqu'au 120°.: & dans le même tems vous aurez le miroir B qui fait un angle de 60°. avec A. ainsi lorsque vous menerez l'index au 60°. de l'arc (qui reelem^{t}. n'est qu'un arc de 30°.) vous aurez B à angle droit avec n. & par consequent vous pouvez observer tous les angles depuis 180 (ou demicercle) jusqu'à 60°.: & même vous pouvez observer des angles suplementaires au delà de 180°. jusqu'a ce que l'objet par derriere est invisible à cause de la tete de l'observateur, qui est entre B & n. un habile hom̄e tourne l'instrument de façon que seulement l'oeil & part du front y est au milieu. Reelement ce sextant n'a pas toutes les avantages, qu'il pourroit avoir avec un apparatus comme celui que je viens de decrire mais en revanche il est plus simple. il porte sa rectification avec lui meme, car tous les angles de 120°. y sont près de deux façons (sur chacun des miroirs horizontaux A & B) ce qui sert à demontrer que l'arc de 60°. est exact. autrement il ne poura pas etre montré par les mêmes divisions: & outre celà on y ajuste le verre pour l'observation par derriere, avec une certitude immancable & sans aucun recours à l'horizon. ordinairement on fait les sextants un peu plus grands que 60°. (120°.) & par consequent on a alors plus de points communs dans les divisions, pour examiner si elles sont exactes.

J'ai fait executer deja tous ces instruments: & tous repondent entierement à mon atteinte. M^{r}. Ramsden est finissant un en cuivre jaune que Je pense apporter avec moi à Paris pour le faire voir à L'acad^{e}. Royale, avant de l'envoyer à M^{r}. de Briqueville, officier du Roy à Brest,[10] qui me l'ordonna depuis long tems. mais ce diable de menteur (Ramsden) m'a entretenu jusqu'à ce moment, sans le finir. Lui est enchanté de ces Sextants, dont je viens de vous parler, à cause de leur simplicité: & dit qu'il entreprendra un bientôt.

Mon voyage à Paris & Flandres a eté retardé jusqu'à ce moment. Par la mort du D^{r}. Knight (le fameux, dont Musshenbroeck parle avec tant

10 Bon-Chrétien, Marquis de Briqueville (1723-1803), who at the time of this letter was "capitaine de vaisseau" and was based at Brest.

d'eloge par ses decouvertes sur Le magnetisme[11]) Je fus nommé pour revoir & certifier les Bussoles de la marine d'Angleterre: qui sont faites selon les principes de feu D^r. Knight; & c'est à moi que Lui les a communiqués par l'etroite amitie que nous avions tous deux.[12] ainsi moi etant Le seul qui les connois, & l'ouvrier M^r. Adams qui les faisoit, venant de mourir aussi depuis peu,[13] Je fus obligé d'attendre & de retarder mon voyage jusqu'à ce que cet affaire soit arrang[ée,] ce que Je crois sera fait en 8 ou 15 jours. aussitôt je parts pour Bruxelles, & de là pour Paris. Je ne m'y arreterai pas Long tems. Car je pense etre de retour ici avant Noel. 72.10.26

N.B. Le miroir S fig. 3me. reçoit les raions reflechis du miroir n, & les reflechit lui meme vers w sans aucun embarras du miroir A, qui pour lors est fort incliné, comm'on le voit marqué par des petits points, & n'empeche aucunem^t. qu'ils passent par chaque coté. Je vous recommende beacoup de faire graver ces figures avec exactitude.

Je suis fache au contraire de ce que vous dites, de ce que Vos Ephemerides ne seront pas en francois! ne croiez vous pas devoir faire present de votre 2^d. vol. du Recueil à M^r. Maskelyne qui vous fit present du dernier vol. du Nautical Almanac que Je vous envoyai? à Dieu mon cher. pensez à Votre santé, & croiez moi votre pour la vie

Magalhaens.

A Monsieur
Monsr. Bernoulli Astronome Royal de Prusse
à Berlin

11 Cf. Pierre van Musschenbroek, *Cours de physique expérimentale et mathématique* (Paris, 1769), *I*, 459.

12 Gowin Knight, FRS (1713-1772) in the 1740s developed a method of making artificial magnets, which he subsequently employed in making much improved mariner's compasses. After successfully demonstrating the merits of these instruments to the Admiralty, he was appointed to certify compasses made according to his method and supplied to the Navy by the instrument maker George Adams. This was the position that now passed to Magellan. A compass bearing Magellan's certification is held by the Mitchell Library, Sydney.

13 The famous instrument maker George Adams died suddenly on 17 October 1772.

72.10.27 *To Louis-Henri Duchesne*[1]
Archives Nationales, Paris, MS T/160 (10-12)

Londres 27 Oct^{bre}. 1772

Monsieur & très cher ami

Je vous ecris encore celle-ci de Londres, oû je me trouve arreté par des acidents imprevûs. Car M^{r}. Adams qui faisoit Les Bussoles pour La marine d'Angleterre vient de mourir cette semaine derniere,[2] & il me faut attendre que cet affaire soit arrangé: ce que J'espere sera fini en peu de Jours. Savoir, il me faut attendre si son fils (qui est jeune encore) sera ou non son sucesseur. Car moi aiant eté nomé (moimeme) pour examiner & certifier ces Bussoles, après la mort de mon bon ami Le Dr Knight, qui me devoilà les principes de Leur construction, & pas aucun autre Le possede (excepté Le feu M^{r}. Adams & son fils pour ce qui regarde La practique) il est necessaire de voir le tour que cet affaire prendra: quoique Je crois que ce sera son fils qui devra continuer cette entreprise.[3] aussitôt que ceci sera fini, je part d'ici ; mais en attendant je voudrois savoir de Vos nouvelles: & particulierem^{t}. si vous etes déjà de Retour à Paris. Car autrem^{t}. j'irai à Bruxelles oû je m'arreterai quelque semaine jusqu'à ce que vous soyez de Retour à Paris, oû je souhaite de Vous avoir lors de mon arrivée. ecrivez donc moi aussitôt que vous Recevrez celle ci à moins que vous ne l'ayez point avant le 6 ou 8 du mois prochain: car alors je vous prierai de m'ecrire à Bruxelles chez Monsg^{r}. Le Duc d'Aremberg: parceque comme je viens de vous dire en cas de ne point recevoir de vos nouvelles ici avant mon depart, J'irai alors en droiture à Bruxelles, & J'y attendrai votre Reponse. nous avons differents objets à discourir ensemble, oû J'ai besoin de votre avis & directions. à Dieu mon cher & bon ami. employez mes dispositions en tout ce que vous fera plaisir, car personne n'en aura plus à Vous marquer toute La gratitude qui vous est

1 Annotated by Duchesne: "Rien a faire attendre".

2 The instrument maker George Adams died on 17 October 1772.

3 Following Adams' death, his business was initially carried on by his widow Ann, and then by his son, George Jr (1750-1795). As Magellan indicates, following the death a few months earlier of Dr Gowin Knight, he had taken over the contract Knight had had with the Royal Navy to certify the improved mariner's compasses made by Adams for the Navy according to Knight's method. On this, see also Magellan to Duchesne, 7 July 1772.

due à votre bon Coeur, comme celui qui Le connoit à fond, & qui est pour La vie 72.10.27

Votre

Magalhaens

A Monsieur
Monsr. Duchesne Secret^{re}.
de M^{r}. Trudaine de Montigny
à Paris

To Jacques-André Mallet[1] 72.12.03

Bibliothèque de Genève, Geneva, Ms suppl. 1654, fol. 21-22

à Bruxelles 3 Dec^{bre}. 1772

Monsieur

Il ne m'a pas eté possible de m'acquiter plutôt de ma promesse; car ce ne fut que dans ce tems ci que J'ai pu sortir de Londres pour faire le voyage de Paris; oû Je compte me trouver dans la quinzaine, avant Noel. ainsi vous pouvez m'adresser à Paris votre certificat aussitôt, pourque Je l'y fasse signer par les membres etrangers de la S^{e}. R^{le}. de Londres que J'y connois; & dans la suite je l'apporterai avec moi, & le presenterai à la Soc. R^{le}. oû je ne manquerai pas de faire tout ce que je pourrai par mes amis, pour effectuer votre election. Mais celle-ci depend plus de l'hazard que du merite du candidat: Car si les autres concurrents ont plus grand nombre d'amis, le nombre de leurs voix l'emportera sur le plus digne, à cause du petit nombre de deux membres etrangers dont l'election est fixée pour chaque année.[2]

1 Annotated: "M^{r}. Magalhaens. X^{bre}. 1772. Rep". See Mallet to Magellan, 19 February 1773.

2 In earlier times, there had been no restriction on the number of foreigners who could be elected to the Royal Society. However, the number of foreign members eventually became unacceptably high and increasingly tight restrictions were imposed from the 1760s onwards, both on the number who could be elected and on the procedures adopted in electing them. In the early 1770s, there were in fact three foreigners, not two, being elected each year; later, a complete freeze on new elections was imposed for several years. Mallet appears never to have been nominated.

72.12.03 C'est à cette cause que, si je me resolve à devenir candidat un jour pour la meme societé, je ne me presenterai pour membre etranger, mais pour y etre admis entre les Nationaux, dont le Nombre n'est point fixé du tout: & selon une des nouvelles regles de la Societé, tout etranger qui aura vecû en Angleterre quelques mois, poura etre reçu entre les membres Nationaux, en paiant la contribution que ceux-ci paient annuellement; savoir 5 guinées en entrant & deux & demie chaque année.[3]

Il est bien honourable, Monsieur, pour l'Astronomie, que les grands & les Etats y emploient des sommes pour batir de beaux & commodes observatoires; mais elle n'en tire souvent autant d'avantages, comme lorsque des particuliers savants & zelés comme vous, les erigent eux memes, & en sont entierement les maitres. C'est presque toujours l'intrigue & la faveur passionée, qui place des genies mediocres dans des places pareilles: au lieu qu'un amateur intelligent scait surmonter les incommodités & defauts d'un observatoir mediocre; mais un astronome mediocre ne fera jamais beaucoup de bon dans le meilleur observatoire du mond. Le plan de votre intendu observatoire me fut communiqué par feu notre aimable D^r^. Bevis:[4] & je le montrai aussi à differentes personnes, qui le trouverent fort bien entendû. M^r^. Alex. Aubert[5] batit aussi à ses depens un petit observatoire, dans la campagne proche de Londres: & peut etre aurai-je quelques choses à vous communiquer de lui dans l'avenir.

J'ecrivis à Monsr. Pennington qui tient l'observatoire qui fut de M^r^. Ludlam à Cambridge,[6] & à M^r^. Hornsby[7] Professeur d'Astronomie à Oxford pour avoir quelques unes des observations correspondantes que vous souhatiez avoir: mais aucun d'eux n'en avoit point. Les seules que J'ai pû obtenir sont les suivantes, que M^r^. Maskelyne me communiqua, faites dans l'Observatoire Royal de Greenwich. Je vous envoye le papier original de la main du même M^r^. Maskelyne, pour plus de satisfaction tant de ma part, que de la Votre.

---

3 Magellan was in fact elected a Fellow on the Society's Home List on 21 April 1774.

4 Dr John Bevis, FRS (1693-1771).

5 Alexander Aubert, FRS (1730-1805).

6 William Ludlam (1717-1788) of St John's College, Cambridge, established an observatory in the college and in 1769 published the observations he had made there during the preceding two years (*Astronomical Observations made in St. John's College, Cambridge, in the Years 1767 and 1768...* [Cambridge, 1769]). When Bernoulli visited Cambridge in 1769, Ludlam had just been appointed to a living in Lincolnshire and had left the observatory in the charge of "M. Pellington", that is, Isaac Pennington, a young Scholar of the College who subsequently rose to become Regius Professor of Physic in the University and Master of St John's.

7 Thomas Hornsby, FRS (1733-1810), Savilian Professor of Astronomy at Oxford.

Je serois charmé d'avoir quelques nouvelles interessantes que vous communiquez sur votre cheri objet l'Astronomie, mais je n'en ai point. M^{r}. l'Abbé Chev^{er}., Biblioth^{e}. de la nouvelle librairie qu'on vient d'ouvrir ici publiquem^{t}., vient de sortir de chez moi, lui est un amateur aussi de l'Astronomie, comme vous pouvez voir par differentes de ses communications à la S. R^{le}. dont il est membre.[8] J'appris de lui que M^{r}. Piggot [Anglois][9] avec M^{r}. Needham,[10] sortirent hic matin d'ici pour Gent, à fin de continuer les observations necessaires pour determiner la latitude & longit. des villes principales de la Flandre Austriaque. Je suis faché de les avoir manqué par un jour; car je n'arrivai qu'hier soir ici. on va etablir ici une nouvelle Academie Imperiale des sciences.[11] Les lettres patentes sont attendues chaque jour, mais je ne sçais encore le detail. du moins J'espere que l'Astronomie n'y sera point oubliée. 72.12.03

Vous aurez vû dans une note du 2^{d}. vol. du Recueil pour l'Astron^{es}. de M^{r}. Bernoulli que j'ai fait quelqu'amelioration sur le quadrant marin. J'ai [...][12] un Memoire à la Soc. R^{le}.,[13] & ai communiqué au dit notre commun ami le substance des avantages de ma construction: peut etre lui en parlera plus au large dans son 3me. vol. Si la passion de pere ne m'aveugle pas, je me flate d'avoir porté cet instrument à une perfection qu'on n'attendoit point. Mes instruments sont à l'abri des ouvriers maladroits: car même etant mauvaisem^{t}. divisés, on y peut reconnoitre le valeur reele des angles qu'on prend avec eux. un octant des miens peut prendre des angles au delà de 120°., & les sextants de même: & la meme invention etant appliqué à des Quadrants Astronomiques, montrera & corrigera leurs defauts, soit originalment de la main de l'ouvrier en les divisant erroneously soit ceux qu'ils aient acquis, par l'alteration qu'ils aient [souferte]. il n'est pas possible de vous donner ici un plus grand detail: peutetre vous en parlerai-je plus au

8 Abbé Jean Chevalier, FRS, sent a number of reports to the Royal Society during the 1750s describing astronomical observations made at Lisbon while he was presbyter of the Congregation of the Oratory there. Having later moved to Belgium, he was in 1772 appointed Librarian at the newly renovated Royal Library in Brussels, formerly the library of the Dukes of Burgundy, that had recently been converted into a public library.

9 Nathaniel Pigott, FRS.

10 Abbé John Turbeville Needham, FRS (1713-1781), Director of the Académie Impériale et Royale des Sciences et Belles-Lettres in Brussels at its foundation.

11 The Académie Impériale et Royale des Sciences et Belles-Lettres of Brussels had its beginnings in 1769 and was granted Letters Patent by the Empress Maria Theresa of Austria on 16 December 1772.

12 Paper damaged.

13 Cf. 72.09.04.

72.12.03 large dans une autre. en attendant, si je puis vous etre utile à quelque chose, disposez de ma bonne volonté, etant avec toute consideration possible.

Monsieur
votre très humble & très obeiss[t]. serviteur
J.H. Magalhaens.

N.B. lorsque vous m'ecrirez à Paris, avant Noel, pretez une seconde envelope sur ma lettre, adressée à Monsr. Trudaine de Montigny Conseiller d'Etat & Intendant des Finances à Paris.

Je compte de retourner en Angleterre au commencem[t]. de Janvier prochain. mon adresse y est à present at n°. 25. Fetter Lane London.

A Monsieur
Monsr. J.A. Mallet, Astronome
Cour de S[t]. Pierre
Geneve

72.12.07 *To Louis-Henri Duchesne*

Archives Nationales, Paris, MS T/160 (10-12)

à Bruxelles 7 Dec[bre]. 1772

Monsieur & très Cher Ami

Je suis arrivé ici depuis 4 jours, & ne suis pas encore certain du jour que je pourrai partir d'ici pour Paris. C'est pour cette raison que Je vous ecris d'avance ces deux mots, pour vous annoncer, que La caisse avec les differents instruments, dont Je vous parlai dans ma precedente Lettre,[1] doit arriver bientôt chez vous; car elle etoit deja à bord du premier vaisseau qui devoit partir pour Calais, avant mon depart de Londres, qui fut le 27 du mois dernier. le mauvais tems m'obligea d'alonger ce voiage plus que je le pensois. J'ai traversé la mer de Douvre à Calais le 30 de Nov[bre].: & M[r]. Caffieri[2] me dit que Le vaisseau avec ma caisse n'etoit pas encore arrivé: mais à present il est très probable qu'il est effectivement arrivé, & que La caisse est deja en chemin pour Paris.

1 See Magellan to Duchesne, 16 October 1772.
2 See Magellan to Duchesne, 17 January 1772.

Lorsqu'elle y arrivera vous n'avez qu'à la ouvrir avec soin, pour en ôter les 3 aulnes de flanelle que vous m'aviez ordonnées,[3] & qui y sont empaquetées dans un pacquet separé. Car etant pour etre usées dans L'hiver, peut etre vous trouverez deja le tems assez froid pour en avoir besoin. il y a aussi un autre paquet avec du tricot fin de Laine, qui est pour M^r^. Le Chev^er^. de Bory. apparement il vous en a deja parlé, ou à Monsr. Trudaine, Là dessus: car il y a deja quelques mois (plus de 3 pour le moins) que Je lui recommendai de parler pour celà, vû qu'il n'etoit pas possible de trouver une autre voye pour L'envoyer. Pour ce qui regarde les autres articles contenus dans cette caisse, qui sont differents instrum^ts^. de Mathematique, & entre autres mon nouveau Quadrant marin,[4] pour Le faire voir à L'Academie avec d'autres choses pour La même &c. je vous prie de les Laisser rester tous empaquetés dans La meme caisse, jusqu'à ce que J'y arrive; car ce sont des choses qui se gatent aisement, lorsqu'on ne les connoit pas bien, ni comment les manier &c. 72.12.07

M^r^. Le Marquiz de Pezay,[5] que vous sans doute connoissez, & qui merite de l'etre par ses excellentes qualités, me pria de parler à Monsr. Trudaine (lorsque Je partis de Londres) pour permetre d'adresser sous son nom une petite caisse, contenant des morceaux d'histoire naturele, savoir morceaux de diferents mineraux, pierres, spars &c. qui sont pour La Comtesse de Monbarré,[6] dame d'honneur de M^de^. Adelaide.[7] Monsr. Trudaine, qui connoit & estime ce digne officier, ne manquera J'espere d'acorder cette faveur. ainsi je vous envoie la lettre ci jointe,[8] que je vous prie cacheter & jeter dans la poste d'Angleterre aussitôt que M^r^. Trudaine aura donné son agrement. elle contient les adresses & directions necessaires pour faire cet envoy de Londres avec sureté, selon qui Je les promis au dit Marquiz.

Je vous prie aussi d'envoier l'autre Lettre ci-jointe à notre estimable ami le D^r^. Sanchez.[9] elle contient un petit billet pour l'Abbé Bourriot,[10] que le

---

3 See Magellan to Duchesne, 16 October 1772.

4 See Magellan to Duchesne, 31 March 1772.

5 Alexandre-Frédéric-Jacques Masson, Marquis de Pezay (1741-1777).

6 Thaïs Françoise Parfaite de Mailly (1737-1819) in 1753 married Marie Alexandre Eléonore Louis César de Saint-Mauris (1732-1794), Count (later Prince) of Montbarey, minister and secretary of state of war. Sometimes the title was written Montbarré.

7 Marie Adélaïde de France (1732-1800) was the 4th daughter of King Louis XV. At first she was known as Madame Quatrième, then as Madame Troisième after her sister's death, and finally as Madame Adélaïde.

8 Letter not found.

9 Letter not found.

10 Note not found. L'abbé Bourriot was a canon at Bazas (Bordeaux) and an enthusiastic astronomer. He contributed to the perfection of achromatic lenses and the spherometer.

72.12.07 D[r]. envoyera par Son domestique plus aisement que vous ne le pouvez faire par les votres, ceux-ci aiant plus àfaire que le premier.

Je fais tout mon possible pour partir d'ici au plutôt. mais je crains ne pouvoir encore cette Semaine. Monsg[r]. le Duc d'Aremberg me fait des acueils les plus honourables & benignes: m'invite à diner souvent avec toute sa famille & personnes de distinction qui lui font compagnie à sa table. Lui va partir bientôt pour Paris avec M[dme]. la Duchesse, & le Prince son ainé, qui va epouser, comme vous savez, la fille du Comte de Lauragais.[11] Je fais monter actuelem[t]. chez le Duc une des nouvelles balances, dont je crois vous avoir parlé l'année derniere. J'y ai fait executer des ameliorations de mon invention qui la rendent très avantageuse. La caisse avoit eté gardée, toute fermée jusqu'à ce que J'arrivai ici, & fut ouverte sous mes yeux. un charpentier ou deux sont actuellement à apreter la charpente necessaire pour la monter sous une arcade de la seconde cour de son palais; oû elle servira, outre l'usage du concierge &c, pour etre examinée, & imitée à l'avantage du public: car sa construction est telle que sans l'usage d'aucun poids & sans d'autre trouble que d'y mettre les choses à peser dans le bassin, on voit tout de suite le total du poids par un index, ou main qui le montre sur un quadrant. Je lui ai expedié aussi environ 12 boisseaux de l'orge de Siberie (dont je vous ai envoyé un petit sac le perimtems[12] dernier) que lui (le Duc) veut faire semer par differents de ses fermiers, à fin de repandre dans le pais ce grain precieux pour l'avantage public. il est bien à souhaiter que tous Les grands eussent de pareils penchants pour le bien des hommes. tout le mond y gagneroit: & le bonheur n'en seroit pas si eloigné comm'on le voit ordinairement de la Societé Civile.

à Dieu mon cher & bon ami, jusqu'au revoir. nous avons quelques objets utiles de quoi nous entretenir. Votre coeur est sensible à la dignité de pareils entretiens: & meme dans le cas de ne pas etre àmeme pour les executer, nous aurons du plaisir à en discourir.

Je suis pour La vie
Mon cher ami
Votre
Magalhaens

N.B. après avoir faite celle-ci Je Reçois une Lettre du D[r]. Sanchez, qui me l'adressa ici.[13] ainsi je n'ai plus besoin de lui ecrire avant mon depart.

---

11 Louis Englebert, later 6th Duc d'Arenberg (1750-1820), married in 1773 at Saint Sulpice, Paris, Louise de Brancas-Villars, Comtesse de Lauragais (1755-1812), daughter of Louis Léon Félicité, Duc de Brancas and Comte de Lauragais (1733-1824).

12 printemps. See Magellan to Duchesne, 31 March 1772.

13 Letter not found.

To Matthew Boulton[1]

73.00.00

Birmingham Central Library, Boulton and Watt collection, MBP 244, 52

Magalhaens's Complim^{ts}. wait on M^{r}. Boulton, and sends him this half dozen of steel bars; which he desires to be sent to Birmingham, in order to be hardned and polished for magnetism. as to the hardness he desires they may be hardned to the utmost possible. but the polishing is no need of being the highest. Magalhaens desires to know what is the expence by the dozen of such workmanship, as he has a great number of them to undergo the same operation: this being but a tryal to ascertain the whole expence. therefore begs that a good guess of it may be send back to him along with these samples when done as above.[2]

Tuesday after noon at n°. 4. old North street Red Lyon Squ^{re}., Holborn.

To
M^{r}. Boulton
at n°. 14 Cannon street
London

1 Annotated "London Magalhens". The letter is undated. That it was written in 1773 has been established by comparison with Magellan to Boulton, 9 September 1773, in which Magellan follows up his request concerning the steel bars.

2 Following the death of Dr Gowin Knight on 8 June 1772, Magellan took over Knight's contract to supply magnets, made according to Knight's method, to the Royal Navy. The steel bars were needed to rebuild the section of Knight's "magnetical machine" that had been destroyed, along with most of Magellan's belongings, in a fire at his house in December 1772 while he was away in Paris.

73.02.17 *From Jacques-André Mallet*

Bibliothèque de Genève, Geneva, Ms suppl. 1654, fol. 47[1]

17 Feb. 1773.
par M^{r}. Deluc.

M^{r}.

J'ai bien reç. au mois du Xbre dern. la lett. que vous m'av. fait l'hon. de m'ecrire[2] & je vous remercie très sincer^{t}. de l'offre oblig^{te}. que vous m'y faites de vouloir m'aider de votre recomm. auprès de vos amis, p^{r}. me faire obten. l'hon. d'etre memb. de la Soc. R. plus^{rs}. raisons m'ont determiné à renvoyer ma presentation p^{r}. une autre ann. p^{r} la quelle je pense que vous voudr. b.[3] me conserver votre bo[nn]e volonté.

J'aurai peutetre occas. d'envoyer dans l'interv. à la Soc. R. quelques papiers ou observat. qui pourroient contribuer en qlque chose à la réussite de cette élection, j'ai d'ailleurs appris, que M^{r}. Deluc mon compatriote etoit sur les rangs cette annee cy & il vaut mieux ne pas nous metre dans le cas de nous nuire reciproquement.[4] C'est ce même M^{r}. Deluc qui part actuellem^{t}. p^{r}. Londres, & qui se chargera de vous remettre ma lettre, c'est un homme de genie, & très habile physicien, qui vient de do[nn]er au public un ouvrage sur le Barometre & Thermom. & sur les modif. de l'atm. qui renferme beaucoup de choses intéressantes & nouvelles.[5] vous vous etes occupé, M^{r}. avec succès de ces memes objets, & vous serez sans d^{te}. bien aise de faire connoiss. avec M^{r}. Deluc. Vous aurez peutetre aussi entendu lire un mèmoire qu'il a envoyé depuis peu à la Soc. Roy. sur un Hygrometre de son invention.[6]

Ce que vous me dites de l'amelioration que vous avez donnée au quad^{t}. des marins me fait tres fort souhaiter de voir votre memoire, qui doit etre fort curieux, il sera sans d^{te}. imprimé dans les Transact, & j'espere, que préablem^{t}. M^{r}. Bernoulli nous en donnera un extrait dans son 3e vol.[7]

---

1 Letter not found. The text is from a draft kept by Mallet.

2 See Magellan to Mallet, 3 December 1772.

3 bien?

4 J. A. Deluc was elected a Fellow of the Royal Society on 10 June 1773. Mallet, however, seems never to have been nominated.

5 Deluc, *Recherches sur les modifications de l'atmosphère: contenant l'histoire critique du baromètre et du thermomètre...* (Geneva, 1772).

6 De Luc read his "Account of a new hygrometer" to the Royal Society on 10 June 1773; it was subsequently published in *Philosophical Transactions*, 63 (1773), 404-60.

7 Magellan read a paper on "A new construction of the sea-quadrant" to the Royal Society on 17 December 1772, but it was never published in the *Philosophical Transactions*. Instead, an expanded version was published separately in Paris as *Description des*

J'ai eté occupé pend^{t}. tout l'été passé à dirig. la constr. de cet observat. 73.02.17
que je vous avois annoncé dans ma dern. lettre,[8] il a eté bati sur le même plan que vous connoissez, & il a assez b. reussi les instrum. y sont placés & en etat de servir depuis le commencem^{t}. de cette annee, quelque belles nuits de Janv. m'ont permi d'observ. l'opposit. de Mars, & je me dispose à observ. celle de Saturne a la fin de ce mois cy mon intention est de ne negliger aucune des observat. qui se presenteront, & d'en calculer les résultats à mesure, je crois qu'il est import^{t}. que l'observ. calcule luimeme ses observat, il est plus en etat que pers. de juger du poids de chacune, & de choisir parmi plusieurs celles qu'il estime les plus sures, d'ailleurs elles couroient gr^{d}. risque de n'etre jamais employées car il me semble qu'on observe beaucoup, & qu'on calcule peu.

Je vous remercie des 2 observ. d'ecl. de sat. de ♃ que vous m'avez procurées, la differ. des merid. de Paris & Geneve que j'en ai conclu est en tems de 14':40'', & c'est exactem^{t}. le même résultat que avois tiré cy dev^{t} par d'autres observ. comparées à celles de M^{r}. Mess.[9] cette differ. des merid. est bcp plus petite que celle qu'on trouve dans la connois. des T. qui est des 17'[10]

Je vous serois bien obligé, M^{r}. de vouloir me procurer à votre loisir les correspond^{tes} à ci deux cy

1772. 22 Xbre, 7: 18: 44. Em. 2^{d} Sat. ♃ bo[nn]e obs. lun. [...][11]

1772. 23 Xb. 4 47 22. Sat. 1^{r}. assez bien quoique dans le crepusc.

Je serai b. charme que vous me donn. qlfois des nouv. de l'observat. & des trav. de M^{r}. Alex. Aub.[12]

M^{r}. Magalhaens
at N°. 25 Fetter Lane.
London.

---

octants et sextants anglois ... (Paris, 1775). The original paper is at the Royal Society, *Letters & Papers*, V, 323.

8 See Mallet to Magellan, 30 March 1772.

9 i.e. Messier.

10 *Connoissance des temps pour l'année... 1773* (Paris, 1771), p. 134.

11 Word or words illegible.

12 i.e. Alexander Aubert. Mallet's draft is unsigned.

73.03.05 *To Louis-Henri Duchesne*

Staatsbibliothek Preussischer Kulturbesitz, Berlin, Sammlung Darmstädter, G2 1770(1)

à Londres 5 Mars 1773
at n°. 4 Old North Str. Red Lion Square

Monsieur & cher ami

Je suis arrivé ici vers la fin de la semaine derniere après un voyage très dangereux sur mer. Celle-ci acompagne une brochure très interessante que j'envoye à M^{r}. Trudaine sur les experiences faites par le D^{r}. Priestly sur differentes sortes d'air.[1] Je vous prie de lui faire souvenir la lettre qu'il promit de me envoyer pour Milord de Shelburn,[2] & sur lequelle je vous ai deja parlé de Bruxelles. aiez la bonté de me dire si la soude des rivieres dont je vous envoiai la description fut goutée: & si M^{r}. La Voisiere a reussi dans l'entreprise pour desaler l'eau de mer par la nouvelle methode.[3] Je me trouve si desorienté par la perte de mes papiers, & de mon logis, que je ne trouve pas assez de quoi entretenir mes amis.[4] le tems me manque: & je ne scai pas de quel coté me tourner le premier. Cependant je espere dans peu de semaines de me remettre comm'il faut dans l'antien train de vie; mais il faudra bien travailler pour celà. à Dieu mon cher & bon ami, souvenez vous de celui qui est sincerem^{t}.

Votre de coeur & pour la vie
Magalhaens

A Monsieur
Monsr. Duchesne
Secretaire de Mr Trudaine de Montigny &c.
à Paris

---

1 Joseph Priestley, "Observations on Different Kinds of Air", *Philosophical Transactions*, 62 (1772), 147-264.

2 William Petty, Lord Shelburne (1737-1805), Priestley's patron and one of the leading British statesmen of the day.

3 Lavoisier was a leading member of a commission established by the Académie Royale des Sciences to investigate making sea water potable, following Magellan's reporting a method of doing this; see Magellan to Duchesne, 4 May 1772.

4 Magellan's lodgings at Fetter Lane had burnt down on 21 December 1772, while he was in Paris.

To Pierre Macquer 73.03.23

Bibliothèque Nationale, Paris, MS fr. 12306, f. 19-20

Londres 23 Mars 1773 at n.° 4
Old N. Str^t^. Red Lyon Square

Mon cher Monsieur

Je n'arrivai ici que vers le commencem^{t}. de ce mois: & le grand desordre de mes petits affaires à cause de l'incendie qui consuma mon Logis,[1] ne m'a pas permis de vous marquer ce que Je trouve à L'egard de la Lunete que vous m'avez ordonnée. premierem^{t}. si vous vous contentez d'une petite d'environ 15 ou 16 pouces de foier à triple objectif de 2 pouces d'ouverture comme celle de M^{r}. Bezout,[2] elle ne vous coutera que <u>10</u> louis d'or ici; & vous pouvez l'avoir tout de suite. Mais si vous la voulez d'environ 3½ pieds, à double objectif d'environ 30 lignes d'ouverture, pour lors il faut la faire faire tout exprès. M^{r}. Dollond me promit de l'entreprendre, par le même prix de <u>10</u> Louis (ou guinées) elle sera en tuiau de mahogany, avec un piedestal &c, au lieu que l'autre de 16 ou 17 pouces a le tuiau de cuivre qui Reçoit au-dedans son piedestal, comme vous pouvez voir chez M^{r}. Bezout qui en a une fort bonne. Vous n'avez qu'à prendre votre resolution, & me la communiquer, en m'envoiant aussi l'argent par Lettre de change ou comme vous voudrez. si vous parlez à M^{r}. Messier, peut etre il aura quelqu'argent aussi à m'envoier & se chargera de tirer La Lettre de Change. N.B. il faudra ajouter audelà des <u>10</u> louis quelques livres de plus [8$^{\#}$ ou 10$^{\#}$] pour paier la caisse, porteur, & lettres &c. ou autrem^{t}. je vous en tiendrai compte dans la suite.

Je vous prie de dire de ma part à M^{r}. Cassini Le fils, que Son Mémoire a eté presenté à la Societé R^{le}. J'en fis la question au D^{r}. Matty, secretaire de cette Societé qui me Repondit dans L'affirmative.[3]

---

1 On 21 December 1772, there was a fire at Magellan's dwelling, 25 Fetter Lane, while he was away in Paris. Following his return to London, he went to live at Old North Street, Red Lyon Square for a few months before going to 12 Nevil's Court, Fetter Lane where he lived until his death in 1790.

2 The mathematician Étienne Bezout (1730-1783).

3 César-François Cassini de Thury or Cassini III (1714-1784), French astronomer and cartographer. No record has been found in the Royal Society's *Journal Book Copy* of Cassini's paper having been read at a meeting of the Society, and neither was it published in the Society's *Philosophical Transactions*.

73.03.23 En toute occasion que je pourrai vous etre bon à quelque chose, faitez usage de mes bonnes volontés, & vous ne manquerez de trouver combien Je suis de Coeur

Mon Cher Monsieur

Votre hum^{le}. & Ob^{t} Serv^{r}.

Magalhaens

A Monsieur
Monsr. Macquer Professeur de
Chymie de l'Acad. R^{le}. des Sciences
Rue S^{t}. Sauveur
A Paris

73.03.23a *To Nathaniel Pigott*

Royal Astronomical Society, London, MSS Pigott, 25

Dear Sir

As soon as I arrived here, did sent to M^{r}. Maskelyne your Letter: and the next time I saw him in the Royal Soc^{y}., he assured me to have already sent the observations he had to you, tho' he could not furnish them all. I have wrote to two acquaintances in Oxford, and Cambridge: and have not yet any answer. besides I applyed to y^{e}. h^{le}. M^{r}. Beauclerc,[1] and just now received from his Astronomer the following.

		equal time	app^{t}. time
Novr. 14	Emers. ♃'s 1st. Sat.	5^{h}.44^{m}.0^{s},6	= or 5^{h}.59'.8''.
Decr. 23	Em. ♃ 1st. Sat.	4.21.43,7	= or 4.21'.41'',6

These observations were made (with a 3½ feet Achromatic of M^{r}. Ramsden's make) with y^{e}. utmost exactness. I believe there could not be an error of half a second. the Long^{de}. of Muswelhill (the place of M^{r}. Beauclerc's observatory) is 29^{s}½ of time west of Greenwich: it was determined from y^{e}. Latitudes & Distance: the distance was taken from Rocque's

1 Hon. Topham Beauclerk, FRS (1739-1780), a wealthy patron of science and literature, had established a private observatory at Muswellhill, London.

Map.[2] I suppose there is not an Error of 2 seconds. I believe the moon's Eclipse was not observed in England, at least I'm certain it could not any where near London. &c.... Signed Is. Dalby. who is the astronomer kept at Muswelhill's Observʸ. by Mʳ. Beauclerc.[3] 73.03.23a

N.B. I have seen the above 3½ f. achrom. of Ramsden the last summer. it is a triple objectif one, exceedly good, at least as the best of Dollond's make. but am sorry that Mʳ. Dalby did not expresse the magnifying power used in these observations. if you think necessary, I'll write to him again for that question.

Mʳ. Ramsden tells me that about ten days ago he did sent a barometer to you: but add that it was directed for Mʳ. Needham.[4] therefore please to speak with Mʳ. Needham, and see if this already sent was realy arrived for him at Ostend: or if he knows any thing about it: that I may send one I have already got from him, to your friend at Dunkerque, according to your desire: but for fear of any mistake, I'll keep it in my hands, till I receive your answer. for I do not account at all, nor do relay on what Ramsden sais. please to tell me also, where shall I apply for the 5½ guineas I have already paid for this Barometer, and for the expenses of yᵉ. box and sending it aboard yᵉ. ship &c.

You was very right (and I also) in thinking that my account of yᵉ. new Octˢ. & Sextants should be undone, and not printed in yᵉ. Royˡ. Socʸ.: the worthy gentleman you know,[5] took it into pieces in yᵉ. most shamefull, tho' artfull, way. is a pity that his examen of yᵉ. said Memoir, is not given to yᵉ. public. it should well expose his ingenuity, as well as his unfair and injust reasoning and partiality. However the Public is not to be mislade by private animosities. I intend to publish the said Memoir; and add such notes, as to make appear the unfair objections. but you need not to say this my intention to nobody in the mean time.[6]

2 John Rocque, *A Plan of the Cities of London and Westminster and Borough of Southwark with the Contiguous Buildings* (London, 1746).

3 Isaac Dalby (1744-1824), a mathematics teacher who was employed by Topham Beauclerk until the latter's death in 1780 to operate his observatory at Muswellhill. He was subsequently employed on the trigonometrical survey to connect the meridians of Greenwich and Paris.

4 John Turbeville Needham (1713-1781), founding Director of the Académie Impériale et Royale des Sciences et Belles-Lettres at Brussels.

5 The Astronomer Royal, Nevil Maskelyne, who favoured the alternative improvements due to Peter Dollond, presented a highly critical account of Magellan's quadrant to the Royal Society on 21 January 1773 (Royal Society, *Journal Book Copy*, vol. XXVII [1771-4], p. 333 ff.).

6 Cf. Magellan's *Description des octants et sextants anglois* (London, 1775).

73.03.23a Please to tell M^{r}. Gordon,[7] when you see him, that I am waiting for his answer about y^{e}. particular he recommended to me; and upon which I have already wrote to him by the post.[8]

Pray my humble Resp^{s}. to M^{dme}. Pigott, & your Son: and to any friends, who honour me with their memory. If I am of any Service, make free use of my dispositions, and believe me for Life

Dear Sir
your most ob^{t}. humble Servt.
Magalhaens

London 23 Mars 1773
at n°. 4. O. N. Str^{t}. R. Lyon Sq^{re}.

A Monsieur
Monsr. Pigott Gentilhomme Anglois
vis à vis La Cour
à Bruxelles

7 William (later Sir William) Gordon, British Minister Plenipotentiary at Brussels, 1765-1777?

8 Letter not found.

73.03.23b *To Johann III Bernoulli*

Öffentliche Bibliothek der Universität Basel, MS L I a 709, f. 41-42

Londres 23 Mars 1773
at n°. 4. O. N. Str^{t}. R^{d} Lyon Square

Monsieur & très Cher ami

Je viens d'arriver depuis peu de Flandres, & Paris, oû je passai Le Noel. aussitôt je passai chez M^{r}. Gordon un des plus fameux Seedsmen de Londres,[1] & lui donnai l'ordre des Semences les plus variées de

1 The nurseryman James Gordon (*ca.* 1708-1780).

l'Amerique [q]ue vous me ordonnez pour L'Academie R[le].; mais ce ne fut qu'hier que J'ai pu les avoir. Lui m'assura qu'elles arriveront dans le tems plus propre pour etre Semées: ainsi J'ai remis La caisse, marquée Acad[e]. R[l]. Berlin, à M[r]. Ernst[2] selon votre avis, qui me paia mon deboursem[t]. & vous l'envoiera au plutôt possible. elle con[ti]ent 111 or 112 sorts de semences differentes, pour lesquels il me char[g]ea £8:14[s]:0 Est. etant des plus completes & rares suites qu'il a jamais faites. On y trouvera dedans la liste detaillée, avec sa quitance. vous y trouverez les Tables pour la Refraction & Paralaxe,[3] que M[r]. Shepherd m'avoit [pro]mis pour vous. à vous dire le vrai, je ne l'ai pas encore vû depuis mon [reto]ur. mais en passant chez le Libraire, & les voiant publiées, je lui dis que [le] D[r]. m'avoit promis une copie pour vous. Lui m'avoua qu'il souhaitoit en [en]voyer à Berlin, mais qu'il n'en avoit point d'ordre &c. Cependant j'ai [...][4] sur moi cet affaire, pour ne pas perdre cette occasion. & je vais ecrire bientôt à lui. J'espere qu'il ne le trouvera pas mal. C'est un gros vol. in folio, tout en tables, dont vous verrez L'utilité. J'y ai mis les les tables de Dodson[5] [4[s].6[d]] N.B. le meme A. a fait d'autres Tables beaucoup plus avantageuses & utiles des logarithmes & nombres &c.[6] mais celles-ci coutent plus d'une guinée & meme 1½. / plus Flora Virginica[7] [7[s]:0] Flora Ameri[cae] Septentrionalis[8] [1[s].8[d]] / un pamphlet de Landen[9] [1[s]:6[d]] le porterage & lett[res] 4[s].6[d]. en tout £9:12[s]:6[d] que J'ai reçus de Mons[r]. Ernst, & dont je lui passai Reçu. 73.03.23b

aussitôt que M[r]. Ferguson publiera l'usage de sa Rotula, je vous envoyerai une copie.[10]

---

2 From the trading house of Smith & Ernst, Great St Helen's, London?

3 *Tables for Correcting the Apparent Distance of the Moon and a Star from the Effects of Refraction and Parallax*, ed. Anthony Shepherd (London, 1772).

4 Word obliterated.

5 James Dodson, *The Calculator, being...Tables for Computation, adapted to Science, Business, and Pleasure* (London, 1747).

6 James Dodson, *The Anti-Logarithmic Canon; being a Table of Numbers, consisting of Eleven Places of Figures, corresponding to All Logarithms under 100,000...* (London, 1742).

7 Gronovius, *Flora Virginica* (2nd ed., Leiden, 1762).

8 Forster, *Flora Americae septentrionalis* (London, 1771).

9 Presumably John Landen, *Animadversions on Dr Stewart's Computation of the Sun's Distance from the Earth* (London, 1771).

10 James Ferguson, *The Description and Use of the Astronomical Rotula...* (London, 1775)?

73.03.23b mes nouveaux Quadrants & Sextants Marins sont approuvés, & estimés par tous ceux qui jugent sans partialité, ni prevention. Ma[is] il en a beaucoup qui s'entrainent & se laissent entrainer par des vües & raisons particulieres. C'est le public à juger, qui a tort. reelem^t. vous m'aviez promis sans faute de inserer dans le 3me. volume de votre recueil le precis de ce que je vous avois communiqué: & ne puis pas etre très content de ne le voir que si vaguem^t. annoncé dans Le second vol.[11]

J'ai eté à Paris peu de tems: & j'y ai vû plusieurs fois M^r. de la Lande & nous avons parlé en bonne harmonie & amitié.

il n'a pas ici aucun Dicionaire Irlandois publié depuis peu. l'ouvrage de feu M^r. Wood sur Homere n'est pas encore imprimé.[12]

Vous n'avez que mettre dans une Caise Les Memoires de Votre Acad^e. viz. de 1767-68-69 & 1770, & les envoier par Hambourg, ou comme vous voudrez adressée à M^r. Robertson[13] Bibliothecaire de la Soc^e. R^le. en Crane Court London: en chargeant dessus tous les deboursem^ts., qui seront ponctuelem^t. paiés par La S^te. R^le.

Je fera[i] tenir bien tôt à M^rs. Valtravers[14] & Maskelyne Votre 2^d. vol.:[15] car je me trouve si acable d'affairs, que je n'ai pas encore pû y penser. Ma Maison c'est à dire mon Logis, qui consistoit de 3 Chambres un bon grenier & Caves en Fetter Lane fut tout consumé dans un incendie avec touts mes instrum^ts. papiers, meubles &c. &c. le meme jour que J'arrivai à Paris, le 21 de decembre dernier. il me faut arranger toute mil choses, qui en sont la consequence: & si ce n'etoit pour ne pas perdre La Saison, je n'executerois aussitot l'ordre de l'Academie.

J'en ai deja parvenû M^r. Valtravers un de ces jours du present que vous Lui faites.

ni M^r. Landen ni M^r. Waring[16] n'ont point publié aucune chose dernierem^t. outre la petite brochure du premier que je vous envoie.

Voilà tous Vos articles Respondus Je crois. à present il ne me reste qu'à vous souhaiter de la Santé, & vous dire que selon moi, & plusieurs autres, vous devriez faire quelques petits voyages, & suspendre vos applications

---

11 Despite Magellan's remonstrances, Bernoulli did not include a more adequate account of his invention in the third volume of his *Recueil pour les astronomes*, when this was eventually published in 1776.

12 The new edition of Robert Wood's *Essay on the Original Genius and Writings of Homer* eventually appeared in 1775.

13 John Robertson, FRS (1712-1776), Librarian at the Royal Society since 1768.

14 Rodolph Valltravers, FRS.

15 i. e. of Bernoulli's *Recueil pour les astronomes*.

16 Edward Waring, FRS, Lucasian Professor of Mathematics at Cambridge.

Literaires. La vie sans Santé, n'est que miserable: & c'est de notre interet de la Rendre autant agreable qu'il est possible. à Dieu mon Cher ami: Croyez moi pour La vie 73.03.23b

mon cher
Votre de Coeur
Magalhaens

N.B. Je n'ai pas pû obtenir encore aucun Compte net de la Vente de vos Livres; M^r. Elmsley se trouve à present fort occupé.

A Monsieur
$Mons^r$. Bernoulli Astronome Royal de Prusse
à Berlin.

To Pierre Macquer 73.04.04

Bibliothèque Nationale, Paris, MS fr. 12306, f. 21-22

Londres 4 Avril 1773

Mon cher $Mons^r$.

Je viens de Recevoir par cette poste avec La lettre de change pour £12: 8^s:11^d Esterl. dont je vous tiendrai compte. Mais je vois par votre Lettre[1] que vous vous avez mepris, ou moi Je me suis mal expliqué.[2] car La Lunette de 10 guinees n'est pas faite par Dollond: quoiqu'à vous dire le vrai cet ouvrier ne les fait point inferieures. Si vous vous donnez La peine de passer chez le $Chev^{er}$. de Bory qui a une de cette espece faite par Dollond, & dans La suite passer chez M^r. Bezout[3] qui a une autre faite par le meme ouvrier (Ramsden) dont je vous parle je crois que vous ne trouverez aucune difference sensible, quoique La premiere (de Dollond) coute 15 guinées, & la Seconde (de Ramsden) ne coute que $seulem^t$. 10 guinées. peut etre que Dollond ait à cette heure rebaissé le prix des siennes: car je n'en ai

1 Letter not found.
2 See Magellan to Macquer, 23 March 1773.
3 The mathematician Étienne Bezout (1730-1783).

73.04.04 pris depuis long tems: mais je ne crois pas qu'il les donne si bas comme le premier. Considerez là-dessus, & donnez votre Resolution

a votre tres h. S^{r}.

Magalhaens

a Monsieur
Monsr. Macquer de l'Acad. R^{le}. des Sciences
Rue de S. Sauveur
à Paris

73.04.29 *To Johann III Bernoulli*

Öffentliche Bibliothek der Universität Basel, MS L I a 709, f. 45-46

Londres 29 Avril 1773
at n°. 4. Old N. St^{t}. R. L. Sq^{re}.

Monsieur & cher ami

Le porteur de celle-ci est M^{r}. Wolfe,[1] dont je vous parlai l'année derniere, qui ne manquera de passer chez vous quelque tems dans cet eté. C'est un vrai digne homme & un de mes meilleurs amis; Je vous Le recommande comme tel à vos bonnes graces; & ne manquerai de prendre toutes celles, dont vous le favourirez, comme faites à moi-même. Lui vous donnera une brochure du D^{r}. Priestley,[2] que je vous prie de presenter de ma part à M^{r}. Margraaf,[3] & de Lui recomender M^{r}. Wolfe, avec lequel il ne manquera de se plaire, aiant acquis une grande connoissance dans Les matieres de Chymie.

---

1 Peter Woulfe, FRS (?1727-1803). It seems that Woulfe did not go to Berlin that summer, after all (see 73.11.27), but somehow Magellan's letter got to Bernoulli anyway.

2 Probably Joseph Priestley, *Directions for Impregnating Water with Fixed Air* (1772).

3 Andreas Sigismund Marggraf (1709-1782), director of the chemical laboratory of the Berlin Academy of Sciences since 1753.

Voici l'article de la Lettre que J'ai reçue de M^r^. le D^r^. Shepherd = I am glad that you had forwarded my intended present (les Tables pour la refraction & paralaxe) to M^r^. Bernoulli.[4] I intended to sent a Letter along with y^e^. same to thank him for y^e^. present he made me of his voyage Astronomique. You'll suply this deffect &c.... In his account of Cambridge he says that I saw Venus in the Meridian before M^r^. Maskelyne at Greenwich:[5] had he said Mercury, it had been true... If he should think it worth his while to mention these Tables (celles dont il vient de Vous faire present) in his next recueil, I could wish he would confine himself to what he finds in the Preface. = Vous aurez [d]onc[6] soin de penser au contenû dans cet article de M^r^. Shepherd. 73.04.29

Voici le compte courant de M^r^. Elmsley, que Je viens de Recevoir dans ce moment.

M^r^. Bernoulli

Cred^r^.		per Contra	
19 Recueil. vol. 1. at 3^s^: --	£2:17^s^:0^d^	16 vol. du Recueil qui restent encore à vendre -----------	2: 8: 0
24 d^o^. -- vol. 2^d^. at 3^s^: ----	3: 12: 0	14 Lettres à vendre -------	1: 2: 6
27 Lettres Astr.	- at 1: 9 -- 2: 2: 9	3 Recueils donnés de presents & 2 lettres* ----------------	:12: 6
	£8: 11: 9	payé à M^r^. Magalhaens ------	4: 3: 0
	paid --- 8: 6: 1		8: 06: 0
Crediteur de --- £	5: 8		

* Les trois Recueil dont il est question sont ceux du vol. 2^d^. qui vous m'avez ordonné de presenter de Votre part à Messrs. Valtravers, Maskelyne, & à la Societé Royale. pour ce qui regarde les deux copies des Lettres Astronomiques, Je ne me souviens plus à present. apparem^t^. une etoit pour La Societé R^le^. &^a^. mais mes papiers etant tous brulés comme je vous l'ai dit dans ma derniere, Je ne puis vous parler exactement de celà.

Vous voyez bien qu'en ostant les livres qui restent encore à vendre du produit de ce qu'on a vendu, & de meme les £4. 3^s^. 1^d^ que Je pris pour Les livres que Je vous envoyai l'année derniere, il ne reste d'argent Liquide entre les mains de M^r^. Elmsley que Seulem^t^. 5^s^: 8^d^. Esterl. jusqu'à ce moment.

4 Cf. 73.03.23b.

5 Bernoulli, *Letters astronomiques*, p. 120.

6 Conjectural reading.

73.04.29 Voilà tout ce que Je puis dire à present, n'ai[ant][7] autres nouvelles interessantes à Vous communiquer, excepté le voyage qu'on vait faire vers Le pole, pour y faire des decouvertes. C'est le Capitain --- je ne puis pas me souvenir à present le nom.[8] M^{r}. Wolfe vous le dira de bouche. M^{r}. Lyons[9] fameux calculateur dont vous avez Le nom sur La solution de differents problems l'acompagne je crois dans La capacité de Astronome ou philosophe &c.

Je vous souhaite une meilleur santé, & vous prie au nom de tous vos amis de l'epargner autant qu'il depend de vous.

J'ai l'honneur d'etre
Mon cher ami
Votre très humble & obeiss^{t}. Serv^{r}.
Magalhaens

P.S. Le livre de M^{r}. Wood sur le caracter & genius d'Homere n'est pas encore imprimé.[10]

A Monsieur
Monsr. Bernoulli, Astronome R^{l}. de Prusse
à Berlin

avec une brochure pour M^{r}. Margraaf.

7 Conjectural reading.
8 In 1773 an expedition under Captain Constantine Phipps (later Baron Mulgrave) set out to find a northern route to India. Blocked by ice to the north of Spitzbergen, the expedition returned without any result.
9 Israel Lyons (1739-1775), calculator for the Board of Longitude.
10 Cf. Magellan to Bernoulli, 23 March 1773, note 15.

To Nathaniel Pigott

73.04.30

Royal Astronomical Society, London, MSS Pigott, 26

Dear Sir

I received yesterday the following observations in a paper which the R^{c}. Francis Wollaston F.R.S. communicated to me, it being a copy of his account inserted in the Phil. Transactions almost ready to be published.[1] they were made at <u>Chislehurst</u> in Kent in his observatory, whose latitude is 51°.24'.33''. and its Longitude 4'.39''. (= 18'',6. in time) east of y^{e}. R^{l}. Observatory at Greenwich. amongst many other observations contained in this paper the following are of y^{e}. number you desire.

Eclipses of Jupiter's Satelites observed with a 3½ foot achromatic refractor of Dollond, with triple objectif, magnifying <u>100</u> times.

1772	app. time	
Aug. 28	8^{h}.28'.14''	first Satel. emers. very doubtful.
Oct. 20	11.14.32	ditto - - - doubtful.
Oct. 20	7.40.15	first Satelite in contact
	7.44.20	- - - occultation

N.B. these two Last observations were made with the same Refractor, magnifying 150 times. the author sais, that they are not to be depended upon within several seconds; and yet, if pursued regularly, might have their use &c.

Upon the whole I dont see any advantage to be had for your purpose from these observations: but I thought fit to send them, as a mark of my care and regard to your commands. I cannot help being surprised at not receiving any answer from you to the query about the Barometer, which M^{r}. Ramsden says to have sent to you, directed to M^{r}. Irwin, Consul at Ostend,[2] for M^{r}. Needham.[3] by these means I know not if M^{r}. Debrou[4] or M^{r}. Needham is to have it; tho' the Later tells me to know nothing of

1 Wollaston, "Astronomical Observations made at Chislehurst in Kent", *Philosophical Transactions*, 63 (1773-4), 67-78.

2 Robert Irvine, British Consul at Ostende, Bruges, &c.

3 Cf. Magellan to Pigott, 23 March 1773.

4 Presumably Captain (later General) Philippe-Joseph De Brou (1732-1796), an engineer in the Austrian Army in Flanders.

73.04.30 any barometer being arrived to Ostende &c. perhaps the whole may be a Tale or a mistake of M^{r}. Ramsden. I thought it should be decided by your answer: and in the mean time do suspend the sending any, either to your friend at Ostende or to M^{r}. Needham at Bruxelles.

I am with the greatest Regard

Sir

your most Obed^{t}. humble Servt.

Magalhaens

London 30 April 1773
at n°. 4. Old N. St^{t}. R. L. Sq^{re}.

P.S. I could not obtain any observations from Cambridge nor from Oxford. D^{r}. Shepherd[5] tells me in a Letter of y^{e}. end of y^{e}. last month[6] to have none at all: and the silence of y^{e}. two other Astronomers makes me think they are in the same circonstance. as to M^{r}. Maskelyne he told me to have send you the few he had, corresponding to yours. if necessary, I'll write to Geneva & Berlin, where I have two friends in the astronomical way.[7] but without any answer from you, I cannot, nor chuse to take any steps further.

A Monsieur
Monsr. Pigott, Gentilhomme Anglois
vis à vis La Cour
à Bruxelles

5 Anthony Shepherd (1722-1795), professor of astronomy at Cambridge.
6 Letter not found.
7 i. e. J. A. Mallet and Johann III Bernoulli respectively.

To Louis-Henri Duchesne

Archives Nationales, Paris, MS T/160 (10-12)

Londres 11 May 1773
at n°. 4 Old N. St. R. L. Sq^{re}.[1]

Mon cher ami

Je vous envoye la Lettre ci jointe pour M^{r}. Le Ch^{er}. de Bory[2] en Reponse aux questions qu'il m'a faites pour M^{r}. de Montigny. il s'y agit de la meilleure construction des hydrometres, ou aerometres, ou pese-liqueurs: J'ai bien vû comm'on s'y sont pris quelques uns de vos savants. M^{r}. L'Abbé Bourriot & le Sucesseur de l'Abbé Nolet[3] ont travaillé sur cet sujet: mais quoique J'ai assez bien compris & leur theorie & leur practique, Je ne puis pas leur donner la preference, ni même croir que leurs instruments puissent jamais etre exacts, ni capables de repondre aux vues qu'on se propose. La route Angloise que J'expose Ce me semble assez intelligiblem^{t}., est l'unique qu'on doit suivre, si l'on veut marcher sur, avec precision, & rendre justice aux parties interessées. Je vous envoie la lettre ouverte pour que vous puissiez La lire, avant La donner.

Peut-etre que M^{r}. L'Abbé Rosier y trouvera un article fort interessant pour son journal:[4] mais je ne veux point qu'il dise par qui Lui a Reçu cette communication.

Vous verrez par La Lettre ci jointe, que M^{r}. l'Abbé Morelet[5] s'est trompé dans le prix des petits horloges, qui coutent une guinée & demie au lieu d'une guinée tout seule comme vous me le marquez dans votre Lettre.[6]

---

1 Red Lion Square.

2 Letter not found.

3 Mathurin Jacques Brisson (1723-1806) succeeded to the position of professor of experimental physics at the Collège Navarre formerly held by the abbé Jean-Antoine Nollet (1700-1770). A paper of his on determining the density of liquids was published by the Académie Royale des Sciences: "Mémoire sur le rapport des différentes densités de l'esprit-de-vin, avec ses différens degrés de force ...," *Histoire de l'Académie Royale des Sciences, avec les mémoires*, 1769 (publ. 1772), Mémoires pp. 433-452. No publication by Bourriot on this subject has been identified.

4 The abbé François Rozier had recently launched his journal, *Observations sur la physique, sur l'histoire naturelle et sur les arts*. It would appear that he did not take up Magellan's suggestion, since there is no article on hydrometers in either of the two volumes that he published in 1773.

5 The abbé André Morellet (1727-1819), economist and man of letters.

6 Letter not found.

73.05.11 J'ai passé ce matin chez Le D[r]. Franklin pour savoir si l'ouvrier vouloit m'attraper: & lui m'a dit que celui acheté par l'Abbé Morelêt couta aussi 1 ½ guinée. Je vais vous envoier la semaine prochaine une caisse par Rouen, oû je mettrai ces 2 petits pendules: & dans 2 ou 3 jours je vous envoierai une autre petite caisse par Calais, qui Contient une Lunete pour M[r]. Maquer,[7] qui m'assura dans une de ses lettres avoir parlé expressem[t]. & arrangé cet affaire avec M[r]. Trudaine: mais j'y ajouterai quelques autres articles, pour epargner autant qu'il est possible le nombre de ces envois.

Je vous prie de mes humbles Respets chez M[r]. Le Marquiz de Mirabeau,[8] & tout ceux qui m'honore de Leur souvenir.

La poste va fermer
à Dieu
votre pour La vie
Magalhaens

7 Pierre Macquer; see Magellan to Macquer, 4 April 1773.

8 Victor Riquetti, Marquis de Mirabeau (1715-1789), French economist and statesman.

73.05.14 *To Pierre Macquer*

Bibliothèque Nationale, Paris, MS fr. 12306, f. 23-24

Londres 14 May 1775 at n.° 4
Old N. Str. R. L. Sq[re].

Mon cher Monsieur

Je viens de faire empacqueter dans uns caisse à l'adresse de M[r]. Trudaine de Montigny, votre Lunette achromatique faite par Ramsden qui couta 10 guinees.[1] elle me paroit une des meilleures que J'ai jamais vue, & ne doute que vous en serez content. Je lui ai fait faire une caisse de Mahogany pour l'empacqueter toute demontée piece par piece, qui est La manière La plus avantageuse pour La transporter en voyage: car j'ai appris par des acidents repetés, que La manière de les porter avec le piedestal audedans etoit sujete à les casser avec les cahos des voitures: comme il arriva à celle du Chev[er]. de Bory, qu'il a eté obligé d'envoier à recommoder ici, & qui vint à Lui

1 See Magellan to Macquer, 23 March and 4 April 1773.

couter presque le double du premier prix, je veux dire, presqu'autant comme 73.05.14
Le premier prix. Cependant La petite caisse ronde (cylindrique) oû Cette Lunette peut etre fermée, avec le piedestal audedans du tube, peut servir pour la Commodité des petits voiages en ville &^{a}. La caisse de mahogany couta $12^{s}:0^{d}$: la depense proportionelle de la caisse d'embalage, à la douane & jusqu'a bord du vaisseau $5^{s}.6^{d}$, 3 postages de Lettres $2^{s}.6^{d}$. Some totale £11:10:0, qui etant deduites de £12:8:11 que vous m'avez envoyées, Restez crediteur de 18^{s}: 11^{d}: ce qui fait en argent de france $21^{\#}.12^{s}$. par l'echange de guinée contre Louis d'or. J'ecrivai Cette meme poste à M^{r}. Messier pour vous rendre ces $21^{\#}.12^{s}$ pour mon compte de l'argent qu'il a reçu pour moi. Mais comme cette lunete difere en quelque chose de celles de Dollond pour ce qui regarde la monture, qui est superieure à celles-ci, & outre celà elle a un oculaire celeste de plus, Je vais vous donner quelqu'idée de ces particularités.

la fig^{re}. 1re. represente la lunete toute montée pour les objets terrestres, en apliquant l'oeil en r. lorsque vous voulez user les oculaires celestes, devissez le tube des oculaires en h n. & metez y à vis un des oculaires celestes, que vous trouverez en a dans le centre de l'axe du piedestal. de l'autre coté vous trouverez un autre oculaire qui grossit plus, il est simple, mais pas aussi claire & distincte que l'oculaire à deux verres. il y a aussi un verre obscur pour

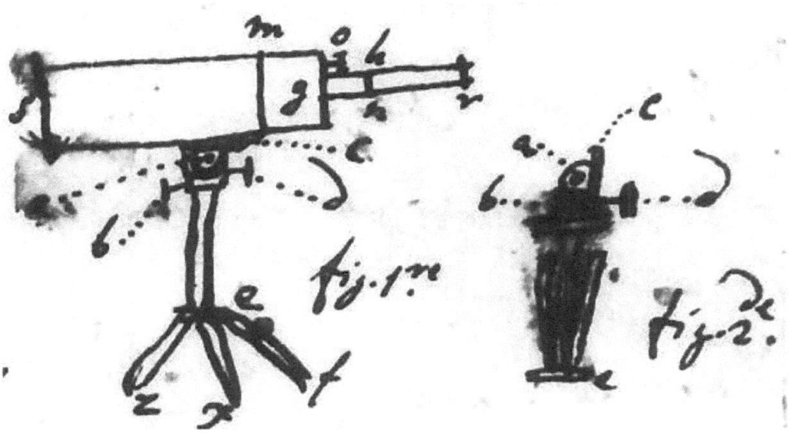

observer le soleil. lorsque vous voudrez empacqueter le piedestal audedans du tube, pour le porter dans la petite boete cylindrique, il faut devisser le tuyau en m. otez la coulisse c qui tient au piedestal. devissez le trepied fxz en e & tournez la à rebours en le vissant de nouveau, comme vous voyez

73.05.14 dans La fig^{re}. 2de. otez la tête de la vis b fig^{re}. 1re: & vissez-le dans le trou b de la fig^{re}. 2de.: tournez la coulisse comme vous La voiez dans La fig. 2de.: mettez le tout dans le gros tuyau: vissez la partie m.o.: & poussez le tuyau des oculaires r.g qui entrera dans le vuide du piedestal: & le tout sera Redui à un seul volume cylindrique, qui pourra etre renfermé dans La caisse, ou etui noir.

La tete o sert pour ajuster avec precision le foier, car en le tournant, il fait mouvoir tres doucem^{t}. le tuyau g.r. des oculaires. Ce tuyau a une croix (+) qui est la marque de son enfoncement. Cette croix doit ruer toujours en haut, lorsqu'on met ce tuyau dans sa place. La tete b fig. 1re. sert pour donner du mouvement tres lentement dans le sens horizontal à la Lunete. l'autre tête d sert pour donner un mouvem^{t}. pareil dans le sens vertical.

Mais lorsqu'on veut mouvoir la Lunete à la fois d'un coup, alors il ne faut que pousser au dehors, ou en bas, ces boutons; & pour lors on les degage de leur engrainure. lorsque la lunete est à peu près adressée vers l'objet, on repousse en arriere la tete de ce vis, & alors on les tourne pour mouvoir lentem^{t}. la Lunete vers l'objet &c.

Vous trouverez dans l'etuy noir La clef de la boete de mahogany; & outre cela une brochure avec l'excellent Memoire du D^{r}. Priestley sur les differentes especes d'air;[2] que Je vous prie d'accepter, pour fournir quelque bon article à la 2de. edition de votre Dictionnaire de Chymie. on le souhaite avec empressem^{t}. tout partout.[3]

J'ai L'honneur d'etre avec une parfaite consideration & toujours prêt a vous Rendre tous Les services dont Je suis capable

Monsieur

Votre très humble & obeiss^{t}. Serv^{r}.

J.H. de Magalhaens

P.S. Je crains ne pouvoir ecrire cette Poste à M^{r}. Messier. Du moins je lui ecrirai la suivante.

a Monsieur
Monsr. Macquer, Professeur de Chymie
de L'Acad. R^{le}. des Sciences
Rue S^{t}. Sauveur
à Paris

2 Joseph Priestley, "Observations on Different Kinds of Air", *Philosophical Transactions*, 62 (1772), 147-264.

3 The second edition of Macquer's *Dictionnaire de chimie* was not published until 1778.

To Nathaniel Pigott 73.06.01

Royal Astronomical Society, London, MSS Pigott, 27

London 1 Jun 1773 at
n°. 4. Old N. St. R. L. Sq^{re}.

Dear Sir and respected friend

I do believe that Ramsden's assertions on the article of your Barometer are true, and at least once he is in the right.[1] it appears to me, as a fact, that he has sent y^{t}. instrum^{t}. tho' in a silly manner, without giving any proper notice to you, or to M^{r}. Irwin at Ostende. I called on him, and shew'd the article of yours relating thereto. but he declares to have sent the instr^{t}. and adds that in a week or 10 days you shall hear of it's being at Ostende or Bruges, as M^{r}. Winkleman of Bruxelles,[2] whom you perhaps know (he being a Merch^{t}. somewhat in y^{e}. Math^{al}. way) has promised to inquire for it, and forward the notice thereof. as to y^{e}. Opera Glass Ramsden acknowledges his fault, and promised to make one this week, which I'll take care to send to you as soon as I can get it, by directing it to M^{r}. Irwin at Ostende. I was tempted to send you one made by Dollond, which I think as good as those of Ramsden, and of an equal aperture as well as price (18 shillings): but feared that the name of Ramsden might be a desirable circumstance to y^{e}. Person.

You are very kind in thinking so favourably of my poor habilities. Was I possessed of any sufficient share for doing some material services to y^{e}. new Academy, then it should be my greatest ambition to obtain a place therein. But unequal as I am to the task, I'll confine my self to the pleasure of being as serviceable as I can in my present situation, or at most as a Corresponding Member, if such are to enter in the present

1 Cf. 73.03.23a and 73.04.30.

2 Two instruments that survive in modern collections, a single-year calendar medal for 1776 signed "Winckelmans a Brussele" and a garden sundial signed "Winckelmans 1777", testify to the presence of an instrument-maker named Winckelmans in Brussels at this period. However, nothing else is known about him. See Victor A. Rasquin, *Dictionnaire des constructeurs belges d'instruments scientifiques (des origines à 1914)* (Brussels, 1996), p. 99.

73.06.01 plan.[3] Very near to y^e^. decline of Life by my age, and never disposed before for public interferings (much Less at this time) by a peculiar taste for retired imployements, where my own choice, freedom, and friends are the only concerned, I should not be able to undergo now any considerable alteration in the place already enjoyed in this happy corner of the world. From time to time I'll continue to make some Little excursions abroad, in order to see my good friends, as far as health and circumstances may allow me that pleasure, and to revive with them those feelings, which serve to keep up the strings of this wearisome Life. Be sure that you shall allways be held amongst them all in the most distinguished Consideration by

Dear Sir
Your most Obed^{t}. humble serv^{t}. & true friend
Magalhaens

A Monsieur
Monsr. Pigott, Gentil-homme Anglois
vis à vis La Cour
à Bruxelles
<u>Flandres</u>

3 Pigott had evidently suggested to Magellan the possibility of his transferring to Brussels and becoming a member of the new Academy of Sciences there. In principle, transferring to Brussels was not essential, since the Academy's statutes allowed for up to six of its twenty-six Ordinary Members to be foreigners; but these positions must already have been spoken for, because the proposal that Magellan should become a member went no further at this time. Magellan's counter-proposal went no further, either, since the statutes did not provide for the election of Corresponding Members. Magellan was, however, elected a full member of the Academy some years later, on 21 November 1785, the limit on the number of foreign members having been relaxed and then abandoned in the meantime. Membership gave him, besides a pension, personal nobility within the Austrian empire, including the right to wear a sword.

To Louis-Henri Duchesne[1]

73.06.08

Archives Nationales, Paris, MS T/160 (10-12)

Londres 8 Juin 1773 at n°. 4
Old N. Str. R. Lyon Sq^{re}.

J'ai, mon cher ami, chez moi depuis quelques jours les deux petites pendules que vous m'avez ordonnées. elles partiront bientot par la voie de Rouen avec quelques barometres pour Son Em^{ce}. Le Card. de Luynes[2] & d'autres choses, sous l'adresse de M^{r}. Trudaine. Je paiai ces deux petites pendules <u>trois</u> guinées, qui par la charge courant de 29 ¾ font 76$^{\#}$. 4½^{s} de France. Le voyage de Banks & Solander vient de paroitre en 3 vol. In 4°. avec des planches de taille douce.[3] j'ai crû ne devoir pas manquer d'acheter un pour la Bibliotheque de M^{r}. Trudaine. je le paiai 3 guinées aussi, qui font autant, c'est à dire <u>152$^{\#}$. ·9^{s}.</u> en tout. je les ai mis dans un pacquet dans une caisse de Livres, qui va partir incessam^{t}. pour M^{r}. Gibert Libraire[4] dans le quai des Augustins à Paris, qui ne manquera pas de vous les envoier aussitôt qu'ils arriveront. La raison pourquoi je ne les ai envoiés en droiture à M^{r}. Trudaine fut, pour ne pas multiplier les envoys. car j'en ai expedié depuis peu deux autres caisses à son adresse; qui sont deja parties & contient differents articles pour plusieurs Messrs. de l'Academie R^{le}. des Sciences. Or comme vous m'avez recommandé de ne pas augmenter le nombre, je n'ai pas cru raisonable d'envoier dans si peu de tems une 3me.

La premiere caisse ci dessus contient les suivants articles. n°. 1. un hydrometre. n°. 2. un livre des droits de la douane d'Angleterre. Ces deux articles sont pour M^{r}. Montigny. ils couterent 98$^{\#}$.12^{s}.

---

1 Annotated by Duchesne: "Rep. le 17 juin. et le 14 juillet." Letters not found.

2 Paul Albert de Luynes (1703-1788), Bishop of Bayeux from 1729 and then archbishop of Sens from 1753, was made Cardinal in 1756. He was also interested in astronomy and physics, and was elected an honorary member of the Académie Royale des Sciences in 1755.

3 John Hawkesworth, ed., *An Account of the Voyages undertaken by the Order of His Present Majesty for making Discoveries in the Southern Hemisphere: and successively performed by Commodore Byron, Captain Wallis, Captain Carteret, and Captain Cook…, drawn up from the Journals which were kept by the several Commanders and from the Papers of Joseph Banks, Esq.* (3 vols., London, 1773). To Cook's chagrin, the voyage around the world that he had so successfully commanded in 1768-1770 was often referred to, as here, as the voyage of Banks and Solander.

4 The bookseller Jean Baptiste Gibert (1727?-1784), who since 1761 had had his shop at the Quai des Augustins, "à la descente du Pont Neuf, Rue des Mathurins", under the sign "*A la belle Image.*".

73.06.08 n°. 3 & n°. 4. est un compas de proportion en ivoire pour M^{r}. le Marquiz de Courtenvaux, avec un Electrometre pour le même. j'avertis cette poste M^{r}. Messier de les lui envoier, & se charger de la distribution de ces articles, & en recevoir l'argent pour ne vous pas embarasser avec ces bagatelles. La valeur de ces 2 articles est 145$^{\#}$. 4^{s}.

n°. 5 sont 3 vol. des Transactions Philosophiques, celles de Philadelphie,[5] & une parcelle de Brochures, le tout pour L'Abbé Rozier. valeur 87.$^{\#}$ 2^{s}.

n°. 7 une parcelle de petites Lunetes pour M^{r}. Messier, dont il a le compte, & deux rasoirs pour M^{r}. Le Chev^{r}. de Bory.

n°. 9 contient une platine de cuivre jaune divisée en pouces, pour M^{r}. L'Abbé Bourriot.

n°. 10 & 11. contient une lunette achromatique, & sa caisse, qui sont pour M^{r}. Macquer, qui m'ecrivit il y a Long tems de la lui envoier sous l'adresse de M^{r}. Trudaine. apparam^{t}. il lui a parlé là dessus.

n°. 12 contient un Livre de Navigation qui est pour M^{r}. le Presid^{t}. de Saron, avec trois regles de bois, qui sont au fond de la caisse: aussi bien qu'un pacquet n°. 13 avec une ou deux vessies de moutarde.

il y a aussi une brochure adressée à M^{r}. de la Voisiere,[6] que vous Lui rendrez de ma part.

le n°. 14 contient deux Livres & quelques brochures que vous envoyerez à notre bon ami Le D^{r}. Sanchez.

vous y trouverez aussi le petit thermometre de poche, que vous m'avez demandé pour L'Abbé Morelet.[7] ce thermometre couta une guinée, qui par le change courant ci dessus, & plus 2$^{\#}$. pour la partie proportionelle des depenses de cette caisse fait 27$^{\#}$. 8^{s}. tournoises. il est marqué n°. 8. M^{r}. Morelet ne doit pas trouver mal de paier le change courant, qui fait 25$^{\#}$. 8^{s}. pour le guinée, & les 2$^{\#}$. pour les depenses en proportion du total de la caisse. pour arranger cette distribution raisonablem^{t}. j'ai dit dans une Lettre à M^{r}. Messier de diviser la depense qui sera chargée pour cette caisse jusqu'à Paris, de la diviser par 70. de me charger 20 parties de ces 70 à mon compte. 17 à M^{r}. l'Abbé Rozier: 10 à M^{r}. Macquer: 8 au D^{r}. Sanchez: 5 à M^{r}. de Montigny: & 15 à M^{r}. le Presid^{t}. de Saron. Celle-ci est la distribution proportionelle, que j'ai crû plus raisonable: aiant egard au poids, & aux autres charges des differents articles. Vous pouvez remettre s'il vous plait les 6 guinees c'est à dire les 152$^{\#}$. 9^{s}. à M^{r}. Messier, qui doit Reçoir[8]

5 The first volume of the *Transactions* of the American Philosophical Society, published in 1771.

6 Lavoisier.

7 The abbé André Morellet (1727-1819), economist and man of letters.

8 recevoir?

differentes autres parcelles d'argent d'autres Messieurs pour m'envoier le tout ensemble par une Lettre de change à deux usances: ce dont j'ai bien besoin, car mes fonds sont epuisés: la somme totale allant au delà de deux mil livres tournoises. il m'a falu emprunter de l'argent pour y suffire. 73.06.08

L'autre caisse que je vous ai expediée la Semaine derniere, sous La meme adresse de M^r^. Trudaine ne contient que 4 sextants marins, & 3 bussoles azimuthales pour M^r^. Le Chev^er^. de Borda: & cependant elle est beaucoup plus grande que la premiere: & par consequant si les deux arrivoient ensemble elle devroit payer plus de la moitié des frais de Calais à Paris.

Il n'y a rien de nouveau: je me Remets à ma derniere Lettre, & Suis pour La vie

tout à Vous

Magalhaens

A Monsieur
Monsr. Duchenne Secretaire
de M^r^. Trudaine de Montigny
a Paris

To Nathaniel Pigott 73.06.25

Royal Astronomical Society, London, MSS Pigott, 28

Dear Sir

The last post was retarded, and consequently I did not receive yours till Wednesday. M^r^. Ramsden promised to give me this day the opera glass: but on calling on him just now he sais cannot finish it till to morrow: and perhaps till the next month according to custom. therefore I went to M^r^. Dollond, where I take this, which appears to me the most distinct of all. I paid it five and twenty shillings, and send the receit along with it.[1]

intend to pass the next week by M^r^. Bird, who lives now near Marybonne,[2] and will forward you his answer about your Quadrant.[3] the

1 Magellan evidently enclosed the Dollond opera glass with this letter.
2 i. e. Marylebone.
3 Cf. 72.00.00.

73.06.25 first opportunity will endeavour to get from Ramsden the answer to your queryes. for although he promised me just now, to write them this night and send them by this same post to you, I cannot believe him to perform it. I will not forget your tubes, and will send the answer about y^{e}. deal rods for pendulum.

I am

dear Sir
your most obed^{t}. h^{ble}. serv^{t}.
Magalhaens

London 25 June 1773

If you had told me the sight of y^{e}. person for whom this opera glass is intended, perhaps I could send one more fitted than this. for as I am short sighted, take always those to be the best, which appear the more distinct on account of a more concave eye glass. and this was the more so, among many others I tryed, many of them would cost but 18 shillings, like those of Ramsden. but thoght the difference of y^{e}. price of no consequence, in regard to the distinctness of this glass.

My complim^{ts}. to Mr Needham. pray tell him that his parcel of books for D^{r}. Morton,[4] as well as the ten[?][5] prints, for the subscriptions I made, are not yet arrived, nor any notice to be had as yet. as to his new magnetometer I do not[6] when the whim will take Ramsden to begin to make it: tho' I have spoke over and over again about it.[7]

A Monsieur
Monsr. Pigott Gentilhomme Anglois
vis à vis La Cour
à Bruxelles

---

4 Presumably Charles Morton, FRS (1716-1799), under-librarian at the British Museum.

5 Reading uncertain.

6 know *omitted.*

7 Ramsden is not known to have made magnetometers or, indeed, any other magnetic instruments apart from some rather unsatisfactory mariner's compasses. Perhaps Magellan who, a year earlier, following the death of Gowin Knight, had taken over Knight's equipment for making artificial magnets, was trying to push Ramsden into making magnetic instruments, presumably using magnetic needles supplied by Magellan and made according to Knight's method?

To Johann III Bernoulli 73.06.25a

Öffentliche Bibliothek der Universität Basel, MS L I a 709, f. 47-48

Londres 25 Juin 1773
at n°. 4 old North Str. R. Lyon Sq^{re}.

Monsieur & très Cher ami

Voila deja 3 ou 4 postes que je comptois pouvoir vous repondre à l'honeur de la Votre du 8 May dernier,[1] sans pouvoir la faire par des accidents imprevûs & meme à present je me trouve avec assez peu de tems pour le faire avant que La poste soit fermée. Je suis charmé que vous aiez trouvé la caise envoyée par Stetin en bon ordre.

Les Tables des ascensions droites furent imprimées par ordre de M^r. Makenzie[2] (qui m'en fit present) pour l'usage de son Equatorial, fait par Ramsden, qui est un Chef d'oeuvre,[3] & qui excede beaucoup celui dont vous avez reçu la description, construit par Dollond. Cette description & l'autre du Sextant je les ai reçues de Dollond pour Les envoyer à Mes amis,[4] & je crois lui avoir dit que je vous en envoyerois. M^r. Shepherd a bien agrée l'envoy des Tables, que j'avois fait:[5] & a pris votre 2^d. vol. du Recueil chez Elmsley, qu'il n'avoit pas encore. & me pria de vous presenter ses complim^{ts}., & en cas de parler de ces tables au public, de faire usage seulem^t. de ce qui est contenu dans l'avant propos, ou preface.

M^r. Wolfe le Chymiste est parti pour France & de là pense aller cette année à Berlin. je lui ai donné une Lettre pour vous[6] avec un detail de ce que disoit M^r. Shepherd pour vous; & aussi de vos comptes avec Elmsley.

---

1 Letter not found.

2 Hon. James Stuart Mackenzie (?1719-1800). Cf. 72.04.07. No copy of the tables to which Magellan refers, which were presumably privately printed, has been located, nor any published reference to them.

3 Ramsden made four small portable equatorial instruments to the same design at this period, for Mackenzie, Mackenzie's brother the Earl of Bute, Joseph Banks, and Sir George Shuckburgh. The design is described and illustrated in C. Vince, *A Treatise on Practical Astronomy* (Cambridge, 1790), p. 152 and plate vi.

4 P. and J. Dollond, *Description and Uses of the New Invented Universal Equatorial Instrument, or Portable Observatory* (London, 1771); idem, *Directions pour l'usage de l'octant de Hadley perfectionné...* (London, 1772).

5 Cf. Magellan to Bernoulli, 23 March 1773.

6 See Magellan to Bernoulli, 29 April 1773.

73.06.25a Combien il en avoit de vendre de vos ouvrages &c. en cas qu'il (Wolf) n'aille pas cette annee non plus à Berlin, ou que cette letre n'arrive chez vous, je vous repeterai dans une autre ce detail, qu'il est impossible à present.

Je vais imprimer mon petit memoir sur les nouveaux Sextants & Octants,[7] dont je vous envoyerai quelques copies. vous y verez La coquinerie des objections faites par Maskelyne, qui par le moien de cette chicane, empeche qu'on l'imprimat dans les Transactions, ce qui ne tourneroit pas à l'avantage de l'invention des Oct^s^. & Sext^s^. de Dollond dont il a epousé de les proner & recommender tout partout. il sçait bien ce qui Lui convient: car sans doute il a ses raisons pour agir de la sorte &c.[8]

laissons là le passé & ne parlons d'omissions, involontaires, je crois.

J'ai deja prevenû le D^r^. Franklin des Remercim^ts^. de Votre Acad. & du present qu'elle envoiera à la Soc^e^. de Philad^e^.

la Sieve d'Erasthotenes est inseré dans le dernier vol. des Transact. Phil. pour 1773.[9] Vous ne m'avez donné encore des ordres de la part de l'Acad. pour vous le prendre. en effet il vaut mieux d'attendre quelqu' occasion, de quelque autre envoy plus volumineux. Car ces petits packets sont souvent perdus à bord des vaisseaux. Comm'il m'arriva deja plusieurs fois.

M^r^. Votre ami qui veut les nombres qui le manquent des Transactions est très mal informé. Car j'ai connu bien feu M^r^. Osborne[10] qui debitoit ces n^os^.; mais ce n'etoient que quelques uns qui par hazard lui tomboient entre Les mains: & jamais ceux qui sont rares depuis longtems, & que les

7 Magellan, *Description des Octants et Sextants Anglois…* (London, 1775).

8 Magellan made the same complaints about Maskelyne publicly in the account of his invention published in Rozier's journal, drawing a parallel with what he saw as Maskelyne's equally biased response to John Harrison's newly perfected chronometer: "En effet cet homme intéressé déchira la méthode de M. H——son, afin de pouvoir établir plus facilement celle des Longitudes par les distances de la lune aux étoiles. Pour remplir cet objet, il rendit nécessaires les Ephémérides marins qu'il compose, & dont il en tire les émolumens. On peut supposer qu'il avoit quelqu'autre projet lucratif en vue lorsqu'il recommandoit si fort l'invention de son protégé [i.e. Dollond], & s'acharnoit si violemment contre le Précis de ce Mémoire, dans lequel on trouve quelque chose de mieux. Peut-être y voyoit-il un obstacle pour établir la réputation de l'instrument de son Favori, dont il vouloit partager les profits qui en devoient résulter" (*Observations sur la physique*, 4 [1774], 132).

9 The article, by Samuel Horsley, actually appeared in the last part of the *Philosophical Transactions* for 1772; see 72.07.21, n. 13.

10 Probably Thomas Osborne of Gray's Inn, "the most celebrated bookseller of his day … who carried on a successful trade from the year 1738 to 1768" (*Literary Anecdotes of the Eighteenth Century* [London, 1812], III, 401).

libraires achetent au plus grand prix pour completer des suites entieres des Transactions. Je viens de prendre une d'hazard par 50 guinées pour un de mes amis de Flandres: & peutetre ne pourrai je avoir une autre aussi tôt. ainsi vous pouvez l'assurer que son affaire est infesable. Je scais qu'on reimprime en Alemagne, les Transact. Philos. (je ne sais pas bien si c'est à Ausbourg[11] ou quel endroit). pour lors il y pourra se pourvoir. 73.06.25a

Je serois fort charmé d'envoier à Messrs. les aut[eurs][12] du Journal, l'universal Catalogue:[13] & en effet je l'avois pris chez moi pour leur envoyer. mais lorsque Je reçus votre Lettre, je n'ai pas trouvé aucun adresse affranchie de la poste: & dans un tel cas celà leur feroit une depense horrible. Car pour l'envoier par des vaisseaux, c'est une voye si ennuyeuse & si incertaine qu'il ne faut pas y penser. Je crois que ces Messieurs pouront l'avoir mieux à bon compte, en s'adressant à une Compagnie qu'il y a à Danzic (à ce qu'on me le dit) par laquelle on fait aller d'ici ces sortes de papiers periodiques & gazetes &c. en s'adressant à quelqu' homme de Lettres à Danzic, il ne manquera de leur procurer cette commodité.

Je ne sais pas des experiences de M^{r}. Conclaud & Mercier. Mais M^{r}. de Luc, Cit^{n}. de Geneve qui se trouve ici me dit que son ami, compatriote, M^{r}. Le Sage a un admirable sisteme demontrée par lequel il fait vo[ir][14] que toute gravitation n'est qu'une impulsion des corpuscules, qu'il appelle .... je ne s[ais][15] pas si ultramontains &c. lorsque cet ouvrage paroitra, que le pauvre Newton & ses partisants seront confondus!...!...![16]

---

11 Augsberg?

12 Conjectural reading.

13 *The Universal Catalogue for the year* ... (3 vols., London [1772-74]).

14 Conjectural reading.

15 Magellan writes "se".

16 The experiments ascribed to Coultaud (sic) and Mercier were published in the *Journal des beaux-arts et des sciences*, June 1769, 389-419 and December 1771, 389-416 respectively, and caused a considerable stir. Both involved comparing the rates of temperature-compensated pendulum clocks, one placed high on a mountain, the other at its base. In opposition to what Newtonian gravitational theory predicted, it was reported that the upper clock ran faster, that is, that the weight of the bob *increased* with elevation. Eventually the reports were shown by the Genevan physicist Georges-Louis Le Sage (1724-1803) and his friend Jean-André Deluc (1727-1817) to be fabrications; see "Lettre de M. Lesage sur la fausseté de deux suites d'expériences...", *Observations sur la physique, sur l'histoire naturelle et sur les arts*, 1 (1773), 249-260. Who was responsible for them remains a mystery; see James Evans, "Fraud and Illusion in the Anti-Newtonian Rear Guard: The Coultaud-Mercier Affair and Bertier's Experiments, 1767-1777", *Isis*, 87 (1996), 74-107. Despite what Magellan says, Le Sage fully accepted Newton's notion of a universal inverse-square gravitational force. However, like older contemporaries such

73.06.25a Elmsly a vendu 19 Recueils. vol. 1. à 3^{s}. = 2.17. plus 24 vol. 2 = 3.12. plus 27 lettres à $1^{s}.9^{d}$.= $2^{£}.7^{s}.3^{d}$. total = $8^{£}$:16: 3. / Il reste à vendre 16 vol. du Recueil = 2. 8^{s}. plus 14 lettres = $1^{£}$. 4^{s}. 6^{d}. plus 3 recueil de present & deux lettres de presents aussi = 12. 6 plus en argent qu'il me paye = $4^{£}$. 3^{s}. 1^{d}. total $8^{£}$. 8^{s}. 1. qui deduits de 8. 16. 3 ci dessus montre que lui (Elmsley) vous est debiteur de 10: 6^{d}.

Celles ci etoient les circonstances de vos comptes dans le [2][17] Avril dernier, lorsque je fis les miennes avec lui (Elmsley). Depuis ce tems là je ne sai pas ce qu'il a vendu, qui peut etre n'est rien du tout. N.B . le vol. de M^{r}. Shepherd fut depuis ce tems là.

tachez de retablir votre Santé & ne vous point appliquer au delà de vos forces. à dieu mon cher.

Je suis
Votre pour la Vie
Magalhaens

A Monsieur
Monsr. Bernoulli Astronome Royal de Prusse
à Berlin

---

as Leonhard Euler, he believed that this force must have a mechanical cause. Moreover, as Magellan reports, Le Sage believed he had discovered the cause. He supposed the world to be filled with "ultramundane" corpuscles travelling at great speeds in all directions. The motions of some of these are interrupted by gross matter, which thus partially screens other matter beyond it from the rain of corpuscles; as a result, the two bits of matter are pushed together. With some plausible auxiliary hypotheses, Le Sage was able to derive a "force" between them that varied as the masses and inversely as the square of their distance apart. See "Physique mécanique de Georges-Louis Le Sage, redigée d'après ses notes...", pp. 1-186 in Pierre Prévost, *Deux traités de physique mécanique* (Paris/Geneva, 1818); also Samuel Aronson, "The Gravitational Theory of Georges-Louis Le Sage", *The Natural Philosopher*, 3 (1964), 51-74.

17 Number partially obscured.

To Louis-Henri Duchesne[1]

73.07.23

Archives Nationales, Paris, MS T/160 (10-12)

Londres 23 Juillet 1773

Mon cher ami

Je vous envoye ici le plan de l'Hospital de Plymouth,[2] que vous donnerez de ma part avec la Lettre ci jointe à Mr. le Comte Carbury.[3] Car c'est lui qui me l'ordonna, à ce que je crois par convention ou desir de Mr. Trudaine. Je vous assure que j'ai eu bien du trouble pour trouver, aborder, & effectuer cet affaire. mais je suis bien aise d'avoir Reussi: Ceci etant un objet interessant, qu'on doit consulter, pour Rebatir l'hotel Dieu.[4] Je vous envoye aussi le Reçu de l'architecte qui l'a fait copier d'après Les desseins originaux: Je l'ai paié comme vous voyez 5 guinées, que vous aurez soin je vous prie de me faire tenir par Mr. Messier.

Je vous ecrirai encore la poste suivante sur les articles que j'ai envoyés par mer, & entre eux vos deux pendules de 3 guinées.[5] mais il est tard & la poste va fermer a present,

adieu

votre de coeur

Magalhaens

A Monsieur
Monsr. Duchesne Secretre.
de Mr. Trudaine de Montigny

1 Annotated by Duchesne: "a joindre".

2 The Royal Naval Hospital, Stonehouse, Plymouth was the prototype of a new concept of hospital planned to avoid propagation of diseases, either through the position of the wards and pavilions or through the system of forced ventilation.

3 Giovanni Battista Carburi (1722-1801?), professor of medicine at Turin, 1754-1770, went to Paris in 1773 as personal physician to Maria Teresa of Savoy, wife of the Comte d'Artois.

4 In 1772 there was a great fire at the Hotel Dieu, the most famous hospital in Paris, that led to its having to be rebuilt.

5 See Magellan to Duchesne, 27 July 1773.

73.07.27 *To Louis-Henri Duchesne*
Archives Nationales, Paris, MS T/160 (10-12)

Londres 27 Juillet 1773
at n°. 4. Old North Str^{t}. R. L. Sqr^{e}.[1]

Monsieur & Cher ami

Je viens d'expedier à la fois par la voie de Rouen differentes Commissions dont j'ai eté chargé par des personnes qui ont parlé à Monsr. Trudaine, ou qui le parleront sur celà par example M^{r}. le Card^{l}. de Luynes pour lequel est une caisse marquée **MM**, n°. 5. avec deux Barometres,[2] une autre avec la meme marque, n°. 4. qui contient des plateaux de verre pour l'electricité, pour M^{r}. le Marquiz de Courtenvaux. Celle de la meme marque n°. 6 contient vos deux petites pendules. une autre marquée n°. 3. contient deux barometres pour M^{r}. Messier & Patû.[3] Celle marquée n°. 2, & une autre qui n'a pas de numero, mais qui a la meme marque commune **MM** contient des instrum^{ts}. pour M^{r}. le Presid^{t}. de Saron: & je vous dirai en confidence que la premiere de ces deux contient la fameuse machine à diviser, dont il n'a pas d'egale en france: ce que je crois etre un des plus grandes services, que j'ai rendu pour l'avancem^{t}. de vos artistes.[4] mais M^{r}. le Presid^{t}. m'a prié de ne pas le divulger: & je vous charge de meme: mais vous pouvez toucher un mot à M^{r}. Trudaine. Car malgré le silence que M^{r}. le Presid^{t}. de Saron souhaite dans cet article, ce n'est que pour l'usage & perfection des instrum^{ts}. de Mathematique qu'il en a faite la depense. Quoique cette machine est forte & de gros metal, elle ne manque pas d'etre fort delicate: ainsi je souhaite, que personne n'y touche, sans que M^{r}. le Presid^{t}. soit present. Lui est la personne la plus attentive & la plus capable pour manier des choses pareilles, que Je connois au monde. en fin je vous Recommende ces deux caisses, & les autres trois aussi, & M^{r}. Messier auquel j'ecris cette poste sera de concert avec vous lorsqu'elles arriveront, en remontant la

1 Red Lion Square.

2 See Magellan to Duchesne, 8 June 1773.

3 André-Claude Patu de Mello (1726-1799) possessed an extensive library of scientific books and also a large collection of physical and astronomical instruments which, however, Lalande is reported to have said, he neither used nor let anyone else use. See *Catalogue des livres de la bibliothèque de feu A. C. Patu de Mello, suivi de la notice d'une collection précieuse d'instruments de physique* ... (Paris, 1799).

4 The machine in question was the first of Jesse Ramsden's circular dividing engines, completed in 1768; see Magellan to J. A. Mallet, 13 October 1775, and Anita McConnell, *Jesse Ramsden (1735-1800): London's Leading Scientific Instrument Maker*(Aldershot, 2007), p. 42.

Seine par bateau, afin d'eviter tout derangem^{t}. acidentel. Je lui ai envoyé le Reçu du Capitain du vaisseau, pour qu'il passe chez vous àfin de l'envoyer par La poste au Receveur des fermes de Rouen, comme vous m'aviez dit; mais ne Sachant pas son nom, j'ai pris cet expedient à fin de prevenir tout equivoque ou meprise. Monsr. le Chev^{er}. de Borda, qui avoit parlé avec M^{r}. Trudaine pour une lunete, qu'il m'ordonna ici, l'aura bientôt par la voie de Calais, dans une caisse marquée **MM** n°. 1 qui est adressée (de même que les autres par Ia la voie de Rouen) à M^{r}. Trudaine de Montigny. lorsqu'elle arrivera, vous l'ouvrirez, & y trouverez les deux lunetes que vous avez ordonnées pour votre ami vis à vis les Tuilleries.[5] mais comme vous n'especifiez la qualité, je vous envoye une avec son piedestal, dont le prix est de quatre guinées. & une autre à tuyaux, de deux pieds de longueur lorsqu'elle est toute etendue. Celle-ci ne coute que deux guinées. de façon que les deux ne coutent que 6 guinées: en tout 152$^{\#}$. 6^{s}., que vous me ferez grace d'envoier avec les autres 6 & les 5 guinées des desseins pour M^{r}. le Comte de Carbury que je vous envoyai la poste derniere.[6] qui en tout font 17 guinées, ou 431$^{\#}$. 5^{s}. selon le change courant. car je me trouve epuisé d'argent pou[r] Ces Commissions: & serois bien aise de me trouver à meme de payer ceux qui m'en ont preté pour ces objets. Je souhaite bien savoir comment a Reussi M^{r}. La Voisiere avec la desalation de l'eau Salée.[7] 73.07.27

dans cette meme caisse de la lunete pour M^{r}. le Chev^{er}. de Borda, est aussi un petit pacquet de 3 livres in 8°. pour M^{r}. le Marquiz de Courtenvaux, & un petit pacquet avec un paire de ciseaux que Son Secretaire, Mr Baptistin,[8] m'a demander d'ici. plus un petit etui de lunete en noir, pour M^{r}. le Presid^{t}. de Saron: & en fin un mape du detroit de Magellan, qui appartient au voyage que Je vous ai envoyé pour M^{r}. Trudaine en 3. vol. in 4°. par la voye de M^{r}. Gibert libraire:[9] mais qui n'etoit encore tout à fait gravé dans ce tems là.

à Dieu mon cher & bon ami. portez vous bien, & croyez moi pour La Vie

votre de coeur
Magalhaens

A Monsieur
Monsieur Duchesne Secretaire
de M^{r}. Trudaine de Montigny
a Paris

5 Not identified.
6 See Magellan to Duchesne, 23 July 1773.
7 See Magellan to Duchesne, 4 May 1772 and 5 March 1773.
8 Not identified.
9 See Magellan to Duchesne, 8 June 1773.

73.08.03 *To Joannes Franciscus Thijsbaert*
University Library, Université de Liège, Ms. 4130

Londres 3 Aoust 1773

Mon cher ami

Celle-ci est produite par le depit oû je suis pour ne pas etre à meme de pouvoir accomplir vos commissions avec l'expedition que Je souhaite, à moins de sacrifier la bonté des instruments à la celerité de les finir plutôt. C'est donc pour epancher mon chagrin, & pour me consoler avec la communication de mon bon ami, que je vais vous dire ce que j'ai effectivem^{t}. deja chez moi & prêt à partir; & l'etat de ce qui me manque. 1: J'ai deja une bonne Lunete achromat^{e}. de 16 ou 17 pouces de foyer: & une autre d'environ 7 pouces, qui est celle que vous m'avez ordonnée dernierem^{t}. pour votre propre usage & ne coute que 4 guinées. elle a aussi son petit oculaire celeste, dont quelques unes de cette espece ne sont pas fournies. pour l'autre, elle en a deux: un simple, & l'autre à double lentille: au lieu que celles de Dollond du meme genre n'ont qu'un seul; coutent 15 guinées, c'est à dire 5 guinées de plus que celle-ci: & ne sont pas meilleurs pour l'effet. J'ai aussi votre Barometre de Montaigne: c'est à dire comme celui de M^{r}. Pigott. J'ai un prisme assez passable: mais je n'ai pas encore pu trouver un autre pareil. J'ai le prisme à fluides, dont je vous parlai. l'autre achromatique que vous avez ordonné, fait par M^{r}. Whitford.[1] J'ai un excellent niveau, avec sa lunete achromatique d'environ 19 pouces de foyer; c'est un excellent instrument. un officier qui l'a vû, fut bien faché que cet instrum^{t}. devoit aller pour etre gardé dans un cabinet de Physique, & non pas pour etre employé dans La Campagne, à prendre le niveau des rivieres & canaux &c. Les deux miroirs concave & convexe, dont le premier est monté dans un piedestal de mahogany, sont deja finis: aussi bien que le telescope Newtonien: & je les aurai dans ma chambre dans deux jours. Le microscope Solaire l'est aussi: & je l'aurai pour ce tems là. Le tellescope gregorien je puis l'avoir à toute heure. J'ai aussi Le livre du Pere Becaria[2] deja chez moi. Vous voiez donc qu'il ne me manque plus que les 7 plateaux, mon octant, & le cube solide de verre. pour ce dernier, j'ai fait deja executer dans les verreries par differents fois plus de 6 ou 7. mais il n'est pas possible d'obtenir un qui soit assez affranchi des filements qui aportent

1 Thomas Whitford, instrument maker at 27 Ludgate Street, St. Paul's, London.
2 Giambatista Beccaria, *Elettricismo artificiale* (Turin, 1772)?

du malheur dans la Refragibilité. je vais encore demain dans une verrerie, 73.08.03
qu'on m'assure aura un pot de verre de la meilleure qualité pour ce tems là: & je serai charmé de reussir. Je pourois avoir des plateaux il y a long tems, faits par M^{r}. Dollond. mais il charge des prix si hauts, que Je me suis soumis à les attendre de l'attelage de Ramsden. en effet M^{r}. Dollond me fit depuis peu 6 plateaux savoir 2 de 12 pouces françois, 2 de 18, & 2 de 15 pouces, qui couterent dix guinées; ce qui est un prix exorbitant; mais qui est compatible au hazard qu'on court de les casser, comm'en effet on en cassa plus de 3 ou 4, pour avoir les 4 de 15 & de 18 pouces. Pour ce qui regarde mon quadrant, il est à demi fait depuis long tems & je pense que ce ne sera qu'un fort peu de tems à le completer. vraim^{t}. ce ne sera cet instrum^{t}. qui fera arreter les autres.

Pour la description de la Machine Elect. de Nairne, elle n'a rien d'extraordinaire dans sa construction, que la justesse, & la grandeur de ses parties. Concevez un Cylindre qui a un pied de diametre ou meme 13 pouces, & environ 20 pouces ou deux pieds de longueur. il est monté à l'ordinaire dans un exieu,[3] & est enduit au dedans avec de la composition rouge, dont parle le D^{r}. Priestley dans son Hist^{re}. de l'Electricité.[4] on lui donne du mouvem^{t}. avec une roue d'environ 2½ de diametre [pour][5] qu'il tourne pas trop vite. on a arrangé dessous ce cy[lindre], deux autres cylindres de verre très forts, couverts de [soie][6] rouge, qui sont dans le sens horizontal; ils ont environ 3 pouces de diametre, & servent, comme de blancards pour y monter le coussinet, oû se fait le frotement du grand cylindre. leur avantage est seulem^{t}. pour avoir de l'electricité negative, en otant la chaine, qui pend du coussinet, & communique avec la terre, lorsqu'on opere sur l'electricité positive. Le Conducteur à du moins 8 pieds de longueur, & un de diametre. il est fait de bois fort leger, & couvert du tin-foil, ou feuille d'etain, qui sert à doubler les bouteilles pour l'experience de Leyde: & il est monté sur un cylindre de verre perpendicular, à peu près comme ceux qui suportent le coussinet. mais ce qu'il y a de particulier, c'est une piece de soye cirée ou plutot vernisée gh qui est fixe sur le bord hi superieur, tout au long du cylindre, & qui couvre le bord I du coussin A. aussi bien que la partie efa du cylindre, qui tourne dans le sens des letres abcdef.

3 essieu?

4 Joseph Priestley, *History and Present State of Electricity* (London, 1767) or, more likely, the French translation of this (Paris, 1771).

5 Words obscured here and in next two lines by seal.

6 Conjectural reading.

73.08.03 l'avantage de cette Soye (qui est un corps non electrique, ou non-conducteur) est d'empecher, que l'electricité, excitée par le frotement du coussinet, ne reentre, ou recule en arriere vers le bord superieur du meme qui est couvert avec de la soye: ainsi cette matiere passe tout au long au tour avec le cylindre, & va etre acumulée dans le grand Conducteur. Le D^r^. Newt[6] donne un Memoire à la Soc^e^. R^le^. oû il remarque cette observation du reflut, ou retour, ou reculement de l'electricité vers le coussinet, & la methode de l'empecher en grand partie. J'ai vû qu'on employe avec sucès des morceaux de cette soye n & n dans les deux

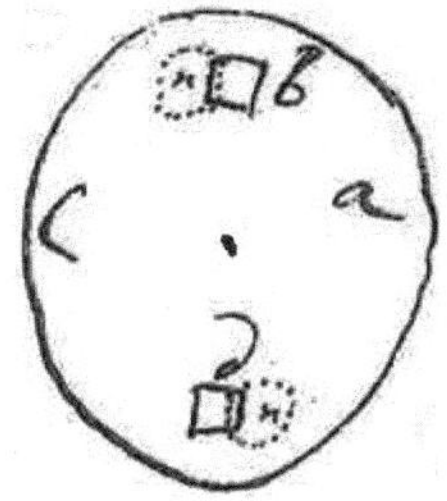

coussinets b & d du plateau abcd, qui tourne dans le sens de ces letres, & se tient flotants du coté que le plateau tourne.

à dieu, mon cher Monsieur, soyez assuré que je ne m'oublie de vos affaires, & que suis pour la vie

Votre de Coeur
Magalhaens.

P.S. mes complim^ts^. à M^r^. le [Prevot][7], & autres amis qui m'honorent de leur souvenir.

A Monsieur
Monsr. L'Abbé Thisbaert, Professeur
au College de l'Université
Coll^e^. Castrum
à Louvain
Flandres

---

6 i.e. Nooth; cf. "An Extract of a Letter from Dr. Nooth to Dr. Franklin, F.R.S. on some Improvements in the Electrical Machine", *Philosophical Transactions*, 63 (1773), 333-9.

7 Conjectural reading; presumably the Abbé Jean-François de Marci, who was provost of St Pierre in Louvain from 1772.

To Louis-Henri Duchesne 73.08.03a

Archives Nationales, Paris, MS T/160 (10-12)

Londres 3 Aoust 1773

Mon cher ami

J'ai fait mettre à bord du premier vaisseau qui partira demain pour Calais une autre caisse marquée **MM** n°. 2, qui contient les 5 cartes biographiques, & les autres 5 historiques que vous m'avez ordonnées par votre derniere du 23 Juillet.[1] vous y trouverez autant de brochures, qui donnent l'explication & l'usage de ces cartes: & outre celà J'y ai ajouté 3 vol. d'un ouvrage, qui fait du bruit ici, sur le contract Social, par le même Auteur à ce qu'on dit, de celui du Sisteme de La Nature.[2]

J'ai cru que La Biblioth^{e}. de M^{r}. Trudaine devoit l'avoir. il ne couta que 9 shillings. Les 10 cartes ci dessus avec leurs brochures couterent 5 guinées. caisse & porteur 2^{s}. 8: depenses à bord du vaisseau & à la douane 4^{s}. 6. total £6: 1: 2. Ce qui par le change courant fait 146$^{\#}$. tournoises, que vous aurez la bonté de donner avec celles de ma derniere lettre[3] à M^{r}. Messier, le tout ensemble fait 414$^{\#}$. 16^{s} tournoises.

Vous avez bien tort de faire des apologies par les embaras que vous dites me donner. il faudroit me considerer d'un coeur peu sensible, & moins reconnoissant pour ne pas me rejouir de pouvoir vous rendre ces petits services ici. vous me n'avez faits, & me faites actuelem^{t}. de bien plus remarquables.

aiez la bonté de me marquer dans le papier de votre lettre la grandeur & figure precise du thermometre cassé de M^{r}. de Montigny: & je ferai faire ici au plutôt un autre, monté dans la plaque que vous souhaitez de metal blanc, ou de l'argent. Mais seulem^{t}. parlez lui sur celà encore. Car l'alteration qui fait L'ivoire n'est pas trop considerable pour ne pas faire usage de cette matiere, qui s'essaya assez bien & est fort propre lorsqu'on la met dans les

---

1 Letter not found.

2 The works in question are no doubt by Paul-Henri Thiry, baron d'Holbach (1723-1789), namely his anonymously published *Système social, ou Principes naturels de la morale et de la politique, avec un examen de l'influence du gouvernement sur les moeurs* (London, 1773) and *Système de la nature ou des loix du monde physique et du monde moral* (London, 1770). The term "contrat social" had been popularized by Jean-Jacques Rousseau in his *Du contrat social, ou principes du droit politique* (Amsterdam, 1762).

3 See Magellan to Duchesne, 27 July 1773.

73.08.03a liqueurs, pour en eprouver la temperature: au lieu que le metal blanc, & même l'argent, se noircissent, & laissent un petit gout desagreable. C'est la raison pourquoi on prefere l'ivoire; quoiqu'en effet on use ici de plaques metaliques aussi, dans des cas pareils. ainsi vous n'avez qu'à me dire encore un mot Là dessus. Mais encore si ce thermometre est pour servir seulem^{t}. dans La poche dans une caisse ou etuy separé, pour lors la grandeur precise n'est pas bien necessaire, car un peu plus ou moins ne gatera rien dans la longueur de la tige &c.

Vous trouverez dans l'autre caisse dont je vous ai parlé dans ma Lettre precedente, marquée **MM** n°. 1. qui contient la lunete pour M^{r}. le Chev^{er}. de Borda, vous y trouverez je diz la carte du detroit de Magellan, qui appartient à l'un des 3 vol. du voyage que J'envoyai, & que vous avez Reçus de M^{r}. Gibert, pour M^{r}. Trudaine.[4] Cette carte n'etoit pas encore finie lors que j'expediai cet ouvrage.

vous verrez par ma Lettre precedente que vos deux petites pendules sont deja parties, & peutetre [arrivées à][5] Rouen. J'ai deja ordonné les 4 de plus, que vous venez d[e] demander dans cette Lettre: & je vous en parlerai de la Reponse [que] je Recevrai là dessus.

il n'est possible de fixer le prix de la machine Pneumatique sans savoir la quantité & qualité de l'apparatus qu'on souhaite. la meilleure sorte de ces machines sont celles qui peuvent servir aussi pour condensoirs. il en a qui coutent 30 & même 40 guinées. Celle qu'on à fait selon les principes de Smeaton pour Le Roy ici,[6] couta beaucoup plus à ce qu'on dit. mais qui aime le solide, sans des ornements, n'a pas besoin d'en avoir de la Sorte.

Je suspendrai de vous envoyer aucune autre chose jusqu'à ce que vous m'avisez de Votre Voyage de Retour. Vous voyez ce qui est parti par Rouen: & qui doit aller par La Seine en remontant jusqu'à Paris: je vous prie d'en avoir soin, etant de concert & intelligence sur celà avec M^{r}. Messier, que J'ai fait mon Comissaire, ou plutot gardien de mes commisions à Paris, en revanche de celles que Je fais ici pour Lui. lorsque vous verrez M^{r}. Bouriot faitez lui une bonne querelle de ne pas m'avoir donné de ses nouvelles depuis très long tems. M^{r}. l'Abbé Rozier est egalem^{t}. criminel à mon egard

4 See Magellan to Duchesne, 27 July 1773.

5 Paper damaged at the end of this and the following two lines; words illegible.

6 The London instrument maker George Adams in 1761 made a magnificent air pump for King George III, the design of which was based on a pump built some years earlier by the engineer John Smeaton (1724-1792); see Alan Q. Morton and Jane A. Wess, *Public and Private Science: the King George III Collection* (Oxford, 1993), p. 247.

depuis long tems (je ne sais pas pourquoi) malgré des affaires & resolutions que je lui demandai. 73.08.03a

à Dieu mon cher & bon ami Je suis pour la vie

Votre de coeur

Magalhaens

A Monsieur
Monsr. Duchenne Secretaire
de Monsr. Trudaine de Montigny
à Paris

To Louis-Henri Duchesne[1] 73.08.27

Archives Nationales, Paris, MS T/160 (10-12)

Londres 27 Aoust 1773

J'ai Reçu, mon cher & bon ami, la mesure du Thermometre qu'il faut faire executer en argent pour M[r]. de Montigny:[2] & J'en ai deja donnés des ordres pour celà. aussitôt qu'il sera fini je l'envoyerai par la premiere occasion, quoique je ne sais pas si je l'aurai aussitôt. Car comme vous me dites d'aller faire un tour en Savoye jusqu'au mois d'Octobre, je ne compte faire aucun envoy avant votre Retour, à moins que vous m'ordonnez le contraire,

Je ferai desormais des envoys beaucoup plus rares car tout au plus il n'y aura qu'une seule caisse capable de contenir les autres, à fin d'éviter la pluralité de ces expeditions. Car à vous dire le vrai presque tous ces Messieurs ne prendent pas gard de me dire de leur envoier leurs instruments adressés à M[r]. Trudaine, par un seul mot peut etre qu'ils lui ont dit, & sans d'autre ceremonie: & il peut bien arriver que M[r]. Trudaine s'en oublie dans La suite de leur avoir dit oui.

Je serois bien Charmé de pouvoir faire un petit tour à Paris lors de votre retour. peutetre M[r]. la Voisiere seroit-il plus à meme de faire aller La machine pour faire L'eau douce de celle de la mer, avec plus de succès

1 Annotated by Duchesne: "Rep. le 20 8e. 1778"; letter not found. "1778" is certainly a mistake; it should be 1773.

2 See Magellan to Duchesne, 3 August 1773.

73.08.27 si je m'y trouvois &c^{a}. Mais je crains ne pouvoir pas me debarasser aussitôt, d'une infinité de petits affaires qui me retient ici.

Au reste aiez La bonté de m'annoncer votre retour, & presenter mes Compl^{ts}. aux amis communs: & particulierem^{t}. à M^{r}. L'Abbé Nicoli,[3] en Lui disant que j'ai deja ecri[t] à M^{r}. L'Abbé Fontana,[4] sur l'objet dont-il me parloit dans La lettre qu'il m'adressa par son moien:[5] mais que Je n'ai pas encore Reçu aucune Reponse.

Au Reste conservez vous, & jouissez autant de bonheur comme vous souhaite Celui qui est pour La vie

tout à Vous

Magalhaens

A Monsieur
Monsr. Duchenne Secretaire
de M^{r}. Trudaine de Montigny
à Paris

3 The abbé Raimondo Niccoli (?-1780), secretary of the Parisian legation of the Grand Duke of Tuscany.

4 Letter not found. Felice Fontana (1730-1805) was a professor at the University of Pisa before being appointed by the Grand Duke of Tuscany to organize his physics cabinet.

5 Letter not found.

73.09.09 *To Matthew Boulton*[1]

Birmingham Central Library, Boulton and Watt collection, MBP 244, 53

London 9 Sep^{ber}. 1773 at n°. 4
old North Str^{t}. Red Lyon Squre.

Dear Sir

I am afraid that you have intirely forgotten my affair of y^{e}. magnetical bars, there being so long a while without hearing no further about them.[2] I told the clerk who brought the samples to me 1^{rst}. that I wanted them

1 Annotated "London 9th: 7^{br}. 1773 / Magalhaens / received 11th. d^{o}. d^{o}. / Answ^{d}. 16th: 8^{br}: 1773."

2 See 73.00.00.

to be harden'd still more, viz as hard as possible. 2^{d}. that I was contented with the polishing of the samples. 3. that I had no objection to y^{e}. prices of ye. work, viz 6 shillings for hardning & polishing the biger sise ones, and 3 or four shill^{s}. for y^{e}. 2^{d}. sise &c. 4. that I wanted to know the prices by which you should be able to furnish me some dozens of the largest sise bars intirely new made I mean, as big and of the same dimensions & shape as the larger sise of my samples. these are the queries I want to know in order to send the Large parcel, which I have by me, intirely spoiled by the fire which burn'd down my Lodgings:[3] and beg you be so kind as to answer by the turn of y^{e}. post. 73.09.09

I have also spoken about impregnating the above samples, before you send them to me, with y^{e}. magnetism. this was only by curiosity's sake, in order to see to what degree of strenght they were impregnated but as to the Large quantity of bars, intended to be sent there for cleaning hardning & polishing them, I don't want, nor desire, that they should be impregnated there with any magnetism.

I am with the greatest regard and true friendship

Dear Sir
Your most obed^{t}. h^{ble}. serv^{t}.
Magalhaens

P.S.
Pray my Compl^{s}. to D^{r}. Small.[4] and will be glad to know if your new fire engine is finished, and works well?

To
Mr Boulton
at Soho
in Birmingham

---

3 On 21 December 1772, while Magellan was in Paris, his dwelling at 25 Fetter Lane was destroyed by fire. The insurance paid £400 for the goods, books, utensils, stock and wearing apparel (Sun Insurance Records, ref. 11936, p. 539). It may have been the heat of the fire that demagnetized the bars used to magnetize compass needles. From early in 1773, after returning from Paris, until nearly the end of that year, Magellan lived at no. 4, Old North Street, Red Lion Square, Holborn.

4 William Small (1734-1775), Boulton's physician and a leading member of the Lunar Society of Birmingham.

73.10.07 *To John Fothergill*[1]

Birmingham Central Library, Boulton and Watt collection, MBP 244, 54

London 7. Oct^{ber}. 1773 at n^{o}.
4. Old North Street Red Lyon Square

Sir

On my speaking this morning with our Common friend, D^{r}. Fothergill[2] about some affair depending on your manufactory, he told me to mention his name, that your friendship towards him, may help the ending of it, of which I'll give here the account.

Some months ago I spoke here in London with M^{r}. Bolton, about the hardning and polishing a good quantity of magnetical Bars, which have been spoiled in the fire, by which my Lodgings were burnt down about a year ago. accordingly I send half a dozen of these bars, of 3 different sises, in order to try what could be done with.[3] Some time after they were sent to me, with y^{e}. prices they should cost to be done in that manner. I found them prety well, only that I wished they should be more hard: and on that account send word back along with them, in order to have them newly hardned if possible: and as to y^{e}. prices I had not objection.

Since that time, which is more than 2 or 3 months ago, I had not any answer: in consequence of which I wrote a letter on y^{e}. subjet to M^{r}. Bolton:[4] and think, either he did not receive it, or has not been able for y^{e}. hurry of business to answer: for I know his politeness, to not think he has done it purposely.

Now as great part of these bars are for y^{e}. Royal Society, and D^{r}. Fothergill is concerned in this object, I told him what has passed. Therefore beg you in his name, to inquire about it viz. about the three pairs of y^{e}. bars I have sent back, and see if they can be tempered much more hard that is to Say, as hard as possible: and at ye same time strait. then order to be polished as before: and please to send them back to me, along with y^{e}. price or expences for each dozen of each size as well as the cost of each dozen of y^{e}. greatest

1 Fothergill evidently passed this letter to Matthew Boulton since it is filed among the latter's correspondence from Magellan and is annotated in his characteristic style: "London 7th: 8^{br}: 1773 / Magalhaens / received 9th. d^{o}. d^{o}. / answ'd 16th. d^{o}. d^{o}.".

2 Dr John Fothergill (1712-1780).

3 See 73.00.00.

4 Magellan to M. Boulton, 9 September 1773.

size of y^e^. three, intirely newly made. for I will want some quantity of new ones of that sise. And when I receive and examine these 3 pairs, I will be able to judge, and to send you the [...][5] Cargo. for I have a great number, which require the same operation. I wait for your answer and am 73.10.07

Sir
Your most obd^t^. h^ble^. ser^t^.
Magalhaens

To
M^r^. Fothergill
at his Manufactory
in Soho
near <u>Birmingham</u>

5 Illegible.

To Nathaniel Pigott 73.10.12

Royal Astronomical Society, London, MSS Pigott, 29

London 12 Oct^ber^. 1773

Dear Sir

This is to assure you of my being about the order of his R. H. Prince Charles.[1] I went to the most famous glass-house we have, and bespoke the man, who is reckoned to be the most clever in drawing thermometrical and barometrical Tubes, & whom I know this many years, he being the only one I employ for these objets. as soon as there is a pot of very fine glass, he will send me word, for I chuse to be present at y^e^. operation, in order to refuse those tubes which have some blemishes without lost of time and have others made to remplace them. And as soon as the whole is ready (varnish and all) I'll send it according to y^e^. directions given by you. But I wish you should warn those who are concerned in these glasses, about the

1 Presumably Prince Charles of Lorraine (1712-1780), long-time Governor-General of the Austrian Netherlands and founder of the Académie Impériale et Royale des Sciences et Belles-Lettres at Brussels.

73.10.12 nature of y^{em}.; as it is meer chance to get tubes of an uniforme diameter, there being no art to make them of a regular bore. That is the reason why spirit-levels are grinded and worked in the inside, with great trouble & expense. M^r. Nearne[2] made here one about 2 years ago, which shew'd the inclination of one minute by y^e. space of 15 lines. I wanted then one very fine, for M^r. Le Marquiz de Courtenvaux of Paris, and endeavour'd to engage him in the task. but he told me that he was not sure of always doing it so nice: and it perhaps should cost me 25 guineas. the same or more difficulty is met with in y^e. $Thermomet^l$. Tubes. only meer chance afords one tube amongst thousands, which on tryal is found of an equal bore through a good lenght: and this is the reason of making the graduation to each thermometer by a good patern. viz. by comparing the espaces gone thro' by the same degree of heat in the regular thermometer which the makers keep as a patern. From hence you see that we may get one thousand tubes made, with all possible care, and perhaps not one will turn out perfect: tho' some times there will be found four or five in a dozen, good enough for use.

I've spoke a thousand times to Ramsden about his experiments on S^t. Paul's: and never could get any positive answer, but promises of looking & searching his memorandums.[3] I have seen of late an engraved plate, where he sets the heights for each fall of y^e. mercury. but this he set aside, saying there were some errors or mistakes, which renders it useless untill corrected. Now he is very busy with M^r. de Luc: and perhaps it is about some scheems of y^e. kind. This gentleman (M. de Luc) has met with y^e. greatest reception in England. indeed he is very clever in y^e. barometrical way: and must have a very extraordinary patience to go thro' 2 large vol. on y^e. subject.[4] if you have this work, you may find there much better Tables, than those we could get from Ramsden. The last told me on Saturday, that he will make now another kind of Barometers, much superior to y^e. former. but I will wait the light of them, for believing his assertions. As soon as I can, I'll call on M^r. Bird to see your $Astron^{al}$. $Quad^t$. before it be sent:[5] and am sure of being an exceeding good $instrum^t$. I have got from y^e. R. Soc^y. the 1^{rst}. part of y^e. Ph^l. Trans. of y^{is}. year for you, and for M^r. Needham:[6] which I'll not fail of sending by the fi[rst op]portunity.[7] My next will carry

2 i.e. Edward Nairne (1726-1806).

3 Cf. 72.00.00.

4 Jean André Deluc, *Recherches sur les modifications de l'atmosphere...* (2 vols., Geneva, 1772).

5 Cf. 72.00.00.

6 John Turberville Needham (1713-1781).

7 Paper damaged.

the Lading bill of [y^{e}.] glasses, above spoken of. Pray my respects to our 73.10.12
comm[on] friends, who may mention me: and whatever may be the case of employing my friendly dispositions, don't deprive me of the pleasure of shewing how much sincerely I am

Dear Sir
Your most obed^{t}. humble serv^{t}.
J.H. de Magalhaens.

A Monsieur
Monsr. Pigott Gentil-homme Anglois
vis à vis La Cour
à Bruxelles
Flanders

To Nathaniel Pigott 73.10.29

Royal Astronomical Society, London, MSS Pigott, 30

Londres 29 Oct^{bre}. 1773

Mon Cher Monsr. & respectable ami

Les deux bills ci-joints vous montreront que le total de votre commission pour son A. R^{le}.[1] est deja acompli, & mis à bord du vaisseau Diligence Cap^{tn}. Tho. Fennley, qui partira le premier pour Ostende. La petite caisse marquee SAR contient les bouteilles avec differents sortes de vernis, dont je vous parlerai bientôt: & celle marquée ASA contient tous les tuyaux de verre, dont vous trouverez le detail de l'autre coté du bill que je paiai à la verrerie, avec le total des depenses, qui font la some de £14:7^{s}:1^{d}½ Esterlin. La caisse marquée **DAR** contient des oignons de fleurs pour S.A.S^{m}. le duque d'Aramberg.[2] Celle marquée **NEE** n°. 1. contient la premiere part des Transactions Philos. de cette annee. une copie pour vous. une autre pour La Bibliotheque Royale & une 3me. pour M^{r}. Needham.

---

1 Prince Charles of Lorraine; see 73.10.12.
2 i.e. Charles-Marie-Raymond, duc d'Arenberg (1721-1778).

73.10.29 l'autre n°. 3 avec la même marque, contient un assortiment de vernis pour M^{r}. l'Abbé Thisbaert de Louvain:[3] & la barrique marquée n°. 2 contient des pommes de terre Howard potatoes,[4] pour M^{r}. de Belen, greffier du conseil à Bruxelles.[5] C'est par delicatesse que j'y ai mis des marques differentes, & que je n'ai pas mis le tout ensemble, pour que le franchisse des droits &c^{a}. ne s'extende qu'aux articles que S. A. R^{le}. voudra. Je ne manquerai d'ecrire cette meme poste à M^{r}. Hennesy, pour qu'il ait soin de ces caisses, qui pourront arriver peutetre plutôt, que vous lui faisiez part de Bruxelles. Au reste je puis vous assurer d'avoir pris toutes les mesures possibles pour bien executer cette commission; pour seconder les vues de votre desir: & pour que le tout fut parfait. Mais vous ne savez peutetre les grandes dificultés pour reussir dans la manufacture des tuyaux gros & epais. on a fait ceux-ci à differentes reprises, & on a gaté plus de la moitié. Ce fut avec bien de la dificulté, qu' n'a pas chargé plus de 2^{s}. par Livre pessante de ces tubes.

A present je vais vous faire le detail de ce qu'il faut observer pour bien vernir, & bien argenter les pieces de metal. L'essentiel de cette manipulation consiste, pour le premier objet de verniser, à mettre le vernis bien egalement etendu sur la surface metalique. Mais c'est bien plus tôt dit, que de la bien faire, malgré toute la simplicité du fait. La premiere fois que j'ai vû cette operation, j'ai crû pouvoir la faire tout de suite. mais je fus bien etonné de la grande quantité de circonstances, qui me faisoient manquer. ainsi il ne faut pas se rebuter, si l'on ne reussir point d'abord: il faut persister avec patience, & recommencer plusieurs fois la meme piece qu'on a gatée. 1^{remt}. il faut que la piece de cuivre jaune, qu'on veut verniser, soit parfaitem^{t}. finie; & qu'on lui ait donné un très beau poli, dans le meme sens; c'est à dire, que si la surface est plane ou oblongue, il faut que les vestiges du poli aillent touts du meme coté dans la longueur de la piece. J'appelle vestiges du poli, ces petites raies, lignes, ou enfoncem^{ts}. presqu'insensibles de la surface, qu'on ne peut distinguer qu'à la loupe. Ce sont ces petits amas de raies paralleles, qui rendent la couleur du metal plus sensible: car plus une

3 Joannes Franciscus Thijsbaert (1736-1825), professor of experimental physics at the University of Louvain.

4 The Howard or Cluster potato was introduced from America by the famous prison reformer John Howard in 1770; it became widely grown, chiefly as a food for stock (Redcliffe N. Salaman, *The History and Social Influence of the Potato* [Cambridge, 1949], p. 162).

5 Presumably the Baron de Beelen described as "Secrétaire-Greffier du Conseil Privé des Domaines et des Finances entre 1761 et 1792" in H. Tarlier, *Almanach de la Cour de Bruxelles ... de 1725 à 1840* (Brussels, 1864), p. 57.

piece de metal est parfaitem^{t}. polie, plus la couleur paroit foncée, comm'on le voit dans l'argent, l'or, &c. ainsi pour que la surface paroisse belle, & egale dans la couleur, il faut que ces amas de petites rayes soyent toutes dans le même sens. 73.10.29

2^{d}. Il faut aussi que la surface de la piece soit bien degraissée, & proprem^{t}. nettoyée: particulierem^{t}. lorsqu'elle fut polie à l'huile. il est donc necessaire de la froter bien alors, dans le même sens, avec de la fleur de l'emery, du tripoly, ou de la pierre pourrie: ou du moins avec de la craie (crayon blanc) bien fin; & dans la suite la bien essuyer, avec un linge bien propre.

3. on choisira la qualité du verniz qui convient mieux à la piece. par example, si la piece n'est pas de la premiere delicatesse, comme une pompe pneumatique &c. on fera usage du vernis marqué **A**, qui est pale, ou autrement du vernis **B** qui est d'une couleur fort foncée. Mais dans le cas de vouloir que la couleur de la piece soit entre les deux extremes, on fera pour lors un melange de ces deux vernis, dont on reconnoitra la couleur, en l'eprouvant sur un morceau de cuivre jaune bien poli &c. Mais si c'est une piece d'ouvrage fin, vg. une lunete de poche, un Quadrant Astronomique, un Theodolite &c. alors on doit employer le vernis marqué **C** qui est pale. C'est avec cette couleur que nos Artistes vernisent leurs instruments en general: quoiqu'il soit quelques fois plus agreable de leur donner une couleur plus rehaussée, Ce qu'on obtiendra en y melant du vernis **D**, ou même l'employant tout seul. N.B. 1. il vaut mieux avoir fait ce melange 3 ou 4 heures auparavant de l'employer, en le secouant bien & le laissant reposer long tems. N.B. 2. lors qu'il y a des divisions dans les instruments, il ne faut pas y mettre dessus les divisions aucun vernis; particulierem^{t}. s'il y a quelqu'alidade ou coulisse qui passe dessus.

4. on allumera du charbon de bois, dans un pot de fer ou dans un rechaud, sur le quel on mettra un trevet, ou 3 pointes de metal pour suporter une planche de metal bien propre, qui soit à la distance de 3 ou 4 pouces de la braise ou charbon allumé. Si l'on met avec le charbon quelques cendres, celà le conservera mieux dans un degré de chaleur presqu'egal. Cette planche de metal doit etre assez grande pour contenir la piece qu'on veut verniser, sans en laisser quelque partie exposée à l'action immediate du feu, ni des vapeurs, ou efluvia, qui en sortent. Mais si la piece est trop grande, & trop grosse pour lui donner un degré egal de chaleur tout partout, vg. si c'est un grand tuyau de Telescope ou Lunete &c. alors on y met dedans un fer chaud, ou même du charbon de bois alumé. Celle-ci est une des operations les plus dificiles. les meilleurs artistes la manquent souvent, & sont obligés de recommencer plusieurs fois.

73.10.29 5. on mettra du vernis qu'on aura choisi (n°. 3) dans la coupe ou tasse, que j'envoye, montée comm'elle l'est, sur son piedestal: & on prendra le vernis avec un des deux pinceaux, que j'envoye aussi, le grand, ou le petit, selon la grandeur de la piece. après avoit trempé, ou mouillé le pinceau dans le vernis, il faut decharger le surplus, en passant 3 ou 4 fois le pinceau contre la plaque de cuivre, qui est croisée à travers sur la tasse que j'envoye. Car le pinceau ne doit etre qu'á peine mouillé de vernis: autrem^t^. il le laisse, comm'estagné, dessus la surface du metal, ce qui la rend eparpillée & fort desagreable.

6. La piece doit etre assez chaude pour recevoir le vernis: mais point plus chaude qu'on ne puisse pas la soufrir à la main. on doit la retirer dessus la planche, oû l'on avoit mise la piece à echaufer, ou mêm'oter la planche ensemble avec la piece. on la tiendra avec un linge propre tandis qu'on la vernise, & non pas avec la main nude. si la piece est ronde, il faut avoir preparé une manche de bois, ou manivelle, pour faire tourner la piece dans le sens de son poliment, tandis qu'on y applique du vernis avec le pinceau. Mais si la piece est unie, on y mettra le vernis dans le sens longitudinal, & selon la direction du poliment. C'est cette operation, qui demande une certaine adresse, ou tour de main, dont on ne peut pas donner la description. il faut mettre le vernis sur la piece, en frottant toujours dans le meme sens, & en Ligne droite, avec une certaine velocité, qui fasse ettendre la couche du vernis bien egalem^{t}., & bien mince; de façon que le vernis soit sec par la chaleur de la piece lors que le pinceau y passe le 2de. fois. & lorsqu'on prend du vernis, il faut toujours decharger le surplus du vernis à chaque puise, sur la traverse qui est dessus la tasse. Lorsqu'on a donné une seconde couche, & quelques fois une 3me., de vernis sur la piece, bien egalem^{t}. tendû, on la remet encore pendant quelques minutes à echaufer, pour que le vernis soit bien sec. après on l'ote, & on la laisse refroidir sans la toucher. Car si on la touche, pendant qu'elle est chaude, il y restent des vestiges des doigts comm'autant de taches.

7. on doit prendre bien gard de ne pas verniser la piece lorsqu'elle est trop chaude car alors le vernis se brule, & reste sur la surface en taches noires. C'est pour celà qu'il faut essayer sur quelqu'endroit, que ne soit pas visible, par example au dessus, ou au dedans de la piece y touchant legerem^{t}. avec le pinceau. Pour le contraire si la piece est froide, on gate tout: car le vernis ne s'attache point à la surface com'il faut; mais forme comme des croutes, qui s'elevent par ci par là, à mesure qu'on les touche avec la main ou le pinceau. C'est encore ce degré de chaleur moyenne, qu'il faut apprendre à force d'experiences, & après avoir gaté plusieurs pieces. &c.

8. Lorsque la piece est gatée, ce qui arrive souvent, c'est à dire lorsque l'operation est manquée, il faut avoir à la main de l'esprit de vin, pour la lever bien avec, à fin d'oter tout le vernis: & s'il est brulé, il faut souvent polie de nouveau la piece pour en ôter les taches. 73.10.29

9 Si les pieces sont ciselées, avec des gravures, feuillages, & endroits sabloneux &c. il faut les preparer par le moien de l'eau seconde, ou aqua fortis assez afoiblie, pour bien nettoyer toute la crasse & hordures & taches de la surface metalique. on peut quelques fois employer de la lessive: & on frote bien toute la surface avec des broches de fil de metal. dans un mot, il faut faire en sorte que toute la surface, sur laquelle on veut mettre du vernis, soit tout à fait de la couleur du metal, sans taches, ni endroits plus obscurs &c. Car sans avoir une surface bien nette, d'une couleur egale, & toute à decouvert, le vernis ne fera que faire paroitre plus les endroits moins eclatants & obscurs, &c.

10 En fin c'est la practique seule, acompagnée de la reflection, qui menera l'artiste à faire cet ouvrage en perfection, prevoiant les acidents, & ecartant les dificultés, qui peuvent l'empecher de bien reussir. N.B. au lieu de l'esprit de vin ci dessus n°. 8, on peut avoir un pot avec des cendres de bois & de l'eau, qu'on echaufe sur le feu pour y lever les pieces gatées, avec cette lessive: & on fait la meme operation aux pieces vieilles, qui ont eté vernisées. Car cette lexive chaude, fait tomber le vernis de la surface du metal, & la laisse dans le même tems fort nette.

11. Pour ce qui regarde la methode d'argenter il n'y a qu'à bien polir, avec la fleur d'emery, la surface du metal. mais s'il y a des gravures qu'on veut faire paroitre noires, vg. la graduation des thermometres, les quadrants des pendules &c. on echaufe la piece, & on y met de la cire noire à cacheter de la plus fine: dont on emporte le surplus avec la pierre pouce. après celà on polit fort nettem[t]. la surface du metal &c. on met la piece ainsi polie & bien lavée sur une planche, qui est croisée à travers sur les bords d'un grand sceau d'eau pure de fontaine. on met, dessus la piece, un peu de la pate d'argent que je vous envoye. & on la frote bien dessus de cette surface metalique, avec la main, ou avec un morceau de cuir mou. Mais nos ouvriers preférent toujours de froter avec leurs mains, qui sont grossieres & caleuses. la pate doit etre melée de tems en tems avec quelques goutes d'eau, qu'on jette dessus, tandis qu'on frote. or comm'elle est composée des sels, avec la solution d'argent très fin, les particules metaliques s'attachent à la surface de cuivre, tandis que les sels sont dissus & fondus avec l'eau. C'est pour celà qu'il faut bien degraisser & polir avec de l'emery humide cette surface. Celà fait on plonge la piece deja argentée dans le sceau d'eau: & si l'on voit qu'il n'y a assez d'argent, on repete l'operation. il est bon de laisser la piece

73.10.29 dans l'eau pendant une demiheure, pour le debarasser de tout sel. on l'ote après: & on l'essuye doucem^{t}. avec un linge bien fin & bien propre: & on la met à echaufer comme dans le n°. 4.: & on y met une couche très unie & mince, du vernis blanc de la bouteille marquée **E**. Mais, au lieu de pinceau, il est mieux d'appliquer le vernis avec un morceau de toille fine de lin, bien propre, qui ait eté un peu use. on le plie 4 ou 8 fois, comm'un plumasseau de cirurgien: & on prend avec lui du vernis blanc qu'on ettend vite, bien egalement, sur toute la surface argentée, en allant toujours du meme coté en ligne droite, par des tranches paralleles. C'est le vernis qui garantira pendant long tems la couleur de l'argent, car sans celà il ne manqueroit pas de noircir au bout d'un certain tems. Je m'etonne com̄'on n'a pas encore pris cette methode pour preserver les pieces d'argenterie, qui ne sert pas à manger: car pour celle-ci, il seroit desagreable d'avoir l'odeur du vernis. Mais il a très peu de monde qui reflechisse sur des objets communs: on se laisse entrainer par la coutume, sans y penser du tout...! N.B. il faut avoir soin de laver bien avec l'esprit de vin les pinceaux, après en avoir fait usage, à chaque fois. autrement ils se gatent, & ne servent bien dans la suite. J'ai lié ceux que j'envoye, à la hauteur, que je crois plus avantageuse pour l'en servir avec moins d'embarras.

Voilà une lettre bien longue & bien mal ecrite. mais je n'ai pas pû faire mieux: et j'espere que Vous saurez m'excuser là dessus en attention à la bonne volonté & disposition avec la quelle j'ai l'honneur d'etre

Mon cher Monsr.
votre très humble & ob^{t}. serv^{r}.
Magalhaens

A Monsieur
Monsr. Pigott, Gentilhome Anglois
vis à vis La Cour
à Bruxelles
Flandres

To Louis-Henri Duchesne 73.11.02

Archives Nationales, Paris, MS T/160 (10-12)

Londres 2 Nov^{bre}. 1773

Mon cher ami

C'est avec bien du plaisir que j'aprends votre hereux retour. Les lunetes que Je vous ai envoyées sont de la meilleure espece dans leur genre, & vous faites fort bien de garder pour vous la meilleur de toutes les 3. A present je comprends ce que vous entendez par lunete de poche, qui n'est pas d'autre chose qu'une achromatique, à peu près comme celles qu'on use au theatre, sans aucun pied, & $seulem^t$. avec un ou deux oculaires. J'aurai soin de vous envoyer deux bien choisies par la premiere occasion.

Je suis bien sensible de ce que vous me dites sur l'envoy des choses d'Angleterre. mais on m'ecrit souvent d'avoir le consentment de M^r. Trudaine pour telle & telle chose. par example M^r. le $Chev^{er}$. de Borda, Monsieur le D^r. Macquer, M^r. le Marquiz de Courtenvaux &c. & j'ai toujours soin de dire dans les lettres d'annonce lors que chaque caisse part d'ici de ne s'oublier de parler de celà à M^r. Trudaine. desormais j'en aurai une autre precaution, de vous en faire part avant de celà pour que vous vous certifiez de leur exactitude.

je vous prie de presenter mes complimts. à M^r. de Montigny, & de lui dire que je ne m'oublie pas de son thermometre.[1] mais qu'il ne m'a pas eté possible de l'arracher jusqu'à present des mains de l'ouvrier, quoiqu'il me le promet chaque semaine. Neanmoins si je ne puis pas l'avoir jusqu'à Samedi, je lui oterai la planche d'argent que je lui ai fourni depuis très long tems, & je le ferai faire par un autre ouvrier.

Je vous chercherai le livre d'edimburg sur le principe mercuriel de Beccher,[2] pour votre ami. c'est un livre dont je n'ai pas entendu parler encore: à moins que ce ne soit un autre nom d'Auteur.

Jusqu'au present je n'ai pas oui dire un mot d'aucunes additions aux laboratoires de poche, si ce n'est un petit traité de mon ami Forster, le meme, qui est à present dans le voyage aux terres Australes. Car il a fait

---

1 See Magellan to Duchesne, 27 August 1773.

2 Johann Joachim Becher (1635-1682) was a German chemist, the father of the idea of phlogiston (afterwards developed by G. E. Stahl), who elaborated the idea that mineral bodies are composed of three "earths", of which the third, the mercurial principle, gives metals their chief characteristics. The work supposed to have been published in Edinburgh has not been identified; see also Magellan to Duchesne, 12 November 1773.

73.11.02 une brochure sur la maniere d'essayer avec d'autres Liqueurs & acides &c.[3] mais on n'a pas suivi (que je sache) cette methode, pour la construction des petites caisses de poche, qui d'ailleurs deviendroient trop grands, & incommodes. Cependant je m'informerai de nouveau, & vous en dirai ce qu'il y en a.

Je voudrois bien pouvoir faire un tour à Paris dans un ou deux mois d'ici: peutetre pourrions nous alors pousser les experiences de M^{r}. La Voisiere, qui ne meritent pas de rester dans l'oubli.

M^{r}. Le Marquiz Griselli de Rosignan,[4] vient de partir d'ici pour Paris. il apporte avec soi une petite caisse adressée à M^{r}. Trudaine de Montigny, avec quelques livres pour Monsr. le Comte Carburi,[5] auquel j'ecris cette meme poste.[6] il y a dedans un petit pacquet avec 3 brochures pour notre bon ami le D^{r}. Sanchez, & un autre pour M^{r}. Messier. Le Marquiz de Rosignan est ami de M^{r}. le Comte Carburi: & lui me promit de donner le petit pacquet à M^{r}. Messier. pour l'autre pour le D^{r}. Sanchez, je vous prie d'en avoir soin qu'il lui soit envoyé. le deboursement pour les livres de M^{r}. le Comte Carburi, fait la some de 57$^{\#}$. 8^{s}. tournois, que je vous prie de donner à M^{r}. Messier pour mon compte, comme j'en avertis aussi à M^{r}. Le Comte.

à Dieu mon cher bon ami: conservez vous, & soiez assure de l'attachement sincere de

votre pour la vie
Magalhaens

Aiez la bonté de donner cette petite lettre à M^{r}. Messier[7]

a Monsieur
Monsr. Duchesne Secretaire
de Monsr. Trudaine de Montigny
à Paris

3 Johann Reinhold Forster, *An Easy Method of Assaying and Classifying Mineral Substances...* (44 pp., London, 1772)? Forster accompanied James Cook on his second voyage of exploration, 1772-5.

4 Francesco Maria Grisella, Marquis of Rosignano (1722-1802), was a gentilhomme of the House of the Kingdom of Sardinia.

5 See Magellan to Duchesne, 23 July 1773.

6 Letter not found.

7 Letter not found.

To Joannes Franciscus Thijsbaert 73.11.05

University Library, Université de Liège, Ms. 2617

Londres 5. Nov^{bre}. 1773

Monsieur & très Cher ami

La Votre du 18 du mois dernier[1] m'a rempli d'une vraye joye, voiant que vous & ceux qui voient vos instruments, les trouvent[2] satisfaisants. je suis faché cependant par le petit acid^{t}. du miroir concave. mais si vous prennez un petit morceau de la feuille d'etain comme celle qui sert pour les bouteilles electriques, & vous y mettez un peu de mercure, dans peu de minutes le mercure y prendra; alors otez la pelicule avec un morceau de coton. metez encore une goute de mercure dessus, & après avoir bien nettoyé la pelicule, mettez ce morceau sur l'endroit qui fut gaté par le frottem^{t}., & vous verrez que celà ne paroit point. mais vous devez avant celà nettoyer ce meme endroit du verre. le meilleur c'est de vous exercer sur un miroir commun dont vous gaterez exprès un endroit. & lorsque vous y reussirez bien, alors vous ne manquerez de bien reussir aussi sur le miroir concave.

J'ai eu l'occasion d'envoyer quelques caisses à Ostende, & j'y ai ajouté pour Vous une petite, marquée NEE n°. 3., dont j'ai deja donné avis à M^{r}. Knuÿt. elle contient differentes especes de vernis jaune, avec un long detail pour le bien savoir appliquer sur les pieces de cuivre jaune. il y a tout l'attirail necessaire, même deux pinceaux arrangés comm'on les use dans cette operation. Vous y trouverez aussi un quart de livre de la pate d'argent, pour argenter les pieces, qui en ont besoin, vg. les plaques des Thermometres, quadrants &c. & aussi une bouteille de vernis blanc, pour couvrir cette argenture, ou quelqu'autre piece argentée de Votre Cabinet, à fin de lui conserver la couleur contre les impressions de l'air. en fin j'y ai ajouté plus d'un tiers d'aune de la soye cirée, dont on fait usage pour l'electricité.

---

1 Letter not found.

2 Magellan ended a line at this point with "trou-", then began the next, as if he had forgotten this, with "trouvent".

 Supposons que ednm soit le coussin d'un cylindre qui tourne

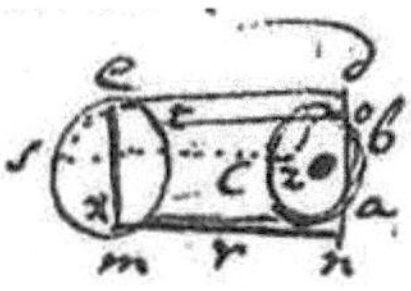

dans le sens aboc. Coupez un morceau de cette soye cirée (ou plutôt vernisée, car c'est propremt. un verniz avec lequel on l'apporte) & appliquez-le au dessous des 3 quarts de la partie du coussin qui frote contre le cilindre. C'est à dire à t.o. & laissez le reste de cette soie de façon qu'elle couvre le dessus du cylindre tandis qu'il tourne. pour lors vous verez, que la machine donne beaucoup plus d'electricité. Car on a observé, que l'electricité, qui est excitée par le frotement du coussin, regurgite en arriere sur le meme coussin. ainsi en mettant de cette soye vernisée sur le bord qui touche en dernier lieu au cylindre, la matiere electrique est forcée d'aller dans le meme sens avec le cylindre, sans reculer. N.B. il est bon de mettre une chaine, ou barre de metal en contacte avec le bord inferieur du coussin, pour laisser courir abondament le fluide electrique de la terre au coussin.

Vous aurez je crois demelé la division du Nonius de votre Barometre. Car il y a 20 divisions en chaque pouce anglois: c'est à dire, chacun est divisé en 20 parties. or le Nonius a 25 de ces parties pour subdiviser. donc $20 \times 25 = 500$. ainsi chaque pouce anglois est divisé reelemt. en 500 parties. Mais comme les nombres y sont mis doubles, c'est à dire, chaque division du Nonius est marqué par 2, il s'en suit, que ces nombres multiples de 2 dans le Nonius, marquent des miliemes ($2 \times 500 = 1000$) malgré que la division reele ne soit que en 500 parties.

Je vous suis bien obligé par le present des cats de mer.[3] & plus encore par les observations & notice que vous m'en donnez. Je les attends à tous moments, & j'espere que Mr. Knuyt ne [...][4] oubliera de m'en donner l'avis. J'en ferai present à la Societé Roya[le][5] & au British Museum.

Soyez bien assuré que je n'aurai pas de peine à quiter les bords de cette Isle, pour aller revoir des amis comme vous, & tous vos Messrs. mais, malgré mon desir, & même quelques petits affaires assez pressants, que J'ai à Paris, je crains ne pouvoir pas sortir d'ici qu'au bout de l'hiver. en

3 Sea-cats (wolf-fish)? Cf. Magellan to Thijsbaert, 25 January 1774.
4 Word illegible.
5 Paper damaged.

attendant je vous prie de presenter mes humbles Respects; sans oublier votre digne Chancelier,[6] qui par son merite, & aimables qualités, ne peut pas manquer d'etre très present dans la memoire de tous ceux qui ont le plaisir de le connoitre. Portez vous bien, & ne prodiguez votre santé, qui est le don le plus precieux, & pas assez connu des humains. C'est le bien publique, qui s'interesse à la votre, aussi bien que l'amitie de celui qui est pour la vie 73.11.05

Votre de Coeur
Magalhaens.

M^{r}. Heussens[7] a montré ici dans le microscope, que le sang n'a pas de <u>globules</u> rouges, mais des petits cercles aplatis, comme des lentilles. pour les voir bien il faut diluer une goute de sang, dans 3 ou 4 fois autant de limphe. il a donné sur celà un memoire à la Societé R^{le}. qui ne manquera d'etre imprimé dans les Transactions Philosophiques, pour la derniere partie de cette annee. peut etre vous fera du plaisir cette nouveauté, pour le montrer à vos eleves lorsque vous traiterez de ces matieres dans votre cours.

A Monsieur
Monsr. L'Abbe Thisbaert, Professeur
dans L'Université de <u>Louvain</u>
au College de <u>Castrum</u>
Flandres

---

6 Joseph de Crumpipen (1737-1809), Chancellor of Brabant and first President of the Brussels Academy of Sciences.

7 i.e. Hewson; cf. "On the Figure and Composition of the Red Particles of the Blood, commonly called the Red Globules. By Mr. William Hewson, F.R.S. and Teacher of Anatomy", *Philosophical Transactions*, 63 (1773), 303-23.

73.11.12 *To Louis-Henri Duchesne*
Archives Nationales, Paris, MS T/160 (10-12)

Londres 12 Nov^{bre}. 1773

Mon cher ami

la lettre ci jointe containt une demi douzaine de feuillets que M^{r}. Baptistin, me pria de lui envoyer sous votre envelope. apparament il vous a parlé de celà: car sans cette supposition, je n'oserais point de la faire.

Je vous prie de faire passer à M^{r}. Messier la parcelle des volumes du journal de M^{r}. l'Abbé Rozier qu'il à fait[1] remettre chez M^{r}. Trudaine, pour les envoyer en Angleterre: car il y en a pour la Societé Royale & des Arts & d'autres: & on est un peu faché de les avoir aussi retardés. M^{r}. Messier aura la bonté de m'envoyer cette parcelle par la premiere occasion: & en cas de ne pas avoir aucune avant Noel, il n'aura qu'à passer chez M^{r}. de la Tour libraire, dans la Rue de S^{t}. Jacques,[2] qui envoye souvent des Livres à M^{r}. Elmsley, Libraire ici à Londres.[3]

Dans ce moment je viens de parler à M^{r}. Elmsley, qui dit de les envoyer tout de suite à M^{r}. Gilbert,[4] libraire, Rue des Maturins, car il doit faire tout de suite un envoy de Livres ici. ainsi vous n'avez qu'à faire un pacquet, & y mettre dessus = à M^{r}. Elmsley. Libraire à Londres, pour le rendre à M^{r}. Magalhaens London. = en cas que M^{r}. Gibert ait deja envoyé sa pacotille, alors envoyez ce pacquet adressé dans la meme maniere, chez M^{r}. Briasson,[5] libraire dans la Rue de S^{t}. Jacques, qui doit envoyer aussi bientôt une autre caisse de Livres au meme M^{r}. Elmsley ici à Londres.

aiez la bonté de dire à M^{r}. de Montigny, que l'ouvrier, etoit ce matin gravant la platine de Son Thermometre. Je fus obligé de le faire faire par un autre ouvrier, voyant que le diable de menteur Ramsden, ne faisoit que me remettre d'un jour à l'autre, sans jamais finir cet ouvrage.[6] aussitot que le Thermometre sera fini, je ne manquerai pas de l'envoyer par La premiere occasion. votre pendule ou pendules n'est pas encore finie: & partira de meme aussitôt que Je l'aurai.

---

1 *sic.*
2 Louis François de la Tour (1727-1807), bookseller and publisher at Paris, 1745-78.
3 The well-known London bokseller Peter Elmsley (1736-1802).
4 i. e. Gibert; see below. See also Magellan to Duchesne, 8 June 1773, note 5.
5 The bookseller Antoine Claude Briasson (1700-1775).
6 Jesse Ramsden was notorious for failing to meet promised delivery dates for instruments he had contracted to supply.

Je ne puis pas decouvrir l'ouvrage sur le principe mercuriel de Becher.[7] apparament il y a quelque meprise dans le titre: ainsi tachez de m'envoyer le vrai titre si vous le pouvez, avec la date de l'edition &c. 73.11.12

a Dieu mon cher ami, portez vous bien, & conservez-vous pour La satisfaction de tous ceux qui vous aiment, entre les quels je me flate de meriter un des premiers rangs.

votre de coeur
Magalhaens

Mes humbles Respects à Messrs. Le Marquiz de Mirabeau,[8] Abbé Nicoli,[9] & tous les autres de son assemblée qui m'honorent de leur souvenir; sans oublier l'Abbé Boudeau[10] & Mr. du Pont.[11]

7 Like Magellan, we have not been able to identify the book that Duchesne was seeking. According to the famous chemist Johann Joachim Becher (1635-1682?), the mercurial principle was one of three universal principles, the component of mineral substances that gave metals their distinctive character.

8 Victor Riquetti, Marquis de Mirabeau (1715-1789), French economist and statesman.

9 The abbé Raimondo Niccoli (?-1780), secretary of the Parisian legation of the Grand Duke of Tuscany.

10 Nicolas Baudeau (or Boudeau) (1730-1792), economist and physiocrat.

11 The physiocrat Pierre Samuel du Pont de Nemours (1739-1817), who in 1768 had succeeded Nicolas Baudeau as editor of the influential journal, *Ephémérides du citoyen ou Bibliothèque raisonnée des sciences morales et politiques.*

To Johann III Bernoulli 73.11.27

Öffentliche Bibliothek der Universität Basel, MS L I a 709, f. 49-50

Londres 27 Novbre. 1773

Monsieur & très cher ami

La Votre du 6 de ce mois m'a causé le plus vif plaisir.[1] Votre retablissement de Santé qui semble aller au mieux, & encore plus l'aisance de votre esprit, qui sembloit acablé par le depart d'une infirmité languissante m'interessent on ne peut plus, pour en resentir toute La joye. Peut etre si vous vous aviez

1 Letter not found.

73.11.27 contenté d'une securité de nom, & d'un defaut d'eclat, comme celui, dont je jouis à mon aise dans ce commencement de ma 52me. annee; vous trouviez vous aussi bien en santé, & aussi robuste comme je me trouve à present. Mais il ne s'agit plus du passé, qui n'a pas du remede. Tachez seulem^{t}. de vous conserver comme vous etes, avec tout le foible de votre corps, & jouissez de toute La force de Votre esprit: c'est Le souhait de tous vos amis.

Je ne sais pas d'autre astronome qui calculat la Table des Asc. droites de M^{r}. Mackenzie, que Je vous envoyai, que Lui meme: il est très capable, & grand amateur de l'Astronomie.[2]

Je ne sais pas bien l'objet de l'attaque de M^{r}. Shepherd par M^{r}. Witchell. apparement il ne trouve pas la publication des grosses Tables pour la Refraction & Paralaxe aussi bonne comm'il l'a proposée lui meme: car c'est M^{r}. Wichell qui en donna les idees, peut etre bien plus compendieuses, & aisées. il fut premié je ne me souviens bien si avec 200 ou 300 guinées par son invention. mais l'offre qu'il fit de les faire calculer à un prix assez modique, fut rejeté; apparament parceque Maskelyne & Shepherd vouloient employer leurs calculateurs, par des raisons qu'ils connoissent mieux que moi ni vous.[3]

M^{r}. Wolf (non pas D^{r}.) que je vous recommendai, a eté en Autriche cet Eté:[4] peutetre qu'il y trouva de quoi s'amuser dans ses pursuites chymiques, sans avoir assez de tems pour aller passer chez vous quelques mois, comm'il le pensoit.

Je pense aller à Paris dans le Printems prochain: & j'y fairai imprimer mon Memoire[5] dont je vous presenterai quelque copie. S'il est vrai que

---

2 Cf. Magellan to Johann III Bernoulli, 25 June 1773.

3 George Witchell (1728-85) of the Royal Naval Academy, Portsmouth, helped compute the tables for the first volume of the *Nautical Almanac* (1767). He was awarded £300 by the Board of Longitude for a method he developed for clearing an observation of the effects of refraction and parallax, a brief account of which was appended, without separate heading, to Maskelyne's description of his own method for clearing an observation, that was published, in 1770, in the *Nautical Almanac* for 1772. (Maskelyne as Astronomer Royal was the person chiefly responsible for the *Nautical Almanac*, while Anthony Shepherd, Plumian Professor at Cambridge and likewise a member of the Board of Longitude, was his principal collaborator.) According to E. G. R. Taylor, Witchell first announced his method to the Board of Longitude in 1764, and the premium he was awarded was for "compiling, correcting and printing 1000 copies of his tables". However, these tables were never printed, being upstaged by Shepherd's *Tables for correcting the Apparent Distance of the Moon and a Star from the Effects of Refraction and Parallax*, published in 1772.

4 The London-based chemist Peter Woulfe (?1727-1803), who regularly spent his summers on the Continent, usually in Paris; see 73.04.29.

5 i. e. his *Description des octants et sextants anglois...* (Paris/London, 1775).

l'interet meut les actions des hommes, on ne peut pas douter que Maskelyne y trouve sien à proner les quadrants de Dollond, & deprimer les miens. Mais ce sera le public à juger de ses raisons. 73.11.27

le compte que je vous envoyai de vos livres avec le libraire Mr. Elmsley, est sur le pied dont il doit vous repondre: car je lui ai donné le resultat du premier prix, & depenses du porterage &c.

Les Barometres de Mr. de Luc ne seront jamais employés, que par des observateurs aussi Soigneux & meme echarnés, que Lui. ils ne sont pas assez portables, ni plus exacts que les bons qu'on fait ici.

Je ne sais rien de la peinture en Cire, ni connois l'ouvrage de Mr. Lambert[6] & du P. Schiffermuller[7] sur les couleurs. Si j'etois un peu plus amateur de cette branche, ou moins paresseux, je me donnerois la peine de les chercher: car je suppose que ces Messrs. ont poussé bien plus loin leurs recherches, que le P. Castel[8] & autres.

Ce fut en Angleterre qu'on fit la sustitution de l'ecorce de Saule au quinquina. voyez les Transact. phil. vol. 53. obs. 33 par Mr. Stone.[9]

Re[...][10] les amitiés de Mr. Valtravers,[11] & d'autres amis qui on[t][12] entendu vos [nouv]elles avec du plaisir.

je communiquerai a Mr. Shepherd votre article qui lui regarde.

Celle-ci va ensemble avec le pacquet des Transact. Phil. de 1772, & la premre. partie du Vol. de cette année. Car on publiera desormais les Transact. de chaque semestre. J'ecris au Bibliothecaire de me donner un ordre en general pour lui prendre les Vol. de ces Transact. aussitôt qu'ils paroitront. Car sans en etre authorisé pour celà, je ne m'en mêle de ces choses.

Vale, neque ama
Magalhaens

P.S. le Mr. Mackenzie ci dessus est the Right honble. James Stuart Mackenzie, frere du celebre favori (ou directeur) du Roy, the Earl of Butte, il est à

6 J.H. Lambert, *Beschreibung einer mit dem calauschen Wachse ausgemalten Farbenpyramide* ... (Berlin, 1772).

7 Ignaz Schiffermüller, *Versuch eines Farbensystems* (Vienna, 1772).

8 Louis-Bertrand Castel, *L'optique des couleurs* (Paris, 1740). The work on colours of Castel, Lambert and Schiffermüller is discussed by Sarah Lowengard, *The Creation of Color in Eighteenth-century Europe* (Gutenberg e-publication, New York).

9 Edmund Stone, "An Account of the Success of the Bark of the Willow in the Cure of Agues", *Philosophical Transactions*, 53 (1763), 195-200. (This is actually item no. 32 in the volume, not 33 as Magellan has it.)

10 Paper damaged.

11 Rodolph Valltravers, FRS (1723-c.1815).

12 Magellan writes "on".

73.11.27 present (M^{r}. Mackenzie) du privy Council de S.M. Britanique. & ce fu[t] pour l'usage de son observatoire portatif, ou Instrum^{t}. Equatorial, que ces tables fur[ent][13] imprimées: dont il me donna quelques copies.

A Monsieur
Monsr. Bernoulli, Astronome R^{l}. &
de l'Acad^{e}. Royale des Sciences de
Berlin

13 Text obscured in binding.

73.12.14 *To Louis-Henri Duchesne*[1]

Archives Nationales, Paris, MS T/160 (10-12)

Londres le 14 Dec^{bre}. 1773

La Lettre ci jointe,[2] mon Cher ami, est pour que vous ayez La bonté de la remettre à notre bon ami M^{r}. l'Abbé Niccoli dans La premiere occasion. mais pour le petit billet je vous prie de l'envoyer par La petite poste, & si M^{r}. Silva[3] ou sa femme vient en consequence de celà, je vous prie de lui donner 24 francs pour mon compte: & de plus 6 francs chaque mois aussi pour mon compte. Ce sont des gens honnetes, que se meurent de faim, & qui n'ont point de resource: je ne puis pas me tenir de compassion pour eux & pour la misere de leur situation. Mais je vous recommende très particulierement de ne rien dire à notre bon vieux le D^{r}. Sanchez ni à M^{r}. Messier qui pourroit sans y penser le lui dire. Car c'est lui qui leur donnoit cette petite charité jusqu'à [...][4] de tems auparavant. Mais je ne sais pas par quelle tracasserie on lui a fait croir que ces gens ne meritoient plus sa faveur. ainsi si le pauvre bon vieux vient à savoir que je leur fais donner cette some de ma poche, il viendra peut-etre à decouvrir quelques

1 Annotated by Duchesne: "Rep. le 26 D^{e}." Letter not found.
2 Letter not found.
3 Polycarpo da Sylva; see Magellan to Duchesne, 18 August 1770.
4 Paper damaged; word illegible.

mauvaises maneuvres ou impostures, qui lui feront du chagrin: & moi je Respecte ce bon vieillard, & connoissant le foible de sa santé, ne voudrois aucunement lui faire trouver du chagrin, meme lorsqu'il seroit le plus juste. 73.12.14

Comme je pense aller dans peu de tems à Paris, je tacherai d'arranger cet affaire charitable, de mon mieux: & vous tiendrai compte de vos deboursements.

Le Thermometre pour M^r^. de Montigny, monté en argent, est deja fini, & bien achevé chez moi.[5] Je n'attends que l'opportunité de le faire partir. La se[maine][6] prochaine Je recevrai deux des petites pendules, je [...] la seconde, car La premiere est deja arrivée : & aussitot je vous les envoyerai, avec Votre petite Lunete de poche que vous m'avez ordonnée, & que J'ai deja chez moi.

à Dieu mon cher & bon ami. presentez mes humbles respects à Monsr. le Marquiz de Mirabeau,[7] & ceux de son assemblee qui m'honnorent de leur souvenir. Conservez Vous, & ne doutez un moment de l'attachement du Coeur de

votre

Magalhaens

P.S. on fait ici une souscription pour publier les voyages manuscripts qui se trouvent dans le British Museum, & qui sont precieux & interessants. C'est Banks & d'autres Messieurs qui s'en melent & qui semble ne manqueront de faire un ouvrage interessant.[8] Je ne leur ai pas encore demandé les circonstances. mais Je le ferai Jeudy [...][9] me direz selon l'avis, si vous jugez [...]

a Monsieur
Monsr. Duchesne Secretaire de
Monsr. Trudaine de Montigny
à <u>Paris</u>

5 See Magellan to Duchesne, 12 November 1773.
6 Paper damaged on this and the following line; words illegible.
7 Victor Riquetti, Marquis de Mirabeau (1715-1789).
8 See also Magellan to Duchesne, 18 January 1774. The project that Magellan describes here seems to have come to nothing.
9 Paper damaged on this and the following line; several words illegible in each case.

73.12.17 *From Jacques-André Mallet*

Bibliothèque de Genève, Geneva, Ms suppl. 1654, fol. 48[1]

Geneve 17 Xbre 1773

M^{r}.

Dans toutes les lett. que vous m'av. f. le plaisir de m'écrire, vous avez b. voulu, m'offrir d'employer vos soins, lorsque je serois dans le cas de faire venir de Lond. de qlque livre ou Instrum. d'Astronomie, je sai M^{r}. que vous vous acquitez de ces commiss. avec aut. de zele que d'intelligence, c'est pquoi je viens m'addresser à vous p^{r} l'achat d'une lunette achromat. à triple objectif, telle que celles de l'obs. de Berlin, de Greenwich, ou celle que vous avez fait construire p^{r} M^{r} le President de Saron en 1769. cette lunette étant un présent que qlqu'un fait à un de mes amis qui freq^{te} bcoup mon observat. je m'interesse par conseq^{t} à la reussité de cette commission, & co[mm]e cette perso[n]e a destiné une certaine somme fixe à la dépense qu'il a intention de faire, nous souhaiterions savoir quel est le juste prix d'une celle lunette, je suppose bien, que ce prix doit varier suivant les differ. ajustemens qu'on peut ajouter à cette lunette, c'est pourq. Je vous prierois M^{r}. de m'informer de toutes les differ^{tes} pièces qu'on peut y joindre & de l'augmentation de prix pour chacune, par ex^{ple}.

1°. Quel est le prix de la lunette simple, avec ses differents assortimens d'oculaires & des mouvemens à vis dans son pieds.

2°. Quel est le prix, si on veut pouvoir lui donner un mouvem^{t} parallatique.

3° Quel est le prix d'un heliometre ou microm object. adapté à cette lunette, avec tout l'assortiment qui lui est nécessaire.

4°. Ne peut-on pas ajuster un reticule de fils qui se coup a angle droit au foyer des oculaires, avec une vis pour faire mouvoir le chassis qui les porte, afin de pouvoir déterminer avec le micromet. object. les differ. d'asc. dr. & de déclin. co[mm]e cela est expliqué dans la lett. astr. de Bernoulli, pag. 91 & suiv.[2] Quel seroit le prix d'un tel ajustement.

Je demande le prix des 3 derniers articles, parce peutetre je ferai moimême la dépense de ces additions, quoique la lunette ne m'appartienne pas, mais il est probable qu'elle restera à l'observat. & je serois bien aise d'augmenter ainsi son utilité, mais quoiqu'il en soit, il faut que la lunette, avec son pied & ses oculaires puisse etre entierem^{t} séparée du micromet.

1 Letter not found. The text is from a draft kept by Mallet.

2 Johann III Bernoulli, *Lettres astronomiques* (Berlin, 1771), pp. 91-4.

c'est a. d. qu'elle ait sa caisse a part & puisse etre transportée oû l'on voudra indépendam[t]. du micrometre. 73.12.17

Comme l'on est fort pressé d'avoir cette lunette, Je vous prie de vouloir bien me repondre le plutot possible aux différ[tes] questions cy des m., & demander à l'ouvrier quand est-ce que l'ouvrage pourroit etre achevé & pret à partir.

Je suis &[a][3]

A Mr Magalhaens at N° 25 Fetter Lane
London

3 The draft is unsigned.

To Matthew Boulton[1] 73.12.22

Birmingham Central Library, Boulton and Watt collection, MBP 244, 55

London 22 Dec[ber]. 1773
at n°. 12 Nevils Court
Fetter Lane

Sir

As I am removed from my lodgings in Red-Lyon Sq[re].,[2] and it is about this time that You wrote, You should have sent to me the whole parcel of my magnetical bars hardned & polished;[3] think necessary to notice the place of my present abode, that no mistake may happen in the direction thereof. Pray don't forget to forward to me the notice of their depart by the post, that I may be ready to receive them: and be carefull to pack them,

1 Annotated: "London 22[d]. X[br]. 1773 / Magalhaens / received the 24th d[o] d[o]. / Answ[d]. 19th. Jan[y] 1774." Boulton's letter not found.

2 Following a disastrous fire at his previous lodgings at 25 Fetter Lane in December 1772, Magellan had been living temporarily in Red Lion Square. He remained at 12 Nevils Court for the rest of his life.

3 See Magellan to M. Boulton, 9 September 1773.

73.12.22 well paper'd, and close enough, to not spoil in the carriage by shaking one against the other.

I am

Sir
Your most ob^{t}. h^{le}. serv^{t}.
Magalhaens

To
M^{r}. Boulton & Co.
at their Manufacture
in Soho
near Birmingham

73.12.27 *From Nathaniel Pigott*

York City Archives, York, Pigott papers, Acc 227.12a[1]

My Dear Sir

I cannot say how much I am obliged to you for all the trouble you are continually taking for me & particularly about his R. Highness's Commission; which must have been a very troublesome one for you: it required all y^{r} good nature to do it in the Carefull and Expeditious manner you have done: the Prince[2] seemed when I sent the parcels, highly pleased with the Tubes and indeed he has good reason to be so; but I fear the Varnish will turn out more a matter of Curiosity than use; let that be as it will, I am not the less obliged to you. I send you inclosed a Draught for £14.7.1½ for the Prince's Commission & £1.5 for an opera glass which, if I am not mistaken, you paid Dollond: making in the whole £15.12.1½. I have in my mind that you have laid out some other money for me, but I cannot say when or for what; if it be so let me know it that I may remit it: now we are about money affairs, I must mention one thing, which I hope,

1 Original not found; text from Pigott's letter-copy book "Mathematical correspondence to 1781". The date is provided by Magellan's reply, i.e. 74.01.18.

2 Prince Charles of Lorraine; cf. 73.10.29.

you will not take amiss: there appears to me no reason why you should be perpetually loosing y^{r}. time for us; neglecting y^{r}. own affairs: laying out y^{r}. money for a considerable time; without any consideration: indeed my good friend that is neither reasonable nor Just: if you would enter into some arrangement in that respect; it would make one more free; for my own part I am ashamed to employ you as I do and as to the Postage of letter I must insist upon being charged with it, as it is quite out of all reason it should be otherwise; do not let the generosity of y^{r}. mind take offence at this proposal it would give me infinite pain to have said any thing that could possibly offend so obliging & so worthy a man. 73.12.27

I dined yesterday with M^{r}. Gordon,[3] who shewed me a letter from you,[4] in which you mention a new construction of a Compound pendulum of your invention.[5] I had scarce time to read the letter & less to consider the thing, but should be glad if you would communicate your Ideas upon this Subject to me; in the mean time I heartily wish you Success as in all you undertake; now my good friend without any further apology, I must beg some further favors of you. in the first place you would oblige me in a very particular manner, if you could procure a satisfactory answer to the following question, being very desirous of obliging the Person who asks it – on Demande la maniere et les procedés de faire de la Colle avec des Pieds de boeuf; comme on la fait en angleterre, and let me know at the same time if there be any Expence attending the above enquiry, as I may charge it to its proper account.

I have sent a considerable number of astrono: observations I made in these parts to M^{r}. Maskelyne for the R. S.[6] and as I have continu'd the same, I must desire you to procure for me the following observ. corresponding to those I have lately made; not from Greenwich, because M^{r}. Maskelyne will send me his but from Oxford or any other good observatory the longitude of which is well Settled: the observations I want are

	1773				
augst.	15. Imm. ♃'s first Sat.	(a)	Sep^{r}.	30	Eclipse of the ☽
22.	first Sat.		oct.	9	Em. first Sat.
24.	Imm. second Sat.			13.	Em. Second Sat.

3 The English Minister in Brussels.

4 Letter not found.

5 Magellan published a description of his new pendulum in his *Notice des instrumens d'astronomie, de géodesie, de physique, &c. faits dernierement à Londres, par ordre de la Cour d'Espagne...* (London, 1780), pp. 210-212.

6 Royal Society.

73.12.27 29. Imm. first Sat
31. Imm. first Sat.
(a) D^{o}. Imm. Second Sat
Novr. 1. occultation of aldebaran
D^{o} Emers. first Sat.
10. Em. first Sat.
Decr. 10 Em. first Sat.
19. Em. first Sat

20 Em. 2^{d}. Sat.
26 Em. third Sat
9re. 2. Imm. third Sat.
21. Em. Second D^{o}.

the sooner you can get the above observ. the more you will oblige me; I would willingly have sent you mine did I think they could be of the least use to you. I made y^{r}. Comp^{ts} to such of y^{r}. acquaintance I have had an opportunity of seeing. adieu my D^{r}. Sir I leave you for fear of missing the Post and am Sincerely & with a very particular regard

Y^{r}. oblig'd humble Servt.

To Mr Magelhaens
n°. 4 Old N. Street Red Lion Square London

73.12.30 *To Johann III Bernoulli*

Öffentliche Bibliothek der Universität Basel, MS L I a 709, f. 51-52

Londres 30 Dec^{bre}. 1773
at n°. 12 Nevils Court Fetter Lane

Monsieur & très cher ami

Aussitot que Je reçus l'honneur de la votre du 14 du courant,[1] je tachai de trouver quelqu'un qui voulusse se charger de debiter les cartes geographiques de Votre Acad^{e}.; car pour moi il ne seroit pas possible de le faire moi-meme. Mais je n'ai pas trouvé personne de Confidence & reputation, qui veuille le faire aux termes que Vous proposer (de 25 per c^{t}.,

1 Letter not found.

les frais jusqu'à Hambourg deduits, ce que je comprends, payés sur la marchandise) Car, selon l'information que M^r^. Elmsley[2] m'a donné, les droits d'entrée pour chaque estampe etrangere, ici à La Douane, est de 4 pences: ce qui vient presqu'au ½ de la valeur. Si les termes etoient de donner un tant per c^t^., & que tous Les frais soient deduits de la vente: ou même qu'on fixat le prix net de chaque carte, comme vous avez fait avec vos Receuils,[3] & qu'on en debitat ici pour Le compte de l'Academie: alors on scauroit à quoi se tenir. Car on computeroit les deboursem^ts^.; & on les chargeroit, avec le profit pour l'avantage du debiteur. Mais meme dans ce dernier cas, J'ai trouvé une bonne dificulté chez un Marchant de Cartes geographiques, auquel je demandai s'il en avoit celles publiées par l'Acad. R^le^. de Berlin pour pouvoir juger sur le debit, & vous informer là dessus. Sa reponse fut, qu'il ne connoissoit d'autres cartes de Berlin, qu'un petit atlas publié il y a plus de 10 ou 20 ans, qui etoit fort mal gravé & qui ne se vendoit plus. Je discourus avec lui, sans Lui ouvrir mon intention: & je trouvai qu'il n'avoit pas aucune grande idee de vos graveurs; & ajouta, qu'il pouvoit avoir Les cartes les plus bellement gravées à Paris, par des prix fort raisonables &c. Comme je ne connois pas non plus les Cartes dont Vous parlez, ni je puis juger par votre Lettre que Celles en question soient plus modernes & meilleures que les anciennes, J'ai laissé Là mon jaseur sans rien dire d'avantage. 73.12.30

J'ai envoyé les Transact. Phil. à M^r^. Le Bibl^re^. de Votre Acad. par La voye d'un jeune Negociant, Nommé M^r^. Zumbrock,[4] & Lui s'envoya adressé avec d'autres choses, à M^r^. Spener,[5] Libraire chez vous qui doit les avoir reçus à cette heure. J'y ajoutai une Lettre pour vous dans Laquelle je vous repondois aux articles de la Votre precedente: une autre pour Le meme Bibliot^re^. dans Laquelle je lui disois de donner l'ordre en general; car c'etoit à cette cause, que je n'ose pas Les prendre, avant de recevoir votre Lettre derniere.

Mon ami M^r^. Wolfe[6] vient d'arriver de sa tournée en Alemagne, sans passer à Berlin. il s'amusa avec Les curiosités de la Saxe, & Autriche: & lors que l'automme arriva, il n'avoit plus de tems que pour retourner en Angleterre. Neanmoins je Vous suis infinim^t^. obligé par vos dispositions

---

2 P. Elmsley, bookseller in London.

3 Bernoulli, *Recueil pour les astronomes* (1771-6).

4 William Henry Zumbrock, merchant at No. 2 Walbrook, London? But see p. 598, where Magellan refers to "M^r^. Frederic Zambruck de Londres".

5 Johann Karl Philipp Spener (1749-1827), for many years director of the Berlin publishing company Haude und Spener, official publisher for the Berlin Academy of Sciences.

6 The chemist Peter Woulfe; cf. 73.11.27.

73.12.30 à Lui rendre service. A present J'ai chez vous (à Berlin) un jeune-homme, fils de mon bon ami M^{r}. Forster, qui se trouve actuelem^{t}. dans un voyage autour du monde avec le Cap^{tn}. Cook, pour visiter Les Terres Australes, & examiner les productions de Hist^{re}. Naturelle de ces endroits Là. [7] Ce garçon se trouve aprentif chez un de vos Marchants, pour s'instruire dans le commerce. Le Nom du March^{t}. est M^{r}. Schlüsser,[8] & celui du garçon est Charles Forster.[9] En cas que vous Le voyez, & lui puissiez etre de quelque service, je vous en serai très obligé, par L'amitié que J'ai pour son Pere, qui est un homme à Talents, & fort distingué en diferen^{ts}. branches de Literature, & par les egards que J'ai pour toute Sa famille, dont il me chargea, en partant pour son Voyage, qui lui prendra peutetre plus de 3 ou 4 ans. C'est un devoir de l'amitié de songer à tout ce qui appartient à ses amis, & c'est avec des sentim^{ts}. pareils, que je me flate de meriter celle dont vous honorez

Mon cher ami
Votre très humble & ob^{t}. Serv^{r}.
Magalhaens

P.S. La Societé R^{le}. d'ici a bien reçu les vol. de vos Memoires,[10] & retourna ses Remerciements &c. je m'y trouvois dans l'Assemblee lorsque le Bibliothecaire les presente &c. & dans la premiere occasion que je verrai le D^{r}. Franklin, je lui demanderai, si celles destinées pour la Societé de Philadelphie[11] Lui furent delivrées, ou non. Dans ma Lettre suivante je vous en parlerai, à moins de m'oublier.

A Monsieur
Monsieur Bernoulli, Astronome R^{l}.
à Berlin
Prusse

---

7 Johann Reinhold Forster (1729-98) and his oldest son, Georg, travelled as naturalists on James Cook's second voyage around the world, 1772-5.

8 Not identified.

9 Carl Reinhold Thomas Forster spent several years in Berlin before joining the Prussian Army in 1778; see Michael E. Hoare, *The Tactless Philosopher: Johann Reinhold Forster, 1729-1798* (Melbourne, 1976), p. 164.

10 i. e. the *Mémoires* of the Académie des Sciences et Belles-Lettres of Berlin.

11 The American Philosophical Society.

To Jacques-André Mallet[1] 74.01.03

Bibliothèque de Genève, Geneva, Ms suppl. 1654, fol. 23-4

Londres ce 3 Janvier 1774
at n°. 12 Nevils Court, Fetter lane

Monsieur

Je n'ai pas reçu l'honneur de la votre du 17 du mois dernier[2] assez tôt pour parler à Mr. Dollond sur les questions que vous souhaitez avoir repondues. Car pour ce qui regarde l'ouvrage de Ramsden dans des lunetes d'un foyer aussi grande que vous en souhaitez, il est inutile d'en discourir avec lui. reelemt. il en a deja fait de cette grandeur, une qui m'a paru très belle; mais il ne faut compter que sur Mr. Dollond; autremt. il faudroit attendre 2 ou meme plus d'années: car Ramsden n'en fait que très raremt.; & peutetre n'a pas fait d'autre que celle que J'ai vûe. Je parlai donc avec Mr. Dollond sur cet affaire: & voici sa reponse.
1. une lunete monté en tuyau de Mahogany à triple objectif de 3 ½ pieds de foyer, avec un piedestal composé du meme bois &c. – – 25 guinees.
2. La meme en tuyau de cuivre jaune, avec l'apparatus pour les mouvements lents; axe long de cuivre jaune, & piedestal comme l'autre – – 36 guin.
3. Si l'on veut qu'elle ait un mouvemt. parallatique, avec un axe petit peut etre d'un pied & demi ou deux pieds – prix – – 45 guinees.
4. Mais si l'on veut seulement avoir ce mouvement parallatique par le moyen de deux blocks comme celle-ci

dont la section transversale fait qu'on peut ajuster le plan x.z dans le sens de celui de l'Equateur; alors elle n'aura pas la fermeté de l'axe du mouvement, comme dans le n°. 3. & la depense n'ira pas au delà de 2 ou 3 guinées au delà de celle du n°. 2.
5. une Lunete comme celle du n°. 2 avec un micrometre objectif, comme celle de Mr. Le Presidt. de Saron, coute – – 50 guin.

---

1 Annotated: "Magalhaens Janvier 1774 Repondu". See Mallet to Magellan, 24 January 1774.

2 See Mallet to Magellan, 17 December 1773.

74.01.03 6. par oû vous voyez que l'addition du micrometre objectif est un affaire de 14 guinées, lorsqu'on l'ajuste à la lunete. mais tout seul, on le fait par 12 guin. il est bien aisé de diviner la raison.
7. Le reticule filaire, dans un chassis &c. pour determiner les differences d'asc. droite & declin., coute environ deux guinées & demie de plus.

N.B. la lunete est toujours separable du micrometre: car celui ci est toujours empacqueté dans une boite separement.
8. en fin une lunete de la sorte du n°. 2 montée dans une machine parallatique, comme celle decrite par de la Lande au n°. 2400 de son Astronomie,[3] avec un cercle vertical, & un de l'Equateur en cuivre jaune, divisés &c. & la monture en mahogany; & un micrometre filaire, est un affaire de 70 guinées. telle est celle que j'ai fait faire ici, & que Je suis [à près] pour l'examiner sur jupiter quelque bonne nuit; car elle est tout à fait finie pour M^{r}. Messier de Paris.

J'ai parlé à M^{r}. Maskelyne sur vos observations correspondentes: & lui me dit qu'il seroit bien charmé de vous en communiquer celles dont vous auriez besoin: & que vous n'aviez qu'à lui ecrire pour cela [...]ture.[4]

Celles que J'ai pû obtenir, des mentionées dans votre lettre pre[...][5] sont comme suit: l'observ. du 1er. satel de Jupiter
23 Dec^{ber}. 1772 at 4^{h}. 21'.41",8 apparent time at Muswellhill
29',4 long. of Muswellhill west of Greenwich
4^{h}. 22'.11,2 app^{t}. time at Greenwich.[6]

L'observation du 22 fut empechée par le tems couvert. Je pense faire un tour en France & Flandres ce printemps prochain: ainsi J'aurai encore le tems de veiller à votre instrument avant mon depart, pourvû que vous ne tardez pas beaucoup à me donner des ordres, & me marquer le Merchant[7] ou personne qui doit fournir ici l'argent de la depense: car je suis reponsable ici pour les ordres que je donne aux artistes & n'aiant assez de fond pour les commissions de mes amis, ils me donnent toujours ici quelqu'un qui le fournisse: ou autrem^{t}. m'envoyent le montant par quelque lettre de change:

---

3 Lalande, *Astronomie*, vol. 2 (2nd edition, Paris, 1771), pp. 793-9, "Description de la lunette parallatique" (paragraphs 2400-2407).
4 Paper damaged.
5 precedente? Paper damaged; but see Mallet to Magellan, 19 February 1773.
6 In the space to the left of these two lines, the following calculation has been entered, probably by Mallet:
9: 16
29
9: 45 de Paris
7 *sic.*

& alors je deposite dans la Banque de Londres cette somme jusqu'à ce que les instruments soyent finis. Car ce fut pour assurer l'argent de mes correspondents le mieux possible, que je me resolus d'ouvrir une compte & caisse avec cette Banque: & J'ai nommé des administrateurs très reputables, pour en avoir soin en cas de ma mort avant que les commissions de chaque ami correspond^{t}. soient acomplies. Car quelques fois J'en avois des somes considerables entre mes mains, qui m'etoient fort desagreables, dans l'aprehension de quelqu'acident sur ma vie. C'est avec ces sentiments & exactitude, que je me proposai de meriter l'estime de ceux qui m'honnorent de la leur: & que Je suis de coeur 74.01.03

Monsieur
votre tr. h^{le}. serv^{tr}.
Magalhaens

A Monsieur
Monsr. Mallet, Professeur d'Astronomie
Cour de S^{t}. Pierre
à <u>Geneve</u>
en Suisse

To Louis-Henri Duchesne[1] 74.01.07

Archives Nationales, Paris, MS T/160 (10-12)

Londres le 7 Janvier 1774
at n°. 12 Nevils Court Fetter Lane

Monsieur & tres cher ami

Je vous ecris un peu plus à Loisir le detail du contenû dans La caisse que j'ai adressée par Calais à M^{r}. Trudaine de Montigny marquée

n°. 1. premierement elle contient deux petites boetes marquées <u>duchenne</u> dessus, qui contient les deux pendules dont je vous parlai.[2] l'une est faite par le meme artiste, & de la meme sorte que les autres que je vous envoyai: & je

1 Annotated by Duchesne: "Rep: le 1. Mars." Letter not found.
2 See Magellan to Duchesne, 14 December 1773.

74.01.07 la payai avec le porterage[3] & petit arrangement £1: 14: 6. l'autre pendule fut faite ici à Londres sous ma direction, & elle differe de l'autre en ce qu'elle marque Les heures distinctement unes des autres, & outre celà elle va plus d'huit jours, par le moien d'une double poulie que je lui y fis ajouter. je me flate qu'on sentira bien aisem^{t}. la superiorité & l'avantage qu'elle a sur l'autre, tant par ces deux circonstances, comme par la perfection de l'ouvrage: & ne me couta que 11½ shillings de plus; c'est à dire £2: 6^{s}: 0^{d} Vous trouverez dans La Caisse de cette pendule la petite Lunete de poche en roussete[4] verte, que vous aviez ordonnée. elle couta £1: 7: 0^{d}

il y a aussi une petite boete, qui contient Le thermometre, monté en argent, & gradué selon les deux echelles de Reaumur & de Fahrenheit, qui est pour Monsr. de Montigny: auquel vous presenterez mes compliments. Je le payai l'argent y compris £2: 1^{s}: 0^{d} .

il y a aussi une caisse triangulaire au dedans de la meme, marquée **BRV^{ille}.** n°. 1. qui contient deux Sextants marins & une lunete pour Un officier de la Marine à Brest;[5] & que vous aurez la bonté de donner à M^{r}. Messier, au quel j'ecris cette meme poste[6] pour s'en charger de celà.

enfin il y a au fond une autre caisse avec deux Sextants marins pour M^{r}. Le Chev^{er}. de Borda. elle est marquée **BOR^{da}.** n°. 1., que je vous prie d'envoyer au dit Chev^{er}. en cas de se trouver a Paris: autrem^{t}. M^{r}. Messier en aura soin jusqu'à ce qu'il Retourne &c. un petit pacquet avec des brochures pour M^{r}. le D^{r}. Sanchez, vous ou M^{r}. Messier aurez La bonté de la lui remettre. & une autre petite boete marquée **Fonde**, M^{r}. Messier en aura soin. elle contient deux petites bouteilles de vernis à cuivre. Vous voyez bien que J'ai renfermé tous ces articles dans une seule caisse, pour ne pas en multiplier Le nombre, comme vous me l'avez averti. M^{r}. Le Comte Carbury[7] ne m'avoit ecrit de vouloir avoir une de ces petites pendules. vous vous arrangerez avec Lui là dessus à l'egard de ces deux: mais au cas de vouloir attendre, vous n'avez qu'à m'en avertir. Car j'aurai au commencement de la semaine prochaine 3 autres finies, qui seront à votre service. Savoir, une qui montre les heures separem^{t}.; mais qui ne va plus de 2 ou 2 ½ jours, comme les autres: & coutera 38 shilling. une autre

3 Portage.
4 Roussette – species of shark, the skin of which was used to cover cases or boxes.
5 Bon-Chrétien, Marquis de Briqueville (1723-1803) who at this time was "capitaine de vaisseau" and was based at Brest; see Magellan to Johann III Bernoulli, 26 October 1772.
6 Letter not found.
7 Giovanni Battista Carburi (1722-1801?); see Magellan to Duchesne, 23 July 1773, n. 3.

comme la 2de. ci dessus, du meme prix de £2: 6s: 0d: & une autre, qui ne s'arretera pas dans les seconds tandis qu'on la monte, & qui aura la verge du pendul corrigée [...][8] les derrangements du froid & du chaud, de façon qu'elle sera très propre pour des observations astronomiques. Je crois que son prix n'ira pas au delà de 3 guinees. peutetre que celle ci sera la plus à propos pour notre bon ami L'Abbé Bourriot. 74.01.07

Pour ce qui regarde les charges proportionelles de ces articles contenus dans la caisse totale, je croirois qu'on pourroit les charger dans la forme suivante, pour les frais depuis Calais jusqu'à Paris. on dividroit la somme totale en 34 parties. La caisse du Chever. de Borda en paieroit 12 de ces parties. l'autre caisse triangulaire autant, c'est à dire douze parties aussi. Les deux pendules 7 parties. la boete marquée **FONde.** deux parties: & le Thermometre en paieroit une seule. Celle-ci est la distribution qui me paroit assez raisonable, d'autant plus que les pendules ne sont point chargées avec la depense de la douane & caisses ici &c. au Reste je me Remets à votre jugement.

& J'espere avoir votre Reponse sur les pendules, c'est à dire sur Le choix de celles qui vous manquent pour vos amis selon que vous me l'aviez dit dans une de vos Lettres.

à Dieu mon cher ami conservez vous & ne me Refusez pas Le plaisir de vous servir si je le puis, etant pour La Vie

mon cher ami

le votre de coeur

Magalhaens

a Monsieur
Monsieur Duchenne, Secretaire
de Mr. Trudaine de Montigny
à Paris

8 Paper damaged; word illegible.

74.01.18 *To Nathaniel Pigott*

Royal Astronomical Society, London, MSS Pigott, 31

London 18 Jany. 1774 at n°. 12
Nevils Court Fetter Lane

My Dear Sir

Yours of 27 Last I received,[1] with the bill for £15.12.1½ which was immediately acquited: but am doubtfull about the 25 shillings for y^{e}. opera-glass being due to me still, unless I have sent to you, or on y^{r}. acc^{t}., two opera-glasses; there being one already paid to me according to my memorandum book. the other little expenses on your account are perhaps some Tubes for barometers, which M^{r}. Needham furnished you from those I did send to him; these were 15 in all, & M^{r}. Needham only paid for 10. if you had the other five, they [came] to 7^{s}:6^{d}. I am very happy to find that y^{e}. Tubes for his R. Highness did please:[2] as I took all possible care to suceed, am highly pleased there with. for this is the chief advantage I seek after, that of employing my time and cares in doing some good offices, and chiefly to those I love & esteem. However some times the Merch^{ts}. with whom I deal, make me some presents as an acknowledgm^{t}. for y^{e}. prefference: and in some articles of common runing they make to me the same allowances, as to any other Tradesman. So that now and then I am defray'd for the wearing of my shoes, besides the pleasure of doing service to my good friends.

This will be enough in answer to your kind remonstrances on the subject: and for inducing you to be free with me, in whatever occurs that I may be of any service either to you, or to your friends. as to Letter & other expenses I make, you have seen that I charge the whole on each bill.

I have removed to a new Lodging, as you see at y^{e}. top of this Letter: this and y^{e}. bad weather hinder'd me from gratifying the curiosity of your friend about the method of making our glue. nor I know how to come at the source, unless I apply to some manufacture on the pretense of some large order, for one or two hundred weight at least of it. if your friend thinks proper to have it sent to any merch^{t}. to dispose of it, there can be not

1 See Pigott to Magellan, 27 December 1773.

2 Cf. 73.10.29.

any Lost: and if any, it must be considered as one expense attending such an information. please to answer on this. 74.01.18

I have corrected my pendulum from the effects of heat and cold, by a very simple method, viz. that of fixing one bar on y^{e}. back of it to y^{e}. wall, kept by hooks or rings, and only fasten'd at y^{e}. botom, at such a distance from y^{e}. top, as to counteract by it's expansions or contractions upwards, as much, as the rod of y^{e}. pendulum (which hangs on it, and is unconnected with y^{e}. wheel work) expands or lessens downwards. I have adjusted it by means of a stove to a very great nicety: & there apply'd a new mechanism to keep it a going whilst it is winding up: but as soon as the winder, or key, is taken out, it ceasses immediately from acting on y^{e}. wheels. it goes 8 day only, in order to avoid a greater number of wheels. the seconds-hand comes out on y^{e}. center of y^{e}. dial; and the hour and minute-hands are carried by the axles of their respective wheels, without any quadrature or additional wheel work to bring their motions to y^{e}. center as usual. there are some other particulars, which render this pendulum more free than others from common errors, & less liable to go wrong: but, what is more remarquable, is the cheapness of y^{e}. whole. for it costs me hardly 6 guineas, instead of 35 that I should be obliged to pay for a good one of Shelton or Elicot,[3] and I hope will go full as well. Besides this I caused some cheap pendulums to be made for common use. they show t[he] seconds, minuts, & hours distinctly, and have but 3 wheels in their real movement. these I have diversified in 3 or 4 ways. those which go but 2 days or 2 ½, cost only 38 shillings. those which go one week, cost 2:6:0^{d}: those which do not stop nor retroceed whilst winding up, and have the rod corrected from the effects of heat and cold, cost but 3 guineas: and I think these will prove very good for poor Philosophers, who love astronomical observations. I have now in my study, where I write this, two of y^{e}. first kind and one of y^{e}. last, which I go to send to Paris. at present I have my work-man (for I got of late a very good one) employed in a new kind of perennial repeater: and if it succeeds as well in practice as it appears in theory, you will see it soon at Bruxelles. N.B. after writing you this, I recollect that M^{r}. Dollond did sent to you an opera-glass, which I paid on your account: and perhaps is this, you pay me now in this account.

3 John Shelton (1697-ca.1774) and John Ellicott (1706-1772), well known makers of astronomical clocks.

74.01.18 I'll endeavour to get as many observations as I can, of those you mention in your Letter: and perhaps will send them along with the Barometer of M^{r}. Needham, together with the speech of Sir John Pringle, when he deliver'd the gold medal to D^{r}. Priestly.[4] it is a very good elaborate one on y^{e}. history of the discoveries on air: and is deliver'd to y^{e}. members of y^{e}. R^{l}. Society: where I'll take yours and another for M^{r}. Needham.

I'll speak to Dollond about your 2 opera glasses: but why did you not wrote to me about them? he cannot be more zealous or ready than I am to oblige you!... the price of Quadrants astronomiques by Ramsden are as follows. those of 12^{inch}. radius, with 2 Tellescopes, well balanced and moveable level &c. 42 guineas. if mounted like those made by Bird, only 36^{g}. those of 18^{inch}. radius 50^{g}.: and those of 24^{inch}. 70 guineas. I bought 3 or 4 years ago one of 12^{inch}. radius of Ramsden, for 25 guineas only: but now you see that the price is so much rised: and depend on that of Birds being not above what he charges to every one. indeed these instrum^{ts}. cost great deal to be made well: no Mathem^{l}. maker becomes rich. M^{r}. Bird is far from being one: and Ramsden still works.

It took into my head to improve Ramsdens barometers and will have soon some made according to my ideas by my work-man, which hope will answer better.[5] but of this another time. Please to pres^{t}. my hu^{ble}. Resp^{cts}. to your Lady & family; and to our common friends; whilst I am with the utmost regard and steem

Dear Resp^{d}. friend
your most obd^{t}. h^{le}. serv^{t}.
Magalhaens

A Monsieur
M^{r}. Pigott, Gentil-homme Anglois
vis à vis La Cour
à Bruxelles

4 Pringle, *A Discourse on the Different Kinds of Air, delivered at the Anniversary Meeting of the Royal Society, November 30, 1773* (London, 1774).

5 Magellan described various improvements to the barometer in his *Description et usages des nouveaux barometres, pour mesurer la hauteur des montagnes, et la profondeur des mines...* (London, 1779).

To Louis-Henri Duchesne

74.01.18a

Archives Nationales, Paris, MS T/160 (10-12)

à Londres le 18 Janv^{r}. 1774
at n°. 12 Nevils Court Fetter Lane

J'ai Reçu, mon cher ami, la votre du 25 du mois dernier,[1] dans Le moment que j'allois mettre à la poste celle que je vous ecrivis il y a 8 jours.[2] premierem^{t}. Je vous Remercie très de Coeur par La bonté de donner pour mon compte le petit argent au pauvre Silva:[3] & Je vous en tiendrai compte, lors que nous arrangerons les notres. vous voyez par ma derniere l'envoy que Je vous fis de votre petite lunete, & des deux pendules, aussi bien que du Thermometre pour M^{r}. de Montigny &c, sur quoi & sur touts les autres articles je me raporte à ma d^{te}. Lettre derniere. A present Je vous annonce d'avoir deja finies chez moi Les 2 autres pendules simples, qui partiront Lors que vous l'ordonnez. J'ai aussi une troisieme garnie de tout Le necessaire pour Correspondre & servir aux observations astronomiques: mais elle coute un peu plus, c'est à dire 3 guinées au lieu de 38 shillings que Les Deux simples coutent. Je la destina pour notre bon ami l'Abbé Bourriot auquel vous aurez la bonté d'envoyer la Lettre ci jointe,[4] qui contient la description de sa pendule.

J'apprends avec plaisir le sort du Pere Boscowic, malgré de desagrement des envieux,[5] & si vous avez occasion de me Recommender à Lui, en attendant que j'aille à Paris, ce que je compte de faire le perimtems prochain, vous m'obligerez beaucoup.

J'ai eté ce matin avec M^{r}. Banks: & lui me dit d'avoir deja plus d'un cent ou meme 150 de Souscripteurs: mais il n'est pas encore arreté, comment on commencera cette edition, qui doit etre très interessante.[6] J'y veillerai, & ne manquerai pas de profiter de l'occasion &c.

---

1 Letter not found.

2 From what follows, it is clear that Magellan means his letter to Duchesne dated 7 January 1774.

3 See Magellan to Duchesne, 14 December 1773.

4 Letter not found.

5 When the Jesuit Order was suppressed in Italy in 1773, Boscovich accepted an invitation from King Louis XV to move to Paris where he was appointed director of optics for the French navy, with a pension of 8,000 livres.

6 See Magellan to Duchesne, 14 December 1773. Magellan's doubts about this project seem to have been well-founded, since it came to nothing.

74.01.18a Je vous suis très obligé par tous vos soins & egards de votre bonne & vraie amitié de Coeur: & ne manquerai pas jamais d'y correspondre de Reste de ma vie jusqu'au dernier soupir, ettant sincerem^{t}. tout à Vous

Magalhaens

a Monsieur
Monsr. Duchenne Secretaire
de M^{r}. Trudaine
à Paris

74.01.24 *From Jacques-André Mallet*

Bibliothèque de Genève, Geneva, Ms suppl. 1654, fol. 48[1]

Geneve le 24 Janv^{r}. 1774.

M^{r}.

J'ai reçu la lettre que vous m'av. fait l'h. de m'écrire, le 3 du c^{t}.,[2] & je vous remercie du détail très circonstancié que vous m'y donnez du prix des Instrumens, Celui qui m'avoit chargé de cette commiss.[3] se détermine p^{r}. le N^{r}. 1, c.a.d.[4] pour la lunette, montée en mahogany, du prix de 25 guin. & moi j'y joins pour mon cpte. le micrometre objectif du prix de 14. Je suppose bien qu'il pourra s'ajuster au N° 1. tout co[mm]e au N° 2. Je suis etonné qu'il y ait une si gr^{de} diff. de prix entre ces 2 Numeros, j'aurois fort souhaité qu'il eut choisi ce N° 2 mais il en trouve le prix trop considerable, au reste je crois que je pourrai faire faire ici une espèce de machine parallatique, ou au moins une monture à mouvemens lents à bcoup meilleur compte, mais aussi moins bien exécutée.

Je souhaiterois aussi que vous y joigniez le reticule filaire dans un chassis que vous mettez à 2½ guin. Cela fera en tout 41½ guin. qui vous seroit payées par Messrs. Mark Liotard, Aubertin & Rivier Négocians de la Cité, c'est à ces Messrs. que je vous prie de remettre les instrumens bien

1 Letter not found. The text is from a draft kept by Mallet.
2 See Magellan to Mallet, 3 January 1774.
3 H.B. de Saussure?
4 c'est à dire.

embalés, dans une même caisse afin qu'il n'y arrive point d'accidens, & le plutot qu'ils seront achevés sera le meilleur. 74.01.24

Je vous prie encore pour profiter de cet envoy de me procurer le nautical almanack p[r] 1774 & celui pour 1775 s'il a deja paru et enfin si cela ne vous donne pas trop de peine d'y joindre quelques graines p[r] le jardin, qui soyent de bonne espèce,
Savoir une once de graine de choufleurs, colliflowers
.... ¾ livre graine de racines jaunes, orange red carots
et vous aurez la bonté de remettre à ces Messrs. Liotard &[a]. ou de m'envoyer un compte de tous ces objets séparés, parce que c'est pour differ[tes] personnes.[5]

A M[r]. Magalhaens
at N°. 12. Nevils Court, Fetter Lane.

5 Mallet's draft is unsigned.

To Nathaniel Pigott 74.01.25

Royal Astronomical Society, London, MSS Pigott, 32

London 25 Jany. 1774 at n°. 12
Nevils Court Fetter Lane

My Dear Sir

Here I send you the only observations I could get, Corresponding to yours. I spoke to M[r]. Dollond about your opera glasses, and he said to have realy sent them to you according to your directions:[1] if you please I'll pay him for; and you may send y[e]. money when you think proper: but Let me know how many you have realy received from him. Please to tell M[r]. Needham, that his barometer is ready and will be sent this week by the first ship. I am waiting for the tryals of M[r]. Nairne previous to the make of his instrument, in order to ascertain the effect thereof, of which I'll acquaint him by my next: and at the same time will send the speach of y[e]. Presid[t]. of y[e]. R[l]. Soc[y]. I took for him and for you, with another I bought

1 Cf. 73.12.27 and 74.01.18.

74.01.25 for M^{r}. Chevalier,[2] it being a very elaborate paper on the discoveries of different kinds of air, which is to be annexed to y^{e}. Philos. Transactions.[3] these pamphlets will probably go along in a small box of fresh seeds of North-American Trees, which D^{r}. Fothergill[4] gave me, and of which I'll make a present to his Highness y^{e}. Duc of Aramberg, for his pepinieres of Enghien.[5] If purchance you see the Duc, you may tell this to him: and that I wish He may order his gardner to gather again in the next season more seeds of the cocombres[6] de Barbarie; for I am affraid those I received from his Highness, will not produce, according to what I was told from the gardner of D^{r}. Fothergill, and three more amongst whom I distributed that parcel; they finding those seeds too light, and almost intirely empty of any germen.

the 18th. instant viz the very same night I wrote my last, I observed Saturn, with the two refractors I have almost ready to be sent to Messrs. Messier & de la Lande: and I could distinguish the appearance of the ring again with both tellescopes. please to tell me, if you have discerned it before that night. I asked Thursday Last M^{r}. Maskelyne in y^{e}. R^{l}. Soc^{y}.; but he said he had not observed that planet.

M^{r}. Priestly has made of Late two great steps.[7] he finds that Electricity decomposes the air. he puts into a small tube ab a tincture of archile, or of tournesol, which is blewish, and leaves the space ae filled

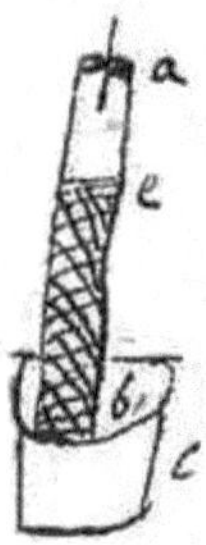

2 Father Jean Chevalier, at this period Keeper of the Royal Library in Brussels.

3 Pringle's essay was published separately as *A Discourse on the Different Kinds of Air, delivered at the Anniversary Meeting of the Royal Society, November 30, 1773* (London, 1774) and later reprinted in his *Six Discourses delivered...when President of the Royal Society on occasion of Six Annual Assignments of Sir Godfrey Copley's Medal* (London, 1783).

4 The well known Quaker physician Dr John Fothergill (1712-1780) maintained a magnificent botanic garden at Upton, near Stratford, and had strong links with the British colonies in North America.

5 The Duc d'Arenberg had an extensive tree nursery at his family estate at Enghien in the Austrian Netherlands.

6 concombres?

7 Priestley gave an account of the experiments here described in the first volume of his *Experiments and Observations on Different Kinds of Air* (London, 1774), pp. 183ff. and 215 ff. respectively.

only with common air. the tube is stoped in a with sealing wax: and there is a wire d at y^{e}. top. then by receiving continually electrical sparks on d, he finds that the air contained in ae is discharged of a certain portion of acid, which changes the tincture into a purplish colour. but after a certain time it discharges no more acid. this air, so deprived of its acid diminishes in its bulk, and is very noxious to animals. hence it appears to be a real decomposition: and that the acid smell we feel when we are about Electrical experiments does not belong to the electric fluid, itself: but to y^{e}. real decomposition of y^{e}. air. and y^{e}. same hapens in y^{e}. Lightining &c. another discovery of Priestly's is that of one species of air, in which a candle burns as well, or better than in common air: and nevertheless it is very noxious. this is the nitrous air saturated with phlogiston, vg. where iron has been kept long while, or hepar sulphuris.[8] from whence we learn, that the burning of a candle is by no means a sure test for ascertaining the good qualities of air. it seems that the lessening of its bulk is the more to depend upon, for on mixing this air with another quantity of the nitrous air, not any diminution hapens in it's bulk. indeed this very same experiment, when made in common air, is very amazing. for instance put two measures of common air into a vessel topside down in mercury, or in water: then add to it one measure of nitrous air. immediately the whole takes a redish colour, and the three measures of these two airs are reduced to about one and $^{4}/_{5}$. so that the addition of one third, not only vanishes, but even reduces the other two to $1^{4}/_{5}$ only. this air so reduced is very noxious to animals. a mice for instance being put into it, dyes instantly. &c. but if this very same air so lessened and noxious, is well beaten with water, it's bulk is still somewhat diminished; and becomes wholsome and innocent for animal respiration. &c. what is become of that $1^{1}/_{5}$, which is lost? was it by y^{e}. addition of some noxious particles that it became so noxious? and that these being attracted by y^{e}. water, it becomes again wholesome? &c. &c. &c...? 74.01.25

Please to comunicate this to your new Academy,[9] to M^{r}. Needham, Duc d'Aramberg, Gordon[10] & other friends.

I am ever
yours
Magalhaens

8 Liver of sulphur, made by boiling sulphur with a solution of potash.

9 i. e. the Académie Impériale et Royale des Sciences et Belles-Lettres at Brussels.

10 William (later Sir William) Gordon, British Minister Plenipotentiary at Brussels, 1765-1777?

74.01.25a *To Joannes Franciscus Thijsbaert*
University Library, Université de Liège, Ms. 4130

Londres 25 Janvier 1774
at n°. 12 Nevils Court, Fetter Lane

Monsieur & très Cher ami

Je vous dois depuis Long tems Remercier pour Le present des meer-cats,[1] que J'ai reçu, & dont je fis present à different amis. Celui que Je donnai à M^r^. Banks, est plein de vie: & quoiqu'on n'y observe des proprietés bien eclatantes pour La meteorologie, il y joue un drole tour, qu'on n'avoit encore observé dans aucun autre poisson, savoir celui de dormir la Nuit sur le Coté comme un homme. on en a eté ici bien charmé avec cette curiosité d'histoire naturelle: même M^r^. Banks me dit, que si l'on pourroit avoir une vingtaine ou trentaine de ces poissons vivants, il voudroit bien Les envoyer dans sa terre (en Lancashire) pour les naturaliser dans quelque marais. en effet si vous croyez que celà soit une affaire de peu de depense, c'est à dire d'environ 3, 4, ou 5 guinées, Je serois bien charmé de Lui procurer cette satisfaction; car il me dit qu'il voudroit faire cette depense à Son compte, pourvû que celà ne fut bien cher: & c'est celà à peu près ce que je jugerois proportioné à un pareil apetit.

Notre D^r^. Priestly vient d'enricher la Phil. N^le^. de quelques nouvelles decouvertes.[2] Je vais vous en rendre part, dans l'idee qu'il Vous sera agreable de les savoir. Il trouve que l'electricité decompose reelem^t^. l'air de son acide. vous savez que cet acide a eté bien controverti, & pas encore bien demontré. Voici l'experience de Priestly.[3] il remplit à demi un tuyau de verre ao. qui est cacheté

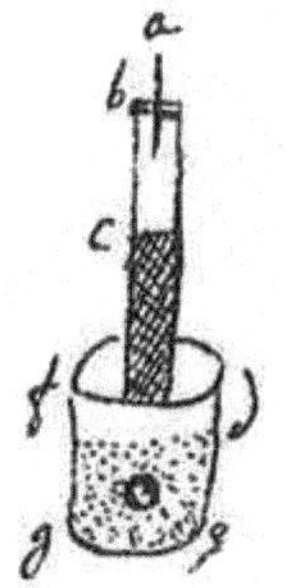

1 Magellan is here referring not to the small mammal now known as the meercat but to a fish, the sea-cat or wolf-fish.

2 Joseph Priestley, *Experiments and Observations on Different Kinds of Air*, vol. 1 (London, 1774).

3 Ibid., pp. 183 ff.

en b avec de la cire d'Espagne avec un fil a de fer ou archal, au milieu: c'est depuis c jusqu'à o, que ce tube est demi-rempli avec la tinture d'archile ou de tourne sol,[4] dont la couleur est bleuatre. l'espace bc est plein d'air seulement. & la vase fdeg qui est demiplein de la meme tinture sert à tenir le tuyau renversé avec l'embochure en bas[.] Ce tuyau a environ une ligne de diametre. Celà fait, il reçoit un grand nombre d'etincelles electriques dans le fil a qui passent a travers l'air vers la tincture: & au bout de quelque tems, cette tinture devient rougeatre. mais s'il change la tinture, sans changer l'air aussi, l'electricité n'y fait aucun changement. ce qui prouve que la portion d'air bc, etant une fois decomposée de son acide, ne peut pas decharger plus qu'elle en avoit. Cet air ainsi depouillé de son acide, est très nuisible pour la vie animale, il est diminué en son volume: & la couleur rougeatre qu'il donne à la tincture bleue, prouve que c'est de l'acide, qui y etoit contenû. delà on connoit pourquoi nous sentons un odeur forte & aciduleux lors qu'on travaille aux experiences electriques. C'est la decomposition de l'air, causée par le feu electrique, & non pas ce feu, qui cause cet odeur. de même dans les endroits frapés du coup de tonnerre, on sent un odeur sulphureux, qui provient de la decomposition de l'air, qui se separe de son acide, par Le feu electrique, & non pas le même feu qui porte cet odeur. 74.01.25a

Le meme Priestly a trouvé une espece d'air, dans le quel une chandelle continue aussi bien alumée que dans l'air commun: & qui neanmoins est très nuisible.[5] C'est l'air nitreux, dont il parla dans le Memoir que Je vous envoyai, lors qu'il est chargé du phlogistique, soit pour y avoir mis long tems des morceaux de fer, ou pour y avoir mis peu de tems du hepar sulphuris.[6] de façon que l'epreuve qu'on connoissoit jusqu'à present pour reconnoitre la bonté de l'air, en y mettant une chandelle alumée, n'est plus une epreuve sans exception: car elle y brule aussi bien dans ce mauvais air, & meme mieux, que dans l'air commun. Voilà deux pas je crois bien avantageux, qu'on a fait de plus dans La Philosophie.

Je souhaite bien, que Vous & vos confreres poursuivissent ces experiences des differentes sortes d'air. L'apparatus qu'il decrit dans son Memoir [est][7] très simple, & aisé:[8] & Je fus très enchanté de le voir operer. il fit devant moi l'experience de l'action de l'air nitreux sur l'air commun.

---

4 i. e. litmus.

5 Priestley, *Experiments and Observations...*, pp. 215 ff.

6 Liver of sulphur, made by boiling sulphur with a solution of potash.

7 Word obscured by seal.

8 Priestley, *Experiments and Observations...*, pp. 6 ff.

74.01.25a Soit abncod un sceau de bois, qui soit

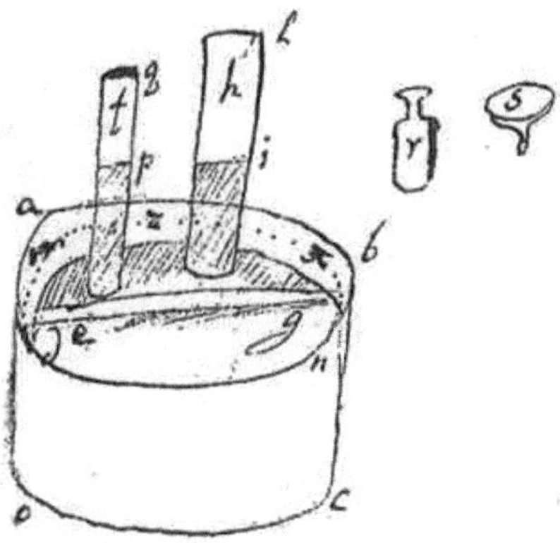

rempli d'eau jusqu'à la ligne marquée par des points m.z.x.: & qu'il y soit une tablete eg dans un coté, environ un pouce au dessous de la surface d'eau mzx: sur laquelle soient les 2 verres f.h. à la renverse. le verre h contient de l'air nitreux depuis l jusqu'à i. prennez le verre f & remplissez-le d'eau, en le sumergeant dans le sceau. tournez le avec le fond en haut, & prennant la fiole r qui est pleine d'air commun, avec l'embouchure en bas, mettez-la dans l'eau, tenant f avec la gauche, & tournez adroitem^{t}. la bouche de r, de façon que l'air entre & monte jusqu'au haut de f. repetez la meme operation, pour que f soit ocupé avec deux mesures d'air commun, mesurées par la phiole r.: prennez l'entonoir S sur la bouche de r, & les tennant au dessous de la surface d'eau avec la gauche, tous deux pleins d'eau, prennez de la droite h, & faitez sortir autant d'air, qui entrera par S, jusqu'à remplir la phiole, qui etoit pleine d'eau. remettez h à sa place: & prennant f, sans en ôter jamais son embouchure du dessous la surface de l'eau, appliquez y celle de r, que vous conservez à la renverse dessous l'eau, pleine d'air nitreux, & en la tournant, faitez entrer cet air dans f. alors vous aurez mis 3 mesures d'air ensemble, savoir deux d'air commun, & une d'air nitreux. vous verrez aussitôt que cette misture prend une couleur rougeatre, & que l'espace que vous aviez marqué plein d'air commun jusqu'à p, au lieu d'augmenter par la 3me. mesure d'air nitreux, devient encore moindre d'environ une 5me. partie. de façon que les 3 mesures de ces deux airs, se reduisent à $1\,^{4}/_{5}$. qu'est elle devenû donc cette $1\,^{1}/_{5}$ mesure d'air? or cet air ainsi diminué de volume est très nuisible. un souri mis dedans, meurt tout de suite. mais si vous foüetez bien cet air nuisible avec de l'eau commune, il diminue encore un peu de volume, mais il n'est alors plus nuisible. Est ce donc que l'air nitreux aporte quelques corpuscules nuisibles à l'air commun? mais s'il y en ajoute quelque chose, pourquoi diminue t-il son volume? Est ce que l'eau atire & prend ces corpuscules nuisibles, & laisse l'air ainsi diminué, plus depuré & tout à fait sain &c.? voilà les problemes utiles à resoudre pour le genre humain. 74.01.25a

mes respects à Mr. le Chancelier[9] & autres amis.

Je suis pour la vie
Votre
Magalhaens

A Monsieur
Monsr. L'Abbé Thisbaert
Professeur dans L'Université
au College de Castrum de
Louvain
en Flandres

9 Joseph de Crumpipen (1737-1809), Chancellor of Brabant and first President of the Brussels Academy of Sciences.

To Gabriel de Bory 74.01.25b

Archives de l'Académie des Sciences, Paris, Dossier Magalhaens[1]

Mon cher Chevalier

je me trouvois jeudi dernier dans l'assemblée de la Societé Royalle de londres, lorsque l'on y presenta l'ouvrage de M. Lavoisier Sur les emanations elastiques qui se degagent de differents Corps.[2] je suis bien faché de ne pouvoir satisfaire ma curiosité sur ce sujet traité par une main si habile comme celle de ce jeune savant, car j'en attends beaucoup de ses grands

1 Letter not found. The text is from a copy in a secretary's hand, headed "Copie de la lettre écrite par M. Magalhaens a M. de Bory le 25 janv: 1774" (where "de Bory" has been written over "Le Chevalier de Borda", *crossed out*). The letter has been published in *Oeuvres de Lavoisier: Correspondance*, ed. René Fric, fascicule II, 1770-1775 (Paris, 1957), pp. 421-4.

2 The Royal Society's *Journal Book*, XXVIII, 266, records the presentation on 20 January 1774 of a copy of Lavoisier's *Opuscules physiques et chimiques* (Paris, 1774).

74.01.25b talents. le titre annonce quelque chose de sistematique, Et je ne doute pas que M. Lavoisier l'ait etabli sur des experiences Et des faits incontestables, car c'est cette methode qui peut uniquement nous amener a developer les misteres de la nature, au lieu que la Theorie sans l'appuy de l'observation ne fait que creer des idées imaginaires qui n'existent hors de la tete des reveurs qui y pensent. mais en attendant que je puisse satisfaire ma Curiosité a l'egard de louvrage de M. Lavoisier, je[3] veux vous Communiquer deux pas nouveaux que M. le chevalier de priestley a fait dernierement dans la phisique experimentale qui me semblent fort interessants. Le premier est la decomposition de l'air en separant son acide si long temps dispute Et pas encore bien demontré avant ce temps, ce qu'il fait par le moyen du feu electrique.[4] il prend un tuyau de Verre d'environ une ligne de diametre, il met un fil de metal a un des deux bouts Et le scelle avec de la Cire d'espagne,[5] cela fait il y met de la teinture d'archile ou de Tournesol qui comme vous savéz est d'un bleu fort (Et peut etre celle de violet fera le meme effet) Et y laisse un espace d'air commun d'environ un pouce Et mettant le bout ouvert dans un Vase avec de la meme Teinture pour que celle du Tuyau ne touche point en bas, il y reçoit des etincelles d'une machine electrique sur le fil de metal qui au bout de quelque temps font prendre une Couleur Rougeatre a la teinture du tuyeau, mais pas au dela d'un Certain degré, Et s'il renouvelle la teinture sans renouveller l'air qui avoit servi auparavant, elle ne changera pas de Couleur. on trouve que cet air diminue de volume dans l'operation, Et qu'il est nuisible pour la respiration animale, car une souris qu'on aura mis dans un Verre dans lequel on ramassera une bonne quantité de cet air depouillé de son acide perira incessament. il est donc demontré quant a moi par cette experience que cet air est depouillé de son acide par le feu electrique, Et que c'est par cette decomposition de l'air, Et non par le meme feu electrique que nous sentons une odeur sulphureuse dans les operations avec la machine electrique, Et dans l'occasion des accidents par le coup de tonnere, Et peut etre c'est cet air decomposé et nuisible qui tue quelquefois les animaux que nous appellons frappés du coup de tonnere quoiqu'on n'y trouve pas aucune blessure.

---

3 An extract commencing at this point and continuing to "...qu'on pourra decider complettement la mauvaise qualité de l'air", written out in the same secretary's hand, is to be found among Lavoisier's papers (Archives de l'Académie des Sciences, Paris, Fond Lavoisier, 1642).

4 J. Priestley, *Experiments and Observations on Different Kinds of Air*, vol. 1 (London, 1774), pp. 183-6.

5 A mixture of lac, wax, chalk and cinnabar that was used for sealing documents.

l'autre experience de ce grand observateur des phenomenes de la nature est celle d'une espece d'air qui etant tres nuisible a la Vie animale, n'etouffe point une chandelle allumée;[6] vous savéz que les mineurs se servoient de cet expedient pour eprouver la qualité de l'air ou ils devoient travailler, mais le chevalier priestley trouva que si l'air nitreux est surchargé de phlogistique, une chandelle allumée y brulera aussi bien Et meme mieux que dans l'air commun, quoique cependant ce meme air nitreux ainsi phlogistiqué soit tres nuisible aux animaux qui ne peuvent y vivre un moment, c'est en mettant du hepar Sulfuré[7] un peu de tems dans un Vaisseau d'air nitreux, Et en y mettant des morceaux de fer fort long temps que cet air acquiert la qualité de conserver la flamme d'une chandelle aussi bien que l'air commun, Et neantmoins il a la mauvaise qualité d'etre tres nuisible a la vie. ainsi ce ne sera desormais par cette epreuve de la chandelle qu'on poura decider complettement la mauvaise qualité de l'air. il semble que l'epreuve la plus infaillible pour connoitre le mauvais air est celle que le meme S. priestley a proposé dans son excellent memoire inseré dans les transactions philosophiques de londres pour l'année 1772[8] qui consiste a y meler de l'air nitreux et observer s'il souffre ou non quelque diminution dans son volume reellement. l'experience qu'il a repeté devant moi est tres frapante, il prit un bocal de verre plein d'Eau qui etoit renversé dans un sceau avec de l'Eau aussi, il y dechargea deux phioles pleines d'air commun, Et marqua l'endroit de l'espace qu'il y occupoit, il y ajouta une troisieme phiole de l'air nitreux, Et ce melange prit tout de suite une Couleur rougeatre, Et au lieu d'augmenter de Volume, il se reduit a un cinquieme de moins de l'espace que les deux mesures d'air commun occupoient auparavant. cet air ainsi diminué de son Volume est tres nuisible a la vie animale, mais si on le bat Et le fouette bien avec de l'Eau commune, il redevient respirable Et perd la mauvaise qualité. il est a remarquer que cette meme quantité d'air nuisible en perd une partie de son Volume dans l'operation de le battre Et fouetter avec l'Eau, mais qu'est il devenu la troisieme partie Et un cinquieme du volume de cet air? car au lieu de trois mesures d'air qu'on mele ensemble, on ne trouve qu'une Et quatre cinquiemes ou Environ, est ce que l'air nitreux introduit quelques emanations nuisibles dans le Volume de l'air commun dans le meme temps en diminuant son Volume? Et que ces emanations sont attirées par l'Eau dans la seconde operation en se dechargeant de l'air 74.01.25b

6 Priestley, *Experiments and Observations on Different Kinds of Air*, vol. 1, pp. 215-7.
7 Liver of sulphur, made by boiling sulphur with a solution of potash.
8 J. Priestley, "Observations on Different Kinds of Air", *Philosophical Transactions*, 62 (1772), 147-264; pp. 212-4.

74.01.25b auquel elles etoient attachées dans la premiere, Et dont elles faisoient part de son Volume? est ce que l'air nitreux reserve, ou qu'il detruit le Ressort de l'air commun en tout ou en partie? et que l'eau le retablit ou quelle lave celles qui etant nuisibles s'etoient attachées a celles de l'air qui n'avoient pas perdu leur Ressort. Voila des problemes que je proposé aux philosophes pour les resoudre sur des Experiences Et observations bien averées.

jai l'honneur d'etre avec tout l'attachement et Egards possibles

Mon cher chevalier
Votre &c
Magalhaens.

P.S. jai observé Saturne avec les lunettes acromatiques que je fais faire ici pour M. Messier et M. de la lande dans la nuit du 18 au 19 de ce mois, Et jai pu croire aussi bien que M. dollond l'anneau de cette planette bien distincte. je souhaiterois savoir dans quelle nuit elle commença de paroitre Et d'etre Vue par vos observateurs, ce soir j'irai encore eprouver ces deux lunettes, mais il ne seroit alors temps de mettre celle ci a la poste.

74.01.25c *To François Rozier*[1]

M. le Docteur Priestley, connu à juste titre par ses Ouvrages, & principalement par ses expériences sur l'air fixe, vient d'en publier de nouvelles, couronnées du plus grand succès. Ce sont deux pas de plus dans la connoissance de l'air qui nous environne, ou de l'air athmosphérique.

Le premier, est la décomposition de l'air commun, au moyen de laquelle, son acide si long-temps regardé comme un problême, & qui n'étoit pas encore démontré, en est séparé par le secours du feu électrique. Il prend un tube de verre d'une ligne environ de diametre; à une de ses extrêmités, est placé un fil de métal, bouchée & scellée avec de la cire d'Espagne.[2] On remplit presque

1 Letter not found. The text is from an extract published in *Observations sur la physique, sur l'histoire naturelle et sur les arts*, 3 (1774), 145-7, under the heading, "Extrait d'une lettre de M. Magalhaens, Correspondant de l'Académie Royale des Sciences; Sur de nouvelles Expériences de M. Priestley, sur l'Air fixe, Datée de Londres le 25 Janvier 1774."

2 The wax commonly used for sealing letters or documents.

entierement ce tube avec la teinture de tournesol ou de violette, l'une & l'autre ont une belle couleur bleue; on a soin cependant de laisser dans ce tube une portion d'air commun, l'espace environ d'un pouce ou de trois quarts de pouce de longueur, & on met l'autre bout dans un vase rempli de la même teinture, pour que celle du tube ne tombe pas. Après avoir préparé cet appareil très-simple, on reçoit des étincelles d'une machine électrique sur le fil de métal qui traverse la teinture du tube. Au bout de quelque temps, on voit que cette teinture prend une couleur rougeâtre, & parvenue à un certain point, cette couleur n'augmente plus. Si on renouvelle l'air, cette couleur rougeâtre augmente. Si on renouvelle seulement la teinture sans renouveller l'air, la teinture ne change pas du tout: on observe que cet air diminue de son volume à mesure que la teinture devient rougeâtre; & que dans cet état, il est très-nuisible à la respiration animale. Une souris mise dans un verre plein de cet air, ainsi diminué de volume, & dépouillé de son acide, a péri presqu'aussitôt. C'est donc un vrai acide que le feu électrique fait dégager de la portion de l'air commun au-dessus de cette teinture, & qui la rend rougeâtre; & c'est une vraie décomposition que le feu électrique opere dans cet air. On voit par-là que l'odeur sulfureuse que nous sentons dans les expériences électriques, ne provient pas du feu ou matiere électrique, mais de la décomposition que ce feu cause dans l'air voisin. On doit en dire autant de l'odeur qu'on ressent assez souvent dans les endroits récemment frappés par la foudre, & peut-être c'est ce même air décomposé & nuisible que tue quelquefois les animaux qu'on dit frappés du tonnerre, quoiqu'on ne trouve sur eux aucune blessure ni contusion. 74.01.25c

Il faut encore observer que si on met dans ce tuyau de verre de l'air inflammable. Le feu électrique, en passant par cet air inflammable, n'en sépare point d'acide, puisque la teinture bleue ne change point de couleur.

L'autre expérience de ce célebre Physicien a pour objet une espece d'air qui, quoique très-mal-sain & très-nuisible à la vie des animaux, n'éteint cependant pas la lumiere d'une chandelle, d'une bougie, &c. On sait que les mineurs se servent d'une lumiere pour éprouver la qualité de l'air dans lequel ils doivent travailler. Mais M. le Docteur Priestley prouve que si l'air nitreux est surchargé de phlogistique, une chandelle allumée y brûlera aussi-bien, & même mieux que dans l'air commun, quoique ce même air nitreux, ainsi phlogistiqué, soit en lui-même si nuisible aux animaux, qu'ils n'y peuvent vivre un seul instant. C'est en mettant de l'*hépar sulphuris* ou *foie de soufre*[3] pendant un peu de temps dans un vaisseau chargé d'air

3 Liver of sulphur, made by boiling sulphur with a solution of potash.

74.01.25c nitreux, ou en y mettant de petits morceaux de fer, & les y tenant pendant long-temps, que cet air acquiert la qualité qui le rend propre à conserver la flamme d'une chandelle, (de même que le seroit l'air commun) quoique sa qualité soit pernicieuse & nuisible à la vie des animaux. Ainsi on ne décidera plus à l'avenir par l'épreuve de la chandelle allumée, de la mauvaise qualité de l'air. Il paroît que l'épreuve la plus infaillible pour connoître cette mauvaise qualité, est celle que M. Priestley proposa dans son Mémoire inséré dans les Transactions philosophiques (ann. 1772)[4] qui consiste à y mêler de l'air nitreux, & à observer si réellement il éprouve ou non quelque diminution dans son volume. L'expérience qu'il a répétée devant moi, est frappante. Il prit un bocal plein d'eau, renversé dans un seau également plein d'eau. Il déchargea dans cette eau deux phioles pleines d'air commun, & marqua l'endroit de l'espace que cet air occupoit. Il y ajouta une troisieme phiole d'air nitreux, & ce mêlange prit tout de suite une couleur rougeâtre. Ce dernier, au lieu d'augmenter le volume de l'air, réduisit cet espace au cinquieme de moins que les deux mesures d'air occupoient, auparavant ce mêlange. Cet air ainsi diminué dans son volume, est très-nuisible à la vie des animaux; mais si on le bat, ou si on l'agite vivement avec de l'eau commune, il redevient alors propre à la respiration, & perd toutes ses mauvaises qualités. On remarque encore que cette derniere opération de le battre, de l'agiter avec l'eau, lui fait perdre de nouveau quelque chose de son volume.

On demande qu'est devenue la troisieme partie avec la cinquieme, & plus du volume de ces trois phioles d'air, puisqu'au lieu de trois mesures d'air qu'on mêle ensemble, on n'en trouve dans la suite qu'une seule & quatre cinquiemes ou environ? Est-ce que l'air nitreux introduit, communique quelques émanations nuisibles dans l'air commun, & que dans le même temps il en diminue le volume? Mais ceci semble contradictoire. Est-ce que ces émanations sont attirées, & pour ainsi dire lavées dans leur frottement avec l'eau? Cela pourroit être vraisemblable; mais comment diminuroient-elles le volume de l'air au lieu de l'augmenter? Est-ce que l'air nitreux resserre ou qu'il détruit le ressort de l'air commun en tout ou en partie, & qu'alors ses parties occupent un espace plus petit qu'auparavant? est-ce que l'eau rétablit ce ressort, ou qu'elle lave les parties nuisibles qui s'étoient attachées à celles de l'air, & qui n'avoient pas perdu leur ressort? Voilà des problêmes à résoudre, dignes du travail des plus zélés Physiciens.

---

4 Joseph Priestley, "Observations on Different Kinds of Air", *Philosophical Transactions*, 62 (1772), 147-252; pp. 211 ff.

To Louis-Henri Duchesne 74.01.25d

Archives Nationales, Paris, MS T/160 (10-12)[1]

a Londres le 25 Janv^{r}. 1774
at n°. 12 Nevils Court Fetter Lane

Celle-ci, mon cher ami, est seulem^{t}. pour vous marquer que La pendule à double poulie, que Je vous envoyai,[2] aura peut etre besoin d'un plus grand poids. Car Je viens d'observer cette faute dans celle que J'ai à present dans ma chambre de la meme construction. ainsi prennez 3 livres de plomb en deux petites barres, & attachez les une à chaque coté du poids de plomb avec 4 vis. Celà remediera à ce defaut sans avoir besoin d'employer quelqu'horloger, qui d'aileurs trouveroit peut etre son compte à la gater ou decrier.

Vale me que ut soles ama
tuus ex corda[3]
Magalhaens

N.B. les autres 3 sont deja finies & pretes à partir dans ma chambre, avec votre avis.

a Monsieur
Monsr Duchenne Secretaire
de M^{r}. Trudaine
à Paris[4]

---

1 A receipt is filed next to this letter, as follows: "1er. Fev^{r}. 1774. M. duchesne | Port d'une caisse venant d'angleterre pesant 150^{l}. | 47#. 17^{s}."

2 See Magellan to Duchesne, 7 January 1774.

3 Farewell and maintain your friendship for me | yours from the heart.

4 Following this letter, as the final item in Duchesne's file of correspondence received from Magellan that is preserved at the Archives Nationales, there is a letter addressed to him signed "Caffieri l'ainé | D^{eur}. de la poste" and dated "Calais ce 17 J^{er}. 1774". It reads as follows:

Monsieur

Jay L'honneur de vous donner avis que je vous envois sous Couvert de M. Trudaine de montigny une Caisse que jay recu de Londres de M^{r}. magalhanns elle est marquée RM n°. 2 & partira dicy Lundy prochain par le Carosse pour arriver a Paris le dimenche suiv^{t}. je vous supplie de vouloir bien m'en accuser la reception, jay deboursé pour port de Londres icy avaries a 10 p % du Cap^{e}. plomb, acquit a Caution porteurs & autres memes frais ... 23. 8. que le directeur de La messagerie ma remboursé & donc il se pourvoira sur vous avec le port. je vous supplie de me renvoyer L'acquit à Caution qui accompagne cette Caisse bien & Duement dechargé, pour m'evitter L'amende de 3000$^{\#}$ porté par les ordonnances.----- Jay L'honneur d'etre avec une parfaitte Consideration

Monsieur

Votre tres humble & tres obeissant Serviteur

The letter is annotated in the same handwriting: "Le numero de L'acquit a Caution est 211. – | Je dis 211."

74.02.00[1] *From Jacques-André Mallet*
Bibliothèque de Genève, Geneva, Ms suppl. 1654, fol. 59[2]

Suite à Mr. Magalhaens

Je suis chargé encore par la même pso[nn]e pour qui est la lun.[3] de m'informer si l'on fait en Anglet. de ces reticules de verres sur les quels on a tracé des parallèles très fines & à une tres petite distance les unes des autres co[mm]e par exple de 1/10 lin. j'en ai reçu de pareils d'un artiste d'Augsbourg nommé Brander,[4] qui trace ces fig. avec une très grde exactitude, & je sai qu'on s'en sert co[mm]e de micrometres soit pour des lunettes soit pour des microscopes. L'usage qu'on voudroit en faire, c'est d'en adapter un à une lunette de 7 à 8 pouces, pour avoir ainsi une lunette de poche avec cette espece de micrometre, pour pouvoir mesurer grossieremt en campagne des hauteurs & des distances inaccessibles, c'est pour un naturaliste qui passe une partie de l'été à courir les montagnes.[5] Il demande si l'on n'a rien en Angleterre qui puisse servir lieu de cela, on pourroit bien ajuster un micrometre filaire à vis, mais on trouveroit cela trop couteux & trop compliqué il ne voudroit pas mettre à une pareille lunette au delà de 4 à 5 guin. Si d'ici à l'envoy de la caisse vous trouviez de rencontre, & à un prix honnete, la bonne édition des Tabl. de Sherwins[6] in octavo, Je vous serois bien obligé de me l'envoyer par la même occasion. Voila Monsieur bien des Commissions que je vous recommande mais surtout la lunette & le micrometre; & que nous serons charmés de recevoir le plutot possible.

Je vous suis b. obl. de l'observation de Muswellhill, je ne manquerai pas d'écrire à Mr. Maskelyne pour les autres que je souhaiterois avoir[7] Presentez lui je vous prie mes Complimens en attendant, & soyez persuadé que Je suis avec un sincere attachemt

Votre tres humble Servitr
J.A. Mallet

---

1 Approximate date established from Magellan to Mallet, 22 February 1774, which seems to respond to matters raised in Mallet's draft, as well as in his letter of 24 January.

2 Letter not found. The text is from a draft kept by Mallet.

3 See Mallet to Magellan, 17 December 1773.

4 Georg Friedrich Brander (1713-1783); see M. Daumas, *Les instruments scientifiques aux XVIIe et XVIIIe siècles* (Paris, 1953), p. 335.

5 H. B. de Saussure; see Mallet to Magellan, 11 March 1774.

6 Henry Sherwin, *Mathematical Tables, contrived after a Most Comprehensive Method...*, 5th ed. (London, 1772). See also Magellan to Mallet, 22 February 1774.

7 See Magellan to Mallet, 3 January 1774.

To Antoine-Laurent Lavoisier 74.02.11

Archives de l'Académie des Sciences, Paris, Fond Lavoisier, 1642[1]

a Londres ce 11 Fevr^{r}. 1774
at n°. 12 Nevils Court Fetter Lane

Je Vous ecris, mon cher Monsr., moins pour vous remercier du present que Je viens de recevoir de votre part par Les mains de M^{r}. Woide,[2] que pour vous feliciter d'une production vraiement philosophique & savante.[3] On en est ici fort content, & moi j'en suis ravi de voir que vous savez si bien vous acquiter des hautes esperances que J'avais conçû de Vos talents. La note que vous avez faite sur la decouverte de M^{r}. Vennel[4] m'a donné beaucoup de satisfaction. Car je m'avois presque brouillé avec l'Abbé Rozier de ne pas avoir fait justice dans ce particulier, attribuant toute l'invention de l'air fixe à M^{r}. Venel, qui n'a jamais connû d'autre air, dans les eaux acidules, que l'air commun comm'il paroit à ne pouvoir pas s'en douter par son meme Memoir, imprimé dans la Collection des Savants Etrangers.[5]

Le D^{r}. Hales avoit trouvé bien la meme chose avant lui, et se trompa de meme.[6] peut etre M^{r}. Vennel ne pensa bien ce qu'il disoit dans son Memoir: car il devroit avouer, que toute sa decouverte n'etoit dans le fond que la meme chose deja avancée par Hales. Vous avez rendu justice à ces deux savants, & au D^{r}. Priestly: & J'en suis très content, aussi bien que tous ceux qui ont lû votre Livre.

A present Je me flate que Vous poursuivrez l'objet de la distilation & desalaison de l'eau de mer par la nouvelle methode; & je ne doute pas que cette entreprise n'acquiere pas quelques degrès de plus grande perfection

---

1 Corrected version of the text published in *Oeuvres de Lavoisier: Correspondance*, ed. René Fric, fascicule II, 1770-1775 (Paris, 1957), pp. 430-1.

2 Presumably the Polish-born Charles Godfrey Woide, FRS (1725-1790), preacher at the Dutch Chapel Royal, St James Palace, London.

3 Magellan has evidently just received a copy of Lavoisier's *Opuscules physiques et chimiques* (Paris, 1774).

4 The chemist Gabriel François Venel (1723-1775).

5 Chapter VI of Lavoisier's *Opuscules* is devoted to Venel's experiments, and includes the following footnote (p. 34): "M. Venel a toujours supposé que le fluide élastique, contenu dans les eaux minérales, étoit le même que l'air de l'atmosphère; on verra dans la suite ce que l'on doit penser de cette opinion." Magellan's reference is to G. F. Venel, "Mémoire sur l'analyse des eaux de selters ou de seltz", *Mémoires de mathématique et de physique, présentés à l'Académie Royale des Sciences par divers savans*, vol. 2 (Paris, 1755), Part 1, pp. 53-79; Part 2, pp. 80-112.

6 Stephen Hales, *Vegetable Staticks* (London, 1727).

74.02.11 entre vos mains.[7] Je pense aller à Paris au bout de mars prochain: & serai très charmé de pouvoir vous aider dans cette entreprise avec les pauvres talents que Je possede. en attendant, conservez moi vos amitiez, & soiez assuré de l'attachem^{t}. sincere & pour la vie

de
Votre
Magalhaens

P.S. presentez mes humbles Resp^{ts}. à Mad^{e}. votre digne Epouse, & reste de votre respect^{ble}. famille

a Monsieur
Monsr. de La Voisiere, de l'Acad^{e}. R^{le}. des sciences
Rue des bons Enfants
a Paris[8]

7 In February 1773 Magellan had reported to Trudaine de Montigny recent English experiments with an apparatus for distilling seawater on board ship. At Trudaine's suggestion, the French Navy had commissioned Lavoisier to follow up the idea. See also Magellan to Lavoisier, 24 November 1774 and 1 July 1775.

8 La france *added in another hand.*

74.02.22 *To Jacques-André Mallet*[1]

Bibliothèque de Genève, Geneva, Ms suppl. 1654, fol. 25-6

à Londres ce 22 Fev^{r}. 1774
at n°. 12 Nevils Court Fetter Lane

Monsieur

En consequence de la Votre du 24 du mois dernier[2] J'eus differentes conferences avec M^{r}. Dollond pour l'affaire de votre Lunete. Mais il n'ose point me la promettre avant l'eté prochain. Ces sortes d'Instruments ne

1 Annotated "Mars 1774".

2 See Mallet to Magellan, 24 January 1774.

sont pas faits sans des ordres particuliers: & il en a tant de promises, & si peu de bon flint-glass pour en faire les objectifs, qu'il ne peut pas fixer precisement le tems: d'autant plus que cette spece[3] d'ouvrage demande un tems clair, & une atmosphere assez pure, pour l'essaier sur les objets celestes. Les deux qu'il a faites pour M^{r}. Messier, & M^{r}. de la Lande (qui font un objet d'environ cent livres Esterlines en semble) ne sont pas encore decidem^{t}. eprouvées. nous les avons essayées plusieurs fois sur jupiter & Saturne: & nous n'avons pas encore eté satisfaits tout à fait. Ainsi, mon cher, il faut attendre patiement pour avoir quelque chose de satisfaisant: car je ne vous envoyerai aucunem^{t}. une Lunete ordinaire, & dont je ne sois pas bien satisfait: la votre devant servir bien differement que celles des gens simplem^{t}. curieux, pour des observations delicates. pour ce qui regarde la monture de votre lunete, il vaudra bien mieux la prendre montée dans un tuyau de metal, & non pas de bois: mais sans aucun piedestal: ce qui reviendra à peu près au meme prix. Car je crois que vous pouvez la faire monter bien moins cherem^{t}. chez vous dans une machine paralatique: qui dans le meme tems lui sert de Piedestal &c. mais nous aurons encore du tems pour discourir là dessus: & je tacherai de me borner à peu près dans les termes de la depense que vous me dites, y compris le micrometre objectif & le chassis filaire &c. 74.02.22

J'ai remis à Messrs. Liotard, Aubertin & Co. un pacquet sous votre adresse, qui contient les deux almanaks nautiques pour 1774 & 75. (prix 7^{s} :0^{d}) / une ounce de la meilleur graine de choufleur - 2^{s} :0^{d} / 12 onces (ou 3/4 de livre) d^{o}. de carotes 1^{s} :6^{d} / & j'y ai ajouté une once de navets jaunes, qui ont quelque chose d'extraordinaire pour la couleur des racines qui sont jaunes après etre bouillies: & d'un gout excellent = 3^{d}. seulem^{t}.

On a reimprimé depuis peu les Tables logarith. de Sherwin. mais on n'est pas encore bien convaincû de leur egalité avec l'ancienne edition, qui est fort rare.[4] C'est par cette raison que je ne vous envoye la premiere.

On fait bien ici ces reticules sur le verre pour les microscopes: mais elles ne valent rien pour les lunetes. Si M^{r}. votre ami de campagne, veut y mettre une guinée de plus, je crois pouvoir lui envoyer une fort jolie lunete de poche, d'environ 7 pouces de foyer, que je fais garnir d'une espece de micrometre filaire pour mon usage: & que J'espere n'ira pas au delà de

3 espece?

4 Henry Sherwin, *Mathematical Tables, contrived after a Most Comprehensive Method…*, 5th ed. (London, 1772). Earlier editions were published in 1706, 1717, 1726 and 1741. See Mallet to Magellan, […] February 1774, in which Mallet requests "la bonne edition" of Sherwin's work.

74.02.22 6 guinees ou 6 1/2. Ce petit micrometre n'aura que 1 fils à coulisse qui se mouvera par une vis, dont je mesurerai sur une base connue, la valeur angulaire de chaque pas: j'y mettrai seulem^{t}. 2 fils aux deux extremités du champ de vision: & je tacherai de practiquer l'idée du Professeur Wilson, en reduissant l'epaiseur des fils par l'applatissement.[5]

N.B. Je vous dis qu'il n'aura qu'un fil à coulisse, c'est à dire mobile: car on en a fit ici avec 4. fils mobiles, dont chaque pair etoit à angle droit, & qui se mouvoient dans le meme tems, de façon qu'on prennoit par la meme position du micrometre les deux diametres (à rectangle) d'une planete. M^{r}. le Presid^{t}. de Saron en a une que Je lui envoyai d'ici, avec son instrument des passages meridiennes. Cette espece de micrometres seroit d'un grand usage pour prendre l'A. droite & declinaison dans un imstrum^{t}. paralatique. mais ils sont difficiles à etre bien executé.

Je crois avoir repondu aux articles de la votre: & s'il y aura quelque chose de votre service, vous pouvez encore me l'ecrire avant que je part pour mon tour de Paris & Flandres, que ne sera pas avant la fin de Mars prochain. Je vois de tems en tems ici M^{r}. de Luc votre digne patriote.[6] il se porte à merveille, & trouve toute l'estime, qu'il merite par ses talents & qualités, entre les Anglois.

J'ai l'honneur d'etre
Monsieur
votre t. h^{ble}. Serv^{r}.
Magalhaens

A Monsieur
Monsr. Mallet, Professeur de
Astronomie Cour S^{t}. Pierre
à Geneve
en Suisse

5 See Alexander Wilson, *Philosophical Transactions*, 64 (1774), 105-7.
6 The Genevan physicist Jean André de Luc (1727-1817) had settled in London in 1772.

To Charles Messier 74.03.00[1]

American Philosophical Society Library, Philadelphia, João Jacinto de Magalhães (Jean-Hyacinthe Magellan) Collection, Mss BM25

Vous savez, mon cher ami, l'attention que J'ai pour que votre Lunete soit des meilleures: & c'est cette raison pourquoi on en retarde tant l'envoy. car le meme Mr. Dollond y prend un particulier interet, & ne veut s'empresser de me la delivrer sans que nous soions bien convaincûs de son excellence. elle m'a parû dans les derniers essais, avoir tant soit peu trop de Lumiere au tour du corps de Saturne, quoiqu'elle le montroit fort bien coupé & net. Notre atmosphere n'a pas eté en etat de bien demeler, si c'etoient des vapeurs ou la disposition acidentelle &c. Car il me paroissoit qu'elle n'avoit pas tant dans les essais precedents. Mr. Dollond ne peut se resoudre à toucher à la Courbure des verres, sans avoir plus de certitude sur celà: & meme quand il l'auroit, ces touches delicates demandent un tems plus favourable. C'est autant pour la Lunete de Mr. de la Lande, auquel vous expliquerez ces raisons, & que J'y veille egalemt. en attendant tachez de m'envoier l'argent de Votre Lunete, ou du moins la plus grande partie: car sans celà je me trouverai fort embarrassé.

Je viens d'envoyer sous l'adresse de Mr. Trudaine, à Mr. Du Chenne une caisse qui contient 5 petites autres dedans. une marquée SAN contient des livres pour Mr. le Dr. Sanchez. mais il ne sçait pas par quelle voye elle arrivera, & vous aurez soin de la lui envoyer, sans lui dire rien, en chargeant sur mon compte la 5me. partie de la depense totale d'ici jusqu'à Paris. Car il m'a dit dans sa lettre de la lui envoyer toute simplemt. à son adresse: mais[2] [...] de trouver, en faisant passer [...] membre syndical des Libraires &c. aussi bien que plus de depense. Les 3 [...] empaquetées en papier brun. qui vont dans le fond, contient des outils de menuiserie pour quelques curieux: celles-ci furent ordonnées par Mr. le Prince Pignatelli[3] & sont pour Mr. le Chever de

1 Letter dated on the basis of the annotation at its head, "recu le 22 de mars", and the further annotations throughout the letter.

2 This is the final word on the first page of the letter. As a result of water damage to the top of the sheet, much of the first line on the next page and the first words in the following two lines have been lost.

3 Juan Joaquín Atanasio Pignatelli de Aragón y Moncayo (1724-1776), Spanish Ambassador in Paris, 1763-72?

74.03.00 Llovera,[4] Tresorier du Roy d'Espagne, chez son Ambassadeur à Paris qui vous les payera à vous pour mon compte, comme je lui ecris par cette meme poste.[5] mes deboursements ici pour ces 3 caisses montent à 279#.4^{s}. tournois:[6] auxquelles vous devez ajouter les 3 cinquiemes de la depense jusqu'à Paris, comme je le lui dis dans ma lettre. Enfin la caisse marquée D.C.H. contient les petits articles suivants. un pacquet pour M^{r}. le Chev^{er}. de Bory, qui doit vous payer pour mon compte 28#.15^{s}.[7] / un autre pour M^{r}. le Chev^{er}. d'Arci,[8] qui doit vous payer 57#.16^{s}.[9] pour mon compte. / un autre pour M^{r}. l'Abbé Rozier. / un autre pour M^{r}. Du Chenne qui doit vous payer sur le compte du Comte Carbury[10] & pour le mien par le contenû dans ce paquet qui sont des medecines &c. la somme de 80#.0^{s} / il y a aussi le Nautical Almanak pour vous, dont M^{r}. Maskelyne vous fait present. / un pacquet pour M^{r}. Macquer, de la part du D^{r}. Lettsom.[11] / un pacquet pour M^{r}. le President de Saron / & un autre pour M^{r}. de la Fonde,[12] ou l'Abbé Bourriot, qui doit vous payer pour mon compte 93#: 8^{s}. tournoises.[13] Reelem^{t}. chacune de ces parcelles doit payer proportionellem^{t}. pour la depense depuis Calais jusqu'à Paris: c'est à dire, on devroit partager la 5me. partie de la depense totale, appartenante à cette 5me. caisse, par les 6 ou 7 parcelles qui y sont contenues, & chacune devroit payer son ecot. mais vous ferez là dessus ce que vous jugerez avec M^{r}. Duchenne. les sommes que je viens de vous annoncer, sont celles de mes deboursements ici: & les additions de la depense proportionelles jusqu'a Paris y devroient etre ajoutées respectivement à chacune.

Vous aurez sans doute deja reçû la caisse que Je vous annonçai dans ma lettre precedente: M^{r}. de Briqueville[14] m'ecrivit deja qu'il alloit

4 Llovera's name has been underlined. He was treasurer of the Spanish Embassy in Paris.
5 Letter not found.
6 Annotation: "recu le 28 avril 1774".
7 Annotation: "recu le 23 mars 1774".
8 i.e. Comte Patrick d'Arcy (1725-1779).
9 Annotation: "paie le 24 avril 1774".
10 See Magellan to Duchesne, 23 July 1773, note 3.
11 The celebrated Quaker physician John Coakley Lettsom (1744-1815).
12 Joseph Aignan Sigaud de la Fond (1730-1810)?
13 Annotation: "recu le 18 avril 1774".
14 Bon-Chrétien, Marquis de Briqueville (1723-1803); see Magellan to Johann III Bernoulli, 26 October 1772.

vous faire remettre les 144#.8^{s}[15] pour avoir la caisse qui lui appartenoit &c. 74.03.00

aussitôt que vous aurez recû cet argent, aiez la bonté de me l'envoyer ensemble par une Lettre de Change; car J'en ai grand besoin.

Je vous ai envoyé la gazete oû l'on avoit mis les elements de votre derniere comete, sous l'envelope que vous m'aviez dit, de M^{r}. de Boynes.[16] J'espere que vous n'avez pas manqué de l'avoir reçu dans son tems.

À dieu, mon cher ami. j'ai encore beaucoup de lettres cette poste. Je serois charmé d'avoir le precis du raport qu'on fit à l'Acad^{e}. de vos travaux sur les nebuleuses, & nombre de vos observations, cometes &c. Car celui que vous m'aviez envoyé il y a 2 ans, fut consumé avec le reste de mes papiers dans mon incendie.

Vale me que ama
tuum
Magalhaens

A Monsieur
Monsr. Messier, Astronome de l'Acad^{e}. R^{le}. des Sciences
Hotel de Cluny,
Rue des Maturins
à <u>Paris</u>[17]

---

15 Annotation: "reçu les 144#.8^{s}. le 7 mars 1774. les frais à la charge de mon ami 18#.15^{s}. voir sa lettre du 31 jan 1773." Letter not found.

16 Pierre-Etienne Bourgeois de Boynes, Secretary of State with the Department of the Navy in Paris.

17 Annotation:

Depense de la caisse monte à	48#	12
P. le Prince Pignatelli	36	0
P. M^{r}. Duchenne et le Doct. Sanches	12	12
[*illegible word*]	6	6

74.03.11 *From Jacques-André Mallet*
Bibliothèque de Genève, Geneva, Ms suppl. 1654, fol. 50[1]

Geneve le 11 mars 1774

J'ai reçu M^r. la lettre que vous m'avez fait le plaisir de m'écrire du 22 Fev.[2] Je suis faché d'apprendre que notre lun. ne puisse pas etre finie de qlque mois mais, il vaut trop mieux, l'avoir plus tard, si c'est un moy. de l'avoir plus parfaite, j'ai une entiere confiance aux soins que vous voudrez prendre p^r qu'elle réponde à la réputation de l'ouvrier, & à l'objet p^r. lequel elle est destinée. C'est b. mon intention, co[mm]e j'ai deja eu l'hon de vous l'écrire, de faire faire ici une machine parallat. pour monter la lunette, lorsqu'on voudra s'en servir pour des observat. astron. mais co[mm]e il est possible que cette lunette ne reste pas touj à mon observat. je souhaiterois bien qu'elle eut son piedestal particulier, aussi simple qu'on voudra, duquel elle put etre séparée p^r la fixer sur la machine parall. lorsqu'on le souhaiteroit, Je conviens qu'un tuyau de metal seroit plus solide qu'un de bois, mais ce dernier fait avec soin ne seroit-il point suffisant S'il n'y a pas d'inconveniens, je le préfére par économie parce que vous vous rappellerez, M^r. que la lunette simple avec son pied, sans micrometre, & sans chassis filaire, ne m'appartient pas, & qu'on n'y destine que 25 guinées.

J'ai communiqué l'article de votre lettre rélatif à la petite lunette de 7 pces de foyer, à mon ami le Prof. Dessaussure,[3] pour qui elle est déstinée, il approuve la maniere dont vous comptez de la faire exécuter, & cela p^r le prix des 6 à 6½ G.[4]

1 Letter not found. The text is from a draft kept by Mallet.
2 See Magellan to Mallet, 22 February 1774.
3 i.e. H.B. de Saussure.
4 The draft is unsigned.

To Nathaniel Pigott 74.03.22

Royal Astronomical Society, London, MSS Pigott, 33

London 22 March 1774

My Dear Sir

I found of late one of my acquaintance, who gave me a particular account of y^{e}. process for making the best glue: and being excited by curiosity I consulted the Dictionaire des Arts & Metiers of y^{e}. last edition 1773 at Paris,[1] where to my great surprize, the very same particulars are so well described as to make unnecessary any further explanation. the whole point consists on having at hand the best materials, as our great tanners have in England, by the quantities of skins, which are damaged or spoiled for making good Leather: and that is the reason why these manufactures are so nasty and offensive by their stink, and roteness.

there were but 3 trials on as many frost day in the last season, by which the compounded pendulum of my clock was set to right: so that in the fourth tryal I could not perceive the least alteration of it's lenght from near the freezing point of y^{e}. thermometer, to the 86 in y^{e}. Fahrenheit's scale. my method was the same described by the Late Ingenious M^{r}. Elicot in the Philos. Transact.[2] it speaks for itself: and am fully convinced that neither heat nor cold, by themselves alone, will affect it's regularity. this clock as it stands now in a mahogany case, does not exceeds *7* guineas.

I directed by the post according to your desire, the opera glass of Dollond to M^{r}. Le Duc D'Ursel,[3] the very same day I received yours of y^{e}. 5th. instant.[4] the Seeds and young trees for Monsr. le Chev^{er}. Turgot,[5] are already sent to Dover, and notice given to M^{r}. Factor & Minett[6] in order to forward them to Calais. but I could not get sooner than yesterday the bill of M^{r}. Gordon by which you see the cost has been 6:8^{s}:8^{d}. Esterl. the beans I bought in another shop, where I am very certain of being genuine Windsor ones, for 4^{s}:0^{d}.

---

1 *Dictionnaire universel des arts et métiers* (Paris, 1773), p. 614.

2 John Ellicott, "A Description of Two Methods, by which the Irregularity of the Motion of a Clock, arising from the Influence of Heat and Cold upon the Rod of the Pendulum, may be prevented", *Philosophical Transactions*, 47 (1751-2), 479-94.

3 Charles, Duc d'Ursel (1717-1775).

4 Letter not found.

5 Étienne-François Turgot (1721-1788), older brother of France's controller general of finance, was a noted agricultural innovator.

6 Fector, Minet & Co. monopolized the merchant services from Dover before 1778.

74.03.22 have here by me a very good achrom. Telescope about 9 inch long, made by Ramsden, which cost 4. guineas. / two fine razors without ornament, price 10^{s}:0^{d}. / Nautical almanack for 1775, for 3^{s}:6^{d} / one dozen pieces of elastic gum at 6^{d} each – 6^{s}:0^{d} [N.B. this is commonly sold at y^{e}. Mathem^{l}. shops in small squares, as large dyes, at 3^{s}:6^{d} each.] these articles together with y^{e}. Duc Ursel's opera glass of 1:5:0 / y^{e}. 7^{s}:6 you mentioned in this last Letter for the barometical tubes: and 3:6 of letters, porters, & messages for the seeds and young trees, which ought to be charged along with the inclosed two bills for the Chev^{er}. Turgot, make the sum of £13:12^{s}:2^{d} Sterl. which you'll please to send at your Leisure. But as to the mode of sending you these small articles now in my hand, I am almost puzled; on account of them making so small a parcel, which however will make as much expences on being sent in y^{e}. usual way with entering it in y^{e}. customhouse &c. as if it was worth twenty times the double. but I will follow your order, whenever you decide about it. the last part of Philos. Transact. for 1773 is not yet printed: and I'll take care of sending it, as soon as published.

the 3½ foot achromatics of Dollond like that at Greenwich, cost 20 guineas in a woden tube. 25^{g}. if mounted in a stand with little Rakwork. 36 guineas, if mounted in brass tube, large rakwork and with a finder. about 43^{g}. if mounted in a paralactic stand. 50 guineas if furnished with one objectif achromatic micrometer: and about 70 guineas if mounted in a paralactic machine, as described in de la Lande's Astronomy,[7] and furnished with a Graham's micrometer:[8] such as I have now one made by him, for M^{r}. Messier of Paris.

I have not yet seen M^{r}. Maskelyne since your Letter came. the first time I see him, will not forget your message.

Pray my most sincere Complim^{ts}. & good wishes to all our worthy friends, and particularly to those who honour me in their memory: and believe me

me Dear Sir
your most obed. h^{le}. serv^{t}.
Magalhaens.

A Monsieur
Monsr. Pigott, Gentil-homme Anglois
vis à vis La Cour
a Bruxelles
Flandres

7 Lalande, *Astronomie*, vol. 2 (2nd ed., Paris, 1771), pp. 793-9 and Plate 176.

8 George Graham, FRS (1673-1757).

To Jacques-André Mallet[1]

74.04.15

Bibliothèque de Genève, Geneva, Ms suppl. 1654, fol. 27-8

Londres ce 15 Avril 1774

Je soufre, Monsieur, une si grande inquietude lorsque Je retarde de repondre à mes amis, que je ne puis pas m'empecher de vous ecrire celle-ci, avant de l'envoy que je pensois vous faire bien avant ce tems, de la lunete à petit micrometre pour M^{r}. Desaussure.[2] plusieurs petits acidents imprevus ont empeché l'ouvrier d'executer cet ordre, malgré tous les soins & applications que j'ai fait pour celà. Mais Je ne manquerai pas de vous l'expedier avant mon depart pour Paris, que J'ai fixé pour le commencem^{t}. du mois prochain. M^{r}. Dollond, co̅me Je vous l'ai marqué, ne sera pas à meme d'entreprendre votre lunete avant mon retour: & Je ne manquerai pas de prendre toutes les mesures, pour que vous soyez bien servi, d'après les avis & circonstances que vous m'avez marquées.

On vient de presenter à la Sociéte R^{le}. un instrument très simple que je crois etre le meme que vous souhaitez pour prendre des angles sans aucun calcul &c.[3] il ne consiste qu'en deux regles bien divisées en pouces, centiemes, & meme plus de pouce &c. qui se meuvent par le moyen d'une charniere, oû est le zero (dans le centre de son axe) de chacune des divisions de ces deux regles, savoir ab & ag. il y a une troisieme regle cf aussi bien divisée, arrangée en coulisse, qui porte deux pinules au centre de chaque coulisse d.e: & de meme il y a une autre en a; ou plutot des petites lunetes d'aproche &c.

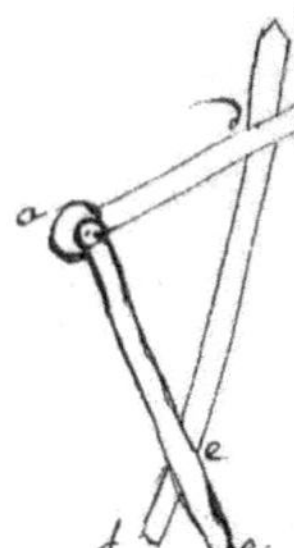

1 Annotated "Avril 1774".

2 i.e. H.B. de Saussure.

3 The Royal Society's *Journal Book* records (*Journal Book Copy*, XXVIII, 43) the reading at the Society's meeting on 24 March 1774 of a description of a "new invented double sector" by Mr Andrew Marr. The paper was not published.

74.04.15 Vous concevez bien que ces regles etant bien divisées & soudivisées, donneront la valeur des cotés de chaque triangle avec beaucoup plus de precision que la planchete, ou table des arpenteurs, & bien plus aisem^{t}. que les Theodolites ordinaires. Si l'on adaptoit ces regles sur un bon niveau: & qu'on montoit une lunete d'aproche dans chaque coulisse, de façon qu'elle puisse avoir du mouvem^{t}. vertical, pour trouver les objets à quelque hauteur au dessus ou desous du plan horizontal, pour lors cet instrument seroit le plus complet de tous ceux qu'on employe dans l'arpentage. Pour le present on ne fait encore ces instruments. ils sont dans leur enfance. peut etre je ferai executer un selon les idees ci dessus, lors de mon retour.

M^{r}. de Luc est très sensible à votre souvenir, & vous prie d'accepter les siens. il n'etoit hier dans la Societé R^{le}. apparem^{t}., il etoit engagé ou en campagne.

J'ai l'honneur d'etre avec toute consideration possible

Monsieur
votre tres humble & obeiss^{t}. Serv^{r}.
Magalhaens

P.S. Je me souviens bien d'avoir vû ici M^{r}. de Saussure, que le D^{r}. Tourton me fit connoitre:[4] & Je le menai, si je ne me trompe pas, chez M^{r}. Lane, pour y voir quelques experiences electriques. presentez lui de très humbles respet^{s}. de ma part.

A Monsieur
Monsr. Jaq. And. Mallet, Profess^{r}. d'Astronomie
Cour de S. Pierre
à Geneve
en Suisse

4 Saussure visited England in 1768. While there, he fell ill and was attended by "son ami le docteur Turton" (J. Senebier, *Mémoire historique sur la vie et les écrits de Horace Bénédict Desaussure …* [Geneva, 1801], p. 48).

To Johann III Bernoulli 74.04.29

Öffentliche Bibliothek der Universität Basel, MS L I a 709, f. 53-54

a Londres le 29 Avril 1774

Monsieur & très cher ami

Les cartes geographiques mentionées dans La Votre du 26 Fevrier[1] sont arrivées de puis peu entre mes mains: & je trouve La plus grande dificulté à en disposer. Car aiant parlé à plusieurs qui gardent des boutiques de ce genre, & qui sont bien reputés, ils ne veulent point se charger d'autre vente que des leurs, malgré le profit de 25 per cent que je leur ai offert sur celles de votre Acad^{e}.:[2] quant à ceux moins reputés, de qui sont des petits marchants, je ne veux pas Les leur confier, de peur que La perte tombe sur l'Academie: & pour ce qui me Regarde, il est très improbable d'avoir aucune occasion pour le total. Seulem^{t}. d'un tems à l'autre, quelqu'ami prendra peutetre quelque carte, comm'en effet deux de mes amis en ont pris 3 ou 4 de ces cartes.

Mais ce qui m'auroit mis dans l'erreur sans La Lettre & prix y-mentionés de M^{r}. Naudé,[3] c'etoit la valeur en argent d'Angleterre que Vous me donniez: car en consultant un Marchand qui connoit votre argent, il y en avoit presque La moitié de difference outre celà Vous me disiez, que l'Academie me faisoit present d'une copie de chaque carte, au Lieu que M^{r}. Naudé ne disoit rien de celà dans sa Lettre, & seulem^{t}. parloit de 25 per cent, que l'Academie m'offroit pour la vente de ses cartes. Ce que je ne puis pas accepter, n'etant point Marchant, ni aiant aucune boutique: ainsi c'est sur ce pied que j'ai parlé à ceux qui en tient, en leur faisant ce meme offre: & cependant je n'ai pas pû y reussir encore.

M^{r}. Maskelyne vous envoye cette copie du Nautical almanac & j'ai cherché par tout les oeuvres de Letherland, mais Je trouve qu'il n'a

---

1 Letter not found.

2 Cf. Magellan to Bernoulli, 30 December 1773, for earlier comments by Magellan on the likely difficulty in selling the Academy's maps.

3 David Naudé, astronomical assistant at the Berlin Academy of Sciences and son of the former member of the Academy, Philipp Naudé the younger (d. 1745). Letter not found.

74.04.29 jamais imprimé aucun traité ou Memoir, excepté dans les Transactions philosophiques.[4]

L'etat complet de vos comptes que vous souhaitez de M^{r}. Elmsly, est celui que Je vous ai envoyé: car je l'eus de sa propre main: ainsi vous pouvez compter Là dessus.

Mon ami, M^{r}. Wolfe va partir d'ici dans peu pour chez vous, & il vous apportera la Memoir qu'il avoit l'année derniere sur l'air fixe, & un autre pour M^{r}. Margraff.[5] dans ce tems là il etoit plus precieux qu'à present etant deja inseré dans Les Transact. Phil. qui sont publiées:[6] mais ce ne fut pas ma faute, que Vous ne l'eusses auparavant. Je compte vous ecrire par Lui, pour Vous le recommander de nouveau, aussi bien que son compagnon, qui est un jeune Chymiste de ma Connoissance.

Quelqu'un m'a dit ici que Vous alliez voyager cet eté, & Je crois que c'est celà qui vous manque le plus, pour dissiper les effets de vos complaintes. Car je sçais bien la force & impression de ces maladies ou incommodités habituelles. rien de mieux que changer d'air, de pais, de maniere de vie, & se donner du mouvement: qui sont les consequences des voyages.

Je souhaite beaucoup savoir sur quelle relation & sur quelle memoire autentique est fondée la route du vaisseau portugais de David Melquer, venû du Japon par le Pole du Nord en 1660, com̄'on le voit exprimé dans le Hemisphere du Comte de Redern, publié en 1762 par Votre Acad^{e}. R^{le}.?[7] Certes le nom Melquer n'est pas portugais du tout. Je viens d'ecrire là dessus à Lisbonne pour me certifier. Mais je doute beaucoup, qu'on en sache rien, par l'ignorance qu'on y a des faits pareils.

Je vais partir dans peu de jours pour Paris, d'ou je retournerai au bout d'un ou deux mois pour Londres.

Celle ci part avec les Transactions Philos. de l'année derniere, que J'adresse à M^{r}. de Merian, Biblioth^{e}. & Direct^{r}. de Belles lettres: & lui sans

---

4 Despite what Magellan says, no article by anyone with a name like this appeared in the *Philosophical Transactions.* Magellan may have been referring to the watchmaker Peter Litherland (1756-1804) of Warrington and later of Liverpool.

5 The noted chemist Andreas Sigismund Marggraf (1709-82), a leading member of the Académie des Sciences et Belles-Lettres in Berlin.

6 No paper by Peter Woulfe (or by anyone else with a similar surname) on fixed air was ever published in the *Philosophical Transactions*!

7 Sigismund Ehrenreich de Redern, *Hemisphere septentrional dresse en 1754 par Mr le comte de Redern, curateur de l'Academie des Sciences et belles lettres pour l'eclaircissement de ses considerations sur le globe execute par ordre de l'Academie* (Berlin, 1762).

doute aura soin de vous la faire tenir. Je vous souhaite une parfaite santé, & suis de Coeur 74.04.29

Mon cher Monsieur
Votre très h^{ble}. & obeiss^{t}. serv^{r}.
Magalhaens

P.S. La Soc^{e}. Royale vient de m'elire en Membre comme les Nacionaux, aiant residé ici depuis le tems requis par l'estatut[8]

A Monsieur
Monsr. Bernoulli, Astronome de l'Acad^{e}. R^{le}. de Prusse
à Berlin

---

8 Magellan was elected FRS on 21 April 1774. The regulation to which Magellan here refers was embodied in a series of resolutions adopted in the 1760s by the Council of the Royal Society but not enshrined in a Statute, that distinguished between Home and Foreign Members, but allowed foreigners who had resided in Great Britain for a period of at least six months to be elected on the Home list.

To Johann III Bernoulli 74.05.16

Öffentliche Bibliothek der Universität Basel, MS L I a 709, f. 55

Londres ce 16 May 1774
at n°. 12 Nevils Court Fetter Lane

Monsieur & cher ami

Enfin de cette fois Vous verez Le porteur de la presente qui est mon bon ami M^{r}. Wolfe,[1] membre de la Soc^{e}. Royale ici, dont je Vous ai parlé tant de fois: & auquel Je vous prie de rendre tous Les bons offices que son merite personnel ne manquera pas de Vous meriter, & que notre amitié vous influira. J'espere qu'avant son arrivée, vous aurez deja reçue la lettre que Je vous ecrivis en Vous envoyant dans le meme tems Le Nautical almanak dont M^{r}. Maskelyne Vous fit present pour L'annee suivante de 1775.[2]

---

1 The chemist Peter Woulfe (?1727-1803).
2 See Magellan to Bernoulli, 29 April 1774.

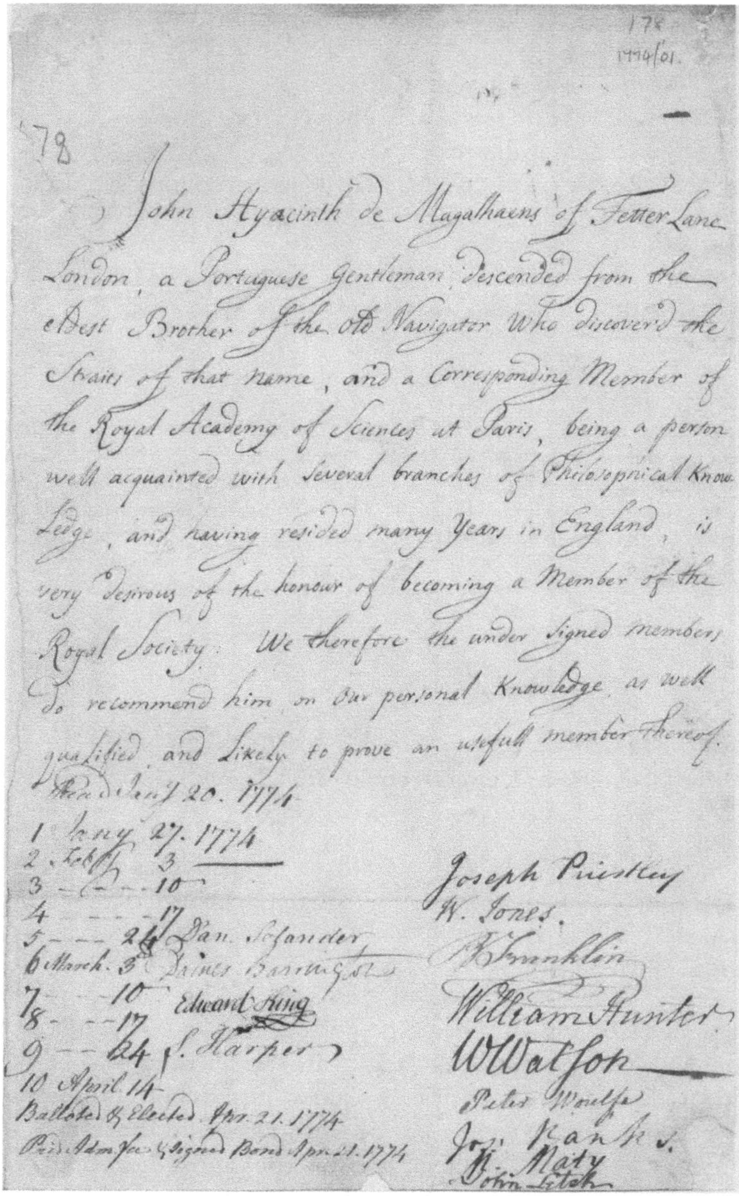
178
1774/01.

178

John Hyacinth de Magalhaens of Fetter Lane London, a Portuguese Gentleman, descended from the eldest Brother of the Old Navigator who discover'd the Straits of that name, and a Corresponding Member of the Royal Academy of Sciences at Paris, being a person well acquainted with several branches of Philosophical Knowledge, and having resided many Years in England, is very desirous of the honour of becoming a Member of the Royal Society. We therefore the under signed members do recommend him on our personal Knowledge as well qualified and likely to prove an usefull member thereof.

Read Jan. 20. 1774
1 Jany 27. 1774
2 Feb 3
3 — 10
4 — 17
5 — 24 Dan. Solander
6 March 3 Daines Barrington
7 — 10 Edward King
8 — 17
9 — 24 S. Harper
10 April 14
Ballotted & Elected Apr. 21. 1774
Paid Adm fee & signed Bond Apr. 21. 1774

Joseph Priestley
W. Jones.
B. Franklin
William Hunter
W Watson
Peter Woulfe
Jos: Banks.
M. Maty
John Letch

Figure 4. Magellan's nomination certificate for election to the Royal Society of London, signed by Joseph Priestley, William Jones, Benjamin Franklin, William Hunter, William Watson, Peter Woulfe, Joseph Banks, Matthew Maty, John Letch, Daniel Solander, Daines Barrington, Edward King and Samuel Harper. As the annotation at the bottom of the certificate indicates, Magellan was elected on 21 April 1774. Courtesy Royal Society of London.

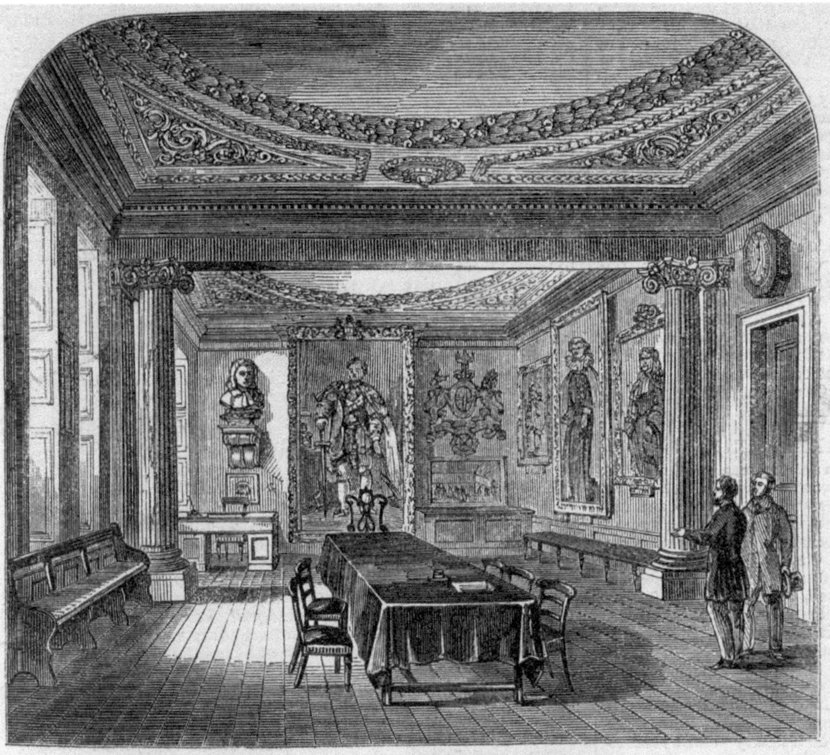

ROOM IN CRANE-COURT, FORMERLY THE MEETING-PLACE OF THE ROYAL SOCIETY.

Figure 5. Meeting room of the Royal Society of London in the Society's house in Crane Court, London, where meetings were held from the time the Society purchased the house in 1710 until 1780, when the British Government provided the Society with free accommodation in Somerset House (*Illustrated London News*, 12 December 1863, p. 597).

74.05.16 Je vais partir incessament pour Paris, oû je m'arreterai jusqu'au milieu de Juillet suivant & de là je pense retourner en droiture chez moi à Londres.

Je vous prie de m'informer en plûtot possible, si vous admetez dans Le recueil de Votre Academie R^{le}. des Memoires des savants etrangers, je veux dire qui ne sont pas membres de Votre Academie. Car un de mes amis ici voudroit envoyer un Memoir sous des objets fort interessants dans La Physique (ou Philosophie) experimentale pour etre inseré dans quelqu'une des collections d'aucune Academie etrangere: & J'ai pensé à celle de la Votre. en cas que ceci soit faisable (& en supposant que La Memoire soit approuvé &c.) je vous prie de me Le faire savoir au plutôt possible, en m'adressant votre Reponse à Paris = chez M^{r}. Sanchez Doct^{r}. en Medecine Rue des Blancs-manteaux =

Comme Je vous ai ecrit assez largement dans ma derniere sur Les articles de la Votre, il ne me reste qu'à Vous assurer de nouveau, combien Je suis avec La Consideration la plus parfaite, & vraie amitié de Coeur

Mon cher ami
Votre très humble & ob^{t}. Serv^{r}.
J.H. de Magalhaens

74.05.16a *To Antoine-Laurent Lavoisier*

Archives de l'Académie des Sciences, Paris, Fond Lavoisier, 1642[1]

a Londres ce 16 May 1774

Monsieur

Le porteur de celle-ci est M. Wolfe, membre de la Societé Royale,[2] & un de mes intimes amis, que Je prends la liberté de Vous recommander, Com'un sujet très distingué par son savoir particulierem^{t}. dans la chymie, cette charmante science qui fait vos delices. Lui vous rendra de ma part la derniere publication qui vient de paroitre du D^{r}. Priestly sur les differentes especes d'air[3]

1 Corrected version of the text published in *Oeuvres de Lavoisier: Correspondance*, ed. René Fric, fascicule II, 1770-1775 (Paris, 1957), pp. 453-4.

2 The chemist Peter Woulfe, FRS (1727?-1803).

3 Joseph Priestley, *Experiments and Observations on Different Kinds of Air* (vol. 1, London, 1774).

& vous vous y trouverez cité deja par votre travail sur ce sujet, avec les egards qui vous sont dûs. 74.05.16a

Je suis bien faché de ne pouvoir encore partir avec cet ami pour Paris, comme je l'avois determiné; mais je ne tarderai qu'une semaine tout au plus: & je garde pour Notre entrevue de vous entretenir d'autres objets. en attendant, conservez moi vos amities & croyez moi pour la vie

Monsieur
Votre très humble & ob^{t}. serv^{r}.
Magalhaens.

à Monsieur
Monsr. de la Voisiere de l'Acad^{e}. R^{le}. des Sciences
Rue des Bons enfants
derriere le Palais Royal
à Paris

To Jacques-André Mallet[1] 74.06.03

Bibliothèque de Genève, Geneva, Ms suppl. 1654, fol. 29-30

Londres ce 3 Juin 1774

N.B. mon adresse à Paris est = Chez M^{r}. Sanchez doct^{r}. en Medecine: rue des Blancs-manteaux =

Je vous tiens parole, Mon Cher Monsr., car Je viens de remettre à Messrs. Liotard & Aubertin, une petite boite avec la lunete de poche, garnie du micrometre nouveau, que vous aviez ordonnée pour M^{r}. de sausseur.[2] Ce ne fut qu'avant hier, que J'ai pû l'avoir finie, & je vais partir dans 3 jours pour Paris, oû J'attendrai de vos nouvelles, avec l'information de ce que vous jugerez de ce petit instrument. Ce fut le premier qu'on a fait de la sorte, & il a falû avoir du recours à des methodes differentes pour reussir. La relation ou communication que le Professeur Wilson fit à la Societé Royale sur sa methode d'appliquer des fils aplatis, au lieu de fils ronds, dans les micrometres, vient d'etre imprimée dans la premiere partie du vol. des Transact^{s}.

1 Annotated "N^{O}. 6 Juin 1774".
2 i.e. H.B. de Saussure.

74.06.03 Philos^{es}., qu'on distribue hier aux Membres de la Societé R^{le}. pour cette annee 1774.[3] Mais aucun ouvrier n'a pas encore executé de ces micrometres ici: & je suspecte qu'ils en seront rebutés. Le detour que J'ai pris est different tout à fait de celui du Professeur: & il m'a reussi à merveille. La petite lunete est arrangée avec son apparatus de micrometre &c. dans une petite boite de mahogany; & j'envoie aussi la caisse noire dans la quelle on peut porter la lunete, toute seule, dans la poche, avec son petit piedestal &c. J'ajustai d'avance avec l'artiste que je ne lui donneroit plus de 6 ½ guinées pour le total: mais J'ai bien vû, qu'il y a perdû, etant obligé de refaire & recomencer de nouveau son ouvrage, par differentes manieres: & je l'ai satisfait de ma poche avec plaisir, par la satisfaction d'avoir fait reussir une machine que Messrs. Dollond & Ramsden regardent comm'impracticable (& peu vantageuse) à ce qu'ils pretendent par la dificulté qu'ils y trouvent: vous y verrez bien combien gagne-ton à employer de ces fils aplatis; car j'ai fait mettre le fil horaire (on soit horizontal) du plus fin que nos meilleurs Artistes employent dans les micrometres, pour faire juger de son epaisseur comparé avec les fils verticaux, qui semblent ne faire que le 5me. ou meme le 8me. partie du premier.

Messrs. Liotard & Co. me payerent $7^{l}.11^{s}.11^{d}$ pour votre compte, sur mon reçû, que je leur donnai. en voici le detail. dans le 22 de Fevrier je vous envoyai le Nautical almanac pour 1774 & 75. prix $7^{s}.0^{d}$ / plus une once de semence de choufleur 2^{s}. / 12 d^{o}. de carotes $1^{s}.6^{d}$ / 1 d^{o}. de navets jaunes 3^{d}.[4] / deux lettres reçues – $1^{s}.8^{d}$ / à present cette petite lunete, avec son le[5] micrometre $6^{£}:16^{s}:6^{d}$ / empaquetage, porteur & different messages que je n'ai pas pû faire moi meme $3^{s}.6^{d}$ / some totale (à moins d'erreur) £7:11:11.

Vous trouverez une espece de description de ce micrometre & son usage, dont je pouvois bien vous epargner la lecture, & à moi la peine de dire ce que vous n'avez pas besoin d'apprendre. mais vous excuserez cette bavardice & me ferez du plaisir en me disant votre sentiment là dessus.

On fit hier l'election des Membres etrangers de la Societé R^{le}.: il n'y avoit que 3 candidats; & tous les trois furent elûs, quoiqu'on refusa un pour membre domestic, comm'on les appelle. Entre les 3 elûs, fut un, M^{r}. le D^{r}. Poissonier que vous connoissez sans doute, Membre de l'Acad^{e}. R^{le}. des sciences de Paris.[6] Lorsque vous vous resoudrez etre notre candidat, je

---

3 Alexander Wilson, "An Improvement proposed in the Cross Wire of Telescopes", *Philosophical Transactions*, 64 (1774), 105-7. Wilson was Professor of Practical Astronomy at the University of Glasgow.

4 See Magellan to Mallet, 22 February 1774.

5 *sic.* The text goes from one line to the next at this point.

6 Pierre-Isaac Poissonnier (1720-1798), Medical Inspector-General for France's ports and colonies and a member of the Académie Royale des Sciences since 1765, was elected FRS

puis vous offrir une voix certaine; car je fus elû le mois dernier (Je veux dire d'Avril) com̄e Membre National (or domestic member) avec une pluralité de 62 voix contre 8 seulement de ceux qui tacherent de cabaler contre moi.[7]

On y lit hier une lettre fort curieuse du D^{r}. Franklin sur l'effet suprenant, que l'huile produit sur l'eau, pour en a[...][8] le mouvement, causé par le vent dans sa surface.[9] Peut-etre [...] vous deja vû ce phenomene dans votre lac. du moins vous pouvez [...] lorsqu'il est agité par les vents. Car en y jetant petit a[...] deux bouteilles d'huile, vous observerez qu'une grande [...] surface devient sensiblem^{t}. calme & tranquile &c. pour [...] du coté du vent. il semble que la repulsion de ces d[...] cause de ce phenomene. l'huile se repond avec une veloc[...] une fort grande etendue de la surface de l'eau, & y fait [...]leurs prismatiques. dans cet etat, l'action de l'air (ou le vent) [...] pas assez de prise sur l'eau pour laquelle il y a une attraction remarquable. discourrez sur cet objet avec M^{r}. de Saussure; & presentez lui mes respects.

M^{r}. Dollond ne me donne encore aucune esperance de pouvoir faire votre lunete dans aucun tems determiné, à cause de defaut ou total dissete qu'il a [...][10] qui soit bon pour cet ouvrage. il y a plusieurs autres qui [...]ssi des ordres pour d'autres lunetes de cette espece; & qu'il ne peut [...] sçait encore dire le tems, celà dependant de l'hazard d'obtenir du bon verre. [...] Je ne m'oublierai pas de mon coté, de le soliciter.

[J]'ai l'honneur d'etre
Monsieur
Votre très h^{ble}. & ob^{t}. Serv^{r}.
Magalhaens

A Monsieur
Monsr. Mallet, Professeur d'Astronomie
dans la Cour de S. Pierre
à Geneve
en Suisse

on 2 June 1774. Also elected that day as Foreign Members were Anton Georg Eckhardt (1740-1810) of The Hague and the Viennese mineralogist Ignace Born (1742-1791). The minutes do not identify the unsuccessful candidate for election on the home list.

7 Magellan was elected a Fellow of the Royal Society on 21 April 1774. The minutes of the meeting do not record votes cast for and against.

8 The edge of the page has been torn away and for several lines the final word or words on the line are missing.

9 Benjamin Franklin, "On the stilling of waves by means of oil", *Philosophical Transactions*, 64 (1774), 447-60.

10 Paper damaged. For several lines, the first word or words on each line are missing.

74.08.02 *To Benjamin Franklin*

American Philosophical Society Library, Philadelphia, Hays Calendar, I, 163[1]

Monsieur

Celui qui vous presentera cette lettre est M^r^. Ludwig,[2] Neveu du celebre M^r^. Ernest de Leipsick,[3] & fils de M^r^. Ludwig savant Medecin de la meme ville, qui J'ai l'honneur de vous recommender comme digne de votre connoissance. Il voyage pour s'instruire avec les savants etrangers, & pense de passer quelque tems à Londres. Le plaisir que vous prennez à recevoir les gens de merite comme M^r^. Ludwig, m'encourage à prendre cette Liberté dans la consideration que vous ne manquerez pas d'en etre content.

Je pense que mon retour en Angleterre sera vers la fin du courant mois: & ce sera alors que J'aurai la satisfaction de vous repeter, combien je suis avec toute consideration possible & vrai attachement de Coeur

Monsieur
votre tres humble & tres ob^t^. serviteur
Magalhaens

à Paris ce 2 Aout 1774

P.S. tous vos amis ici me prient de vous presenter leurs respects; & en particulier M^r^. Le Marquiz de Courtenvaux, M^r^. Le Roy,[4] & M^r^. de Bourg.[5] Ce dernier parle d'aller faire un tour en Angleterre ce Automne. Tous vos petites Commissions furent rendues & delivrés au plutot possible après que J'ai pû les avoir ici: ce qui ne fut point aussitôt comme mon arrivée à cause du grand retard du vaisseau, qui aportoit ma male.

A Monsieur
M^r^. Le D^r^. Franklin
Craven street near Charing Cross
in the Strand
London

1 Published in *The Papers of Benjamin Franklin*, vol. 21, pp. 270.

2 Christian Ludwig (1749-1784), German doctor, son of Christian Gottlieb Ludwig (1709-1773), also a doctor in Leipzig.

3 Presumably Johann August Ernesti (1707-1781), professor of theology at the University of Leipzig.

4 Jean Baptiste Le Roy, for many years Franklin's chief supporter among the French electrical investigators.

5 Barbeu Dubourg, editor of the two-volume French edition of Franklin's *Oeuvres* (Paris, 1773).

To Johann Castillon[1]

Si je ne supposois que votre Correspondant ici vous donnera les nouvelles courantes, je ne manquerois pas de vous en annoncer quelques-unes qui sont venues à ma connoissance; par exemple, celle du Mercure fulminant, que M. *Bayeu* a fait, en dissolvant le mercure dans l'esprit de nitre, le précipitant par l'alcali fixe, le desséchant, & y ajoûtant partie égale de souffre. Ce mêlange détonne furieusement, avec un certain degré de chaleur, à peu près comme l'or fulminant.[2]

On fait ici une lunette[3] de quatre pieds de diametre, que M. *Trudaine de Montigny* fait construire généreusement pour l'Académie. On pense que le tout coûtera au delà de 12 ou 15 mille francs. Elle est composée de deux verres concavo-convexes fort épais, & sera remplie d'esprit de vin: il y en aura une seconde pour réunir le foyer, qui sera de deux pieds de diametre, aussi remplie d'esprit de vin. Le foyer commun sera d'environ 8 pieds: & l'on espere que la face[4] de son foyer sera supérieure à tout ce qu'on connoit de ce genre. On croit pouvoir faire les premiers essais à la fin de ce mois.

M. *de Lavoisiere* avoit réduit en charbon les diamans exposés au foyer d'une grande Lunette massive de *Tschirnhausen*;[5] ou du moins il y avoit produit une croûte charbonneuse; & il avoit trouvé des émanations élastiques dans cette opération, comme celles qu'on appelle à présent *air fixe*. Il trouva aussi une quantité considérable de ces émanations dans la calcination des cailloux & du quartz. Vous savez que ce laborieux Académicien a donné dernierement un fort bon Traité sur cette matiere, qui fait un gros Volume *in 12*;[6]

1 Letter not found. The text is from an extract published in *Nouveaux mémoires de l'Académie Royale des Sciences et Belles-Lettres, Berlin* (1774; publ. 1776), Histoire pp. 29-31, under the heading, "Extrait d'une Lettre de M. Magellan à M. le Professeur de Castillon, de Paris le 18 Septembre, 1774". The Italian mathematician Jean François Mauro Melchior Salvemini de Castillon, or Castiglione (1709-1791), was professor of philosophy and mathematics at the University of Utrecht before becoming Director of the Mathematics Section of the Berlin Academy of Sciences.

2 Pierre Bayen (1725-1798) made numerous experiments on the precipitates of mercury at this period; see Partington, *A History of Chemistry*, III, 394-9.

3 In the Royal Society of London's copy of the publication, "lunette" has been crossed through and "lentille à brûler" written in, in an 18th-century hand.

4 In the Royal Society's copy, "face" has been altered to "force", in the same 18th-century hand.

5 Ehrenfried Walther von Tschirnhaus (1651-1708), physicist and geometer, built several large burning mirrors. One of these, a concave mirror of polished copper, 3 ft in diameter, was purchased by the Duke of Orleans and used by several generations of French chemists.

6 A.-L. Lavoisier, *Opuscules physiques et chimiques* (Paris, 1774). The work usually cited is an octavo edition.

74.09.18 & que le Docteur *Priestley* cite délà dans la collection qu'il a publiée de ses expériences sur cette matiere.[7]

M. le Duc *de Chaulnes* a fait arranger une carcasse de barres de fer pour monter en forme d'une espece de cabane sur la plaine, lors de quelque orage. Il est déterminé de s'y trouver & de faire descendre le tonnerre des nuées avec son cerf-volant qui m'a paru fort ingénieux. Je l'ai fait souvenir du sort du Professeur *Richmann* de Pétersbourg;[8] mais il ne paroit aucunement appréhender le danger. Il a fait construire une excellente machine électrique, qui lui coûte plus de 100 guinées, & qui donne des étincelles à plus de 9 ou 10 pouces de distance. La derniere fois que je le visitai au Château de Vincennes, il se préparoit pour un travail en grand sur le phosphore de l'urine, & me montra du sel microcosmique, dans une masse transparente, comme le plus beau crystal.

M. *Baumé* montre à l'Académie du mercure calciné, ou précipité *per se*, moyennant un bouillonnement continuel d'environ deux ans. Il s'est formé en petits crystaux rougeâtres; & même il y a de petites lentilles applaties de la même couleur, mais opaques. Il assure que le mercure étoit pur, sans aucun souffre.

Mr. *Bayeu* a fait voir que les marbres colorés de *Campan* dans les Pyrénées sont composés du vrai *schiste* (espece d'ardoise,) & de terre calcaire. Tout le monde savoit déjà l'existence de celle-ci dans le marbre; mais on ne supposoit pas que la matiere argilleuse du *schiste* entrât dans sa composition.

M. *Aublet* a découvert que le vrai *ambre-gris* n'est qu'une substance végétale, c'est à dire, le résidu résineux du suc laiteux d'un arbre de la Guyane, produit par la simple évaporation. Cet arbre s'appelle *Couma*: il donne une grande quantité de suc, lorsqu'il est entamé par une hache, ou par quelque autre accident. C'est sur les eaux des rivieres, après les grandes pluies, que ces amas de suc concret arrivent quelquefois par le courant des rivieres. M. *Aublet* fait imprimer actuellement son Ouvrage sur les Plantes de la Guyane: il y a 392 Planches.[9] La derniere est celle de cet arbre avec les fruits: mais l'Auteur n'a pas pu voir les fleurs pour déterminer la Classe du Systeme de *Linné*. Cet Ouvrage paroîtra en Décembre prochain. J'ai actuellement sous les yeux la derniere Planche: & toutes les autres sont entierement gravées. Jusqu'à présent on n'avoit que des conjectures sur l'origine de l'ambre gris.

---

7 J. Priestley, *Experiments and Observations on Different Kinds of Air*, vol. 1 (London, 1774).

8 The St Petersburg academician G.W. Richmann (1711-1753) died when the pointed conductor he had erected from the window of his room was struck by a bolt of lightning while he was taking a measurement from it.

9 J. Aublet, *Histoire des plantes de la Guiane Françoise...* (Paris, 1775).

To Joseph Banks[1] 74.10.08

British Library Add. 33977, f. 38[2]

Dear Sir,

I cannot help rejoicing that foreigners do justice to your merit and endeavour to acknowledge it in the most durable maner. Monsr. Aublet[3] put in my hands two new plants intended as a present to y^{e}. R^{l}. Society, which I have forewarded already along with other things of y^{e}. kind to D^{r}. Fothergill. One of them is called by your name Bankia:[4] and the Author gives along with the plant (dry'd), two descriptions viz. one in latin, of which I send here a copy; and another much more extensive in french. Altho' I am to set out to England in about a fortnight, thought proper however to forward to you this notice, as a mark of the interest I take in whatever thing regards your person, being with the utmost regard & steem,

Dear Sir
your most Obd^{t}. obliged Servt.
J.H. de Magellan

Paris 8 october 1774
at the Presid^{t}. Saron's Hotel, rue de l'Université.

P.S. M^{r}. Aublet's Natural History of Guyane will be published in december next.[5] He has many other new plants of y^{e}. Iland of France[6] &c. which he told me should give willingly to any curious Naturaliste, who may chuse to publish them, without any further expense, but that of y^{e}. drawings, which he requires to be made under his own inspection.

---

1 Marked "M^{r}. Magellhaen Octr. 20 –74".

2 Published in *Scientific Correspondence of Sir Joseph Banks, 1765-1820* (London, 2007), I, 67-8.

3 Jean Baptiste Christophe Fusée Aublet (1720-1778).

4 There is a later annotation, presumably by Daniel Solander: "this seems never to have been published. DS." It was certainly not published in the *Philosophical Transactions*, despite both the specimens and Aublet's descriptions having been presented as Aublet intended to the Royal Society at its meeting on 10 November 1774 (Royal Society, *Journal Book Copy*, XXVIII, 118-9).

5 Aublet, *Histoire des plantes de la Guiane Françoise…* (Paris, 1775).

6 Aublet had spent nine years, 1752-62, in the Île de France (Mauritius) in the service of the French East India Company, and had established a botanic garden there.

74.10.08 His direction is = dans La Rue du Paon près celle de St. Victor, Fauxbourg St. Marceau =

To Joseph Banks Esqr. F.R.S.
New Burlington Street
London

Bankia (Borbonensis)

Pentandria, monoginia

Cal. Perianthicum minimum, turbinatum, quinque deutatum.
Cor. Petala quinque, minima, calice longiore, viridia, ovata, concava intra vivisuras calicis disco infra germen inferta
Stam. Filamenta quinque, brevia, intra Petala visco inserta. Anthere ovatæ biloculares
Pist. Germen inferum, subrotundum, calicis fundo aductum, disco coronatum, stylus brevis, acutus, stigma carnolum rotundum.
Per. Orupa olivae formã.
Sem. nucleus rugosus, ovatus, acutus, bilocularis, singulum loculamentum ab una parte, suturis utrimque prominentibus, quandoque unum aboritur.

Bankiæ

Arbor trunbo cylindrico 40 aut sexaginta pedali ad basim costis seu radicibus quinque & amplius complanatis, latè supra turam protensis, suffulto ad immitatem ramosissimo, ramis oppositis, erectis, & declinatis, crassissimis, undique spasis, ramulis oppositis, foliosis, cylindriacis
Folia opposita, ovata, dentata, globra, rigida & quasi sessilia.
Folia arboris juvenis oblonga, lanceslata, glabra, crassa, leviter dentata, sessilia & venis rubris notata.
Flores panniculati, terminales & axilares. Ramuli panniculi oppositi ad basim, squemata muniti
Fructus similes olivæ.
Species huius generis fructu ovato aut oblongo ab incolis distinguitur è trunco & ramis huius arboris succus gommosus, limpidus, & flavus defluit.
Florebat Martio & Aprili.
Frutescebat mense junii & Julii.
habitat in Sylvis insulæ Franciæ
Nomen incolarum Bois d'olive; et Borbonense, Bois rouge.

From Jacques-André Mallet 74.11.04

Bibliothèque de Genève, Geneva, Ms suppl. 1654, fol. 51[1]

Geneve le 4 N^{bre}. 1774

J'ai reçu M^{r}. Il y a deja qlqtems les 2 paquets que vous m'avez fait le plaisir de m'addresser[2] & je n'aurois pas tardé si longt. à vous en donner avis, si je n'eusse toujours attendu de voir Mr Dessaussure p^{r}. lui demander comment il trouvoit sa lun. je n'ai point encore pu lui en parler avec un peu de detail, ne l'ayant vu qu'en courant. il m'a paru qu'il en etoit tres content, seulem^{t} il l'auroit souhaité un peu moins volumineuse, p^{r} la poche, mais cela etoit difficile. je l'ai examinée avant que de la lui remettre & je l'ai trouvée bien nette & tres bo[nn]e, ce fil applati me paroit avoir parfaitemt reussi, & je regarde cette invention co[mm]e tres utile car qd on prend des haut. de petite étoil. qui ont peu de lumiere, elles sont totalem^{t} cachées derriere les fils ordinaires & il peut y avoir une inexactit. de plus^{rs} second^{s}.

Je compte que vous etes actuellem^{t} de retour à Lond. & j'espere que vous me donerez bientot des nouvelles de notre lunette de 3½ pds. que nous sommes tres impatiens d'avoir. Je souhaiterois en même tems savoir le prix d'une machine parallatiq. en bois de la gr^{deur} de celle décrite dans l'astron. de M^{r}. de la L.[3] avec les deux cercles de metal divi, les niveaux &c. & toutes les autres choses necessaires, je suppose même que pour une lun. co[mm]e celle là qui peu pesante, il faut une vis p^{r}. pouvoir faire mouv. lentem^{t} la lunette, sans quoi il seroit à craindre qu'elle ne se mût par sauts cette machine serviroit de pied à notre lunette, mais je voudrois cepend^{t}. qu'on put l'en séparer, p^{r} y pouvoir appliq. d'autres lunettes, si on le souhaitoit. Marquez moi donc je vous prie, a quoi reviendroit la lun. sans piedestal avec son tuyau de cuivre, garnie du micromet. objectif. & du petit reticule dans un chassis mobile. le tout monté sur la mach. parall. precéd^{te}, Suiv^{t} le prix, nous pourrons peutetre faire faire cette nouv. machine. Dites moi aussi quelle augmentation de prix feroit un piedestal de bois le plus simple possible, pour y placer la lunette lorsqu'on l'oteroit de dessus la mach. parall.[4]

A M^{r}. Magalhaens
at N°. 12 Nevils Court, Fetter Lane.

---

1 Letter not found. The text is from a draft kept by Mallet.
2 See Magellan to Mallet, 3 June 1774.
3 Lalande, *Astronomie*, vol. 2 (2nd edition, Paris, 1771), pp. 793-9.
4 The draft is unsigned.

74.11.24 *To Antoine-Laurent Lavoisier*

Archives de l'Académie des Sciences, Paris, Fond Lavoisier, 1642[1]

Londres ce 24 Nov^{bre}. 1774
at n°. 12 Nevils Court Fetter Lane

Mon Cher Confrere

Je viens d'envoyer un panier avec quatre douzaines de notre biere appellée porter à bord du vaisseau l'Aimable Henriette, capit^{ne}. Nicolas Mignard, qui le debarquera à Rouen: & je vous envoye ici le Reçu du Capitain, en vous priant de charger quelqu'un dans cette ville là pour le recevoir, & vous l'envoyer par quelque barque en montant la riviere jusqu'à Paris. Lorsque celà sera arrivé, je vous prie d'envoyer une douzaine de ces bouteilles de porter à M^{r}. le Presid^{t}. de Saron rue de l'Université une autre douzaine a M^{r}. Duchenne:[2] & vous ferez des autres ce que vous voudrez car c'est pour vous en faire gouter de cette bierre porter que je l'envoye principalement. Je souhaiterois que M^{r}. Trudaine de Montigny y prend du gout pour cette bierre: car elle est fort corroborante & très saine: & je pourrois Lui envoyer d'ici autant qu'il voudroit, en cas que cette boisson s'accordat bien avec son estomac. Il ne manqueroit d'en sentir un tres bon effet. elle est très nourrissante, & n'echaufe point comme le vin le fait, à ceux qui ont la poitrine peu forte et pas robuste.

J'ai deja jetté des vues pour decouvrir la vraye methode de brasser cette biere; & un de mes bons amis m'a promis de me chercher cette information bien averée.

Mais J'ai voulû aussi vous faire gouter des vins de mon pais, dont on fait un grand usage ici à Londres. ainsi j'ai fait embarquer aussi dans le meme vaisseau sous la meme adresse de M^{r}. Trudaine, un autre panier cacheté, contenant 78 demi-bouteilles de vin Portugais. dont voici le detail. il y a 26 demi-bouteilles cachetées en noir, qui contient du vin blanc du Porto. vous garderez 8 de ces demi bouteilles: envoyerez 6 à mon vieux ami le D^{r}. Sanchez dans la Rue des blancs manteaux: & les 12 restantes sont moitié pour M^{r}. le Presid^{t}. de Saron: et l'autre moitié pour le Marquiz de S. Simon,[3] auquel je promis aussi de lui faire gouter

1 Corrected version of the text published in *Oeuvres de Lavoisier: Correspondance*, ed. René Fric, fascicule II, 1770-1775 (Paris, 1957), pp. 463-5.

2 i.e. Louis-Henri Duchesne.

3 Maximilien-Henri, Marquis de Saint Simon (1720-1799), French writer and translator.

de ce vin: & le meme Presid^{t}. de Saron les lui fera tenir. il y a aussi 26 demi-bouteilles de vin blanc de Carcavellos: qui seront distribuées de meme que les precedentes: celles ci sont marquées avec un cachet en cire verte. de plus il y a 13 demi bouteilles cachetées avec de la cire jaune qui contienent du vin blanc de Lisbone. vous en prendrez 5 de celles-ci, & envoyerez les 8 à M^{r}. de Presid^{t}. de Saron, moitié pour lui & moitié pour le Marquiz de S. Simon. Enfin il y a 13 cachetées avec de la cire rouge, qui contiennent du vin rouge du Porto tel que nous le buvons ici à Londres: & vous en disposerez de la meme maniere: 5 pour vous, & 8 pour Messrs. de Saron & de S. Simon. 74.11.24

Vous aurez deja vû dans le voyage du Cap^{tn}. Phips[4] un petit recit ou precis de la maniere employée par lui & M^{r}. Irwin[5] pour desaler l'eau de la mer: vous y observerez qu'il ne donne pas la figure de la vraie machine à distiller comme celle que Je vous ai communiquée. il paroit qu'on a voulû en cacher encore à la connoissance du Public la vrai construction de cet alambique, dont l'usage est d'une si grande consequence pour toute sorte de distilation en grande. Je vous prie de faire cette observation à M^{r}. Trudaine. J'apporterai avec moi une copie de ce voyage pour sa bibliotheque, & je vous montrerai ce detour adroit dans la figure qui y est gravée.

Je pense partir à la fin de la semaine prochaine, & de vous embrasser bientôt à Paris. mes Respects à M^{de}. votre Epouse, à Mlle., & à toute votre famille & amis & suis avec toute considération & amitié de Coeur

Mon cher Monsieur
Votre trés h^{ble}. & obeiss^{t}. serv^{r}.
Magellan.

N.B. 1°. J'ai fait attacher à chaque panier (à celui de la biere, & à l'autre avec du vin) l'adresse de M^{r}. Trudaine de Montigny; & non pas la votre, croiant que cela donneroit plus d'aisance à le retirer du Vaisseau vous là arrangerez celà avec M^{r}. Trudaine &c.

4 Constantine Phipps (later Baron Musgrave), *A Voyage towards the North Pole* (London, 1774) includes an appendix (pp. 205-21), "Account of Doctor Irving's Method of obtaining Fresh Water from the Sea by Distillation".

5 Dr Charles Irving, ship's surgeon on Phipps' ship *Racehorse*. In February 1773, well before Phipps sailed for the Arctic, Magellan had reported to Trudaine de Montigny recent English experiments with an apparatus for distilling seawater on board ship. At Trudaine's suggestion, the French Navy had commissioned Lavoisier to follow up the idea. See also Magellan to Lavoisier, 1 July 1775.

74.11.24 N.B. 2do. le Mr. Duchenne ci dessus nommé pour une douzaine de biere porter, est le secretaire en second de Mr. Trudaine, auquel je suis redevable de beaucoup d'egards & services depuis la mort de mon feu ami Mr. Jars.[6]

a Monsieur
Monsr. De la Voisiere. Fermier genal.
Rue neuve des bons Enfants
à Paris

6 Antoine Gabriel Jars (1732-1769).

74.11.28 *To Jacques-André Mallet*

Bibliothèque de Genève, Geneva, Ms suppl. 1654, fol. 31-32

Londres 28 Novbre. 1774
at no. 12 Nevils Court Fetter Lane

Je viens de recevoir, Monsieur, La Votre datee du 4 de ce mois,[1] avec un grand plaisir: & suis charmé que Mr. Dessaussure soit content de ce premier essai du micrometre à fils plats, qui etoit l'unique moien d'etre apliquer à des oculaires d'un foyer aussi petit. J'ai aporté une Lunette de 15 pouces de foyer à Paris, qui etoit fournie d'un micrometre pareil, & fut destinée pour l'usage de Mr. Messier dans son observatoir de la Marine, pour montrer, ce qu'on pourra faire avec cette espece de fils dans Les micrometres. Mon retour à Londres n'est que depuis le commencemt. de ce mois, pour quelques petits affaires pressants, mais il me faut encore partir dans peu de jours pour Paris, à fin d'achever quelques objets, qu'il m'a falu laisser incomplets. J'ai vu plusieurs fois Mr. Dollond & j'ai appris que Votre Lunette sera finie vers le commencemt. de Janvier: qui sera à peu près le tems de mon retour ici: soyez assuré que j'y preterai toute attention possible pour qu'elle soit digne de vous. il y a 3 ou meme 4 de mes amis

1 See Mallet to Magellan, 4 November 1774.

qui veuillent autant de Lunettes pareilles à la Votre: & si je ne trouve cette premiere assez nette pour le ciel, j'attendrai une des autres pour vous en donner la meilleur: car aucun d'eux n'etant pas Astronome de profession, peut bien se contenter pour des objets terrestres d'une Lunette qui soit moins parfaite. La justice & amour de l'Astronomie exige cette attention de ma part: mais il ne faut pas que vous le disiez: car on n'envisage toujours la justice que par egard à soi meme. 74.11.28

Le prix que M^r^. Dollond demande pour une Machine parallatique comme celle decrite par M^r^. de la Lande, est 40 guinées: ce qui reelem^t^. est trop. Mais il n'est proprement qu'un opticien: & tout le reste de l'ouvrage est fait par des ouvriers qu'il employe pour celà. Celle qu'il fit dernierem^t^. pour M^r^. Messier couta à peu près le meme. C'etoit une Lunete comme La Votre, avec un micrometre à fils: & fut paye 70 guinées. Un de mes amis qui a une excellente Lunete de cette espece à Paris, voyant ce prix trop grand, a pris le parti de le faire executer par un ouvrier François, qui croit pouvoir le faire par presque la moitié: quoique Je pense qu'il ne le fera jamais à ce marché là. Au reste J'ai ici des ouvriers que J'employe à la journée, qui sont fort capables d'executer, ce qu'on leur ordonne: & je pense faire faire une machine de ce genre, mais differente de celle de M^r^. de la Lande: qui aura plus d'avantages, & ne coutera pas tant. Celle de M^r^. Messier, qui est suivant le plan de celle de M^r^. Lalande, ne peut pas aller jusqu'au pole: & meme il y a quelques petits defauts, qui lui ont donné du degout. Je me suis presque querelé dans cette occasion là avec M^r^. Dollond: qui enfin me dit qu'il prennoit sur soi le soin de la faire aussi parfaite qu'il le pourroit: & cependant il y a des defauts, que M^r^. Messier a trouvés bien desagreables. C'est par cette raison, que Je ne compte plus de me tenir aux promesses de M^r^. Dollond: mais de faire executer ces choses à ma façon.

La Lunete avec le tuyau en Cuivre, comme J'ai ordonné La Votre, un oculaire terrestre, deux cellestes, un troisieme avec un croix de fil, avec un trouveur & verres noirs &c. coute 26 guinées, selon la reponse que le meme Dollond me fit à Paris pour un autre de mes amis, qui en vouloit une de cette espece. Le micrometre objectif fait une difference de 12 à 14 guinées en cas d'etre achromatique. Je vous prie d'y reflechir sur cette espece de micrometre: car M^r^. l'Abbé Rochon,[2] & d'autres personnes intelligentes soutiennent, que ces micrometres ne donnent pas une mesure constante, & dependent beaucoup de la disposition de la pupile de l'oeil.

2 Alexis Marie Rochon (1741-1817) was an authority on optical instruments, noted especially for developing a prismatic micrometer.

74.11.28 M[r]. Dollond charge environ 5 guinées pour le piedestal simple de Mahogany pour suporter La lunete: & s'il y a un axe long de metal avec le mouvement lent à vis pour l'horizont, & pour le sens vertical, je suis sur qu'il charge encore plus. Mais le cas est qu'il ne veut pas donner des details dans les prix de ses instruments: & je crois qu'il vaudra toujours mieux de ne pas prendre de Lui que ce qui est essentielem[t]. de Son resort, c'est à dire la Dioptrique achromatique.

Je n'ai pas entendu parler ici d'aucun Tellescope plus grande que 12 pieds de foyer. il appartient à M[r]. Short, frere de feu M[r]. Short celebre & savant Artiste dans ce genre.[3] Ce Short vivant y a fait quelques improvements assez plausibles. savoir il l'arrange en sorte que deux observateurs voyent à la fois le meme objet assez distinctement l'un regardant à la Newtonienne, & l'autre à la Gregorienne. Outre cela il a trouvé le moien de le faire grossir prodigieusement, mais pas tant comme il le dit: tout au plus 10.000 fois, & non pas 50.000. Il n'est pas vrai que l'objet soit egalement eclairé, comm'il le pretendoit. J'ai vû à Paris le fameux Telescope de D. Noel, avec le D[r]. Priestly une soirée ensemble avec le Duc de Chaulnes. M[r]. de la Lande en a donné les dimensions dans la Connoissance des Tems de l'année suivante. je crois qu'il a 22 ou plus de pieds de foyer.[4]

La Lune vûe par ce Tellescope est d'une netteté & detail prodigieux. mais en voyant Jupiter, je ne l'ai pas trouvé assez distinct; pas meme aussi bien qu'avec un autre bien plus petit. D. Noel dit qu'il n'etoit encore arrangé comm'il faloit pour voir Jupiter assez distinct. &c.

Vale me qua ama
tuum
Magellan

A Monsieur
Monsr. J.A. Mallet Astronome
dans La Cour de S. Pierre
à Genève
en Suisse

3 Thomas Short, younger brother of the famous telescope maker James Short (1710-1768).

4 *Connoissance des temps pour l'année commune 1775* (Paris, 1774), p. 338. As Lalande described it, the principal mirror of the telescope constructed by Dom Noël, 'Opticián du Roi au Château de la Meute [i.e. La Muette]', was 22 pouces in diameter and had a focal length of 24 pieds 4 pouces; observations of terrestrial objects indicated that the telescope enlarged things 437 times.

To Joannes Franciscus Thijsbaert 75.00.00

University Library, Université de Liège, Ms. 4130

Mon cher Monsieur

Directions pour obtenir les avantages
de la nouvelle invention sur les
quadrants marins.

Après avoir bien entendû l'usage de l'octant de Mr. Hadley (ou Quadrant marin) en suivant Les instructions du traité que Je vous envoye, composé par Mr. Ludlam,[1] vous n'avez qu'à substituer l'apparatus de mon invention, aux miroirs horizontaux, de l'octant commun que je vous envoye de même. mais pour eviter toute sorte de meprise, observez que La figure premiere, du traité de Ludlam, represente un octant ordinaire, avec l'addition d'une petite Lunete 01, qui est montée sur le pilier A: au lieu que l'octant que je vous envoye n'a pas de lunete, mais une pinule en A, & une autre en 02: la premiere sert pour observer par devant, & la seconde pour les observations par derriere. mais lorsqu'il y a une Lunete adaptée à cet instrument, pour lors on s'en sert de la même, dans un endroit ou dans l'autre. Je n'ai pas jugé de vous acheter un octant garni d'une Lunete, à cause d'avoir une que je vous envoie avec mon apparatus, qui peut etre appliquée sur ce quadrant dans l'endroit A. comme vous le verrez. Mais je ne vous conseille d'en faire aucun usage, qu'après de vous avoir bien exercé à tous les usages de cet octant, sans d'autre secour que l'oeil simple, & sans aucune Lunete. Car autrement vous vous trouverez très embarassé dans La practique: & ce n'est que peu à peu que vous deviendrez maitre de cet instrument. Je vous conseille aussi de vous exercer au commencement à prendre des angles terrestres c'est à dire des distances angulaires soit verticales ou horizontales. Car, pour prendre des distances des astres, & des hauteurs du soleil, ou d'autre planete &c. il faut deja avoir acquis un peu d'habitude. Mr. Ludlam parle bien au long de l'usage qu'on peut faire de cet instrument sur terre & je vous conseille de vous procurer un horizon artificiel, comme celui que l'auteur decrit à pag. 20. Ce que vous pouvez vous faire faire chez vous. Je n'ai pas taché de Vous envoyer un Octant des meilleurs, à cause du prix qui seroit du moins 8 ou 9 guinées, pour l'octant tout seul: & comme vous ne le voulez avoir que pour en expliquer La nature de cet admirable instrument, j'ai crû que celuici suffiroit.

1 William Ludlam, *Directions for the Use of Hadley's Quadrant ...* (London, 1771).

75.00.00 Je suppose donc que vous avez etudié votre octant, & connoissez toutes ses proprietés par le traité de Ludlam: à present il s'agit de connoitre les avantages de ma nouvelle invention. ôtez les deux miroirs horizontaux H1. & H2., avec tout l'atirail qui les tient par dessous, aussi bien que les verres rouges qui sont en L.2. (fig^{re}. 1re. du Traité de Ludlam), & la platine de cuivre sur laquelle ces deux miroirs, & les verres rouges sont mis. prennez l'apparatus, qui est dans la petite boete, representé par cette figure-ci: & mettez les deux vis, qui sont dessous, dans les deux trous, oû etoit le miroir H2 de l'octant (toujours la meme fig^{re}. 1re. de Ludlam), & la pinule 02.: de façon que la plaque qR (fig. 12 ici) soit presqu'en

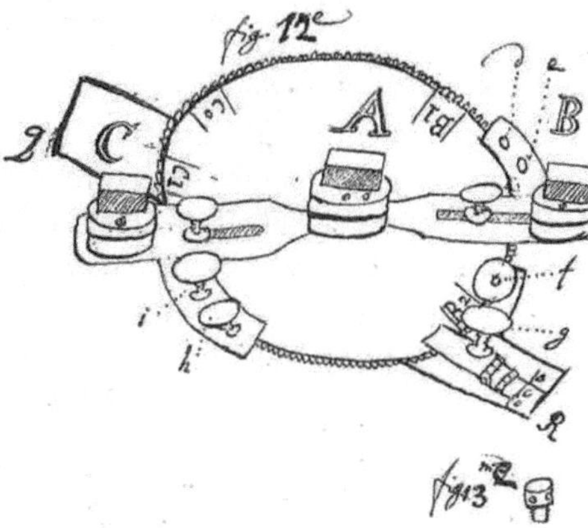

contact avec l'index MN, lorsqu'il est sur le 90°. du limbe; particulierem^{t}. tachez d'aprocher le bout R assez bien de l'index: & reserrez par les deux ecroux, ces deux vis assez fermes dans le corps de l'octant; & pour lors votre instrument sera prêt pour resoudre les Problemes suivants, qui n'ont eté jamias resolus par aucun autre instrument, qu'on sache jusqu'à present.

Avant de commencer avec les Problemes, il sera necessaire de vous prevenir que le plan, dans lequel se meuvent les trois miroirs ABC de cet apparatus, doit etre bien parallele au plan de l'instrument: ce que vous trouverez aisem^{t}. par la coincidence des images, l'une par vision directe, à travers de la partie diaphane d'un des miroirs; & l'autre par vision de reflexion sur la partie etamé du même miroir. Car si par example vous trouvez cette coincidence dans le miroir B, lorsqu'il est dans la position representée dans la figure 12 (ci-jointe) avec l'index à 90°.; & que dans la suite vous menez l'index à 0°.: & tournant le cercle dentelé, par le moien de la manivelle du pignon f, vous le faites retroceder 45°., c'est à dire, jusqu'à ce qu'il devienne parallele aussi au même miroir I (fig. 1re.) de l'index, pour avoir la coincidence des deux images: Je dis donc que si dans ce cas les deux images coincident parfaitem^{t}. bien, il est clair que le plan, oû se fait ce mouvement (le cercle dentelé de l'apparatus) est parfaitement parallele au plan de l'instrument, dans lequel se fait le mouvement de l'index. 75.00.00

N.B. en cas que par le mouvement du voyage il y ait quelqu'alteration: vous trouverez 3 ecrous, comme celui representé par la fig. 13me. qui servent à remettre ce parallelisme: & dans le bord de la boete, est un morceau de fil d'acier, qui sert à tourner la tête de ces ecrous qu'on appelle à caprestan. Vous ne devez pas supposer que les octants nouveaux sont aussi compliqués comme cet apparatus. J'en ai fait executer en differentes sortes, bien plus simples que celle-ci. Mais J'ai jugé de vous envoyer cet apparatus, pour generaliser cette methode: puisque par son moyen on peut donner toutes ces avantages à un octant construit dans la methode ordinaire, comme celui que je vous envoye. C'est ce meme apparatus, que J'ai presenté, & montré, l'annee derniere, à l'Academie Royale des Sciences de Paris: & je suis bien charmé de pouvoir vous l'envoyer (le meme identique) pour votre cabinet de l'université &c. Vous voiez bien à present pourquoi je vous disois dans une de mes lettres, que ceci ne feroit pas retarder l'envoi des autres instruments. Car il estoit tout fait depuis l'année derniere: & je n'ai fait que seulem^{t}. le faire nettoyer & polir par L'ouvrier &c. entrons en matiere.

Je viens de vous dire que la coincidence des deux images prouve le parallelisme du miroir horizontal H1 avec celui de l'index I. Je crois que vous en trouverez aisem^{t}. la demonstration par les loix de la catoptrique, aussi bien que de l'autre circonstance, qui fait La base du Quadrant de Hadley; savoir que l'angle formé par les deux miroirs (lorsqu'ils ne sont pas paralleles) n'est que la moitie de la distance angulaire des deux objects, l'un observé par des raions directs à travers de la part diaphane du miroir horizontal H1; & l'autre vû par la double reflexion sur le miroir I & la partie etamé du miroir horizontal. Mais si vous vous en doutez, jetez les yeux sur cette fig. 14. Soit eb le miroir de l'index: & tr celui qu'on appele

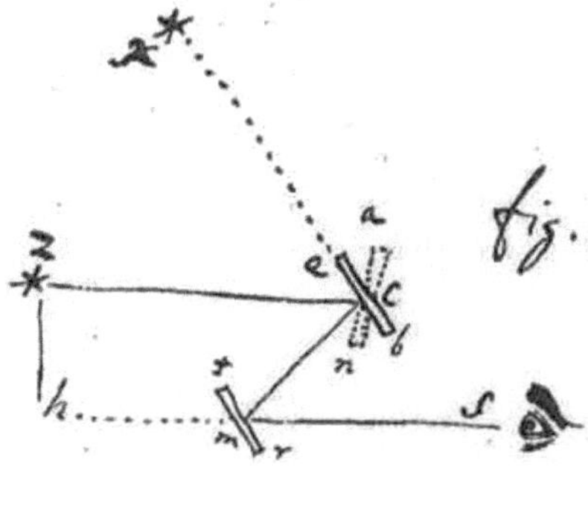

fig. 14

75.00.00 horizontal, qui est demi etamé & demi diaphane. il est evident que l'oeil voiant par la partie diaphane l'objet h doit voir dans le meme endroit l'objet z, lorsque le miroir eb est parallele à tr. car l'angle d'incidence doit etre toujours egal à celui de reflexion: ainsi l'angle externe ecz, doit aussi etre egale à bcm: & de la meme maniere les autres deux angles alternes rms & tmc. ainsi l'image de l'objet z sera reflechie de c à m, & de m à s: & coincidera avec l'image du raion direct h.s: & par consequence l'oeil s vera z sur h. Donc, si le parallelisme des deux miroirs donne la coincidence des deux images; il est evident que l'inverse de cette proposition est egalem^{t}. vraie: c'est à dire, que la coincidence des deux images, preuve le parallelisme des deux miroirs. N.B. lorsque la distance zc ou hs est infinie, c'est à dire que le raport de la ligne perpendiculaire entre zh n'est pas sensible, pour lors l'image reflechie du meme objet z tombera precisement sur l'image de vision directe du meme objet, sans le moindre paralaxe. V^{e}. Ludlam §. 187. 195.

A present supposons qu'un autre objet x doit etre observé, & qu'on veut avoir son image reflechie, qui tombe precisement sur celle de l'objet h. il est bien evident, que dans ce cas il faut mettre le miroir eb dans une telle situation, que l'angle incident soit egale à l'angle de reflexion. on voit que zcx est egale à m.c.b: ainsi il faut que la moitié du miroir eb avance la moitie en avant, tournant de b jusqu'à n: & que l'autre moitié recule autant; depuis e jusqu'à a: parce que c'est seulement dans cette position, que l'angle externe de la coincidence xca sera egale à l'autre angle externe de reflexion mcn: & alors le raion reflechi de c tombera, sous la meme inclinaision qu'auparavant, sur m; & de là entrera l'oeil dans la meme direction de h.s & par consequent l'objet x paroitra dans le meme endroit de h.

donc il est demontré que la distance angulaire de deux objets z (qui à une grande distance se confond avec h) & x c'est à dire que l'angle xcz est le double de l'angle de l'inclinaison des deux miroirs ncb (egale à ncm) &c. C'est à cette cause que les octants n'ont qu'un arc de 45°. pour mesurer 90°.: car chaque demi-degre du limbe vaut un degré entier dans l'observation. De même Les Sextants, n'aiant que seulem^{t}. 60°. dans le limbe, servent neanmoins pour observer jusqu'à 120°. &c^{a}. Celle-ci est la maniere la plus simple & aisée, que J'ai trouvée pour faire connoitre la theorie de ces instruments; aucun auteur que je sache ne l'a pas rendue aussi aisée. Voyez com̄ent le fameux Smith l'explique dans son Traité de l'Optique Chap. 12 du Livre 3me. Le Pere Pezenas a traduit cet ouvrage de l'Anglois en langue Francoise.[2]

2 Robert Smith, *Cours complet d'optique, traduite de l'anglois … avec des additions considerables par L. P. P. [i. e. Pezenas]*, 2 vols (Avignon, 1767).

C'est le meilleur ouvrage qu'on connoit sur cette matiere. allons à mes Problemes nouveaux.[3] 75.00.00

Probleme 1er. Mesurer, par l'observation de devant, des angles plus grands, que La Valeur de l'arc de l'instrument.

Solution. pour plus de commodité je vous ajoute ici les deux figures suivantes. la fig^{re}. 15me. montre un Octant

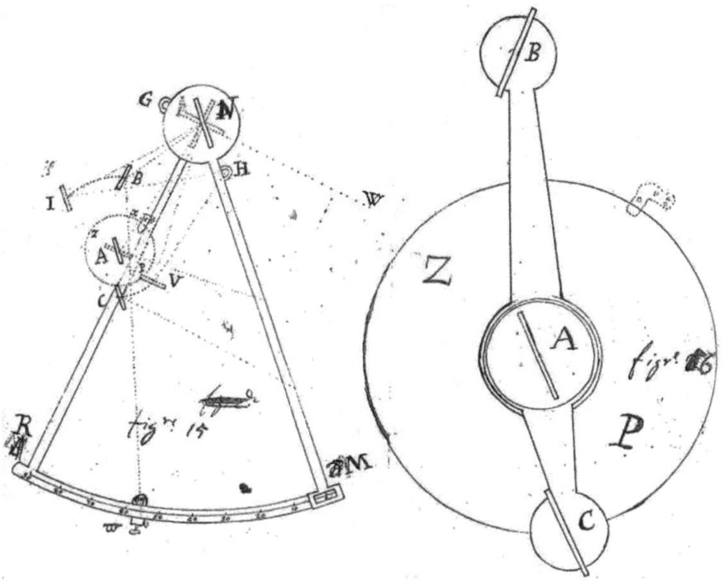

garni de mon apparatus, mais dont on voit seulem^{t}. les lignes qui sont necesaires pour La demonstration. La fig^{re}. 16 montre ce meme apparatus dans une echelle plus grande, sans aucun mechanisme pour les mouvements. Voici la solution du probleme. Menez l'index ou alidade NM. (fig. 15) à 90º.: faitez B parallele a N par la methode connue de faire coincider les deux images d'un meme objet (vz. le soleil, lune, etoile &c) l'une par vision directe, & l'autre par vision reflechie. remetez l'index à 0º.: faitez A & C paralleles à N. Celà fait tournez le Cercle ZP. sans toucher à la position respective de ces miroirs, jusqu'à ce que B arivant en I soit parallele à N. & pour lors le miroir

3 What follows summarizes Magellan's account in his *Description des octants et sextants anglois, ou quarts de cercle à reflection...* (Paris, 1775), pp. 5ff. This suggests that the letter was written at about the same time as this work was published, i.e. *ca.* 1775, and is the basis of our approximate dating of it.

75.00.00 C se trouvera en V, avancé de 45°. au delà de l'endroit oû il etoit auparavant. & par consequence il seroit parallele à N si l'alidade ou index auroit recule d'autant (45°.) & seroit dans la position NW. ainsi en observant sur le miroir qui est alors en V, on pourra mesurer tous les angles au dela de 90°.; tandis que ceux au desous de 90°. peuvent etre observés sur le miroir B. on voit dans cette figure la direction des raions de chaque miroir, marquée par des points jaunes. J'ai fait differentes experiences avec le meme apparatus, que je vous envoye, appliqué au même octant, & je trouve qu'on peut observer de la sorte des angles jusqu'à 162°. & meme plus, par l'observation de devant. Voici la practique de ce probleme. metez l'alidade au commencem^{t}. du limbe. C'est à dire à 0°. prennez la piece z ci marquée & mettez la dans

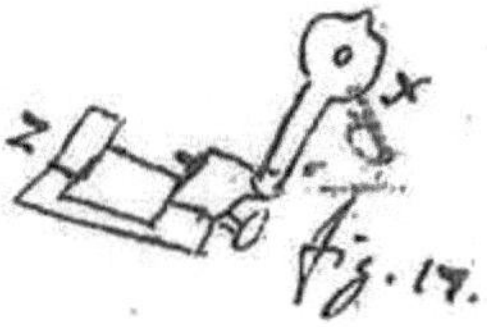

le bras NM vers le milieu, à l'endroit oû vous trouverez un A gravé sur le bois. alors si vous regardez sur le miroir A, & voiez quelqu'objet à travers de sa partie diaphane, vous verrez aussi le même objet dans la partie qui est etamée. mettez la piece z (fig. 17) un peu plus bas à l'endroit marqué C_1 sur le bras NM: & faitez autant sur le miroir C (fig. 15 & 12) que vous aurez soin de faire reculer, en tournant le bouton ou pignon h jusqu'à ce que le bras ou piece qui le porte est sur la marque gravée C_1 sur le cercle dentelé. Mais je m'oubliai de vous dire que pour faire coincider les deux images sur le miroir A, vous devez tourner d'un coté ou de l'autre le pignon f (fig. 12): & aussitot que vous trouvez cette coincidence, vous devez alors serrer la vis g. qui fait arreter le cercle. à present pour trouver la coincidence des images sur le miroir C, vous devez tourner seulement le pignon h: & lorsque la coincidence sera faite, vous devez serrer la vis I, qui fait attacher le bras qui porte ce miroir, au cercle commun. Celà fait, menez l'alidade ou index, à 90°. au bout du limbe vers R (fig. 15.) mettez la pinule z (fig. 17) dans l'endroit du limbe ou il est un B_1 gravé sur le bois: & pour lors faites la coincidence des deux images sur le miroir B qui doit tomber sur la gravure B_1 qui est gravée sur le cercle dentelé. le mouvement de ce miroir B, est executé par le pignon d (fig. 12): & lorsque vous aurez trouvé la coincidence des images, vous serrerez la vis e. Celà fait remettez l'alidade à 0°.: deserrez la vis g: & tournez le pignon f. jusqu'à ce qu'aiant mis la pinule z (fig. 17) à l'endroit H de la fig. 15 & regardant par elle (par son trou) vous trouvez la coincidence des images sur le miroir B: ce qui n'arrivera, jusqu'à ce qu'il ait reculé 45°. N.B. vous trouverez gravé sur le bois du bras NM fig. 15. un B_2 pour marquer l'endroit de la pinule pour cette derniere operation. Lorsque vous trouverez cette derniere coincidence, vous serrerez la vis g. & pour lors l'instrument est ajusté pour observer par devant, jusqu'à 90°. sur le miroir B qui pour lors se trouvera en I (fig. 15.) & depuis 90°. jusqu'au delà de 160°. sur le miroir C qui pour lors se trouvera en V. fig. 15. mais pour observer sur ce dernier il faut mettre la pinule un peu plus haut, sur la marque C_2. qui est au dessus du rond de l'endroit de la pinule ordinaire. 75.00.00

N.B. 2°. Vous trouverez dans cet apparatus une piece, comme celle marquée ci dessus dans la fig^e^. 17.: mais la pinule, est mise sur une autre piece plus petite, qui sert pour l'attacher au cercle dentelé, dans le cas du probleme 2^d^. A present que je vous ai expliqué, je crois que bien nettement tout le mechanisme de cette machine que j'appelle mon nouveau apparatus, je ne ferai que vous donner simplement la Solution des problemes suivants; sans rien dire de la methode de faire usage de la machine: car l'operation est toujours la même en tous les cas.

Probleme 2^d^. Ajuster le miroir A à angle droit avec le miroir N (fig. 15.), par un seul objet, & sans aucun recours à l'horizon: pour pouvoir faire les observations par derriere; c'est à dire, avec le dos tourné à un des objets.

N.B. vous avez deja vû par le traité de Ludlam, qu'on ne peut pas ajuster le second miroir pour observer par derriere, que par le moien de l'horizon de devant & celui de derriere. V^e^. o d°. Ludlam §. 79 & suivants: or il est impossible de faire cet ajustement dans la nuit, lors qu'on veut observer la distance de la lune à quelqu'etoile pour en deduire la longitude selon la nouvelle methode: & même de jour il y a bien des occasions, oû l'horizon est couvert avec du brouillard: & d'ailleurs on ne doit jamais faire des observations sans rectifier chaque fois l'ajustem^t^. de son instrument. Car il change & soufre des alterations par des accidents ce qui cause autant d'erreurs. &c.

Solution. repetez le probleme precedent: & le miroir A aura reculé deux fois 45°., qui est la position qu'il doit avoir pour faire des observations par derriere.

Un de nos artistes[4] fait une marque sur le limbe de l'instrument qui est à la distance de 90°. d'une autre. il attache une queue au miroir A & après l'avoir mis parallele à N tourne à la main cette queue: & pretend qu'il est ajusté. Cette methode qui est purem^t^. mecanique, ne vaut rien:

4 Peter Dollond.

75.00.00 car il est dificile de remettre la queue precisem^{t}. sur les points marqués auparavant. outre celà il faut que le centre du mouvement soit precisement concentrique à l'angle de 90°. qui y est marqué & l'on sçait combien celà est dificile de s'en assurer: & soit que ces marques soient alterées ou changées par quelqu'acident, ou qu'elles ne soient bien prises, tout doit contribuer à des erreurs dans La practique. Cependant malgré la superiorité & evidence de ma methode, il s'est t[r]ouvé un malhereux tracassier, vray coquin, qui prefere en public cette methode incertaine à la mienne, qui est indubitable. Mais la raison est qu'il a de l'interêt à preferer l'autre, etant bien payé pour celà par les profits dont L'ouvrier partage avec lui.[5]

N.B. lorsque vous voudrez resoudre ce probleme avec l'apparatus que je vous envoye, laissez rester serrée la vis g (fig. 12): menez l'alidade à 90°. desserez la vis e: & tournant le pignon d, faites avancer le bras qui porte le miroir B jusqu'à l'endroit marqué B_2 sur le cercle dentelé: ajustez la coincidence des images par le pignon d en regardant par la pinule que vous aurez mise en ω fig. 15: & après celà serrez la vis e. remettez l'alidade à 0°., deserrez la vis g, & tournant le pignon f, faites reculer le cercle entier de 45°., jusqu'à ce que vous trouvez la coincidence des images sur B: & alors serrez la vis g. Car alors aura le miroir A reculé deux fois 45°. Q.E.D.

Probleme 3. Reconnoitre, tant sur mer que sur terre la valeur precise de l'arc ou Limbe de l'instrument, sans employer qu'un seul instrument.

Solution. Repetez trois fois le probleme precedent: & pour lors Le miroir A aura parcourû tous les 360°. du cercle entier. Si l'arc est justement de 45°., le miroir A se trouvera au bout de la derniere operation parallele à N avec l'alidade à 0°.: mais s'il y a quelque difference, en la divisant par 8, le quotient montrera l'erreur qu'il y a, en plus ou en moins, dans l'arc total du Limbe.

Probleme 4. Reconnoitre la valeur precise de chaque 30°. 20°., ou autre nombre aliquot quelconque du Limbe de l'instrument.

Solution. Mettez les deux miroirs CB ou BA, à l'angle (fig. 15) que vous souhaitez examiner, selon la methode du Probleme premier. parcourrez le limbe entier d'un bout à l'autre avec cet angle, tournant alternativement l'index, & le cercle ZP & divisez la difference trouvée au bout de vos

5 Magellan's complaint is directed against Nevil Maskelyne, the Astronomer Royal, who reported to the Royal Society most unfavourably on Magellan's instrument; see Royal Society of London, *Journal Book Copy*, vol. XXVII (1771-4), pp. 333ff.

operation, en cas qu'il en ait quelque une, par le nombre des operations memes. Cette correction vous servira pour dresser une table, selon laquelle vous devez corriger vos observations. N.B. On voit par ce Probleme, qu'un Octant fort mal divisé, peut neanmoins servir à faire des observations correctes; pourvû qu'on employe avec adresse & pacience les attentions que cette nouvelle methode exige. 75.00.00

N.B. 2do. Ce Probleme & le precedent doivent preceder dans la practique les deux premiers, dont la certitude depend de l'exactitude de la grandeur & des divisions du Limbe.

Je crois que jusqu'à present on n'a jamais pensé à avoir des Instruments, qui pussent se corriger d'eux memes, devenant non $seulem^{t}$. capables d'anoncer leurs defauts, mais de servir à les corriger sans d'autre secours.

Probleme 5 Trouver les erreurs de division, ou ceux causés par quelqu'acident, dans les Quadrants astronomiques.

Solution. fig. 15. Ajoutez un verre semblable à N, au bout de l'alidade ou Lunete mobile du Quadrant, sur le centre de son mouvement: & un apparatus comme celui representé dans la fig. 16, au coté du Quadrant Astronomique. Examinez par le Probleme 3me. la valeur de son limbe, & par le 4me. celle de son divisions. & s'il en a des defauts, dressez en une table, qui servira dans la suite pour corriger vos observations &c.

Probleme 6. diviser un arc ou cercle donné.

Solution. Prennez un arc egal autant qu'il est possible au raion du cercle: & examinez par les problemes 5 & 3 sa precise valeur. dans la suite cherchez les subdivisions de cet arc par la methode du Probleme 4. Le premier Sextant etant divisé, passez par la meme methode, ses divisions aux Les autres sextants: & tout le cercle sera $entierem^{t}$. divisé.

Voilà je crois bien assez de proprietés dans cette invention pour La rendre estimable. Je ne crois pas que de mon vivant elle soit d'un [...]se[6] assez ettendu. mais je me contente d'avoir rendu ce Service à la posterité, sans aucun avantage personel.

Aussitôt que J'aurai un peu de loisir, je mettrai la derniere main à un petit traité ou Memoir sur cet objet, que je rendrai public: & je ne m'oublierai

6 Paper damaged, part of word missing.

75.00.00 pas de vous en envoyer une ou deux copies; car je m'y ettendrai bien plus sur les particularités de cette construction.

J'ai l'honneur d'etre pour la Vie

Mon cher Monsieur
Votre vray ami & très affectioné Serviteur
Magalhaens

à M^{r}. l'Abbé Thisbaert

Post scriptum

Je ne vous envoye le traité de Ludlam comm'un ouvrage parfait: tout au contraire je le trouve très prolixe sur des points qui ne valent rien; par example les methodes qu'il donne pour trouver l'index error, qui ne sont pas necessaires. Car on ajuste à chaque fois le miroir horizontal, avec l'index precisem^{t}. sur le 0°.: & alors il n'aura point d'index error. Mais sur des autres matieres & questions importantes, il passe trop legerment. quelques fois il radote, quelquefois il parle confusement, & mêle une chose avec l'autre. Comme par example lorsqu'il dit que les observations avec les octants ou Sextants de Hadley ne sont pas assez exactes (v^{e}. §. 183 &c.). Car lorsque celà arrive ce n'[est pas][7] la faute de l'instrument, mais de ce que les miroirs ne sont [pas] bons, ou que les sens de l'observateur ne peuvent raporter precisement dans la meme point visuel les deux objets qu'on compare &^{a}. Mais malgré tous ces defauts je crois que de tous les petits traités que je connois sur cette matiere, celuici de Ludlam est le plus clair, & plus à la portée des Lecteurs. à vous dire le vrai, il n'y a pas un seul traité assez bon & complet sur ce sujet, qui d'ailleurs merite bien d'en avoir un fait par quelqu'habile homme. vous trouverez dans la pinule de mon apparatus, deux vis qui servent à donner du mouvent à la Lunete lorsqu'on l'employe, à fin de la m[...][8] de façon que l'axe de vision soit parallele au plan de l'instrument. de même vous trouverez deux fils paralleles au foyer du tuyau des oculaires celestes, c'est à dire, du tuyau qui renverse [ces][9] objets. ils servent pour le meme effet. Voyez là dessus un Memoir de l'Astronome Royal, M^{r}. Maskelyne, qui est dans les Transactions philosophiques pour l'année 1772. pag. 99.[10]

7 Paper damaged, ends of two lines missing.
8 Paper damaged, part of word missing.
9 Conjectural reading, paper damaged.
10 N. Maskelyne, "Remarks on the Hadley's Quadrant, tending principally to Remove the Difficulties which have Attended the Use of the Back-Observation, and to Obviate the Errors that might Arise from a Want of Parallelism in the Two Surfaces of the Index-Glass", *Philosophical Transactions*, 62 (1772), 99-122.

To Nathaniel Pigott 75.01.02

Royal Astronomical Society, London, MSS Pigott, 34

Paris 2 Jany. 1775. at M^r^. le Presid^t^.
de Saron's, Rue de l'Université

Dear Sir

I received on my arrival here, but very few days ago, your favour of y^e. 12th. last month:[1] and am sorry for y^e. word mention'd to M^r. Needham on your account. There was at that time a Circonstance, in which I found inconveniency by y^e. scantiness of my purse, which I am now and then reduced to, by the pleasure of doing service to my many friends abroad. But being arrived here, have already received some part of what was owing to me: and need not make use of your offer till my arrival at Bruxelles, which hope may be about y^e. end of y^e. next month.

I had a talk with M^r. Messier, about your becoming a Corresponding Member of y^e. R^l. $Acad^y$. of sciences here: and think of no difficulty for carrying it into execution, provided you do qualify your self, by sending some <u>Memoir</u> or <u>Memoirs</u> to be presented to y^e. R^l. $Acad^y$., which always is previously required from any Candidate. No doubt but you have materials enough at hand: and it is no matter on <u>what subject</u> you do it. I should be very glad you may send something before I set out from hence: that I may have an opportunity of speaking to my friends, as accidentally, which you may easily conceive do better, than otherwise. But Let it be as it may, you can depend on my interest and friends. we (Messier & I) are of opinion that you must do better in datting from Louvain or other abode than Bruxelles, where y^e. R^l. $Acad^y$. has already two Correspondents, viz M^r. Needham & M^r. Chevalier.[2] otherwise there will be a great objection to have more in the same place. I don't know if at present is any vacant place for Correspondents. but will inform my self about it on Saturday next, which will be the first assembly after Christmas. you know that there is a certain number of Corresponding members, as well as in the other Classes of this $Acad^y$.: and if there are many Candidates for any vacancy, the more one makes himself known & y^e. sooner, so much the

1 Letter not found.

2 Needham had been named correspondent of Buffon in 1768, while Chevalier, Keeper of the Royal Library in Brussels but previously based in Lisbon, had been named correspondent of J.-N. Delisle in 1753.

75.01.02 better.[3] I'll not forget y^{e}. <u>Connoissance</u> des tems for y^{e}. current year (775) and if you Like it, will add the profond treatise of my good friend M^{r}. du Sejour on y^{e}. comets that could be apprehended to hurt y^{e}. earth: which is just now published.[4] it is about 4 french livres cost. in the mean time Sincerely & heartly am

Dear Sir
your most obd^{t}. h^{ble}. Servt.
Magellan

Pray my h^{ble}. resp^{s}. to M^{de}., and friends.

A Monsieur
Monsr. Pigott Gentilhomme Anglois
vis à vis La Cour
à Bruxelles

3 Pigott's candidature was successful, he being named correspondent of Messier on 12 June 1776.

4 Achille-Pierre Dionis du Séjour, *Essai sur les comètes en général, et particulièrement sur celles qui peuvent approcher de l'orbite de la terre* (Paris, 1775).

75.02.08 *From Jacques-André Mallet*

Bibliothèque de Genève, Geneva, Ms suppl. 1654, fol. 52[1]

8 Fev. 1775.

Après ce que vous m'avez marqué dans votre derniere lettre datée du 28 Nov^{re}.[2] touchant le micrometre objectif dont j'avois envie, & après avoir consulté dernierem^{t} Mr Messier, qui pense com̄e vous sur cet instrum^{t}, j'y renonce dans la crainte de faire une dépense inutile & assez considerable & je me contenterai d'un micrometre filaire ordinaire, mais com̄e j'ai deja un tel micrometre que Dollond m'a fait autrefois p^{r} ma lun. ach. de 10 p^{d} & qu'il semble assez inutile d'en avoir 2 dans un même observatoire, je desirerois de pouvoir l'ajuster à la nouvelle lunette, & cela ne me paroit

1 Letter not found. The text is from a draft kept by Mallet.

2 See Magellan to Mallet, 28 November 1774.

pas difficile, je n'ai qu'a vous envoyer les dimensions exactes du porte micrometre, vous les trouverez à la fin de ma lettre.[3] Si vous pensiez que cela n'est pas praticable, ou qu'au moins il seroit necessaire que l'ouvrier eut ce porte micrometre, je vous prierois de me l'écrire tout de suite, peutetre aurai-je à la fin du mois une occasion p[r] vous le porter à Paris où je crois que vous serez encore, mais je preferer ois bien qu'il ne fut pas besoin de vous l'envoyer, parce que je ne pourrois faire aucun usage de mon micrometre, jusqu'à ce qu'il fut revenu. 75.02.08

C[om]bien couteroit l'addit. d'un tel micrometre

Je renonce aussi à la machine parallatiq. de Dollond, dont le prix est exhorbit. je crois que je la ferai faire à Paris par Mr Meniers[4] que M[r] De la L. me recomande bcoup, & qui en a deja fait une p[r] M[r] le Pres. de Saron, on me marque qu'il y a fait plus[s] changem utiles. je lui écrirai de vous voir avant votre depart, parce qu'il n'aura point la lunette p[r] l'ajuster à la machine, & qu'il faut par conseq[t] qu'il en sache bien toutes les dimensions.

Il me faut necessairem[t] un pied à la lun. indépendam[t] de la mach. parall. parce que je suis b. aise que cette mach. puisse aussi servir p[r] d'autre lunett. si l'occasion s'en presente, ainsi je vous prie M[r]. de faire faire un pied. de bois le plus simple possible, afin qu'il coute peu, il suffit qu'il soit solide & qu'on puisse donner facilem[t] à la lun. tous les mouvemens necessaires. J'espere que desque vous ferez de retour à Londres, vous pourrez me donner de bonnes nouvelles de cette lunette, que nous attendons avec b. de l'impatience, mais cepend[t] tres resignés à attendre un peu plus longtems pour l'avoir plus parfaite.

Je souhaiterois que vous y joignissez une de ces lunettes de nuit qui servent à chercher les cometes, je crois qu'elles ont communem[t] 1 pied de long, ne grossissent pas beaucoup, mais ont une ouverture & un champ assez large. j'en ai une achr. de 3 pied qui est d'une clarté etonnante, & qui peut servir à cet usage mais le champ n'est pas assez grand.[5]

A Mr Magalhaens à Paris
Membre de la S.R. de L.
& Corresp de celle de Paris

---

3 The draft does not include the dimensions promised.

4 The instrument maker Pierre-Bernard Mégnié ("Mégnié le Jeune") (c.1751-1807)?

5 The draft is unsigned.

75.02.22 *To Samuel More*[1]

Royal Society of Arts, London, RSA/AD/MA/100/10/436

à Paris le 22 Fevrier 1775

Dear friend

Here I send you the proposal of a new member for our Society, which I have signed with M^{r}. Messier & M^{r}. Gaucher:[2] he is proposed for a domestic or paying member: and his entry-money will be paid after his election or when I arrive there. I know the idea of these gentlemen is only for the honour of having their names in our List: but think that the entry viz the two guineas pay enough for the printing of their names. Besides he proposes to send some prints of his engraving. those he has shown to me are not despicable. M^{r}. Gaucher intends to send more of his own to y^{e}. Society, and perhaps his own bust, which he has by himself very well done.

If you think it necessary, make another proposal and some of our friends will no doubt sign it on your application. When M^{r}. Godefroy is elected by y^{e}. Society, please to direct your Letter of notice by y^{e}. post as soon as possible: for he wants to add the qualification = de la Societé des Arts d'Angleterre = to a large print now finishing.[3] his direction is = to M^{r}. Godefroy, Graveur, rue des Francs bourgeois, porte S^{t}. Michel, à Paris = receive the thanks of M^{r}. Gaucher & mine for your souvenir and present of y^{e}. small print, untill he writes himself, which he said, will do on my going back to London. I hope it will be before the end of y^{e}. next month. my complim^{ts}. to M^{rs}. More and friends. am sorry for y^{e}. sickness of M^{r}. Arbuthnot.

---

1 Annotated "M^{r}. Magellan on M^{r}. Godefroy for a Member". Samuel More (1726-1799) served as Secretary of the Society of Arts from 1770 until his death in 1799; see George Mercer, "Mr More of the Adelphi", *Journal of the Royal Society of Arts*, 127 (1979), 96-103, 173-9, 237-44.

2 The engraver Charles Etienne Gaucher (1740-1804) was first a pupil of Basan and afterwards of J. P. Le Bas. He himself had been proposed for membership of the Society of Arts by Magellan and elected on 27 February 1771; however, he only ever paid one subscription, so that by 1775 his ability to propose new members must have been in question.

3 Francis Godefroy (1748-1819) was elected a member of the Society on 15 March 1775. According to the Minute Book, he was proposed by Magellan and seconded by Dr Johnson and Mr Songa; it would thus appear that More had indeed found it necessary to "make another proposal". Like Gaucher, Godefroy only ever paid one subscription.

If any thing [...][4] be here of your service, don't deny me the pleasure of shewing how much Sincerily 75.02.22

I am
Dear Sir
Your most obed[t]. hl[e]. Serv[t].
Magellan

Pray mend my name in the new list of Members. Since no body spels it right: let it be so as in the list of y[e]. Royal Society. viz.
= Magellan, John Hyacinth de ........F.R.S. &c.

To
M[r]. More, Secretary
to y[e]. Society for y[e]. Encouragement
of Arts, &c.
in the Adelphi London

---

4 Paper damaged, word illegible.

To Nathaniel Pigott 75.02.26

Royal Astronomical Society, London, MSS Pigott, 35

Paris 26 Feby. 1775 at y[e]. President
de Saron's, Rue de l'Université

Dear Sir

I received yours of y[e]. 16 instant[1] yesterday in y[e]. assembly of y[e]. R[l]. Academy, by y[e]. hands of M[r]. Messier. We both bespoke M[r]. Le Monnier to propose you as a Corresponding Member, which he is very willing to do. but you must send us at least your Observations now with these of y[e]. Saturn's occultation, and some others of y[e]. most interesting you have

---

1 Letter not found.

75.02.26 done to ascertain the positions of y^{e}. Towns in Flanders &c.[2] please to send a good account of y^{e}. instrum^{ts}. there employ'd, and y^{e}. circomstances thereof &c. as soon as you do this, you'll be put into y^{e}. List of Candidates: and in the next election I have no doubt but we will succeed, tho' there are some others waiting for y^{e} same correspondence.[3] Please to direct your observations to M^{r}. Messier as soon as possible: and put over them a second cover directed to M^{r}. Trudaine de Montigny Conseiller d'Etat & Intend^{t}. general des Finances à Paris = and do the same with your letter or Letters to me whilst I am here beg y^{e}. favour of y^{e}. inclosed paper to Professor Thisbaert.[4] tell him that I intend to write to him very soon about y^{e}. queries of his Letter as soon as I can get them answer'd: and add also the method of sending me his Letters during my stay here: which I hope will be but 3 weeks or there abouts.

M^{r}. Messier will send you y^{e}. corresponding observations you desire from him. but you must send first your corresponding ones to those you desire: and let this be your rule for y^{e}. time to come, it being desagreable to any observer the sending his own observations without having in his hands those for which they are required. but to be more explicit in a matter in which I am not concerned, and of course have no personal interest, suppose that I have a set of observations. if they are good so that I am confident of their exactness, I should not be pleased to send them that any other may be able to correct his own by mine, without my knowledge; which has been often the case with Master Maskelyne,[5] as I have been told by the Late M^{r}. Short,[6] who knew him better than many others. and if I am doubtfull about the goodness of my own observations; when I see the others, I can give some excuse for y^{e}. difference; or not send them at all &c Surely you are not like Maskelyne: but it is better to follow a general rule with every one.

our Phil^{rs}. are very well with y^{e}. Newtonian Laws of Gravitation and have Laught at all Bertier's David's &c. heavy heads. I hear that y^{e}. Piemont

---

2 Pigott evidently did as Magellan urged because Pierre-Charles Le Monnier (1715-1799), *pensionnaire astronome* at the Académie Royale des Sciences in Paris, presented his observations to the Académie on 22 March 1775. See also Magellan to Pigott, 31 March 1775. On Pigott's astronomical survey of Flanders, see A. McConnell and A. Brech, "Nathaniel and Edward Pigott, Itinerant Astronomers", *Notes & Records of the Royal Society*, 53 (1999), 305-18.

3 Pigott was named correspondent of Messier on 12 June 1776.

4 The paper included in this edition as 75.00.00?

5 Rev. Nevil Maskelyne, Astronomer Royal.

6 James Short, FRS (1710-1768), the well known telescope-maker and astronomer.

or Turin academy has undertaken to try the controverted experiments upon y^e^. Alpes; but know nothing of y^e^. success.[7] 75.02.26

I am with y^e^. great steem of a heartly friend

Dear Sir
your most Ob^t^. Ser^t^.
Magellan.

N.B. have you received la Connoissance des Tems for 1775? & y^e^. Book of M^r^. du Sejour?[8] do tell me if you have? or not?

A Monsieur
Monsr. Pigott, Gentil-homme Anglois
à Louvain
Flandres

---

7 Joseph-Etienne Bertier (1702-1783) and Jean-Pierre David (1737-1784) were among those who took up most enthusiastically reports published in 1769 and 1771 over the names respectively of Coultaud and Mercier, that purported to controvert Newton's gravitational theory by showing, by means of pendulum experiments performed in the French Alps, that gravity actually *increased* with altitude – hence Magellan's reference to "heavy heads". In 1773, the experiments were shown by Georges-Louis Le Sage to be fraudulent. Bertier, however, persisted with experimental claims of his own before finally retracting them in 1777. See James Evans, "Fraud and Illusion in the Anti-Newtonian Rear Guard: The Coultaud-Mercier Affair and Bertier's Experiments, 1767-1777", *Isis*, 87 (1996), 74-107.

8 Achille-Pierre Dionis du Séjour, *Essai sur les comètes en général…* (Paris, 1775).

To Jacques-André Mallet[1] 75.03.05

Bibliothèque de Genève, Geneva, Ms suppl. 1654, fol. 33-4

Paris ce 5 Mars 1775
Chez M^r^. le Presid^t^. de Saron
Rue de l'université

Monsieur

Comm'il seroit desnecessaire de vous envoier ici votre micrometre j'ai differé de repondre à la votre du 8 Fevrier,[2] jusqu'à ce que je verrois la construction que M^r^. Messier vouloit donner à la machine parallatique pour

---

1 Annotated: "10 mars 1775".

2 See Mallet to Magellan, 8 February 1775.

75.03.05 Mr. le Presidt.[3] car il est faux que cette machine soit faite. C'est seulemt. un petit modele que l'ouvrier a fait en bois. J'ai discouru avec lui, & je pense qu'il la fera bien; car j'ai vû de son ouvrage dont je suis content. Il vient de sortir de chez moi dans la minute pour me faire voir le different modelle, qui reponderoit aux idees dont je lui parlai il y a quelques jours. C'est que je voudrois qu'il y eut quelque mechanisme pour ajuster l'axe de la machine à l'axe du monde. Car il est malheureux de ne pas etre maitre d'arranger son insrument à la fantasie, & d'en pouvoir corriger les ecarts sans appeller un ouvrier. Je suis satisfait de ce dernier arrangement.

Soit ab la lunette qui est dans un chassis m. Ce chassis est au bout de l'axe du mouvement vertical de la lunete, qui est l'axe cd: il porte au bout c le cercle des declinaisons: & je lui ai dit la precaution necessaire qu'il faudra prendre pour qu'il tiene bien ferme à cet axe: hz est l'axe de l'Equateur qui porte l'index zt sur le demicercle immobile xtw pour les ascensions droites. Le chassis m aura des vis de rapel, pour remettre la lunete bien perpendiculaire (l'axe optique de la lunette) à l'axe de mouvement cd: or la piece en qui tient ferme à l'axe polaire doit avoir aussi une coulisse pour que cet axe td soit perpendiculaire à l'axe hz: quoiqu'il pourroit bien etre ajusté comm'il faut par l'artiste. Mais outre celà il faut avoir aussi le pouvoir d'adapter l'axe hz à la hauteur du pole en cas de changement de latitude, ou de derrangement de l'instrument. ainsi il faut que le bout z soit en etat d'etre hausé ou rabaissé à volonté par la vis s: & pour y reussir il faut aussi que le colet n puisse tourner sur les points des deux grosses vis fr. Vous voyez par cette esquise que votre instrument doit avoir toutes les qualites requises. il pourra etre mis parallele à son axe pour observer la marche des astres les plus près du Pole car on pourra le tourner dans tout le sens dans cette position, excepté la portion ocupée par le soutien du piedestal; & on pourra le mettre sens-dessous-dessus pour le verifier &c.

A present il s'agit que vous prenniez votre resolution là dessus. dites moi si vous voulez que je fasse venir ici votre lunette, en comptant sur l'honneur de Mr. Dollond. Car je ne l'ai point vue, & je ne pourrai pas l'examiner avant le milieu du mois prochain. L'ouvrier sans doute fera beaucoup mieux le piedestal en l'aiant ici, pour ajuster le chassis m qui doit la porter. &c.

Pour ce qui regarde le micrometre vous ne pouvez pas eviter de faire ajuster chez vous un bout de tuyau qui enchassera proprement dans celui de la Lunette. il vaut mieux de ne pas vous embarasser de celà pour le present. C'est une bagatelle, dont il ne faut charger Dollond, qui n'est lui meme

---

3 i. e. Bochart de Saron.

plus qu'un bon opticien. tout le reste est par les mains de ses ouvriers, qui quelques fois font de la besongne sans propos &c. que vous aiez une bonne Lunette est & doit etre toute la demande à esperer de Mr. Dollond. si vous attendez jusqu'à mon arrivée à Londres vous aurez un pied de Lunette pour la votre & je vous envoyerai aussi la Lunette de nuit. Seulement dittez moi le prix. les meilleurs de Dollond coutent, si je ne me trompe pas 3 guinées. mais il en a d'un moindre prix. repondez moi s.v.p. sous une 2de. envelope à Mr. Trudaine de Montigny, Conseiller d'etat & Intendt. general des finances à Paris: oû je serai jusqu'au bout de ce mois. 75.03.05

Je suis
Monsieur
Votre t. hle. serr.
Magellan

A Monsieur
Mr. J.A. Mallet, Astronome
Cour de S. Pierre
à Geneve
Suisse

75.03.15 *From Jacques-André Mallet*
Bibliothèque de Genève, Geneva, Ms suppl. 1654, fol. 53[1]

Geneve 15 Mars 1775.

J'ai reçu Monsieur votre lett. du 5 de ce mois[2] avec la descript. de la mach. parallatiq. dont je suis très satisfait, il y aura je pense des niveaux p^r la pouvoir placer bien horizontale, & une vis de rapelle p^r pouvoir faire tourner lentem^t l'axe polaire. car autrem^t je craindrois qu'il fut difficile de faire suivre à la lunette le parallele d'un astre. Je souhaiterois donc que M^r. Meniers[3] se mit tout de suite à cet ouvrage, & je vous prie de lui faire servir mon intention.

Pour ce qui regarde la lunette, quoique j'aye grande idée de l'habilité & de l'honneur de Mr Dollond, je crois cepend^t qu'il est plus sur dattendre votre retour à Londres qui nest pas eloigné d'ailleurs je serois b. aise de recevoir la lunette avec son pied en même tems pour la remettre ainsi toute ajustée à son proprietaire & il me paroit qu'il ne sera pas difficile à l'ouvrier de Paris de faire le chassis sans avoir la lunette, surtout lorsque vous pouvez lui en donner exactem^t les dimensions, ou les lui envoyer à votre arrivée à Londres, je vous prie de ne pas oublier que le pied que vous ferez faire à Lond. doit etre fort simple.

Quant à la lunette de nuit, j'en ai une de 3 pieds avec le tuy. de mahogany qui me coute 4 guinées, elle est excellente, mais j'en trouve le champ trop petit, il n'est je crois que d'env. 1½ deg.. Bernoulli parle dans ses lettres astron. p. 156. d'une lun. de nuit d'un pied qu'a Mr Messier, qui coute 1 guinée chez Dollond,[4] c'est de cette espece que j'en désirois une, mais si en y mettant de plus on l'a beaucoup meilleure, à la bo[nn]e heure, je souhaite seulem^t de la netteté & un champ large, il suffit que le tuy. soit de bois.[5]

A Mr Magellan à Paris

---

1 Letter not found. The text is from a draft kept by Mallet.
2 See Magellan to Mallet, 5 March 1775.
3 Pierre-Bernard Mégnié? See Mallet to Magellan, 8 February 1775.
4 Johann III Bernoulli, *Lettres astronomiques* (Berlin, 1771), p. 156.
5 The draft is unsigned.

To Jacques-André Mallet[1]

75.03.31

Bibliothèque de Genève, Geneva, Ms suppl. 1654, fol. 35

Je viens de recevoir la reponse de M^{r}. Dollond de Londres du 24. Mars dans la quelle il me dit que la Lunette pour M^{r} l'Astronome que Je lui avois ordonnée avant tout autre (car il a à present 5 de plus à faire de ce genre pour mes autres amis) etoit finie. elle a un tube d'oculaires qui grossit d'environ 40 fois pour les ocultations des Etoiles, un autre qui grossit environ 85 fois, garni d'une croix filaire: & deux autres encore pour le ciel, sans aucun fil, dont l'une grossit 85, & l'autre 150 fois, outre le tube des oculaires terrestres. le tuyau est en cuivre, comme de celle de Mr Messier: & si cette Lunette est assez bonne pour les observations de Astronomie, comme je l'espere, je conseil très fort M^{r}. Mallet de ne pas la ceder à personne: car M^{r}. Dollond me dit ouvertement qu'il ne peut pas faire aucune pour le present, manquant entierement de bon verre (flint glass): de façon que je pense de renvoier l'argent qu'on m'avoit remis deja pour quelques unes des 5 lunettes ci dessus de mes autres amis, à moins qu'il ne veuillent pas attendre la destinée de l'hazard, puisque ce n'est qu'au hazard qu'on rencontre un bon morceau, peut etre en cent livres Esterlines d'emplatte de ce meme flint-glass. peut etre c'est à cette cause qu'il (Dollond) vient de hausser les prix de ces Lunettes. Car il m'envoye dans cette meme Lettre une liste imprimé des prix de ses ouvrages, & j'y vois qu'il met à 28$^{£}$:7^{s}:0^{d} les memes qu'il vendoit jusqu'à present à 25 guinées. Le prix qu'il me marque pour celle que j'ai destinée à M^{r}. Mallet, decrite ci dessus est de vingt huit guinées. Je pense partir dans la quinzaine d'ici (ou environ) & j'aurai le tems de recevoir reponse plus claire de M^{r}. Mallet sur son piedestal. l'addition pour ajuster l'axe, je dis l'inclinaison de l'axe polaire suivant la latitude à quelques degres de difference va à 5 louis d'or de plus. je voudrois savoir s'il le veut? ou non? sans cette qualité, la machine parallactique ne coutera que 25 louis d'or. M^{r}. Megnier[2] n'est pas encore retourné de la campagne. j'y ai passé deux fois depuis que J'ai reçu la lettre de M^{r}. Mallet: & ne suis sorti de Paris qu'hier à 8 du soir, pour venir coucher ici. j'y retournerai ce soir meme, mais il me faut encore une reponse positive de M^{r}. Mallet sur ce point. aussitôt que j'arriverai à Londres je ferai faire le piedestal qu'il souhaite tout simple pour la Lunete; à moins qu'il ne

1 Annotated: "De Paris du 31 Mars 1775. Arrivée dans une lettre à M^{r} dessausure"; headed in Magellan's hand, "Voici l'Article pour M^{r}. Mallet." Letter to Saussure not found.

2 Pierre-Bernard Mégnié (c. 1751-1807)? See also Mallet to Magellan, 8 February 1775.

75.03.31 change d'opinion. Car s'il la garde pour son usage comme je le lui conseil, il n'en aura pas besoin. Son piedestal parallatique aura des niveaux, & une vis de mouvement lent pour suivre le cours des astres. Par le catalogue de Dollond les lunettes de nuit en mahogany d'un pied coutent 27 shillings. celles de [2] pieds[3] coutent 2 guinées & demie. Le Nautical Almanak pour l'annee prochaine ne paroitra que le mois suivant, suivant l'avis que j'ai reçu de londres. Je demande deux mots de reponse sous couvercle ou 2de. envelope à M^r^. Trudaine de Montigny Conseill^r^. d'Etat & Intend^t^ General des Finances, tandis que Je suis à Paris.

Votre de coeur
Magellan

3 The paper has been torn away at the seal and the missing words (or their general sense) written in around the tear, presumably by Mallet. The number written in here remains difficult to read, but see Mallet to Magellan, 7 April 1775.

75.03.31a *To Nathaniel Pigott*

Royal Astronomical Society, London, MSS Pigott, 36

Paris ce 31 Mars 1775

Dear Sir

I have receiv'd the favour of your Letter datted y^e^. 13th. Instant[1] with your Observations, which were presented to y^e^. R^l^. Acad^y^. by M^r^. Lemonnier,[2] and your name added to y^e^. candidates for the Correspondence. I am sorry you have not send them sooner. Your election should come whilst I am here. think however that it cannot fail of success.[3] M^r^. Messier and

1 Letter not found.

2 Pierre-Charles Le Monnier (1715-1799) presented Pigott's observations to the Académie Royale des Sciences on 22 March 1775. Pigott later presented his observations to the Royal Society, where they were read at the meeting on 9 December 1775 and afterwards published in the *Philosophical Transactions*, 66 (1776), 182-95. A further paper, including new determinations supplied by Messier and Johann III Bernoulli of the meridians of Paris and Berlin respectively, appeared in *Mémoires de l'Académie Impériale et Royale des Sciences et Belles-Lettres de Bruxelles*, 3 (1780), 171-6.

3 Magellan's expectations were not fulfilled so soon; though Le Monnier proposed Pigott as a Corresponding Member immediately after presenting his observations to the Académie, he was not elected until 12 June of the following year.

self have engaged our friends, and as soon as a vacancy hapens surely you'll have a very good interest in your side. Mr. Lemonnier knows your habilities in ye. Astronoml. capacity chiefly; and is very willing to serve you, notwitstanding your dismissing or forgeting the former acquaintance & Correspondence with him, misleaded perhaps by some Little brouillon where intrigues & tales had more weight then they deserved. Mr. Messier was a little mortified by the imputation you mistook in my Letter on account of ye. communication of Astronoml. Observations, without sending before hand your own.[4] it was not Mr. Messier, who did dictate me the article on the subject. but I did from my self open to you the sentiments of those I have hear'd speaking on the matter many a time: and thought proper to unbosom them to you, as a true friend. in general we don't see on our affairs, but what we are capable of doing to others, and often expose ourselfs to censure, and to be mistrusted in the same points in which others really do as we never should, by not being aware of other's doings. it is enough on the subject; and hope you'll see the friend, not the censor. 75.03.31a

My little work is not yet printed.[5] it is above a forthnight the printer disapoints me: and fear I will not be able to go to Bruxells before the midle of ye. next month.

Pray don't forget to send any other observations you can make to Mr. Messier: and put a Second cover to the letter, directed to Mr. Trudaine, as I have told of mine; who is very willing to help and protect science and those who apply themselfs to any branch thereof.

If on any occasion you write to Chevr. Turgot,[6] it will be not amiss to speak about your election into ye. Rl. Acady., but it is not necessary to write only for that purpose.

the Connoisse. des Temps for the ensuing year is not yet out. if I can get it before my depart, will not be forgoten.

I wish you pursue the ocultations of ye. stars by the moon, and for that purpose the nautical almanak is as good if not better than the Connoissance des Tems, this being a Copy of ye. other in that article. it is a pity that practical observers do not care them more than they do.

---

4 Cf. 75.02.26.

5 Magellan, *Description des octants et sextants anglois...* (Paris, 1775). The work was also printed in London in the same year.

6 Étienne-François Turgot (1721-1788), a member of the Académie Royale des Sciences in Paris, not his more famous younger brother, France's controller general of finance at this time.

75.03.31a Pray my respects to Mad[e]. & familly, as well as to all our common friends: and sincerely am

Dear respected friend
your true one
Magellan

P.S. I have set your abode at Louvain for the reasons I told you in one of my Letters.[7] Traveler as you are at present, you can adopt any abode you please.

a Monsieur
Mons[r]. Pigott, Gentil-homme Anglois
Vis à vis La Cour
à Bruxelles
Flandres

---

7 See Magellan to Pigott, 2 January 1775.

75.04.01 *To Pierre Macquer*
Bibliothèque Nationale, Paris, MS fr. 12306, f. 25-27

Pour M[r]. Le D[r]. Macquer[1]

Mons[r]. Ed[d]. Delaval, gentil-homme Anglois,[2] membre Savant de la Societé R[le]. de Londres, a montré par une suite d'experiences, dans un Memoire, qui fut communiqué par M[r]. de Castillon à l'Acad[e]. R[le]. des Sciences de Berlin, que les changements des couleurs qui se trouvent dans les corps, soit du regne animal, soit du mineral, ou soit vegetal, y sont produites par La differente grosseur ou dimension à laquelle sont reduites les particules visibles de ces corps. De façon que ce Monsieur a montré que La Loi etablie par Le Chev[r]. Newton sur les couleurs causées dans les corps diaphanes, comme l'air, le verre, & l'eau par cette cause de la differente grosseur ou dimension des particules de ces memes corps,

---

1 Magellan's letter was enclosed with a letter he sent to someone else, not identified, to be passed on to Macquer.

2 Edward Hussey Delaval (1729-1814).

subsiste egalement dans les autres corps qui ont une couleur permanente. 75.04.01
M^r^. Delaval est le premier qui appliqua [les] loix Newtoniennes à ces corps derniers, & qu'elles sont en effet les memes dans les uns & dans les autres. Le Memoire cidessus sera tres probablement inseré dans le Volume des Memoires de l'Acad^e^. de Berlin, qui doit paroitre aux Paques de l'année prochaine (1776).[3]

Ce Savant m'a montré un grand nombre d'experiences contenues dans son Memoire: & Je me rappelle entre autres d'une singularité, qui est plus sensible dans les fluides. Car pour passer d'une couleur à une autre extreme suivant la suite ou gradation Newtonienne, ils ne le font qu'en donnant toutes les nuances des couleurs intermediaires. C'est à dire, ne passant pas du rouge par exemple au vert, sans donner premierement le violet, en suite le bleu, après de quoi on apperçoit le vert &c. Mais ce que je trouve de plus singulier encore, c'est que dans le cas de ne pas donner les couleurs intermediaires; alors ce fluide devenoit diaphane avant de passer à la couleur extreme du bleu; comme pour montrer le saut (saltus latin), on remarque sa soumission à la loy qu'il ne pouvoit pas suivre.

Une Lettre que Je viens de recevoir de ce Philosophe, datée le 12 du mois dernier,[4] contient quelques articles, dont je crois interessant d'informer l'Academie. = Je vous devoilerai à present (dit M^r^. Delaval) le paradoxe chymique que Je vous ai montré dans votre dernier retour à Londres. Vous Savez qu'on dit du bleu de Prusse qu'il n'est pas assujetti à l'action des acides. Cette opinion, aussi bien que plusieurs autres relativement à cette substance, est tout à fait erronée.[5] Car J'ai mis quelque huile de vitriol sur le bleu de Prusse. La misture devint immediatement chaude, & jetta une vapeur volatile & sulphureuse, qui marquoit la separation qu'y etoit faite du Phlogistique par l'action de l'acide vitriolique. En laissant cette misture quelque tems, j'ai trouve que tout le bleu de Prusse etoit dissou, par l'huile de vitriol; & etoit devenu d'une couleur brune pale, epais, & opaque, qui se deposa au fond du vaisseau. la huile de vitriol surabundante nageoit au

3 Delaval, "Recherches expérimentales sur la cause des changements de couleur dans les corps opaques naturellement colorés", *Nouveaux Mémoires de l'Académie des Sciences et Belles-Lettres [de Berlin]*, 1774 (published 1776), 154-94. Delaval subsequently published a more extended account of this work, *An Experimental Inquiry into the Cause of the Changes of Colours in Opake and Coloured Bodies…* (London, 1777); its significance is discussed by Alan E. Shapiro, *Fits, Passions, and Paroxysms: Physics, Method, and Chemistry and Newton's Theories of Colored Bodies and Fits of Easy Reflection* (Cambridge, 1993), pp. 230ff.

4 Letter not found.

5 Delaval described numerous experiments on Prussian blue in the paper published in Berlin, pp. 176ff.

75.04.01 dessus du melange. Elle avoit acquis une couleur brune foncé, un odeur particulier avec d'autres circonstances, qu'il seroit long à detailer ici.

Par Les experiences qui suivent, il paroit que le bleu de Prusse est soluble dans une huile bien forte de vitriol; mais qu'il n'est pas soluble dans la meme huile de vitriol, lorsque celle-ci est diluée avec de l'eau.

Après avoir separé la huile de vitriol surabondante qui surnageoit dans le vase, J'ai mis quelque partie de cette solution dans un verre: oû j'ai melé de l'eau commune. Le melange fut fait, & resta pendant fort peu de tems sans aucune autre couleur mais dans l'espace d'une minute, il devint bleuatre: & quelques minutes après le total prit un fort beau bleu. de façon que le bleu de Prusse, qui avoit eté dissous dans l'huile de vitriol, l'abandonna, lors qu'on y mit de l'eau, & enfin tomba au fond du vaisseau, sous La meme couleur ordinaire qu'il avoit auparavant. Celle-ci etoit la Liqueur, que je vous ai montrée, & qui fut nommé le paradoxe chymique.

Notez que si l'on ajoute de l'alkali fixe à cette solution du bleu de Prusse, l'acide en est atiré, & il reprend immediatement la Couleur bleu qu'il avoit avant La solution. Vous pouvez communiquer cette experience à M^r^. Macquer de ma part, en l'assurant de la grande Consideration & Regard que J'ai pour ce Savant & si vous voulez, vous etes le maitre de La comuniquer à vos amis.

Le D^r^. Higgins dont vous appreciez dignement les talents,[6] a fait une experience, qui prouve plus completement l'existence de l'alcali dans le Tartre, independament de l'ignition (brulement) qu'aucune autre qu'on ait faite en France ou ailleurs. Il ajouta de l'eau au Tartre crud, & le mit en digestion dans une cave froide. le melange devint fort putride: & aiant evaporé la moitié de cette Liqueur, elle devint parfaitement alcaline. En y ajoutant de l'acide vitriolic afoibli, il en resulta une forte & violente effervescence, qui fit sortir la liqueur en ecume blanche en partie hors du vaisseau. Cette meme Liqueur change le sirop de violetes en vert: & donne toutes les marques d'un fort alcali. J'ai vu moimeme Cette experience: & Je frequente Les Lectures de cet habile chymiste.[7]

Je vous ai dit dans ma derniere que l'alkali caustique precipitoit la chaux de l'eau de chaux. Je ne Reponds pas neanmoins du meme effet, si l'on y

---

6 The Irish chemist Bryan Higgins (1737 or 1741-1818) opened a school and laboratory of practical chemistry in London in 1774. Priestley and Magellan were both enrolled in 1775. Higgins later accused Priestley of plagiarism, which gave rise to a confrontation between the two.

7 Magellan at this time still held Higgins in great consideration, but this situation changed with the dispute between Priestley and Higgins. From that moment, Magellan started to call Higgins a charlatan; see, for example, Magellan to Macquer, 24 November 1775.

employera de l'alkali caustique mineral. De fait je ne pretends point avoir etudié à fond cette philosophie aerienne. Mais Le D^r^. Higgins a developé toute cette matiere de l'air fixe avec une grande Clarté. Il a montré par un grand nombre d'experiences delicates, que l'air fixe n'est autre chose plus que l'air commun, chargé de Phlogistique: & que l'air appellé inflamable, est une vraie vapeur acide, chargée aussi de Phlogistique. Lorsque vous serez de retour chez vous, Je vous montrerai cette matiere dans Le plus grand Jour. 75.04.01

Communiquée ce 1er. Avril 1775
par Magellan

From Jacques-André Mallet 75.04.07

Bibliothèque de Genève, Geneva, Ms suppl. 1654, fol. 54[1]

7 Avril 1775

Mr Dessaussure m'a remis, Monsieur, l'article de votre lettre qui me regarde,[2] j'apprends avec plaisir que la Lun. est finie, & j'attends avec impatience votre arrivee à Londres pour l'examiner je ne puis pas me passer du piedestal que je vous ai demandé parce qu'il faut que la lunette puisse se separer de la machine parallatique, & que ma commission est de faire venir une lunette avec son pied, celui qui l'a demandée n'y a voulu consacrer que 25 g. peutetre en mettra-til 28, puisque Dollond a augmenté ses ouvrages, & c'est moi qui payerai le piedestal, sans rien dire c'est prquoi je souhaite qu'il soit très simple.

Je prendrai la lun. de nuit de 2 pieds, pourvu que le champ soit aussi large que celui de la lun. d'un pied.

Pour ce qui regarde Mr Meniers,[3] je consens à cette augmentation de 5 louis d'or p^r. l'addition relative à l'inclinaison de l'axe polaire, quoiqu'elle me paroisse assez chère relativem^t au total, je ne crois pas qu'il faille faire cette économie, surtout à cause des changem^s qui peuvent arriver au pied

1 Letter not found. The text is from a draft kept by Mallet.
2 See Magellan to Mallet, 31 March 1775.
3 Pierre-Bernard Mégnié?

75.04.07 de la machine en bois. Je souhaite que M^{r}. Meniers vueille se mettre à l'ouvrage le plutot possible & qu'il me donne avis aussitot qu'il sera achevé.

Je suis Mon cher Monsr avec le plus sincere attachem^{t}

Votre très affect. Servit.[4]

A Mr Magellan à Paris,
sur l'enveloppe à Mr Trudaine de Montigny
Conseill^{r} d'Etat, Intend^{t} Gen. des Finan^{s}.

---

4 The draft is unsigned.

75.04.08 *To Nathaniel Pigott*

Royal Astronomical Society, London, MSS Pigott, 37

Dear friend here I send you the certificate of y^{e}. R^{l}. Acad^{y}., by which you see that your Observations will be printed in the Academy's collection.[1] as soon as there is a vacancy your election will be proposed by M^{r}. Le Monnier, and I have no doubt of succeeding well.[2] I have submitted to y^{e}. Acad^{y}. my Little Treatise, and there were named 3 Commissaries for the revisal of it.[3] this will hinder my setting out till the week after Easter on account of the little vacancies of y^{e}. Acad^{y}. and then I'll have the pleasure of seeing you at Bruxelles with other friends to whom I beg you my heartly Compl^{s}.

Your sincere friend & serv^{t}.
Magellan

Paris this 8 April 1775.

---

1 Despite what Magellan says, Pigott's observations were never printed in the Academy's *Mémoires de mathématique et de physique, présentés...par divers savans.*

2 Le Monnier had in fact proposed Pigott as a Corresponding Member on 22 March 1775, immediately after presenting his observations to the Academy of Sciences. However, Pigott was not elected until 12 June 1776, Messier then being designated as his correspondent within the Academy; *Index biographique de l'Académie des Sciences, 1666-1978* (Paris, 1979), p. 417.

3 Magellan, *Description des octants et sextants anglois...* (Paris, 1775). The three commissioners appointed by the Academy of Sciences to review the work were Borda, Bezout and Bory; they reported favourably on 29 April.

To Jacques-André Mallet[1] 75.04.10

Bibliothèque de Genève, Geneva, Ms suppl. 1654, fol. 36-37

Paris ce 10 Avril 1775

Monsieur

Je vous ai ecrit deux mots dans une Lettre à M^{r}. de Saussure,[2] & peut etre en recevrai je cette semaine votre reponse à ce que je vous demandois. Cependant comme mon tems pour partir d'ici s'aproche de fort près, je crois devoir m'assurer d'avantage contre quelqu'acident dans l'incertitude si par hazard l'absence de M^{r}. de Saussure vous empechoit de recevoir la note ci dessus. Son contenû consistoit principalment dans ces deux points, qui n'etoient pas bien decidés dans la Lettre derniere que vous m'avez ecrite. Savoir

1^{0}. Si le piedestal parallactique dont je vous ai fait la description doit avoir, ou non l'apparatus ou ajustement Polaire dans son axe, à fin de le mettre exactement à la Latitude du Lieu dans le quel on en voudra faire usage.

Dans ce cas l'ouvrier dit ne pouvoir pas la faire pour moins de 30 Louis d'or. Mais si vous vous contentez de l'avoir (cette machine) sans cet ajustement, & seulement appropriée ou adaptée à la hauteur du Pole de Geneve, comme sera celle que l'ouvrier va entreprendre pour M^{r}. Le Presid^{t}. de Saron, qui sera adaptée à la Latitude de Paris; alors le prix ne sera que de 25 Louis d'or. il est donc necessaire de me dire explicitement quelle est votre Volonte là dessus?

2°. Vous aurez aussi la bonté de nommer la personne qui doit recevoir & payer ici cette machine pour vous l'envoyer. Car comme je vais partir dans la semaine après Pacques ou tout au plus dans la suivante il faudra que l'ouvrier sache à qui elle doit etre delivrée & de qui il doit recevoir son argent. M^{r}. Menier[3] n'est pas encore revenû de son pais, oû il alla pour se marier & arranger des affaires de famille. J'ai eté chez lui differents fois. il y a actuellement quelques ouvriers. mais sans qu'il arrive ici on ne peut pas conclure ni prendre aucune arrangement.

Mon petit ouvrage est à la derniere feuille: & il faut attendre jusqu'à la rentrée de l'Academie, qui est actuellement en vacance jusqu'à la semaine

1 Annotated: "Rec. le 15 av 1775".

2 See Magellan to Mallet, 31 March 1775. Letter to Saussure not found.

3 Mégnié?

75.04.10 apres Paques, pour avoir le raport des commissaires que J'ai demandés pour l'examiner. ainsi J'aurai le tems de recevoir ici votre reponse avant mon depart. Mes Resp^{s}. à M^{r}. Saussure, & que je ne m'oublierai pas de lui envoyer bientôt son instrument. la Connoissance des Tems pour 1776 va paroitre dans peu de jours. & les Ephemerides jusqu'à 1784 vient d'etre publiées. Voyez si vous en voulez?

Je suis de Coeur
Monsieur
Votre très h^{l}. serv^{r}.
Magellan.

P.S. mettez une seconde envelope à votre reponse, sous l'adresse de M^{r}. Trudaine de Montigny Intend^{t}. Genl. des Finances à Paris.

A Monsieur
Monsr. Mallet Professeur d'Astronomie
Cour de S. Pierre
à Geneve
Suisse

75.04.17 *From Jacques-André Mallet*

Bibliothèque de Genève, Geneva, Ms suppl. 1654, fol. 55[1]

Geneve le 17 Avril 1775

J'ai bien reçu, Monsieur, votre derniere lettre inserée dans celle de Mr Dessaussure,[2] & j'espere qu'actuellem^{t} vous avez reçu ma réponse qui est du 7 de ce mois.[3] vous y verrez que pour avoir la machine parallatique plus parfaite je veux bien y consacrer les 5 louis de plus p^{r}. l'apparatus relatif au changem. d'inclin. de l'axe polaire, mais j'espere que M^{r}. Meniers[4] se chargera de l'encaisser solidement pour qu'il n'y arrive point d'accident dans le transport. quand l'ouvrage sera achevé, il les remettra à Messrs.

---

1 Letter not found. The text is from a draft kept by Mallet.
2 See Magellan to Mallet, 31 March 1775. Magellan's letter to de Saussure not found.
3 See Mallet to Magellan, 7 April 1775.
4 Mégnié?

Banquet & Pache Negocians, Cul de Sac de l'étoile,[5] qui lui payeront les 30 Louis, j'aurai soin de les prevenir ladessus; & comme M^r^. Meniers pourroit souhaiter d'avoir qlque assurance du payem^t^ je joins ici un billet que je vous prie de lui remettre & qui lui servira de lettre de change.[6] 75.04.17

vous pourriez, Monsr. lui remettre un expl. de la Con. d. T.[7] 1776 & des nouvelles Ephemerides p^r^ 10 ans,[8] & il joindroit tout cela dans la même caisse.[9]

A Mr Magellan à Paris
sous enveloppe à Mr Trudaine de Montigny
Conseill^r^ d'Etat Intend^t^ gener^l^. des Finances

5 See Magellan to Duchesne, 30 November 1770, note 12.
6 At the bottom of the page, Mallet has written as follows: 'Je prie Mess^s^ Banq^t^ & Pache de payer à M^r^. Meniers faiseur d'Instrumen. de mathemat. la som̃e de trente Louis neufs soit 720 L. de [Fr.] lorsqu'il lui remettra la caisse qui doit contenir la machine que je lui ai commandée. Geneve. Jacques André Mall'
7 *Connoissance des Temps.*
8 Lalande, *Éphémérides des mouvemens célestes, pour le méridien de Paris ... contenant les dix années de 1775 à 1784* (Paris, 1774).
9 The draft is unsigned.

To Jacques-André Mallet 75.05.24

Bibliothèque de Genève, Geneva, Ms suppl. 1654, fol. 38-39

Enghien près de Bruxelles
ce 24 Mai 1775

Monsieur

Enfin me voici en chemin pour retourner chez moi à Londres oû je tacherai de vous faire expedier au plutôt possible votre Lunette. J'ai parti de Paris Lundi dernier avec Son Altesse le Duc d'Aramberg, qui m'a amené à cette maison de campagne pour voir la culture qu'il y fait donner aux plantations d'arbres & plantes curieuses, dont je lui a envoyé une bonne partie de semences. Je ne resterai que très peu de jours, d'autant plus qu'il va partir pour ses etats en Alemagne; & je pense etre de retour à Londres dans la quinzaine. J'ai attendu tous les jours M^r^. Megnié à Paris pour lui

75.05.24 donner votre billet:[1] mais il n'etoit encore arrivé, malgré les esperances que sa soeur me donnoit à chaque fois que j'y allois, disant qu'il seroit de retour dans la semaine &c. Je lui ai ecrit une lettre pour fixer le tems dans le quel il voudroit s'obliger à finir votre machine parallatique: lui me repondit qu'en trois mois, de façon qu'en ne comptant ce mois-ci, vous l'aurez finie à Paris au bout du mois d'Aoust, en cas qu'il soit fidelle à sa parole. Dans cette supposition vous devrez recommender quelqu'autre de vos amis à Paris de veiller à l'execution de cet instrument que je ne doute soit bien fait & d'usage, car je suis fort content de l'ouvrage de ce jeune artiste. Je crois vous avoir dit que c'etoit une fausseté le raport d'un instrument pareil fait par lui pour M^r^. le Presid^t^. de Saron. Ce n'etoit qu'un projet, & il ne l'a pas encore commencé; quoique M^r^. le Presid^t^. seroit bien aise de l'avoir. C'est le malheur de ces ouvriers Francois de donner beaucoup dans le faste & parade. Il vient de ce[2] marier dans son pays (à Dijon), & me disoit dans sa letre qu'il s'attendoit d'etre reçu Academicien dans l'Academie de son Pays. je lui ai repondu sechement à cet article, en le faisant souvenir que des titres Academiques n'ajoutent rien au vrai merite intrinsique: & que s'il ne feroit pas de bons ouvrages, qu'il n'y auroit pas des gens sensées, qui lui ordonassent de travailler pour l'egard de la parade de son titre d'Academicien &^a^. C'est la malhereuse disposition de cette Nation de ne penser qu'à la parade & à faire figure dans le monde, sans payer plus d'attention au reel & essentiel des affaires &c.

J'ai fait arranger à Paris le petit instrument destiné pour M^r^. le Professeur de Saussure. Ce n'etoit que le simple changement de la platine qui porte la Lunette. J'employai le meilleur Artiste de Paris pour cet objet de si peu d'ouvrage, & si aisé à faire: & cependant j'eus le desagrement de voir la platine entiere gatée & que l'instrument ne pouvoit produire son effet par la maladresse de l'artiste. ainsi je l'ai pris avec moi, pour le faire racomoder à Londres, & l'envoyerai au plutôt possible par la meme voye que vous avez deja reçu la petite lunette, avec les ephemerides & les deux exemplaires de mon Traité sur les Octants que Je vous ai promis, un pour vous & l'autre pour M^r^. de Saussure.

Un Baronet Anglois, Jeune Astronome, mais dont les connoissances & les talents meritent bien le nom d'extraordinaires, va faire un voyage en Italie. Il est fourni des plus excellents instruments pour toute sorte d'observations. jamais voyageur a entrepris un pareil voyage en particulier aussi bien preparé que lui: & je suis sur que vous serez enchanté de faire sa

1 See Mallet to Magellan, 17 April 1775.
2 se?

connoissance. il porte un excellent instrument equatorial ou observatoire portatif, qui est un bijou; par lequel il determine les latitudes à peu de secondes (environ 28"). il y a fait adapter des pieces fort ingenieuses. une bussole de declinaison, qui peut etre adaptée à cet instrument. une autre pour l'inclinaison (deeping needle), qui est très bien faite & charmante. une petite pendule qui frape les demi seconds, d'une construction singuliere & d'un très petit volume. deux barometres pour prendre les hauteurs des montagnes, & une autre quantité d'instruments philosophiques. le tout est si bien arrangé qu'il les porte tous dans sa chaise de poste. Vous me saurez très bon gré de cette connoissance. je lui ai donné une lettre de recomendation pour vous:[3] & Je souhaite que vous jouissiez de tout plaisir philosophique d'une telle connoissance. Son nom est Sir George Shuckburg.[4] il est membre de la Societé Royale de Londres, & très respecté par tous les connoisseurs dans cette branche de Science &c. 75.05.24

Vale neque ut soles ama
tuum
Magellan.

P.S. si par hazard vous souhaitez mieux d'avoir votre machine parallactique executée à Londres, je crois pouvoir la faire faire à peu près par le meme prix par les ouvriers particuliers que j'employe; quoique Dollond dise ne pouvoir pas le faire par le meme prix.

N.B. Je vous renvoye le billet que vous aviez destiné pour Mennier,[5] par la raison que Je vous ai marqué, de ne pas etre arrivé à Paris avant mon depart. Car je n'ai pas crû devoir le laisser entre d'autres mains, que les siennes. Vous pouvez le lui envoyer (ou un autre) par la voye que vous jugerez à propos.

A Monsieur
Monsr. Mallet, Professeur d'Astronomie
Cour de S. Pierre
à Geneve
Suisse

---

3 Letter not found.
4 Sir George Shuckburgh (1751-1804).
5 Mégnié.

75.06.23 *To the Duke Charles-Marie-Raymond d'Arenberg*[1]

Archief Van Arenberg, Enghien/Edingen, Belgium

Monseigneur

La Lettre dont V. A. S^{me}. m'a honoré du 13 du courant,[2] authorise mon obeissance, mais elle ne sauve pas tout à fait mon inquietude, ni diminue La Sensation qu'excite sur moi très souvent Le Souvenir de l'etat actuel de Sa Persone & de toute Sa Famille.[3]

J'arrivai ici le 12 du courant, comme je l'avois projecté dans ma derniere lettre,[4] & je portai tout de suite celle de V. A. S^{me}. qui contenoit La consultation sur l'incommodité de Mad^{e}. La Princesse, à M^{r}. Le Comte de Belgioioso[5]; il etoit déjà couché, se trouvant encore mal convalescent de sa maladie. J'y retournai le matin suivant, & le trouvai bien mieux que je ne l'esperois, mais encore fort foible & pale. Il promit de faire avertir bientot le Medecin & d'en faire expedier à V. A. S. La reponse, que je suppose arrivera avant celle-ci; car aiant eté deja deux autres fois pour Le voir (mais sans le trouver) J'y ai appris neanmoins que Le D^{r}. Hunter[6] avoit eté avec Lui le jour après que Je Lui delivrai La Consulte.

J'ai vu dimanche dernier M^{r}. Arbuthnot dans sa compagnie & nous avons bien causé sur V. A. S^{me}.[7] Il me dit devoir aller en Flandres dans quelque tems cet Eté ou l'automne, pour amener un de ses enfants à une

1 Charles Marie Raymond d'Arenberg (1721-1778), fifth Duke of Arenberg, Prince de Ligne and 11th Duke of Aarschot.

2 Letter not found.

3 One of the Duke's younger sons had died while Magellan was visiting the Duke; see Magellan to Sanches, 23 June 1775.

4 Letter not found.

5 Ludwig Karl Graf von Barbiano-Belgiojoso [Ludovico Carlo Maria di Barbiano, Conte di Belgioioso] (1728-1801) was Austria's Envoy extraordinary and Minister plenipotentiary in Great Britain, 1770-82. He was later Vice-Governor of the Austrian Netherlands, 1783-7.

6 William Hunter (1737-1783).

7 John Arbuthnot (1729-1805) of Mitcham, author of *An Inquiry into the Connection between the Present Price of Provisions, and the Size of Farms* (London, 1773), and a good friend of Magellan's. See also "Mémoire de M. Jean Arbuthnot, Ecuyer Anglois, Membre de la Société Royale, & de celle pour l'encouragement des Arts de Londres, sur les principes & construction de sa Charrue; Communiqué à l'Académie Royale des Sciences de Paris, par M. Jean-Hyacinthe de Magalhaens, Gentilhomme Portugais, résidant à Londres, Membre de la Société Royale de cette Ville, & Correspondant de la même Académie Royale des Sciences de Paris", *Observations sur la physique, sur l'histoire naturelle et sur les arts*, 4 (1774), 284-297.

pension particuliere du coté d'Anvers. Je serai charmé de savoir le tems du Retour de V. A. S[me]. à Bruxelles ou à Enghien, pour tacher d'arranger ce voyage de façon que cet honnet-homme puisse avoir le plaisir de presenter ses Respects à V. S. S[me].: peut-etre dans un cas pareil m'efforcerai-je de l'acompagner, pour ma propre satisfaction. Mais quoiqu'il en soit de l'effet de ce souhait, je me tiens & je compte fermement sur l'assurance que V. A. S[me]. me donne dans cette Sa Lettre, savoir, que, s'il y aura quelqu'occasion dans laquelle mon vrai attachement de Coeur puisse etre de Service à Sa Personne ou à Sa Famille, j'obtiendrai La satisfaction de pouvoir montrer par l'effet, combien je suis vraiement & avec le plus profond Respect 75.06.23

Monseigneur
de V. A. S[me]
Serviteur tres obeiss[t]. & tres obligé
Magellan

à Londres ce 23 June 1775

To Ribeiro Sanches[1] 75.06.23a

Österreichische Nationalbibliothek, Vienna
A.N. Ribeiro Sanches Korrespondenz, codex 12714, ff. 151-152v

Londres ce 23 June 1775

Meo Am[o] e S[r]. e m[to]. do Cor[am].

Aqui cheguei felizm[te]. em 12 do Corrente, & começo a despachar uma boa quantid[e]. de commissoins q me esperavão, particularm[te]. a das agulhas de marear p[a]. as quais ainda vim à bom tempo. a jornada de Bruxelles foi m[to]. boa & tãobem a de Louvain: Mas o duque de Arenberg perdeo um de seos filhos, que era o seo benjamin: & toda a sua familia se achou na

1 Annotated by Sanches: "Resp[a]. a 3 de Julho" [Answered 3 July] and "+ Pharm. Edmb. 1774 [= *Pharmacopeia Collegii Regii Medicorum Edinburgensis* (Edinburgh,1774)] + New Dispensatores [= M.S. Leury, *The new Dispensatory, containing the Elements of Pharmacy, the Materia Medica, &c.* (3rd ed., London, 1770)] + Glass two leters upon the Inoculation [= Thomas Glass, *A Letter ... to Dr. Baker on the Means of Procuring a Distinct and Favourable Kind of Small-pox, etc.* (London, 1767) and *A Second Letter ... to Dr. Baker on Certain Methods of Treating the Small-pox during the Eruptive State* (London, 1767)]".

75.06.23a maior consternação. eu tãobem participei huma grande parte & lhes dei mostras sinceras da minha gratitude e dos meos affectos: pois quando chegou a noticia do perigo em q o Princepe se achava, me offreci à lhe ir assistir á sua enfermidade, e certamente o executaria se dentro em dois dias não viesse a noticia da sua morte. o bom Duque me fez a honra de me assegurar que nisto tinha visto a sincerid^{e}. dos meos affectos, & que delles faria uso &^{a}. como d'hum amigo verd^{o}. na verd^{e}. o tal princepe era de hum excellente caracter, eu oconheci pessoalm^{te}. na viage & demora q fis há dois ou 3 annos em Bruxellas: & sem embargo da distancia q era em Moravia, onde elle estava enfermo e onde elle morreo, q dista coiza de 300 legoas de Bruxel[las] eu effectivam^{te}. me metteria immediatam^{te} à caminho. Mas elle acabou na flor ou botão de flor da idade, pois não tinha mais de 18 an[nos] & se acha descançado fóra dos perigos e enfados, amofinaçoens vilhacadas & patifarias desta vida buz-na.

Achei o nosso Abbe Chev^{r}. com saude & com a Sua Biblioth^{a}.: he geralm^{te}. estimado (& o merece) & deveria viver mais contente do q em Lisboa no tempo presente. M^{r}. Needham cada vez mais gordo continuando as suas propensoens Physico-religionarias. agora vou mandar-lhe a Historia dos Americanos por M^{r}. Adair, q pretende serem descendentes ou pelomenos terem ideias de Religião Mozaica & conservarem ainda Laibos da vida dos Patriarchas &^{a}. o livro he dos mais mal escritos & menos bem arranjados que eu tenho visto em Inglez mas não importa. talvez que delà nasça algũa dissertação sobre este bem importante objecto, donde algum Cagot tirará bellas asneiras p^{a}. fazer perder o tempo a outros tais como elle.

Aqui travalha-se ainda sobre o ar fixo. O D^{r}. Highins pretende q não he mais doq o ar commum impregné de Phlogistique.

Recebi carta (ou p^{a}. melhor dizer achei carta) do D^{r}. Wolfe de Danzic, que se Recomenda à V.M. & me pede suas noticias. he hum [hom]em[2] de bem & honrado amigo. diz-me q tem feito alteraçoens [...]o sistema consideraveis pela nova materia do ar fixo. eizaqui outro artigo da sua carta que he datada de 17 Fevr^{o}. passado = Je me fache que les Anglois & les François se disputent le quel des deux a donné la grande verole à l'isle d'otahitie. Ce prejugé merite d'etre censuré. la verole, la grande & la petite, est si ancienne que toute la generation animale. la propreté est le preservatif dela premiere. Les Turcs n'en savent presque pas, & les Otaheitiens s'en mocquent, sachant que les bains les en gueriront bientôt. les chretiens sont cochons par modestie & par principes de religion. Toute pourriture se communique. il n'est pas necessaire de la faire venir de l'Amerique & de l'Arabie. J'ai gueri la verole

2 Conjectural reading.

une infinité de fois par l'abstinence, les bains, & la boisson aqueuse & 75.06.23a
copieuse, sans mercure l'effet du ☿ est de gater toute nourriture pour un certain tems, pendant le quel la boisson aqueuse peut laver la Lymphe: & j'ai observé mille fois, que le ☿ ne guerit pas la verole, si le malade n'amaigrisse point. = &^{a}. Ora aqui tem V.M. a opinião deste Philosopho. Verd^{a}. ou falsa, ella sempre provará q he bem aventurado aquelle que como eu, come, bebe, e f-de, sem ter necessid^{e}. dos Senhores Medicos.

brevem^{te}. lhe mandarei alguns livrinhos novos com os q V.M. me deo na sua Lista. Conserve-se V.M. , & viva quieto no seio da sua familia adoptiva: de lhe minhas saudades, & aos mais amigos se he q os há: & mande fare buz-re tuto questo mondo buz-ne.

Vale me que ut soles ama

a Monsieur
Monsr. Sanchez D^{r}. en Medecine
Rue des Blancs-manteaux
à Paris

London 23 June 1775

My dear friend

I arrived here safely on the 12th of this month, & started dispatching a good quantity of commissions that awaited me, particularly that of the marine compasses[3] for which I am still in good time. The visit to Brussels was very good & also that to Louvain: but the Duke of Arenberg lost one of his sons, his Benjamin,[4]: & I found the whole family in the greatest consternation. I also participated a great deal & gave them sincere demonstrations of my gratitude and my affections: because when the news of the Prince's danger arrived, I offered to go and assist him in his infirmity, and I would certainly have done so if it was not that the news of his death came two days later. The good Duke did me the honour of assuring me that in this he had seen the sincerity of my affections,

3 Until his death in June 1772, Dr Gowin Knight, who had developed a greatly improved form of mariner's compass using tempered steel needles magnetized by a method he had invented, was responsible for certifying all compasses supplied to the Royal Navy. Following Knight's death, Magellan took over this contract; see his account in "Notice des Instrumens d'astronomie, de géodesie, de physique, &c. faits dernierement à Londres, par ordre de le cour d'Espagne ...", in his *Collection de différens traités...* (London, 1780), p. 216.

4 i.e. his sixth child, Prince Charles Joseph, who died on 28 May 1775, aged 20. The dead prince was not the Duke's youngest son, however, as Magellan's referring to him as "Benjamin" implies, since he had a younger brother, Prince Louis Marie (1757-1795). Magellan was obviously misinformed.

75.06.23a & that he would make use of them &c. as from a true friend. In fact the prince was of an excellent character, I knew him personally from the visit that I made two or 3 years ago to Brussels: & in spite of the distance to Moravia, where he was ill and where he died, that is about 300 leagues from Brussels, I would of course have started the journey immediately. But he passed away in the flower or flower button of age, since he was no more than 18 years old & is now beyond the perils and annoyances, afflictions, knaveries & villainies of this buzzona[5] life.

I found our Abbé Chevr.[6] in good health & with his Library: he is generally esteemed (& he deserves it) & should live more content than in Lisbon at the present time. Mr. Needham,[7] fatter and fatter, continues his physico-religious propensities. Now I am going to send you the History of the Americans by Mr. Adair,[8] who pretends they are descendants of or at least have ideas of the Mosaic Religion & still conserve traces of the life of the Patriarchs &c. The book is one of the worst written & least well arranged that I have seen in English, but never mind. Maybe it will give rise to some dissertation about this important subject, whereby some Cagot[9] will come up with nice follies to make others like him waste their time.

Here they are still working on fixed air. Dr. Highins[10] sustains that it is no more than common air impregné de Phlogistique.

I received a letter (or better said, I found a letter) from Dr. Wolfe from Danzic, who recommends himself to you & asks me for news of you.[11] He is a good man & an honourable friend. He says that he has been making considerable alterations [...] system[12] for the new matter of the fixed air. Here is another extract from his letter dated 17th Febr. last: "Je me fache que les Anglois & les François se disputent le quel des deux a donné la grande verole à l'isle d'otahitie.[13] Ce prejugé merite d'etre censuré. La verole, la grande & la petite, est si ancienne que toute la generation animale. La propreté est le preservatif de la premiere. Les Turcs n'en savent presque pas, & les Otaheitiens s'en mocquent, sachant que les bains les en gueriront bientôt. les chretiens sont cochons par modestie & par

5 Most probably a modified form of *buggerona*, probably in Venetian or northern Italian dialect, whereby the letter "z" was used instead of "g" = cheating, deceitful. (We thank Dr Dino Bressan, formerly of the University of Melbourne, for this explanation.)

6 João (or Jean) Chevalier (1722–1801) was a priest of the Congregation of the Oratorians. In 1760, upon the persecution of the Oratorians by the Portuguese Prime Minister, the Marquis of Pombal, he was forced into exile in a country village, but a year later he went to Paris and then to Brussels where he was at this time librarian at the Royal Library.

7 John Turberville Needham (1713–1781).

8 James Adair, *The History of American Indians; particularly those nations adjoining to the Mississipi, East and West Florida, Georgia, South and North Carolina, and Virginia, etc.* (London, 1775).

9 Cagot = outcast, pariah.

10 Bryan Higgins (1737?- 1820).

11 Letter not found.

12 Stained paper.

13 Tahiti.

principes de religion. Toute pourriture se communique. Il n'est pas necessaire de la faire venir de l'Amerique & de l'Arabie. J'ai gueri la verole une infinité de fois par l'abstinence, les bains, & la boisson aqueuse & copieuse, sans mercure l'effet du ☿[14] est de gater toute nourriture pour un certain tems, pendant le quel la boisson aqueuse peut laver la Lymphe: & j'ai observé mille fois, que le ☿ ne guerit pas la verole, si le malade n'amaigrisse point." &c. Now here you have the opinion of this Philosopher. True or false, it will always prove that blessed is the one like me who eats, drinks, and f---ks, without needing medical gentlemen. 75.06.23a

Soon I will send you a few new little books with those you mentioned in your list. Look after yourself, & live peacefully within your adoptive family: give them my regards, & other friends if there are any: & fare buz-re tuto questo mondo buz-ne.[15]

Vale me que ut soles ama[16]

14 ☿ = mercury.

15 fare buz[zera]re tuto questo mondo buz[zero]ne = hold all this deceitful world in contempt. See note 5 for the spelling.

16 Farewell and love me as you are accustomed to do.

To Antoine-Laurent Lavoisier[1] 75.07.01

Mon cher confrère et très-cher ami,

Je vous écrirai celle-ci a mesure que je lis votre ouvrage,[2] L'avertissement est admirable.

1 Letter not found. The text given here was published in *Oeuvres de Lavoisier*, volume 4 (Paris, 1868), pp. 762-5, and reprinted in *Oeuvres de Lavoisier: Correspondance*, ed. René Fric, fascicule II, 1770-1775 (Paris, 1957), pp. 492-5.

2 In February 1773 Magellan had reported to Trudaine de Montigny recent English experiments with an apparatus for distilling seawater on board ship. At Trudaine's suggestion, the French Navy commissioned Lavoisier to follow up the idea. Magellan and a naval officer, Deslette, were also involved in the work (see the report to Turgot published in *Oeuvres de Lavoisier*, IV, 758-60). Magellan's comments set out in the present document relate to a MS that Lavoisier had sent him, that was eventually published anonymously in 1781 under the title, *Nouvelle construction d'alembic pour faire toute sorte de distillation en grand avec le plus d'économie dans l'opération et le plus d'avantage dans le résultat, en deux parties; la première comprenant son application à la distillation des eaux-de-vie; et la seconde à la dessalaison des eaux de mer à bord des vaisseaux*. The MS, in Lavoisier's hand, survives. It has a marginal annotation: "Donné une expédition à M. Magalhaens le 17 mai 1775 et publié". The text as published conforms to that of the

75.07.01 n° 4. «La solution de ce probleme …elle interesse l'etat… c'est d'elle que dépend la chute ou le succès de *presque* tous les etablissements.» Otez le mot *presque*, et specifiez en disant: *tous les etablissements de ce genre*. «C'est elle qui *peut* établir la balance entre cette *branche* de commerce… C'est elle qui *peut* rompre l'équilibre, la concurrence, ou qui peut les rétablir, etc…».

n° 5. Il faut amplifier un peu plus ici sur les avantages généraux de cette invention relativement à la navigation

n° 6. Ajoutez la transition, par exemple: «Entrons en matière,» ou «Nous allons en développer les principes, etc…».

n° 8. Otez ces mots a la fin: «On a été même dans les fabrications en grand jusqu'à faire passer à travers un courant d'eau continu». Ces mots sont ici mal à propos. Il faut commencer le n° 9 disant: «Mais on n'en a pas fait ici une bonne application; car l'eau froide qu'on renouvelle, etc… ». Il ne faut pas parler ici d'un courant, parce que, s'il est un courant, il emportera l'eau déjà chaude. L'application du courant doit venir lorsque vous traiterez de la rechute ou circulation de la vapeur condensée sur la surface du liquide, etc. qui viendra plus bas dans le n° 11.

n° 11. Il faut le diviser en deux, en ôtant ce qui regarde la circulation ou rechute de la vapeur, qui se refroidit de proche en proche: et faisant voir que même dans le cas d'avoir un courant d'eau froide (ce qui est très rare, et impossible d'avoir dans presque toutes les fabriques distillatoires) alors cette eau froide, en touchant le chapiteau, y cause la circulation de la vapeur dont vous parlez dans le n° 8 ci-dessus.

n° 13. Mauvaise expression de 50 pour l'eau froide, et de 10 pour la chaleur du tuyau! il faut mettre du moins 25 ou même 30 pour la chaleur du tuyau du serpentin. Voici une notice qui vient à propos, ou du moins mérite d'être mise en note. Suivant les experiences du Dr Black le degré de la chaleur de la vapeur est souvent 1012. Or l'eau bouillante n'etant que 212 (toujours dans l'echelle de Fahrenheit), on voit que les 800 degrés doivent opérer à echauffer le tuyau avant que la vapeur soit réduite en masse, comme l'eau bouillante: et ces 800 degrés doivent être réduits à zéro par le refroidissement de l'eau qui est en contact avec la Surface extérieure du chapiteau et du serpentin etc, avant que cette réduction ait place. Avez-vous un bon thermometre chimique, c'est à dire à mercure avec un fort long col? Faites cette experience, mettez-le dans un chapiteau où l'on évaporera de l'eau, mais sans toucher le fluide qui bouillonne, et remarquez à combien de degrés il monte au delà

MS, Magellan's suggestions being ignored. The actual authorship of the text is uncertain. Lavoisier was obviously involved and Magellan clearly saw him as responsible for it, but the editors of Lavoisier's *Oeuvres* note that the clumsy title is uncharacteristic of him.

du point de l'eau bouillante (212 degrés de Fahrenheit): bien entendu qu'il ne faut pas avoir d'eau réfrigérente sur ce chapiteau; et qu'il faut rabattre le contact de l'air froid dans la partie du tuyau hors du chapiteau. Mais, pour revenir, il faut changer les chiffres 50, 10, 40, etc. de ce n° 13. 75.07.01

Explication de la figure 1, planche 1ere, n° 7. «Cette eau est conduite du réservoir E par un tuyau de cuir h' h'», *ajoutez* «à chacune des deux machines distillatoires: elle se répand… »

n° 24. Ajoutez à la fin après les mots: *soit de le revêtir d'un enduit léger de maçonnerie*, «ou en y faisant une espèce de cloison de maçonnerie, de façon que la communication entre F''' et B fig. 2, soit au-dessus de la ligne tddt de la figure 3; car alors l'action du feu qui sortira par la cheminée B'B'AB, ne pourra toucher aucunement la paroi R'R de la chaudière.»

n° 28. Souvenez-vous de faire changer les chiffres de cette figure, c'est à dire le 7 en 2: et de la citer ainsi: fig 4) dans le n° 8 ci dessus.

N.B. Il faut relire soigneusement cette description ayant examiné les figures chaque fois qu'on les citera, après en avoir corrigé les nombres dans les planches. Autrement on citera à faux les figures par des nombres faux, et il y aura une grande confusion.

Il faut ajouter encore un mot dans l'introduction du dessalement de l'eau de mer sur les deux objets, savoir: celui de fournir de l'eau douce à l'équipage dans des voyages longs; celui d'éviter dans quelques cas pressants le retard d'aller prendre de l'eau, soit pour se trouver trop éloigné des pays amis, soit pour ne pas manquer quelque coup important dans les expéditions maritimes. Enfin on peut encore s'épargner de la place pour d'autres objets, soit des munitions ou des marchandises, en ne prenant qu'une petite quantité de barriques d'eau, qui seront remplies d'eau douce par la machine, à mesure qu'on en distille, etc…

Cahier de la distillation de l'eau de mer

n° 12. M.M. fig. 9 represente des *bancs* (non pas des *barres*), etc.

Il faut ajouter à ce *cahier* c'est à dire à la fin de cette introduction pour la *distillation de l'eau de mer* deux mots en disant «qu'on va entrer dans l'explication des figures contenues dans la planche 2 et suivantes, dont le lecteur concevra encore mieux le sens s'il veut se donner la peine de lire ce qui se trouve de commun dans la machine qu'on a representée dans la planche 1ere destinée a la distillation des eaux-de-vie sur mer.».

De même il faudra faire une espèce de transition ou passage, au bout du premier discours sur les eaux de vie, disant: «Qu'on va parler de l'application de cette machine distillatoire pour servir à bords des vaisseaux pour dessaler l'eau de mer: et que dans la suite on expliquera en détail toutes les pièces representées dans les quatre planches etc.».

75.07.07 *To Ribeiro Sanches*[1]

Österreichische Nationalbibliothek, Vienna
A. N. Ribeiro Sanches Korrespondenz, codex 12714, 155-156v

7 Julho 1775

Meo Am^{o} e S^{r}. e m^{to} do Cor^{m}.

Agora Recebo neste instante a sua estimada e dezejada carta do 3 do Corr^{te}., a q vou Responder bem picado da sua pulha sobre as arreitacoens do estado prezente. Senão estivera tão depressa por ser tarde e ainda não ter jantado, havia de extender-me em hua Longa invectiva pelo gostinho vindicativo ou invejozo como de q^{m}. ja não pode, & se alegra de augmentar o n^{o}. dos companheiros. Pois saiba q hontem á noite depois de bem cansado & suado, por ter estado hum dia mui quente & abafadiço, fui fazer uma vizita a antiga Nely, depois de varios queixumes a q satisfis dizendo q [tinha][2] estado auzente estes 3 años nas Indias orientais, lhe montei nas ancas (mas por diante), & me confe[ssou][3] q havia a very Long while q não tinha sido t[ão][4] bem forn- -cada.

Mandarei os Livros q dezeja com mais 4 ou 5 q achei novos.

o am^{o} N[ee]dham[5] ja tem a critica de q V.M. fala. eu avi sobre a sua meza, e elle gostava bem della.

não tenho tanto susto por Chevalier. elle está bem, e não lhe falta senão o persuadir-se q o está.

me alegro q V.M. pas[seie][6] por esse fauxbourg & faça um pouco de exercicio, q lhe [serve][7] de alivio p^{a} o corpo & espirito.

eu vou em 15 dias pass[ar] 3 ou 4 com o D^{or}. Priestley & Mylord Shelburne q me invitou. falaremos de ares fixos. elle está p^{a}. imprimir um 2do. vol. so[bre][8] a mesma materia. tem feito m^{tas}. exp^{a}. novas, & achado rezultado[s] importantes. Como a distancia à Bath he de 3 ou 4 legoas talve[z] q eu dê lá uma chegada, e verei o D^{r}. La Cour. veja V.M. se quer

1 Annotated "R^{a} a 20 Julho" [Ans. 20 July].
2 Conjectural reading.
3 Obliterated margin. Conjectural reading.
4 Conjectural reading.
5 Obliterated margin. Conjectural reading.
6 Obliterated margin. Conjectural reading.
7 Obliterated margin. Conjectural reading.
8 Obliterated margin. Conjectural reading.

algua coiza pª. elle. eu ainda não me incontrei [com][9] o D^or^ Seq^ra^. Ainda q o busquei varias vezes no Caffe &ª. mas sei q passa bem 75.07.07

tenho saude & arr-to bem a pezar das [s]uas pragas. tudo vai bem & melhor cada vez, porq sei tomar partido, & tiralo de tudo, com os polvilhos da Philª. martiniana q he a minha Madre Celestina. em effº. não tenho de q me queixar: antes varios acontecim^tos^. vierão a propozito pª. melhor navegar.

o Conde Mushkin [P]usskin partio pª. Russia provavelm^te^. pª. não tornar, pelo- [menos ess]a[10] foi a sua intenção, pois se venderão [os t]rastes de Caza, como q^m^. não faz conta de vir.

hontem foi a ultima Seance da Soc^e^. R^l^., onde se lêrão novas experiencias sobre o calor q os animaes podem suportar sem grande daño: por ellas se prova q Boerhaave se enganou[.] homens & animaes podem suportar até 220, 230 & mesmo 250 gr. de Fahrenheit, sem padecer danno na organização e funçoens, pois ao mesmo tempo o principio vital tem a faculdade de conservar os fluidos animais pouco mais do q 120 ou 130 gráos. são experiencias de hum novo Medico daqui, q he mui capas & applicado, não me lembra agora o nome (cuido q he Irlandez ou Escossez) eu o conheço devista. ao mesmo tempo os ovos metidos neste Lugar das experiencias se endurão &ª. M^r^. Jean Hunter fez semilhantes experiencias sobre o frio q os animaes podem sofrer: & achou q o mesmo principio vital tem a faculdade de conservar os fluidos do corpo em calor m^to^. superior a[o] do ambiente do corpo animal vivente. &ª. Voilà les plus g[ran]des nouvelles Litteraires. Je suis après quelque chose de [...][11] sur la bussole. hontem de manhan fui ensayar uma sobre a Thamise, que apezar dos empuxoens dos remos & grandes balanços, apontou sempre & governou admiravelm^te^. ao Norte.

Vale atque me ut soles ama

a Monsieur
Monsr. Sanchez D^r^. en Medecine
Rue des Blancs-Manteaux
à Paris

9 Obliterated margin. Conjectural reading.
10 Stained paper. Conjectural reading.
11 Illegible.

75.07.07 7 July 1775

My Dear friend

I just received at this very moment your esteemed and desired letter of the 3rd of this month,[12] to which I shall answer very angrily because of your mockery about the libidinous excitations of my present state. Were I not in such a great hurry due to the late hour and not having had dinner, I would write a long invective against the vindictive or envious pleasure of one who can't do it any more and is happy to increase the number of those like him. So you may know that last night, tired and sweaty, due to the very hot and suffocating day, I went to pay a visit to old Nelly, after various complaints that I justified by saying that I had been absent for three years in the East Indies, I jumped on her hips (and so on), & she confessed that for a very long while she had not been so well fornicated.

I'll send the books you wish together with a further 4 or 5 new ones.

our friend Needham already has the critique that you mention.[13] I saw it on his table, and he really likes it.

I'm not so worried for Chevalier. He is well, and he only needs to be persuaded that he is so.

I'm happy that you walk by that fauxbourg and do a little exercise that may give you some relief in the body and in the spirit.

In 15 days I'm going to spend 3 or 4 days with Dr. Priestley & Milord Shelburne[14] who invited me. We'll talk about fixed airs. He is about to print a 2nd volume on the same matter.[15] He has been doing many new experiments, & has obtained important results. As the distance to Bath is 3 or 4 leagues maybe I shall call there and see Dr. La Cour.[16] See if you want anything to send to him. I haven't yet met Dr. Sequeira[17] though I looked for him several times at the Cafe &c. but I know that he's well.

I'm healthy & I belch well[18] in spite of your imprecations. Everything is going well and better all the time, because I know how to take the best part from

12 Letter not found.

13 Not identified.

14 Priestley's patron William Petty-Fitzmaurice, 1st Marquess of Lansdowne and Earl of Shelburne (1734-1805).

15 Joseph Priestley, *Experiments and Observations on Different Kinds of Air*, vol. 2 (London, 1775).

16 Philippus (or Philip) de la Cour, alias Abraham Gomes Ergas (?-1786), physician of Portuguese Jewish origin, graduated from Leiden and established himself in Bath and later in London, attracting a very elegant clientele.

17 Isaac Henriques Sequeira (1738-1816), Jewish physician of Portuguese origin, studied at Bordeaux and Leiden and started practising in London in 1771. He became attaché to the Portuguese Diplomatic Mission in London.

18 Magellan uses the word "arroto", which ordinarily means "to burp" or "to belch". However, it is clear from the sequence of letters to Sanches that he is using it as code or slang for the pleasure involved in having sexual intercourse.

everything, with a sprinkle of the martinian philosophy[19] that is my Mother Celestine.[20] In fact I have no reason to complain: instead, various events promise an even better navigation. 75.07.07

Count Mushin Pusskin departed for Russia probably for good, at least that was his intention since his household effects were sold, as if he has no intention of returning.[21]

Yesterday was the last seance of the Royal Society, where new experiments were described about the heat that animals can stand without great damage.[22] By these it is proved that Boerhaave[23] was wrong. Men and animals can stand up to 220, 230 & even 250 degr. Fahrenheit, without suffering damage in their organization and functions, since at the same time the vital principle has the faculty of maintaining the animal fluids at little more than 120 or 130 degrees. These are experiments of a new local Doctor, who is very able & diligent. I don't recall his name (I think he is Irish or Scottish).[24] I know him by sight. At the same time eggs put in the place where the experiments take place last &c. Mr. John Hunter did similar experiments on the cold that animals can endure: and he found that the same vital principle has the faculty of maintaining the body fluids at a much greater heat than that surrounding the living animal body &c.[25] Voilà les plus grandes nouvelles Litteraires. Je suis après quelque chose de [...] sur la bussole. Yesterday morning I went to try one on the Thames that in spite of the thrusts of the oars & great oscillations, always pointed out & directed admirably to the North.

Vale atque me ut soles ama[26]

19 i.e. the philosophy of Martin Luther.

20 Madre Celestina (or Mother Celestine) is a character of Spanish and Portuguese literature of the 16th and 17th centuries who does marvellous things by means of witchcraft.

21 Contrary to Magellan's suspicions, Count Aleksei Semenovich Musin-Pushkin continued as Russian ambassador in London until 1779.

22 Charles Blagden, "Further Experiments and Observations in a Heated Room", *Philosophical Transactions*, 65 (1775), 484-94. Blagden's paper was read at the Royal Society, as Magellan reports, on 6 July 1775.

23 Blagden refers to Boerhaave, *Elements of Chemistry*, vol. 1, p. 275.

24 Blagden in his paper and a preceding one on the same subject (*Philosophical Transactions*, 65 [1775], 111-23) credits the experiments he describes to Dr. George Fordyce (1736-1802), a Scot who trained in medicine at Edinburgh and Leiden before setting up practise in London. Fordyce was elected FRS in 1776.

25 John Hunter, "Experiments on Animals and Vegetables, with respect to the Power of producing Heat", *Philosophical Transactions*, 65 (1775), 446-58. Hunter read his paper to the Royal Society on 22 June 1775.

26 Farewell and love me as you are accustomed to do.

75.07.14 *To Antoine-Laurent Lavoisier*

Archives de l'Académie des Science, Paris, Fond Lavoisier, 1642[1]

Monsieur

J'ai gardé tout exprès pour cette poste le compte que je vais vous donner de la coquinerie du Capitain, qui a les deux caisses à son bord pour les apporter à Rouen. C'est le Capitain de l'aimable Henriette si je ne me trompe pas dans le nom: mais comme vous avez reçu par la poste derniere, le billet signé de sa main en François pour la caisse avec l'appareil de verre pour l'air fixe, & pour l'autre avec un plateau de verre pour M^r^. Brisson,[2] il n'y peut avoir aucune meprise tant sur son nom, comme sur celui de son vaisseau. lors que J'ai vu la charge de deux louis de fret (48^l^.) pour deux petites caisses qu'un autre n'oserait pas charger au delà de 12^l^., je me portai moi meme â la Bourse: & je lui demandai de me rendre mes caisses ou de rabaisser l'exorbitant prix qu'il y avoit mis. il refusa de faire ni l'un ni l'autre: disant que le courtier (Broker en Anglois) l'avoit fixé de la sorte. dans ce cas je m'adressai alors au Broker, qui me repondit l'avoir jugé par l'information du Capitain. Je le fis appeller, car il n'etoit pas loin, & lui soutint que les caises en question etoient assez grandes pour meriter le prix chargé: Je lui protestai de me plaindre à ceux qui pouvoient le faire rendre raison de son tort en France (car j'ai vû qu'ici Je n'aurois d'autre parti que d'augmenter des frais) & il s'a mocqué de cette commination. Ainsi Je crois devoir vous rendre part de ce procedé, dans la consideration qu'un mot de la part de votre Bureau au Receveur ou Directeur du Bureau de la Douane à Rouen, sera suffisant pour faire redresser ce tort, lorsque ce coquin y arrivera. Car c'est abuser & meme avilir le caractere François d'agir si ouvertement contre la practique & coutume honete des gens de commerce dans des cas comme le present. Il est bon de vous avertir que ce vaisseau partira de ce port chargé des Marchandises d'ici de Londres qui n'ont pas d'entrée en France: & que les depeches & meme les Reçus qu'il donne pour les marchandises mises à son bord, son datés d'ostende, avec une fausseté

1 Corrected version of the text published in *Oeuvres de Lavoisier: Correspondance*, ed. René Fric, fascicule II, 1770-1775 (Paris, 1957), pp. 496-7.

2 Presumably the naturalist and experimental physicist Mathurin-Jacques Brisson (1723-1806).

la plus effrontée Le Reçu que Je vous ai envoyé, & l'autre pareil que J'envoyai aussi la poste precedente à M^r^. Brison en font evidence. 75.07.14

J'ai l'honneur d'etre avec toute consideration possible

Monsieur
Votre très h^ble^. & très obeiss^t^. serv^r^.
Magellan

à Londres ce 14 Juillet 1775.

a Monsieur
Monsieur La Voisier de l'Acad^e^. R^le^. des Sciences
Rue neuve des Bons Enfants
à Paris

To Antoine-Laurent Lavoisier[1] 75.07.14a

Archives de l'Académie des Sciences, Paris, Fond Lavoisier, 1642[2]

Mon cher confrère,

J'ai cru devoir vous envoyer cette lettre, ainsi ecrite sans d'autres articles, pour que vous puissiez faire usage pour punir le coquin de ce Capitain à Rouen, si par hazar il vous reste le moien de le faire par votre Bureau ou par quelque autre moien. C'est voler d'une maniere la plus criante que d'agir comme ce coquin l'a fait dans l'affaire du fret de ces deux caisses. Si vous faites quelque chose à ce faquin pour le punir, je serai charmé de le savoir. il le merite fortement & sans pitié.

---

1 The note is on a scrap of paper, undated and unsigned, and gives no indication of addressee, but the handwriting is Magellan's and the contents indicate that it accompanied his letter to Lavoisier dated 14 July 1775.

2 Published in *Oeuvres de Lavoisier: Correspondance*, ed. René Fric, fascicule II, 1770-1775 (Paris, 1957), p. 497.

75.07.14b *To the Duke Charles-Marie-Raymond d'Arenberg*
Archief Van Arenberg, Enghien/Edingen, Belgium[1]

Monseigneur

Faute d'une bonne occasion je fus obligé de retarder jusqu'à present d'envoyer à V.A. S^{me}. Les crayons qu'Elle souhaitoit avoir pour Son portefeuille. M^{r}. Nedham gentil-homme Anglois,[2] qui passera quelque tems en Flandres & qui merite d'etre connu de V.A.S., veut bien s'en charger. Il apporte aussi un petit Livre, qui contient La traduction françoise de l'ouvrage de feu D^{r}. Gregory d'Edimburg,[3] dont j'ai envoyé à V.A.S. L'original il y a près d'un an & demi. La traduction ne l'égale point de beaucoup près: mais Les avis paternels qui y sont contenus, renferment des sentiments si delicats sur Le beau sexe, & des idées si precieuses & touchantes pour La conduite des jeunes-filles, que j'ose prier Mad^{me}. La Duchesse[4] & Sa Belle-fille M^{de}. La Princesse,[5] de vouloir bien destiner ce petit Livre (après l'avoir Lu attentivement, & l'avoir jugé digne de Leur approbation) pour que La jeune Princesse votre petite fille,[6] y prenne Les premieres Leçons, lors qu'elle arrivera à l'age de commencer à Lire, & de recevoir Les premieres idees de sa conduite.

L'attention de mon esprit ne Laisse pas d'accompagner V.A.S^{me}. dans son voyage à Arenberg, quoique je n'aie pas obtenu L'honneur de L'y accompagner personellement; Et Sans l'aprehension d'etre importun ou fastidieux je ne manquerois pas d'y envoyer plus souvent Le tableau de mes voeux & des souhaits de mon Coeur pour La tranquilité de celui de V.A., malgré tout le mauvais coloris de mon pauvre jargon.

Le meme M^{r}. Nedham porte pour V.A.S. un petit mape-monde, dans Le quel j'ai crayoné en rouge La route que mon ami M^{r}. Forster & son fils

1 The Arenberg archive also includes a copy written in a different hand of part of this letter, from "Le meme M^{r}. Nedham..." to "... dans son pais", headed: "Extrait D'une Lettre de M. L'abbé Magellan De Londres Le 14 Juillet 1775".
2 John Turberville Needham (1713–1781).
3 *Legs d'un père à ses filles* (London, 1774), French translation by "le Prof. Delapierre" of John Gregory, *A Father's Legacy to His Daughters* (London, 1774).
4 Louise-Marguerite von der Marck, Comtesse de La Marck et de Schleiden, Duchesse d'Arenberg.
5 Louise Antoinette de Brancas-Villars (1755-1812), Comtesse de Lauraguais, daughter of Louis Léon Félicité de Brancas, Duc de Lauraguais, Duc de Brancas and Elisabeth Pauline de Gand-Vilain de Mérode de Montmorency, Princesse d'Isenghien.
6 Princess Pauline d'Arenberg, born on 2 September 1774.

ont faite autour du monde,[7] avec le Cap^{tn}. Coock[8] dans ce dernier voyage: ils m'ont ecrit du Cap de Bonne Esperance par un vaisseau qui partoit pour L'Angleterre Le jour suivant de leur arrivée; La lettre est datée du 22 Mars dernier:[9] & on les attend ici dans peu de jours, car ils ne devoient y rester au Cap, qu'environ un mois pour se refraichir d'un si Long voyage. il n'y eut une seule personne à bord de ce vaisseau qui mourut de maladie. on attribue cette preservation à l'usage frequent qu'on y pratiquoit du Sour-Krout,[10] plat alemand que V. Alt^{e}. connoitra assez bien dans ses campagnes militaires, & à l'usage du wort,[11] c'est à dire de la liqueur faite avec le malt, ou grain preparé, dont on fait La biere. C'est une espece de petite biere toute fraiche avant de fermenter. Voici l'explication des chifres que j'y ai crayoné ce matin avec le craion rouge, ensemble avec M^{r}. Daines Barington frere du Secretaire d'Etat de Notre Marine, qui a voulû m'aider avec les notices qu'il savoit outre celles contenues dans ma Lettre.[12] V. A. S. ne doit pas prendre cette route comm'exacte. C'est seulement un à peu près. En partant d'Angleterre il y a plus de 3 ans, ce vaisseau arriva au Cap de Bonne Esperance en (a) avec l'autre vaisseau du Cap^{n}. Fourneaux.[13] de là ils sont allés chercher le Cape de la Circoncision, qu'ils ne purent point trouver après y avoir croisé plus de 3 jours dans ce parage.[14] il monturent en droiture au Sud jusqu'à (6) 66^{d}. 10^{m}. de Latitude. ils y trouverent uniquement de la glace. ils tournerent à l'Est par c, d e, entre 50 & 60^{d} de latitude, c'est à dire par environ les 55^{d}. du sud, par f jusqu'à la celebre isle de Otahitee marquée par g (ou environ) dans la mape. C'est à peu près alors que le Cap^{tn}. Cook se separa du Cap^{tn}. Fourneaux qui est deja de retour il y a environ un an. Mais Coock arrivant à la Nouvelle Zelland (h) prit la route tant soit peu à l'Est jusqu'à (i) la hauteur de 66^{d}. 30^{s}. du sud; oû il trouva des montaignes de glace convertes de Niege &^{a}. de là il tourna à K au Nord jusqu'à 54^{d} vers 75.07.14b

7 Johann Reinhold Forster (1729–1798) and his son, Johann Georg Adam Forster (1754-1794) accompanied Captain James Cook on his second voyage of discovery, 1772-5.

8 i.e. Cook.

9 Letter not found.

10 sauerkraut.

11 The liquid extracted from the mashing process during the brewing of beer or whisky.

12 Daines Barrington, FRS (1727-1800).

13 Tobias Furneaux (1735-1781). For this second voyage, James Cook commanded HMS *Resolution*, while Furneaux commanded its companion ship, HMS *Adventure*.

14 Jean-Baptiste Charles Bouvet de Lozier (1704-1786) reported his discovery on 1 January 1739 of a very high and iced land that appeared through the mist. He believed it was a promontory of the Antarctic mainland and called it "Cape Circumcision", but doubts remained concerning his report. Cook was instructed to find Cape Circumcision.

75.07.14b l'Est. il parcourut un peu de ce parallele jusqu'à L: & de là il monta presque droit au Pole jusqu'à 71^{d}. 10 du Sud à (m). Il n'y trouva que de la glace, & il paroit par le journal qu'il y trouva aussi quelque terre car il lui a donné le nom de Nouvelle Georgie Australe:[15] il retourna vers le Nord jusqu'à n, pour voir la terre qu'on dit avoir eté decouverte par Juan Fernandes:[16] mais il n'y trouva rien alors il mit vers le Nordwest jusqu'à l'Isle de Packe, autrement de Paris, ou suivant autres de Roggewein (marquée par 7) qui est à 28^{d} de lat. du Sud & 106^{d} à l'ouest de Londres.[17] Cette Isle est habitée par des gens policés, & il y fut Reçu fort bien. C'est un excellent climat, fertil & delicieux. on en peut voir La description dans le voyage de Roggewein qui se trouve dans la Collection des voyages de Harris en Anglois, faite par le D^{r}. Campbell.[18] De là il prit à la meme route de Norwest jusqu'aux Marquizes de Mendoza (marquées 8), qui sont habitées par des gens assez amis & policés, vers les 10^{d}. de lat. du Sud.[19] dans la suite il retourna[20] à Otahitee, marquée (p.g.) oû il fut Reçu tres amicablement, & oû il s'arreta pour quelque tems considerable. + de là ils partirent en droiture au Cap de Horn: decouvrirent une Isle nouvelle (n) vers les 54^{d}. du Sud & une autre (x) vers les 59^{d}. de Latitude du meme Sud[21] & enfin retournerent au Cap de Bonne Esperance d'oû on les attend tous les jours.

+ Je me suis oublié de dire qu'en venant de l'Espirito Santo[22] (r) vers s ils decouvrirent plusieurs Isles, particulierement une grande en S, qu'ils appellerent Nova Calidonia[23] environ 7 degrés avant d'arriver à la Nouvelle Zelande.

M^{r}. Forster fils me marque qu'il ne sera pas necessaire de chercher plus Les Terres Australes, car elles n'existent plus que dans les tetes des reves creux de la façon qu'on en parle dans quelques Auteurs. en effet une si grande

15 i. e. South Georgia.

16 Juan Fernández Islands, discovered by the homonymous Spanish explorer and navigator (c.1536–c.1604), situated about 667 km off the coast of Chile in the South Pacific Ocean.

17 The first European to record a visit to Easter Island (Île de Pâques) was the Dutchman Admiral Jacob Roggewein (or Roggeveen).

18 John Harris (ed.), *Navigantium atque itinerantium bibliotheca: or, a Compleat Collection of Voyages and Travels...* (2 vols, London, 1705 and subsequent editions); Magellan is probably referring to the revised edition by John Campbell "with large additions, and continued down to the present time" (London, 1764).

19 Marquesas Islands (Las Islas Marquesas de Mendoza), French Polynesia.

20 Word repeated as the page was changed.

21 The East Scotia Ridge, an active back-arch spreading centre located to the west of the South Sandwich island arc in the South Atlantic Ocean.

22 The largest island of the New Hebrides (Vanuatu).

23 New Caledonia, named in honour of Scotland (i.e. Caledonia).

quantite de glaces sembre[24] annoncer des terres: mais elles sont si près du pole & elles sont dans des climats si desavantageux pour la population civile qu'il n'y a pas de tentation à les habiter: Lui & Son Pere, tous les deux destinés pour Les Recherches philosoph[es]. & d'histoire Naturelle, portent une collection nombreuse de Plantes nouvelles jusqu'au Nombre de 260, & plus de 200 animaux nouveaux en y comptant Les poissons. La montre Marine faite par Kendal[25] à l'imitation de celle de Harrison a prouvé à merveille, en donnant très bien La Longitude du vaisseau.[26] Ce meme jeune voyageur, avoit deja vû dans si peu de tems le voyage precedent ecrit par Hauksword,[27] & il paroit en etre fort malcontent. il assure que les habitants de La Nouvelle Zelande ne sont pas aussi mechants comm'on les decrit dans cet ouvrage &[a]. on dit qu'ils avoient pris abord du vaisseau un des habitants d'Ottahitee pour l'amener en Europe; mais il leur a paru si stupide au bout de quelque tems, que Le Capit[n]. Cook l'a fait amener à Terre dans son pais. 75.07.14b

Il est tems de finir Cette Longue Lettre, en Reiterant les vrais sentiments de Coeur, avec lesquels J'ai l'honneur d'etre avec un profond Respect

Monseigneur

de V.A. S[me]

Serviteur tres h[ble]. & très obligé

J.H. de Magellan

à Londres ce 14 Juillet 1775

---

24 semble?

25 The watchmaker Larcum Kendall (1721-1795), whom the Board of Longitude appointed to make copies of John Harrison's marine chronometers. Cook took Kendall's so-called K1 on his second voyage and found it to perform extremely well.

26 John Harrison (1693-1776) invented the marine chronometer for determining longitude at sea.

27 John Hawkesworth, *An Account of the Voyages undertaken … for making Discoveries in the Southern Hemisphere and performed by Commodore Byron, Captain Wallis, Captain Carteret, and Captain Cook…* (3 vols, London, 1773).

75.07.24 *To Ribeiro Sanches*[1]

Österreichische Nationalbibliothek, Vienna
A. N. Ribeiro Sanches Correspondenz, codex 12714, 157-158v

Londres 24 Julho 1775

Meo Am^{o} e S^{r}. e m^{to}. do Cor^{am}.

Por crista q estou picado como uma cebola, com a sua metafora de fogo ja em borralho. á manhan pelas 10^{h}. está ja dada apalavra p^{a}. me despicar á sua saude pois assim o dezeja. tomára saber onde tinha lá ainda guardada a sua expressão de Riba-Coa ...! logo q Li a sua carta fui fóra: compreilhe o Kirkland; e antes de entrar em caza, passei por hũa certa tal, com q^{m}. estava arrufado por me ter faltado á pitada; apertou-me a mão, dizendo-me de mansinho q á manhan viria ver-me. pois olha lhe disse [...], eu ja não estava m^{to}. teo amigo. mas vem q t'o hei de meter até a arreigada por conta de fulano. What you say? I dont understand you? ... nothing at all, sais I. dont fail, and I'll wait for you about ten in y^{e}. morning. he hũa das nossas, isto hé, como as nossas q o fazem por gosto & sem interesse. em effeito eu ainda lhe não dei a valia de um seitil. mas tenho lhe feito algum bom serviço, alem de malhadura de q ella gosta mais q tudo. he aquella mesma de q^{m}. eu lhe contei, q na prim^{a}. vez q a colhi de jeito, me perguntou no meio da obra: and now what shall you think of my readiness? vim p^{a}. caza um pouco quente (pois fez hoje calma) & com sede. fiz uma tijela de migas cuido q se escreve meigas. a saber vinho (branco) do Porto com agoa & assucar, q he bebida q lá se não conhece na sua França, nem em Riba-Coa tão pouco, e puz me a escrever-lhe esta. Mas não a acabarei senão á manhan depois de ter dezempenhado o meo brio com a minha reconciliada, q se meixe á portugueza, & q nunca me disse ora basta, q ja estou cançada. antes parece q tem pundonor em não se contentar até escorrepichar o molho.

estou tentado a mandar-lhe um arratel do nosso assucar, q aqui chamão Lisboa q he do q eu uzo, mas q he feito aqui. he m^{to}. menos branco q o outro. mas sabe a assucar. o q eu gostei em Paris, & delle comprei [o] inverno passado um pouco, chamado castonada, não tem doce. parece adulterado. digão o q quizerem. aqui tudo he caro (ainda q o assucar desta casta não he senão a 6 pence q fazem [...][2]0 Reis) mas não ha tanta adulteração, nem a metade, como em Paris.

---

1 Annotated: " recue a 5 Ag^{o}." [received 5 August], "resp^{a}. a 11 7bre." [answered 11 September], and "about the inkpot".

2 Illegible.

Tabem tenho [h]um tinteiro p^{a}. mandar-lhe. he de vidro grosso mas cristalino. & tem um bico unicam^{te}. onde se mette a penna. deste modo, não toma mais tinta do q a necessaria. eiz aqui a sua figura. 75.07.24

a he a tapadoira, de vidro m.^{to} bem grinded, de modo q

não deixa passar o ar dentro do Reservoir, ou corpo do tinteiro. c he um bico, como o de hum candieiro onde se mete a penna b. ora p^{a}. o encher de tinta abre-(se) a rolha a poem-se o dedo no buraco c com um bocadinho de papel, p^{a}. não o sujar com a tinta (o dedo): logo q está cheio, poem-se-lhe a tapadoira a, q se aperta algum tanto. e então tira-se o dedo do buraco c. Como o ar não pode entrar por a a tinta não sahe por c, e fica suspensa até que se gasta seja à escrever, ou pela evaporação, a tinta q esta no bico c, porq então entra uma bolha de ar, & outra tanta tinta desce p^{a}. encher o bico c. se se inclina o tinteiro de a até b, então pode ainda descer mais tinta até encher todo o bico c. Ora como o ar não pode comunicar com a tinta q está dentro, não ha mais evaporação do q a da superficie da tinta q está no bico c. qd^{o}. a tinta faz sedimento, então, pode lavar-se o tinteiro em abrindo a rolha a. esta rolha he feita como hũa tassa, p^{a}. nella se metter obreias &^{a}. Ora voume a val de lençoes. ....

Meu amigo está V.M. servido. pudera escrever um tomo inteiro, se quizesse relatar-lhe todos os particulares, q deixo p.^{a} outra occasião.

V.M. engana-se com a pedra do edificio sobre as bussolas. não he coisa q possa durar m^{to}. tempo. ha m^{tos}. pretendentes, & m^{tas}. circonstancias q não prometem duração. mas vamos em q^{to}. vamos as m^{as}. correspondencias de commissão, essas sim q podem sempre ficar, & sempre dão p^{a}. os gastos.

não sei a sua ideia de aljibeiras providas. he termo vago. p^{a}. mim hũ shilling sempre foi assaz de provizão, pois basta p^{a}. me satisfazer o dezejo. V.M. là me ouvio como arranjei à Portuguesa os cem Luizes das destilaçoens da agoa do mar. Senão imprimisse pouparia a metade, pois tanto me custou a minha portuguesada. o leit(e) bem coado fica em viagens idas & vindas, & jogos de independencias, pois he necess^{o}. m(os)trar q não se necessita, p^{a}. alcançar por outros caminhos o q sempre se nega a q^{m}. faz ver q depende, q espera, & q está necessitado &^{a}. &^{a}. tal he a vida buzarona. ou p^{a}. milhor dizer tal a fazem os individuos humanos.

vou partir domingo 30 do Corrente p^{a}. Calne onde vive o D^{r}. Priestley, & Mil^{d}. Shelburne. talvez q chegue a Bath, & veja La Cour.

28 de Julho. Agora fiz embalar nova caixa p^{a}. Paris da qual M^{r}. Messier sera o distribuidor. elle mandará à V.M. um paquete q contem o seg^{te}. Kirkland on childbed fever – 3^{s}./ Pharmac. d'Edimburgo, encadern^{a}. – 4^{s}.3^{d}. Transac. Phil. Vol. 65 part. 1^{a}. – 7^{s}. 6^{d} ainda q he mui piqueno p^{a}. o

75.07.24 preço. / Farr on consonption – 1^s./ obser. on y^e. caract. of a Physician 2^s. 5^d. / grant on gaol pestilential fever – 3^s. / Chandler on y^e. Cataracta. 2^s.6^d./ Whithers on y^e. abuse of Medicine – 4./ Aikin on Hospitals. 1^s. 6^d. / Med^{al}. Transact. o principio do 3° tomo – 1^s. 0^d. / animadvers. on kin-cough – 1^s. 6^d./ dimsdale on small pox – 2^s. 6^d./ duncan coment. n°. 9 & n°. 10 – 3^s./ Lind on sea-men – 5^s. 0^d. / Heartly on y^e. Physical theory of mind by D^r. Priestley 5^s. 0^d. / e demais disto o tinteiro de chrystal q assima descrevi. despezas respectivas até Paris, segundo a observação das encomendas das outras vezes – 4^s. 6^d. Pois p^a. evitar embaraços, mando o calculo feito sobre os gastos, à fim q M^r. Messier não tenha mais do q carregar toda a soma na minha conta. total £ 2.11.7 q em dinh^o. de frança faz 61$^{\#}$.18^s.

Rogo a V.M. q tenha a bond^e. de mandar a inclusa á petite poste, exceto se o Ivan houver de passar pela Rue Copeau.

Saud^{es}. a M^e. Jeane, Toinete, Marie & autres amis se he q os ha. estou mui cançado de trabalhar com as encomendas desta Remessa & de escrever p^a. esta posta.

Vale me que ut soles ama

N.B. nao he possivel achar a carta do D^r. glass sobre o small-pox. a impressão está exhausta, & pode ser q não se reimprima. Porem V.M. porq não a manda pedir peremptoriam^{te}. por hũ escrito a M^r. Le Roy? Olhe q he boa paciencia somitega de lhe deixar là tão longo tempo os seos Livros ...! q tem V.M. q perder ou q ganhar em sofrer semilhantes imparraçoens?

London 24 July 1775

My dear friend

For goodness' sake I am piqued as an onion, with your metaphor of fire in the ashes. Tomorrow by 10 o'clock the word is given to take revenge, to your health, since you so wish. I wonder where you had kept your expression from Riba-Coa[3] ...! As soon as I read your letter[4] I went out. I bought you the Kirkland;[5] and before returning home, passed by a certain one, with whom I was angry because she had missed a meeting; she shook my hand, saying gently that tomorrow she would come to see me. Look, I said, I am no longer your friend. But come so that

3 A far inland region of Portugal, high up the Douro River.

4 Letter not found.

5 Thomas Kirkland, *A Treatise on Child-bed Fevers* (London, 1774).

I may go into you up to the root on behalf of the other fellow. What do you say? I don't understand you? ... nothing at all, says I. Don't fail, and I'll wait for you about ten in y[e] morning. She is one of ours, that is, like ours who do it for the fun & without interest. In fact I have not yet given her a penny. But I have done her some good service, besides the screwing that she loves more than anything. She is the one I told you about, who, the first time I took her properly, asked me in the middle of the affair: and now what shall you think of my readiness? I came home a bit hot (since today was a calm day) & thirsty. I made a bowl of soaked bread soup, that is white Port wine with water and sugar, a beverage that is not known in your France, nor in Riba-Coa either, and started writing this letter. But I won't finish it before tomorrow after having satisfied my pride with my reconciled one, who shakes like a Portuguese & has never said to me enough, I'm tired. Instead she seems to make a point of honour of not being content before draining the last drop. 75.07.24

I am thinking of sending you one pound of our sugar which I use, that here is called Lisbon but is made here. It is much less white than the other. But it tastes like sugar. The one I tried in Paris, & bought a little of last winter, named costonada,[6] has no sweetness. It seems adulterated. Whatever they say, here everything is expensive (though sugar of this kind is just 6 pence that makes [...]0 reis[7]) but there is not so much adulteration, not even half that in Paris.

I also have an inkpot to send you. It is made of thick but crystalline glass, & has a single beak where the pen is introduced. In this way it takes no more ink than necessary. Here is its figure. a is the stopper, of well grinded glass, such that it does not let the air into the

reservoir, or inkpot body. c is a beak, like that of a lamp where the pen b is introduced.

Now to fill it with ink the stopper a is opened, a finger is put in the hole c with a little bit of paper, so as not to stain it (the finger) with ink: as soon as it is full, the stopper a is put in and somewhat fastened. Then the finger is taken out of the aperture c. As the air cannot enter through a the ink doesn't come through c, and remains suspended until it is consumed either through writing or by evaporation, because then a bubble of air enters, & as much ink comes down to fill the beak c. If the inkpot is inclined from a to b, then more ink can come down till the beak c is completely filled. As the air cannot communicate with the ink that is inside, there is no more evaporation than that from the surface of the ink that is in the beak c.

---

6 A type of brown sugar.

7 *Real* (pl. *reis*) = Portuguese unit of currency. From account books kept at the Physics Cabinet of the University of Coimbra, where some notes on the acquisition of instruments made in London at this time are found, the equivalence to English currency is: 1 guinea = 3,740 reis, or £1 = 3,400 reis approximately.

75.07.24 When the ink produces sediment, the inkpot can be washed by opening the stopper a. This stopper is made like a cup for wafers &c. Now I am going to bed.

You are served my friend. I wish I could write a whole volume to tell you all the particulars, but I leave that for another occasion.

You are mistaken concerning the compasses as the foundation stone. It cannot last a long time. There are many competitors, & many circumstances that don't promise a long duration. But as we go on in the mean time the commissions from my correspondences will always stay, & they cover expenses.

I don't know about your idea of well provided pockets. It is a vague expression. To me one shilling has always been enough provision, since it is enough to satisfy my desire. You heard how I arranged in the Portuguese way the hundred Louis from the distillations of seawater. If I had not printed it I would have saved half, such was the cost of my portuguesada.[8] Milk is well skimmed by running to & fro, & by attitudes of independence, because it is necessary to show that one is not in need, in order to obtain by other means what is always denied to those who show that they depend, expect & need it &^{c}. &^{c}. Such is the buzarona[9] life or to put it better, so individual humans make it.

I'm leaving on Sunday the 30th of this month for Calne where D^{r}. Priestley lives, as well as Milord Shelburne. Maybe I shall go to Bath & see La Cour.

28th of July. Now I made up a new box to be arranged and sent to Paris of which M^{r}. Messier will be the distributor. He will send you a package containing the following: Kirkland on childbed fever[10] – 3^{s}. / Pharmac. of Edimburg,[11] hard bound – 4^{s}.3^{d}. Transac. Phil. vol. 65 part. 1st. – 7^{s}.6^{d} though it is too small for the price./ Farr on consumption[12] – 1^{s}. / obser. on y^{e}. caract. of a Physician[13] 2^{s}.5^{d}. / Grant on pestilential fever[14] – 3^{s}.-/ Chandler on y^{e}. Cataracta[15]. 2^{s}.6^{d}./ Whithers on y^{e}. abuse of Medicine[16] – 4./Aikin on Hospitals[17]. 1^{s}.6^{d}. / Med^{al}. Transact.[18] the

---

8 Bravery in Portuguese style. Much of Magellan's memoir, *Nouvelle construction d'alambic pour faire toute sorte de distillation en grand*, was printed in 1775. However, due to unforeseen circumstances, the work was not issued until 1781.

9 See Magellan to Sanches, 23 June 1775.

10 See above, note 5.

11 *Pharmacopœia Edinburgensis ...* (London, 1753).

12 Samuel Farr, *Aphorismi de marasmo, ex summis medicis collecti* (Altenburg, 1774).

13 Probably J. Gregory, *Observations on the Duties and Offices of a Physician* (London, 1770).

14 W. Grant, *An Essay on the Pestilential Fever of Sydenham* (London, 1775).

15 Samuel Chandler, *A Treatise of a Cataract, its Nature, Species, Causes and Symptoms, with a Distinct Representation of the Operations by Couching and Extraction, etc.* (London, 1775).

16 Thomas Withers, *Observations on the Abuse of Medicine* (London, 1775).

17 John Aikin, *Thoughts on Hospitals ... with a Letter to the Author by Thomas Percival* (London, 1771).

18 *Medical Transactions, published by the College of Physicians in London*, 3 vols. (London, 1768).

beginning of the third vol. 1^{s}.0^{d}. /animadvers. on kin-cough[19] – 1^{s}.6^{d}. /dimsdale on small pox[20] – 2^{s}.6^{d}. / Duncan coment.[21] no. 9 & no.10 – 3^{s}. / Lind on seamen[22] – 5^{s}.0^{d}. / Heartly on y^{e}. Physical theory of mind by Dr. Priestley[23] 5^{s}.0^{d}. / and apart from this the crystal inkpot that I described above. Expenses to Paris, according to the previous orders – 4^{s}.6$^{d.}$ To avoid embarrassment, I send the calculation of the expenses, in order that M^{r}. Messier has no more to do than put the total on to my account. Total £ 2.11.7 which in French money makes 61$^{\#}$.18^{s}. 75.07.24

I ask you to send the included à petite poste, unless Ivan[24] passes by Rue Copeau.

Regards to M^{e}. Jeane, Toinete,[25] Marie[26] & autres amis if any are there. I'm tired of working on the orders of this remittance & of writing this letter.

Vale meque ut soles ama[27]

N.B. It's not possible to find the letter of D^{r}. Glass about small-pox.[28] The edition is exhausted, & maybe it will not be reprinted. However why don't you ask for it through writing to M^{r}. Le Roy?[29] Look how good your avaricious patience is in leaving your books there for such a long time ...! What have you to lose or gain in suffering such troubles?

---

19 Thomas Kirkland, *Animadversions on a Late Treatise on the Kink-Cough: To which is annexed, an Essay on That Disorder* (London, 1774).

20 Thomas Dimsdale, *The Present Method of Innoculating for the Small-pox. To which are added, some Experiments, Instituted with a View to Discover the Effects of a Similar Treatment in the Natural Small-pox ...* (2nd edition, London, 1767).

21 *Medical and Philosophical Commentaries* [by A. D.] (Edinburgh, 1773 onwards).

22 J. Lind, *An Essay on the Means of Preserving the Health of Seamen* (London, 1774).

23 Joseph Priestley, *Hartley's Theory of the Human Mind, on the Principle of the Association of Ideas, &c* (London, 1775).

24 Ivan (or João) Pernelet was possibly studying surgery with the help of Sanches.

25 Marie Jeanne (Joana as Magellan often calls her) Pernelet was Sanches' housekeeper. Antoinette (Toineta, as Magellan sometimes calls her) Pernelet, a cousin of Marie Jeanne, became Sanches' housekeeper when Marie Jeanne died.

26 Possibly the little girl adopted by Sanches at that time.

27 Farewell and love me as you are accustomed to do.

28 Probably Thomas Glass, *A Letter ... on the Means of Procuring a Distinct and Favourable Kind of Small-pox. And the Use of Cold Air and Cold Water in Putrid Fevers* (London, 1767).

29 Bookseller in the rue St. Jacques in Paris.

75.08.00 *To [François Rozier]*[1]

Vous avez peut-être appris que le Capitaine Cook étoit déja arrivé de son voyage autour du monde; c'est une méprise de ceux qui l'ont avancé. Cet infatigable Marin est seulement au Cap de Bonne-Espérance, d'où je viens de recevoir une lettre de M. Forster, datée du 22 Mars dernier.[2] Le Capitaine Fourneaux[3] arriva jusqu'au 67e. degré de latitude au Sud, mais le Capitaine Cook a pénétré jusqu'au 71e degré 10'' du Sud, & passé jusqu'au 106e degré 30''. à l'Ouest de Greenwich. Il n'y a pas de terre ferme dans cette position; mais il a trouvé plusieurs Isles, dont quelques-unes ont jusqu'à 80 lieues de longueur. M. Forster apporte environ 260 plantes nouvelles, & 200 espèces d'animaux inconnus, en comprenant les poissons dans ce nombre. Voici comment ce célèbre Naturaliste s'explique dans sa lettre: "Notre voyage a été si heureux, que je ne crois pas qu'on fasse de nouvelles découvertes géographiques sur les prétendues terres australes. Nous pouvons nous flatter d'avoir fait de notre côté une collection considérable, quoique nous n'ayons pas eu, à beaucoup près, les grandes ressources de M. Banks.... J'ai parcouru la relation publiée par M. Hawkesworth,[4] & au premier coup-d'oeil elle m'a paru remplie de faits exagérés & très-peu qui soient bien exacts.... Il ne faut pas se faire une idée trop désavantageuse de nos nouveaux Zélandois, quoiqu'ils ayent massacré 11 Anglois de la chaloupe l'*Aventure*, & mangé le Capitaine François Marion avec vingt-

---

1 Letter not found. The text is from an undated extract published in *Observations sur la physique, sur l'histoire naturelle et sur les arts*, 6 (1775), 131-4, in the issue for August 1775, under the heading, "Extrait d'une Lettre de M. de Magellan, relative au nouveau voyage autour du Monde fait par le Capitaine Cook, & à quelques expériences sur l'Air fixe". An Italian translation of the article was published in *Scelta di opuscoli interessanti tradotti da varie lingue*, 12 (Milan, 1775), 3-11.

2 Letter not found. J. R. Forster accompanied Cook as naturalist on his second voyage of exploration, 1772-5.

3 Tobias Furneaux (1735-1781) commanded *Adventure*, consort of Cook's ship *Resolution* during Cook's second voyage of exploration, 1772-5. The two ships became separated and pursued the latter stages of their journeys independently.

4 J. Hawkesworth, *An Account of the Voyages undertaken by the Order of His Present Majesty for making Discoveries in the Southern Hemisphere...* (3 vols, London, 1773). Volumes 2 and 3 are devoted to an account of Cook's first voyage around the world, 1768-70, when he was accompanied by Joseph Banks. A copy of the work awaited Cook at Capetown on his way back to England in 1775; he was not pleased by the many inaccuracies he found in it (see J. C. Beaglehole, *The Life of Captain James Cook* [London, 1974], p. 439).

trois hommes de son équipage.[5] Je puis vous assurer que ce sont de bonnes gens qui ne traitent point mal les Etrangers. Je me suis trouvé presque seul au milieu d'eux, & ne puis que me louer du traitement que j'en ai reçu. On retrouve chez eux les témoignages d'honnêteté avec lesquels une Nation, qui n'est point civilisée, peut vous accueillir." 75.08.00

J'ai appris par d'autres du même voyage, que la montre marine faite par M. Kendal, à l'imitation de celle de M. Harrison, avoit eu le plus grand succès, donnant la longitude de ce vaisseau avec la plus grande précision.[6]

On n'a pas perdu, pour cause de maladie, un seul homme de l'équipage pendant un voyage d'un si long cours. On attribue cet heureux effet à l'usage salutaire du *sourkrout*[7] (ou chou porté à la fermentation acide), & à la boisson du *Wort* (ou malt de bière), qu'on distribuoit à l'équipage.[8] Je ne sais si on a fait également usage d'eau imprégnée d'air fixe. J'avois, dans le tems, préparé pour ces voyageurs les choses nécessaires à cet effet.[9] Cette méthode étoit la plus expéditive alors, mais aujourd'hui on a préparé à Londres un appareil en verre ou en crystal, au moyen duquel on fait ce mêlange d'air fixe avec l'eau avec beaucoup plus d'aisance, cependant elle exige plus de tems, & il faut deux ou trois heures pour que l'eau soit bien chargée d'air fixe.

Le Docteur Highius[10] démontre, à ce que l'on prétend, par une grande variété d'expériences, 1°. que l'air fixe est composé d'air commun & de phlogistique, mais que souvent d'autres matières entrent dans sa composition; 2°. que toutes les fois que le phlogistique est mêlé avec l'air commun, alors cet air est ce qu'on appelle *air fixe*. (Voyez la note rapportée à la fin de ce Mémoire[11]). Le Docteur Priestley, dans une lettre que je

---

5 Hawkesworth devotes several pages (vol. 2, pp. 388-95) to the cannibalism of the New Zealanders. The crew of one of *Adventure*'s boats were killed and eaten by Maori people at Queen Charlotte Sound, as were the French navigator Marc-Joseph Marion du Fresne and his crew in 1772, as Cook and his companions learned when they reached Capetown.

6 Cook carried with him on his second voyage a copy made by Larcum Kendall of John Harrison's famous chronometer, in order to test the performance of Harrison's invention under operational conditions.

7 i. e. Sauerkraut, used successfully as an antiscorbutic.

8 David Macbride had advocated the use of wort as an antiscorbutic in his *Experimental Essays …* (London, 1764).

9 Cook made no mention of any such apparatus in his account of the measures taken to preserve the health of his crew (*Philosophical Transactions*, 66 [1776], 402-6).

10 Bryan Higgins (1737?-1818). Magellan may be referring to his *A Syllabus of the Discourses and Experiments…* (London, 1775), or to the course of lectures summarized in this work that both Magellan and Priestley attended.

11 No such note appears at the end of the memoir.

75.08.00 viens de recevoir de lui,[12] dit autrement, c'est-à-dire, qu'il ne pense pas comme le Docteur Highius, sur la nature de cet air, & il avoue qu'il ne sait rien de certain là-dessus. Il ajoute que ses expériences postérieures lui ont donné des résultats singuliers, & qu'il prépare un second ouvrage qui contiendra des faits peut-être encore plus singuliers que ceux rapportés dans le premier volume.[13] Revenons à la manière de penser de M. Highius, & voici en abrégé quelles sont ses assertions. 1°. Si on met le pyrophore de M. Homberg,[14] (composé, comme on le sait, avec l'alun & le froment brûlé & réduit en charbon) dans une phiole au-dedans d'un récipient, l'air s'impregne du phlogistique qui s'échappe du pyrophore & devient air fixe. 2°. Lorsque le charbon brûle, le phlogistique qui s'en dégage, mêlé avec l'air commun, fait l'*air fixe*. 3°. Une chandelle qui brûle dans un récipient, rend l'air qui est dedans *air fixe*. 4°. La calcination des métaux où le phlogistique se dégage du corps métallique dans un vaisseau fermé, en rend l'air *fixe*. 5°. Le foie de soufre qui consiste dans l'union de l'acide vitriolique avec le phlogistique & l'alkali fixe, étant exposé à l'air, est décomposé, il ne reste qu'un *tartre vitriolé*, c'est-à-dire, l'acide vitriolique & l'air fixe; mais l'air dans lequel le phlogistique est reçu, devient *air fixe*. 6°. Dans la fermentation végétale, c'est l'air & le phlogistique qui s'en dégage, c'est-à-dire, l'*air fixe*. 7°. Les peintures, huiles & matières odorantes qui, comme on le sait, se déchargent de leur phlogistique, rendent l'*air fixe*. 8°. Le phosphore de Kunkel[15] (ou d'urine), étant décomposé dans l'air lentement, rend cet air *fixe* par le phlogistique qui s'en dégage. 9°. La terre calcaire réduite en chaux, exhale son phlogistique, en remplit l'air voisin, & par conséquent l'air qui environne les fours à chaux, est phlogistiqué, ou est *air fixe*. 10°. Enfin, lorsqu'on mêle de la limaille de fer avec le soufre, & qu'on pétrit ce mêlange avec de l'eau, le phlogistique en sort, & ce qui reste est l'acide vitriolique chargé de fer; mais l'air qui reçoit le phlogistique devient *air fixe*.

On voit que dans toutes les matières dont on vient de parler, il n'y a rien de commun que le phlogistique; & que toutes les fois qu'il se dégage de ces matières & qu'il se mêle avec l'air, alors cet air devient fixe. On voit encore

12 Letter not found.

13 The second volume of Priestley's *Experiments and Observations on Different Kinds of Air* was published later that year.

14 Among the numerous discoveries of the chemist Wilhelm Homberg (1652-1715) was a powder spontaneously inflammable in air that became known as "Homberg's pyrophorus".

15 The German chemist Johann Kunckel (or Kunkel) (1630?-1703) was often credited with the discovery of phosphorus.

dans l'expérience de M. de Laval sur le *bleu de Prusse*,[16] que le fer réduit en petites particules, étant phlogistiqué (ce qui fait le fond du bleu de Prusse), étant une fois dissous par l'huile de vitriol, prend une couleur brune; mais aussi-tôt qu'on y mêle de l'eau, celle-ci s'empare de l'acide, & alors le vrai bleu de Prusse reprend sa couleur primitive. De même dans le foie de soufre, si l'on n'y ajoutoit pas de l'eau, ou si l'air n'avoit pas d'humidité, il ne se décomposeroit pas. En y ajoutant de l'eau, le phlogistique se dégage, & se mêlant avec l'air, le rend *air fixe*. Qu'arrive-t-il dans la calcination des métaux? A l'air libre, le phlogistique s'empare de l'air & le rend *fixe*; au lieu que si on travaille dans des vaisseaux fermés, alors il n'y a que la petite portion du phlogistique qui trouve de l'air libre pour s'unir à lui, ce qui fait que la calcination n'agit que sur une des plus petites parties. Tout le reste ne se calcine point. L'expérience du Pere *Beccaria* de Turin, vient ici fort à propos.[17] Ce Physicien prit de la limaille d'étain & la mit dans une phiole de verre, qu'il fit cacheter hermétiquement avec un grand flacon de verre, en fondant les deux embouchures à la lampe. Il marqua avec un lime l'endroit où ces deux flacons se trouvoient en équilibre. Il mit le petit flacon avec la limaille sur une flamme d'esprit-de-vin, & observa toujours qu'il y avoit une petite pellicule de calcination proportionnée à la quantité d'air du flacon, c'est-à-dire, plus grande lorsque le flacon étoit plus grand. Le poids total étoit toujours le même (des deux flacons ainsi sellés hermétiquement), tant avant qu'après la calcination; mais le flacon d'air pesoit plus après, car il ne faisoit pas équilibre avec l'autre lorsqu'on les présentoit devant la marque faite à la lime. Suivant les expériences du Docteur Priestley & de M. Lavoisier, cet air contenu dans le grand flacon, étoit un *air fixe*, & on voit en même-tems, suivant la théorie du Docteur Highius, que c'étoit le phlogistique qui s'échappoit de la partie calcinée de l'étain qui rendoit cet air, *air fixe*; car c'est le phlogistique qui, par la séparation du métal, le rend calciné. Enfin, qu'on considère ce qui arrive dans l'eau de chaux. La terre calcaire qui y est, se trouve privée de son air & de son phlogistique par la calcination; qu'on y ajoute de l'air fixe, d'abord il s'unira avec sa partie terreuse, elle se précipitera & deviendra trouble, &c. 75.08.00

16 Edward Delaval, "Recherches expérimentales sur la cause des changements de couleur dans les corps opaques naturellement colorés", *Nouveuax Mémoires de l'Académie des Sciences et Belles-Lettres [de Berlin]*, 1774 (published 1776), 154-94; pp. 176ff.

17 Beccaria described his experiment in a letter to Lavoisier dated 12 November 1774, an extract from which was published in *Observations sur la physique, sur l'histoire naturelle et sur les arts*, 4 (1774), 453.

75.08.25 *To Ribeiro Sanches*[1]

Österreichische Nationalbibliothek, Vienna
A.N. Ribeiro Sanches Correspondenz, codex 12714, ff. 149-150

Londres 25 Ag^{o}. 1775

Am^{o} e S^{r}. e m^{to} do Cor^{am}

Aqui cheguei depois de 4 ou 5 dias. passei coiza de 10 com M^{d}. Shelburne, 6 com o D^{r}. Priestley q imprime actualm^{te}. um 2do. vol. sobre as suas novas descubertas de differentes especies de ar: fui à Bath, onde o D^{r}. La Cour me Recebeo e reteve em sua caza 4 dias. falamos m^{to}. de V.M. elle achou o seo centro. ganha din^{o}. mais q algum dos outros mas com modo de gente & sem se abaixar au petit gain. não sahe antes das 10^{h}. ainda q seja p^{a}. hum Princepe. elle trata o Princepe de Anhalt q ali está com sua Esposa. eu tinha feito conhec^{o}. com elle em caza de M^{d}. Shelburne, onde o Princepe veio passar alguns dias e invitou-me à hir jantar com elle & tãobem o D^{r}. La Cour foi invitado. mas elle não pode ou quiz hir. o certo he q estava resfriado, & apenas sahia a ver raram^{te}. algum doente. nunca sahe senão nũa cadeirinha ou chaise à porteur. elle em fim está no seio de Abraham. poucos dias antes tinha ido fazer uma vizita a Wales onde gastou mais de hua semana ou 15 dias com sua M^{er}. Mas teve not^{a}. em Oxford de q o Princepe de Anhault lhe vinha recomendado pela Duqueza de Bristol (cuido q est he o seo nome) & veio logo p^{a}. Bath. ficamos m^{to}. mais amigos q d'antes. quer estar em correspondencia com̃igo. ordenoume Livros, um Barometro & thermometro &^{a}. eu intendo hir jantar á manhan com o D^{r}. Sequeira: & não deixarei de lhe escrever no principio da semana q vem, pois tenho aqui um conselheiro do gran Conseil de Malinas q me foi Recomendado, & dom^{o} vou com elle ao Campo. logo q vier ficarei com mais tempo p^{a}. outros objectos.

Jantei dom^{o}. passado com meo amigo Forster, q veio com o Cap^{tao}. Cook do mar do Sul onde descubrio novas ilhas desconhecidas. elle (Forster) me diz q o galico se acha de tempo immemorial em varias ilhas onde nunca chegárão Europeos. Assim q parece q esta enfermid^{e}. foi gerada em varias partes, sem communicação de hũas com outras. Eu quiz aproveitar a ocazião do D^{r}. le Begue du Presle p^{a}. mandar à V.M. estas Regras, ainda

1 Annotated "Recu a 9 de 7^{bro}." [Received 9 September] and "R^{a} a 9 de 8^{bro}" [Answered 9 October]. Letter not found.

q me acho acabrunhado com m^{to}. a q responder nesta posta. Saud^{es}. a M^{a}. 75.08.25
J^{e}. Toinete & mais amigos se he q os ha.

Vale me que ama

a Monsieur
M^{r}. Sanchez D^{r}. en Medecine
Rue des Blancs-manteaux
à Paris

London 25 August 1775

My Dear Friend

I arrived here 4 or 5 days ago. I spent approximately 10 days with Milord Shelburne, 6 with D^{r}. Priestley who is presently printing a 2nd vol. on his new discoveries of different species of air:[2] I went to Bath, where D^{r}. La Cour received me and kept me at his place for 4 days. We spoke much about you. He has found his niche. He earns more money than anyone else but in a proper manner & without lowering himself au petit gain. He doesn't leave home before 10 a.m. even for a Prince. He takes care of the Prince of Anhalt[3] who is there with his wife. I had met him at Milord Shelburne's house, where the Prince spent a few days and invited me to dinner as well as D^{r}. La Cour. But he could not come or didn't want to. The fact is that he had a cold, & only rarely went out to see a patient. He only leaves on a chair or chaise à porteur. At last he is in the peace of Abraham.[4] A few days before, he visited Wales, where he spent more than a week or a fortnight with his wife, but he heard the news at Oxford that the Prince of Anhalt was recommended to him by the Duchess of Bristol (I think this is her name)[5] & promptly returned to Bath. We have become much friendlier than before. He wants to keep up a correspondence with me. He ordered books, a barometer & thermometer &^{c}. I intend to go and dine tomorrow with D^{r}. Sequeira: & I shan't fail to write to you early next week, since I have here a counsellor of the Grand Council of Malines[6] who was recommended to me, & on Sunday I will go with him to the countryside. As soon as I come back I will have more time for other matters.

2 The second volume of a series of three entitled *Experiments and Observations on Different Kinds of Air*, by Joseph Priestley (London, 1774-77).

3 Probably the Anglophile Leopold III Frederick Franz (1740-1817), Prince of Anhalt-Dessau, who is known to have visited England on several occasions.

4 Cf. *Genesis*, *14*: 15, where it is written that God said to Abraham, "Thou shalt go to thy fathers in peace; thou shalt be buried in a good old age".

5 Elizabeth Chudleigh, Countess of Bristol (1720-1788), wife of Augustus John Hervey, third Earl of Bristol, and also Duchess of Kingston as a result of a bigamous marriage to the second Duke of Kingston.

6 Possibly Godefridus Gasparus Tackoen (1720-1788), lawyer and counsellor at the Grand Council of Malines (Mechelen), the Supreme Court of the Low Countries; see Magellan to Sanches, 15 November 1776.

75.08.25 I had dinner last Sunday with my friend Forster,[7] who returned with Capt. Cook from the South Seas[8] where he discovered new unknown islands. He (Forster) tells me that syphilis has been found since time immemorial in various islands where Europeans have never landed. So it seems that this infirmity originated in different parts of the world, without communication among them. I wanted to take the occasion of Dr. le Begue du Presle[9] to send you these lines, though I find myself overwhelmed with much to answer in this post. Regards to Me. Je. Toinete & other friends[10] if any are there.

Vale me que ama[11]

---

7 Johann Reinhold Forster (1729-1798) went as naturalist with James Cook on his second voyage around the world, 1772-5, accompanied by his son Georg (1754-1794).

8 James Cook (1728-1779), the famous English explorer, returned to England at the end of his second voyage around the world on 30 July 1775.

9 Achille-Guillaume Le Bègue de Presle (c. 1735-1807), French physician, a friend of Jean-Jacques Rousseau and Antonio Ribeiro Sanches.

10 See Magellan to Sanches, 24 July 1775, note 25.

11 Farewell and love me.

75.09.00 *From Jacques-André Mallet*

Bibliothèque de Genève, Geneva, Ms suppl. 1654, fol. 56[1]

A Mr Magellan à Londres.

Septembre 1775

Vous jugez bien mal de moi, en concluant de mon Silence, que je ne me souhaite plus d'avoir la lun. achrom. de Dollond.[2] Soyez sur au contraire que si j'eusse changé d'avis, je me serois très pressé de vous l'écrire, pour vous demander la permiss. de degager ma parole, & qu'agir autrement auroit eté très malhonnete. Ce n'est pas non plus la faute de la poste, car j'ai b. reçu en son tems la lettr. que vous m'av. f. le plais. de m'écrire d'Enghien le 24 May.[3] Pourquoi donc ne vous ai-je répondu, c'est qu'il m'a paru que votre lettre ne demandoit aucune réponse. vous me dites, <u>me voici à chemin pr retourn. à Londres oû je tacherai de vous faire expedier au plut. possib. votre lun.</u> vous m'aviez deja marqué de Paris qu'elle etoit finie. Je n'avois donc rien à

---

1 Letter not found. The text is from a draft kept by Mallet. This letter is clearly of a later date than Magellan to Mallet, 1 September 1775, to which it is the answer.

2 See Magellan to Mallet, 1 September 1775.

3 See Magellan to Mallet, 24 May 1775.

faire qu'à attendre votre arr. à L. & la nouvelle du départ de la Lun. aussi connoissant M[r] votre exactitude, etois-je fort surpris de ne recevoir aucune nouvelle, Quoiqu'il en soit Je suis b. aise d'apprendre que cette L. est entre vos mains, & je vous prie de vouloir bien me l'envoy. le plutot possible, en y joignant le piedestal commun que je vous avois demandé. 75.09.00

Quant à la machine parallatique, il est vrai que dans votre lettre vous m'offrez de la faire faire à Londres, c'est le seul article qui exigeat une rep. & si je ne vous en ai point fait, c'est que j'attendois tj[s] quelque nouv. de M[r]. Meniers,[4] vis à vis duquel je me suis cru en qlq. façon engagé, je n'en ai point eu jusque à présent, je ne sais point s'il a travaillé ou non à la machine, ou s'il ne compte plus sur moi, ainsi Monsieur pour vous répondre, Je vous dirai que je prefere bien à avoir cette machine faite à Londres sous votre direction, car qlque bo[nn]e opinion que j'aye de M[r]. Meniers, il est François, ce n'est pas une recommandation en fait d'exactitude. mais la chose ne peut pas avoir lieu sinon que 1° Je ne sois sur d'etre entierem[t] dégagé avec M[r] Meniers, & 2° que le prix ne surpasse pas les 30 Louis que j'avois consacré à cela; car je ne comptois d'abord y mettre que 25 L. Je suppose la machine faite sur le dessin que vous m'avez envoyé, & sur le modele & les dimens. que Mr Meniers devoit y donner. Je voudrois savoir si Mr Men. a commencé la machine pour moi, si je lui écris, il ne manquera pas de dire que oui, Ne pourriez vous point Monsieur savoir cela de quelque maniere, mais en attendant que tout cela soit décidé je ne voudrois pas que l'envoy de la Lun. en souffrit du retard, ainsi je vous prie M[r]. de la remettre avec son piedestal simple aussi tot qu'il sera achevé à M[rs] [Naville] &c. les même negocians dont vous vous etes deja servi une fois qui sont chargés depuis un an ou deux de vous rembourser vos depenses.[5] vous pourrez mettre dans le même ballot le petit instrum. de Mr Dessauss.[6] qui a ete extremem[t] faché de l'accid[t] qui l'a empeché de le recevoir plutot, il auroit fort desiré l'avoir dans un voyage qu'il a fait dernierem[t] dans les montagnes.

Je vous remercie bcoup de ce que vous m'avez procuré la connoiss[ce] de M[r]. le chev. Shuckburg, j'ai eu un gr[d] plais. à voir ses instrumens je souhaiterois qu'il fut passé à Geneve dans une autre Saison, parce que vivant à la Campagne je n'ai pu le voir que rarem[t], il a fait sur nos montag. vois. qlques observat. avec ses barometres.[7]

---

4 Mégnié.

5 While the name of the London agents to whom Mallet here refers is not clear, it is certainly different from the names he had specified earlier; see Mallet to Magellan, 24 January 1774.

6 i.e. H.B. de Saussure.

7 The draft is unsigned.

75.09.01 *To Jacques-André Mallet*[1]
Bibliothèque de Genève, Geneva, Ms suppl. 1654, fol. 40-41

Londres ce 1. Sep^{bre}. 1775
at n°. 12 Nevils Court Fetter Lane

Monsieur

Si je ne savois par experience la falibité[2] des postes, même de celles qu'on suppose les plus exactes & les plus regulieres, je ne manquerois de conclure par votre silence que vous aviez perdu la patience & ne vouliez plus la Lunette achromatique de M^{r}. Dollond que je vous ai ordonnée; ou du moins que vous aviez changé d'avis, & vouliez vous en passer. dans ce dernier cas je vous assure que je n'en serois point faché du tout. Car celle-ci est la derniere que M^{r}. Dollond peut fournir jusqu'à ce que le hazard nous donne du flint-glass assez bon pour des ouvrages pareils: & il y a long tems qu'on en trouve une grande disete, ou pour mieux dire une entiere disette. Quoiqu'il en soit pour le fait du motif qui cause votre silence, voici a peu près ce que je vous ecrivis dans ma derniere lettre du 26 May dernier[3] qui fut mise à la poste de Bruxelles, oû je me trouvois avec son Altesse le Duc d'Arenberg, qui m'y mena à son retour de Paris, avant du mien en Angleterre. 1°. Je vous envoyai dans cette lettre ci-dessus le billet que vous m'aviez fait tenir à Paris pour la valeur de 25 louis a peu près, que M^{r}. Megnier[4] devoit toucher en faisant le piedestal paralactique que vous souhaitiez pour votre Lunette. J'avois gardé ce billet entre mes mains en attendant le retour à Paris de cet artiste, qui se trouvoit absent dans sa patrie (si je ne me trompe pas Dijon): mais il s'y laissa rester de semaine en semaine à cause de son mariage & d'autres affaires: & n'etoit pas de retour lorsque je partis de Paris le 22 Mai. ainsi j'ai cru ne devoir pas le laisser à personne, mais de vous l'envoyer, comme je le fis en effet dans ma lettre ci-dessus. 2°. je vous disois qu'aussitôt que j'arriverois en Angleterre, & que je pourrois avoir cet instrument, c'est à dire cette lunette finie entre mes mains, je ferois faire le piedestal commun que vous souhaitiez. 3°. je crois vous avoir dit aussi, que je tenterois de faire reussir ce meme piedestal pour pouvoir servir de machine parallatique. mais je ne me souviens pas de cette

1 Annotated: "Septembre 1775".
2 falibilité?
3 See Magellan to Mallet, 24 May 1775.
4 Mégnié.

particularité. Car je n'ai pas gardé de copie de la lettre en question. 4°. enfin: que j'avois fait faire une petite alteration ou changement dans l'instrument destiné pour M^r. le Prof^r. dessauseur,[5] mais que l'artiste de Paris l'avoit manqué: & qu'il faloit le faire racommoder de nouveau à Londres. &^a. 75.09.01

A present je vous dis de plus que j'ai deja chez moi la Lunete finie de Dollond: que je payai £29.8^s.0^d Esterl. & que si vous ne la voulez pour vous entierement, c'est à dire pour vos observations astronomiques, independement de la volonté d'aucun autre, alors je veux la garder & la passer a quelqu'autre Astronome, pour le bien & service de l'Astronomie. Car peut etre M^r. Dollond ne fera pas une autre de sa vie: attendu la detresse oû l'on est de bon-flint depuis long tems. M^r. de la Lande a vendu celle qui je lui avois procurée à une autre personne: & à present il ne peut pas avoir une de cette espece, ni meilleur ni moins bonne, malgré toutes les applications & deligences qu'il a faites. M^r. de Luc qui s'avoit chargé de cette commission aussi que moi, reçut la meme reponse & refus de M^r. Dollond, tout à cause de ne pas pouvoir avoir un bon morceau de flint-glass.

J'ai fait un tour à Bath, & Oxford après mon retour à Londres, d'oû je ne suis revenu que la semaine derniere. je vais faire executer le piedestal pour cette lunette que je vous destinois suivant mes idees, & recommoder ou arranger de nouveau le petit instrum^t. destiné pour le Professeur Dessasseur, auquel vous presenterez, je vous prie, mes Resp^s.: & en atte[ndant][6] je pourrai avoir votre resolution là dessus suivant que je viens de vous detailler dans cette Lettre. N.B. je n'eus la lunette finie qu'à la fin de Juillet: je fis mon tour à Bath & Oxford dans la suite. Je vis dans ce dernier endroit le beau observatoire qu'on y batit. il y a deja des insrum^ts. excellents. & excede deja celui de Greenwich dans l'instrument des passages, & dans le secteur au Zenith, qui tous les deux sont superieurs.

Je suis avec toute consideration possible
Monsieur
Votre très h^ble. serv^r.
Magellan

A Monsieur
Monsr. Mallet Professeur d'Astronomie
Cour de S. Pierre
à Geneve
Suisse

---

5 i.e. H.B. de Saussure.
6 Paper damaged.

75.09.06 *To Benjamin Wilson*

British Library, Add. MS 30094, f. 154[1]

London 6 Sept^{ber} 1775
Nevils Court Fetter Lane

Dear Sir

Here I send according to your desire the article of the letter I received from y^{e}. R^{d}. M^{r}. Beccaria, Professor at Turin, dated 1. Sept^{ber}. 1772,[2] in answer to my inquiry about y^{e}. practical method for making his curious experiment already published in y^{e}. Philos^{l}. Transact. of y^{e}. R^{l}. Soc^{y}. of London vol ... pag ...[3]

The occasion for my inquiry was the want of success, my good friend the Abbé Bourriot, canon of y^{e}. Cathedral of Bazas in Gascoigne, had in repeating this experiment at Paris, after I wrote to him about it. But I am as yet very sorry that something is still wanting in the method hereafter prescribed by the Author to have any success in it; since I am sure I did take all proper care to perform the tryals I made of late without better answering my expectations. I have already acquainted Professor Beccaria with y^{e}. whole, desiring him to favour me with fuller instructions about this experiment: and I'll not fail on communicating them to you as soon as I happen to have his answer thereto.

I am

Dear Sir
Your most obed^{t}. h^{ble}. serv^{t}.
J.H. de Magellan.

---

1 Wilson published a lightly edited version of this letter in the "Additions" he inserted in the second edition of his *A Series of Experiments on the Subject of Phosphori, and Their Prismatic Colours: in which are discovered, some New Properties of Light* (London, 1776), pp. 100-01. In doing so, Wilson reported that he had witnessed the experiments to which Magellan refers in the letter and had asked him to confirm in writing their lack of success. Magellan would thus have written the letter knowing that Wilson intended to publish it.

2 Giambatista Beccaria (1716-1781), professor of physics at Turin. Beccaria's letter not found. Following Beccaria's death, Magellan sent the letters he had received from him to Count Prospero Balbo in Turin, at Balbo's request, for inclusion in the comprehensive edition of Beccaria's writings that Balbo was preparing (see Magellan's annotation on a letter from F.U.T. Aepinus published in *Gentleman's Magazine*, 54 [1784], 563-4). However, Balbo's proposed edition was never published.

3 Magellan left spaces with rows of dots in which to insert the volume and page details of Beccaria's paper. This was a brief note from Beccaria, "Letter ... to Mr. John Canton F.R.S. on his new Phosphorus receiving several Colours, and only emitting the same", *Philosophical Transactions*, 61 (1771), 212.

The article of Profess^{or}. Beccaria's letter runs as follows. 75.09.06

... Metto nelle scattolle di ottone fosfforo elettissimo: nella stanza buia metto tali scattolete in centro ad anelli di ferro molto massicci, & quasi roventi; & cosi spoglio il fosfforo di ogni Luce: Copro le scattolete cosi coperchi di ottone al buio: e cosi coperte, una persona le reca al sole, dove il fosfforo bee la Luce attraverso al vetro colorato per alcuni secondi. Apro al buio le scattolete, che mi si recano cosi chiuse: e allora vedo il fosfforo a rendere una luce tinta di quel colore, che ha il vetro incastrato nel coperchio. Perciochè la Luce, che passa per tali vetri é pure composta; eperò la Luce che rende il fosfforo é solamente tinta di quel colore, che passa più copioso per i cristalli. La differenza, se bene mi sovengo, é piu discernevole usando un coperchio di colore rosso intenso, ed uno di colore giallo. Ho sperimentato lo stesso con diamanti fosfforici.

y^{e}. Translation in to English.

I put the most selected phosphorus in small boxes of brass. these are put into thick rings of iron, made red-hot, in a dark room: by which means the phosphorus is discharged of y^{e}. whole light it had received before. the boxes are then cover'd with caps of metal, each a piece: and a second person carries them to the Sun-shine, where y^{e}. phosphorus absorbes it's beams during some seconds, thro' the colour'd glass (it is meant, with which every box remains cover'd, after its cap of brass has been taken off). I discover them as soon as they are brought to me in y^{e}. dark closet. and then I perceive the phosphorus gives a light tinged with y^{e}. same colour of y^{e}. glass which was over it. Altho' the light transmitted through the glasses is mixed with other colours, that however given by the phosphorus is only tinged by the colour which passes more copiously thro' the respectif glass. the difference, if I recolect well, is more distinctly perceived when a high-red glass covers the top of y^{e}. box, and likewise when a yellow-glass is imployed. I have found the same result in making such experiments with phosphoric diamonds.

N.B. M^{r}. Magellan desires to have the English corrected in case this might be printed: but at the same time wishes that no changement be made any ways in the sense, nor any part of y^{e}. whole be left out.[4]

---

4 Wilson published the extract from Beccaria's letter, in both the original Italian (following the second version, set out below) and an improved English translation, in the second edition of his *Series of Experiments on the Subject of Phosphori*, pp. 95-7.

75.09.06 Metto[5] nelle scattolette di ottone fosforo elettissimo; nella stanza buia metto tali scattolette in centro ad anelli di ferro molto massicci, e quasi roventi, e cosi spoglio il fosforo di ogni luce: copro le scattolette con coperchi di ottone al buio; e cosi coperte, una

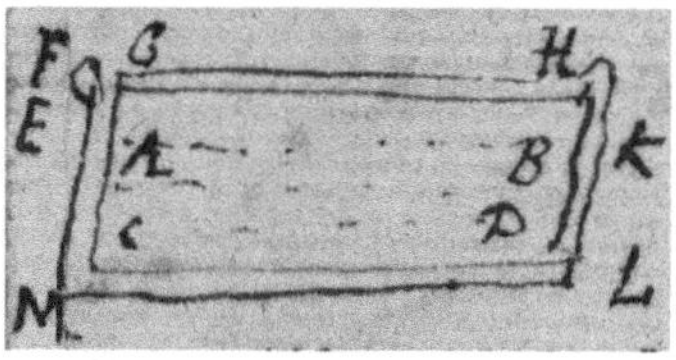

persona le reca al sole, dove il fosforo bee la luce attraverso al vetro colorato per alcuni secondi; Apro al buio le scattolette, che mi si recano cosi chiuse; e allora vedo il fosforo a rendere une luce tinta di quel colore, che ha il vetro incastrato nel coperchio – Perciochè la luce, che passa per tali vetri é pure composta; eperò la luce che rende il fosforo é solamente tinta di quel colore, che passa più copioso per i cristalli – La differenza, se bene mi sovengo, e più discernevole usando un coperchio di colore rosso intenso, ed uno di colore giallo. Ho sperimentato lo stesso con diamanti fosforici. ABCD fosforo. AMBL scattoletta di ottone. GTE, HIK coperellio di ottone. GH vetro colorito incastrato nel coperihio[6]

Mondovi 1. 7bre. 1772

Giambatista Beccaria F.S.P.[7]

To

M^r^. Wilson F.R.S.

Great Russel Street

Facing y^e^. British Museum

London

---

5 In the manuscript, everything to this point is in Magellan's handwriting. What follows, which repeats the earlier extract in Italian with additions and some variant spellings, is in a different hand, but on the same single folded sheet of paper.

6 *sic.* The lettering is as given, even though not all the letters used appear on the diagram. The lettering and spelling were corrected and the diagram drawn more clearly, as follows, in the version published by Wilson:

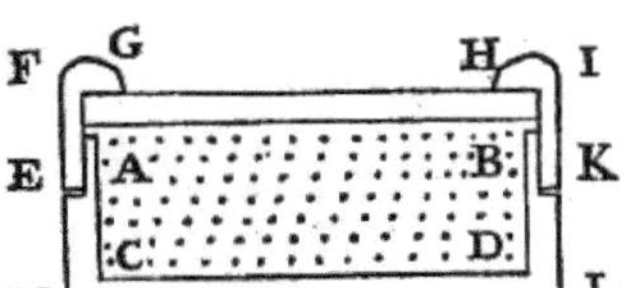

7 Frater scholarum piarum. Beccaria was a member of the Piarist order of priests. Though he was based in Turin, Mondovi was his home town.

To Ribeiro Sanches[1]

75.09.16

Österreichische Nationalbibliothek, Vienna
A.N. Ribeiro Sanches Correspondenz, codex 12714, 161-162v

Londres 16 Sept^{bro}. 1775

Meo am^{o} e S^{r}. e m^{to}. do Cor^{am}.

Começo a escrever esta em Reposta à de V.M. de 7 do Corr^{te}. Recebida esta manhan, de caza do D^{or}. Seq^{ra}. com q^{m}. jantei hoje, & a q^{m}. Li a sua magana Letre Conimbricense. Como elle me comunicou hũa do D^{r}. La Cour q ultimam^{te}. Recebeo, quero copiar-lhe aqui o artigo q me diz Respeito, p.^{a} q V.M. julgue do humor & pachorra deste comum e estimado amigo; pois q vejo pela de V.M. o m^{to}. q se interessa no seo bien etre.

"divide my thanks with M^{r}. Magellan for the trouble you both share in my trifling commissions. I sincerily wish to make better returns than empty words. The box I Received this morning (hũa caixa em q lhe mandey varios livros) but have not yet had time to open it. What strange beings! how anxious yesterday to have it! to day possession wipes off curiosity. is it not the same through Life? indeed it is. We in every stage of Life are meer children of fancy. enough. this heavenly storms of thunder Lightning and rain I believe would make Foot western &c. quit their jock and spritly wit, to moralise with me. still you are not to suppose me in a very grave disposition. no my friend. it means no more nor less than to prolong our conversation, whilst I am writing to you I please myself; if it answers the same purpose whilst you read, exegi monumentum ... &^{a}.

Aqui tem V.M. o humor do nosso D^{r}. La Cour.

eu tenho ja todas as suas encomendas de V.M. q mandarei pela primeira occazião. a saber os livros e o tinteiro como o q V.M. deo ao Conde Carburi. Cada um custa 2 shillings q carregarei pois V.M. o quer na sua Conta.

Tenho estado varias vezes com o nosso Enviado q he hum vray galant homme. bom juizo & excellente natural. vive na melhor vida domesti[ca][2] com sua Mulher q he uma fidalga de Guimaraens e q não tem nada das

---

1 Annotated "Resp^{a}. a 9 de 8^{bro}." [Answered 9 October]. Letter not found.

2 Here and elsewhere in this letter, words or parts of words have been lost through staining of the paper or obliteration of the margins.

75.09.16 basofias da Corte. agora lhe deo uma menina, depois do primº. parto q foi macho, ambos vivem: mas o filho está mal dos dentes q lhe sahem, doença q apenas se conhece em Portugal, e q nestes paizes do norte faz grande estrago nas Crianças. O D^{r}. Sequeira passa bem. hoje he 6ª frª. 30 de 7^{bro}. e cuido q á manhan hirei jantar com elle, exceto se for pª. caza do Amigo Arbuthnot onde passo com gosto & satisfação a maior p^{te}. dos Domingos q me ficão Livres. eu lá estive domingo passado e a segunda frª. por isso talvez q me deixe ficar o q vem, e então hirei jantar no sabº. com Sequeira. de outra sorte não teria tempo pª. hir na Carrossa de Mitcham q parte ás 4 horas da tarde neste tempo. Não me admira q V.M. se ache logrado por Castriot[o] & por todas as sevandijas do mundo, pois q V.M. se deixa dar ao Logro, [&] ainda não tem aprendido a considerar todo o hommem como um verdº. vilhaco emq^{to}. não acha por obra q elle o não hé. esta deve ser a regra geral pª. viver no mundo: mas ainda q eu a conheço mui bem [pela] propria experiencia, não me gabo de a practicar toties quoties.

V.M. me deixa em aflição com os disgostos em q se acha como V.M. diz por cauza do Sangue. ainda em cazo q V.M. tivesse filhos, não devia affligir-se por não ter mais pª. gastar com elles. mas não os tendo, nem parentes q se achem privados do physico necessario pª. o sustento da vida, porq rezão se agonia V.M. com coizas q não pode Remediar! nada ha senão o physicam^{te}. necessario pª. a propria subsistencia ou Recuperação da doença q deva darnos aflição. Mas em cazo q V.M. tenha algũa occurrencia de necessid^{e}. pª. si próprio, não me faça a afronta de m'a encobrir. em todo o tempo pode tirar á vista sobre mim por 15 ou 20 guineos; e tãobem mais, pois tenho amigos pª. me valer & pª. acudir a q^{m}. devo. Porem eu sei construir toda esta barafunda. q^{tos}. conhecem a V.M. estão certos de o fazerem infeliz com as suas choradeiras pª. lhe chuparem o sangue: & V.M. corta por si próprio & pelo necessario pª. o seo bien etre & conservação, som^{te}. pª. lhes dar pª. as suas superfluidades. eiz ahi o q se segue de possuir um coração maviozo como as Mains da nossa Terra. ora basta de Sermão

[V.]M. não faz se não rir-se das minhas proezas: & como já não arr—ta, não quer crer q os outros ainda q hum pouco idózos o possão fazer. estou Rezoluto na primª. vez q for à Paris à levar comigo hũa Ingleza, pª. lhe dar uma prova de facto á sua vista diante de testemunhas: pois a dizer-lhe a verd^{e}. o não poderei fazer limpamente com hũa franceza. tanto q aqui cheguei, cuidei renovar de 10 annos pelo menos: e ainda q me acho m^{to}. mais gordo contra o meo dezejo (mas descanço de corpo e alma sempre fazem engordar) não deixo por isso de dar o meo furo com toda a perfeição.

a triste nova do Princepe de Arenberg, q recebeo hum tiro de espingarda no rosto e peito andando á caça, me tem dado m^ta^. pena. estive quazi à partir p^a^. Bruxelles. porem Recebi noticia de q vai melhor, e q não ficará cego como se temia. 75.09.16

Vale me que ama.

Hoje escrevo em Resposta ao D^r^. le Begue du P[resle] de proposi[to] [...] amigo M^r^. de Malsherbes talvez q faça algũa coiza por elle. iterum vale. mes amities & bons souhaits à M^e^. Jeane, sa cousine, sa fille, &^a^.

a Monsieur
Monsr. Sanchez, D^r^ en Medecine
Rue des Blancs-manteaux
á Paris

London 16 September 1775

My Dear Friend

In response to yours of the 7th of this month, received this morning,[3] I start writing this from D^r^. Sequeira's home with whom I dined today, & to whom I read your lustful Coimbra Letre. As he communicated to me one from D^r^. La Cour that he received lately, I want to copy here to you the article that refers to me, so that you may judge the humour and patience of this common friend: since I see from yours how much you care about his bien etre.

"divide my thanks with M^r^. Magellan for the trouble you both share in my trifling commissions. I sincerely wish to make better returns than empty words. The box I received this morning (a box in which I sent him various books[4]) but have not yet had time to open it. What strange beings! How anxious yesterday to have it! Today possession wipes off curiosity. Is it not the same through Life? Indeed it is. We in every stage of life are meer children of fancy. Enough. This heavenly storms of thunder lightning and rain[5] I believe would make Foot western &c.[6] quit their jock[7] and spritly[8] wit, to moralise with me. Still you are not to suppose me in

3 Letter not found.
4 Note inserted by Magellan.
5 The autumn of 1775 was excessively wet in the south of England and Wales.
6 The actor, satirist and wit Samuel Foote (1721-1777) and the comic actor Thomas Weston (1737-1776), who often appeared with Foote.
7 jocund?
8 sprightly?

75.09.16 a very grave disposition. No my friend. It means no more or less than to prolong our conversation, whilst I am writing to you I release myself; if it answers the same purpose whilst you read. exegi monumentum[9] … &c.

Here you have the humour of our Dr. La Cour.

I have all your orders already that I will send at the first opportunity. That is the books and the inkpot like the one you gave to Count Carburi.[10] Each one costs 2 shillings that I will charge since you want it on your account.

I have been a number of times with our Envoy[11] who is a vrai galant homme. Good sense & excellent nature. He lives in the best domestic life with his wife who is a noblewoman from Guimaraens[12] and has none of the vanities of the Court. She gave him a little girl, after the first delivery that was a boy, both alive: but the boy suffers from his teeth falling out, an illness hardly known in Portugal but which in these northern countries does much harm in children. Dr. Sequeira is well. Today is Friday 30th September and I'm thinking of having dinner tomorrow with him, unless I go to my friend Arbuthnot[13] where I spend with pleasure and satisfaction most of my free Sundays. I was there last Sunday and Monday and so maybe I shall stay here on the next, and then I will go and have dinner with Sequeira. Otherwise I would not have time to go in the Mitcham coach[14] that leaves at 4 p.m. at this time of year. I am not surprised that you feel cheated by Castrioto[15] & by all vermin of this world, because you let yourself be cheated, & have not learned to consider every man as a true villain until you find by his deeds that he is not. This must be the general rule to live in the world: but though I know it very well by experience, I don't praise myself as practising it toties quoties.[16]

You leave me in affliction with the troubles you have due to the blood as you say. Even if you had children, you shouldn't be worried by not having more to spend on them. But since you don't, nor relatives physically unable to earn their

---

9 I have set up a monument.

10 See Magellan to Sanches, 24 July 1775. Giovanni Battista Carburi (1722-1801?), professor of medicine at Turin, 1754-1770, went to Paris in 1773 as personal physician to Maria Teresa of Savoy, wife of the Comte d'Artois.

11 Luiz Pinto de Sousa Coutinho (1735-1804) was Portuguese Minister in London, 1774-88.

12 Catarina Micaela de Sousa César e Lencastre (1749-1824) was married by proxy to Luiz Pinto de Sousa (see n. 11) while he was in Brazil, and later went with him to London. She was a very cultivated woman and made the Portuguese legation for a time one of the most remarkable meeting centres of those who cultivated the arts and sciences.

13 John Arbuthnot of Mitcham, farmer and author of books on farming.

14 The Mitcham coach left every day from the Spread Eagle Inn in Gracechurch Street; see *The Compleat Compting-House Companion or Young Merchant and Tradesman's Sure Guide* (London, 1763).

15 Félix António Castrioto, Portuguese journalist.

16 As often as occasion demands.

living, why do you worry about things you can't remedy? Nothing except what is physically necessary for subsistence or recuperation from illness should bother us. But if you should find yourself in need, don't do me the injury of hiding it from me. At any time you can have 15 or 20 guineas on my account and more, because I have friends to help me and to help those to whom I am indebted. However I know how to deal with all this mess. All those who know you are certain to make you unhappy with your whimpering for your bien etre & maintenance, just to give them part of it for their excesses. This is what comes of having a soft heart like our country's mothers. Now enough sermons 75.09.16

You just laugh at my prowess: & as you don't belch[17] any more, you don't believe that others though a little old can do it. Next time I go to Paris I have decided to take with me an English girl, to prove it to you under your own eyes before witnesses: since I truly can't do it properly with a French woman. As soon as I arrived here, I felt rejuvenated by 10 years at least: and in spite of getting much fatter than I would wish (but quietness of body and soul always make one fatter), I don't fail to give my shot perfectly well.

The sad news of the Prince d'Arenberg, who was shot in his face and chest while hunting, brings me much sorrow. I almost decided to go to Brussels. However I had news that he is better, and will not become blind as feared.

Vale me que ama[18]

Today I write in answer to D^r^. le Begue du Presle on purpose [...] friend of M^r^. Malsherbes[19] maybe he will do something for him. iterum vale.[20] Mes amities & bons souhaits à M^e^. Jeane, sa cousine, sa fille, &^c^.

17 See Magellan to Sanches, 7 July 1775, note 19.

18 Farewell and love me.

19 Guillaume-Chrétien de Lamoignon de Malesherbes (1721-1794) was appointed Secretary of State of the Maison du Roi, shortly after Louis XVI succeeded to the French throne in 1774.

20 Once again, farewell.

75.09.29 *To Johann III Bernoulli*

Öffentliche Bibliothek der Universität Basel, MS L I a 709, f. 56-57

Londres 29 Sept^{bre}. 1775 at n°. 12
Nevils Court Fetter Lane

Monsieur & très cher ami

Je viens d'arranger tous mes comptes avec M^{r}. Elmsley, entre lesquels se trouvoient la vente de vos Recueils & lettres astronomiques.[1] en voici le Resultat.

16 Recueils pour les Astronomes vol. 1. en feuilles pour etre vendus à 3^{s}. chacun --------	2. 8^{s}. 0^{d}
24 lettres Astronomiques à 1^{s}. 9^{d}. ----------------	2. 2. 0
24 Recueils vol. 2^{d}. à 3^{s}. -------------------------	3. 12. 0
3 Recueils vol. 1. à 3 -----------------------------	9. 0
3 Lettres astronomiques à 1/9 -------------------	5. 3
	8. 16. 3

Depenses que Je vous fis en 9 octobre 1772 suivant l'avis --	4. 3. 1
2 recueils & deux lettres Astronomiques pour des presents, suivants votre avis ---------------------------------	12. 6
pour brocher ces 5 articles -------------------------------	.10
postages de differentes lettres & petits messages que Je n'ai pas pu faire par moi meme --------------------------	3. 6
10 Copies de lettres Astronomiques qui ne sont pas encore vendues ---	17. 6
3 Recueils qui ne sont pas encore vendus -----------------	9. 0
cette lettre. payera ici à la poste --------------------------	1. 0
Some --------	6. 7. 5
qui etant deduite de -------	8. 16. 3

Vous avez donc entre mes mains & à vos ordre

La somme de £2. 7. 10.

que Je vous remettrai par La voye que vous m'indiquerez.

---

1 Respectively Bernoulli's *Recueil pour les astronomes* and his *Lettres astronomiques*.

Je vous ai envoyé le petit ouvrage de Kean[2] en Novembre dernier par la Voye de Mr. de Castillon,[3] auquel je chargeai le prix de 1s. 6d, que vous lui devez payer. Je vous envoye la liste suivante pourque Vous fachiez de trouver chez vos observations, ou celles de vos amis & correspondents, les observations correspondentes en quelque part que ce soit de l'Europe, pourvû qu'on donne aussi la position de l'endroit de chaqu'observation correspondente. Celles-ci sont pour determiner la position de differentes endroits villes &c. de Flandres & du Brabant, oû ces observations furent faites par Mr. Pigott qui donnera les siennes si on les lui demande.[4] Je vous prie de nous faire cette besogne pour le service de la geographie du Pays en question, oû il y a beaucoup d'erreurs dans les cartes &a. 75.09.29

Je vous prie de faire avertir Mr. de Castillon, que Je n'ai pas encore reçu les livres du Journal qu'il me dit vouloir envoyer à Londres, & que J'attends sa Reponse à la Lettre que Je lui ecrivis sur l'echange de livres par le prix de Libraire contre les siens (& je dis autant pour les votres) savoir lettre de Pekin sur la langue Chinoise & Egyptiene avec 28 planches in 4o.[5] à 6s. / Kuhnius de æquationibus cubicis. in 4o.[6] à 3s. / Exhortations d'un Quaquer à Mr. de Voltaire. in 8o.,[7] à 2s. 3d. Journal de Rozier.[8] 18 vol. in 12mo. & 36 in 4o. C'est un journal de Physique, Philosophie Nles. &c. mechanique &a. à 2s. par vol. il faut prendre la collection entiere.

Vous savez que le Captn. Cook est arrivé de son grand voyage au tour du monde.[9] il a decouvert plusieurs isles nouvelles. Mr. Forster & son fils[10] qui l'accompagnerent, vont publier les plantes nouvelles & animaux. le meme Forster qui est mon intime ami, publiera bientot le journal de ce

2 Presumably T. Kean, *A New and Easy Method of finding the Longitude at Sea, with like Accuracy that the Latitude is found* (London, 1774).

3 Johann Castillon (Giovanni Salvemini Castiglione) (1704-91), director of the Berlin Observatory.

4 Pigott, *Philosophical Transactions*, 66 (1776), 182-95.

5 *Lettre de Pékin, sur le génie de la langue chinoise et la nature de leur écriture symbolique, comparée avec celle des anciens égyptiens; ... par un père de la Compagnie de Jésus* [i.e. Pierre Cibot] (Brussels, 1773).

6 H. Kühn, *Tentamen de aequationibus cubicis quibuscunque perfecte resolvendis opus posthumum* (Danzig, 1771).

7 *Exhortations à Mr. de Voltaire, avec le parallele entre lui and Mr. Rousseau* (London, 1774). The half-title of this work represents it as part of a series, "Les Vrais Quakers".

8 *Observations sur la physique, sur l'histoire naturelle et sur les arts*, ed. François Rozier (1771-).

9 Cook arrived home from his second voyage around the world on 30 July 1775.

10 Johann Reinhold Forster (1729-98) and his oldest son Georg (1754-94).

75.09.29 voyage oû il aura plus de 80 planches & mapes. on publiera aussi bien tot les observations Astronomiques faites dans ce voyage.[11]

Portez vous bien mon cher ami, & soyez assuré du sincere attachement de Coeur de

Votre t. h Serv^{r}.
Magellan

On demande les observations correspondentes qui suivent.

1773	
August.	15. Immers. du 1er. Satelite de Jupiter
	22. Im. 1er. Sat.
	24. Im. 2^{d}. Sat
	29. Im. 1er. Sat.
	31. {Im. 1er. Sat. +
	{Immers. du 2^{d}. Sat. +
Octobre.	9. Emers. du 1er. Sat. +
	13. Em. du 2^{d}. Sat. +
Nov^{bre}.	1. Em. du 1er. Sat.
	10. Em. du 1er. Sat. +
Dec^{bre}.	10. Em. du 1er.Sat. +
	19. Em. du 1er. Sat. +
1774	
Janv.	11. Emers. du 1er. Sat.
Fev^{r}.	3. Em. du 1er. Sat.
	18. Em. du 2^{d}. Sat. +
Juillet	24. Imm. du 2^{d}. Sat.
	26. Im. du 1er. Sat. +
Aout.	25. Im. du 2^{d}. Sat. +
	27. Im. du 1er. Sat. +

11 In 1776 the Forsters published a preliminary account of the botany of the voyage, *Characteres generum plantarum*; two years later, the father published his definitive account of the natural history and anthropology of the voyage, *Observations made during a Voyage round the World, on Physical Geography, Natural History and Ethnic Philosophy*. In the meantime, there was much controversy over the publication of the official journal of the voyage. As Magellan's letter makes clear, the elder Forster expected to do this as well; later, there was a proposal that he and Cook should do it jointly. Eventually, however, the Admiralty decided that Cook would do it alone, Forster being forbidden to publish any account of his own until after Cook's had appeared – whereupon Forster turned his journals over to his son, who was not bound by the Admiralty interdiction. His very successful *Voyage round the World* (London, 1777) beat Cook's official account into print by six weeks.

Oct^{bre}. 5. {Imm. du 3^{d}. Sat. 75.09.29
{Im. du 1er. Sat.
14. Im. du 2^{d}. Sat.
21. {Im. du 1er. Sat.
{Im. du 2^{d}. Sat.

1775
Fev^{r}. 15. Emers. du 1er. Sat.
19. Em. du 2^{d}. Sat. +
22. Em. du 1er. Sat. +
Mars 1. Emers. du 1er. Sat. +

N.B. les observations marquées avec la croix + sont celles dont on a le plus grand besoin. mais on sera aussi bien aise d'avoir les autres qui ne sont point marquées, pour les comparer avec celles observées à Paris. on demande la Longitude & position de l'endroit oû l'on aura fait chaqu'observation.

A Monsieur
Monsr. Bernoulli, Astronome R^{l}.
membre de l'Acad^{e}. R^{le}. des Sciences
à Berlin
Prusse[12]

12 Bound with Magellan's letter, at f. 58, is a list, in Bernoulli's hand, of observations of the immersions and emersions of Jupiter's satellites on some of the dates in question, recorded at Tyrnau, St Petersburg, Greifswald, Uppsala, Stockholm, Marseilles, Milan and Berlin.

75.09.30 *To the Duke Charles-Marie-Raymond d'Arenberg*

Archief Van Arenberg, Enghien/Edingen, Belgium

Monseigneur

Il s'en a falu très peu que je ne sois pas le porteur de cette Lettre pour mitiguer l'inquietude continuelle oû je me trouve depuis la malheureuse nouvelle du desastre du Prince.[1] Je me suis porté tout de suite chez M^{r}. Le Comte Belgioioso,[2] mais il ne savoit plus que moi. enfin j'ai reçu une Lettre de M^{r}. Tackoen de Malines,[3] qui me dit que le Prince ne perdra pas La vue. Voilà Monseig^{r}. deux des plus grandes epreuves que La Providence a voulu faire de V.A.S^{me}. en fort peu de tems. puissent La bonne Philosophie & la Religion soutenir son Courage, calmer les violentes agitations de Son Esprit, & Lui rendre La paix & tout Le bonheur pour Le reste de Ses Jours.

J'avois eu L'honneur d'Ecrire à V.A.S^{me}. par un gentil-homme Anglois de mes intimes amis qui partit le 2 de ce mois. mais il tomba malade en chemin: il vient de retourner à Londres; & m'a remis ma Lettre.

M^{r}. L'Eveque d'Iprès[4] m'avoit ecrit[5] que Mad^{e}. La Princesse avoit obtenû le plus grand retablissement de Sa Santé avec les eaux de Spa, & que V.A.S^{me}. avoit par là vû couronés Ses Soins avec un hereux succès, dont je felicitois V.A.S. dans un des articles de la meme Lettre.

J'ai passé quelque tems à Bowood avec Mylord Shelburne,[6] qui m'a ordonné de presenter ses Respects à V.A.S., à Mad^{e}. La Duchesse & aux Princes. Il partit après le commencement d'Aout pour Ses Etats en Irlande, qu'il m'a dit etre habités par les plus grands paresseux & faineants du monde. Il y a fait passer des Agriculteurs de differentes Provinces d'Angleterre, & il est determiné à leur donner tout l'encouragement possible, pour reveiller ces faineants de leur inaction. La Lettre dont V.A.S^{me}. m'a honoré en date

1 Louis Engelbert Pierre Marie Joseph Augustin (1750–1803), future 6th Duke of Arenberg, had damaged his eyes in an accident while out hunting.

2 Ludwig Karl Graf von Barbiano-Belgiojoso [Ludovico Carlo Maria di Barbiano, Conte di Belgioioso] (1728-1801), Austrian Minister plenipotentiary in London.

3 Letter not found. Probably Godefridus Gasparus Tackoen (1720-1788); see Magellan to Sanches, 25 August 1775.

4 Ypres.

5 Letter not found.

6 William Petty, 2nd Earl of Shelburne and 1st Marquess of Lansdowne (1737-1805), British Prime Minister in 1782-3, was Joseph Priestley's chief patron. He had his seat at Bowood House near Calne in Wiltshire, and owned extensive estates in Ireland.

du 24. du mois dernier,[7] me fait imaginer que les habitants d'Arenberg se trouvent à peu près dans la meme situation. Ne seroi-t-il pas possible de les exciter de cette lethargie par des methodes pareilles? 75.09.30

J'ai eté dimanche & lundi derniers chez M^r^. Arbuthnot: & je l'ai laissé encore dans le doute si j'irois à Bruxelles avant Lui; car ce fut à mon Retour que je trouvai la Lettre de M^r^. Tackoen; suivant Les apparences, M^r^. Arbuthnot laissera son voyage en Flandres pour le perintems[8] prochain.

Je ne m'oublie pas de l'horloge pour Everlê.[9] J'en ai fait deja quelques essais. il sera très simple, mais fait suvant[10] ma methode & n'aura pas besoin d'etre monté en plusieurs années. On y lira mon nom dessus: & comm'il survivra mon existence, il temoignera encore lorsque je ne serai plus, le devouement de Coeur, la Reconnoissance, & le profond Respect avec le quel J'ai l'honneur d'etre

Monseigneur

de V. A. S^me^

Serviteur tres h^ble^. & très obligé

J.H. de Magellan

at n°. 12 Nevils Court Fetter Lane.

7 Letter not found.

8 printems?

9 The château de Heverlee, near Leuven, was one of the houses of the Duc d'Arenberg.

10 suivant?

75.10.13 *To Jacques-André Mallet*[1]

Bibliothèque de Genève, Geneva, Ms suppl. 1654, fol. 42-43

Londres ce 13 Octobre 1775

Monsieur

Ce n'est pas pour avoir jugé mal de vous, mais pour avoir acquis un peu de connoissance practique du monde, que je vous ai supposé ennuyé d'attendre votre Lunette depuis un si long tems qu'elle fut ordonnée.[2] Je viens de l'entrer dans mon livre sur votre compte; & je ne perdrai pas un moment à faire executer le piedestal simple que vous souhaitez: mais encore à condition que la lunette restera entre vos mains, & que vous l'employerez dans votre observatoire. ainsi vous pouvez vous expliquer là dessus avec ce Monsr. votre ami dans ces termes: car je croirois faire un grand tort à d'autres astronomes, en envoyant cette lunette qui pour le present est l'unique à avoir de ce genre, pour quelqu'un qui ne l'employat avantageusement à des observations interessantes. Je suis sur que ce piedestal ne sera pas aussi cher comme ceux que fait Dollond, qui charge 7½ g pour chacun separement, & Je compte y faire adapter une piece à charniere, comme celle que j'avois intention d'ajouter à celle de M^r^. le Presid^t^. de Saron, dont vous avez un mot à page 90 des Lettres Astronomiques de M^r^. Bernoulli.[3] Je vais ecrire à Paris pour savoir, si M^r^. Menier[4] a entrepris le piedestal paralatique, & alors je ne douterai de le faire executer ici, si lui ne l'a pas fait, comme je le crois. Je suis bien de votre opinion que les François malgré tout le ton qu'ils prennent & les airs qu'ils se donnent, ne sont pas comparables à nos simples artistes anglois. J'eus l'adresse de faire passer en France une excellente machine à diviser, faite par Ramsden,[5] & pour laquelle on m'a blamé ici assez fort, car quelques uns la vouloient à tout prix, mais l'artiste n'a pas voulu la leur ceder. Vous trouverez dans le Journal de Rozier à page 147 du mois

---

1 Annotated: "de Londres. Magellan. 24 octobre 1775".

2 See Magellan to Mallet, 1 September 1775, and Mallet to Magellan, September 1775.

3 Johann III Bernoulli, *Lettres astronomiques* (Berlin, 1771), p. 90.

4 Mégnié.

5 See Magellan to L.-H. Duchesne, 27 July 1773.

de Fevrier 1773,[6] un precis de cette machine, quoique l'auteur s'a trompé disant que le traçoir etoit mis en action avec le talon, au lieu que c'est la main qui meut le traçoir. J'ai actuellement un cercle divisé par cette machine, de 39,4 lignes (mesure de France) de diametre. il est divise en degrés & quarts de degré avec un Nonius de 30 divisions, ce qui donne des demi-minutes: & qui fut fait pour avoir une epreuve de la bonté de cette machine avant de l'envoyer en France. Si vous prennez la proportion de 7: 22, la peripherie sera 10,319 – & puisqu'il y a 43200 demiminutes, chacun n'occupera que l'espace d'une cinque millieme partie de pouce ou $^{238}/_{100000}$. on ne trouve pas un seul minute de difference, en parcourant toute la peripherie avec l'alidade qui contient le Nonius. Or c'est avec cette meme machine (dans laquelle travailloient les eleves, & meme quelques fois la femme de Ramsden) qu'on a divisé dernierem^{t}. à Paris, pour M^{r}. Le Chev. de Borda, un Instrument, qu'il trouva si defectueux qu'il fut obligé de me l'envoyer ici pour etre redivisé de nouveau: car il y trouva plus de 3 minutes de difference en quelques endroits. Ainsi j'ai perdu toute esperance sur les artistes François: & Je vais envoyer à Paris cette epreuve de la meme machine, pour qu'on y voit que le defaut procede de la main du mauvais artiste, & non pas de la machine. 75.10.13

Mais retournant à votre Lunete. elle a 42½ dans le Clair du diametre de l'Objectif: & elle paroit excellente. il y a un oculaire terrestre, & 4 celestes, deux de ces derniers ont des croix filaires: Dollond les a mis (ces deux derniers) dans des tuyaux qui s'ajustent à la main. Si vous croyez que vos ouvriers ne mettront bien ces oculaires dans le tuyau à pignon oû les autres 3 oculaires sont mis, je vous les ferai changer ici. Car pour vous le dire en passant, je menage autant qu'il est possible les idées de Dollond, même lorsqu'elles ne me plaisent pas, pour mieux obtenir, ce dont lui seul peut me fournir, c'[est] à dire de cette espece de Lunettes Astronomiques, quoique pour le present on n'en peut pas avoir plus jusqu'à ce que nos verreries recouvrent la maniere de faire du bon flint-glass. Il a encore une autre chose, c'est que cette lunette merite d'avoir une caisse de mahogany comme celle de M^{r}. de Saron & de M^{r}. Messier, qui coutera environ une guinee à peu près. Car pour le present Je ne l'ai

6 "Précis d'une machine inventée et exécutée par M. Ramsden... pour diviser toutes sortes d'instrumens", *Observations sur la physique, sur l'histoire naturelle et sur les arts et métiers*, 1 (1773), 147-8.

75.10.13 pas reçue que dans une packing caise de sapin. Vous me ferez avis sur ces articles au plutôt: & Je suis

Monsieur
Votre très h^{ble}. serv^{r}.
Magellan

P.S. voici une liste d'observations dont on demande les Correspondantes.[7] Je vous prie de me fournir celles que vous en aurez.

A Monsieur
Monsr. Mallet Professeur d'Astronomie
dans La Cour de S. Pierre
à Geneve
Suisse[8]

---

7 List not found.

8 The file includes a sheet (f. 49) headed 'Extraits de differ^{tes} lettres de Mr Magellan', as follows:

24 Janv^{r}. 1774	guin.
1°. La lun. de 3¼ p. avec tuyau de bois, & pied de bois	25
2°. La lun. avec tuyau de cuivre, axe de cuivre, & vis p^{r}. les mouv^{s} lent	36
3°. La préced^{te} avec mouv^{t} parallatiq. l'axe de 1½ ou 2 p^{ds}	45
4°. Celle du N°. 2 avec un micrometre objectif	50
5°. La lun. avec tuyau de cuivre, montée sur une machine parall. avec cercle vertical, & de l'equateur divisés, & micrometre filaire	70

Nov^{bre}. 1774.
Une lun. de 3½ p^{d}. tuyau de cuivre, 1 oc. terrestre, 2 célest. un autre avec croix de fil, un trouveur, verres noirs &c. sans piedestal coute 26 guin.

31 Mars 1775.
La Lun. est finie et coute 28 guin.

Il y a	1 tube p^{r} les occult. grossit	40 f^{s}.
	1 tub	85 garni d'une croix
	1	85
	1	150
	1 tube d'ocul. terrestre.	

1^{r} Sept. 1775
Il a eu la lunett. finie en Juillet 1775. & il l'a paye 28 guin.

Octob. 1775
La lun. lui paroit excellente, elle a 42½ lig. dans le clair du diam. de l'objectif
Il y a un ocul. terrest. & celest. dont 2 avec des croix de fil
Il faut une caisse qui coute une guinée.

To Ribeiro Sanches[1] 75.10.27

Österreichische Nationalbibliothek, Vienna
A.N. Ribeiro Sanches Correspondenz, codex 12714, ff. 159-160v

Londres 27 Out^{bro}. 1775

Meo am° e S^{r}. e m^{to}. do Cor^{am}.

Dilatei esta depois de ter Recebido a de V.M. de 9 do Corr^{te}. por esperar uma do D^{r}. Seq^{ra}. q me disse queria escrever a V.M. sobre os banhos de Russia, mas como elle a não mandou até gora, não dilato esta mais à fim de não pôr V.M. em alguã aprehensão pela minha situação corporal & espiritual. ambas são boas como d'antes. Mas V.M. deixame suspenso e admirado com o q parece inculcar na sua sobre a mutualid^{e}. dos nossos interesses; pois cuidei q ambos tinhamos bem entendido o q tantas vezes temos repetido, desde o principio das nossas conversaçoens: e q o q eu tenho ou posso ter, he, & deve ser de V.M. todas as vezes q necessitar. a esta antiga declaração não havia mais q ajuntar senão q não quero nem sofro q V.M. corte por sim, e q diminua, em vez de acresentar nos adminiculos & Circunstancias Commodas q a sua id^{e}. e pouca saude requerem: & por isso na minha carta precedente indiquei a soma q podia tirar sem receio & á vista sobre mim em toda a occazião. fiquemos pois intendidos sobre esta materia daqui por diante, sem q seja necessario fazer novas repetiçoens; pois bem sabe q so pelas não fazer, puz de parte as do Celebre Breviario Romano com toda a sua admiravel latinidade.

o amigo Sequeira pôz sua carroça nova, m^{to}. boa & decente. nesta terra he hũ traste essencialm^{te}. necess°. a qualquer medico. elle passa bem, e cuido q cada vez adianta em practica & conhecim^{tos}. o nosso Ministro o Cav^{ro}. Pinto (q he sobrinho do ultimo Mestre de Malta falecido) faz delle m^{to}. apreço, & m^{tas}. outras pessoas distintas. os seos livros hirão antes do fim do mez ou pelo menos no meio do q vem, pois em escrevendo esta ultima palavra me lembro de outras coizas q ainda não estão prontas.

o desastre do Duque de Arenberg sucedeo em seo f°. mais velho, o q cazou com M^{le}. Lauragais. M^{r}. Gordon Seo amigo intimo, foi o q descarregou sobre elle. teve um grão em hũ olho, & dois no outro. Comtudo espera-se que não ficará inteiram^{te}. cego. o tal Gordon (Ministro desta Corte á de Bruxellas) q eu conheço m^{to}. bem & hé de hũ excellente caracter, tem

1 Annotated "R^{a} a 2 de 9^{bro}" [Received 2 November].

75.10.27 sido goardado á vista p[a]. não se vingar sobre si mesmo á Ingleza em se dezenbaraçando da vida propria. Se acazo M[r]. Gordon não atira áquella perdiz, o Duque mesmo de Arenberg lhe atiraria, & ficando mais em façe ao Princepe seo filho, lhe faria maior dãno, & sem duvida o mataria, pois sempre carrega m[to]. mais do ordinario a sua espingarda. de modo q o mal ainda pudera ser m[to]. mais lamentavel.

tive not[as]. de lisboa onde vejo q Alvares vai ás nuvens & cedo tera habito de Christo &[a].

folgo q o Pr[ce]. Pignatelli não largasse os ossos na sua empreza de Don Quixote, ou p[a]. melhor dizer, á Sebastianista portuguez[a].

Saud[es]. á fam[a]. adoptiva.

Vale & me ama

a Monsieur
Monsr. Sanchez D[r]. en Medecine
Rue des Blancs-Manteaux
à Paris

London 27 October 1775

My Dear Friend

I delayed this letter after having received yours of the 9th of this month[2] since I waited for the one that D[r]. Sequeira told me he wished to write to you about the Russian baths, but as he did not send it till now, I won't delay this one longer in order not to cause you any apprehension about my corporeal and spiritual situation. Both are good as before. However you left me anxious and surprised with what you seem to point out in yours about the reciprocity of our interests; since I supposed we had well understood what we have so many times repeated from the beginning of our conversations: and that what I possess or may possess in the future is and must be yours, any time you need it. To this old declaration there is nothing to add except that I don't want or suffer your decreasing instead of increasing the assistance & comfortable circumstances that your age and poor health require: and so in my preceding letter I indicated the sum that you can draw on me without apprehension in specie & at any time.[3] Let us remain clear about this matter from now on without need for further repetitions; because you know that precisely in order not to make them, I put aside the celebrated Roman Breviary with all its wonderful Latinity.

2 Letter not found.

3 See Magellan to Sanches, 16 September 1775.

Our friend Sequeira has acquired a new, very good and decent carriage. In this country that is an essential tool for any physician. He is going well and I think he advances more and more in practice and in knowledge. Our Minister, Chev[er]. Pinto (who is a nephew of the last deceased Master of the Order of Malta) regards him very highly, as do many distinguished persons. Your books will go before the end of the month or at least in the middle of the next, since on writing this last word I remember other things that are not yet ready. 75.10.27

The Duke d'Arenberg's disaster happened to his eldest son, the one who married M[le]. Lauragais.[4] It was M[r]. Gordon,[5] his close friend, who fired the shot that hit him. He had a pellet in one eye and two in the other. However it is hoped that he won't become completely blind. That M[r]. Gordon (Minister of this Court to Brussels), whom I know very well and who is of excellent character, has stayed out of sight by not taking revenge on himself in the English way by doing away with his own life. If by chance M[r]. Gordon had not shot at that partridge, the Duke himself would have done so &, standing more in front of his son, would have caused greater damage & undoubtedly would have killed him because he always charges his rifle more than normal so that the harm could have been still much more regrettable.

I had news from Lisbon where I see that Alvares rises up to the clouds and sooner or later will have the habit of Christ &[a].[6]

I'm glad that Prince Pignatelli[7] didn't leave his bones at his Don Quixote-like enterprise or, to say it better, in the Portuguese sebastianist way.[8]

Greetings to the adoptive family.

Vale & me ama

---

4 See Magellan to Duchesne, 7 December 1772.

5 William (later Sir William) Gordon, British Minister Plenipotentiary at Brussels, 1765-1777.

6 i.e. will be made a Knight in the Portuguese order of chivalry, the Order of Christ. On Alvares, see Magellan to Sanches, 20 April 1762, note 21.

7 Juan Joaquín Atanasio Pignatelli de Aragón y Moncayo (1724-1776), Spanish Ambassador in Paris, 1763-72? The Quixote-like enterprise has not been identified.

8 Young King Sebastian of Portugal (1554-1578) led an army to conquer the north of Africa but was defeated at the battle of Alcacer-Kibir (Morocco) and was never seen again. He became a symbol of quixotic chivalry.

75.11.13 *To Antoine-Laurent Lavoisier*

Archives de l'Académie des Sciences, Paris, Fond Lavoisier, 1642[1]

Mon cher ami

Voici le sommaire des Lectures du cours de chymie du D^{r}. Highins.[2] J'ai cru vous faire plaisir en vous l'envoyant, car en effet c'est un abregé tout à fait extraordinaire du plus grand charlatan qu'on a jamais vû;[3] mais qui reelement est un homme de beaucoup de merite.

Je suis on ne peut plus mortifié de ne pas avoir de vos nouvelles, & ne croiois pas vous meriter cet oubli. ni le tems ni la distance n'effaceront jamais mon souvenir, ni ma gratitude. Je soufre pour ne pas savoir de Vos nouvelles ni de Mad^{e}. votre digne & aimable Epouse philosophique, ni de M^{r}. votre Pere, Mlle. votre Tante, Messrs. Paulze Pere[4] & fils &^{a} &^{a}. &^{a}.

dites moi si votre machine de verre pour l'air fixe est arrivée sans acident? si le coquin du Capitain a eté puni?[5] si le traité sur la desalaison de l'eau de mer a eté imprimé, ou du moins si l'on pense de l'imprimer bien tôt?[6] &^{a}. &^{a}. &^{a}.

J'envoye regulierem^{t}. à M^{r}. Trudaine toutes les postes des feuilles de l'ouvrage du D^{r}. Priestley c'est à dire de son 2^{d}. vol. sur les nouvelles decouvertes de differ^{tes}. especes d'air &^{a}.[7] & je ne doute pas qu'il vous les communiquera à mesure: dans le meme tems j'envoye autant de feuilles pour le D^{r}. Gibelin que le meme D^{r}. Priestley pria dans une lettre M^{r}. Trudaine, d'affranchir pour la poste & de les envoyer à Aix en Provence, à fin qu'il puisse avancer au plutot la traduction de ce second volume[8]

1 Corrected version of the text published in *Oeuvres de Lavoisier: Correspondance*, ed. René Fric, fascicule II, 1770-1775 (Paris, 1957), pp. 504-6.

2 Bryan Higgins, *A Syllabus of the Discourses and Experiments...* (London, 1775); see J. R. Partington, *A History of Chemistry*, vol. 3 (London, 1962), p. 727.

3 Magellan's abuse was doubtless prompted by the accusations of plagiarism that Higgins had levelled at Joseph Priestley; see Priestley's *Philosophical Empiricism, containing Remarks on a Charge of Plagiarism respecting Dr. H-s, interspersed with Various Observations relating to Different Kinds of Air* (London, 1775).

4 Mme Lavoisier's father, Jacques Paulze.

5 See Magellan to Lavoisier, 14 July 1775.

6 See Magellan to Lavoisier, 1 July 1775. The report in question was not published until 1781.

7 J. Priestley, *Experiments and Observations on Different Kinds of Air*, vol. 2 (London, 1775).

8 The French edition of Priestley's work, *Expériences et observations sur différentes espèces d'air*, was published in Paris in 1777. The translator was identified as "M. Gibelin, Docteur en Médecine, Membre de la Société Médicale de Londres"; he was presumably the Jacques Gibelin (1744-1828) who later (1787) produced a French-language abridgement of the *Philosophical Transactions*.

Je vous prie de mes h[bles]. respects à toute votre Famille, & à tous nos bons amis communs: voyez si je puis vous rendre quelque service, & comptez sur le sincere attachement 75.11.13

de Votre
Magellan

à Londres ce 13 Nov[bre].
1775 at n°. 12 Nevils Court Fetter Lane.

From Jacques-André Mallet 75.11.13a

Bibliothèque de Genève, Geneva, Ms suppl. 1654, fol. 57[1]

A Mr Magellan

Geneve 13 Nov. 1775

M[r].

J'ai reçu il y a peu de jours la lett. que vous m'avez fait le plais. de m'écrire.[2] Je suis b. aise d'apprendre, que la lunette va enfin bientot etre en route pour Geneve, car je pense que le piedestal sera bientot fait, je le souhaite le plus simple possible. Quant à la caisse, je trouve bien qu'il convient qu'elle en ait une de mahogani. & relativem[t] à la maniere d'ajuster les oculaires je ne comprends pas bien de quoi il est question, mais je suis sur de trouver ici un ouvrier très intellig[t] & très habile.

voila je crois M[r] ce que vous me demandez dans votre lettre, & vous trouverez cij[3] après les observat. de Satellit. que vous désirez. J'ai l'[4]

1 Letter not found. The text is from a draft kept by Mallet.
2 See Magellan to Mallet, 13 October 1775.
3 cijoint? The promised observations are not appended to the draft.
4 The draft breaks off in mid-sentence, and is unsigned.

75.11.17 *To Ribeiro Sanches*[1]

Österreichische Nationalbibliothek, Vienna
A.N. Ribeiro Sanches Korrespondenz, codex 12714, ff. 163-164v

Londres 17 Nov^{bro}. 1775

Meo Am^{o}. e S^{r}. e m^{to}. do Cor^{am}.

A de V.M. de 2 do Corr^{te}. me deo m^{to}. gosto, e serîa este m^{to}. maior se as suas molestias se achassem diminuidas. Aqui andou um andasso (palavra da nossa beira) quazi geral, com tosse & pela maior p^{te}. fevre. parecia epidemia: mas pouco perigo tem avido. eu tao be[m] tive um pouco: & curei-o com dieta, & agoa quente com assucar.

Aqui achará V.M. na ult^{a}. pagina 2 ou 3 perguntas de M^{r}. Mickle, tradutor de Camoens, q brevem^{te}. sera publicado; pois som^{te}. lhe faltão duas ou 3 paginas do Prefacio. Se V.M. pode commodam^{te}. dar algũa Reposta a estas perguntas, mademas logo q for possivel pela posta seg^{te}.

Pelo prim^{o} Navio p^{a}. Lisboa partirá o masso com 12 N^{os}. do Med^{al}. Commentaries of Edimburg, remetido à Florencio Teixeira de Azevedo, Mercador, q mora (ou morou) na rua dos [...]stos Lisboa. p^{a}. entregar ao S^{r}. D^{r}. João Henriques Ferreira q os mandará buscar & pagará os gastos do fréte &^{a}. V.M. là dará avizo a seo Irmão [p]^{a}. q lhe dê esta Noticia. ocusto he 16^{s}. 6^{d}. q carregarei na [co]nta de V.M.

Ja comprei o Wallace's system of Midwifry 4°., Butter puerperal fevers, & harwood on classic-editions. eu m'encontrei com este ulti[mo] Autor uma destas noites em Caza de meo amigo Paradise, q V.M. vio [em] Paris com o Irmão do Conde Carbury[.] pareceo-me homem m^{to}. instruido [no]s AA. Classicos; & de hum genio [...] & aplicado. Não he possivel achar Home's Methodus Materia Medica. cuido q V.M. se enganou no titulo: pois ninguem conhece que o seo principia ou Instituts de Medecine.

Cuido q o Sketches of y^{e}. Hist^{y}. of Men é composto por um tal Myl^{d}. Kaimes. mas não conheço nada de particular sobre o Autor. Se tiver ocazião, e não me esquecer, eu me informarei.

o D^{r}. Seq^{ra}. passa bem: e ainda me não deo a sua Carta: cuido q jantarei á manhan com elle. La Cour passa bem no seo Centro de Bath.

Sinto a aflição do D^{r}. le Begue, & o desastre do princepe de Arenberg ainda m^{to}. mais pelo conhecim^{to}. pessoal do Sojeito, & obrigaçoens q devo

1 Annotated "R^{a}. a 27 9^{bro}." [Answered 27 November]. Letter not found.

a toda a familia. miseraveis humanos! eu escrevi a o Abb^{e}. Chevalier p^{a}. 75.11.17
saber not^{as}. do Princepe: cuido q me Responderá.

q o Princepe Pignatelli torne à Paris, & se arrange mais feliz do q o anno passado!

Aqui fiz conhecim^{to}. com o Princepe Orloff, com q^{m}. jantei em caza de Myl^{d}. Shelburne. disse-me q o fosse ver, o q fiz ante-hontem. reteveme a jantar com elle & sua comitiva. he desembaraçado, maganão do alto. grande penetração, vivacid^{e}., e engenho. faleilhe em mandar á Acad^{a}. de Petersbourg um exemplar do meo tratado. não som^{te}. isso, me disse elle, mas quero accompagnalo de uma carta m^{a}. p^{a}. o Director q he meo amigo & foi meo oficial (não sei se general ou q?) e deo me logo a sua adresse, q he M^{r}. Damachnew, Directeur de l'Acad^{e}. des Sciences, gentilhomme de la Chambre &^{a}. Cuido q V.M. là o verá em Paris. he très engageant par une franchise, naive, & plaine: c'est lui qui se dit très ignorant lui meme, mais qui a une très vaste connoissance de la Physique & Chymie, & Metaphy[si]que Moderne &^{a}. &^{a}. esta manhan despachei um caixote d'instrumen^{tos}. p^{a}. o Princepe Gallitzin à la Haye, q m'os ordenou por via de hum amigo commum: & eizm'aqui introduzido com S^{res}. Russos: q provavelm^{te}. não darâõ perda, mas proveito. Saud^{es}. á fam^{a}. adoptiva & a cada hum della, & dos mais amigos se he q os ha verdadeiros.

Vale & me ama

P.S. Cada semana imagino poder mandar p^{a}. Paris varios Instrum^{tos}.: mas não acabo de os ver acabados. Com elles partirão os Livros de V.M.

iterum vale.

A Monsieur
Monsr. Sanchez D^{r}. en Medecine
Rue des Blancs-manteaux
á Paris

London 17 November 1775

My Dear Friend

Yours of the 2nd of this month[2] gave me great pleasure, and it would be even greater if your diseases were lighter. There has been an almost general infection (andasso as they say in our Beira) with cough & in most cases fever. It looked like an epidemic: but without much danger. I also had it a little & cured it with diet & hot water with sugar.

2 Letter not found.

75.11.17 Here you will find on the last page 2 or 3 questions from Mr. Mickle,[3] translator of Camoens, which will be published soon; because only two or 3 pages of the Preface are missing. If you can easily give answers to these questions, send them to me as soon as possible by the next post.

In the first ship to Lisbon the package containing 12 issues of Medal. Commentaries of Edinburg[4] will be sent, addressed to Florencio Teixeira de Azevedo,[5] Merchant, who lives (or lived) at rua dos [...]stos Lisbon to handle for Dr. João Henriques Ferreira,[6] who will send for them & pay the cost of the freight &c. Tell your brother to give him this news. The cost is 16s. 6d. that I will charge on your account.

I have already bought Wallace's system of Midwifry[7] 4°., Butter puerperal fevers,[8] & Harwood on classic-editions.[9] I met this author one night at the home of my friend Paradise,[10] whom you saw in Paris with the brother of Count Carbury.[11] He seemed a very learned man in the classical authors; & of a [...] and diligent spirit.

I can't find Home's Methodus Materia Medica.[12] I think you have the wrong title since nobody knows anything but his principia or Instituts de Medecine.[13]

I suspect that Sketches of ye. Histy. of Men[14] is by Milord Kaimes. But I know nothing particular about the author. If I have an opportunity, and I don't forget, I will obtain the information.

---

3 The following questions are written in a hand other than Magellan's, presumably Mickle's:

> "Were the Portuguese Governors of India arbitrary, or did they govern by any certain & established system or code of laws? If they did, what was that system? Could a copy of the commission of any one of these Governors be procured?
> N.B. It is the first Governors, those about the time of Camoens, who are enquired after."

William Julius Mickle published a translation of "The Lusiad" of Camoens: *The Lusiad; or the Discovery of India: An epic poem,* translated from the original Portuguese with an introduction and notes by William Julius Mickle (Oxford and London, 1776). Many editions of this translation were published in the following decades.

4 *Medical and Philosophical Commentaries by a Society in Edinburgh* (London, 1773).

5 Unidentified person.

6 Probably José (not João) Henriques Ferreira, physician and member of the Lisbon Academy of Sciences, who was a nephew of Ribeiro Sanches.

7 Robert Wallace Johnson, *A New System of Midwifery* (London, 1769)?

8 William Butter, *An Account of Puerperal Fevers, as they appear in Derbyshire, and some of the counties adjacent* (London, 1775).

9 Edward Harwood, *A View of the Various Editions of the Greek and Roman Classics* (London, 1775).

10 John Paradise (1743-1795), English linguist with a remarkable knowledge of several languages, was elected F.R.S. in 1771. His house was open to literary men and leading personages of his time, including Samuel Johnson and Joseph Priestley.

11 See Magellan to Sanches, 16 September 1775.

12 Francis Home, *Methodus materiæ medicæ* (Edinburgh, 1770).

13 Francis Home, *Principia medicinæ* (Edinburgh, 1762).

14 Henry Home, Lord Kames, *Sketches of the History of Man* (2 vols., Edinburgh, 1774). The work was initially published anonymously, but Kames's authorship was later acknowledged.

Dr. Sequeira is well and didn't give me your letter yet. I will have dinner with him tomorrow. La Cour is going well in his Centre at Bath. 75.11.17

I understand Dr. le Begue's affliction,[15] & the disaster of Prince Arenberg[16] still much more because of my personal knowledge of him, & the obligations I owe to all the family. Miserable humans! I wrote to Abbé Chevalier[17] for news of the prince: I hope he will answer.

May Prince Pignatelli[18] return to Paris, & be happier than last year!

I made the acquaintance of Prince Orloff,[19] with whom I had dinner at Milord Shelburne's home. He told me to go and see him, which I did the day before yesterday. He kept me to have dinner with him & his companions. He is an unaffected scoundrel of great acuteness, liveliness and talent. I mentioned I would be sending to the St. Petersburg Academy a copy of my treatise. Not only that, he told me, but I want to send with it a letter of mine to the director who is my friend & was my officer (I don't know if general or ?) and he gave me his address, he is Mr. Damachnew, Directeur de l'Acade. des Sciences, gentilhomme de la Chambre &c.[20] I think you will see him in Paris. He is très engageant par une franchise, naïve, & plaine: c'est lui qui se dit très ignorant lui meme, mais qui a une très vaste connoissance de la Physique & Chymie, & Metaphy[si]que Moderne &c. &c. This morning I dispatched a box of instruments to Prince Gallitzin[21] at The Hague, who ordered them through a common friend: & here I am amongst Russian gentlemen: that probably will give no loss, but profit. Regards to the adoptive family & to each one of them, & to other friends if there are any true ones.

Vale & me ama[22]

P.S. Each week I think I can send to Paris various instruments: but I don't see them finished. Your books will go with them.

iterum vale[23]

---

15 Possibly the "contagious cold" that Magellan in a later letter reports Le Begue du Presle to be suffering from; see Magellan to Sanches, 1 January 1776.

16 See Magellan to Sanches, 16 September 1775.

17 Letter not found.

18 Juan Joaquín Atanasio Pignatelli de Aragón y Moncayo (1724-1776), Spanish Ambassador in Paris, 1763-72?

19 Grigory Grigorievich, Count Orlov (1734-1783), military officer and lover of Catherine II, Empress of Russia. He planned with his brother Aleksei Grigorievich the coup d'état that overthrew Tsar Peter III and made Catherine Empress.

20 Sergei Domashnev (1743-1795) was Director of the St. Petersburg Academy of Sciences, 1775-83.

21 Dmitri Alekseevich Gallitzin (or Golitsyn) (1734-1803), Russian Ambassador at The Hague, was deeply interested in science.

22 Farewell and love me.

23 Again farewell.

75.11.20 *To Joseph Priestley*[1]

Dear Sir,

Among the many important discoveries for which the philosophical world is indebted to you, chiefly in that new and extensive branch of natural philosophy respecting *different kinds of air*, a very striking one is the exhibiting, in the form of this fluid, many solid bodies, and most of the known acids. It is with the most pleasing astonishment that I have always beheld that experiment, by which any unprejudiced mind must be convinced, that atmospherical air, even the purest, and the fittest for animal respiration, is produced by heat from a mixture of nitrous acid with any dephlogisticated earth, as *red lead*, *chalk*, &c. For, after having extracted from this mixture all the air that fire can expel, which is a prodigious quantity, when you repeatedly add fresh nitrous-acid to the residuum, you get a fresh quantity of this purest air, till all the earthy substance disappears.

This, however, being a very tedious process, when carried on with the most scrupulous attention, it came to my mind, that it would be sufficient to examine at the end of the first process, whether the residuum from the above mixture contained any part of the nitrous acid which had been put to it. To clear up this doubt, I kept for that purpose, with your approbation, the very same residuum of that process of the kind which we made a few days ago, to shew this wonderful kind of air to his Highness Prince Orloff,[2] and with it I made the following experiments.

I put into a large phial a weak blue tincture of archil, and after mixing it well, I poured two thirds of it into two smaller phials, in one of which I put a good quantity of the said residuum, and into the other as much of the dried mixture of red lead, with nitrous acid. The blue colour of this last phial disappeared in a few seconds, leaving the liquor almost limpid and transparent; but the other tincture, with the residuum out of which the air

---

1 Letter not found. The text given here was published as an appendix to Priestley's *Experiments and Observations on Different Kinds of Air*, vol. 2 (London, 1775), pp. 379-81, under the heading "A Letter from Mr. Magellan to Dr. Priestley, on the Subject of Dephlogisticated Air". The letter was evidently a last-minute addition to the volume, which was issued in December 1775, since it is not listed in the table of contents, this having presumably been printed already. A French translation appeared in the translation of Priestley's work, *Expériences et observations sur différentes espèces d'air*, Tome III (Paris, 1777), pp. 258-62.

2 Prince Gregory Orlov (1734-1783), former favourite of the Empress Catherine the Great of Russia.

had been expelled, shewed no change of colour, when compared with the remaining quantity of the tincture left in the large vessel. 75.11.20

I repeated this morning the same experiment with the tincture of turnesole,[3] and found the same effect, with this only difference, that the tincture turned reddish in the glass, which contained the dried mixture of red lead with nitrous acid, whilst the other kept its blue colour.

This seems to evince, with the greatest certainty, that the nitrous acid in this experiment is entirely set free by the action of the fire, in the form of air, and being at the same time combined with some part of the earthy matter, becomes respirable air. It is remarkable that this air has no acid in it, as may be concluded from the effect of shaking it with the above tincture in a phial, which I did: for it does not change the blue colour; whereas if the same operation is made with fixed air, it is changed almost instantaneously into a very decided reddish colour, as is well known.

Now since the air produced from the mixture of earth with nitrous acid, not only does not discover the least acidity, but proves to be the purest and the most wholesome for animal respiration, it plainly demonstrates that either air is not an *element*, or acid is not one, as some chuse to assert: since nitrous acid is reduced into air, together with the earth, in the above experiment, without leaving behind any acidity to impart it to the air which comes out of it. As to myself, I should rather think there is a transmutation of *elements* into one another, if such *name* may be used in this case; for we see by the above experiment, that acid and earth are transmuted into air, and by the experiments of Mr. *Godfrey*, published in 1747,[4] it seems that *water* is convertible into earth.

If you think the above may be any elucidation or confirmation of your experiments on this subject, you are at liberty to make what use of them you please.

Dear Sir, Yours, &c.

J.H. De Magellan.

London, 20th Nov. 1775.

3 i. e. litmus.

4 [Godfrey, Ambrose], *A Curious Research into the Element of Water; containing Many Experiments on that Fluid Body...* (London, 1747).

75.11.24 *To Pierre Macquer*

Bibliothèque Nationale, Paris, MS fr. 12306, f. 29-30

Londres le 24 Nov^bre^. 1775

Mon Cher Confrere

Celle-ci est pour vous prevenir de la part du D^r^. Priestley sur l'article de l'air fixe que vous avez commencé à travailler pour la 2de. edition de votre Dictionaire, que tout le monde attend avec impatience.[1] le 2^d^. vol. de l'ouvrage du D^r^. Priestley sur cette matiere, va paroitre incessament:[2] & j'en ai deja envoyé par la poste la plus grande partie à M^r^. Trudaine de Montigny,[3] qui ne laissera pas de vous le communiquer si vous lui en parlez: du mois[4] M^r^. lavoisiere qui sçait l'anglois l'aura sans doute vû, & pourra vous en donner des nouvelles. je lui envoye Cette meme poste la feuille V.[5]

Vous y verrez la fausseté que le D^r^. Higgins en vrai charlatan a decidée que l'air fixe n'etoit que de l'air commun, combiné avec du Phlogistique. en effet j'en donnai les arguments du D^r^. Higgins dans une lettre que j'ecrivis à M^r^. le Chev^r^. de Bory,[6] & qui peut etre il ne manque de communiquer à l'Acad^e^.: mais lorsque J'ai consulté le D^r^. Priestley là dessùs, J'ai vû le sophisme du D^r^. Higgins: & vous verrez par ce dernier ouvrage du D^r^. Priestley, qu'il n'avoit jamais adopté ce raisonement: & que les experiences montrent le contraire. Car l'air phlogistiqué, n'est pas acide, ni peut etre presqu'entierem^t^. absorbé par l'eau comme l'air fixe l'est. tout l'equivoque du D^r^. Higgins fut, qu'en phlogistiquant quelque quantité d'air commun, l'air fixe en fut degagé: de façon que l'eau de chaux en fut precipitée: mais le reste de cet air n'etoit que phlogistiqué; n'etoit pas acide; & ne peut pas aucunement etre absorbé par l'eau &^a^. Vous trouverez un grand nombre de belles experiences nouvelles dans Cet ouvrage. Je suis très faché que vous ne puissiez lire l'original, & soyez sujet aux meprises des traducteurs.

Vale & me ama

Magellan

Je vous prie de donner ce petit billet ci-joint à M^r^. Le Chev^r^. de Borda.[7] mes Respects à tous nos amis communs.

---

1 The second edition of Macquer's *Dictionnaire de chimie* was not published until 1778.

2 Priestley, *Experiments and Observations on Different Kinds of Air*, vol. 2 (London, 1775).

3 Letter not found.

4 moins?

5 Letter to Lavoisier not found.

6 Letter not found; but see 75.08.00.

7 Note not found.

To Antoine-Laurent Lavoisier[1] 75.12.00

Archives de l'Académie des Sciences, Paris, Fond Lavoisier, 1642[2]

Monsieur & cher ami

Voici une espece d'Apologie du Dr. Priestley contre les pretentions du charlatan Higgins.[3] Je viens de vous envoyer depuis quelques jours le 2d. vol. de l'ouvrage du Dr. Priestley[4] dont il Vous fait un cadot: ce volume n'est pas encore publié mais J'ai eu l'adresse d'en obtenir des premiers pour profiter de l'occasion d'une Caisse qui partit à l'Adresse de Mr. Trudaine. Mr. Duchenne (ou Mr. Messier) aura soin de vous la remettre lorsqu'elle arrivera à Paris.

Il ne m'a pas eté possible de trouver les deux Livres sur le Nitre dont vous me parlez dans la lettre de Mde. votre Epouse:[5] & Je suis fort assuré que sans un extraordinaire hazard, il ne sera pas possible de les trouver du Tout.

Vale & me ama
Magellan

P.S. Mes hbles. resps. à Mde. votre Epouse & à Mlle.
iterum vale

Je viens d'envoyer cette meme poste les dernieres feuilles de l'ouvrage ci dessus à Mr. Trudaine.

---

1 The letter is undated and gives no indication of addressee, but the contents indicate that it was addressed to Lavoisier and was written a few days after Magellan's letter of 15 December 1775 to Marie Lavoisier.

2 Corrected version of the text published in *Oeuvres de Lavoisier: Correspondance*, ed. René Fric, fascicule II, 1770-1775 (Paris, 1957), pp. 507-8.

3 Joseph Priestley, *Philosophical Empiricism, containing Remarks on a Charge of Plagiarism respecting Dr. H-s, interspersed with Various Observations relating to Different Kinds of Air* (London, 1775).

4 Joseph Priestley, *Experiments and Observations on Different Kinds of Air*, vol. 2 (London, 1775).

5 See Magellan to Marie Lavoisier, 15 December 1775.

75.12.12 *To Ribeiro Sanches*[1]

Österreichische Nationalbibliothek, Vienna
A.N. Ribeiro Sanches Korrespondenz, codex 12714, 165-166v

12 Dezembro 1775

Meo Am^{o}. e S^{r}. e m^{to}. do Cor^{am}.

A de V.M. me alegra m^{to}. pois diz q passa melhor. eu passo bem, mas sinto um principio de defluxo e porisso hirei á cama antes das 11, q he o meo Remedio junto com a dieta, e agoa quente com assucar. ontem á noite estive com o D^{r}. Priestley, & ambos bebemos á saude de V.M. um bom copo de Madeira. á manhan cuido hirei jantar com o nosso Enviado, Cav^{ro}. Pinto (q he home̅ m^{to}. de bem & de excellente coração.) e ajustaremos hũ dia p^{a}. hir ver as experiencias do D^{r}. Priestley, q sera talvez sabado proximo. elle jantou com Mil^{d}. Shelburne 5^{a}. fr^{a}. passada, & eu (q o vi na 4^{a}. precedente, jantando com elle) lhe disse q na 6^{a}. fr^{a}. devia hir jantar com Mylord. olhe se se engana, me disse elle, pois eu estou invitado p^{a}. a 5^{a}. & talvez q a sua invitação seja do mesmo dia. Certam^{te}. não, lhe respondi. oh! pode ser q tenha lá outros Ministros Estranjeiros: q^{to}. a mim eu mais estimaria hir com V.M. na 6^{a}. fr^{a}., pois gosto mais da socied^{e}. filozofica do q da ceremonioza. por aqui vera V.M. a qualid^{e}. do sojeito. na verd^{e}. acompanhia da 6^{a}. fr^{a}. não era mais do q eu, Arbuthnot, & Young. eu disse isto mesmo à Mylord Shelburne q^{do}. estavamos á meza, q gostou m^{to}. deste trait.

Eu tãobem recorri aqui ao nosso Ministro p^{a}. as noticias da India q M^{r}. Mickle dezeja. elle escreveo duas ou 3 vezes a hũ seo bom amigo na Torre do Tombo: & está picado de ainda não ter resposta. nem talvez q não terá alguma.

os livros q V.M. me enco̅mendou estão ja em hũ masso dentro do caixáo q vai partir p^{a}. essa terra. o Commũ amigo M^{r}. Duchenne (q ja veio p^{a}. Montigny, e talvez sera ja chegado a Paris) terá cuidado de lhos mandar &^{a}. Porem a obra de Cullen sobre os afogados ainda não sahio ao publico. Com tudo imagino poder mandar-lhe uma copia, pois o Livreiro m'a prometeo. o Rush sur l'histoire Naturelle dela medecine des Indiens, foi embargado aqui, onde se queria fazer uma nova edição, por ordem do mesmo A, & de Franklin seo intimo amigo; por ser necess^{o}. elucidar melhor alguns pontos, &c^{a}. emq^{to} estas negras disputas com os Americanos não se acabarem,

1 Annotated "Resp^{a}. a 1 Janr^{o}. 1776. Rép^{a}. dia de anno novo" [Answered on 1 January 1776, New Year's Day]. Letter not found.

não ha esperança de q este livro se publique. Comtudo se V.M. mandar as duas Regras no fim desta, ao D^{r}. du Bourg, na rua Coupeau, Fauxbourg S^{t}. marceau, elle lhe emprestará a copia q o D^{r}. Rush lhe mandou por minha via. o D^{r}. Dubourg he homen de idade, de hũ excellente caracter, & m^{to}. meo amigo. 75.12.12

Tãobem a Dissertação sobre a Medecina dos Lapoens, não he livro q se ache aqui. eu cuido q o D^{r}. Le Begue o vio citado no Medical Commentaries, mas he livro estranjeiro a Inglaterra.

agora tenho de responder a uma carta do Princepe Galitzin sobre a electricid^{e}., mas esperará p^{a}. a posta seg^{te}.

tive carta de Chevalier, q não passa bem. corpo piqueno, lisboeta delicado, e expatriado! nos outros beiroens passamos bem sobre todas estas dezavantagens.

Jantei sabd^{o}. com Sequeira, q passa bem, e ocupado. elle se recom̃enda. o D^{r}. La Cour está como o peixe n'agoa, nas suas assembleas de Bath.

o Mizeravel Letch está as portas do ultimo extremo a q a sua fraca Cabeça o conduzio. tem comido & dispendido nestes 4 ou 5 annos, com as fantazias de sua bella mulher outras tantas mil livras esterlinas. não sei qual sera a ultima scena, q necessariam^{te}. he chegada, pois tudo está esgotado. não penso nelle sem affliҫão: esteve aqui comigo ha 3 dias. vê bem os seos passados erros mas não ha remedio ...!

Vale atque iterum vale

Magellan's Compl^{s}. to his worthy friend D^{r}. Dubourg. wishes to know, if he has received the parcell of M^{r}. Williams, sent by Magellan thro' y^{e}. hands of M^{r}. Messier. M^{r}. Williams is on going to grenadoes: we have here good news of D^{r}. Franklin, who is now at y^{e}. head of y^{e}. Congress in America. Magellan begs the favour of D^{r}. Dubourg Lending the work of D^{r}. Rush on Y^{e}. medecine of y^{e}. Indians, to his good friend D^{r}. Sanchez, who will return it very safely: and his own compl^{s}. to M^{rs}. Dubourg and com̃on friends.

A Monsieur
Monsr. Sanchez D^{r}. en Medecine
Rue des Blancs-manteaux
a Paris

75.12.12 12 December 1775

My Dear Friend

Your letter[2] fills me with happiness because you say you are feeling better. I am well, but feel a cold starting and for that I shall go to bed before 11, that is my medicine together with diet, and hot water with sugar. Last night I met D^{r}. Priestley, & we both drank to your health a good glass of Madeira. Tomorrow I shall have dinner with our Envoy, Chevalier Pinto[3] (who is a good man and of excellent heart) and we shall arrange a day to see D^{r}. Priestley's experiments, that will probably be next Saturday. He dined with Milord Shelburne last Thursday, & I (who saw him the preceding Wednesday, dining with him) told him that on Friday I would have dinner with My Lord. Look if you are wrong, he told me, because I am invited for Thursday & perhaps your invitation is for the same day. Certainly not, I answered. Oh! Maybe he has other foreign Ministers: as for me I should prefer to go with you on Friday, because I prefer the philosophical to the ceremonial company. So you see the quality of the person. In the event, the company on Friday was no one else but me, Arbuthnot[4] & Young.[5] I told this same story to Milord Shelburne when we were at table, and he much liked this trait.

I also asked our Minister for the information about India that M^{r}. Mickle desires. He wrote once or twice to a good friend of his at the Torre do Tombo:[6] & is annoyed at not having had a response. Perhaps he won't ever have one.

The books you ordered are already packed in the box that will leave for France. Our common friend Duchenne[7] (who came to Montigny,[8] and perhaps has already arrived in Paris) will take care to send them to you &^{c}. However the work by Cullen on drowned[9] is not yet published. Nevertheless I imagine I can send you a copy, because the librarian promised it to me. Rush sur l'histoire Naturelle de la medecine des Indiens[10] was held up here, where a new edition was to be published, by order of the same Author, & of Franklin[11] his intimate

---

2 Letter not found.

3 Luiz Pinto de Sousa Coutinho (1735-1804) was Portuguese Minister in London, 1774-88.

4 John Arbuthnot of Mitcham, farmer and author of books on farming.

5 Arthur Young (1741-1820), writer on agriculture.

6 The headquarters of the National Archives of Portugal.

7 Louis-Henri Duchesne (1737-1793), secretary of Trudaine de Montigny, France's Intendant Général des Finances.

8 Presumably Montigny-Lencoup, a commune in the modern department of Seine-et-Marne, not far from Fontainebleau, where Trudaine de Montigny had his château.

9 William Cullen, *A Letter to Lord Cathcart ... concerning the Recovery of Persons Drowned and Seemingly Dead* (London, 1776).

10 Benjamin Rush, *An Oration delivered February 4, 1774, before the American Philosophical Society, containing an Enquiry into the Natural History of Medicine among the Indians in North America ...* (Philadelphia, 1774).

11 Benjamin Franklin (1706-1790).

friend because of the need to better elucidate some points, &c. Until these disputes with the Americans are finished, there is no hope of this book being published. However if you send the two lines at the end of this to Dr. du Bourg,[12] at rue Coupeau, Faubourg St. Marceau, he will lend you the copy that Dr. Rush sent him through me. Dr. Dubourg is an aged man of an excellent character, & a very good friend. 75.12.12

Also the dissertation about the Medicine of the Lapps is not found here. I think Dr. le Begue saw it referred to in the Medical Commentaries,[13] but it is a book foreign to England.

Now I have to answer a letter from Prince Galitzin about electricity, but it will wait for the next post.

I had a letter from Chevalier,[14] who is not well. Small body, delicate Lisboner, and expatriate! We from Beira[15] rise well above all these disadvantages.

I had dinner with Sequeira on Saturday, he is well and busy. He sends greetings. Dr. La Cour is like a fish in the water, in his assemblies at Bath.

Miserable Letch[16] is on the verge of the ultimate extreme to which his weak head led him. He has eaten & spent, these 4 or 5 years, with the fantasies of his beautiful wife, many thousands of pounds sterling. I don't know what the final scene, that necessarily will arrive soon since everything is exhausted, will be. I don't think of him without distress: he was here with me 3 days ago. He recognises his past errors but there is no cure ...!

Vale atque iterum vale[17]

Magellan's compliments to his worthy friend Dr. Dubourg. Wishes to know if he has received the parcel of Mr. Williams,[18] sent by Magellan thro' ye. hands of Mr. Messier. Mr. Williams is on going to grenadoes:[19] we have here good news of Dr. Franklin, who is now at ye. head of ye. Congress in America. Magellan begs the favour of Dr. Dubourg's lending the work of Dr. Rush on y.e medicine of ye. Indians,[20] to his good friend Dr. Sanchez, who will return it very safely: and his own compls. to Mrs. Dubourg and common friends.

---

12 Jacques Barbeu-Dubourg (1709-1779), French medical doctor and botanist.

13 *Medical and Philosophical Commentaries* (Edinburgh, 1773 onwards).

14 Letter not found.

15 A province in the central part of Portugal that at the time embraced both Magellan's home town, Aveiro, on the coast, and Penamacor in the interior, where Sanches was born.

16 John Letch, F.R.S. (?-1781) was one the members who proposed Magellan to the Royal Society.

17 Farewell and again farewell.

18 Possibly Jonathan Williams, Jr., Benjamin Franklin's grand-nephew, in a letter to whom, in March 1776, Franklin mentions that in a letter of September 1775 Williams had proposed a voyage to the West Indies.

19 Possibly the island of Granada, in the West Indies.

20 See note 10.

75.12.15 *To Marie Lavoisier*

Archives de l'Académie des Sciences, Paris, Fond Lavoisier, 1642[1]

Madame

La Philosophie avec toute sa grande suite n'est que trop glorieuse de trouver qu'une Dame si charmante & si respectable à tous egards s'occupe de ses interets: & Je me suis presqu'enorgueilli du devoir honorable que votre Lettre m'impose, pour vous remercier de la part qui me touche de ce bonheur.

J'ai eté frapé par la nouvelle inattendue du decès de M^r^. Lavoisiere Pere, que J'ai laissé en si bonne santé en Juin dernier, & que J'etois si fort obligé d'aimer par ses qualités & par ses bonnes dispositions à mon egard. La bonne nouvelle de la disposition de Melle. m'en soulage en grande partie, avec celles de M^r^. Paulze[2] & reste de toute votre respectable famille; auxquels severally je vous prie de presenter mes h^bles^. respects.

La brochure ci jointe[3] contient une petite idée de la maniere de faire le salpetre en Amerique, à page 46. J'ai deja fait quelques diligences pour les livres que vous me marquez: mais sans en trouver jusqu'à ce moment un seul. Je continuerai encore à faire des recherches nouvelles; quoique je crains ne pas avoir meilleur succès, etant des livres d'une date ancienne:[4] & je ne manquerai pas de vous envoyer aussitôt ce que je pourrai decouvrir.

M^r^. le D^r^. Priestley a eté fort content du tonelage & mensuration de l'air fixe par Le Duc de Chaulnes. il va partir lundi pour passer le Noel avec sa famille, & reviendra à Londres vers le milieu de Janvier: Il m'a prié de vous presenter ses très humbles resp^s^.: & a donné des ordres pour me fournir les feuilles qui manquent de son 2^d^. vol. sur l'air,[5] aussitôt qu'elles seront tirées. il vient de finir l'impression d'une espece de defense de ces decouvertes

1 Corrected version of the text published in *Oeuvres de Lavoisier: Correspondance*, ed. René Fric, fascicule II, 1770-1775 (Paris, 1957), pp. 532-3.

2 Mme Lavoisier's father, Jacques Paulze.

3 Not identified.

4 The works in question are identified in Magellan to Lavoisier, 16 January 1776.

5 Joseph Priestley, *Experiments and Observations on Different Kinds of Air*, vol. 2 (London, 1775).

centre le charlatan du Dr. Higgins, dont Je compte vous envoyer la poste suivante une copie.[6] 75.12.15

J'ai l'honneur d'etre avec tout respect & devouement possible

Madame
Votre très hble. & très obt. servr.
Magellan

à Londres ce 15 Decbre. 1775

6 Commencing in 1774, Bryan Higgins (*c.*1741-1818) gave lectures in chemistry at his house in London. In 1775 he accused Priestley of stealing experiments on air from him and other chemists. Priestley in response published his *Philosophical Empiricism, containing Remarks on a Charge of Plagiarism respecting Dr. H-s, interspersed with Various Observations relating to Different Kinds of Air* (London, 1775).

To the Duke Charles-Marie-Raymond d'Arenberg 75.12.19

Archief Van Arenberg, Enghien/Edingen, Belgium

Monseigneur

Je me represente très Souvent Le triste etat dans Lequel V. A. Sme. doit se trouver avec toute Son Auguste Famille depuis Long tems: & Je voudrois bien pouvoir, au prix de ma vie, en en diminuer L'amertume. Mais il n'y a que La Saine philosophie qui puisse fournir des Lenitives dans des circonstances pareilles: & il faut esperer que Le tems viendra à bout d'assoupir L'aigreur de la sensation à force de s'habituer à La Resignation du Sort ou plutot de la Providence. Les informations que j'ai Reçu par differentes personnes, me font croire, que Mr. Le Prince regagnera La vue:[1] & c'est toujours quelqu'avantage que de retrancher une grande partie de son malheur.

J'ai pris La liberté d'envoyer à V. A. Sme. une machine de verre pour faire chez Soi & en tout tems, des eaux minerales comme celles de Spá: il n'a pas Long tems qu'on l'a inventée ici comm'on le voit par La description

1 The Duke's son Louis Engelbert, the future 6th Duke of Arenberg, had been blinded in an accident while out hunting.

75.12.19 inserée dans les dernieres Transactions Philosophiques de Londres.[2] On en a deja fait un grand nombre pour L'usage de differentes personnes avec un extraordinaire succès. Cette machine est dans une caisse à l'adresse de V.A.Sme. que J'ai deja fait partir pour Ostende, recommendée au Soin de Mr. Hennessy[3] auquel j'ai fait avis par La poste derniere. On y trouvera une description écrite de ma main pour la practique & usage de cette machine; & je me flate que le valet de Chambre Mr. Causson, ou autre domestique quelconque de V.A.S. pourra aisement se mettre au fai[t] avec un peu d'attention à l'instruction que J'envoye.

J'y ai ajouté un petit miroir de toilette pour Mde. La Princesse, que Je prie V.A.S. de Lui faire tenir. C'est de ces miroirs de la verité, faits de crystal pur, qui rendent l'image sans d'autre couleur brune, comm'on le voit aisement en Le mettant à coté d'un autre miroir ordinaire. il y a aussi un petit pacquet de Semences de la palme à huile de la Cote de Mosquitos,[4] pour Mr. Michau de Louvain:[5] parceque ces semences demandent des serres chaudes pour produire.

Il y a aussi un pacquet de Livres imprimés, & 2 petites caisses qui contient deux pendules pour deux amis: & je prie V.A.S. de les confier à Mr. l'abbé Needham, auquel j'ecrirai la poste suivante pour en avoir soin.[6]

Comm'on assure que les eaux artificielles, faites avec cette espece de machine philosophique, produisent les memes effets, que les naturelles de Spa & d'autres endroits, J'ai pensé que celà pourroit etre de service pour Mde. La Princesse d'Arenberg ou pour Made. La Duchesse, ou quelqu'autre de Sa Famille. ainsi je n'ai pas voulu perdre cette occasion pour temoigner

---

2 John Mervin Nooth, "The Description of an Apparatus for impregnating Water with fixed Air; and of the Manner of conducting that Process", *Philosophical Transactions*, 65 (1775), 59-66. Magellan introduced some modifications to this apparatus; see his *Description of a Glass Apparatus, for Making Mineral Waters ...* (London, 1777), and also his "Lettre ... contenant la description d'un appareil de crystal, pour faire, en peu de minutes & à peu de frais, des eaux minérales semblables à celles de Pyrmont, Spa, Saltzer, &c.," *Observations sur la physique, sur l'histoire naturelle et sur les arts*, 9 (1777), 280–87.

3 Patricius Hennessy (1741-1784), head of the Ostende trading company, Patricius Hennessy & Co.; see Jan Parmentier, "The Sweets of Commerce: The Hennessys of Ostend and their Network in the Eighteenth Century", in David Dickson, Jan Parmentier and Jane Ohlmeyer, eds, *Irish and Scottish Mercantile Networks in Europe and Overseas in the Seventeenth and Eighteenth Centuries* (Ghent: Academia Press, 2007), pp. 67-91.

4 Part of the eastern coast of Central America, today shared between Nicaragua and Honduras.

5 Jean Joseph Michaux (1717-1793), professor of botany and director of the botanic garden at Louvain.

6 Letter not found.

mes Soins & mon Zele pour tout ce qui regarde le Service de S. A. Sme., etant 75.12.19
constament de Coeur & pour La vie penetré de Ses bienfaits & attentif aux sentiments avec les quels J'ai l'honneur d'etre avec un profond Respect

Monseigneur
de V. A. Sme
Serviteur très hble. & très obligé
Jean Hyacinthe de Magellan

à Londres ce 19 Decbre. 1775

Cette Machine pour faire les eaux minerales artificielles dans tout tems & dans quelqu'endroit qu'on souhaite de les avoir, est très commode, & utile: car suivant les observations & experiences repetées qu'on en a faites de ces eaux artificieles, elles ont les memes vertus medicinales, que les eaux minerales, dont on est obligé aller faire usage à des grandes distances avec beaucoup de trouble & d'inconveniences: au lieu qu'on peut avoir de ces eaux chez soi, toutes fraiches, dans fort peu de tems, à l'heure qu'on veut, & presque sans aucune depense. Voici l'usage de cette machine.

Prennez le vaisseau de verre A, mettez-le dans le plat de bois B pour eviter de mouiller la table avec quelques goutes d'eau, qui pourront tomber. otez le globe à queue de verre C: & remplissez le verre D avec de l'eau. remettez le verre C sur le verre D: & posez-les tous deux sur le piedestal de bois E. mettez alors un peu d'eau dans le vase A, jusqu'à la hauteur fg, c'est à dire jusqu'à ce que le fond soit couvert d'eau. Celà fait jettez toute

75.12.19 La quantité d'acide vitriolique (autrement nommé huile de vitriol) dans l'eau du vase A. Prennez la mesure de fer blanc h, pleine de terre calcaire, autrement de craion blanc, ou même de marble en poudre; & jettez cette poudre dans la la[7] misture qui se trouve dans le verre A. nettoyez bien avec un linge l'embouchure on, de façon qu'il n'y reste de la poudre, pour ne pas coler ou cementer ensemble avec l'autre verre. mais faitez celà assez vite; car aussitôt que la poudre calcaire tombe sur l'eau melée avec l'acide vitriolique, il y a une considerable fermentation. Prennez donc tout de suite le globe de verre D, avec l'autre C qui est sur lui, & mettez-le Sur le verre A, de façon que tous les trois verres restent, comme la figure les represente.

On verra tout de suite, que l'air fixe qui sorte de la fermentation, monte par la petite soupape de verre, qui se trouve en on, & passe à travers l'eau qui est en D, dont une quantité egale monte par la queue de verre m, jusqu'a vaisseau C qui etoit vuide.

On laissera rester cette eau ainsi en fermentation (c'est à dire ainsi couverte d'air fixe, qui se degage de la fermentation dans le verre A) pendant 3 ou 4 heures; au bout desquelles, on peut en faire usage, en ouvrant le trou z, pour oter l'eau du vaisseau D. Mais l'eau qui aura monté dans le vaisseau C, ne se trouvera pas impregnée ou mineralisée: ainsi il faut la oter avant d'ouvrir le bouchon z. pour cet effet on met le doit sur le trou x du verre C: & on l'ote de dessus du verre D, pour avoir l'eau minerale qui est en celui ci toute pure.

Enfin lorsqu'on voudra avoir de l'eau minerale calibeate, ou autrement ferrugineuse, comme celle des eaux de Spa, Pyrmont &ª. il n'y a d'autre trouble qu'à mettre dans le vaisseau D, un peloton ou deux, de petits fils de fer bien mince. Car l'eau aussitôt qu'elle est impregnée avec l'air fixe, agit fortement sur le fer, & en prend les vertus.

N.B. 1°. Si au lieu de 3 ou 4 heures, on laisse rester l'eau plus long tems vg. 5 ou 6 heures, elle sera alors plus fortement impregnée &ª.

2°. Le verre g est la mesure de l'acide vitriolique qu'on doit employer à chaque fois. mais un peu plus ou moins, ne fait pas une grande difference. il faut vuider dans la suite le verre A, en jettant ce qui est dedans, & repeter la meme operation ci dessus en employant les memes quantités de nouveau, pour avoir de cette eau minerale autant de fois qu'on en voudra.

3°. on envoye avec cette Machine un verre quarré plein d'acide vitriolique, & un grand pot de poudre calcaire; qui fourniront assez de matiere pour un grand nombre de fois.

Le tout de la part du plus fidel Serviteur
Magellan

7 *sic.*

To Joseph Banks 76.00.00[1]

British Library Add. 33977, f. 92

Magellan called to present his Compl^{s}. to Mr Banks. he has received a letter from the author of Table Academique, of which 3 vol. are already published, and contains all the matters treated of in y^{e}. collection of y^{e}. R^{l}. Acad^{y}. of Sciences and many others,[2] to which a list of all y^{e}. Members and Correspondents is to be added &^{a}. the Author desires to know 1°. the Christian name of M^{r}. Banks 2.º the year, place and county of his birth &^{a}. and 3°. the titles of his works printed or ready for being printed, their format & number &^{a}. &^{a}.

these particulars the author endeavours to get in order to mention them in the above list. If M^{r}. Banks pleases rather to write to y^{e}. Author (than to send the above queries to Magellan) the direction is so =

à M^{r}. l'Abbé Rozier
Place S^{te}. Genevieve
à Paris

then please to add an outside cover directed

à Monsieur
Monsr. Trudaine de Montigny
Intend^{t}. general des Finances &^{a}.
à Paris

this outside cover is a kind of Frank to y^{e}. Author.

---

1 In the bound sequence of Banks' correspondence, this letter is placed among letters received by him in late March and early April 1779. Given the dates of publication of the volumes referred to in the letter, however, it would appear that the letter must have been written in late 1775 or early 1776, and we have dated it accordingly. It must in any event have been written before Trudaine de Montigny's death in August 1777. The text has been published in *Scientific Correspondence of Sir Joseph Banks, 1765-1820* (London, 2007), I, 195-6, as a postscript to Magellan to Banks, 6 March 1779.

2 F. Rozier, *Nouvelle table des articles contenus dans les volumes de l'Académie Royale des Sciences depuis 1666 jusqu'en 1770...*, 4 vols (Paris, 1775-6). The first three volumes included a "Table des articles contenus dans les volumes de l'Académie", respectively sections A-D, E-L and M-S. The fourth volume, evidently in preparation at the time this letter was written, included section T-Z and also a "Table des noms des académiciens et de leurs correspondans, depuis 1666 jusqu'en 1770 inclusivement, dans laquelle on trouve le catalogue des ouvrages qu'ils ont fait imprimer, et la liste de leurs mémoires ou observations qui sont consignés dans les volumes de l'Académie". Since Banks was not elected a correspondent of the Paris Academy of Sciences until 1772, it would seem that Rozier's request that Magellan transmitted to him was based on wrong information.

76.01.01 *To Ribeiro Sanches*[1]

Österreichische Nationalbibliothek, Vienna
A.N. Ribeiro Sanches Korrespondenz, codex 12714, f. 518-519v

Ravensbury 1 de Jan^{o}. 1776

Am^{o} e S^{r}. e m^{to} do cor^{am}.

Em 12 do passado escrevi a V.M., e agora faço esta de caza do nosso bom amigo q V.M. não tem a dita de conhecer pessoalm^{te}., João Arbuthnot, onde vim passar estes dois dias com m^{to}. gosto. A mulher & cunhada estão engagées p^{a}. hir esta noite ao baile; & provavelm^{te}. eu e elle repetiremos a scena de ha 2 ou 3 annos com os piquenos, hum dos quais ja me disse q se lembrava da outra, e esperava hoje a Repetição.

Eu mandei à V.M. hũa parcella de Livros, com outras encomendas p^{a}. outros amigos: & M^{r}. Messier ou M^{r}. Duchenne terão cuid^{o}. de os fazer entregar a V.M. eizaqui a Lista.

Aitken Essays & cases – 4^{s}. / Prestwitch on poisons – 6 / Hutchison's Moral Phil^{y}. 2 vol. 4°. bound – £1.4^{s}.0^{d}./ Walace' System of Midwifery -£1.1^{s}.0^{d}./ Harwood on various editions of classics – 3^{s}./ Butter on puerperal fevers - 2^{s}.6^{d}./ eleven numbers of Med^{al}. Commentaries, para o seo am^{o}. de Lisboa – 16^{s}.6^{d}./ porteur à bord du vaisseau – 8^{d}./ Duncan's heads of lectures – 6^{d}./ Cullen's first Lines (very scarce and only to be had from his pupils) – 4^{s}. 6^{d}./ white on sea water – 1^{s}.0^{d}. Hamilton's midwifery (a good new book from Edimburg) – 4^{s}.6^{d}. Cullen on drownded (sent by y^{e}. poste)[2] – 1^{s}.6^{d}./ Leak on child-bed fever – 4^{s}.[3] / despezas[4] &c. – 6^{s}.0^{d}.

---

1 Annotated: "Recue a 9 Janr^{o}. " [Received 9 January].

2 Someone (Sanches?) has noted "dobra" [duplicate].

3 The following has been written at this point, then deleted: "Lovet electrical Philos^{er}. – 3^{s}.0^{d}./ Pringle on ye Torpedo – 1^{s}.6^{d}. / the 11th. number of Med^{al}. comment. – 1^{s}.6^{d}. / Priestley's 2^{d}. vol on air – 6^{s}.0^{d}". The works in question are: Richard Lovett, *The Electrical Philosopher: Containing a New System of Physics ...* (Worcester, 1774); Sir John Pringle, *A Discourse on the Torpedo, delivered at the Anniversary Meeting of the Royal Society ... Nov. 30, 1774* (London, 1775); A. Duncan (ed.), *Medical and Philosophical Commentaries. By a Society in Edinburgh* (London/Edinburgh, 1773-1795), no. 11; and J. Priestley, *Experiments and Observations on Different Kinds of Air*, vol. 2 (London, 1775).

4 "Respectivas de Caixa, porteurs, d'Alfandega" [respecting Box, porters, Customs] deleted.

demais achará V.M. hum embrulho q contem outro tinteiro de cristal do valor de 2^{s}.0^{d}. semilhante ao q o Conde Carbury lhe cubiçou: & com elle vai hũ bocado do nosso Rhuibarbo inglez, secado & preparado por hũ Mercador, amigo meo, de Milão; o qual tem tomado o Rhuibarbo a sua conta, isto he por seo divertimento, tem feito & procurado fazer por via de outros repetidas experiencias: & tem achado q he tão bom & produz tão bons effeitos como o melhor ruibarbo q vem da china. elle tem hũ livro de entrada sobre esta raiz. tira cada mez hũa ou duas raizes da terra de 5 ou 6 annos; apara-as, seca-as á hũa certa distancia do fogo &^{a}. e assenta todas as circunstancias q achava & q vem á sua noticia. Elle tem mandado m^{ta}. semente p^{a}. Italia & varias plantas ja crecidas: & diz q com o tempo sera um ramo de Com̄ercio q o estado podera poupar; pois produzira assaz p^{a}. o seo gasto &^{a}. 76.01.01

Os artigos assima fazem a soma de £ 5.15^{s}.8^{d}. Esterlinas, q em dinh^{o}. de França fazem 132$^{\#}$. 14^{s}. tournoises, e q se lhe pezarem no escritorio ou na fantazia, V.M. poderá dar a M^{r}. Messier, ou M^{r}. Duchenne por m^{a}. conta, alem da parte proporcional dos gastos de Calais até Paris q o mesmo Sujeito lhe dirá.

O am^{o}. du Presle me dizia na sua ultima q se achava com o andaço catarral, e não sabia como V.M. estava, por não poder sahir p^{a}. ver V.M. eu tãobem aqui fui atacado duas ou 3 vezes com este andaço q tem sido geral, & tem despachado alguns p^{a}. o outro mundo donde ninguem torna, sinal de q vale mais do q o prezente. Porem tenho-me curado com o meo methodo, matando-o á fome, & afogando-o com agoa quente e assucar antes de hir á cama. Ultimam^{te}. senti fortes gripes & piles. estou bem das prim^{as}. com a abstinencia de vinho & cerveja, pois não erão senão procedidas de hum principio inflamatorio: e o mesmo remedio me tem feito mitigar as almorreimas, q ainda não estão totalm^{te}. sans.

Hontem pela manhan antes de partir p^{a}. aqui passei por caza do D^{r}. Seq^{ra}., q ja sahe de caza, mas ainda sente alguns restos do seo dezastre. eu fui p^{a}. jantar com elle no sabado precedente á semana do Natal isto he em 16 do passado, q^{do}. o criado me disse q elle estava de cama. Subi assima & então soube como na 3^{a} ou 4^{a} fr^{a}. precedente, tinha sido entalado por hum carro contra a parede ao voltar na rua. cahio por terra, & todos o derão por morto. Sangrado não deitou sangue da prim^{a}. picada. mas sim qd^{o}. o picárão no outro. metido na cama continuou m^{to}. mal, vomitando tudo, & nenhũ Remedio lhe valia. conservou sempre o juizo, & vendo-se sem esperanças, pedio hũ copo de agoa de neve, como ultima despedida da vida. logo q a bebeo sintio hũ alivio consideravel. foi continuando, & com ella pode reter algũa coiza no estomago, de modo q em 4°. dia em q o vi,

76.01.01 ja estava assentado numa Cadeira: e como digo ja se acha capaz de sahir no seo Coche a fazer algũas vizitas. elle me pedio q lhe presentasse Suas Lembranças & aff^{os}.

Sab^{o}. passado jantei com nosso am^{o}. M^{r}. Paradise, q V.M. lá vio com Lascaris, & me pedio lhe presentasse os seos Resp^{os}. tãobem alli jantou M^{r}. Lytton de Bath, m^{to}. amigo do D^{r}. La Cour q me deo boas not^{as}. delle e Sua familia.

na 4^{a}. fr^{a}. jantei com o nosso Enviado, q hé daquella classe da qual mui poucos Portuguezes de distinção se podem gloriar. elle passa bem, mas sua espoza, magra, moça, com sua febrinha devez em qd^{o}. não parece q se dá bem com este clima. dá me pena, pois merece mais feliz dispozição. ja tem dois meninos (macho e femea): esta he a mais velha & naceo 2 ou 3 dias depois de chegar a Falmouth, q não foi pouca ouzadia de se meter ao mar em tal estado. mas he Portugueza, e isso basta!

Ora ja q falamos em Portuguezes, quero-lhe dar copia de hũa lembrança q achei na mão do Sec[re]tario do nosso Ministro, rapaz de bom propozito; com condição q V.M. a entre no Seo Livro, & a le[ia] da m^{a}. p^{te}. ao D^{r}. le Begue p^{a}. credito da Patria. eila aqui: = Lembrança communicada pelo dez[^{r}] Joze de Siabra e Sylva, sous-secretario de Estado, ao Tenente general Simão Frazer. No Arc[hi]vo da Torre do Tombo, onde se conservão os originais dos Negocios do Reino de Portugal, se acha a copia ou original do Perdão q El Rey D. Afonso 3°. deo ao P. Fernando da villa de Arouca, pelas culpas de ter dormido com sette Irmans, onze afilhadas, quinze comadres, hũa sua thia, e mais cincoenta mulheres, de quem teve cento & noventa & sette filhos; à Saber quarenta fêmeas & cento & cincoenta & sette machos. = Agora V.M. dezafie lá da minha parte toda a França & resto da Europa, se são capazes de mostrarem um factum autentico desta casta.

Vale & me ama.

Saud^{es}. à M^{a}. Jeane, & Resto da familia adoptiva; & mais amigos se he q os ha.

iterum vale. m^{tos}. annos & bons, Saude; paz, & felicid^{e}.

A Monsieur
Monsr. Sanchez D^{r}. en Medecine
Rue des Blancs-Manteaux
à Paris

Ravensbury 1 Jany. 1776 76.01.01

My Dear Friend

On the 12th of last month I wrote to you,[5] and now I do it from the home of our good friend whom you don't have the pleasure of knowing personally, John Arbuthnot, where I came with pleasure to spend this couple of days. His wife & sister-in-law are engagées to go tonight to the ball; & probably I and he will repeat the scene of 2 or 3 years ago with the children, one of whom told me he remembered the other one, and was expecting the repetition today.

I sent you a parcel with books, and other orders to other friends: & M^r. Messier or M^r. Duchenne will have to handle them for you. This is the list:

Aitken Essays & cases[6] – 4^s. / Prestwitch on poisons[7] – 6 / Hutchison's Moral Phil. 1°. & 2 vol 4°.[8] bound – £1.9^s.0^d./ Walace' system of Midwifery[9] – £1.1^s.0^d./ Harwood on various editions of classics[10] – 3^s / Butter on puerperal fevers[11] – 2^s.6^d./ eleven numbers of Med^al. Commentaries, for your friend of Lisbon[12] – 16^s.6^d./ porteur à bord du vaisseau – 8^d./ Duncan's heads of lectures[13] – 6^d./ Cullen's first Lines (very scarce and only to be had from his pupils)[14] – 4^s. 6^d / white on sea water[15] – 1^s.0^d. Hamilton's midwifery (a good new book from Edinburg)[16] – 4^s.6^d. Cullen on drownded[17] (sent by y^e. poste) – 1^s.6^d./ Leak on child-bed fever[18] – 4^s. / expenses &c – 6^s.0^d.

---

5 See Magellan to Sanches, 12 December 1775.

6 John Aitken, *Essays and Cases in Surgery* (London, 1775).

7 John Prestwich, *Prestwich's Dissertation on Mineral, Animal and Vegetable Poisons ...* (London, 1775).

8 Francis Hutcheson, *A System of Moral Philosophy* (2 vols 4°, London/Glasgow, 1755).

9 Robert Wallace Johnson, *A New System of Midwifery, etc.* (London, 1769).

10 Edward Harwood, *A View of the Various Editions of the Greek and Roman Classics ...* (London, 1775).

11 William Butter, *An Account of Puerperal Fevers ...* (London, 1775).

12 A. Duncan (ed.), *Medical and Philosophical Commentaries. By a Society in Edinburgh* (London/Edinburgh, 1773-1795). Sanches's friend was D^r. João Henriques Ferreira; see Magellan to Sanches, 17 November 1775.

13 Andrew Duncan, *Heads of Lectures on the Institutions of Medicine* (Edinburgh, 1775)?

14 The first of the four volumes of William Cullen's *First Lines of the Practice of Physic*, based on his lectures at the University of Edinburgh, was not published until 1777. However, unauthorized copies of his lectures had been circulating among his students for many years.

15 Robert White, *The Use and Abuse of Sea Water impartially considered ...* (London, 1775).

16 Alexander Hamilton, *Elements of the Practice of Midwifery* (London, 1775).

17 William Cullen, *A Letter to Lord Cathcart ... concerning the Recovery of Persons Drowned and Seemingly Dead* (London, 1776).

18 John Leake, *Practical Observations on the Child-bed Fever ...* (London, 1775).

76.01.01 In addition you will find a package containing another crystal inkpot costing 2^{s}.0^{d}. analogous to the one that Count Carbury envied,[19] & with it goes a bit of our English rhubarb, dried & prepared by a merchant, a friend of mine, from Milan, who has been taking Rhubarb on his own account, that is for fun. He has done & tried to do repeated experiments through others: & has found that it is as good as & produces as good effects as the best that comes from China. He has a logbook on this root. Every month he takes one or two roots 5 or 6 years old from the earth, chips them, dries them at a certain distance from the fire &^{a}. and takes note of all the circumstances found that come to his notice. He has been sending a lot of seeds to Italy & various grown plants: & says that given enough time it will be a branch of commerce that the state could keep out of; because it would produce enough to meet the need &^{a}.

The articles above amount to the sum of £ 5.15^{s}.8^{d}. Sterling, which in French currency makes 132$^{\#}$.14.^{s} tournoises, and if that becomes a burden in your office or in your mind, you can give to M^{r}. Messier or M^{r}. Duchenne on my account, as well as the proportional part of the expenses from Calais to Paris that the same person will tell you.

Our friend du Presle told me in his last letter[20] that he had a contagious cold and didn't know how you were, because he couldn't go out to see you. I also suffered two or 3 times from this same infection that has been general, & has dispatched a few to the other world from whence nobody comes back, a sign that it is worthier than the present one. However I have been curing it with my method, killing it with hunger & drowning it in hot water with sugar before falling into bed. Lately I felt strong gripes & piles. I have recovered from the former by abstinence from wine & beer, since they were merely caused by an inflammatory principle: and the same medicine has alleviated my haemorrhoids that are not yet completely well.

Yesterday morning before coming here, I called in to D^{r}. Seq^{ra}.'s home. He already goes out, but still feels some troubles from his accident. I was going to dine with him on the Saturday before Christmas, that is the 16th of last month, when the servant told me that he was in bed. I climbed the stairs & then I learnt how on the preceding Tuesday or Wednesday, he got stuck between a carriage and the wall as he turned into the street. He fell down, & everybody thought him dead. On being bled, he didn't bleed at the first sting but he did when they picked elsewhere. In bed he continued very bad, vomiting everything, & no medicine was effective. He always kept his mind, & feeling no hope, he asked for a glass of snow water, as a last farewell to life. As soon as he drank it, he felt considerable relief. He continued & with it he could retain something in the stomach, so that on the fourth day, when I saw him, he was sitting on a chair: and as I say he is

19 See Magellan to Sanches, 16 September 1775.
20 Letter not found.

now able to go out in his coach to pay a few visits. He asked me to present his regards to you. 76.01.01

Last Saturday I dined with our friend Paradise,[21] whom you saw with Lascaris[22] & who asked me to present his respects to you, also dined there with Mr. Lytton of Bath,[23] a good friend of Dr. La Cour[24] who gave good news of him and his family.

On Wednesday I had dinner with our Envoy,[25] who is of that class of which few Portuguese of distinction can claim. He is well but his wife, thin, young, sometimes with a little fever, does not seem to stand this climate well. it makes me sad, because she deserves a happier disposition. She has two children already (male and female): the latter is the elder & was born 2 or 3 days after arriving at Falmouth, it took no little courage to go to sea in that condition. But she is Portuguese, and that is enough! Now that we talk of Portuguese, I want to give you a copy of a note I found with the secretary of our Minister, a young man of good character; on condition that you put it in your book and read it for my part to Dr. le Begue on our country's credit. Note communicated by Justice Joze Siabra e Sylva,[26] Under-secretary of State, to Lieutenant General Simão Frazer.[27] In the Archive of Torre do Tombo, where the originals of the Affairs of the Kingdom are kept, the copy or the original is found of the Forgiveness that King Afonso the 3rd gave to Father Fernando of the town of Arouca, found guilty of having slept with seven Sisters, eleven stepdaughters, fifteen women friends, an aunt of his, and another fifty women, by whom he had one hundred & ninety-seven children, forty females & one hundred & fifty & seven males. = Now challenge on my behalf all France & the rest of Europe, whether they are capable of showing an authentic factum of this kind.[28]

Vale & me ama.[29]

Regards to Ma. Jeane & all the adoptive family, and friends if any are there.

iterum vale. Have many and good years, health; peace, & happiness

---

21 John Paradise (1743-1795); see Magellan to Sanches, 17 November 1775.

22 Not identified.

23 Not identified.

24 Dr. Philipus La Cour.

25 Luiz Pinto de Sousa Coutinho (1735-1804), Portuguese Minister in London, 1774-88.

26 José de Seabra e Silva (1732-1813), Portuguese statesman who held important positions under the governments of King José and Queen Maria I.

27 Simon Frazer (or Fraser) (1726-1782), Colonel in the British Army, was sent to Portugal in 1762 as brigadier-general and later lieutenant-general in the Portuguese army during the war against Spain.

28 Sanches in his Journal, on 2 February 1776, repeats this story and writes the following comment: "Procreador Notavel e Desavergonhado" [remarkable shameless procreator].

29 Farewell and love me.

76.01.04 *To Charles Messier*

American Philosophical Society Library, Philadelphia, João Jacinto de Magalhães (Jean-Hyacinthe Magellan) Collection, Mss BM25

Londres 4 Janvier 1776

Mon cher ami

Si celle-ci arrive encore à tems ajoutez aux livres que Je vous demandois, & qui doivent venir dans une caisse, comme je l'ai avisé à M^{r}. le D^{r}. le Begue du Presle, avec plusieur autres, ajoutez je vous prie, la <u>Chymie raisonée & practique</u> de Mr. Beaumé,[1] pour moi & marquez moi le prix, aussi bien que des autres livres &^{c}. je vous ecrirai au large la poste suivante. à present il ne m'est pas possible. mes amitiés à M^{r}. de Sejour,[2] son Pere, M^{r}. & M^{de}. le Blanc, Abbé [...], & à tous les amis de la Acad^{e}.

Vale & ama
tuum Magellan

N.B. à M^{r}. Le Monnier,[3] que les observations sur la variation de l'aimant en differentes parages de la mer du Sud &^{c}. suivant le voyage de Cook, seront bien tot publiées par ordre de l'Amirauté,[4] qui a tous les papiers des Astronomes de ce Voyage dernier, & dont il n'est pas possible d'avoir copie avant leur publication.

iterum vale.

A Monsieur
Monsr. Messier, Astronome de la
Marine, membre de l'Acad^{e}. R^{le}. des Sciences
Hotel De Clugni rue des Maturins
<u>à Paris</u>

---

1 A. Baumé, *Chymie expérimentale et raisonée* (Paris, 1773).

2 The mathematician and astronomer Achille-Pierre Dionis du Séjour (1734-1794).

3 Pierre Charles Le Monnier (1715-1799).

4 See W. Wales and W. Bayly, *The Original Astronomical Observations, made in the course of a Voyage towards the South Pole, and round the World, in His Majesty's Ships the Resolution and Adventure, in the Years 1772, 1773, 1774, and 1775* (London, 1777), which includes (pp. 367-85) a table headed "Azimuths of the Sun's Center, taken with an Azimuth Compass; together with the Altitudes of his Lower Limb, taken at the same time, with Hadley's Sextant, for determining the Variation of the Magnetic Needle ...".

To Antoine-Laurent Lavoisier 76.01.16

Archives de l'Académie des Sciences, Paris, Fond Lavoisier, 1642[1]

Londres ce 16 Janvier 1776

Monsr. et cher ami

Voici une gazette, oû vous trouverez un article de Philadelphie sur le salpetre. Les deux livres, savoir John Mayow Tractatus quinque Physico medici Oxon. 1674,[2] & Nat[al]. hist[y]. of Nitre by W[m]. Clarke London 1670,[3] ne peuvent pas etre trouvés malgré toutes les diligences, que J'ai faites moi-meme, & par le moien de differents colporteurs. La lettre de M[de]. Lavoisier[4] parloit d'un Extrait des papiers Anglois du mois d'Avril 1763; Mais je n'ai pas pu obtenir aucune idée ici de cet Extrait. Si ces livres se trouvent dans la Bibliotheque du Roy, ou dans quelqu'autre Bibliot[e]. à Paris, il ne vous sera pas dificile d'en obtenir la lecture & d'en faire les extraits necessaires. Mais s'ils ne se trouvent pas chez vous, & vous les croyez fort importants pour vos vues, faitez me le savoir: car Je ferai des recherches ici pour Les decouvrir dans quelque Bibliotheque de Londres ou en Oxford: & je Vous en ferai faire des extraits.

Je me suis informé sur la methode qu'on employe dans les Indes Orientales pour obtenir une si prodigieuse quantité de Salpetre, comm'on aporte en Europe chaque année. On m'assure qu'il y a des gens en forme de Compagnies pour ramasser des plastrons d'oû l'on extrait le salpetre: que ces gens suivent d'après les devastations que les differents peuples font des villages & peuplades detruites pas les guerres, famines, & malheurs, qui sont assez frequents dans les indes, par lesquels acidents il y a plusieurs villes & vilages abandonés & deserts tout à fait, oû la Corruption & pouriture des cadavres des animaux aussi bien que des gens memes remplit par la dissolution de leurs parties constituentes, les murailles avec ces concretions salines &[a]. & que par là on ramasse cette prodigieuse quantité de salpetre que nous importons en Europe.

---

1 Corrected version of the text published in *Oeuvres de Lavoisier: Correspondance*, ed. René Fric, fascicule III, 1776-1783 (Paris, 1964), pp. 537-8. The letter is unaddressed but the contents make plain that it was intended for Lavoisier.

2 John Mayow, *Tractatus quinque medico-physici...* (Oxford, 1674), the work in which Mayow famously wrote of a "nitro-aerial spirit" that took part in chemical reactions.

3 William Clarke, *The Natural History of Nitre: or, A Philosophical Discourse of the Nature, Generation, Place, and Artificial Extraction of Nitre, with Its Vertues and Uses* (London, 1670).

4 Letter not found; but see what was evidently Magellan's response, dated 15 December 1775.

76.01.16 Ne pourroit-on pas faire quelque chose de pareil, dans les bois & montagnes ou champaignes eloignées des grandes villes de quelques lieues, oû l'on seroit obligé d'aporter tous les cadavres des Hospitaux, & des gens pauvres qu'on ensevelit à present dans les cimiterres au dedans des villes avec au prejudice des vivants? une police pareille feroit grand honneur au Ministere.

Mes h^bles^. respects à M^de^. Lavoisiere & a Mlle. votre tante à M^r^. Paulze Pere[5] & fils: de meme à tous nos amis communs. ditez à M^r^. Le Roy que J'ai reçu sa lettre, & peut etre lui repondrai-je dans cette meme poste.

Vale & me ama
tuum[6]
Magellan

5 Mme Lavoisier's father, Jacques Paulze.
6 Farewell & love me | yours.

76.01.18 *To the Duke Charles-Marie-Raymond d'Arenberg*
Archief Van Arenberg, Enghien/Edingen, Belgium

Londres ce 18 Janvier 1776

Monseig^r^.

Je conçois bien que l'objet de cette lettre n'a pas besoin d'etre presenté à l'attention de V. A. S^me^.: mais J'espere de sa bonté qu'elle m'excusera ces reflections des necessaires.

Je vois par la lettre de V. A. S. de 6 de ce mois,[1] que les remedes internes ont endomagé la Santé du Prince: & je suis on ne peut plus mortifié, que les Messieurs oculistes aient voulu faire le charlatan dans ce cas vis à vis du Prince. Car sa complainte n'est autrement que locale.[2] il ne s'agit que de retablir chirurgicalement, je veux dire par La maneuvre de l'operation manuelle ce qui etoit possible d'arranger, & laisser le reste au cours ordinaire des remedes externes; mais encore ceux-ci doivent etre très doux

1 Letter not found.
2 The Duke's son, the future 6th Duke of Arenberg, had damaged his eyes in an accident while out hunting.

& sans aucune autre tendence qu'à prevenir l'acès de l'air & l'action de l'ambient. il faut ne pas appliquer des corrosifs ni des espiritueux, qui ont la qualité de coaguler les humeurs des yeux: & absolument se garder de troubler l'economie animale avec aucune sorte de remedes internes. tout au plus si le corps est caustif,[3] il ne faut que prendre quelque purgatif leger, & le moins irritant qu'il sera possible. Dans un mot l'erreur des humains le plus dangereux est de s'eriger audessus du rang oû la Nature les a mis: & c'est le cas de ces pretendus Savants, mais vrais charlatans, qui abusent de la confidence des autres. Si ce M^r^. l'Abbé des Monceaux est de cet avis, & poursuit cette methode à la bonne heure: mais si V.A.S. voit qu'il prend une autre route, Je la prie de l'envoyer à traiter des affaires de son Breviaire.[4] C'est une pretention la plus insolente que de pretendre connoitre le mechanisme secretoire des humeurs, qui remplissent Les differentes cavités des yeux, ou de s'entêter à les ameliorer. Si les tuniques ou membranes des yeux du prince sont offensées, & l'operation manuelle n'a pas reussi à les arranger comm'il faut pour transmetre en propre ordre la lumiere qui fait la sensation de la vision: si l'organe qui Reçoit les impressions est endomagé, il n'est pas possible par aucun remede interne de l'arranger. Ce n'est qu'à la poursuite des operations de l'economie animale qu'on doit se tenir pour reparer ces defauts: & il n'y a rien de plus à faire, que seulement empecher en quelque sorte s'il est possible, les suites du derangement que l'ambient, ou la lumiere y peuvent causer: ce qui après un si grand cours de tems passé depuis l'acident, ne peut à present exiger aucun procedé extraordinaire. 76.01.18

Je demande mil pardons a V.A.S. par n'avoir melê de la medicine oculaire: mais mon vrai attachem[t]. de Coeur pour Le Prince & pour tout ce qui Regarde la personne & Famille Auguste de V.A.S., ne me permet pas d'etoufer ou cacher mes sentiments: & Je veux faire ce peu qui est en mon pouvoir pour satisfaire les emotions de ce meme attachem[t]. de Coeur, & vivera devouement eternel, avec lequel J'ai l'honneur d'etre avec le resp[t]. le plus profond

Monseigneur

de V.A.S.

Serviteur le plus humble & obligé

Magellan

P.S. Que V.A.S. accepte les h[bles]. Respects de M[r]. Arbuthnot qui a très fortem[t]. senti tant de malheurs &[a].

3 There is no word "caustif" in French. Magellan probably had in mind the English word "costive" which translates not as "costif" or "caustif" but as "constipé".

4 Abbé des Monceaux (or Desmonceaux) (1734-1806) was author of *Traité des maladies des yeux et des oreilles…* (Paris, 1786), and other works on diseases of the eye.

76.01.18a *To Ribeiro Sanches*[1]

Österreichische Nationalbibliothek, Vienna
A.N. Ribeiro Sanches Korrespondenz, codex 12714, 169-170v

Londres 18 Jan^{o}. 1776

Meu am^{o}. e S^{r}. em^{to} do Cor^{am}.

Pela minha ultima de 2 do Corr^{te}. terá V.M. visto q nos Livros q mandei a V.M. não vai o Kirkland por serem já partidos m^{tas}. semanas antes de Receber a carta de V.M. do 1. deste mez. brevem^{te}. (em 4 ou 5 semanas) mandarei outras encomendas p^{a}. Paris & com ellas partirá este Livro, em cazo de não ter antes disso not^{as}. de V.M. ter achado o seo. Com elle partirá a Lusiada de Camoens[2] em Inglez[3] m^{to}. bem traduzida, com um prefacio ou Introdução de 167 paginas, mui filozofica, mui Politica, Historica & Poetica de q estou certo q V.M. gostará. à pezàr das suas grunhiduras là está o nome de V.M. na Lista dos suscritores. porem como elle vai à Petersbourg em vez de à Paris, póde V.M. consolar-se q todos cuidarâõ ser outro do Seo nome q se acha em Rússia, e não V.M. mesmo.

Estou picado com a sua maganeira de me dizer graças sobre o silencio das minhas Nelys, Betys, Polys &^{a}. &^{a}. &^{a}. pela crista d'um galo, q na prim^{a}. vez que tornar a Paris, o hei de fazer trocer as linhas! olhe o q he a inveja! pois saiba q'ainda não ha 20 minutos, q a minha Bety m'o mamou até a arreigada, e me protesta q a pezar dos meos 54, ainda a sirvo melhor q hum seo unico amante de menos de 30 de idade. Mas V.M. q já não arr-ta não só constroe as minhas modestas reticencias, como verd^{as}. fraquezas, & completos cumprim^{tos}. das suas profecias, mas tão bem o silencio do pobre am^{o}. Duchenne sobre a parcella dos livros, da qual elle ainda não

1 Annotated: "recebida a 30 Jan^{ro}" [received 30 January] and "Resp^{a} a ambas a 5 Fevereyro" [Answered both 5 February]. Letter not found.
Annotated on the envelope "A Mr Messier 400#| resp^{a} a esta | Que lhe mando 400 | Armstronge of Fevers | Que a sua conta recebida na carta ~~não~~ he err[a]^{da} e que todas apare[cem]" [To Mr. Messier 400# | reply to him | That I send him 400 | Armstronge of Fevers | That his account received | in the letter is wrong | that everything appears]. The book referred to here was probably John Armstrong, *Medical Essays* (London, 1773).

2 Altered by Sanches to "os" instead of "a", meaning that the title should be "The Lusiads" and not "The Lusiad" as Magellan wrote. Cf. Magellan to Sanches, 27 February 1776.

3 Sanches has inserted an asterisk at this point, leading to his marginal annotation: "ha outra antiga que vi em Londres" [there is another old one that I saw in London]. The older translation of Camoens' poem to which he refers was possibly the one by Richard Fanshaw (London, 1655).

tinha noticia alguma quando V.M. lhe perguntou por ella: pois eu não lhe escrevi com o detail du contenu dans les deux caisses senão em 9 deste Mez; sabendo q o Navio se tinha dilata[do] m^{to}. mais alem do tempo prefixo em q prometeo de dar á véla p^{a}. Calais o q visto e bem considerado devemos concluir 1° q Duchenne não podia dar a V.M. algũa noticia dos seos Livros; & 2° q V.M. como q^{m} já não arr-ta, deitou suspeita de mal visto, ou mal entendido, ou mal dezejado sobre aquelle amigo, q sabe viver, & tãobem servir q^{do}. pode. 76.01.18a

Temos depois 15 dias m^{ta}. neve nesta terra. Cuido q me acho em Serra d'Estrella. os caminhos estão cubertos della: ao pé da caza de nosso am^{o}. Arbuthnot estavão 3 carros enterrados em neve: e o moço do Correio tãobem là deixou a sua carreta, & tomou a mala das Cartas em hũ Cavalo, q lá fez passar por hum caminho de pé. eu cuidei seg^{da}. f^{ra}. passada q a carruagem em q vim p^{a}. Londres ainda q a 6 cavalos tãobem ficaria no caminho: e estou vendo q talvez não acharei á manhan carruagem p^{a}. lá tornar como costumo, pois sempre vou depois de m^{tos}. mezes là passar o sabb^{o}. á noite, e o domingo; e torno na 2^{a}. fr^{a}. pela manhan, & ás vezes na 3^{a}. fr^{a}. Com tudo agora neste momento Recebi uma carta do nosso honrado amigo o D^{r}. Wolfe de Danzic, o q me faz cuidar q algũas malas de correio de fóra chegárão. em Canterbury houve uma inundação: e as gazetas não falão senão de acidentes por cauza desta neve & máo tempo. o d^{o}. D^{r}. Wolfe me pede o recomende na sua Lembrança, & lhe agradeça o prezente q V.M. lhe fez do seo Tratado sobre a hist^{a}. da maladie venerienne. o D^{or}. Hunter deo á Socied^{e}. Real uma Memoria p^{a}. provar o mesmo objeto. elle se prevaleceo das autorid^{es}. q V.M. cita na sua Dissertação, mas não sei se a citou, ainda q estou certo q elle a leo. elle me mostrou uma data, pela qual Arias montano tinha sido agregado em 1507 (ou coiza semilhante pois não me lembro já) á faculd^{e}. de Medicina de Salamanca. a carta do D^{r}. Wolf est[4] do 26 Dez^{bro}. passado. elle me d[iz][5] "les doubles tierces & les simples tierces ont fait beaucoup de ravage: actuellement nous avons les fievres quartes très rebelles. La cruauté des Prussiens a fait mourrir plusieurs milliers de travailleurs à leur Fortresse près de Marienwerder & à leur canal de Bromberg. Beaucoup de coups de baton, point de logement, mauvaise nourriture, leur ont donné une Epidemie terrible. Il est fort curieux de reflechir sur le circuit des maladies febriles. à Warsovie dans une practique extensive, Je ne me souviens pas d'avoir vû une fievre quarte, &

4 Magellan uses a French word instead of a Portuguese one, in accordance with the letter he has received.

5 Stained paper.

76.01.18a fort rarement des tierces simples, pendant l'espace de 15 ans & plus. C'étoit tout des fievres bilieuses remittentes, ou meme des continues. Depuis 5 à 6 ans la scene a changé. Je ne vois que rarement les fievres continues, & les remittentes changent toujours en intermittentes, qui pourtant ne sont pas moins mechantes. la même chose s'observe à Warsovie. Cette année-ci le plus du monde a eté emporté par des desynteries qui manquoient rarement de suivre la fievre, quoiqu'on avoit eté purgé à plusieurs reprices" Ora veja V.M. là se o mundo & os climas não vem a ser tãobem caducos como nos outros!

vale & me ama.

Saudes. á fama. adoptiva & amigos se hé q os há.

a Monsieur
Monsr. Sanchez Dor. en Medecine
Rue des Blancs-manteaux
à Paris

London 18 January 1776

My Friend and very Dear Sir

Through my last of the 2nd of this month[†6] you may have seen that in the books I sent you, Kirkland's[7] isn't included since many weeks passed before receiving your letter on the 1st of this month.[8] Soon (in 4 or 5 weeks) I will send the other parcels to Paris, & this book will go with them, in the absence of news that you have found your own. With it I will send the Lusiad of Camoens in English[9] very well translated, with a preface or Introduction of 167 pages, very philosophical, very political, historical & poetical which I am certain you will enjoy. In spite of your grumblings, your name figures in the list of subscribers. However as it goes to Petersbourg instead of Paris, you can be assured that everyone will presume that it is someone else with your name who is found in Russia, and not yourself.

I'm cross about your roguery in making jokes about the silence of my Nelys, Betys, Polys &a. &a. &a. By the crest of a cock, on my next journey to Paris I'll make you change those lines! Look how envy is! Well, know that not more than

6 Magellan is probably referring to his letter of 1 January 1776.

7 Thomas Kirkland, *A Treatise on Child-bed Fevers* (London, 1774).

8 Letter not found.

9 Luis de Camoens, *The Lusiad: or, the Discovery of India*, translated with an introduction and notes by William J. Mickle (London, 1776). Sanches has inserted an asterisk at this point, leading to a marginal annotation; see note 3 above.

20 minutes ago, my Bety sucked it up to the root, and tells me that in spite of being 54, I serve her better than her only lover aged less than 30. But since you don't belch any more,[10] you not only misconstrue my modest reticences for true weaknesses & complete vindications of your prophecies, but also the silence of our poor friend Duchenne about the parcel of books, of which he didn't yet have any news when you asked for it because I only wrote to him with the detail du contenu dans les deux caisses on the 9th of this month,[11] knowing that the boat was delayed much more after the scheduled time of its promised sailing to Calais: from which seen and well considered we must conclude 1°. that Duchenne could not give you any news of your books; 2°. that you, as you don't belch any longer, let fall over that friend, who knows how to live, & serves so well whenever he can, a suspicion that he was frowned upon or unappreciated or unwanted. 76.01.18a

We have had a lot of snow for 15 days in this country. It seems that I am at the Estrela Mountains.[12] Roads are covered with it: near the house of our friend Arbuthnot there were 3 carts buried in snow: and the postman also left his wagon & took the letter-bag on a horse, and managed to get through along a footpath. I thought that the coach I used to come to London would also be stuck on the way, even though pulled by 6 horses: I am afraid that I may not find a coach to go again tomorrow as usual, since I have been going there for many months to spend Saturday nights and Sundays; and I come back on Monday morning, & sometimes on Tuesday. However I have now received at this very moment a letter from our honourable friend Dr. Wolfe of Danzic,[13] which makes me think that some mail has arrived from abroad. In Canterbury, there was a flood: and the magazines talk only of accidents because of this snow & rough weather. The said Dr. Wolfe asks me to remember him to you, & to thank you for the present you made him of your treatise on the history of the maladie venerienne.[14] Dr. Hunter read to the Royal Society a memoir to prove the same object.[15] He took advantage of the authorities that you list in your dissertation but I don't know if he quoted it, though I am sure that he read it. He showed me a date on which Arias Montano had been aggregated in 1507 (or something similar since I don't remember it any more) to the Faculty of Medicine of Salamanca.[16] Dr. Wolf's letter is of 26th Dec^ber^. last. He says to me: "les doubles tierces & les simples tierces ont fait beaucoup de

10 See Magellan to Sanches, 7 July 1775, note 19.

11 Letter not found.

12 A mountain range in Portugal, behind Coimbra.

13 Letter not found.

14 Sanches, *Dissertation sur l'origine de la maladie vénérienne*... (Paris, 1765).

15 John Hunter read a paper on the origin of the Lues Venerea to the Royal Society on 14 December 1775 that was never published; see Royal Society, *Journal Book Copy*.

16 Benito Arias Montano (1527-1598) was a Spanish scholar better known for his oriental studies but who also studied medicine and contributed the preface and perhaps somewhat more than this to a medical treatise, Francisco Arceo's *De recta curandorum vulnerum ratione* ... (Antwerp, 1574). The date Magellan gives for his aggregation is clearly wrong.

76.01.18a ravage: actuellement nous avons les fievres quartes très rebelles. La cruauté des Prussiens a fait mourrir plusieurs milliers de travailleurs à leur Fortresse près de Marienwerder & à leur canal de Bromberg. Beaucoup de coups de baton, point de logement, mauvaise nourriture, leur ont donné une Epidemie terrible. Il est fort curieux de reflechir sur le circuit des maladies febriles. à Warsovie dans une practique extensive, je ne me souviens pas d'avoir vû une fievre quarte, & fort rarement des tierces simples, pendant l'espace de 15 ans & plus. C'étoit tout des fievres bilieuses remittentes, ou meme des continues. Depuis 5 à 6 ans la scene a changé. Je ne vois que rarement les fievres continues, & les remittentes changent toujours en intermittentes, qui pourtant ne sont pas moins mechantes. la meme chose s'observe à Warsovie. Cette année-ci le plus du monde a eté emporté par des desynteries qui manquoient rarement de suivre la fievre, quoiqu'on avoit eté purgé à plusieurs reprices". Now see if the world & climates are not as transitory as we ourselves!

Vale & me ama

Regards to the adoptive family & friends if any are there.

76.01.26 *To Joannes Franciscus Thijsbaert*

University Library, Université de Liège, Ms. 4130

Londres le 26 J[anvier 1776][1]

Monsieur & très cher ami

Votre Lettre[2] m'a remis ma tranquilité pour La Caisse que Je croiois perdue: mais je voudrois encore savoir dans quel jour & dans quel mois elle fut envoyée pour Strasbourg. Car le billet que Vous m'avez envoyé du Commissionaire de M^r^. Poullet, n'a pas aucune date: & peut etre cette caisse a eté envoyée de Louvain depuis fort long tems, & se trouve arretée quelque part en chemin. Je vous prie donc d'examiner encore ce point pour me tranquilizer entierement.

1 Paper damaged, part of date cut off. The missing text has been restored on the basis that Priestley's book, referred to in the letter, was published in December 1775, and Magellan's phraseology suggests that he is writing soon after it appeared, i.e. in January 1776 rather than the following June or July.

2 Letter not found.

J'ai donné des ordres positives pour deux machines au lieu d'une seule, 76.01.26 qui feront une belle apparence dans Votre Cabinet. une de ces machines contiendra le mouvement de la Terre au tour du Soleil, avec toutes les phases de la Lune, le tout monté sur un beau piedestal, ou plutot sur une espece de plateforme, ornamentée &ª. L'autre montrera tous [les mouvem]ents des planetes autour du Soleil, entre les quels [sera la Ter]re[3] avec son satelite (la lune) mais ce dernier n'aura pas les phases, qui se trouvent dans l'autre. elle sera egalem^t^. & pareillem^t^. montée comme l'autre & fera son pandant. le prix de ces deux machines sera de 55 guinées, & si vous voulez qu'on les couvre ici avec un chassis de verre, enchassé en mahogany, alors celà coutera 5 ou 6 guinées de plus. Je crois que ces deux machines feront une meilleure parade que l'Orrery tout seul: & elles contient le total: mais lors qu'on veut faire entrer le tout dans une seule machine, alors elle devient beaucoup plus chere. Ces deux machines seront executées par Adams en deux mois d'ici toutes completes. Je donne la preference à cet artiste sur Martin, parce que son ouvrage est beaucoup superieur pour l'estabilité & arrangement des pieces. au lieu que celui de Martin malgré tout le bon coup d'oeil, ne merite point de confidence. Son Quart de Cercle astronomique que Vous aurez vû à Liege en est La preuve, qui selon l'opinion de notre ami Pigott,[4] ne vaut absolument rien du tout, & je crois qu'il s'entend dans ces matieres.

L'argent que J'ai actuellem^{t}. entre mes mains de Votre compte, est £50. 12^{s}. 1^{d}. ainsi vous voyez bien qu'il y manque très peu de chose pour l'emplete entiere. mais comme vous me parlez d'y mettre jusqu'à 70 Livres, Je ferai executer ces deux orreries avec un peu plus de magnificence, & lui ferai faire des boetes & chassis de Mahogany avec des bons verres &^{a}.

Je vous envoyerai bientôt l'ouvrage de Priestly 2^{d}. Vol. sur l'air[5] oû il a des choses fort interessantes: une espece d'air 4 ou 6 fois plus pur que celui de l'atmosphere. un autre qui devient corps solide (come des petites coquilles d'oeufs) lorsqu'il touche dans l'eau &^{a}. il a fait dernierem^{t}. des experiences pour demontrer sa decouverte sur l'usage de la Respiration: savoir qu'elle sert pour decharger le trop de phlogistique de nos corps. le sang des veines des animaux etant plus obscur que celui des arteres, devient rubicond clair lorsqu'on le met dans l'air dephlogistiqué: & le sang arterieux mis dans l'air phlogistiqué, devient foncé & noiratre (meme à

3 Paper damaged, words and letters missing.
4 Nathaniel Pigott.
5 J. Priestley, *Experiments and Observations on Different Kinds of Air*, vol. 2 (London, 1775).

76.01.26 travers de la limphe & d'une membrane). lorsqu'on fait ces experiences & on voit ces effets, on doit etre convaincu que cette operation de l'economie animale (la respiration) n'est que seulem^{t}. pour decharger la surabondance du phlogistique, qui d'aileurs deviendroit tres nuisible, & nous tueroit par suffocation s'il ne se dechargoit avec l'expiration. mes Respects aux amis.

Vale & me ama
tuum[6]
Magellan

A Monsieur
Monsr. L'Abbé Thisbaert
Doyen du Chapitre de S. Jacques & Principal du College Royal
à Louvain
en Flandres

6 Farewell & love me | your.

76.02.02 *To Ribeiro Sanches*

Österreichische Nationalbibliothek, Vienna
A.N. Ribeiro Sanches Korrespondenz, codex 12714, 171-172v

2 Fev^{o}. 1776

Am^{o}. e S^{r}. m^{to}. do Cor^{am}.

Estou com grande cuidado por V.M., causado do frio terrivel q aqui temos ha mais de 3 semanas. o thermom^{o}. tem estado à 10 gráos de Fahrenheit; & dizem chegára à 6 gráos, o q he frio m^{to}. terrivel. tudo gelado. eu vi hontem passar a Tamise de hũa banda á outra por sima do caramelo. V.M. q he tão sensivel ao frio, não pode deixar de padecer com este, q dizem he tãobem mui grande em Paris. eu passo de saude, sem deixar de hir todos os domingos a passalos com o meo amigo Arbuthnot ao campo. ja hontem á noite começou a desgelar um pouco: mas as ruas estão cheias ainda de gelo.

a semana passada se lêo uma descuberta do D^{r}. Priestley sobre a respiração animal. a saber q o seo uzo hé som^{te}. p^{a}. descarregar o excesso ou surplus do phlogistico. tomou sangue das veias de animaes q he de cor escura, meteo-o em uma bexiga, e pola em um vazo cheio de ar dephlogisticado. Com este ar tomou a cor rubicunda e esbranquicada do sangue das arterias. ao contrario este ultimo metido em ar phlogisticado, tomou uma cor denegrida. desde Hipocrates até o prezente ninguem soube dizer o verdº. uzo da respiração. huns q era pª. aquentar outros pª. esfriar o sangue. outros pª. atenuar &ª. &ª. este ar dephlogisticado do qual V.M. achará a descrição no 2do. vol. de Priestley q eu lhe mandei, virá talvez à ser um grande Remedio nas fevres & doenças podres. 76.02.02

diga-me como passa. Saud^{es}. á familia adoptiva.

vale & me ama

não sou mais largo porq estou cançado de escrever hoje pª. a posta. a vista começa a turbarse & he necessº. descançala.

iterum vale.

quarta frª. passada jantei (como quazi sempre faço neste dia em cada semana) com o nosso Enviado e sua Esposa. são sujeitos dignos da maior estimação. ninguem ha q os não ame, se os conhece. paga visitas a toda a gente q o procura &ª.

iterum atque iterum vale

a Monsieur
Monsr. Sanchez Doct^{r}. en Medecine
au Bureau du Cuivre,
Rue des Blancs-manteaux dans le Marais
à Paris

2 Feby. 1776

My Dear Friend

I am very concerned for you, due to the terribly cold weather that we have had here for more than 3 weeks. The thermometer has been at 10 degrees Fahrenheit & it is said that it was at 6 degrees, which is a very terrible cold.[1]

1 January 1776 was one of the coldest Januarys that England has experienced since records began to be kept in the mid-seventeenth century. Day after day, temperatures were well below freezing. Unusually heavy falls of snow closed many roads for long periods and the River Thames was frozen over for some time.

76.02.02 Everything is frozen. I saw yesterday that one could cross the Thames from one bank to the other over the frozen snow. You, who are so sensitive to the cold, cannot but suffer from this one that is said to be also very great in Paris. I am in good health, without refraining from going to spend every Sunday with my friend Arbuthnot in the countryside. Last night it started thawing a little, but the streets are still full of ice.

Last week a discovery of D^r^. Priestley's about animal respiration was read.[2] That is that its use is just to discharge the excess or surplus of phlogiston. He took blood from the veins of animals, which is of dark colour, put it in a bladder and placed it in a vase full of dephlogisticated air. With this air it acquired the red and whitish colour of arterial blood. On the other hand, this last, put in phlogisticated air, took the darkened colour. From Hippocrates to the present time no-one was able to tell the true use of respiration. Some said that it was to warm, others to cool the blood, still others to attenuate it &c. &c. This dephlogisticated air, of which you will find the description in Priestley's 2nd vol.,[3] which I have sent to you, will perhaps become an important medicine for fevers & putrid diseases. Tell me how you are. Regards to the adoptive family

vale & me ama[4]

I don't continue because I am tired of writing for the post. My sight is beginning to get troubled & it's necessary to give it some rest.

iterum vale[5]

Last Wednesday I had dinner (as I do almost every week on this day) with our Envoy[6] and his wife. They are worthy of the greatest esteem. There is no one who, knowing them, doesn't love them. He pays visits to every one who seeks him out &c.

iterum atque iterum vale[7]

---

2 Priestley read his memoir, "Observations on Respiration and the Use of Blood", at the Royal Society on 25 January 1776; see Royal Society, *Journal Book Copy*, vol. XXVIII. It was subsequently published in the *Philosophical Transactions*, 66 (1776), 226-248.

3 J. Priestley, *Experiments and Observations on Different Kinds of Air*, vol. 2 (London, 1775).

4 Farewell and love me.

5 Again farewell.

6 Luiz Pinto de Sousa Coutinho (1735-1804), Portuguese Minister in London, 1774-88.

7 Again and again farewell.

To Ribeiro Sanches 76.02.23[1]

Österreichische Nationalbibliothek, Vienna
A.N. Ribeiro Sanches Korrespondenz, codex 12714, ff. 179-180v

Londres

Meu am^{o}. e S^{r}. e m^{to}. do Cor^{am}.

Vejo q V.M. aplaude com maganeira ratinha a suspeita de q lhe faço companhia nas exequias da parsilla quâ quondam Achyles eramus. Mas a melhor prova de q se engana, he q não me pico do remoque: e vou continuando com o póte á fonte, como as nossas michélas de Coimbra sem mais embaraço, que um sorrizinho seco por debaixo da mantilhinha azul, ou do marôto de baeta porsima da tiricana parda. Parece-me q as estou vendo pela travessa de quebra costas da nossa Athenas Lusitana. o q lhe gavo he a prezunção da nobre Ordem de S^{to}. André, como se algum da nossa terrinha, desde o Fernandes remendão até o fidalgo coroado pudesse provar q não participa de tal privilegio por fas, ou por nefas ...! mas são vapores de Riba-Côa ...!

Não pude ver Quarta fr^{a}. o nosso Enviado; era a da cinza & não quiz fazer a hypocrizia de ganhar uma indigestão com o bacalháo, ou pratos quadragessimaes, não hirei domingo á quinta de nosso amigo Arbuthnot; e então hirei jantar com o Chev^{r}. Pinto a q^{m}. direi q V.M. se engana com seo Pai ou thio em vez d'elle mesmo: pois elle veio a Coimbra m^{to}. depois de mim, q fui novato no anno de 1734.

Escreverei brevem^{te}. ao Princepe Pignatelli: se V.M. o vir antes disso, presente-lhe os meos Resp^{os}.

V.M. está mui devoto, & não falla senão em rezar por contas, pedindoas a q^{m}. as não tem depois de m^{tos}. annos. a dizer-lhe a verd^{e}. estou com m^{ta}. pressa p^{a}. hir á cama q passa da meia noite, & virá o D^{or}. Priestley ás 8 horas p^{a}. hirmos almoçar com seo Cunhado daqui hũ quarto de legoa; donde hiremos todos 3 a Bow, p^{a}. ver o effeito da nova maquina a fogo q elle executou na sua fundição, em Shropshire.

em outra occazião lhe mandarei os gastos e contas q dezeja.

---

1 Given Magellan's remark in the course of this letter about Ash Wednesday, the date of the letter could be 22, 23 or 24 February, since in 1776 Easter Sunday was on 7 April and Ash Wednesday therefore on 21 February. Most probably Magellan wrote on 23 February since otherwise he would either have referred to Ash Wednesday as "yesterday" or to the following Sunday as "tomorrow".

76.02.23 Recebi a lettra de 300#. q V.M. me mandou: & não tenho animo de agradecer-lhe a continuação do prez^{te}. porq não tenho ideia das suas superabundancias & superfluid^{es}. e assim tãobem porei ponto aqui por basta.

Peço a V.M. q mande o seo criado comprar-me hum Livro broché, intitulado = Recueil de Memoires & d'observations sur la fabrication du Salpetre = este livro foi impresso depois pouco tempo, cuido q por ordem da Acad^{a}. R^{l}. das Sciencias &^{a}. he bom q elle seja som^{te}. broché & não encadernado. Ponha-lhe simplesm^{te}. hum sobrescrito = To M^{r}. Robert Bigg.

at y^{e}. gen^{al}. Post office

London =

Mas não ajunte carta, nem nada mais escrito no sobrescrito, nem dentro delle. Isto he encomenda de hũ amigo a q^{m}. dezejo servir, & que tem pressa de [v]er esta obra. V.M. me dirá em carta pelo correio separadam^{te}. o q^{to}. custou &^{a}. Porem se o seo criado não souber fazer esta encom̃enda, com presteza, peço q avize da m^{a}. p^{te}. o D^{or}. Begue du Presle; pois sem duvida não deixará de o fazer. Certam^{te}. o seo criado será m^{to}. besta, se achar nisto algũa difficuld^{e}.

vive vale & ama

tuum Mag^{n}.

Saud^{es}. á familia adoptiva, com tanto q não haja vomitos, nem engulhos &^{a}. Cá direi a minha Neli o q^{to}. V.M. prezume em sua perda.

a Monsieur
Monsr. Sanchez Doct^{r}. en Medecine
Rue des Blancs-manteaux
à Paris

London

My Dear Friend

I see that you applaud with a petty joke the suspicion that I keep company with you in the exequies of pars illa quâ quondam Achyles eramus.[2] But the best proof that you are wrong is that I don't get angry at the remark, and continue to go with the jug to the fountain, like our whores of Coimbra with no more embarrassment than a little smile under the blue mantilla or the [rustle] of coarse fabric over brown burel. It seems that I see them in the quebra costas lane[3] of our

2 Adapted from Petronius Satyricon, 129.I: "that part where once we were an Achilles", i. e. in past erotic adventures. Sanches's letter not found.

3 A steep and narrow street leading from the lower part of Coimbra to the upper town. The name means "Back break".

Lusitanian Athens.[4] What I praise in you is the presumption that, as in the noble Order of St Andrew,[5] anyone in our little country, from the patcher Fernandes to the crowned nobleman, could prove that he doesn't take part in that privilege by hook or by crook…! But these are the spirits of Riba-côa[6] …! 76.02.23

On Wednesday I couldn't see our Envoy;[7] it was Ash Wednesday & I didn't want to partake in the hypocrisy of getting indigestion from the cod or Lenten dishes. I shan't go on Sunday to the farm of our friend Arbuthnot and shall have dinner with Chevalier Pinto, to whom I'll say that you are wrong about his father or uncle rather than himself, since he came to Coimbra long after me who was a novice in the year 1734.[8]

I shall write soon to Prince Pignatelli:[9] if you see him before I do, give him my respects.

You are very devout, & don't speak except in praying by beads,[10] asking those who haven't had them for many years to tell you the truth. As a matter of fact I am in a hurry to go to bed since it is past midnight & D^r^. Priestley will come at 8 o'clock for us to go to have lunch with his brother-in-law[11] a quarter of a league from here, from where the 3 of us will go to Bow, to see the effect of the new fire machine that the latter made in his foundry in Shropshire.

On another occasion I shall send you the expenses and accounts you wish.

I received the promissory note of 300#. that you sent me: & I don't have the courage to thank you for the continuation of the present,[12] because I have no idea of your oversupplies & surpluses and so I also stop here.

I ask you to order your servant to buy me a book broché entitled = Recueil de Memoires & d'observations sur la fabrication du Salpetre[13] = this book was

---

4 Coimbra, where Magellan lived for many years.

5 A Masonic order of the Scottish rite of such low prestige that virtually anyone could belong to it.

6 A region in the east-central part of Portugal, the men of which took particular pride in their provincial origins, seeing themselves as more virile and trustworthy than city-dwellers. Sanches spent part of his youth in the nearby town of Guarda.

7 Luiz Pinto de Sousa Coutinho (1735-1804), Portuguese Minister in London, 1774-88.

8 Possibly a reference to Sanches's having sent Pinto de Sousa in 1758 an extract from a work by Bartolomeo Intieri on the conservation of grain; see Magellan to Sanches, 15 April 1777.

9 Juan Joaquín Atanasio Pignatelli de Aragón y Moncayo (1724-1776), Spanish Ambassador in Paris, 1763-72?

10 i.e. saying the rosary.

11 John Wilkinson (1728-1808), the famous ironmaster and industrialist. Priestley married Wilkinson's sister Mary.

12 These 300#. were the regular amount that Sanches had promised to give Magellan in return for the (unknown) favour Magellan undertook for him at the risk of his own life, leading him to flee Portugal in 1763.

13 [Académie Royale des Sciences, Paris], *Recueil de memoires et d'observations sur la formation & sur la fabrication du salpetre …* (Paris, 1776).

76.02.23 printed a short time ago, I suppose by order of the R^{l}. Acad^{y}. of Sciences &^{a}. It's good that it's just broché & not hardbound. Just put on the envelope =

To Mr. Robert Bigg
At y^{e}. gen^{al}. Post office[14]
London =

But don't attach a letter, or any thing else written on the envelope, or inside. This is an order for a friend whom I wish to assist, & who is impatient to [see] this work. You will tell me in a letter by separate post how much it cost &^{c}. However if your servant is not able to do this task readily, I ask you to tell D^{r}. Begue du Presle on my part since he undoubtedly will do it. Surely your servant is a fool if he finds some difficulty in this.

vire vale & ama
tuum Mag^{n}.[15]

Regards to the adoptive family and may there be no vomits or nausea &^{c}.
I shall tell my Nelly how much you presume at her loss.

---

14 Robert Bigg was a sorter at the General Post Office; see *Court and City Calendar* (London, 1762).

15 Flourish, farewell and love me | your Magellan.

76.02.27 *To Ribeiro Sanches*[1]

Österreichische Nationalbibliothek, Vienna
A.N. Ribeiro Sanches Korrespondenz, codex 12714, 167-168v

Londres 27 Fev^{ro}. 1776

Meu am^{o}. e S^{r}. e m^{to}. do Cor^{am}.

Vou replicar ás duas de V.M. de 3 & 14 do Corr^{te}. q me tirárão do susto em q [o] gelo passado me poz por sua cauza, sabendo o q^{to}. lhe he contrario; pois pelo q me Respeita, nenhũ abalo me deo, "porq q^{m}. come, bebe, & fo-e, com todo o mal pode."

Jantei 6^{a}. fr^{a}. passada com D^{r}. Sequeira, q se Recomenda em amigo. o Secret^{o}. do nosso Ministro tãobem ahi jantou, & elle sera o Medico da

---

1 Annotated: "Resp^{a}. a 13 Mars 1776" [Answered 13 March 1776]. Letter not found.

familia. a Mulher do mesmo, q he hũa verd^{a}. Senh^{a}. Portugueza, muito 76.02.27
amavel, disse á meza n^{a}. 4^{a}. fr^{a}. passada (q he o dia q me fica Livre, e q destinei p^{a}. vêlos) q estimava m^{to}. o D^{or}. Sequeira & q havia de tomalo p^{a}. lhe assistir & á sua familia: & sem duvida o Secret^{o}. (ainda rapaz, mas sezudo & de bom natural) lhe falou na materia, pois ambos se levantárão da meza & forão falar sós p^{a}. o seo Cabinete. eu á manham me hei de achar com o D^{or}. Seq^{ra}. à jantar em caza do nosso Enviado, & saberei o q se passa.

desde antes do Natal q tomei a venêta de hir todos os sabb^{os}. até à 2^{a}. fr^{a}. p^{a}. caza do nosso bom amigo Arbuthnot, ainda o mais rigoroso frio da passada gelada me não embarassou a viagem. tomei por minha conta arranjar-lhe os seos Livros, & em 3 liçoens q dei a seo filho mais velho lhe fiz comprehender a lingoa portugueza, de modo q com a ajuda do dicionario poderia bem continuala, se tivesse curiozid^{e}. assaz. Mas he rapaz de 15 a 16 anos, pouco activo e molacheirão,[2] de q o Pai se entristece: mas o 2do. q he de 13 ou 14 de id^{e}. mostra grande applicação. Estuda na celebre escola de westminster, & sempre se acha no prim^{o}. lugar do seo banco. aqui tenho os Livros p^{a}. mandar a V.M. seg^{do}. a sua carta: e partirão pela prim^{a}. occazião. as suas contas estão como suas. eizaqui as q tenho no meo Livro. Se as q mandei na carta passada erão differentes, foi erro da copia, ou do Leitor. Aitkin – 4^{s}./ Prestwich 6^{s}/ Hutchinson 1$^{£}$. 4^{s}./ valace 1$^{£}$. 1^{s}./ harvood 3^{s}. Butter 2^{s}.6/ p^{a}. Coimbra 11 Med^{al}. Comentaries 16.6 custo do porteur 8^{d}. (& não 8^{s}.)/ Duncan's Lines 6^{d}/ Cullen Lect. 4^{s}.6^{d}/ White 1^{s}.6^{d}/ Hamilton 4^{s}.6^{d}/ Culen on drownded 1^{s}.6^{d}/ leak 4^{s}./ Lovet 3^{s}./ Pringle on Torpedo 1^{s}.6^{d}/ one Med^{al}. Commentaries 1^{s}.6^{d}/ Priestley on air 6^{s}. (& não 6^{d}.). dois tinteiros à 2^{s} cada hũ fazem 4^{s}./ gastos respectivos de caixa & duana, & até bordo do Navio 6^{s}./ total £5.15^{s}.8^{d}.

Cá puz no meo canhenho o dinh^{o}. q V.M. entregou por m^{a}. conta à Messier, q como ja disse, estava em depozito p^{a}. quando V.M. se quizer servir delle: & tãobem ja disse a V.M. q tirasse sobre mim á vista qd^{o}. lhe for necessario até 20 ou 25 Luizes. e fique isto dito de hũa vez p^{a}. sempre. eu não tenho q Replicar sobre o dito de Quevedo viz. q he necess^{o}. retener p^{a}. tener: mas som^{te}. advirto q deve ser sem faltar ao seo necess^{o}., nem ao seo bien etre physique.

V.M. está enganado com as suas ideias ferrenhas sobre o Rhubarbo. a experiencia mostra tudo o contrario: e na verd^{e}. este clima & este terreno não differem daquelle da Tartaria em q esta planta crece, mais do q a Borgonha differe de Portugal, onde bons & fortes vinhos nacem & se fazem.

---

2 molageirão?

76.02.27 Li ao nosso Ministro o articulo encomiastico q V.M. faz do nosso Hercules de Arouca, de q elle gostou muito, & se rio bem com o N.B. da sua christandade velha pela graça de S.M.F^{ma}.

terá V.M. o Kirkland q ja aqui tenho enpaquetado p^{a}. o seo embrulho com o Armstrong Essays.

Mas terá V.M. a pulha de q a Lusiada não he os Lusiadas. isto seria pôr o Poeta m^{to}. fóra dos Eixos da Epopeia: & M^{r}. Mickle ao contrario quer q elle seja o prim^{o}. depois de Homero & Virgilio q atinou com a verd^{a}. Epica.

V.M. não achará no 2do. vol. de Priestley quazi nada sobre o uzo da Respiração. este objecto está n'uma Memoria q elle com̃unicou ultimam^{te}. á Soc^{e}. R^{l}. m^{to}. depois de estar impresso o d^{o}. 2do. vol. a sua observação he nova, & fundada sobre experiencias notaveis. nem Kaw Boerhaave nem ninguem antes delle (Priestley) acertou com este achado.

Sir John Pringle vive com saude, & he falso o ter morrido.

D^{r}. la Cour passa bem, & me mandou um grande cumprim^{to}. na carta de Sequeira, por hum barometro q daqui lhe enviei por sua ordem a Bath. Sempre está de perninha, e não escreve senão por whim ao d^{o}. Sequeira segd^{o}. a venêta q lhe deo depois de certo tempo.

as instituiçoens de Medecina por Culen de q ha noticia até agora, não são mais do q o First Lines q mandei: e este he o q eu anunciei com aquelle nome.

ainda não pude colher o q se segue [...][3] do capitulo 2°.: o q mandei a V.M. foi por via de hum discipu de Culen, q se desfez delle. talvez q cedo venhamos a ter algũa coiza mais delle impressa.

nem no meo livro, nem mesmo na copia da conta q V.M. me manda, não se acha o Leak mais do q hũa vez, em lugar das duas q V.M. diz: pelo q fique V.M. com o epitheto de podão aritmetico, até q prove o contrario.

Pelo q toca aos mexilhoens de Aveiro, bem sabe q são mariscos, e q fazem arr-tar ainda aos q como eu se achão nos 54. ainda hontem fui ver a m^{a}. antiga Nely, q está mui gorda e cazada. disse-me maravilhas do marido, q não lhe prohibe ver os seos antigos amigos. he hũ original dos mais exoticos q jamais se virão. disse-lhe q vinha das West-Indies [e] q me lembrei de a hir ver. agradeceo-me m^{to}. a vizita, & foime ainda divertir em caza da vizinha.

Saud^{es}. ao Cav^{ro}. Brito q se regale lá com as suas marionettes de Paris. eu cá meti o seo nome na Lista dos suscritores do Camoens: mas isso não obriga a tomalo se o achar caro.

Saud^{es}. a M^{a}. Joana & fam^{a}. adoptiva. espero q nos veremos neste verão q vem: mas ainda não tenho nada de bem determinado. Corre a agoa por

3 Stained paper.

onde vai o rio, ainda q os dezejos sempre vão pela congosta dos affectos. 76.02.27
veja lá q expressão esta de riba-coa ...!

vale atque iterum vale

a Monsieur
Monsr. Sanchez D^or^. en Medecine
Rue des Blancs Manteaux
à Paris

London 27 Feby. 1776

My Dear Friend

I shall answer your two letters of the 3rd & 14th of this month,[4] which freed me from the concern to which the last cold had given rise on your account, knowing how harmful it is to you; as for me, it didn't trouble me, "because he who eats, drinks, & fu-s, can stand anything bad".

I dined last Friday with D^r^. Sequeira, who recommends himself as a friend. Our Minister's[5] secretary also dined there, & he will become the family doctor. The Minister's wife, who is a true Portuguese lady, very pleasant, said at table last Wednesday (which is my free day, and which I assigned to seeing them) that she had a great esteem for D^r^. Sequeira & that she wanted to have him assisting her & her family: & no doubt the secretary (still a young man, but wise & of good nature) talked to him about the matter, since both left the table & went to his office to talk alone. Tomorrow I shall meet D^r^. Sequeira at dinner at our Envoy's house, & I shall know what is going on.

Since Christmas I have been in the habit of going every Saturday till Monday to our good friend Arbuthnot's house; even the most rigorous cold weather of the last frost didn't prevent me going. I took on the task of setting his books in order, & in 3 lessons to his son I made him understand the Portuguese language, in such a way that with the help of the dictionary he could well continue, if he had enough curiosity. But he is a boy 15 or 16 years old, not very active & aimless, which saddens his father: but the 2nd who is 13 or 14 years of age shows great enthusiasm. He is at the Westminster school, & is always the first in his class. I have here the books to send you according to your letter: and they shall go on the first occasion. Your accounts are as you have them. These are the ones I have in my book. If those I sent in the last letter were different, it was a mistake in copying or of the reader.[6] Aitkin– 4^s^. / Prestwich 6^s^. / Hutchinson £1^£^.4^s^. / valace £1.1^s^. /

4 Letters not found.

5 Luiz Pinto de Sousa Coutinho (1735-1804), Portuguese Minister in London, 1774-88.

6 Except for the inkpots, these items are all listed (and identified) in Magellan to Sanches, 1 January 1776; but the last few, having been listed, are there then crossed out.

76.02.27 harvood 3^{s}. / Butter 2^{s}. 6 / to Coimbra 11 Med^{al}. Commentaries 16.6. cost of porteur 8^{d}. (& not 8^{s}.) duncan's Lines 6^{d}. / Cullen lect. 4^{s}. 6^{d}. / White 1^{s}. 6^{d}.[7] / Hamilton 4^{s}. 6^{d}. / culen on drownded 1^{s}. 6^{d}. / leak 4^{s}. / lovet 3^{s}. / Pringle on Torpedo 1^{s}. 6^{d}. / one Med^{al}. commentaries 1^{s}. 6^{d}. / Priestley on air 6^{s}. (& not 6^{d}.) two inkpots at 2^{s}. each make 4^{s}. / expenses of packing & customs, & till taken on board 6^{s}. total £5.15^{s}.8^{d}.

I took note in my book of the money you gave to M^{r}. Messier on my account, that, as I said before, will be on deposit for you to use when you want it: & I also said that you could draw when you find it necessary up to 20 or 25 Louis. And that this be said once and for all. I don't need to replicate the aphorism of Quevedo who said that it is necessary to save in order to possess,[8] but just warn that this should not be done by going without that which is necessary to one's bien etre physique.

You are wrong in your stubborn ideas about Rhubarb. Experience shows quite the contrary: and in fact this climate & this land don't differ from those of Tartary, where this plant grows, more than Burgundy differs from Portugal where good & strong wines are born & made.

I read to our Minister the encomiastic article of yours about our Hercules of Arouca,[9] which he much enjoyed & laughed heartily at the N.B. of your old Christian situation by the grace of Her Most Faithful Majesty.[10]

You will have the Kirkland that I have already here packed to send with Armstrong's Essays.[11]

But you will joke that the Lusiad is not the same as the Lusiads.[12] This would be to put the poet far from the Epics: & M^{r}. Mickle on the contrary wants him to be the first after Homer & Virgil who succeeded with a true Epic.

You will find scarcely anything in Priestley's 2nd vol. about the use of respiration. This material is in a memoir he communicated recently to the R^{l}. Soc.[13] long after the said 2nd vol. was published. His observation is new & supported by remarkable experiments. Neither Kaw Boerhaave[14] nor anyone before him (Priestley) achieved this finding.

Sir John Pringle lives in good health, & it is false that he has died.

---

7 In his letter of 1 January 1776, Magellan gave the price of this work as 1^{s} 0^{d}, which is probably what he ought to have written here since then the total shown would be correct.

8 Francisco Quevedo (1580-1645), Spanish moralist writer.

9 i. e. the priest from Arouca referred to in Magellan to Sanches, 1 January 1776.

10 Sanches was Jewish but had been baptized a Christian, and so was considered an "old Christian" by the grace of the Queen.

11 Thomas Kirkland, *A Treatise on Child-bed Fevers* (London, 1774) and John Armstrong, *Medical Essays* (London, 1773). See Magellan to Sanches, 18 January 1776.

12 See Magellan to Sanches, 18 January 1776, note 2.

13 Joseph Priestley, "Observations on Respiration and the Use of Blood", *Philosophical Transactions*, 66 (1776), 226-248.

14 Abraham Kaau Boerhaave, *Perspiratio dicta Hippocrati per universum corpus anatomice illustrata...* (Leiden, 1738)?

Dr. la Cour is going well, & sent me a great compliment in the letter to Sequeira for a barometer that I sent him on his order to Bath. He is handicapped, and doesn't write except by whim to Sequeira according to the caprice that he had some time ago. 76.02.27

the institutions of Medicine by Culen, which is known now, are nothing else than the First Lines that I sent: and this is the one I announced with that name.[15]

I could not get the next [...][16] of the 2nd chapter: what I sent you was through a [student][17] of Culen's, who got rid of it. Maybe we shall have soon something of him in print.

Neither in my book, nor even in the copy of the account you sent me, is Leak found more than once, instead of twice as you say: So you should keep the epithet of clumsy arithmetician until you show proof to the contrary.

In what concerns Aveiro's mussels, you know well that they are shellfish and that they make belch[18] those who, like me, are in their 54th year. Even yesterday I went to see my old Nely, who is very fat and married. She told me marvels about her husband, who doesn't forbid her to see her old friends. He is an original, one of the most exotic ever seen. I told her that I had come from the West Indies [and] that I came to see her. She thanked me for the visit and took me to have fun at her neighbour's.

Regards to Chev. Brito.[19] May he have fun with his Paris marionettes. I put your name on the list of subscribers to the Camoens but that doesn't compel you to take it if you find it expensive.

Regards to Ma. Joana & the adoptive family. I hope we shall meet next summer: but I have nothing well settled yet. Water runs towards the river, even though wishes always go via the narrow lane of the affections. Look at this expression from Riba-coa[20] ... !

vale atque iterum vale[21]

---

15 William Cullen, *Institutions of Medicine, Part 1: Physiology* (Edinburgh, 1772). The first volume of Cullen's *First Lines of the Practice of Physic*, likewise based on his lectures at the University of Edinburgh, was published in 1777, but unauthorized copies of his lectures had been circulating among his students for many years.

16 Paper damaged

17 Paper damaged. Though the word is illegible, its meaning can be inferred; see note 16 above.

18 See Magellan to Sanches, 7 July 1775, note 19.

19 Joseph Caetano Jeronimo de Brito, whose letters appear in the correspondence of Ribeiro Sanches. It seems that Brito spent some months in Paris at this time, returning to Portugal towards the end of the year; see Magellan to Sanches, 17 December 1776, note 5.

20 A region in the east-central part of Portugal, the men of which took particular pride in their provincial origins.

21 Farewell and again farewell.

76.04.02 *To Antoine-Laurent Lavoisier*

Archives de l'Académie des Sciences, Paris, Fond Lavoisier, 1696[1]

Cher Confrere & très cher ami

j'ai vu en fin la belle & excellente pompe Pneumatique faite par Nairne sur le principe de celle de M[r]. Smeaton:[2] elle est très aisée à travailler: & Je l'ai vu rarefier l'air dans le recipient du moins jusqu'à la 3469me. fois. Le Doct[r]. Priestley me dit hier soir qu'elle alloit encore beaucoup plus loin (à 7 ou 8 mille fois). n'aiant plus d'un seul bassin son prix est de 35 guinées. mes h[bles]. resp[s] à M[de]. votre Epouse, à Mlle. Messrs. Paulze &[a].

tuus
Magellan

Ce 2 Avril 1776

1 This note appears on the first page of an offprint of Priestley's paper, "Observations on Respiration, and the Use of the Blood", *Philosophical Transactions*, 66 (1776), 226-48, inscribed "Mr Lavoisier from the Author". It is published in *Oeuvres de Lavoisier: Correspondance*, ed. René Fric, fascicule III, 1776-1783 (Paris, 1964), pp. 561-2.

2 Edward Nairne, "An Account of Some Experiments made with an Air-pump on Mr. Smeaton's Principle...", *Philosophical Transactions*, 67 (1777), 614-48.

From Nathanael Wolf[1]

Avis touchant un Portrait de Copernic, *présenté à la* Societé Royale, *par M. le Dr.* Wolf *de* Danzic: *extrait d'une Lettre du dit Dr. à M. de* Magellan, *Membre de la Soc. R.*

Danzic, le 7 Avril, 1776.

Monsieur et très cher ami,

Le capitaine qui vous délivrera celle-ci, vous apportera en même tems la copie d'un portrait du fameux Copernic, que je vous prie de présenter à votre illustre Societé Royale, comme un temoignage de mon dévouement et respect pour cet illustre Corps.[2] Le possesseur de l'original, dont le portrait fut copié avec la plus grand attention est M. le Chambellan Hussarzewski.[3] Il a déja refusé 100 ducats, et ne le cédera pas sa vie durant pour aucun prix: mais il me le veut léguer après sa morte. C'est à dire très probablement après la mienne: de sorte qu'il n'y a rien à espérer de ce coté-ci. Nous avons un portrait de Copernic dans la grande Eglise de Thorn,[4] dans une espéce de mausolée, érigé une trentaine d'années après la mort de cet homme célebre, par un médecin de la ville, qu'on dit avoir été un de ses parents.

Hartknoch a fait graver ce portrait, et l'a inseré dans sa Chronique de la Prusse.[5] Notre original a été comparé sur le lieu avec celui du mausolée, et on a trouvé les traits du visage absolument les mêmes: mais l'habillement est très different. Celui de Thorn le represente devant l'autel en fonction de prêtre. Dans le notre il est habillé en pélisse, avec les cheveux de la tête plus soigneusement arrangés et coupés, comme en habit de parade. Le pinceau est sûrement Italien du vieux tems en question, comme on trouvera en regardant d'autres de ce tems et les comparant avec celui-ci. Par exemple, on sait que les peintres dans ce vieux tems, même Raphael, ne savoient pas encore donner aux yeux cette vivacité, que le plus médiocre peintre depuis plus d'un siecle, ne manque pas de donner à ses peintures. Je ne dis pas

1 Letter not found. The text is from *Philosophical Transactions*, 67 (1777), 33-7, where the original French version is published in parallel with an English translation, as a running footnote. The precedence is inverted here, to reflect the language in which the letter was actually written. The heading was presumably constructed by the Royal Society's editor. The published article is annotated: "Read Dec. 7, 1776." There is no valediction.

2 The portrait, an oil on panel, 41.9 by 33 cm, is still in the possession of the Royal Society (ref. no. P/0027).

3 Tomasz Hussarzewski (1732-1807), professor of history at the University of Vilnius.

4 Torun, Poland.

5 Christopher Hartknoch, *Alt- und neues Preussen* (Frankfurt/Leipzig, 1684), p. 371.

76.04.07 que les vieux peintres n'approchent pas plus de la nature avec leurs yeux tranquiles sans feu; mais ils déplaisent à notre siecle, qui ne veut que des yeux petillants. Au moins c'est la marque d'une ancienneté plus grande qu'un siecle et demi. L'inscription fait voir que le peintre etoit Italien. Il y a deux siecles qu'on a cessé de peindre sur du bois.

L'histoire de notre original est la suivante. Il a été conservé dans la collection des ducs de Saxe Gotha, dans le chateau du même nom, toujours avoué comme un véritable original: on prétend même prouvé comme tel par les archives; ce qui est fort croyable, vû que le prince evêque de Warmie, qui l'a obtenû du duc defunt, etoit trop circonspect et trop bon connoisseur pour s'en laisser imposer. Enfin c'est après les paroles du prince evêque, que je fais cette assertion. Ce fut environ l'année 1735 qu'etant à Gotha, il remarqua ce portrait dans la gallerie du duc. Les preuves qu'on lui fit de son authenticité, augmenterent son envie de le posséder. Il l'obtint enfin par un vol qu'il fallut faire à la Cathédrale de Warmie,[6] où se trouvoit un très ancien portrait d'un des ancétres du duc de Saxe Gotha, qui etoit jadis evêque de Warmie, et dont le portrait manquoit dans la collection des portraits de sa famille. On fit donc un échange des deux originaux. Le prince evêque Grabowski[7] enfin a laissé celui de Copernic à son favori M. Hussarzewski.

La copie est absolument la même que l'original: excepté que j'ai fait coller trois rêgles sur le derriere de la table pour empecher son courbissement. Depuis que ma copie a été tirée, M. Hussarzewski, croiant bien faire, a fait renouveller son original, c'est à dire, y mettre du vernis: et actuellement il est si bien le même que la copie, qu'on peut a peine s'appercevoir de la moindre différence. Mon peintre est M. Lorman de Berlin,[8] assez célebre dans son art.

An Account of a Portrait of Copernicus, presented to the Royal Society by Dr. Wolf of Dantzick: extracted from a Letter of his to Mr. Magellan, F. R. S.

Dantzick,
April 7, 1776.

Dear Sir,

The captain who will deliver this to you, will also put into your hands a copy of an original portrait of the famous Copernicus, which I beg you will present to the Royal Society, as a testimony of my devotion and attachment to that respectable

6 The cathedral at Frombork, Poland, seat of the Prince Bishops of Warmie (Warmia).

7 Adam Stanislaw Grabowski, Prince-Bishop of Warmia (1698-1766). Copernicus lived in the Castle of Olsztyn, capital of Warmia.

8 Friedrich Anton August Lohrmann (1735-1800).

body. The original, from which it is copied with the greatest accuracy, is in the possession of the Chamberlain Hussarzewscki, who has already refused one hundred ducats for it, and will not part with it at any rate during his life, but intends to bequeath it me after his death; for which reason there is no probability of my ever possessing it, as he is likely to survive me many years. We have a portrait of Copernicus in the great church at Thorn in a kind of mausoleum, erected about thirty vears after the death of that great man, by a physician of that town, who is said to have been one of his relations. 76.04.07

Hartknoch has inserted a print taken from this portrait in his Chronicles of Prussia. Our original has been compared with that of the mausoleum, and the features of the face are found to be perfectly similar, but there is a great difference in the dress. That at Thorn represents him kneeling before an altar, in the attitude of a priest officiating; in ours he is cloathed in fur, with his hair more carefully dressed, and as it were in a habit of ceremony. The painter of it was certainly one of the old Italians, as will appear by comparing it with other works of those masters; for instance, it is known that the painters of those times, and even Raphael, never gave to the eyes that brightness which the most indifferent artists within this century never fail to express in their portraits: not but what the serene and seemingly inanimated countenances of the former artists came nearer to nature than the sparkling eyes which are now all the fashion. This however is a proof that the portrait is at least one hundred and fifty years old; the inscription shews that the painter was an Italian; and it must further be observed, that it is now two centuries since they left off painting on wood.

The history of this portrait is as follows. It was formerly in the collection of Saxe Gotha, where it was always considered as an original, which is even said to appear from the archives of that court, and is the more probable, as the prince-bishop of Warmia, who obtained it from the late duke of Saxe Gotha, was too good a connoisseur and too cautious to be deceived in this respect. That bishop being at Gotha in the year 1735, observed this portrait in the gallery of that palace; the proofs that were produced of its authenticity made him very desirous to acquire it. He at length obtained it by a kind of theft which it was necessary to commit on the cathedral of Warmia, in which there was a very old portrait of one of the ancestors of the dukes of Saxe Gotha, who had been bishop of that see, and whose picture was wanting in the duke's collection of the portraits of his family. An exchange was accordingly made of the two originals, and the bishop has since bequeathed that of Copernicus to his favourite Mr. Hussarzewski.

The copy I herewith send you is very exact in every respect, except that I have glued three wooden lists on the back of the board to prevent its warping. After my copy was taken, the proprietor of the original thought proper to have it repaired and varnished, and they are now so much alike that it is scarce possible to discern any difference. The name of my painter is Lorman of Berlin, an artist of some reputation.

76.04.07a *From Nathanael Wolf*

Royal Society of London, Letters & Papers, VI, 236[1]

Je joins ici mes observations astronomiques de Danzic, dont je me flate qu'elles seront acceptables, comme les premieres, faites dans cette ville, avec de bons instruments, & les seules presque faites après Hevelius.[2] Il est important de constater exactement la situation de l'Observatoire de ce grand Homme. Son Observatoire etoit à 300 pas au N.E. de ma chambre. Je pense, comme la plus part de mes observations donnent plus que les Tables, que la difference en longitude avec Greenwich sera une minute & demie plus que celle qu'on a adoptée. M^r^. Maskelyne en jugera: en lui presentant mon papier, aiez la bonté de lui dire que Je suis toujours rempli de plus grand respect pour son merite.

J'ai eté plus hereux cet hiver-ci à prendre des hauteurs correspondentes, que l'année passée, de sorte que j'ai pû mieux veiller à la marche de ma pendule. Je la croiois inalterable dans sa marche par le changement de la temperature. Feu M^r^. Short qui l'avoit comparée assez long tems avec la sienne me l'avoit assuré: & je ne m'en suis pas apperçu ni à Varsovie ni à Dirscham. peut etre que l'emplacement de la pendule, la defendoit du trop grand changement de la temperature, & que j'attribuois ses irregularités à l'inexactitude de mes observations. Mais cet hiver-ci les disparitions de Syrius m'ont donné toujours le meme resultât que les hauteurs correspondentes du Soleil: & aujourd'hui je ne doute plus que 2 degrés de Reaumur donnent une seconde de difference par jour, & meme un peu plus. Quoique le thermometre cette année-ci a eté bien long tems entre 9 & 16 degres de Reaumur au dessous du zero, il n'a jamais baissé dans mon observatoire au dessous de 7½: & la pendule etoit presque toujours envelopée dans une pelisse. Voilà une decouverte bien desagreable. Vous me ferez un grand plaisir d'en parler à M^r^. Maskelyne: peutetre qu'il aura observé quelque chose d'approchant: mais son observatoire est inebranlable, au lieu que le mien ne manque pas de trembler plus ou moins avec le mouvement des carrosses qui passent

1 The MS is in Magellan's hand, and is accompanied by an English translation headed "Extract of a Letter from Dr Wolf to Mr Magellan dated Danzig April 7. 1776. together with a set of astronomical observations", and also by a 5-page document with Wolf's observations, this being marked "Dr Wolfs Astronomical Observations | Read March 13". The Royal Society's *Journal Book* confirms that Wolf's report was not read until 13 March 1777.

2 The famous Danzig astronomer Johannes Hevelius (1611-1687).

par la rue, à peu près comme celui du P. Hell[3] à Vienne. Vous verrez par 76.04.07a
l'examen de la pendule depuis le Solstice tous ces changements. Je ne marque pas l'etat du thermometre qu'àprès chaque observation, de sorte que le jugement porté sur ma pendule, ne doit pas etre pris en toute rigueur. Vous serez surpris peut etre que le nombre de mes observations soit si petit pendant 3 mois. C'est pourtant tout ce que le beau tems a permis de faire, & je peux dire que J'ai manqué très rarement, quand le tems a [...][4] un tant-sort-peu favorable. C'est à Danzic à peu près comm'à Londres.

Je vous embrasse de tout mon Coeur, mon cher ami, & ne cesserai qu'avec la vie d'etre

&^{a}.

Signé Dr Wolf.

---

3 The Jesuit astronomer Maximilian Hell (1720-1792), director of the Vienna Observatory from 1756.

4 Word obscured by repair to paper.

To Ribeiro Sanches[1] 76.04.16

Österreichische Nationalbibliothek, Vienna
A.N. Ribeiro Sanches Korrespondenz, codex 12714, ff. 173-174v

Londres ce 16 Abril 1776

Meo am^{o}. e S^{r}. e m^{to}. do Cor^{am}.

A ultima de V.M. me fez conhecer q não se achava bem: mas já pela Reposta q tive do D^{r}. Du Presle soube q V.M. se achava melhor, & mesmo q até tinha passado já por sua caza. Com a primavera canta o cuco, & tudo vai melhor com tão boa muzica. Mandei em hũa caixa p^{a}. Paris um paquete dos Livros seg^{tes}. q serão entregues em caza de V.M. a saber Kirkland on child bed fever 3^{s}. 0^{d}./ Musgrave on Nerves – 2^{s}. 6^{d}./ Armstrong Essays 1^{s}. 6^{d}./ Chalmer's on South Carolina fevers 5^{s}. 6^{d}./ Percival Essays the 3^{d}. vol. 5^{s} 0^{d}./ Philos^{al}. Transact. vol. 65 part. 2/ Pringle's speech – 1^{s}. 6^{d}./ Levison on blood – 2^{s}. 6^{d}./ Rutty on Edimburg & London Dispensary's 3^{s}. 0^{d}./ Camoens translated into English, bound and charmingly arrayed 1$^{£}$. 6^{s}. 0^{d}.

---

1 Annotated "Resp^{a}. a 27 Mayo" [Answered 27 May]. Letter not found.

76.04.16 Eu faria a ceremonia de mandar tãobem hum p^{a}. o Cav^{ro}. Brito (a q^{m}. V.M. me recomendará Saudozo); mas sendo Livro caro, não quero q caia na matraca por ceremonia. Se elle o quizer não tem mais do q hir a caza de M^{r}. Ruault Livreiro na rua da Harpa [a] q^{m}. mandei 3 exemplares p^{a}. dispor delles a q^{m}. os quizer.

Taobem Lá mandei ao D^{or}. Begue du Presle um grosso vol. em 4°. halfbound, q contem os dois tomos do D^{r}. Smith sobre as cauzas da riqueza das Naçoens. na carta q escreverei na posta seg^{te}. ao mesmo D^{or}. du Presle, lhe pedirei q empreste este Livro a V.M. p^{a}. se divertir com as longas ideias politicas, economicas, civis, mechanicas, & comerçantes deste D^{or}. Pedagogo: e depois de V.M. o Ler á sua vontade o mandará p^{a}. a logea do mesmo Livreiro Ruault, onde elle o vendra a q^{m}. o quizer; pois he Livro Caro, & sobre materias q nem V.M. nem eu não podemos endireitar nem retrocer p^{a}. o bem ou p^{a}. o mal.

O D^{or}. Sequeira passa m^{to}. bem de saude e de officio: & da mesma sorte o D^{r}. La Cour. Quanto a mim cada vez vou engordando mais, & já me enfado de tanta gordura. estou vendo q não poderei hir dar hũa volta este verão a Paris nem fóra de Inglaterra, pelos m^{tos}. empecilhos q me occorrem cada dia. Porem divirtome, como sempre. Cada sabado vou á caza de campo do meo amigo Arbuthnot, onde fico até a 2^{a}. fr^{a}. pela manhan em q torno p^{a}. Londres. todas as 4as. fr^{as}. tenho as destinadas p^{a}. hir jantar com o nosso Enviado e sua digna esposa, ambos de feição portugueza, excellente caracter & amaveis. tenho actualm^{te}. aqui comigo o filho mais velho do nosso Arbuthnot, q he meo companheiro continuo aonde quer q vou. Seo pay quer q elle saiba o Portuguez, e q aprenda comigo a puxar pela vida. Cuido q o destina p^{a}. a Companhia das Indias orientaes, onde o Portuguez pode ser de bom serviço. he bom rapaz (coiza de 15 anos) mas mui preguiçozo. o Irmão 2do. q he de 13 à 14 anos, entrou o seo nome já no serviço da Marinha, onde hum de seos thios he actualm^{te}. comodore.

Eiz aqui como tacitis senescimus annis, de dia em dia e de cuidado em cuidado se nos-passa a vida; e em q^{to}. ella assim corre sem lhe sentirmos o pezo, deixemo-la hir na boa hora.

deme V.M. not^{as}. suas e da familia adoptiva: veja se por lá há algum milagre do cordão do nosso Padre S. Francisco neste mayo, emq^{to}. lhe Lembrarem as zurraduras dos nossos heroes da alimaria ardente pelos viçosos campos de Sendelgas, & veigas do Mondego.

vale & me ama

a Monsieur
Monsr. Sanchez, D^{or}. en Medecine
Rue des Blancs-manteaux
à Paris

London 16 April 1776 76.04.16

My Dear Friend

Your last letter[2] showed me that you were not well: but by the answer of Dr Du Presle[3] I realized that you were better, & you had even called at his home. With spring the cuckoo sings, & every thing goes better with good music. I sent in a box to Paris a parcel with the following books that will be delivered to your place, that is Kirkland on child bed fever[4] 3s. 0d. / Musgrave on nerves[5] – 2s. 6d. / Armstrong Essays[6] 1s. 6d. / Chalmer's on South Carolina fevers[7] 5s. 6d. / Percival Essays[8] the 3rd vol. 5s. 0d. / Philos.al Transact. vol 65. part 2. / Pringle's Speech[9] – 1s. 6d. / Levison on blood[10] – 2s. 6d. / Rutty on Edimburg & London Dispensary's[11] 3s. 0d. / Camoens translated into English, bound and charmingly arrayed[12] 1£. 6s. 0d.

I would have the politeness to send one also to Chev. Brito (to whom you would do me the favour of passing on my regards); but its being an expensive book, I don't want to be jeered at for being excessively polite. If he wants it, he just has to go to Mr. Ruault's bookshop in the rue de la Harpe[13] to whom I sent 3 copies for him to dispose of to those who want them.

I also sent to Dr. Begue du Presle a thick vol. in 4°. half-bound, containing the two tomes of Dr. Smith about the causes of the wealth of Nations.[14] In the letter I shall write in the next post to Dr. Du Presle, I shall ask him to lend you this book for you to get pleasure from the long political, economical, civil, mechanical, & commercial ideas of this Pedagogue Dr.: and after you have read it, if your wish

---

2 Letter not found.

3 Letter not found.

4 Thomas Kirkland, *A Treatise on Child-bed Fevers* (London, 1774).

5 Samuel Musgrave, *Speculations and Conjectures on the Qualities of the Nerves* (London, 1776).

6 John Armstrong, *Medical Essays* (London, 1773).

7 Lionel Chalmers, *An Account of the Weather and Diseases of South-Carolina* (2 vols, London, 1776).

8 Thomas Percival's *Philosophical, Medical, and Experimental Essays* (London, 1776) was generally seen as the third volume of his *Essays Medical and Experimental…* (2nd ed., 2 vols, London, 1772-3), and this was made explicit in the 3-volume edition that appeared under the latter title in 1777.

9 Sir John Pringle, *A Discourse on the Attraction of Mountains, delivered at the Anniversary Meeting of the Royal Society, November 30, 1775* (London, 1776).

10 Gumperz Levison, *An Essay on the Blood…* (London, 1776).

11 John Rutty, *Observations on the London and Edinburgh Dispensatories …* (London, 1776).

12 Luis de Camoens, *The Lusiad: or, the Discovery of India*, translated with an introduction and notes by William J. Mickle (London, 1776).

13 Nicolas Ruault, publisher and bookseller in Paris.

14 Adam Smith, *An Inquiry into the Nature and Causes of the Wealth of Nations* (2 vols, London, 1776).

76.04.16 you can send it to the shop of the same bookseller Ruault, where he will sell it to whoever wants it; since it is an expensive book, & on matters that neither you nor I can put right or wrong in either the good or bad direction.

Dr. Sequeira is very well in his health and in his job: & also Dr. La Cour. As for me, I'm getting fatter, & I'm tired of so much fat. I'm afraid that I can't go for a tour to Paris or abroad, because of the many impediments that arise every day. However I please myself, as always. Each Saturday I go the country house of my friend Arbuthnot, where I stay till Monday morning when I come back to London. I have all Wednesdays set to have dinner with our Envoy and his honourable wife, both of a Portuguese nature, of excellent character & pleasant. I have at present here with me the eldest son of our Arbuthnot, who accompanies me wherever I go. His father wants him to know Portuguese, and to learn from me how to thrive in life. I imagine that he destines him to the East India Company, where the Portuguese may be of good service. He is a good boy (about 15 years old) but very lazy. His brother, who is 13 or 14 years old, has already signed on for service in the Navy, where one of his uncles is now Commodore.

Here is how tacitis senescimus annis,[15] from day to day and from concern to concern life passes us by; and while it runs on without our feeling its weight, let it go in good time.

Give me news of you and the adoptive family: see if there is some miracle of the cord of our Father St Francis[16] in this coming month of May, while you remember the braying of our heroes with the burning loins by the blooming fields of Sendelgas & plains of the Mondego.[17]

vale & me ama

---

15 Ovid, *Fastes*, VI, 771: "we all grow old in silence".

16 The coarse cord worn by St Francis of Assisi to hold his garments together was reputed to have wrought several miracles.

17 Sendelgas is a village a short distance down the Mondego River from Coimbra where there was a convent with many young nuns. Magellan seems to be teasing Sanches about his lack of sexual activity by reminding him of the student festivities at Coimbra in May each year, at the end of the academic year. Ribbons symbolizing the various university degrees were ceremonially burnt and many of the students, no doubt with fire in their loins, serenaded the young nuns at the Sendelgas convent and displayed their gallantry with passion and sighs by constructing sonnets or rhymed verses from phrases offered to them by the nuns, at the same time as they savoured the sweet delicacies that the nuns had made for them.

To Alessandro Volta 76.04.25

Istituto Lombardo, Milan, Racc. fr. Volta[1]

Pour Mr. Volta
De la part de l'Auteur
& par le soin de son très h^{ble}. $serv^{r}$.
Magellan

P.S.

La machine de l'electricité continuelle ou perenne vient d'etre executée ici avec succès: même on les fait fort petites pour la poche d'environ 3 pouces de diametre. On peut voir dans l'histoire de l'Electricité du D^{r}. Priestley une experience Je crois de M^{r}. Wilkie,[2] qui aiant fondu de la resine dans un vase de metal, elle donnoit de l'electricité toutes les fois qu'on les separoit. Quoiqu'il en soit, C'est un phenomene fort remarquable, que la machine de M^{r}. Volta donne en perfection, & avec un avantage très agreable.[3]

---

1 The sheet on which this note was written was originally the initial blank page of a copy of Volume 2 of Joseph Priestley's *Experiments and Observations on Different Kinds of Air* that Magellan sent to Volta on Priestley's behalf, but has now been removed from the book. The main body of the letter (i.e. the postscript) was first published in *Aggiunta a lettere voltiane inedite*, ed. Z. and A. Volta (Como, 1883-4) and was later republished in the National Edition of Volta's *Epistolario*, vol. 1, p. 124. Subsequently the book in which the letter was inscribed was located among the family papers by some of Volta's descendants, and the sheet with the letter on it was removed and presented to the Istituto Lombardo (where it has not, however, been given a definite item number in the collection of Volta's papers). In 1966 the gift inscription that precedes the postscript was published in *Aggiunte alle opere e all' epistolario di Alessandro Volta*, pp. 78-9. The date of the whole is uncertain; Volta's editors were unable to do more than to suggest that it was written after 25 April 1776 but before 6 August 1778. Since the postscript constitutes Priestley's initial response to Volta's letter to him of 10 June 1775 announcing the invention of the electrophorus (published in *Scelta di opuscoli interessanti*, 9 [Milan, 1775], 91-107), it should in fact be dated prior to Priestley's own letter to Volta of 25 April 1776, not after this as Volta's editors have done.

2 i. e. Johan Carl Wilcke (1732-1796).

3 Magellan's response on Priestley's behalf to Volta's electrophorus was typical in its doubting the originality of the experiment but acknowledging the particular effectiveness of Volta's experimental arrangement. The experiment with resin melted in a metal cup to which Magellan refers was published in Wilcke's *De electricitatibus contrariis* (Rostock, 1757); Wilcke seems to have performed it together with his friend and teacher, F.U.T. Aepinus, who published his own description of it in his *Tentamen theoriae electricitatis et magnetismi* (St Petersburg, 1759). Subsequently, as Magellan notes, Priestley had given an account of it in his *History and Present State of Electricity* (London, 1767), pp. 232-4.

76.04.25 M^{r}. le D^{r}. Priestley a eté fort content de cette experience; & m'a prié de le faire savoir à M^{r}. Volta par cette occasion; se trouvant pour le present très occupé sur d'autres branches de la Physique, differentes de l'electricité pour le faire de sa main.[4]

J.H. de Magellan

4 As Volta would have learned from the book in which this note was inscribed, Priestley was by this time deeply immersed in the experiments on gases on which his fame now chiefly rests.

76.05.21 *To the Duke Charles-Marie-Raymond d'Arenberg*

Archief Van Arenberg, Enghien/Edingen, Belgium

Monseigneur

Celle-ci est bien moins pour complimenter V.A.S^{me}. sur Son retour & de Son Illustre Famille au charmant sejour d'Everlê,[1] que pour Les y accompagner avec mes souhaits pour Leur repoz d'Esprit, & pour ce douce oubli, qui fait le bonheur de la vie humaine beaucoup plus souvent que La jouissance des plaisirs Les plus vifs.

Les dernieres notices que J'obtins de Paris par La voye de l'Abbé Nicoli,[2] me marquoient que Monsr. Le Prince s'avoit retabli en santé, avant son depart. C'est gagner l'objet principal: & on attendra le reste de la vigueur de son age, tandis que La compagnie La plus aimable & affectueuse, ne manquera pas de Lui adoucir Le desagrement de l'incom̃odité des yeux; puisque La Providence a eté assez genereuse en Lui destinant une infinité d'autres satisfactions bien plus precieuses & Solides.[3]

Je ne me suis oublié, Monseig^{r}., de la commission d'un horloge pour l'Escalier d'Everlé. Mais j'avouerai avec Repentence quoique sans en rougir, que mon devoir de reconnoissance pour V.A.S^{me}. & pour tout ce

1 The château de Heverlee, near Leuven, one of the houses of the Duc d'Arenberg.

2 Abbé Raimondo Niccoli (?-1780), secretary of the Parisian legation of the Grand Duke of Tuscany.

3 The Duke's son Louis Engelbert, the future 6th Duke of Arenberg, had been blinded in an accident while out hunting.

qui La Concerne, se trouve tout à fait degeneré en attachement Le plus sympathique, de meme qu'on observe L'amitié degeneré en amour entre les deux sexes. Après l'acident du Prince je me trouvai tombé dans une telle despondence,[4] que je me lassai de suivre les projets formés sur ce sujet. J'envoyerai cependant en peu de tems une de mes pendules ordinaires, qui pourra servir en attendant une meilleure jusqu'à ce que Je reprende la poursuite de mes idées.[5] 76.05.21

Il est très probable que Je ne pourrai tenir cet eté ici; sans y aller en personne, pour revoir de mes yeux, les objets que je porte sur Le Coeur depuis Long tems, malgré une infinité de petits embarras, & d'engagements, qui tachent de m'en empecher. Le Dr. Priestley vient de Sortir de ma chambre, & me pria de presenter à V.A.Sme. ses plus humbles Respects. My lord Shelburne m'en a parlé plusieurs fois de meme avec Les plus dignes sentiments.

Je viens de Lire dans les co͞mentaires d'Edimburg que l'air fixe appliqué à deux cancers, avoit soulagé les malades.[6] Celà m'a fait souvenir de cette pauvre dame de la connoissance de V.A.Sme. près d'Enghien.[7] J'en ai parlé tantot au Dr. Priestley: Rien peut empecher de tenter ce secours, qui a deja eté appliqué en d'autres cas, avec succès.

J'ai l'honneur d'etre avec le plus vrai attachemt. senti de Coeur, & profond respect

Monseigneur

de V.A.Sme.

Serviteur le plus hble. & obligé

Jean Hyacinthe de Magellan

à Londres ce 21 Mai 1776

---

4 Magellan evidently has in mind here the English word "despondency".

5 A description of this clock was later published; see Magellan to F. Rozier, April 1781. There Magellan revealed that it took much more time than he expected to perfect the instrument.

6 There is a brief report in *Medical and Philosophical Commetaries, by a Society in Edinburgh*, 1 (1773), 271-2 on the successful use of fixed air in the treatment of certain cancers, that may be what Magellan is referring to.

7 See Magellan to the Duke of Arenberg, 23 July 1776.

76.05.29 *From Charles Messier*[1]

I have observed, since the 14th of May, a belt of a fainter light on the body of Saturn, opposite to the part of the ring behind the planet. It is pretty broad, and almost as distinct as those of Jupiter. It was with a very good achromatic of three foot and a half, made by Mr. Dollond, that I discovered this appearance. I wish you would communicate it to the astronomers, because those who are furnished with better instruments may, perhaps, see some inequalities in this belt of Saturn, and so the time of the planet's revolution on its axis may be better ascertained than it is at present. Mess. John and James Cassini seem to have been the only astronomers who discovered this phenomenon[2] about the end of the last century.[3]

---

1 Letter not found. The text is from *Philosophical Transactions*, 66 (1776), 543, where it was published under the heading, "A Belt on the Disc of Saturn described in an extract of a Letter from Mr. Messier, F.R.S. to Mr. Magellan, F.R.S. Dated Paris, May 29, 1776." An annotation indicates that it was read to the Royal Society on 6 June 1776.

2 J.D. Cassini, "Observations nouvelles sur Saturne", *Histoire de l'Académie Royale des Sciences*, 1715, Mém. pp. 41-9?

3 See also Royal Society of London, AP.4.13, which is a sheet on which is pasted a drawing of Saturn that Magellan has evidently cut off the bottom of Messier's letter to him. The drawing is labelled, in Messier's hand: "Figure renversée comme la donne l'instrument". Above it appears the valediction of Messier's letter: "à Dieu cher ami porté vous bien et aime moi toujours, Messier" and the date, "Paris 29 Mai 1776". The following is written above and below the drawing, in Magellan's hand:

> The present appearance of Saturn with it's belt, drawn by M^{r}. Messier Astronomer at Paris, in a Letter written the 29 Mai 1776 to M^{r}. Magellan F. of the Soc^{y}.
>
> N.B. M^{r}. Messier mentions in his Letter, that the best time to observe the appearance, is in y^{e}. twilight.
>
> It appears by the 59 vol of y^{e}. Philos^{al}. Transactions at page 454 that M^{r}. Messier observed once before, that Saturne has some belts round its body. it was the 28 March 1766 that he observed them at Colombe, a country seat of y^{e}. Marquiz de Courtenvaux near Paris; with an achromatic refractor of 12 feet, made by M^{r}. Dollond.

To Ribeiro Sanches[1] 76.06.16

Österreichische Nationalbibliothek, Vienna
A.N. Ribeiro Sanches Korrespondenz, codex 12714, ff. 175-176v

Londres 16 Junho 1776

Meo Am°. e Sr. e mto. do Corm. vou escrever esta emqto. V.M. lá estará ouvindo a Sta. Missa, pois he Domingo pela manhan, tempo q me acho ociozo e só no meo quarto, donde talvez não sahirei q ás 4 ou 5 da tarde pa fazer algum passeio, tendo feito mtos. todos estes dias pa. acompanhar e servir (servindo-me) ao nosso commum amigo Espanhol enxertado em Italiano o Dr. Cazimiro Ortega, q aqui se acha, & partirá pa. Hollanda 3a. fra. q vem. fico encarregado das suas Comissoens botanicas & cientificas, de q sempre me rezultará algum interesse. V.M. ahi o verá de Retorno antes q parta pa. Espanha, o q não fará senão no outono.

Se V.M. vir o Dor. Le Begue diga-lhe q cá recebi a sua carta. e q lhe mandarei os Livros q me pede, como tãobem os 3 Musgraves q V.M. me ordena: mas não me lembra algũ outro em q V.M. falasse, excéto a continuação do Medical commentaries q mandarei á medida q sahirem. V.M. lá receberia uma piquena brochura de Grant on Cough dedicada ao Dr. la Cour: & ultimamte. outra de Musgrave sobre worm-fever, q mandei juntar com outra copia ao Dr. du Presle, & q V.M. lhe pedirá em cazo q elle se esqueça.

Me alegro mto. q V.M. comece a sahir, & q va passar ao Fauxbourg St Antoine neste tempo quente. ainda q nós o não temos aqui. tem chovido frequentemte., & o tem feito tãobem esta manhan. mas o tempo está excellente, como as nossas primaveras de riba-Coa.

Sahio aqui hũ tratado de von Rosenstein sobre as doenças das crianças traduzido do Sueco. he bom livro, ainda q não cuido tem couzas novas: já o tenho pa. lho mandar aqui no meo quarto.

O Dor. Seqra. passa bem. está Medico do nosso Embor. & mto. estimado da nossa Embaixatrice. elle me contou ter curado pelo metodo de Musgrave differentes cazos de hũa tosse horrivel & convulsiva, com o topico da agoa quente aplicada com boa esfregação no ventre &a. acompanhada de algũ remedio apparente, q não fizesse mal, por dentro à fim de cubrir e encubrir o Empiricismo.

Na Sociede. Real se lerão ultimamte. varias experiencias notaveis & crueis sobre os animais, pa. conhecer q os nervos não servem senão de conductores do influxo do cerebro pa. os membros do corpo: & q a irritação

1 Annotated "respa. a 1 de Julho" [answered 1 July]. Letter not found.

76.06.16 das partes os mete em movim[to]., sem ajuda dos nervos. a inspiração nos bofes he o melhor socorro p[a]. os afogados: & a bronchotomia (cuido q este he o nome da abertura do Larinx) pode ser uzada com proveito em semilhante cazo.

Admirou-me q o Conego Chevalier não informe a V.M. das pensoens q se tem destinado aos acad[os]. de Bruxellas, entre os quais elle não deixará de ter a sua, com M[r]. Needham &[a]. mas não me admira q hũ lisboeta, perfumado com as fidalguias Portuguezas não se ache piando como o pintainho pelas honras passadas, q certam[te]. não hão de tornar, pois já não existem os q nellas figuravão & fazião figurar. ainda não estou decidido a sahir este verão fóra desta ilha. se o fizer não será mais q até Flandres.

Eu mandei certam[te]. a V.M. o Chalmers on South Carolina fevers. veja V.M. se o D[or]. du Presle o goardou ou não, e avizeme. o preço das Phil[al]. Transact[s]. he de 7[s]. 6[d]. & o Discurso de Pringle 1[s]. 6[d].

o D[or]. La Cour virá brevem[te]. fazer hũa vizita aos am[os]. de Londres. elle está cortezão completo isto he com todos os principios despoticos nos cascos.

Musgrave publicou depois de pouco tempo hũ tratadinho latino de seo Pai (ou de seo thio) sobre a gota. senão fosse tão antiquado, havia de mandalo a V.M.

Fui a Oxford havera coiza de 3 semanas com o D[or]. Ortega, q queria ver esta universid[e]. ainda vestida á monastica com Labios de Seiscentista: & passando pela rua onde se vende a Magnesia do D[or]. Glass, entrei na botica ou logea: e perguntei por elle: ali me disserão q o D[or]. em Medecina discipulo de Boerhaave ainda vive (cuido q me disserão em Exiter) mas o Irmão, autor da magnesia, he morto. e eiz aqui tem V.M. a Reposta da pergunta q me fez varias vezes sobre a existencia deste seo condiscipulo.

Ora vou acabar esta com máo humor. Hũa das minhas Nelis a q[m]. fiz saber hontem q ficarîa hoje em caza isto hé q não hiria passar como costumo o Domingo com o meo amigo Arbuthnot, não me veio vizitar. a vilhacona ou fez q não entendeo, ou não se quiz dezengager d'outro engagement: por crista q m'o ha de pagar. São quazi duas depois do meio dia: e ella não apparece. estou picado vou vestirme p[a]. lhe hir por os cornos com outra comadre: e ao depois talvez q va jantar com o amigo Paradise emq[to]. por cá se arr-ta, não se me dá nada do q vai nem do q vem. diga lá a Messier q va bugiar com as suas choramingas do seo corpo Acad[o].: q eu hirei q[do]. m[to]. quizer à Paris; pois não temo nem espero nada dos q cahem nem dos q sobem mas certam[te]. as suas michellas de Paris são a mais maldita raça do universo. as poucas q vi nesse anno quazi inteiro q ahi assisti, sempre se punhão sobre a borda da cama p[a]. q as fo[...]sse em pé. olhe q gosto!

bem hajam as Inglezas q não sabem f-der senão de costas, sobre a cama estendidas como hũa ran. 76.06.16

vive, vale & me ama

P.S. Diga ou faça dizer à Messier q passe por caza de Ruault Libraire, & q lá verá na gazette Ingleza London-Chronicle o annuncio das faxas ou bandas de Saturno, q elle quis se publicassem. o mesmo avizo foi impresso em outras gazetas, e eu o comũniquei á Socied[e]. Real. &[a]. iterum vale

Saud[es]. a M[a]. Jeane, Toinete, Marie, le mari, & tout autre ami s'il y en a, qui merite ce nom.

A Monsieur
Monsr. Sanchez D[or]. en Medecine
Rue des Blancs-manteaux
à Paris

London 16 June 1776

My Dear Friend

I'm going to write this while you may be attending Holy Mass, since it's Sunday morning, a time when I'm idle and alone in my room, which perhaps I shan't leave before 4 or 5 in the afternoon to go for a walk, having gone many times these days to accompany and to serve (serving me also) our common friend, Spanish grafted on Italian, D[r]. Cazimiro Ortega,[2] who is here & will leave for Holland next Tuesday. I shall be in charge of his botanical & scientific commissions, which will certainly pay some interest. You will see him back there before he returns to Spain, which he won't do before autumn.

If you see D[r]. Le Begue, tell him that I have received his letter[3] and shall send him the books he asks for as well as the 3 Musgraves[4] that you want: but I don't remember any others you have mentioned except the continuation of the Medical Commentaries,[5] which I'll send as soon as it is published. You would have received a little brochure by Grant on cough[6] dedicated to D[r]. La Cour, & recently another one by Musgrave about worm-fever,[7] which I sent together with another copy to D[r]. du Presle, & that you will ask him for in case he forgets. I am very pleased that you begin to go out & pass along the Fauxbourg St Antoine in

2 Casimiro Gomez de Ortega (1741-1810), Spanish physician and botanist who studied at Bologna before becoming a professor in Madrid.

3 Letter not found.

4 i.e. three copies of Samuel Musgrave, *Speculations and Conjectures on the Qualities of the Nerves* (London, 1776); see Magellan to Sanches, 13 September 1776.

5 See Magellan to Sanches, 24 July 1775.

6 William Grant, *A Short Account of the Present Epidemic Cough and Fever* (London, 1776).

7 Samuel Musgrave, *An Essay on the Nature and Cure of the So-called Worm-Fever* (London, 1776).

76.06.16 this warm weather, though we don't have it here. It has often been raining & also this morning. But the weather is excellent, like our springs in riba-Coa.[8]

A treatise by von Rosenstein translated from Swedish about children's diseases[9] was published here. It's a good book, though I think it has no novelties; I have it already with me in my room to send to you.

D^{r}. Sequeira is well. He is our Ambassador's physician & is greatly prized by our Ambassadress.[10] He told me he had cured by Musgrave's method different cases of horrible & convulsive cough, with the topical remedy of hot water rubbed on the abdomen &^{c}. along with a fictitious medicine that would not do harm inside, in order to cover and hide the empiricism.

At the Royal Society several remarkable & cruel experiments on animals were read that were intended to show that the nerves serve only as conductors of the influx from the brain to members of the body, & that irritation of the parts sets them in motion without the help of the nerves.[11] Inspiration into the lungs is the best help for the drowned, & bronchotomy (I think this is the name for opening of the larynx) can be used with success in such cases.

I was surprised that canon Chevalier didn't inform you about the pensions that are reserved to the academicians of Brussels, among which he certainly has one himself, along with Mr. Needham &^{c}. but I'm not surprised that a Lisboner, perfumed with Portuguese nobilities, peeps like a young chicken for past honours that will certainly not come back since those who were awarded them & made the awards are no longer there. I haven't yet decided whether to leave this island in the summer. If I go, it won't be farther than Flanders.

I'm sure I sent you Chalmers on South Carolina fevers.[12] Please see if D^{r}. Du Presle kept it or not, and tell me. The price of the Phil^{al}. Transact^{s}. is 7^{s}.6^{d}. & the Discourse of Pringle 1^{s}.6^{d}.

D^{r}. La Cour will soon come to visit his friends in London. He is a perfect courtier, that is, he has all the despotic principles in his head.

Musgrave published a short time ago a little treatise by his father (or uncle) about gout.[13] Were it not so old fashioned, I would send it to you.

I went to Oxford 3 weeks ago with D^{r}. Ortega, who wished to see that university still dressed in monastic style with marks of the 17th century, & when

8 See Magellan to Sanches, 23 February 1776, note 21.

9 Nicolaus Rozen von Rosenstein, *The Diseases of Children, and their Remedies*, trans. Anders Sparrmann (London, 1776).

10 See Magellan to Sanches, 27 February 1776.

11 A paper by a Mr. Crookshanks, or Cruikshanks (who was introduced by Dr John Hunter), was read at the meeting of the Royal Society on 13 June 1776, giving an account of experiments made on nerves of living animals. The paper was not, however, published in the *Philosophical Transactions*.

12 Lionel Chalmers, *An Account of the Weather and Diseases of South-Carolina* (2 vols, London, 1776); see Magellan to Sanches, 16 April 1776.

13 William Musgrave, *De arthritide primogenia et regulari* (London?, 1776). Samuel Musgrave was the son of William Musgrave (1655?-1721), physician and antiquary at Exeter.

passing along the street where the Magnesia of Dr. Glass[14] is sold, I went into the pharmacy or shop and asked for him. There I was told that the Dr. in Medicine and disciple of Boerhaave still lives (I think they said in Exiter[15]) but that his brother, the author of the Magnesia, is dead.[16] And that is the answer to your question about the existence of this colleague of yours. 76.06.16

I am finishing this letter in a bad temper. One of my Nellys to whom I gave notice that I would stay at home, that is, that I wouldn't go as usual to spend Sunday with my friend Arbuthnot, didn't come to visit me. The knave either misunderstood or didn't want to disengage from another engagement; on my honour she will pay for that. It is almost two in the afternoon, and she hasn't come. I am cross[.] I'll get dressed to make her a cuckold with another woman, and then maybe I'll go and have dinner with my friend Paradise[.] As long as one belches[17] I don't care about what goes or comes. Please send Messier to hell about his business, with his lamentations about his Academic body: because I shall go to Paris only when I want to, since I don't fear or expect anything from those who fall or from those who rise; but certainly his Parisian michellas[18] are the most damned race in the universe. The few I saw that whole year when I was there always placed themselves at the edge of the bed in order to be f-ked while standing. How pleasant! Thanks be to the English women who don't know how to f-k except lying on their back, on the bed lying down like frogs.

vive, vale & ama.[19]

P.S. Tell Messier or send him a message to pass by Ruault Libraire, & that he will see there, in the English gazette London Chronicle,[20] the announcement of the stripes or bands of Saturn that he wanted to have published. The same announcement was printed in other papers, and I communicated it to the Royal Society. &c.[21]

iterum vale.[22]

Regards to Mme. Jeane, Toinete, Marie, le mari, & tout autre ami s'il y en a, qui merite ce nom.

---

14 Samuel Glass, author of *An Essay on the Magnesia Alba, wherein its History is Attempted ... and Use of It Recommended* (Oxford, 1764).

15 Thomas Glass (1709-1786) studied under Boerhaave at Leiden, graduating MD in 1731. He subsequently practised as a physician in Exeter for many years.

16 Samuel Glass died in February 1773.

17 See Magellan to Sanches, 7 July 1775, note 19.

18 whores.

19 Flourish, farewell and love me.

20 In the *London Chronicle*, 6 June 1776, there is an extract of a letter from Paris dated 29 May, reporting the announcement by Messier of his observation of "an obscure belt on the body of Saturn opposite the part of the ring behind the planet...".

21 Messier, "A Belt on the Disc of Saturn described in an Extract of a Letter...", *Philosophical Transactions*, 66 (1776), 543; see 76.05.29.

22 Farewell again.

76.06.21 *To Louis-Henri Duchesne*
Wellcome Institute Library, London[1]

Londres ce 21 Juin 1776

Mon cher ami

Il y a un Siecle que Je n'ai pas de vos nouvelles, & je ne sçais pas si les changements du Ministere de M^{rs}. Turgot & Malsherbes, ont fait quelqu'un dans ceux qui nous interessent.[2] Je ne sçais non plus si M^{r}. Trudaine est deja de retour à Paris, & si Mad^{e}. Trudaine & le Comte Carburi[3] sont deja revenus en France du Voyage d'Italie. aiez la bonté de me dire si vous avez reçu les articles que Je vous ai envoyés dernierem^{t}. dans le commencem^{t}. de Avril dernier. Et dites moi aussi si Monsr. Trudaine veut avoir encore les autres 12^{lb}. de Thé de la meilleure sorte. Car comme vous savez Je ne lui envoyai que seulem^{t}. 12^{lb}. au lieu des 24^{lb}. qu'il m'avoit ordonnées, pour attendre l'occasion de l'avoir plus parfait.

Pourtez vous bien, mon cher ami, & soyez assuré de l'attachem^{t}. pour la vie

de Votre
Magellan

A Monsieur
Monsr. Duchenne Secretaire
de Monsr. Trudaine de Montigny
à Paris

---

1 Letter transcribed in 1977 but now missing.
2 Malesherbes resigned from the Ministry in April 1776. A month later, on 12 May 1776, Turgot was dismissed from his position as Controller General.
3 Giovanni Battista Carburi (1722-1801?); see Magellan to Duchesne, 23 July 1773, note 3.

To Ribeiro Sanches[1] 76.07.12

Österreichische Nationalbibliothek, Vienna
A.N. Ribeiro Sanches Korrespondenz, codex 12714, ff. 177-178v

Londres 12 Julho 1776

Meo Am^{o}. e S^{r}. & m^{to}. do Cor^{am}.

A de V.M. escrita dos seos Jardins de diante & detras das portas de S^{to}. An^{to}. me mostra o seo humor assanhado contra os mexilhoens d'Aveiro, o qual certam^{te}. mudaria se os tivesse p^{a}. cear, pois q por experiencias indubitaveis fazem arr-tar de toda a sorte. Tãobem aqui temos tido ventos & chuvas nestes dias, q fazem o tempo fresco como as nossas primaveras: & particularm^{te}. entre as pernas das minhas bettys, Nelys, & fannys, onde ainda devez em q^{do}. me divirto com o bello jogo do toque em boque, jogo de q nem o nome se usa em toda a Ribacoa. Ora vamos adelante & por diante á portugueza deixando o trazeiro p^{a}. os becos f-tos[2] dos Italianos. Leia V.M. como e q^{do}. quizer por seo divertim^{to}., e dê o livro a Ruault m^{to}. q^{do}. lhe parecer. eu lá escrevi o correio passado a o D^{or}. Begue du Presle queixando-me deste magarefe de Ruault, q não se dignou darme hũa Reposta desde q lhe mandei varias encomendas, & parcellas de dinh^{o}. a Receber q importão mais de mil Livras (tournoises).

Agora tirei sobre elle por coiza de 800$^{\#}$., & fiz avizo nas formas. mas estou Receando q não me pague a Letra a q^{m}. a Levou. Verd^{a}.m^{te}. esta g^{te}. quer-ser servida, & não querem dar hũ passo por outrem. assim vai & hirá este mundo buz-ne![3]

agora tive carta do am^{o}. Chev^{er}. q parece melancolico. coitado ainda não achou tempo de se endurecer aos vaivens da fortuna: e ainda parece q pia pela patria como o pintainho pela mãi!

A um amigo me avizou q hũ dos da nossa Terra a q^{m}. foi prezentado o meo Tratado sobre os Octantes, dera sinais de querer procurar-me o retorno á Patria &^{a}. &^{a}. Que lhe parece a V.M. de tentação p^{a}. hũ estoico como eu, q renegou toda a Filosophia de Pangloss, & q não segue senão a do grande Martinho ...! Vou escrever hũa grande bazumbaya a este Fidalgão, & oferecer-lhe os meos serviços cá pelos arrabaldes do Norte, donde lhe mandarei q^{tas}. encomendas for servido, se me mandar o dinh^{o}. p^{a}. ellas.

---

1 Annotated "Resp^{a} a 18 de Julho" [answered 18 July]. Letter not found.
2 Becco fottuto – cheater?
3 buzzerone; see Magellan to Sanches, 23 June 1775, note 5.

76.07.12 Aqui está o amigo Needham tão gordo, & tão encasquetado das suas classes da vitalid[e]. Como d'antes: fui mostrar-lhe domingo passado a planta movente i.e. q continuam[te]. move hũa ou outra das suas folhas piquenas, commo se fosse u[m co]rninho de caracol. Chama-se Burrum Chundally: he nativa (habitat seg[do]. a fraze do grande Linneo) nas bordas do Ganges perto de Dacca &[a]. o am[o]. Needham fez suas Remarques, & lá lhe servirá p[a]. apojar as suas classes das differentes vitalidades.

na prim[a]. occazião mandarei a V.M. os 3 Musgraves q ja aqui tenho no meo quarto. mas faltão me not[as]. do am[o]. Duchenne, q he o meo guia.

hirei falar outra vez com o Mercador sobre os livros q mandei p[a]. seos sobrinhos. provavelm[te]. estão empatados na Alfandega, ou mais certam[te]. na Meza Censoria. elles devião procuralos & falar não só com florencio Teixeira, mas buscalos nestas duas partes. Saudades á familia adoptiva.

vive, vale, & me ama

A Monsieur
Monsr. Sanchez Doct[or]. en Medecine
Rue des Blancs-Manteaux
à Paris

London 12 July 1776

My Dear Friend

Your letter[4] written from your gardens in front of and behind the gates of St. Anthony[5] shows me your furious humour against Aveiro's mussels that would certainly change if you had them for dinner, because by unquestionable experiences they make one belch in every way.[6] We also these days have had winds & rain that make the weather cool like our springs; & in particular between the legs of my Bettys, Nellys, & Fannys, where from time to time I still have fun with the nice game of touch and bite, a game of which not even the name is used any longer in all Riba Coa.[7] Let us go on leaving the bum to the f--ing[8] cuckolds of the Italians. Read how and when you want at your pleasure, and give the

4 Letter not found
5 The Porte de St Antoine was one of the gates through the walls that used to surround Paris, and was located at the eastern end of the rue St Antoine next to the Bastille, not far from the rue des Blancs-Manteaux where Sanches lived. The structure was demolished in 1778.
6 See Magellan to Sanches, 7 July 1775, note 19.
7 See Magellan to Sanches, 23 February 1776, note 6.
8 Conjectural reading.

book to Ruault whenever you like. I wrote to D^{r}. Begue du Presle by the last post[9] 76.07.12
complaining about this butcher Ruault, who didn't deign to give me an answer after I sent him several orders, & parcels to the value of more than one thousand Pounds (tournoises).

Now I have drawn on him for something like 800$^{\#}$., & gave notice of it in the forms. But I'm afraid that he won't pay on the promissory note to the person who took it. In fact these people want to be served, & don't want to take a single step further for others. So this cheating world goes and will go on!

I had just now a letter from our friend Chevalier[10] who seems melancholic. The poor man hasn't yet found the way to toughen himself against the ups and downs of fortune: and it seems that he still chirps for his country like a poult for its mother!

A friend gave me notice that someone from our country to whom my treatise on octants[11] was presented showed signs of wanting me to return to the motherland &^{c}. &^{c}. What do you think of such temptation for a stoic like me, who has abjured all the philosophy of Pangloss[12] & doesn't follow any but that of the great Martin[13]...! I'm going to write a great and profound acknowledgment to this gentleman & offer my services in these northern districts, from whence I'll send all the orders I am asked for, if he sends the money for them.

Our friend Needham is as fat & as obsessed with his classes of vitality as before.[14] Last Sunday I showed him the moving plant, that is, the plant that continuously moves one or another of its small leaves, as if it was the little horn of a snail. It is called Burrhum Chundally:[15] it is native (habitat following the term of the great Linnaeus) to the banks of the Ganges near Dacca &^{c}. Needham made his Remarques,[16] & that will serve to support his classes of the different vitalities.

9 Letter not found.

10 Letter not found.

11 Magellan, *Description des octants et sextants anglois...* (London, 1775).

12 The tutor in Voltaire's *Candide*, who maintained, despite overwhelming evidence to the contrary, that everything that happened was for the best, in this best of all possible worlds.

13 i.e. Luther.

14 Needham's celebrated (but ultimately flawed) microscopical investigations had led him to the view that all effects in nature arise from the combination of two antithetical principles, action and reaction, that living organisms are composed of particles organized by a vital force that acts as the formative principle in all vital change, and that there is a scale of living beings of increasing levels of vitality as action predominates over resistance. See J. Roger, *The Life Sciences in Eighteenth-Century French Thought* (Stanford, 1997), pp. 399ff.

15 The moving plant *Hedysarum gyrans* was formally described by Carl von Linné jr. in *Supplementum plantarum systematis vegetabilium* (1781). As part of the description, Linné recorded the plant's common name as "Buram Chadali. Bengal".

16 Possibly Needham's article, "Remarques sur l'analogie, & le passage du vegetal au vital", pp. 239-261 in L. Spallanzani, *Nouvelles recherches sur les découvertes microscopiques et la génération des corps organisés ...* (London/Paris, 1769).

76.07.12 At the first opportunity I'll send you the 3 Musgraves that I already have here in my bedroom.[17] But I'm still missing news of our friend Duchenne who is my guide.

I'll go and talk to the merchant about the books I sent to your nephews. Probably they are stuck in Customs, or more certainly at the Censorial Table. They should look for them & not only speak with Florêncio Teixeira[18] but search for them in these two places. Regards to the adoptive family.

vive, vale & me ama.[19]

---

17 i. e. three copies of Samuel Musgrave, *Speculations and Conjectures on the Qualities of the Nerves* (London, 1776); see Magellan to Sanches, 13 September 1776.

18 See Magellan to Sanches, 17 November 1775.

19 Flourish, farewell and love me.

76.07.23 *To the Duke Charles-Marie-Raymond d'Arenberg*

Archief Van Arenberg, Enghien/Edingen, Belgium

Monseigneur

Le papier ci joint,[1] contient la methode practique de secourir les malhereux qui soufrent des cancers. Je me souviens bien comment V. A. S^me^. & M^de^. La Duchesse retournerent tristes l'année derniere de chez la pauvre dame près d'Enghien, qui est affligée de ce malhereux acident. J'avois Lu deja plus d'une fois la nouvelle du soulagement que l'application de L'air fixe procure dans cette maladie: mais ce ne fut que hier matin en revenant à Londres de chez M^r^. Arbuthnot, que J'eus une Longue conversation avec un cirurgien,[2] qui actuellem^t^. applique ce remede avec succès. Je m'empresse donc de le mander ici en detail à V. A. S^me^., sachant bien qu'Elle a toujours à Coeur tout ce qui peut etre utile & soulager Les pauvres humains.

M^r^. Arbuthnot me dit qu'il compte aller dans peu de tems dans Les Paiz bas: & s'il l'execute je ne manquerai pas de l'acompagner, pour avoir La honneur de Revoir V. A. S^me^. & Son auguste Famille. M^r^. Needham qui se trouve ici m'en a donné des nouvelles du Retablissement en santé de M^r^. Le Prince.[3] C'est deja un grand point degagné: La Jeunesse & le tems feront Le reste.

---

1 See below.

2 Possibly the "Mr. Hurdman" to whom Magellan refers in his letter to Condorcet, 11 March 1777.

3 The Duke's son, the future 6th Duke of Arenberg, had damaged his eyes in an accident while out hunting.

J'ai envoyé La traduction Angloise du grand Camoens Portugais pour La Bibliotheque de V.A.S^{me}.[4] si Elle ne fait pas usage de cette Langue dans Sa lecture, elle ne sera pas ignorée peut etre des Princes qui Seront Ses successeurs. 76.07.23

J'ai l'honneur d'etre avec le plus vrai attachement de Coeur & ame & avec le plus profond Respect

Monseigneur

de V.A.S^{me}.

Serviteur le plus h^{ble}. & le plus obligé

Jean Hyacinthe de Magellan

à Londres ce 23 Juillet 1776.

Maniere d'appliquer l'air fixe aux Cancers: ce qui a procuré en peu de tems une cessation des douleurs, & une grande diminution dans Le Cancer.[5]

Prennez deux grandes vessies a & b: Liez l'embouchure de chacune à un tuyau deverre cm (ou à un morceau de pipe à fumer), avec les ficelles marquées ns, ns.: & coupez le fond de la vessie f.g: de façon qu'elle fasse comme une espece de manche pendente, comm'on le voit dans La figure.

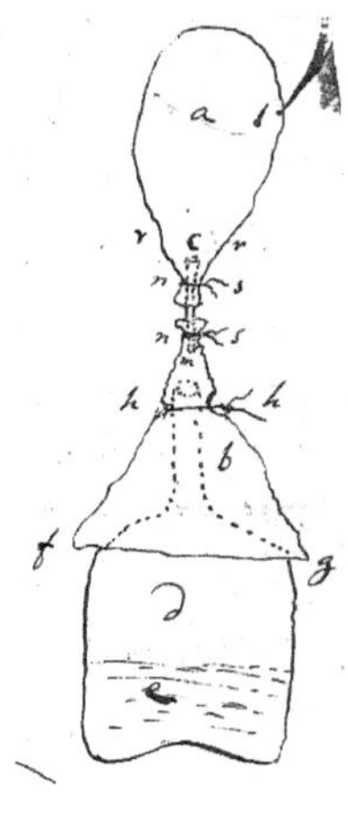

4 *The Lusiad; or, The Discovery of India. An Epic Poem*, trans. William Julius Mickle (Oxford, 1776).

5 Magellan also sent a copy of this report to Jacques Rozier, who promptly published it, with minor editorial adjustments, in his *Observations sur la Physique...*, August 1776, pp. 132-135. Rozier added to Magellan's report further reports on the treatment of cancers with fixed air, together with comments of his own.

76.07.23 Prennez une grande bouteille d.e: mettez y des petits morceaux de craye, ou terre calcaire, ou meme du marble blanc concassé comme du poivre: ajoutez y un peu d'eau de fontaine, de façon que la craie, ou morceaux de marble concassé, soient couverts par l'eau. jettez y un peu d'huile de vitriol. aussitot celà fera une grande effervescence. C'est par cette raison qu'on laissera les deux tiers de la bouteille vuides, pour avoir assez d'espace, capable de contenir l'ecume de l'effervescence, sans qu'elle entre dans la vessie. Liez alors la vessie b au cou de la bouteille avec la ficelle hh. dans ce cas l'air fixe qui se degage de cette effervescence, passera par le tuyau mc: & remplira la vessie a.

N.B. avant d'employer ces vessies, il faut les rendre assez moles, en les trempant pendant quelques minutes dans l'eau: dans la suite on les essayera avec un Linge parce qu'autrement elles seront fort glissantes.

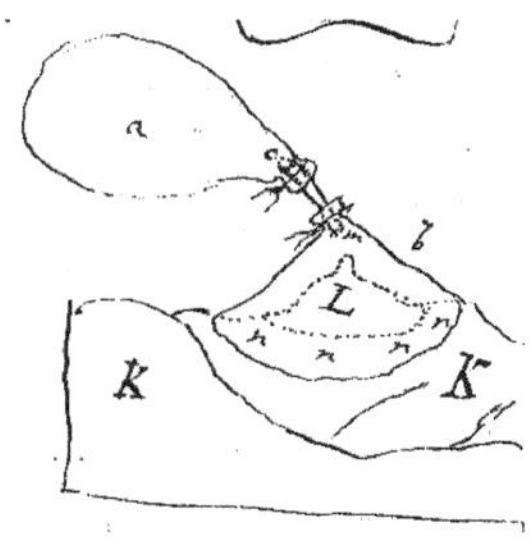

A present soit Le malade couché dans son lit, ou si l'on veut assis sur une chaise. Soit KK le corps du malade, que je suppose couché: & soit L la mamelle avec le cancer. Aussitot que la vessie a sera pleine d'air fixe, prennez-la avec la main par rr, de façon qu'en la pressant entre les doigts, par dessus du tuyau c, l'air ne puisse pas s'en echaper. Deliez la ficelle hh: & appliquez la manche de la vessie fg ainsi panachée, tout au tour du mamellon & du cancer L. Le malade meme ou La femme de chambre, pressera avec ses mains, mais doucement, toute l'extremité panachée de la vessie nnnn, contre le corps du malade, & tout au tour du cancer. Dans cette situation on pressera un peu La vessie a, pour que l'air fixe sorte par le tuyau cm, & soit repandu tout à l'entour du cancer, sans s'echaper.

Dans peu de tems on verra que le volume d'air fixe diminue considerablement, etant absorbé par le cancer.

on devroit avoir 2 ou 3 pairs de vessies arrangées comm'on l'a dit ci dessus en parlant de celles a & b, liées ensemble à un morceau de tuyau de verre, ou de pipe à fumer, pour mettre sur le col de la bouteille, & profiter l'air fixe, qui continuera à se degager dans la bouteille, tandis qu'on fait usage des premieres.

On peut faire toute cette operation dans une demi-heure: & il sera bon de la repeter du moins deux fois par jour. 76.07.23

un cirurgien Anglois traite actuellem[t]. par cette methode un cancer terrible, qui empechoit depuis long tems le pauvre malade de reposer ni dormir, à cause des vives douleurs qu'il soufroit. Il a fort peu de jours qu'il a commencé ce traitem[t]., & le cancer est deja reduit à 12 pouces de circonference, au lieu de 16 pouces qu'il avoit auparavant. Les douleurs ont cessé entierem[t]. & le miserable peut dormir fort à son aise. on espere que la continuation le guerira entierement. Mais en cas même que celà n'arrive point, c'est deja avoir beaucoup gagné, que de faire evanouir le cruel tourment des douleurs qu'il soufroit, & de lui concilier le someil. J'ai eu une longue conversation hier avec le cirurgien: qui m'a dit son nom & demeure, & meme promit de venir me voir ce matin. apparem[t]. il fut empeché par quelqu'acident; & quoique Je me suis oublié son nom, Je ne manquerai pas de le trouver, & Je m'informerai de la suite de cette cure. J'ai Lu qu'on avoit aussi appliqué ce Remede del'air fixe à cette terrible maladie des cancers dans quelqu'autre paiz: mais je ne me souviens oû: & n'aiant pas sçu la methode de l'appliquer, J'ai suspendu mon jugement. à present ce fait recent me donne l'occasion de mettre à meme de la practiquer tous ceux qui voudront & auront de ces malades en des circonstances pareilles. hereux si par cette attention Je deviens l'instrument pour soulager L'humanité dans une maladie si terrible & si malhereuse! fait à Londres ce 23 Juillet 1776

J.H. de Magellan

76.07.29[1] *To Jacques-André Mallet*[2]

Bibliothèque de Genève, Geneva, Ms suppl. 1654, fol. 44-45

Detail des depences faites avec La Lunette achromatique & petit Sextant, envoyés à Geneve.

1775

26 Juillet

Je payai à M^{r}. Dollond une Lunette achromatique de 3½ pieds de foyer, à triple Objectif, que Je lui avois ordonnée dans l'intention de l'invoyer à M^{r}. Mallet, Profess^{r}. d'Astronomie à Geneve, qui m'en avoit ordonné une de cette espece, avant plusieurs autres de mes correspondents qui souhaitent en avoir des pareilles. Le grande tuyau de cette Lunette est en cuivre jaune: elle a un tuyau d'oculaires terrestres, & quatre Oculaires celestes; dont il y a deux à croix filaire, pour observer le passage des astres. Mais elle n'avoit pas aucun autre equipage, comm'il paroit par le Reçu originel ci-joint de M^{r}. Dollond, qui est pour la somme de - - - £29:8^{s}:0^{d}

M^{r}. Mallet m'avoit marqué, qu'il vouloit avoir une monture, ou piedestal, simple, fait ici pour cette Lunette; quoiqu'il avoit ordonné une monture parallatique à Paris. En consequence de celà, je fis faire ici un piedestal très simple en mahogany qui couta - - - 4:14:6.

Après avoir examiné la difficulté reele, ou pour mieux dire, l'impossibilité, oû se trouvent depuis long tems nos Opticiens pour faire des Lunettes de cette espece, pour n'avoir plus de bon flint-glass: & que le meme M^{r}. Dollond, auquel j'avois ecrit de Paris pour avoir plusieurs autres lunettes pareilles pour differents amis, ne pouvoit en faire plus de cette espece: j'ecrivis à M^{r}. Mallet, que Je comptois toujours lui envoyer celle-ci, qui probablem^{t}. seroit la derniere à avoir de cette espece: mais qu'il faloit me donner sa parole d'honneur, qu'elle n'appartiendroit point à aucune autre personne, comm'il me l'avoit fait entendre dans une de ses lettres.[3] M^{r}. Mallet ne me repondit pas un mot sur cet article: & je trouvai d'ailleurs par des informations sures, que ce n'etoit pas Lui (M^{r}. Mallet) qui seroit le proprietaire de cette Lunette, mais un autre gentil-homme,[4] par lequel elle devoit etre payée. Voiant donc que mon objet de ne pas ceder cet instrument, que pour L'avantage & service de l'Astronomie, à quelqu'un qui l'employat dans un observatoire regulier, devenoit frustré; Je me suis consideré comme

1 Document dated from date given in the body of the letter.

2 Annotated: "Cpte de Magellan p la lunette de Dessaussure".

3 See Magellan to Mallet, 13 October 1775.

4 i.e. H.B. de Saussure.

degagé de l'arrangement en question, & j'ai gardé cette Lunette entierem[t]. à ma disposition. Par consequence Je lui ai fait faire les equipages & ajustements que J'ai jugés à propos & suivant ma fantaisie; sans cependant me determiner à la refuser à M[r]. Mallet, en cas que Je fus convaincû, qu'elle seroit uniquement employée à l'usage de l'Astronomie:[5] & Je continuai dans cette idée jusqu'au 29 Juillet 1776, ce qui fut Le tems, dans lequel J'ai crû devoir n'envoyer plus aucun instrument à Geneve; aiant vû qu'on n'y savoit pas distinguer proprement entre la bonté essentielle & utilité des Instruments, & leurs agrements apparents. En effet sans l'interposition de M[r]. Alex[e]. Aubert, qui m'a assuré que cette lunette ne seroit employée qu'au service de l'Astronomie, & qui m'a prié de m'oublier du passé, en consideration de notre ancienne amitié &[a]. Je n'aurois point changé dans ma resolution 76.07.29

Il est evident d'après l'exposé que Je viens de faire, que l'equipage & articles suivants, ne furent point ordonnés par M[r]. Mallet; mais que Je les ai fait faire de mon accord; parce que J'etois le maitre d'arranger ma lunette, comme Je voudrois. Je commençai par rejetter le piedestal ordinaire presque tout à fait; car je n'en conservai que les trois jambes de mahogany & j'ai fait faire l'equipage qui suit.

Une planchette double avec des charnieres fortes, qui forme le chapiteau du piedestal, sur le quel il y a une bonne monture de metal, avec le mouvement vertical, & horizontal: les cercles de ces deux mouvements, sont bien divisés; & chacun a un Nonius. Il y a un coin de mahogany avec des vis necessaires, pour etre arreté oû on le voudra, à fin d'elever la planchette superieure, pour que ce plan devienne parallele à celui de l'Equateur, suivant La Latitude de l'endroit. J'ai fait mettre 4 tourillons de metal, avec des tetes à vis, à distances egales du centre de pesanteur de la Lunette, à fin d'y poser son corps en tout sens, & de reconnoitre par là La ligne de Collimation, ou axe optique. Il y a deux manches à jointe universelle, que nous appellons D[r]. Hook's joints, pour les mouvements lents de la Lunette: des pignons doubles, pour la faire mouvoir d'un coté ou de l'autre: & 3 gros vis aux bouts inferieurs des jambes, pour mettre le total de niveau &[a]. Cet ouvrage couta - - - £12:17:6

Deux niveaux à boule d'air, avec des chassis & montures correspondentes, l'un pour niveler le chapiteau du piedestal: & l'autre pour etre employé sur les tourillons de la Lunette, à fin de trouver le zero du cercle vertical, c'est à dire à fin de connoitre à quel degré de le cercle corresponde la ligne de collimation, lorsqu'elle est parallele à l'horizont. Ils couterent, y compris le mauvais acident de deux verres de cassés - - £2:12:

5 "1776" written in left margin.

76.08.06 *To Johann III Bernoulli*

Öffentliche Bibliothek der Universität Basel, MS L I a 709, f. 59-60

Londres ce 6 Aout 1776

Monsieur & très cher ami

C'est avec un très grand plaisir que Je reçois la votre du 15 Juillet[1] avec l'annonce des Livres & brochures que vous m'envoyez par La voye de M^{r}. Deker. Quoique Je ne suis pas trop content de l'exactitude de Son correspond^{t}. de Hambourg: car il envoye apparem^{t}. sans propre mesure les choses qui Sont destinées pour ici. il y a deja une caisse de Livres avec le 17 & le 18 vol. (12 exemplaires de chacun) du Journal de M^{r}. de Castillon[2] qui sont absolum^{t}. perdus, faute d'envoyer ici avis de la voye, vaisseau & nom du Cap^{tn}. par Lequel ce pacquet ou caisse fut envoyée. Je vous prie donc d'en parler là dessus à M^{r}. Deker & à M^{r}. de Castillon; & d'ecrire s'il l'en faut là dessus à son Correspond^{t}. à Hambourg. Vous sentez bien combien il est desagreable & meme prejudiciable de faire pareilles ommissions & par consequent des pertes pareilles.

Lorsque votre envoye arrivera ici entre mes mains soyez assuré que Je ne manquerai pas de faire & executer de mon mieux tout ce que vous desirez. Pour ce qui regarde Le Manuscrit Je me flate que Vous y donnez des ordres positives sur sa destination mais si celà est quelque chose astronomique il seroit bon que vous ecriviez vous meme Là dessus à M^{r}. Maskelyne, qui est un fort mauvais sujet, & sans sa faveur on ne fait rien dans La Branche Astronomique dans La Societé Royale. J'ai eu avec lui une demelée sur mon invention de la construction nouvelle des octants & Sextants, qui ne sera jamais racom̃odée; parceque Je l'ai mis devant les yeux du Public, & J'y ai rendu compte assez intelligible de la mauvaise humeur & hônetetè de le Monsieur.[3]

---

1 Letter not found.

2 Presumably the *Journal littéraire* edited by Castillon and others and published in Berlin, commencing in 1772.

3 Cf. Magellan's *Description des octants et sextants anglois* (London, 1775). In an end-note, Magellan describes how the first instrument built according to his design was "executed perfectly well" by Ramsden, and how Maskelyne had seen it at Ramsden's in Magellan's hands "& peut-être l'avoit-il vu plusieurs autres fois, tandis qu'on y travailloit". "Cette seule circonstance", he went oñ, "suffit pour montrer la mauvaise foi des objections faites contre la construction pratique de cet instrument, tandis que l'Auteur étoit absent de Londres."

Mr. Speiner[4] m'a remis ici une de vos listes[5] pour un Astronome de Cadix que J'ai envoyé deja.[6] J'ai eu occasion de lire une de ces Listes: & en effet je crois que Vous avez fait quelques meprises & avez manqué plusieurs noms qui meritent mieux d'y etre que d'autres reelemt. C'est un objet trop vast que parler de l'etat actuel d'une science & de ceux qui s'y employent. il faudroit voyager beaucoup, & pour ainsi dire dans le meme tems par toutes les parts de l'Europe. par example Je ne vois pas pourquoi vous y metez le Dr. Solander qui n'est plus Astronome que son compagnon de voyage Mr. Banks: & tous les deux Je ne sais pas qu'ils s'adonnent à l'Astronomie mais bien à la botanique &a.[7] Je crains je le dis que cette entreprise n'est pas tout aisée à bien executer. Si je pourrai vous y rendre service, [je ne][8] manquerai pas certainement. 76.08.06

Le Dr. Sheperd[9] ne travailla pas beauc[oup de][10] quelque tems. Je n'ai pas encore vu l'observatoire de Richmond:[11] m[ais j'ai][12] eté tout exprès à Oxford pour voir celui qui s'y batit.[13] Pour [...][14] regarde celui de Mr. Aubert Je le crois le plus parfait & le [...]garai[15] de l'Europe, quoique le batimt. ne soit pas grand [ou les][16] Instrumts. du plus grand raion.[17] C'est un très digne homme que [...][18] Aubert. Monsr. de Luc travaille actuellemt. à perfectioner [... Hy]grometre: & a arreté ceux qui se

4 Johann Karl Philipp Spener (1749-1827); cf. 73.12.30.
5 Bernoulli, *Liste des astronomes connus, actuellement vivans* (Berlin, 1776).
6 In his *Liste*, Bernoulli lists five astronomers at Cadiz, as follows: "Vinc. Doz; Tofinno; Varillas; Velasque; d'Ulloa". There is a note concerning "Tofinno" (i. e. Vicente Tofiño de San Miguel y Wanderiales), saying that he might be somewhere else in Spain.
7 Daniel Carl Solander (1736-1782) and Joseph Banks had travelled as naturalists on James Cook's first voyage around the world. Both were primarily botanists, as Magellan states. Solander did, however, while at Tahiti make observations of the transit of Venus (*Philosophical Transactions*, 61 [1771], 412).
8 Conjectural reading; paper damaged.
9 Anthony Shepherd, Plumian professor of astronomy at Cambridge.
10 Conjectural reading
11 King George III maintained a private observatory at Richmond (later called the Kew Observatory), where Stephen Demainbray (1710-1782) was employed as astronomer.
12 Conjectural reading.
13 The Radcliffe Observatory at Oxford commenced construction in 1772 under Thomas Hornsby (1733-1810), Savilian professor of astronomy and the first Radcliffe observer.
14 Paper damaged.
15 Paper damaged.
16 Conjectural reading
17 Alexander Aubert (1730-1805) had an extremely well-equipped observatory at Loampit Hill, near Deptford.
18 Paper damaged.

76.08.06 faisoient suivant sa pre[mier cons]truction, qui certès n'etoient pas encore bien parfaits.[19]

Je suis etonné que les pendules ne soient pas encore arrivées chez vous. elles furent adressées d'ici à M^{r}. Caspar, Keuffmann à Hambourg[21] par la voye de M^{r}. Frederic Zambruck de Londres[21] dans le vaisseau Lady Christine Cap^{tn}. Heinrick Nickes: & cette caisse devoit etre adressée à Messrs. Haude & Spener à Berlin. Je vous prie de tacher d'avoir des nouvelles de celà & de vous adresser par lettre un cas de defaut au dit M^{r}. Keuffmann à Hamburg.[22]

Voilà le tems de fermer à la poste

Vive, vale & ama
tuum
Magellan[23]

A Monsieur
Monsr. Bernoulli Astronome R^{l}.
& de l'Acad^{e}. R^{le}. des Sciences de Prusse
Berlin

19 The Swiss-born Jean André De Luc (1727-1817) took up residence in England in 1773, and in that year gave an account to the Royal Society of a new hygrometer he had developed (*Philosophical Transactions*, 63 [1773], 404-60). As Magellan reports, De Luc soon realised the deficiencies of the instrument and abandoned it for one of an entirely different design (see his *Idées sur la météorologie* [London, 1786], I, 38).

20 Possibly Caspar Voght (1707-1781), Hamburg merchant and later senator. Magellan would appear not to have known the word "Kaufmann" or that it means "merchant" in German.

21 See 73.12.30, note 4.

22 Magellan eventually managed to track down the missing items himself; see 76.10.12a.

23 Enjoy life, farewell & love | your | Magellan.

To Joannes Franciscus Thijsbaert

University Library, Université de Liège, Ms. 2617

Londres ce 9 Aout 1776

Monsieur & très cher ami

Si vous etes content de vos orreries,[1] Je le suis aussi bien, de ce que vous le soyez: car c'est mon plus grand plaisir, que de rendre service à qui s'entend & en connoit La valeur. Vous n'avez rien à faire avec Adams pour le payement de ces instrum[ts].; c'est moi qui les ai payés, com̄e toujours, avant de les envoyer.

M[r]. Needham se trouve ici, & partira pour Bruxelles vers le 20me. de ce mois. il a vu quelque experience que Je lui ai montré du D[r]. Priestley, & il a eté si frapé de l'effet, qu'il me pria de lui procurer un appareil complet pour les Repeter. Ce seroit un objet interessant pour vos Cours de Physique. Cet apparatus ne va pas au delà de 2 ou tout au plus 3 guinées: or Je compte aller vers la fin du mois prochain en Flandres & par consequence Je pourrai vous faire cet exercise des differents especes d'air. si vous voudrez J'envoyerai d'avance cet appareil.

on vient de traduire en françois le 2de. vol. du D[r]. Priestley:[2] & on l'imprime par souscription. parlez à vos amis, & voyez si vous faites quelque nombre de Souscripteurs. ce sera en 2 Vol. en 12mo. prix 6[lb]. de France. Le Traducteur, qui est le D[r]. Gibelin, mon ami, vient de m'ecrire cette meme poste là dessùs.[3]

M[r]. Needham apportera pour la Biblioth[e]. de Votre Université, un exemplaire de l'ouvrage posthume du celebre Mathem[n]. Robt. Simson, relié en marroquin.[4] C'est un present que J'obtins du Comte Stanhope qui fit à ses depens cette belle edition, & ordonna de relier tout pour l'envoyer de present à differents endroits, sans qu'aucun exemplaire fut vendu; ce qui rend ce present beaucoup plus estimable. lorsqu'il arrivera chez vous il faudra m'ecrire une Lettre de remercim[t]. que Je puisse montrer au dit Comte Stanhope, pour sa satisfaction &[a].

---

1 See Magellan to Thijsbaert, 26 January 1776.

2 Joseph Priestley, *Expériences et observations sur différentes espèces d'air*, Tome II (Paris, 1776).

3 Letter not found.

4 Robert Simson, *Opera quaedam reliqua*... (Glasgow, 1776).

76.08.09 Je vous envoye par la meme voye le Memoire de M^{r}. Delaval, sur les preparations des couleurs fluides, telle qu'il se trouve imprimé dans le dernier vol. de l'Acad^{e}. de Berlin.[5] il sera fort interessant de faire ces preparations, & d'en montrer l'effet dans votre Cours de Physique

mes Resp^{s}. à M^{r}. le Prevot[6] Chancelier[7] & Messieurs qui m'honorent de leur souvenir

Vive, Vale & ama
tuum
Magellan[8]

A Monsieur
Monsr. L'Abbe Thisbaert
Proffess^{r}. de Philos^{e}. & Principal
du College Royal
à Louvain
Flandres

---

5 Edward Delaval, "Recherches expérimentales sur la cause des changements de couleur dans les corps opaques naturellement colorés", *Nouveaux mémoires de l'Académie Royale des Sciences et Belles-Lettres [de Berlin]*, 1774, Mém. pp. 154-94.

6 The Abbé Jean-François de Marci, provost of St Pierre in Louvain from 1772.

7 Joseph de Crumpipen (1737-1809), Chancellor of Brabant and first President of the Brussels Academy of Sciences.

8 Enjoy life, farewell & love | your | Magellan.

76.08.13 *To Horace Bénédict de Saussure*

Bibliothèque de Genève, Geneva, Archives de Saussure, 10, f. 181

Londres ce 13 Aout 1776

Monsieur

Je vous assure sur mon honneur, que je ne suis pas faché contre vous, parceque vous differez de mes opinions, soit d'après votre propre jugement, soit d'après celui de quelques autres. Mais tandis que celà subsiste, je ne puis, sans manquer à moi-meme, rien changer dans les resolutions, que je vous ai annoncées par ma lettre precedente.[1] J'ai l'honneur d'etre avec bien de la consideration 76.08.13

Monsieur
Votre tres h^{ble}. & tres obeiss^{t}. serv^{r}.
Magellan.

A Monsieur
Monsr. de Saussure, Professeur en Philosophie
à Geneve
en Suisse

1 Letter not found; but see Magellan to Mallet, 29 July 1776.

From Jacques-André Mallet 76.09.05

Bibliothèque de Genève, Geneva, Ms suppl. 1654, fol. 58[1]

A M^{r}. Magellan à Londres

Je suis b. fac. M. C. M.[2] de n'avoir pas pu rep. plut. à la lettre que vous m'av. fait le plais. de m'écr. il y a qlques mois.[3] J'etois absent alors, & j'appris ensuite que Mr Dessaussure[4] a qui vous aviez écrit par le même courier [vous] avoit répondu relativem^{t} au piedestal de la lunette, qui étoit le seul article de votre lettre qui exigeoit de moi une réponse.[5] Tout cela etant arrangé, rien ne devoit retarder davantage le depart de cette lunett. que nous attendons depuis si longt^{s}. c'est moi qui en soufre, elle est destinée à mon observatoire. j'ai le désir & l'espérance de la posseder depuis le commencem^{t} de 1774 qu'elle fut commandée à Dollond. mais surtout depuis votre lettre du mois de 7bre de l'ann. derniere[6] oû vous me marquiez que vous aviez chez vous la lunette finie depuis le mois de Juillet.

1 Letter not found. The text is from a draft kept by Mallet.
2 Mon Cher Monsieur.
3 Letter not found.
4 H. B. de Saussure.
5 Letters not found. Mallet's comment makes it highly unlikely that he is referring here to Magellan's letter to him of 29 July 1776.
6 Magellan to Mallet, 1 September 1775.

76.09.05 si j'eusse prévu que ce piedestal eut pu etre la cause d'un si long retard, je vous aurois bien prie de l'expedier sans cela. Enfin actuellem[t] qu'il est fini, que vous avez l'addr. du negocian a qui le paquet doit etre remis, & par qui il doit vous etre payé, J'espere M. C. M que rien ne vous empêchera de nous expedier le plutot possible, si cela n'est deja fait, cette excellente lunette, etant bien persuadé par la connoiss. que j'ai de votre exactitude ordinaire, que tous ces retards n'ont pas dépendu de vous.[7]

Je suis de tout mon coeur avec un sincere attachem[t]

M. C. M.
Votre tres affect. Serv.
Mall.

5e 7bre. 1776

7 Is this letter an indirect response to Magellan to Mallet, 29 July 1776?

76.09.10 *To Antoine-Laurent Lavoisier*

Archives de l'Académie des Sciences, Paris, Fond Lavoisier, 1642[1]

Londres ce 10 Sep[bre]. 1776

Monsieur & très cher confrere

Votre excellente Pompe Pneumatique est deja parti d'ici pour Calais, & j'ai deja fait avis à M[r]. Caffieri Directeur de la poste dans le meme port,[2] pour vous l'envoyer, adressée à M[r]. Trudaine, en vous chargeant ses deboursements. J'ai eu cette pompe chez moi dès le 11 Juillet jusqu'au 20 Aout, pour voir si la secheresse des valves avoit lieu. Je l'ai essayée ce jour en presence du comte [Mamelli],[3] que peut etre vous connoissez. il

1 Corrected version of the text published in *Oeuvres de Lavoisier: Correspondance*, ed. René Fric, fascicule III, 1776-1783 (Paris, 1964), pp. 568-72.
2 Charles Augustin Caffieri (c.1739-?).
3 Not identified.

est grand ami de l'Abbé Fontana, & fort habile: & nous avons jugé que 76.09.10
l'exhaustion se faisoit jusqu'à la douzieme mille fois (12,000). Car apres avoir fait l'exhaustion, nous avons plongé dans le vase de mercure (au dedans du recipient) la gauge nm de façon que l'embouchure m etoit au dessous du mercure x. alors en laissant entrer l'air, le mercure a monté presque jusqu'à v, laissant seulement l'espace vn rempli d'air, qui n'etoit au delà de la $^{1}/_{12000}$ partie de la capacité totale de la gauge n.m.

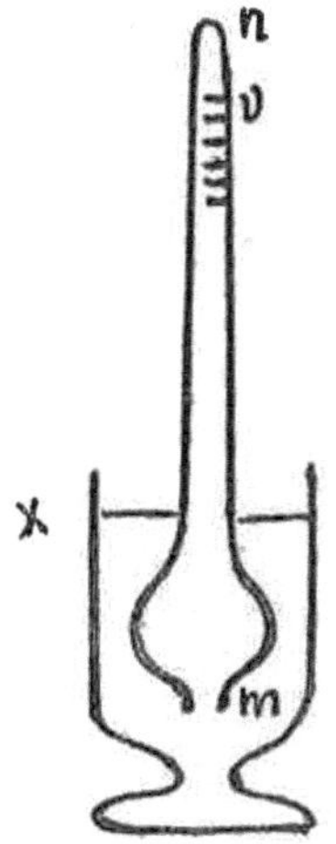

J'observai neanmoins que le bassin du grand robinet ne le renfermoit pas entierement, ainsi J'ai fait faire cette correction, & cette operation fit le retard que cette machine a eu jusqu'au 2 du courant mois, que Je l'ai essayé encore de nouveau chez l'artiste: & je l'ai trouvé egalem^{t}. bonne. Vous verrez la gauge que J'ai ordonné de laisser avec le mercure dans le petit tuyau: & J'espere qu'il arrivera encore à Paris dans la meme position, qui est à peu près à la 10,000 partie. Lorsque vous voudrez faire un très bon vuide, il faut n'employer pas du cuir; mais mettre seulem^{t}. le recipient sur le grand plateau, sans aucune chose plus que de l'huile à l'entour. il faut le couvrir (ce recipient) avec la plaque de metal, au milieu de laquele passe la verge, qui tient à son bout le tuyau de la gauge; en y mettant de l'huile au tour du recipient cylindrique, dans l'endroit oû il touche la plaque ronde. il est bon d'oindre les endroits oû touchent les bouts couverts de cuir qui sont au dedans de la vis de la gauge torricellienne, & dans tous les autres endroits, oû il peut avoir quelque soupçon ou crainte de l'air. Le meilleur de cette construction, est que toutes les pieces qui composent la pompe, peuvent se degager les unes des autres, par des vis. de façon qu'on peut la demonter, examiner toutes les valves, le piston &^{a} y remettre du cuir nouveau dans le piston; remplacer celui des valves, &^{a}. mais quant à ce dernier des valves, au fond du cylindre, ou corps de la pompe, l'artiste a crû qu'il valoit mieux y mettre de la soye vernissée: & vous en trouverez un morceau de cette soye dans la caisse n°. 2. il faut le mettre toujours avec la surface la plus unie vers l'endroit oû il touche le metal. Cependant Je crois que le cuir de gant de femme est du moins aussi bon que la soye cirée. J'ai fait faire au meme Artiste une relation ou Directoire, que Je vous envoye cette meme poste sous l'adresse de M^{r}. Trudaine, pour ne pas doubler ou tripler le postage de cette lettre. il faudra que vous l'etudiez attentivem^{t}. avant de demonter la pompe.

76.09.10 Le corps de la pompe entiere est renfermé dans la caisse marquée n°. 2. il faut en defaire les vis qui la renferment du coté d'en haut, & dans la suite les vis d'une des surfaces laterales, sans faire usage de marteau.

La caisse marquée n°. 1 contient deux ou 3 autres caisses plus petites, qui renferment le recipient cylindrique, pour l'exhaustion la plus forte: & le recipient tres epais pour condenser, avec les vis de bois pour l'y arreter & rafermir au corps de la pompe. vous y trouverez 2 clefs pour visser & deviser les parties de la pompe. la plus petite ne sert que pour la gauge de la condensation. il y a aussi dedans, la manche du pignon qui fait monter & descendre le piston. Cet axe fait un tour & demi: il est très aisé à travailler: un enfant peut faire aller & venir ce piston avec la manche. il y a aussi la piece de soye cirée, et quelque cuir. 2 tourne-vis. la gauge Torricellienne. le goblet de bois, oû l'on met du mercure, pour le faire monter dans le tuyau de cette gauge. Il y a un petit colier d'yvoire dans cette forme qu'on met au bout inferieur

de cette gauge, & qui surnage dans le mercure pour bien determiner le zero de la mesure de cette meme gauge. enfin il y a 2 petites gauges, deja divisées & marquées depuis 500 jusqu'à 4,000.

Il y a dans la meme caisse un paquet de bas de soye, dont un pair blanc est pour vous, dans l'idée que la nouveauté vous fera plaisir. le tissu de cette espece de bas est tout à fait singulier: car si l'on y fait un trou, la maille ne se defait pas, comme dans les bas ordinaires. J'en ai vû, oû l'on avoit fait exprès un trou avec un couteau, & ils avoient deja eté lavés 3 ou 4 fois, sans que ce trou devint plus grand, ni qu'aucune maille se defit. on pretend que ces bas durent pour 3 ou 4 des ordinaires, ainsi les paye-t-on presque le double, savoir 18 shilling le pair. il y a un autre pair noir pour M^{r}. le Chev^{r}. de Bory avec un 2^{d}. dedans, que Je vous prie de lui donner: & enfin un autre pour M^{r}. Du Chenne[4] secretaire de M^{r}. Trudaine.

Vous enfin trouverez dans la meme caisse deux sextants, chacun dans sa caisse ou boite de Mahogany. ils sont pour M^{r}. de Briqueville, Cap^{tn}. de Vaisseau du Roy à Brest[5] auquel j'ecris cette meme poste pour les faire chercher chez vous, en vous faisant remettre la valeur que Je vous

4 Louis-Henri Duchesne.
5 See Magellan to Johann III Bernoulli, 26 October 1772.

prie de garder pour mon compte & de me l'envoyer avec l'argent de cette pompe par la voye de Mr. [Herries],[6] ou comme vous voudrez. l'argent que Mr. de Briqueville doit vous mander pour mon compte est 681l. 12s. tournois. 76.09.10

J'espère que vous escuserez ce trouble, en consideration que ce sont des objets de service pour la marine de France, & pour vos savants. vous verrez que J'ai fait un rabaissement dans votre compte pour les frais respectifs du porterage & du poids de ces sextants. C'est a dire d'autant de moins que J'ai jugé qu'ils peuvent avoir augmenté les frais de cette caisse jusqu'à Paris.

Outre les deux caisses ci dessus, il y a une troisieme, qui contient differents articles, particulieremt. les livres qu'on envoye de present pour l'Acade. Royale, & pour plusieurs de ses membres &a. J'ai profité de cette occasion pour l'envoyer sous la meme adresse: sachant bien que Mr. Trudaine est tout porté à favoriser les savants de l'Acade. &a. Mais la distribution des articles de cette caisse vous sera je crois enuyeuse: ainsi Je vous prie d'en donner la liste à Mr. Messier qui y a aussi quelque chose pour lui; ou autremt. envoyez la à Mr. Ruault[7] ou à quelqu'autre libraire, ou personne capable d'en faire la distribution, & d'en recevoir la quote part des depenses & frais de cette caisse, comme vous verrez detaillés dans la liste que Je vous envoie sous l'adresse de Mr. Trudaine ensemble avec la votre. Car les articles contenus dans cette 3me. caisse vous doivent paier la 3me. partie des frais depuis Calais jusqu'à Paris, & en outre ceux que J'ai faits ici pour elle jusqu'à bord du vaisseau, qui montent à 13s. 6. vous devez recevoir ces deux parcelles pour votre compte, car elles vous appartiennent, cette derniere etant mise dans votre compte & l'áutre etant payé par vous à Mr. Caffieri, qui doit vous charger pour toutes les 3 caisses.

J'ai fait passer à Mr. Henry[8] l'article de votre lettre derniere. Son adresse est = a Mr. Henry Apothicaire à Manchester. Angleterre = vous pouvez lui ecrire par la poste. N.B. Mr. Caffieri de Calais doit vous faire avis du depart & arrivée des 3 caisses à Paris, & doit vous charger pour ses deboursemts.: & vous aurez soin de lui faire avoir son acquit de Mr. Trudaine pour les 3 caisses &a.

---

6 There were at least two London merchants named Herries at this time, Herries & Co. of 4 St Mary Axe and Michael Herries of Philpot Lane.

7 Nicolas Ruault, publisher and bookseller in Paris.

8 i.e. Thomas Henry.

76.09.10 vous ne me dites pas pour quelle rue vous avez demenagé.

Mes h^{bles}. $resp^{s}$. à M^{de}. Lavoisiere, M^{le}. votre Tante, & à $Mess^{rs}$. Paulze Pere & fils & Mlle. mes $Compl^{s}$. à tous nos amis co͞muns qui se souviennent de moi: & croyez moi pour La Vie

Mon cher $Mons^{r}$.
Votre très h^{ble}. $serv^{r}$.
Magellan.

P.S. Je compte aller faire un tour en Hollande & Flandres mais Je serai de retour avant la fin d'octobre.

A Monsieur
$Mons^{r}$. De la Voisiere, Fermier gen^{al}.
& Membre de l'$Acad^{e}$. R^{le}. des sciences
Rue des Bons Enfants
à Paris

76.09.12 *To John Walsh*[1]

Smithsonian Institution Libraries, Washington, MSS 945A

M^{r}. Magellan presente ses h^{bles}. $resp^{s}$. à M^{r}. Walsh: & s'il a quelque chose de son Service en Hollande ou en Flandres M^{r}. Mag^{n}. y va faire un tour de 3 ou 4 Semaines, dans peu de jours, & ne manquera pas d'etre charmé d'y pouvoir rendre quelque Service à M^{r}. Walsh.

Le Bill ci joint est celui de la Valeur des Journaux reçus par M^{r}. Walsh, qu'il aura la bonté de faire payer à son plaisir.

Ce 12 Sep^{tbre}. 1776: at n°. 12 Nevils Court Fetter Lane London.

To
John Walsh Esqr.
Chesterfield Street.

1 John Walsh, FRS, MP (1726-1795) was a cousin of the Astronomer Royal, Neville Maskelyne. Letter not found.

To Johann III Bernoulli 76.09.13

Öffentliche Bibliothek der Universität Basel, MS L I a 709, f. 61-62

Londres ce 13 Sep^{bre}. 1776

Monsieur & très cher ami

J'ai Reçu les livres que vous m'avez annoncés dans votre lettre du 15 Juillet.[1] j'ai envoyé les trois presents pour Messrs. Maskelyne, Shepherd & Soc^{e}. R^{le}., & vous remercie de celui que vous me faits avec toute reconnoissance. J'ai lû attentivem^{t}. La lettre destinée pour La Societé Royale avec le livre manuscrit: & comme je suis vrai ami, je me suis obligé d'y faire deux objections, parceque je connois ce qui doit ecriver selon les circonstances, & je ne voudrai jamais soufrir que mes amis s'exposent à quelque desagrement. il y faut, mon cher ami, supprimer absolum^{t}. le premier & le dernier article. le premier, parceque tout Candidat ne s'y presente jamais, que par un certificat signé de 3 membres etrangers & autant par des membres domestiques.[2] Votre merite est fort connû, & Je ne doute pas que si vous m'envoyez un Certificat signé par 3 membres etrangers de la Soc^{e}. dont je crois qu'il y a à Berlin plus en nombre, ou du moins vous pouvez le faire signer à Paris par vos amis: alors dis-je, si vous me l'envoyez ainsi signé, J'aurai soin de le faire signer ici par 3 ou plus de mes amis, & vous serez admis Candidat dans les formes, sans les quels aucune personne n'est jamais elue dans La Societé. Voici la forme de ce certificat. Notez qu'il faut mettre le nom de bateau & de famille à la place N. & ses qualités de Professeur, ou Astronome Royal &c^{a}. après.

N. of Berlin &^{a}. a gentleman well versed in N. (par example in Astronomical & Mathematical Knowledge &^{a}.) as it appears by his publications, particularly by his Treatise in N. N. &^{c}. &^{c}. &^{c}..... being desirous of becoming a Fellow of y^{e}. Royal Society of London, we whose names are

1 Letter not found.

2 Bernoulli had evidently expressed a desire to become a member of the Royal Society. He was eventually nominated, in 1787, in accordance with the procedures that Magellan here describes, and Magellan was one of his signatories. By that time, however, the Society had further tightened its procedures for electing foreign members and imposed a strict limit on the total number of foreigners who could be Fellows at any one time. At a ballot held on 3 April 1778, 23 foreign candidates were put forward but only 13 were elected; Bernoulli was not one of them.

76.09.13 underwriten, do recommend him, on our personal knowledge, as highly deserving that honour, and Likely to become an usefull member.

N. ....
N. ....
N. ....

Pour ce qui regarde le dernier article de votre lettre, il ne peut pas avoir le moindre succès dans La Societé R^{le}. & outre celà ce ne serviroit que de vous exposer. Ce corps est gouverné par une commitée ou conseil de 20; dont la moitié est renouvellée chaque année. Ce corps ne donne jamais rien d'encouragement lucratif: tout au plus ce qu'on pourroit esperer, ce seroit qu'il le feroit imprimer. Mais j'en doute beaucoup, parce que cet ouvrage ne regardant que les Calculateurs des Tables Astronomiques, la Societé n'a rien à faire avec celà. Si vous concevez que ces tables peuvent etre d'un grand Secours immediat pour la Navigation: alors je vous conseillerois de vous adresser au Board de la longitude. Peut etre même que cela etant pour les calculateurs des ephemerides, ces Tables y seroient bien reçues & qu'il y auroit une Recompense à esperer. Car c'est le seul corps qui donne des Recompenses sur cet objet.[3] Mais pour celà il faudroit avoir gagné l'amitié de M^{r}. Maskelyne & du D^{r}. Shepherd[4] & de M^{r}. Hornsby[5] Astronome de Oxford, qui sont tous trois Membres de ce board of Longitude. Je sçais que Maskelyne n'est pas de mes amis; & il seroit assez de soupçoner que je m'y interessois, pour y faire toute l'opposition. dans un cas pareil de vouloir s'adresser au Board de la Longitude, il vous faudroit ecrire en droiture à Maskelyne &^{a} &^{a}. pour tacher de l'y interesser &^{a}. en fin il faudroit arranger cet affaire plus de soin & avec grandes mesures &^{a}. mais pour La Societé Royale, il faut absolum^{t}. ne pas penser aux 2 articles dont je viens de parler: & y renoncer entierem^{t}. tel est mon opinion, & mon avis, en vrai ami.

Je vous prie de dire à M^{r}. de Castillon que le Vol. 16 s'es[t] trouve double: & qu'il faut m'envoyer la douzaine de copies du Vol. 18 du Journal au plutôt.[6] Peut-etre pourroit on envoyer encore une demi-douzaine de

3 Bernoulli had apparently sought a premium from the Royal Society for the tables he had developed which, following Magellan's advice, he subsequently submitted to the Board of Longitude and for which he did indeed receive a gratuity (see 78.01.30). The tables were eventually published by the Board as *A Sexcentenary Table, exhibiting at sight the result of any proportion where the terms do not exceed 600 seconds or 10 minutes, with precepts and examples* (London, 1779).

4 Antony Shepherd (1721-1796), Plumian professor of astronomy at Cambridge.

5 Thomas Hornsby (1733-1810), Savilian professor of astronomy at Oxford.

6 Probably Magellan is here referring to the *Journal littéraire* edited by Castillon and others and published in Berlin from 1772. See also 76.08.06.

chacun des autres vol. Car le libraire me dit qu'il lui restent peu de copies completes. Mais il faut recomender à ce drole correspondent de Hamburg, qui envoye ce qu'il doit envoyer, en droiture à moi meme en me donnant avis par la poste dans la forme acoutumée: car cette caisse fut adressée à un tel fou qui se trouve ici, & qui a voulu la retenir pendant plusieurs jours (plus de 8 ou 10 jours, selon que je fus informé) seulem^{t}. parceque ce compere Hamburgeois y avoit mis une demi douzaine de feuilles pour ce Monsieur qui est un vrai insolent. il faut parler decidément là dessus à M^{r}. Deker, car ce n'est pas celle-ci la maniere d'agir avec personne, ni de faire des affaires. Je vous prie de recommender ce point à M^{r}. de Castillon: & celà m'epargnera le trouble & la depense du postage d'une autre lettre. 76.09.13

J'ai l'honneur d'etre avec le plus vrai attachem^{t}. de Coeur

Mon cher ami
Votre
Magellan

A Monsieur
Monsr. Bernoulli, Astronome
Royal, & de l'Acad^{e}. R^{le}. des sciences
à Berlin

To Ribeiro Sanches[1] 76.09.13a

Österreichische Nationalbibliothek, Vienna
A.N. Ribeiro Sanches Korrespondenz, codex 12714, ff. 181-182v

Londres 13 Sep^{bro}. 1776

Am^{o}. e S^{r}. e m^{to}. do Cor^{am}.

Ha mais de hum Seculo ao q me parece q não tenho n^{as}. directas de V.M. emq^{to}. o D^{or}. Le Begue ou o meo am^{o}. Messier me dão algũas, vou passando sem tão dezagradavel sensação; mas q^{do}. estas faltão, não posso

1 Annotated "recebida 4^{a} fr^{a} 18 7^{bro}" [received on Wednesday 18 September] and "Resp^{a}. a 19 de 7^{bro} para Bruxelles" [answered 19 September to Brussels]. Letter not found.

2 Provinces in central Portugal, from where both Magellan and Sanches came.

76.09.13a descançar. aqui soube q V.M. estivera molestado, q se achava melhor, q tornava a estar amorrinhado & doentiço q são das nossas expressoens da Beira alta & baixa &^{a}.

ha ja alguns dias q mandei uma caixa p^{a}. Paris a M^{r}. de Lavoisiere Fermier general & da Acad^{a}. das sciencias, onde vai um paquete de Livros p^{a}. V.M. & outro p^{a}. o D^{or}. Le Begue.

[l]á avizei a este ultimo do conteûdo no seo: agora direi os q se contem no de V.M. 1°. Gibbon's Hist^{y}. of y^{e}. Fall of y^{e}. Roman Empire 4°. £1:1:0/ 2. Phil^{al}. Transact. vol. 66 p. 1. = 7^{s}. 6^{d}./ trez exemplares de Musgrave on Nerfs = 7.^{s} 6^{d}./ 4. Med^{al}. Inquiries [v]ol. y^{e}. 5th. = 6^{s}. 0^{d}/ 5. Percival's Essays vol. y^{e}. 3th. = 5^{s}. 0^{d}./ 6. Med.^{al} Comentaries n°. 12 = 1^{s}. 6^{d}./ 7. Rozen on Children. = 5^{s}. 0^{d}.

Como os am^{os}. Du Presle & Messier tem livros no mesmo caix[ão] elles terão cuid^{o}. na parcella de V.M.: e talvez q M^{r}. de Lavoisiere mand[e] tudo a M^{r}. Messier p^{a}. fazer a distribuição, & carregar os gastos proporcionalm^{te}. &^{a}.

vejo todas as semanas o nosso Ministro, e Sua amavel Esposa, q vem de lhe dar agora a 3^{a}. criança, h[ũa] menina. Alli se acha actualm^{te}. o senhor de Murça q vem destinado p^{a}. Ministro em algũa destas Cortes do Norte. jantei lá com elles 4^{a}. fr^{a}. passada, e me pareçe ter m^{to}. bom proposito & capacid^{e}. ontem jantei com o Embaix^{or}. do Imperio, onde achei o general Pelgrini: & là lhe ajustei uma partida , p^{a}. elle e p^{a}. o S^{r}. de Murça e o nosso Ministro hirem ver em particular o Museo Britanico. se não me achasse ocupado com outros objetos menos ociozos, não me ficaria hũ instante q estes Senhores Estranjeiros me não ocupassem. Com tudo faço tenção se estiver em Londres 3^{a}. fr^{a}., hir assistir lhes na vizita do Museo. Mas cuido q no mesmo dia á noite partirei com o meo bom am^{o}. Arbuthnot p^{a}. Hollanda, e com os dois filhos mais velhos q lá vai deixar n'uma Acad^{a}. p^{a}. aprenderem o Hollandez & Fran[cez] Lingoas uteis a ambos, um dos quais ja está alistado no serviço de Marinha, & o outro he destinado p^{a}. o Commercio da India Orien[tal].

o amigo Sequeira q passa bem & frequenta o nosso Ministro de q^{m}. he Medico da familia, me pedio presentasse [a V.M.] os seos affetos, e q lhe mandasse algũa ideia do methodo e aplicação da tintura de Cantaridas. se V.M. a achou em algum livro, tenha a bond^{e}. de dar o nome, e bastará p^{a}. não lhe dar mais trabalho.

Eu faço tenção de vir com Arbuthnot por Bruxellas, onde verei o nosso melancolico Chevalier. M^{r}. Needham aqui esteve coiza de hum mez este verão: mas já tornou p^{a}. Bruxellas. está gordo & sempre o mesmo no Resto do seo porte.

Aqui soube q tirárão da Lista o pobre Barros, da lista digo dos Correspond^{es}. da Academia Real. verd^{a}.m^{te}. elle se faz morto não só p^{a}. as academias, mas p^{a}. os amigos: e isto depois de m^{tos}. annos. assim vai a vida humana ...! 76.09.13a

eu passo bem de saude. occupo-me com as minhas maquinas, Relogios, instrumentos &^{a}. &^{a}. Mas cá suspeitei depois de 8 ou 15 dias q as pragas de V.M. começavão a cumprirse; pois n'uma vizita da minha Nely, ella me disse: ora não te canses hoje, eu virei á manham: e o peor foi q no dia seg^{te}. não foi melhor servida q no precedente. fiquei picado do máo sucesso, & me lembrei das suas profecias praguentas & agouradas, q parece ameação o cumprir-se. Com tudo estou esperãdo pelo inverno, q faz enrigecer as fibras. Porem eu cada vez engordo mais apezar do meo dezejo. Não sei q lhe faça. como, bebo, & durmo bem: & não tenho cuid^{os}. nem afliçoens q me consumão.

goze V.M. saude & Repouso igual ao menos, como lhe dezejo. dê m^{as}. Saud^{es}. á fam^{a}. adoptiva, e aos mais amigos se hé q os ha.

Por agora não ha novid^{es}. Literarias. este tempo das Fer[ias] he o tempo morto da Inglaterra. eu serei de Retorno à Londres an[tes] do fim de 8^{bro}.: Espero ter not^{as}. de V.M. em Bruxellas.

vale

a Monsieur
Monsr. Sanchez D^{or}. en Medecine
Rue des Blancs-manteaux
à Paris

London 13 September 1776

My Dear Friend

It's more than a century, it seems to me, without direct news from you. As long as D^{r}. Le Begue or my friend Messier gives me some, I carry on without such an unpleasant feeling; but when these miss on news, I can't relax. I knew that you'd been sick, that then you were better, then again you were weakened & slightly ill, to use our expressions from Beira alta & baixa[2] &^{c}.

Some days ago I shipped a box to Paris to M^{r}. de Lavoisiere, Fermier general & of the Academy of Sciences, in which there is a packet of books for you & another for D^{r}. Le Begue.

3 Letter not found.

76.09.13a I gave notice to him of the content of his:[3] now I tell you what are in yours. 1° Gibbon's Hist^{y}. of y^{e}. Fall of y^{e}. Roman Empire[4] 4°. £1:1:0 / 2. Phil^{al}. Transact. vol. 66. p. 1.[5] = 7.^{s}.6^{d}./ three copies of Musgrave on Nerfs[6] = 7^{s}.6^{d}./ 4. Medical Inquiries[7] [v]ol. y^{e}. 5th. = 6^{s}.0^{d}./ 5. Percival's Essays[8] vol. y^{e}. 3rd. = 5^{s}.0^{d}. / 6. Med^{al}. Commentaries[9] no.12 = 1^{s}.6^{d}. / 7. Rozen on children[10] = 5^{s}.6^{d}.

As our friends Du Presle & Messier have books in the same box, they will take care of your parcel: and maybe that Mr. de Lavoisiere will remit everything to M^{r}. Messier to do the distribution & to assign the expenses proportionally &^{c}.

I see our Minister[11] every week, and his kind wife who has just given him a 3rd child, a little girl. At his house now is the Earl of Murça[12] who is on his way to some of these northern Courts. Last Wednesday I had dinner with them, and he seemed to have good manners & capacity. Yesterday I had dinner with the Ambassador of the Empire;[13] where I found General Pelegrini:[14] & I formed a party of him and Murça and our Minister to go and visit the British Museum in particular. If I was not busy with other less frivolous matters, I wouldn't have a single moment that these gentlemen wouldn't take from me. However, if I'm in London next Tuesday, I intend to accompany them on the visit to the Museum. But I'm afraid that, the same night, I'll leave with my good friend Arbuthnot for Holland, with his eldest sons whom he will leave there in an Academy to learn the Dutch & French Languages, which will be useful to both, one of them being already enlisted in the naval service, & the other being intended for the East India Company.

---

4 Edward Gibbon, *The History of the Decline and Fall of the Roman Empire* (6 vols., London, 1776-88). The first volume of this work was published in February 1776 and sold out almost instantly. A second edition was published in June, and it was probably a copy of this that Magellan sent to Sanches. Subsequent volumes of the work were not published until the 1780s.

5 Part 1.

6 Samuel Musgrave, *Speculations and Conjectures on the Qualities of the Nerves* (London, 1776).

7 *Medical Observations and Inquiries, by a Society of Physicians in London* (6 vols., London, 1757-84), the 5th volume of which was published in 1776.

8 Thomas Percival's *Philosophical, Medical, and Experimental Essays* (London, 1776) was generally seen as the third volume of his *Essays Medical and Experimental...* (2nd ed., 2 vols, London, 1772-3).

9 *Medical and Philosophical Commentaries* [by A. D.] (Edinburgh, 1773 onwards).

10 Nicolaus Rozen von Rosenstein, *The Diseases of Children, and their Remedies,* trans. Anders Sparrmann (London, 1776).

11 Luiz Pinto de Sousa Coutinho (1735-1804), Portuguese Minister in London, 1774-88.

12 Manuel Guedes de Miranda, Earl of Murça, entered on a diplomatic career under the Marquis de Pombal. He was counsellor at the London embassy and then Minister in Denmark.

13 Ludovico Barbiano di Belgioioso (1728-1801), Austrian Ambassador in London.

14 Probably General Karl Clemens Count Pellegrini, the military engineer who designed the massive Terezín Fortress in Bohemia.

15 A substance obtained from the beetle known as the Spanish fly, that can help men keep

Our friend Sequeira who is well & visits our Minister where he is the family physician, asked me to present his affections to you, and wants you to send him some idea of the method of application of the cantharides tincture.[15] If you found it in a book, please give the name of this, and that will be enough to save you more trouble. 76.09.13a

I intend to pass with Arbuthnot through Brussels, where I'll see our melancholic Chevalier. Mr. Needham was here for about a month this summer, but he has already returned to Brussels. He is fat & always the same in the rest of his behaviour.

Here I heard that poor Barros was taken off the List, the list of Correspondents of the Royal Academy.[16] Truly he makes himself dead not only to the academies but to friends even of many years. Thus goes human life ...!

I'm in good health. I'm busy with my machines, clocks, instruments &c. &c. But I suspected after 8 or 15 days that your curses began to be fulfilled, since in a visit to my Nelly, she told me: don't tire yourself today, I'll come by tomorrow: and the worst was that the next day she was not better served than the preceding day. I was made cross by the ill success, & remembered your cursed and foreboded prophecies, that seem to threaten being fulfilled. However I am waiting for winter that strengthens the fibres. Contrary to my wish, I'm getting fatter and fatter. I don't know what to do. I eat, drink & sleep well: & I have no troubles or afflictions that consume me.

May you have the same health & rest that I wish you. Give my regards to the adoptive family, and to the other friends if there are any there.

For now there is no literary news. This period of holidays is the dead time in England. I'll be back in London before the end of October: I hope to receive news from you in Brussels.

vale[17]

---

a longer erection but that is toxic in larger doses.

16 The Portuguese astronomer José Joaquim Soares de Barros (1721-1793) had been a corresponding member of the Paris Academy of Sciences since 1757, but on 12 June 1776 his name was removed from the list of members.

17 Farewell.

76.09.17 *To Johann Albrecht Euler*[1]

St Petersburg Branch, Archives Department, Russian Academy of Sciences, f. 1, op. 3, Nr. 62, ll. 252-253ob

Monsieur,

Il y a peu de jours que j'ai envoyé d'ici les articles suivants, qui sont destinés pour etre presentés au corps Illustre de votre Acad^e. Imp^le.: & par cette raison je crois necessaire de vous en faire cet avis. 1°. un vol. in fol. contenant les Observations Astronomiques, faites dans l'Observatoire Royal de Greenwich:[2] & 2°. La premiere partie du Vol. 66 des Transactions philosophiques: tous les deux de la part de la Société R^le. de Londres. 3°. un vol. in 4°. relié en maroquin, contenant les oeuvres posthumes de Mathem^c. par feu Robert Simson, Profess^r. de Glascow;[3] que M^r. le Comte Stanhope presente à votre Acad^e. Imp^le. des sciences; & dont j'espere que vous m'accuserez la reception pour le participer avec les remerciem^ts. de l'Acad^e. à ce genereux Seig^r., qui a fait à ses depens cette belle edition, & qui l'a fait reliée toute, sans permetre qu'un seul vol. fut vendû, pour la distribuer par les Acad^es. & Savants les plus distingués de l'Europe. 4°. Enfin j'y ai ajouté un petit Traité que j'ai publié l'année derniere à Paris, sur les Instrum^ts. à Reflection, pour Observer sur Mer,[4] que je vous prie de vouloir bien presenter avec les autres trois; malgré son peu de merite, & seulem^t. en consideration de son utilité pour les marins.

Ces quatre articles furent empaquetés dans une caisse, qui contient une des pendules Astronomiques que j'ai fait faire ici depuis quelques années,[5]

1 Annotated: "reçu le 10 Octobre 1776 & lû à l'Académie le 17 du meme mois". See *Protokoly zasedanii konferentsii Imperatorskoi Akademii Nauk...*, III, 262.

2 N. Maskelyne, *Astronomical Observations made at the Royal Observatory at Greenwich, from the Year 1765 to the Year 1774* (London, 1776).

3 R. Simson, *Opera quaedam reliqua ...* (Glasgow, 1776).

4 Magellan, *Description des octants et sextants anglois...*(Paris/London, 1775).

5 In his *Notice des instrumens d'astronomie, de géodesie, de physique, &c faits ... par ordre de la Cour d'Espagne* (1780), p. 203, Magellan states that all the pendulums in the collection had been made according to Graham's design. The first had been made by Grignion, "& je la fis arranger pour les observations astronomiques, en y adaptant les penduls de mon invention, dernièrement perfectionnés, dont la longueur n'est point sensiblement changée par les variations de la température de l'atmosphere".

adressée à S. Exc[e]. le gen[al]. Baur[6] au quel j'ecrirai la poste suivante,[7] pour le prevenir là dessùs. J'ai envoyé un grand nombre de ces pendules nouvelles à mes amis en differents endroits de l'Europe, qui les trouvent excellentes, malgré la modicité du prix, qui n'excede 4, ou 4 ½ guinées, au lieu des 35 & meme 50 guinées, qu'il faut payer pour les pendules astronomiques chez les artistes les plus fameux. 76.09.17

Un de nos plus grands libraires à Londres, m'a parlé un de ces jours, disant qu'il souhaitoit debiter ici les Commentaires de votre Acad[e]. Imp[le]., pourvû qu'on les lui passe au prix marchant, qu'on dit monter à 7½, ou 8 shillings par vol. Son nom est M[r]. Nourse dans le Strand; il est fort connû par son bon caractere & par les editions de plusieurs ouvrages. Il prendra d'abord une suite complette des Novi Commentarii, qui sont, je crois, 18 ou 19 vol.: & de plus une une[8] seconde suite des memes après le 3me. vol. exclusive. Il voudroit aussi une demi-douzaine d'exemplaires des Lettres philosophiques de M[r]. Euler, adressées à une dame ou Princesse Allemande.[9] il faudra bien examiner & collationer chaque volume avant de l'envoyer; & voir qu'il soit complet & sans fautes. Cet objet etant relatif aux interets de votre Illustre Acad[e]., j'ai crû que vous n'auriez pas de la difficulté à parler là dessùs à la personne qui a la surintendance de ces affaires: en cas que ce soit un de vos libraires, il feroit bien d'envoyer ici un catalogue des livres de Petersburg avec cette commission; parce qu'on ignore ici pour la plupart les publications qu'on fait chez vous.

Faites moi l'honneur, je vous prie, de presenter mes h[bles]. respects & mes ofres aux membres de votre Illustre Academie, pour tout ce qui peut etre de leur service dans ce pais. ma connexion avec la plupart de nos Savants, & l'employ que je fais continuellem[t]. des artistes les plus habiles en toute sorte des instruments de Physique & d'Astronomie, m'ont affranchi la grande correspondance qui j'entretiens avec beaucoup de gens de lettres dans la plupart de l'Europe, aux quels je ne laisse pas d'etre souvent serviable. Je serois bien flaté, Monsieur, si vous & vos Savants Academiciens voudroient

6 General F.W. von Baur (or Bauer) (1731-1783) was a military engineer recruited by the Empress Catherine to overhaul the administration of the Russian army.

7 Letter not found.

8 *sic.*

9 L. Euler, *Lettres à une princesse d'Allemagne sur divers sujets de physique et de philosophie* (3 vols, St Petersburg, 1768-72).

76.09.17 egalem^{t}. employer les bonnes dispositions avec les quelles j'ai l'honneur d'etre avec la plus grande consideration & respect

Monsieur
Votre très h^{ble}. & obeiss^{t}. serv^{r}.
Magellan

à Londres ce 17 Sep^{bre}. 1776
au n°. 12 Nevils Court Fetter Lane

P.S. La poste precedente fut fermée avant que je puisse finir cette lettre. Ce retard a donné lieu à la nouvelle suivante, qui fera plaisir à votre Acad^{e}. M^{r}. Walsh nous a fait voir hier, le 19 de courant (Sep^{bre}.), l'eteincelle du coup electrique dans le gymnotus, ou Anguille de Surinam; on trouve actuellem^{t}. 4 de ces poissons à Londres. Notre Presid^{t}. le Chev^{r}. Pringle, & plusieurs autres Membres de la Soc^{e}. R^{le}., avec d'autres gentil-hommes, s'y trouvoient. nous avons vû 10, ou même 12 fois cette eteincelle, en faisant passer le coup à travers du metal, oû il y avoit une petite separation: on avoit soupçonné que ce poisson le gymnotus & la Torpille etoient electriques i.e. operoient par vertu electrique. M^{r}. Walsh est le premier qui l'ait demontré. personne avoit jamais vû le feu ou eteincelle electrique dans ce poisson, même plusieurs philosophes croioient que c'etoit une action mechanique.[10]

---

10 Walsh seems not to have published his observation of an electrical spark from the gymnotus, but a report based on information supplied to J.B. Le Roy by Magellan and Sir John Pringle was published in Rozier's journal (*Observations sur la physique, sur l'histoire naturelle et sur les arts*, 8 [1776], 331-5). Walsh's observation seems to have been well-known at the time but a generation later, Faraday was unable to find a published source to back up a reference to it by Humboldt; see M. Faraday, *Experimental Researches in Electricity* , vol. 1 (London, 1839), p. 101.

To the Duke Charles-Marie-Raymond d'Arenberg

Archief Van Arenberg, Enghien/Edingen, Belgium

à Londres 27 Sep^{bre}. 1776

Monseign^{r}.

Il y a près d'un mois, que Je comptois partir pour Hollande, avec mon ami M^{r}. Arbuthnot, qui y va amener deux de ses enfants les plus agés, pour une Acad^{e}. oû ils apprendront l'Hollandois & François, deux langues avantageuses pour Les objets auxquels ils sont destinés. L'un aïant deja entré dans La liste du Service de la Marine & l'autre etant determiné de passer aux Indes Orientales.

Mais M^{de}. Arbuthnot vient d'acoucher d'un onzieme enfant, & celà a retardé notre depart. J'ai reçu hier une Letre là dessus, oû cette digne consorte d'un tel Epoux (V. A. S^{me}. connoit bien Les deux) me dit qu'elle ne lui donne pas la permission pour partir avant que l'Enfant soit baptisé & Je ne doute point que V. A. S. trouvera qu'elle a raison. En consequence de celà nous partirons d'ici le 8 du mois prochain, allant en droiture à Amsterdam, nous reviendrons par la Haye, & de la à Bruxelles pour presenter nos humbles Respects à V. A. S^{me}. Mais nous serons obligés de nous depecher le plutôt possible, aîant tous les deux des affaires pressants au commencem^{t}. du mois de Novembre à Londres.

J'ai envoyé le commencem^{t}. de ce mois une pendule des plus simples de ma construction pour Everlé,[1] ou je la fixerai à mon arrivée, car je ne manquerai pas d'aller à Louvain voir mes amis les Professeurs & le Chancellier.[2] ainsi J'ai prié M^{r}. Hennessy de Ostende,[3] de l'envoyer en droiture par Louvain.

Lorsque M^{r}. Needham fut ici à Londres il me montra une petite Liste de Semences que V. Al^{e}. S^{me}. voudroit avoir; savoir du Tulipier, des 4 especes de Magnolia, du Chenne de Lacomb, & du Pin de Weymouth. Mais il n'a pas eté possible de trouver, que seulem^{t}. des semences de ce dernier, le Pin de Weymouth. Pour les autres il n'y a pas de Semences à avoir. Cependant on peu[4] avoir des plantes en assez bonne quantité. Celles du Tulipier coutent environ une demi courone chacune, a peu près de la hauteur de deux pieds.

---

1 The château de Heverlee, near Leuven, one of the houses of the Duc d'Arenberg.

2 The mathematician Jean François de Marci (?-1791), chancellor of the University of Louvain in 1772.

3 See Magellan to the Duke of Arenberg, 19 December 1775, note 3.

4 peut?

76.09.27 Les Magnolias dès 5 jusqu'à une guinée & les chennes de Lacombe coutent environ 5 shillings, lorsqu'ils sont de 4 à 5 pieds de hauteur.

Le bon vieillard Gordon vive encore, quoique malatif[5] & le fils qui demeuroit en ville, avec une grande boutique de graines, vient d'avoir sa maison brulée:[6] cependant il en a sauvé beaucoup, & a deja ouvert une autre boutique dans La meme Rue.[7]

J'ai l'honneur d'etre
Monseigneur
de V. A. S^me^.
serv^r^. très h^ble^. & tres obligé
Magellan

---

5 maladif.

6 The nurseryman James Gordon (*ca.* 1708-1780); see Magellan to Johann III Bernoulli, 25 January 1771, note 8.

7 A seed shop in Fenchurch Street. See Magellan to Johann III Bernoulli, 16 April 1771, note 4.

76.10.11 *To the Duke Charles-Marie-Raymond d'Arenberg*

Archief Van Arenberg, Enghien/Edingen, Belgium

Monseigneur

Mon depart[1] est encore retardé jusqu'au 16 de Ce mois, à cause des affaires de mon ami Arbuthnot & celà m'a donné le tems de recevoir de M^r^. Le Comte de Belgioioso[2] un petit paquet de la part d'une Dame dont je suis oublié le nom, adressé à S. A. S^me^. Madame la Duchesse d'Arenberg, qui contient quelque commission ou achat fait ici dans le pays. J'ai mis ce paquet dans une caisse, avec une Lunette que Le general Pelegrini[3] m'a recommandé de lui faire arranger ici, & un autre petite parcelle pour M^de^. la Comtesse d'Epinoi,[4]

---

1 See Magellan to the Duke of Arenberg, 27 September 1776.

2 Ludwig Karl Graf von Barbiano-Belgiojoso [Ludovico Carlo Maria di Barbiano, Conte di Belgioioso] (1728-1801), Austrian Minister plenipotentiary in London.

3 The Austrian general Karl Clemens Graf von Pellegrini (1720-1796), Director-General of Engineering and Fortifications and Supreme Director of the Academy of Engineers.

4 Presumably Maria Franziska von Liechtenstein (1739-1821), wife of Prince Charles Joseph Lamoral de Ligne (1735-1814), one of whose titles was Prince d'Épinoy. The couple usually spent winter each year in Brussels.

dans une caisse, que J'ai adressée à Mr. Hennessy à Ostende[5] pour l'envoyer à V.A.Sme. à Bruxelles: & Je lui fais avis cette poste sur cet objet. J'ai ajouté deux livres dans la meme caisse pour un de mes amis, que je lui delivrerai moi-meme à mon arrivée, si V.A.Sme. aura La bonté de me Les faire garder. 76.10.11

Je sui deja si honteux d'avoir si fort procrastiné mon depart, à cause des affaires de Mr. Arbuthnot, que je vais demain pour Rester chez lui jusqu'au moment de notre depart. J'y ai couché cette nuit derniere. Ce matin avans 6 heures Je le trouvai dans son cabinet avec son fermier, ou chef du labourage, pour lui donner ses ordres, pour semer du froment &a &a. nous entrames La Chaise de poste tout de Suite après dejeuner (vers les 7h.) pour venir à Londres: Il est allé à ses affaires, & moi je suis rentré chez moi pour faire à V.A.S. cet avis.

Je dinai mercredi dernier avec Le Comte d'Acunha,[6] qui est arrivé ici pour Retourner à Lisbone. Il m'a donné des nouvelles de V.A.Sme. & de Sa Famille, & que Tous se trouvoient à Heverléz[7] actuellement: Nous aurons bientot ici 5 Ministres de Portugal, savoir celui qui vient de la Haye le Comte d'Acunha, celui qui va lui suceder Mr. Orta,[8] celui qui va à Danemarque, Mr. Guedes,[9] celui qui va à Petersbourg Mr. Naples,[10] & en fin celui qui reside à Londres Le Chevr. Pinto.[11] Messrs. Orta & Naples ne sont pas encore arrivés, mais on les attend tous les jours. C'est un nombre qui rarement se rencontre de la meme Cour & dans le meme endroit.

Les Chiens du Roy d'Angleterre ont fait un grand carnage dans les fous Americains, qui veulent ne pas etre esclaves.[12] ils ont pris Long Iland[13] vis

---

5 Patricius Hennessy (1741-1784), head of the Ostende trading company, Patricius Hennessy & Co.; see Magellan to the Duke of Arenberg, 19 December 1775, note 3.

6 D. José Vasques Álvares da Cunha (1734-1812), 2nd Count of Cunha and nephew of the great diplomat to King João V, D. Luiz da Cunha.

7 Heverlee (or, in French, Héverlé or Heverlé), a family seat of the d'Arenbergs.

8 Francisco José Horta Machado (1746-1817).

9 Manuel Guedes de Miranda, Earl of Murça, entered on a diplomatic career under the Marquis de Pombal. He was counsellor at the London embassy and then Minister in Denmark.

10 José de Nápoles Tello de Meneses was nominated Minister Plenipotentiary to the Court of St. Petersburg in 1776 but this nomination was never effective, perhaps because of the sudden illness of King José followed by his death early in 1777.

11 Luiz Pinto de Sousa Coutinho (1735-1804), Portuguese Minister in London, 1774-88.

12 The Declaration of Independence was approved by the American revolutionary Congress on 4 July 1776 and proclaimed by George Washington to his troops in New York City on 9 July. After that, crowds in many cities tore down all signs representing royalty.

13 The American revolutionary forces suffered a heavy defeat in the Battle of Long Island, New York, on 27 August 1776.

76.10.11 à vis de New-York, & tué 3 ou 4 mille Americains. Voilà des nouvelles bien consolantes pour l'humanité ...!

J'ai l'honneur d'etre
Monseigneur
de V. A. S^{me}.
$Serv^{r}$. très h^{ble}. & très obligé
Magellan

à Londres Ce 11 Octobre 1776

A Son Altesse S^{me}.
$Monseig^{r}$. Le Duc d'Arenberg
Flandres
à Bruxelles

76.10.12 *To [Charles Messier]*

American Philosophical Society Library, Philadelphia, João Jacinto de Magalhães (Jean-Hyacinthe Magellan) Collection, Mss BM25

Londres ce 12 Octobre 1776[1]

Mon cher ami

Je viens d'apprendre la nouvelle funeste de la mort de M^{de}. de Montigny. la lettre ci jointe[2] contient les sensations de mon ame dans ces tristes circonstances. Vous aurez la bonté de la donner à M^{r}. Trudaine lorsque vous le jugerez à Propòs.

Je vous ai envoyé un pair de bas de Soye noire de la nouvelle manufacture d'ici; & je ne sçais pas si je vous l'ai dit deja dans quelqu'une de mes lettres. Ce pair de bas furent empaquetés dans une caisse $d'intrum^{ts}$. pour M^{r}. La Voisière, adressée à M^{r}. Trudaine $\&^{c}$. mais excusez moi Je viens de revoir mon livre, & Je trouve que Je vous ai ecrit le 10 Sep^{bre}.[3] ainsi il est probable que je vous en ai deja parlé dans cette lettre là; car cette caisse partit d'ici le $commencem^{t}$. du meme mois.

1 Annotated "rep. le 3. 9^{b}." Letter not found.
2 Letter not found.
3 Letter not found.

Donnez moi des nouvelles de M[r]. Le Comte Carburi.[4] J'ai continué à Lui envoyer toute cette année les brochures periodiques qu'il m'avoit ordonné avant son voyage en Italie. il ne m'a pas ecrit depuis son retour: je serai charmé d'apprendre de ses nouvelles. 76.10.12

à dieu mon cher ami. portez vous bien & croiez moi pour la vie

votre de coeur
Magellan.

P.S.

Il y a plus de 4 ou 5 Semaines que je suis pret à partir pour La Hollande & Flandres, pour y faire un tour avec un de mes amis. mais c'est à sa cause que je ne suis pas encore parti: cependant, celà sera avant 3 ou 4 jours sans faute.

Si vous voulez m'ecrire à Bruxelles, j'y trouverai votre lettre; car j'y passerai avant la fin de Novembre, qui est le tems que j'ai fixé pour mon retour à Londres. Si vous y ecrivez, adressez votre lettre à la poste Restante: autrem[t]. adressez chez moi à Londres, car je laisse ordre de me garder soigneusem[t]. mes lettres &[c].

Mes resp[s]. & amitiés à tous nos amis communs.

Vale[5]

4 Count Giovanni Battista Carburi, physician to Maria Teresa of Savoy; see Magellan to Louis-Henri Duchesne, 23 July 1773.

5 Farewell.

To Johann III Bernoulli 76.10.12a

Öffentliche Bibliothek der Universität Basel, MS L I a 709, f. 63

Londres ce 12 Octobre 1776

Monsieur & très cher ami

J'ai cherché avant hier M[r]. Isaac Hankins Browne[1] selon l'avis de votre lettre datée le 28 Sep[bre].[2] & Je trouve qu'il est encore à Spá. Probablem[t].

1 Presumably Isaac Hawkins Browne (1745-1818), later a long-serving member of parliament. What Bernoulli's connection was with Browne is unknown, but probably they met while Browne was making an extended continental tour after completing his studies at Oxford in 1773 – a tour from which, it would appear from Magellan's letter, he had not yet returned to England.

2 Letter not found.

76.10.12a il ne viendra jusqu'au milieu ou à la fin de Novembre: & Je ne manquerai pas alors de le chercher & lui delivrer votre manuscript;[3] car Je retournerai avant la fin de Novembre à Londres, après le petit tour que Je vais faire immediatem^{t}. [e]n Hollande & Flandres. J'aimerois bien que vous me donnasses une [i]dée precise du travail qu'on epargne par l'usage de vos Tables. Car je n'ai pas eu occasion de faire grandes calculs de ce genre, & n'ai pas Le tems de faire la comparaison entre les operations ordinaires que les calculateurs sont en usage de practiquer, & celles qu'on pourra abre[g]er en employant vos Tables. Car je voudrois donner une idée bien [n]ette de vos Tables & leurs avantages à M^{r}. Browne, & meme à ceux de ma Connoissance, qui pourroient en quelque maniere directe [o]u indirectem^{t}. favourir & seconder votre application au Board de la Longitude &^{c}. à mon Retour nous parlerons des autres articles de votre lettre, sur la Liste des Astronomes,[4] & les Instrum^{ts}. de l'Observatoire de M^{r}. Aubert,[5] que Je ne douterois pas preferer à aucun de ceux que Je connois, tant pour les instrum^{ts}., position & arrangement, que pour la comodité de l'observat[eur][6] en la considerant tout Seul & sans assistant, comm'un curieux Astronome &^{a}.

Je vous Remercie de la Communication des articles pou[r] M^{r}. de Castillon.[7] Je lui ai ecrit, & Je me flate que vous aurez deja l[e] pendule Astronomique &^{a}. Car J'ai decouvert oû la Caisse etoit arreté[8] & hereusem^{t}. le meme Cap^{tn}. du vaisseau qui se trouvoit ici pour le 2de. fois n'etoit pas encore parti. J'ai ecrit la dessus à M^{r}. de Castillon &^{a}.

Vive, vale & ama
tuum
Magellan[9]

---

3 The manuscript of Bernoulli's *Sexcentenary Table, exhibiting at sight the result of any proportion where the terms do not exceed 600 seconds or 10 minutes, with precepts and examples* (London, 1779).

4 For Magellan's criticism of Bernoulli's *Liste des astronomes connus, actuellement vivans* (Berlin, 1776), see 76.08.06.

5 Alexander Aubert (1730-1805) had an extremely well-equipped observatory at Loampit Hill, near Deptford.

6 Conjectural reading; text obscured by binding.

7 Johann Castillon (1704-1791), director of the Berlin Observatory.

8 See 76.08.06.

9 Enjoy life, farewell & love | your | Magellan.

To [...] Wingate 76.10.15

Wellcome Institute Library, London[1]

Sir

I have received an account from Paris viz from the Author of y^{e}. Journal de Physique,[2] by which it appears that he sends regularly every mont one number of this work for M^{r}. Walsh according to my direction & order,[3] by the way of M^{r}. Robert Bigg.[4] Besides that I have called lately on the same M^{r}. Bigg, and find that he receives but one single cahier of this journal every month. these two circonstances evince clearly that M^{de}. des Saints[5] does not send any number of the said work: and of course my account with M^{r}. Walsh is very right.

I have given orders already to send no more to M^{r}. Walsh by the post after December next: and so you'll take care that M^{rs}. Desaint do not fail on sending it after December next.

I am
your m^{t}. obed^{t}.
Magellan

London this 15 Oct^{ber}. 1776

To M^{r}. Wingate
at M^{r}. Nourse's Bookseller
facing Catherine Street
Strand
London

1 Letter transcribed in 1977 but now missing.
2 François Rozier.
3 See Magellan to J. Walsh, 12 September 1776.
4 Bigg was an official at the General Post Office in London.
5 Desaint & Saillant were engravers and printers at Paris. Following Desaint's death, his widow took over the running of the business.

76.10.31 *To Ribeiro Sanches*[1]

Österreichische Nationalbibliothek, Vienna
A.N. Ribeiro Sanches Korrespondenz, codex 12714, ff. 183-184v

a Leyde 31 Outubro 1776

Am^{o}. e S^{r}. e m^{to}. do Coração

Bem considero q as dilaçoens e incertezas desta minha jornada aos Paises baixos (não imagine q falo dos femeninos; pois destes ultimos ainda me despacho & me avîo assaz bem) podem ter sido a cauza de me faltarem noticias de V.M. depois de Longos tempos: e tãobem me Lizonjeo de q acharei carta de V.M. em Bruxellas, como lhe pedi na minha ultima. Com tudo não posso socegar, sem lhe escrever a prezente. a comunicação de ideias com os verd^{os}. amigos q conhecemos como taes, faz o maior gosto duravel, & dá a maior satisfação & a sensação mais agradavel q conheço na vida.

Eu esperei por nosso amigo Arbuthnot em Inglaterra mais de 5 ou 6 semanas, p^{a}. virmos ambos a este paiz & tornar pelo de Bruxellas p^{a}. Londres. Porem vendo q os seos negocios & embaraços de familia (sua Mulher lhe deo ultimam^{te}. o undecimo filho, e todos são vivos, & sadios &^{a}.) lhe não consentião o pôr-se a caminho, tomei a Rezolução de hir p^{a}. sua caza, como de ordin^{o}., no Sabado 12 de outubro, & lhe disse q eu sem falta partiria na semana seg^{te}., ou com elle, ou sem elle: & em effeito lá fiquei toda a semana, pois eu estava envergonhado de me ter despedido tantas vezes de outros amigos, sem acabar de partir. Quando chegou a 5^{a}. fr^{a}., vendo q elle entrava na sua post-chaise p^{a}. hir a Londres, lhe pedi q tomasse um Lugar na carruage de Harwich; pois eu certam^{te}. partiria no dia seguinte: & q se elle quizesse tomar mais dois Lugares p^{a}. seos dois filhos, em cazo de elle não poder hir, q eu os traria comigo &^{a}. Quando tornou á noite, disse-me q tinha tomado 3 lugares no Coche de Harwich, & q me ficaria obrigado se eu tomasse cuid^{o}. dos rapazes. Partimos pois de Londres na 6^{a}. fr^{a}. 18 de 8^{bro}. pela meia noite. chegamos a Harwich pelas 4 da tarde do dia 19. embarcamos na manhan do dia 20. chegamos com a mais feliz passagem à Helvootsluyle pelas 8 do dia 21. no mesmo dia viemos dormir aqui à Leyde: & no dia 22 fui metter os dois rapazes na Academia ou Escola, q dista coiza de 5 milhas, em Noordwyk. no mesmo dia fui dormir já só & dezembaraçado dos piquenos Arbuthnots, a Harlem. na Manhan

1 Annotated "recebida a 4 9^{bro}" [received 4 November] and "resp^{a} a 7 9^{bro}" [answered 7 November]. Letter not found.

do dia 23 cheguei a Amsterdam, onde fui m^{to}. bem Recebido de varios 76.10.31
sujeitos, curiosos da Physica experimental, & em particular hũ thio do mais velho dos rapazes, pois este mais velho naceo da 2da. Mulher q meo amigo Arbuthnot teve (a prezente he a 3^{a}., & della tem nacido os dez filhos m[ais] novos). Este sujeito he um dos principaes mercadores de Amsterdam. tem todo o comercio de Russia &^{a}. a sua caza chama-se Pye Rich & Wilkieson: este ultimo he o seo nome. Este cavalh^{o}. não permitio q eu jantasse uma só vez na Estalage. Recebi carta do Princepe Gallitzin da Haya, q me esperava sem falta, & q de proposito deixava a caza de Campo onde se achava com a Princeza, perto de Breda, p^{a}. me vir esperar na Haya. estive até 26 em Amsterdam. no 27 fui com um amigo a Harlem p^{a}. ver o D^{or}. van Marum, q inventou uma nova maquina Electrica, na qual o frotement hé unicam^{te}. feito por meio do azougue puro. No dia 28 parti p^{a}. Utrech, onde fiz conhecim^{to}. com varios Professores dos quais Recebi hum acolhim^{to}. m^{to}. honrado. parti no dia 29 á noite, & cheguei hontem pela manhan á la Haye: onde jantei com o Princepe Gallitzin, q me disse fixasse hũ dia p^{a}. lá me achar com os Professores Allaman & Gaubius, q dezejavão ver as novas experiencias das differentes especies de Ar &^{a}. parti pois ontem á noite p^{a}. Noordwyk àfim de dizer àDeos aos piquenos Arbuthnots, & ver se estavão contentes. hoje pela manhan os deixei, & vim aqui à Leyde outra vez, p^{a}. ver os ditos Professores dos quais fui Recebido com os maiores agrados; particularm^{te}. o Profess^{r}. Allaman, q não sabe como me obrigar com novos favores. fomos ver o Horto-botanico & o Museo, a sala dos Instrum^{tos}. Philosoph^{os}., & delà viemos por caza do Prof^{r}. Gaubius, com o qual ajustamos um rendez-vous p^{a}. sabado q vem (depois de amanhan) em caza do Princepe Galitzin à la Haye pelas 11 da manhan, & lá jantaremos todos com elle: onde lhes farei varias experiencias sobre differentes especies de Ar, pelo methodo do D^{r}. Priestley, q he o mais simples & mais perfeito. Gaubius me falou em V.M. com m^{ta}. estimação, & da mesma sorte o Princepe q^{do}. hontem jantei com elle. eu alli achei (em caza do Princepe) hum cirurgião, q veio ultimam^{te}. de Petersburgo, cujo nome me esqueceo já, mas elle se achará lá sabado q vem. elle fez a operação do cancro do peito á Senhora Muschkin Pushkin em poucos minutos, & com bom sucesso, & me disse q provavelm^{te}. tornará p^{a}. Londres com seo Marido p^{a}. a Embaixada de Inglaterra, o q m^{to}. estimarei.

eu conto de partir outra vez para la Haye á manhan, & provavelm^{te}. partirei na 2da. fr^{a}. (4 de Nov^{bro}. em q completo 54 de id^{e}.) p^{a}. Roterdam, onde não me dilatarei mais de hum dia. de lá vou a Antuerpia & a Malines, onde um amigo, conselheiro au grand conseil, bom vivant & q me tem perseguido p^{a}. lá o ver, me espera. de Malines à Bruxellas & à Louvain. tenho o Professor Thibaert, q ali me espera com ancia. emfim espero q antes do fim

76.10.31 de Nov^{bro}. me terei restituido á minha Inglaterra, único lugar onde Respiro descançado, & sem susto politico, a pezar do rapido progresso com q vejo q o sistema caminha ao despotismo. Porem espero q não chegará à praticar-se perfeitam^{te}. no pouco q já agora me pode restar de vida.

Tenho gostado m^{to}. da Hollanda, & lhe restituo o bom conceito de q eu a tinha privado na minha estimação, & até ás mesmas Hollandezas faço a mesma justiça; pois huma q provei á vontade, achei q se meixia tãobem & com tanto donaire & com tanto dengue como se fosse huma portugueza da Beira ou da Serra de Estrella.

Peço a V.M. q não me deixe sem carta & novas suas em Bruxellas à la poste restante.

Os D^{ores}. Sequeira & La Cour ficarão bem, q^{do}. deixei Inglaterra. eu me despedi do prim^{o}., pois o D^{r}. La Cour está longe, e elle me pedio de novo q soubesse de V.M. a verd^{a}. receita p^{a}. fazer a tinctura de Cantharides, & o methodo da sua applicação, como tãobem os seos effeitos & cazos em q sucede bem &^{a}. veja V.M. se pode sem molestia dar-me um paragrafo na sua carta sobre esta materia.

não sei se disse já a V.M. na minha ultima (da qual me esqueci de fazer nota) q o S^{r}. Cunha, Ministro q aqui foi na Haya, se achava em Londres de retorno p^{a}. Portugal, onde os negocios da sua caza & familia pedião a sua prezença. eu jantei com elle & com seo filho em caza do Cav^{ro}. Pinto antes de sahir de Londres, cuido q em 9 do corrente: pareceo-me homem de bom caracter, pacifico, moderado, & não deixa de ver claro em varias materias importantes. esperava-se em Londres um tal S^{r}. q he do Algarve, & vem p^{a}. Enviado a Haya. alem de hũ S^{r}. Napoles q he de Viseu & vai p^{a}. Enviado a Petersburgo.

o Princ^{e}. Gallitzin me disse ontem as diligencias inuteis as mais extraordinarias, q tinha feito p^{a}. alcançar hũa boa quantid^{e}. de Terra de Mafra p^{a}. ma[ndar] p^{a}. Petersburgo. Elle falou nisto á Emperatriz: & até em nome della se pedio ao Marquez de Pombal q mandasse & permitisse o sahir a d^{a}. terra &^{a}. respondeo q sim, mas q tudo era uma charlataneria &^{a}. ainda atégora não foi possivel o alcançala. Coitado de q^{m}. he disgraçado …! pobre Barros …! q^{to}. a mim eu tãobem creio q essa terra de mafra não he mais do q um absorbente, com terra calcaria. mas q^{m}. sabe o q poderia suceder a favor do pobre inventor …! Saud^{es}. á familia adoptiva. tenha V.M. saude, vida & descanço.

vale

a Monsieur
Monsr. Sanchez, Doct^{r}. en Medecine
Rue des Blancs-manteaux
à Paris

Leyden 31 October 1776 76.10.31

My Dear Friend

I think that the delays and uncertainties of this journey of mine to the Low Countries (don't imagine that I'm talking about the feminine ones; since of these I still despatch & look after them rather well) must be the cause of having been without your news for such a long time and I also flatter myself that I will find a letter from you in Brussels, as I asked for in my last.[2] However I can't be calm without writing the present one to you. Communication of ideas with true friends whom we know to be such creates the most lasting pleasure, & gives the highest satisfaction & most agreeable sensation that I know in life.

I waited for our friend Arbuthnot in England more than 5 or 6 weeks, in order to come to this country together & return through Brussels to London. However, seeing that his businesses & family obstacles (his wife lately gave him his eleventh child, and all are alive & well &c.) didn't allow him to get under way, I took the resolution of going to his house, as usual, on Saturday 12th October, & told him that I would depart without fail the next week, with or without him: & in fact I stayed there all the week, because I was embarrassed at having said good-bye so many times to other friends, without departing. When Thursday arrived, seeing that he got in his post-chaise to go to London, I asked him to book a place in the Harwich coach; because I would for sure depart the next day: & that if he wanted to book two additional places for his two children, if he couldn't go, I would take them with me &c. When he came back that night, he told me that he had booked 3 places in the Harwich coach, & would be obliged if I took care of the boys. We departed from London on Friday 18th October at midnight. We arrived at Harwich at 4 in the afternoon of the 19th. We embarked in the morning of the 20th and arrived after the happiest journey at Helvootsluyce[3] at 8 on the 21st. The same day we came to stay here in Leyden: & on the 22nd I took the two boys to the academy or school, that is about 5 miles distant, in Noordwyk. On the same day I went to stay at Harlem, alone already without the Arbuthnot children. In the morning of the 23rd I arrived at Amsterdam, where I was very well received by various persons curious about experimental physics, & in particular an uncle of the eldest of the boys, since this eldest was born of the 2nd wife of my friend Arbuthnot (the present one is the 3rd, & the ten youngest were born of her). This man is one of the most important merchants in Amsterdam. He has all the commerce of Russia &c. His house is called Pye Rich, & Wilkieson:[4] the last being his name. This gentleman didn't allow me to have a single dinner at the inn. I received a letter from Prince Gallitzin[5]

2 See Magellan to Sanches, 13 September 1776.

3 Helvoetsluis (now Heellevoetsluis), a fortified town and naval port in western Holland.

4 Pye Rich & Wilkieson were an Anglo-Dutch banking firm in Amsterdam. Henry Pye Rich later withdrew from the company.

5 Prince Dimitrii Gallitzin (1738-1803), Russian ambassador at The Hague and a great enthusiast for experimental philosophy.

76.10.31 from The Hague, who waited for me without fail, & who would leave his country house where he was with the Princess, near Breda, to come specially and wait for me at The Hague. I was in Amsterdam until the 26th. On the 27th I went with a friend to Harlem to see Dr. van Marum, who has invented a new electrical machine in which the frotement is produced just by means of pure quicksilver.[6] On the 28th I departed for Utrecht, where I made the acquaintance of various professors from whom I received a very respectful reception. I departed on the night of the 29th & arrived yesterday morning at The Hague, where I had dinner with Prince Gallitzin who asked me to fix a day to meet Professors Allaman[7] & Gaubius,[8] who were desirous to see the new experiments on the different kinds of air &c. I departed therefore last night to Noordwyk in order to say goodbye to the Arbuthnot children, & to see if they were content. This morning I left them & came here to Leyden again to see the said professors, who received me with the greatest satisfaction; particularly Professor Allaman, who doesn't cease to oblige me with new favours. We went to see the Arboretum & the Museum, the Philosophical Instruments room, & then came by Prof. Gaubius's house, with whom we arranged a rendez-vous on the following Saturday (the day after tomorrow) at Prince Galitzin's house in The Hague at 11 in the morning; & there we all shall have dinner with him: where I shall make several experiments about different kinds of air, by the method of Dr. Priestley, that is the simplest & most perfect. Gaubius mentioned you with great esteem, as did the Prince when I dined with him yesterday. There (at the Prince's house) I found a surgeon, who arrived lately from Petersburg, whose name I now forget, but he will be there next Saturday. He did the breast cancer operation on Lady Muschkin Pushkin[9] in a few minutes, & with success, & told me that she probably will return with her husband to the Embassy in England, which I shall very much appreciate.

I intend to go again to la Haye in the morning, & probably I'll depart on Monday (4th November when I turn 54) for Roterdam, where I shan't stay more than one day. From there I'll go to Anthwerp & to Malines, where a counsellor au grand conseil and bon vivant[10] who has been asking me to see him there, awaits me. From Malines to Brussels & to Louvain, where I have Professor

---

6 Martinus van Marum (1750-1837) gave lectures in experimental physics at Haarlem, where he was subsequently director of Teyler's Museum. Magellan is probably referring to the electrical machine he built in 1774, with shellac plates and mercury as the rubbing agent, described in M. van Marum, *Verhandeling over het electrizeeren…* (Groningen, 1776).

7 Jean-Nicholas-Sébastien Allamand (1713-1787), professor of philosophy at Leiden from 1749, where he specialized in natural history. He also became responsible for the lectures in experimental physics following the death of Petrus van Musschenbroek in 1761.

8 Hieronymus David Gaubius (1705-1780), professor of medicine at Leiden from 1734, renowned for his lectures in pathology.

9 Wife of the Russian ambassador to London, Count Aleksei Semenovich Musin-Pushkin.

10 Probably Godefridus Gasparus Tackoen (1720-1788), lawyer and counsellor at the Grand Council of Malines (Mechelen), the Supreme Court of the Low Countries; see Magellan to Sanches, 25 August 1775 and 15 November 1776.

Thibaert[11] who is waiting anxiously for me. I hope that before the end of November I shall at last be back in my England, the only place where I breathe quietly & without political fear, in spite of the rapid march that I see the system making towards despotism. However I hope this will not get perfectly into practice during the little that now remains of my life. 76.10.31

I have enjoyed Holland very much, & I regain the good concept I had of it in my private esteem, & I even do the same justice to the Dutch women; since one I tried at ease, I found moved as well & with such grace & with such coquetry as though she was a Portuguese of Beira or of the Serra da Estrela.

I beg you not to leave me without a letter & your news in Brussels à la poste restante.

D^{rs}. Sequeira & La Cour were well when I left England. I said farewell to the former, D^{r}. La Cour being far away, and he asked me again to get from you the true recipe to make the tincture of cantharides, & the method for its application, as well as its effects & cases where it succeeds &^{c}.[12] Please see if you can without trouble include a paragraph in your letter about this matter.

I don't know if I told you in my last letter (of which I forgot to take note) that M^{r}. Cunha, the former Minister at The Hague,[13] was in London on his way back to Portugal, where the affairs of his house & family demanded his presence. I dined with him & his son at Chevalier Pinto's house[14] before leaving London, I think on the 9th of this month: he seemed a man of character, peaceful, moderate, & he sees clearly in several important matters. A certain man from the Algarve was expected in London, who comes to be Envoy to The Hague,[15] besides a M^{r}. Napoles[16] who is from Viseu & goes as Envoy to Petersburg.

Prince Galitzin told me about the most extraordinary but unfruitful attempts he had made to obtain a good quantity of earth from Mafra[17] to send to Petersburg.

---

11 Joannes Franciscus Thijsbaert (1736-1825), professor of experimental physics at the University of Louvain.

12 See Magellan to Sanches, 13 September 1776, note 15.

13 D. José Vasques Álvares da Cunha (1734-1812), 2nd Count of Cunha, nephew of the eminent D. Luiz da Cunha, was Secretary (1764), then Minister Plenipotentiary (1766), and finally Portugal's Envoy Extraordinary at The Hague until 1776.

14 Luiz Pinto de Sousa Coutinho (1735-1804), Portuguese Minister in London, 1774-88.

15 Cunha was succeeded as Portugal's Envoy Extraordinary at The Hague by Francisco José Horta Machado who was from Faro in the Algarve.

16 José de Nápoles Tello de Meneses was nominated Minister Plenipotentiary to the Court of St. Petersburg in 1776 but this nomination was never effective, perhaps because of the sudden illness of King José followed by his death early in 1777. The first Portuguese Minister to Russia was Francisco José Horta Machado, who transferred from The Hague in 1778.

17 The Portuguese astronomer José Joaquim Soares de Barros had told Gallitzin, and also Sanches, that a calcareous earth known as *terra de Mafra* from the town of Mafra, 20 miles from Lisbon, had been found to have healing properties against cancer. Gallitzin asked the Marquis de Pombal in the name of the Empress, Catherine the Great, to arrange for some to be sent to her.

76.10.31 He talked about this to the Empress: & it was actually in her name that the Marquis de Pombal was asked to order & allow the exportation of the said earth &c. He consented, but in the end everything was charlatanry &c. Till this moment it was not possible to get it. Unfortunate poor fellow. ...! Poor Barros[18] ...! As for me I think that this earth of Mafra is nothing else but an absorbent, calcareous earth, but who knows what could happen to the poor inventor ...! Regards to the adoptive family, may you have good health, life & peace.

Vale.[19]

18 See Magellan to Sanches, 13 September 1776.

19 Farewell.

76.11.15 *To Ribeiro Sanches*[1]

Österreichische Nationalbibliothek, Vienna
A.N. Ribeiro Sanches Korrespondenz, codex 12714, ff. 185-186v

Bruxellas 15 Novbro. 1776

Meo amo. e Sr. e mto. do Coração

Aqui cheguei hontem com bom sucesso, e me acho hospedado no Palacio de S.A.Sma. o Duque de Arenberg, donde escrevo esta depois de Receber as duas de V.M. de 19 de Setembro & 7 do corrente, q mto. mto. estimei. Pois ainda q o estado da sua saude, fraqueza ou debilide. de compleissão, falta de apetite, e pouco effeito do Rhuibarbo me fazem uma bem triste sensação; consolo-me em ver q não lhe falta nem se lhe enfraquece o animo = "nec sperat in festis, nec nutovit secundis, alteram sortem bene preparatum pectus ... fugoces Labuntur anni: nec pictas moram rugis instanti que senectae afferet. Sed omnes una manet nox, perpetua dormiende: & calcanda semel via Lethi." = Mas deixemos ideias melancolicas. basta q nos-saibamos someter ao nosso fado sem Reluctancia, nem renitencia inutil: mas não nos-occupemos da scena, q não tem nada de galante. vamos ás suas cartas, q o nosso bom Conego Lisboeta me entregou. Coitado ...! falta-lhe não sei q? & faz-me pena q não sabe contentar-se do prezente, q, como hontem lhe disse, he sem comparação mais do q o q me faz feliz: pois sem hũ só Real

1 Annotated by Sanches: "Respa a 1 Xbre pa Londres" [Answered 1 December to London]. Letter not found.

de Renda certa, nem conezia, nem pensão, vivo, gozo, f-do, & viajo Ledo e contente, mandando fare buz-rare tuto questo mondo fotuto. &^{a}. 76.11.15

Agradeço a adresse q V.M. deo a M^{r}. Klingstedt, q verd^{a}.m^{te}. tem pecado no artigo de puntualid^{e}. vis à vis de moi.

espero saber os porques Relativos ao amigo Duchenne; pois he necess^{o}. saber e conhecer com q^{m}. se trata na vida.

Eu vi varias vezes o Medical Museum. nunca achei nelle grande interesse scientifico: q^{do}. vir o D^{or}. Sequeira, q espero seja antes do fim da semana q vem, lhe perguntarei q conceito faz deste jornal, q em effeito he como todos os demais, videlicet, a catch penny of Booksellers.

Esteja descançado q não faltarei em mandar a V.M. a continuação do Jornal Medico de Edinburg assim q sahir pois goardo sempre uma nota, no meo Livro, do ultimo Numero q Remeto a V.M.

As Lectures de Culen nunca forão impressas (excéto sobre a Mat^{a}. Medica): meo & nosso amigo Paradise assim m'o assegurou, depois q veio da sua jornada a Edimburgo, na qual lhe encomendei que m'as procurasse a todo o preço.

Cuido q V.M. define bem o amigo Ortega. Eu certam^{te}. o estimaria m^{to}. mais, se elle não tivesse vivido nove annos em Italia. em geral um caracter Italianado pode bem ser de grande utilid^{e} p^{a}. si proprio; mas rarissimam^{te}. p^{a}. os outros.

ora gavo-lhe os fumos da sua Riba-côa! O Balsamão vai descarregar-se no Baroza, & este no Doiro q termina a Beira-baixa, a qual por antiphrase todo o mundo sabe q^{to}. he mais sublime do q a Beira-alta. mas isto são disputas q lá arranjarei com o Cav^{ro}. Pinto q^{do}. chegar a Londres. no entretanto dê V.M. Lá uma vista á geographia da Serra de Estrella, p^{a}. q nos-entendamos melhor.

Ja vi & li o Tratadinho de M^{r}. Fuss sobre os achromaticos. pouco dis q todos não saibam: & nenhũa utilid^{e}. por elle se tirará na pratica.

Agradeço a noticia dezejada pelo D^{r}. Seq^{a}. sobre a tinctura das cantaridas, a q^{m}. a communicarei sem falta.

Tenho gozado não sei porq, de hũa boa parte da Benção da Tia velha de V.M.: e o q mais hé ainda por cá escorre m^{to}. bem a calda das pevides de cabaço, apezar dos 55 q comecei no mesmo dia em q deixei a Haya

as ideias de V.M. sobre o movim^{to}. do mercurio no Barometro, são assaz analogas as q eu tenho; mas sobre outro principio, a saber o pezo absoluto da atmosphera, que crece ou diminui segundo a maior ou menor quantid^{e}. de vapor, q se acha dissolvido nella: a cuja dissolução depende da quantid^{e}. do fluido electrico, q causa a atração ou repulsão necessaria p^{a}. produzir este effeito.

76.11.15 fiquei intimam^{e}. agradado & cativo do bom coração do Princepe Galitzin. falamos sem Rebuço, & com o coração nas mãos. cultivarei a sua amizade com todos os meos affectos. pois os merece. os Prof^{ores}. Gaubius & Aleman ficarão admirados & satisf^{os}. das Exper^{as}. q lhes mostrei sobre differ^{es}. especies de ar. o prim^{o}. he frio, mas boa cabeça, & ao menos he bom negativam^{e}.: o segundo tem m^{to}. de bom hom̃em; & he m^{to}. cordial. ficaremos em boa correspondencia. Eu não vi o Castrioto, nem o procurei. as ideias q delle me deo o Conde D'Acunha antes de partir de Londres me confirmárão na q eu delle formo depois de m^{tos}. años. à saber q he hũ hom̃em vitando. Taobem tive a fortuna de não me encontrar na Haya com o Bahia, q ainda he peor do q o outro.

O S^{r}. Orta ainda não chegou a Londres: mas sei q ja ali chegou hum parente do nosso digno Ministro o Cav^{ro}. Pinto, hom̃em instruido, sem prejuizos, & de hũ excellente caracter, o qual foi seo ouvidor em Mato grosso. O gosto de conhecelo & tratalo servirá tão bem a apressar o meo Retorno a Londres.

Eu sahi da Haya em 4 do corrente, depois de jantar com M^{r}. Joel, Enviado de Dinamarca, q he hũ amavel sojeito, mui amigo do Cav^{ro}. Pinto, do qual eu lhe trouxe uma carta particular, q elle não quiz fiar da posta. tinha-me despedido do Princepe Galitzin na noite preced^{e}., a q^{m}. deixei quazi pela meia noite, tendo a passado quazi toda em conversação aberta so por so. fui dormir na mesma noite a Rotteradam, onde passei dois dias com o Consul Inglez meo antigo amigo, & outros da mesma nação com q fiz conhecim^{to}.: no dia 7 parti p^{a}. Dort por agoa: caminhei a pé 3 milhas até Newveer: tomei barco até Laage Swallow onde dormi. no dia 7 tomei a posta pelas 4 da manhan, & cheguei a Breda, antes q as portas se abrissem. entrei immediatam^{e}. no coche p^{a}. Antuerpia onde pernoitei. era 6^{a}. fr^{a}. e não achei jantar senão peixe salgado. maldito bigotismo Austriaco! quis Deos q o Estalajadeiro em Antuerpia me tomou pelo q eu não sou (protestante) e me deo a cear um bocado de vitella fria no meo quarto. eu me achava com hũa febrinha pela indigestão: mas pela manhan, estava são & arr-tado como hũ franciscano da Ponte. tomei a carroça p^{a}. Malinas, onde jantei com o meo amigo M^{r}. Takoen, conselheiro ao grand Conseil. negro jantar de peixe duro e indigesto. elle devia hir na mesma noite à Louvain, & eu o acompanhei na sua carruage. fui pouzar no Coll^{o}. Real com o com̃um amigo o Profess^{or}. Thisbaert. no domingo, fui vizitar o Duque de Arenberg, q se achava com toda a Familia à Evre, menos de hũa milha de Louvain. Jantei com elle, q me Recebeo com toda a cordialid^{e}.: abracei o Princepe cego, de q^{m}. tenho alguas esperanças ainda q fracas: pois parece q o cristalino tem formado hũa cataratta, & talvez q a operação de lha abater, ou extrahir, lhe deixe alguma vista ainda q confuza. no mesmo dia partiram todos pª. Bruxellas, por cauza da assembleia dos Estados do Brabante q ainda dura. gastei os dias 11 & 12 em arranjar & Repetir as varias Experiencias sobre differentes especies de ar ao Professor Thisbaert, que as fará no curso da Physica, & me ficou sumamᵉ. obrigado. no dia 12 á noite veio o filho do Conselheiro Takoen de proposito buscar-me na sua Carruage, pª. hir jantar em Malines com seo pai no dia segᵗᵉ.; o q fizemos. dormi em sua caza no dia 13; & parti pª. Bruxellas no dia 14. onde cheguei pelo meio dia. vim buscar o Duque de Arenberg, q ordenou um Leito no mesmo quarto, em q seo defunto e amado filho, o Princepe Carlos, habitava: e onde agora escrevo esta Longa carta, emqᵗᵒ. a companhia da sua Assemblea se acha ceando; pois eu não ceio ordinariamᵗᵉ.: & alem disso não gosto destas grandes, esplendidas, mas insignificantes, insipidas & imphilosophicas companhias. hoje fiz as preparaçoens necessarias, n'um quarto de Mʳ. Needham pª. á manhan lhe mostrar as experiencias principaes sobre differentes especies de Ar. o Secretario & outros da nova Acadª. ali se acharão, & se tiverem algum pouco de miolos, não deixarão de ficar convencidos da importancia deste novo ramo da Physica experimental. a deos meo amº. do Coração q he mais de meia noite passada. meos affectos á familia adoptiva e a os verdᵒˢ. amigos, entre os quaes o Dʳ. du Presle. 76.11.15

vive, vale, atque me ama[2]

a Monsieur
Monsʳ. Le Doctʳ. Sanchez &ª
Rue des Blancs-manteaux
a Paris

---

2 A note has been added at the end in another, unidentified hand, as follows:
[S.] Dᵒʳ. 2ª fª. escrevi à vm de[s]endolhe que o Sʳ Magalhaens não tinha chegado mas no mesmo dia soube que estava ja em Louvania, e chegou aqui 5ª fª. com mº. boa saude, e deo mᵗᵒ. gosto a todos os seus amigos: e hoje assisti as experiencias que fez para mostrar as diversas qualidades de ar com as machinas inventadas por Mʳ Prestley: que mᵗᵒ. folguei de ver.
Estimarei q VM passe com saude: os nevoeiros me tem hum pouco carregado a cabeça mas vamos passando de pé.
[Dʳ., on Monday I wrote to you telling that Mʳ. Magalhaens had not yet arrived but on the same day I learned that he was already in Louvain, and he arrived here on Thursday in very good health and gave much pleasure to all his friends: and today I attended the experiments, which I very much enjoyed, that he performed to show the different kinds of air with the machines invented by Mʳ. Priestley.
I shall be pleased if you are in good health: the fogs have made my head a little heavy but we keep going on.]

76.11.15 Brussels 15 November 1776

My Dear Friend

I arrived here in good order and am accommodated in the palace of His Most Serene Highness the Duke of Arenberg, where I'm writing from after receiving your two letters of the 19th of September & the 7th of the current month,[3] that I was very pleased to get. Because though your health, with weakness or debility of complexion, loss of appetite and rhubarb having litle effect makes me feel unhappy; I'm glad to see that you don't let your spirit sink or weaken = "nec sperat in festis, nec nutovit secundis, alteram sortem bene preparatum pectus ... fugoces Labuntur anni: nec pictas moram rugis instanti que senectae afferet. Sed omnes una manet nox, perpetua dormiende: & calcanda semel via Lethi."[4] = But let us leave melancholic ideas. Enough that we know how to submit to our fate without reluctance or useless obstinacy: But let us not devote ourselves to the scene, which has nothing graceful about it. Let us come to your letters, that our good canon from Lisbon[5] handed to me. The poor ...! He lacks something & it makes me sorry that he's not happy with the present which, as I told him yesterday, is without comparison more than what makes me happy: since without a single penny of assured rent, or sinecure or pension, I live, enjoy myself, f—k, & travel joyful and happy, wishing to fare buz–rare tuto questo mondo fotuto.[6] &ª.

I thank you for giving the adresse to Mr. Klingstedt,[7] who truly has very much missed in punctuality vis à vis de moi.

I hope to know the whys relative to our friend Duchenne; since it's necessary to know and understand with whom one deals in life.

I saw the Medical Museum several times.[8] I never found a great scientific interest in it: when I see Dr. Sequeira, which I hope will be before the end of next week, I'll ask him what he thinks of this journal, that in fact is like any other, videlicet,[9] a catch penny of Booksellers.[10]

3 Letters not found.

4 He does not hope in festive days nor did he nod in favourable times a breast well prepared for the other lot ... the fleeting years slip by nor will righteousness bring a delay to wrinkles and advancing age. But a single night awaits all for unending sleep and the path of death must be trodden once. (Mixed quotations from Horace, Odes I and II.)

5 João (or Jean) Chevalier (1722–1801).

6 send cheating this entire f...king world. See also Magellan to Sanches, 23 June 1775.

7 Baron Timothy de Klingstedt (1710-1786) from St Petersburg visited England in 1772, seeking information about the latest improvements in agriculture.

8 *The Medical Museum: or a Repository of Cases, Experiments, Researches, and Discoveries* (3 vols, London, 1763-4). Part of a 4th volume was published in 1781.

9 that is to say.

10 A book intended merely to sell.

Don't worry lest I miss sending you the continuation of the Medical Journal of Edinburgh[11] as soon as it is published, because I always keep a note, in my book, of the last issue that I sent you. 76.11.15

The Lectures by Culen have never been printed (except about the Materia Medica[12]); my friend and ours, Paradise, assured me of that when he came back from his journey to Edinburgh, where I asked him to find them for me at all cost.

I think you well define our friend Ortega. I would certainly praise him much more, if he had not lived for nine years in Italy.[13] In general an Italian character can well be of great use for oneself; but rarely for others.

I praise your vanities of your Riba-coa![14] The Balsamão flows into the Baroza,[15] & this into the Douro that limits Beira-baixa, which by antiphrasis everyone knows how much more sublime this is than Beira-alta. But these are disputes that I'll have with Chevalier Pinto[16] when I arrive in London. In the mean time, have a look at the geography of Serra da Estrela, in order that we can better understand each other.

I have already seen & read the little treatise of Mr. Fuss about achromatics.[17] He says little that others have not already known: & it will bring nothing useful to practice.

Thank you for the news about the tincture of cantharides that Dr. Sequeira wanted,[18] I shall communicate it to him without delay.

I don't know why, but I've been enjoying a good part of the blessing of your old aunt: and what is more, the sauce of lust[19] still drains very well, in spite of my beginning my 55th year on the day I left The Hague.

Your ideas about the motion of the mercury in the Barometer are rather similar to mine; but due to a different principle, which is the absolute weight of the

---

11 Probably *Medical and Philosophical Commentaries* [by A.D.] (Edinburgh, 1773 onwards).

12 William Cullen, *Lectures on the Materia Medica, etc.* (London, 1773). Cullen's *Institutions of Medicine, Part 1: Physiology* had, in fact, been published in Edinburgh in 1772, while unauthorized copies of his lectures had been circulating among his students for many years.

13 Casimiro Gomez de Ortega (1741-1810), Spanish physician and botanist who studied at Bologna before becoming a professor in Madrid.

14 A region of east-central Portugal, near where Sanches grew up.

15 The small stream Balsemão flows into the left margin of the Baroza River that flows into the left margin of the Douro River. Balsemão is a tiny old village by the stream of the same name, which was the origin of the title of Viscount of Balsemão given to Luiz Pinto de Sousa by Queen Maria I in 1801.

16 Luiz Pinto de Sousa Coutinho (1735-1804), Portuguese Minister in London, 1774-88.

17 Probably Nicolas Fuss, *Instruction détaillée pour porter les lunettes de toutes les différentes espèces au plus haut degré de perfection don't elles sont susceptibles tirées de la théorie dioptrique de Mr. Euler le père, et mise à la portée de tous les ouvriers en ce genre...* (St Petersburg, 1774).

18 See Magellan to Sanches, 13 September and 31 October 1776.

19 The expression that Magellan uses, *pevides de cabaço*, is unknown in contemporary Portuguese. However, its meaning is clear from Magellan's letter to Sanches of 17 December 1776, where he uses the same phrase again.

76.11.15 atmosphere that rises or falls according to the greater or smaller quantity of vapour that is dissolved in it, the dissolution of which depends on the quantity of the electric fluid that causes the attraction or repulsion necessary to produce this effect.

I was deeply pleased & captivated by the good heart of Prince Gallitzin. We talked frankly & with heart in hand. I shall cultivate his friendship with all my sympathy, since he deserves it. Profs. Gaubius & Alaman were surprised & pleased with the experiments that I showed them about the different kinds of air. The former is a cold man but with a good brain, & at least is good negatively: the latter is a good man; & is very cordial. We'll keep in regular correspondence. I didn't see Castrioto[20] or look for him. The ideas transmitted about him by Count Acunha[21] before leaving London confirmed the idea I've had about him for many years. That is, that he is a man vitando.[22] I also had the fortune of not meeting Bahia[23] at The Hague, who is even worse than the other.

Mr. Orta[24] has not yet arrived in London: but I know that a relative of our honourable Minister Chevalier Pinto arrived there, a learned man, without prejudice, & of excellent character, who was his auditor in Matogrosso.[25] The pleasure of making his acquaintance & dealing with him contributes to hastening my return to London.

I left The Hague on the 4th of this month after having dinner with Mr. Joel, Envoy of Denmark,[26] who is a pleasant man, a good friend of Chevalier Pinto from whom I brought a private letter that he didn't want to trust to the post. I had said goodbye the night before to Prince Gallitzin, whom I left at almost midnight, having passed almost all that night in open conversation. Just the two of us. I stayed that same night in Roterdam, where I spent two days with the English Consul, an old friend, & others from the same nation that I met: On the 7th I departed for Dort[27] by ship: I walked 3 miles to Newveer:[28] and embarked for Laage Swallow[29] where I stayed. On the 7th I took the stage-coach at 4 in the morning[30] & arrived at Breda before the gates were open. I immediately got into

20 Félix António Castrioto, Portuguese journalist.

21 D. José Vasques Álvares da Cunha (1734-1812), 2nd Count of Cunha.

22 A man to avoid.

23 Felis (or Felix) António Bahia was a not very honest Portuguese citizen living in Paris.

24 Francisco José Horta Machado (1746-1817).

25 Pinto de Sousa's cousin, Luís de Albuquerque e Melo Pereira e Cáceres (1739-1797), was his judicial assistant when he served as Governor of the Brazilian province of Matogrosso; see Magellan to Sanches, 11 July 1777.

26 Christian Sehested Juel was Danish envoy in the Netherlands from 1773 to 1778. Thanks are due to Mr. Henrik Stissing Jensen, archivist from Rigsarkivet, Denmark, for help with this identification.

27 The English name for the Dutch town of Dordrecht.

28 Probably, Nieuwe Merwede, via the road named Nieuwe Merwedeweg.

29 Lage Zwaluwe.

30 Magellan's journey from Laage Swallow to Antwerp via Breda was presumably on 8 November, which was a Friday.

the coach to Antwerp where I stayed that night. It was Friday and I couldn't find anything for dinner except salted fish. Damned Austrian hypocrisy! Fortunately the innkeeper in Antwerp took me for what I am not (Protestant) and gave me a piece of cold veal for dinner in my room. I had a little fever due to indigestion: but by morning I was well & belching like a Franciscan of the Bridge.[31] I took the car to Malines, where I dined with my friend Mr. Takoen,[32] counselor at the grand Conseil. An awful dinner of harsh and indigestible fish. He had to go that same night to Louvain, & I accompanied him in his carriage. I stayed in the Royal College with our common friend Prof. Thisbaert.[33] On Sunday, I visited the Duke of Arenberg, who was with all his family at Evre, less than a mile from Louvain. I had dinner with him, who received me with complete cordiality: I embraced the blind Prince, for whom I still have some hopes though faint, since it seems that the crystalline has developed a cataract, & perhaps the operation to remove it may leave him with some sight though confused. On the same day they all departed for Brussels, due to the Assembly of the states of the Brabant that is still going on. I spent the 11th & the 12th arranging & repeating the various experiments on the different kinds of air for Professor Thisbaert, who will do them in the physics course & was most obliged to me. On the night of the 12th Counsellor Takoen's son arrived on purpose to fetch me in his carriage, to have dinner in Malines with his father the next day, which we did.[34] I stayed at his house on the 13th & left for Brussels on the 14th, where I arrived the same day. I came to look for the Duke of Arenberg, who ordered me a bed in the same room, where his deceased and beloved son, Prince Charles, lived: and where I am now writing this long letter, while the company from his Assembly are having supper, since I ordinarily don't have supper: & besides I don't like these grand, splendid but insignificant, insipid & non-philosophical companies. Today I made the necessary preparations in a room of Mr. Needham's in order to show him tomorrow the main experiments about the different kinds of air. The secretary & other members of the new Academy[35] will be there, & if they have a few brains, they will be convinced of the importance of this new branch of experimental physics. Goodbye my beloved friend since it's past midnight. My love to the adoptive family and to the true friends, including Dr. Du Presle. 76.11.15

vive, vale, atque me ama.[36]

31 Possibly a reference to a procession of the Franciscans held annually in Coimbra since medieval times, that by the 18th century had degenerated into a hedonistic display, to such an extent that the Bishop of Coimbra found it necessary to ban it.

32 Godefridus Gasparus Tackoen (1720-1788) lawyer and counsellor at the Grand Council of Malines (Mechelen), the Supreme Court of the Low Countries.

33 Joannes Franciscus Thijsbaert (1736-1825), professor of experimental physics at the University of Louvain.

34 The son of Godefridus (see letter to Sanches, 25 August 1775) was Henricus Arnoldus Josephus Tackoen (1751-?) also a lawyer in the Great Council of Mechelen.

35 The newly-formed Brussels Academy of Sciences, of which Needham was director.

36 Flourish, farewell and love me.

76.11.27 *To the Duke Charles-Marie-Raymond d'Arenberg*
Archief Van Arenberg, Edingen/Enghien, Belgium

Ostende 27 Nov^bre^. 1776.

Monseigneur

C'est mon devoir par des titres aussi justes que multipliés, de ne pas sortir du Continent sans Reiterer a V. A. S^me^. Le[1] sentiments de ma gratitude & de mon attachement, qui ne finiront qu'avec ma Vie. Je viens d'arriver à ce port depuis peu de minutes: aussitot je me portai chez M^r^. Hennessy pour savoir ce que etoit devenue La petite caisse contenant La parcelle pour Son Altesse M^de^. La Duchesse, La Lunette du Gen^al^. Pelegrini, & les autres deux articles.[2] Le Navire ne fait que commencer à decharger. J'y allai moi meme, & J'y ai vû encore La caisse en question; que Je priai Le Commis de M^r^. Hennessy de faire depecher au plus vite pour Bruxelles. Ce fut la circonstance de l'enlevement des matelots, qu'on fit à Londres pour Les vaisseaux de guerre,[3] qui causa un si grand retard à la sortie de ce vaisseau pour la destination. Et voilà comment le malheur de la guerre s'etend non seulement aux generaux quoiqu'etrangers, mais aux Dames Les plus distinguées & Respectables.

Je viens on ne peut plus enchanté des belles manieres du R^d^. P.D. Mann,[4] & encore plus etonné de son grand savoir en plusieurs branches Les plus utiles. C'est un personnage des plus Respectables & d'un merite le plus distingué & le plus avantageux. Connoissances profondes, Jugement solide, & genie transcendant, avec des vues pleines de cette humanité & interet public, qui doivent Le rendre très cher, non seulement à tous ceux qui Le connoissent en particulier, mais encore plus à Ceux qui par leur Rang se trouvent à La tete des Affaires publics. Je ne puis laisser cet objet sans signifier à V. Al^e^. S^me^. combien je suis faché que ce grand homme ne soit plus connû, & que Votre Souveraine[5] ne sache pas evaluer la perte actuelle que l'Etat fait à ne pas l'employer dans quelque place proportionnée à son Caractere, oû il puisse deployer ses talents & ses vertues pour Le bien de l'Eglise & de l'Etat. quelle perte irreparable si on tarde à Le faire, & qu'on Le laisse Sortir pour d'autre paiz, oû peut etre on n'a pas besoin d'un sujet

1 *sic.*
2 See Magellan to the Duke of Arenberg, 11 October 1776.
3 The American War of Independence had broken out a few months before this.
4 Theodore Augustus Mann (1735-1809) was an English Catholic priest and a member of the Imperial Academy of Sciences and Belles-Lettres of Brussels. He was *Secrétaire perpétuel* of the Academy, 1787-1794.
5 Maria Theresa (1717–1780), Empress of Austria.

de cette cathegorie ...! Je tremble pour ce pais, comm'un bon patriote: car V.A.S[me]. & tout ce qui Regarde sa famille, me fait prendre à Coeur les avantages de l'endroit oû Elle a fixé sa demeure. V.A.S[me]. fera le plus grand service à cet Etat, en employant son attention & son interet à cet objet, que je crois d'une extreme consequence 76.11.27

Que Le bonheur le plus complet de V.A.S[me]. & de toute son Auguste Famille couronne Les souhaits constants de celui qui a L'honneur d'etre avec un Respect Le plus profond & attachement sans bornes

de V.A.S[me].

Servit[r]. le plus h[ble]. & le plus obligé

Jean Hyacinthe de Magellan.

A S.A.S[me]. Le Duc d'Arenberg &[a]
à Bruxelles

P. S. Je me suis oublié de dire que je vais partir ce soir par le paquebot pour Douvre.[6] le tems est admirable: & apparem[t]. je serai à Londres vendredi Soir, ou J'attendrai constament L'honneur des ordres de V.A.S[me].

6 Dover.

To Joseph Priestley[1] 76.11.30

A Letter from Mr. J. Hyacinth de Magellan F.R.S. *to* Dr. Priestley, *concerning Experiments on Phosphori, his Method of impregnating Water with Fixed Air, and his Eudiometer.*

London, Nov. 30, 1776.

Dear Sir,

I arrived but last night from my tour through Holland and Flanders; and will not delay to acquaint you with a piece of philosophical intelligence, which I am sure will be no less agreeable to you than it was to me when I

1 Letter not found. The text given here was published as an appendix to Priestley's *Experiments and Observation on Different Kinds of Air*, vol. 3 (London, 1777), pp. 376-80.

76.11.30 first heard it. On my coming to Leyden, I visited the celebrated professor Allamand,[2] who, with great kindness, shewed me his large collection of philosophical instruments, among which is the best set I ever saw of white and coloured glasses, crystals and diaphanous substances, nicely cut into prisms, and other shapes for dioptrical experiments. This put me in mind of the unsuccessful attempts made at Paris by my friend Abbé Bourriot, and here in London by myself and Mr. B. Wilson, to repeat that curious experiment of Father Beccaria, professor at Turin, related in vol LXI of Philos. Trans.[3] viz. that the sulphureo-calcareous phosphorus, gives the same coloured light, when brought into a dark room, which it received through a coloured glass exposed to the sun.

I told professor Allamand how puzzled I had been at not being able to succeed in that experiment, which I attributed to not employing so good glasses, as either those of professor Beccaria, or such as those I had before me. He answered, that one of his experiments was nearly the same with that of Father Beccaria: for "Having exposed," said he, "a prepared[4] piece of the Bononian calcareous phosphorus, to the coloured rays of the sun, after their being separated by one of my prisms, and looking to it in the dark, I found the phosphorus gave the colour of the separated rays to which it had been exposed." Two days after I met again with the same professor Allamand at the Hague, where he came purposely with professor Gaubius,[5] by invitation of Prince Gallitzin, Ambassador from the Empress of Russia,[6] and one of the most intelligent lovers of Natural Philosophy of his rank, in order to see some of your experiments that I was to shew on different kinds of air, having sent, before I set out from England, the necessary apparatus to the Prince for that purpose.[7] There I introduced the same topic about the above experiment on the Bononian phosphorus: and professor Allamand appealed to Mr. Hemsterhuis, who was present, and is a very learned gentleman, residing at the Hague, well

2 Jean Nicolas Sébastien Allamand (1713-1787), professor of natural philosophy at Leyden.

3 Beccaria, "Letter...to Mr. John Canton F.R.S. on his new Phosphorus receiving several Colours, and only emitting the same", *Philosophical Transactions*, 61 (1771), 212. See also Magellan to B. Wilson, 6 September 1775.

4 The following footnote is included at this point: "Although the preparation of this calcareous matter is said to be kept as a secret, it is thought to consist, but in a kind of calcination with some sulphureous substance: and on this account it is nearly the same thing as the phosphorus employed by professor Beccaria."

5 Hieronymus David Gaubius (d. 1780), professor of pathology at Leyden.

6 Prince Dmitri Alekseevich Gallitzin (or Golitsyn) (1738-1803), Russian ambassador at The Hague.

7 See also Magellan to N. Pigott, 3 February 1777.

versed in philosophical experiments.[8] This gentleman had been present also at those of professor Allamand: he asserted that he had seen the very same phenomenon: and added a circumstance, which confirms my idea, that its success depends upon the good quality of the glass employed, and perhaps of a certain degree of strength in the solar rays, when it is tried. For Mr. Hemsterhuis, observed that, after some time, when the distinct colour seen on the phosphorus, began to weaken, the phosphorus turned to a yellowish cast, as when it was exposed simply to the light of the sun, without any separation of its coloured rays. Besides this, I have in my possession an authentic letter from Italy,[9] by which it appears that a young person of the first rank, with two noblemen his tutors, before whom the said experiment of professor Beccaria was repeated, had seen the same phenomenon, and that the colours of the phosphorus, when brought to the dark room, were distinct enough to enable them to guess, without any previous knowledge, what was the true colour of the glass, through which it had been exposed to the sun. 76.11.30

Now this appears to me, and I think also to every body else, to be a matter of *fact*, which in philosophical matters is as much to be depended upon when seen once, as if it was observed every day.

I am sorry to have seen, by a printed letter of the said professor Beccaria, that he gave it almost up, as if he had been mistaken in his experiment, by taking the shadow, or diluted darkness of the phosphorus, as a decided colour.[10] This he says was in consequence of Monsieur Zannotti, president of the Academy of Bologna,[11] himself and others not being able to see the same phenomenon. But that circumstance related by Monsieur Hemsterhuis, just now mentioned, may fully account for our unsuccessful attempts: and I hope soon to have a proper quantity of the said sulphureo-calcareous phosphorus, prepared by myself, which I shall send to professor Allamand desiring him to repeat with it this experiment, with his good apparatus, which I doubt not but will have the desired effect, and farther confirm the truth of F. Beccaria's observation.

I am,
Dear Sir,
Your most obedient, &c.

8 Probably Frans Hemterhuys (1720-1790), Dutch archaeologist and philosopher.

9 Letter not found.

10 Giambatista Beccaria, *A Letter … to Mr. Wilson, concerning the Light Exhibited in the Dark by the Bologna Phosphorus…* (London, 1776), p. 5.

11 Francesco Maria Zanotti (1692-1777).

76.11.30 P.S. I hope soon to have an opportunity of explaining to you in person two contrivances of mine, which I hope may be useful ones.[12] One relates to *Mr. Parker's glasses* for impregnating water with fixed air,[13] which, in the common method of using them, requires six or eight hours for each process. To avoid this delay, I use a double set of the middle and upper vessels, with another *stand*, so contrived with a convex bottom, that they may be easily agitated while they rest upon it. As soon as one of these middle vessels is properly supplied with fixed air, I take out that and the upper vessel belonging to it, and placing them upon the stand, I shake the water very well till it has absorbed the air, while the other vessels are receiving their supply of fixed air from the lower vessel, which is common to them both. Then letting out the residuum of fixed air, by separating the upper vessel from the middle one, I replace it on the lower vessel, that it may get a fresh supply of fixed air, while I shake the other set, which is now replenished. This operation I repeat three or four times; and by this means can fully impregnate the water contained in both these vessels in less than half an hour.

The other contrivances I want to shew you are two new *eudiometers* to measure the degree of salubrity of the air in different places. One of them, which I reckon the better, being the more simple, and the neater of the two, is according to the original idea of your experiments on this subject: but in neither of them do I make use of any *cock*, both on account of its being difficult to be made, and likewise subject to be out of order. Those of Mess. the Chevalier Landriani[14] and the Abbé Fontana,[15] seem to be liable to this inconvenience. The experiments already made in most parts of Italy, by the former of those gentlemen, with his own eudiometer, deserve the greatest praises; and it is to be wished that philosophers would more generally apply themselves to this interesting subject of inquiry.

---

12 Magellan, *Description of a Glass Apparatus, for making Mineral Waters .. together with the Description of some New Eudiometers ...* (London, 1777).

13 Described in Joseph Priestley, *Experiments and Observations on Different Kinds of Air*, vol. 2 (London, 1775), pp. 298-303.

14 Marsilio Landriani, *Observations sur la physique*, 6 (1775), 315-6; also M. Landriani, *Ricerche fisiche intorno alla salubrità dell'aria* (Milan, 1775).

15 Felice Fontana, *Descrizioni ed usi di alcuni stromenti per misurare la salubrità dell'aria* (Florence, 1774).

To Ribeiro Sanches[1]

76.12.17

Österreichische Nationalbibliothek, Vienna
A.N. Ribeiro Sanches Korrespondenz, codex 12714, ff. 187-188v

Londres 17 Dec^{bro}. 1776

Meo Am^{o}. e S^{r}. e m^{to}. do Cor^{am}.

Ainda q não tenho nada particular q avize a V.M., & q me acho com um milhão de coizas p^{a}. arranjar hoje & nos dias seg^{tes}. [...][2] dezembaraçar-me p^{a}. hir passar o Natal no campo com Mylord Shelburne q me invitou q^{do}. jantei com elle & mais o nosso Enviado & seo Primo (ouvidor q foi de Matogrosso) domingo fez 8 dias: com tudo não posso Resistir ao prurito de Replicar a carta de V.M. de 27 Nov^{bro}. Cheguei felizm^{te}. a Londres e sinto q V.M. perca o amigo Brito coitado de q^{m}. tem raizes ...! não há ninguem q as não tenha; porem as minhas não são tais q me impessão a minha Liberd^{e}. Cantabat vacuus coram Latrone viator. não tenho q perder pelo decreto de S.M. Fidelissima & por isso não torno; o q em effeito não deixaria de fazer igualmente, ainda q tivesse m^{to}. q perder.

Não me dá cuid^{o}. a divida de Klingstedt ella não he mais q 50 shillings. Q pague ou q não pague nada importa, pois com este nada de perda ficarei aprendendo &^{a}.

Se me picar com a sua incredulid^{e}. nas pevides de cabaço, mandar-lhe-ei huma certidão das minhas Nelys, bettys & mais michellas Londinenses: e se isso não bastar sou capaz de o hir à fazer trocar as Linhas &^{a}. &^{a}. &^{a}.

Admira-me a sua acrimonia com o Senhor grego olhe q novid^{e}. q o enganasse! p^{a}. q vivemos-nós neste mundo senão para sermos o joguete do engano, da vilhacada, da ignorancia, & da superstição ...? estava capaz de dizer a V.M. q mente, senão fosse cazo de tirarmos pelas espadas, quando ouço dizer-me q nunca ninguem o enganou mais do q uma vez. olhe q turra lá da Serra de Estrella ...! Pois eu gavo-me que tenho sido enganado mais de 30 mil vezes: ainda não ha 8 dias q estive esperando por Nely mais de duas horas, com m^{to}. p^{a}. escrever, & sem poder escrever nem Ler, com o diabo das pevides de cabaço nos miolos, q nem um so periodo querião dictar-me p^{a}. carta nem p^{a}. Ler &^{a}. &^{a}. e emfim ella não veio. perdi o meo tempo & m^{a}. pàciencia: fui ao seo Lodging & tinha saido com outro amigo. veio no outro dia fizemos as pazes com gosto Reciproco: & p^{a}. dizer-lhe a

1 Annotated "Recue a 22 X^{bre}" [Received 22 December], "R^{a}. a 30 Xbre" [Answered 30 December] and, on the envelope, " Mr de Magalhaens 1776".

2 Illegible.

76.12.17 verd^{e}. não creio q esta sera a ultima nem a penultima vez q sera necessario Renovar as pazes. eiz aqui o q passa a q^{m}. arr-ta. Mas ella tem aquillo tão apertadinho, tão gostozinho, & move-se com tal geito q não posso deixar de lhe perdoar todas as suas infidelid^{es}.

Não he possivel q V.M. visse o nosso Enviado em Coimbra. elle estudou depois do meo tempo. Elle me disse q estivera em correspondencia com V.M., e q Recebera livros ou não sei q por via de V.M. &^{a}.

ainda não tive carta de M^{r}. Le Begue, e lhe farei o q ordenar porq o amo. não ha novid^{es}. Medicas nem phil^{as}. Interessantes p^{a}. V.M. eu lá mandei algũas á Acad^{a}. de q Messier lhe dará conta, e q não valem grande coiza. aqui se diz q o Rei da terrinha morrera: mas o nosso Enviado ainda não tinha algũa noticia em 4^{a}. fr^{a}. passada. á manhan saberei, se tem algũa mais, pois conto hir jantar com elle.

o Journal de Medecine não vale nada. he hũ catch penny. [...] Le Begue lhe mostrará o 13°. N^{o}. do Med^{al}. Commentaries de Edimburgo, q lhe mandei a elle pela posta. o de V.M. hirá pela prim^{a}. occazião. Não chore o Conego Lisboeta de Bruxellas. elle não he obrigado a hir de tarde á Livraria: não lhe falta nada senão o contentar-se com o q tem, & renunciar ao cuidados patriot[icos]. hũ dos Sujeitos q me tinha escrito sobre o eu tornar p^{a}. Portugal & que queria arranjar este affaire, me escreveo ultimam^{te}. tão seco q me fez rir bem deveras. o tolo cuidava q eu havia de estimar o fazer semilhante asneira, como a de me hir entre arabios, envejosos & pretinhos de guiné ...!

hontem me fez aqui outro amigo invitar. fui jantar com elle. propoz-me um emprego, q eu cuido poderia conseguir, sem renunciar ao Santo Catholicismo, & q vale perto de 350 ou 400 Livras Esterlinas por anno mas q me tiraria a minha Liberd^{e}. pessoal, & me sojeitaria a trinta mil marotos & vilhacos. deilhe os agradecimentos, pois na verdade he am^{o}. sincero; mas disselhe q hũ bocado de beef no meo quarto, correspondendo cõs meos amigos, & fazendo livrem^{te}. o q queria, valia mais do q aquelles com̃odos pecuniarios.

Recados á familia & amigos.

vive, vale, & ama

tuum

ora veja! estou depressa p^{a}. hir jantar com Sir John Pringle, mas ainda enchi a folha de fio a pavio.

iterum vale

a Monsieur
Monsieur Sanchez D^{r}. en Medecine
Rue des Blancs-manteaux
à Paris

London 17 December 1776 76.12.17

My Dear Friend

Though I have nothing in particular to report to you, & I am busy with a million things to arrange today & to dispose of in the next few days in order to go and spend Christmas in the country with Mylord Shelburne, who invited me when I had dinner with him & our Envoy & his cousin (who was district judge of Matogrosso)[3] eight days before last Sunday: I can't resist the wickedness of replying to your letter of 27 Nov^ber^.[4] I arrived in London happily and I'm sorry you have lost your friend Brito[5] with whom you had roots ...! There is no one who doesn't have them; however mine are not such as to impede my freedom. Cantabat vacuus coram Latrone viator.[6] I have nothing to lose by the decree of His Most Faithful Majesty should I not return; which indeed I wouldn't do in any case, even if I had much to lose.

I don't care about Klingstedt's debt,[7] it is no more than 50 shillings. It doesn't matter whether he pays or not since I shall learn from this insignificant loss &ª.

If I get angry with your incredulity regarding my lust, I will send you a certificate from my Nelys, Bettys & other whores londinensis: and if that is not enough, maybe I will go and make you change your opinion &ª. &ª. &ª.

I'm surprised at your acrimony towards the Greek gentleman. How strange that he would cheat you! Why do we live in this world except to be the plaything of deceit, of villainy, of ignorance, & of superstition ...? I might tell you that you lie, would it not lead to our resolving it by sword, when I hear you say that no one cheated you more than once. Look at the dispute about the Serra da Estrela ...! Well, I pride myself that I have been cheated more than thirty thousand times: even eight days ago I waited for Nely for more than two hours, having a lot to write, & without being able to write or read with such a devil of lust in my brain,[8] so that I did not even finish a letter or reading anything &ª. &ª. &ª. and after all that, she didn't come. I wasted my time & my patience: I went to her lodging & she had gone with another friend. She came the next day and we made peace with reciprocal pleasure: & to tell the truth I don't believe that this will be the last nor the last but one time that we will need to renew the peace. Is this what happens to someone who belches?[9] But she has that thing so narrow, so delicious, & she moves with such skill that I cannot but forgive all infidelities.

---

3 Chevalier Pereira; see Magellan to Sanches, 15 November 1776, note 26.

4 Letter not found.

5 Joseph Caetano Jeronimo de Brito whose letters appear in the correspondence of Ribeiro Sanches. From these we learn that Brito had been in Paris but returned to Portugal at this time. See also Magellan to Sanches, 27 February 1776, note 20.

6 "The penniless traveller will sing in the presence of the highwayman"; or: "he who has nothing has nothing to lose" (Juvenal, Sat. X, v. 22).

7 See Magellan to Sanches, 15 November 1776, note 8.

8 See Magellan to Sanches, 15 November 1776, note 20.

9 See Magellan to Sanches, 7 July 1775, note 19.

76.12.17 It isn't possible that you saw our Envoy[10] in Coimbra. He studied after my time. He told me that he had been corresponding with you and had received books or something else through you &ª.

I have not yet had a letter from Mr. Le Begue. I will do what he orders since I love him. There is no interesting medical or philosophical news for you. I sent some to the Academy that Messier will report to you, that are not worth much. Here it is said that our motherland's King has died: but last Wednesday our Envoy had as yet no news. Tomorrow I'll know if he has any further news, since I intend having dinner with him.[11]

The Journal de medecine[12] is worth nothing. It is a catch penny. Mr. Le Begue will show you the 13th no. of the Med^al. Commentaries of Edimburgh,[13] that I sent to him by post. Yours will go at the first opportunity. Don't cry for the Lisboner Canon of Brussels.[14] He doesn't have to go to the library in the afternoon: he lacks nothing except being content with what he has & renouncing troubles in the motherland. One of the people who had written to me about my returning to Portugal & who wanted to arrange this affair, wrote to me so roughly recently that I had a good laugh. The fool thought that I should like doing such a stupid thing as to go and set myself among envious Arabs & Negroes from Guinea ...!

Yesterday another friend invited me over. I had dinner with him and he proposed a job that I think I could get without having to renounce holy Catholicism, & that is worth almost 350 or 400 pounds sterling a month. But it would take away my personal freedom, & would submit me to thirty thousand rogues & knaves. I thanked him because he is truly a sincere friend, but told him that a piece of beef in my room, corresponding with my friends & doing freely what I wished was worth more than all those pecuniary comforts.

Compliments to the family & friends.

vive, vale, & ama

tuum

Look! I am in a hurry to have dinner with Sir John Pringle, but still I filled the sheet from beginning to end.

iterum vale

---

10 Luiz Pinto de Sousa Coutinho (1735-1804), Portuguese Minister in London, 1774-88.

11 The rumour that Magellan had heard was not correct, although concerns about the health of Portugal's King José I were fully justified. The King had the first of several seizures on 12 November 1776 and on 27 November he transferred power to Queen Maria. He died on 24 February 1777.

12 Possibly *Journal de médecine, chirurgie, pharmacie*, ... published in Paris since 1758 and renewed in 1776.

13 *Medical and Philosophical Commentaries* [by A. D.] (Edinburgh, 1773 onwards).

14 João (or Jean) Chevalier.

To the Duke Charles-Marie-Raymond d'Arenberg[1] 76.12.22

Archief Van Arenberg, Enghien/Edingen, Belgium

Monseigneur

Cette Lettre contiendra un objet des plus importants pour le bien-etre des particuliers & de l'Etat: & quoiqu'elle soit en[2] mauvais françois & un peu Longue, Je suis bien persuadé que V.A.S^me^. ne s'ennuyera pas de sa Lecture.

Lorsque Je passai par Neuport,[3] Je trouvai que Le R^d^. P.D. Mann[4] etoit d'opinion, que La division des Fermes en petites portions, etoit d'un avantage beaucoup superieur à celui des grandes fermes. Il m'a parû que V.A.S^me^. panchoit[5] aussi à cette opinion: & comme Je savois que Messrs. Arbuthnot, & Young avoient pensé beaucoup sur ce sujet, avoient suivi, & meme demontre, que Les grandes fermes etoient beaucoup plus avantageuses que les petites; je proposai d'etudier les arguments & les principes, sur lesquels Ils etablissoient cette doctrine, aussitôt que J'aurois opportunité de Le faire. Ce fut dimanche derniere que J'eus une bonne occasion, me trouvant seul dans La Librerie Chez le même ami Arbuthnot avec le Traité, qu'il publia il y a 2 ou 3 ans,[6] sur le sujet, devant mes yeux. Je suis parfaitem^t^. convaincû que ces deux Messieurs ont raison: mais il n'est pas possible de renfermer, dans les bornes d'une Lettre, les arguments & les inductions demonstratives, sur les quelles Leur doctrine est fondée.

---

1 There is a copy of this letter, with extensive annotations by Theodore Mann, at Archives de l'Académie Royale de Belgique, Brussels, at Nr. 000390: Pièces de Son Altesse Mgr le duc d'Aremberg, de M. Magellan et de D. Mann sur les Grandes-Fermes en 1777. There are some differences between the copy and the original manuscript. The copy was published in Claude Bruneel, *L'hostilité à l'égard des grands fermes: un aspect du populationnisme dans les Pays-Bas Autrichiens* (Publication no. 93, Centre Belge d'Histoire Rurale, Louvain-la-Neuve, 1990), pp. 69-81.

2 *sic.*

3 Nieuport/Nieuwpoort in western Flanders. Magellan would have passed through the town on his way back to England at the end of his trip to the Low Countries the previous autumn.

4 Theodore Augustus Mann (1735-1809) was an English Catholic priest who was Prior of the Carthusian monastery at Niewpoort. In 1774 he moved to Brussels to become a member of the newly founded Académie des Sciences et Belles-Lettres de Bruxelles.

5 penchoit?

6 [John Arbuthnot], *An Inquiry into the Connection between the Present Price of Provisions, and the Size of Farms, with remarks on Population as affected thereby....* (London, 1773).

76.12.22 Il faudroit remonter jusqu'aux premiers resorts de la Nature & du Coeur humain, qui font agir les hommes en general, dans toutes leurs entreprises & travaux journaliers; pour pouvoir juger solidement de l'effet que telle ou telle circonstance, produriot[7] sur eux.

Reelement vouloir circonscrire un fort grand nombre de personnes, en general, sans presque aucune distinction, ni attention à leurs dispositions & capacités naturelles, sous une uniformité de vie & d'occupation, comme seroit celle des fermiers à petites fermes pour gagner leur entretien & de leurs familles, est une idée si exotique & singuliere, qu'elle est tout à fait dans le gout monacal. elle est entierem^{t}. repugnante aux passions humaines, qui sont La force motrice de nos actions. Non, Monseigneur! il faut que les hommes puissent avoir toujours un champ très étendû, pour y repaitre leurs fantaisies; à fin que leur ambition puisse les aiguilloner au travail, & à s'occuper de leurs projets.

Il seroit fort redicul de vouloir à present, que les hommes agissent en vraies machines, comme les pauvres marionets de ces vassaux dans le tems du Systeme feodal. Ce n'etoient pas les querelles des gros seigneurs, qui avoient aneanti tout idée d'Agriculture dans ces tems-là. C'etoit que chacun de cette espece d'esclaves se bornoit uniquement au simple necessaire; sachant bien qu'il ne pouroit monter jamais à une autre classe superieure, par tous les efforts de son industrie sur Les champs. Reduire à des petites fermes tout Le terrein d'une Province, ou d'un Royame,[8] fut-il le plus fertile du monde entier, ne manqueroit pas d'avoir actuellement Les memes consequences.

Chaque fermier retreci dans sa petite ferme, n'est dans le fond qu'une espece de moine (à la famille près), qui en doit participer necessairement la nonchalance: il doit se borner par principe au simple necessaire; & celà d'autant plus que ses travaux deviendront plus penibles pour obtenir la subsistance physique de sa personne & de sa famille. Toute La somme des petits profits qu'il gagnera, même aux depens de toute son industrie & de son application continuelle, ne le mettront jamais à son aise, ni ceux de sa famille. Ainsi il ne fera rien au delà de ce qui suffira pour son entretien: & par consequence cet agriculteur ne fera que rester tout au plus au même niveau, & comm'endormi dans le train ordinaire de sa tache; sans faire aucun effort extraordinaire, qu'il sçait ne lui apportera aucun avantage assez sensible: il n'aura pas Le Courage de rien entreprendre pour pousser ses Recherches & ses travaux au delà du train ordinaire. L'Etat assurement

---

7 produiroit?
8 Royaume.

n'y gagnera rien, & moins encore le petit fermier. Tout au contraire une ou deux mauvaises racoltes, le deperissement de ses bestiaux, par une maladie epidemique, ou par acident, ou enfin une maladie dans sa famille ou en Lui meme, suffiront pour Le ruiner de fond en comble; ou pour le moins il se degoutera du miserable metier de la petite ferme qui ne peu le secourir; tandis que ceux d'une autre profession, ne souffrent ordinairement tant, par cette espece d'acidents, qui sont au delà des bornes de la volonté humaine. C'est alors L'Etat qui, dans ce cas, soufrira une double perte reele. 76.12.22

C'est bien tout autrement vis à vis du grand fermier dans des cas pareils. La somme des memes profits, qui divisés separement entre plusieurs petits fermiers, ne leur procureroient pas un avantage considerable; etant reunis entre les mains d'un seul, Le mettront, Lui & sa famille, dans un etat d'aisance fort distinguée. Il s'enrichera à vüe d'oeil: il aura abondament de quoi entretenir sa femme & ses enfans avec un peu d'eclat: & ceux ci deviendront à leur tour autant de citoiens aisés & utiles à l'Etat, chefs de famille hereuses, qui seront autant d'exemples pour exciter de l'emulation dans ceux, qui connoissent la source de leur bien etre; & qui, par consequence, tacheront d'exercer Leur industrie & d'entreprendre avec ardeur les memes moyens & la meme profession, dont ils voient la preuve incontestable de la bonne fortune qu'elle apporte à ceux qui s'y appliquent avec soin & assiduité. Au lieu que Les petites Fermiers & Leurs familles, au mieux aller, c'est à dire dans le cas impossible, ou du moins improbable, de ne pas avoir jamais de mauvaises racoltes, ni d'autres malheurs; ils resteront tous dans une mediocrité, qui n'excitera jamais de l'envie, ni l'emulation de ceux qui les connoissent, & qui ont observé le peu d'avantage, que toute Leur industrie & travaux Leur ont produit.

C'est la meme Somme considerable des profits du grand fermier, qui Le mettra à même d'exciter ses talents, employer son industrie, & redoubler ses efforts extraordinaires, pour reparer Les malheurs & acidents facheux, qui necessairement font la ruine du petit fermier. C'est le grand fermier qui se trouvera en etat de poursuivre ses vües d'agriculture, avec d'autant plus d'ardeur, qu'il esperera de reparer dans la Racolte suivante, les pertes qu'il a essayées dans la mauvaise qu'il vient de soufrir. S'il a perdu beaucoup par le deperissement ou epidemie mortelle des Bestiaux; il a de l'argent & du credit, suffissants pour les remplacer: & quoiqu'il faille les acheter à un haut prix, il espere d'en etre bien dedomagé avec des interets & meme de l'usure la plus honnete. Enfin si ce sont des maladies personelles ou de sa famille, c'est lui seul Le grand fermier, qui trouvera avec son argent de quoi payer du monde ou un bon substitut, pour supléer à son assistance personnelle dans Le management de sa ferme, ou pour Le soulagement de

76.12.22 sa famille, qui ne languira par defaut du traitement necessaire, comme il arriverá,[9] dans un cas pareil, à celle du petit fermier.

Mais pourquoi faut-il recourir au developement especulatif des actions humaines…? Parlons des faits: & mettons là dessùs la decision de cette controverse. Quel est Le Pays ou La Province comme dans l'Europe, ou meme dans Le monde entier, oû l'on trouve tout Le terrein assez divisé en petites portions ou entre petits fermiers, qui soit dans le meme tems dans un etat florissant? oû les particuliers soient à leur aise; le Souverain assez puissant; & l'Etat assez nerveux pour soutenir son independance, sans etre à la merci de ses voisins? Si par hazard il y a quelque Province à peu près dans le cas en question: c'est à dire, qui soit pleine presque tout à fait de petits-fermiers, est-elle jamais celle oû regna l'aisance des particuliers, & sur Laquelle le Souverain puisse dependre pour les subsides de l'Etat? … Tout au contraire ne voit-on pas qu'elle est toujours la plus foible, la plus en proye à la misere par La moindre mauvaise racolte, & la plus abondante en ces petites querelles & ces dissensions chagrines, qui font la ruine reciproque de ses habitants? parce qu'en fin, ce n'est pas l'homme à son aise, qui est le plus mechant, generalement parlant. Malhereuses especulations …! pourquoi faut-il nous tourmenter la fantaisie avec des theories fondées sur des suppositions gratuites …! pourquoi faut-il nous promener dans les espaces chimeriques de notre imagination? tandis que nous avons des faits incontestables, pour nous eclaircir dans La route qu'il faut suivre. Mais c'est une Epidemie universelle de l'esprit humain, que de vouloir rafiner toujours dans les objets de son especulation: & comme tout homme est naturellem^{t} paresseux pour aprofondir & bien analyser les faits, & les causes qui les produisent; aussitôt qu'il rencontre un motif, tant soit peu especieux, & hors du commun qu'il voit practiquer, alors il s'y arrete tout court, & le prend pour un principe indubitable, pour en deduire toute sorte de consequences, meme Les plus absurdes & insoutenables.

Mais dit-on (quoique je ne conçois pas sur quelle authorité); ceux qui auront grandes fermes, en laisseront une grande partie sans culture. Je suis etonné d'une objection si chimerique! assurement elle est de pure theorie; car on n'a jamais vû un cas pareil, a moins que ce fermier en question ne soit pas un vrai sot, ou un imbecille: & alors le meme arriveroit, si la ferme seroit la plus petite. Autrement, ce seroit la meme chose que dire: le marchand sera charmé de gagner trois pour cent, au lieu de 20 & 30 qu'il pourra avoir, pour la meme marchandise …!

---

9 *sic.*

Comment donc, Messieurs les Especulateurs, avez vous deja rendu les hommes assez indifferents à leurs propres interets, qu'ils se contentent avec peu, lorsqu'ils peuvent avoir beaucoup ...! nous voilà tous philosophes ...! l'age d'or va revenir...! 76.12.22

Assurem^t^., Messieurs, vous parlez sans avoir examiné à fond ce que vous dites. Je suppose que vous avez vû de grandes fermes peu cultivées: Mais avez vous examiné, si c'etoit, parce que la mauvaise qualité du terrein ne valoit pas La peine de la culture? ou parce que La mauvaise situation n'assuroit pas un debouché favorable & profitable des fruits de cette terre, pour le cultivateur? ou parce que des Loix genantes empechoient de les transporter aileurs? ou en fin parce que le fermier ne pouvoit pas entirer un parti avantageux, à cause de vos etablissements mal entendus, de vos principes destructifs, & de vos privileges Seigneuriaux & barbares; qui, pour assourir, exclusivement du peuple, un amusement sauvage & sache de tuer en parade triumphante (oh que les hommes sont rediculs ...!) des bettes qui n'ont pas de defense, & qui se fuient de vous, obligez les grands & petits fermiers à voir leurs champs devastés, & leurs semailles ravagées, par ce nombre immense d'animaux nuisibles, qui ne trouvent pas de quoi subsister dans l'enorme étendue de vos parks & forets, tout à fait inutiles au genre humain ...!

Mais allons: mettons Les choses à termes égaux. Qui sera à même de bien cultiver une terre ingrate, & de faire égouter les terres trop humides par des grands frais: comme par exemple celles d'Enghien? Le petit fermier, ou un Duc d'Arenberg? Car un grand fermier feroit le même, s'il etoit maitre d'Enghien, & avoir de l'argent avec de l'intelligence. Sera-t-il le petit fermier, qui aura du courage, pour risquer son peu d'argent à faire des essays, quelques fois inutiles; à changer la methode ancienne par une meilleure, à adopter des instruments couteux, mais plus à propos; & à employer des graines & semences plus produisantes & choisies, mais plus cheres? Sera-t-il le petit fermier, qui pourra depenser gros pour marler[10] sa ferme de 9 en 9 ans, pour l'engraisser avec du bon fumier, pour l'entourer de bonnes hayes, pour former de bonnes plantations d'arbres &^a^ &^a^ ...? assurement non.

Detaillons encore un peu; & l'erreur paroitra d'avantage. Comment veut-on que le labourage soit fait? à la beche, ou à la charrue? que l'on prendre l'un ou l'autre parti, n'est-il pas certain, que s'il faut le travail de 4 beches ou d'une charrue, pour 60 Arpens,[11] dans un certain nombre de jours; il est evident, qu'on sera forcé d'employer 40 beches, ou 10 charrues

10 marner?

11 An *arpent* was an ancient measure of area equivalent, according to the region, to 14 to 21 hectares.

76.12.22 pour faire le meme labour, dans un tems egal, sur 600 Arpens de terre semblable? Mais il y aura toujours cette difference; que si un boeuf, ou deux personnes de la famille du petit fermier, vient à manquer par une maladie, ou par quelqu'accident, il lui restera la moitié de son terrein, ou même tout entier, sans labourage: au lieu que le gros fermier ne manquera jamais de monde, ni de bestiaux, ni d'argent (à soi, ou prèté) pour suppléer à des acidents pareils.

Qu'on me dise encore, qu'est ce que fera le petit fermier avec un pair de boeufs; & meme il lui faudra deux pairs, selon que son sol sera plus dificile à labourer. Certès ces animaux peuvent faire un labourage double de celui de sa ferme. n'est ce pas une perte reele que ce defaut d'employ? De meme ce fermier aura besoin d'autant d'especes d'instrumens & outils d'Agriculture, que le grande fermier. Voilà donc des avances d'argent qui produiront beaucoup plus chez le grand fermier que chez le petit; car une grande partie se gate plus par la rouille & acidents, que par leur employ effectif. ainsi proportion faite, le petit fermier a besoin de plus d'argent respectivement à ses profits, que celui qui travaille en grand.

Mais dit-on en fin: les petits fermiers augmenteront la population d'avantage, que les grands fermiers: & c'est la population, qui fait le puissance de l'Etat. Cette objection est aussi fausse, qu'elle est absurde. Un Roy avec trente mille hommes bien armés, & bien nourris, battera à plate couture un autre avec soissante, ou même cent mille faineans, mal armés, & mourants de faim. D'aileurs je viens de montrer, que le meme travail champestre demande le même nombre de mains: ainsi il est faux que les grandes fermes occupent moins de monde que Les petites. Pour ce qui regarde les bestiaux, il est certain, que le grand fermier aura besoin d'un moindre nombre de charrues en proportion que l'autre. Mais il y a bien d'autres profits, qui induiront le grand fermier à avoir autant & même plus de bestiaux, proportion faite à ceux du petit fermier. il les employera à d'autres travaux, soit à la charette &[a]. ou il en engraissera pour Le boucher, ce qui raporte de grands profits.

Cependant J'ose dire de plus que cette objection est tout à fait absurde, par defaut de principe. C'est l'aisance des individus, c'est le bonheur personel, qui sont la mere & le vrai pere de la population. Or ces deux circonstances ne sont pas compatibles, du moins par la plupart, avec le petit fermier. Celui-ci à peine peut-il se tenir au niveau de ses egaux. Qui sont ceux, qu'on voit perir Les premiers, dans une grande disete? les petits ou les gros fermiers? Le petit fermier n'a qu'un foible soutien en soi meme & en son petit Revenû, produit de ses travaux dans sa petite ferme: au Lieu que l'ouvrier du grand fermier, depend de la grande masse de son maitre.

Car il est sur de gagner du pain, sans en courir les risques. & aussitot que ses enfants sont capables du moindre employ, ils y seront admis dans une immense varieté d'emplois proportionels à leur age qui se trouvent dans La grande ferme; tandis que les enfants du petit fermier seront entierement à sa charge, & ne feront presque rien qui le dedomager des frais de leur entretien. Ainsi voit-on tout par tout, que la proportion des gens mariés & des garçons, entre les simples Laboureurs qui n'ont que Leur travail pour leur Subsistence, est incomparablement plus grande, que celle entre les possesseurs de terre, ou fermiers mediocres. 76.12.22

Il est de fait, que Le Souci pour L'avenir, detruit, dans les etats mediocres, cet oubli paisible, qui fait toujours un bonheur très reel entre ceux, qui s'abandonnent à La Providence, sans d'autre soutien que leur propre industrie. Ainsi Le laboureur journalier, ne fait pas dificulté d'epouser une femme & devenir pere d'une famille, qui, comme je viens de l'observer, ne Lui coutera pas si cher à entretenir, que si c'etoit sur une petite ferme à Lui-meme.

Enfin pour faire mieux Sentir toute l'absurdité de l'opinion que je combate actuellement, il ne faut que comparer La manufacture des grains & toute l'Agriculture (qui en effet n'est qu'une vraie manufacture des productions de la terre) avec une autre manufacture quelleconque. Pourra-t-on soutenir serieusement qu'un ouvrier tout seul, peut faire, à aussi bon marché & en si grande quantité, un ouvrage quelconque; comme lorsque ce même ouvrage est fait en grand dans une, ou plusieurs manufactures? Si par exemple le même homme alloit deterrer la mine, fondre le cuivre, le purifier & le tourner en laiton par le melange de la calamine; le reduire en fil, le couper par morceaux pour faire des epingles, leur aiguiser les pointes, ajouter les têtes, les polir & colourer avec l'étain &ª.; pourroit cet ouvrier, dis-je, vendre chaque epeingle par moins de 6 solds la piece: & pourroit-il en fournir autant, comm'une de nos dames modernes employent dans un Seul jour? Qu'on applique cette idée à l'Agriculture en general, & l'on verra toute l'absurdité de vouloir la faire aller à petits pas cerrés; comme dans l'enfence[12] du monde; ou comme dans une Isle deserte, oû quelque malhereux auroit echapé d'un naufrage, qui ne pourroit pas se prevaloir des avantages de la societé des autres hommes, ni du benefice des Arts, qui sont toujours poussés à la plus grande perfection, & à la plus grande utilité publique, lorsqu'on les cultive en grand.

Y a-t-il rien de plus singulier, que vouloir retrêcir le grand Art de la Agriculture, en la faisant poursuivre toute isolé par des particuliers, qui seront privés de l'avantage d'etre conduits par un directeur, dont les interets

12 enfance.

76.12.22 reunis font son propre avantage; & qui par consequence les conduira beaucoup mieux, que si chaque branche, ou partie du travail, qui doit tendre au même bout, se trouvoit dependant des varietés & fantaisies de chaque petite tête, qui ne travailleroit pas le concert avec les autres, & sous une seule direction?

De façon que Le bois de la charrue, le fer qu'on y employe, les clous, les roues & la charpente du chariot; que dis-je? le cuir même des souliers du petit fermier, Le drap de son habit, enfin les plats mêmes de sa cuisine, & tout le Reste dont il fait usage &^{a}. doit être fait en grand dans une manufacture, pour que le tout soit non seulement à plus bon marché; mais aussi dans la plus grande quantité possible: cependant La plus essentielle de toutes les manufactures, celle des fruits de la terre, doit etre faite simplement en migniature, & celà par des petits ouvriers les plus gênés en tout sens, & les moins pourvûs de ces charmes d'un avenir hereux (ou du moins plus aisé) qui sert de premier moteur à tous les travaux & entreprises humaines...! Que Les hommes sont inconsequens, même dans les affaires les plus communs de la vie civile...! &^{a}. ...! &^{a}. ...!

Voici, Monseigneur, Les idées qui me sont venûes, sur l'objet qui fait celui de cette Lettre. Hereux! si elles sont aussi utiles, que je les crois vraies: & encore plus si elles peuvent eclaircir, sur ce sujet, quelqu'un qui ait en son pouvoir La direction des affaires publics, pour prendre La route la plus avantageuse à l'humanité. Si V.A.S^{me}. jugeroit à propòs d'en faire extraire une copie, corrigée dans le langage; Je serois charmé que le R.P.D. Mann fut à meme de la Lire.[13]

J'ai L'honneur d'etre avec un Resp^{t}. le plus profond, & un attachem^{t}. le plus inviolable & Reconnoissant

Monseigneur

de V.A.S^{me}.

Servit^{r}. le plus h^{ble}. & le plus obligé.

Jean Hyacinthe de Magellan.

à Londres 22 Dec^{bre}. 1776

---

13 See the Duke d'Arenberg to T.A. Mann, 26 January 1777, in Bruneel, op. cit. (note 1). Mann, like the Duke, was very much opposed to Magellan's ideas. He set out his response to Magellan in a memoir that he read at the Brussels Academy's session of 20 April 1780 that was subsequently published as "Mémoire sur la question: Dans un pays fertile & bien peuplé, les grandes fermes sont-elles utiles ou nuisibles à l'état en général?", *Mémoires de l'Académie Impériale et Royale des Sciences et Belles-Lettres de Bruxelles*, 4 (1783), 200-222. The subject was also developed by Mann in another paper entitled "Mémoire sur les moyens d'augmenter la population & de perfectionner la culture, dans les Pays-Bas Autrichiens", *ibid.*, pp. 161-196.

To Nathaniel Pigott 77.00.00

Royal Astronomical Society, London, MSS Pigott, 44

Dear Sir,

I know of no remedy, in regard to your Barometers, but to have patience. I called there, after I came to Town, at M^{r}. Ramsden: he send me answer that they were in hand: but could not say when should be done.

I have asked the price of the hanging paper, like your Sample: and find it costs 7^{s}. per piece. each piece contains 12 yards. it should be better to have a larger, at least wider sample, in order to not mistake the drawing &^{c}.

I have ready for sending to you the Memoirs of y^{e}. R^{l}. Acad^{y}. containing that on y^{e}. Nebuleuses,[1] and the epitome of De la lands Astronomy,[2] as well as the Phil^{al}. Transact. vol. 67. p.1.[3] See if you want any thing else from here, in order to send all together according to your direction at Bristol.

I wish you should apply yourself to Maskelyne for y^{e}. observations you want made in England. You know that we are not on very friendly terms this long since.

My resp^{s}. to your Lady & sons: and heartly am

Dear Sir
ever yours
Magellan

Nathanael Pigott Esqr.
in Great Frampton near Cow Bridge
Glamorganshire

1 Messier, "Catalogue des nébuleuses et des amas d'étoiles, que l'on découvre parmi les étoiles fixes, sur l'horizon de Paris", *Histoire de l'Académie Royale des Sciences, avec les mémoires*, 1771 (Paris, 1774), 435-61.

2 Presumably Lalande, *Abrégé d'astronomie* (Paris, 1774).

3 Magellan's comment gives an indication of the year, at least, in which this letter was written. Volume 67 of the *Philosophical Transactions* is for the year 1777; Part I was probably issued relatively early in the year. The letter was evidently written after Magellan's letter to Pigott of 5 August 1777, since at the time that was written, Magellan had only just sent Pigott's barometers to Ramsden.

77.01.28 *To Ribeiro Sanches*[1]

Österreichische Nationalbibliothek, Vienna
A.N. Ribeiro Sanches Korrespondenz, codex 12714, 153-154v

Londres 28 Jan^{o}. 1776[2]

Meu am^{o}. e S^{r}. e m^{to}. do Cor^{am}.

A carta de V.M. de 30 de Dez^{bro}. me deo m^{to}. gosto, q seria m^{to}. maior se a saude de V.M. fosse melhor. mas meo amigo do Cor^{am}. tal he a nossa sorte, q ou nos vamos hir acabando por pedaços, ou somos surprendidos qd^{o}. menos o cuidamos. verd^{a}.m^{te}. este ultimo parece o mais dezejavel, pelo menos aos q não sabem sojeitar-se tranquilam^{te}. ao seo fado. Porem o Filosopho não se embaraça nem de hũ nem do outro: e como não depende do seo arbitrio q a cortina cubra a scena neste ou naquelle passo; deixa hir a comedia até o ponto em q elle se vê forçado de Retirar-se. Bem sei q o estado de sofrer dores, cançasso, despondencias, tedios &^{a}. &^{a}. he um estado mui dezagradavel & triste. Porem isto mesmo serve a suavizar as negras ideias do termo a q corremos. Quem sabe a aflição q passa pelo animo de q^{m}. morre no auge da felicid^{e}.! imagino q o seo horror deve ser maior doq pensamos. Porem qd^{o}. a enfermid^{e}., as dores, & a nossa má situação em q nos achamos, nos-fazem sentir todo o incom̃odo da nossa existencia, he sem duvida q a ideia do fim dellas nos-consola em vez de nos-afligir. Vamos pois adelante caballeros: & se a sympatia ou [a] amizade sentida pudesse fazer algua diminuição, ou pelo menos Re[pa]rtição em comum do q hé pessoal, esteja na certeza q não passara tão mal, como passa.

Aqui se diz q Espanha vai tomar parte do q possuimos perto do Rio de la plata: & que Inglaterra não nos dará o minimo socorro. na verd^{e}. q nos-importa possuir mais ou menos terreno, se com elle não se augmenta a felicid^{e}. dos vasallos?

Hontem á noite estive com o nosso Enviado alguns minutos, homem estimavel & digno do maior apreço. sua Esposa não se dá bem com os ares deste clima: mas não imagino q tenha animo p^{a}. sahir delle auzentando-se

---

1 Annotated: "R^{a}. a 6 Fevr a [com tresentas]" [Recd 6 Febr [with three hundred]].

2 Magellan has written the wrong year – a common enough mistake made by people writing letters in January. Internal evidence makes it clear that 1777 is intended. In particular, Magellan passed through Brussels in November 1776, whereas in November 1775 he was in London; the letter on the size of farms that he mentions is undoubtedly his letter to the Duke d'Arenberg of 22 December 1776; and he is known to have spent Christmas 1776 at Bowood with Lord Shelburne.

de q^{m}. ama como verd^{a}. Portugueza. o Chev^{r}. Horta, q vai Ministro p^{a}. a Haya, & verd^{a}.m^{te}. he sojeito de excellentes partes, foi dar uma volta pelas principais p^{tes}. desta Ilha. o Princepe Gallitzin a q^{m}. escrevi sobre a sua bella disp[ozi]ção & conhecim^{tos}. na physica experimental, lhe expressou ja por minha via [o] alvoroço com q o espera p^{a}. se occupar destes objetos dignos de todo o homem q pensa, & mil vezes mais divertidos, q todas as danças & futilidad^{es}. das Cortes & suas etiquetas. O Princepe Gallitzin me faz certam^{te}. m^{ta}. honra, & eu lhe sou unido de coração pelo excelente caracter honrado & bem intentionado q nelle conheci. 77.01.28

Eu fui passar os dias da festa a Bowood com Mylord Shelburne, q a isso me invitou. São coiza de 92 milhas, q fiz em hũ dia das 4 da manhan até as 6 ou 7 da tarde, pe[la] diligencia Ingleza, especie de post-chaise, q merece bem o d^{o}. nome, em lugar q as diligencias de França, são propriam^{te}. negligencias. alli h[a] quazi todos os dias compagnia nova de gente da prim^{a}. Nobreza. eu q não [sou] de tal gosto, pouco depois de jantar sahia p^{a}. o meo quarto, ou hia visitar o meo bom amigo D^{r}. Priestley, q mora alli perto. V.M. brevem^{te}. verá impressa uma carta minha, q ali escrevi, sobre uma nova construção de vidros em q se faz a agoa mineral artificial. estive com Mylord até dia de Reis, no qual tornei p^{a}. Londres: onde faço execu[tar] os instrum^{tos}. q varios amigos me pedem, q me servem de divertim^{to}. e q me fazem viver em esta especie de independencia philosoph^{a}. assaz estimado dos bons, & assaz livre das tracassarias dos máos. hoje Recebi uma carta humanissima do Duc d'Arenberg, que se queixa de eu lhe não ter escrito. Porem foi falta da posta, pois antes de partir p^{a}. Bowood, i.e. antes de dia de Natal eu lhe escrevi uma carta de duas folhas bem cheias, sobre a questão Agro-politica; a saber, q as grandes farmes são mais avantajozas p^{a}. o Estado, do q as piquenas. Quando eu passei por Bruxellas en Nov^{bro}. observei que a opinião geral, era pelas piquenas. isto hé que os grandes labradores, os quais cultivão grande quantid^{e}. de campos, são perjudiciais ao Estado & á agricultura: e q o melhor he q cada labrador cultive pouca terra &^{a}. ora nosso amigo Arbuthnot & M^{r}. Young escreverão sobre esta mat^{a}. & seguirão a opinião contraria. eu prometi de examinar esta questão: & li com cuid^{o}. o Tratado de Mr Arbuthnot, n'um dia q passei só na sua livraria o hanging day. i e o domingo. Considerei na mat^{a}. & achei outros m^{tos}. argumentos à seo favor q expendi n'uma carta ao mesmo Duque; pois he ponto interessante p^{a}. o bem do publico & dos individuos. Mostrei esta carta, antes de a fechar, ao nosso Ministro q quiz copiar os argumentos della, & cuido q [por] esta Rezão ficou retardada. Sinto q não deixei copia della; pois cuido q [as] Rezoens a favor da Agricultura em grande, são decisivas.

77.01.28 diga ao D[or]. Begue du Presle q cedo lhe escreverei, & mandarei os livros q me pede: com estes mandarei a V.M. algum Livro novo, se o ouver e q valha a pena.

minhas saudades á Familia adoptiva, & mais amigos se hé q os há.

vive vale, & ama

tuum

M.

Não sei se disse a V.M. já, q o amigo Somborski, Capelão da Embaixada da Russia em Londres trouxe consigo, no outono passado 7 ou 8 mossos Russos por ordem da Imperatrix, p[a]. aprenderem a Agricultura Ingleza. Eu lhe fiz fazer conhecim[to]. com nosso amigo Arbuthnot, & brevem[te]. hirão todos morar em hũa caza perto da farme deste galant-home [p[a].] ali verem os melhores principios practicos, mas fundados na mais solida [teo]ria. ao depois hirão viver alguns annos em differentes Provincias de Inglate[rra] p[a]. se instruirem nos differentes methodos &[a]. antes de tornar p[a]. a Russia.

iter[um]

a Monsieur
Monsr Sanchez, Doct[r]. en Medecine
Rue des Blancs-manteaux
à Paris

London 28 January 1776[3]

My Dear Friend

Your letter of 30th Dec[ber]. gave me much pleasure,[4] which would be even greater if your health were better. But, my dear friend, that is our destiny, that either we see ourselves finishing piece by piece, or we are taken by surprise when we least expect it. Truly the latter seems the more desirable, at least for those who don't know how to submit to their fate. However the Philosopher does not get embarrassed with one or the other: and as it does not depend on his will that the curtain covers the scene in this or that moment; he lets the comedy continue till the point when he is forced to retire. I know that the state of suffering pains, fatigue, disesteem, tedium, &[a]. &[a]. is a very unpleasant & sad state. Yet this serves to soften the dark thoughts about the end towards which we are running. Who knows the affliction that passes over the spirit of the one who dies in the height of happiness! I imagine that his horror must be greater than we think. Nevertheless when infirmity, pain, & the poor situation we are in make us feel all

3 See previous note.
4 Letter not found.

the discomforts of our existence, there is no doubt that the idea of their ending brings us comfort instead of grief. Let us go adelante caballeros:[5] & if sympathy or friendship experienced, or at least a sharing in common of that which is personal, could cause any relief, be sure that you won't go on as badly as you do. 77.01.28

Here they say that Spain is going to take over some of what we possess near Rio de la plata:[6] & that England will not render us the least support. Indeed, what good is it for us to possess more or less land, if with it the subjects' happiness is not increased?

Last night I was for a few minutes with our Envoy,[7] a respectable man & worthy of the highest esteem. His wife does not get on well with this climate: but I don't imagine she has the courage to separate from him whom she loves as a true Portuguese woman. Chevalier Horta[8] who will be Minister at The Hague & is truly a man of excellent parts, came to tour the principal sights of this island. Prince Gallitzin,[9] to whom I wrote about his good interest & knowledge in experimental physics, expressed to him through me the enthusiasm with which he awaits him in order to devote himself to these objects worthy of any man who thinks, & a thousand times more pleasant than all the dances & futilities of courts & their etiquette. Prince Gallitzin certainly does me much honour, & I feel very close to him at heart for the excellent, honourable & well disposed character that I knew him as.

I went for the feast days to Bowood with Mylord Shelburne, who invited me. It is a matter of 92 miles, that I made in one day from 4 in the morning till 6 or 7 in the afternoon, in the English stage-coach, a kind of post-chaise that well deserves the name, whereas the stage-coaches of France are rather negligence.[10] There you find almost every day a new company of people of the first rank. I am not of that sympathy and shortly after dinner went to my room, or went to visit my friend D^r^. Priestley, who lives nearby. You will shortly see printed a letter of mine that I wrote there, about a new construction of the glasses to make artificial mineral water.[11] I stayed with Mylord until Epiphany, when I went back to London: where I had the instruments executed that various friends ask me for, which amuses me and makes me feel, living this kind of philosophical independence, greatly esteemed by the good ones & far enough away from the wickedness of the bad. I received today a very kind letter from the Duke d'Arenberg,[12] who complains

5 Spanish expression meaning: "let's go on, gentlemen".

6 The Portuguese colony of Sacramento, on the La Plata River in what is now Uruguay, was exchanged with Spain by the Treaty of Santo Ildefonso (1777) together with some other territories, for the island of Santa Catarina on the south coast of Brazil, which the Spanish had captured late in 1776.

7 Luiz Pinto de Sousa Coutinho (1735-1804), Portuguese Minister in London, 1774-88.

8 Francisco José Horta Machado (1746-1817).

9 Dmitri Alekseevich Gallitzin (or Golitsyn) (1734-1803), Russian Ambassador at The Hague.

10 Magellan plays with the words, criticizing the French transports. In fact, stage-coach in French is "diligence", the opposite of "negligence", which is the same word as in English.

11 Magellan, *Description of a Glass Apparatus for Making Mineral Waters...* (London, 1777).

12 Letter not found.

77.01.28 that I have not written to him. However this was due to there being no post, since before I left for Bowood, that is before Christmas, I wrote him a letter of two full sheets about agro-political matters; that is, that large farmes are of greater advantage to the State than small ones.[13] When I passed through Brussels in November I noticed that the general opinion was in favour of the small: that is, that the great farmers who cultivate large fields are prejudicial to the state & to agriculture, & that it is better that each farmer cultivates small fields &ª. Our friend Arbuthnot & Mr. Young will write on this matter & will follow the opposite opinion. I promised to examine this matter: & I read with care Mr. Arbuthnot's treatise[14] on a day I passed at his library, the hanging day, i.e. the Sunday. I considered the matter & found many arguments in its favour that I expounded in a letter to the same Duke, since it is an interesting point for the good of the public & of the individuals. I showed this letter, before closing it, to our Minister, who wanted to copy its arguments, & maybe for this reason it was delayed. I'm afraid I didn't keep a copy of it; but I think that the reasons in favour of large-scale agriculture are decisive.

Tell Dr. Begue du Presle that I shall soon write to him, & shall send the books he asks for: with these I shall send you some new books if there are any worth having.

My regards to the adoptive Family, & any other friends if any are there.

Vive, vale & ama
tuum
M.[15]

I don't know if I had already told you that our friend Somborski,[16] Chaplain of the Russian Embassy in London, brought with him, last autumn, 7 or 8 Russians by order of the Empress,[17] to learn about English agriculture. I introduced them to our friend Arbuthnot, & soon all of them will go to live in a house near the farme of this galant-home in order to see the best practical principles, but supported by the most solid [theo]ry. Then they will go to live for a few years in different provinces of England to get instruction in the different methods &ª. before returning to Russia.[18]

ite[rum][19]

---

13 See Magellan to the Duke d'Arenberg, 22 December 1776.

14 John Arbuthnot, *An Inquiry into the Connection between the Present Price of Provisions, and the Size of Farms...* (London, 1773).

15 Flourish, farewell and love your M.

16 Andrei Samborskii was the long-serving priest of the Russian Embassy's church in London, 1765-1779.

17 Catherine the Great, Empress of Russia 1762-96.

18 The seven students remained with Arbuthnot for some two years and later went to Arthur Young's farm at Bradfield, Suffolk. Some of them went elsewhere, including to Robert Bakewell's famous farm at Dishley Grange near Loughborough. Most returned to Russia in 1784.

19 Once again.

To [Gabriel de Bory] 77.01.31

Archives de l'Académie des Sciences, Paris, Fond Lavoisier, 1642[1]

Londres ce 31. Janvier 1777

Mon cher Chevalier et très cher ami

Il y a quelque tems que je vous ai annoncé[2] quelques expériences faites avec les nouvelles pompes pneumatiques, comme celle que j'envoyai d'ici à M. de la Voisiere, qui paroissoient surprenantes. A présent je vais vous rendre compte de celles qu'on fit chez M. Nairne, Membre de la S. R^{le}. et un de nos 1ers. Artistes, qui exécuta aussi une un peu mieux arrangée pour Mylord Shelburne, et dont le prix ne sera peut-être moins de 45 ou 50 guinées. Je me trouvai chez lui mercredi dernier (28 de ce mois) avec le D^{r}. Priestley, le Ch^{er}. Shuckburg que vous y avez vu à Paris, M. Volaston,[3] le D^{r}. Orme,[4] le D^{r}. Kaye[5] et M^{r}. Camton.[6] Ce fut vers l'une heure après midi qu'on a commencé les expériences suivantes, qui prouveront, je crois, combien toutes celles qu'on a faites jusqu'à présent avec la machine pneumatique sont Eloignées d'être bonnes à compter là dessus; et qui demontrent que la seule jauge, pour juger de l'exhaustion ou condensation, faites par ces machines à piston, est la nouvelle inventée par le même M. Smeaton, et connue sous le nom de Jauge à poire.[7]

1 Letter not found. The text given here, a corrected version of the text published in *Oeuvres de Lavoisier: Correspondance*, ed. René Fric, fascicule III, 1776-1783 (Paris, 1964), pp. 588-90, is a copy in another hand that was apparently prepared for Lavoisier. No indication is given of the person to whom it was originally addressed. M. Fric indicates that it was addressed to Lalande, but offers no basis for this; both the salutation and Magellan's reference to this letter in his letter to J. A. Euler, 27 June 1777, suggest that it was more likely addressed to the Chevalier de Bory.

2 Letter not found.

3 Wollaston? Possibly Francis Wollaston, FRS (1731-1815).

4 Possibly Dr David Orme (c1728-1812), MD Edinburgh 1749, Senior Licenciate of the Royal College of Physicians and man-midwife extraordinary at the City of London Lying-in Hospital? (See *Gentleman's Magazine*, 82 [1812], 489.)

5 Dr (later Sir) Richard Kaye, FRS (1736-1809)?

6 Canton? Possibly William Canton, who succeeded his father John Canton, FRS (1718-1772) as proprietor of the Spital Square Academy, London; or his brother Thomas Canton, who was one of the executors of Dr Orme's estate.

7 Smeaton described his "pear" gauge – so called on account of the shape of its glass bulb – in his "Letter ... concerning some Improvements ... in the Air-Pump", *Philosophical Transactions*, 47 (1751-2), 415-24. Experiments similar to those here described by Magellan was reported by Edward Nairne, "An Account of Some Experiments made with an Air-pump on Mr. Smeaton's Principle; together with Some Experiments with a Common Air-pump", *Philosophical Transactions*, 67 (1777), 614-48.

77.01.31 J'en ai envoyé deux de ces jauges avec la machine pneumatique de M. de la Voisiere. Si vous ne les avez pas vues, il pourra bien dire à l'Academie quelle est leur forme et leur usage. On commença a travailler vers l'une heure après midi. Le Barometre, se trouvoit à 29 pouces 85 centiemes de pouce anglais.

1re. Expérience. La pompe étant bien nettoyée, sans aucun huile ou de l'eau, ni aucun cuir, le Récipient étoit si bien ajusté sur le plateau de la Machine, qu'il ne laissoit pas entrer de l'air, Etant cementé au dehors avec un cement composé de la cire jaune, de la résine, colophane et du Suif de mouton. On travailloit la pompe 10 minutes continuellement en chaque expérience. Il y avoit à côté une Jauge Torricellienne, une autre ordre. au dedans du Récipient, et une troisieme Jauge à poire de Smeaton. La Jauge Torricellienne monta à 29 pouces 75 centiemes. La Jauge ordre. fit descendre le mercure a $^{2}/_{40}$ de pouce, c'est à dire le Mercure dans le tube n'étoit que $^{2}/_{40}$. au dessus de celui dans la tasse: et la Jauge de Smeaton montra 600 fois d'exhaustion.

2de. Expérience. On ouvrit le couvercle de Récipient et on y introduit un cuir trempé dans l'huile et suif, comme celui qu'on employe dans les plateaux des pompes ordres. Alors la grande Jauge Torricellienne montra 29 p. 70 centiemes. La Jauge ordre $^{5}/_{40}$ de pouce; et la Jauge de Smeaton 32000. fois d'exhaustion.

3me. Expérience. On ôta le cuir: on travailla la pompe le même tems. La Jauge Torricellienne montra 29 pouces 75': La petite Jauge $^{1}/_{40}$: et la Jauge de Smeaton ne montra que 900. fois d'exhaustion.

4me. Expérience. On ouvrit le couvercle du Récipient; on y introduit un petit Cylindre de bois de buis, environ un pouce en diametre et 3 ou 4 pouces de longueur. On travailla la pompe le même tems. La Jauge Torricellienne montra 29 p. 70''. La petite Jauge ordre. $^{4}/_{40}$ de pouce: et la Jauge de Smeaton 16000. fois d'Exhaustion.

5me. Expérience. On ôta le cylindre de buis, et on y introduit un cercle de cuir blanc sans aucun apprêt, i.e. comme on l'achette dans les boutiques pour l'usage des pompes. Alors la Jauge Torricell. monta à 29 p. 60". la commune descendit à $^{7}/_{40}$ et la Jauge de Smeaton montra au delà de 32000. fois d'exhaustion.

6me. Expérience. On ôta ce même cuir: on le dessécha bien vis-à-vis du feu; et on l'introduit dans le Récipient: après avoir travaillé le même tems, la Jauge Torricell. montra 29. 75". la Jauge ordre. montra $^{3}/_{80}$ et la Jauge de Smeaton ne montra que seulement 400. fois d'Exhaustion.

7me. Expérience. On a pris une autre pompe, dont la construction est selon les principes de Smeaton, mais différemment arrangée que celle destinée pour Mylord Shelburne. On y adapta le Cylindre à sec, avec le

même cement résinieux à l'entour par le dehors. On la travailla toute séche: 77.01.31
et on trouva que le petit Jauge ord^re^. avoit descendu jusqu'à $^{3}/_{40}$ de pouce; mais la Jauge de Smeaton ne montra que seulement 300. fois d'Exhaustion.

8me. Expérience. On ouvrit le couvercle du Récipient, et on y introduisit le même cuir rond de l'expérience 5me. et 6me.; mais après l'avoir bien trempé dans l'eau mêlée avec de l'esprit de vin parties égales. Dans ce cas, la Jauge ord^re^. n'a pas descendu qu'à $^{22}/_{40}$ de pouce: et la Jauge de Smeaton n'a montré que 450. fois d'Exhaustion.

On voit d'après ces expériences que la Jauge de Smeaton est la seule qui peut mesurer la quantité réelle de l'Expansion du fluide qui se trouve dans le Récipient de la Machine pneumatique, lorsqu'on la fait travailler, et que les autres Jauges, dépendantes de la pression de ce même fluide, ou de l'air extérieur sur le Mercure, ne sont aucunement propres à donner une idée assez precise de l'exhaustion ou pression. M. Nairne publiera dans les Transactions une suite fort considérable de ses expériences sur ce sujet. Il nous a dit qu'en mettant dans une de ses pompes nouvelles un grain et demi d'Eau, le reste de la pompe étant bien séche, il ne trouva que 4000 fois d'exhaustion: Et dans la suite en pesant l'eau, il trouva qu'on avoit évaporé un grain: c'est à dire, que les deux tiers de cette eau étoient évaporés. Une autre fois il y mit une bonne quantité d'Eau: et alors la pompe ne fit l'exhaustion qu'environ 80 fois. D'où l'on doit conclure que un peu d'humidité fait une grande expension dans le fluide au dedans du Récipient; mais au contraire une grande quantité d'eau empêche de s'y faire une Exhaustion considérable.

Je vous prie de communiquer ces Expériences à l'Académie R^le^. et en particulier à M. le Président de Saron avec mes plus affectionnés respects et de même à Mad^e^. Si l'Abbé Rozier voudra imprimer ces nouvelles, elles sont, je crois, bien interessantes.

Vive, vale et ama[8]
Signé, Magellan.

8 Enjoy life, farewell and love.

77.02.00 *To Gabriel de Bory*[1]

Le Révérend Pere Beccaria, Professeur de Physique à Turin, fort connu par ses découvertes, fit une expérience très-curieuse sur le phosphore sulphuro-calcaire, dont on trouve la description dans les Transactions Philosophiques de Londres. Il remplit de petites boëtes de métal avec ce phosphore, & après les avoir dépouillées de toute lumière, par le moyen d'une chaleur très-forte dans une chambre obscure, il les couvrit avec des verres de différentes couleurs; il les exposa ensuite aux rayons du soleil, & les découvrant dans l'obscurité, il trouva que le phosphore rendoit la même couleur du verre qui l'avoit couvert. Cette expérience est également insérée dans les Transactions Philosophiques, année 1771, ou du moins dans le volume suivant.[2] Je me trouvai à l'Assemblée de la Société Royale, lorsqu'on en fit la lecture, & je la communiquai à mes amis, habitans dans différentes Villes de l'Europe. M. l'Abbé Bourriot, aujourd'hui Chanoine de la Cathédrale de Bazas,[3] répéta cette expérience, il ne réussit point, & elle me fit part de son peu de succès. J'écrivis alors au R. P. Beccaria, pour lui demander la marche qu'il avoit suivie pour son expérience; sa réponse fut assez détaillée pour ne plus se tromper.[4] Je la communiquai de nouveau à plusieurs de mes Correspondans; mais de retour à Londres, après le séjour assez long que je fis à Paris en 1775, j'appris que M. Wilson, Membre de la Société Royale, avoit beaucoup travaillé sur les phosphores, & que son travail l'avoit conduit à tirer des conséquences différentes, comme on le voit par le Traité qu'il publia sur ce sujet. (1)[5]

Je ne manquai pas, à la première occasion, de lui communiquer la vraie méthode du Pere Beccaria; je me chargeai de préparer les boëtes de métal; nous fîmes les expériences avec le plus grand soin, & nous ne parvînmes pas

1 Letter not found. The text is from an undated version published in *Observations sur la physique, sur l'histoire naturelle et sur les arts*, 9 (1777), 153-5, in the issue for February 1777, under the heading, "Lettre de M. de Magellan, de la Société Royale de Londres, à M. de Bory, de l'Académie Royale des Sciences de Paris, Sur une Expérience faite avec le Phosphore".

2 Beccaria, "Letter ... on his new Phosphorus receiving several Colours, and only emitting the same", *Philosophical Transactions*, 61 (1771), 212.

3 See Bourriot to Magellan, 17 December 1770.

4 Letters not found.

5 Benjamin Wilson, *A Series of Experiments relating to Phosphori...* (London, 1775; 2nd ed. London, 1776). There is a footnote at this point: "Voyez dans le Volume du second Semestre, année 1776, page 72, l'extrait & la suite des expériences de M. *Wilson*, sur *les Phosphores & sur les couleurs prismatiques qu'ils offrent dans l'obscurité.*" See *Observations sur la physique, sur l'histoire naturelle et sur les arts*, 8 (1776), 72-80.

à voir aucune couleur sur aucune des boëtes où le phosphore étoit recouvert par des verres colorés. Nous n'y découvrîmes qu'un peu plus d'ombre, plus ou moins foncée, suivant l'épaisseur & l'obscurité du verre, sans distinguer la plus légère nuance colorée. M. Wilson publia notre expérience,[6] & quoique j'aye exprimé mon doute sur la cause de notre peu de succès, soit en l'attribuant à quelque défaut dans l'appareil ou dans la manipulation, cependant, le Pere Beccaria témoigne dans sa Lettre imprimée, que je n'ai parlé ainsi que par partialité pour M. Wilson, & qu'il n'est pas content de ma manière à rendre hommage à la vérité.[7] 77.02.00

On voit par un article de la Lettre du P. Beccaria, imprimée & dattée du 29 Mai 1776, qu'il commence à douter de sa propre expérience; sans doute, parce qu'il l'a répétée sans trouver les mêmes apparences phosphoriques, vues à la première fois. Il y dit qu'il peut s'être trompé, en prenant la différente graduation de la lumière, ou la nuance de l'ombre, pour une vraie couleur, quoique foible; puisque, dit-il, ni M. l'Abbé Bourriot, ni M. Wilson, ni M. Magellan, & sur-tout, ni M. Zanotti, Président de l'Institut de Bologne, n'ont point vu ce phénomène comme je l'avois aperçu la première fois.[8]

M. Allamand, célèbre Professeur de Physique à Leyde, & très-habile dans la pratique des expériences physiques, m'assura, lors de mon passage dans cette Ville au mois de Novembre dernier, qu'il avoit reçu les rayons colorés, après les avoir séparés avec un bon prisme sur le phosphore préparé de Bologne (*qui est une Pierre calcaire, dont la préparation consiste à ce que l'on dit dans une manipulation chymique avec du soufre* (1)[9]), & que ce phosphore, examiné ensuite dans l'obscurité, lui avoit rendu la même couleur. Ce célèbre Professeur ajouta, que M. Hemsterhuis, savant distingué de la Haye, avoit été présent à ses expériences. Peu de tems après, je me rencontrai avec ces deux Physiciens & M. le Professeur Gobins, qui me confirmèrent le fait chez M. le Prince Gallitzin, Ambassadeur de Russie, en Hollande. Les momens que ce Seigneur n'emploie pas aux affaires du cabinet, sont entièrement consacrés à l'étude de la Physique, & sa maison est le rendez-vous général de ceux qui s'occupent des Sciences.

---

6 Wilson, *A Series of Experiments...*(1776), pp. 94-117, especially pp. 94-101.

7 Giambatista Beccaria, *A Letter ... to Mr. Wilson, concerning the Light Exhibited in the Dark by the Bologna Phosphorus...* (London, 1776). Beccaria in this work nowhere directly accuses Magellan of partiality, though this may perhaps be read into his pointedly twice referring to Magellan as Wilson's "friend".

8 Ibid., p. 5.

9 The editor has inserted a footnote at this point, explaining how these phosphors are prepared.

77.02.00 Il est donc évident, que des circonstances particulières se sont opposées à la réussite des expériences. L'appareil de M. Allaman, est composé des plus beaux verres que j'aie jamais vus, & il est plus que probable, que ceux dont se sont servis MM. Wilson, Zanotti, Bourriot & moi, & même le Pere Beccaria dans ses dernières expériences, n'avoient pas la perfection nécessaire pour réussir complètement. Il seroit à desirer, pour le bien de la Physique, que ceux qui ont de bons instrumens, répétassent ces expériences.

77.02.03 *To Nathaniel Pigott*

Royal Astronomical Society, London, MSS Pigott, 38

London 3 February 1777
n°. 12 Neville Court Fetter Lane

Dear Sir

I owe an answer to your kind Letter of y^{e}. 29th october[1] which I found waiting for me when I came home on y^{e}. begining of Dec^{ber}.: and, as I went to Wiltshire few weeks after to pass there the holy days, So many Little things have called for my attention after my second return home, that I have been oblidged to retard this indespite of my wishes. M^{r}. Needham at Bruxelles was long time waiting for his books, sent from here at his depart: for want of proper care at Ostende, they had been carryed to Bruges: and that may be the reason, why your friend had not received the book, when he wrote to you: but certainly he had received it before I arrived at Bruxelles: and M^{r}. Needham corrected, by my means, the mistake he had done in charging him with the cost, which M^{r}. Needham, after my information, was to return to him back. So that now it is you who owns to me the 18^{s}: 9^{d} of the cost of this book and two Letters: which you'll send when convenient.

I found the Duke of Arenberg with his Family at Eure, when I arrived at Louvain the 9th. Nov^{ber}. last after my Ramble thro' Holland. they came the tenth to Bruxelles, where I went the 14th. and stay'd above a week with them. The Prince is quite blind as to any use of Life: hardly can he

1 Letter not found.

seen more than to distinguish light from darkness. there is however a 77.02.03
kind of Cataract before the pupil of each eye: which when removed by the operation, may chance to restore his sight, if the retina is not hurted, which nobody can guess. He is full of spirits, and gay enough, submiting himself to his fate: but I could not help feeling with the rest of that amiable respectable family for such a happy an acid^{t}.

I have shewn the experim^{s}. on air[2] to different curious both in Holland & Flandres, with admiration of every sensible beholder. Professor Allamand and the famous D^{r}. Gaubius[3] went purposely to y^{e}. Haye, to see those I made to the Prince Gallitzin,[4] to whom I sent before hand a proper Apparatus from London: and have now many others order'd from different quarters.

M^{r}. Nairne has shewn here some very curious experiments in the air-pump, of his making, according to the principles of M^{r}. Smeaton.[5] it appears by them that the pear-gage invented by the same Smeaton, is the only one to rely upon, in order to know the true degree of exhaustion, or expansion of the fluid, contained in the receiver. when the whole is very dry, without any Leather or any wood in the inside, this gage shews but about 600 times of exhaustion: but if there be put in a piece of leather or of wood, even dry wood to all appearance, it will shew an expansion above 30000 times, which is a wonderfull thing, and is beiond conception: but such it is.

I beg my humble resp^{s}. to M^{rs}. Pigott and Son and believe me for Life

Dear Sir
your most obed^{t}. h^{ble}. serv^{t}
Magellan

Nathaniel Pigott Esqr.
at Stow in the Would

---

2 A repetition of some of Joseph Priestley's experiments, as described in the first two volumes of his *Experiments and Observations on Different Kinds of Air* (London, 1774-5); cf. 76.11.30 for a published reference by Magellan to his demonstration.

3 Respectively Jean Nicolas Sébastien Allamand (1713-1787), professor of natural philosophy, and Hieronymus David Gaubius (d. 1780), professor of pathology at the University of Leyden.

4 Prince Dmitri Alekseevich Gallitzin (or Golitsyn) (1738-1803), Russian ambassador at The Hague, devoted much of his time to experimental philosophy.

5 Nairne, "An Account of Some Experiments made with an Air-Pump on Mr. Smeaton's Principle; together with Some Experiments with a Common Air-Pump", *Philosophical Transactions*, 67 (1777), 614-48. Smeaton had described his "pear-gage" in his "Letter... concerning Some Improvements made by Himself in the Air-Pump", *Philosophical Transactions*, 47 (1751-2), 415-28.

77.02.04 *To the Duke Charles-Marie-Raymond d'Arenberg*
Archief Van Arenberg, Enghien/Edingen, Belgium

Monseigneur

Tout est pour Le mieux, suivant l'opinion de Pope[1] & Leibnitz:[2] & J'en suis convaincû pour cette fois. A peine avois-je fini La Lettre que J'eus L'honneur d'adresser à V.A.S^{me}. La veille de Noel;[3] que je partis pour Bowood près de Bath, oû je passai Les jours de ces fetes avec Myl^{d}. Shelburne qui m'y invita avec d'autres amis.[4] Je laissai ma lettre recommandée au soin de quelqu'un, qui s'en oublia; & ce fut la cause de son retard: ma[is] ce fut aussi par cette faute que je gagnai La Lette[5] gracieuse dont V.A.S^{me}. se daigna de m'honorer, avant Sa seconde du 27 Janvier;[6] & qui devroit mettre Le comble au bonheur dont je jouis au nombre de ceux qui se sont devoués de Corps & d'ame au service de V.A.S^{me}., si je n'aurois pas Reçû d'avance une si grande quantité de preuves eclatantes & reiterées de la grande bonté de Son Coeur, & des grandes qualités de Son Ame.

Je suis très sensible au souvenir de M^{r}. Le Colonel de Brequin,[7] & je suis très convainqû[8] des avantages de sa sonde, tant pour examiner le fond des Rivieres, que de celui des côtes, & bas fonds des mers incon-

1 Cf. the final lines of Epistle 1 of the *Essay on Man* (1733-4) by Alexander Pope (1688-1744):
Safe in the hand of one disposing Pow'r,
Or in the natal, or the mortal hour.
All Nature is but Art, unknown to thee;
All chance, direction, which thou canst not see
All discord, harmony not understood,
All partial evil, universal good:
And, spite of pride, in erring reason's spite,
One truth is clear, whatever is, is right.

2 Gottfried Wilhelm von Leibniz (1646-1716). Fundamental to his metaphysical system were the propositions that everything in the world is linked together in a grand, mutually compatible harmony, and that in creating the world, God, being good, inevitably created the best of all possible worlds. It follows that everything that happens in the world is for the greater good of the whole.

3 See Magellan to the Duke of Arenberg, 22 December 1776.

4 Bowood was Lord Shelburne's country seat.

5 Lettre?

6 Letters not found.

7 Jean-Baptiste Brequin de Demange (1712-1785), mathematician and engineer. He married Nicole de Chedville (1731-?), the Empress Maria Theresa's lady of honour.

8 convaincû?

nûes.[9] J'envoyai deja cet article copié de la Lettre de V.A.S^{me}., à M^{r}. de Castillon, Auteur de l'excellent Journal Litteraire de Berlin,[10] qui peut etre s'inserera dans quelque volume, avec l'annonce de l'anemometre ingenieux & singulier du même M^{r}. de Brequin, qui fera sans doute un present interessant au public, en donnant le dessein de Sa construction, & des changements qu'il a faits aussi à La machine pour battre les pilotis. Je pense qu'il connoit celle de cette derniere espece, inventée en Angleterre, dont on a fait souvent usage dans differents endroits. Mais si M^{r}. de Brequin ne la connoit pas, j'aurai le plaisir de lui envoyer une planche, oû elle est bien representée: & peutetre sera à peu près la meme chose qu'il aura inventé; car il n'est pas impossible de se rencontrer en fait de mechanique, lorsqu'on cherche les memes objets. 77.02.04

M^{r}. Arbuthnot m'a prié de presenter sa Reconnoissance à V.A.S^{me}. par Le souvenir dont Elle se digne de l'honnorer. Il ne manque assurem^{t}. de desirs pour aller faire un tour en Flandres & Hollande, oû il seroit charmé d'aller voir ses enfants,[11] & dans Le meme tems examiner L'etat d'Agriculture que M^{r}. Hopa, Deputé de Amsterdam à La Haye,[12] a consulté avec Lui depuis mon Retour. Mais La grande multiplicité d'affaires & soins de sa nombreuse famille, ne Lui Laissent pas assez du tems à epargner pour suivre son penchant. Il m'a deja parlé plus d'une fois de ce voyage projeté pour le perintems[13] prochain. Je lui ai promis de l'y acompagner: mais Je doute beaucoup qu'il l'execute effectivement.

Ce fut encore La semaine derniere que J'ai trouvé M^{r}. Young[14] chez l'ami commun M^{r}. Arbuthnot: &, en Leur disant d'avoir Reflechi Moi-meme sur Leur opinion relativement aux grandes & petites fermes, pour La quelle Je croiois avoir trouvé d'autres raisons puissantes, ils me Repondirent aussi bien assurés du fait, que Je ne puis point concevoir, que la question devienne problematique, sans supposer un defaut, ou meme une erreur essentielle dans Les Loix du paiz, oû Le contraire soit practiqué. Cependant Lorsqu'il s'agit des faits, il n'y a pas d'argument qui puisse tenir; pourvû que ces faits soient attribués à leurs causes reeles, dont La complication fait souvent qu'on

9 Magellan gave notice of Brequin's sonde in *Observations sur la Physique*, 1 (1773), 64; in vol. 2 (1773), pp. 336-7, there appeared a "Réponse de M. le Colonel & Ingénieur De Brequin, aux réflexions de M. Magalhaens, sur la Sonde que le premier a inventée".

10 Castillon was the principal editor of the *Journal littéraire, dedié au Roy … par un société d'académiciens* that was published in Berlin from 1772.

11 Arbuthnot's children were attending school in the Netherlands.

12 Adrian Hope (1709-1781), a member of both the Dutch Parliament and the Amsterdam City Council?

13 printemps.

14 The agricultural reformer Arthur Young (1741-1820).

77.02.04 prend Le change, en attribuant le succès à celle dont il ne depend point. Je conçois fort aisement que chaque contrée a des circonstances relatives à sa position, qui font varier à l'infini l'application de la Solution generale du Probleme. Mais quelconque soit Le cas oû les hommes se trouvent, il sera toujours vrai que leur disposition suivra les memes panchants, que leur sont implantés par leur constitution naturelle. La passion pour leur interets personels sera toujours Le premier moteur de leurs actions: vouloir l'aneantir & la supprimer par des Loix genantes, & pas des limitations economiques authorisées par le gouvernement, est une espece de fanatisme politique, aussi dangereux & aussi inutile comme celeci[15] des institutions monacales. Tout au contraire Laissez agir les hommes librement: & le tout se metra aussitôt de niveau, de soi-meme. Ceux qui auront des Talents se distingueront des mediocres: ceux qui n'en auront point, Leur serviront par besoin pour l'ouvrage manuel: & la Liberté que tous auront, pour agir Selon leur genie, fera briller un grand nombre de ceux que par l'esclavage de la classe oû ils sont nés, se trouvent ensevelis pour jamais, sans aucun avantage pour eux memes, ni pour L'Etat.

Je suis bien d'acord, Monseig^{r}. que le monde ira toujours son train, malgré les rêves honnetes du philosophe. Mais que Le philanthrope se rejouisse du moins dans ses rêves, puisqu'il n'y a pas du mal a bien rêver, tandis qu'on vois tant de malheur à notre reveil Sur La face du globe que nous habitons. Peut etre Les Tourbillons qui environnent le notre, ont une marche plus hereuse tant pour le physique, que pour Le moral & le politique. Je n'en sçais rien: mais il est plus doux de le croire, ou de le supposer. Cependant pour ce qui Regarde celui, dont je connois ce qui vient fraper sur mes sens, Je suis convaincû qu'il n'y a rien de mieux que de bien Rêver pour son bien-etre & celui des autres; sans se trop facher, neanmoins lors que ces beaux rêves ne sont point du tout accomplis. J'attends avec impatience les rêves savants de D. Mann: & soit en Rêvant, ou bien eveillé, Je m'aplaudirai toujours de l'honneur d'etre avec un Resp^{t}. sans borne, & un attachem^{t}. de coeur sans Rêve

Monseigneur

de V. A. S^{me}.

Serv^{r}. tres h^{ble}. & très obligé

Magellan.

Londres ce 4 Fev^{r}. 1777[16]

---

15 celle-ci.

16 The date appears on the top right of the last page of the letter.

To Ribeiro Sanches[1] 77.02.18

Österreichische Nationalbibliothek, Vienna
A. N. Ribeiro Sanches Korrespondenz, codex 12714, ff. 191-192v

Londres 18 Fevr°. 1777

Am°. e Sr. e mto. do Coração

V. M. esqueceo-se endossar a Lettra de cambio q por isso não pode ser recebida nem negociada aqui: e por isso a mando aqui: V. M. não tem mais do q assinar o Seo nome nas costas della.

Aqui tem V.M. as nossas contas pois q não descança sem ellas:

1776	£ s d
2 Abril 1 Camoens __	1. 6: 0
Rutty on dispensaries ____	3. 0
Kirkland ____________________	3. 0
Musgrave _________________	2. 6
Armstrong essay _____________	1. 6
Chalmers____________________	5. 6
Philosal. Transact. vol. 65. p. 2 ___	7. 6
Discourse of ye. Presidt. ________	1. 6
Levison on blood _____________	2. 6
9 Mai ----- grant cough ______	1. 0
11 Julho – Musgrave on worm fever_	. 6
9 Agosto – Gibbon Roman histy.	1: 1: 0
Phil. Transact. vol. 66. p.1 ______	7: 6
3 Musgraves on nerves ________	7: 6
Medal. Inquiries vol. 5__________	6. 0
Percival Essays vol. 3._________	5. 0
Medal. Comentaries vol. 12 ______	1. 6
Rozen______________________	5. 0
Duncan's Coment. n°. 13________	1. 6
	______
	5: 9. 6

1 Annotated "Respa. a 27 Fevr°" [Answered 27 February] and "com a lettra de cambio de 300#" [with the bill of exchange for 300#]. Letter not found.

77.02.18 Estas 5:9.6 valem em dinh^{o}. de França 131$^{\#}$. 8^{s}. V.M. pagou por mim à Ruault 96$^{\#}$. logo restava a dever 35$^{\#}$.8^{s}. isto entenda-se não contando as 300$^{\#}$. q V.M. mandou o anno passado.

Aqui esteve agora M^{r}. Somborski, o Capellão Russiano q cuido V.M. conhece, & e elle se Recomenda affectuozo a V.M. Não sei se ja disse a V.M. q elle trouxe 7 ou 8 mossos Russos comsigo no Outono ultimo, por ordem da Imperatriz, p^{a}. aprenderem aqui a melhor agricultura, a fim de a hirem ensinar & practicar em Russia. Eu o introduzi a o meo amigo Arbuthnot, q como V.M. sabe he o melhor agricultor da Inglaterra, & da Europa inteira: e tem ajustado de hir passar com elles, isto he, M^{r}. Somborski tem ajustado de tomar uma caza ao pé da farme de M^{r}. Arbuthnot, onde hirá morar com os 7 ou 8 Russos toda a primavera & verão seguinte, a fim de tomarem as ideias mais proprias dos principios da Agricultura: ao depois hirão huns p^{a}. uma, outros p^{a}. outras provincias de Inglaterra p^{a}. aprenderem a practica q se requer nos differentes terrenos, & circunstancias Locaes &^{a}. pois tendo conhecido os verd^{os}. principios, facilmente podem ao dep[ois][2] julgar das Rezoens q fazem variar a practica nos terrenos q são differentes &^{a}.

espero q V.M., ou o D^{or}. Begue du Presle me mandem pela prim^{a}. posta ou Correio, o Livro q pedi na minha ultima com a adresse à M^{r}. Roberto Bigg at y^{e}. gen^{al}. Post Office. London =

Eu là mandei ao D^{r}. Le Begue du Presle o d[iscur]so [l]ido do Presid^{t}. da Socied^{e}. R^{l}. sobre a preservação da Saude dos marinheiros: e cuido q elle o tera communicado a V.M. pela prim^{a}. occazião mandarei a V.M. as Tranzaçoens q vem de sahir do prelo com mais algum livrinho se o D^{or}. Sequeira o aprovar. Eu jantei com elle na semana de entrudo em caza do nosso Enviado: elle me disse q tinha actualm^{te}. uma cura quazi feita com applicação do mercurio a espasmos, ou não sei q, a qual havia de communicar a V.M. em hua carta.

Saud^{es}. á familia adoptiva

vive, vale & ama

tuum

M.

a Monsieur
Monsieur Sanchez Doct^{r}. en Medecine
Rue des Blancs-manteaux
à Paris

2 Conjectural reading.

London 18 February 1777 77.02.18

My Dear Friend

You forgot to endorse the bill of exchange that can't be drawn or negotiated here:[3] and therefore I send it enclosed: you have only to sign your name on the back.

Here are our accounts since you won't rest without them.

1776	£ s d
2 April 1 Camoens_______	1. 6: 0
Rutty on dispensaries ________	3. 0
Kirkland __________________	3. 0
Musgrave ________________	2. 6
Armstrong essay ____________	1. 6
Chalmers __________________	5. 6
Philos^{al}. Transact. vol. 65. p.2^{d}__	7. 6
Discourse of y^{e}. Presid^{t}. _______	1. 6
Levison on blood[4] __________	2. 6
9 May – grant cough ______	1. 0
11 July Musgrave on worm fever[5]	. 6
9 August Gibbon Roman hist^{y}.	1: 1: 0
Phil. Transact. vol. 66. p.1_____	7: 6
3 Musgraves on nerves _______	7: 6
Med^{al}. Inquiries vol. 5 _________	6. 0
Percival Essays vol. 3_________	5. 0
Med^{al}. Commentaries vol.12 ____	1. 6
Rozen[6] ____________________	5. 0
Duncan's Comment. no. 13[7] ______	1. 6
	______
	5: 9. 6

These 5:9.6 are worth in French money 131$^{\#}$: 8^{s}. You paid 96$^{\#}$ to Ruault[8] for me. Thus there remains 35$^{\#}$. 8^{s}. This is without taking into account the 300$^{\#}$. that you sent last year.

Mr. Somborski, the Russian Chaplain[9] whom I think you know, was here just now, & he sends affectionate greetings. I don't know if I have already told you

3 The bill of exchange to the value of 300$^{\#}$ sent by Sanches corresponds to the annuity that Sanches had undertaken to send to Magellan every year.

4 For the works listed to this point, see Magellan to Sanches, 16 April 1776.

5 For this and the previous item, see Magellan to Sanches, 16 June 1776.

6 For the items sent on 9 August, see Magellan to Sanches, 13 September 1776.

7 See Magellan to Sanches, 17 December 1776.

8 Bookseller in Paris.

9 See Magellan to Sanches, 28 January 1777.

77.02.18 that he brought 7 or 8 young Russians with him last autumn, by order of the Empress, in order that they learn here the best agriculture, to teach & practice it in Russia. I introduced him to my friend Arbuthnot who, as you know, is the best farmer in England, & in all Europe: and has arranged to stay there with them, that is, Mr. Somborski has arranged to rent a house near Mr. Arbuthnot's farm, where he will live with the 7 or 8 Russians during all of the next spring & summer, in order to acquire the most proper ideas of the principles of Agriculture: then some of them will go to one, others to another of the provinces of England to learn the practice required in the different soils & local circumstances &^c^. Once they know the true principles, they can easily judge the reasons why the practice varies in different soils &^c^.

I hope that you or D^r^. Begue du Presle will send me by the first post or mail, the book I asked for in my last letter, addressed to Mr. Roberto Bigg[10] at y^e^. gen^al^. Post Office London =

I sent D^r^. Le Begue du Presle the speech read by the President of the Royal Society about the preservation of the health of seamen:[11] and I assume he has communicated it to you. At the first opportunity I'll send you the Transactions that have just been published together with another little book if D^r^. Sequeira approves it. I had dinner with him in carnival week at our Envoy's house:[12] he told me that he had at present an almost complete cure with the application of mercury to spasms, or I don't know what, which he would communicate to you in a letter.

Regards to the adoptive family

vive, vale, & ama
tuum[13]
M.

---

10 In the 4th edition of the *Court and City Calendar: or Gentleman's Register* for 1762, Robert Bigg is listed as "Window man on the By-days" at the General Post Office in London. He had doubtless been promoted since then to a more senior position.

11 Sir John Pringle, *A Discourse upon some Late Improvements of the Means for Preserving the Health of Mariners, delivered at the Anniversary Meeting of the Royal Society, November 30, 1776* (London, 1776).

12 Luiz Pinto de Sousa Coutinho (1735-1804), Portuguese Minister in London, 1774-88.

13 Flourish, farewell & love your.

To Joannes Franciscus Thijsbaert 77.02.25

University Library, Université de Liège, Ms. 4130

Londres ce 25 Fevrier 1777

Monsieur & très cher ami

Je viens de vous envoyer par La voie de notre bon ami M[r]. Needham, un fort petit paquet dans une caise que Je lui ai envoyée à Bruxelles avec des verres pour Les experiences sur differentes especes d'air. Ce petit paquet ne contient qu'une phiole de Chrystaux de cuivre, & une boite avec deux douzaines de boules de pith: ce qui vous fournira deux experiences nouvelles dans vos lectures, qui sont fort agreables & nouvelles. 1°. prennez un morceau de feuille d'etain, aussi large qu'une ou plutôt deux cartes à jouer. prennez de ces crystaux, que vous casserez s'il est necessaire sur un papier, & etendez-les au long sur le bord de la feuille d'etain; mais faites celà assez vite; tournez-le tout dans un rouleau que vous plierez aux deux bouts, tout de meme comme vous verrez le petit rouleau de feuille d'etain que Je vous ai envoyé dans le meme paquet. alors vous mettrez ce rouleau sur une brique ou sur une table, & en moins d'une minute, vous verrez que le rouleau se gonfle de lui meme, & qu'il creve avec une vraie effervescence chaude. meme il detonne & on y voit des ettincelles reeles de feu. Cet effet depend de l'action des crystaux du cuivre (qui ont eté formés par la solution de ce metal dans l'acide nitreux) sur la feuille d'etain: & c'est par cette raison qu'il faut se depecher, lorsqu'on les met sur la meme feuille & qu'on en fait le rouleau &[a]. Ce phenomene fut decouvert, il n'y a pas long tems en Angleterre, par un Chymiste Charlatan, nommé le D[r]. Highins. Vous pouvez voir ce memoire dans le vol. 63 des Transact. Philos. pour 1773 à page 137.[1]

Pour ce qui regarde les petites boules de pith elles servent à une experience fort jolie, & qui fait surprendre. elle est de l'invention de M[r]. Bryant, Anglois,[2] qui s'est adonné avec empressement aux phenomenes de l'electricité. Prennez un gros goblet de verre, en le

---

1 Bryan Higgins, "Actual Fire and Detonation produced by the Contact of Tin-foil, with the Salt composed of Copper and the Nitrous Acid", *Philosophical Transactions*, 63 (1773), 137-41.

2 Possibly Dr. Wiliam Bryant (1730/1-1786) who described the electric eel but we have not found reference to the experiment described by Magellan.

77.02.25 tennant avec la main au dehors un peu au dessous de son embouchure. Faitez adapter avant celà une pointe de metal au bout du conducteur de votre machine electrique. faitez travailer cette machine de façon que le conducteur soit assez chargé d'electricité: & appliquant l'interieur de votre gros goblet au point du conducteur, tandis que la machine tourne, continuez de l'y soutenir, en le tournant doucement, de façon que la pointe de metal touche successivement tout l'interieur, c'est à dire la surface interieure du goblet. or dans ce cas vous savez que ce goblet sera chargé au dedans avec de l'electricité positive, & que la surface exterieur se dechargera par votre main & bras, pour devenir negative, comme dans La bouteille de Leide: ainsi il faut avoir soin de ne pas approcher trop votre doigt de l'embouchure du goblet, parce qu'alors vous recevrez le coup electrique, & la surface interieur du verre ou goblet sera dechargée. lors que vous croyez qu'il est assez chargé, tournez-le vite sur les boules de pith, que vous aurez mises ensemble dans un tas, sur une table, ou sur un plat de metal: & vous verrez que ces petites boules entrent dans un mouvem^{t}. perpetuel, sautant de haut en bas sans s'arreter presque: & celà pendant long tems, quelques fois ce mouvement, qui semble volontaire, dure 6, 7, & meme jusqu'à 9 heures. Lorsqu'il se ralentit, vous n'avez qu'aprocher votre main, & toucher l'exterieur du goblet, qui est renversé, pour leur donner de la vitesse. vous concevez bien que le mouvement des petites boules, très legeres en elles memes, n'est que celui de l'electricité qui se decharge par leur moien. la table ou plateau n'etant pas electrisé, permet que les boules soient attirées en haut, par la surface interieure du goblet[.] aussitôt qu'elles y sont chargées d'electricité, elles sont repulsées en bas, & s'y dechargent dans la table ou plateau: la meme operation recomence de nouveau, & ainsi la dance des boules dure, tandis qu'il y a plus d'electricité dans l'interieur du goblet, que dans la table ou dans le plateau. N.B. si l'on deseche bien ces boules au feu, alors il n'y aura pas assez de dance: car dans ce cas, le defaut d'humidité, fait que les boules ne sont bien affectées par l'electricité.

J'ai eté presque tempté à vous envoyer plus de ces bouteilles très delicates dans le fond avec des tuyaux en forme de S, pour les experiences sur differents airs. Car si vous avez poursuivi à travailler dans ces experiences, vous en aurez sans doute plusieurs decassées. M^{r}. Le Prince de Gallitzin[3] m'a deja envoyé demander d'avantage, & je viens de lui envoyer une ou deux douzaines. Je crois que J'ai mis dans votre petit

3 Russian ambassador at The Hague and an enthusiastic experimental philosopher.

paquet, une petite brochure que J'imprime actuellem[t]. sur ma construction nouvelle pour faire les eaux minerales avec de l'air fixe, en peu de minutes, & en plus grande quantité. J'ai completé mon eudiometre ou instrument pour reconnoitre la salubrité de l'air respirable; & je vais finir sa description, qui sera imprimée dans la meme brochure avec des figures dans la meme planche.[4] si vous le souhaitez, je vous en envoyerai un de ces instruments. 77.02.25

Je vous prie de mes complim[s]. à Messrs. tous nos amis comuns: & en particulier à M[r]. Le Chancelier,[5] à M[de]. & Messrs. la Mere, Pere & fils, oû nous avons soupé la veille de notre depart. ditez-moi, comment se plait la jeune Chanoinesse dans son Noviciat. Je m'oublie toujours leur nom de famille, mais je ne m'oublierai jamais leurs personnes, Leur etant si obligé par leur politesses & bon accueil &[a].

vive, vale, & ama
tuum
Magellan[6]

A Monsieur
Monsr. L'Abbé Thisbaert, Doyen de
S. Jacques, Principal du College Royal, & Profess[r].
dans L'Université de
Louvain
Flandres

4 Magellan, *Description of a Glass Apparatus, for Making Mineral Water;... together with the Description of Some New Eudiometers...* (London, 1777).

5 Joseph de Crumpipen (1737-1809), Chancellor of Brabant and first President of the Brussels Academy of Sciences.

6 Enjoy life, farewell & love | your | Magellan.

77.03.00 *To Johann Castillon*

Öffentliche Bibliothek der Universität Basel, MS L I a 709, f. 67[1]

[...]

Je vous prie de dire à M^r^. Bernoulli que je lui ai acheté déjà le petit livre House-Wife de susanna Carter,[2] qui partira avec les vôtres: que je ne sais rien de Sir Howkins Brown;[3] seulement un ami commun, M^r^. Poor,[4] vint chez moi, il y a plus de 2 mois, pour avoir le même livre que je devois délivrer au dit M^r^. Brown. Je l'ai vu depuis, et il m'a dit qu'il avoit rempli ce que M^r^. Brown avoit desiré qu'il fît.

[...][5]

1 Letter not found. The text is an extract included by Castillon in a letter to Johann III Bernoulli, 25 March 1777, that Castillon introduced as follows: "Je vous rends graces de la lettre de M^r^. Magellan que vous m'avez envoyée. Voici la copie de l'article qui vous regarde dans la mienne." See also Magellan to J. Bernoulli, 14 March 1777.

2 Susannah Carter, *The Frugal Housewife; or, Complete Woman Cook* (London, 1772).

3 Isaac Hawkins Browne; see Magellan to J. Bernoulli, 12 October 1776, note 1.

4 The mathematician Edward Poore, FRS (1745-1803)?

5 Castillon added to Bernoulli: "Si vous ecrivez bientôt à M^r^. Magellan, je vous prie de lui dire que je m'acqui[ter]ai exactement de ses commissions. Vous pouvez me faire tenir ce que vous [v]oulez lui envoyer; je le mettrai dans mon ballot. La Société de Londres donnant à l'Académie ses Transactions, la Commission n'[a] pas attendu le présent des observations de Greenwich pour lui offrir ses nouvelles Tables. On les envoya dès qu'elles parurent, et M^r^. Magellan m'a écrit qu'il les avoit présentées."

77.03.11 *To the Marquis de Condorcet*[1]

Bibliothèque de l'Institut de France, Paris, MS 876, no. 35

Londres ce 11 Mars 1777

Monsieur

Les premieres tentatives pour La Cure du Cancer par l'application de l'air fixe, furent faites par M^r^. Hurdman, cirurgien Anglois,[2] dont je m'avois

1 Marie Jean Antoine de Caritat, Marquis de Condorcet, well-known French mathematician, philosopher and politician.

2 Not identified.

oublié le nom lorsque Je les annonçai à l'Acad^{e}. R^{le}.: il fut obligé de sortir de Londres pour des affaires pressants: & à son retour il vint chez moi, pour m'apprendre qu'au bout de 20 ou 25 jours les douleurs & l'elargissement de la plaie du cancer avoient repris leur cours ordinaire, sans que l'application continuée de l'<u>air fixe</u> produisit le meme effet qu'au commencement. De celà il faut conclure, qu'on manque encore de connoitre l'obstacle qui rendit sans effet la continuation de ce remede, dont le commencement avoit donné des esperances si flateuses. 77.03.11

Lorsque Je fus à Bruxelles en Novembre dernier, M^{r}. Creschini,[3] gentilhomme de la Chambre du Prince Charles, me pria d'instruire M^{r}. Imerbaut[4] cirurgien, qui traitoit le cancer d'une dame sa parente, ou meme Je crois que c'etoit sa propre femme: Je lui ai expliqué & montré tout le detail de cette application: & Je le priai de me laisser savoir l'effet que ce Remede produisit. Jusqu'au present [je n']ai pas[5] reçu aucune nouvelle: & Je crains que Le succes [n'a pas] repondu à notre attente.

M^{r}. Hurdman m'a promis de tenter encore de nouveau ce remede sur le premier malade qu'il aura de ce genre: & qu'il tachera de trouver la cause d'oû provient cet obstacle qui empeche la continuation des memes bons effets que ce Remede produit dans les premiers tems.

Soyez assuré que Je ne manquerai pas d'annoncer à l'Acad^{e}. & au public, une decouverte si interessante à l'humanité, si par bonheur elle vient à etre connue avec certitude & aussi complete qu'il le faut.

J'ai l'honneur d'etre avec l'attachement le plus vrai, & parfaite reconnoissance

Monsieur
Votre très h^{ble}. & obeiss^{t}. serv^{r}.
Magellan

A Monsieur
Monsieur Le Marquiz de Condorcet
Secretaire perpetuel de l'Acad^{e}. R^{le}. des Sciences
à l'Hotel de la Monnoye
à Paris

3 Jean Henri Creskens?

4 Himmelbauer? There was a surgeon of this name who was surgeon to Their Royal Highnesses, 1785-92.

5 Paper damaged, words missing at the end of this and the following line.

77.03.11a *From Johann Castillon*

Royal Society of London, Letters & Papers, VI, 246[1]

Nous vimes ici le 26 Fevrier une superbe Aurora Boreale. Elle se leva très peu au dessus de l'horizon du coté du Nord, & fut accompagnée d'une lumiere très vive, qui s'etendoit sur tout le Ciel. À 8^h. 21', tems vray, ce trait paroissoit sortir d'Arcturus, & passoit par dessus les Gemaux[2] & la tete du Taureau jusqu'à l'horizon occidental. Quelques minutes après ce trait quita Arcturus, & passa un peu vers le Sud. à 8^h. 46', il s'etendoit de l'Orient à l'Occident, passant par Jupiter & entre Aldebaram[3] & les epaules d'Orion. à l'Orient il commençoit par le point de l'horizon qui est audessous d'Arcturus, & à l'Occident il touchoit un point de l'horizon à 20°. environ de l'ouest du coté du Sud. À 8^h. 53', Il commençoit & finissoit aux memes points de l'horizon, & passoit par le dos du Lion, entre Jupiter & Procyon, & par dessus l'epaule orientale d'Orion, oû il resta jusqu'à 9^h. passées. À 9^h. 19'. Le trait des memes points de l'horizon passoit par dessus La queue du Lion, entre Castor & Pollux, par dessus le Ceinturon d'Orion, jusqu'au 62°. de l'Equateur, oû il se courboit. Peu après la lumiere commença à s'affoiblir vers le milieu. À 9^h. 31'. on ne la vit presque plus. Sa direction etoit par la queue & le coeur du Lion, la tete de l'Hydre, Syrius, & Procyon, & par dessus la nebuleuse dans l'epée d'Orion, jusque vers le 60°. de l'Equateur, oû elle se courboit fortement vers l'horizon occidental. Quelques minutes après le trait disparut entierement; mais l'Aurore boreale, qui etoit toujours restée près l'horizon au Nord, parût jusqu'après le lever de la Lune, qui arriva ici à 9^h. 41'. Le milieu, tant de l'Aurore que du trait, se trouva dans le meridien Magnetique. L'Academie vous prie de faire part de ces Observations à vos Astronomes, & de lui communiquer les leurs.

---

1 The text is in Magellan's hand and is headed: "Copy of part of a Letter from Monsr. de Castillon, Member of y^e. R^l. Acad^y. of Sciences of Berlin & F.R.S. to M^r. de Magellan Fellow of the R^l. Soc^y., dated 11th. March 1777, Containing the account of an extraordinary Aurora Borealis: communicated by the same at y^e. desire of the Said R^l. Acad^y. of Berlin." It is accompanied by an English translation in another hand.

2 Gémeaux?

3 *sic*. The spelling is corrected in the translation.

To Johann Albrecht Euler[1] 77.03.14

St Petersburg Branch, Archives Department, Russian Academy of Sciences, f. 1, op. 3, Nr. 63, ll. 24-25^{ob}

Londres 14 Mar. 1777

Monsieur

Ce n'est pas par negligence que j'ai retardé cette reponse à la Commission dont vous m'avez honnoré, pour l'Acad^{e}. Imperiale des Sciences.[2] Mais voulant repondre avec precision à tous les articles; il y en avoit, qui demanderont du tems pour obtenir la decision. Je fis un extrait de votre lettre, pour la presenter dans le premier conseil de la Soc^{e}. R^{le}., qui a la direction entiere de ces affaires: on y ordonna tout de suite, qu'on fourniroit à votre Acad^{e}. Imp^{le}. tous les volumes des Transactions qui vous manquent depuis l'année 1767, comme vous le marquez, y compris le Vol. 57 & suivants, que la Soc^{e}. R^{le}. prie le corps de votre Acad^{e}. vouloit bien accepter de leur part.[3] On trouva par le registre du Bibliothecaire, que le Vol. 59 avec les 3 suivans vol. des Transactions Phil. avoient eté delivrés à quelqu'un pour le remettre chez vous: on aura de la difficulté à les retrouver. Mais, que ces 4 vol. paroissent ici, ou non; celà n'empechera point, que le total soit complet, & prêt à partir par les premiers vaisseaux pour St. Petersburg.

Pour ce qui regarda l'Abridgement des Transact., je me suis informé chez le libraire qui les vend (car ils ne fut point imprimé par ordre de la Société R^{le}.) & je trouve qu'il n'en a plus que les 10 tomes, que vous avez d'imprimés; quoiqu'on les fait relier souvent en 11, & même en 13 vol.[4]

Les Acta ou Commentaires de Votre Acad^{e}. Imperiale, qui se trouvent dans la Bibliotheque de la Société R^{le}. de Londres, commencent dans l'année 1726, vol. 1, qui fut imprimé en 1728; & finissent par le vol. XVI pour l'année 1771, qui fut imprimé en 1772. Tous les suivants manquent

---

1 Annotated: "reçu le 26 Mars 1777 & lû à l'académie le 27 du même". See *Protokoly zasedanii konferentsii Imperatorskoi Akademii Nauk...*, III, 295.

2 Euler's letter not found.

3 The Council of the Royal Society agreed at its meeting on 29 January 1777 to furnish the Academy of Sciences in St Petersburg "with all volumes of the Phil. Trans. which have been published since the year 1767 inclusive; such as are not in the house to be purchased at the Society's expence" (Royal Society, *Council Minutes Copy*, vol. 6, p. 317).

4 *Philosophical Transactions and Collections, to the End of the Year 1750, abridg'd and dispos'd under General Heads...* (London, 1705-56). Though nominally a 10-volume set, the collection was often bound up in more volumes than this.

77.03.14 dans cette collection. Vous devez en savoir le nombre, & aurez la bonté de nous les envoyer, suivant l'accord genereux de votre Acad^{e}. Imp^{le}.[5]

A present il s'agit d'arranger cet envoy reciproque. Je suis convaincû d'après des experiences reiterées, qu'il ne faut jamais laisser aux soins des libraires les commissions litteraires, dans lesquelles ils n'ont pas un interet bien considerable pour leur commerce. Ce sont des mercenaires, ou pour mieux dire des vraies sangsues, qui ne cherchent que succer la substance pecuniaire des gens de lettres, sans avoir le moindre egard pour l'avantage de ces derniers.

Je prendrai donc sur moi le soin de faire empaqueter les Vol. des Transactions Philos. qui vous manquent: & je les delivrerai au marchand que vous me nommerez ici: Soit M^{r}. Baxter,[6] ou un autre quelconque. Et je vous prie de faire bien empaqueter dans une caisse les vol. de vos Commentaires pour la Société R^{le}; & de les delivrer à Messrs. Robt. & John Hay & Co., Merchan^{s}. Anglois à S^{t}. Petersburg, aux quels j'ecris cette meme poste[7] pour vous rembourser de cette depense; de meme que le Marchand d'ici que vous nommerez, me remboursera de la mienne, suivant l'avis que vous lui ferez en lui donnant vos ordres là dessus.[8]

Si vous souhaitez quelque chose de ce pais-ci, je Vous l'envoyerai très volontiers, dans la meme caisse avec les Transact. pour l'Acad^{e}.:[9] & je profite de même de vos offres obligeants, en vous priant d'ajouter dans la même caisse les 3 articles suivants pour mon compte, & dont les memes Messrs. Hay & Co. vont avoir ordre pour vous rembourser, ensemble avec la depense de la caisse & emballage des Commentaires ci dessus. Savoir. 1°. le tom XV des vos Commentaires nouveaux pour 1770: & ceux qui seront sortis après l'année 1772: car j'ai le vol. XVI pour 1771, sans le vol. precedent, & sans les suivants (après 1771). 2°. Je manque les Tables ou planches 7me. & 9me. du Tome 3me. des vieux Commentaires imprimé en

5 There is an annotation at the bottom of the page at this point: "N.B. Je fus obligé d'ouvrir cette lettre pour corriger les nombres: & de la cacheter de nouveau".

6 Alexander Baxter, a member of the Russia Company since 1752, was appointed Russian Consul-General in London in 1773.

7 Letter not found.

8 The arrangement suggested by Magellan runs counter to what had been envisaged by the Council of the Royal Society, which in agreeing to supply missing issues of the *Philosophical Transactions* had resolved that Euler "be requested to transmit the future publications of the Academy of Petersburgh to the Society thro' the hands of P. Elmsley Bookseller in the Strand".

9 The Academy resolved to ask Magellan to obtain (1) a complete set of the *Nautical Almanac*; (2) the three volumes it lacked of Thomas Pennant's *British Zoology* (London, 1768-70; 4th ed. 1776-7); and (3) Pennant's *Genera of Birds* (Edinburgh, 1773). See *Protokoly*, III, 295.

1732: ainsi je vous prie de tacher de me les envoyer, s'il est possible. 3°. enfin je vous prie d'y ajouter aussi, *2* ou 3 exemplaires des lettres de M^{r}. Euler, votre digne Pere, à une Princesse d'Allemagne,[10] pour autant de mes amis, qui me les demandent. Vous aurez la bonté d'ajouter ces 3 articles dans la meme caisse avec les Comment^{s}. pour la Soc^{e}. R^{le}. & prendrez vos deboursem^{t}. du meme M^{r}. Hay & Co. comme je viens de vous l'annoncer. Lorsque celà sera expedié, je serai charmé d'en savoir le detail par la poste. 77.03.14

Pour eviter le trouble d'une autre lettre à M^{r}. l'illustre Euler, votre Pere, permettez que j'ajoute ici ces deux mots, que je vous prie de lui participer. Savoir: qu'on a fait toute diligence possible dans la Société R^{le}. pour trouver son Memoire sur la Theorie de la lune: mais on a eu la mortification de ne pas le trouver.[11] on a taché de s'informer des papiers de feu M^{r}. Murdock,[12] pour voir si ce Memoire s'y trouvoit: & on n'a pas trouvé aucune notice. S'il etoit possible d'avoir une copie du brouillon que M^{r}. Euler aura laissé chez soi de ce precieux Memoire la Société Roy^{le}. seroit bien flatée de l'avoir, & moi de le [...][13] presenter en son nom.

Je serai fort curieux de savoir si la pendule du gen^{al}. Baur[14] a la justesse que d'autres pareilles en differens endroits de l'Europe ont montré. Si je saurois qu'elle seroit portée à votre observatoire, je ne manquerois pas d'envoyer en sa place une autre avec l'echapment à repos, qui pour la plupart reussissent encore mieux. Mais je me suis circonscript dans la some de l'ordre du general: & ce qui est encore plus drole, son commissaire ici, l'a pris si fort à la lettre, qu'il a refusé d'en paier les depenses concomitantes de la caisse & embalage, sans un ordre nouveau du meme general, qui apparem^{t}. s'a oublié de le donner jusqu'ici.

Nous avons actuellem^{t}. des pompes pneumatiques d'une grande perfection. on a poussé le vacuum jusqu'a la 32000me partie, je puis meme assurer l'avoir sû jusqu'à la 64000me partie. C'est à dire: que l'air au dedans du recipient, etoit rarifié jusqu'à trente deux mille fois de son volume. C'est avec ces pompes qu'on a decouvert dernierem^{t}. l'erreur des experiences qu'on a faites jusqu'à present avec les jauges ordinaires: qui montrent seulem^{t}. la pesanteur especifique du fluide rarifié. au lieu que celles à poire, de l'invention de M^{r}. Smeaton, montrent l'etat actuel de la rarefaction du

---

10 Leonhard Euler, *Lettres à une princesse d'Allemagne sur divers sujets de physique et de philosophie* (3 vols, St Petersburg, 1768-72).

11 The missing work was in the hands of Anthony Shepherd, professor of astronomy at Cambridge; see Magellan to J. A. Euler, 27 June 1777.

12 The mathematician Patrick Murdoch, FRS (d. 1774).

13 Paper damaged, word or words missing.

14 See Magellan to J. A. Euler, 17 September 1776.

77.03.14 fluid au dedans du recipient.[15] Lorsqu'on reseche bien tout l'interieur du recipient, alors l'expansion ne passe pas au delà le 600 ou 700 fois, mais aussitot qu'on y met un morceau de cuir, quoique sec en apparence; ou du bois sec: alors le meme effet repandit: & le bois, ou le cuir devient plus leger. N.B. vous concevez bien, qu'il ne faut pas mettre aucun cuir, ou matiere humide au dedans, dans le cas ci dessùs. Ces pompes coutent 35 guineas. Communiquez je vous prie cette nouveauté à l'Acad^e^. Imper^le^. de ma part; en cas qu'on ne l'ait pas encore communiqué d'aileurs.

Vous avez actuellem^t^. ici un cirurgien très habile, & très avide pour acquerir toute sorte d'instructions. il travaille jour & nuit en suivant presque tous nos cours scientifiques. Je l'introduirai un de ces jours à la connoissance du D^r^. Priestley. Son nom est M^r^. Thode.[16] peut etre le connoissez vous en personne.

Le 3me. vol. de ce Philosophe (Priestley) sur differentes especes d'air, paroitra la semaine prochaine, il se traduit actuellem^t^. à Paris. les deux premiers sont deja traduits & publiés en françois.[17] C'est une branche nouvelle de Physique qui promet beaucoup, & qui a deja ouvert un champ trop vast à nos connoissances. J'ai envoyé d'ici un grand nombre des instruments necessaires pour celà, à plusieurs savants en differents endroits de l'Europe.

Mais il faut finir dans le bon ton des anciens.

Vive, vale, & ama.
tuum
Magellan[18]

A Monsieur
Monsr. Jean Albert Euler
directeur des Etudes au Corps Imp^l^. de Terre
& Secretaire des Conferences de l'Acad^e^.
Imp^le^. des Sciences
à S^t^. Petersbourg

15 Smeaton described his "pear" gauge – so called on account of the shape of its glass bulb – in his "Letter ... concerning some Improvements ... in the Air-Pump", *Philosophical Transactions*, 47 (1751-2), 415-24. The work here described by Magellan was reported by Edward Nairne, "An Account of Some Experiments made with an Air-pump on Mr. Smeaton's Principle; together with Some Experiments with a Common Air-pump", *Philosophical Transactions*, 67 (1777), 614-48; the principle in question was simply to remove all sources of moisture from the space being evacuated.

16 Not identified.

17 Joseph Priestley, *Experiments and Observations on Different Kinds of Air*, vol. 3 (London, 1777); *Expériences et observations sur différentes especes d'air* (3 vols, Paris, 1777).

18 Enjoy life, farewell, & love | your | Magellan.

To Johann III Bernoulli 77.03.14a

Öffentliche Bibliothek der Universität Basel, MS L I a 709, f. 64-65

Londres 14 Mars 1777

Monsieur & Cher ami

Quoique Je vous envoyai deux mots par la voye de M^r^. de Castillon la poste derniere,[1] Je crois necessaire de vous dire d'envoyer 3 ou 4 du premier vol. de votre Receuil pour Les Astronomes,[2] dont le tout est tout à fait vendu; & il restent assèz des autres, qui peut etre demandent d'etre completés avec le premier.

Je vous ai dit dans la Lettre de M^r^. Castillon ce qui se passa sur vos Tables manuscriptes, quoique sans les nommer & par des termes qu'il ni personne ne peut entendre. Vous y verrez, que Monsr. Browne[3] a fait prendre de ma main les Tables par M^r^. Poor,[4] qui est un bon ami & vrai honnet homme. Sans doute elles sont presentées au Board of Longitude & vous devriez ecrire à vos amis Maskelyne, D^r^. Sheperd M^r^. Hornsby &^a^. sur cet affaire, en leur faisant concevoir la valeur & l'usage interessant de Ces Tables.[5]

J'envoyerai bientot d'ici quelque caisse de livres à M^r^. de Castillon, & j'y ajouterai la brochure frugal house wife.[6]

De meme M^r^. de Castillon doit m'envoyer bient[ôt] quelques nombres du Journal:[7] & vous pouvez allors envoyer les exemplaires ci dessùs du premier vol. de Votre Recueil.

Je me trouve à present trop acablé de petits aff[ai]res qui ne me Laissent pas assez du tems. ainsi je finis en Vous assurant combien Je suis pour La vie

Mon cher ami
Votre tres h. & ob^t^. serv^r^.
Magellan

aiez la bonté d'envoyer les lignes dans la page suivante à M^r^. de Castillon.[8]

---

1 Magellan to J. Castillon, March 1777?

2 Bernoulli, *Recueil pour les astronomes* (1771-6).

3 Isaac Howkins Browne; see Magellan to J. Bernoulli, 12 October 1776, and Magellan to J. Castillon, March 1777.

4 The mathematician Edward Poore, FRS (1745-1803)?

5 Bernoulli's tables were eventually published as *Sexcentenary Table, exhibiting at sight the result of any proportion where the terms do not exceed 600 seconds or 10 minutes, with precepts and examples* (London, 1779).

6 Susannah Carter, *The Frugal Housewife; or, Complete Woman Cook* (London, 1772).

7 Presumably the *Journal littéraire*; see 76.08.06.

8 Page not found.

77.03.21 *To Ribeiro Sanches*[1]

Österreichische Nationalbibliothek, Vienna
A.N. Ribeiro Sanches Korrespondenz, codex 12714, ff. 193-194v

Londres 21 Março 1777

Am°. e S^r^. e m^to^. do Coração

Se as infirmidades dependessem do nosso arbitrio ou dos nossos amigos, teriamos Rezão p^a^. nos quixar & p^a^. nos afligir; mas sendo, como são, decretos do Fado; não nos-fica outro dezafogo q sofrer a p^te^. q he inevitavel, procurando atenuala com os Lenitivos da socied^e^. affectuoza dos amigos, & tambem com o divertimento comico das Loucuras humanas de q as Ruas hãbitadas são o theatro, & das quais a conversação da vida civil (ainda a dos mais estupidos, ou dos mais discretos) he o Recitativo. Por estas Rezoens q V.M. deve confessar bem fundadas & philosophicas, não posso aprovar q viva separado da convivencia civil, & q não Receba visitas inuteis dos importunos; pois em as tomando como medecina moral p^a^. o seo divertimento, ficará sentindo menos a fatalid^e^. das suas incom̃odidades. a attenção dividida, ainda mesmo com um pouco de amofinação, faz huma certa suspenção ou pauza nas nossas sensaçoens, as quais q^do^. variadas não são tão insofriveis. o Quevedo chamava inferno aquelle sempre sempre do matrimonio: e cuido q he texto de authoridade tão canonica como a do Direito Eclesiastico

Recebi a Lettra de cambio por 300^#^. q foi paga puntualm^te^. e taobem o Livro das Memorias sobre o Nitro de q quizera saber o preço p^a^. o carregar na Conta do Sujeito q m'o pedio.

Ainda não tive occazião p^a^. me informar de M^r^. Zimlin Somborski sobre o Quali da Russia: e não deixarei de dar p^te^. ao Cav°. Pringle, se o erro for essencial. Mandarei o worship of Christ brevem^te^.: se he p^a^. o seo uzo, gavo-lhe o bom gosto de buscar semilhante Leitura. o meo amigo Priestley, q he hommem de bem, bom philosopho, honrado, sincero, & de grande penetração, tem-se declarado contra a espiritualid^e^. & immortalid^e^. da alma, q deve perecer com o corpo; mas q deve ser resuscitada com elle por milagre de J.Chr°.: diz q he realm^te^. material, & q a espiritualid^e^. não pode ser defendida pelas Escrituras Santas &^a^. &^a^. Com tudo isso assenta & cré a Revelação, & q as mesmas Escrituras & evangelhos forão inspirados

1 Annotated "R^a^. a 3 de Abril" [Answered 3 April] and "Que o livro sobre o nitro custa 5^#^" [that the book about nitre costs 5^#^]. Sanches' letter not found.

por Deos, exceto naquelles artigos q os escritores das mesmas Escrituras ajuntarão da sua cabeça &ª. &ª. Ora veja lá V.M. como tudo isto se combina em hũa cabeça tão excellente & judicioza: & o q he mais em hum animo tão humano, amavel, & cheio de bons sentimentos ...! tal he o espirito humano ...! Elle está imprimindo hũa obra metaphysica sobre esta materia, & sobre a Liberdade das açoens humanas: e o q mais hé Li com m^to^. gosto a prim^a^. folha q achei della ja impressa sobre a sua meza uma destas noites ...! 77.03.21

Mostrei ao nosso Enviado o artigo da carta precedente de V.M. elle a leo toda: & rio assaz com as nenias sobre a suposta falta das minhas arrtaçoens. Elle me disse = ora pois diga ao nosso amigo, q visto ter tão boa lembrãça dessa da m^a^. caza q lá conheçeo em Coimbra em 1716 ou 1717; tãobem se lembrara melhor de me ter escrito a mim proprio (cuido q a Vienna, ou Malta, & Portugal) e me ter mandado de prezente o Livro do Abb^e^. fulano (esquece-me o nome) sobre o methodo de prezervar o grão &ª. q eu ainda conservo a Sua carta, e ainda a encontrei com gosto hũ destes dias entre os meos papeis: q eu o estimo & sempre estim[arei][2] & lhe dezejo todo obem e Saude &ª. =

Coitada da M^a^. Joana. Tosse, catharro, febres, annos, mao humor, & o q mais he, falta daquillo q V.M. não pode dar-lhe; são outros tantos artigos p^a^. estar melancolica, & fazer outro tanto aos q a vem. mas he p^a^. admirar q não queira tomar Remedios: q^m^. sabe se ella crê tão pouco nelles como eu, ou se ella está ja tão doutora como V.M. ...? Delhe minhas saud^es^. & á prima, sobrinha & primo &ª.

vive, vale & ama
tuum
M.

Hontem me disserão na praça q Elrei D. Jozé morrera: & q seo neto fora ja aclamado, o qual tinha cazado pouco tempo antes com sua thia. ainda não tive ocasião de me informar com maior certeza: nem nisso me intereço assaz p^a^. hir daqui 2 milhas p^a^. o saber. vivo indifferente p^a^. tudo o q lá se passa naquelle paiz. dezejo bem a todo o genero humano: & vi[ve]rei aqui no meo cantinho em esquecim^to^. politico de q^to^. por là se passar.

a Monsieur
Monsr. Sanchez Docteur en Medecine
Rue des Blancs-Manteaux
à Paris

2 Conjectural reading.

77.03.21 London 21 March 1777

My Dear Friend

If infirmities depended on our will or our friends', we would have reason to complain & to be afflicted, but being, as they are, decrees of fate, there is no relief except to suffer the part that is inevitable, seeking to attenuate it with the palliative of the affectionate society of friends, & also with the comic amusement at the human follies of which the inhabited streets are the theatre, & of which the conversation of civil life (even of the most stupid, or the most discreet) is the recitative. For these reasons, which you must confess to be well established & philosophical, I cannot approve of your living separated from civil society, & of your not receiving useless visits of annoying people; since taking these as moral medicine for your amusement, you will feel less the fatality of your disturbances. Divided attention, even with a little grief, makes a certain suspension or pause in our sensations, which when varied are not so unbearable. Quevedo[3] called hell the always and again of matrimony: and I take it as a text of as much canonical authority as ecclesiastical law.

I received the bill of exchange for 300#.[4] that was paid punctually and also the book of memoirs about nitre[5] of which I wanted to know the price to charge for it on the account of the man who asked me for it.

I have not yet had opportunity to get the information from Mr. Zimlin Somborski[6] about the quali[7] in Russia: and I shan't fail to pass it on to Chevalier Pringle, if the mistake is essential. I will send soon the worship of Christ:[8] if it is for your use, I praise your taste in looking for such reading. My friend Priestley, who is a man of virtue, a good philosopher, honourable, sincere & of great discernment, has declared himself against the spirituality & immortality of the soul, which must perish with the body but would be resuscitated with it through the miracle of Jesus Christ: he says that it is actually material, & that its spirituality cannot be defended by the Holy Scriptures &c. &c. In spite of all this, he is definite & believes in the revelation, & that the same scriptures & gospels were inspired by God, except for those articles that the writers of these scriptures added out of

3 The Spanish moralist writer Francisco Quevedo (1580-1645) was opposed to matrimony. He himself married against his will at 54 but the union lasted only three months.

4 The annuity that Sanches paid to Magellan; see Magellan to Sanches, 23 February 1776 and 18 February 1777.

5 [Académie Royale des Sciences, Paris], *Recueil de memoires et d'observations sur la formation & sur la fabrication du salpetre ...* (Paris, 1776); see Magellan to Sanches, 23 February 1776.

6 Andrei Afanes'evich Samborskii was the long-serving priest of the Russian Embassy's church in London, 1765-1779. Magellan may be making a joke about Samborskii's becoming a man of the soil (in Russian, *zemlya*); see Magellan to Sanches, 18 February 1777.

7 mortality.

8 "Theosebes", *Letters on the Worship of Christ, addressed to the Rev. George Horne ...* (London, 1776)?

their own heads &c. &c. Now consider how all this combines in such a judicious & excellent head: & what is more in a spirit so humane, kind, & full of good feelings ...! Such is the human spirit ...! He is printing a metaphysical work about this matter[9] & about the freedom of human actions: and of which I read with pleasure the first printed page that I found on his table one of these nights ...! 77.03.21

I showed our Envoy[10] the extract from your last letter.[11] He read it all, & laughed greatly at your obsession with my supposed lack of belching.[12] He told me: "please tell our friend, that since he has such a good memory [...] of my house that he knew in Coimbra in 1716 or 1717, he also would remember having written to me (I think to Vienna, or Malta, & Portugal) and having offered me the book of Abbé so-and-so (I forget his name) about the method of preserving grain &c.: that I still keep his letter, and have found it with pleasure one day in the midst of my papers: that I praise & always will praise him & extend to him all good wishes and health &c."

Poor Mª. Joana.[13] Cough, catarrh, fevers, age, bad temper & what else, lack of what you cannot give her, are as many reasons to be melancholic, & to make those who see her the same. But it's surprising that she doesn't wish to take any medicines: who knows if she believes in them as little as I do, or if she is already a doctor as you are ...? Give her my regards & also to the cousin, niece &c.

vive, vale, & ama[14]

tuum

M.

Yesterday I was told in the market-place that King D. José had passed away[15] & that his grandson who had married his aunt had been acclaimed. I have not yet had opportunity to become informed with more certainty: nor am I interested enough to go 2 miles from here to check it. I live indifferent towards everything that happens in that country. I wish all humankind well: & I shall live here in my little corner in political forgetfulness of which happens there.

---

9 Joseph Priestley, *Disquisitions relating to Matter and Spirit* (London, 1777).
10 Luiz Pinto de Sousa Coutinho (1735-1804), Portuguese Minister in London, 1774-88.
11 Letter not found.
12 See Magellan to Sanches, 7 July 1775, note 19.
13 Maria Joana Pernelet, Sanches' house-keeper.
14 Flourish, farewell and love me.
15 King José I died in Lisbon on 24 February 1777.

77.04.00 *To Joseph Priestley*[1]

Je ne puis mieux employer, Monsieur, les instans de loisir dont je jouis, depuis quelques jours, dans votre voisinage, qu'en vous communiquant quelques méthodes que j'ai imaginées, & dont il me semble que l'usage pourra être d'une grande utilité au Public. C'est avec grand plaisir que j'ai observé, que ma façon de penser se rapportoit à la vôtre dans presque toutes nos idées philosophiques, & sur-tout en ce point: de regarder avec indifférence toute découverte, quelque ingénieuse qu'elle puisse être, s'il n'en résulte pas un véritable avantage pour l'humanité.

Parmi le grand nombre de découvertes contenues dans vos Ouvrages philosophiques, la méthode de produire, avec le secours de l'art, des eaux minérales, semblables à celles de Pyrmont, de Saltzer, de Spa, &c., dont les vertus dépendent de l'air fixe, dont elles sont impregnées; & l'idée de trouver le méchanisme d'un instrument qui puisse marquer en tout tems & en tous lieux, le degré de salubrité de l'air que nous respirons, sont, sans doute, de la plus grande importance à l'humanité. Le succès avec lequel la première de ces deux découvertes a été employée, est actuellement connu par-tout, & les observations intéressantes que le Chevalier Landriani a faites au sujet de la seconde, en parcourant presque toute l'Italie avec son *Eudiomètre*, démontrent clairement la vérité de mon assertion.

Aussi-tôt que votre Mémoire[2] sur la manière de faire des Eaux de Pyrmont, m'est parvenu, en 1772, j'en envoyai aussi-tôt un grand nombre d'exemplaires dans toutes les parties de l'Europe, où j'entretiens une correspondance littéraire; & telle a été depuis long-tems ma coutume, toutes les fois que j'ai eu connoissance de quelque découverte utile. Je fis alors quelques légers changememens dans votre méthode, afin de procurer plus de facilité à faire passer l'*air fixe* dans l'eau. Ce changement fut indiqué dans la Traduction Françoise de votre Mémoire qu'on a faite à Paris (1),[3]

1 The text is a French translation of the first part only of Magellan's *Description of a Glass Apparatus for Making Mineral Waters...*, which is presented as a letter to Priestley dated 3 January 1777. It was published, undated, in *Observations sur la physique, sur l'histoire naturelle et sur les arts*, 9 (1777), 280-87, in the issue for April 1777, under the heading, "Lettre de M. Magellan, de la Société Royale de Londres, au Docteur Priestley, Contenant la Description d'un Appareil de crystal, pour faire, en peu de minutes & à peu de frais, des Eaux minérales semblables à celles de Pyrmont, Spa, Saltzer, &c.".

2 Priestley, *Directions for Impregnating Water with Fixed Air, in order to communicate to it the Peculiar Spirit and Virtues of Pyrmont Water...* (London, 1772).

3 Editor's note: "Voyez cette Traduction dans le Journal de Physique, Août 1772." See "Manière d'imprégner l'Eau d'air fixe, & de lui communiquer les propriétés de l'Eau de Pyrmont...", *Observations sur la physique, sur l'histoire naturelle et sur les arts*, August 1772.

d'après l'exemplaire que j'avois envoyé à M. Trudaine de Montigny, ce grand 77.04.00
Promoteur de la Philosophie naturelle, dont vous avez fait mention, page 268 de votre deuxième volume sur les différentes sortes d'air.[4] Quelque tems après, M. Blunt imagina une machine qui rendoit encore plus aisée cette opération, & j'en envoyai une de sa façon, à Turin, au Marquis de Rozignan, ce judicieux amateur de toutes les expériences philosophiques, qui est actuellement Ambassadeur du Roi de Sardaigne, à Berlin.

Une autre construction d'une machine de crystal, pour remplir le même objet, fut donnée par le Docteur Nooth dans le 65e volume des Transactions Philosophiques;[5] mais cette machine étant encore assez impairfaite, fut encore perfectionée par M. *Parker*, de la manière dont vous avez rendu compte dans votre susdit Ouvrage, page 293 & suivantes.[6] Ces machines ont été envoyées, en très-grand nombre, dans différentes parties du monde, & même jusques aux grandes Indes; & il est de notoriété, que beaucoup de personnes se sont très-bien trouvées de l'usage de ces eaux artificielles & *acidules*.

J'ai néanmoins reconnu, qu'en suivant le procédé tel qu'il est indiqué dans les explications imprimées, envoyées avec ces machines, l'opération avoit l'inconvénient d'être fort longue, puisqu'il falloit quatre, & même souvent six ou sept heures pour obtenir une suffisante quantité d'eau impregnée d'*air fixe*. C'est ce que j'éprouvai encore plus sensiblement le mois de Novembre dernier, étant chez Son Excellence le Duc d'Aremberg, que je dois regarder comme mon *Mécène*, après toutes les faveurs que j'ai reçues de lui. Connoissant que ce généreux Prince étoit doué, autant qu'aucune personne de son rang, de toutes les qualités les plus propres pour encourager toutes les découvertes utiles à l'humanité, je lui envoyai de Londres une des machines, perfectionnée suivant la dernière méthode; mais lorsque je voulus l'essayer, à mon arrivée chez lui, je sentis combien la lenteur de l'opération étoit désagréable. Je réfléchis alors sur ce qu'on pourroit faire pour éviter cet inconvénient. A la fin, j'imaginai l'appareil

4 Priestley, *Experiments and Observations on Different Kinds of Air*, volume 2 (London, 1775), p. 268, in noting the publication of a French translation of his pamphlet only a few weeks after the English version appeared, attributed this to "the laudable zeal of Mr. Trudaine, for promoting all philosophical and useful improvements". In doing so, Priestley repeated the acknowledgment of Trudaine's zeal already published with the French translation of his pamphlet.

5 John Mervin Nooth, "The Description of an Apparatus for impregnating Water with fixed Air; and of the Manner of conducting that Process", *Philosophical Transactions*, 65 (1775), 59-66.

6 Priestley, *Experiments and Observations on Different Kinds of Air*, volume 2 (London, 1775), pp. 293-303.

77.04.00 suivant, dans lequel, par l'addition de quelques pièces, l'opération est si raccourcie, que non-seulement elle s'achève en peu de minutes, mais encore on obtient, par ce procédé, deux fois autant d'eau impregnée d'*air fixe*, que par le procédé connu jusqu'à présent.

Description de l'Appareil.

Voyez (*figure* 1) une des machines, perfectionnée par M. Parker, laquelle est marquée A, B, C. Cette machine est placée sur un plateau de bois, pour éviter que l'eau qui pourroit se répandre pendant l'opération, ne tombe sur la table. Le vaisseau du milieu marqué B, a un collet, lequel est fermé par un bouchon de crystal, marqué V. Ce bouchon est composé de deux parties percées de petits trous, pour laisser passer l'air à travers, & entre ces deux parties, doit être ménagé un intervalle, dans lequel est contenue une petite lentille qui doit agir comme une soupape, en laissant passer l'air de bas en haut, mais empêchant son retour de haut en bas. La totalité du collet du vaisseau du milieu B, doit entrer dans le goulot du vaisseau inférieur A, & le boucher hermétiquement.

Le vaisseau supérieur C, se termine dans son extrémité inférieure, en forme de tube. Ce tube doit être courbé à-peu-près comme une retorte, pour empêcher que l'*air fixe* ne puisse pénétrer dans le vaisseau supérieur C. Avant d'avoir atteint la surface de l'eau contenue dans le vaisseau du milieu B, ce tube qui termine le vaisseau C, doit aussi boucher hermétiquement le goulot du vaisseau mitoyen B, & l'extrémité supérieure du vaisseau C, doit avoir un bouchon bien ajusté à son goulot, qui doit être percé dans le milieu. Le vaisseau supérieur C, doit être fait pour recevoir la moitié de la liqueur qui peut être contenue dans le vaisseau du milieu B, & l'extrémité de son tube recourbé ne doit pas descendre plus bas qu'à la moitié du vaisseau du milieu B.

Outre ces trois vaisseaux A, B, C, (*figure* 1) il faut en avoir deux autres, tels que G & H, (*figure* 2.) Ces deux vaisseaux G & H, qui ne sont que des vaisseaux de rechange, doivent être parfaitement semblables aux vaisseaux B & C, de la figure 1. Ces deux vaisseaux de rechange doivent être placés en attente sur un pied de bois marqué F, (*figure* 2.) Outre toutes les parties parfaitement semblables dans les vaisseaux B & C, & leurs vaisseaux de rechange G & H, chacun des deux vaisseaux A, B, (*figure* 1), & le vaisseau G de rechange avec le vaisseau B, doivent avoir des ouvertures (*m, n, l,*) bouchées hermétiquement avec des bouchons de crystal ajustés, afin de pouvoir s'en servir au besoin, ainsi qu'il sera mentionné ci après.

Le pied de bois marqué K, (*figure* 3), doit avoir, à son extrémité supérieure, une forte pièce de crystal *x*, comme une espèce de virole, laquelle aura été ajustée de manière à fermer hermétiquement avec les goulots inférieurs des vaisseaux B & G, & cette espèce de virole, ou de goulot de crystal, sera exactement cimentée sur le pied de bois K. La forme de ce pied de bois, qui doit être plat en-dessous, afin d'avoir une assiette solide, se concevra facilement à la vue de la figure 3. La figure 4 représente un large entonnoir de crystal, qui puisse entrer dans l'ouverture supérieure du vaisseau A, (*figure* 1.) La figure 5 représente une petite phiole de crystal, pour servir à mesurer la quantité d'acide vitriolique à employer, & la figure 6 représente une petite mesure d'étain, pour la quantité de craie ou de marbre pilé à employer dans chaque procédé. 77.04.00

Explication du procédé de l'opération.

Commencez par soulever tout ensemble les vaisseaux B & C, (*figure* 1), ensuite séparez les vaisseaux supérieurs C & H des deux vaisseaux du milieu B & G, remplissez alors, d'eau bonne à boire, les deux vaisseaux du milieu B & G, & recouvrez-les ensuite avec leurs deux vaisseaux supérieurs C & H.

Prenez quinze ou seize measures d'eau avec la petite phiole P, (*figure* 5) versez-les dans le vaisseau inférieur A, remplissez ensuite la même phiole P, d'huile de vitriol que vous verserez sur l'eau que vous avez mise dans le fond du vaisseau A.

Vous aurez soin de commencer par mettre dans l'embouchure du vaisseau A, l'entonnoir de verre *q*, (*figure* 4) & ensuite, vous introduirez la craie ou le marbre pilé, par cet entonnoir, qui est nécessaire pour empêcher que la matière calcaire, en touchant au collet du vaisseau, ne vint à s'y attacher, & à y former ensuite un mastic pendant l'opération, ce qui empêcheroit de pouvoir séparer le vaisseau du milieu du vaisseau inférieur. Il faut avoir soin de ne pas remuer le vaisseau inférieur pendant qu'on y introduit la substance calcaire, car il se formeroit une effervescence tellement subite, qu'elle entraîneroit la plus grande partie des matières préparées dans le vaisseau A. Si même cela venoit à arriver, il faudroit ouvrir le bouchon (*m figure* 1) afin de prévenir la casse du vaisseau, en donnant du jeu à l'effervescence; & malgré cela, si l'effervescence avoit gagné à un certain excès, il seroit nécessaire de jetter tout le contenu; de bien laver le vaisseau avec de l'eau fraîche, & recommencer l'opération, parce que les matières en effervescence, s'attacheroient au col des vaisseaux, & les mastiqueroient

77.04.00 ensemble; mais si on répand les matières calcaires avec l'attention de ne pas remuer le vaisseau inférieur A, dans lequel se fait le mêlange, (& dont on feroit peut-être bien, pour plus grande précaution, de charger de plomb le plateau de bois sur lequel il est établi), alors, l'effervescence sera peu considérable dans le commencement, & on aura le tems de continuer aisément l'opération.

Aussi-tôt que la pierre à chaux, ou le marbre en poudre, auront été introduits dans le vaisseau A sur l'eau, & l'acide vitriolique qu'on y avoit versé auparavant, il faudra retirer l'entonnoir (*q, figure* 4) & prendre les deux vaisseaux B & C qui étoient en attente sur le pied de bois (K, *figure* 3) & les placer immédiatement tels qu'ils sont sur l'embouchure du vaisseau A qui contient la préparation.

Alors, *l'air fixe* désengagé de la base calcaire par l'effort de l'acide vitriolique, s'élèvera à travers la soupape dans le vaisseau B. Lorsque cet *air fixe* est arrivé jusqu'au haut du vaisseau du milieu B, il déloge une partie de l'eau à boire dont on a rempli, avant l'opération, le vaisseau du milieu, & cette eau est forcée à monter à travers le tube recourbé dans le vaisseau supérieur, en proportion égale au volume de l'air fixe qui vient remplir le vaisseau du milieu. Si l'opération est bien conduite, le vaisseau supérieur C sera bientôt rempli d'eau, & le vaisseau mitoyen B, sera rempli moitié d'eau, moitié d'air fixe, ce qui se connoîtra aisément lorsqu'on verra les petites bubes d'air montant dans le tube recourbé que forme l'extrémité inférieure du vaisseau C; ce qui arrivera en deux ou trois minutes.

Aussi-tôt que cet effet se présente, soulevez tout ensemble, avec précaution, les deux vaisseaux B & C, & placez-les tels qu'ils sont sur le pied de bois ou pied d'attente, (K, *figure* 3, alors, prenez de dessus le pied F, (*figure* 2) les deux vaisseaux de recharge G & H, & placez-les doucement sur le vaisseau A, tandis que l'opération se fait dans les vaisseaux G & H, comme elle vient de se faire dans les vaisseaux A & B; secouez alors vivement les deux premiers vaisseaux qui sont sur le pied d'attente K, en soutenant le tout dans vos mains. Les parties d'air fixe & d'eau qui se trouvent alors dans le vaisseau du milieu sans être mêlées, se confondront ensemble par l'effet de l'agitation, & l'eau absorbera la plus grande partie de *l'air fixe*, ce qui se connoîtra lorsqu'on verra que l'extrémité du tube recourbé qui étoit dans l'air, se trouvera recouvert par l'eau qui aura monté, en proportion de ce qu'elle aura absorbé d'air.

Il sera suffisant d'agiter l'eau de cette manière pendant deux ou trois minutes. Cela fait, détachez le vaisseau supérieur C, de manière que l'eau qui y reste puisse tomber dans le vaisseau B, & que l'air absorbé puisse s'échapper; ensuite, ôtant ces deux vaisseaux de dessus le pied d'attente K, placez-les, joints ensemble, sur le pied de rechange F: pendant ce tems-là, les deux vaisseaux de rechange G & H se trouveront dans le même état que les premiers A & B, c'est-à-dire, que le vaisseau mitoyen G, sera rempli moitié d'eau, & moitié d'air fixe, & le vaisseau supérieur H, sera rempli de l'eau que l'introduction de l'air fixe aura forcé d'y monter. Prenez alors ces deux vaisseaux de rechange; placez-les comme vous avez fait les premiers sur le pied d'attente K, & replacez immédiatement les deux premiers B & C, sur le goulot du vaisseau A, afin de les remplir encore d'air fixe, tandis que vous agiterez l'eau dans les vaisseaux de rechange qui sont sur le pied d'attente K, de la même manière que vous avez fait déjà dans les premiers, pour mêler l'eau avec l'air dans le vaisseau mitoyen G. Lorsque l'effervescence commence à se rallentir beaucoup dans le vaisseau A, on peut la ranimer en y donnant un petit coup avec la main, de manière que la substance calcaire qui s'est amoncelée au fond du vaisseau en se remêlant avec l'acide vitriolique, puisse lui fournir encore de l'air à dégager. Cependant, lorsqu'il arrive que le tout est épuisé, & qu'il ne s'exhale plus d'air, on peut rajouter de la matière calcaire ou de l'huile de vitriol, & enfin, si cela n'est pas suffisant, on peut reverser encore de l'eau sur le mêlange. Ces additions peuvent se faire dans le vaisseau inférieur A, en soulevant les deux vaisseaux supérieurs, & les plaçant sur le pied de rechange F. Dans cette circonstance, on doit toujours avoir grand soin de faire usage de l'entonnoir de verre (*q*), afin d'éviter l'inconvénient du mastic qui se formeroit au col des vaisseaux. 77.04.00

Lorsque cette opération a été répétée alternativement, trois ou quatre fois au plus, avec chaque assortiment de vaisseaux, en laissant ainsi échapper à chaque fois *l'air fixe* qui n'aura pas voulu s'incorporer avec l'eau, & y en introduisant de nouveau, en les remettant ainsi alternativement sur le vaisseau inférieur A, qui contient le mélange; l'eau contenue dans les deux vaisseaux mitoyens B & G, sera suffisamment saturée, & sera d'un gout plus agréable que les eaux naturelles de Pyrmont, & de Seltzer, lesquelles non-seulement ne contiennent pas autant *d'air fixe* que nos eaux artificielles en peuvent absorber, mais en outre, ont un goût salin assez désagréable, qui, loin de contribuer à leurs vertus médicales, peut être, au contraire, nuisible dans des maladies compliquées. Ces eaux artificielles ne sont pas moins lympides que l'eau ordinaire, quoiqu'elles aient absorbé de l'air fixe en une proportion presqu'égale à leur volume. Par cette méthode, le procédé ne durera guerres plus d'un quart-d'heure, & fournira le double de la quantité d'eau qu'on obtient par le procédé ordinaire. En ouvrant le bouchon (*l* ou *n*), on peut en tirer la quantité qu'on veut boire immédiatement; autrement, il vaut mieux la laisser dans la machine, sans quoi si on la laissoit exposée

77.04.00 au contact de l'air ordinaire, elle deviendroit insipide, & perdroit, ainsi que les eaux naturelles, au bout d'un certain tems, la saveur *acidule.* Ces eaux artificielles peuvent néanmoins se conserver assez long tems dans des bouteilles bien bouchées, & renversées de bas en haut. En général, elles sont tellement semblables aux eaux naturelles les plus acidules, qu'on peut les amener au point de mousser comme le vin de Champagne. M. Warltire a produit cet effet en tenant *l'air fixe* comprimé sur la surface de l'eau dans le vaisseau du milieu, mais pour cela, il faut que les vaisseaux soient très-forts, sans quoi ils seroient sujets à se casser.

On peut aisément ajoûter à ces eaux, une qualité ferrugineuse, en y mettant deux ou trois paquets de *fil de fer à ressorts*, ou deux ou trois onces de petits clous. Ces eaux, suivant l'expérience de M. *Lane*, dissolvent le fer assez bien pour en être saturées dans l'espace de quelques heures;[7] & suivant Sir John Pringle, on peut y ajoûter huit ou dix gouttes de *tinctura Martis cum spiritu salis*, afin de les faire ressembler plus parfaitement aux véritables eaux de Pyrmont.[8]

Il n'est pas douteux qu'on ne puisse faire en Médecine, un usage avantageux de ces eaux artificielles, car non-seulement on peut y faire dissoudre les différens sels qu'on trouve dans plusieurs eaux minérales, renommés par leurs différentes vertus; mais encore on peut s'en servir comme de véhicule *ou de collature* qui, en donnant du ton & du ressort à l'estomac, rendroit les drogues qu'on seroit obligé de faire passer dans certains cas, plus supportables, & moins *nauséabondes.*

Je finirai par vous observer qu'on peut également faire passer *l'air fixe* dans le vin, dans la bière, dans le cidre, & dans quelque liqueur que ce soit; lors même que la burre *est passée*, on pourroit en rappeller le goût par cette méthode, mais cette agréable & piquante saveur, où ce goût *acidule*, communiqué par l'air fixe, & qui est si sensible avec l'eau pure, le seroit à peine dans le vin ou les autres liqueurs, qui ont un goût décidé par elles-mêmes (1).[9]

---

7 T. Lane, "On the Solubility of Iron in simple Water, by the Intervention of Fixed Air", *Philosophical Transactions*, 59 (1769), 216-27.

8 Pringle, "A Discourse upon some late Improvements of the Means of Preserving the Health of Mariners", delivered at the Royal Society on 30 November 1776; in his *Six Discourses...* (London, 1783), p. 173.

9 Editor's note: "Pour completter les détails de ces Machines, nous croyons devoir ajouter celle de M. le Duc de Chaulnes." There follows (pp. 287-91) an account of the method devised by the Duke to saturate water with fixed air.

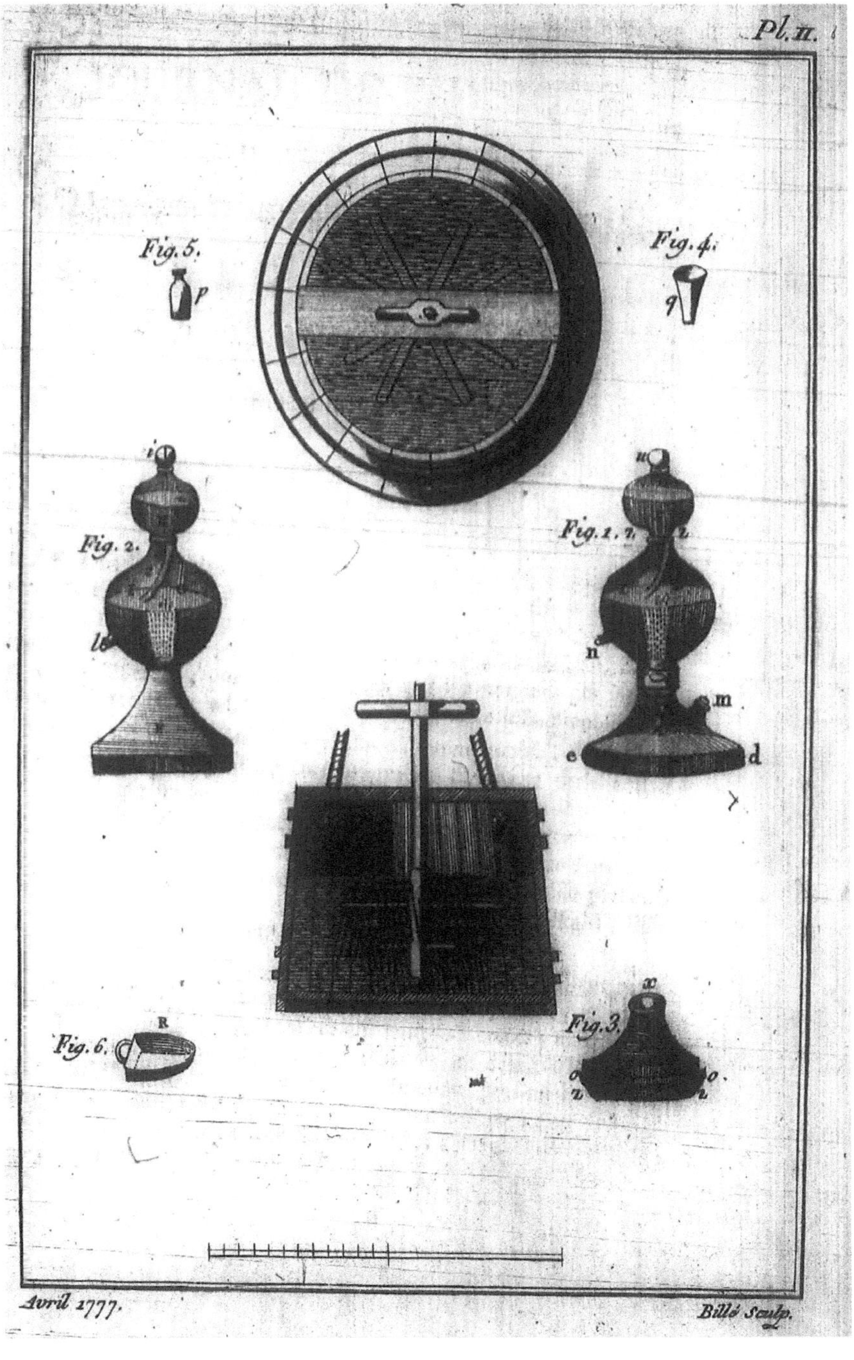
Pl. II.
Fig. 5.
p
Fig. 4.
q
i
Fig. 2.
u
Fig. 1.
n
m
e
d
x
Fig. 3.
R
Fig. 6.
Avril 1777.
Billé Sculp.

77.04.15 *To Ribeiro Sanches*[1]

Österreichische Nationalbibliothek, Vienna
A.N. Ribeiro Sanches Korrespondenz, codex 12714, ff. 195-196v

Londres 15 Abril 1777

Am^{o}. e S^{r}. e m^{to}. do Coração

A de V.M. de 3 do Corr^{te}. me deo m^{to}. gosto como sempre, não obstante o seo warth, warth, espera, espera. lembrei-me q não [te]ndo q responder um tal Espanhol, contra uma censura q se lhe fez, [to]mou a pena na mão, & fez um grande catalogo de todas as faltas [gr]amaticaes, oratorias, & orthograficas do escrito do seo adversario. Se não fora por imitar um tal exemplo, havia de dizer-lhe q warth não se escreve com h depois do t em allamão: & q ainda posso fazer & dizer da ist es, em vez de wart, wart. Mas V.M. ja não sabe senão do q passou, & dos ourelos das meias das nossas tiricanas de Coimbra.

Tãobem aqui temos varias noticias daquella terra como as de V.M., ainda q menos explicitas. Comtudo affirma-se q passão de mil & 400 os prezos q ainda se acharão vivos p^{a}. sahir dos calaboiços. Mas q tal seja publico & conhecido, e q não se faça um exemplo publico p^{a}. escarmento dos ministros futuros, mostra q aquella terra não pode ser habitada por q^{m}. conheceo & soube prezar a sua Liberd^{e}.

o meo am^{o}. Priestley foi p^{a}. sua caza, em Calne. sua M^{er}. se acha perto do Pa[e] & lá ficará até o outono. elle não passa bem. tem uma queixa n[as] tripas, uma especie de colica seca: & não he mui attentivo á sua propria conservação: trabalha m^{to}., pouco alimento, m^{ta}. app[li]cação e meditação &^{a}. &^{a}.

elle dezejava ter dois livrinhos d[q][2] este correio escrevi ao D^{r}. le Begue, p^{a}. m'os mandar pela posta [...][3] adresse de M^{r}. Roberto Bigg at y^{e}. gen^{al}. Post Ofice M. london. C[om] um M antes do London: & sem algũa escritura de maõ junto com o Livro. Se acazo M^{r}. le Begue não estiver na terra, & V.M. puder mandalos buscar pelo seo criado, ficar-lhe ei m^{to}. obrigado, pois são Relativos ao estado de sua Mulher, q elle ama, e ella o merece. o titulo hé = Cathechisme sur les acouchements pour usage des sages femmes. = o tit^{o}. do Segundo hé = observations sur les pertes de Sang. par M^{r}. le Roux =.

---

1 Annotated "Reçue a 20 Avril" [Received 20 April], "Resp^{a}. a 28 Mayo" [Answered 28 May] and "aqui a conta e n^{o} de livros que me mandou" [here the account and number of books he sent me]. Sanches' letter not found.

2 Conjectural reading.

3 Illegible.

O nosso Ministro, q he dos raros daquella terra, & talvez o único q ella tem produzido depois de m^{tos}. seculos, tanto pela capacid^{e}. como pelo coração, diz q V.M. deve estar equivocado com outro sojeito em lugar seo; pois não seria facil esquecerse de lhe ter mandado no anno de 1758 pouco mais ou menos hum extracto de hũa obra impressa d'um tal M^{r}. Entieri, sobre a conservação dos grãos, trigo &^{a}. com o modo q se pratica na Russia p^{a}. secar estes mesmos grãos &^{a}. o q tudo foi remetido a Portugal ao Morgado de Balsamão &^{a}. &^{a}. talvez q isso fosse mandado pela via de Gonçalo Xavier, q eu sei tem, & teve grande correspondencia, q ainda dura com este Fidalgo. 77.04.15

alegreime com os elogios do gallico. não faltou q^{m}. com elle risse ás cangalhadas p^{a}. o exprimir á portugueza, não porq eu o mostrasse na sua carta; pois sou fiel; mas porq me apanhou a carta, e a lêo por si mesma, com silencio, mas com gostinho, q he nat^{al}. das nossas portuguezas, emq^{to}. eu & o marido nos-sorriamos por [b]aix[o...][4] da capa.

agora mandei o n°. 14 dos Com̃ent^{os}. de Edimburgo ao D^{r}. du Presle pelo correio: q não deixará de os emprestar a V.M. até q chegue o q lhe mandarei com os seos livrinhos &^{a}.

minhas saud^{es}. a Marie Jeane & fam^{a}. adoptiva. q passem bem, & fação o seo demenagement com satisfação, & sem m^{ta}. fadiga, se he possivel; pois sei q^{to}. custa, & q$^{to.}$ se perde, & se destroe &^{a}. &^{a}.

ha tempoq não vi o D^{r}. Seq^{ra}. mas sei q passa bem & da mesma Sorte o De la Cour, q dizem virá a Londres brevem^{te}. vizitar os conhecidos, p^{a}. ao depois tornar p^{a}. o seo Centro de Bath.

vive, vale, & ama
tuum
M.

P.S. Se se acha embaraçado com despezas & gastos, & podem servir de alivio 50 ou 100 luizes, tire sobre m[im] á vista sem mais avizo, nem ceremonia; pois posso pagar no mesmo instante: e não esteja là a sofrer dores de barriga como as nossas tiricanas de Sendelgas, q são achacadas do ventre logo q provão das pivides de cabaço, mas não o querem confessar por medo do q dirão as más lingoas.

a Monsieur
Monsr. Sanchez, Doct^{r}. en Medecine
Rue des Blancs-manteaux
à Paris

4 Torn paper.

77.04.15 London 15 April 1777

My Dear Friend

Yours of the 3rd of this month[5] gave me much pleasure as always, in spite of your warth, warth,[6] wait, wait. I remembered that a certain Spaniard, having nothing to answer against a reproach made to him, took pen in hand & made a large catalogue of all the grammatical, oratorical & orthographical mistakes in the writing of his adversary. If it were not to imitate such an example, I would tell you that warth isn't spelled with h after the t in German: & that I still can do & say da ist es[7] instead of wart, wart. But you don't speak but of the past & of the tops of the stockings of our country girls of Coimbra.

We have also had here several reports from that land like the ones you have had, though less explicit. However it is said that the number of prisoners found alive to leave jail, number more than one thousand & 400.[8] But that such news is public & known, but that a public example is not made as a warning to future ministers, shows that those who have known & experienced their freedom cannot live in that country.

My friend Priestley went to his home in Calne. His wife is near her father & there she will stay till the autumn. He is not well. He has an intestinal complaint, a kind of dry colic: & he is not very attentive to his own well-being: he works a lot and eats little, with much diligence and meditation &c. &c.

He wished to obtain two little books [for which] I wrote this post to Dr. le Begue,[9] to send them by the post to the address of Mr. Robert Bigg at ye. genal. Post office M. London.[10] With an M before London: & without any handwriting with the book. If by mischance Mr. le Begue is not there & you can get them through your servant, I should be very grateful, because they are related to the condition of his wife, whom he loves and she deserves it. The title is = Cathechisme sur les acouchements pour usage des Sages femmes.[11] = the title of the Second is = Observations sur les pertes de Sang. par Mr. le Roux =.[12]

Our Minister,[13] who is among the rare ones of that country, & perhaps the only one it has produced for many centuries, as much for his capacity as for his

---

5 Letter not found.

6 warten = to wait, in German.

7 There it is.

8 Large numbers of political prsoners were released following the downfall of the Portuguese Prime Minister, Pombal.

9 Letter not found.

10 See Magellan to Sanches, 18 February 1777, note 10.

11 Anne Amable Augier du Fot, *Catéchisme sur l'art des accouchements pour les sages-femmes de la campagne fait par ordre et au dépens du gouvernement* (Paris, 1775).

12 Laurent Charles Pierre Le Roux, *Observations sur les pertes de sang des femmes en couches et sur les moyens de les guérir* (Dijon/Paris, 1776).

13 Luiz Pinto de Sousa Coutinho (1735-1804), Portuguese Minister in London, 1774-88.

heart, says that you must be mistaken, taking another person for him; since it would not be easy to forget that you sent him in about 1758 an extract of a work of a Mr. Entieri[14] about the conservation of grains, wheat &c. with the manner practised in Russia to dry these same grains &c. which was all forwarded to Portugal to the Morgado de Balsemão[15] &c. &c. Maybe that was sent through Gonçalo Xavier,[16] whom I know has & had a large correspondence that still continues with this nobleman. 77.04.15

I was happy with your eulogies of syphilis. There were many who cackled to express it, in a Portuguese way of speaking; not because I showed her your letter, since I am faithful; but because she took the letter, and read it herself, in silence but with the pleasure which is natural in our Portuguese ladies, while her husband & I smiled behind our sleeves.

I have now sent no. 14 of the Edinburgh Commentaries[17] to Dr. du Presle by mail, that he will certainly lend to you until the arrival of the one I'll send you with your little books &c.

My regards to Maria Jeane & the adoptive family. May they continue well, & make their demenagement with satisfaction & without too much fatigue, if possible;[18] since I know how much it costs, & how much is lost & destroyed &c. &c.

It is some time since I last saw Dr. Sequeira but I know he is well & also Dr. De la Cour, who is said to be coming to London soon to visit his acquaintances and then returning to his Centre in Bath.

vive, vale, & ama[19]
tuum
M.

P.S. If you are concerned with expenditures & payments, & 50 or 100 Louis can be of relief, draw on my account without any notice or ceremony, because I can pay it in the same instant: and don't go suffering belly-aches like our country girls of Sendelgas, who become sick in the belly as soon as they feel lust but don't want to confess it for fear of gossip.[20]

---

14 Bartolomeo Intieri, *Della perfetta conservazione del grano* (Naples, 1754), a work sometimes erroneously attributed to F. Galiani.

15 i. e. to Pinto de Sousa, who was at the time the first-born son or heir to Balsemão.

16 Gonçalo Xavier de Alcáçova Carneiro (1712-after 1774) was a Portuguese doctor who corresponded with Ribeiro Sanches.

17 *Medical and Philosophical Commentaries* [by A. D.] (Edinburgh, 1773 onwards).

18 Sanches and his household were about to move house.

19 flourish, farewell and love me.

20 The country girls from the Coimbra region (including Sendelgas) were known as "tricanas" after their special way of dressing.

77.04.22 *To Louis-Henri Duchesne*[1]
Smithsonian Institution Libraries, Washington, MSS 945A

Londres ce 22 Avril 1777

Voici, mon cher ami, une graine particuliere qui fera du plaisir à Vos amis botanistes d'Italie. elle est de cette espece de riz, qui croit dans les terres hautes de la Chine, sans avoir besoin d'etre summergé dans l'eau comme le riz ordinaire. on le seme dans la terre Labourée come du bled: & elle produit abondament. il y a long tems que J'ai entendu parler de cette espece de riz mais on n'en pouvoit avoir la Semence. & on s'attend que si on vient à bout de le multiplier assez pour le cultiver en grand dans les climats chauds de l'Europe, cette espece de grain deviendra une objet très utile & avantageux. il faudra tremper cette graine 24 heures dans l'eau avant le semer: la semaille se fait toujours dans le printems. & meme il seroit prudent de garder la moitié pour le printems prochain; car la saison fera deja trop avancée en Italie lorsqu'elle y arrivera. on ne devroit pas epargner la depense de l'envoyer par la poste, car il s'agit d'un objet interessant pour le bien public.

Il a bien long tems que Je n'ai pas de vos nouvelles. J'ai visité le neveu de M^r^. le marquiz de Cordon, Ambass^r^. du Roy de Sardaigne dans cette Cour,[2] qui m'aporta une de vos Lettres,[3] & Je ne manquerai pas de lui rendre tous les services dont Je serai capable. Je connois beaucoup son oncle, qui est très aimable aussi bien que Mad^e^. La Marquize qui est un ornement à son sexe. Ce fut alors que J'ai appris que vous etes secretaire actuelle d'une des dames de France,[4] & apparement celà vous attache à la Cour assez pour ne pas avoir trop de tems pour Vos amis anciens. Je voudrois cependant savoir si vous n'etes plus attaché à M^r^. Trudaine, & si je puis vous adresser encore, comm'auparavant les differents articles de nos amis communs &^a^.

Quoiqu'il en soit vous pouvez bien compter que Je ne cederai pas les droits de l'amitié & de l'attachement sur le compte de vos occupations nouvelles, & que vous pouvez toujours compter que Je suis pour La Vie

Mon cher ami
votre de Coeur
Magellan

---

1 Annotated "Rep: le 19 May". Letter not found.
2 Victor Sallier de la Tour, Marquis de Cordon, Sardinian Ambassador in London, 1774-84.
3 Letter not found.
4 Duchesne had acquired the position of secretary to the Comtesse de Provence, sister-in-law of Louis XVI.

Donnez moi des nouvelles de Mr. le Comte Carburi.[5] Je lui ai fait depuis long tems une Commission pour l'Italie: & Je n'ai pas reçu de ses nouvelles depuis celà. est-il malade, ou est-il allé faire quelque voyage? 77.04.22

A Monsieur
Monsr. Duchesne, Secretaire de Made. de France
Hotel de Tallard rue des Enfans rouges
à Paris

5 Giovanni Battista Carburi (1722-1801?).

To Ribeiro Sanches[1] 77.05.20

Österreichische Nationalbibliothek, Vienna
A. N. Ribeiro Sanches Korrespondenz, codex 12714, ff. 197-198v

Londres 20 Maio 1777

Meu amo. e Sr. e mto. do Coram.

Ha mto. tempo q não Recebi notas. de V.M.: dezejo q não seja por cauza de Molestia. Talvez q se ache actualmte. embaraçado com a sua mudança de caza, q não he pouca matraca. eu passo bem de animo e corpo, sem novide. ou mudança q valha huma figa; como qm. ve a festa dos toiros, em distancia bastante pa. não levar cornada nem azurriada. Aqui se diz q o Palacio do Sr. Inquizidor mor se está preparando em Lisboa, pois o Sr. Cardeal Patriarca foi advertido pa. o despejar. he pois pa. esperar q o Sr. Principe D. Jozé sahira do seo Retiro do Bussaco, & vira exercitar o seo honrado cargo pa. honra & gloria da nossa Sta. Fé Catha. Romana. dizem q fora necessario tirar o medalhão com a Figura do Carvalho q estava no pedestal da Estatua Equestre do Rey defunto, porq lhe atiravão continuamte. com lama da Rua. Veja V.M. o q s[ão][2] os nossos heroes do tempo prezente ...! ha dois dias q os aplausos chovião sobre esse grande Sujeito; e agora já lhe atirão com

1 Annotated "Ra a 28 Mayo" [Answered 28 May] and "falla da conferva" ["speaks about conferva"]. Sanches' letter not found.
2 Conjectural reading.

77.05.20 lama ao focinho, mas focinho da sua imagem ...! Elle está na sua quinta, debaixo do seo Parreiral de Pombal, esgravatando os dentes depois de ter tantas façanhas executado a seo salvo. acho-lhe graça ao dizerem q lhe hão de pedir conta do q fez sofrer a tanta gente. Em governos tais, onde o Rei he Lei viva, não ha mais disputa sobre o q o Ministro faz, em dizendo q assim o ordenou seo amo.

mas falemos do q nos importa. Mandei a V.M. com outros Livros pª. o Dr. Begue du Presle, os Livros seguintes, & ja dei a elle avizo delles todos. os q pertencem a V.M. são os seguintes. Philal. Transact. vol. 66. p. 2 & o discurso do Presidte. = ambos 9:s0d worchip of Christ, cuido q 1s. 0d/ Medal. Commentaries vol. 13. – 1s. 6d/ o vol. 14 hirá pela primª. occazião, pois tãobem ja sahio./ Saunders on Antimony. cuido q 3s. ou 2s. 6d. V.M. lá verá o preço na primª. pagina./ o 3º. vol. de Priestley sobre differs. especies de ar. – 6s. 0d/ & cuido q he tudo o q ha de novo. a obra de Robertson sobre a America sera publicada esta semana em 2 vol. 4º. eu a Mandarei a V.M. pois cuido q me disse a queria tanto q sahir.

não ha novides. Litterarias q valhão hum corno. o Capitão Cook escreveo do Cabo de Boa Esperança, q alli achára a notª. de q outro Capitão de Mar e guerra Hollandez, vendo-se fortemente atacado de escorbuto na sua gente em hũa viagem longa, fizera recolher aquella especie de limos verdes, ou conferva q se produz á roda dos navios da pte. de fora, e q hé mto. amargoza, e a deo em forma de Sallada á sua gente: e com isso a salvou do estrago do escorbuto. Aqui ainda vive hũ cirurgião q se achou na celebrada expedição de Mylord Anson, sobre a Náo chamada o C[enturi]ão[3] se não me engano; & dizem q elle com outros camaradas fizerão uso da mesma Salada da dª. conferva ou Limos verdes, por meio da qual se salvárão dos estragos do Escorbuto.

Aqui vi hũm destes dias o Dr. Sequeira, q passa bem, & me pedio o recomende a V.M. &ª.

vejo frequentemte. o nosso Enviado, q he o fidalgo mais capaz, cheio de conhecimtos. uteis, & o mais proprio pª. o Ministerio q talvez hoje existe entre os Nossos: se as coisas fossem como devião hir, elle ja aqui não estaria; pois tenho por certo q nenhũ o igoala de qtos. por lá estão governando o bôlo. Quanto a mim sentirei a sua auzencia qdo. daqui se for: mas parece-me q não ha por lá qm. tenha miolos [pª.][4] o conhecer.

Estou cuidando no pobre Barros, na esperança q mude de fortuna. não tenho notª. de Chevalier, isto he se vai ou não cedo pª. a sua dezejada Patria,

3 Stained paper; conjectural reading.
4 Torn margin; conjectural reading.

pela qual tanto piava como pato pela agoa do charco. Se V.M. por là tiver notas. mais miudas & gostozas, não deixe de m'as communicar. Saudes. a Ma. Joana. 77.05.20

vale

a Monsieur
Monsr. Sanchez, Doctr. en Medecine
Rue des Blancs-manteaux
à Paris

London 20 May 1777

My Dear Friend

It's a long time since I had news from you: I hope it's not due to illness. Maybe you are troubled at present with your moving house, which is no small thing.[5] I am well of spirit and body, without novelty or change that is worth a penny; like one who sees the bullfight at enough distance to avoid either a wound from a horn or derision. Here it is said that the palace of the Master Inquisitor is being prepared in Lisbon, because the Cardinal Patriarch was warned to evacuate it. It is then to be hoped that Prince D. José will leave his retreat of Bussaco,[6] & come to perform his honourable office to the honour & glory of our Holy Roman Catholic faith. They say that it was necessary to remove the image of Carvalho[7] that was on the pedestal of the equestrian statue of the deceased king[8] because street mud was continually being thrown at it. You see how our present-day heroes are ...! Two days ago applause rained down on this great person; and now mud is thrown at his snout, or rather at the snout of his effigy...! He is at his estate, under his arbour in Pombal,[9] picking his teeth after having done so many things unmolested. I find it funny when it is said that he will be called to account for what he made so many people suffer. In such governments, where the king is the living law, there is no more dispute over what the minister does, than saying that his master ordered it.

But let us talk about what matters to us. I sent with other books to Dr. Begue du Presle the following books for you, & I have given him notice of them all. Those

5 By the time Magellan next wrote to Sanches, on 27 June 1777, the latter had a new address.

6 A mountain near Coimbra where there was a Capuchin monastery and a summer house of the King.

7 The deposed Prime Minister, Sebastião José de Carvalho e Melo, Marquis of Pombal.

8 The equestrian statue of King José I in Lisbon that was erected by Pombal and also had his effigy on it.

9 Pombal is a little castle town south of Coimbra.

77.05.20 that belong to you are the following: Phil^{al}. Transact. vol. 66. p. 2 & the discourse of the President[10] = both 9^{s}: 0^{d} /; Worchip of Christ,[11] I think that is 1^{s}:0^{d} / Med^{al}. Commentaries[12] vol. 13.- 1^{s}. 6^{d} / vol. 14 will go at the first opportunity since it is already published./ Saunders on Antimony.[13] I think it is 3^{s}. or 2^{s}.6^{d}. you will see the price on the first page./ the 3rd vol. of Priestley about different kinds of air.[14] – 6^{s}. 0^{d}. / & I suppose that's all that's new. Robertson's work about America[15] will be published this week in 2 vol. 4°. I'll send it to you because I think you told me you wanted it as soon as it was published.

There are no literary novelties worth a pin. Captain Cook wrote from the Cape of Good Hope[16] that he found news there of another Captain, a Dutchman, whose crew was severely affected by scurvy due to a long journey, and he ordered that the kind of green algae or conferva that grows on the sides of ships, which is very bitter, be picked, and gave it in the form of salad to his people: and with that he saved them from the damage of scurvy. Here there still lives a surgeon who was with the famous expedition of Milord Anson, on the vessel called the Centurion if I am not mistaken; & it is said that he with other comrades made use of the same salad of the said conferva or green algae, by which they were saved from the damage of scurvy.[17]

Some days ago I saw D^{r}. Sequeira, who is well & who asked me to pass on his greetings to you &^{c}. &^{c}.

I often see our Envoy,[18] who is the most capable gentleman, full of useful knowledge, & the most proper to be a Minister who can be found today among our people: if things were as they should be, he wouldn't be here any longer, since I'm absolutely sure that no-one equals him among those who are ruling the roost over there. As for me, I'll feel his absence when he is gone: but it seems to me that there is no-one there with the brains to acknowledge him.

---

10 Sir John Pringle, *A Discourse upon some late improvements of the means for preserving the health of Mariners, delivered at the anniversary meeting of the Royal Society ... Nov. 30, 1776* (London, 1776).

11 "Theosebes", *Letters on the Worship of Christ, addressed to the Rev. George Horne ...* (London, 1776)? See Magellan to Sanches, 21 March 1777.

12 *Medical and Philosophical Commentaries* [by A. D.] (Edinburgh, 1773 onwards).

13 William Saunders, *Observationes de antimonio ejusque usu in morbis curandis* (London, 1773).

14 Joseph Priestley, *Experiments and Observations on Different Kinds of Air,* vol. 3 (London, 1777).

15 William Robertson, *The History of America* (2 vols., London, 1777).

16 Captain James Cook spent six weeks at Capetown in October-November 1776, at the commencement of his third, fatal voyage of exploration in the Pacific Ocean.

17 George Anson (1697-1762) sailed around the world in 1740-44, raiding the Spanish colonies in South America during the War of the Austrian Succession and capturing a Spanish treasure galleon off the Philippines. While rounding Cape Horn and subsequently, Anson's squadron suffered appalling losses from scurvy, but there is no mention in the official account of the voyage of their using conferva as an antiscorbutic.

18 Luiz Pinto de Sousa Coutinho (1735-1804), Portuguese Minister in London, 1774-88.

I'm worried about poor Barros,[19] and hope that his luck changes. I have no news from Chevalier,[20] that is, if he is going soon or not to his desired motherland for which he chirped so much, like a duck for the water of the pond. If you have more detailed & tasty items of news there, don't fail to report them to me. Regards to Mª. Joana. 77.05.20

vale.[21]

19 See Magellan to Sanches, 13 September 1776.
20 See Magellan to Sanches, 17 December 1776.
21 Farewell.

To Jean-Baptiste Le Roy 77.06.20

Bibliothèque de l'Institut de France, Paris, MS 876, no. 36

Londres ce 20 June 1777

Mon cher Monsieur & mon cher Confrere

Vous n'avez rien à changer sur l'avantage que vous avez ploidé des garde foudre plutôt que garde tonnerre (car ce n'est pas le bruit, mais le feu de la foudre qui fait tout) à cause de celle qui tomba dans un coin du batiment à Purflet. On a lû hier à la derniere seance de la Soc^e^. R^le^. (qui entra en vacance jusqu'à Novembre) le raport des commissaires, qui examinerent le local.[1] Ce fut une discontinuité des pieces metalliques qui causa ce petit acident. On avoit supposé que les crampons entre les goutieres de plomb se touchoient, ou du moins que celà n'empecheroit que le feu electrique continuat sa route atravers le metal: mais il arriva que dans une distance d'environ de 8 pouces, apparement la pluie y contribua, la matiere sauta d'un morceau à l'autre, & fit eclater en morceaux un coin de la picure & briques intermediaires &ª.

Mais je compte vous envoier un detail plus precis & minutieux, par lequel vous & tout le monde verra demonstrativement que sans le metal qu'il avoit dans le toit, il y auroit peut etre le plus de malheur possible.

1 The committee's report, signed by W. Henley, T. Lane, E. Nairne and J. Planta, was read at the Royal Society's meeting on 19 June 1777, while a dissenting report was read by Benjamin Wilson; see *Philosophical Transactions*, 68 (1778), 236-42.

77.05.20 Cependant M[r]. Wilson lut à la Societé une espece de protestation, comme votre Nolet dejadis faisoit, contre les Conducteurs: & tous les gens qui n'entendent pas la matiere ne manqueront de s'ecrier contre cette precieuse decouverte. teles sont malheureusem[t]. les tetes les plus nombreuses de la sote race humaine...!

Je n'ai pas ete bien cette semaine: & me trouve encore assez fatigué pour ecrire d'avantage pour le present.

Dites de ma part à M[rs]. Bailly & de Bory que J'ai reçu leurs Lettres,[2] & ai deja taché de chercher l'information qu'ils desirent &[a] &[a]

Vale & ama
tuum
Magellan[3]

P.S. Dites je vous prie de ma part ou faites savoir, à Messrs. Jeaurat, de la Lande, Boscowich, & Messier, que J'ai envoyé dernierement dans une caisse à M[r]. Trudaine de Montigny les almanaks Nautiques pour l'année 1780 qui vient de paroitre, & qui leur sont envoyés par l'indication de M[r]. Maskelyne. M[r]. Duchenne secretaire de M[r]. Trudaine en aura soin selon l'avis que je lui fais aujourd'hui. Dites à M[r]. Lemonnier que les observations Astronomiques & sur la bussole & montres marines dans les derniers voyages de Cook & de Fourneaux, vient de paroitre, & seront bientot en vente prix d'une guinée.[4] C'est un vol. in 4°. qui est fort interessant. Je lui envoyerai s'il en veut avoir.

iterum vale.[5]

N.B. il y a aussi dans ce meme caisse un almanak Nautique pour M[r]. Bailly pour la meme annee 1780: dont vous lui donnerez avis &[a].

A Monsieur
Monsr. Le Roy, de l'Acad[e]. R[le]. des Sciences
aux galeries du Louvre
à Paris

2 Letters not found.

3 Farewell & love | your | Magellan.

4 W. Wales and W. Bayly, *The Original Astronomical Observations made in the Course of a Voyage towards the South Pole and round the World ... in the Years 1772, 1773, 1774 and 1775* (London, 1777).

5 Farewell again.

To Johann Albrecht Euler[1] 77.06.27

St Petersburg Branch, Archives Department, Russian Academy of Sciences, f. 1, op. 3, Nr. 63, ll. 67-69ob

Londres 27 Juin 1777
at n°. 12 Nevils Court Fetter Lane

Monsieur

La vaisseau Anne Captn. J. Hunter, qui partira incessament de cette Riviere pour St. Petersburg, a une caisse à son bord marquée I.A.P. n°. 1. qui contient les articles suivants selon l'ordre de la Lettre dont vous m'avez honnoré en date du 5 Mai dernier.[2]

1. Les vol. 57, 58, 59, 60, 61, 62, 63, 64, 65, & la partie 2de. du vol. 66 des Transact. Philoses. de Londres, avec une copie des Nouveaux Estatutes[3] & le discours du Presidt. pour l'année derniere;[4] le tout est destiné pour l'Acde. Imple. comme present de la pt. de la Societé Rle. de Londres:[5] vous savez que le vol. 66 partie premiere partit d'ici l'annee derniere dans une caisse de Genal. Baur,[6] qui vous l'a fait remettre pour votre Acade.

2. Les Observations Astronomiques & sur l'aiguille &a. faites dans les dernieres voyages à la Mer du Sud,[7] que la Haute Commission de la Longitude envoye de present à Mr. Euler votre digne Pere, auquel je vous prie de presenter aussi mes hbles. resps. J'ai scû par Le Dr. Shepard, qui en est membre (Of the Board of Longitude), que le livre de Mr. Euler sur les mouvemts. de la Lune, Lui avoit eté presenté.[8] Et Mr. Thode, dont je vous parlai dans ma lettre precedente,[9] s'est chargé de vous apporter une

---

1 Annotated: "reçu le 10 Juillet 1777 & lû le même jour à l'Académie."

2 Letter not found.

3 *Diplomata et statuta Regalis Societatis Londini, pro scientia naturali promovenda* (London, 1776).

4 J. Pringle, *A Discourse upon Some Late Improvements of the Means for Preserving the Health of Mariners, delivered at the Anniversary Meeting of the Royal Society, Nov. 30, 1776* (London, 1776).

5 Receipt of the works listed was reported by Euler to the St Petersburg Academy of Sciences on 14 August 1777 (Old Style); see *Protokoly zasedanii konferentsii Akademii Nauk...*, vol. 3, p. 316.

6 See Magellan to J. A. Euler, 17 September 1776 and 14 March 1777.

7 Presumably W. Wales and W. Bayly, *The Original Astronomical Observations made in the Course of a Voyage towards the South Pole and round the World in ... the "Resolution" and "Adventure" in the Years 1772, 1773, 1774 and 1775* (London, 1777).

8 L. Euler, *Theoria motuum lunae...* (St Petersburg, 1772). See also Magellan to J. A. Euler, 14 March 1777.

9 See Magellan to J. A. Euler, 14 March 1777.

77.06.27 collection des publications de la meme Haute Comission comme presentes pour Mr. votre Pere. C'est le libraire Nourse qui me donna cette notice: & vous l'aurez avec plus de precision à l'arrivée du meme Mr. Thode.

3. Vous Trouverez dans La meme Caisse deux exemplaires d'une Lettre Françoise venue de Pekin, sur la pretendue identité des caracteres chinois, & des Hierogliphes Egyptiens,[10] dont je vous prie de presenter une de ma part pour la Bibliotheque de votre Acade. Imperiale; & de garder l'autre pour La Votre, avec un exemplaire de mon petit Traité sur les octants,[11] & un autre d'une critique sur Voltaire, dont l'auteur anonyme, qui est de mes amis, m'a fait present.[12]

4°. Je vous envoye les 14 vol des Nautal. almanak qui ont parû jusqu'a present, avec quelques Tables pour les acompagner,[13] qui couterent £2:11:6.

5°. Pennant vol. 3 = 6s:0d: d°. supplement aux premiers trois vol. = £1:1:0: & en fin le dernier vol. sur the crustatious animals = 12s:0d.[14] Le petit traité du meme Naturaliste intitulé Genera of Birds,[15] ne se trouve plus, que par un hazard extreme.[16]

6°. Il n'est pas possible de trouver qu'egalement par un hazard extreme, les Tables de Sherwing de la 4me. edition.[17] J'ai trouvé neanmoins celles de l'edition 3me. faite par le meme Gardiner,[18] & qui sont très rares, & estimées. J'ai crû profiter de l'occasion; & je vous les envoye. Cependant si vous ne trouvez bien de les avoir, vous n'avez qu'à me les renvoyer sans façon: car je les garderai avec plaisir, ne les aiant moi même. elles coutérent, à cause de sa rareté, 17s:6d.

7°. C'est sous la meme Condition que Je vous envoye la Table general de l'Acade. Rle. des sciences, qui est un Livre moderne très utile; ce sont 4 vol.

---

10 *Lettre de Pékin, sur le génie de la langue chinoise et la nature de leur écriture symbolique, comparée avec celle des anciens égyptiens; ... par un père de la Compagnie de Jésus* [i.e. Pierre Cibot] (Brussels, 1773).

11 Magellan, *Description des octants et sextants anglois* (London, 1775).

12 Possibly *Exhortations à Mr. de Voltaire, avec le parallele entre lui and Mr. Rousseau* (London, 1774); see Magellan to J. Bernoulli, 29 September 1775.

13 *Tables requisite to be used with the Nautical Ephemeris for finding the Latitude and Longitude at Sea* (London, 1767).

14 T. Pennant, *British Zoology*, 4th ed. (4 vols, London, 1776-7).

15 T. Pennant, *Genera of Birds* (Edinburgh, 1773).

16 The St Petersburg Academy had requested a complete set of the *Nautical Almanac* and the works by Pennant; see Magellan to J. A. Euler, 14 March 1777, note 9.

17 *Sherwin's Mathematical Tables...*, 4th ed., carefully revised and corrected by William Gardiner (London, 1761).

18 H. Sherwin, *Mathematical Tables...*, 3rd ed., revised and corrected by William Gardiner (London, 1741).

In-4°. brochés.[19] Leur prix est £2: 8: 0. La table des Transactions Philosophiques, que feu D^{r}. Matty avoit presque finie avant sa mort, sera bientôt imprimée:[20] & ce fera un autre objet bien precieux dans le même genre. 77.06.27

8°. Je vous envoye une pendule Astronomique d'echapem^{t}. à repos, faite suivant mes ideés,[21] que Je me flate repondera à vos vuës; & fera un bon service. elle couta £5: 15: 6 & la caisse £1: 11: 6: & en la comparant avec celle du Gen^{al}. Baur, vous ne manquerez pas de juger, que la petite difference du prix y correspond avec avantage.

9°. La caisse pour embaler tous ces articles, & travail de l'embaleur, coutérent 7^{s}: 6^{d}: messages & lettres 6^{s}:10^{d} N.B. Il vaut mieux lorsqu'on ecrit à Londres, ecrire sur une grande feuille de papier, & fermer dans la meme: car les commis de notre poste chargent toujours autant pour l'envelope, ou pour tout autre petit morceau de papier qu'ils y peuvent appercevoir dedans, comme si elles etoient autant de lettres distinctes. Le porterage à la douane, loquet, searchers, porterage wharfage. &^{a}. 7^{s}: 8^{d}: bateau, lading bills & broker 6^{s}: 6^{d}. La Somme totale fait £16: 11^{s}: 6^{d} Esterlines[22] que vous aurez la bonté de faire payer pour mon Compte à Messieurs Robt. & J^{o}. Hay, auxquels j'envoye cette meme poste le lading bill selon la coutume; car puisque ces Messrs. doivent vous rembourser pour la depense des livres dont Je vous parlai dans ma lettre precedente, J'ai crû qu'il etoit plus à propos de leur remettre en droiture moimeme cette caisse, que d'aller incommoder M^{r}. Baxter,[23] qui d'aileurs ne vous est pas connû en particulier.

Les nouvelles pompes dont je vous parlai[24] sont à peu près la meme chose, ou pour mieux dire sont dans le fond les memes de l'invention de M^{r}. Smeaton, dont vous trouverez une description dans les Transact. Philos. il y a environ 15 ou 20 années.[25] Mais celles-ci sont beaucoup mieux executées: & les experiences qu'on a faites avec elles ont demontré une grande meprise sur l'evaluation du vuide, qui semble aler[26] jusqu'à 32 & meme à 64 mille fois: c'est à dire, que la portion du fluide qui se trouve expandu dans le recipient,

19 F. Rozier, *Nouvelle table des articles contenus dans les volumes de l'Académie depuis 1666 jusqu'en 1770...* (4 vols, Paris, 1775).

20 P. Maty, *A General Index to the Philosophical Transactions, from the First to the End of the Seventieth Volume* (London, 1787).

21 Magellan described his improvements in his *Notice des instrumens d'astronomie, de géodesie, de physique, &c. faits ... par order de la Cour d'Espagne* (London, 1780), p. 209.

22 A note is attached to the letter at this point: "£16.11.6 @ 48es assignatum: R^{l}. 82.87–".

23 Alexander Baxter, Russian Consul-General in London.

24 See Magellan to J. A. Euler, 14 March 1777.

25 J. Smeaton, "Letter ... concerning some Improvements ... in the Air-Pump", *Philosophical Transactions*, 47 (1751-2), 415-24.

26 aller?

77.06.27 a augmenté autant de fois dans son volume. Ces pompes sont travaillées in vacuo, je veux dire le piston qui fait sortir & rarefier l'air qui reste dedans, travaile toujours dans un tuyau, d'oû l'air peut sortir, mais oû pas aucun air peut entrer. J'ai donné une relation des experiences etonnantes de ces pompes à l'Acad^e^. R^le^. des sciences, par lesquelles il est evident que ce n'est pas l'air qui s'expand jusqu'à un si haut degré; mais l'humidité qui s'y trouve dedans: car si on travaille ces pompes à sec, employant du cement resineux au lieu des cuirs mouillés, alor elles ne vont au delà de 600 fois ou environ. M^r^. de la Place lut un Memoire à la rentrée de cette Acad^e^. (de Paris) sur ce sujet:[27] & le chev^r^. de Bory qui y tient ma correspondance, m'ecrivit dernierem^t^. que ma lettre sur ce sujet soit inserée dans le journal de l'Abbe Rozier (qui est une des plus excellentes publications periodiques de nos jours) avec une planche qui representera la pompe de cette espece que J'ai envoyée d'ici à M^r^. de la Voisiere mon ami, Membre de la meme Academie.[28] Vous aurez aussi dans le vol. suiv^t^. des Transact. Philos^es^. le recit de ces nouvelles & suprenantes experiences.[29]

Nous avons à present ici un ridicul tapage philosophique, qui peut etre arrivera chez vous, & dont je souhaite pour le bien de la Philosophie & du Public, que vous preveniez votre Academie. Dans l'année 1772 la Soc^e^. R^le^. deputa quelques membres pour consulter sur les conducteurs electriques qu'on devoit mettre aux magazins à poudre (de canon) à Purflet, environ 15 miles de Londres.[30] M^r^. Wilson, qui a fait tant de tapage avec les phosphores,[31] fut de cette commitée, & il dissentit des autres sur la circumstance seule de que ces conducteurs metalliques finissoient emoussés & non pas une pointe aiguisée, comme tout homme qui connoit un peu

27 The *procès-verbaux* of the Paris Academy of Sciences record that on 9 April 1777 Laplace read a paper "Sur la nature du fluide qui reste dans le récipient de la machine pneumatique". However, this was never published and no trace of it remains: see *Dictionary of Scientific Biography*, vol. 15, p. 392.

28 Magellan is evidently referring to his letter of 31 January 1777 that was probably addressed to the Chevalier de Bory. Magellan's expectations notwithstanding, it seems never to have been published.

29 Edward Nairne, "An Account of Some Experiments made with an Air-pump on Mr. Smeaton's Principle; together with Some Experiments with a Common Air-pump", *Philosophical Transactions*, 67 (1777), 614-48.

30 The committee comprised Henry Cavendish, William Watson, Benjamin Franklin, Benjamin Wilson and the Society's assistant secretary, John Robertson. Its report, accompanied by a dissenting report by Wilson, was published in *Philosophical Transactions*, 63 (1773-4), 42-8.

31 Shortly before this, Wilson had engaged in a vigorous public dispute with Joseph Priestley and others over the behaviour of phosphori. See B. Wilson, *A Series of Experiments on the Subject of Phosphori*, 2nd ed. (London, 1776).

l'electricité doit juger qu'ils doivent etre. Cependant ce M[r]. qui veut savoir & raisoner mieux que les autres, protesta contre: les conducteurs furent mis en consequence pointûs, quoique pour le contenter il y a toute apparence qu'on les emoussa un peu. Voilà que dernierement vient un coup de tonnerre à la faveur de ce criailleur, faute d'attention de ceux qui [preferent] à cet affaire. Car la maison des assemblées de Purflet est dans une coline: & a un parapet de muraille autour du toit. Ce parapet est couvert de pierres de taille, qui sont raffermies par des crampons de fer, qui y sont plombés: & le conducteur qu'on y a mis au milieu du Toit est à la distance de 46 pieds du coin du parapet, qui avance le plus hors de la coline. Si l'on auroit fait une communication metallique entre les crampons de ces pierres du parapet & le conducteur, tout seroit bien: mais on n'y prit pas garde en 1772: & voilà que dernierem[t]. vient un nuage chargé d'electricité, & etant attiré par le montage, & peut etre par les conducteurs qui ne sont pas bien pointus, passa plus près du crampon près de ce coin, y dechargea un coup, qui trouvant à la distance de 7 pouces le coin d'une plance du plomb des goutieres, y sauta tout de suite; & se dechargea sans faire dans la suite le moindre domage: & seulement en sautant ces 7 pouces fit eclater quelques briques, laissant un trou d'environ 13 pouces à cause des briques qui sauterent &[a]. Voici donc que les avantages des conducteurs metalliques sont prouvés par ce meme acident, car la foudre passa en silence par la goutiere de plomb, sans faire le moindre domage: mais le criailleur Wilson, qui veut se faire un nom, fait du tapage & sans doute ecrira par toute L'Europe en decriant les conducteurs sans penser au mal qu'il fait par là a public ignorant dans la physique.[32] le papier finit, & moi aussi en me signant avec toute consideration possible 77.06.27

Monsieur
Votre t. h[ble]. serv[r].
Magellan.

A Monsieur
Monsr. Jean Albert Euler
Secretaire de l'Acad[e]. Imp[le]. des Sciences
à S[t]. Petersburg

32 A new committee established by the Royal Society, to consider the circumstances of the accident at Purfleet, reported along the lines indicated by Magellan in his description of what happened. Wilson once again filed a dissenting report and subsequently received great publicity for his spectacular attempts to provide experimental support for his views. See R.W. Home, "Points or Knobs: Lightning Rods and the Basis of Decision Making in Late Eighteenth Century British Science", in P. Heering, O. Hochadel and D. Rhees, eds, *Playing with Fire: Histories of the Lightning Rod* (Philadelphia, 2009), pp. 97-120.

77.06.27a *To Ribeiro Sanches*[1]

Österreichische Nationalbibliothek, Vienna
A.N. Ribeiro Sanches Korrespondenz, codex 12714, ff. 199-200v

27 Junho 1777

Meu am^{o}. e S^{r}. e m^{to}. do Cor^{am}.

Apenas ha meia hora q fechei uma carta p^{a}. Euler f^{o}., sobre uma pendula astronom^{a}. q me pedio lhe mandasse das minhas; e estava tão acabrunhado q me fui deitar sobre a cama: pois ainda não estou totalm^{te}. restabelecido da influenza q por aqui andou de dores nas tripas, tosse & febre. vai senão quando uço[2] bater na porta, levanto-me, abro, e acho hũa das minhas Nelis, q sabe ser o dia do correio proprio p^{a}. me achar. valha-te mil diabos, não sabes q ainda não estou muito bom. no matter; I don't care to be disapointed.... good by. no, no sayd I: come in, and Let us see what we can do. fomos p^{a}. sima da cama, & foi tal a tezão q ella mesma disse: oh oh, I see you wanted to cheat me. acabamos a nossa obra, foi-se ella embôra: & eu achome tão refrescado & bom q me puz logo á banca. escrevi outra carta ao Mercador a Petersburgo a q^{m}. mando a caixa com a pendula & livros: chegou o comiss^{o}. da posta com a carta de V.M. de 23 do corrente: e eizme aqui são como hum pero, respondendo-lhe immediatam^{te}. como q^{m}. se acha aliviado & restabelecido. Veja lá o q he arr-tar ...! e ainda mais o descar-gar ...!

hontem jantei com o D^{or}. Sequeira em caza do nosso Emb^{or}. q nos deo boa sopa portugueza, com bom prezunto do melgaço, & payo d'alentejo, arroz, & marmelada fina &^{a}. tão bem esteve molestado a semana passada: & da mesma sorte fui achar meo am^{o}. Arbuthnot, sua M^{er}. & as duas filhas mais velhas no domingo passado, q fui estar com elle no campo. foi andaço geral. o D^{r}. Seq^{ra}. me disse q tivera m^{tos}. doentes desta maneira. n'um dia em q eu estava mal, veio elle ver-me por acazo, aconselhou-me isto & aquilo &^{a}. pois olhe; sabe V.M. q hei de fazer? hei de jantar uma pouca de vaca fria, em pedacinhos com hũa salada q ja mandei buscar. isso he uma parvoice, me disse elle não não respondi eu; pois eu sinto um particular appetite p^{a}. isso. Comi a sellada & vaca: & logo me achei m^{to}. melhor. e tãobem apostarei q se V.M. entrasse neste quarto havera 3 quartos de hora,

1 Annotated "recebida a 3 de Julho" [received 3 July] and "R^{a} a 24 Julho" [Answered 24 July]. Letter not found. The envelope is annotated in a different hand: "port de letre 2$^{\#}$ 10^{S}" [carriage of letter 2$^{\#}$ 10^{S}].

2 ouço?

havia de mandar a minha Nely coas más horas ...! pois eu sei agora por experiencia, q faria uma grande parvoice ...[!] 77.06.27a

mandarei a V.M. o dic^{o}. de Medecina q não me parece grande [cou]za,[3] pela prim^{a}. occazião, & procurarei as addicoens de Dimsdale &^{a}.

a tomada da Ilha de Sta. Catherina ha de parir quatrinos p^{a}. Espanha, com q^{m}. o nosso Min^{o}. me disse q já estavamos com tregoas, e [de][4] ajustes de estado &^{a}. eu já lhe Repeti o bom ditto do Pr^{e}. Pignate[lli] q o nosso paiz hia entrando pela boca da Espanha. elle sabe [...][5] de Longe, & imagino q não deixa de lhe ver geito, mais cedo ou m[ais] tarde. q^{to}. a mim eu não vejo q isso seria uma grande mal p^{a}. a Nação: pois mal por mal, assaz mal temos nos estado, & assaz mal estamos & estaremos ainda.

lá mandei a V.M. n'um masso por via de M^{r}. du Presle, Robertson Hist^{y}. of America. 2 vol. 4°. £ 1.16:0^{d}/ Culen's Institutes 3^{s}./ o 4$^{me.}$ & 5me. cahier do med^{al}. commentaries 3^{s}./ Mudge on inoculation 2^{s}. 6^{d}./ e demais um tinteiro de vidro, q he da melhor invenção q ainda vi. tapa-se com o dedo o buraco a, e enche-se de tinta pelo outro z q ao depois se fecha com a sua rolha. neste tinteiro nunca se tira mais tinta

com a penna do q he necess^{o}. pelo buraco a. e por isso nunca cahem borroens: e a tinta nunca se evapora por estar sem outro contacto com o ar senão no buraquinho a q he mui piqueno & se o calor faz q o ar em sima se dilate, nunca pode transbordar, como sucede nos outros tinteiros q a V.M. mandei.

vou responder á sua antecedente de 26 de Maio. não estou agora p^{a}. beatisses por isso não lhe falo nas contas: ainda [es]to[6] meio arr-do ...

o artigo da nossa terra he excellente. o S^{r}. Pinto e sua portugueza Espoza rirão bem com elle: mas nem elles nem eu, não nos cancaremos com m^{tas}. vias sacras, jejuns, penitencias, novenas &^{a}. &^{a}. &^{a}.

eu là mandei a M^{r}. le Begue a dissertação de hulme: esta practica necessita confirmação. o Quid facient Leges, ubi sola pecunia regnat he um achaque tão antigo como Petronio, & tão moderno como vemos. isto aqui vai dos diabos. A America desmembrada, sera o unico asylo p^{a}. os espiritos philos^{os}., tanto q a paz se fizer: pois he necess^{o}. q em fim se faça ao menos q^{do}. esta grande massa de dinheiro em Inglaterra começar a fazer falta sensivel.

---

3 Obliterated margin. Conjectural reading.
4 Obliterated margin. Conjectural reading.
5 Obliterated margin. Illegible.
6 estou?

77.06.27a minhas saud^es^. a M^a^. Joana á Prima & á sobrinha. sinto q o marido va de mal em peor. ao menos espero q podera continuar o oficio de marido, q he a unica & geral consolação femenina emq^to^. dura.

vale atque iterum vale.

N.B. não quero q V.M. venda os seos livros. conserve-os ao menos p^a^. compagnia. q^m^. ficar q feche a porta, & disponha do q quizer & do q achar.

iterum vale.

a Monsr.
Monsr. Sanchez, Doct^r^. en Medecine
Cimitiere de S. Jean au bout de la Rue de la Verrerie
à Paris

27 June 1777

My Dear Friend

Half an hour ago I sealed a letter to Euler fils[6] about one of my astronomical pendulum clocks that he asked me to send him; and I was so distressed that I went to lie on the bed, because I'm not yet completely recovered from the influenza that came through here accompanied by belly-aches, cough & fever. Suddenly I heard knocking on the door; I get up, open it, and find one of my Nellys who knows that the mail day is the right one to find me. Help me a thousand devils, don't you know that I'm not yet well? No matter; I don't care to be disappointed.... good by. no, no sayd I: come in and Let us see what we can do. We went over to the bed, & the excitement was such that she said: oh, oh, I see you wanted to cheat me. We finished our job, she went away, & I feel so refreshed & well that I sat myself at once at the table. I wrote another letter, to the merchant in Petersburg to whom I'm sending the box with the pendulum clock & books,[7] the mailman arrived with your letter of 23rd of this month:[8] and here I am, sound as a bell, answering you immediately as someone who feels relieved & recovered. You see what it is to b-lch[9] and even to unload! Yesterday I dined with D^r^. Sequeira at our Ambassador's house who gave us good Portuguese soup, with good cured ham from Melgaço & smoked pork sausage from Alentejo,[10] rice & fine quince jam &^c^. He was also ill last week: & in the same condition I found my friend Arbuthnot,

6 See Magellan to Johann Albrecht Euler, 27 June 1777.
7 Letter not found.
8 Letter not found.
9 See Magellan to Sanches, 7 July 1775, note 19.
10 Respectively a small fortified town in northern Portugal famous for its cured ham, and a southern province of Portugal renowned for its pork sausages.

his wife & the two eldest daughters last Sunday when I went to visit him in the country. It was a general disease. D^r^. Sequeira told me he had many patients in this condition. One day when I was ill he came to see me by chance and advised this & that &c. Do you know what I'll do? I'll have a little cold beef for dinner, in small pieces with a salad that I've already asked for. That's folly, he told me. No, no I answered; because I feel a special appetite for that. I ate salad & beef & at once felt much better. And I also bet that if you had entered this room 3 quarters of an hour ago, you would have sent my Nelly away because I was feeling bad ...! But I know now by experience, that that would have been a silly thing to do ...! 77.06.27a

I'll send you at the first opportunity the Dictionary of Medicine,[11] which doesn't seem a great work, & I'll look for Dimsdale's additions[12] &c.

The conquest of Santa Catarina Island[13] will produce quadruple benefits for Spain, with whom our Minister[14] said we already have a truce and are in negotiation &c. I repeated to him the saying of Prince Pignate[lli][15] that our country was almost being swallowed by Spain. He knows [...] at a distance, & I imagine he won't fail to agree with it, sooner or later. As for me, I don't see that as a great harm to the nation since, harm for harm, we have been bad enough, & so we are & will be.

I sent you in a package through Mr. du Presle, Robertson Histy. of America[16] 2 vol. 4° £1.16.8d / Culen's Institutes[17] 3s. / the 4th. & 5th. cahier of the Medal. Commentaries[18] 3s. / Mudge on inoculation[19] 2s. 6d./ and also a glass inkpot that is of the best invention I have seen. One stops up with a finger the hole a and fills it with ink through the other z that is then closed with its stopper. In this inkpot

one never takes more ink than necessary through the hole a and, because of that, blots never fall out: and the ink never evaporates due to being in contact

11 George Motherby, *A New Medical Dictionary* (London, 1775).

12 Thomas Dimsdale, *The Present Method of Inoculating for the Small-pox. To which are added Some Experiments instituted with a View to Discover the Effects of a Similar Treatment in the Natural Small-pox ...* (2nd edition, London, 1767)? Several editions and translations of this work were published in the following years.

13 The Island of Santa Catarina, off the south coast of Brazil, was the scene of battles between Portugal and Spain from the time the colony of Sacramento (in modern Uruguay) was established by the Portuguese in 1680. Late in 1776, Spain occupied the island, but in the following year the Treaty of Santo Ildefonso returned the island to Portugal in exchange for Sacramento and some other territory.

14 Luiz Pinto de Sousa Coutinho (1735-1804), Portuguese Minister in London.

15 See Magellan to Sanches, 17 November 1775.

16 William Robertson, *The History of America* (2 vols., London, 1777).

17 William Cullen, *Institutions of Medicine. Part I: Physiology* (Edinburgh, 1772); see Magellan to Sanches, 4 July 1777.

18 *Medical and Philosophical Commentaries* [by A. D.] (Edinburgh, 1773 onwards).

19 John Mudge, *A Dissertation on the Inoculated Small-pox* (London, 1777).

77.06.27a with air only through the little hole a which is very little, & if heat makes the air expand, it can never spill over, as happens in the other inkpots that I sent you.[20]

I'll answer your letter of 26th May.[21] I'm not in the mood for bigotry and then I'm not talking about the accounts; I'm still half bel-ed.

The article[22] about our country is excellent. Mr. Pinto and his Portuguese wife found it amusing but neither they nor I will get tired from many stations of the cross, fasts, penances, nine-day-prayers &c. &c. &c.

I finally sent Mr. le Begue the dissertation of hulme:[23] this practice needs confirmation. The Quid facient Leges, ubi sola pecunia regnat[24] is an ailment as old as Petronius, & as modern as we see. Here things are going to hell. A separated America will be the sole asylum for philosophical spirits, as soon as peace is achieved: since necessarily this will finally happen, at least when the great lack of money in England begins to be felt.

My regards to Ma. Joana, and to her cousin & her niece. I'm sorry her husband is getting worse and worse. At least I hope he can continue the function of a husband, which is the sole & general feminine consolation as long as it lasts.

Vale atque iterum vale.[25]

N.B. I don't want you to sell your books. Keep them at least for company, whoever remains behind will close the door & dispose of what you want & decide on.

iterum vale[26]

---

20 See Magellan to Sanches, 16 September 1775 and 1 January 1776.

21 Letter not found.

22 Not identified.

23 Nathaniel Hulme, *A Treatise on the Puerperal Fever …* (London, 1772).

24 What will the laws achieve where money alone rules (Petronius, *Satyricon*, Ch. XIV).

25 Farewell and again farewell.

26 Farewell again.

To Ribeiro Sanches[1]

Österreichische Nationalbibliothek, Vienna
A.N. Ribeiro Sanches Korrespondenz, codex 12714, ff. 201-202v

Londres 4 Julho 1777

Am^{o}. e S^{r}. e m^{to}. do Coração

Antes q V.M. me peleje pela f-ria da carta passada, vou mandarlhe aqui a dezejada conta, q certam^{te}. não acabarei de escrever, se a m^{a}. Neli torna hoje pelo vezo. em 2 Abril 1776 mandei a V.M. 1 Camoens .1$^{£}$. 6^{s}. 0^{d}. Rutty on London & Edimburg dispensaries – 3^{s}. 0^{d}. / Kirkland on child bed fevers – 3./ Musgrave on nerfs 2^{s}. 6^{d}./ Armstrong essay – 1^{s}. 6^{d}./ Chalmers on Caroline fevers – 5^{s}. 0^{d}./ Percival essays – 5^{s}./ Phil^{al}. Transact. vol. 65 p. 2^{d}. – 7^{s}. 6^{d}./ Pringle's discourse – 1^{s}. 6^{d}./ levison on blood – 2^{s}. 6^{d}.

9 de Agosto Gibbon's History of y^{e}. fall of y^{e}. Roman Empire £1.1.0/ Phil. Transct. vol. 66. p. 1. – 7^{s}. 6^{d}. Med^{al}. Inquiries vol. 5th. – 6^{s}./ Percival Essays vol. 3^{d}. – 5^{s}./ Med^{al}. Commentaries n.° 12 – 1.6^{d}/ Rozen – 5^{s}./ Med^{al}. Coment. n°. 13 – 1^{s}. 6^{d}.

no prim^{o}. de Maio 1777 – Phil. Transact. vol. 66. p. 2^{d} – 7^{s}. 6^{d}. discourse – 1^{s}. 6^{d}./ worship of christ – 1^{s}. 0^{d}/ Sounders on Antimony não sei se 3^{s}./ o 3°. vol. de Priestley on air 6^{s}./ 14th N^{o}. of Med^{al}. Commentaries – 1^{s}. 6^{d}/ Culen's Physiology – 3^{s}./ Robertson Hist^{y}. of America – 1. 16. 0/ Mudge on inoculation – 2. 6^{d}/ Med^{al}. Com. n°. 15 – 1. 6^{d}

Agora vou mandar a V.M. na semana proxima & pela mesma via Culen's Practice of Physic vol. y^{e}. 1^{rs}. 6^{s}./ Motherby's dictionary – 1$^{£}$. 10^{s}. 0^{d}./ griffith on ethick fever 2^{s}./ – expenses of these different parcels about 7^{s}. or 8^{s}. Total £11: 3^{s}: 6^{d} qui valent 268$^{\#}$. pour la compte desquelles J'ai reçu 96$^{\#}$. que vous avez payé pour moi a Ruault selon votre Lettre de 17 Julho 1776. plus 300$^{\#}$. dans le 11 Fevrier dernier: e emfim 5$^{\#}$. du Traité sur le [nitre][2] que V.M. me mandou pela posta total 400$^{\#}$. Livras tournoises donde vem q eu sou devedor depois do d^{o}. Abril do anno passado de 132$^{\#}$. q M^{r}. le Begue du Presle pagará do dinh^{o}. q tem recebido por minha Conta, cujo Resto elle mandara p^{a}. Inglaterra pela via q ja lhe indiquei.

Ora já estará V.M. satisfeito e descançado com as nossas contas de bugalhos. Cuido q bem se lembrara da cantiga = o velho mais a velha jogarão

---

1 Annotated "Resposta a 25 Julho" [Answered 25 July]. Letter not found.
2 Illegible. Conjectural reading.

77.07.04 a bugalhinha: mas o velho de maroto, meteo-lho na buraquinha. isto sim q são cantigas das nossas Marias lá pelos alqueves de Serra d'Estrella.

Hontem pela manhan fui almoçar com o D^{or}. La Cour, q veio visitar alguns amigos à Londres. tinhamos jantado no dia antecedente na Taverna do Adelphi, & là bebemos á sua saude. está todo encasquetado de Fidalgos, q certam^{te}. o tem invitado ás suas terras. partio hontem ás duas da tarde p^{a}. o duque de Northumberland & nos deo uma ladainha de gente de primeira Classe, q absolutam^{te}. o querem ver nas suas terras & cazas de Campo. sua Mulher não pode vir almorçar comnosco, achando-se molestada: mas não coiza q impedisse a continuação da viagem. o D^{or}. Seq^{ra}., q tinha jantado com nosco na quinta fr^{a}. na Taverna, foi obrigado a partir na noite seg^{te}. pela posta p^{a}. Britel[...]ston coiza de 20 milhas de Londres, chamado por hum dos seos doentes, q lá se achava tomando banhos de mar, e q se tinha achado mal. o D^{r}. La Cour me falou m^{to}. em V.M. e me pedio q lhe dissesse como conservava fielm^{te}. o seos affectos &^{a}. &^{a}. Respondilhe porq não lhe escrevia duas Regras ao menos: disseme q certam^{te}. o faria [bre]vem^{te}., & assim o recomendou a Miss Davies, um[a] bella young Lady, sua companheira de viagem, q com elle passa g^{de}. parte do anno em Bath (cuido q não he da nossa casta) e ella me prometeo q lho avia de lembrar, e q^{do}. mais não fosse q ella mesma escreveria em Seo nome. visto isto está V.M. em risco de Receber carta escrita por mãos de neve. olhe não lhe pegue o fogo q semelhantes Neves fazem atear n'um coração Portuguez.

o nosso Ministro de Balsamão vai dar uma volta à Bath com sua Espoza: hirá passar alguns dias com Mylord Shelburne à Bowood, onde o mesmo Mylord me invitou tãobem: mas eu duvido q me possa dezembaraçar de Londres p^{a}. ir esperdiçar o tempo &^{a}.

vive, vale & ama
tuum &^{a}.

aqui temos not^{a}. q as hostild^{es}. com Port^{al}. estão suspensas, & q se entra em Tratado. eu vejo bem no q acabará, & como tudo ira pouco mais ou menos: mas por lá & por cá, más fadas há.

iterum vale.

a Monsieur
Monsr. Sanchez, D^{r}. en Medecine
Cimitiere de S^{t}. Jean Rue de la Verrerie
à Paris

London 4 July 1777 77.07.04

My Friend and very Dear Sir

Before you quarrel with me due to the fury of the last letter,[3] I'm going to send here the desired account, that I certainly shan't finish writing if my Nelly returns today for the usual. On the 2nd of April 1776 I sent you[4] 1 Camoens 1$^{£}$.6^{s}.0^{d}. Rutty on London & Edimburg dispensaries – 3^{s}.6^{d}./ Kirkland on child-bed fevers – 3./ Musgrave on nerfs 2^{s}.6^{d}./ Armstrong essay – 1^{s}.6^{d} / Chalmers on Caroline fevers – 5^{s}.0^{d}./ Percival essays – 5^{s}./ Phil^{al}. Transact. vol 65. p. 2^{d}. – 7^{s}.6^{d}./ Pringle's discourse – 1^{s}.6^{d}./ Levison on blood – 2^{s}.6^{d}

9th of August[5] Gibbon's History of y^{e}. fall of y^{e}. Roman Empire £1.1.0/ Phil. Transact. vol. 66. p.1. 7^{s}.6^{d} Med^{al}. Inquiries vol. 5th – 6^{s}/ Percival Essays vol. 3^{d} – 5^{s}/ Med^{al}. Commentaries no.12 – 1.6^{d}/ Rozen – 5^{s}/ Med^{al}. Comment. no. 13 – 1^{s}.6^{d}.

On the first of May 1777[6] – Phil. Transact. vol. 66 p.2^{d}. – 7^{s}.6^{d}/ discourse – 1^{s}.6^{d} / Worship of Christ – 1^{s}.0^{d}/ Saunders on Antimony I don't know if it is 3^{s}. / the 3rd vol. of Priestley on air 6^{s}. / 14th No. of Med^{al}. Commentaries – 1^{s} 6^{d}/ Cullen's Physiology – 3^{s}./ Robertson Hist^{y}. of America – 1.16.0 / Mudge on inoculation – 2.6^{d} / Med^{al}. Com. no.15 – 1.6^{d}

Now I am going to send you next week & by the same way Cullen's Pratice of Physic.[7] vol. y^{e}. 1st. – 6^{s}. / Motherby's Dictionary[8] – 1$^{£}$.10^{s}.0^{d} / Griffith on ethick fever[9] – 2^{s}. / – expenses of these different parcels 7^{s} or 8^{s} Total £11: 3^{s}: 6^{d}. qui valent 268#. pour la compte desquelles j'ai reçu 96#. que vous avez payé pour moi a Ruault[10] selon votre Lettre du 17th July 1776.[11] plus 300#. dans le 11 Fevrier dernier: and at last 5#. du Traité sur le nitre that you sent me by post[12] total 400.# livres tournoises whence I am in debt since April last year an amount of 132# that M^{r}. le Begue du Presle will pay out of the money he has received on my account, the remainder of which he will send to England in the way I have already told him.

Now you may be content and calm about our accounts of trifles. I hope you will remember well the song = the old man and the old woman played the "bugalhinha":[13] but the naughty old man introduced it in the little hole. These are the true songs of our Marias there in the lands of Serra d'Estrella.

3 Letter not found.

4 See Magellan to Sanches, 18 February 1777.

5 Ibid. Magellan fails to note here single books sent on 9 May and 11 July 1776 that he included in the list in the earlier letter.

6 See Magellan to Sanches, 20 May and 27 June 1777.

7 William Cullen, *First Lines of the Practice of Physic*, vol. 1 (Edinburgh, 1777).

8 George Motherby, *A New Medical Dictionary* … (London, 1775).

9 Moses Griffith, *Practical Observations on the Cure of Hectic and Slow Fevers* … (London, 1776).

10 Bookseller in the rue de la Harpe, Paris.

11 Letter not found.

12 [Académie Royale des Sciences, Paris], *Recueil de memoires et d'observations sur la formation & sur la fabrication du salpetre* … (Paris, 1776); see Magellan to Sanches, 21 March 1777.

13 A boy's game.

77.07.04 Yesterday morning I had lunch with Dr. la Cour, who came to visit some friends in London. We had lunch the day before at the Tavern of the Adelphi,[14] & there we drank to your health. He is obsessed with noblemen, who certainly have been inviting him to their properties. He left yesterday at two p.m. to the Duke of Northumberland & gave us a rigmarole of first-class people who absolutely want to see him in their estates & country houses. His wife could not come and have lunch with us, being ill: but not such a serious problem as would impede the continuation of their journey. Dr. Sequeira, who had dinner with us on Thursday at the Tavern, had to leave the next night by the stage-coach to Britel[...]ston,[15] about 20 miles from London, called by one of his patients who was already having sea baths, and who felt bad. Dr. La Cour spoke a great deal about you and asked me to tell you how he kept faithfully your sympathy &c. &c. I answered why wouldn't he write you at least two lines: he told me that he would certainly do so soon, & so he recommended it to Miss Davies, a beautiful young lady, his travel companion, who spends a large portion of the year with him in Bath (I think she is not of our race) and she promised me that she would remind him and in the last resort would write herself in his name. This said you run the risk of receiving a letter written by hands of snow. Be careful not to catch the fire that such snows ignite in a Portuguese heart.

Our Minister de Balsamão[16] is going on a trip to Bath with his wife: he will spend a few days with Milord Shelburne at Bowood, where the same Milord invited me also: but I doubt that I can leave London to waste my time &c.

vive, vale & ama

tuum[17] &c.

Here we have news that hostilities against Portugal are suspended, & that they enter into a treaty.[18] I see well where it will finish, & how everything will go more or less: but both there & here, evil fairies exist.

iterum vale[19]

14 A group of neoclassical buildings in London between the Strand and the Thames, laid out in 1768-72 by the four Adam brothers and hence called Adelphi (from the Greek for brothers).

15 The paper is torn but we conjecture that Magellan is referring to Brighthelmstone, an ancient designation of the town of Brighton. In the middle of the eighteenth century, this small fishing village became a fashionable place where rich people went to take sea baths that supposedly had curative properties. However the distance from London is not 20 miles as Magellan says but nearer to 60 miles.

16 Luiz Pinto de Sousa Coutinho (1735-1804), Morgado de Balsemão, Portuguese Minister in London.

17 Flourish, farewell and love. your.

18 Late in 1776, Spain occupied the Portuguese island of Santa Catarina, off the south coast of Brazil, but in the following year the Treaty of Santo Ildefonso returned the island to Portugal in exchange for the Portuguese settlement of Sacramento (in present-day Uruguay) and other territory.

19 Farewell again.

To Ribeiro Sanches 77.07.11

Österreichische Nationalbibliothek, Vienna
A.N. Ribeiro Sanches Korrespondenz, codex 12714, ff. 203-204v

11 Julho 1777

Am^{o}. e S^{r}. e m^{to}. do Coração

hoje fiz meter a bordo do Navio p^{a}. Calais o Dicion^{o}. de Motherby; o prim^{o}. vol. completo da practica de Culen: & o tratadinho de Griffith sobre a Hectica: tudo vai remetido ao D^{r}. Begue q o mandará a V.M.

4^{a}. fr^{a}. passada sahio daqui p^{a}. Hollanda, e de lá por Flandres p^{a}. Paris o Cav^{ro}. Pereira, primo do nosso Enviado nesta Corte, acompanhado do seo secretario Cypriano Ribeiro Freire, rapas de boa feição e juizo modesto, q absolutam^{te}. me pedio duas Regras p^{a}. V.M., debaixo de palavra q V.M. o não ha de hir vizitar, nem pagar-lhe vizitas &^{a}. e q de outra sorte queria antes ficar privado do gosto de ver a V.M. a q^{m}. depois longo tempo venera &^{a}. Se por acazo o companheiro, não cuide q tem a affabilid^{e}. & mais qualid^{es}. do primo q aqui fica.

Minhas saud^{es}. á Familia adoptiva a q^{m}. dezejo perf^{a}. saude & felicid^{es}. N.B. o Cav^{ro}. Pereira foi corregedor & teve outros semilhantes cargos emq^{to} o Cav^{ro}. Pinto esteve em Mato grosso por governador. ambos forão de Portugal p^{a}. a America: mas o Ch^{er}. Pereira ficou ainda là servindo depois do Primo: fez m^{to}. bom dinh^{o}., comprou fazendas & fará uma caza distinta da do irmão, pois este he filho segundo &^{a}. Veio ver aqui o Primo: tem estado aqui alg[umas ve][1]zes: vai agora ver a Hollanda, Flandres, & França até Paris: de là tornará aqui, & hirá p^{a}. o seo Cadoz de Portugal, comer boas peras pigarças, & bom prezunto de Lamego Lá p^{a}. o pé do Pezo da Regoa &^{a}. &^{a}.

Vive & vale, atque iterum vale

a Monsieur
M^{r}. Sanchez D^{r}. en Medecine
Cimitere de S^{t}. Jean Rue de la Verrerie
à Paris

1 Stained paper. Conjectural reading.

77.07.11 11 July 1777

My Dear Friend

Today I sent aboard the ship to Calais the Dictionary of Motherby,[2] the first complete vol. of the practice of Cullen,[3] & the little treatise of Griffith about Hectics:[4] everything is addressed to Dr. Begue who will send it to you.

Last Wednesday Chevalier Pereira, cousin of our Envoy in this Court,[5] left for Holland and from there through Flanders to Paris, accompanied by the latter's secretary Cypriano Ribeiro Freire,[6] a young man of good figure and modest judgement who begged me to write two lines to you, given that you won't go and visit him or pay him visits &c. and that otherwise he would prefer to be deprived of the pleasure of seeing you whom he has revered for a long time &c. lest his companion doesn't care to have the kindness & further qualities of his cousin, who is staying here.

My regards to the adoptive family to whom I wish perfect health & happiness. N.B. Chevalier Pereira was Corregidor & had other similar posts while Chevalier Pinto was in Mato Grosso as governor.[7] Both went from Portugal to America: but Chevalier Pereira continued to serve there after his cousin left. He made a large amount of money, bought farms, & will build a mansion separate from his brother's, since he's the second son &c. He came here to see his cousin: he has been here for some time. He's now going to Holland, Flanders & France as far as Paris: from there he'll return here & then go to his burrow of Portugal, to eat good speckled pears & good cured ham from Lamego near Pezo da Regoa[8] &c. &c.

Vive & vale, atque iterum vale[9]

---

2 George Motherby, *A New Medical Dictionary …* (London, 1775).

3 William Cullen, *First Lines of the Practice of Physic*, vol. 1 (Edinburgh, 1777).

4 Moses Griffith, *Practical Observations on the Cure of Hectic and Slow Fevers …* (London, 1776).

5 Luís de Albuquerque e Melo Pereira e Cáceres (1739-1797) was Governor and Captain General of the Captainship of Mato Grosso in Brazil, 1772-88, and a cousin of Luiz Pinto de Sousa Coutinho (1735-1804), the Portuguese Minister in London.

6 Cipriano Ribeiro Freire (1749-1824) was a protégé of Pombal who took diplomatic jobs, being appointed secretary to the Portuguese legation in London in 1774 in charge of commercial and industrial matters. After Luiz Pinto de Sousa ceased to be the Portuguese minister in London, Freire was appointed chargé d'affaires. In 1791 he was minister resident in the United States and later minister plenipotentiary.

7 See Magellan to Sanches, 15 November and 17 December 1776.

8 A town southeast of Porto, not far from the Douro River, famous for its cured hams.

9 Flourish, farewell and again farewell.

To the Marquis de Condorcet

Bibliothèque de l'Institut de France, Paris, MS 876, no. 37

Londres 21 Juillet 1777

Monsieur

Je trouvai enfin dimanche dernier Le Dr. Solander[1] chez soi, après 3 ou 4 fois que j'avois allé chez lui. il me dit n'avoir pas fait aucune experience avec la Phytolacca:[2] & me recommenda de parler avec Mr. Justamond, cirurgien qui epousa la fille de feu Dr. Matty, que vous avez connû sans doute.[3] J'y fus tout de suite. Il m'assura ne pas avoir fait usage de cette plante: mais il a un especifique ou remede particulier avec lequel il a fait des guerissons notables sur des cancers. il m'en a montré dans de l'esprit de vin, qu'il avoit extrait sans l'application du couteau, mais simplement avec son Remede. il dit que si le patient qui a le cancer sans etre encore ouvert, lui envoyera une information bien detaillée de son etat actuel, il pourra lui envoyer d'ici le remede avec la direction pour l'appliquer. mais que si le cancer est ouvert, alors il faudroit venir ici. Il m'assura que Mr. Minors, cirurgien dans l'Hopital de Midlessex[4] ici à Londres, avoit gueri tout à fait un cancer dans la levre, avec l'application reiterée de l'air fixe. je vous prie de communiquer cette notice à l'Abbé Rozier pour l'inserer dans son journal, comme interessante à encourager l'application de ce remede nouveau. Ajoutez y, que me trouvant hier soir dans l'assemblée du Chevr. Pringle, Mr. Wedenberg, Dr. en Medecine Suedois,[5] qui s'y trouvoit aussi, nous assura qu'en Alemagne on avoit obtenû la cure parfaite d'un autre cancer avec l'application de l'air fixe.[6] Le Chevr. Pringle dit n'avoir pas entendû d'aucun essay fait avec la Phytolacca, qui produissit du bon succès.

---

1 The naturalist Daniel Carl Solander, FRS.

2 *Phytolacca Americana*, still used as a homeopathic medicine in the treatment of tonsilitis.

3 John Obadiah Justamond (1737-1786), surgeon at the Westminster Hospital from 1770 and author of *An Account of the Methods Pursued in the Treatment of Cancerous and Scirrhous Disorders, and Other Indurations* (London, 1780). Justamond was elected FRS in 1775.

4 Isaac Minors, surgeon at the Middlesex Hospital, 1753-79.

5 Adolphus Fridericus Wedenberg (1743-1828).

6 An extract from Magellan's letter reporting the claims of Minors and Wedenburg was published in *Observations sur la physique, sur l'histoire naturelle et sur les arts*, 10 (1777), 163, in the issue for August 1777.

77.07.21 Le meme Dr. Wedenberg dit qu'il savoit qu'on en avoit fait application (de la Phytolacca) en Suede aussi; mais sans succès. Le Chevr. Pringle dit que les Poultices ou Cataplasme des carotes, appliquées sur les cancers ouverts, avoient produit le bon effet d'adoucir l'acrimonie & les douleurs &a.

J'ai connû assez particulierement ici feu Mr. Guy,[7] qui avoit un Remede particulier pour les cancers, & qui fit des guerissons parfaites, & extraordinaires. Mais il faut avouer que ce meme remede manqua aussi quelques fois de produire son bon effet. Le Roy de Suede envoya un medecin expressement à Londres, qui s'appelloit Bierchen[8] pour examiner les effects[9] de ce Remede, & etudier la practique de l'appliquer. par ordre du meme Roy il fut fait un agreement ou contract par lequel Mr. Guy s'obligeoit de fournir une certaine quantité de ce remede chaque année au dit Mr. Bierchen, en lui paiant 5000£ Esterlines d'une fois: ce qui fut executé: & il le fournit tandis qu'il vecût. Le Dr. Wedenberg me dit hier soir qu'on avoit certainement fait plusieurs cures parfaites des cancers en Suede avec l'aplication de ce remede. Mais après la mort de Mr. Guy, on n'a pas pû obtenir la quantité estipulée de ce remede. Le Dr. Gahn, Medecin Suedois, est venû ici il y a 4 ou 5 ans à Londres, pour retablir l'observance de ce contract: mais il n'a pû reussir. Car Mde. Guy la veuve, avoit fait un nouveau contact[10] avec un autre cirurgien &a. Outre celà le fils ainé[11] du même feu Mr. Guy se trouvoit alors en France &a. Enfin pour completer ce que Je sçais de cet affaire: Ce Mr. Guy fils, est Revenû à Londres; je le rencontrai il y a 6 ou 10 [...][12] lui m'invita d'aller le voir, car j'ai eté plusieurs fois e[n com]pagnie du vivant de Son Pere, qui frequentoit quelques amateur[s ... su]jets philosophiques, & les invitoit souvent chez soi. Ce Son fils me donna son adresse; mais lorsque J'allai pour le voir, ce qui fut depuis plusieurs mois, il avoit changé de demeure: & je ne l'ai rencontré depuis, ni sçû de lui. Le Dr. Wedenberg me dit hier soir, que le Ministre Envoyé de Suede à cette Cour avoit ici à Londres ordre de sa Cour pour lui faire un procès, & l'obliger à fournir la quantité estipulée du Remede par son pere &a.

---

7 Richard Guy (?-1767), author of *An Essay on Scirrous Tumours and Cancers* (London, 1759).

8 Pehr af Bjerkén (1731-1774).

9 *sic.*

10 contract?

11 Melmoth Guy, who published a work in the same year on the treatment of cancer, *A Select Number of Schirrous andCancerous Cases Successfully Treated without Cutting, by a Peculiar Remedy of Melmoth Guy, Surgeon* (London, 1777). For the relationship between Richard and Melmoth Guy, see J.A. Paris and J.S.M. Fonblanque, *Medical Jurisprudence*, vol. 3 (London, 1823), pp. 153-9.

12 Paper damaged, text missing at the end of this and the next two lines.

Ce sont toutes les informations que j'ai pû obtenir, relatives à l'objet dont vous me parlez: Je serai enchanté que vous y trouvez quelque chose d'utile: & Je suis avec toute consideration possible 77.07.21

Monsieur le Marquiz
Votre tres h[ble]. & obligé serv[r].
Magellan

N.B. M[r]. Justamond me dit avoir essayé aussi lui-même, l'application de l'air fixe; mais qu'il n'avoit pas eu un si bon succès que M[r]. Minors. ainsi vous voiez, qu'il n'a pas de remede infailible pour tous les cas.

A Monsieur
Monsr. Le Marquiz de Condorcet
Secretaire de l'Acad[e]. R[le]. des Sciences
à Paris

To Nathaniel Pigott 77.08.05

Royal Astronomical Society, London, MSS Pigott, 39

London 5 Augt. 1777

Dear Sir

Your Barometers are arrived: I have sent them to Ramsden, with a copy of y[e]. article of your letter relating there to, and charge M[r]. Woodward to seek for them, in order to avoid his further lyes to me, he being about 7 or 9 pounds owing still to me since these 2 or more years; which fear of my requesting to be paid, would very probable make him still more averse to finish your work. as soon as he does it, I'll forward by sea (or by land if you please) the said Barometers & tubes &[a]. together with a small parcel of music which was found in the cases from Bruxelles: and also the figure of the plant of M[rs]. Pigott, to whom I beg my respectfull comp[s]. as well as to your sons. the plant is but a kind of Borrago: and cannot be well ascertained without seeing the flower in perfection &[a].

I have wrote to M[r]. Messier with your queries about the corresponding observat. to yours &[c]. which I copied for him in the same letter, and gave him your present Direction at Frampton &[c].

77.08.05 the waterman Woodward is astonished, you have not received his letter about your plate. he performed your Commission, and send it on y^{e}. 11th. past month, by the James & Son's wagon to Bristol, as you order'd him: and he himself put a letter to you in the post, on the same day.

I have order'd also the vol. of French Memoires for 1772 as M^{r}. Ed^{d}. desires,[1] besides that other vol. with y^{e}. account of ye nebuleuses: because it is not possible to have that Memoire alone.[2]

I congratulate you & M^{rs}. Pigott for ye good success of your chevalier, and must conclude in a hurry by assuring you that I am heartly

your's
Magellan

N.B. J'eus la curiosité d'examiner la parcelle de musique: & Je trouve qu'elle est imparfaite. savoir elle commence a page 17 jusqu'à page 30. après il vient la page 39 jus à page 73.

ainsi il y manquent les premieres 5 arietes: & depuis l'ariete 12 jusqu'à l'ariete 18: c'est en tout 9 arietes qui y manquent. crus pouver ecrire à celui qui vous a envoyé cette musique de vous la completer, si vous la souhaitez. &^{c}. car je ne crois pas qu'on les ait dérobées à la Duane de Londres.

iterum vale[3]

Nathaniel Pigott Esqr.
at Frampton near Cowbridge
in Glamorganshire

1 i. e. Pigott's son Edward (1753-1825), who was also active as an astronomer.

2 Messier's "Catalogue des nébuleuses et des amas d'étoiles, que l'on découvre parmi les étoiles fixes, sur l'horizon de Paris" had appeared in the volume of the *Histoire... avec les mémoires* of the Paris Academy of Sciences for 1771 (published in 1774), pp. 435-61.

3 Farewell again.

To Nathaniel Pigott 77.08.08

Royal Astronomical Society, London, MSS Pigott, 40

London 8 Augt. 1777

Dear Sir

your plate-chest, or case, has ben left at the ofice of y^{e}. wagons at Bristol, which I named in my Last Letter.[1] it is directed to M^{r}. Maskeline wine-Merch^{t}. at Bristol: and it appears by the Receit of y^{e}. goods there deliver'd, which has been sent to y^{e}. office at London, that this chest or case is received at y^{e}. corresponding office in Bristol &^{a}. so that you have nothing else to do, but give your orders accordingly. the chest as I aprehend, has your name upon it.

the fitest microscopes for y^{e}. Lady you mention, are those called flower-microscopes, as here represented. the eye is applied to a: and the object is applied to b. these are of different prices. viz. from 6^{s} to 12^{s} or 15^{s}. they go in a little case, as a pair of spectacles.

the 3 Barometers are come, and are sent to Ramsden, with a note of y^{e}. particulars, transcribed from your letter.

I have wrote to M^{r}. Messier for y^{e}. observations and longitudes you want, and gave your direction in the country, to him. M^{r}. Jeaurat[2] is now in London: will return soon to Paris, and I will charge him again with the same astron^{al}. commission.

I have given already the order for y^{e}. French Memoirs for 1771, a case of books being almost ready to be sent here for me. now I see by your last letter that it is the same of 1772 I had order'd before: because it is this, and not that of 1771, that contains the nebuleuses.[3] had you expressed your self so in the preceeding letter, you should not be exposed

---

1 See Magellan to N. Pigott, 5 August 1777.

2 Edme-Sébastien Jeaurat (1724-1803), astronomer at the Observatoire in Paris.

3 Magellan has made a mistake here, since Messier's famous catalogue of nebulae appeared in the volume for 1771; cf. Messier, "Catalogue des nébuleuses et des amas d'étoiles, que l'on découvre parmi les étoiles fixes, sur l'horizon de Paris", *Histoire de l'Académie Royale des Sciences, avec les mémoires*, 1771 (published 1774), pp. 435-61.

77.08.08 to have now 2 vol. of y^e^. French Memoirs, instead of one. I'll give this post a conter-order: and if the case is not yet sent to Calais, the wrong will be mended. N.B. Messier's Memoir of y^e^. Nebuleuses cannot be had separate.

I received a letter from him (Messier) lately, where he tells me to have made very assiduous observations on y^e^. spots of the sun, since January last. he has there observed a very great number, once he there rekoned 50 at once. he will make a very interesting Memoir on this subject, after he has continued to observe till the end of the year.[4] Should it not be attributed, this unaccountable weather of this Summer, at least partialy, to such kind of circumstances in the first Luminary?

Vale.
tuum Magellan[5]

Nathaniell Pigott Esqr.
great Frampton, near Cowbridge
Glamorganshire

4 Messier announced the same intention in a paper read to the Académie des Sciences on 21 June 1777 (*Histoire de l'Académie Royale des Sciences*, 1777, Mém. p. 464), but the promised memoir did not appear, being perhaps rendered superfluous by Lalande's memoir on the same subject (*ibid.*, 1778, Mém. pp. 393-424).

5 Farewell | your Magellan.

To Ribeiro Sanches[1]

77.08.15

Österreichische Nationalbibliothek, Vienna
A N. Ribeiro Sanches Korrespondenz, codex 12714, ff. 205-206v

15 Agosto 1777

Meo am°. e Sr. e mto. do Coram.

O amigo Freire me entregou os dois exemplares da sua dissertação, edição de Hollanda, dos quais mandei logo um por via de hum filho do Medico Irlandez q morreo em Lisboa, & cuido q se chamava vale, o qual partio pª. Edimburgo, segd°. sua Excª. me disse, q por elle mandava outro impresso a Mr. Robertson. Eu tive cuidado de escrever na capa, defronte do frontespicio do seo Livro estas palavras = Mr. Magellan prend la liberté de Laisser cette brochure chez S. Exce. le Chevr. Pinto, pour etre envoyée de sa part à Mr. Robertson: celle-ci etant une edition plus complete, que celle qu'il Lui envoya sur le meme sujet, il y a quelques années à Edimburg. Peut etre que Mr. Robertson après l'avoir Lû avec un peu d'attention, y trouvera des faits, qui meritent quelque place à coté de son assertion page 464 du vol. 1er. de son Excellent Histoire de l'Amerique, edition de Londres. ce 10 Augt. 1777 = eu não estou em correspondencia com este celebre Historiador, nem se me dá nada disso. o caracter geral dos Escocezes não he dos q eu amo. Secos, almas de páo, finos dessimulados, & verdados. Italianos da grande Bretagna. Jantei quarta frª. passada com o Cavro. Pinto: Sua Espoza vai partir pª. Portugal nos principios ou meio de Setembro. ella vai c[om][2] seo Primo q V.M. lá vio, o Cavro. Pereira. fazem tenção de passar no fim deste mez por Bowood, caza de campo de Mylord Shelbur[ne] q lá os invitou (he perto de Bath, & dista daqui 80 & tantas milhas) e eu faço tenção de là me achar tãobem por esse tempo [ten]do[3] sido invitado tãobem, pª. lá passar alguns dias, o q farei por gozar de varias novides. Philosas. do nosso infatigavel Priestley, que he seo Biblioth°. & mora ao pé de Bowood.

não tenho q repetir sobre o ajustado. eu quero & insisto em q V.M. não se encurte no necessario & com̃odo, segundo os seos annos & achaques. quando tiver necesside. tire sobre mim: & emqto. ouver, não deixarei de concorrer. fique isto dito de hũa vez pª. sempre.

---

1 Annotated "Recebida a 21" [Received on 21] and "Respª. a 28" [Answered on 28]. Letter not found.
2 Obliterated margin. Conjectural reading.
3 Obliterated margin. Conjectural reading.

77.08.15 Aqui me dizem q M^{r}. Trudaine vem de morrer, depois de se ter dimitido dos seos cargos, o q creio; pois este retiro de negocios habituaes, faz sempre hũa impressão n'um corpo achacado & fraco. não ha de faltar quem sinta a sua morte: & eu sou do numero. mas já estou calejado: e não m'embaraço dos sucessos da vida.

eu là escrevi ao D^{r}. Le Begue sobre o Dicion^{o}. de Motherby: & cuido q elle o procurará e mandará a V.M.

não sabe q^{to}. tenho rido com o terço, cantado pelos nossos soldados dizem q a S^{ra}. Rainha nunca assina decreto nem ordem, sem ter a mão cuberta com relíquias dos Santos. he o meio mais seguro p^{a}. em tudo ter o ceo à seo favor. ditozo governo ...! tão bem me disserão (na praça, pois bem sabe q em caza de Ministros, nunca ha novid^{es}. q se digão) q tinha sahido uma pasquinada da qual eizaqui o sentido. = o Rey hé hum santo: a Rainha hũa Beata: o Ministro nem ata, nem dezata: mal por mal, antes o Pombal.= em fim cá estaremos de longe vendo moscas por corda, & toiros pelos ares da patria, até que tudo entre pela boca dentro da Espanha. Parece-me q estou vendo cá de longe o D^{or}. Sonsinho, com as suas contas na mão, pescoço torcido ajudando ás missas, & correndo as vias sacras de Lisboa com todas as Senhoras Fidalgas, minhas Senhoras ...! queira D^{s}. q lhe não pegue fogo ao Sambenito; pois a Inquizição & a Fradaria cedo tornaráõ ao seo Zenith.

os Banhos de Dominicetti são universais. i.e. de Russia, Alemanha &^{a}. &^{a}. &^{a}. pois elle os dá de toda a sorte & por todos os modos. tem caldeiras a ferver, q comunicão com canudos de metal, donde ha registros p^{a}. varios quartos, cameras, cabinetes &^{a}. &^{a}. ali, ou os doentes estão assentados em hũ nicho, & todos cercados de vapor: ou som^{te}. uma p^{te}. do corpo, ou hũ so membro está exposto ao vapor: ou taobem o corpo todo deitado no Leito, recebe o vapor &^{a}. &^{a}. em fim a varied^{e}. he infinita & certam^{te}. engenhoza. elle mistura varias ervas, & drogas nestes vapores: mas o beneficio certam^{te}. vem principalm^{te}. do vapor quente & aquozo &^{a}. eu cuido q este homem tem gasto mais de dez mil livras Esterlinas na construção destes banhos.

eu estou vendo q o S^{r}. Pinto não ficará aqui longo tempo; pois sua Espoza q se vai p^{a}. caza não he bom sinal. Certam^{te}. elle hé homem de mais juizo, capacid^{e}., instrução & honra q todos q^{tos}. sahirão daquella terra: & hum tal homem n[ão][4] he dos q semilhante terra merece ter empregados.

O amigo S[eq^{ra}.][5] prometeo q havia de escrever. elle disseme q não tinha sucesso com a tintura das cantaridas. agora está aqui um rapaz

4 Obliterated margin. Conjectural reading.
5 Conjectural reading.

de 10 à 12 ann[os][6] q se acha como tolhido da cintura pª. baixo; tem-me lembrado se as unturas e esfregaçoens das quentaridas lhe serião uteis. elle he filho de hum acunhado de Arbuthnot. seo pai tem grande riqueza, e gastará tudo por ver seo filho restabelecido. elle foi nacido e criado em Hollanda: cuido q tem o systema nervozo relaxado. q opina V.M. sobre semilhante cazo ...? 77.08.15

Saudes. á Familia adoptiva.

vale & ama

tuum &ª.

a Monsieur
Monsr. Sanchez Dor. en Medecine
Cimitiere St Jean, Rue de la Verrerie
à Paris

15 August 1777

My Dear Friend

My friend Freire[7] handed over to me the two copies of your dissertation, edition of Holland,[8] one of which I sent at once through a son of the Irish Doctor who died in Lisbon & I think was named Vale,[9] who departed for Edinburgh, since His Excellency told me that he would send another printed work to Mr. Robertson through him. I was careful to write on the cover opposite the frontispiece of your book these words = Mr. Magellan prend la liberté de laisser cette brochure chez Son Excellce. le Chevr. Pinto, pour etre envoyée de sa part à Mr. Robertson: celle-ci etant une edition plus complete, que celle qu'il lui envoya sur le meme sujet il y a quelques années a Edimburg. Peut etre que Mr. Robertson après l'avoir lû avec un peu d'attention, y trouvera des faits, qui meritent quelque place à coté de son assertion page 464 du vol. 1er. de son excellent Histoire del'Amerique, edition de Londres. Le 10 Augt. 1777.[10] = I am not in correspondence with this famous

6 Obliterated margin. Conjectural reading.
7 See Magellan to Sanches, 11 July 1777, note 6.
8 Sanches, *Dissertation sur l'origine de la maladie vénérienne ...* (new ed., Leyden, 1777). There had been earlier editions in English and French, the earliest in 1751; a second Leyden printing followed in 1778, with a different frontispiece.
9 Unidentified person.
10 Sanches argued that syphilis had not come to Europe from America but had arisen locally. The Scottish historian William Robertson in his *History of America* (London, 1777) cited, on the page to which Magellan refers, an earlier (1765) edition of Sanches' dissertation, but criticized both his evidence and his conclusions.

77.08.15 historian, nor do I care to be. The general character of the Scottish people is not to my liking. Rough, wooden souls, refined hypocrites & true Italians from Great Britain. I dined last Wednesday with Chevalier Pinto: his wife leaves for Portugal at the beginning or middle of September. She is going with her cousin whom you saw over there, Chevalier Pereira, and they intend to spend the end of this month at Bowood, the country property of Milord Shelburne who invited them – it's near Bath, & is 80-odd miles from here – and I intend to be there also at that time, having been invited as well to spend a few days there, which I'll do for the pleasure of several philosophical novelties from our indefatigable Priestley,[11] who is his librarian & lives near Bowood.

I have no need to repeat what is arranged. I want & insist that you don't do without things that are necessary & useful, given your age & ailments. When you need to, draw on me: & as long as I have something, I won't fail to contribute.[12] This is said once and for all.

I was told here that Mr. Trudaine has just died,[13] after resigning his offices I believe, because this retreat from customary business always causes some shock in an ill & weak body. There are many who will regret his death & I'm among them; but I'm experienced and don't get upset by the events of life.

I wrote to Dr. Le Begue about Motherby's Dictionary:[14] & I imagine he'll get it and send it to you.

You don't know how much I've laughed at the rosary's being sung by our soldiers. It's said that the Queen never signs any decree or order, without having her hand covered with relics of saints. It's the safest way to have the heavens on her side in everything. Fortunate government ...! I was also told (in the street, because you know well that in a minister's house there is never any news to be heard) that a lampoon has been circulating, the meaning of which is the following = the king is a saint, the queen is a bigot, the minister leaves everything without solution: between the present and the past evil, choose Pombal. = after all, here we'll be far away, looking at disorder until everything goes into the mouth of Spain. It seems to me that I see from afar Dr. Sly, with his beads in his hand, neck twisted, serving at mass & running the stations of the cross in Lisbon with all the noble dames, my ladies ...! May God will that the sanbenito[15] is not set on fire, since the Inquisition & the monks will soon return to their zenith.

---

11 i. e. Joseph Priestley.

12 See also Magellan to Sanches, 16 September and 27 October 1775.

13 Trudaine died on 5 August 1777.

14 George Motherby, *A New Medical Dictionary* ... (London, 1775).

15 The garment worn by penitent (yellow) or impenitent (black) heretics who were condemned by the Inquisition to be burnt at the stake.

Dominicetti's baths[16] are universal, that is Russian, German &c. &c. &c. since he gives them of any kind & by any method. He has boiling caldrons that communicate through metal tubes, with taps to several rooms, chambers, cabinets &c. &c. Patients are either seated in a niche with all or only a part of the body surrounded by vapour: or just one member is exposed to the vapour: or again all the body, lying on a bed, receives the vapour &c. &c. After all, the diversity is infinite & certainly ingenious. He mixes several herbs & drugs in these vapours: but the benefit certainly arises mainly from the hot & humid vapour &c. I suppose this man has spent more than ten thousand pounds sterling on the construction of these baths. 77.08.15

I guess that Mr. Pinto won't stay here for long, since his wife's going home is not a good sign. Certainly he is a man of better judgement, capability, instruction & honour than all who came out that country: & such a man is not among those that the country deserves to have employed.

Our friend Sequeira promised me he would write. He told me he had no success with the cantharides tincture. Now there is a young boy 10 to 12 years old who is completely paralysed from the waist down.[17] I wonder if the ointments and rubbings of the cantharides could be useful to him. He is the son of a brother-in-law of Arbuthnot's. His father is very wealthy, and will spend anything to see his son recovered. He was born and brought up in Holland: I suppose his nervous system is relaxed. What do you say about this case ...?

Regards to the adoptive family.

vale & ama

tuum[18] &c.

---

16 Bartolomeo Dominiceti, an Italian doctor of Venetian origin who established himself in Chelsea, in London, with a clinic for vapour bath treatment. He is said to have treated over 16,000 patients, including the Duke of York. He published books on the subject; see, for example, *A Plan for extending the Use of Artificial Water-baths, Pumps, Vaporous and Dry Baths, Fumigations and Frictions* ([London], 1771).

17 The son of the Amsterdam merchant Thomas Wilkieson; see Magellan to Sanches, 13 January 1778.

18 farewell & love | your &c.

77.08.28 *To Nathaniel Pigott*

Royal Astronomical Society, London, MSS Pigott, 41

London 28 Augt. 1777

Dear Sir,

I am sorry to tell you, that your writing to the ingenious Scoundrel[1] in the most earnest manner will avail nothing; you must wait till his whim of complying with your request comes to his head: and God knows, if it will be soon or late. I remember well to see often letters directed to him on his desk, and upon the chimney, without even being open, till they were either lost or destroied; and on asking the reason, was told, he knew or guessed the contents, and had no mind to see them &^{a}.

I wrote some while ago to Messier, with the list of ye corresponding observations you want, & position of the places there marked. now have given a copy to M^{r}. Jeaurat,[2] with the additional list of those observations mentioned in your last letter. he will set out for Paris to day or to morrow, and have no doubt, but he will do what you desire. M^{r}. Jeaurat has given 13 pages of y^{e}. latitudes & longitudes, of all y^{e}. different places he could find, in his Connoissance des Tems for 1779[3] (you may have a copy of this book, if you please): and told me will undoubtedly insert in the next those you offer'd of y^{e}. Low Countries and Flanders, made and averred by your self, if you send them as soon as possible to him (I'll have means of forwarding them by the post without expense).[4]

M^{r}. Jeaurat does not want the elements, nor the calculations at large. he will take your word for them: and only please to send simply the results in this manner, for instance

Noms des Lieux	Difference des Meridiens		Latitude ou Hauteurs
	en tems	en degrés	de Pole
Ostende	0°.2'.13''. os.[5]	0°.33'.13''	51°.13'.55''.

1 i.e. Jesse Ramsden; cf. Magellan to N. Pigott, 5 August 1777.

2 Edme-Sébastien Jeaurat (1724-1803), astronomer at the Observatoire in Paris, who was visiting England at the time.

3 The table actually appeared on pp. 224-37 of the *Connoissance des temps* for 1780, which was published in 1777 (the volume for 1779 having been published a year earlier).

4 The *Connoissance des temps* for 1781 included a more extensive list of places for which positions are given than did the previous year's volume, the list now extending to 17 pages. It is not clear, however, whether any of the added data were due to Pigott.

5 ouest?

this sample is taken from the above mentioned Connoissance des tems for 1779, and is marked with a * both in the latitude & longitude which means, that both have been taken by some members of the R^l^. Acad^y^. of Paris. if you compare these positions with yours, you will see if they agree: and if not I think you should send your observations made at Ostende, to prove that yours are right. But you'll do as you please. 77.08.28

a deeping needle like that in ye R^l^. Society by Nairne,[6] costs 30 guineas. a variation compass like that in y^e^. same place costs 20 pounds. a Raingage as that in the same place, which is the only one to be depended upon, costs about 9 or 10 pounds according as the copper pipe is bigger or smaller.

Pray my compl^s^. & Resp^s^. to M^rs^. Pigott and sons. I'll go on wednesday next to Bowood to pass some days at Mylord Shelburne, where I'll be often with D^r^. Priestley[7] at my return to Town will see if your Ramsden has done any thing for you

vale & ama
tuum
Magellan[8]

N.B. I have given your direction both to M^r^. Messier and to M^r^. Jeaurat, in order to send directley to you in Glamorganshire their observations.

Nathanael Pigott Esqr.
at Great Frampton, near Cow-bridge
in Glamorganshire

6 All three instruments are described in Henry Cavendish, "An Account of the Meteorological Instruments used at the Royal Society's House", *Philosophical Transactions*, 66 (1776), 375-401. The dipping needle is described in more detail in Nairne, "Experiments on Two Dipping-Needles... executed for the Board of Longitude", *Philosophical Transactions*, 62 (1772), 476-80.

7 Priestley was at this period employed as librarian by William Petty, Earl of Shelburne, whose family seat was at Bowood in Wiltshire.

8 Farewell & love | your | Magellan.

77.09.11 *To Nathaniel Pigott*

Royal Astronomical Society, London, MSS Pigott, 42

Bowood Park near Calne
in Wilts - this 11 Sep^{ber}. 1777

Dear Sir,

I intend to return to London the next week, and will do what I can about your instruments. indeed I have order'd M^{r}. Woodward to call now and then on that sad Liar,[1] in order to put him in mind of them: Because if I meddle with it, am affraid it will be worse, as he is indebted to me of 7 or 8 guineas, which I reckon as lost, and which he will think I am endeavouring to recover by any orders or instruments. He was so unfair as to give me a bill of some work as unpaid, of which I shewd to him his own receit for &^{c}. he is a <u>vitandus homo</u>.[2] I have sent the 22 Augt. a fine pair of peble spectacles,[3] and the flower microscope[4] to M^{de}. la Comtesse de Wegnante Rue du Long chariot à Bruxelles, dans une caisse de Livres, adressée à M^{r}. l'Abbé D. Mann,[5] who has already told me in a letter, he will send every thing there contained to the respective directions.

Your letter to M^{r}. Jeaurat is already gone to Paris, and will certainly be deliver'd. I gave him, before his depart, a fair copy of the corresponding observ^{s}. you want, mentioned in two of your last letters:[6] and no doubt he will endeavour to get them together with M^{r}. Messier, who has already your direction at Frampton house, to send the answer directed by the post to you.

---

1 i.e. Jesse Ramsden.

2 a man to be avoided.

3 Spectacles with lenses made of colourless, transparent rock crystal. The price stated by Magellan in the postscript is very high, since George Adams in the price list appended to his *Essays on the Microscope* (1787) gives "The best double-jointed silver spectacles... with Brazil pebbles... £1.16.0".

4 Either a common botanical microscope or (perhaps more likely in view of the comparatively low price that Magellan quotes) a simple botanical magnifier with two or three choices of lens that swivel into position.

5 Theodore Augustus Mann (1735-1809) was an English Catholic priest who lived in Brussels for many years, where in 1774 he became a member of the newly founded Academy of Sciences and Belles-Lettres. He was *secrétaire perpétuel* of the Academy, 1787-1794. Mann's letter to which Magellan refers has not been found.

6 Letters not found.

I have forgot what you call a Claude Lorraine? is it not a convexe dark mirror?[7] what you mean by rulers of Inacka wood?[8] are they flat, or round? what use are they for? let me know this and you'll be served. 77.09.11

as to the dark glasses tell me if you like better a modern scheem I have seen and seems superior to any dark glasses. it consists of two pieces of glass, one very dark and the other very light coloured as here by the margin.

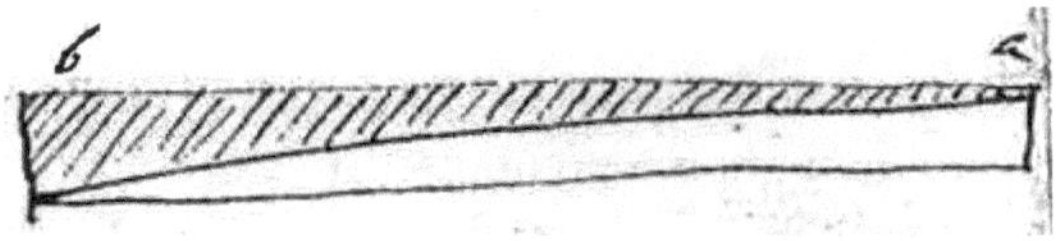

now such 2 glasses fitted in a brass frame give the best advantage you can wish, you being the master of receiving the solar rays with y^e^. desired strengh, according as you receive them towards *a* or towards *b*. I do not [know][9] how much this costs, but I have seen this contrivance, dont recollect whether of Nairne's or of Dollond.

mes h^bles^. resp^s^. to M^rs^. Pigott & sons: and am

yours
Magellan

I'll not forget the vol. of savans etrangers with your observations.[10]

iterum vale[11]

the flower microscope cost one guinea the pebble spectacles in silver, with a fine casse did cost £2. 14^s^. 0^d^

Nathanael Pigott Esqr.
at Great Frampton
near Cow-Bridge
Glamorganshire

7 A Claude Lorraine glass is "a somewhat convex dark or coloured hand-mirror, used to concentrate the features of a landscape in subdued tones". The earliest printed use of the term recorded by the *Oxford English Dictionary* is 1789!

8 Not identified.

9 Editorial addition.

10 Pigott's observations appear not to have been published in the *savants étrangers*, i.e. the *Mémoires ...présentés par divers sçavans* of the Paris Academy of Sciences.

11 Farewell again.

77.09.21 *To Ribeiro Sanches*

Österreichische Nationalbibliothek, Vienna
A. N. Ribeiro Sanches Korrespondenz, codex 12714, ff. 207-208v

Bowood Park 21 Sep^{bro}. 1777

Am^{o}. e S^{r}. e m^{to}. do Coração.

Escrevo esta da Caza de campo de Mylord Shelburne, coiza de 90 milhas de Londres & perto de Bath, onde vim passar estes poucos dias depois de 5 do Corr^{te}. por invitação deste Fidalgo, q me faz m^{ta}. honra. Eu aqui esperava despedirme da S^{ra}. D. Catherina Espoza do nosso Embaixador q devia passar por qui a semana passada, com seo marido e Primo Miguel Pereira, p^{a}. daqui hir a Bath, & de là a Falmouth, pois está resolvida a partir p^{a}. Portugal com seo Primo, por cauza da pouca saude q goza neste clima, & tambem por cauza da sua caza em Balsamão q necessita de arranjamento &^{a}. Porem tive carta desta Sr^{a}. q me diz ter cahido enfermo seo Espozo com hũa febre, e q não sabia q^{do}. se porîa a caminho. visto isto tornarey p^{a}. Londres depois de a manhan: e em effeito já teria partido 5^{a}. fr^{a}. passada, pois não fazia tenção de estar auzente de Londres mais de 15 dias. Porem Mylord me propoz tão cortez & affectuozam^{te}. de esperar por seos filhos p^{a}. vellos antes de partir, visto q elles devem chegar hoje aqui de Devonshire onde tem estado este verão a tomar os banhos de agoa salgada, q me pareceo incivilid^{e}. o não me dilatar mais estes 4 ou 5 dias. Em effeito esses dois rapazes (o maior não tem mais de 12 annos) são dotados de hum amavel caracter, e eu lhes quero tanto como se fossem meos.

Procurarei as 16 paginas q faltão de seo Culen's Physiol^{y}.

Ja tinha q^{m}. me fizesse o instrumento [p^{a}.][1] servir de dentes; mas á vista da carta de V. M., suspendi a ordem.

Antes de sah[ir][2] de Londres copiei com toda a exactidão o avizo de V. M. p^{a}. a cura daquelle menino tulhido, filho de M^{r}. Wilkieson, cunhado de nosso amigo Arbuthnot pela sua prim^{a}. ou 2da. Mulher. Eu lhe mostrei a carta & lhe mandei a tradução do artigo em questão: qd^{o}. chegar à Londres saberei o q tem rezolvido. q elle siga, ou não, esse remedio; & que elle ache ou não o bom effeito q se espera: sempre isto servirá de lhe mostrar q tomamos a peito os seos interesses, p^{a}. nos-ficar obrigado.

Antes de sahir de Londres mostrei ao Cavlh^{o}. Pereira & ao nosso Embaix^{or}. os artigos respectivos da carta de V. M. como tãobem ao am^{o}. Freire: & todos me pedirão significasse a V. M. o seo agradecim^{to}. &^{a}.

---

1 Obliterated margin. Conjectural reading.
2 Obliterated margin. Conjectural reading.

O Princepe Pignatelli cá me escreveo sobre differentes instrumentos; e eu lhe respondi a tudo, com os affectos q lhe devo. pela sua carta vi, q se perdeu a q eu escrevi a Madrid no tempo da sua heroica expedição, tal ou quazi tão bem sucedida como a do nosso Don Quixote D. Sebastião. 77.09.21

a agoa deitada sobre pedras d[e] seixos ardentes, não faz outro effeito senão o reduzir-se em vapores. estes são os q opérão sobre o corpo humano q a elles está exposto. mas estes mesmos vapores devem necessariamente condensar-se pouco a pouco, tanto q os seixos se esfrião. Ora os vapores dos banhos de M^{r}. Domenicetti sendo continuam^{te}. renovados pela ação da grande caldeira ou alambique q ferve devem produzir m^{to}. maior effeito. tenha V.M. paciencia nesta minha contradição ao q V.M. parece opinar, relativam^{te}. á superiorid^{e}. dos vapores Russos: b[em] sabe q eu sou candido e sincero: & q não sei fazer comprimentos falsos em materias philosophicas. nos vapores de agoa quente, quer seja ou não reduzida nelles por seixos ardentes, nunca se acha ar fixo, sendo q elles não são mais do q agoa reduzida [a] particulas minimas. Se V.M. visse o effeito destes vapores de Dominicetti, q passados & conduzidos a hũa distancia mui consideravel por canudos de metal, podem fazer ferver hũa panella, q nelles poem sumergida & rodeada: de modo, q som^{te}. se emprega hũa caldeira de agoa fervendo, a qual estando bem cuberta & com̃unicando por varios canudos, pode fazer ferver & cozinhar, sem outro fogo, hũa ou 2 duzias de guizados, havia de ficar pasmado. este methodo tem a vantagem, de poder levar os vapores onde o Medico dezeja, & applicalos som^{te}. a hũa ou duas p^{tes}. do corpo, onde se necessita o seo effeito. Mas o D^{r}. Dominicetti não he o único q assim os applica. eu sei q varios Medicos delles usam com bom sucesso.

Tenha V.M. a bond^{e}. de cortar o pedaço [… des]ta[3] carta, & mandalo ao amigo D^{r}. Begue du Presle, [a] q^{m}. escrevi daqui o correio passado. Minhas saud^{es}. á Familia adoptiva. goze V.M. sempre melhor saude, & sirva-se dos meos affectos.

vale.

Coitado do Dieudonné. faz-me pena o seo estado, e a aff[lição] das duas Primas. mas q remedio! São pençoens desta triste negregada vida, q sem embargo disso tanto se ama e se dezeja faça-lhe V.M. a todos meos affectuozos sentimentos, q lhes dezejo do Coração. deixe endireitar o olho á coifa da piquena. q aprenda a gan[har] o pão, & q tenha boa sorte!

iterum vale.

---

3 Stained paper.

77.09.21 O Cav^{ro}. Pinto, sua Epoza, & seo primo aqui chegárão ao jantar. o Cav^{ro}. Pinto se acha [...][4] em 3 dias, eilo aqui em jornada. me pede lhe dê suas lembranças & mais o [Primo.][5]

a Monsieur
[Monsr.] Sanchez D^{or} en Mede[cine]
[Cimitier]e de S^{t}. Jean, au bout de la [Rue de la] Verrerie
à Paris

Bowood Park 21 September 1777

My Dear Friend

I write this letter from the country house of Milord Shelburne, about 90 miles from London & near Bath, where I came to spend these few days after the 5th of this month by invitation of this Nobleman, who does me great honour. I expected to say goodbye here to Mme Catherina, the wife of our Ambassador,[6] who was to call here last week with her husband and her cousin Miguel Pereira, and then to go to Bath & therefrom to Falmouth since she has decided to leave for Portugal with her cousin, due to the poor health she suffers in this climate, and also due to her house in Balsamão needing arrangements &^{a}. However I had letter from this lady saying that her husband was taken ill with a fever, and that she didn't know when she would start on her way.[7] So I'll return to London the day after tomorrow: and indeed I would have left last Thursday, because I did not intend to be away from London more than 15 days, but My Lord proposed so courteously & affectionately that I wait to see his two sons before leaving, since they will arrive today from Devonshire where they have been sea bathing this summer, that it seemed an incivility not to delay another 4 or 5 days. In fact, these two boys (the elder is no more than 12 years old) are endowed with a kind character, and I love them as much as if they were mine.

I'll look for the 16 missing pages in your Culen's Physiol^{y}.[8]

I already had someone making the instrument to serve as teeth; but on reading your letter,[9] I suspended the order.

Before leaving London, I copied with all accuracy your advice regarding the cure of that paralysed boy, the son of M^{r}. Wilkieson, brother-in-law of Arbuthnot by his first or 2nd wife.[10] I showed him the letter & sent him a translation of the article in question: when I arrive in London I'll know what he decided. Whether

4 Obliterated margin.

5 Conjectural reading.

6 Catarina Micaela de Sousa César de Lencastre, wife of Luiz Pinto de Sousa Coutinho (1735-1804), Portuguese Minister in London.

7 Letter not found.

8 William Cullen, *Institutions of Medicine. Part I: Physiology* (Edinburgh, 1772); see Magellan to Sanches, 27 June 1777.

9 Letter not found.

10 See Magellan to Sanches, 15 August 1777.

or not he follows this remedy; & whether or not he finds the good effect that is hoped for, this in any case serves to show him that we have his interests at heart, in order that he remains obliged to us. 77.09.21

Before leaving London I showed to Chev[er]. Pereira & our Ambassador as well as to our friend Freire the various articles of your letter: & all asked me to express their thanks to you &[a].

Prince Pignatelli wrote about different instruments; and I answered everything with the sympathy I owe him.[11] Through his letter I saw that the one I wrote to him to Madrid at the time of his heroic expedition that succeeded almost as well as that of our Don Quixote D. Sebastião was lost.

Water poured on very hot pebbles doesn't do anything else than turn into vapour. This is what operates on the human body that is exposed to it. But these same vapours must necessarily condense slowly as the pebbles cool. Now as the bath vapours of M[r]. Domenicetti are continually renewed by the action of the big caldron or alembic that is boiling, they must produce a much greater effect. Be patient about my contradicting what you seem to think concerning the superiority of the Russian vapours: you know that I am innocent and sincere, & that I don't pay untrue compliments in philosophical matters. In hot water vapours, being generated or not by hot pebbles, there is never <u>fixed air</u>, since they are no more than water reduced to minimal particles. If you saw the effect of these vapours of M[r]. Dominicetti, which, transported & conveyed to large distances through metal tubes can make a pot boil, and that only a single boiling water caldron, well covered & communicating through several tubes, can make <u>one</u> or <u>2</u> dozen stews boil & cook without additional fire, you would be astonished. This method has the advantage, of allowing the vapours to be taken wherever the Doctor wishes, & applied only to the one or two parts of the body where its effect is needed. But D[r]. Dominicetti is not the only one to apply them. I know that several physicians use them with good success.

Please cut off this part of this letter & send it to our friend D[r]. Begue du Presle, to whom I wrote by last post. My regards to the adoptive family. I wish you always better health, & make use of my affection.

vale.[12]

Poor Dieudonné.[13] I'm sorry for his state and the affl[iction] of his two cousins. But what can be done! They are costs of this sad unlucky life that nevertheless one loves and desires so much. Give them all my affectionate sentiments, that I

---

11 Letters not found. They must have been written some time before this, if our identification of Juan Joaquín Atanasio Pignatelli de Aragón y Moncayo (1724-1776), Spanish Ambassador in Paris, 1763-72, is correct. Such a dating is by no means impossible, however, given that Magellan made very similar remarks about Pignatelli's expedition, almost two years earlier; see Magellan to Sanches, 27 October 1775.

12 Farewell.

13 Pierre Dieudonné, partner of Antoinette Pernelet, cousin of Sanches' housekeeper Marie Jeanne Pernelet and also a member of Sanches' household.

77.09.21 wish them from my heart. Pay attention to the girl's coif. May she learn to earn a living, & have good luck![14]

iterum vale.[15]

Chev^{r}. Pinto, his wife and cousin arrived at dinner time. Chev^{r}. Pinto recovered his health in 3 days, here he is on a journey. He asks me to send his regards as does his cousin.

---

14 Magellan is probably referring here to Maria Dieudonné, the young daughter of Pierre Dieudonné and Antoinette Pernelet, and urging Sanches to pay attention to her upbringing.

15 Farewell again.

77.09.30 *To Ribeiro Sanches*[1]

Österreichische Nationalbibliothek, Vienna
A.N. Ribeiro Sanches Korrespondenz, codex 12714, ff. 209-210v

Londres 30 Sep.^{bro} 1777

Meo Am^{o}. e S^{r}. e m^{to}. do Cor^{am}.

Aqui mando a V.M. a informação inclusa sobre a doença do filho mais velho de M^{r}. wilkieson. V.M. conhece mui bem q^{to}. me tocão no Coração as affliçoens dos bons amigos: e por isso me perdoará este trabalho, q dezejo não seja incompativel, nem onerozo á situação de V.M. Eu estive sabado passado em Knightsbridge chez D^{r}. Kelly, q he o Dominicetti Inglez, e applica m^{to}. propriamente os banhos, tanto quentes, de vapor, & mornos, como os de agoa fria. O rapaz tem boas cores, & não parece doente em nada senão na falta de movimento nas pernas & no espinhaço: Pois em o pondo assentado em hũa cadeira de braços he necessario atalo nella & p^{a}. assim dizelo, suspendelo com uma correia por baixo dos braços, & seguralo com outra pela barriga, porq de outra sorte elle não se pode sustentar, & cairia da cadeira abaixo. eu fiz uma tradução exacta dos Remedios q V.M. aconselhou na sua carta, e a entreguei a seo Pai antes de partir p^{a}. o campo. Agora elle conta de trazer o filho p^{a}. sua caza, & mudar de methodo, fazendo hua consulta de bons medicos p^{a}. esse fim: e por essa Rezão elle me trouxe esta manhan a conta ou informação incluza, p^{a}. q V.M. veja se tem algũa coiza q alterar no methodo curativo q aconselhou. Agradecerei m^{to}. q V.M. me faça Reposta com a brevid^{e}. q commodam^{te}. puder.

---

1 Annotated "R^{a}. a 6 d'8^{bro}." [Ans. 6 October]. Letter not found.

fui 6ª. frª. passada á caza de campo de Mylord Stanhope q he daqui 23 milhas, pª. ver a utilissima experiencia q seo filho Mylord Mahon ali fez, sobre o methodo de edificar as cazas, sem q possão ser consumidas por algum incendio. Tudo consiste em formar uma boa argamaça (mortier) de 1ª parte de areia: duas p^tes^. de cal: & 3 partes de feno cortado do comprimento de 3 polegadas. tanto as paredes, como o solho da caza he bem cuberto & formado com esta argamaça, com os pedaços ordinarios de ripas &ª. tudo bem cuberto de argamaça, de modo q não fique algũa greta por onde o ar entre até a madeira &ª. o mesmo plancher he posto sobre esta sorte de stratum d'argamaça & páo; sobre a qual se mete areia solta, de modo q não fique algũa gréta pª. o ar passar. hũa caza, feita nesta forma, foi cheia de fagots: lancouse-lhe o fogo: ardêrão as logeas com tal força, q os vidros de hũa janella forão derretidos: no entretanto varias pessoas se achavão no sobrado: e eu tão bem ali me achei, & comemos de la crême à la glace: em fim poz-se-lhe o fogo por fóra, & nada bastou pª. q ella fosse abrazada. o sobrado som^te^. se tostou um bocado: de modo q apenas uma oitava p^te^. das taboas tomou fogo. 77.09.30

Agora partirá pª. Paris o filho mais velho de M^r^. Forster, a q^m^. darei uma carta de recom̃endação pª. M^r^. le Begue du Presle. he bom sujeito, mui moderado mui capaz. Este he o filho de Forster q fez a viagem ao Mar do Sul com seo pai, mas he mui diferente delle na prudencia [&] socego; decente sem ser teimoso, falando com acerto e moderado. elle se achou tão bem prezente á dª. experiencia de Mylord Mahon.

minhas Saud^es^. á familia adoptiva. goze V.M. saude & quietação: & viva contente & sem cuidados da vida q certam^te^. não vale [a est]ima[2] q della tomamos.

vale

O D^r^. Musgrave, q achei tão bem na caza de campo com Mylord Mahon, alem de mª. outra gente distinta &ª. me pedio q o puzesse á Sua obedª. me falou com affeto em V.M. disse-me q examinára hum doente cancerado a q^m^. se applicava a terra de Mafra em Paris &ª. eu ja o enculquei pª. a consulta q se deve fazer sobre o filho de M^r^. Wilkieson.

o D^or^. Sequeira se recomenda tãobem: e me diz q escreverá a V.M. na semana q vem

vale.

A Monsieur
Monsieur Sanchez Doc^r^. en Medecine
Cimitiere de S. Jean Rue de la Verrerie
au Marais
à Paris

2 Conjectural reading.

77.09.30 London 30 September 1777

My Dear Friend

Here I enclose the information about the illness of the eldest son of Mr. Wilkieson.[3] You know very well how my heart is touched by the affliction of my good friends: and so you'll forgive doing this work, which I don't want to be incompatible with or onerous to your situation. Last Saturday I was in Knightsbridge, chez Dr. Kelly,[4] who is the English Dominicetti[5] and applies the baths very properly, whether hot, vapour or tepid, as well as the cold-water ones. The boy has much colour, & doesn't look ill in any way except in the lack of motion in his legs & spine: since in seating him in an armchair it is necessary to tie him to it &, so to speak, to suspend him with a leather strap under his arms, & to hold him with another round the belly because otherwise he can't hold himself up & would fall off his chair. I made an exact translation of the medicaments you advised in your letter,[6] and handed it to his father before leaving for the country. Now he intends to bring his son to his house & to change the treatment, taking medical advice to this end: and for this reason, this morning he brought the enclosed account or information, so that you can see if there is anything to modify in the curative method you advised. I will appreciate it if you answer me as soon as you comfortably can.

Last Friday I went to the country house of Milord Stanhope,[7] 23 miles away, to see the most useful experiment that his son Milord Mahon[8] made there, about a method of building houses that can't be destroyed by any fire. Everything consists in making a good mortar (mortier) of 1 part of sand, two parts of lime & 3 parts of hay cut into pieces 3 inches in length. The walls as well as the wooden floor are well covered & formed with this mortar, with the ordinary batten pieces &c. all covered with mortar, in such a way that no crack remains whereby the air contacts the wood &c. The same plancher is put over this sort of stratum of mortar & wood; on which loose sand is put, in such a way that no crack remains through

3 See Magellan to Sanches, 15 August 1777.

4 Christopher Kelly M. D., a Licentiate of the Royal College of Physicians, established in Knightsbridge a clinic for treatments using water baths.

5 Bartolomeo Dominiceti, the Venetian doctor who introduced water-bath treatment in the 1760s.

6 Letter not found.

7 Philip Stanhope (1714-1786), FRS, second Earl Stanhope. Educated in Utrecht and Geneva, he acquired a love for mathematics, the Greek language and democratic principles. He arranged the publication and distribution at his own expense of a posthumous edition of the works of the mathematician Robert Simson (1776). Priestley dedicated to him the 3rd volume of his *Experiments and Observations on Different Kinds of Air* (1777).

8 Charles Stanhope (1753-1816) second son of Philip Stanhope, who became Lord Mahon when his elder brother died at Geneva, was a Fellow of the Royal Society and an enthusiastic experimental philosopher.

which the air can pass. A house made in this manner was filled with fagots and set on fire: the ground floor burned so strongly that the window glasses melted: in the meantime several persons were on the first floor: I was also there, & we had de la crême à la glace: finally, a fire was set outside, but nothing would set the house ablaze. The floor was just a little toasted, in such way that only one-eighth of the wooden boards caught fire.[9] 77.09.30

Mr. Forster's eldest son,[10] to whom I will give a letter of recommendation to Mr. Le Begue du Presle, is leaving for Paris. He is a good person, very moderate and very able. This is Forster's son who made the trip to the South Seas with his father, but he is very different from him in prudence & quietness; decent but not stubborn, speaking with judgement and moderate. He was also present at the said experiment of Milord Mahon.

My regards to the adoptive family. May you enjoy good health & quietness, & live happily & without concerns about life, which certainly is not worth the esteem we ascribe to it.

vale[11]

Dr. Musgrave,[12] whom I also found at the country house with Milord Mahon, as well as many other distinguished people &c. asked me to remember him to you and spoke with affection of you. He told me that he examined a cancerous patient to whom Mafra earth[13] was applied in Paris &c. I have told him about the consultation that will be held about the son of Mr. Wilkieson.

Dr. Sequeira sends greetings also & tells me that he will write to you next week.

vale.[14]

9 Mahon described his technique in *Philosophical Transactions*, 68 (1778), 884-94.

10 Georg Forster (1754-1794), who accompanied his father, Johann Reinhold Forster (1729- 1798), when he went as naturalist with James Cook on his second voyage around the world, 1772-5.

11 Farewell.

12 Samuel Musgrave (1732-1780), FRS, physician and classical scholar.

13 See Magellan to Sanches, 31 October 1776, note 17.

14 Farewell.

77.10.07 *To Johann Albrecht Euler*[1]

St Petersburg Branch, Archives Department, Russian Academy of Sciences, f. 1, op. 3, Nr. 63, ll. 80-81^{ob}.

Londres 7 Octobre 1777

Monsieur

Tous les articles mentionés dans votre lettre du 1 Aout,[2] que Je reçus à la campagne, me sont arrivés: & tous sont delivrés, à l'exception de la medaille pour le D^{r}. Morton & Maskeline,[3] ce que je ne manquerai pas de faire à leur retour. Je vous prie de retourner mes humbles remerciem^{ts}. à Votre Illustre Acad^{e}. pour le present qu'elle m'a fait de Sa Medaille & jeton.[4] Les presents pour La Societé R^{le}. sont deja delivrés au Bibliothecaire, ils seront presentés dans la premiere assemblée au commencem^{t}. de Novembre: & elle ne manquera de faire les remerciements. Enfin la lettre que Vous m'avez envoiée avec la votre, fut delivrée au Portier de Mylord [Harwhich][5] qui promit de l'envoyer immediatement, au Monsieur à qui elle etoit adressée, qui se trouve constament à la campagne à quelques milles de Londres. Les amis qui m'avoient demandé les vol^{s}. des commentaires de Votre Ac^{de}., ont trouvé, que votre corps a fort mal choisi son libraire. Car il faut que ce M^{r}. vous trompa d'une maniere ou de l'autre, pour donner les mêmes livres à un prix si bas aux Hollandois, que les memes livres achetés en Holande chez un libraire, ne coutent pas autant comme ce qu'il vous chargea pour les miens, sans compter les depenses de la caisse, lettres, fret &^{a}. Le vol. pour l'annee 1773 ne couta qu'11 shillings en Hollande, au lieu que 4 Roubles dont il chargea. les vol. 19 & 20 font 16^{s}.4^{d}: & ceux à 3 roubles, font 13^{s}.6^{d}, sans compter les depenses. Je sçais bien par experience que les Marchands libraires ne manquent jamais de tromper s'ils le peuvent: mais en attendant le debit des ouvrages de Votre Academie soufre; car c'est une erreur de croire qu'on ne se passera souvent de ce qu'on ne pourra pas avoir avec une depense raisonable.

1 Annotated "reçu le 20 Octobre 1777". See *Protokoly zasendanii konferentsii Imperatorskoi Akademii Nauk*..., III, 335-6.

2 Letter not found.

3 In March 1777, the St Petersburg Academy decided to present a medal and *jeton* to each of its corresponding members, among whom Morton and Maskelyne were numbered; see *Protokoly*, III, 297-8.

4 Magellan was not yet a member of the St Petersburg Academy and his name was therefore not on the initial distribution list for the Academy's medals. A decision had evidently subsequently been made to cultivate him.

5 Harwich? i.e. the prominent politician Wills Hill, Lord Harwich?

Mylord Mahon vient de faire une experience devant beaucoup de monde qui est de la plus grande utilité Publique.[6] C'est la maniere qu'il a trouvé de faire batir des maisons, avec des materiaux les plus communs à bon marché, & même en bois, qui ne peuvent etre consumées par aucun incendie. Je fus tout exprès à Chevening, environ 23 ou 24 milles de Londres, pour en voir le succés. il y fit batir une maison en bois avec le mortier & methode de son invention; on remplit le rez chaussé avec des fagots: on y mit le feu, qui fut si furieux, que les verres d'une fenetre fondirent comme de la Cire. tandis que cette chaleur bruloit au dessous, beaucoup de monde y monta à son etage les uns après les autres. J'en fus du nombre. nous n'y trouvames aucun Chaleur sensible. même nous y fumes servis avec de la creme à la glace. en suite on mis une grande charpente de bois à coté de la maison, qu'on remplie du fagots & qu'on alluma en suite. Le feu fut horrible: mais il ne prit pas à exterieur de la maison. enfin on mit plusieurs fagots au Escalier, qui etoit en bois avec du mortier: mais il n'en brule plus que la surface exterieure, jusqu'à la profondité d'un dixieme de pouce. M[r]. [...] Samborski,[7] chapelain de l'Ambassade Russe ici à Londres s'y trouva aussi pour voir cette experience: & il m'a dit qu'il communiqueroit la methode chez vous. Reelement elle est très aisée, & consiste essentiellem[t]. en ce que le total des murailles & de la plancher soit bien solidem[t]. rempli avec du mortier, qui ne casse point ni fait des crevasses pour donner accés à l'air: ainsi les flames ne pouvant toucher (par example le plancher) que seulem[t]. dans une seule surface, n'en brulent pas d'avantage, que jusqu'à une petite profondeur &[a]. Mylord Mahon m'a dit qu'il publieroit une brochure sur ce sujet;[8] qui assurem[t]. devroit être traduite & communiqué en tout paie. 77.10.07

L'affaire de la foudre qui toucha la maison des assemblées à Purflet[9] a eté portée plus loin. Wilson, ce meme Wilson pour lequel Votre Acad[e]. Imp[le]. a envoyé sa medaille,[10] a voulû faire du tapage là dessùs, & employa toute sa fourberie pour imposer non seulement le Public, mais aussi le Roy

6 C. Mahon, "Description of a Most Effectual Method of Securing Buildings against Fire", *Philosophical Transactions*, 68 (1778), 884-94. Mahon reports that he demonstrated his experiments before "the President and some of the Fellows of the Royal Society, the Lord Mayor and Aldermen of the City of London, the Committee of City Lands, several of the foreign ministers, and a great number of other persons". Magellan was evidently one of those in attendance.

7 Father Andrei Samborskii was the long-serving priest of the Russian church in London (1765-79).

8 No such pamphlet appears to have been published, though the promise of a separate publication was repeated in the report in the *Philosophical Transactions*.

9 See Magellan to J. A. Euler, 27 June 1777.

10 Wilson had been a corresponding member of the St Petersburg Academy since 1765, and so received one of the medals distributed by the Academy to its members.

77.10.07 & Famille Royale, qui à ce qu'on dit, a deja fait rabattre les conducteurs pointus de la maison de la Reine. A present que sa charlatanerie & son imposture vient d'etre decouvertes, il ne pourra pas manquer de s'attirer tout le mepris des honnetes gens: & si le conseil de la Societé R^{le}. en prendra connoissance, il faudra l'expulser du nombre de ses membres, comm'un imposteur indigne d'y avoir son nom. Il obtint permission du Roy de monter un grand Appareil electrique aux depens du Bureau, apparement de Purflet, pour demontrer par des faits la verité de sa theorie.[11] il forma un conducteur monstrueux, avec des tambours militaires mis en forme presque circulaire, les unes après les autres, & couverts avec la feuille d'etain: il avoit environ 150 pieds de longueur & environ 2 de diametre. C'etoit proprement un magazin d'electricité, qu'on y mettoit, en tournant une ou deux machines electriques ordinaires. Ce conducteur, ou magazin fut suspendu par des cordons de soie dans la grande Sale du Pantheon. il fit arranger une petite maison, pour representer celle de Pourflet, qui etant tirée par un cordon, attaché à un grand poids, la faisoit mouvoir avec une grande velocité, courant sur une coulisse, qui passoit par dessoùs du grand conducteur, ou magazin qui representoit une nuée chargée d'electricité. Tout le monde sçait qu'en presentant une pointe metallique à un magazin ou conducteur chargé d'electricité, elle toute se decharge insensiblem^{t}. sans laisser echaper aucune etincelle &^{a}. au lieu que si l'on y presente une boule de metal, meme à une distance plus grande que celle oû se trouvoit sa pointe, alors elle attire une etincelle très forte, sans se decharger insensiblem^{t}. comme dans le cas precedent. Malgré cette verité incontestable, notre charlatan Wilson trouva le moien de faire paroitre le contraire: car en mettant une petite boule sur la maison (modelle de celle de Purflet) elle passoit vitement sous le bout du grand magazin des tambours & n'attiroit l'etincelle; au lieu que si elle avoit une pointe de metal, alors elle recevoit une forte eteincelle. Je me trouvois en campagne lors que ces experiences se faisoient à Londres. J'en etois etonné de ce que mes amis m'[ecrivirent] là dessus. Je reçus bientôt avis que Mylord Mahon, avoit publiquement confondu la coquinerie de Wilson:[12] mais personne n'avoit pas encore decouvert la construction que sa fourberie avoit employé, pour montrer un effet tout à fait contraire à ce qu'on connoit dans cette matiere. en fin je retournai à Londres le 24 du mois dernier; & j'appris qu'il y auroit de la

11 B. Wilson, *An Account of Experiments made at the Pantheon, on the Nature and Use of Conductors* (London, 1778).

12 See *London Evening Post*, 16 September 1777. See also R.W. Home, "Points or Knobs: Lightning Rods and the Basis of Decision Making in Late Eighteenth Century British Science", in P. Heering, O. Hochadel and D. Rhees, eds, *Playing with Fire: Histories of the Lightning Rod* (Philadelphia, 2009), pp. 97-120.

compagnie des gens instruites au Pantheon le jeudi soir, 2 du courant mois. 77.10.07
Je m'y trouvai avec un grand nombre de curieux. M^r^. Henley, membre de la Soc^e^. R^le^., fit charger le magazin: & e[n][13] presentant une boule de metal, elle reçut une grande eteinelle, au lieu qu'une pointe n'en reçut aucune, quoiqu'à une moindre distance. Celà fait, M^r^. Nairne, autre membre de la Soc^e^. R^le^. y presenta une pointe metallique isolé sur une manche de verre couvert de verniz: alors cette pointe au lieu de decharger l'electricité, attira une grande eteinelle, au lieu qu'une boule, mise dans la meme manche isolante, n'en attira gueres quoiqu'on l'approchat de plus près. M^r^. Nairne prit alors un grand ruban de feuille d'etain qu'il mit dans la pointe isolée: dans ce cas la pointe dechargea en silence l'electricité sans aucune eteincelle, au lieu qu'une boule de metal mise à la meme manche isolante, mais dans ce dernier cas communicant avec la terre par le ruban de feuille metallique, recevoit l'eteincelle electrique même à une plus grande distance que la pointe: car celle-ci ne peut pas recevoir aucune eteincelle qu'en la metant fort soudain à une très petite distance du conducteur. Tout le monde vit alors en quoi consistoit la fourberie de Wilson. en effet en mettant le ruban d'etain (qui etoit fort grand, & touchoit à l'autre bout avec la terre près la cheminée) sur la petite maison lorsqu'elle avoit un conducteur pointû, il n'y avoit point d'eteincelle en passant par dessous le conducteur: mais en y mettant une boule de metal, alors elle attiroit toujours une grosse eteincelle, meme en passant à une plus grande distance du conducteur ou magazin electrique des tambours &^a^. En effet en examinant la communication metallique, que ce charlatan Wilson avoit mis sur la petite maison, elle consistoit dans un fil de metal qui finissoit dans un anneau, & cet anneau glissoit au long d'un fil horizontal tandis que la maison glissoit sur le chassis en coulisse, de façon que la velocité de son mouvement faisoit que l'anneau n'etoit pas en contact avec le fil horizontal, dans le moment que la pointe passoit par dessous du conducteur. outre celà la petite maison etoit faite de bois desseché dans le four (baked wood) & etoit bien vernisée: ainsi elle isoloit parfaitement bien la boule ou la pointe metallique qui etoit sur la petite maison. Cependant lorsqu'il arrivoit le contraire, c'est à dire lorsque l'anneau du fil de communication touchoit l'autre fil horizontal, dans ce cas le charlatan ne reussissoit pas dans son experience: mais ceci n'arrivoit que très rarement. Dans une matiere dont les consequences sont si importantes pour le bien du Public, on ne peut pas trop divulger les faits. Ceux-ci nous apprennent que les conducteurs metalliques, pour preserver de la foudre, doivent etre non seulement bien pointus mais avoir une bonne communication metallique avec la terre:

---

13 Paper damaged, word partly illegible.

77.10.07 autrement ils ne meritent pas le nom de conducteurs. L'imposture de Wilson etant decouverte, il est necessaire de la communiquer tout par tout, pour que le monde philosophique ne se laisse pas tromper par des fourberies des jueurs[14] de goblets comme Wilson. Vous aurez la bonté, de communiquer ces faits à Votre Acad^e^. Imp^le^., & j'en garantis la veracité aiant eté temoin oculaire de ce que Je viens de [referir]. J'avois eu soin de delivrer moi même au sieur Wilson la medaille que vous m'avez envoyée pour Lui, avant le 2 du courant de peur qu'il arrivat ce qui en effet arriva, raison pourquoi je pense de m'eloigner le plus possible de sa communication.

Vale & ama
tuum Magellan[15]

A Monsieur
Monsr. J.A. Euler
Secretaire de l'Acad^e^. Imper^le^. des Sciences
à S^t^. Petersburg
Russie

---

14 joueurs?
15 Farewell & love | your | Magellan.

77.10.17 *To the Marquis de Condorcet*
Bibliothèque de l'Institut de France, Paris, MS 876, no. 38

Londres 17 Oct^bre^. 1777

Monsieur

Je me flate que vous me faites La justice de croir que Je n'ai pas retardé cette Reponse que par des obstacles insurmontables. Le tems d'eté est celui, dans lequel il est le plus difficile de trouver du monde en ville: Je fus aussi à la Campagne quelques semaines, & à mon Retour, je n'ai pû trouver la derniere information sur vos questions, qu'avant hier.

L'air fixe employé par M^r^. Minors,[1] est celui dont le D^r^. Priestly a parlé dans son ouvrage,[2] adoptant Le meme nom (air-fixe) que d'autres lui avoient donné avant Lui: c'est cet air, qui sort des pierres & terres calcaires

---

1 See Magellan to Condorcet, 21 July 1777.
2 Joseph Priestley, *Experiments and Observations on Different Kinds of Air* (London, 1774).

(comme la craie, le marble &c[a].) par l'action de l'acide vitriolique, delayé avec de l'eau commune. Le Recit que J'en ai donné, & qui se trouve dans le Journal de Rozier, sur son application aux cancers, est assez detaillé pour le concevoir dans La practique. J'ai parlé moi-meme avec Messieurs Minors fils & Pere[3] en differens fois: ils m'ont assuré d'avoir gueri parfaitement bien un cancer dans la levre d'un Chimney-sweeper, je crois que vous les appelez Ramoneurs en françois. il faisoit degager l'air fixe dans une jarre de verre: & le malade y mettoit dessus la vapeur (ou air fixe) qui sortoit la levre avec le cancer. Cependant M[r]. Minors, le Pere, m'avoua qu'il n'avoit pas encore fait aucune autre cure avec ce remede: qu'il avoit trouvé toujours en differents autres cas oû il l'avoit appliqué, que les douleurs & le mauvais odeur des cancers disparoissoient au commencement mais qu'au bout de quelques jours les douleurs retournoient, malgré la continuation de ce remede: à l'exception du cas ci dessus du ramoneur, qui fut tout à fait gueri. 77.10.17

On dit ici, que Le Roy de France a acheté le secret d'un remede efficace contre les cancers. Si celà est, je vous prie de m'envoyer le detail, pour le communiquer dans ce pais ci: car en particulier Je l'ai promis (de m'en informer) à M[r]. Minors, & de lui faire part de la reponse.

J'ai cherché tout par tout M[r]. Guy:[4] & je trouve enfin qu'il s'est retiré de Londres, sans laisser notice du lieu de sa demeure. Le libraire qu'imprima dernierement son livre sur les cures du cancer par sa medecine,[5] me pria bien fortement de lui dire le resultat de mes recherches pour reavoir son argent; car il ne lui a payé encore la depense de cette publication. Vous voyez bien par là le peu qu'on peut s'attendre de ce Monsieur. On m'assure en outre, que ce Remede de Guy n'est point generalement assuré: & que plusieurs fois il a manqué de guerir le cancer &[a].

J'ai l'honneur d'etre avec toute consideration possible & vrai attachement de Coeur

Monsieur
Votre très h[ble]. & ob[t]. ser[r].
Magellan

A Monsieur
Monsr. Le Marquiz de Condorcet,
Secretaire de l'Acad[e]. R[le]. des Sciences

3 Isaac Minors was a surgeon at the Middlesex Hospital, 1753-79. When he retired, he was succeeded by his son, Daniel Minors, also a surgeon, who practised at the Middlesex Hospital, 1779-1801.

4 See Magellan to Condorcet, 21 July 1777.

5 Melmoth Guy, *A Select Number of Schirrous and Cancerous Cases Successfully Treated without Cutting, by a Peculiar Remedy of Melmoth Guy, Surgeon* (London, 1777).

77.11.07 *From Jan Henrik van Swinden*
Leiden University Library, MS B.P.L. 755, no. 197[1]

Agréez Monsieur la liberté que je prends de vous présenter les deux ouvrages ci joint.[2] Le premier vous est du, comme à un savant distingué, que je suis charmé de pouvoir assurer de mon respect: et l'autre vous l'est en outre comme une marque de ma reconnoissance pour les belles observations faites à Chatham en Janv 1776, que vous avez bien voulu me faire parvenir par M Dentan.[3] Vous verrez Monsieur que je me suis empressé d'en faire usage. Je serois charmé que ces ouvrages Vous parussent dignes que vous les fissiez connoitre en Angleterre.

Je vous ai beaucoup d'obligations Monsieur du bel appareil pour l'Air fixe &c que vous m'avez fait parvenir il y a 9 mois. Je m'en suis servi avec grand succes, et il est encore en bon etat, car quoique j'aye fait un grand nombre d'experiences, j'ai eu le bonheur de ne casser aucune bouteille meme des plus minces. J'ai prié mon ami M. d'Entan, de vouloir bien me procurer par vous Monsieur un Eudiometre, et quelques nouvelles bouteilles decrites dans le 3e Tome de Priestley, ainsi que le Pamphlet que vous avez publié sur le 1er de ces sujets.[4]

Je serois bien flatté Monsieur que vous voulussiez me permettre de m'informer quelquesfois aupres de vous de ce qui se fait sur les sciences en Angleterre, où je n'ai aucune relation parmi les savans. J'en avois eu avec feu M. le Secretaire Maty à qui j'avois meme envoyé il y a 5 ans quelque chose pour la societé: mais cela est resté je crois parmi ses papiers: et d'autres occupations, surtout mon grand travail sur les aiguilles aimantées, qui a

1 Original not found; text from file copy retained by van Swinden (and numbered by him N°. 256). Published in David Willemse, "Suite d'un voyage aux Pays-Bas: João Jacinto de Magalhães (1722-1790) et ses rapports avec Jean Henri van Swinden (1746-1823)", *Arquivos do Centro Cultural Português*, 7 (1974), 225-278.

2 The works in question were Van Swinden's *Dissertation sur la comparaison des thermometres* (Amsterdam, 1778) and his *Observations sur le froid rigoreux du mois de Janvier 1778* (Amsterdam, 1778); see Van Swinden to Magellan, 22 May 1778.

3 Magellan sent observations on a "great cold" experienced at Chatham, made by the surgeon Samuel Simmons, to Van Swinden; see Magellan to Van Swinden, 5 May 1778. Pierre Gédéon Dentan or Dentand (1746-1780) from Geneva accompanied J. A. De Luc during some of his expeditions in the Swiss Alps. He later settled in Haarlem.

4 Van Swinden is probably referring here to Magellan's *Description of a Glass Apparatus for Making Mineral Waters Like Those of Pyrmont, Spa, Seltzer, etc. ... together with a Description of some New Eudiometers, or Instruments for Ascertaining the Wholesomeness of Respirable Air* (London, 1777).

été couronné à Paris,[5] m'ont empéché de m'en informer aussi souvent que je l'aurois fait sans cela. 77.11.07

J'ai l'honneur, &ca.

Franecker 7 Nov 1777.
à M. Magellan à Londres

5 "Recherches sur les aiguilles aimantées, et sur leurs variations regulieres, qui ont partagé le prix proposé pour l'année 1777", *Memoires présentés à l'Academie Royale des Sciences par divers sçavans*, vol. 8 (Paris, 1777).

To Johann Albrecht Euler[1] 77.11.14

St Petersburg Branch, Archives Department, Russian Academy of Sciences, f. 1, op. 3, Nr. 63, ll. 90-91ob

Londres ce 14 Novbre. 1777

Monsieur

J'ai eu l'honneur de presenter à la Societé Rle. les differents ouvrages que l'Acade. Imple. de Petersburg lui a gracieusemt. envoyés, & le Secretaire Mr. Planta, me dit hier dans l'assemblée, qu'il me donneroit un de ces jours la lettre de remercims. que le Presidt. & Conseil de la meme Societé ont ordonné d'ecrire pour cet effet. J'y ai trouvé un Monsr. Groote, que je ne connoissois auparavant que de Vue & auquel je priai, en consequence des offres qu'il me fit, de faire passer à Petersbourg avec peu de frais quelques petits pacquets que J'envois à l'avenir pour Votre Acade., en commençant par le Vol. 67 part 1. de nos Transactions Philosophiques, que J'ai deja entre mes mains, & qui sera acompagné de la lettre du Secretaire ci dessùs, & de quelques exemplaires d'une petite brochure que Je viens d'imprimer sur les eudiometres, ou Instrumts. pour reconnoitre la Salubrité de l'air respirable.[2]

Je ne vous dirai rien sur la belle decouverte de Mr. Achard de Berlin, qui a trouvé la vraie formation des cristaux & pierres precieuses;[3] mystere

1 Annotated: "reçu le 26 Novembre 1777."

2 Magellan, *Description of a Glass Apparatus, for Making Mineral Waters; ... together with a Description of some New Eudiometers, or Instruments for Ascertaining the Wholesomeness of Respirable Air* (London, 1777).

3 See F. Achard, *Bestimmung der Bestandtheile einiger Edelgesteine* (Berlin, 1779), and especially the appendix (pp. 122-8).

77.11.14 de la nature, que les plus grands philosophes n'avoient pu deceler jusqu'à present. Vous devez en etre completem^{t}. informé de cette invention, peut etre par la meme voye par laquelle je reçus cette notice. J'ajouterai seulem^{t}. qu'on va repeter ici chez une Personne de rang & fortune, cette même experience après le detail que je lui ai communiqué sur cet objet: & je me flate que, malgré l'incredulité d'un de nos bons chymistes qui la considere fort douteuse, nous aurons le plaisir de la voir reussir. on travaille deja à completer tout l'appareil, qui est assez simple, mais qui demande assez de la justesse pour en attendre un bon effet.[4]

Monsr. Walsh membre de la Soc^{e}. R^{le}. & du Parlem^{t}., qui comme Je crois vous avoir dit une autre fois, fut le premier à faire paroitre l'eteincelle electrique dans le coup ou secousse, donnée par le gymnotus ou Anguile de Surinam,[5] nous montra hier à midi une autre phenomene assez singulier du meme poisson, qui semble demontrer qu'il a un 6me. sens, dont l'espece humaine, malgré toute la preeminence dont elle jouit entre les animaux, ni aucun autre connû de ceux ci, ne partagent point. M^{r}. Walsh fit voir à une nombreuse compagnie de gentilhommes, entre lesquels se trouvoit une grande partie du corps diplomatiques, que J'avois invité, que ce poisson est sensible aux matieres conductrices de l'electricité, lorsque celles-ci se trouvent dans le cas de l'etre actuellement, mais seulem^{t}. dans le premier moment. en voici l'experience. on prepara un sceau assez large, avec de l'eau dans la temperature d'environ 75°. de Fahrenheit, qui est celle qu'on conserve constament pour preserver la vie de ce poisson, actuellem^{t}. l'unique en vie dans l'Europe, ou du moins chez nous ici. on mit les bouts de 4 fils de metal dans le sceau d'eau, & on fit passer les autres 4 bouts sur une table, qui etoient munis chacun d'une boule de metal, submergeé dans un gros verre a boir, presque plein d'eau. On mit alors l'anguile dans ce sceau: & elle s'y promenoit sans faire le moindre cas des 4 boules, ni des fils de metal, dont elles renfermoient les bouts, qui y etoient plongées. alors M^{r}. Walsh prit un demicercle metalique, dont il plongea les deux bouts à la fois dans les deux

4 The attempts to repeat Achard's experiment did not succeed, and for several years thereafter Magellan in his correspondence with Johann III Bernoulli unsuccessfully exhorted Achard to provide further details of what he had done.

5 See Magellan to J.A. Euler, 17 September 1776. John Walsh succeeded in obtaining a spark from the *gymnotus* in 1775 but never published a formal account of the experiment. However, J.-B. Le Roy reported Walsh's success, quoting in an article published in Rozier's journal (*Observations sur la physique*, 8 [1776], 331-5) from a letter he had received from Walsh on the subject, and also presenting further details of Walsh's experiments that he drew from a letter he had received from Magellan (letter not found). See also Marco Piccolino, *The Taming of the Ray: Electric Fish Research in the Enlightenment from John Walsh to Alessandro Volta* (Florence, 2003), especially pp. 26-30.

goblets qui contenoient les boules des deux bouts des fils de metal, dont les autres deux bouts, avec leurs boules, etoient plus près de l'anguile. dans ce moment le poisson fit un saut & s'agita avec violence: en plongeant le meme cercle de communication dans les autres deux verres dont les extremités correspondoient aux fils près du poisson, il continuoit à se donner du mouvement; & ainsi on le faisoit promener à volonté tout au tout en avant ou en arriere comm'on le vouloit. Monsr. Walsh sera plus explicite dans les circonstances de ce phenomene, dont il a fait un Memoire que Je crois sera inseré dans le volume suivant des Transactions Philosophiques.[6] Ainsi je me contente de faire cet annonce simplem^{t}. & sans aucune broderie de conjectures, pour l'information & amusement de Vos Confreres. 77.11.14

Je me souviens de vous avoir dit un peu des charlataneries philosophiques de Wilson, le peintre,[7] car je me confirme de plus en plus chaque jour que cet homme merite d'etre rayé de la liste des Philosophes, quoiqu'il a encore son nom dans celle des membres de la Societé Royale, qui par une adhesion, peut etre trop forte, à l'ancienne institution de ne pas entrer dans aucune dispute, ne semble se donner aucun mouvement pour le chasser de son corps.[8] Reelement si l'on reflechit tant soit peu sur les soins etudiés, qu'il a employés pour gagner l'approbation des ignorants, & pour faire paroitre aux yeux du vulgaire le contraire des phenomenes de la Nature, en employant de l'artifice le plus reflechi, on me fera, j'espere, la grace de trouver qu'il ne merite pas moins que tous les epithetes dont je le charge. Vous aurez sans doute lû deja l'annonce historique de la foudre qui frapa le 13 Mai dernier la grande chambre des assemblees a Purflet. Le Journal de l'Abbé Rozier en a un article très simple & assez exact.[9] il n'y manque que seulement de dire 1°.

---

6 No such memoir was published. Walsh's experiment was, however, described by Cavallo in his *Complete Treatise on Electricity* (London, 1795), vol. 3, pp. 309-11, and mentioned in the correspondence of other observers besides Magellan; see Piccolino (2003), pp. 202-3.

7 See Magellan to J.A. Euler, 7 October 1777. As well as being an enthusiastic natural philosopher, Benjamin Wilson was a well-known portrait painter who enjoyed the King's patronage.

8 Magellan's hopes notwithstanding, Wilson was never expelled from the Royal Society. However, his reputation in scientific circles was badly damaged by the lightning-rod affair; see R.W. Home, "Points or Knobs: Lightning Rods and the Basis of Decision Making in Late Eighteenth Century British Science", in P. Heering, O. Hochadel and D. Rhees, eds, *Playing with Fire: Histories of the Lightning Rod* (Philadelphia, 2009), pp. 97-120.

9 *Observations sur la physique, sur l'histoire naturelle et sur les arts*, 10 (1777), 137-9. The article concludes as follows:

> Dans le Courier de l'Europe, No. 6, il est fait mention d'une Lettre de M. Wilson à la Société Royale de Londres, dans laquelle on lit ces mots: "On ne peut tirer aucun avantage de ces conducteurs, qui repugnent, suivant moi, aux vrais principes de la

77.11.14 que le nuage qui dechargea le coup, parut rester immobile sur la maison: & 2°. que s'il avoit quelque mouvement c'etoit du coté d'oû etoit le vent, savoir du Sudouest; de façon qu'il a toute l'apparence que ce nuage etant repoussé par la pointe du grande conducteur, fit la culbute, & dechargea sur le crampon de fer, qui etoit de l'autre coté au Nordest, & qui l'attira de preference: parcqu'en effet les experiences electriques montrent, que les pointes repoussent les corps legers (les nuages sont de ce genre) chargés d'electricité, tandis que les boules ou corps emoussés les attirent, & les font eclater avec force. Mais que ces memes pointes decharg[ent][10] insensiblement la charge electrique des corps denses, oû il y a de l'acumulation d'electricité [...] homme conversant dans cette matiere, connoit bien tout celà. mais Wilson a voulû imposer au vulgaire; & par consequent a pris tous les moyens opposés à ceux de la nature pour produire des effets contraires. Il a fait un conducteur en forme de fer à cheval, composé de tambours couverts de feuille d'etain & suspendus par de la soye, qui assurement ne sont pas analogues assez pour representer un nuage. outre celà il y adapta une boule de metal, qui ne se trouve point dans les nuages. il a fait executer un modele representant la grande maison de Purflet: mais trouvant qu'il ne reussissoit, etant fait en bois ordinaire, il fit reduire ce bois dans le four, & ensuite il le vernisa 3 fois; ce qui assuremt. n'est pas le cas d'une maison de briques ni de pierre. il y adapta un conducteur de sa façon avec des interstices à fin d'y faire eclater les eteincelles electriques. il y mit un conducteur pointu, & prepara un autre avec une boule de 3/10 de pouce de diametre. or il est clair qu'une pointe est toujours une pointe egalement aigue dans le modelle, & dans le conducteur reel: mais une boule de 3/10me. est tout à fait differente d'une de 3 ou 4 pouces pour le moins de diametre, comm'il les conseille. enfin, contraire au phenomene de Purflet, il donna une velocité de 9 pieds par second à ce modele dans le meme tems que le nuage à Purflet etoit, ou paroissoit presqu'immobile: mais il faut penser, que le mouvement angulaire de 9 pieds par second, à une distance de 2 pouces telle qu'etoit celle du modelle vis à vis du conducteur, donne une velocité angulaire plus grande infiniment, que celle d'un boulet tiré par un canon, lors qu'on la considere mesurée à la distance de quelques pieds entre le vrai nuage chargé d'electricité, & le conducteur d'une maison. Cependant avec tous ces desavantages, ceux qui connoissent la matiere, ont vû qu'en

---

Physique, & aux vrais intérêts de la Société en général". Cette assertion étonnante a donné lieu à la Lettre dont on vient de publier un extrait, pour empêcher le Public d'être trompé sur un article aussi important, & pour l'engager à avoir recours à un préservatif, dont on est redevable au célèbre M. Franklin, dont les travaux & les occupations ont toujours eu pour but général le bien de l'humanité.

10 Paper damaged, letters or words missing from the end of this and the following line.

mettant une boule d'un pouce & demi de diamettre sur le modele, meme en lui donnant toute cette velocité tout à fait opposée & impossible dans la nature des choses, elle etoit frapée à la distance d'environ 6 pouces, tandis que la pointe ne l'etoit qu'a bien moins d'un & demie: & que si l'on mouvoit lentement le modelle, quoique fourni avec la petite boule de $^{3}/_{10}$ de pouce, elle etoit toujours frapée, tandis que le conducteur point[û] ne l'etoit jamais. Il est impossible de vous decrire tous les autres detours que le charlatan prit pour imposer la populace. Il avança que le coup de tonnerre à Purflet n'etoit qu'un vrai coup lateral: & pour le prouver, il y avoit dans le Pantheon des grandes bouteilles de Leide, sur des plaques de fer blanc: il y mettoit des morceaux de fer interrompus qu'il faisoit communiquer avec le dischargeur ou manche à discharger. alors il chargeoit les deux jarres ou grandes bouteilles, par un fil de communication avec la machine electrique: & en les dechargeant, alors on voioit paroit des eteincelles à chaque interruption metallique. J'allai un jour tout exprès avec Mylord Viscount Mahon pour voir cette charlatanerie. Mylord dit au supleant de Wilson "comment donc, ne voyez vous pas, que ce feu vient de la communication avec la machine, & non pas de la charge des bouteilles". pardonnez moi Mylord, repliquat-il, laissez moi faire les experiences de Mr Wilson: & dans la suite, vous ferez les votres. en effet après avoir chargé les bouteilles avec le fil de communication, Mylord monta sur un piedestal isolé; ota le fil de communication, & en dechargeant les bouteilles comm'auparavant, il n'y parut la moindre eteincelle entre les pieces metalliques. Le papier finit, & le tems aussi pour envoier cette lettre à la poste. Nous atendons que le charlatan publie ce qu'il a à dire sur ses experiences d'imposition;[11] & aussitôt il y aura plus d'un honnet homme qui exposera ses fourberies au public. 77.11.14

J'ai l'honneur d'etre avec toute consideration

Monsieur
votre très h^ble^ & ob^t^. ser^r^.
Magellan

à Monsieur
Monsr. Jean Albert Euler
Secretaire de l'Acad^e^. Imper^le^. des Sciences
à S^t^. Petersburg
Russie

11 B. Wilson, *An Account of Experiments made at the Pantheon, on the Nature and Use of Conductors* (London, 1778).

77.11.14a *To Ribeiro Sanches*[1]

Österreichische Nationalbibliothek, Vienna
A.N. Ribeiro Sanches Korrespondenz, codex 12714, ff. 211-212v

Londres 14 Nov^{bro}. 1777

Meo am^{o}. e S^{r}. m^{to}. do Cor^{am}.

Quarta fr^{a}. passada; indo p^{a}. caza do nosso Enviado, encontrei com o comis da posta q me entregou a dezejada carta de V.M. de 5 do corrente. fui andando & lendo ao mesmo tempo, não sem rizada de quando em quando, de modo q uma das minhas Bettis de fronte de cuja janella passei, me perguntou What good news? bem pode V.M. considerar q não me esqueci de ler o artigo q respeitava ao nosso Ministro, q me deo a agradavel noticia de ser chagada sua Espoza com feliz sucesso a Lisboa dentro de 5 dias de Falmouth menos alguas horas. elle me leo p^{te}. da carta do Primo, Miguel Pereira q V.M. ahi vio: & diz q acha menos boas as nossas Portuguezas, depois q por cá vio as Inglezas.

Tem V.M. rezão no q diz das nossas santas e devotas personagens. tem boa vontade p^{a}. fazer o bem, mas parece q não sabem como. Com tudo lá virá a illuminação da graça do Spirito Santo, q lhes mostrará o santo caminho do Ceo, logo q fizerem m^{tas}. novenas, & correrem m^{tas}. vias sacras, com os exercicios de S^{to}. Ignacio le Chev^{er}. de la Vierge &^{a}. &^{a}.

Acho graça ao S^{r}. Klingstedt com as promessas do S^{r}. Domashenof. Quem sabe se elle julga q o diploma da correspondencia (q eu tenho já sem diploma) vale mais do q os 50 shillings das loupes, ou vidros engyscopicos, q elle não se lembrou de pagar desde 4 de agosto de 1775? q^{to}. mais vivemos, mais novos phenomenos conhecemos da raça humana!

Brevem^{te}. mandarei a V.M., por via do amigo du Presle, algũas brochuras q valem a pena, ainda q não tenhão nada de extraordinarias; & com ellas hirão as paginas q faltão, & já tenho, na obra de Culen.

Aqui me deo noticias de V.M. o Livreiro Elmsley, e por isso pude esperar ategora sem escrever; q na verd^{e}. he como o remedio de godemiches das nossas freiras, com os quaes ellas se cossão com hum gosto passajeiro, mas q Logo lhes faz comichar de novo o pa[rra]meiro,[2] por lhe faltar o verd^{o}. humido radical, q as satisfaça plenam^{te}. segd^{o}. a ordem das Leis da

1 Annotated "Recebida a 20 9bre" [Received 20 November] and "Resp^{a} a 29 X^{bro}" [Answered 29 December]. Letter not found.

2 Partly illegible. Conjectural reading.

propagação. Lembr[o-me][3] agora por este nome o do tribunal Italiano q o tem pª. estender o imperio da Sᵗᵃ. fé. admiro o refinamento dos nossos mysticos & legisladores ortodoxos; q sempre dérão preferencia aos termos, e expressoens fornicatorias, pª. alegorizar os altos officios & mysterios da Religião. as nossas portuguezas tem a mais alta devoção com a conceição da virgem Maria, com a encarnação do Verbo eterno. o homem regenerado pela graça he Santo; Jeˢ Chrº. he o Espozo das beatas: e os despozorios das santas visionarias, celebrados aos thalamos de glorias celestes, tem dado mᵗᵒ. q entender ás noviças de Sendelgas. e educandas de Sᵗᵃ. Clara de Coimbra. mas deixemonos de freirices pª. q não nos suceda algum desterro da corte. 77.11.14a

Aqui mostrou hontem Mʳ. Walsh, q o gymnotus, ou Enguia electrica de Surinão, possûe um 6º. sentido q nenhũ de nós goza elle conhece qᵈᵒ. os conduttores metallicos estão em acto de o serem. dois fios de metal, por exemplo de ferro, cujas pontas estejam metidas de hua parte na agoa da celha onde a tal enguia está, não lhe dão o minimo cuidado. mas logo q as outras duas pontas, ainda q mᵗᵒ. distantes, & fóra da sua vista, são postas em contacto, no mesmo instante a enguia dá um pulo & se move com velocidᵉ. de hũa parte pª. a outra.

Mʳ. Achard de Berlin descubrio ultimamᵗᵉ. a verdª. formação dos cristaes e pedras preciosas, q atégora ninguem pode achar. elle meteo terras alkalinas em hũ vazo de agoa bem impregnada com <u>ar fixo</u>. o fundo deste vazo era poroso, & a deixava filtrar por uma camada de areia em pó, pª. em fim filtrar de novo por outro fundo semilhante. no fim de <u>10</u> semanas elle achou verdadeiros cristaes duros e diaphanos, por baixo do fundo exterior. em deitando no vaso <u>cal martial</u>, os cristaes q se formárão, tinhão uma bella côr como o rubi. Esta he certamᵗᵉ. uma das mais bellas descubertas do nosso seculo.

aceite V.M. saudozas memorias do secretº. do Nosso Enviado, & particulares affetos deste ultimo, q sempre fala em V.M. com singular affeição e estimação. há dias q não vi Sequeira tanto q o vir, lhe perguntarei se já acabou a carta pª. V.M.

minhas sa[udᵉˢ.][4] á familia adoptiva, q talvez hirei ver na primavera do anno q vem.

vale

a Monsieur
Monsieur Sanchez, Dʳ. en Medecine
Cimiterre de Sᵗ. Jean Rue de la Verrerie
à <u>Paris</u>

---

3 Conjectural reading.
4 Partly illegible. Conjectural reading.

77.11.14a London 14 November 1777

My Dear Friend

Last Wednesday: when I was going to our Envoy's house,[5] I met the postman who handed me your letter of the 5th of this month that I'd been waiting for.[6] I went on walking and reading at the same time, not without laughing from time to time in such a way that one of my Bettys in front of whose window I passed asked me: What good news? Well you can be sure that I didn't forget to read the article concerning our Minister, who gave me the pleasant news of the successful arrival of his wife in Lisbon within 5 days less a few hours of leaving Falmouth. He read me part of the letter of his cousin, Miguel Pereira, whom you saw there & who says he finds our Portuguese women not so good after seeing the English ones here.

You're right in what you say about our saintly & devoted Portuguese women. They wish to do good, but they don't know how. However the grace of the Holy Spirit will come and show them the holy way to heaven, as long as they say many prayers & run through many stations of the cross, with the exercises of St. Ignatius, le Cheveer. de la Vierge,[7] &c. &c.

I find Mr. Klingstedt[8] funny about the promises of Mr. Domashenof.[9] Who knows if he thinks that the diploma of correspondence (which I already have without diploma) is worth more than the 50 shillings for the lenses, or engyscopic[10] glasses, for which he forgot to pay since 4 August 1775? The more we live, the more new phenomena of the human race we come across!

Soon I'll send you, through our friend du Presle, some worthy though not very extraordinary brochures; & with them will go the missing pages, which I already have, from Cullen's work.[11]

Here I had news from you via the bookseller Elmsley,[12] and so I could wait till now without writing; that is truly like the remedy of the dildoes of our nuns, with which they scratch themselves in a momentary pleasure but that soon make

---

5 The Portuguese Minister in London was Luiz Pinto de Sousa Coutinho (1735-1804).

6 Letter not found.

7 Saint Ignatius of Loyola, the founder of the Jesuit order, had declared himself "Knight of the Virgin" and, in accordance with the traditional vigil of arms, spent a whole night with sword drawn, ready to defend the divinity and purity of the Virgin Mary. See M. Capefigue, *Les fondateurs des grands ordres religieux: Saint Ignace de Loyola et les Jésuites* (Paris, 1865), p. 14.

8 Baron Timothy de Klingstedt was Chief Auditor of the Russian Admiralty at Archangel. Concerning his debt to Magellan, see Magellan to Sanches, 15 November 1776, note 8.

9 Sergei Domashnev (1743-1795), Director of the St Petersburg Academy of Sciences, 1775-83. Whatever promises he may have made to Klingstedt, the latter was never appointed a corresponding member of the Academy, as Magellan was; see B. L. Modzalevskii, *Spisok chlenov Imperatorskoi Akademii Nauk 1725-1907* (St Petersburg, 1908).

10 microscopic.

11 See Magellan to Sanches, 21 September 1777.

12 Peter Elmsley (1736-1802), bookseller in the Strand.

them rub their genitals again, for lack of the true radical humour that would fully satisfy them according to the laws of propagation. This reminds me now that this is the name of the Italian tribunal that is so called to extend the empire of the holy faith.[13] I admire the refinement of our mystics & orthodox legislators who always gave preference to fornicatory terms and expressions, to allegorize the high offices & mysteries of religion. Our Portuguese women have the highest devotion to the Conception of the Holy Virgin, to the incarnation of the Eternal Word. The man regenerated by grace is a saint; Jesus Christ is the spouse of bigots; and the weddings of the saintly visionaries, celebrated in the nuptial bed of celestial glories, have given the novices of Sendelgas[14] and the pupils of St. Claire of Coimbra a lot to think about. But let us refrain from nun-like behaviour in order not to be exiles from court. 77.11.14a

Yesterday M^r^. Walsh[15] showed here that the gymnotus, or electric eel of Surinam, possesses a 6th sense that none of us has. It knows when metallic conductors are functioning as such. Two metal leads, for example of iron, whose points are plunged from one side into the water of the tub where the said eel is, don't cause it the least worry. But as soon as the other two ends, though very distant & out of its sight, are put in contact, in the same instant the eel jumps & moves rapidly from one place to another.[16]

M^r^. Achard of Berlin discovered recently the true formation of crystals and precious stones, which until now no one could find out.[17] He put alkaline earths in a vase with water well impregnated with fixed air. The bottom of this vase was porous, & he let it be filtered by a layer of powdered sand, to filter it finally by another, similar bottom. In 10 weeks he found true hard and diaphanous crystals beneath the external bottom layer. On pouring martial lime[18] into the vase, the

13 The Sacred Congregation for the Propagation of the Faith, the department of the Roman Curia that is charged with the supreme direction and administration of the missionary activity of the Church.

14 A village near Coimbra where there was a large convent.

15 John Walsh (1725?-1795), FRS, published several papers on electric fish in the *Philosophical Transactions*, earlier in the 1770s.

16 Walsh never published an account of this remarkable experiment. Brief reports were later published by J. Ingenhousz, *Vermischte Schriften phisisch-medizinischen Inhalts* (Vienna, 1782), pp. 275-6, and T. Cavallo, *A Complete Treatise on Electricity* (4th ed., London, 1795), vol. 3, pp. 309-11. Thereafter, however, it seems to have slipped from view. Its significance is discussed by Marco Piccolino, *The Taming of the Ray: Electric Fish Research in the Enlightenment from John Walsh to Alessandro Volta* (Florence, 2003), pp.201-4.

17 Franz Carl Achard, "Copie d'une lettre ... contenant la découverte qu'il a faite sur la formation des cristaux & des pierres précieuses", *Observations sur la physique, sur l'histoire et sur les arts*, 11 (1778), 12-14. Doubts subsequently arose concerning Achard's "discovery".

18 Martial = pertaining to or containing iron. Hence martial lime = calcium oxide containing iron impurities.

77.11.14a crystals that were formed had a beautiful colour like ruby. This is certainly one of the most beautiful discoveries of our century.

Please accept the hearty remembrances of the secretary of our Envoy[19] & the special affection of the latter, who always speaks of you with singular friendship and esteem. I haven't seen Sequeira for days and as soon as I see him, I'll ask him if he has already finished his letter to you.

My regards to the adoptive family, whom I'll perhaps see in the spring of next year.

vale[20]

---

19 Cipriano Ribeiro Freire (1749-1824).

20 Farewell.

77.12.08 *To James Watt*

Birmingham Central Library, Boulton & Watt collection, Box 22/B3/1[1]

London 8 Dec^{ber}. 1777
at n°. 12. Nevils Court Fetter Lane

Sir

I have here in London one of my Correspondents abroad, who by the information I gave him of y^{e}. superiority of your new fire engines, des[ires] to have a fuller Knowledge of the particulars, which you alone can better communicate and give in detail to him. for this purpose he intends to go to Birmingham in order to speak with you about the matter, and I intend to company him there, provided we may know that you are to be there at Soho. Beg therefore a line by the Return of the post, that we may be sure of finding you at home: and heartly am

Sir
Your most h^{ble}. Servt.
Magellan

---

1 Letter not found; MS is a clerk's transcription.

To Ribeiro Sanches[1] 78.01.13

Österreichische Nationalbibliothek, Vienna
A.N. Ribeiro Sanches Korrespondenz, codex 12714, ff. 189-190v

Londres 13 Jan^{o}. 1777[2]

Meo Am^{o}. e S^{r}. e m^{to}. do Coração

Aqui cheguei no dia 6 do Corr^{te}., tendo estado estas festas correndo a posta por Salisbury, Bath, Bowood, oxford & Birmingham. achei a estimada Carta de V.M. de dia de Natal com outras muitas, & foi a prim^{a}. q Li pelo gosto q as suas sempre me dão, ainda q já não espero novas de arr-taçoens &^{a}. Vejo q V.M. está triste & solitario, q disso he a verd^{a}. cauza. a vida humana he p^{a}. a Socied^{e}.: e o descanço q a sua privação nos-offerece he bem amargado dobradam^{te}. pela melancholia q isso nos causa. os desgostos, e as mesmas amofinaçoens q a Socied^{e}. nos causa, ainda os trabalhos e embrulhadas, são as sombras da pintura, q fazem relevar as cores riantes da scena da vida humana. Aqui vai a incluza p^{a}. o S^{r}. Duque de Lafoens. pague V.M. p^{a}. a Cabra, no porte da posta pois não sei onde móra, nem onde adresser esta carta. eu lhe fiz prezente de hũ Camoens Inglez no anno de 76, e lhe escrevi duas cartas, q ambas imagino se perderão, uma foi a Malta e outra a Vienna. a falta de Reposta, nem chuz, nem buz me faz imaginar q elle as não Recebeo. agora espero q esta lhe hirá ás mãos. Se quizer Responder, m^{to} bem: e senão, lá vai p^{a}. o Cadoz de Portugal, onde reina o esquecim^{to}. de q^{to}. por cá fica ou se passa &^{a}. Quando vir o nosso Abb^{e}. Chev^{r}. dê lhe minhas Lembranças, e q estimarei saber q^{do}. faz a sua jornada p^{a}. o paraizo da Gloria Portugueza, onde me parece estou ouvindo a estas horas o harmoniozo Canto do Terço á noite, pela boca de mil soldados & soldadas &^{a}. as boquinhas d'alfenim das nossas Lisboetas fazendo carinhas com o rozario na mão, & os olhinhos no amante rebuçado

---

1 Annotated "R^{a}. no 2° Fevr° 1778" [Received 2nd February 1778].

2 The annotation by Sanches suggests that Magellan has written the wrong year – a common enough mistake made by people writing letters in January – and that the letter was written in January 1778, not 1777. This is confirmed by a comparison of this letter to Magellan's letter to Sanches dated 10 February 1778 in which Magellan's comments about the Duke of Lafoens are clearly a continuation of those made here, and by Magellan's news that Chevalier is going to Portugal, something he could not have safely contemplated doing in January 1777, while Pombal was still in power.

78.01.13 ao canto do Santo terço &ª. &ª. Não escreverei ao Barros, sem saber onde lhe dirigir a carta pelo correio: mas escrever-lhe-ei por Navio com hũa pª. o D^{or}. Franzini, ou o D^{r}. Ciera q ambos o conhecem e são seos amigos. ao Re[s]to elle não necessita de saber q eu o dezejo servir, & q sempre o servirei de Coração. alegro-me q tenha com q viver independente.

os seos Reparos na Lithurgia de Williams sam bem fundados: mas he necessº. guiar o povo segundo as suas nhenhices de hymnos, evening praiers &ª. &ª.

o S^{r}. Pinto agradece cordialm^{te}. a Lembrança de V.M.: elle não ama Ceremonias. Recebe os affectos, e sabe corresponder com outros sem affectação nem etique[ta]

Mandarei o Rol dos Livros em outra ocazião, q agora não tenho te[mpo]

Ora veja là q desgosto! por lhe terem furtado alguns livros ...! Se fosse em outro tempo, e q lhe furtassem a mulher, ou lhe cor[tas]em os c-lhoens como ao pobre Abaelardo: isso sim q era p[ª.] sintir. Porq não tem V.M. a pachorra do D^{r}. La Cour a q[^{m}.] vi em Bath, todo ocupado com os seos Duques, Marquezes, Mylords, & Myladies &ª. &ª. lá me falou m^{to}. em V.M. mas isso de escrever, senão são Receitas q dão dinheiro, nem elle nem o D^{or}. Sequeira não intendem nada. so nós outros os Consules das Naçoens e de todos os filhos da p-ta q ha na Europa, e as vezes de todas as putas honradas, e por honrar, somos os q sabemos escrever cartas, e pagar pª. a Cabra dos Correios & galpurreas d'Alfama &ª.

o rapaz de M^{r}. Wilkieson, q estava tulhido da cinta pª. baixo, Recobrou quazi totalm^{te}. com os banhos de vapor, & sem outra applicação. se a cura durar, será de admirar.

não ha nada por ora de interessante: se acazo sahir algũa coiza com jeito não me esquecerei de manda la a V.M., ao menos pª. lhe dar materia a ralhar e mandar far buzerare questi bechi f-tuti &ª. a obra metaphisica do nosso D^{r}. Priestley Recebe aceitação: m^{tos}. a comprão. são dois vol. em 8º. no primº. prova, ou pretende provar q tudo he material e nada espiritual: & q isto he evidentem^{te}. demonstravel pela escritura Santa. na 2da. p^{te}. sobre a necessid^{e}. philosophica, mostra q não ha liberd^{e}. algua nas nossas açoens &ª. elle he crente & profitente acerrimo da Revelação. diz q em morrendo cada hum, logo cessa toda a sua existenciâõ: mas q em vindo o tempo do milagre da Ressurreição, então todos tornarâõ á vida, do mesmo modo material q antes tinhão &ª. &ª. e tudo isto elle o diz provar evid^{e}.m^{te}. pela Sta. Escritura. ate V.M. lá estes negalhos. em hũa Biblia manuscrita do Vaticano está este Disticho. = Hic Liber est inquo quaerit sua dogmata quisque: et inquo reperit dogmata quisque sua. a bella tradução ingleza diz

= here every sect do for their tenets look: and each one finds his tenets in this book. 78.01.13

Vale & ama
tuum &ª.

a Monsieur
Mons[r.] Sanchez, Doct[r]. en Medecine
Cimitere de S[t]. Jean Rue de la
Verrerie
au Marais
à Paris

London 13 January 1777[3]

My Dear Friend

I arrived here on the 6th of this month, having spent the holidays chasing the post through Salisbury, Bath, Bowood, Oxford & Birmingham. I found your longed-for letter on Christmas Day[4] with many others, & it was the first one I read on account of the pleasure that yours always give me, though I don't expect news of be--chings &[c].[5] I see you are sad & lonely, that is the true cause of it. Human life is meant for society: and the quietness that its privation offers us is embittered twice over by the melancholy that it causes. The sorrows and the same vexations to which society gives rise, even the troubles and confusions, are the shadows of the painting that bring to life the bright colours of the scene of human life. Concerning the enclosure for the Duke of Lafoens,[6] please pay for the cost of the post because I don't know where he lives or where to address this letter to. I made him a present of an English Camoens[7] in the year 76 and wrote him two letters[8] that I imagine were lost, one to Malta and the other to Vienna. The lack of response, not a single word, makes me think that he didn't receive them. Now I hope this one will get to his hands. If he wishes to answer, very well: if not, it comes down to his hiding in Portugal, where forgetfulness reigns about what happens elsewhere &[c]. When

3 See note 2 above.

4 Letter not found.

5 See Magellan to Sanches, 7 July 1775, note 19.

6 Enclosure not found. D. João Carlos de Bragança (1719-1806), second Duke of Lafões, filled important public, military and diplomatic roles during the reign of Queen Maria I. He was the founder of the Royal Academy of Sciences in Lisbon in 1779 and its first President.

7 Camoens, *The Lusiad; or the Discovery of India: An epic poem*, translated from the original Portuguese with an introduction and notes by William Julius Mickle (Oxford and London, 1776).

8 Letters not found.

78.01.13 you see our Abbé Chevalier[9] give him my regards, and say that I should like to know when he is making his journey to the paradise of Portuguese glory, where I seem to hear the harmonious singing of the rosary at night in the mouths of a thousand soldiers & their female companions &c, the little mouths of our touchy Lisbon ladies making grimaces with the rosary in their hands & their little eyes hidden by the corner of the holy rosary fixed on their lovers &c. &c. I shan't write to Barros[10] by the post without knowing where to send the letter: but I'll send him a letter by ship with one to Dr. Franzini[11] or Dr. Ciera[12] since both know him and are his friends. Besides he doesn't need to know that I wish to help him, & that I always will with my heart. I'm glad he has means to live independently.

Your remarks on the liturgy of Williams[13] are well grounded: but it's necessary to lead people according to their sanctimonious hymns, evening prayers &c. &c.

Mr. Pinto thanks you cordially for your compliments: he doesn't like ceremony. He receives compliments and corresponds with others without presumption or etiquette.

I'll send the list of the books on another occasion, since I don't have time now.

How terrible! Being robbed of some books ...! If it was in another time, and if your wife was taken, or your b—ls were cut off as happened to poor Abelard,[14] then you would really feel sorry. Why don't you have the patience of Dr. La Cour whom I saw in Bath, fully occupied with his Dukes, Marquises, Milords, & Miladies &c. &c. He spoke much about you, but in what concerns writing, if it is not prescriptions that earn money, neither he nor Dr. Sequeira intend doing anything. Only we, the consuls of the nations and of all the sons of a bitch who exist in Europe, and sometimes of all honest whores and those who will be, are those who know how to write letters and to pay for the postage & the [exorbitance][15] of Alfama[16] &c.

Mr. Wilkieson's boy,[17] who was paralyzed from the waist down, recovered almost completely with vapour baths & without any other treatment. If the cure is lasting, it is to be wondered at.

9 João (or Jean) Chevalier (1722–1801). See also Magellan to Sanches, 10 February 1778.

10 The Portuguese astronomer José Joaquim Soares de Barros (1721-1793).

11 Miguel Franzini (?-1810), Venetian mathematician invited by the Marquis of Pombal to go to Portugal to teach at the newly reformed University of Coimbra.

12 Miguel Antonio Ciera (fl. 1770), Italian engineer from the Piedmont, went to Portugal in the middle of the century to help in the demarcation of the boundaries between the Spanish and Portuguese colonies in South America. He was professor of mathematics at the Royal College of Noblemen in Lisbon and later professor of astronomy at the reformed University of Coimbra.

13 David Williams, *A Liturgy on the Universal Principles of Religion and Morality* (London, 1776)?

14 Peter Abelard (1079-1142) who, following his love affair and marriage with Heloïse, was castrated at her uncle's instigation.

15 Conjectural meaning.

16 The borough of Lisbon where the General Post Office was located at the time.

17 See Magellan to Sanches, 30 September 1777.

There is nothing of interest for now: if anything worthwhile comes to light I shan't forget to send it to you, at least to give you something to scold and send far buzerare questi bechi fotuti &c.[18] The metaphysical work of our Dr. Priestley receives acceptance: many buy it. It is in two volumes in 8o.[19] In the first volume he proves, or intends to prove, that everything is material and nothing is spiritual: & that this is evidently demonstrable from Holy Scripture. In the 2nd part, about philosophical necessity, he shows that there is no freedom at all in what we do &c. He is a believer & tenaciously professes the revelation. He says that whenever anyone dies, he immediately ceases all existence: but when the miracle of resurrection comes, then all return to life, in the same material way as they were before &c. &c. and all this he says he proves from Holy Scripture. Tie these pieces of thread. In a manuscript Bible in the Vatican is this distich = Hic Liber est inquo quaerit sua dogmata quisque: et in quo reperit dogmata quisque sua.[20] The beautiful English translation says = here every sect do for their tenets look: and each one finds his tenets in this book. 78.01.13

Vale & ama

tuum &c.[21]

18 Send cheating these fucking cuckolds. See Magellan to Sanches, 23 June 1775, note 5.

19 Magellan is referring to Joseph Priestley's *Disquisitions relating to Matter and Spirit* (London, 1777). This was a single-volume work; however, Priestley's *The Doctrine of Philosophical Necessity Illustrated*, published in the same year, was described as an appendix to the former work and the two were often sold together.

20 This epigram is due to the Protestant theologian Samuel Werenfels (1657-1740), who wrote it in this manner: "Hic liber est in quo sua quærit dogmata quisque, Invenit et pariter dogmata quisque sua." See S. Werenfels, *Opuscula theologica, philosophica et philologica* (Basel, 1718), p. 859.

21 Farewell and love me. Your &c.

To the President and Council, Royal Society of London 78.01.22

Royal Society of London, Miscellaneous Manuscripts, 3. 16

Gentlemen,

In consequence of the notice I received from the Clerk of the Royal Society, relating to a mistake in the Bond given by myself for [...][1] contribution money of Dor. Nathanael Mathew Wolf of Danzic, lately elected home-member of this Socy., I transcribed the two first articles

1 Paper damaged, illegible.

78.01.22 of y^{e}. 3^{d}. chapter of our Statutes; and I sent them to him by the Post. I received of late his answer,[2] with a Letter directed to Sir John Pringle on this Subject, which I have deliver'd. In his letter to me, D^{r}. Wolf desires to represent to this Council, that, finding him-self engaged at present in considerable expenses, building an Astronomical Observatory, and purchasing instruments, for the Philos^{al}. Society (Societé Physicale) at Danzic, it becomes highly inconvenient, if not impossible, to him to pay the 26 guineas at once in lieu of ye yearly contribution: and begs the Council of the R^{l}. Society, whose aim is to enlarge and contribute towards the progress of Science, may please, either to remit the whole of his contribution, Since it is in Their power to do so, of which power being exercised there have been several instances: or to accept a new Bond properly drawn up, binding some person to pay, instead of him, the said annual contribution; since the words and the very spirit of the above article, seem to exclude only the accepting of a Bond from any newly elected member of the Society, who lives beyond 60 miles from London, but by no means the Bond of any other person who lives nearer, and oblidges himself to pay for the same.[3]

This Statute is intended, as it seems, simply to secure the Society from the trouble and further expence of sueing, at a greater distance, for recovery in case of failure; which end is fully answer'd by a second person's Bond, if Living within the said distance from London. And if the ready money of 26 guineas paid by a second person for the contribution composition of any member, is always accepted by the R^{l}. Society; it is but common sense (tho' perhaps a mercantile one) to say, that the Bond of a second person living within the required distance must be equaly acceptable.

Now, Gentlemen, Let me be allow'd, as one who has at heart the honour and the interests of this Society, equally as well as any other who may ever boast of the same feelings; Let me be allow'd, I say, to offer the following circumstances to your Consideration. 1^{rst}. The newly elected member, Doct^{r}. Wolf, is very well known in France, Germany, Poland, and even in England, as a learned gentleman, and of an uncommon, even very extraordinary merit as a philosopher, who is spending almost all he

2 Letter not found.

3 The Statute in question required Fellows elected on the Society's home list – and it is most unclear why Wolf was elected on the home rather than the foreign list – prior to their admission, either to pay the sum of 26 guineas in lieu of all future contributions, or to execute a bond rendering themselves liable to pay at the rate of one shilling per week to the Society so long as they continued members. What Magellan is here arguing against is the further proviso, that Fellows whose usual place of abode was more than sixty miles from London did not have the option of executing a bond, but were required to pay the 26-guinea composition fee.

possesses for the service and progress of Science. A worthy Member of this present and Last Council, can give testimony, that I have consulted him long while ago, at the request of Doct^r^. Wolf, about the building of the above Observatory. 2^dy^. His election not only has been noticed to him by the Secretary of the Soc^y^.; but his name has been already inserted in the printed Lists for the anniversary election of this year, besides it's entry in the minute books of y^e^. Society. These Lists were sent to all the members in or near London: and of course this fact is already known to every one into whose hands they hapen'd to fall afterwards. 78.01.22

This being premised, I needs must apprehend, that, if both the favours above requested, are denied to a gentleman so circumstanced as he is, some disagreable reflections may be expected to be cast upon the whole Body of this Society; as the object for such a denial, bears but the only appearance of a pecuniary one: and if so, I cannot help thinking, but that such motives shall naturally be recorded, at least in the minutes of the above Philos^al^. Society of Danzic, not to the great honour of this of London. Indeed should the evil genius of pecuniary interest ever become the sole and leading interpreter of our Statutes, instead of those liberal and generous principles suitable to true Philosophers; the epithet Philosophical, when taken by, or given to, such a Society, must either be then a nugatory, or an ironical one.[4]

I am with the greatest due regard and Respect

Gentlemen
your most Obed^t^. h^ble^. Servt.
John Hyacinth de Magellan

London 22 January 1778
at n°. 12 Nevils Court Fetter Lane

To the President and Council of the Royal Society
Crane Court
London

4 Magellan later apologised for the intemperate tone of these remarks; see Magellan to the President and Council, Royal Society of London, 23 March 1778. They surely can have done Wolf's cause no good, and the apology, written the day before the Council of the Society was to meet, is a transparent attempt to undo the damage done. It was, however, to no avail, Council instructing the Secretary to write to Wolf "informing him that he cannot be permitted to give bond for the payment of contributions, but must cause payment of the sum of twenty six Guineas in lieu thereof, and further informing him that if the said sum shall not be paid on his behalf into the hands of the Treasurer of this Society on or before the first day of Novr. next, his name will be out of the printed lists of the Society". It would seem that Wolf paid up, because his name continued to appear on the Society's annual printed lists of members.

78.01.30 *To Johann III Bernoulli*

Öffentliche Bibliothek der Universität Basel, MS L I a 709, f. 65-66

Londres 30 Janv^{r}. 1778

Mon cher ami

Je vous annonce avec plaisir que Le Bureau de la Longitude vous a fait un present de quatre vingt guinees pour l'attention de celui de vos Tables,[1] quoiqu'on ne fait pas usage ici du calcul decimal. C'est ce que le D^{r}. Shepherd[2] m'a assuré hier a L'Assemblée de la Societé R^{le}. Vous n'avez qu'envoyer ordre pour recevoir cette argent & pour vous le remettre. Si vous m'indiquez La Voye, & n'avez pas d'autre pour celà ici, je vous ferai avec plaisir ce service. J'ai reçu les 40 copies de votre second cahier des nouvelles litteraires,[3] avec les ephemerides pour 1779 & 1780.[4] Je vous remercie pour celui-ci, dont j'ai deja fait present à quelqu'un: & si vous m'envoyez 2 ou 3 exemplaires de chacune, peut etre trouveroient des acheteurs. pour les cahiers, je les ai mis chez Elmsley.[5] J'ai examiné votre compte: & Je ne trouve encore aucun argent liquide pour vous remettre. Des petits envoys de livres coutent trop de frais, & il est impossible de tirer aucune avantage. lorsqu'il y a plusieurs livres, alors la depense respective n'est si sensible. Voici par example cet envoye, qui vous coute de fret & à la douane entre 10 & 12 shillings, ce qui fait environ 3 pence de plus pour cahier; ce qui est une grande charge.

Je vous prie de demander positivement à M^{r}. Speiner, Libraire chez Vous, ce qu'il a fait avec les exemplaires de l'ouvrage posthume de Simpson, que Je lui consegnai ici pour envoyer en Suede, Danemarque, & en d'autres endroits, comme des presents de la part du Comte Stanhope, qui le fit imprimer à ses depens.[6] Je l'ai vû dernierem^{t}. & il s'est plaint, que plusieurs

1 Bernoulli's *Sexcentenary Table, exhibiting at sight the result of any proportion where the terms do not exceed 600 seconds or 10 minutes, with precepts and examples*, published in the following year by the Board of Longitude.

2 Anthony Shepherd, Plumian professor of astronomy at Cambridge and a member of the Board of Longitude.

3 Between 1776 and 1779, Bernoulli issued six cahiers entitled *Nouvelles littéraires de divers pays; avec des supplémens pour la liste et le nécrologe des astronomes.* The final cahier included a new printing of his original *Liste des astronomes connus, actuellement vivans*, first published in 1776, together with a new title page that now usually appears at the front of bound sets of the series, that describes the whole work as a supplement to Bernoulli's earlier 3-volume *Recueil pour les astronomes.*

4 The *Berliner astronomisches Jahrbuch für das Jahr n* began with the volume for 1776.

5 The London bookseller Peter Elmsley.

6 Robert Simson, *Opera quaedam reliqua* ... (Glasgow, 1776).

de ces presents n'etoient pas delivrés aux personnes aux quelles ils etoient destinés: Car ces memes personnes ont ecrit, du moins quelques unes, en declarant ouvertement n'avoir rien reçû. Je suis bien mortifié d'avoir confié rien à un sujet si peu soigneux & si peu exact, & dont Les correspondents sont a peu près de la meme qualité. Je rougis pour Lui avoir confié ces presents. C'est le malheur de croire que .... &c. 78.01.30

Je vous prie de presenter mes Resp[s]. à M[r]. de Castillon le Pere. Je n'ai pas de nouvelles de lui depuis très long tems. le journal du Nord, qui etoit le titre de celui qu'on alloit imprimer à Paris, comme continuation du de Berlin, ne paroit pas encore.[7] Ditez Lui (à M[r]. de Castillon Pere) que J'attends ses ordres pour lui envoyer l'argent qui se trouve entre mes mains de la vente du Journal &c. dont il reste encore quelques exemplaires, sans avoir d'acheteurs.

Avez vous vû des christaux formés par la nouvelle decouverte de M[r]. Achard?[8] Je souhaiterois bien avoir quelques uns, pour les comparer avec ceux que J'espere faire ici avec un appareil, qui va etre finis selon sa methode, pour Le meme effet.

Vale & ama
tuum
Magellan[9]

Si par hazard vous y rencontrez le Comte Donhoff,[10] qui a passé plusieurs mois à Londres dernierement, dittez lui que J'attends de ses nouvelles depuis Long tems: &a

A Monsieur
Monsr. Bernoulli, Astronome Royal,
& membre de l'Acad[e]. R[le]. des Sciences de
Berlin

7 The promised journal, apparently a Parisian continuation of the *Journal litteraire*, edited by Castillon and others, that had been published in Berlin for several years, seems never to have been published.

8 Achard, "Copie d'une lettre ... contenant la découverte qu'il a faite sur la formation des cristaux & des pierres précieuses", *Observations sur la physique, sur l'histoire et sur les arts*, 11 (1778), 12-14. More formal reports were published as the final section (pp. 347-350) of Achard's paper, "Chemische Untersuchung verschiedener Edelgesteine", *Neue philosophische Abhandlungen der baierischen Akademie der Wissenschaften*, 1 (1778), 219-350, and as an appendix to Achard's *Bestimmung der Bestandtheile einiger Edelgesteine* (Berlin, 1779), pp. 122-8.

9 Farewell & love | your | Magellan.

10 A Polish (or East Prussian?) nobleman who in 1764 had married the daughter of the third Earl of Tankerville and was later the Prussian ambassador in London.

78.01.30a *To James Watt*[1]

Birmingham Central Library, Boulton & Watt collection, Box 22/B3/2

London 30 Jany. 1778
at n°. 12 Nevils Court Fetter Lane

Sir

I have the pleasure of announcing to you & M^{r}. Boulton, that the exclusive privilege has begun to be requested by very good way of such Person as to give great hopes of obtaining it in France. I have seen an answer sent by the Last post to my friend here, about this matter, which encourages me to forward this notice already to you both. In a few days I will be able to say some thing more decisive about this affair: and will not fail of communicating it to you. Very probable it will be required that either of you should take a trip to Paris: but of this I will acquaint you in time.

I have been informed that you have invented and made or caused to be made, long while ago, a new kind of micrometer for measuring small angles, as the diameters of Planets &^{a}., composed of one or two small prisms, or angular wedges of glass, sliding along the axis of a refracting tellescope. I wish to know, if this is a fact, about what time, the said micrometer was done? and in whose possession it is now, if existing? or at least I wish to know, if you recolect to have spoken of it to any of our curious gentlemen in Astronomical pursuits?

I assure you this my asking such particulars, proceeds not from any iddle curiosity of myne; but that this inquiry is in consequence of the friendly concern I take for the ascertaining your earlier claim to such an invention.[2] I beg you an answer as soon as possible on this matter; and hope you'll allways find that I sincerely am

Dear Sir
Your most obed^{t}. Servt. and friend
Magellan

---

1 Letter not found; MS is a clerk's transcription. Annotated "Mr Magellan Jany 30th 1778".

2 See Watt to Magellan, 2 February 1778. Magellan's question was no doubt motivated by his learning that a similar device had just been developed by the Abbé Rochon; see M. Daumas, *Scientific Instruments of the 17th and 18th Centuries and Their Makers* (London, 1972), p. 271.

From James Watt[1] 78.02.02

Birmingham Central Library, Boulton and Watt Letter books, vol. 3, 2

Feby. 2^{d}. 1778

Mr Magellan

I thank you for information of French privelege.[2] Mr B. is now in London with whom you may confer

About ten years ago I invented a micrometer composed of a prismatical glass placed beyond the object glass of a telescope[3] – It was divided down the middle and when the two parts lay in the same plane all the rays were equally bent and no effect produced but when one glass was turned so as to make a greater angle with the object glass than the other then a parcel of the rays were bent more than others

I have mentioned it to several people in London and doctors Wilson & Irvine at Glasgow have both seen it – The one I made was imperfect and is long since destroyed – I regarded it as a trifle of no moment to me who have no desire to figure away in the memoirs of academies otherwise might produce better claims to y^{t} honour – I thank you for your attention to see Justice done to my inventions and to oblige men to give to Cæsar those things which be of Cæsar

---

1 Original not found; text taken from Watt's letter-copy book.

2 See Magellan to Watt, 30 January 1778.

3 Watt's micrometer, designed in 1770-71, was what we would call a rangefinder, and was used to measure distances in his canal surveying work. In 1772 he invented what is considered the first micrometer to measure small distances precisely. There is one on display at the Science Museum, London, that is attributed to Watt.

78.02.10 *To Ribeiro Sanches*[1]

Österreichische Nationalbibliothek, Vienna
A.N. Ribeiro Sanches Korrespondenz, codex 12714, ff. 213-214v

Londres 10 Fevr°. 1778

Meo Am°. e Sr. e mto. do Coram.

Neste momento acabo de escrever ao nosso Barros, com o sobrescrito q V.M. me indicou na sua Carta do prim°. do Corrente. Se elle quizer Responder, e eu lhe puder servir de algũa coiza, póde estar certo q o hei de fazer com mto gosto.

Recebi o dinh°. q me Remeteo o Dor. le Presle, e puz em credito na Conta de V.M. as 402#.

Brevemte., isto hé antes do fim desta semana, mandarei a V.M. alguns Livritos, com outros ao Dr. Le Begue. entre elles mandarei as Leis dos gentios da India oriental, onde V.M. achará mto. q admirar, & mto. de q se rir desta raça humana, q tanto presume. particularmte. o artigo ou Capitulo 20 sobre o caracter & obrigaçoens do Sexo feminino he notavel: pouco mais ou menos he o q o gentios modernos, & os patriarcas antigos julgavão. a saber, q a mulher não he senão pa. fazer filhos, & pa. servir ao homem &a. &a.

V.M. não diz chuz nem buz sobre a minha carta q mandei dentro da sua pa. o Duque de Lafoens. isto parece silencio de politico Secretario de Estado. a mim basta-me saber se este fidalgo recebeo, ou não, a minha carta. mas q elle me responda, ou não me responda, hé coiza pa. mim muito indifferente. eu mandeilhe cortezmte. um exemplar da tradução do Camoens pelo Conde Scarnafiz, isto he junto com outro pa. o d°. conde, q então estava em Vienna, Embaixador do Rey de Sardenha e agora se acha em Paris no mesmo Cargo. elle Respondeo-me q tinha mandado entregar o d°. Livro ao Sr. Duque de Lafoens; porem elle não se dignoudar-me a mim Reposta, ás cartas q lhe escrevi pelo Correio, à Vienna & a Malta. agora pa. me tirar da duvida se acazo as duas minhas Cartas se terião perdido, mandei esta terceira na de V.M.; e tenho por certo, ainda q V.M. o não diga, q ella lhe foi entregue. Se este Senhor gosta de passar por fidalgo Portuguez na minha estimação, nenhuma pena isso me dá. eu vivo independente, mto. querido & estimado não so de mtos. hom̃ens de bem e filosofos q eu estimo mto. mais do q todos os fidalgos da Europa, mas tãobem de mtos. e mtos. sojeitos tanto e mais altos do q elle, pelos seos nascimentos e Nobreza hereditaria, igoalmte. q adquirida.

A vista disto esteja V.M. socegado neste ponto, pois nada se me dá disso: e somte. ajunto q me alegrarei de ter comprado tão barato como o preço da

1 Annotated "Ra a 23. Fev°" [Answered on 23 February]. Letter not found.

tal obra do Camoens, o conhecimento e clara demonstração do mui pouco q esse sujeito merece os encomios q lhe dão. Mas deixemos lá esse senhor de guiné, q va lá p^{a}. os seos semilhentes da nossa terra bem aventurada ...! 78.02.10

Não deixarei de mandar o Dicion^{o}. de Motherby ao D^{or}. Martens, em direitura a Vienna pela prim^{a}. occazião.

Quando V.M. vir o S^{r}. Huerta, q me Recomendo affectivo. he mosso honrado.

ainda não vi Sequeira. mas sei q passa bem.

informar-me-ei das institutes de Medicine de Culen; e as mandarei, se forem differentes da sua Physiologia, q ja mandei a V.M.; e h[e] uma brochura de 255 paginas em 12mo., ou piqueno 8°. Veja-lhe V.M. o titulo. Estas institutes são certam^{te}. as lectures q V.M. vê citadas. mas eu me informarei de novo: e tãobem da 2da. edição de Macbride.

em chegando o S^{r}. Doria verei o q V.M. me manda sobre as folhas da Practica de Culen. Eu vou mandarlhe outro exemplar completo da Physiologia de Culen. mande-me V.M. pela prim^{a}. ocazião o q V.M. diz q tem imperfeito.

alegrome q Chevalier va, como o pato p^{a}. a agoa, p^{a}. a sua suspirada Lisboa: q D^{s}. o ajude a cantar o terço, & a correr a via sacra. antehontem encontrei com um Inglez q chegava do Porto, e me dizia q lá vîra ordenar dois mil clerigos de uma vez. que bello espectaculo ...! ...! ...! ora tenha V.M. melhor saude, como dezejo: e alegre-se q lá vem a Primavera e o verão com o tempo quente q talvez lhe faça ainda sahir os corninhos ao sol, como o bemaventurado caracol q fututus futuit segundo as grandes descubertas dos S^{res}. Naturalistas modernos.

minhas saud^{es}. á familia adoptiva, q espero visitar no verão q vem; pois faço tenção ir ao continente; & certamente darei hũa volta por Paris, excéto se algum acid^{e}. m'o impedir.

não há nada de novid^{es}. interessantes q eu saiba na Literatura. o nosso Enviado passa bem: cuido q poderei hir jantar como costumo, á manhan com elle. He o unico fidalgo Portuguez q por cá tenho visto por fóra com merecim^{to}. distinto, & bom coração. Mylord Shelburne perdeo o filho mais novo; dentro de 24 horas morreo com hũ ataque nas tripas. Mylord está inconsolavel. eu & todos os q o conheciamos da mesma sorte.

Vale me que ama

a Monsieur
Monsr. Sanchez, D^{r}. en Medecine
Cimitere S^{t}. Jean
Rue de la Verrerie
à Paris

78.02.10 London 10 February 1778

My Dear Friend

At this moment I have just finished writing to our Barros,[2] with the address you indicated in your letter of the first of this month.[3] If he wants to reply, and if I can be of some use, you can be sure that I'll do it with much pleasure.

I received the money that Dr. le Presle sent me, and I entered the 402# as credit in your account.

Soon, that is before the end of this week, I'll send you a few little books, with others for Dr. Le Begue. Among them I'll send the Laws of gentiles from east India,[4] where you'll find much to admire & much to laugh at about this human race, which is so presumptuous. In particular, the article or chapter 20, about the character & obligations of the feminine sex, is remarkable: it is more or less what modern gentiles & ancient patriarchs thought. That is, that the woman is for nothing but giving birth to children & serving man &c. &c.

You say nothing about my letter that I sent inside yours to the Duke of Lafoens.[5] This seems like the silence of a political Secretary of State. To me it is enough to know if this gentleman received my letter or not. But whether he answers, or doesn't answer, is an indifferent matter. I courteously sent him a copy of the translation of Camoens[6] through Count Scarnafiz,[7] that went together with another for the said count, who was in Vienna at the time as Ambassador of the King of Sardinia and is now in Paris in the same position. He told me he had delivered the said book to the Duke of Lafoens who, however, didn't care to answer the letters I wrote to him by post, to Vienna & to Malta. Now to leave me in no doubt, in case my two letters were lost, I sent this third one inside yours; and I'm sure that, even if you don't say so, it was delivered to him. If this gentleman likes to pass for a Portuguese nobleman in my esteem, it doesn't make me sorry. I live independently, loved & esteemed not only by many men of good will and philosophers whom I like more than all the noblemen of Europe, but also by many persons as high as himself or higher by their births and nobility, hereditary as well as acquired.

In view of this you may be calm at this point, because I don't care a penny. And I only add that I'm glad I have bought as cheaply as that work of Camoens

2 Letter not found.

3 Letter not found.

4 Probably *A Code of Gentoo Laws, or, Ordinations of the Pundits...*, translated by Nathaniel Halhed (London, 1776).

5 See Magellan to Sanches, 13 January 1778.

6 Camoens, *The Lusiad; or the Discovery of India: An epic poem*, translated from the original Portuguese with an introduction and notes by William Julius Mickle (Oxford and London, 1776).

7 Filippo Maria Giuseppe Ottone, Conte di Ponte di Scarnafigi, was Ambassador of the Kingdom of Sardinia in London in the early 1770s, before transferring to Vienna and then to Paris.

the knowledge and clear demonstration of how little this man deserves the eulogies that he is given. But let us leave aside this gentleman of Guinea, may he go to his fellows in our blessed country ...! 78.02.10

I shan't forget to send the Dictionary of Motherby[8] to Dr. Martens,[9] directly to Vienna, at the first opportunity.

When you see Mr. Huerta,[10] please remember me to him heartily. He is an honourable young man.

I haven't yet seen Sequeira. But I know he is well.

I'll inquire about the Institutes of Medicine of Cullen; and shall send them if they are different from his Physiologia that I already forwarded you; it is a brochure of 255 pages in 12° or small 8°.[11] Please check the title. These institutes are certainly the lectures that you see quoted. But I'll inquire again: and also about the 2nd edition of Macbride.[12]

As soon as Mr. Doria[13] arrives, I'll look into what you ask me about the pages of Cullen's Practice. I'll send you another complete copy of Cullen's Physiologia. Send me at the first opportunity the one you say is imperfect.

I was pleased that Chevalier is going, like the duck to the water, to his greatly desired Lisbon:[14] may God help him to sing the rosary, & to run the stations of the cross. The day before yesterday I met an Englishman who arrived from Oporto, who told me that he saw two thousand clergymen at a time ordained there. What a splendid show ...! ...! ...! May you have better health, as I wish: and be happy that spring is arriving and summer, with the warm weather that perhaps may still make your horns reach out to the sun, like the blessed snail that fututus futuit[15] according to the great discoveries of modern naturalists.

My regards to the adoptive family, whom I hope to visit next summer since I plan to go to the Continent & certainly will pass through Paris, unless some accident prevents me from doing so.

---

8 George Motherby, *A New Medical Dictionary* (London, 1775).

9 Carl von Mertens, medical doctor of Vienna, author of *Observationes medicæ de febribus putridis, de peste, nonullisque allis morbis* (Vienna, 1778).

10 A Spanish-like way of referring to Francisco José Horta Machado (1746-1817)? See Magellan to Sanches, 28 January 1777.

11 The work Magellan describes here is William Cullen's *Institutions of Medicine. Part I: Physiology* (Edinburgh, 1772), a different work from the definitive edition of his lectures on medicine published under the title *First Lines of the Practice of Physic*, the first volume of which appeared in 1777.

12 David MacBride published *Experimental Essays upon Different Subjects* (London, 1764), and later *Experimental Essays on Medical and Philosophical Subjects* (London, 1767) which, despite being much enlarged and having a different title, was described as a second edition of the earlier work. Another edition, described as the third but in fact the second under the new title, appeared in 1776 and it may be this that Magellan was referring to.

13 Unidentified person.

14 See Magellan to Sanches, 13 January 1778.

15 Fucked, it fucks.

78.02.10 There is no interesting news in the literature that I know of. Our Envoy[16] is well: I presume I can go and have dinner with him tomorrow, as usual. He is the only Portuguese nobleman whom I've met abroad with distinguished merit & a good heart. Milord Shelburne lost his youngest son; in 24 hours he died of an intestinal fit.[17] Milord is inconsolable. I & all who knew him are the same.

Vale me que ama[18]

---

16 Luiz Pinto de Sousa Coutinho (1735-1804), Portuguese Minister in London.

17 Shelburne's son William Granville Petty-Fitzmaurice (his second son by his second wife, Lady Sophia Fermor) died on 28 January 1778.

18 Farewell and love me.

78.02.20 *To [...]*[1]

Vous me demandez, Monsieur, ce que c'est que le nouveau micromètre de M. Maskeline, ma réponse est bien simple. M. Maskelyne a mis dans l'ame de l'objectif d'une lunette les mêmes prismes achromatiques, dont M. l'Abbé Rochon, de l'Académie Royale des Sciences de Paris, avoit parlé il y a près d'onze ans. En Février 1767, cet Académicien y a lu, dans une des Séances, son Mémoire sur l'application des prismes achromatiques aux lunettes.

Ce Mémoire a été imprimé l'année suivante 1768, à Paris, ainsi qu'on peut le voir à la page 41 des Opuscules Mathématiques de cet Abbé.[2]

Il est certain que cet Ouvrage a été débité sur-le-champ à Londres, par nos Libraires qui ne manquent pas d'acquérir promptement toutes les nouveautés d'un certain genre, quand ile croyent pouvoir les débiter ici avec succès. M. Maskeline n'a présenté à la Société Royale de Londres, ce qu'il appelle son invention, qu'à la fin de Décembre 1777. Il est vrai qu'il

---

1 Letter not found. The text is an extract published in *Observations sur la physique, sur l'histoire naturelle et sur les arts*, 11 (1778), 451-3, in the issue for May 1778, under the heading, "Extrait d'une Lettre de M. Magellan à M. de ***, du 20 Février 1778, Sur le Micromètre de M. Maskelyne."

2 A.M. de Rochon, *Opuscules mathématiques* (Brest, 1768). In addition to giving the wrong place of publication for Rochon's work, Magellan also gave the wrong page reference. The relevant paper is on pp. 51-66 of Rochon's book, while the specific idea of using an achromatic prism is introduced on p. 59.

l'a accompagnée de deux attestations dans lesquelles on affirme qu'il a fait travailler au micromètre dans l'année 1776, c'est-à-dire, long-tems avant que la Société Royale ait reçu de M. l'Abbé Boscovich un Mémoire sur la même matière. 78.02.20

Vous voyez, Monsieur, que la priorité de date, est toute en faveur de M. l'Abbé Rochon. Ce Savant lit un Mémoire en 1767, il le fait imprimer en 1768; & ce n'est qu'en 1777 que M. Maskelyne communique ses idées à la Société Royale. La conséquence est aisée à tirer. Mais en supposant contre toute probabilité, que M. Maskelyne n'ait eu aucune idée, & qu'il n'ait même pas entendu parler de ce que M. l'Abbé Rochon avoit proposé en 1767, peut-on croire qu'il n'ait eu aucune connoissance du micromètre à prisme, inventé & fait, il y a plus de dix ans, par M. James Watz,[3] Ecossois, d'un génie rare & fort connu dans ce pays? Il est l'inventeur des nouvelles machines à feu, dont les effets sont prodigieux, puisqu'elles ont sur les anciennes l'avantage d'économiser sur le chauffage 75 pour 100, c'est-à-dire, les trois quarts de la dépense.

M. Wilson, Professeur de Mathématiques à Glascow, le Docteur Guillaume Irwinc,[4] de la même Ville, connoissent l'existence de ce micromètre. M. le Docteur Lind, de la Société Royale, m'a dit, il y a fort peu de tems, en avoir vu les effets.

C'est donc un fait incontestable & de toute notoriété, que M. l'Abbé Rochon & M. Watz ont appliqué aux lunettes, chacun de leur côté, & vraisemblablement sans se l'être communiqué, le premier des prismes achromatiques, & le second des prismes pour mesurer des distances angulaires.[5]

Il est à remarquer que les recherches de M. l'Abbé Rochon portent principalement sur l'application de la double réfraction du crystal de roche, du crystal d'Islande à la mesure des angles. Les constructions qu'il a imaginées à cet égard, s'appliquent également aux prismes achromatiques de verre qu'il a proposés dès l'année 1767 pour mesurer de plus grands angles, & non-seulement cette application ne lui a pas échappé, mais il a fait voir que le mouvement le long de l'ame, ne pouvoit convenir qu'au crystal de roche & à celui d'Islande, qui ne mesure que de petits angles, & qu'on ne peut pas l'appliquer avec le même avantage aux prismes de verre. La construction qui convient à ces derniers est le mouvement circulaire de

---

3 i. e. James Watt.

4 i. e. William Irvine.

5 Following Maskelyne's presentation to the Royal Society, Watt's friends pressed him to claim priority in the invention of the prismatic micrometer, a model of which Watt said he had constructed in 1770 or 1771 and given to Professor John Anderson of Glasgow University (J. P. Muirhead, *Life of James Watt* [London, 1858], pp. 231-9).

78.02.20 deux prismes achromatiques l'un sur l'autre, & cette construction a été donnée à l'Académie Royale des Sciences, le 24 Février 1776, en présence de M. le Duc de Wirtemberg.[6]

Le Comité de la Société Royale a décidé, que le Mémoire de M. Maskelyne seroit imprimé dans le Volume des Transactions Philosophiques qui va paroître.[7] Mais cela ne peut porter aucun préjudice au droit de M. l'Abbé Rochon, parce qu'au commencement de chaque Volume, la Société Royale met toujours une protestation, par laquelle elle assure ne donner aucune approbation ni authenticité décidée à ce que les Volumes renferment. Chaque Auteur reste seul responsable de la vérité ou de la fausseté de ce qu'il avance.

6 i.e. Württemberg.

7 N. Maskelyne, "Account of a new Instrument for measuring small Angles, called the prismatic Micrometer", *Philosophical Transactions*, 67 (1777), 799-813. Maskelyne's paper was read to the Royal Society on 18 December 1777. An appendix to the paper, pp. 813-5, publishes the attestations to which Magellan refers, signed by Peter Dollond and Alexander Aubert, affirming respectively that in April 1776 Maskelyne had commissioned Dollond to construct a micrometer according to his new design, and that in May 1776 Aubert had seen the finished product. Maskelyne's paper was immediately preceded by Boscovich's description, read to the Society on 19 June 1777, of a similar device, the first invention of which he attributed to Rochon though without giving a precise reference to Rochon's work: R. Boscovich, "Account of a new Micrometer and Megameter", *Philosophical Transactions*, 67 (1777), 789-98. Maskelyne was apparently not aware how much earlier Rochon had suggested using an achromatic prism and must have been under the impression from Boscovich's report that he had done so within the previous few months, for he claimed the entire invention for himself in the following terms (p. 812):
"A paper from the learned Abbé Boscovich was read before this Society the ninth [sic!] of last June, describing a similar contrivance as an invention of the Abbé Rochon, in which the Abbé Boscovich himself also claims some share; I therefore desire to acquaint this Society, that I communicated this invention to Mr. Dollond, and had it executed by him; and also shewed the instrument itself, so executed, to my esteemed friend Alexander Aubert, esq. fellow of this Society, a gentleman very well qualified to judge of things of this nature, above a twelve-month before the communication of the Abbé Boscovich's paper, as will appear from their written attestations... I doubt not that the following attestations of Mr. Aubert and Mr. Dollond will sufficiently prove my title to this invention...".
Delambre, *Histoire de l'astronomie au dix-huitième siècle* (Paris. 1827), pp. 645-52, discussed the respective claims of Rochon, Boscovich and Maskelyne to the invention – he was evidently unaware of Watt's claim, which at that time had still not been published – and gave the credit to Rochon on the basis that he was both the first to conceive the idea and the first to demonstrate it publicly, in several presentations to the Académie Royale des Sciences between January and April 1777.

To Matthew Boulton[1] 78.02.28

Birmingham Central Library, Boulton and Watt collection, Box 22/B3/3

M^r^. Magellan's Comp^s^. to M^r^. Boulton: and sends the Copy of that part of y^e^. Letter he Received from Count Chastnet,[2] about the privilege exclusif, and the machine to be made at Moeres.[3] He desires M^r^. Boulton to consider very attentively all the different objects in view: and to send a word to M^r^. Magellan.

1°. about the circumstances to be expressed in the privilege, caracterising his fire engine?

2°. about what time M^r^. Boulton intends to set out for Paris, thro' Moeres & Burgues[4] &^a^.?

3°. what in[5] the 2^d^. name of M^r^. Boulton, by which he is to go to speak to Comte d'Aronville[6]?

4°. If M^r^. Boulton thinks that some settlement may be made at Moeres to his advantage, of any kind of Manufactures &^a^.

London this 28 Febr^y^. 1778

at n°. 12 Nevils Court Fetter Lane

---

1 Letter not found; MS is a clerk's transcription. Annotated "Mr Maggelhans London Feb 28 1778".

2 Antoine Hyacinthe Anne, Comte de Chastenet de Puységur (1752-1807), was married to Antoinette Louise Claire Ricouart d'Hérouville, daughter of the Comte d'Hérouville with whom Magellan was negotiating on behalf of Boulton and Watt.

3 Les Moeres, a village in France 13 km east of Dunkerque, close to the frontier with Belgium.

4 Bergues? The copyist would appear to have misread Magellan's writing at this point. Bergues is the closest town to Les Moeres, 10 km to the south-west on the road from Dunkirk to Amiens and Paris.

5 is?

6 i.e. Hérouville.

78.03.03 *To Matthew Boulton and James Watt*[1]

Birmingham Central Library, Boulton & Watt collection, Box 22/B3/4

London 3 March 1778
at n°. 12 Nevils Court Fetter Lane

Sir

I received just now a Letter from the chief director of the mines in Prussia,[2] who desires to have 4 of your fire engines for the following effects: after the accounts I have spread of them thro' my acquaintances abroad. they are to be as follow.

1. to rise the water 155 feet high, at the rate of 26 cubic fee of water in each minute, English measure.[3]

2. to rise water 420 feet high at the rate of 47½ cubic feet in each minute[4]

3. to rise ditto 348 feet high, at the rate of ditto, viz. 47½ feet in each minute[5]

4. to rise do. 93 feet high, at the rate of 11 cubic feet in each minute.[6]

all english measure.

Now it is desired to know, if you'll undertake this considerable jobb, in what time you can furnish them and at what price? and if they are dearer than the common fire engines, it must be settled on what terms you oblige your self for the advantages thereof.

2dly. if you chuse to take into Consideration an exclusif priviledge in all the Prussian dominions, I think I can contrive to get it, as soon as the first machine is at work.

3 please to send me an answer as soon as possible, with all possible distinctness, that no blunder or mistake may happen in an affair of such a consequence.

4 If you can not undertake the whole or any of this machines, I have orders to bespeak the old kind of fire engines. but I'll wait for your peremptory answer.

---

1 Letter not found; MS is a clerk's transcription. Annotated "Magelon 3d March 1778".

2 Karl Friedrich Anton, Baron von Heinitz (1724-1802). See Magellan to J. Watt, 21 March 1778.

3 Annotated "21 Inch Cyer pump 8 inch".

4 Annotated "pump 11 inch Cylr 47 inch".

5 Annotated "11 inch pump Cylr 43".

6 Annotated "5 inch pump Cylr 10 inch or 7 inch pump & short 15 inch Cylr".

5 I am desired to send also an estimate of the transport to Hambourgh: if you have any idea of it, please to mention it in your Letter to me 78.03.03

6 as to the payment of these machines, not one is to be deliverd in London without ready money according to the answer I intend to send, after I receive yours to this Letter. However if you think otherwise, please to tell it to me in your Letter.

I heartly am

Dear Sir
Your most obed^{t}. h^{ble}. ser^{t}.
Magellan

Pray my comp^{s}. to M^{r}. Watt and family, your Lady first of all.

I am writing this instant to Paris about the other affair.[7]

To
Messrs. Boulton & Watt
Soho
Birmingham

7 See Magellan to M. Boulton, 28 February 1778.

To James Watt[1] 78.03.12

Birmingham Central Library, Boulton and Watt collection, Box 22/B3/5

London 12 Mars 1778
at n°. 12 Nevils Court Fetter Lane

Sir

I wrote to M^{r}. Boulton about 4 of your new fire engines which are wanting in the mines of Prussia:[2] I gave him the quantity of water and the depth of each: and it is near a week or more that I wait for answer,

1 Letter not found; MS is a clerk's transcription. Annotated "Monsr. Magellan 12 March 1778".

2 See Magellan to M. Boulton and J. Watt, 3 March 1778.

78.03.12 which I dont wonder as you must make your Calculations before, not only to know the expense, but also about what time you can promise with certainty to do the whole. I beg you there fore to be as carefull and expeditious as possible, in giving me your answer, that I may forward it to my friend abroad.

Now this very moment arrives a gentleman from the province of Bretagne in France,[3] who is recommended to me from good friends, and wants also one of your new fire engines for a Coal mine there, which is about 400 feet deep and furnishes about 40 cubic feet french measure per minut.

this gentleman intends to go to speak to you at Birmingham about this matter: but for the reasons that M^{r}. Boulton has no doubt communicated to you about obtaining the privilege in france before hand, I see no necessity of taking that trouble: and the more so as M^{r}. Boulton told me that he intended to Return to London very soon, viz in a few days. Beg therefore a line of answer both about the said 4 engines for Prussia and about the time M^{r}. Boulton will be able to come and be spoke here in London.

I am Sir

Your most obed^{t}. h^{ble}. Servt.
Magellan

Please to tell me the proper direction for you, as I forgot the name of the spot where your house stands: and am afraid there may be another of your same name in Birmingham.

To
M^{r}. J^{s}. Watt
at his new fire engine manufactory
near Birmingham

3 Joseph Jary; see Magellan to M. Boulton, 23 March 1778.

To Johann Albrecht Euler[1] 78.03.13

St Petersburg Branch, Archives Department, Russian Academy of Sciences, f. 1, op. 3, Nr. 63, ll. 100-101

Londres 13 Mars 1778

Monsieur

J'ai bien soupçoné dans ma lettre derniere,[2] que son Excellence Musin Pousthkin[3] n'avoit pas bien compris ce que je lui demandai sur la lampe de nuit, que M^{r}. Domasthnev[4] souhait avoir. J'ai eté avec son Exc^{e}. dernierement, & lui m'indiqua l'espece de cette lampe, qui n'est qu'une sorte de Clepsydre à l'eau, & ne coute qu'environ une guinée & demie. Ainsi je vous prie de dire à M^{r}. de Domasthnev, que je ne manquerai pas de la lui envoyer par la premiere occasion.

Je profiterai de la premiere occasion pour vous envoyer aussi un bon livre très precieuse par son objet, que je viens de recevoir de mon ami l'Abbé Mann, Membre de l'Acad^{e}. Imp^{le}. de Bruxelles. Il fut envoyé ici pour examiner les deux methodes de garantir les maisons d'etre brulées par aucun incendie. Je l'ai accompagné à Chevening chez Mylord Mahon, où il a vû & examiné avec soin sa methode qui est la plus commode etant très peu dispendieuse, & pouvant etre executée par tout ouvrier ordinaire qui soit tant soit peu soigneux à suivre les ordres. J'ai eté de meme avec lui chez Monsr. Hartley,[5] & il a eu toute l'information possible: nous avons vû des essais chez tous les deux en mettant le feu de nouveau, tant dans une partie de la maison faite exprès à Chevening, que dans l'autre de M^{r}. Hartley à Wimbledon: & M^{r}. Mann retournant à Bruxelles imprima la relation de ce qu'il avoit examiné, avec toute clarté & detail necessaire, pour seconder les vuës du Ministre, qui lui avoit donné cette commission.[6]

---

1 Annotated: "reçu le 30 Mars 1778".

2 Letter not found.

3 Alexei Semenovich Musin-Pushkin, Russian ambassador to England, 1766-7 and 1770-79.

4 S.G. Domashnev (1743-1795), the unpopular Director of the St Petersburg Academy of Sciences, 1775-83.

5 The MP David Hartley (1731-1813) received government funding to pursue large-scale experiments in a purpose-built house at Putney Heath, near Wimbledon, to test the ideas on protecting buildings from fire set out in his pamphlet, *An Account of the Method of Securing Buildings and Ships against Fire* (London, 1774).

6 T.A. Mann, *Mémoire sur les diverses méthodes inventées jusqu'a present, pour garantir les edifices d'incendie* (Brussels, 1778). Mann's report was also published in *Observations sur la physique, sur l'histoire naturelle et sur les arts*, 12 (October 1778), 249-73.

78.03.13 Je tacherai donc de profiter de quelqu'un qui partira pour Petersburg afin de vous remettre cette brochure: & je vous prie de prendre les mesures necessaires pour qu'elle soit rendu publique en Russe & distribuée en sorte qu'on puisse profiter de cette invention utile.

Cependant Mylord Mahon publiera bientot lui meme un ouvrage sur sa methode,[7] où il n'est pas douteux qu'il y donnera beaucoup plus de lumieres & eclaircissements utiles: ainsi il sera assez pour le bien public de donner à present dans votre gazete ou quelqu'autre ouvrage periodique le plus populaire chez vous, un extrait, ou precis de ce que l'Abbé Mann dit sur la methode de Mylord Mahon dans cette brochure: & aussitôt que le propre ouvrage de Mylord sera publié ici, je tacherai de vous en envoyer un exemplaire, pour qu'il puisse etre traduit en Russe & repandu tout par tout &^{a}. Vous me ferez bien d'honneur en annonçant la main de qui vous avez reçu cette communication: car tout mon ambition est d'etre connû comm'un de ceux qui ont le bien de l'humanité à coeur.

Je ne sçais pas si je vous ai annoncé le remede nouveau du D^{r}. Hulme, ou pour mieux dire la nouvelle methode d'appliquer l'air fixe à differentes maladies, avec un grand succès, comm'il paroit par les observations de pratique, qu'il vient d'imprimer.[8] il l'a appliqué dans les maladies de la pierre dans la vessie, des [ardues] ou sables dans les reins, dans l'escorbut, meme dans la goute, & dans les fievres hetiques ou de consumption &^{a}.

Le remede consiste en 15 grains de Sel de Tartre en 3 ounces de l'eau [dis]tilée,[9] ou meme de l'eau commune: aussitôt que le malade avale ce boisson[,] soit avaler tout de suite 3 ounces de la meme eau distilée, ou commune, melée avec environ 20 goutes d'esprit de vitriol foible. Cet esprit de vitriol & le sel de tartre sont les memes decrits dans la pharmacopée de Londres. on repete ce remede 4 fois par jour: & les effets ont eté surprenans. Vous voyez bien que l'acide vitriolique agissant immediatem^{t}. sur l'alkali du tartre en doit degager dans l'estomach toute l'air fixe qui y etoit contenû, & il operera son effet au dedans du corps &^{a}. Ayez la bonté de faire repandre chez vous cette nouvelle utile au genre humain.

Vale & ama
tuum
Magellan[10]

---

7 No such separate work appears to have been published; however, Mahon did publish an account of his experiments in the *Philosophical Transactions*, 68 (1778), 884-94.

8 Nathaniel Hulme, *A Safe and Easy Remedy proposed for the Relief of the Stone and Gravel, the Scurvy, Gout, &c....* (London, 1778).

9 Paper damaged at the end of this and the following line.

10 Farewell & love | your | Magellan.

To Ribeiro Sanches[1] 78.03.13a

Österreichische Nationalbibliothek, Vienna
A.N. Ribeiro Sanches Korrespondenz, codex 12714, ff. 215-216v

Londres 13 Março 1778

Meo Am°. e Sr. e mto. do Coração

Em resposta á de V.M. de 23 do passado, q me deo mto. gosto, devo repetir sem ser parente de Sancho Pança q = diz a trempe á sertan: guardate lá q me não sujes = olhe qm. fala em não arr- -tar ...! O Sr. Duque de Lafoens fica restituido ao gráo da sua antiga estimação: mas o Sr. Fidalgo da Ilha da madeira ficou como se tal não fosse. entregou os papeis q V.M. & o Dor. Le Begue me mandárão por elle, em caza do nosso Enviado; & não foi possivel saber onde morava, nem q he feito delle. ouvi dizer q a sua prima. mulher era franceza; & tinha morrido de esmorecida, vendo-se como degradada naquella Ilha: estimarei q não suceda outro tanto a esta Sra. Preston. eu renego dos q não arreitão senão pa. c-nos estranjeiros. Aqui vive hũ seo irmão frade cuido q franciscano, q se retirou da fradaria. Porem não tive ideias delle, q me fizessẽ buscalo, nem visitalo, não obstante q elle me fez a honra de buscarme.

Aqui mando a Lista dos Livros, q remeti ultimamte. à Mr Le Begue, & de q ainda lhe não dei pte., sabendo q não chegarâõ lá estes 15 ou 20 dias. O Livro pa. Mr. Martens foi remetido a Vienna pela via de Bruxellas. e custou 1£.14s. com caixa & remessa

Mr. Le Begue dará a V.M. um par de mouchettes excellente. cortão papel como huma boa tizoira: & tem um scraper dentro: de modo q em o levantando, depois de cortar o murrão; e tornando-o a abaixar, o murrão q estáva cortado dentro delle, passa para o interior da caixa, e não pode cahir fóra ainda q se abra o espivitador. Com elle achará V.M. a tizoira & clippers pa. cortar as unhas. e alem disso uma ode ironica q aqui sahio sobre a patente destas mouchettes, cujo inventor Mr. Pinchbeck, homem ignorantissimo, mas engenhoso, & de hũa original stupidity, disem ser admitido frequentemente á prezença do Rey &a. q ama os niknaks & gimbkrabs deste mecanico. he cheia de drolery & satyr.

bom he q V.M. se divirta com Livros. as nossas Brazileiras quando estão enfadadas com os filhos, chamão a pretinha. "o paulinha toma

1 Annotated "Respa a 24 Marco 1778. no meu livro pag 227.6" [Answered 24 March 1778. In my book page 227.6]. The "book" referred to is Sanches' diary. Sanches' letter not found.

78.03.13a daqui o menino & vai-lhe brincar com a piquinha p^{a}. o divertir, q está m^{to}. impertinente". eu q^{do}. já não arr-tar, hei de procurar outro tanto ao menos o pôr-lhe a mão por sima não ha de deixar de me divertir. faça V.M. outro tanto, & não se lhe dará q lhe tenhão furtado alguns dos seos Livros.

me dá gosto q a nossa terra esteja tão bem defendida pelo nosso S^{to} Antonio de Lisboa, q como bom Soldado montará a cavalo p^{a}. guiar os nossos exercitos de frades e freiras ao campo da batalha. estou em duvida se elle quererá admitir os clerigos no seo exercito: e segundo noticia o numero não he piqueno. Hum Inglez q encontrei ha poucos dias, e q tinha chegado do Porto, me disse gozára da scena de ver ordenar dois mil clerigos de hũa assentada.

O Exc^{mo} Sr Pinto rio muito com a historia da velha Marqueza e Seo bidet: passa bem & cada vez se faz mais amavel e estimado de q^{tos}. o conhecem. não temos sujeito semilhante, tão instruido, tão capaz, & tão agradavel, caracter suave, bom juizo & longa vista.

aqui me disse o D^{or}. Seqr^{a}. (q se recomenda a V.M.) q ja la havia grande vigor na Sta. Inquizição: e q se tinhão prezo por ella varios sujeitos, mesmo officiaes militares, por lerem o Montesquieu, as obras de Russeau, Voltaire &^{a}. &^{a}. logo me lembrei do nosso Sonsinho, como lá suará com as suas palpitaçoens do Coração, sem embargo das S^{ras}. Duquezas & Fidalgas, minhas S^{ras}. &^{a}.

aqui se diz & he bem certo, q o gen^{al}. Effingham fizera hum grande dezacato dentro do nosso Tejo, dourado nas suas areias .....o cazo foi q quiz tomar por força os marinheiros de outro navio Inglez mercante, p^{a}. a sua Náo de guerra: estes fugindo p^{a}. terra, elle os fez seguir com tiros, & matou algum ou alguns portuguezes na praya p^{a}. onde elles fugirão &^{a}. &^{a}. Estes hereges Inglezes não tem medo do nosso S^{to}. general Ant^{o} de Lisboa: mas lá virá o tempo em q elles serão castigados nos infernos pelas suas herezias & pecados: e isso nos dará certam^{te}. uma grande utilidade e divertim^{to}., vendo-os fazer carinhas, emq^{to}. os diabos os assarem no espeto.

Vale &^{a}. Saud^{es}. á Famil^{a}. adopt^{a}

Lembrança das nossas Contas

os Artigos de Livros &^{a}. q mandei a V.M. desde Avril de 76 até Julho de 77 montavão à ....268$^{\#}$.4^{s} = 11$^{£}$:3^{s}:6^{d}

20 Ag^{o}. Phil. Transact. p.1. & Rowley's advice	8. 6
[...][2] Nov^{bro}. withers on chronic deseases, & white on y^{e}. bile	4. 6
728 Med^{al}. Coment. n°. 16 & Bureau on erysipelas	2. 6

2 Illegible.

By a private hand thro' D^{r}. du Presle, holme's 78.03.13a

on the new medicine	2.
[?]6[3] Fev^{o}. Med^{al}. Coment. n°. 17 ----------------------------	7. 6
+ Richardson on Oriental Literature ------------------	3. 6
+ gentoo Laws --	6. 6
Monro's Letters and essays -----------------------	4. 0
+ Culen's Physiology este vai em Lugar do q V.M. tem imperfeiro, & q V.M. me mandará p^{a}. eu o retornar ao Livreiro	
+ Priestley on Matter & Spirit 2 vol. 8°. ____________	8. 0
+ Wilson on dysentery ________________________	2.
uma tizoira, & um instrumento particular p^{a}. cortar as unhas dos pés ---	6. 6
hum espivitador de nova invenção, q corta admiravelm^{te}. e q não deixa cahir o murrão q tem dentro ________	8. 6
+ Hunter, a seg^{da}. parte do seo tratado sobre os dentes __	5. 0
despezas respectivas até bordo do Navio ______	4. 6
Soma total incluindo a prim^{a}. parcella assima (isto hé as £11.3.6)	
	14. 11. 0
q em dinh^{o}. Francez faz	349$^{\#}$.4^{s}. tournois.
Por contra recebi no livro sobre o Salpetre	4$^{\#}$.
777. em 22 de Ag^{o}. por via de M^{r}. du Presle ________	120
778. em 6 de Fev^{o}. por via do mesmo ___________	402
	566$^{\#}$

a Monsieur
Monsr. Sanchez, Doc^{r}. en Medecine
au Cimiterre de S^{t}. Jean, Rue de la Verrerie
au Marais
à Paris

3 Partly illegible.

78.03.13a London 13 March 1778

My Dear Friend

Answering your letter of the 23rd of last month,[4] that gave me much pleasure, I must repeat, though not a relative of Sancho Panza, that = says the trivet to the frying-pan: be careful not to soil me =.[5] Look at you who speak about not belching...![6] The Duke of Lafoens remains at the level of his old esteem:[7] but the nobleman from the island of Madeira[8] remained as if he were not one. He delivered the papers that you and Dr. Le Begue sent me through him, to our Envoy's house;[9] & it wasn't possible to find out where he lived, nor where he is. I heard that his first wife was French & had died of depression, feeling exiled in that Island: I hope that the same doesn't happen to this Mrs. Preston. I hate those who don't feel lust except for foreign [...][10]. A brother of his lives here, a friar – I presume Franciscan – who retired from the friarship. However I had no impression of him that made me look for him, or visit him. Nevertheless he did me the honour of looking for me.

Here I send the list of the books that I shipped recently to Mr. Le Begue, & that I didn't give you notice of, knowing that they wouldn't arrive there for the next 15 or 20 days. The book for Mr. Martens[11] was addressed to Vienna via Brussels, and cost 1£.14s.0 with box & shipment.

Mr. Le Begue will give you an excellent pair of mouchettes.[12] They cut paper like a good pair of scissors & have a scraper inside, so arranged that on raising it after cutting the match and lowering it again, the match inside that was cut passes to the interior of the box, and can't be dropped outside even if the snuffer is opened. With it you'll find scissors & clippers to cut one's nails; and also an ironic ode[13] that came to light here about the patent on these mouchettes, whose inventor Mr. Pinchbeck,[14] a very ignorant but ingenious man & of an original stupidity, is said to be frequently received by the King &c. who loves the niknaks & gimbkrabs[15] of this mechanic. It's full of drollery & satire.

4 Letter not found.

5 Aphorism expressed by Sancho Panza to Don Quixote in Cervantes' famous story, that had become a very common saying.

6 See Magellan to Sanches, 7 July 1775, note 19.

7 Magellan had a low opinion of the Duke; see Magellan to Sanches, 10 February 1778.

8 Not identified.

9 The Portuguese Minister in London was Luiz Pinto de Sousa Coutinho (1735-1804).

10 Cuckolds?

11 See Magellan to Sanches, 10 February 1778.

12 Snuffers.

13 Malcolm MacGregor (i.e. William Mason), *Ode to Mr. Pinchbeck, upon his Newly Invented Candle Snuffers*, a fifth edition of which appeared in 1777.

14 Christopher Pinchbeck (1710?-1783) possessed great mechanical ingenuity and took out several patents (including one in 1776 for the snuffers, which continued to be made until the twentieth century when snuffers began to go out of use).

15 Nick-nacks & gimcracks.

It's good that you're amused by books. When our Brazilian women are tired of their children, they call the black girl: "Look Paulinha take the child away & play with his little penis to amuse him, since he's very impertinent". When I cease belching, I'l look for the same thing – at least, having a hand put on it won't fail to please me. Do the same, & you won't care that some of your books have been stolen. 78.03.13a

I'm pleased that our country is so well protected by our St. Anthony of Lisbon, who as a good soldier will ride his horse to lead our armies of monks and nuns to the battlefield.[16] I doubt if he will admit the clergymen into his army: and according to the news the number is not small. An Englishman I met a few days ago, who had arrived from Oporto, told me how he enjoyed the sight of two thousand clergymen being ordained at one go.[17]

His Excy. Mr. Pinto laughed very much at the story of the old Marquise and her bidet:[18] He is well & more and more kind and esteemed by those who know him. We have no similar person, so educated, so able & of so pleasant a character, calm, with good judgement & long sight.

Here I was told by Dr. Sequeira (who sends his greetings) that there was again great vigour in the Holy Inquisition: and that several persons had been arrested, even military officers, because they read Montesquieu, the works of Rousseau, Voltaire &c. &c. I immediately remembered our Sly,[19] how he will sweat with the palpitations of his heart, notwithstanding the Lady Duchesses & noblewomen, My Ladies &c.

Here they say & it is sure, that General Effingham[20] has made a great insult in our Tagus, golden in its sands .... The situation was that he wished to take the

---

16 Saint Anthony was symbolically appointed a Captain in the Portuguese army in the reign of King Pedro II and was later promoted to Lieutenant Colonel under Queen Maria I.

17 See Magellan to Sanches, 10 February 1778.

18 The story told by Sanches in a previous letter was most likely a malicious anti-Jesuit joke that was circulating at the time. It referred to an anecdote described by Charles-Germain de Saint-Aubin (1721-1786) and his brothers in their *Livre de caricatures tant bonnes que mauvaises*, also known as *Le livre des culs*, based on caricatures collected between 1740 and 1770. The work included an irreverent drawing of the Marquise de Pompadour, "toute nue et faisant bidet, et entourée de robes noires et de bonnets carrés prosternés à ses pieds" with the title: "Ne blâmons personne, ils ont sans doute de bonnes raisons. Les jésuites aux pieds de Mme Pompadour".

19 Cf. Magellan to Sanches, 15 August 1777 ("Dr. Sly"). Magellan may be referring to Sanches' friend Dr J. Alvares, whom he did not like; see Magellan to Sanches, 1 September 1778 ("Sly Alvares") and Magellan to Sanches, 20 April 1762.

20 Magellan has got the name wrong. Captain George Keith-Elphinstone (1746-1823) caused a diplomatic incident between the Portuguese and English Courts when he took over an English merchant ship, anchored in the Tagus estuary, that he thought was trafficking arms to the American colonies, then went ashore and entered into the city of Lisbon chasing its crew, killing some of them as well as some local inhabitants who happened to be passing by. The incident was patched over when the British Government formally apologized to the Portuguese authorities, and the ship and the Captain were released.

78.03.13a sailors of another English merchant vessel to his warship by force: when they ran to the shore, he fired at them, & killed some Portuguese on the beach where they had run to &^{c}. &^{c}. These heretic English have no fear of our General S^{t}. Anthony of Lisbon: but the time will come when they will be punished in hell for their heresies & sins: and that will bring us great profit and fun, seeing their grimaces, while the devils roast them on the spit.

Vale[21] &^{c}. Regards to the adoptive Family

Reminder of our Accounts

the articles of books &^{c}. that I sent you from April 76
to July 77 amounted to .... $268^{\#}.4^{s}$ = $11^{£}:3^{s}:6^{d}$

20 Aug. Phil. Transct. p.1. & Rowley's advice[22] 8 . 6
[...] Nov^{ber}. withers on chronic deseases[23] & white on y^{e}. bile[24] 4 . 6
778 Med^{al}. Commen. no. 16 & Bureau on erysipelas[25] 2 . 6
By a private hand thro' D^{r}. du Presle, holme's on the new medicine[26] 2
[...]6 Feby. Med^{al}. Comment. no. 17 ______________________ 7 . 6
+ Richardson on Oriental Literature[27] ________________ 3 . 6
+ gentoo Laws[28] ________________________________ 6 . 6
Monro's Letters and essays[29] ____________________ 4 . 0
+ Culen's Physiology[30] this goes to replace the one you have
in an imperfect state & that you will send me to return to the bookseller
+ Priestley on Matter & Spirit 2 vol. 8°[31] __________________ 8 . 0
+ Wilson on dysentery[32] ____________________________ 2 .

21 Farewell.

22 William Rowley, *Medical Advice, for the Use of the Army and Navy, in the Present American Expedition...* (London, 1776).

23 T. Withers, *Observations on Chronic Weakness* (York, 1777).

24 William White, *An Essay on the Diseases of the Bile, more particularly Its Calculous Concretions, called Gall-stones* (York, 1771; 2nd edition, London, 1777).

25 James Bureau, *An Essay on the Erysipelas or that Disorder commonly called St. Anthony's Fire* (London, 1777).

26 Possibly Nathaniel Hulme, *Oratio de re medica cognoscenda et promovenda ...Cui accessit via tuta et jucunda calculum solvendi in vesica urinaria inhærentem; ab historia calculosi hominis confirmata* (London, 1777).

27 John Richardson, *A Dissertation on the Languages, Literature, and Manners of Eastern Nations* (Oxford, 1777).

28 See Magellan to Sanches, 10 February 1778, note 4.

29 D. Munro, ed., *Letters and Essays on the Small-pox and Inoculation...* (London, 1778).

30 William Cullen, *Institutions of Medicine. Part I: Physiology* (2nd edition, Edinburgh, 1777).

31 See Magellan to Sanches, 13 January 1778, note 12.

32 Andrew Wilson, *An Essay on the Autumnal Dysentery* (2nd edition, London, 1777).

one scissors, & one special instrument to cut toenails _______	6 . 6	78.03.13a
one snuffer of new invention, that cuts admirably and		
doesn't let the match inside fall out ______________________	8 . 6	
+ Hunter, the sec^{d}. part of his treaty about teeth[33] ___________	5 . 0	
respective expenses until on board the ship _________	4 . 6	
Total sum including the first parcel above (that is the £11. 3. 6)	£14. 11. 0[34]	
which in French money makes	349.$^{\#}$4.^{s}. tournois.	
On the other hand I received in the book about Salpetre[35]	4.$^{\#}$	
	_________	
777. on <u>22nd</u> Augt. through M^{r}. du Presle _____________	120	
778. on the 6th Feby. through the same ______________	402	
	___________	
	566$^{\#}$[36]	

33 John Hunter, *A Practical Treatise on the Diseases of the Teeth, intènded as a Supplement to the Natural History of Those Parts* (London, 1778).

34 Magellan has made a mistake with the addition. The figure here should be £14. 17. 0.

35 [Académie Royale des Sciences, Paris], *Recueil de memoires et d'observations sur la formation & sur la fabrication du salpetre …* (Paris, 1776); see Magellan to Sanches, 23 February 1776.

36 Magellan seems to have made another mistake in his arithmetic here; the total of sums received should be 526$^{\#}$. At the end of the Reminder of the Accounts, there is the following annotation in Sanches' handwriting; it is difficult to reconcile his numbers with those given by Magellan:

"lhe escrevo a 22 Abril
300$^{\#}$. não devem entrar aqui
devo 334
recebeo 220
----
112 + devo"

[I'll write to him on 22 April
300$^{\#}$. should not enter here
I owe 334
he received 220
----
112 + I owe]

The 300$^{\#}$ to which Sanches here refers corresponds to the allowance that he paid regularly to Magellan.

78.03.21 *To James Watt*[1]

Birmingham Central Library, Boulton and Watt collection, Box 22/B3/6

London 21 March 1778
at n°. 12 Nevils Court Fetter Lane

Sir

I have wrote a Letter to M^r. Boulton on the 3^d. instant month,[2] Inquiring him about 4 of your fire-engines, the director of the Prussian mines, who is a particular friend of mine, wants to have there: and indeed I cannot but be highly surprised, for not getting till now any answer at all about the same. I wrote also to you few days after, mentioning the same request:[3] and am equally disapointed in the same maner. Indeed I must suspect either that you both are far distant from your respective Houses, or that some unhappy extraordinary acident has happen'd to you both. in both suppositions I am very concerned: and wish at all events to get some kind of information about your well fare; and altho' I have inquired at M^r. Mathews's compting house two days ago, I was not able to get there any other information, but that M^r. Boulton will not be able to come so soon to Town, as he promised me he should do, the last time I have seen him. Beg you therefore that if this Letter reaches to your hands, and you are alive, to give me an answer about the above affair, in which I am so concerned, as I promised my friend the Baron Heinitz, now Secretary of Estate of y^e. King of Prussia, and general Intend^t. of y^e. Mines in His Majesty's dominions,[4] that I should very soon forward to him the answer I should obtain from you Both.

I have also Received a Letter from Paris, by which it appears that every thing is already in the most favourable situation, for obtaining the exclusif

1 Letter not found; MS is a clerk's transcription. Annotated "Monsr. Magellan 21 March 1778".

2 See Magellan to M. Boulton and J. Watt, 3 March 1778.

3 See Magellan to J. Watt, 12 March 1778.

4 Karl Friedrich Anton, Baron von Heinitz (1724-1802) graduated in mining engineering. After travelling in France and England in 1776-7, he was appointed a member of the Cabinet of Frederick the Great of Prussia and head of the Department of Mining and Metallurgy. He was elected to membership of learned societies both in and outside Germany, even of the American Philosophical Society in Philadelphia.

Privilege, and settling all other matters about Moeres[5] to your most ample and andvantageous desires, and expectations. 78.03.21

There is also arrived in Town one Mr. Jary, of whom I spoke to you in my Last letter, who comes from ye. neighbourood of Nantes, and wishes to alter an old fire engine in a Coal mine somewhere not very far from Nantes, by putting, instead of it, one of your new ones. He told me this morning that he will set out on Tuesday next to Birmingham; and desired to have a Letter of Recom̄endation for Mr. Boulton, which I have promised to him,[6] as he has been Recom̄ended to me as an honest man by Mr. Montaudouin, one of the most capital Merchs. at Nantes,[7] who was in England the Last Summer.

Lastly I have some thing more particular to speak about, with Mr. Boulton, which I dont chuse to trust by writing; altho' it may not happen to be of Consequence to you Both. however as a plain dealer, and possessing a good honest heart, I think right to take that step towards you both. beg there fore a Line of answer by the return of this post, and some kind of certainty about the time that I may see here Mr. Boulton &a.

I am Sir

with ye. utmost Regard and Sincere heart
your most obedt. Servt.
Magellan

To
Mr. Watt, or in his absense to
Mr. Boulton
at their new fire engines Manufactury
near Soho
at Birmingham

5 See Magellan to M. Boulton, 28 February 1778.

6 See Magellan to M. Boulton, 23 March 1778.

7 Jean-Gabriel Montaudouin de La Touche (1722-1780), merchant at Nantes and author of reforming works about agriculture and commerce.

78.03.23 *To the President and Council, Royal Society of London*

Royal Society of London, Miscellaneous Manuscripts, 3. 17

To the Presid^{t}. and Council of the Royal Society of London

Gentlemen

I had the honour of addressing to You a Letter on the 22 Jany. last,[1] concerning the contribution money of an elected member of this Society, in which I now apprehend to have used some free expressions, which, notwithstanding the justice of y^{e}. case I pleaded for, may perhaps be thought unseasonable, and even improper, altho' they were intirely hypothetical, and by no means applicable to the proceedings of this Respectable Council; nor any ways directed to It, but simply to any other, that, in the course of human affairs, may hapen to exist of that description, amongst the innumerous events, which Lay within the wide compass of possibility.

I wish therefore, Gentlemen, that, if this Council is of such an opinion (which I hardly can suspect from the very context of my letter) I wish, I say, that I had not then wrote them: and I add further, that if any kind of offence has been taken by, or if they have been misconstructed even to the least degree of disrespect to You all, I thereby heartly declare, that such never was my intention, and sincerily beg pardon for the same. I am with y^{e}. greatest due Regard and Respect

Gentlemen
your most Obed^{t}. h^{ble}. Servt.
John Hyacinth de Magellan

London 23 March 1778

To the Presid^{t}. and Council
of the R^{l}. Society of London
in Crane Court

---

1 See Magellan to the President and Council, Royal Society of London, 22 January 1778.

To Matthew Boulton[1] 78.03.23a

Birmingham Central Library, Boulton and Watt collection, MBP 244, 56

London 23 March 1778

Dear Sir

The bearer is Monsr. Jarry,[2] a gentleman who came lately from France, and was strongly recommended to me by a person of a very known and respectable caracter. His main object is to agree with you in order to exchange one of the old fire engines into that of your new construction, to work in a mine of which he is the Director. He desires to be recommended to you for this purpose, as far as it may be practicable: and I wish you may do and agree with him as far as it may be usefull, and advantageous to you both; and will always consider all the kindness you'll be so good as to favour him with, as really done to

Dear Sir
Your most obd^t^., obliged servant
Magellan

P.S. I am waiting this long while with much concern for your answer about the four fire engines of yours, now desired in Prussia.[3]

To
M^r^. Boulton
at his Manufactory
in Soho
near Birmingham

1 Annotated: "M Magellans recommend^g^. of Mr Jaring 1778".
2 Joseph Jary (1739-?), director of a colliery near Nantes.
3 See Magellan to M. Boulton and J. Watt, 3 March 1778.

78.03.24 *From James Watt*[1]

Birmingham Central Library, Boulton and Watt Letter books, vol. 3, 7-9

Mar 24th [1778][2]

Mr Magellan

Last night I wrote to you mentioning this reason why we had not sooner acknowledged the favour of yours of the 3^{d} & 12th. ins^{t}.[3] Calculating at the rate of ten strokes p^{r} minute each 7½ feet long the following Engines will answer your purpose

1. Pump 8 inches diam^{r}. 155 feet high, Cyl^{r}. 21 inches diam^{r}. 62 pounds coals
2. Pump 11 inches diam^{r}. 420 feet high, Cyl^{r}. 47 inches diam^{r}. 307 d^{o}.
3. Pump 11 inches diam^{r}. 348 feet high, Cyl^{r}. 43 inches diam^{r}. 250 d^{o}.
4. Pump 7½ inches diam^{r}. 93 feet high Cyl^{r}. 16 inches diam^{r}. with 4 f^{t} stroke. 24 d^{o}

Any of these are capable of giving occasionally at the rate of 12 or more strokes p^{r} minute – We are ignorant whether you mean to use coals or wood, or if coals of what quality they are, have therefore computed it at a medium of this country coals –

As to the article of price we can only say that exclusive of our premium, our Engine will not be dearer than common ones which do the same quantity of work, which will be very plain to you if you consider that the same pumps, House, Working beam and all their furniture will do in one case as well as in the other, that we have no pumps carried up to the house top nor cistern there, nor so expensive an apparatus for the injection – that our cylinder is of a smaller diam^{r}. & the Boiler very considerably less – on the other hand the expense of the condenser and steam case will counterbalance the savings, as to profits on the Workmanship we never charge any.

We will bind ourselves that our engine do considerably more work with [...][4] work than any other

As we have already mentioned to you upon another occasion it would not suit us at all to have any thing to do with the Engines without an

1 Letter not found; text taken from Watt's letter-copy book.

2 *editorial addition.* The year has been established as 1778 from the position of the letter-copy in the sequence in the book in which it is found, and also from the reference in the first sentence to Magellan's two letters dated 3 March and 12 March of that year.

3 See Magellan to M. Boulton and J. Watt, 3 March 1778, and Magellan to J. Watt, 12 March 1778.

4 Word obscured by binding.

exclusive privelege first granted to us throughout his Prussian Majesty's dominions for the term of 14 years at least, this article we wish to be fully understood i. e. y[t] we can take no step in the affair untill that privilege is granted. 78.03.24

We cannot begin to make the Engines before July next and we suppose that we cannot get them ready for setting up before the next spring. That you may better understand what we have said about the prime cost of Engines we beg leave to repeat what we have formerly told you upon that subject – We give drawings and directions to the Founders, Smiths, Carpenters, Masons &c. and superintend the execution of the whole work – our Employers receive and pay the bills of the several workmen and the whole expense of execution. – As to our own profits they must depend entirely upon the advantage the Engine shall be of to our Employers and cannot be determined until we be more fully informed of many particulars especially concerning the patent – We must therefore beg you to procure us answers to the following queries.

1. In what part of Prussia are these Engines to be erected and how near to any large town

2. Is it a hilly or plain country, and can fresh water be had upon the surface to supply the boiler & condenser

3. What are the building materials of the country and how deep under ground lies the surface of the rock or other good foundation.

4. Is it Mines of Coal or Metals.

5. Which is the cheapest fuel there wood or pit coal and what is their prices p[r] quintal of English weight at the places.

6. Is the water in any or all of the mines to be drawn quite up to the surface or is to run off at levels or Galleries; and if so, at what depths below the surface are these galleries.

7. Whether are wooden or iron pump trees to be used, & if iron where are they to be made

8. Whether they have Engineers upon the spot who are used to common fire Engines and can erect the buildings according to drawings.

9. Are the Mines the immediate property of his Majesty or farmed out.

10. If no water can be had for condensation than that from the mine: queritur[5] whether it be sufficiently free from vitriolic acid not to corrode the iron pumps of the condenser, as in that case they must be made of Brass

11. Are the shafts or pits already sunk on which the Engines are to be erected, are they perpendicular or not & what are their dimensions & form.

---

5 Quaeritur – the question is raised.

78.03.24 12. Ground plans of the works about the mouths of the shafts will be wanted and those which cannot be removed must be marked. If the ground is not flat a section of it should be sent passing through the pit and the proposed situation of the Engine house.

As to the freight to Hamburgh you will be able to learn it better at London – The common freight from Bersham foundry where our goods are cast to London is about 15/– per ton – some few things must be made in this neighbourhood which may be shipped from Hull. You know M^{r}. Wilkinson's[6] prices for cylinders and condensers is 30/– p^{r} Cw^{t}.[7] and for plain pipes unbored 18/– p^{r} Cw^{t}.

---

6 John Wilkinson (1728-1808), the "father of the south Staffordshire iron trade".

7 Hundredweight.

78.03.27 *To Johann III Bernoulli*

Öffentliche Bibliothek der Universität Basel, MS L I a 709, f. 71

Londres 27 Mars 1778

Monsieur & Cher ami

Aussitôt que Je reçus Votre Lettre derniere,[1] je cherchai le D^{r}. Sheperd, auquel Je crois que Vous devez principalem^{t}. la gratification qui vous fut accordée par Le Bureau de la Longitude:[2] & par son avis je me presentai dans la premiere seance, qui hereusem^{t}. fut tenue dans La meme semaine; car elles ne sont que rares. Mylord Sandwich Secretaire d'Etat au Department de la marine ou Amirauté,[3] ordonna, après de m'avoir demande si j'avois vos ordres pour recevoir cet argent, au Secretaire d'expedier au plutôt & avec le moindre delay possible le billet pour ce payement. Le Secretaire prit mon adresse & promit de le faire; mais Je n'ai pas encore Reçu aucun avis de lui. Je parlai de nouveau au D^{r}. Sheperd, qui me conseilla d'ecrire un mot au meme Secretaire: Je l'ai deja fait le 21 du courant: mais point d'avis encore. Ce Secretaire est reconnû comm'un homme très Lent; & vous meme savez, que sans l'avis que Je vous ai fait, vous n'en sauriez encore rien du tout,

---

1 Letter not found.

2 See Magellan to Bernoulli, 30 January 1778.

3 John Montagu, 4th Earl of Sandwich, First Lord of the Admiralty, 1771-1782.

tandis qu'on m'a dit, que cette notice est deja inserée dans un des volumes derniers du Journal des savans. 78.03.27

Il faut donc avoir patience, & attendre que le paresseur Ibbetson, c'est le nom du Secretaire en question, veuille bien passer & faire signer l'ordre pour que Votre argent me Soit payé. aussitôt que cet ordre sera entre mes mains soiez assuré que je tacherai de le realiser en argent comptant: car Je pense que ces ordres du Bureau de la Longitude sont comme ceux du Bureau de la marine & de l'ordonance, qui ne sont payables tout de suite mais selon leurs dates: & je sçais qu'il faut disposer de ces derniers par l'intermede d'un broker, pour les reduire sans delay en argent comptant. J'ai deja demande sur cet article le meme D^r^. Sheperd; mais il me dit qu'il n'en savoit rien. Aussitôt que J'aurai l'ordre passé & signé entre mes mains, je vous en ferai part par La poste.

Je crois vous avoir dit dans quelqu'une de mes Lettres precedentes qu'il n'y avoit pas encore d'argent entre les mains de M^r^. Elmsley, provenant de la vente de vos ouvrages, au delà de quelques shillings, provenans de la vente qu'on en aura faite depuis de Juin dernier, qui fut le tems dans lequel J'ai saldé mon compte avec M^r^. Elmsley jusqu'à ce tems là.[4]

M^r^. Spener[5] ne veut pas donner compte des livres que Je lui ai fait remettre ici lors qu'il est venû à Londres pour M^r^. de la Grange[6] disant de ne pas les avoir reçûs. J'ai passé tout exprès chez le D^r^. Forster, qui se souvient bien, aussi bien que son fils,[7] de la parcelle que j'envoyai pour M^r^. de la Grange chez eux, consigné au soin de M^r^. Spener, qui se trouvoit alors chez eux à Londres. Ils sont prets à me donner ce temoignage pour confondre cet homme, qui ne sçait pas se souvenir des commissions qu'on lui a donés, ni de ce qui concerne son caractere. Peut etre en a-t-il fait autant avec les differens exemplaires que Je lui ai delivrés ici de la part de Comte Stanhope pour differens savans dans L'Allemagne.[8] Dieu sçait s'il les a vendu &^a^. &^a^.

aiez la bonté d'envoyer cet autre page à M^r^. de Castillon.[9]

Vale & ama
tuum
Magellan[10]

---

4 See Magellan to Bernoulli, 30 January 1778.

5 Bookseller in Berlin.

6 Presumably the mathematician Joseph Louis Lagrange (1736-1813), who worked at the Berlin Academy of Sciences from 1766 to 1787.

7 Presumably Johann Reinhold Forster and his son Georg.

8 See Magellan to Bernoulli, 30 January 1778.

9 Magellan's letter to Castillon not found.

10 Farewell & love | your | Magellan.

78.04.17 *To Matthew Boulton*[1]

Birmingham Central Library, Boulton and Watt collection, Box 22/B3/7

London 17 April -78
n°. 12 Nevils Court Fetter Lane

Dear Sir

This being the time, rather beyond what you told in your Last, that you intended to come to London, and not seeing or hearing of you, induces me to write this, in order to acquaint you at Large, that every thing relating to the affair of Moeres and it's Consequences, appears every time I Receive any Letter, better and better calculated for producing the advantageous ends we aim at, in regard to you and to the owner.[2] I beg you therefore to give me any idea, if you can, of the time you'll be able to take a trip to the Continent.[3]

I have already got an answer about the passeport, in case of any warlike declaration: and it will be in waiting at Ostende for you, as this must be the port, where you must go first, in Case a war declares before that time.[4]

Pray give me a line, by the post, unless you be at the eve of coming to London.

and believe me sincerily
dear Sir
Your most obd^{t}. h^{ble}. Ser^{t}.
Magellan

P.S.
I have not yet Received any Letter from Berlin about the 4 fire engines.[5]
Pray my Comp^{s}. to M^{r}. Watt, and to M^{r}. Jary[6] in case you see him &^{a}.

N.P. I have Received two Large Letters from France about the affair of Moeres, since I have seen you, that I am certain will give a full Satisfaction. as soon as you come to London, dont fail of calling on me, or sending a Line, and I'll call upon you with them &^{a}

To
M^{r}. Boulton
at his new fire engine Manufactory
near Soho
Birmingham

1 Letter not found; MS is a clerk's transcription. Annotated "Magellan April 17th 1778".
2 i.e. the Comte d'Hérouville.
3 See Magellan to M. Boulton, 28 February 1778.
4 France was about to go to war with Britain in support of the American revolutionaries.
5 See Magellan to M. Boulton and J. Watt, 3 March 1778.
6 See Magellan to M. Boulton, 23 March 1778.

To the Comte d'Hérouville 78.04.17a

Archives Nationales, Paris, MS Marine, G110, ff. 151-2

Londres 17 Avril -78
n°. 12 Nevils Court Fetter Lane

Monsieur

Je me flatois pouvoir Vous annoncer à present le jour que Mr. Boulton devoit partir pour Flandres: mais il n'est pas encore arrivé, comm'il me l'avoit promis, à Londres: & Je viens de lui ecrire de nouveau là dessus.[1] Vous pouvez Monsieur, compter sur mon exactitude & sur mon ardeur pour conclure cet affaire à votre satisfaction, que je ne manquerai pas de le pousser autant que j'en serai capable. En tout cas si la guerre vient à etre declarée, il sera necessaire etre muni d'un passeport, pour entrer en France: et c'est pour cet objet particulieremt. que Je suis impatient pour voir & parler avec Mr. Boulton. Car ce sera assez de le trouver à Ostende, pour passer à Dunkerque & de là à Moeres: mais je vous ecrirai bientot sur cet affaire dans une des postes suivantes.

Je suis infiniment obligé à l'attention, Monsieur que Vous avez eu La bonté de marquer pour la famille de mon ami Mr. Arbuthnot. C'est un vrai galant-homme de beaucoup de merit, qui a soufert très fort par les suites de cette malheureuse guerre de l'Amerique. Il avoit une manufacture des toiles peintes les plus parfaites peut etre qu'on a jamais vû dans l'Europe: cette entreprise l'obligeoit à donner de credit pour des sommes considerables: & le grand nombre de banqueroutes entre nos marchans à Londres, lui a causé des domages très considerables: de façon qu'il sera obligé de se tour[ner][2] entierement pour l'agriculture: oû il est fort eminent, & generalemt. reconnû pour tel. Ce seroit une des plus belles acquisitions pour un Etat quelconque, si l'on pouvoit acquerir un tel sujet à s'y fixer, tel comme Mr. Arbuthnot.

J'envoye ci dedans le Compte courant entre moi & Mr. Le Comte Chastenet,[3] qui me dit dans sa derniere Lettre de m'adresser à vous, Monsieur, qui aurez la bonté de vous y preter, en attendant qu'il puisse s'arranger avec le Chevr. de Borda & les autres amis, pour lesquels une bonne partie de ces instrumens sont destinés. J'ecris donc cette meme post

1 See Magellan to M. Boulton, 17 March 1778.

2 Word partly obscured by binding.

3 Account not found. Antoine Hyacinthe Anne, Comte de Chastenet de Puységur (1752-1807) was the Comte d'Hérouville's son-in-law.

78.04.17a à Mr. le Dr. Le Begue du Presle[4] de passer chez vous si par hazard il a deja reçu quelqu'autre argent de la part de ces Messieurs, au delà des 15 guinees & 4 shillings qui lui furent envoyées par Mr. Bassel de Chateaubourg, & dont j'ai fait deja le rabat dans ce compte; alors cet argent qu'il aura reçu de plus devra etre deduit des 3946lb. tournoises, qui est le reste de mon credit, ou ballance à mon faveur.

La derniere caisse contenant les 3 derniers instrumens circulaires, fut arretée ici dans le batiment Anglois, oû elle alloit partir pour Boulogne: il m'a falu la oter, & l'envoyer à Douvre: mais mon correspondant m'ecrivit, qu'il n'etoit pas permis de l'envoyer par le pacqueboat à Calais: ainsi je fus obligé de l'envoyer à Ostende: & J'ai ecrit à Mr. Romberg[5] de la faire remettre au plutôt à Paris à Mr. le Dr. le Begue du Presle. Dans le cas que Mr. le Comte ait donné des ordres pour la destination de ces 3 derniers instrumens circulaires, Mr. le Dr. Le Begue s'y conformera tout de suite.

J'ai vû hier matin Mylord Shelburne[6] & ai dejeuné avec Lui: il se porte à merveille; & sachant votre amitié reciproque, J'ai cru que cette notice vous seroit agreable.

J'ai l'honneur d'etre

Monsieur le Comte
Votre très hble. & obeisst. servr.
Magellan

A Monsieur
Monsr. Le Comte d'Herouville
Lieutenant general des Armées du Roy
rue du Fauxbourg St. Honoré no. 6.
à Paris

4 Letter not found.
5 Reading uncertain; name partly obscured by binding.
6 The English statesman William Petty, Earl of Shelburne (1737-1805).

From Matthew Boulton 78.04.22

Archives Nationales, Paris, MS Marine, G110, ff. 190-1

Dear S^r^.

I recv^d^ yours of ye 17th[1] & have now the pleasure to tell you that Mr Watt & my self both propose to be in London next week & we will take the first day after our arival to pay our respects to you in Nevil Court & will then come to some conclusion about the business of the Moeres

I find my self & Mr Watt so overwhelmd with business that it would be very inconvenient for either of us to go to Paris at this Moment & therefore I wish [...][2] could contrive to do the business without or
Q? if Monsieur Le Comte d'Herouville could not meet us at Dunkirk. or
Q? if the priviledge could not be sent over by the packett to London & Mr Watt & my self could execute in your presence, & before a Notary Publick, any agreement or instrument which the Comte may think a proper security for our performance of our agreement. or
Q? if we were to write a Letter direct to the Comte & therein recapitulate all that hath already passed between us & after explaining our selves fully request that he would honor us with his remarks, objections & proposals. By this means we shall be more likely to prevent any misunderstanding & the Count will then have all our proposals under our hand. But these things we shall take your advice in & not proceed contrary to it.

If you thought the Count would come or send any proper person impowerd to act for him & meet Mr Watt or me at Dunkirk you may then write to him p^r^ next post because Mr Watt otherwise cannot stay in London longer than 7 or 8 days & you might then receive an answer before he goes to Cornwall but that point I submitt to your better judgment & remain with great regard

Your much obliged & most obed^t^ humb^e^ Servt
Matt^w^. Boulton

1 See Magellan to M. Boulton, 17 April 1778.
2 Paper damaged, word missing.

78.04.22 Soho 22d April 1778

you may fully depend on seeing us next week

We are come to terms with Monsieur Jary[3] which are favourable for him

To
Mr Magellan
No 12 Nevils Court
Fetter Lane
London

---

3 See Magellan to M. Boulton, 23 March 1778.

78.04.24 *To the Comte d'Hérouville*

Archives Nationales, Paris, MS Marine, G110, ff. 153-4

Londres 24 Avril -78

Monsieur

La Lettre ci-jointe de Mr. Boulton,[1] montrera que Je ne me repose point sur les affaires qui vous regardent. Je vous prie donc de reflechir sur son contenû: & de me participer votre resolution sur les differs. articles qui y sont contenûs.

La caisse contenant les 3 derniers instrumens circulaires pour Monsr. Le Comte Chastenet de Puy Segur, fut envoyée, comme Je vous ai marqué dans ma precedente, de Douvre pour Ostende, [à][2] fin de pouvoir etre renvoyée de là à Paris; parcequ'il n'etoit possible de la faire passer en droiture de Douvre à Calais, à cause de la defense qu'on y a d'envoyer aucune marchandise en France, par le paqueboat aux lettres, qui est le seul qui va & vient entre les ports de Douvre & Calais. J'ai ordonné de l'adresser d'Ostende pour Paris au Dr. Le Begue du Presle: & je suppose que comme

---

1 M. Boulton to Magellan, 22 April 1778?
2 Word obscured by binding.

cette caisse entrera par la Flandre en France on n'aura pas difficulté à son entrée de la part de la Douane. Si cependant il y arrivoit quelqu'une, Je me flate que vous donnerez les ordres & instructions necessaires au meme Dr. Le Begue du Presle, pour la surmonter. 78.04.24

J'attends, Mr. le Comte, avec impatience votre reponse au contenû dans cette lettre de Mr. Boulton & vous pouvez etre bien persuadé, que je n'oublierai rien de ma part, pour faire reussir cet affair, relatif à Moeres, ai[ant][3] bien conçu combien celà vous regarde: & je me flate que vous trouverez en toute occasion de vous rendre service que Je suis sincerement & de Coeur

Monsieur
Votre très hble. & très obligé serviteur
Magellan

P.S.
Après des distributions d'un si grand nombre de Proposals, comme J'ai fait distribuer à la rentree de la Bourse de Londres, entre les 2 & les 3 heures de plusieurs jours de poste (qui sont les mardis & les vendredis) dans lesquels il vient un nombre beaucoup plus grand de monde à la Bourse: Je n'ai pas eu aucune autre application que celle d'un tel Monsr. de Fonmore, ecrite dans un mauvais françois & disant qu'il comprenoit la raison de publier ces proposals, que selon lui etoit pour ne pas avoir quelque bon mechanicien &a. & ajoutant qu'il en connoissoit un fort habile qui voudroit entrer en convention pour y etre employé &a. J'ai crû qu'il ne valoit pas la peine de vous envoyer [d']ici[4] cette lettre. Si cependant vous le souhaitez, je vous l'envoyerai par la poste.

A Monsieur
Monsr. Le Comte d'Herouville
Lieutenant General des Armées du Roy
rue du Fauxbourg St. Honoré no. 6.
à Paris

3 Word partly obscured by binding.
4 Word partly obscured by binding.

78.04.28 *To Ribeiro Sanches*[1]

Österreichische Nationalbibliothek, Vienna
A.N. Ribeiro Sanches Korrespondenz, codex 12714, ff. 217-218v

Londres 28 Abril – 78

Am^{o}. e S^{r}. e m^{to}. do Coração

Hoje Recebi com m^{to}. gosto a de V.M. de 24 do Corr^{te}.: mas dame pena q V.M. não desabafe comigo! q he o q tem q o aflige? faltalhe o necess^{o}. p^{a}. a sua subsistencia: se assim he, digamo, & repartiremos ambos do q por ca ouver. Ja lhe disse q não quero, nem aturo, nem sofrerei jamais, q V.M. se encolha, & se negue o q necessita na situação doente, & dezarrei[ma]da em q se acha. Quanto ao mais mande tuto il mondo far-se buz-re.

Tenha V.M. a bondade de agradecer m^{to}. m^{to}. e muitissimo da minha p^{te}. a esse cavalheiro, homem de letras, penetrante, pontilhudo, delicado, escrupulozo &^{a}. &^{a}. &^{a}. pela pouca habilid^{e}. da sua impostura à meo Respeito. assim he q eu as quero & as estimo bem grossas & q toda agente as possa ver de longe sem trabalho nem canseira. eu não vi nem jamais falei com o D^{or}. Williams senão huma ou duas vezes. todos os q me conhecem, sabem m^{to}, bem q eu não vou nem tenho algũa connexão com o d^{o}. Clerigo da Igr^{a}. universal. fui là a prim^{a}. vez por curiozid^{e}. & a segunda por acompanhar o Emb^{or}. do Imperador, o Conde de Belgioiose m^{to}. meo amigo (& com q^{m}. á manhan hei de jantar, pois me escreveo hoje hũ bilhete de invitação) o S^{r}. Ageno, Rezid^{e}. do Estado de Genova, & o nosso amavel Cav^{ro}. Pinto Ministro Enviado de Portugal, q todos 3 tiverão a curiozid^{e}., como eu tinha tido, de hir ver, & ouvir este D^{r}. Williams na sua Capella universal. Pelo q torno a dizer, estou m^{to}. obrigado a esse S^{r}. Gensfoutre por não saber inventar contra mim huma impostura menos facil de pôr em Limpo. Se elle dissesse, q me tinha visto sobre a minha Nely, ou fanny ou outra das minhas conhecidas; então sim q poderia eu ter algum embaraço p^{a}. me rebuçar desta fraqueza, da qual eu dezejára bem q V.M. ainda pudesse ser culpado a estas horas. Diga-lhe V.M. q eu vivo aqui nesta terra de Liberd^{e}., porq sou homem & não escravo, como todos os gens-foutres como elle & seus semilhantes

1 Annotated "Resp^{a} a 7 Mayo" [Answered 7 May]. Letter not found.

A guerra não me dá embaraço nenhum. Se eu tivesse ido p^{a}. a America, 78.04.28
como eu queria haverá 10 ou 12 annos, mas q V.M. me aconselhou de não fazer tal; então sim q me havia de achar agora Capitão ao menos de algũa Companhia d'Americanos.

ainda não desespero de hir ralhar com V.M. neste verão ou no outono q vem, à Paris. eu não sou Inglez nem Francez. sou verd^{o}. Cosmopolita, & je m'en fous de tout.

Quanto ás nossas contas tudo está excellente. [V.M.][2] deve segundo o meo Livro 35$^{\#}$. mas seg^{o}. o seo 52$^{\#}$. Certam^{te}. não podia deixar de ser assim; pois a sua aritmetica nunca foi senão contra si mesmo.

não fale V.M. ao S^{r}. Lafoens no preço do Camoens, nem a Klingsted nos 50 shillings. isso são perdas q eu sofro com gosto p^{a}. melhor conhecer os hommens. o d^{o}. Klingsted não sabe declinar q por sui, sibi, sé. veja (ainda me lembro dos Nominativos!) a sua carta q V.M. me mandou, era p^{a}. dizer-me q o Presid^{e}. da Acad^{a}. de Petersburgo me havia de fazer seo Acad^{o}., coiza de q ainda não tenho noticia, nem creio até q o veja: e em segundo Lugar p^{a}. q lhe visse em q dia tinhão sahido brancos huns bilhetes da Lotaria; pois suspeitava, q o seo correspondente de Londres lhe tinha vendido bilhetes ja sahidos em branco &^{a}. esta desconfiança faz ver a casta do homem. eu lhe Respondi com civilid^{e}. & lhe fiz a sua Comissão, q custou hũ shilling de mais alem dos 50 atrazados. Faço-lhe hũa cruz. e está di[to.]

Esta carta foi escrita com hũa penna de aço de q uzo ja ha 3 ou 4 semanas. poupame o trabalho de a aparar: e ainda q he dura, não se me dá disso pelo costume. q^{do}. for à Paris hei de Levar algumas, & darei huma à V.M. se a quizer. Saud^{es}. á familia adoptiva.

Vale & ama tuum &^{a}. &^{a}.

a Monsieur
Monsr. Sanchez D^{or}. en Medicine
Cimiterre de S^{t}. Jean de la Verrerie
au Marais
à Paris

2 Stained paper. Conjectural reading.

78.04.28 London 28 April –78

My Dear Friend

I received today with much pleasure your letter of the 24th of this month:[3] but I'm sorry you don't open your heart to me! What troubles you? Are you in need regarding your subsistence? If so, tell me, & we shall both share what we find here. I've told you that I neither want, nor can stand, nor shall suffer any more that you should hold back & deny yourself what you need in the situation of illness & helplessness that you're in. As for the rest, send tuto il mondo farse buz-re.[4]

Have the kindness to much, much and very much thank on my part that gentleman, that penetrating, punctilious, delicate, scrupulous &c. &c. &c. man of letters, for the little intelligence of his calumny about me. This is the way I like them, big enough to be seen from far away without much work or fatigue. I haven't seen or spoken to Dr. Williams except once or twice. All who know me are aware that I don't go to or have any connection with the said clergyman of the universal church. I went there for the first time out of curiosity & the second time to accompany the Ambassador of the Emperor, Count Belgioioso,[5] a good friend (& with whom I'll dine tomorrow, since he sent me an invitation card today), Mr. Ageno,[6] the Resident of the State of Genoa, & our kind Chever. Pinto, Minister Envoy of Portugal, because all three were curious, as I had been, to see & hear this Dr. Williams at his universal chapel.[7] Therefore I repeat, I am very obliged to that Mr. Gensfoutre[8] for not being able to invent against me a calumny less easy to rebut. If he had said that he saw me over my Nelly or Fanny or any other of my acquaintances, then yes I might have some trouble in excusing myself for this weakness, of which I wish you could still be guilty now. Tell him that I live here in this country of freedom because I am a man & not a slave, like all the Gensfoutres like him and his fellow creatures.

The war doesn't give me any embarrassment. If I'd gone to America, as I wished to do 10 or 12 years ago, but you advised me not to; then yes I would be now captain at least of some company of Americans.

---

3 Letter not found.

4 Send everybody to hell.

5 Ludovico Count Belgioioso (1728-1801), Envoy and Minister Plenipotentiary of the Holy Roman Empire.

6 Francesco Maria Ageno (1727-1788), ambassador of the State of Genoa to London from 1760 to 1781 with interruptions. His house was a meeting point of London culture.

7 David Williams (1738-1816) opened a chapel in 1776 in Margaret Street, Cavendish Square, in London, that followed a non-Christian, deistic liturgy that Williams had devised. See also Magellan to Sanches, 13 January 1778.

8 Gensfoutre (Jean-foutre or Jeanfoutre) = jackass, incapable or incompetent. The person who invented the calumny has not been identified.

I still don't despair of going and reprimanding you this summer or next autumn in Paris. I'm not English or French. I'm truly cosmopolitan, & je m'en fous de tout. 78.04.28

As for our accounts everything is excellent. [You] owe according to my book 35#. But according to yours 52#. Certainly it couldn't be otherwise, since your arithmetic always went against yourself.

Don't speak to Mr. Lafoens about the price of the Camoens,[9] or to Klingsted about the 50 shillings.[10] Those are losses that I suffer with pleasure for knowing men better. The said Klingsted doesn't know how to decline except by sui, sibi, sé.[11] (See, I still remember the nominative!) His letter that you sent me was to tell me that the President of the Petersburg Academy was going to make me one of their academicians, a thing of which I have not yet had news nor shall I believe until I see it: and in the second place to ask me to see on what day some lottery tickets had come out blank, because he suspected that his correspondent in London had sold him tickets that had already come out blank &c. This suspicion shows the kind of man he is. I answered with civility & added the due commission of one shilling to the 50 owing. I put a cross on it. And that is it.

This letter was written with a steel pen that I've been using for 3 or 4 weeks. It spares me the task of trimming it: and though it's hard, I don't notice this in using it. When I go to Paris I'll take a few, & will give you one if you wish. Regards to the adoptive family.

vale & ama tuum &c. &c.[12]

---

9 See Magellan to Sanches, 10 February 1778.

10 Baron Timothy de Klingstedt was Chief Auditor of the Russian Admiralty at Archangel. He had owed Magellan money for some time; see Magellan to Sanches, 15 November 1776, note 8, and subsequent letters.

11 Of himself, for himself, with himself.

12 Farewell and love. your &c. &c.

78.05.01 *To the Comte d'Hérouville*

Archives Nationales, Paris, MS Marine, G110, ff. 155-6

Londres 1 Mai –78

Monsieur

J'ai l'honneur de vous informer que Messrs. Boulton & Watt arriverent [...][1] Ville hier soir: ils m'en firent part tout de suite, je les visitai le meme soir, & nous arretames cette après dinner pour arranger ce qui concerne à la machine à feu qu'ils sont prets à eriger à Moeres.[2] Ils sont actuellement ecrivant les articles aux quels ils veuillent s'obliger:[3] & en attendant je vais vous donner [...] idees relatives à l'affaire en question & aux circonstances dans lesquelles ils se trouvent actuellem[t]. disposés pour seconder autant qu'il est en leur pouvoir vos vues. 1°. Ces Messrs. se trouvent à present si fortem[t]. engagées à satisfaire au grande nombre de machines qu'on leur demande de tout part, qu'ils ne peuvent pas se porter eux memes ([ou] du moins un d'eux) au delà de l'endroit oû la machine doit etre erigée. 2do. Je [...] par les memes raisons que M[r]. Boulton ne pourra pas à present, s'engager dans aucune manufacture par lui-meme. Cependant si quelque personne voudra l'entreprendre, il ne manquera pas d'y prendre toute la part, qui sera consistente avec les Loix: car en effet un Anglois ne peut pas aller etablir lui meme une manufacture dans un autre paiz, sans se denaturaliser, & rester soumis à des amendes très enormes. Mais si par exemple il y avoit quelque manufacturier, qui finiroit des boutons, ou quelqu'autre objet de commerce, M[r]. Boulton peut le fournir avec les parties de l'ouvrage demi fini, ou meme tout à fait complet à l'exception de la derniere operation &[a]. Ce qui peut faire [une] branche de manufacture très considerable pour la France &[a].

Vous verrez, M[r]. le Comte, par le papier qui accompagnera celle-ci, que Messrs. Boulton & Watt ne veuillent voir le moindre profit ni avantage pecuniaire pour eux memes dans la construction de cette machine, ou de tout autre nombre de machines que vous souhaiterez avoir à Moeres: & vous n'aurez qu'à faire la dep[ense] de la maison, materiaux & ouvrage, qui sera fait avec les ouvriers & artistes. Si votre Agent en peut trouver qui l'executent selon les desseins qui ces Messieurs fourniront, ils en seront charmés: mais en cas d'avoir besoin des pieces travaillées en Angleterre,

---

1 Here and elsewhere throughout the letter, brackets indicate that a word has been obscured (or partly obscured) by binding at the end of a line.

2 Les Moeres, near Dunkirk.

3 See Boulton and Watt to Magellan, 2 May 1778.

ils sont si eloignés de vouloir aucune chose, qui puisse etre soupconée 78.05.01
d'interet pecuniaire pour eux, qu'ils s'offrent à donner la liste de touts les articles & pieces necessaires pour la machine à un commissaire quelconque qu'on voudra leur indiquer.

L'interet ordinaire qui revient à ces Messieurs pour chacune des ces nouvelles machines, est le 3me. de l'epargne qu'on fait avec elles, comparées à la depense de celles de force egale mais de la construction ordinaire, pendant 22 ou 23 années, & celà leur vaut avec quelques unes jusqu'[à] 700 livres Esterlines par an. en effet une, qu'ils vont batir à Chesswater en Cornwall, & pour [la]quelle les proprietaires de cette mine leur ont deja offert 500 livres Est. par année montera selon les articles de leur conven[...] à plus de 2100 livres Est. d'epargne par an. Quelconque puissent etre les profits d'epargne que la machine ou machines erigées ou à eriger à Moeres, Messrs. Boulton & Watt n'en veuillent aucune chose que seulem^{t}. le privilege exclusif, que vous vous obligés de leur obtenir: pourvû que ce soit general pour tout le Royaume de France & ses domaines en quelque part du Monde: pourvû que ce privilege soit effectivem^{t}. rendu valable dans toutes les Provinces & dependences: & que s'il sera necessaire que vous le faisiez enregistrer dans les Parlem^{s}. & Chancelleries ou Tribunaux d'oû il pourra dependre pour la validité. Vous voyez Monsr. Le Comte que ces Messieurs comptent tout à fait sur votre honneur, & que c'est sur ce seul avantage du privilege exclusif qu'ils entrent & font cet engagement. Mais il faudra bien s'exprimer sur le sens de leur proposition. Ils ne pretendent aucun privilege pour manufacturer par eux memes ces nouvelles machines à feu: car il leur est indifferent par qui elles seroient travaillées. Ce qu'ils pretendent avoir c'est le profit du 3me. de l'epargne que leurs machines [fer]ont: ainsi il faut que leur privilege exclusif les met en etat d'empecher qu'aucun autre puisse eriger ces machines sans leur consentement & direction. Ainsi ils n'empechent aucunement le travail & les materiaux de France: meme ils voudront toujours employer [les] ouvriers du pais plutot que les etrangers: Car en effet il n'est point permis d'envoyer [les] ouvriers d'ici pour un autre pais: & ce n'est que seulem^{t}. un Ingenieur ou directeur pour faire eriger la machine selon les idees & desseins de ces Messieurs, qu'ils seront obligés d'envoyer d'ici, ce qui est permis. Et cet Ingenieur sera payé par vous, selon l'agrement [d'a]ccord qu'on fera là dessus. Je vous prie M^{r}. le Comte mil pardons par le desordre de mes idées. mais je veux tacher de faire comprendre bien le sens des expressions de ces Messieurs. L'ingenieur ou directeur de cette machine ou machines à eriger en Moeres, peut etre [un] sujet quelconque qu'il vous plaira. il pourra faire batir la maison, & arranger tout le [...]res de l'ouvrage de la machine. un ingenieur du meme paiz de Moeres pourra aisem^{t}. [et]re chargé de tout celà. Mais pour

78.05.01 y faire mettre la derniere main, & ajuster les pieces plus particulieres & qui necessitent de plus d'attention; alors il faudra que ces Messieurs envoyent quelqu'un qui soit au fait de ce mechanisme: & c'est ce dernier ingenieur qu'il faudra envoyer d'Angleterre, qui, trouvant deja toutes les choses finies & pretes, pourra faire monter la machine en peu de semaines. Le voyage & salaire de cet ingenieur ne pourra pas aller au delà de 40 ou 50 guinées.

Lorsque Messrs. Boulton & Watt seront informés exactement de la quantité d'eau qu'il faudra elever à une certaine hauteur, dans un certain tems, alors ils peuvent resoudre la grandeur & force qui doit avoir leur machine: &, en considerant [les] prix ordinaires tant des materiaux que des ouvriers en Angleterre, ils pourront dire [à] peu près l'argent que celà couteroit ici. Mais vous voyez bien, Monsr. le Comte, que cette depense ne sera pas la meme si la machine sera faite pour la plus grande partie à Moeres meme. Il y a toute l'apparence que la depense respective de la machine deja erigée à Moeres, pourra donner une idée de celle qui chaque machine de Messrs. Boulton & Watt coutera: ce qui ne sera pas possible d'etre evalué que sur les lieux memes, en considerant l'ouvrage qu'il y a à faire & les prix tant des materiaux que des ouvriers dans l'endroit meme &a.

C'est tout ce que J'ai à dire pour vous aider à saisir les disposi[tions] & propositions de ces Messieurs sur le present affaire. La lettre ci-jointe est je crois assez explicite sur ces points.[4] Vous n'aurez, Mr. le comte qu'à les considerer, & à me donner les ordres qu'il vous plaira, & que Je me ferai toujours le plus grand plaisir d'executer avec empressement, etant avec tout respect, & attachemt.

Monsieur Le Comte
Votre très hble. & tres obeisst. Serviteur
J.H. de Magellan

P.S.
Il y a une maniere d'evaluer la depense de ces nouvelles Machines: elles coutent à peu de chose près la meme somme d'argent que les machines ordinaires; mais l'avantage en est qu'elles ne demandent qu'un tiers du charbon pour produire un effet egale.
N.B. quoique ces Messrs. parlent seulement d'un effet double, produit par ces nouvelles machines, il y en est effectivemt. pour la plupart triple & meme quadruple: mais ils aiment à s'annoncer en moins, pour avoir la satisfaction de faire d'avantage au delà de leur engagement.

4 It would seem that Magellan delayed sending his letter until he was able to enclose with it the letter from Boulton and Watt, dated the next day (= 78.05.02), in which they set out the terms they were proposing to Hérouville.

From Matthew Boulton and James Watt 78.05.02

Archives Nationales, Paris, MS Marine, G110, ff. 157-8[1]

London May 2d 1778

Sir

As by your obliging interposition, we perceive, the affair with Mr. le Comte D'Herouville concerning the exclusive privilege is in a proper train, & as we wish to preserve the good Opinion of ye Comte & to prevent misapprehensions, which may possibly arise among people who speak different languages, we therefore beg leave to recapitulate our sense of ye conversation we have had with you and with Monsr. le Comte Chastenet de Puysegur[2] and the more particularly as you mention the having sent to France Mr Boulton's last letter to you[3] which was wrote in a hurry and only intended for your own private inspection.

1st. We propose to furnish Monr. le Comte D'Herouville with all necessary designs and directions for building a Fire engine of such powers as he shall chuse upon the Moere[4] near Dunkirk according to our improved method; and also to provide him with all the cast and hammered iron work and all the Brass and Copper work necessary for the same. – We shall also find a proper Engineer to superintend the erection of the Engine

2do. The Comte D'Herouville is to pay the original cost of the Articles we furnish without any profits or pecuniary advantages charged on our part – We shall previous to the erection of the Engine give him an account of the prices pr pound of the several articles and shall give full liberty to the Comte or his Agents to procure wherever he pleases any of them they may think they can have cheaper provided that they get them of the proper quality and it would be highly agreable to us that the Comte D'Herouville would get every thing done in France that he can[.][5] Monsr. Le Comte D'Herouville is also to pay the workmen & Engineer employed in the Erection.

1 Watt's file copy of this letter, with minor differences in wording and punctuation, also survives, in one of his letter-copy books now at the Birmingham Central Library (Boulton and Watt Letter books, vol. 3, 25-6).

2 Antoine Hyacinthe Anne, Comte de Chastenet de Puységur (1752-1807); see Magellan to M. Boulton, 28 February 1778.

3 Boulton to Magellan, 22 April 1778?

4 i. e. Les Moeres.

5 *editorial addition.*

78.05.02 3d. We will guarantee that such Engine as we shall erect for the Comte shall do more than double the work, with the same quantity of fuel, that is now done by his present Engine or any other steam Engine in France. And if we shall not be able to make good this promise we become bound to reimburse Le Comte D'Herouville for the expence of all the materials made according to our directions and shall take them to our own use.

4th. In every such erection in his country we take for our trouble and invention the value of one third part of the annual savings in fuel which our Engine makes when compared with any one of the common Construction and it is an actual fact, that the proprietors of some of our Engines have agreed to pay us £500 Sterling per annum in lieu of ye third of ye savings for so long as they shall use the Engine; But as the Comte D'Herouville has been so obliging as to undertake to procure for us an exclusive privilege for 15 years for the sole use and application of our inventions and improvements in steam Engines throughout the dominions of France and to procure the same to be registered in ye several parliaments of France where it may be necessary – In consideration thereof we do relinquish all claim of pecuniary profit from him so far as relates to this Engine or the Draining of ye Moere

As we are strangers to the Laws of France we hope that Monsr. Le Comte D'Herouville will take care that the privilege he obtains for us be sufficient to protect the right it is intended to give us, against all invaders.

We beg the favour of Monsr le Comte to transmit to us his observations upon ye above proposals so that we may clearly understand each other, nothing being more our wish than to merit the Continuance of his friendship and good offices

We mentioned in our last letter to you[6] that we could not at present have ye honour and pleasure of waiting upon the Comte D'Herouville at Paris but should be glad to meet him or his Agent at Dunkirk, where the plans may be settled as soon as he pleases – We should be glad to have an answer to this article in course of post as our stay in London will not exceed ten days and our J. Watt goes directly into Cornwall which is 300 miles farther from Paris – We are, Sir,

Your most Obliged Servants
Mattw. Boulton
James Watt

Mr Magellan
Nevill's Court
Fetter lane

6 See M. Boulton to Magellan, 22 April 1778.

To Jan Henrik van Swinden

78.05.05

Leiden University Library, MS B.P.L. 755, no. 256[1]

Londres 5 Mai -78

Monsieur

L'honneur de Votre Lettre datée le 9 Nov[bre]. dernier[2] ne fait qu'arriver à mes mains depuis peu de jours, avec les deux traitez que vous avez daigné de m'envoyer de Votre Composition. J'y ai admiré un travail si bien suivi, & des vuës si vastes en raprochant des principes les plus certains, que je me flate de voir le vrai philosophe bientot Maitre d'entrevoir l'avenir, & devoiler une infinité de phenomenes, qui passent jusqu'à present pour unintelligibles à La raison humaine: C'est assurement à force d'etudier La nature dans les effects qui tombent sous l'observation, qu'on pourra en suivant vos pas, reconnoitre leurs vraies causes.

Je vous ai envoyé une petite brochure que J'ai publiée dernierem[t]. sur les Eudiometres pour l'indagation des qualités de l'air plus ou moins propres pour La Respiration animale, selon la quantité du phlogistique qu'il peut renfermer:[3] & Je vous envoyai un Memoire de M[r]. Nairne sur les meprises qu'on a eu à l'egard du vuide de la pompe pneumatique:[4] & Je serai charmé que vous y trouviez quelque chose d'utile.

J'ai envoyé à M[r]. Dentan[5] un de mes eudiometres fait par notre celebre artiste Parker: mais je crois n'avoir pas envoyé aucun autre verre de ceux du 3me. Vol. du D[r]. Priestley;[6] apparem[t]. M[r]. Dentan ne m'en a pas parlé; ou, s'il l'a fait, je m'en suis oublié, aiant perdu Sa Lettre. En cas que cela soit,

---

1 Published in David Willemse, "Suite d'un voyage aux Pays-Bas: João Jacinto de Magalhães (1722-1790) et ses rapports avec Jean Henri van Swinden (1746-1823)", *Arquivos do Centro Cultural Português*, 7 (1974), 225-278.

2 Van Swinden's surviving draft of this letter is dated 7 November 1777; see Van Swinden to Magellan, 7 November 1777.

3 Magellan, *Description of a Glass Apparatus, for Making Mineral Waters; ... together with a Description of some New Eudiometers, or Instruments for Ascertaining the Wholesomeness of Respirable Air* (London, 1777).

4 E. Nairne, "An Account of some Experiments made with an Air-pump on Mr. Smeaton's Principle; together with some Experiments with a Common Air-pump", *Philosophical Transactions* 67 (1777), 614-48.

5 Pierre Gédéon Dentan or Dentand (1746-1780); see J.H. Van Swinden to Magellan, 7 November 1777, note 3.

6 J. Priestley, *Experiments and Observations on Different Kinds of Air*, vol. 3 (London, 1777).

vous n'avez qu'à me reiterer vos ordres, & pouvez etre assuré d'etre obeï avec tout empressem^{t}. possible.

Je suis bien faché que les Memoires des Savans etrangers paroissent si retardées dans La Collection de l'Acad.^{e}. R^{le}. des Sciences de Paris: car celà me privera de connoitre sitôt les principaux points de Votre travail sur Les aiguilles aimantées.[7] Vous connoissez sans doute celui, qu'a fait le Professeur Lous de Copenhague là dessùs: il en imprima un petit traité en 1773.[8]

Lorsque vous voudrez communiquer quelque chose à La Soc^{e}. R^{le}. de Londres, je serai bien flaté d'en etre chargé.

Sans doute vous avez entendu parler de la decouverte de M^{r}. Achard de l'Acad^{e}. R^{l}. de Berlin, sur la formation des cristaux, moienant L'acide volatile de l'air fixe.[9] Cet Academicien vient de m'envoyer deux cristaux parfaits de sa façon, l'un cal[c]aire (Spateux) sur lequel agit l'acide nitreux, & l'autre parfaitement dur, jusqu'à pouvoir couper le verre, comme les vrais quartz. tous les deux sont diaphanes & beaux, mais particulierem^{t}. le dernier.

Ce même Academicien pretend que les conducteurs pour garantir de la Foudre, seront plus avantageux, s'ils ont un plateau rond au sommet, au lieu d'une pointe.[10] Mais je crains qu'il n'ait pas bien observé les circonstances de ses experiences: car J'ai repeté les mêmes avec Le D^{r}. Ingenhousz un de ces jours: & nous trouvons tout Le Contraire: apparem^{t}. M^{r}. Achard a employé des pointes qui n'etoient pas bien aigues.

En toute occasion oû je pourrai vous etre bon à quelque chose dans ce pais-ci, Je vous prie de me croire sincerem^{t}. & de Coeur

Monsieur
Votre très h^{ble}. & obligé Serv^{r}.
Magellan

---

7 J.H. Van Swinden, "Recherches sur les aiguilles aimantées, et sur leurs variations régulières", *Mémoires de mathématique et de physique, présentées à l'Académie Royale des Sciences [de Paris] par divers savans...*, 8 (1780), 1-576. Van Swinden's paper shared the prize awarded by the Académie in 1777.

8 Carl Christian Lous, *Tentamina experimentorum ad compassum perficiendum ...* (Hafniae, 1773).

9 Achard, "Copie d'une lettre ... contenant la découverte qu'il a faite sur la formation des cristaux & des pierres précieuses", *Observations sur la physique, sur l'histoire et sur les arts*, 11 (1778), 12-14.

10 See Magellan to J. Banks, 12 December 1778.

P.S. Les experiences du grand froid que Je vous communiquai, furent faites par M^r^. Simmons cirurgien,[11] à Chatam,[12] dans l'année 1776. Je suis mortifié d'avoir ommis son nom lorsque je les envoiai à M^r^. Dentan. 78.05.05

A Monsieur
Monsr. J. H. van Swinden,
Profess^r^. de Philosophie dans
L'université de Franeker en Frise Hollande

11 Samuel Simmons, FRS (1750-1813). No publication embodying these observations has been identified.

12 Chatham?

From the Comte d'Hérouville[1] 78.05.17

Birmingham Central Library, Boulton and Watt collection, Box 22/B3/8

Je suis veritablem^t^. faché que les occupations de M^rs^. Boulton & Watt ne Leur permettent pas de venir en France:[2] Leur privilege Exclusif, tel qu'ils l'ont demandé est entierem^t^. expedié & mis en regle; mais comme il y est expressem^t^. stipulé qu'ils ne pourront en jouir, qu'après avoir demontré par l'experience de leur machine Les avantages qu'ils en promettent, il est fort à craindre que Ces Messieurs, faute de faire l'epreuve de leur machine, ne perdent leur privilege; & qu'il ne se trouve quelques autres personnes, les quelles aiant connoissance des machines de Messrs. Boulton établies en Angleterre, ne viennent se presenter en France à leur place, & n'obtiennent le meme privilege, que celui qui leur est expedié: car M^rs^. Boulton & Co. ne doit pas croire que notre gouvernem^t^. se content d'une épreuve qui seroit faite en Angleterre. J'ai crû meme de devoir pas en parler, dans La Crainte que cette proposition n'arretât meme l'expedition de son arret. Si Ces Messrs. persistent à ne point venir en France pour faire leur epreuve & jouir de leur privilege, Ils ne pourront en tirer aucun avantage qu'en donnant leur procuration à quelqu'un Soit Anglois, Soit François, qui vienne ici les

1 Extract headed "Article de la Lettre de M^r^. d'Herouville à M^r^. de Magellan datée le 17 Mai -78; sur l'affaire de Messrs. Boulton & Watt". MS is a clerk's transcription. Annotated "M^r^. Magellan May 22^d^ 1778".

2 See Magellan to d'Hérouville, 1 May 1778.

78.05.17 remplacer. Je crois que s'ils se determinent à prendre ce parti, Je pourrai Leur trouver ici quelqu'un de notre Nation qui pourra traiter avec eux du profit de leur machine à feu sur le pied dont ils conviendront ensemble: Mais ils doivent savoir qu'il y a fort peu de ces sortes de Machines employées en france aujourd'hui; & que celles qui sont sur les mines de Charbon depensent fort peu, parceque Le Charbon y est à fort bon marché.

Quant à celles que M^r^. Boulton proposoit d'établir aux Moeres, Je ne suis pas dans ce moment-ci déterminé à le faire; parceque Je suis depuis peu en marché avec une Compagnie Hollandoise pour Lui vendre les Moeres 170,000 guinées; & que ces Messrs. paroissent preferer les moulins à vent aux Machines à feu; Mais comm'il est ordonné par l'arret du Conseil portant Privilege exclusif en faveur du S^r^. Boulton qu'il fera son épreuve aux Moeres ou à Paris, et que Je voudrois établir à Paris une machine à feu pour l'elevation des Eaux, Je lui demande, en place des machines qu'il m'avoit promis d'etablir aux Moeres, de m'en fournir une à Paris, qui puisse elever les eaux de la Riviere de Soixante pieds de France, & fournir par jour au moins deux cents pouces d'eau: chaque pouce est de 72 muids, ou de huit pieds cubes d'eau, ce qui fait pour chaque muid 280 pintes, que Je crois egales a 70 gallons d'Angleterre.[3] Je voudrois savoir en meme tems, combien l'etablissement de cette Machine couteroit (le frais de construction seront supposés aussi chers à Paris qu'à Londres) & quelle quantité ou poids de Charbon elle consumeroit par jour.

Comme peut etre quelqu'un de ce pais-ci, qui auroit connoissance de la present affaire pourroit s'offrir à M^r^. Boulton pour en traiter avec Lui, Je vous prie de le prevenir de ne rien conclure à ce sujet avec personne, sans ma participation & sans mon avis; parceque surement Je ne permettrai point qu'il soit trompé. J'oubliois de vous dire que les circonstances actuelles politiques en doivent nullem^t^. inquieter M^r^. Boulton. Si La guerre survenoit,[4] je lui procurerois tous Les passeports nécessaires, et il entreroit en France par Ostende. Comme dans ce moment-ci Je ne puis pas etre absolument Maitre de mes actions,[5] Je ne puis Lui promettre de me trouver à Dunkerque ou ailleurs, que suivant Les evenemens publics qui arriveront.

---

3 The *muid* was a unit of capacity, the value of which varied considerably between different parts of France.

4 War between Britain and France was expected to break out at any time—as in fact it did, shortly afterwards, when the French came to the aid of Britain's rebellious American colonists.

5 Comte d'Hérouville was a general in the French army, so that his services were likely to be required in the event of war.

To Ribeiro Sanches[1] 78.05.19

Österreichische Nationalbibliothek, Vienna
A.N. Ribeiro Sanches Korrespondenz, codex 12714, ff. 219-219v

Londres 19 Mai – 78

Meo Am°. e S^{r}. e m^{to}. do Cor^{am}.

Certam^{te}. eu cuidei q ja me tinha bem explicado com V.M. sobre o nosso meo e teu, p^{a}. q V.M. não achasse novid^{e}. no q lhe repeti na m^{a}. carta precedente. isto hũa vez dito, não perciza reiteraçoens: Mas em cazo q eu o não tivesse dito a V.M., certam^{te}. eu sintiria m^{to}. Se V.M. imaginasse q eu me avia nunca de esquecerme da carta q V.M. me escreveo à Estampes qd°. eu vinha da Arrochella p^{a}. Paris, depois de ter escapado ás garras do Pombal.

tenho feito por obrigar o Marquez de Villa alegre & cuido q não irá descontente de mim.

Ja tenho o tratado de Wilkes sobre a hydropsia q partirá com os outros artigos cuido q dentro de 15 ou menos dias

Mas estou picado pelos remoques ácerca do errar fogo. Como se isso não sucedesse aos maiores heróes da arma alta. Pois se me picar, goarde-se q não lha caia em caza o raio da vingança & diga-o là da m^{a}. p^{te}. a M^{a}. Joana, q talvez haja de pagar pelo pato &^{a}. qd°. eu for a essa terra; o q ainda não posso determinar.

o Restabelecim^{to}. da Sta. Inquizição em Goa, se cumprir com as ordens reaes, q são o de fazer todos os processos com abu[ndancia] & publicadas &^{a}. não será tão máo como d'antes. Mas pelo q toca ao novo Codigo das Leis, não lhe falta mato & tojo q cortar, & montijos p^{a}. aplanar.

alegrome q a piquena Maria se meta a debuxar, mas será m^{to}. mais sabia antes de m^{to}. tempo em fazer os originais, Como o escutor[2] Paulo Campi me disse em Roma, hindo-o eu vizitar. Sua Mulher estava prezente, e dizendo-me q ella pintava, ajuntou: eu vizitando-a um dia & vendo as suas pinturas, lhe propuz de fazermos ambos de concerto o trabalho sobre os originaes: o q ajuntado démos neste novo officio com^{to}. melhor gosto q d'antes. o Campi tinha qd°. m'o disse perto de 80 anos: & qd°. cazou assima de 45 ou 50.

o Cardamin ou Cardamomo he hũa flor da India, pois de là vem, & se uza delle na medecina. eu Ja tive da sua semente, & cuido q a mandei p^{a}. o Jardim botan°. de Madrid, ao amigo D^{r}. Ortega.

---

1 Annotated "Recue a 24 Mayo" [Received 24 May].
2 escultor.

78.05.19 estou depressa p^{a}. hir fazer uma mistura de metais p^{a}. a corréção dos pendulos dos Relogios astron^{os}. q aqui se me ordenárão p^{a}. as nossas demarcaçoens do Brazil entre Port^{al}. & Espanha.

o Cav^{ro}. Pinto tem estado mal com fevre: mas está restabelecido: e hontem á noite o deixei, escrevendo p^{a}. a posta de hoje

vale & ama

tuum

Magellan.

A Monsieur
Monsr. Sanchez D^{or}. en Medecine
Cimitere de S^{t}. Jean de la Verrerie
au Marais
à Paris

London 19 May –78

My Dear Friend

I thought surely I had well explained to you about our mine and yours, so that you wouldn't find anything new in what I repeated to you in my last letter.[3] My having made this previous comment, there is no need for reiteration: but in case I haven't said it to you, certainly I would very much regret it if you imagined that I would forget the letter that you wrote to Estampes[4] when I came from La Rochelle[5] to Paris, after having escaped Pombal's tyranny.

I've been working to oblige the Marquis of Villa Alegre[6] and I suppose he won't be unsatisfied with me.

I already have Wilkes' treatise about hydropsy[7] that will go together with the other articles, I hope within the next 15 days or less.

But I'm annoyed with the insinuations about missing fire, as if this wouldn't happen to the greatest heroes of the high weapon.[8] If you tease me, beware of

3 See Magellan to Sanches, 28 April 1778.

4 The small town of Étampes (or Estampes), 56 km SW of Paris?

5 In 1762 or 1763, Magellan fled from Portugal and the oppressive regime of Pombal. He probably fled by sea to the French port of La Rochelle (Arrochella in Portuguese) and travelled from there to Paris via Étampes.

6 Probably Don Andrés Marabel y Silva (1707-1789), who appears in *Memoriales de Titulos Nobiliarios e Hidalgos para obtener Facultad y consignar Renta de Viudedad – Siglos XVII, XVIII y XIX* (Madrid, 1989), pp. 71, 130.

7 Richard Wilkes, *An Historical Essay on the Dropsy* (London, 1777).

8 Sanches had possibly been teasing Magellan about his sexual prowess.

the thunderbolt of revenge striking you, & tell Mme. Joana on my behalf that she may have to pay for the duck[9] &c. when I visit you: I cannot yet fix when this will be. 78.05.19

The re-establishment of the Holy Inquisition in Goa,[10] if the royal orders are executed, which are to perform all processes with [abundance] & published &c. won't be as bad as before. But so far as the new code of laws is concerned, it doesn't lack bush & gorse to cut, & hills to level.

I'm glad that little Maria[11] is starting drawing, but she will be much wiser to make originals soon, as the sculptor Paulo Campi[12] told me when I visited him in Rome. His wife was present and, saying that she was a painter, he added: when I visited her one day & saw her paintings, I proposed to her that we both work on originals: once adjusted to this, we started this new job with much more pleasure than before. Campi was nearly 80 when he told me that:[13] & when he married he was over 45 or 50.

Cardamin or cardamom[14] is a flower from India, since it comes from there, & it is used in medicine. I had its seed already, & I suppose I sent it to the botanical garden of Madrid, to our friend Dr. Ortega.[15]

I'm in a hurry to go and make a mixture of metals to correct the pendulums of the astronomical clocks that I was ordered to arrange for the demarcation of Brazil between Portugal & Spain.

Chevalier Pinto has been ill with fever but he has recovered: and last night I left him writing for today's post.

vale & ama
tuum
Magellan.[16]

---

9 An old Portuguese expression meaning "to play the fool".

10 Suppressed by Pombal in 1774, the Inquisition in Goa was restored by Queen Maria I to prevent any paganization of the local Christians. In 1812 it was finally abolished under pressure from England.

11 Maria Dieudonné?

12 Presumably the well-known sculptor Pietro Paolo Campi (1668-1764).

13 This implies that Magellan met Campi in the late 1740s. However, he probably did not visit Rome until some ten years after this, during the "philosophical tour" he undertook after leaving the Augustinian monastery in Coimbra. It seems, therefore, that he significantly underestimated Campi's age at the time he met him.

14 Cardamom is a spice native to the East Indies and China, also used in medicine as a stomachic.

15 Casimiro Gomez de Ortega (1741-1810), Spanish physician and botanist who studied at Bologna before becoming a professor in Madrid; see Magellan to Sanches, 16 June 1776.

16 Farewell and love | your | Magellan.

78.05.22 *From the Boulton & Watt company*[1]

Birmingham Central Library, Boulton and Watt Letter books, vol. 3, 29-30

May 22d[2]

Mr. Magellan

By your obliging communication (this day) of the Comte d'Herouville's letter[3] we are sorry to find we are not to have the pleasure of meeting him at Dunkirk – Our JW has delayed his journey to Cornwall till now principally on that account, but tho' we both wish to have the honour of a conference with Mr. le Comte he must deny himself that pleasure as he cannot in justice to our employers absent himself from their Business the time of a Journey to Paris – but as we are desirous of availing ourselves of the Comte d'Herouville's kind offer our Mr Boulton will endeavour to arrange his business so as to meet Monsieur Herouville at paris in 10 days from this date[4] and will then personally settle & confirm all that hath past between us – At present it is sufficient to mention that we will with equal willingness erect an Engine at Paris as at the Moere[5] on the terms contained in our last – No person in France has lately made any application to us except le Sieur Joseph Jary concessionaire du Roi pour les mines du Nord en Bretagne[6] – to whom we have made an offer of erecting an Engine upon similar terms to what we offer to our customers in Britain – but upon the condition that our offer shall not be binding upon us without we obtain an exclusive privilege for France – in all this transaction we have not mentioned the Comte d'Herouville's name nor given him any hints by which he could guess at the channel by which we were to obtain the privilege –

JW.

---

1 Original not found; text taken from the company letter-copy book.

2 That the letter was written in 1778 is inferred from the position of the letter-copy in the bound volume of copies kept by the company, and from the annotations on the copy of the letter to which it is a response.

3 See Comte d'Hérouville to Magellan, 17 May 1778.

4 See Magellan to M. Boulton, 29 May 1778.

5 Les Moeres, d'Hérouville's property near Dunkirk.

6 See Magellan to M. Boulton, 23 March 1778.

From Jan Henrik van Swinden 78.05.22a

Leiden University Library, MS B.P.L. 755, no. 256-7[1]

22 Mai 1778

Monsieur

J'ai été très flatté d'apprendre par votre Lettre du 5 de Mai,[2] que mes traités du Thermometre et de l'hyver de 1776[3] ne vous ont pas deplu. L'approbation d'un juge comme vous, est ce que je puis desirer de plus favorable. Je souhaite que vos Physiciens veuillent me juger avec la meme faveur et la meme complaisance, et que mes ouvrages puissent se repandre un peu dans vos contrées. Je souhaiterois pouvoir publier quelque jour mes recherches complettes sur les rudes hyvers de ce siecle.[4]

J'ai très bien receu dans le temps votre traité sur les Eudiometres:[5] et je l'ai lu avec la plus grande satisfaction; il est si clair, et si methodique qu'on ne sauroit après l'avoir lu, trouver de difficulté dans l'usage et le maniement des instrumens memes, comme je l'ai eprouvé après avoir receu l'Eudiometre que vous m'avez envoyé: cet instrument est commode et très exact. C'est un vrai service que vous avez rendu à la Physique en l'inventant. L'examen de l'air par moyen de l'Eudiometre est important, et d'une utilité directe: aussi M Senebier de Geneve[6] me marque-t-il, qu'il joint journellement cette observation à ses autres observations meteorologiques. Je compte qu'il me communiquera ses resultats. Il seroit à souhaiter que tous les meteorologistes en fissent de meme. Cela pourroit servir à faire voir si l'air en tant que tel, est reellement en certain tems par certain vent, certaine température &c^c^, plus sain qu'en d'autres: et cela seroit surtout utile en tems de maladies epidemiques, pour determiner, si celles

1 Original not found; text from file copy retained by van Swinden (and numbered by him N°. 259); annotated "à Mr Magellan à Londres". Published in David Willemse, "Suite d'un voyage aux Pays-Bas: João Jacinto de Magalhães (1722-1790) et ses rapports avec Jean Henri van Swinden (1746-1823)", *Arquivos do Centro Cultural Português*, 7 (1974), 225-278.

2 See Magellan to Van Swinden, 5 May 1778.

3 Van Swinden's *Dissertation sur la comparaison des thermometres* (Amsterdam, 1778) and his *Observations sur le froid rigoreux du mois de Janvier 1778* (Amsterdam, 1778).

4 No such work appears to have been published.

5 Magellan, *Description of a Glass Apparatus, for Making Mineral Waters; ... together with a Description of some New Eudiometers, or Instruments for Ascertaining the Wholesomeness of Respirable Air* (London, 1777).

6 Jean Senebier (1742-1809).

78.05.22a ci sont reellement produites par des miasmes repandus dans l'air: car, si cela est, l'air doit etre alors moins diminué par l'Air nitreux qu'en d'autre temps. Je tacherai des que je le pourrai d'imiter l'assiduité de M. Sennebier sur ce point.

Le Memoire de M. Nairne[7] est très curieux, interessant, et contient nombre de grandes vues qu'il seroit à desirer qu'on suivit plus loin. Quel est donc ce fluide si imperceptible, et dont cependant la force de pression est si grande, qui se developpe dans le vuide? Je ne sais si M. de La Place à Paris n'a pas fait quelque chose la dessus: du moins je me rappelle avoir lu quelque part mais je ne sais plus où, qu'il avoit lu à l'Academie un Memoire, et fait des Exp. sur le fluide qui reste dans le Recepteur après l'extraction de l'Air.[8]

J'aurai soin de remettre au Prince de Gallitzin le montant de ce que vous avez bien voulu debourser pour l'excellent Eudiometre que vous m'avez envoyé. Je compte me rendre à la Haye dans cinq semaines.

Mon Memoire que l'Academie de Paris a couronné sur les Aiguilles aimantées, est entierement achevé d'imprimer, selon ce que M. Le Roy[9] m'a marqué il y a huit jours: mais il se passera encore quelque tems avant que le VIII vol des Sav. Etrangers, où il doit paroitre, soit publié. Je suis mortifié de ce retard.[10] Je connois très bien l'ingénieux ouvrage de M. Lous,[11] et j'en ai beaucoup profité. Mon Memoire est tres long, et contient près de 300 Experiences; j'ai taché de ramener ce qui concerne les aiguilles aimantées à des formules algebriques, et je crois y avoir plus ou moins reussi. La 2e partie contient les Loix sur les variations de l'Aiguille que j'ai deduites de plus de 40000 observations que j'ai faites, et de celles que mon Frere[12] a faites à la Haye. Je desirerois fort que les Physiciens voulussent

---

7 E. Nairne, "An Account of some Experiments made with an Air-pump on Mr. Smeaton's Principle; together with some Experiments with a Common Air-pump", *Philosophical Transactions* 67 (1777), 614-48.

8 Laplace read a paper on this subject at the Easter meeting of the Académie Royale des Sciences on 9 April 1777. It was not published, however, and no trace of the text remains.

9 In the Leiden University Library (collection B.P.L. 755) there are 21 letters from Le Roy – presumably the physicist and member of the Paris Academy of Sciences, Jean Baptiste Le Roy (1719-1800) - to Van Swinden.

10 The volume including Van Swinden's prize-winning memoir was not published until 1780; see J.H. Van Swinden, "Recherches sur les aiguilles aimantées, et sur leurs variations régulières", *Mémoires de mathématique et de physique, présentées à l'Académie Royale des Sciences [de Paris] par divers savans...*, 8 (1780), 1-576.

11 Carl Christian Lous, *Tentamina experimentorum ad compassum perficiendum ...* (Hafniae, 1773).

12 Simeon Pieter Van Swinden (1755-1835), a lawyer at The Hague.

s'appliquer un peu à cet objet. J'ai lieu d'esperer, que quelques personnes en France, et en Suisse, et peut etre en Italie, formeront conjointement avec mon frere et moi une espece de societe sur ce sujet. Nous ferons tous des observ. sur L'aiguille à toutes les heures du jour: et Nous comparerons nos observations. J'ai trouvé que l'aiguille est sujette dans sa variation diurne à 8 mouvemens differens, dont 4 sont reguliers. J'ai discuté cela avec soin. Vous sentez bien que je connois ce que M. Canton a publié la dessus dans le LI vol. des transact.[13] Graham a beaucoup travaillé la dessus, mais n'a publié que des resultats très genéraux:[14] ses papiers existent-ils encore, et y auroit-il moyen d'y avoir acces: sans cela son travail admirable est perdu pour la posterité. Je voudrois que quelqu'un en Angleterre voulut se charger de semblables observations: ce n'est qu'en les multipliant en differens endroits, qu'on peut se flatter de parvenir à quelques Loix générales. 78.05.22a

L'Academie de Baviere a courronné pour la moitié du Prix, mon memoire sur l'Analogie de l'Electricité et du Magnetisme.[15]

Depuis à peu près dix ans, l'aimant fait un des grands articles de mes travaux. Il a paru bien des traités sur ce sujet: Mais je n'ai pu me les procurer tous. Il en a paru entr'autres deux en Angleterre que je desirerois fort d'avoir et que je n'ai pu acquerir quelque soin que je me sois donné pour cela: Voici le titre du premier Calculations and Tables relating to the attractive Virtue of Loadstones; publié en 1729. L'auteur est Lord Paisley. On en trouve un extrait Phil. Trans. N°. 412 p. 245.[16]

Le 2e. est: An account of an attempt to ascertain Longitude at Sea, by an exact Theory of the Variation of magnetical needle, &^{c}. by Zachar Williams. Ce traité a paru en 1755 en 4°, Anglois et Italien. On en trouve un extrait détaillé dans le Journ. Brit. du D^{r}. Maty Tome 16 p. 33 seqq.[17]

---

13 John Canton, "An Attempt to account for the regular diurnal Variation of the horizontal magnetic Needle; and also for its irregular Variation at the Time of an Aurora Borealis", *Philosophical Transactions*, 51 (1759-60), 398-445.

14 George Graham, *Philosophical Transactions*, 33 (1724-5), 96-107; idem, *Philosophical Transactions*, 45 (1748), 279-80.

15 Van Swinden's memoir, awarded a gold medal by the Königliche baierische Akademie der Wissenschaften in Munich, occupies the first volume of his *Recueil de mémoires sur l'analogie de l'électricité et du magnetisme* (La Haye, 1784).

16 James Hamilton, Lord Paisley (later 7th Earl of Abercorn), *Calculations and Tables relating to the Attractive Virtue of Loadstones, etc.* (London, 1729). The summary to which Van Swinden refers was published in *Philosophical Transactions*, 36 (1729-30), 245-50.

17 Zachariah Williams, *An Account of an Attempt to Ascertain Longitude at Sea, by an Exact Theory of the Variation of the Needle ...* (London, 1755). Published with an Italian translation by Giuseppe Baretti.

78.05.22a Oserois-je vous prier M. de voir s'il y auroit moyen de me procurer ces deux Ouvrages: excusez la façon libre dont j'en agis, mais je sais combien vous aimez à obliger. On trouvera surement dans les Bibl. du Museum et de la Societé bien des traités sur l'aiman, publies dans le siecle passé et plus tot que je ne connois que de nom, ou point du tout. J'ai le dessein de publier quelque jour une histoire du Magnetisme[18] et j'ai deja rassemblé des materiaux très considérables. En 1772 j'ai publié un essai d'une Theorie mathem. sur l'aimant dont j'ai fait présenter dans le tems un exempl. à la Societé.[19] Je ne sache pas que les journaux anglois en ayent rendu compte, et je ne crois pas que cet ouvrage soit beaucoup connu en Angleterre.

Le Pr. de Gallitzin m'a parlé et ecrit plus d'une fois sur les belles Exper. de M. Achard pour la production des cristaux.[20] Nous touchons à un monde nouveau, dont on n'avoit gueres d'idée il y a bien peu d'années.

Je ne doute pas que les conducteurs pointus ne soyent les meilleurs: il me semble que le rapport qui se trouve la dessus dans les Transact. le prouve evidemment,[21] ainsi que le travail de M. Le Roy.[22] Le votre et celui du Dr Ingenhouz ajoutent un nouveau poids à ce sentiment.

Si je croiois que quelque dissertation de ma part put ne pas deplaire à la Societé je serois fort tenté de lui envoyer quelque chose: mais il est je crois assez difficile de trouver des occasions.

Si vous voulez bien m'honorer de tems en tems de vos lettres vous m'obligerez infiniment: ayez s.v.p. la bonté de me les addresser selon l'addresse ci jointe.[23] Je serois bien charmé M. de pouvoir vous etre utile dans nos quartiers: vous pourriez vous servir de moi librement, et je serois très flatté de pouvoir executer Vos ordres. Je suis &c

---

18 No such work was ever published.

19 J.H. Van Swinden, *Tentamina theoriae mathematicae de phaenomenis magneticis* (Leiden, 1772).

20 Achard, "Copie d'une lettre ... contenant la découverte qu'il a faite sur la formation des cristaux & des pierres précieuses", *Observations sur la physique, sur l'histoire et sur les arts*, 11 (1778), 12-14.

21 Since the report of the committee established by the Royal Society in 1777 was not published until 1779, Van Swinden must be referring here to the report of the Society's earlier committee on lightning rods, "Report of the Committee appointed by the Royal Society, to consider of a Method for Securing the Powder Magazines at Purfleet", *Philosophical Transactions*, 63 (1773-4), 42-47.

22 J.B. Le Roy, "Sur la forme des barres ou des conducteurs métalliques, destinés à préserver les édifices des effets de la foudre, en transmettant son feu à la terre", *Histoire de l'Académie Royale des Sciences* (1773), Mém. pp. 671-686.

23 Address not shown.

To Matthew Boulton[1] 78.05.29

Birmingham Central Library, Boulton and Watt collection, MBP 244, 57

M^{r}. Magellan's Comp^{s}. wait on M^{r}. Boulton: and, in spite of all his pains and trouble, he has take to put him self ready for setting out to morrow morning, he is very sorry to find, that it is absolutely out of his power so to do: but M^{r}. Boulton may take the places for wednesday morning for Dover without any any other word: for M^{r}. Mag^{n}. will be then ready to set out without fail.[2]

this friday 29 May -78.

---

1 Annotated "Magellan 29th May 78".

2 Magellan had evidently undertaken to go with Boulton to Paris to assist him in his negotiations with the Comte d'Hérouville; see Boulton and Watt to Magellan, 22 May 1778.

To Gabriel de Bory[1] 78.06.00

J'ai l'honneur de vous prévenir, Monsieur le Chevalier, que le bureau des longitudes établi par le Parlement, vient d'accorder 500 livres sterlings à M. Mudge, pour l'encourager à continuer ses travaux sur les montres marines.

On sait assez combien la perfection de ces montres est importante pour la navigation, & en conséquence pour le bien public; ainsi, je pense que

---

1 Letter not found. The text is from the version published in *Observations sur la physique, sur l'histoire naturelle et sur les arts*, 11 (1778), 537-43, in the issue for June 1778, under the heading, "Lettre de M. Magellan à M. le Chevalier de Bory, de l'Académie Royale des Sciences, Relativement à la Montre marine de M. Mudge, Elève du célèbre Graham, & l'un des plus habiles Horlogers de ce siècle."
The text exists in two different states. In what was evidently the original print-run of the journal, the issue for December 1778 included, on p. 455 (i.e. the same page as Magellan's "Lettre sur un Enfant monstrueux" [see Magellan to Rozier, 18 August 1778]), "Errata pour le Mémoire de M. Magellan, inseré dans le Cahier de Juin 1778, page 358 (*sic*)", listing two corrections to be made on p. 538. In other copies we have seen, the corrections have already been made and the "Errata" statement does not appear. Details of the corrections are noted at the appropriate points below.

78.06.00 l'académie sera bien aise d'être informée, avec quelque détail, de ce que M. Mudge a fait en ce genre.

La montre marine de cet excellent artiste, ayant été mise en épreuve à l'observatoire royal de Greenwich, pour que l'astronome du roi (M. *Maskelyne*) en examinât la marche; elle a été avec tant de précision, depuis le 12 novembre 1776, jusqu'au 12 de février 1777 (dernier jour dont j'ai pu avoir les observations), que pendant cet intervalle de temps, elle n'a varié qu'infiniment peu. En effet, elle ne s'est écartée du temps moyen, que d'environ deux tiers de seconde par jour; car au bout de ces trois mois ou de 93 jours, il n'y a eu en tout qu'une différence d'1',1'',8 en avance, sur le temps moyen. Les plus grands écarts qu'on y a remarqués, n'ont été, & cela qu'un seul jour, que de 2'', & en huit autres jours, que de 2'' en avance. Ensuite, les variations en retard, n'ont été que d'1",46 dans un jour, & que d'1'',2 en deux autres jours.

M. Mudge avoit porté cette montre auparavant, c'est-à-dire, en juin 1776, de Londres à Oxford dans une chaise de poste, d'où elle étoit revenue de même, au commencement de novembre. Or, après un transport de 116 milles, pour l'aller & le retour; elle a donné la longitude entre Londres & Oxford à 1",6 près en prenant un terme moyen entre les deux longitudes qu'elle a données, & en arrivant à Oxford, & en revenant à Londres. Car l'ayant mise avant de partir à la seconde sur une excellente pendule à secondes du docteur Heberden,[2] dont la maison est à 31'' de l'observatoire de Greenwich, elle a donné en arrivant à Oxford 5'....3'',5 de différence entre ces deux méridiens, qui est la même, à une demi-seconde près, que celle qui est marquée dans les tables de l'*astronomie* de M. de Lalande.[3]

En revenant à Londres au mois de novembre, elle a donné pour différence entre la maison du docteur Héberden, & Oxford 41...3'',8 qui, jointe à la différence de 31'' qu'il y a entre cette maison du docteur & l'observatoire de Gréenwich, fait 5....1'',8. Or, cette quantité ne diffère pas d'une seconde de la vraie longitude, entre Gréenwich & Oxford, qui par les observations astronomiques les plus exactes, ne va pas au-delà de 5'....1''.[4] Car, en prenant une longitude moyenne entre ces deux différentes longitudes; on a 5'...2'',6, qui ne s'éloigne de la véritable différence de longitude, de ces deux endroits, que d'1''6, comme je l'ai déjà dit.

---

2 The London physician William Heberden (1710-1801).

3 J. J. Lalande, *Astronomie*, 2nd ed. (Paris, 1771-81), vol. 1, Table 1.

4 "5'.... 18''" in the first printing (but given as "5'.... 12''" in the statement of Errata).

Pendant tout le temps que cette montre est restée à l'observatoire d'Oxford, qui a été de 133 jours, elle a été avec tant de justesse, qu'elle n'a pas retardé d'1 ,22 sur le temps moyen.[5] 78.06.00

Je n'ai pas encore été à portée d'examiner par moi-même, comment elle est construite, je sais seulement qu'elle est de la forme des montres ordinaires; quoique beaucoup plus grosse, ayant 5 pouces anglois de diametre, & son cadran 4½; & qu'on la remonte par dessous, comme les montres angloises, une fois toutes les 24 heures: il n'est pas inutile d'observer que cette opération n'arrête en rien la marche. Au reste, voici le principe de sa construction, tel qu'il m'a été communiqué par une personne d'un rang distingué, très-versée dans ces matieres, & qui connoît parfaitement l'intérieur de cette montre.

Le balancier fait ses vibrations d'une maniere tout-à-fait indépendante du rouage; il reçoit de la force motrice une impulsion nouvelle, & toujours égale à chaque vibration, moyennant l'action alternative de deux petits ressorts, placés de chaque côté des palettes; & c'est uniquement ces petits ressorts que le rouage remonte alternativement, pendant que le balancier est tout-à-fait dégagé de leur action.

Ce balancier a deux ressors spiraux (1),[6] tandis qu'il n'y en a qu'un dans les montres ordinaires, afin que ces deux ressorts produisent une action plus réguliere dans les deux sens où se font les vibrations. Il y a un thermometre métallique pour que le *momentum* d'action de ces ressorts ne soit pas plus grand dans une température que dans une autre. Ainsi, par cette disposition générale de la machine, il ne peut y avoir de variation, à proprement parler, ni par les inégalités du rouage, ni par les variations dans la fluidité des huiles, ni par le frottement des roues contre les pignons, ni par celui des pivots de ces mêmes roues; ni enfin, par les différentes intensités d'action des ressorts spiraux, action qui est renouvellée à chaque vibration, pour restituer la quantité de mouvement perdue dans la précédente.

Une autre observation que la même personne m'a fait faire, c'est l'attention qu'a eue M. Mudge, dans l'application du thermometre destiné

5 "par jour sur le tems moyen" in the first printing.

6 Editor's note: "L'horloge marine de M. le Roy, l'aîné, remise à l'académie des sciences à la fin d'août 1766, pour le prix que cette compagnie avoit proposé sur les longitudes pour 1767, & qui a remporté le prix de la même académie en 1769 & 1773, sur le même sujet, a deux ressorts spiraux, comme celle de M. Mudge. La construction marine de M. le Roy, qui étoit déjà connue de plusieurs académiciens en 1767, a été rendue publique par l'impression en 1769. Nous avons cru qu'il n'étoit pas inutile de faire ici cette observation, pour rendre à chacun ce qui lui appartient."

78.06.00 à entretenir l'égalité dans l'action des ressorts spiraux de cette espèce de montre, malgré les variations de température de l'athmosphere, attention qui semble avoir échappé jusqu'à présent à tous les Artistes. Soit MW, (*voyez la figure 1*). Le ressort régulateur du balancier d'une montre marine, que je suppose de la longueur de 19 lignes, ou de 19 parties égales, d'une longueur quelconque; que sa longueur jusqu'en *s* soit celle où les vibrations soient telles qu'elle soit réglée sur le temps moyen, dans la température de 55 deg. du thermometre de Farhenheit, ou de 10 de celui de Réaumur. Soit *der* & *wen*, deux thermometres métalliques composés de deux lames d'acier & de cuivre, placées l'une sur l'autre en sens contraire, c'est-à-dire, celle d'acier en dehors de celle de cuivre, en *der*, celle de cuivre en dehors de celle d'acier en *wen* (1);[7] soit enfin, le curseur *sanr* mobile autour du centre *a* qu'on suppose ici (pour rendre la figure plus intelligible), mobile dans un sens opposé à celui qu'il doit avoir, devant tourner autour du centre du spiral (2).[8]

Supposons actuellement qu'une chaleur de 10 degrés du thermomètre de Réaumur, ou de 15 de celui de Fahrenheit affoiblisse ce ressort, en sorte qu'il soit néccessaire de la raccourcir d'$^1/_{18}$, c'est-à-dire, jusqu'en *t*, pour avoir des vibrations d'une égale durée à celles d'auparavant; il est évident que s'il survient un degré de froid également éloigné de la température moyenne de 55 degrés, c'est-à-dire de $55^\circ - 20^\circ = 35^\circ$ de Fahrenheit; il ne suffira pas d'augmenter la longueur du ressort d'$^1/_{18}$, ou jusqu'en *z*; mais qu'il faudra l'allonger dans une plus grande proportion, ou jusqu'en *m*.

Ainsi, le levier *an* sue lequel agit le thermometre métallique *wcn* doit être au levier *ar*, sur lequel agit l'autre thermometre métallique *der*, en raison inverse des deux longueurs du ressort spiral, c'est-à-dire, comme 18 – 1.18 :: 18.*x*, afin que le curseur *s* au lieu de s'arrêter en *z*, puisse aller jusqu'en *m*, pour un degré de froid également éloigné de la température moyenne. C'est faute de cette attention, qu'on a si souvent échoué dans la

7 Editor's note: "Ce thermometre métallique est celui de Harrison, très-ingénieusement imaginé, & qu'il a adapté à la montre marine à laquelle le parlement d'Angleterre a accordé les 400 & tant de mille livres qu'il avoit promises pour la découverte des longitudes à la mer."

8 Editor's note: "Il n'y a dans la figure qu'un ressort spiral, quoique, comme nous l'avons dit, il y en ait deux dans la montre. On a dit que le petit levier ou curseur *rsna*, n'est que pour faire entendre comment l'auteur s'est proposé de rendre les variations des longueurs de ces ressorts spiraux (relatives aux différentes températures), proportionnelles à ces mêmes longueurs; car son action ne pourroit pas se faire de cette maniere, le curseur devant tourner autour du centre du ressort spiral, afin que dans ses différens mouvemens il reste toujours concentrique à la partie de ce ressort, qui joue entre ses chevilles, &c."

correction des inégalités des montres marines, causées par les différentes 78.06.00
forces de leurs ressorts, relatives aux différentes températures de l'athmosphère: mais il faut avouer que cela n'étoit pas possible à moins de les avoir construites, selon les principes de M. Mudge, c'est-à-dire, en sorte qu'elles soient tout à fait affranchies des effets provenans dans le rouage, des changemens dans l'épaisseur ou dans la fluidité des huiles; qui sont très-considérables (1);[9] car toute cette exactitude dans la marche proportionnelle du curseur ne produiroit aucun effet avantageux dans la construction ordinaire, les changemens dans la température agissant souvent sur la force motrice (les ressorts) d'une maniere, & sur les huiles d'une autre, tellement qu'elles exigeroient respectivement un mouvement contraire dans le curseur. Il faut pourtant dire qu'il arrive quelquefois que ces changemens produisent sur la force motrice & sur les huiles, des effets qui se compensent. Enfin, je ne dois pas oublier que M. Mudge a non-seulement cherché à donner le plus d'égalité possible à l'action (2)[10] qui entretient le mouvement du balancier, mais encore que cet habile artiste a tâché de mettre une très-grande différence de force entre celle qui est nécessaire aux vibrations du balancier, & cette force entretenante du mouvement de la machine: car ces forces sont entr'elles comme 1 à 300 dans sa marche; tandis que dans celle de M. Harrison, elles ne sont que comme 1 à 82, ou à 83. Mais, loin de s'en tenir là, M. Mudge est occupé actuellement à faire deux autres montres marines, qui seront éprouvées sur mer, selon le dernier acte du parlement, sur les longitudes, & dans lesquelles la proportion sera encore plus grande, comme d'1 à 350 ou même davantage.

Je joindrai à cette notice de la montre marine de M. Mudge la description d'un mécanisme fort ingénieux, qu'il a imaginé pour son excellence le comte de Bruhl, ministre de son altesse l'electeur de Saxe auprès du roi d'Angleterre. Le mérite de cette montre consiste principalement dans sa généralité; car on peut par cette méthode employer un rouage assez simple,

9 Editor's note: "On voit assez par cette description de la montre de M. Mudge, qu'il a employé un remontoir, comme l'auroient fait auparavant M. Rivaz, M. Harrison & d'autres. Ainsi, toutes les horloges & montres marines, construites de cette maniere, ainsi que celles qui ont un échappement libre ou à détente, sont susceptibles des mêmes avantages, relativement à l'action de ce thermometre métallique sur le ressort spiral."

10 Editor's note: "Il est important de se rappeller, pour mieux entendre ce qui est dit ici, que le premier moteur n'a d'action dans cette montre qu'en remontant tour à tour deux petits ressorts (tandis que le balancier est entièrement dégagé de leur action), par l'entremise desquels il agit sur les palettes de ce balancier, & que c'est de cette maniere que ce premier moteur en entretient les vibrations."

78.06.00 & dont l'exécution ne demande que la main d'un ouvrier ordinaire, pour produire un mouvement périodique quelconque, qui aura toute l'exactitude qu'on voudra, tel, par exemple, qu'il donnera à un millionieme & même à plus de deux millioniemes près, la révolution demandée. J'ai l'original de ce dessin devant les yeux, fait de la main de M. Mudge, qui est daté de l'année 1766, il a été déjà mis en pratique; cependant, il n'est presque connu de personne. Je me fais un vrai plaisir de le communiquer à l'académie comme une invention utile.

Supposons, par exemple, qu'on veuille représenter le moyen mouvement de la lune par le mouvement d'une pendule. La maniere d'y parvenir, & qui est connue de tout le monde, est celle qui est représentée dans la figure premiere. On a une roue *aa* qui fait son tour en 24 heures, cette roue porte un pignon *b* de huit, qui engraine dans une roue *cc* de 45, cette derniere porte de même un pignon de huit *d* qui engraine dans la roue *ee* de 42, au centre de laquelle est portée la figure de la lune.

Or, il est évident que la roue *aa*, faisant un tour en 24 heures, la roue *ee* fera le sien en 29 jours 53125, c'est-à-dire, en 29 jours 12 heures & 45 minutes; car $^{42}/_{8} \times {}^{45}/_{8} = {}^{1890}/_{64} = 29{,}53125$; mais le vrai mouvement moyen de la lune, n'étant que de 29 jours 12 heures 4 minutes,[11] & un peu plus de 3 secondes & moins de 4; on voit que ce mouvement du rouage est un peu plus lent que le vrai mouvement moyen de la lune, d'environ 57 à 56 secondes; il faut donc faire en sorte que le mouvement périodique de ce rouage aille plus vîte de la $^{57}/_{2551500}$ partie, ou tout au plus de la $^{56}/_{2551500}$ partie du total de la révolution. Car dans la période ci-dessus du 29e jour 12 heures 45 minutes, il y a en effet 2551500 secondes.

Voici la mécanique que M. Mudge a employée pour avoir, avec précision, la révolution requise. Il conserve les deux pignons *b* & *d*, *figure* 2, de même que les roues *cc* & *ee*, & il ajoute la vîtesse

$^{1}/_{1515\ 15/19}$ au pignon *b* de la roue *aa* $^{1515\ 115}/_{19}$ étant le quotient de 86400 (nombre de secondes en 24 heures, divisé par 57.

Pour produire cet effet, il met la roue *aa*, *figures* 2 & 3, sur un canon qui tourne librement sur l'arbre ou tige *oo*, fixé sur la plaque *pp*. L'arbre de la petite roue g, tourne dans le même plan que la roue *aa* entre des plaques, comme à l'ordinaire; mais on ne les a pas représentées dans la figure, non plus que celles des autres petites roues, afin de rendre cette figure plus intelligible. Cette roue g engraine dans le pignon *f*, qui est fixé sur la tige *oo*. L'arbre de la roue g porte une vis sans fin *n*, qui engraine dans la roue

11 44 minutes *is clearly intended.*

h, & celle-ci a une autre vis sans fin sur son arbre qui engraine dans la roue *k*. Cette roue *k* porte au bout de son arbre un pignon *l*, qui engraine dans la roue *m*, au centre de laquelle se trouve le pignon *b* qui engraine dans la roue *cc*, & le pignon *d* de cette roue engraine dans la roue *ee*, qui porte à son centre la lune RR. Il faut donner actuellement les nombres de cette espèce de rouage: les voici. 78.06.00

Premier cas pour accélérer le mouvement total, dans la proportion de $^{57}/_{2551500}$.

Pignon *f*, de	19
Roue g, de	45
Vis sans fin *n*, de	1
Roue *h*, de	21
Vis sans fin *i*, de	1
Roue *k*, de	45
Pignon *l*, de	3
Roue *m*, de	60

Second cas pour accélérer le mouvement total, dans la proportion de $^{56}/_{2551500}$.

Pignon *f*, de	4
Roue g, de	45
Vis sans fin *n*, de	1
Roue *h*, de	15
Vis sans fin *i*, de	1
Roue *k*, de	42
Pignon *l*, de	7
Roue *m*, de	45

Les deux pignons *b* & *d*, aussi bien que les roues *cc* & *ee*, ont les mêmes nombres que dans la figure seconde; savoir,

Le pignon *b* fixé sur la roue *m*, de . . .	8
La roue *cc*, de	45
Le pignon *d*, de	8
Et la roue *ee*, de	42

On a donc, par cette méthode, la révolution de la roue accélérée de $^{57}/_{2551500}$ dans le premier cas, parce que $^{3}/_{60} \times {}^{1}/_{45} \times {}^{1}/_{21} \times {}^{19}/_{45} = {}^{57}/_{2551500}$, & dans le second cas, de $^{56}/_{2551500}$, puisque $^{7}/_{45} \times {}^{1}/_{42} \times {}^{1}/_{15} \times {}^{11}/_{45} = {}^{28}/_{1275750} = {}^{56}/_{2551500}$, pendant que les autres roues ont le même mouvement respectif, & que la roue *aa* fait sa révolution en 24 heures.

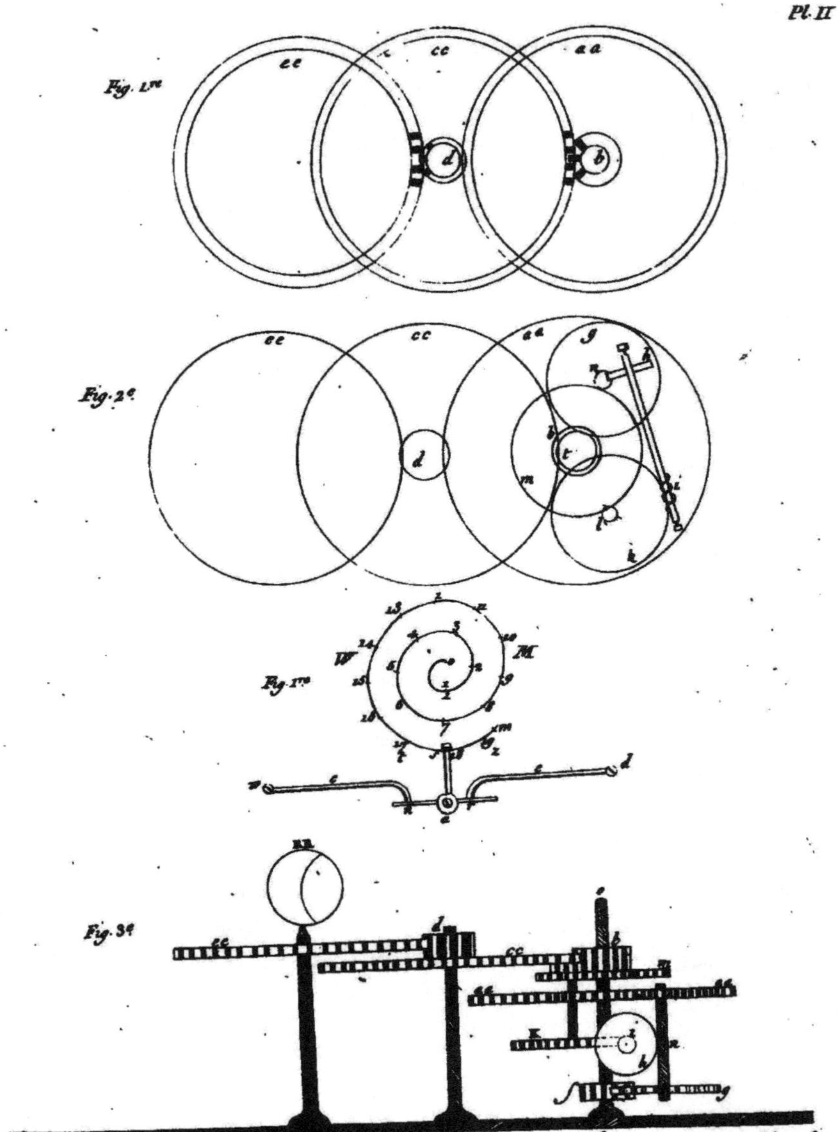
Fig. 1re
Fig. 2e
Fig. 1re
Fig. 3e
Juin 1778

To Benjamin Franklin 78.06.20

Annenberg Rare Book & Manuscript Library, University of Pennsylvania, Philadelphia[1]

My very dear & most respected friend

I cannot help writting these Lines, as a heartly farewell, loaded with all faithfull wishes and deep concern for your well-fare and of every one who has the happiness of being personally engaged in your glorious Cause, that of your own mother country.[2] I am extremely sorry, that my Little position in the world allows me but dry wishes to offer, and still more that the too early date of my age does not permit any personal service on my part, towards so nob[l]e an undertaking. But I hope, that I shall never see the time of saying, victrix causa diis placuit, sed victa Catoni:[3] and will return home to our once agreable, but now ungratefull Land, well pleased after observing the good prospect and favourable appearances on your side.

I have not yet received any Letters from you to your phil[al]. Friends, who are as anxious as possible to see them: but I hope you'll not forget, if possible, to send them this evening.

Please to honour me with some place amongst your most sincere heartly friends, and ever depend on the truest regard & esteem my Dear Sir

of your most obed[t]. devoted serv[t].
Magellan

Paris this 20th. June 1778

my directions are

at the Count d'Herouville's
rue du Fauxbourg S[t]. Honoré n°. 6.
à Paris

and

at n°. 12 Nevils Court Fetter Lane
London.

1 Published in *The Papers of Benjamin Franklin*, vol. 26 (New Haven: Yale University Press, 1987), pp. 663-4.

2 Franklin was at this time living in Paris as Plenipotentiary of the American colonies during the American War of Independence. Magellan had been visiting Paris but was about to return to London.

3 "If the victor had the gods on his side, the vanquished had Cato" (Lucan, *Pharsalia*, I, 128).

78.06.20 Please not to forget, that the Presid^{t}. de Saron[4] and all his family and relations, are of the most sincere and honorable caracter. He is one of the most ingenious Philosophers I know, very good Mathematician and very clearly headed-man. I know him these many years: and every one who is acquainted with him, cannot fail of doing him this justice.

A Monsieur
Monsr. Benjamin Franklin
à Passy
pres de Paris

4 Bochart de Saron, President of the Parlement of Paris.

78.06.20a *To Ribeiro Sanches*[1]

Österreichische Nationalbibliothek, Vienna
A.N. Ribeiro Sanches Korrespondenz, codex 12714, ff. 220-222v

Paris 20 Junho 1778

Meo Am^{o}. e S^{r}. e m^{to}. do Cor^{am}.

Não posso deixar de dar-lhe huã vaya ou pulha em bom portuguez de Riba Coa, pela maganeira da Sua Sta. benção ao Lumiar da porta da sua corte. Se eu Soubesse q V.M. ainda arr-tava, havia de ficar-lhe agradecido, na ideia de q esta terîa algum parentesco com a do Cordão do Seraph^{o}. P. S^{t}. Francisco, tal e qual os seos abrazados filho do Convento da Ponte de Coimbra a possuem p^{a}. felicid^{e}. das nossas Michéllas & tiricanas do Mondego, q saudozo e sombrio La vai murmurar com o grande Oceano (mas em silencio, como as beatas devotas da nossa terra) sobre as divertidas scenas q se passão entre os Sinceirais & Lamedas areosas q adornão as suas margens. Parece-me q estou vendo a V.M. crecendo-lhe a agoa na boca ao Ler estas Regras com as Lembranças do passado. mas de q serve á gente o alembrar-se do passado se tudo passa se não he entristecer-se & magoar-se....!

1 Annotated "recue a vinte & un / a 21 Juin" [received on twenty one/ on 21 June].

Pois regala-te papo q já foste farto; e voltemos folha. 78.06.20a

Lá achei nos sacos 260 ∆[#]. tournoises q V.M. me deo esta manhan, pois todas ellas são suas, vendo q ás bond[es]. antigas de V.M. (não exceptuando as presentes) devo o poder agora cobralas [a] outros. Deixei-as ao S[r]. Tronchin Fermier gen[al]., ou p[a]. melhor dizer ao seo Banqueiro; pois elle não está na terra. e deixarei o Recibo ao bom amigo comum Le Begue du Presle p[a]. q lho entregue quando vier, afim de me fazer dar em Londres o seo valor. q este dinheiro esteja aqui, ou lá, pouco importa; pois a maior p[te]. q he meo, está e estará sempre as ordens de V.M. a q[m]. de direito pertence.

Goze V.M. da saude, e satisfação q lhe dezejo, e sem embargo dos seos ralhos na escada contra a Divindade humana Americana, semilhantes ao dos nossos dirty fowning, coutty dogs de Londres, esteja na Certeza q sou Like a true-born English man.

P.S. Two kisses to the pretty ever your's Mary, and as many to her mother and aunt.

iterum atque iterum vale.
Magellan

a Monsieur
Monsr. Sanchez D[or]. en Medecine
Cemitere de S[t]. Jean, Rue de la Verrerie
à Paris

Paris 20 June 1778

My Dear Friend

I can't refrain from hooting at you or answering you with a scoff in good Portuguese from Riba Coa,[2] for the joke of your holy blessing at the door-step of your court. If I knew that you still bel-ed,[3] I'd be grateful, in the hope that this had some connection with that of the cord of the Seraphic Father St. Francis[4] such as his inflamed sons of the convent at the Coimbra bridge possess, to the

2 A region in the east-central part of Portugal, the men of which took particular pride in their virility.

3 belched; see Magellan to Sanches, 7 July 1775, note 19.

4 The coarse cord worn by St Francis of Assisi to hold his garments together was reputed to have wrought several miracles.

78.06.20a happiness of our whores & countrywomen of the Mondego[5] that flows, nostalgic and shady, to whisper with the large ocean (but in silence, like the pious bigots of our country) about the funny scenes that take place among the willow woods & sandy lanes that adorn its borders. I imagine seeing your mouth watering from memories of the past as you read these lines. But if everything passes away, what is the point of our remembering the past, unless to get sad & hurt ...!

Be pleased that your stomach has been full; and let us turn the page.

I found there in the bags 260 Δ#. tournoises[6] that you gave me this morning, which are all yours, seeing that I owe to your old benevolence (not to mention the present one) my now being able to charge them to others. I left them with Mr. Tronchin Fermier genal.,[7] or better with his banker because he is not in town; and I'll leave the receipt with our good common friend Le Begue du Presle to give to him when he arrives, so that he will give me its value in London. That this money is here or there doesn't matter, since most of what is mine will always be at your disposal to whom it rightly belongs.

May you have good health and the satisfaction that I wish you, and in spite of your scolding on the stairs against the human American divinity,[8] similar to that of our dirty fowning,[9] coutty[10] dogs of London, be sure that I am like a true-born English man.

P.S. two kisses to the pretty ever your's Mary,[11] and as many to her mother and aunt.

iterum atque iterum vale.[12]
Magellan

---

5 The Franciscans had been established in Coimbra since the 13th century. In the 17th century they occupied a new convent near the bridge over the Mondego, the river that flows by Coimbra, that was often referred to as the Convent of St. Francis of the Bridge.

6 i.e. 260 livres tournoises exactly, the delta symbol meaning there were no pence added?

7 Jean-Robert Tronchin (1702-1788), banker at Lyon, had been a Fermier Général in Paris since 1762.

8 Possibly a reference to the description of Benjamin Franklin attributed to the French finance minister Turgot, "Eripuit coelo fulmen sceptrumque tirannis" [He seized lightning from the heavens and the sceptre from tyrants], that was illustrated by Fragonard in a famous allegorical print first published in 1778.

9 fawning?

10 gouty?

11 Maria Dieudonné. See Magellan to Sanches, 21 September 1777, note 14.

12 Again and again farewell.

From Nathanael Wolf 78.06.30

Royal Society of London, Letters & Papers, VII, 87[1]

Notre ami (M^{r}. Bernoulli) arriva le moment, oû l'eclipse alloit commencer. Nous observames alternativement avec le micrometre Objectif. en voici le resultat.

June 26. La pendule etoit plus tard que le tems moien de 6''.

June 25. Abt. L. ⊙ 35°. – 7^{h}. 49'. 40''. – 4^{h}. 14'. 10''. midi
0^{h}. 2'. 0'' : plus tard de 6''. que le tems moien.
Diam. ⊙. 31'. 32'' jusqu'à 31'. 40''. je prends 31'. 36''.

June 24. 5^{h}. 10'. 25''. Commencement manqué, & conclû des autres phases. temps de pendule.

	12'. 0''.	deja bien avancée	
	21. 32	partes lucidae	27''. 21''.6
	27. 19	----------	25. 5.1
	32. 35	----------	23. 57.7
	41. 0	----------	22. 38.4
	49. 21	----------	21. 44.8
6^{h}.	4. 7.	----------	22. 44.2
	11. 47.	---------	24. 59.2
	18. 36.	---------	27. 17.9
	28. 44	La fin	

Donc la plus grande obscuration, à 5^{h}. 49'. 34''. de 9'. 52''.
Distance des Centres 22'. 17''.

---

1 MS is in Magellan's hand and is headed: "Observation de l'Eclipse du Soleil faite à Danzic, le 24 Juin 1778 par Messrs. Nathanael Matthew Wolff Doct^{r}. en Medecine, membre de la Soc^{e}. R^{le}. de Londres, & Jean Bernoulli Astronome Royal de Prusse & Membre de l'Acad^{e}. R^{le}. des sciences de Berlin Communiquée par M^{r}. Magellan qui l'eut dans une Lettre du meme M^{r}. Wolf, datée le 30 de Juin dernier, dont voici l'extrait." MS is annotated: "Account of an Eclipse of the Sun, by M^{r}. Matt. Nath. Wolfe | Read Feby 11th. 1779." The Royal Society's *Journal Book* records that Magellan presented this report on the same day as Maskelyne reported Ludlam's observation of the same eclipse at Leicester.

78.07.14 *To Luíz Pinto de Sousa Coutinho*[1]

Arquivo Histórico Ultramarino, Lisbon, CU-Reino, Cx 258

Excell^{mo}. S^{r}.[2]

Em consequencia das Observaçoens & repostas q V. Exc^{a}. me fez a honra de Communicar, relativas aos Instrumentos ordenados p^{a}. o Serviço da Corte de Portugal, tomei todos os meios q me parecem mais proprios p^{a}. a sua plena execução como vou ponderar a V. Exc^{a}. nos artigos seguintes.

1. As Agulhas magneticas p^{a}. determinar a variação, estão já m^{to}. adiantadas, p^{a}. dar outra forma á sua Construção; mas quando assim o não fosse, a circunstancia de terem hũ circulo vertical com suas divizoens p^{a}. determinar os angulos azimutaes, pode certam^{te}. ser mui vantajoza em varias occazioens, onde falta o quadrante Astronomico. Alem disso a Ordem q Recebi da Corte de Espanha p^{a}. Instrumentos semilhantes aos desta Lista de Portugal, expressam^{te}. aponta 6 Theodolites: e imagino q os Astronomos Portuguezes não serâõ descontentes de possuirem instrumentos com as mesmas propriedades aos quaes ajuntarei, Se V. Exc^{a}. assim o aprovar, outras tantas cadeias de arpentage, como na d^{a}. ordem de Espanha se determina.
2. Os Relogios de aljibeira q se estão fazendo actualm^{te}. são eizatam^{te}. conformes á ideia expressada nestas Observaçoens & Reflexoens; mas o Cazo de estar 4 ou 5 annos sem se alimparem, não pode deixar de cauzar lhes deterimento, pela inspissação do oleo nos pivotes do Regulador & ultimas rodas. Com tudo eu terei cuidado p^{a} q se dêm instrucoens, com as quaes qualquer curiozo, sem ser relojoeiro, & sem desmontar todas as peças, poderá facilm^{te}. remedear em grande parte este incoveniente.
3. Lizonjeo-me q as pendulas a meio Segundo, q se estão actualm^{te}. fazendo, satisfarâõ aos Requizitos apontados, particularm^{te}. se se fizer attencão ás instruçoens q as devem acompanhar.
4. Ja dei ordens necessarias a M^{r}. Dollond p^{a}. q os tubos & pedestais tenhão as tarraxas precizas p^{a}. q occupem menos espaço na embalage.

1 This letter was annexed to Luiz Pinto de Sousa's letter to the Minister Ayres de Sá e Mello, dated the same day, mentioned in Pinto de Sousa's Oporto Copybook, of which the original is at the Arquivos Nacionais da Torre do Tombo in Lisbon.

2 Annotated on the left of the first page: "Exc^{mo}. S^{r}. Luiz Pinto de Souza".

Alem disso ajustei com este Artista pª. q lhes puzesse, à cadahum destes oculos, o q chamamos rack-work; à fim de q os movimentos, vertical & horizontal, pª. conservar o objeto no meio do Campo do instrumento, possão ser os mais suaves: oq he requizito importante pª. não arriscar o verdº. instante das ocultaçoens e immersoens dos satelites de Jupiter. Esta adição do rack-work fará augmentar de quatro guineos o preço destes oculos. 78.07.14

5. O mesmo Dollond entreprendeo de executar os 5 Semicirculos de meio palmo & de duas polegadas de radio: & ambos são coiza de pouco custo

6. Tenho ordenado as 5 Lunettes d'epreuve, q forão omitidas por esquecimento na primeira ordem, e serão feitas, como se ordena, apropriadas aos quadrantes astronomicos. Estas lunettes são mui tediozas no ajustamento pª. q o eixo dioptrico fique equidistante de cada Lado: e custarâõ 6 ou 7 guineos alem do preço de cada Quadrante Astronomico.

7. Devo dizer a V. Excª., q examinando com attenção o quadrante astronomico de segunda mão, q tinha achado já feito, descubri nelle tais defeitos nas divizoens do Limbo, & no ajustam^to^. da Rectificação, q me vi obrigado a rejeitalo inteiram^te^. e já dei ordem à M^r^. Sisson, pª. fazer outro em Seo Lugar.

8. Posso assegurar a V. Excª q se trabalha actualmente em todos os Instrumentos ordenados; pois não me esqueço de vizitar frequentemente os Artistas, recommendando-lhes toda a presteza. Mas como he mais vantajozo pª. a perfeição da obra & cōmodid^e^. dos Artistas, q as mesmas peças se fação consecutivamente, e q as differentes operaçoens não sejão interrompidas com as de outro genero; necessariam^te^. as dilaçoens devem exceder os nossos dezejos.

9. Em fim permitame V. Excª. de observar, q na ordem q recebi da Corte de Espanha pª. 6 Series ou Colleçoens de Instrumentos, pouco mais ou menos como os ordenados pela Corte de Portugal, não só se achão os 6 theodolites, com as suas cadeias de Arpentage, mas tão bem 1º. 6 barometros portativos, sem duvida pª. tomar as alturas dos terrenos & das montanhas. 2º 6 barras magneticas, para retocar ou reimpregnar as agulhas com o magnetismo: oq certam^te^. he necessario em varias occazioens. 3º. o numero dos thermometros he dobrado: isto he doze em Lugar de seis; e com rezão, pela grande fragilid^e^. destes instrumentos, q quebrão com facilid^e^. 4º. seis micrometros pª. os oculos achromaticos. 5º. Seis agulhas de marear: eu faço tenção mandar das mesmas q se uzão na Marinha de Inglaterra, as quaes todas são revistas e assinadas por minha mão, segundo a ordem do Bureau da Navegação, sendo eu o unico a q^m^. o D^or^. Knight, seo inventor & meo intimo amigo, já

78.07.14 falecido, communicou os principios da sua construção. 6°. em fim os seis Quadrantes Astronomicos devem ser de dois pés de radio: mas esta circunstancia os fará menos com̃odos pª. o transporte.

N.B. Parece dezejavel, q os Astronomos Portuguezes não deixem de se achar providos igoalmente, com as ventagens de semilhantes instrumentos.

Fico m^{to}. pª. servir a V. Excª. com a mais pronta Vontade, & sou com a maior estimação & mais distinto Respeito

Excell^{mo}. S^{r}.
De V. Excellencia
servo mais Reverente & Venerador
João Hyacintho de Magalhaens

a Londres 14 de Julho 1778

Veja V. Excª. o P. S. da outra parte

P.S.

Quazi q me esquecia dizer a V. Excª. q ainda não ha nesta terra Relogios ou reguladores maritimos [time-Keepers] feitos por M^{r}. Harrison, q possão ser comprados. Este famoso Artista he já falecido; & seo filho me disse ultimam^{te}. na Socied^{e} Real, q ainda não tinha acabado os Instrumentos necessarios pª. os executar; e q o primeiro, já estava apalavrado pª. a Imperatriz da Russia. Quanto ao Regulador maritimo de M^{r}. Mudge, doqual eu dei a informação q sahio impressa no mez passado em Paris no Jornal do Abbade Rozier, & que sem duvida parece o mais perfeito de todos, ainda se não completou todo o ensayo pª. ser absolutam^{te}. premiado, não obstante q o Governo já fez hum prezente de quinhentas Livras Esterlinas à M^{r}. Mudge pª. o animar a proseguir. Em fim os Reguladores do charlatão Arnold, não valem a pena de q nelles se fale. Pelas Observacoens Astronom^{as}. feitas na ultima viage de Cook ao Mar do Sul; impressas por ordem do Almirantado, se lê a pag. 230, q hum dos Reguladores de Arnold, sem maior distancia doq até o cabo da Boa Esperança, deo hũ erro em Longitude de 29 gráos menos quazi quatro minutos. Agora V. Excª julgará facilm^{te}. se o tal Arnold offereceria pª. os exames da Tal viagem, por ordem do Almirantado, algum dos seos Reguladores, q elle não julgasse o mais perfeito!

As a consequence of the observations & answers I had the honour of receiving from Your Excell^y^. relative to the instruments ordered for the service of the Court of Portugal, I used all means that seemed proper to their completion as I will describe in the following articles.

1 The magnetic needles to determine the variation are already very advanced, taking into account that I wish to give a different shape to their construction; but even if they were not, the circumstance of having a vertical circle with divisions to determine the azimuth angles can certainly be very advantageous on several occasions where the astronomical quadrant is missing. Besides, the order I received from the court of Spain for similar instruments to the present list for Portugal expressly mentions 6 theodolites and I imagine that the Portuguese astronomers would not be unhappy to possess instruments with the same properties; to which I will add, if Your Excell^y^. approve, the same number of arpentage chains[3] as is specified in the order of the Court of Spain.

2 The pocket time pieces that are being made now are exactly as expressed in these Observations & Reflections; but in the event they are 4 or 5 years without cleaning this cannot but cause them damage, due to the thickening of the grease at the pivots of the regulator & the last wheels. However I will take care to give instructions so that any amateur, without being a watch maker & without dismantling all the parts, can easily in great part overcome this inconvenience.

3 I flatter myself that the half-second pendulum clocks that are being made will satisfy the defined requirements, especially if attention is paid to the accompanying instructions.

4 I have already given orders to Mr. Dollond that the tubes and pedestals should have the required screws in order to occupy less room in the packaging. Besides I arranged with this artist that he apply to each of these glasses what we call rack-work, in order that the vertical & horizontal movements made to keep the object in the field of the instrument be the most gentle, which is an important requirement to avoid missing the true instant of the occultations and immersions of Jupiter's satellites. This addition of the rack-work will increase the price of these glasses by four guineas.

3 Arpentage (from the French) means the measuring of land areas in *arpent*, an old unit of area; hence surveying in general.

78.07.14 5 The same Dollond decided to make the 5 semicircles in half a hand[4] & two inches in radius & both don't cost much.

6 I have ordered the 5 Lunettes d'Epreuve that were forgotten in the first order & will be made appropriate to the astronomical quadrants, as requested. Adjusting these telescopes so that the dioptric axis stays equidistant from each side is very tedious, and they will cost 6 or 7 guineas apart from the price of each astronomical quadrant.

7 I must say to Your Excell^y. that on examining carefully the second-hand astronomical quadrant I had found already made, I discovered such imperfections in the divisions of the limb & in the adjustment of the rectification that I had to reject it entirely and have given an order to Mr. Sisson to make a new one in its place.

8 I can assure Your Excell^y. that work is proceeding on all the ordered instruments, because I don't forget to visit the artists frequently, urging them to press ahead at full speed. However, since it is better for both the perfection of the work & the convenience of the artists that the same parts are done consecutively and that the different operations are not interrupted by others of a different kind, necessarily the delays will exceed our wishes.

9 Lastly allow me to observe that in the order I received from the court of Spain for 6 series or collections of instruments more or less like those ordered by the court of Portugal, not only are found the 6 theodolites with their arpentage chains, but also: 1° 6 portable barometers, undoubtedly to take the height of lands & mountains, 2° 6 magnetic bars to touch or re-impregnate the needles with magnetism as is certainly necessary on various occasions, 3° the number of thermometers is double, that is twelve instead of six; and with reason due to the great fragility of these instruments, which break easily. 4° six micrometers for the achromatic glasses. 5° six compass needles: I intend sending the same needles as used in the English Navy, that are all reviewed and signed by my hand according to the order of the Bureau of Navigation,[5] I being the only one to whom Dr. Knight its inventor & my close friend already deceased communicated the principles of their construction. 6° lastly the six astronomical quadrants should be of two foot radius, though this circumstance will make them less easy to transport.
N.B. It seems desirable that the Portuguese Astronomers be provided equally with the advantages of similar instruments.

---

4 A "hand" ["palmo"] is an old Portuguese measure of length equal to 8 inches.
5 i. e. the Board of Longitude.

I stay to serve Your Excell[y]. with the most ready will & I am with the greatest esteem & high respect 78.07.14

Your Excell[y].

Your Excellency's reverent & devoted servant

João Hyacintho de Magalhaens

London 14th July 1778

May Your Excell[y]. see the P.S. on the other side of the page

P.S. I almost forgot to tell Your Excell[y]. that there are not yet in this land Time pieces or maritime regulators [time keepers] done by Mr. Harrison that may be bought. This famous artist is already dead & his son told me lately at the Royal Society that he had not yet finished the necessary instruments to make them & that the first was already promised to the Empress of Russia. As to the maritime regulator of Mr. Mudge of which I gave the information that was published last month in Paris in Abbé Rozier's Journal[6] & that undoubtedly seems the most perfect of all, it has not yet completed its trial so as to be fully rewarded, in spite of the fact that the government already gave a present of five hundred pounds sterling to Mr. Mudge to encourage him to prosecute the work. Finally the regulators of the charlatan Arnold[7] don't deserve to be mentioned. By the Astronomical Observations made during Cook's last voyage to the south Seas, printed by order of the Admiralty, it can be read on pg. 230 that one of the regulators by Arnold, after a distance less than that to the Cape of Good Hope, gave an error in longitude of 29 degrees less about four minutes. Now Your Excell[y]. will easily judge whether the Arnold so named would offer for examination on such voyage, by order of the Admiralty, any of his regulators that he did not consider the most perfect!

6 "Lettre de M. Magellan à M. le Chevalier de Bory de l'Académie Royale des Sciences, relativement à la Montre marine de M. Mudge ...", *Observations sur la physique*, (1778), p. 537; see 78.06.00.

7 The watchmaker John Arnold (1736?-1799).

78.07.17 *To the Comte d'Hérouville*

Archives Nationales, Paris, MS Marine, G110, f. 170

Londres 17 Juillet 1778

Monsieur

Ce n'est que depuis une semaine & demie, que Je suis de retour à ce pais-ci, oû je trouve bien d'affaires, relatifs pour la plupart à mes amis, qui m'occupent tout à fait. Je ne dois cependant pas retarder d'avantage à vous exprimer les sentimens de mon obligation & gratitude par toutes les bontés, dont vous avez bien voulû m'honnorer, & dont je ne m'oublierai jamais dans ma vie. Je suis bien sur, Monsr. le Comte, que s'il ne tenoit qu'à mes souhaits, il n'y auroit point de bonheur, qui ne fut constament à l'entour de vous: mais, outre celui de l'aimable Compagnie de Mad^{e}. la Comtesse, & des deux Anges, Mlle. votre fille & sa cousine, dont vous jouissez, & auxquelles je vous prie de presenter mes humbles respects, certes je ne sçais pas qu'on en puisse rien desirer d'avantage, & à celà près J'ai pour moi l'opinion de Mylord Shelburne, qui fut enchanté d'apprendre de vos nouvelles.[1] Quant au Colonel Barré[2] je m'attends qu'il revienne de la campagne, pour le bien gronder par sa paresse à Vous ecrire, ou pour mieux dire, à ne pas vous ecrire depuis si long tems.

J'ai trouvé à mon retour M^{r}. Boulton dans cette Capitale, & je lui ai rendu le copie du privilege avec toutes vos reflexions sur cet affaire. Il est parti dans peu pour Birmingham, d'oû il promit de m'envoyer sa resolution & reponse aux questions proposées: Je le crois disposé à embrasser vos avis, que je conçois bien les plus sensés & avantageux. Cependant la multitude de ses affaires, augmente de jour en jour; & ce sera la seule raison pour retarder les mesures de sa part. Je viens de recevoir avis du Ministre d'Etat au departement des mines de Prusse, qui me parle de deux commissaires, qui arriveront ici bien tot pour examiner l'effet de ces machines nouvelles, & prendre avec M^{r}. Boulton les arrangemens necessaires:[3] Ce que je vais lui ecrire ce meme jour par la poste.

Je suis, & peut etre vous aussi, dans le doute du destin de notre ami le Comte Chastnet & de ses compagnons de voyage. Je vous prie de m'en donner des nouvelles, lorsque vous en saurez de certaines; car je m'y interesse

1 Lord Shelburne and the Comte d'Hérouville were on friendly terms; see Magellan to d'Hérouville, 17 April 1778.

2 Colonel Isaac Barré who fought in Canada during the Seven Years' War, in the 1750s and 1760s?

3 See Magellan to M. Boulton, 21 July 1778.

de Coeur.[4] Permettez moi de vous prier de remettre le billet ci joint dans la page suivante à M^r^. le Marquiz de Puy Segur;[5] & de vous persuader du sincere attachement avec lequel J'ai l'honneur d'etre pour la vie 78.07.17

Monsieur le Comte
Votre très humble & très obligé serv^{r}.
Magellan

P.S. J'ai delivré la lettre chez l'Abbé Gonetti; mais Je n'ai pas encore pu le voir.

A Monsieur
Monsr. Le Comte d'Herouville
Lieutenant Genl. des Armées du Roy
rue du Fauxbourg S^{t}. Honoré n°. 6
à Paris

4 Antoine Hyacinthe Anne, Comte de Chastenet de Puységur (1752-1807), was a naval officer and the Comte d'Hérouville's son-in-law.

5 On the second half of the folded sheet on which this letter is written, backing on to the address, is the note to the Marquis de Puységur to which Magellan here refers; see 78.07.17a.

To the Marquis de Puységur[1] 78.07.17a

Archives Nationales, Paris, MS Marine, G110, f. 171[2]

Pour M^{r}. Le Marquiz de Puy Segur de la part de Son très h^{ble}. serviteur Magellan.

On demande à la Societé d'Emulation de Paris[3] le vrai sens de la 6me. condition pour la serrure de combinaison, & que cet eclaircissem^{t}. soit

1 Jacques-François-Maxime de Chastenet, Marquis de Puységur (1716-1782), father of Antoine Hyacinthe Anne, Comte de Chastenet de Puységur.

2 This letter was sent as an enclosure with Magellan's letter of 17 July 1778 to the Comte d'Hérouville, being written on the second half of the folded sheet on which the latter document was written.

3 Modelled on the Society of Arts in England, the "Société Libre d'Emulation pour l'encouragement des arts, métiers et inventions utiles" was founded in 1776 in Paris with numerous wealthy and powerful backers, including Puységur. However, it survived for only a few years before ceasing activity soon after 1780. See R. Hahn, *The Anatomy of a Scientific Institution: the Paris Academy of Sciences, 1666-1803* (Berkeley/Los Angeles, 1971), pp. 110-11. Louis-Henri Duchesne was a founding member of the society.

78.07.17a entré dans le livre des minutes de la meme Societé. le voici. "quelque soit le nombre, l'ordre & la forme des parties, il est indispensable, que, sans demonter la serrure, on puisse lui faire executer indifferement telles de ces combinaisons, qu'il plaira choisir au moment, oû l'on voudra fermer ou ouvrir, & sans aucuns preparatifs." le mot ouvrir pris dans le sens ordinaire, donne une idée, qui detruiroit toute la bonté d'une serrure à combinaison; car s'il etoit possible d'arranger les combinaisons interieures de la serrure, avant de l'ouvrir, sans le demonter; alors un etranger, qui atraperoit le zero des combinaisons, c'est à dire, la position oû elles n'empechent le pene pour glisser, n'auroit aucune difficulté pour la ouvrir. au lieu que si cette faculté d'arranger les combinaisons, est practicable seulement avant de fermer la porte, alors le maitre pourra les varier à chaque fois qu'il voudra fermer sa porte, & par consequent la maniere de l'ouvrir sera aussi variée, & aussi differente à chaque fois comm'on le voudra: & on ne pourra la ouvrir sans la combinaison, sous laquelle elle fut fermée par cette fois là, à moins de la casser, ou de la demonter. On demande une declaraison decidée par la Societé sur cet article; car un de mes amis veut devenir candidat, en cas de pouvoir concevoir le sens de cet article, qui à la simple enonciation semble destructif de l'objet qu'on se propose &ª.

On fait actuellemᵗ. les deux charrues & le hou que Mʳ. le Marquiz souhaite avoir d'ici. ce dernier instrument est bien moindre qu'aucune des charrues: & à l'envoyer tronqué, il n'en peut resulter aucun avantage pour l'epargne, ni pour le transport.

Ce 17 Juillet 1778

Londres at nº. 12 Nevils Court Fetter Lane.

To Anne-Robert-Jacques Turgot[1]

78.07.17b

Biblioteca do Museu Nacional de Arte Antiga, Lisbon
Fundo Francisco Sousa Viterbo, Cx nº 1

Londres ce 17 Juillet 1778.

Monsieur,

J'allois vous ecrire sur le tricot veluté, lorsque Je Reçois l'honneur de votre Lettre datée du 9 du Courant.[2] Je me flate qu'à present M^{r}. le D^{r}. Le Begue du Presle vous aurá envoyé vos paquets, car la Caisse doit etre arrivée deja à Paris. Cependant com'il y a des instrumens pour M^{r}. Bezout de l'Acad^{e}. R^{le}. des Sciences, il sera aisé d'en savoir de ses nouvelles par sa voye.

Je vous prie de signifier à M^{r}. L'Abbé Sigorgne,[3] toute mon estime, avec Le desir de lui Rendre tous les services, dont Je serai Capable. Mais puisqu'il est curieux d'observer quelques phenomenes d'Astronomie, il vaudra mieux posseder une pendule qui y soit propre. Celles qui sont faites pour cet objet, & dont J'ai fourni un grand nombre pour differentes parts de l'Europe, oû elles ont prouvé excellentment bien, ont les circonstances suivantes. 1° elles frapent les seconds à repos, ou com'on le dit en Anglois dead beats. 2.° elles ne s'arretent pas lorsqu'on les monte. 3.° elles vont un mois, au lieu d'8 jours & 4.° on les emballe dans une caisse quarrée de mahogany, qui a environ 2 pieds de Long, & qui sert à les couvrir affixées contre une muraille &^{a}. A present j'y vais faire ajouter des palets de rubi dans l'échapement, sur une pour un de mes amis, qui l'a voulû avoir ainsi. Sans Cette derniere circonstance (que je me flate ne coutera pas au dela de 15 ou 20 shillings) leur prix total a eté de 7 ½ guinees, c'est-à-dire sept luis d'or & demi. Si M^{r}. L'Abbé veut de celles-ci, Je serai charmé de le savoir; car à la verité Les simples telles que Je vous ai envoyé par environ 17 ecûs tournois, ne sont pas propres à l'Astronomie. Quant à la verge J'en fais faire toujours de bois à fibres droites & paralleles, comme le Sapin de la Finlande. Car alors toutes les expansions, ou contractions se font lateralement, & non pas en longitude: au lieu que le bois de mahogany, ou d'acajû, n'a que des fibres tortues & transversales &^{a}. ce qui n'est point du tout avantageux. Apropòs Je suis bien faché de ne pas vous avoir fait ajouter, à votre pendule, un reveil: vous ne concevez peut etre pas, combien celà est utile. J'en ai une de ces pendules constament chez moi; je le mets à l'heure de mes rendez vous, de

1 Anne Robert Jacques Turgot (1727-1781), French politician and economist, France's Controleur Général des Finances, 1774-6.

2 Letter not found.

3 Pierre Sigorgne (1719-1809), professor of philosophy at the Collège du Plessis in Paris.

78.07.17b la poste, de sortir &c^{a}. &c^{a}. Celà vaut mieux qu'une douzaine de domestiques, pour me rappeler le moment de mes affaires.

Alons à parler du tricot. 1.° on le fait tout dans une piece, & non pas façoné comm'il faut l'appliquer: ainsi c'est au tailleur à le couper dans le sens du poil veluté qu'on le souhaitera. Voyez si cela vous convient? L'Artiste me dit qu'une piece de bon tricot blanc à velour capable de vous faire la Camisole, Selon les mesures & modeles que vous m'avez remis, coutera environ 5 guinées ou 5 & demie. 2°. pour ce qui regarde Les bas de soye velutée blancs, on demande si l'on veut avoir le poil du velour au dedans vers la chere, ou au dehors? Ayez donc la bonté de me Repondre là dessùs.

Mylord Mahon a poussé bien Loin ses Recherches sur les conducteurs pointûs & emoussés: il a decouvert une Loi nouvelle, c'est-à-dire, que Je crois personne avoit encore reconnû dans l'electricité, & ce qui vaut mieux c'est que les conducteurs pointûs y sont encore plus fortement hereux pour en eviter Le danger. Il a eté ce matin chez moi, & m'assura qu'il alloit bientôt publier son travail là dessùs, qui assurement est de la plus grande utilité publique.[4]

Les gazetes annoncent ici la mort du pauvre J. J. Rousseau. Je l'avois laissé en bonne santé, selon que J'ai pu voir, le 23 du mois dernier à Hermenonville.[5] du moins il y est mort en paix, dans les bras de ses amis, & Monseigneur Votre Archevecque bigot[6] & ... n'aura rien à demeler sur La Sepulture de son cadavre. oh que la race humaine est meprisable ...! puisqu'elle produit des individus tels qu'on place audessus des autres ...! Je vous assure que je me Rejouis Souvent de ma retraite dans Ce pais Philosophique, oû je n'ai rien que craindre, que seulem^{t}. des Loix, auxquelles aucun honet-homme ne doit jamais se Refuser ....

Vale & ama

tuum Magellan[7]

Celle ci est la clause la plus energique & courte du bon tems antien à propòs mes Resp^{s}. à l'Abbé Fontana. on m'en demande ici de son arrivée pretendue, tous les jours.

a Monsieur
Monsieur Turgot ancient Controleur general
Rue de l'Université Fauxbourg S^{t}. Germain
à Paris

4 Charles Stanhope, Lord Mahon, *Principles of Electricity, ... together with an Analysis of the Superior Advantages of High-pointed Conductors etc.* (London, 1779).

5 Ermenonville. Rousseau died on 2 July 1778.

6 Christophe de Beaumont, Archbishop of Paris, had issued a formal condemnation of Rousseau's book, *Émile*.

7 Farewell & love | your Magellan.

To Matthew Boulton[1] 78.07.21

Birmingham Central Library, Boulton and Watt collection, Box 22/B3/9

London 21 July –78

Dear Sir

I Received a Letter from Prussia,[2] announcing that two gentlemen, Counselors of ye. Bureau of ye. Mines, are coming to England, agreable to the information I communicated to ye. Secretary of State of that department, from you, relating to ye. intended 4 fire engines of yours, to be erected there.[3] I give then this notice to you in order to be ready to answer to their Inquiries, and chiefly to have ready a table of ye. expenses for erecting your fire engines as I told you, of different sises &a. You may take my word for it, that no business will ever be done with foreigners in a blind way, as they will alwais term it. Why can you not make a List (which certainly you must have already) of all the pieces necessarily employed in your engine? Can there be any difficulty in setting a certain price each? at least more or less about the value, when worked in the Country? Cant you judge of ye. expense for the work men, timber, &a. &a.? Certainly when any foreigner hears, that you can not fix any price, he will think that you mean to charge more and more &a. every one likes to take his measures, and have a good guess of his expenses in any undertaking whatsoever. As to the premium for your invention, I think it should be better, that you should fix also a Certain sum for each machine, according to it's Sise: and if any other consideration should have place by the circumstances of ye. Country, or of the time &a. it may be at any time added to, or deducted from, the fixed terms &a. the two above gentlemen will come directed to me, and I will not fail of giving you notice of yeir. arrival.

Pray tell Mr. Keir that I have sent his Chymical dictionary of Macquer[4] to Mr. Elmsley:[5] but as he had already sent to him the parcell of books he had order'd, he directed my porter to carry the parcell of the Dictionary to your ware-house somewhere in Trinity Lane &a. which he assured me to

1 Letter not found; MS is a clerk's transcription.

2 Letter not found.

3 See Magellan to M. Boulton and J. Watt, 3 March 1778.

4 P. Macquer, *A Dictionary of Chemistry…*, translated by James Keir (London, 1771). Keir lived in Birmingham and was closely associated at this period with the Boulton and Watt manufactory at Soho.

5 The London bookseller Peter Elmsley.

78.07.21 have done immediately, and that the commissioner or keeper of that warehouse did promise to send the said parcell with all possible speed.

Pray dont forget the affair of France, [an]d[6] the answers to y^{e}. Queries and remarks of y^{e}. Count de Herouville

I am with y^{e}. Greatest regard

Dear Sir
Your most obed^{t}. h^{bl}. ser^{t}.
Magellan

To
M^{r}. Boulton
at his Fire Engine Manufacture
near Soho
Birmingham

6 Paper damaged.

78.08.18 *To [François Rozier]*[1]

J'ai été ce matin avec M. l'Abbé Fontana, (18 août 1778), voir un monstre singulier, dont la femme d'un négociant de cette ville vient d'accoucher, deux heures après avoir été délivrée d'un premier enfant bien formé & constitué comme l'est un enfant ordinaire. Le second, ou plutôt le monstre par excès, n'a qu'un seul corps, surmonté de deux cols & de deux têtes. Son sexe est féminin. Ce qu'il y a réellement de singulier, c'est que dans ce seul corps il y a deux cœurs, dont l'un beaucoup plus grand & plus gros que l'autre. Il a en outre deux especes de diaphragme & deux estomacs; mais, après une certaine longueur, les deux intestins se réunissoient en un seul. La grandeur & la perfection des membres de ce monstre, sont en tout semblables à ceux des autres enfans qui viennent à terme. Nous l'avons examiné attentivement chez M. Sheldon,[2] un des plus célebres anatomistes de cette ville.

1 Letter not found. The text is from the version published in *Observations sur la physique, sur l'histoire naturelle et sur les arts*, 12 (1778), 455, in the issue for December 1778, under the heading, "Lettre sur un Enfant monstrueux; par M. de Magellan, de la Société Royale de Londres".

2 John Sheldon (1752-1808) had a private anatomical theatre in Great Queen Street, London, where he gave lectures in anatomy. He was elected FRS in 1784.

From Matthew Boulton 78.08.22
Archives Nationales, Paris, MS Marine, G110, ff. 172-3[1]

Soho near Birmingham 22[d] Aug[t] 1778

M[r] Magellan

Dear S[r].

I recv[d] the several papers you deliverd to me relating to our obtaining an exclusive privilege in France for the Fire Engine of our invention, and I beg the favour of you to assure the Count d'Herouville that We are fully sensible that He has done all that was possible towards expediting the affair; & that he has fulfilld with honour the part which he undertook. It remains for us to make the required Trial in France, and upon this subject I beg leave to submit to the consideration of the Count the following observations. – –

1st. The superiority of our Engines over those of the old construction can be ascertained only by a comparison. But as there is no old Engine at Paris with which the comparison can be made, We fear that the erection of an Engine in that City would not be effectual towards obtaining a second Arrêt de Conseil granting an exclusive priviledge.

2[c]. As the Count d'Herouville is going to sell the Moeres we cannot make that place, the place of trial, unless the new proprietors should apply to us for that purpose which if they should do, we shall be glad to accomodate the[m][2] with our Engine.

Monsieur Joseph Jary – Concessionnaire du Roi pour les mines du Nord pres de Nantes en Bretagne, has applied to us to erect one of our

1 Boulton's letter is accompanied by a French translation in an unknown hand. There are two different drafts of Boulton's letter in the Boulton & Watt papers, Birmingham Central Library, MBP 244, 58. Both differ significantly from the version finally sent. One seems to be an original draft and is more complete; it is undated but is annotated "Copy of answ[r] to M[r] Magellan relating to Count D'Heroville". The text is given in note 10 below. The other appears to be a copy made more recently of an earlier draft: it is headed "Magellan. J.H. de. / Autograph draft. /[1778.] [?. Oct.] M.B. [?Soho] – J.H. de Magellan [?London] 6 pp. folio. / Endorsed 'Copy of answr. to M[r]. Magellan relating to Count D'Herouville'." There is also a copy of the letter that was sent, with some differences in punctuation, in the Boulton and Watt letter books, vol. 3, 97-8, at the Birmingham Central Library, dated however "Soho | End of Dec[r]. 1778 or 1 Jan[y]. 79".

2 Paper damaged, letter missing.

78.08.22 Engines at his Coal mines near Nantes in place of an old one already erected there, and for this purpose he has come to England.[3] We think this would give us a good opportunity of ascertaining immediately the superiority of our Engine, as there is already one of the old construction upon that spot, with which the Comparison of ours might be made. But fr[om][4] the Copy of the Arret de Conseil which we have receiv[d][5] from you, it appears that the Trial is required to be mad[e][6] either at Paris, or the Moeres. If the Count d'Herouvil[le][7] could obtain permission for us to make this Trial in Bretagne near Nantes and if the trial so made would be effectual towards the obtaining a second Arrêt de Conseil, We could then without loss of time proceed to erect an Engine for M^r^. Jary. We have not thought it proper to make M^r^. Jary acquainted with any of the measures that have been taken towards obtaining an exclusive priviledge, but we have delayd giveing him a decisive answer to his proposition till we knew whether his Engine can be made the subject of the Trial required by the Arret of Conseil. We therefore beg the favour of the Count d'Herouville to adjust that matter & if it can be done to acquaint us that we may be able to give our answer to Mr Jary who waits impatiently for it. But if it be required that some Trial should be made at Paris conformable to the words of the Arrêt then, if no better occasion should offer itself of erecting an Engine in that City, I would send to Paris a Model of the Engine which should be capable of working in the same manner as the large Engines although the Cylinder be not more than 3 or 4 Inches in diameter & it is so much more difficult to make models work than larg Engines that no man hath ever been able to make a Model of a Common Fire Engine work with half the Vigour or with double the quantity of Fuel (in proportion to ye work done) of those on a larger scale & the difference between the effects produced by our Model & one of the old construction would be too evident to leave the least doubt

I also beg you will acquaint the Count d'Herouville that from our being so much engaged all this Year in erecting upward of Twenty (very large ones) of our Beam or reciprocating new Engines, that we have never yet had time to compleat the Wheel or Circular Engine & we therefore

3 See Magellan to M. Boulton, 23 March 1778.
4 Paper damaged, letters missing.
5 Paper damaged, letter missing.
6 Paper damaged, letter missing.
7 Paper damaged, letters missing.

wish our exclusive prividedge to be confin'd to the other sort viz the Beam Engine of which alone we shall make the proposed Trial at least for some time. 78.08.22

Lastly I beg you to assure the Count that we only wait for his Commands to execute the part promised by us towards him, & that I return him my sincere thanks for his endeavours & good disposition towards us, not only in promoteing my interest in the Fire Engine but also in other branches of my Manufacture. In answer to the proposed Questions concerning the effects of our Circular Engine, please to acquaint the Count that as we have not yet had leasure to carry it any further into execution, than was sufficient to convince ourselves of the truth & practicability of our principles, We cannot speak with the same certainty & precision as we can do of the reciprocating Engine, but that there is not the least doubt of its being applicable to any purpose in which a circular Motion is required, such as Mills for Grinding Corn, or laminateing Metals or working of Sugar Mills, or for every purposes in every Manufacture where Water mills are applicable & that these Engines may be made of any degree of power that may be required; and that as to the expence of keeping them going, one Man would be sufficient to manage an Engine, and a power equal to the strength of Two strong Horses exerted for a day, may be obtaind for the expence of one Hundred weight or one Quintel of Coals p^{r} diem (viz 112^{lb}) I must defer till some other occasion answering the question relating to the articles & prices of my Manufacture, I only beg that you will please to acquaint the Count that it consists of whatever the Fashion of the Time requires to be made of Steel Silver Copper plated wth. Silver or Gilded en or Moulu Buttons, Jappand wares, Architectal ornaments both of Tin paper & artificial Stone, with other articles such as paintings, a specimen of which I will send, in the first goods I send to Monsieur Granchez[8] au petite Dunkirk a Paris with directions to be deliverd to the order of the Count, & beg the Count will do me the honor to accept a specimen of my new manner of painting[9]

I am sorry it hath not been in power to send you any reply, to the papers you gave me, before now, but I am overwhelmed with Business at home, & have been obliged to make several journeys into different parts of England, where we are erecting Engines. pray make the best appology

8 See Magellan to Louis-Henri Duchesne, 2 October 1770, note 3.

9 Boulton had recently developed a method of copying pictures.

78.08.22 you can to Count d'Herouville for me & believe me to be with great regard

Dear S[r]
Your faithfull & Obliged Friend & Servant
Matt[w]. Boulton[10]

---

10 The draft version of the letter reads as follows:

I have read the several papers you delivered to me relating to our obtaining an exclusive privilege in France for the Fire-Engine of our new invention, and I beg the favour of you to assure the Count d'Herouville that we are fully sensible that he has done all that was possible towards expediting the affair, and that he has fulfilled with honour the part which he undertook. It remains for us to make the required Trial in France, and upon this subject I beg leave to submit to the consideration of the Count the following observations:

1. The superiority of our Engine over those of the old construction can be ascertained only by a comparison. But as there is no old Engine at Paris with which the comparison can be made, we fear that the erection of an Engine in that city would not be be [*word repeated in going from one line to the next*] effectual towards obtaining a second Arrêt de Conseil granting an exclusive privilege.

2. As the Count d'Herouville is going to sell the Moeres, we cannot make that the place of Trial, unless the new proprietors should apply to us for that purpose, which if they should do, we shall be glad to accomodate them with our Engine.

3. M. Jery Concessionaire &c has applied to us to erect one of our Engines at or near Nantes in Bretagne in place of an old one already erected there, for this purpose he has come to England. We think this would give us a good opportunity of ascertaining immediately the superiority of our Engine as there is already one of the old construction with which the comparison of ours might be made. But from the copy of the Arret de Conseil which we have received from you, it appears that the Trial is required to be made either at Paris or the Moeres. If the Count d'Herouville could obtain permission for us to make this Trial in Britaine, and if the Trial so made would be effectual towards the obtaining the second Arret de Conseil, we could then without loss of time proceed to erect an Engine for M. Jery. – We have not thought proper to make M. Jery acquainted with any of the measures that have been taken towards obtaining an exclusive priviledge, but we have delayed giving him a decisive answer to his propositions till we knew whether his Engine can be made the subject of the Trial required by the Arret de Conseil. We therefore beg the favour of the Count d'Herouville to adjust that matter, if it can be done, & to acquaint us of his success that we may be able to give our answer to M[r] Jery, who waits impatiently for it. [*The second version of the draft lacks anything similar to the following two paragraphs*]

I also beg you will acquaint the Count d'Herouville that we have never yet had time to compleat the circular Engine from being so much engaged in the erection of Beam Engines, and we therefore wish our exclusive privilege to be confined to the other sort, of which alone we shall make the proposed trial, at least for some time.

Lastly I beg you to assure the Count that we only wait for his commands to execute the part promised by us towards him, and that I return him my sincerest thanks for his endeavours and good disposition to promote my interest not only in the fire Engine but also in other branches of my Manufacture.

I am
D[r] Sir
Yours

I have enclosed an Answer to the Questions proposed.
[*The passage that follows is written out separately but marked to be included at this point*]
In answer to the proposed Questions concerning the effects of the circular Engine, please to acquaint the Count that as we have not yet had leisure to carry it any further into execution, than was sufficient to convince us of the truth of our principles, we cannot speak with the same certainty & precision as we can do of the reciprocating Engine, but that we have not the least doubt of its being applicable to any purpose in which a circular motion is required, as mills for grinding corn, or rolling metals; that these Engines may be made of any degree of power that may be required; and that as to the expence of keeping them going, one man would be sufficient to manage an Engine, and a power equal to the strength of two horses may be obtained by means of 112 pounds of Newcastle coals per diem

I must defer till some other occasion answering the question relating to the articles and prices of my Manufacture. I beg however that you will please to acquaint the Count that it consists of whatever the Fashion of the Time requires to be made of copper, steel, silver-plated or gilded, with other articles such as paintings, I have sent one of these paintings (Penelope) to Mr Granchar, petit Dunkerque à Paris, with directions to be delivered to the order of the Count, of whom I beg that he will do me the honour to accept this specimen of my manufacture.
[*What follows is on a separate sheet*]
Question 1. La machine &c

Our circular machine will certainly be applicable to the grinding of corn or any other purposes to which water or wind-mills are commonly applied.
Question 2^{d}. Quel seroit ce son emplacement.
Question 3me. Combien feroit elle mouvoir de meules.
Question 4me.

Answer. As this machine has not yet been compleatly executed we cannot give a decisive answer to these questions, but only will observe that the Machine would require a room 18 feet by 12, besides the room requisite for the millstones; and further, that they may be made of any required degree of power, and that the quantity of fuel requisite to produce an effect equal to that of two horses is 112^{lb} of Newcastle castle [*evidently "coals" is intended; beneath "112^{lb} is written "one quintal"*].
Question 6. One person, even an old man, would be sufficient to manage the machine itself.
Question 7. The machine is certainly very capable of being applied to mills for rolling metal & either the circular or reciprocating Engine would be fit for forges.

78.08.26 *From John Walsh*

American Philosophical Society Library, Philadelphia, Fabroni papers, B/F113

M^r^. Walsh presents his Compliments to M^r^. Magellan and is very happy to find that the Abbé Fontana and M^r^. Fabroni together with Himself will honour his Roof with Their Company on Saturday next – The Gymnotus[1] and the moving Plant[2] will exert their Powers to entertain such Guests

Warfield, Berkshire
26th. Augt. 1778.

---

1 South American fish, the shocks delivered by which Walsh had demonstrated to be electrical.
2 Mimosa.

78.09.01 *To the Comte d'Hérouville*

Archives Nationales, Paris, MS Marine, G110, f. 176

Mon très cher Comte & très resp^ble^. ami

Le papier ci-joint contient la reponse que J'ai reçu dernierement de M^r^. Boulton sur les points en question que vous m'avez donnés pour Lui.[1] Je me flate que vous avez à cette heure arrangé completem^t^. l'affaire de Moeres,[2] & que vous vous trouvez à meme de savoir avec precision L'avantage qu'il y aura à profiter des machines à feu de M^r^. Boulton, soit à l'egard des eaux de Moeres, soit à celles de l'entreprise pour Paris.[3] En toute occasion oû je pourrai vous servir d'ici pour cet affaire, ou pour toute autre, Je me flate, que vous ne me refuserez pas le grand plaisir que J'aurai toutes les fois qu'il me sera possible de vous marquer ma vive reconnoissance par les bontés dont vous m'avez comblé: & Je vous prie de presenter ces mêmes dispositions de ma part avec mes très h^bles^. resp^s^. à

---

1 M. Boulton to Magellan, 22 August 1778?
2 Les Moeres, near Dunkirk.
3 See d'Hérouville to Magellan, 17 May 1778.

M^{de}. La Comtesse, à M^{le}. d'Herouville, & à Mlle. de Mondion,[4] etant pour le reste de la vie avec la plus grande consideration & parfait attachement 78.09.01

Monsieur
Votre très h^{ble}. & très obligé serviteur
J.H. de Magellan

à Londres ce 1 Sep^{bre}. 1778

P.S. J'attends avec impatience le reponse de M^{r}. Le Marquiz de Puy Segur sur la question relative au prix proposé par la Societé d'Emulation:[5] & Je voudrois savoir aussi s'il approuve qu'on lui envoye les 3 instruments d'Agriculture qu'il voudroit avoir, en me servant de la voye de Flandres: car dans les circonstances presentes je ne connois pas aucune autre moins longue, & plus certaine.

S'il y a quelques nouvelles de M^{r}. Le Comte Chastnet de Puy Segur,[6] vous savez bien combien elles me seront agreables, suivant le vrai attachem^{t}. de Coeur que je lui ai voué. Mylord Shelburne[7] se porte bien: il est actuelm^{t}. du cote de la mer en Hampshire pour prendre des bains d'eau salée.

A Monsieur
Monsr. Le Comte d'Herouville
Lieutenant gen^{al}. des Armées du Roy
rue du Fauxbourg S^{t}. Honoré n°. 6
à Paris

4 The Comte d'Hérouville's niece? See Magellan to d'Hérouville, 17 July 1778.

5 See Magellan to the Marquis de Puységur, 17 July 1778.

6 Antoine Hyacinthe Anne, Comte de Chastenet de Puységur (1752-1807), the Comte d'Hérouville's son-in-law. A naval officer, he was currently at sea with the French fleet; see Magellan to d'Hérouville, 17 July 1778.

7 William Petty, Earl of Shelburne (1737-1805).

78.09.01a *To Ribeiro Sanches*[1]

Österreichische Nationalbibliothek, Vienna
A.N. Ribeiro Sanches Korrespondenz, codex 12714, ff. 223-224v

Londres 1 Sep^{bre}. – 78

Meu am^{o}. e S^{r}. e m^{to}. do Cor^{ão}.

Agradeço m^{to}. a nova bem fofa q me dá da sua honrada agregação á Soc^{e}. R^{l}. de Medecina, com tanto q V.M. me faça a m^{ce}. de là nunca meter o pé; não só pelo dissabor de ouvir m^{ta}. patarata & m^{ta}. fofisse sobre o caroço mais empedernido da ignorancia; mas porq a consequencia será Resfriar-se & ter mais q gemer neste inverno ao canto da sua cheminé. Já aqui tenho no meo quarto o livro de Foyer com as addiçoens de Baynard, q he classico no seo genero, e he assaz raro. ámanhã terei o Anderson sobre a Industria, q imagino não pode ser tão bom como o Espanhol Campomanes, q aqui vi, e são 6 vol. onde se achão coisas mui acertadas & de bom Juizo. Ja ordenei um tal Charlton q imprimio um tratadinho em Oxford sobre os banhos frios, & cuido chegará no fim desta semana. na seguinte espero ter prontas varias encoñendas, q farei partir logo pela via de Flandres, & o commum amigo o D^{r}. le Begue du Presle terá cuid^{o}. de mandar a V.M. os artigos q lhe pertencerem.

Tãobem aqui chegou hü Senhor Bezerra estudante da Universid^{e}. de Coimbra, q fugio por não hir morar na Santa Inquizição, visto ser taréco q sabe francez & Inglez, com liç[ões]de Voltaire, Rousseaux, e outros autores mais espirituosos q espirituais. eu jantei com elle em caza do nosso Enviado p^{a}. q^{m}. trouxe cartas: & todos estamos bem servidos do caminho q lá tomão as coizas naquella terra bem aventurada. aqui me disserão q o Sonsinho do Alvares se faz tonto, ainda mais do q na realid^{e}. hé: e q assim será percizo q todos se portem p^{a}. escapar da perseguição dos fanaticos & furbos beatoens. mas voltemos folhas, & deixemos ao mar q ronque, visto cá estarmos ao Largo saons & Salvos.

Hoje foi já o prim^{o}. dia q fiz fogo no meo quarto, pois mudou o tempo quazi de repente p^{a}. bem frio: comtudo hirei jantar á manhan ao campo com alguns amigos daqui [a] coiza de 7 ou 8 milhas, onde acharei entre outros o moderno Cassandro, o D^{r}. Price, Autor q V.M. conhece, & que todos, até os mesmos da Classe Ministerial, estimão pela sua mod[estia] & sinceridade.

1 Annotated "R^{a} a 24 7^{bro}" [Answered 24 September]. Letter not found.

a triste nova da morte do Duque d'Arenberg me deo tanto abalo como pudera, se o não tivesse deixado em est[ado] feliz & contente com toda a sua familia em Enghien no dia 30 de Junho. o termo da vida he indispensavel, mas q este aconteça [no] tempo da felicid^{e}., deixa menos impressão nos q ficamos, do q q^{do}. isso sucede no tempo da tristeza. o Sabio olha com indiferença para o fim q elle sabe lhe compete por lei da sua existencia limitada: mas q elle seja forçado por força da sua infelicid^{e}. a sahir escalabrado como q^{m}. se despenha por entre rochedos p^{a}. o abismo onde se vai sumir, isso sim q dá m^{to}. q sentir a q^{m}. vê a tragedia. 78.09.01a

[v]i haverá 15 dias o D^{r}. Sequeira q passa bem. o La Cour veio perto de Londres haverá hũ mez; porem não o vi.

minhas saud^{es}. á familia adoptiva.

tenha a bond^{e}. de mandar o recado abaixo ao D^{r}. Le Begue du Presle, cortando-o desta carta

Vale & ama
tuum
Magellan

a Monsieur
[Monsr. Sanchez,] D^{or}. en Medecine
[Cimiterre de] S$^{t.}$ Jean, Rue de la [Verrerie][2]
à Paris

London 1 September –78

My Dear Friend

I thank you very much for the pleasant news of your aggregation to the Royal Society of Medicine,[3] provided you promise me that you will never set foot there; not only out of disgust at listening to a lot of silly things & much boasting about the most hardened kernel of ignorance, but because the consequence will be your catching a cold & having more to wail about at the corner of your chimney. I have

2 Paper cut, removing part of the address (see note 19). Missing words inferred from the address given on previous letters.

3 A society created in Paris in 1778 under the impulse of Lassone and Vicq d'Azir with the aim of counteracting the conservatism of the Faculty of Medicine of Paris. However, Magellan was not well informed on this since he here attributes the same conservatism to the new society. He recognized his mistake in his letter to Sanches of 6 November 1782.

78.09.01a already in the room with me the book by Foyer with the additions of Baynard,[4] which is a classic of its kind and very rare. Tomorrow I'll have Anderson's about Industry,[5] that I imagine to be not as good as the Spanish Campomanes,[6] that I saw here, and is in 6 vol. where very wise & judicious things are found. I've already ordered a certain Charleton who published a little treatise in Oxford on cold baths,[7] & I expect it will arrive at the end of the present week. Next week I hope to have several orders ready that I'll send at once through Flanders, & our friend Dr. le Begue du Presle will take care to send you the articles belonging to you.

Also arrived here is a Mr. Bezerra,[8] a student of the University of Coimbra, who ran away so as not to fall into the holy Inquisition's hands, because he is a naughty man who knows the French & English languages, who had lessons from Voltaire, Rousseaux, and other authors rather more spirituous than spiritual. I dined with him at our Envoy's home[9] for whom he had brought letters: & we are all aware of the way things happen in that blessed country. Here I was told that Sly Alvares[10] makes a fool of himself, more than he really is: and this is how everyone has to behave in order to escape the persecution of the fanatic & roguish hypocrites. But let us turn the pages, & let the sea snore, since we are afar, safe & sound.

Today I made a fire in my room since the weather changed almost suddenly, becoming very cold: however tomorrow I'll go and have dinner with some friends in the country 7 or 8 miles away, where I'll find among others the modern Cassandra,[11] Dr. Price,[12] an author whom you know & whom all, even those of ministerial class, esteem for his modesty & sincerity.

The sad news of the death of the Duke d'Arenberg[13] brought me such grief as it could, had I not left him in a happy & contented state with all his family in Enghien last 30th of June. The termination of life is unavoidable, but when it happens in

---

4 Probably *Ψυχρολουσια revived: or, the History of cold bathing both ancient and modern. In two parts. The first ... by Sir John Floyer. The second ... by Dr. E. Baynard ...* (fifth edition, London, 1722).

5 J. Anderson, *Observations on the Means of Exciting a Spirit of National Industry; chiefly intended to promote the Agriculture, Commerce, &c. of Scotland* (Edinburgh, 1777).

6 Pedro Rodríguez Campomanes y Sorriba, *Discurso sobre el fomento de la Industria popular* (Madrid, 1774)?

7 Rice Charleton, *An Inquiry into the Efficacy of Warm Bathing in Palsies* (Oxford, 1770).

8 João Paulo Bezerra de Seixas (1756-1817) had a degree in law from Coimbra University. He later held several positions in Brazil and was Portuguese Representative in the United States (1801), in the Republic of Batavia (1802-1809), and then in Russia.

9 The Portuguese Minister in London was Luiz Pinto de Sousa Coutinho (1735-1804).

10 Sanches' friend Dr J. Alvares; see Magellan to Sanches, 13 March 1778, note 19.

11 Prophet.

12 Richard Price (1723-1791), non-conformist minister and writer on morals, politics and economics.

13 Duke Charles-Marie-Raymond d'Arenberg died on 17 August 1778.

a time of happiness, it leaves less impression on those of us who remain, than when it happens in a time of sadness. The wise man looks with indifference to the end that he knows belongs by law to his limited existence: but when due to unhappiness he has to leave it wounded as if he has fallen over rocks to the abyss into which he'll disappear, that gives those who see the tragedy a lot to regret. 78.09.01a

I saw Dr. Sequeira a fortnight ago and he's well. La Cour passed near London a month ago but I didn't see him.

My regards to the adoptive family.

Please send the message to Dr. Le Begue du Presle, detaching it from this letter[14]

Vale & ama[15]

tuum

Magellan

---

14 A part of the sheet, including, on the reverse, part of the address, has been cut away, no doubt as a result of the message to which Magellan refers being detached.

15 Farewell and love.

To Johann III Bernoulli 78.09.08

Öffentliche Bibliothek der Universität Basel, MS L I a 709, f. 72

Londres 8 Sep^{bre}. –78

Mon cher Monsieur

Celle-ci est pour vous annoncer d'avoir Reçu les 84 livres Esterlines (ou 80 guinées) que le Bureau de la Longitude vous a presentés comm'une gratification pour les Tables que vous Leur avez envoyées, & qui seront publiées à leur depens pour Le bien public.[1] J'ai appris par notre ami commun Mr. le Dr. Wolfe de Danzic,[2] que vous aviez parti pour La Russie.[3]

---

1 See 78.01.30.

2 Dr Nathanael Wolf.

3 Bernoulli left Danzig for Königsberg and St Petersburg on 27 June 1778, and arrived back in Berlin on 21 October; see his *Reisen durch Brandenburg, Pommern, Preussen, Curland, Russland und Pohlen, in den Jahren 1777 und 1778* (6 vols, Leipzig, 1779-80). While in Danzig, he observed with Wolf an eclipse of the Sun on 24 June 1778. Their observations were reported by Magellan to the Royal Society of London on 11 February 1779; see Wolf to Magellan, 30 June 1778.

78.09.08 Cependant je vous avance cette nouvelle, pour que vous m'envoyez (tout de suite, & aussitot que vous la Recevrez à votre Retour) ordre de vous Remettre cet argent, en me disant par quelle voye? & comment? Soit qu'on vous garde cette Lettre jusqu'à Votre Retour; Soit que vous la Receviez dans votre voyage, vous n'avez qu'à donner vos ordres, & vous serez obei. Je vous recommende en particulier d'ecrire vos remercimens à M^r^. le D^r^. Sheperd parce que c'est Lui principalement qui a fait voir le merite & l'utilité de ces Tables dans Le Bureau de la Longitude: & Je crois que sans sa faveur & bonne amitié pour vous, on ne reussiroit point à rien obtenir.

Je me flate qui Vous Reviendrez en bonne Santé chez Vous, & que vous me conserverez toujours l'idée que Je me flate de vous meriter, savoir celle d'etre pour La Vie

Monsieur
votre très h^ble^. & fidel Serviteur
Magellan

N.B. Je n'ai pas touché votre argent que Samedi dernier: & je m'empresse de vous faire cet avis aujourd'hui par la premiere poste pour Berlin. Les depenses que J'ai faites pour recevoir cet argent ne sont pas grandes. C'est à dire pas au delà de 2 guinees & demie, ou environ.

78.09.08a *To Ribeiro Sanches*[1]

Österreichische Nationalbibliothek, Vienna
A. N. Ribeiro Sanches Korrespondenz, codex 12714, ff. 225-226v

Londres 8 de Sep^bro^. 78

Meo Am^o^. e S^r^. e m^to^. do Coração

Neste correio escrevo ao D^r^. Mertens, & ao Negociante por q^m^. lhe mandei o seo caixotinho de Livros. espero q se achará em algum canto dos Commissarios, & q não deixará de o receber o mais cedo q for possivel.

1 Annotated "resp^a^ a 24 7^bro^" [answered 24 September]. Letter not found.

Ainda q não Recebesse esta de V.M. por conta do Mertens, sempre tinha intenção de lhe escrever por fazer o gosto a hũ moribundo, q he parente de V.M., cuido q seo sobrinho, filho do Irmão q V.M. tem na Jamaica. Este sujeito chama-se Framengo, fez hũa boa fortuna; mas cahio numa tisica. partio ja tarde p^{a}. a Europa, e não he possivel q daqui passe, como dezejava p^{a}. Bordeos. o D^{or}. Sequeira me falou nelle, e me disse q teria gosto de verme, sabendo q eu estava [t]ão intimo com seo thio: fomos hontem pela manhan a Hackney, onde o achamos assentado nũma cadeira, hũ verd^{o}. esqueleto, quazi sem poder falar duas palavras juntas. Disse-me q seo Pai se achava m^{to}. bem forte e robusto ainda q de 76 ou 77 annos de idade là na Jamaica. O D^{or}. Sequeira me disse no caminho q este sujeito estava m^{to}. rico, e q tinha feito o seo testamento, em q talvez se lembrasse de V.M. 78.09.08a

Elle me disse tinha tido correspondencia com V.M., e q fazia tenção passar por Paris p^{a}. o ver. Porem está bem dezenganado q não tem q esperar senão o acabar aqui: a febre he continua: o D^{or}. contou 120 pulsaçoens por minuto; e a debilid^{e}. he extrema. Eu faço tenção tornar a visitalo: basta q seja coiza de V.M., e q lhe sirva de algũa consolação o ter algũas noticias suas, q eu lhe communicarei com gosto: pois não posso deixar de sentir com os q padecem: e faz-me maior impressão quando vejo algũ q tem os olhos assaz abertos p^{a}. ver o ultimo golpe com q se lhe acaba a scena da vida.

Não sei se disse a V.M. q aqui tinha chegado n'uma Náo Ingleza o govern^{or}. de Macáo, q he hũ tal S^{r}. Saldanha. Infelizm^{te}. deixouse apoderar de hũa hypocondria q lhe absorbe o juizo, & parece telo perdido. tudo são aprehençoens sem rima nem rezão: e está como hũ aturdido. diz q não dorme; nem toma gosto de tratar com ninguem. o nosso Ministro tem feito tudo o q pode p^{a}. o socegar: & lá foi hoje, ao q creio, morar p^{a}. o bairro ao pé de Sua Excellencia. Cuido q hirei a manhan jantar com este ultimo & verei, se o tal governador se acha melhor, com a sua mudança de quartel.

Não sei q^{m}. me disse q o S^{r}. Duque de Lafoens estava p^{a}. cazar em França antes de partir p^{a}. Lisboa. q seja com maior felicid^{e}. do q o fidalgo q V.M. sabe!

Coitada da Toinete: caro lhe tem sido o seo cazamento. Quanto mais valera o gozar sem ficar encangada p^{a}. toda a vida. A grande Ninon de l'Enclos foi hũa das maiores filosofas q jamais vio o mundo. Ouvi q hũ grande Senhor Inglez elevou da mesma sorte hũa sua filha unica q tinha; & q elle foi o 2do. vol. da grande Ninon.

78.09.08a Cá mostrei a historia da sua carta ao Excelente hommem de bem q aqui gozo: e todos vimos do agoamento q a Irman do seo ratinho arremangado, teria com as mãos diante dos olhos ....!

vale & ama

tuum

M.

a Monsieur
Monsr. Sanchez, Doct^{or}. en Medecine
Cimiterre S^{t}. Jean. Rue de la Verrerie
à Paris

London 8 September 78

My Dear Friend

In this mail I'm writing to D^{r}. Mertens & to the dealer through whom I sent your little packing-box with books.[2] I hope that it will be found in some corner of the Commissaries, & that he'll receive it as soon as possible.

Even had I not received this letter of yours on account of Mertens,[3] I intended writing to you to please one who is moribund, who is your relative, I guess your nephew, son of the brother you have in Jamaica. This man is named Framengo; he made a good fortune but got phthisis. He departed too late for Europe, and it's not possible for him to leave here, as he wished, for Bordeaux. D^{r}. Sequeira mentioned him to me and said that he would like to see me, knowing that I was very close to his uncle. We went yesterday morning to Hackney, where we found him sitting on a chair, a true skeleton, almost unable to speak two words together. He told me that his father was very well, strong and robust there in Jamaica though 76 or 77 years old. D^{r}. Sequeira told me on the way that this man was very wealthy, and that he had made his will, in which he may have remembered you. He said he had corresponded with you, and intended to visit Paris to see you. However he is quite mistaken since he can but wait for his end here. The fever is continuous, the D^{r}. counted 120 heart beats per minute, & his weakness is extreme. I intend to visit him again. It's enough that it's a matter to do with you, and that it may serve as some consolation for him to have news from you, that I'll transmit with pleasure since I can't refrain from being sorry for those who suffer, and I'm very impressed when I see someone who has his eyes open enough to see the last blow coming that ends the scene of life.

2 Letters not found. See Magellan to Sanches, 10 February and 13 March 1778.
3 Letter not found.

I don't remember if I told you that an English ship arrived here with the governor of Macao, who is a certain M^r. Saldanha.[4] Unfortunately he let himself be seized by a hypochondria that absorbs his mind, & he seems to have lost it. Everything is apprehensions without rhyme or reason: and he is like a confused man. Apparently he doesn't sleep and doesn't care to interact with anyone. Our Minister[5] has done everything to calm him & he went, I think, to live in his Excellency's neighbourhood. I imagine I'll have dinner with the latter tomorrow & I'll see if the said governor feels better for the change of quarters. 78.09.08a

I don't know who told me that the Duke of Lafoens was getting married in France before departing for Lisbon. May he find greater happiness than the gentleman you know![6]

Poor Toinete: her marriage has cost her dear. It would be better for her to have fun without having the yoke for life. The great Ninon de l'Enclos[7] was one of the greatest philosophers the world has ever seen. I heard that a great English gentleman educated his only daughter in the same way & that he was the 2nd vol. of the great Ninon.

I showed the story in your letter[8] to the excellent man whom I have the pleasure of dealing with: and we all saw the anxiety that the sister of your angry little mouse would have, with her hands in front of her eyes ...!

vale & ama
tuum[9]
M.

---

4 Diogo Fernandes Salema e Saldanha was Governor of Macao in 1767-1770 and a second time in 1771-1777.

5 Luiz Pinto de Sousa Coutinho (1735-1804).

6 There is no record of the Duke of Lafoens having married at this time.

7 Anne (called Ninon) de Lenclos (1615-1705) was a renowned courtesan in seventeenth-century France who later hosted a famous Parisian salon. She was also a writer who, in *La coquette vengée* (1659) and elsewhere, defended the possibility of living a good life in the absence of religion.

8 Letter not found.

9 Farewell and love | your.

78.10.09 *To Ribeiro Sanches*[1]

Österreichische Nationalbibliothek, Vienna
A.N. Ribeiro Sanches Korrespondenz, codex 12714, ff. 229-230v

9 out^bro^. – 78

Meo am^o^ e S^r^. e m^to^. do Cor^am^.

A de V.M. de 23 & 24 do passado me deo m^to^. gosto, não obstante o remoque de chamar modesta à q a V.M. escrevi no 1°. do mesmo mez. Porem p^a^. lhe abater o seo triunfo, lhe direi (ainda q á minha custa) q me acho com hũ redondo Leicenso verolé na pelle q serve de capa ao Padre de tuti I Santi. q estou entre as mãos de hũ cirurgião, & q tomo pirolas mercuriaes, alem da applicação do enguento do mesmo nome. Porem o q mais hé (e aqui espero q V.M. me fará uma carranquinha d'enraivado) q não sei qual foi das trez Nelis, a q me deo esta galantaria; pois as duas q vi depois, não parecem terem nada deste brinquinho; & pelo q toca á terceira ainda não tive animo de a hir vizitar; nem se lá for, lhe peleijarei; por ser hũa galante creatura, bella, alta, svelta, meixedira, & dotada das mais bellas coxas, & peitinhos q V.M. nunca vio na sua vida. Falei com meo amigo João Hunter, cuidando q visto não ser na glande, onde tinha saido o tal fruncho, talvez não seria venereo. Porem elle me disse q isso não fazia ao cazo; & me aconselhou a aplicação do Ung. Mercurial, com o qual tudo vai ás mil maravilhas, & espero q antes de 15 dias, hirei visitar outra vez as minhas Nelis e dar á saude de V.M. meia duzia de furos &^a^.

não foi possivel achar o Carlton sobre os banhos frios. mas achei hũ tal Henry q tem algũas coizas boas. tenho ja encommendada uma these de hum tal Athol, defendida en Edimburgo, q ainda não pude descubrir; mas q tenho boas esperanças. hũ amigo da Profissão Medica, me disse q no Tratado das agoas minerais de M^r^. Lucas de Dublin, se acha hũ bom capitulo sobre esta mat^a^. dos banhos frios.

o Anderson hé assaz estimado; mas não tem a Reputação de 1ra. classe: & o certo hé q poucos se achão na deste genero de conhecim^os^.

assim he q o Bezerra he filho de hũ minhoto, mas elle he Lisboeta, & tem m^to^. boa mezada p^a^. viver aqui com satisfação, & mesmo boa figura. mas he rapaz, & não tem a alma solta. Suspira por tornar ao ninho: & lá lhe achará o gosto no Cabo: pois aquella terra não he p^a^. mais do que edere, bibere [...]re.

1 Annotated "Sem contas, nem livros" [Without accounts or books] and "Resp^a^ a 18 Octobre" [Answered 18 October]. Letter not found.

ainda não vi, nem se me dá de ver as gazetas & jornais encyclopedicos, á castriotica. não he mato, donde saia coelho. 78.10.09

não me lembro de falar a V.M. em outro Cassandra, senão no D^{r}. Price, por ter prognosticado a verd^{e}. q se experimenta, sem ser acreditado nem crido.

Alegrome q o am^{o}. Chevalier vá p^{a}. comer a sua pensão do Padroado do S^{r}. Duque de Lafoens, de q^{m}. não tive mais outra noticia do q a q V.M. me dá.

Cuido q pelo correio passado lá teria V.M. a grande maravilha da carta do D^{r}. Sequeira. elle me mandou pedir a sua adresse; mas eu não estava ontem em caza q^{do}. elle passou p^{a}. ver-me: e assim não sei de certo se emfim escreveo a V.M., como tantas vezes prometeo sem effeito.

Sinto q la petite Marie, começa a mostrar q he franceza. assim q ella achar o seo suit heart, não a retenha V.M., pois he empreza impossivel. q lá se avenha.

não tornei a ver o seo sobrinho Salomão Nunes por alcunha o Framengo, mais do q hũa só vez; mais por effeito da humanid^{e}., sabendo q teria gosto de ouvir noticias de V.M. do q por comprimento da nossa amizade. Cuido q o D^{or}. Sequeira lá terá falado delle na sua carta. pobre moço! não tenho a minima esperança (nem elle tãobem a tem) de q escape desta. o seo estado não lhe permetra de hir a Bath: e alem disso elle está em Hackney q he hũ ar & hũ sitio excellente, à 3 milhas, ou 4 de Londres.

o govern^{or}. de Macao está melhor: & já disse hũa vez á Sua Exc^{a}. Pinto, q talvez o mercurio lhe tinha hido á cabeça. por isso q^{do}. lhe li o artigo da carta de V.M., sua Exc^{a}. disse = olhe o magano! q advinhou talvez donde vinha o mal! = Agora chegou seo Primo Miguel Pereira de Portugal: e quarta fr^{a}. jantei com ambos. no mesmo Paquebot veio hũ Irmão do tal govern^{or}. de Macao, provavelm^{te}. p^{a}. o conduzir p^{a}. caza: e na verd^{e}. necessita de conductor. eu não o vi, nem o conheço assaz p^{a}. o vizitar &^{a}.

Saud^{es}. á familia adoptiva e q vamos todos vivendo, como melhor pudermos emq^{to}. não entrarmos na Região do descanço Sempiterno. o Amigo Needham, aqui me escreveo n'uma carta como o Duque d'Arenberg Recebera com m^{ta}. pied^{e}. os S^{tos}. Sacram^{tos}. da Sta. Madre Igreja, & tivera o Nuncio á sua cabiceira. Creio q lhe daria huma boa carregação de S^{tas}. Indulgencias. olhe q boa consolação ...! p^{a}. q^{m}. pensa sem peocupação, nem asnatica imbecilid^{e}....!

a Monsieur
Monsr. Sanchez Doct^{r}. en Medecine
Cimitere S^{t}. Jean, Rue de la Verrerie
à Paris

78.10.09 9 October –78

My friend and very Dear Sir

Your letter of the 23rd & 24th of last month[2] gave me much pleasure, in spite of the remark in calling modest the one I wrote to you on the 1st of the same month. However to cut down your triumph, I'll tell you (though at my expense) that I find myself with a large furuncle verolé on the skin that covers the Padre de tuti i Santi.[3] That I'm in the hands of a surgeon & am taking mercurial pills, as well as applying the unguent of the same name. But what is more (and now I expect you'll make a grimace of anger), I don't know whom of the three Nellys gave me this gallantry, since the two I saw afterwards don't seem to have anything to do with this little present, & in what concerns the third I've had no spirit yet to visit her; nor will I dispute with her if I go there because she is a handsome creature, beautiful, tall, svelte, lively, & endowed with the nicest thighs & breast that you've ever seen in your life. I spoke with my friend John Hunter,[4] supposing that its not being at the glans where the said furuncle came out, perhaps it would not be venereal. But he told me that that wouldn't be relevant; & advised the application of the mercurial unguent, with which everything goes very well; & I hope that in less than a fortnight, I'll again visit my Nellys and have half a dozen copulations to your health &c.

It wasn't possible to find the Charleton about cold baths,[5] but I found a certain Henry[6] that has some good things. I've already ordered a thesis by a certain Athol, defended in Edinburgh,[7] that I couldn't find yet; but I have good hopes. A friend in the medical profession told me that a good chapter about this matter of the cold baths is to be found in the treatise on mineral waters by Mr. Lucas of Dublin.[8]

The Anderson[9] is well esteemed but doesn't have the reputation of being 1st class: & what is certain is that there are few works of that quality.

So Bezerra[10] is the son of a man from Minho;[11] but he is from Lisbon & has a good allowance to live here satisfactorily & even keep up a good appearance. But he is young, & does not have his soul free. He is anxious to go back to his

2 Letter not found.

3 testicle.

4 John Hunter FRS (1728-1793), the distinguished surgeon.

5 Rice Charleton, *An Inquiry into the Efficiency of Warm Bathing in Palsies* (Oxford, 1770); see Magellan to Sanches, 1 September 1778.

6 Not identified.

7 Samuel Byam Athill, *Exercitatio therapeutica, ehibens observationes quasdam de usu aquæ frigidæ externo* (Edinburgh, 1778).

8 C. Lucas, *The Theory and Uses of Baths* (Dublin, 1772), pp. 11-36. Magellan later also sent Sanches a copy of Lucas's earlier work, *An Essay on Waters* (3 vols, London, 1756); see Magellan to Sanches, 5 January 1779.

9 J. Anderson, *Observations on the Means of Exciting a Spirit of National Industry...* (Edinburgh, 1777); see Magellan to Sanches, 1 September 1778.

10 See Magellan to Sanches, 1 September 1778.

11 Province in the north of Portugal bordering Galicia.

nest: & in the end will find pleasure since in that country they do no more than edere, bibere [...]re.[12] 78.10.09

I haven't yet seen, nor do I care to see, the encyclopedic gazettes & journals of the Castrioto type.[13] It's not a bush from which rabbits come.

I don't remember having mentioned to you any other Cassandra except Dr. Price,[14] because he foretold the truth, which is subject to experiment, even without being believed.

I'm glad that our friend Chevalier is going to eat his pension through the patronage of the Duke of Lafoens, of whom I've had no more news than yours.

I imagine that by the last post you would have received the marvellous letter of Dr. Sequeira. He asked me your address; but I wasn't at home when he called to see me and so I don't know for sure if he wrote to you at last, as he has promised so many times without effect.

I'm sorry that la petite Marie[15] begins to show that she's French. As soon as she finds her suit heart,[16] you shouldn't restrain her since that is an impossible undertaking. Let her deal with it herself.

I haven't seen your nephew Salomão Nunes nicknamed Framengo again[17] more than once; more out of humanity than to fulfil our friendship. I suppose Dr. Sequeira may have mentioned him in his letter. Poor young man! I haven't the slightest hope (nor has he also) that he may escape this. His state won't allow him to go to Bath: and he is in Hackney which has an excellent air & is an excellent place, 3 or 4 miles from London.

The governor of Macao is better:[18] & I told His Excellency Pinto[19] once that perhaps the mercury has gone to the head. So when I read him the article of your letter, he said = look at the funny one! perhaps he foretold where the evil comes from! = Now his cousin Miguel Pereira has arrived from Portugal, and on Wednesday I dined with both. In the same Paquebot a brother of the governor of Macao arrived, probably to take him home: and he truly needs a guide. I haven't seen him, nor do I know him well enough to visit him &c.

Regards to the adoptive family and may we all keep on living the best we can while we wait to enter the region of eternal rest. Our friend Needham told me in a letter[20] how the Duke d'Arenberg piously received the holy sacraments of the holy mother church, & had the nuncio at his bed.[21] I believe he would have given him a good load of holy indulgences. Look what a great consolation ...! For those who think carelessly or with foolish imbecility ...!

12 To eat, to drink, to ....
13 Félix António Castrioto was a Portuguese journalist.
14 See Magellan to Sanches, 1 September 1778.
15 Maria Dieudonné? See Magellan to Sanches, 21 September 1777.
16 sweetheart?
17 See Magellan to Sanches, 8 September 1778. Sanches' father was also nicknamed Framengo.
18 See Magellan to Sanches, 8 September 1778.
19 Luiz Pinto de Sousa Coutinho (1735-1804), Portuguese Ambassador in London.
20 Letter not found.
21 Duke Charles-Marie-Raymond d'Arenberg died on 17 August 1778.

78.10.12 *From Matthew Boulton*[1]

Birmingham Central Library, Boulton and Watt Letter books, vol. 3, 79-80

Redruth[2] – 12 Oct 78

To
Mr Magellan / No 12 Nevils court Fetter lane London

Advised him of the arrival of self wife & Jerry[3] and that Jerry returns to Birmm. in a few days in order to finish his affairs there & then to go to France directly – Mr Jary hath been so very impatient about his Engine that I have been obliged to tell him what hath passed between us and le Comte d'Hernoville[4] as I wish to avoid giving him cause to say that we have used him ill by keeping him in suspence and therefore I have lately been very open with him. Mr Jary is now determined to go to Paris and to urge the Comte to get the word Paris or Moer in le'arret de council changed to Le Mine de Nord en Bretagne and by that means our trial and his engine would not be delayed but brought to a crisis in a few months – he says he can confirm to the count all we have said about the Engine and expedite the alteration proposed in the arret de Conceil and if Cte. Hl. were not at Paris could get it done by his own interest – but having high opinion of the Comtes honour we cannot do any thing without his approbation and as you have already taken a friendly part in this business hope you will complete it by going directly to Paris at our expence and try to bring the matter to a speedy conclusion agreeable to the conditions contained in my last,[5] which presume you have sent to Comte d Hl. I have not time to prepare a letter for the comte but you may depend upon having it next post & also a bill for £40, which I beg you will place to our account as a part payment for the trouble and expence you have been at – I fear the war[6] may have prevented the letters last sent you from coming to the hands of the Comte & think the only sure way of coming to a conclusion is you going over.

M. Boulton.

---

1 Original not found; text taken from the company letter-copy book.
2 Cornwall.
3 Joseph Jary.
4 i.e. Hérouville.
5 M. Boulton to Magellan, 22 August 1778.
6 France had gone to war with Britain in support of the American revolutionaries.

To Luiz Pinto de Sousa Coutinho[1] 78.10.13

Arquivo Histórico Ultramarino, Lisbon, AHU, CU-Reino , Cx 48, Pasta 37

Exc^{mo}. S^{r}.

1 Em satisfação ao dezejo de V. Exc^{a}. & á minha promessa, darei por escrito a noticia q já lhe participei em outra occazião, não só sobre as Maquinas à fogo ou p^{a}. melhor dizer à vapôr, q aqui se empregão & tãobem em varias p^{tes}. da Europa p^{a}. esgotar as minas & Levantar as agoas a qualquer altura; mas sobre os seos modelos.

Destas maquinas temos actualm^{te}. duas especies em Inglaterra. A mais antiga de q tãobem quazi todas as outras Naçoens do Norte se servem á imitação deste paiz, produzem todo o seo effeito pela ação do pezo da atmosphera, em consequencia da condensação do vapor, q se faz resfriar repentinamente dentro de hũ cylindro de ferro, com hũ esguicho de agoa fria. Porem as maquinas modernas não empregão, mais do q a ação do mesmo vapor, cuja condensação se faz em vazo differente do cylindro: donde se seguem duas grandes vantagens. 1^{a}. o não perder tempo à recuperar o calor perdido no cylindro, com o esguicho da agoa fria: & a segunda o empregar hũ agente m^{to}. mais forte do q o pezo do ar. De modo q as ventagens das modernas sobre as antigas são quazi de 75 por cento. isto he, se hũa das maquinas antigas gastão 400 milreis de carvão por mez, ou por anno, p^{a}. elevarem hũa certa quantid^{e}. de agoa à hũa tal altura: hũa das modernas, cuja construção não he mais dispendioza, produzirá quazi os mesmos effeitos, sem gastar mais do q cem mil reis de carvão no mesmo tempo. Monsr. Wat, q se acha associado com M^{r}. Boulton de Birmingham, ambos meos conhecidos & amigos, he o inventor destas maquinas modernas: & tem privilegio excluzivo, corroborado por hũ Acto do Parlamento p^{a}. elles sós serem os construtores dellas: elles em effeito já tem erigido hũ numero consideravel, com satisfação universal, em varias provincias d'Inglaterra; & aqui ao pé de Londres se achão quatro desta especie. hũa (piquena) em Bow: outra maior em Shadwell; outra em Richmond; & a quarta q brevemente se acabará em Chelsea, meia Legoa somt^{e}. distante de Londres, ou p^{a}. melhor dizer na extremid^{e}. desta Capital. No Verão passado me veio aqui recomendado de Nantes, M^{r}. Jarry, empresario das minas de Bretanha, com o projecto de obter hũa

1 Published by Donzília Pires, "Uma carta de João Hyacintho de Magalhaens", *Publicações do Museu Nacional da Ciência e da Técnica*, N°. 3 (Coimbra, 1973), pp. 125-140.

78.10.13 destas novas maquinas, e com hũa Carta q lhe dei pª. Mr. Boulton foi fixar-se a Birmingham pª. as observar, & se recapacitar das miudezas da sua construção: e segdo. o mesmo Mr. Boulton me disse, haverá 4 semanas vindo a Londres, o dº. Mr. Jarry ainda lá se achava pª. concluir este negocio. o Barão de Heinitz, q V. Excª. conheceo haverá dois annos em Londres, e q se acha hoje Ministro de Estado do Rey da Prussia na direção das minas, me escreveo q queria 4 destas maquinas; &, sobre a Reposta q lhe dei depois de alcançar a Resolução de Mrs. Wat & Boulton, elle me assegura na carta q a V. Excª. mostrei hũ destes dias, q o Barão d'Eschen & seo Irmão, ambos conselheiros dª. Direção, virião a Londres pª. este objeto, pedindo-me quizesse assistir lhes com as minhas informaçoens & avizos opportunos pª. melhor conseguirem os seos intentos. Os artigos q Mrs. Boulton & Wat requerem: são 1º. hũ privilegio exclusivo pª. q ninguem possa fazer destas maquinas, nos dominios & paiz sujetos no mesmo Sobrano, senão elles mesmos, ou seos agentes: & 2do. q lhe darão a 3ª. parte da despeza q com a sua maquina se poupar, comparandoa com qualquer outra das antigas que produza o mesmo effeito.

2 Pelo q Respeita a obter hum modelo perfeito destas novas maquinas, he coiza q ninguem, senão os mesmos Mrs. Wat & Boulton, pode fazer; mas eu sei q elles nunca o fizerão, nem fazem tenção de fazer este favor por valor ou interesse algum, segundo elles mesmos me tem assegurado repetidamente. Eu estou certo q nenhũ artista de caracter se atreverá a entreprender hum tal modelo; mto. mais estando certo q sem lhe dar as devidas proporçoens, esta maquina não produzirá o seo effeito. Eu mesmo q conheço mto. bem o seo mechanismo, não só pella ter visto em Birmingham, onde fui no Inverno passado com meo bom amigo o Conde Chastnet de Puy Segur, q aqui se achou quazi até a Primavera, pª. a ver trabalhar, e tomar as informaçoens com toda a eização; mas porq tenho visto esta maquina toda desmontada em pedaços, por cauza de hũ acide.; não me atrevêra a assegurar q um modelo feito segundo as medidas proprias della, estaria no cazo de dar completa & seguramente ideia da sua construção, com tal exação, q se pudesse fazer hũa grande sem arriscar mto.. A rezão q tenho pª. esta minha duvida, he porq apezar de todo o cuidado, não he possivel, ao menos moralmente, deixar de faltar ou de se enganar nas proporçoens das partes mais piquenas da maquina; & muitas vezes, muito pouco q nellas se erre, basta pª. q as q se fazem pelo modelo, não produzão o dezejado effeito. Alem disto deve considerar-se q a ação proporcional de hũ modelo, teria tão demaziada velocidade (nestas maquinas) q faria tudo em pedaços: e por esta rezão os modelos q se fazem das antigas pª. os Cabinetes de Physica, são alterados de modo q os coups de piston, ou

golpes do embolo, sejão consideravelmente retardados: donde se segue q q^{m}. fizesse executar hũa destas maquinas por hum dos tais modelos, teria uma maquina sũmam^{te}. imperfeita e defeituoza. 78.10.13

3 Eu ja disse a V.M., q tenho hũ bom artista, q me promete de fazer um destes modelos, que mostrará os effeitos das maquinas antigas por 75 guineos elle he sujeito m^{to}. capaz, e tem feito varios q forão aplaudidos. M^{r}. Allamand, Professor de Phil^{a}. experimental, e meo bom amigo, na Universid^{e}. de Leyde, teve hũ destes modelos, feito pelo mesmo sujeito, de q ficou bem satisfeito. Porem a rezão q acabo de mostrar sobre os modelos das maquinas modernas, milita tão bem à resp^{o}. das antigas.

4 Concluirei esta carta com dizer a V. Exc^{a}., com aquelle affetto q sempre conservarei á Patria até o ultimo da Vida, q p^{a}. lá estabelecer hũa destas maquinas com a maior segurança possivel, e ainda com a menor despeza (pois os gastos de alterar, mudar, e emendar a construção defeituoza de semilhante maquina, podẽ montar mais alem do dobro e do triplo do seo custo effectivo) Seria m^{to}. melhor mandar a Inglaterra algum engenheiro habil; mas q seja sujeito practico, e não unicam^{te}. theoretico; por exemplo algum moço q tenha já aprendido a trabalhar na ereção dos nossos moinhos, & em outras manufacturas; e melhor se fosse educado na nossa fundição &^{a}. Este passando em Birmingham 3 ou 4 mezes, ou em outra qualquer parte onde se achão trabalhando estas maquinas, se acharia em estado de tornar p^{e}. Portugal tão bem instruido q pudesse fazelas erigir & construir com toda a segurança e acerto. O exemplo q assima citei a V. Exc^{a}. de M^{r}. Jarry, mostrará facilmente, q este methodo he o mais certo p^{a}. obter o dezejado fim.

Fico m^{to}. p^{a}. Servir a V. Exc^{a}. e Sou com o mais Rev^{e}. Afecto
de V. Exc^{a}.
Servo mais humilde e obrigado
João Hyacintho de Magalhaens.

Londres, 13 Out^{bro}. – 78.

To
His Excellency the Portuguese Ambass^{or}.
Southadley Street
<u>London</u>

78.10.13 Dear Sir

1 Giving satisfaction to your wish and my promise I shall give in writing the news that I announced to you on another occasion, not only about the fire engines or, better said, steam engines that are used here as well as in several parts of Europe to drain mines and raise water to any height; and about models of them.

Of these engines we have presently in England two kinds. The oldest, of which almost all the other northern nations make use in imitation of this country, produce all their effect by the action of the weight of the atmosphere as a consequence of the condensation of vapour that is suddenly cooled within an iron cylinder by means of a cold water jet. However the modern ones use no more than the action of the same vapour whose condensation is produced in a different chamber from the cylinder: from which two great advantages follow. 1. Not losing time to recuperate the heat lost in the cylinder from the cold water jet & secondly, using a much stronger agent than the weight of the air.

Therefore the advantages of the modern ones over the older are of almost 75 percent: that is, if one of the older engines spent 400 thousand reis of coal per month, or per year, to raise a certain quantity of water to a certain height, one of the modern ones, the construction of which is more expensive, will produce the same effects but spending less than one hundred thousand reis of coal in the same time. Mr. Watt, who is associated with Mr. Boulton of Birmingham (both of them my acquaintances and friends) is the inventor of these modern engines & has the exclusive privilege confirmed by an Act of Parliament in order that they are their sole builders: in effect they have already built a considerable number with general satisfaction in several provinces of England; and here near London there are four of this kind. One (small) is in Bow, another bigger one in Shadwell, another in Richmond, and the fourth will soon be finished in Chelsea, just half a league from London or, better put, on the edge of this capital. In the past summer Mr. Jarry, contractor of the mines in Brittany, came to me recommended from Nantes with the project of obtaining one of these engines and, with a letter to Mr. Boulton that I gave him, he went to settle in Birmingham to observe them and to be informed of the details of their construction; and according to what the same Mr. Boulton told me 4 weeks ago on coming to London, the said Mr. Jarry was still there to finish this business. The Baron Heinitz[2] whom you met two years ago in London and who is now the king of Prussia's Minister for the direction of mines wrote to me saying that he wanted 4 of these engines; and on the basis of the answer I gave him following the resolution of M^{rs}. Watt and

2 Friedrich Anton von Heinitz (1725-1802), Prussian inspector general of mines and later Minister under Frederick the Great. The Baron von Eschen mentioned by Magellan was the mining entrepreneur Friedrich Sigismund Waitz von Eschen (1745-1808).

Boulton, he assures me in the letter that I showed Your Excellency some days ago that the Baron d'Eschen & his brother, both counsellors of the said ministry, would come to London for this purpose, asking me to assist them with appropriate information and advice to better achieve their aims. The terms that M^rs^. Boulton & Watt demand are: 1. An exclusive privilege that no one is allowed to build these engines in the domains and country of the said Sovereign except themselves or their agents, 2. that they will receive one-third of the expense that is saved with their engine in comparison with any of the old ones that produce the same effect. 78.10.13

2 As for obtaining a perfect model of these new engines, this is something that the same M^rs^. Watt & Boulton can make, but I know that they have never made any nor have the intention of doing this at any price as they have assured me repeatedly. I am sure that no artist of character will dare to undertake such a model and am sure, too, that without the due proportions this engine will not produce its effect. Even I who know its mechanism very well, not only because I saw it in Birmingham where I went last winter with my good friend the Count Chastnet de Puy Segur[3] who was here almost until spring to see it working and get the information with all exactitude, but because I have seen this machine disassembled in its parts due to an accident, would not dare to assure that a model made according to the same proportions would be such as to give a complete and sure idea of its construction with an accuracy such that a full-sized one could be made without great risk. The reason I have for this doubt is that in spite of all care it is not possible, at least morally, to overlook some proportions of the minor parts of the machine; and often little errors are enough for the engines made from the model not to produce the desired effect. Besides, we must consider that the proportionate action of a model would reach such high speed (in these machines) that it would fall apart and for this reason the models of the older ones that are made for physics cabinets are changed in such way that the coups de piston or piston strokes are considerably slowed down, from which it follows that he who would make one of these machines from one of those models would have a very imperfect and faulty machine.

3 I already told Your Excellency that I have a good artist who promises me to make one of these models to show the effect of the old machines for 75 guineas; he is very able and has made several that have been praised. Mr. Allamand, Professor of experimental physics at the University of Leiden and my good friend,[4] had one of these models made by the same person and he was very pleased. However the reasoning I just mentioned about models of the modern machine applies as well to the older ones.

4 I will finish this letter by saying to Your Excellency, with the same affection for the motherland that I will keep until the end of my life, that in order to set up one of these machines with the best possible safety but still at less expense

3 See Magellan to M. Boulton, 28 February 1778.

4 See Magellan to Sanches, 31 October 1776.

78.10.13 (since the cost to change, alter and modify the defective construction of such a machine could amount to more than double or triple its effective cost), it would be much better to send to England an able engineer but a practical and not just a theoretical one; for instance a young man who has already learned to work in erecting our windmills and other manufactures and better still if he has been educated in our foundries &c. This man if he spent 3 or 4 months in Birmingham, or anywhere else where these machines are working, would return to Portugal so well instructed that he would be able to see them built or erected with every safety and skill. The example that I gave before of Mr. Jarry will show easily that this is the more effective method to achieve the desired aim.

I am ready to serve Your Excellency and am with the most respectful affection
Your Excellency's most humble and obliged servant
João Hyacintho de Magalhaens

London, 13 Oct^{ber}. – 78

To
His Excellency the Portuguese Ambass^{or}.
Southadley Street[5]
London

5 South Audley Street?

78.10.15 *To Johann III Bernoulli*

Öffentliche Bibliothek der Universität Basel, MS L I a 709, f. 73

Londres ce 15 Oc^{bre}. – 78

Monsieur & cher ami

Cette Lettre vous sera delivrée ensemble avec une petite brochure que Je viens de publier ici, sur la mort de feu J.J. Rousseau,[1] par mon ami M^{r}. George Forster, que Vous connoissez bien de nom, etant Le meme, qui

1 A.G. Le Begue du Presle, *Relation ou notice des derniers jours de Mons. Jean Jacques Rousseau…* Avec une addition relative au même sujet, par J.H. Magellan (London, 1778).

ensemble avec son Pere firent Le voyage à la mer du Sud & au tour du Monde avec Le Cap^{n}. Cook: dont il a deja imprimé une excellente Relation, que vous aurez sans doute vû avec plaisir.[2] M^{r}. Forster a beaucoup de Talents & possede toutes les qualités estimables, pour meriter votre amitié: & Je suis bien assuré de son caractere que vous n'aurez occasion de vous repentir des faveurs que Vous lui accorderez en fait de Literature, ou d'introduction à vos amis: pour Les quelles Je vous serai aussi redevable comme si je les reçusse moimême.

J'attends avec impatience votre Reponse à la Lettre que Je vous ai ecrite le 8 Sep^{bre}.,[3] & que vous ne manquerez pas de trouver à Votre retour, qui selon l'article de la Lettre de M^{r}. Euler, ecrit de Votre main,[4] doit etre dans ce mois d'octobre, à peu près lorsque M^{r}. Forster arrivera Chez Vous.[5]

Je me flate que Vous pourrez me donner quelque detail de la Comete, dont M^{r}. Lexel semble avoir determiné La periode à 5 années.[6] J'ai annoncé cette decouverte à quelques uns de ma connoissance qui aiment l'astronomie & ils m'en demandent souvent des nouvelles ulterieures.

J'ai l'honneur d'etre avec un vrai attachement de Coeur

Mon cher ami
Votre très h^{ble}. & ob^{t}. serv^{r}.
Magellan.

---

2 George Forster's father, Johann Reinhold Forster, had expected to write the narrative of Captain James Cook's second voyage of discovery. However, Cook decided to write it himself, and Forster was ordered by the British Admiralty not to write a competing account. Forster, while obeying the letter of the law, set his son to writing it instead, and the result was published a few weeks before Cook's as *A Voyage round the World in his Britannic Majesty's sloop Resolution, commanded by Captain James Cook, during the years 1772-5* (2 vols, London, 1777).

3 Magellan to J. Bernoulli, 8 September 1778.

4 Letter not found.

5 Bernoulli returned to Berlin from his journey to St Petersburg on 21 October 1778. George Forster travelled to Germany in the same month in search of help for his father, who had been imprisoned for debt; he never returned to England.

6 A. J. Lexell, 'Tentamen astronomicum de temporibus periodicis cometarum et speciatim de tempore revolutionis cometae, A. 1770 observati', *Acta Academiae Scientiarum Imperialis Petropolitanae*, 1 (1777), 332-69.

78.10.16 *To Matthew Boulton*[1]

Birmingham Central Library, Boulton and Watt collection, Box 22/B3/10

London 16 Oct^{ber}. –78

Dear Sir

I just now received y^{r}. favour of y^{e}. 12 instant,[2] and think necessary not to delay my answer. I should at any time be ready to oblidge you, but am Certain, that my presence at Paris will not be necessary at all, for the object you have in view. the Privilege was already passed, and signed, and nothing is wanting but to comply with the required forms on your side. I do believe that M^{r}. Jarry will do all that is necessary in the Case: and I'll give him a Letter to M^{r}. de Heroville, who by the by I am affraid is in some bad estate of health, as no answer I had as yet to my Letters, since I came from abroad. He is a very old gentleman, and perhaps glad to drop all kinds of business. Let then M^{r}. Jarry come to me here, and I will furnish him with all the necessary informations and particulars relating to this object.

I beg you however not to forget sending, by him, one picture of your new manufacture, to M^{de}. d'Herouville, to whom I promised it in your name:[3] and this will introduce M^{r}. Jarry much better into M^{r}. de Herouville's house. For you know that the introduction under y^{e}. protection of y^{e}. fair sex hardly fails of success. I should not forgive myself if I was to put you now at y^{e}. unnecessary expenses of my jorney to Paris, without any real necessity for it: and besides I am so much Loaded with different important objets at this moment, that I either must wrong your affair at Paris or those already under my …[4] in London.

I have nothing to add for the present, but to wish you health and happiness; and ever to believe me

Dear Sir
Your most obed^{t}. h^{ble}. serv^{t}.
Magellan

P.S. your Letter to M^{r}. d'Herouville[5] was forwarded by the next post; viz the 31 August.

---

1 Letter not found; MS is a clerk's transcription. Annotated "M^{r}. Magellan Octr 16th 1778".

2 M. Boulton to Magellan, 12 October 1778.

3 See M. Boulton to Magellan, 22 August 1778.

4 Magellan here leaves out a word in going from one page to the next.

5 See M. Boulton to Magellan, 22 August 1778.

my friendly respects to Mr. and Mrs. Watt if they are as I think with you in Cornwall. 78.10.16

Pray don't forget what I have recommended to you about ye. new machines. make by all means a guess of ye. expenses of each, according to it's dimensions: and fix a premium in ready money, or in annual payments during your patent &ca. without any regard to the dearness of ye. fuel abroad. believe me, that no foreigner relishes to take things in the dark. if you settle with your partner a reasonable profit upon each machine of such, or such force; it will be much easier to understand, and to bring to a certain end every undertaking. I was applied to for information about this subject of late by a foreign Minister, to whom his Court gave orders for inquiring about: and he was satisfied with my only information, which he sent to his court. You may well think that the merits of your machines were not wanting due justice from me in this ocasion, as in any other soever. I wish you should send me a little sketch of ye. powers of your machines. viz. how many cubic feet they can rise water to a certain heigh for instance to <u>one foot</u> high, with a pound weight of coals: or something like this, that I hear'd from you once; but of which I have lost the memorandum I did made at that time.

To
Mr. Mathew Boulton
at <u>Redruthe</u>
in <u>Cornwall</u>

78.10.16a *From Matthew Boulton*[1]

Birmingham Central Library, Boulton and Watt Letter books, vol. 3, 81-2

Redruth.[2] 16th. Octr. 1778

To Mr. Magellan

Dr Sr. I had the pleasure of writing to you a few days ago from this place[3] since which I and Mr. Watt have had some farther conversation with Mr. Jary and in consequence thereof he seems very desirous of promoting all in his power, our interest in france as he is now convinced that he cannot obtain one of our Engines until we have obtained the exclusive privilege in France and upon farther reflection upon this subject I think Mons. Jary may be made very useful in obtaining it. In the first place he can certify that he hath seen many of our machines actually at work and that they are very much superior to the common Engine. 2dly. that there is no probability of their being introduced into france in a perfect state but by our means – 3dly. that it will be an accommodation to himself & the public if any privilege in France is confirmed as the Engine may be sent over in 3 ms. & 4th. that Bretagne is a more proper place to make the trial of comparison in, than Paris, because there are two Engines actually at work & the great expence of erecting an Engine at Paris will thereby be avoided. We have therefore drawn up a request in our name & Mr. Jary hath drawn up another in his own name to the conseil d'etat praying that the essai indiqué dans l'arret du 14 Avril soit fait sur la machine a feu qui sera etabli sur les mines de charbon du Nort en Bretagne. We intend to give him a letter to you[4] and another to the Comte d'Herouville & have desired him to lay before you and also before the comte our request & all of his papers relative to the subject. For be assured we will not take any measures that shall not be approved by you & the Count as we rely on you and him only, to conduct us through this affair. I am the more sollicitous to have this matter brought to a speedy conclusion from my apprehensions that if it is not, another party (whom you will reddily guess at) will by their pretensions prevent France from availing herself of all these improvements

1 Original not found; text taken from the company letter-copy book.
2 Cornwall.
3 See M. Boulton to Magellan, 12 October 1778.
4 Letter not found. It appears that it was not delivered; see Magellan to M. Boulton, 6 November 1778.

which M[r]. Watt has lately made or of the true rudiments of the principles of the machine – My principal reasons for wishing you to go over to france arose from the apprehensions that my letter (I sent you 2 m[o]. ago with hopes of your sending it to the Comte)[5] was by some means interrupted in the conveyance as you know that in war time there is no certainty of conveying Letters to such men as the Comte d'Herouville – But if you know of any safe an[d] sure means of conveying letters to the Comte then I think there is no necessity for you going over to paris unless when you see M[r]. Jary and all his paper – If you then think you can secure the point by going with him I will gladly pay you for your trouble and all the expences consequent thereon. I have given orders to Mess[rs]. Matthews & Barton[6] to pay you any money you are in disbourse for your last journey for or on our account in any other respect and likewise to pay to you any money you may think necessary for another journey provided when you have seen M[r]. Jary an[d] all his papers that you think you can promote the affair by going to Paris – Pray excuse the freedom with which I treat you & believe me most sincerely 78.10.16a

Your obliged friend & servant
M Boulton

If the commissioners from Prussia should arrive[7] and think it necessary to see us, I wish you would recommend it to them to come to this place where they will see our Engines in perfection & more curious mines than in any part of England – I shall stay a few weeks longer and should be glad to hear from you.

5 M. Boulton to Magellan, 22 August 1778. Magellan did forward the letter and it was received.

6 Matthews was Boulton's financial agent in London and Barton his associate.

7 See Magellan to M. Boulton, 21 July 1778.

78.10.23 *To Comte Joseph Velasco del Pico*[1]

Archief Van Arenberg, Enghien/Edingen, Belgium

Londres 23 Oc^{bre}. 78

Mon Cher Monsieur

Je vous envoye cette lettre pour S.A.S. que Je vous prie de Lui Lire dans quelqu'occasion moins embarassante Sans Compagnie. Les expressions avec lesquelles, J'ai dit dans cette Lettre, d'avoir fini mon addition à La brochure sur la mort de Jean Jacques Rousseau[2] (dont j'ai envoyé hier par la voye d'Ostende trois douzaines à S. Altesse) sont les suivantes, que mes amis ont trouvées ici pleines de sentiment: & Je vous les copierai en cas que S.A. ait la curiosité de les connoitre d'avance.

page. 44. «Je ne cache point que je fus surpris d'apprendre aussitôt la mort de M^{r}. Rousseau; mais je n'eus pas cette affliction tranchante que J'aurois sentie, s'il fut mort au milieu de ses malheurs passés. Ah! l'on est assez hereux de quiter La vie, avant de quitter le bonheur. Et pour ce qui regarde la mort, elle n'est qu'une suite de notre organisation, une Loy de la Nature, à laquelle il faut se soumettre sans murmurer. Au reste, tous les avantages de la vie ne meritent pas le regret de la perdre.»

J'ai perdu un si grand nombre d'amis & de personnes qui m'etoient cheres depuis que je pense plus à l'essentiel des affaires humaines; que J'ai eu occasion de reflechir souvent sur le sort de notre être, & sur les vrais motifs de la douleur acablante que nous éprouvons lors que nous les perdons. Il semble que nous accueillons ces afflictions avec une espece d'empressement... quelques fois nous y mettons une espece d'honneur, croiant que c'est un devoir, un vrai tribut à la memoire de nos amis. nous nous trompons assurement... c'est nous-mèmes que nous pleurons & non pas nos amis; car c'est nous seuls qui essayons la privation de ce bien etre, dont nous jouissions pendant leur vie: & c'est cette triste sensation, qui forme la douleur de les avoir perdus.

---

1 Joseph Velasco del Pico (1738?-1794?) was apparently born in Luxemburg. He was an officer in the service of Austria in Prague until 1773, when he became secretary to the Duke of Arenberg with whom he shared an involvement with Freemasonry. See Jules Mersch, "Des francs-maçons Luxembourgeois au 18me siècle", *Biographie nationale du pays de Luxembourg depuis ses origines jusqu'à nos jours* (XVIIIe fascicule, Luxembourg, 1970), pp. 345-478.

2 Jean Jacques Rousseau died in Geneva on 2 July 1778. Magellan is referring to his appendix to Le Bègue du Presle's *Relation ou notice des derniers jours de Mons. Jean Jacques Rousseau* (London, 1778).

Peu àprès la mort de Mr. Rousseau, J'essayai une autre perte qui me fût encore bien plus sensible; parce qu'outre les sentimens generaux d'admiration & de Respect qu'on Lui portoit, j'en avois par devers moi de gratitude & d'attachement personnels, soutenûs depuis plusieurs années. Mon affliction fut extreme: elle a eté commune a des Personnes les plus Illustres & Respectables, auxquelles je m'interesse par les devoirs les plus sacrés de la Reconnoissance: & à un grand nombre d'autres, dont le bonheur dependoit presqu'intieremt. de Sa Protection. (S. A. Sme. feu le Duc d'Arenberg[3]) J'eus alors plus que jamais besoin de faire des Reflexions serieuses, sur ces separations cruelles qui nous désolent, pour tacher de reconnoitre les principaux motifs qui occasionent cette vive douleur, que nous éprouvons dans des occasions pareilles; & pour décéler en fin, combien notre moi y joue le plus grand role à notre insçû.» 78.10.23

«En Considerant que la mort est une condition attachée à notre existence, je me suis dit: pourquoi nous plaindre... Ah! ne nous imaginons point de changer l'ordre de l'Univers par d'inutiles clameurs! Soyons Sages. Respectons les decrets de la Toute Puissance: & ne raportons point à nous Seuls tous les objets de la Creation. Non, me suis je dit: je ne payerai plus un tribut avili par l'amour de moi-meme, à la memoire de ceux qui me furent chers. En les perdant j'ai de moins autant de liens qui m'attachoient à la vie. Mais, tandis que je leur survivrai, je sacrifierai mes attentions & mes homages à tout ce qu'ils ont laissé de plus cher: & je cultiverai, encore au delà du tombeau, leur amitié, par mon attachement & par mes services.»

«L'on m'excusera, si l'on veut, d'avoir mis si peu de suite dans mes idées, & d'avoir tant parlé de moi meme & de la douleur dont je suis acablé, dans ces lignes que je viens d'ajouter à cette Relation. C'est qu'en écrivant j'ai suivi les impulsions de mon Coeur: & quand il est emû, on le sçait, l'esprit a peu à faire: la main copie aveuglement le sentiment de l'ame. Je ne fais point d'apologie: & je n'efface rien; parce que ce n'est pas un traité que J'ecris, mais un épanchement de Coeur, que je laisse echaper, pour adoucir ma douleur.» Fin.»

Les deux mots que J'ai mis dans une Note, en obseque de feu Monsgr. Le Duc, iront peut etre plus loin à la posterité, à coté du Recit d'un Rousseau, qu'un eloge funebre fait tout exprès à sa Memoire. Quoiqu'il en soit pour le fait, j'ai suivi Mon penchant en philosophe & à ma maniere, c'est à dire comme je fus en etat de la faire.

---

3 The Duke Charles-Marie-Raymond d'Arenberg died at Enghien on 17 August 1778.

78.10.23 Je vous serai bien obligé, si vous me voulez informer sur l'etat de la Caisse que J'envoyai à feu Mons^{gr}. Le Duc dans les derniers jours de sa vie, mais qui n'etoit pas encore arrivée lors que vous m'avez ecrit votre Lettre derniere.[4] car il y avoit des Livres pour l'Abbé Mann, & pour le Secret^{e}. de l'Acad^{e}. Imp^{le}.;[5] & Je souhaite savoir s'ils les ont Reçus; car leur Silence me laisse indecis de ce qu'ils sont devenus.

Je vous prie d'emploier ma bonne volonté en tout qui sera de votre Service. dites moi si son A.S.[6] a deja fait l'operation de la cataracte & donnez moi des nouvelles de toute Son Auguste Famille.

Je suis de Coeur & pour La vie

Monsieur

votre très h^{ble}. serv^{tr}

Magellan.

A Monsieur
Monsr. Velasco, Secretaire de
Son Alt^{e}. S^{me}. Le Duc d'Arenberg
à Bruxelles

4 Letter not found.
5 J. Des Roches (?-1787) was *secrétaire perpétuel* of the Academy of Sciences in Brussels.
6 i. e. the new Duke, Louis-Engelbert d'Arenberg.

To the Duke Louis-Engelbert d'Arenberg[1]

Archief Van Arenberg, Enghien/Edingen, Belgium

Monseigneur

J'ai pris La liberté d'envoyer à V. A. S^me^. une petite caisse par la voye de M^r^. Hennessy à Ostende,[2] contenant trois ou quatre douzaines d'une brochure, que Je viens de publier ici sur les derniers jours du pauvre Jean Jacques Rousseau. J'y ai fait une addition, dans La quelle je n'ai pû m'empecher de finir sans exprimer Les circonstances, oû Je me trouvois en l'écrivant.[3]

Ma plume n'y a fait que copier Les sentimens du Coeur: & V. A. S. le connoit assez bien, pour ne pas me refuser L'honneur & les croire bien sinceres.

Je souhaiterois que cette brochure fut distribuée dans l'Auguste Famille de V. A. S. & entre ses bons amis. J'ai adressé moi-meme La 4me. douzaine à ceux que Je connois, & J'espere que Mr. Velasco[4] aura La bonté de la faire distribuer ou delivrer Respectivement.

Mon ami Le Dr. Inghenhouz[5] m'a dit un de ces jours, chez le Conte de Belgioioso,[6] que S. A. S. vouloit qu'il fut inoculer La jeune Princesse, sa fille; Je me flate que desormais il n'en aura aucun dans Son Auguste Famille qui ne veuille pas se prevaloir de ce grand secour, pour s'affranchir d'une maladie si fatale.[7]

J'ai L'honneur d'etre avec le plus profond Respect & plus vrai attachement de Coeur,

Monseigneur

de V. A. Sme.

Servr. très hble. & très obligé

Jean Hyacinthe de Magellan.

---

1 Louis Engelbert Pierre Marie Joseph Augustin (1750–1803) was the second son of Charles-Marie-Raymond d'Arenberg and his wife Louise Margaret von der Marck. He became the 6th Duke of Arenberg at his father's death on 17 August 1778.

2 Patricius Hennessy (1741-1784), head of the Ostende trading company, Patricius Hennessy & Co.

3 Jean Jacques Rousseau died at Geneva on 2 July 1778. Magellan arranged the publication, in London, of Le Bègue du Presle's *Relation ou notice des derniers jours de Mons. Jean Jacques Rousseau* (London, 1778), to which he added, as an appendix, an account of his own visit to Rousseau shortly before he died. In this, he made reference to his own emotions on learning of the death of Duke Louis' father.

4 Joseph Velasco del Pico was secretary to the Duke of Arenberg. See 78.10.23.

5 Jan Ingenhousz or Ingen-Housz (1730-1799).

6 The Austrian Ambassador in London.

7 Inoculation against smallpox had been a common practice in Britain since the 1720s, but less common elsewhere in Europe. King Louis XV of France had died of smallpox only four years before this, in 1774.

78.10.27 *From Johann Albrecht Euler*[1]

St Petersburg Branch, Archives Department, Russian Academy of Sciences, f. 1, op. 3, Nr. 55, l. 94

à St. Petersbourg d. 27 Octobre 1778

Monsieur

J'ai l'honneur de Vous notifier, que l'Académie Impériale des Sciences Vous a reçu au nombre des ses Associés externes.[2] C'est un témoignage public du cas distingué et de l'estime parfaite qu'Elle fait de Vos travaux et que Vous meritez à tant de tîtres.

Je m'applaudis d'en être dans ce moment l'interprète et je faisis avec un véritable plaisir cette Occasion de Vous exprimer, Monsieur, la joye que Votre aggregation à notre Academie m'a fait.

Dès que le Diplome Académique sera signé et expedié je le remettrai avec la première Occasion;[3] je me récommande à vôtre bienveuillance, et j'ai l'honneur de me dire avec une consideration distinguée

Monsieur
votre très-humble & très obeissant
Serviteur & Confrère
Jean Albert Euler

à M^{r}. de Magellan à Londres

1 Annotation: "Copie".

2 Magellan was elected a foreign member of the St Petersburg Academy of Sciences on 28 September 1778 Old Style.

3 Diploma not found.

Bill of Exchange[1] 78.10.27a

Istituto Lombardo, Milan, Racc. fr. Volta, 53[2]

Londra a *27 Ottobre 1778* Per *P 193.8.10 da s 115 f. B*^{co}.
A due Mesi data Pagate per questa *mia* prima di Cambio *all'ordine delli SS*^{i}*. Ant*^{o}*. e Bart*^{meo}*. Songa: Perze Centonovantatrè soldi otto, e denari dieci da Soldi Centoquindeci fuori banco* – – – – – – – – – –
valuta *da' Medesimi* – – – – – – – – – – – che passarete *in Conto* – – – Secondo L'aviso a Dio

Jean Hyacinthe de Magellan

All' Illmo*. Sig*^{r}*. D*^{n}*. Alessandro Volta Di Como*
prima
in Genova

London 27 October 1778 For p *193.8.10 at 115 genovese soldi*[3]

In two months pay on this *my* first bill of exchange *to the order of Messrs. Antonio and Bartolomeo Songa: one hundred and ninety three pezze eight soldi and ten denari at the rate of one hundred and fifteen genovese soldi*
currency *by the same*[4] and pass it *to account* as per the banking agreement between us

God be with you

Jean Hyacinthe de Magellan

To the most illustrious Mr. Alessandro Volta of Como
first
in Genoa

---

1 See Magellan to A. Volta, 27 October 1778 (i.e. 78.10.27c). For a discussion of 18th-century bills of exchange, see Edward Hatton, *The Merchant's Magazine* ... (London, 1719), pp. 205-11, 259.

2 Published in *Epistolario di Alessandro Volta*, vol. 1, pp. 292-3. Two bankers' documents signed by Magellan. The documents are printed forms with spaces that have been filled in; the sections that have been filled in are printed here in *italics*. On the first document, a note has been added in another hand: "Accett^{a}. con Sua de 17: 9bre" [accepted with his of 17 November]. The second has two bankers' annotations on the reverse.

3 Lire and soldi "fuori banco" were the money of Genoa previous to its annexation to Sardinia.

4 i.e. in accordance with the currency fixed by the same Messrs Songa.

78.10.27a Londra a *27 Ottobre 1778* Per *P 193.8.10 da s 115 f. B*[co].
A due Mesi data pagate per questa *Mia* Seconda di Cambio (non avendolo con la prima) *all'ordine delli SS*[ri]*. Ant*[o]*, e Bart*[meo]*. Songa: Perze Centonovantatrè, Soldi Otto, e denari dieci da soldi Centoquindeci fuori banco* – – – – – – – – valuta *da Medesimi* – – – – – – – che passarete *in Conto* – – – – – Secondo L'Aviso a Dio

Jean Hyacinthe de Magellan

All' Illmo*. Sig. D*[n]*. Alessandro Volta di Como*
2do.
in Genova

London 27 October 1778 For p *193.8.10 at 115 genovese soldi*

In two months pay on this *my* second bill of change (not having done it with the first) *to the order of Messrs. Antonio and Bartolomeo Songa: one hundred and ninety three pezze eight soldi and ten denari at the rate of one hundred and fifteen genovese soldi* currency *by the same* and pass it *to account* as per the banking agreement between us

God be with you.

Jean Hyacinthe de Magellan

To the most illustrious Mr. Alessandro Volta of Como
second
in Genoa

To Jan Henrik van Swinden[1] 78.10.27b

Leiden University Library, MS B.P.L. 755, no. 297

Mr. Magellan presente ses Comps. à Mr. Van Swinden, & lui envoye cette petite brochure qu'il vient de publier.[2] il n'a rien de nouveau assez interessant en Philosophie; à moins que l'efronterie enragée du charlatant Wilson ne soit un phenomene assez singulier: car malgré la decision de tous les bons philosophes, & de toutes les experiences faites avec sincerité, & sans tour de main comme les siennes, il est evident & demontrable que les Conducteurs electriques pour garantir de la foudre, doivent etre pointûs, & pas emoussés, ou arrondis en boule comm'il le pretend: & ce qui est singulier, il ecrit & tache de faire ecrire tout par tout, d'avoir convai[n]cû & montré tout le Contraire par son escamoutage.[3]

Vale & ama
tuum
Magellan[4]

le 27 Octbre.–78

---

1 Annotated at end: "Rec. 9 Novembr. 1778." Published in David Willemse, "Suite d'un voyage aux Pays-Bas: João Jacinto de Magalhães (1722-1790) et ses rapports avec Jean Henri van Swinden (1746-1823)", *Arquivos do Centro Cultural Português*, 7 (1974), 225-278. Willemse has, however, read the date of the letter as 24 October 1778.

2 *Relation ou notice des derniers jours de Mons. Jean Jacques Rousseau; circonstances de sa mort; et quels sont les ouvrages posthumes, qu'on peut attendre de lui*, par Mons. Le Begue de Presle, avec addition relative au même sujet par J.H. de Magellan (London, 1778).

3 Benjamin Wilson, *An Account of Experiments made at the Pantheon, on the Nature and Use of Conductors* (London, 1778), in which he argued against the recommendations of the Royal Society's committee on lightning rods embodied in its report, *Philosophical Transactions*, 68 (1778), 236-8.

4 Farewell & love | your | Magellan.

78.10.27c *To Alessandro Volta*

Archivio di Stato di Milano, Fondo Studi, parte antica, cart. 134[1]

Londres 27 Oc^{bre}. –78

Monsieur

Le Navire Hollandois nommé Catherine, dont le Cap^{tn}. est Valkert Hendrick qui vient de partir pour Genes, vous apporte 3 Caisses, marquées de Votre nom, dont voici le Contenû. La machine d'Achard, ou pour mieux dire de ma construction pour repeter ses experiences[2] montée sur une petite table de mahogany avec deux Cylindres, l'un pour servir, & l'autre pour y supleer en cas d'acident £ 4. 2. 6 / Coin de mahogany avec bassin de porcelaine pour des experiences avec du mercure, à la façon du D^{r}. Priestley = 3^{s}. 10^{d}. / maquineta pour passer des verres par dessous l'eau 14^{s}. 6^{d} / 3 petites bouteilles & 3 verres avec deux Couvercles de bois &^{a}. 5^{s}. 6^{d} / deux cassettes de mahogany pour des experiences en grand avec du mercure, à la façon de l'abbé Fontana £1: 5^{s}: 4^{d} / un beau oval avec des tiroirs au dessous, des jambes à vis pour le mettre de niveau £5. 14. / piece de mahogany assez epaisse pour servir de tablette dans le meme ovale 6. 6. / lampe & trepied pour fair de l'air dephlogistiqué en compagnie &^{a}. 2^{s}. 0^{d} / 3 verres grands, comme Recipiens, pour garder grandes quantités d'air nitreux à la fois 6^{s}. / une livre & demie de tubes pour faire des communications, en les pliant à la lampe &^{a}. 4^{s}. 6^{d}. / 6 supports de fer blanc japané de brun, pour soutenir des phioles en digestion 6^{s}. 0^{d} / 8^{lb}. de fluor spatosus 7^{s}. 6^{d} / les deux caisses 11. 10. / 3 suites de jares petites & grandes pour le mercure 15^{s}. 0^{d} N.B. je n'ai pas le tems de vous copier le compte des verres de Parker qui monte à £17. 7^{s}. 6^{d} que Je vous envoyerai sous la couverte de M^{r}. le Comte de Firmian.[3] Car la poste va fermer dans peu de tems: & il faut vous avertir d'avoir tiré sur vous pour La Valeur de cet envoy: dont la somme faite à la hate, monta à £32: 7^{s}: 2^{d}. mais je viens de trouver deux meprises, que je rectifierai dans la premiere occasion. Je fis assurer la valeur de les 3 caisses par £4. 4^{s}. 0^{d} la polizza[4] couta 6^{s}: 4, & le total monta à £36: 17: 6. qui par le

---

1 We are grateful to Dr Alessandra Ferraresi, Università di Pavia, for drawing this letter to our attention.

2 See Magellan to Volta, 9 April 1779.

3 Carlo Conte di Firmian (1718-1782), liberal Lombardy politician and Maecenas of innovative philosophers such as Volta.

4 Insurance contract.

Change courant, donna pezze 193: 8: 10^{d} pour les quelles Je passai lettre de Change à payer en deux mois de ce jour, à faveur des memes Messrs. Songa, à la quelle vous ferez honneur avec votre aceptation.[5] 78.10.27c

le tems presse si fort qu'a peine ai je Celui de fermer cette lettre pour empecher qu'elle ne soit retardée pour La poste suivante. Je serai plus long dans la lettre suivante.

Vale & ama
tuum
Magellan[6]

A Monsieur
Mon Sieur Alexandre Volta
Professeur de Philosophie
à Como
près de Milan

---

5 See bill of exchange to Volta of the same day.

6 Farewell & love | your | Magellan.

To Ribeiro Sanches[1] 78.10.30

Österreichische Nationalbibliothek, Vienna
A.N. Ribeiro Sanches Korrespondenz, codex 12714, ff. 231-232v

Londres 30 Out^{ro}. – 78

Meo Am^{o}. e S^{r}. e m^{to}. do Cor^{am}.

Ainda não pude ver o D^{r}. Seq^{ra}. depois q Recebi a m^{to}. estimada de V.M. de 18 do Corr^{te}., p^{a}. saber de seo sobrinho e tirar mais algũa informação. O q me parece, he q V.M. lhe escreva a elle se puder; pois quando mais não seja, sempre lhe dará gosto, & a especie de consolação q qualquer no seo estado, ainda pode Receber.

---

1 Annotated "R. a 5 de 9^{bro}" [Answered 5 November]. Letter not found.

78.10.30 acho-lhe graça ás Rapaduras do tacho q V.M. supoem me hão de ficar do prez^{te}. da minha Nely. pois esteja certo q se engana, pois já estou, e estarei bom, & capaz p^{a}. outra semelhante. Tomára eu q V.M. a visse hũ destes dias q me veio fazer vizita: q^{to}. chorou, & protestou q não era ella q^{m}. me fizera aquelle cadot; tomára q V.M. visse as provas oculares q me fez, levantando o docel das alegrias, & abrindo, com dois dedinhos de marfim, os beicinhos da mais linda gruta das Parcas, cercada da mais bella fillegrana d'oiro q V.M. nunca tal vio na sua vida.

Sinto q V.M. se acha com catarro, febre, & tosse; mas q^{to}. ás suas purrias economicas, não conserve mais tempo o coração de puta de Riba-Coa: & tudo hirá ás mil maravilhas. Como quer V.M. exercitar o atributo da omnipotencia, q só compete ao Padre eterno? faça o q puder, sem incommodo do seo fizico necessario: e ria-se do resto desta vida buzarona.

Já aqui tenho o Livro de Lucas, q diz algũa coiza com geito sobre os banhos de agoa fria, & sobre o seu uzo &^{a}. Porem ainda não pude obter a these de Edimburgo. o sujeito já esta em Inglaterra, & dizem q poucas milhas de Londres: mas ainda não pude velo, nem falar a q^{m}. o visse. he rapaz, & pouco pode dizer, mais do q repetir o q os outros disserão. eu aqui tenho varias theses de Edimburgo q lá lhe mandarei, & V.M. com o D^{r}. Begue du Presle as repartirão entre si.

o D^{r}. Price, he o q escreveo on Civil Liberty, mostrando o perigo Politico da Ruptura com as nossas Americas. ninguem o crêo, e agora [to]dos[2] vem os effeitos q elle predisse. por isso lhe chamo o bom Cassandro. elle he Ministro, dissidente, m^{to}. amigo do D^{r}. Priestley & tãobem meo.

Seg^{da}. fr^{a}. proxima vai partir p^{a}. Lisboa o Saldanha, q foi gov^{or}. de Macáo, juntam^{te}. com seo Irmão e Sobrinho q o vierão buscar. elle acha-se restabelecido da sua passada doença. o Nosso Ministro, vai mandar o menino (q he o mais velho de 4 annos, q tinha ficado com elle em Londres) p^{a}. sua Mãy ao mesmo tempo, isto he no mesmo paquette em q partirâõ estes fidalgos Saldanhas, acompanhado de hũa criada & hũ criado. deste modo ficará só sem familia, mais do q os criados & o Secretario. Verd^{a}.m^{te}. he vida triste p^{a}. q^{m}. está como elle: e ainda q tenho sũmo gosto de aqui o ver empregado, tomára, q elle tãobem fosse p^{a}. sua caza, fazer bem aos seos dependentes &^{a}. e não viver assim auzente, gastando o seo e a sua vida a servir grandes ingratos & mal Reconhecedores do q se lhes faz.

2 Partly illegible.

de politicas nem sei nem quero saber. Deixalos lá quebrar as cabeças: q os Portugueses, ou os espanhoes, ou os Inglezes, ou outros quaesquer fiquem ou não sevandijados pouco me importa. 78.10.30

Vale.

a Monsieur
Monsr. Sanchez, D^or^. en Medecine
Cimitere S^t^. Jean, rue de la Verrerie
à Paris

London 30 October –78

My Dear Friend

I haven't yet been able to see D^r^. Sequeira since I received your much appreciated letter of the 18th of this month,[3] to learn about your nephew and get some more information. What seems right to me is that you must write to him if you can, because at least that will give him some pleasure & the sort of consolation that anyone in his condition can still receive.

I find funny the scrapings of the pan that you suppose will stay from the present from my Nelly.[4] Well, be sure that you're wrong, since I already am and will be well, & capable of a similar one. I wish you could have seen her one of these days when she paid me a visit: how she cried & protested that it was not her that had given me that cadot;[5] I wish you could have seen the visual proofs she made to me, lifting the dossal of joys & opening, with two little ivory fingers, the small lips of the most beautiful grotto of the Parcae, surrounded by the prettiest golden filigree that you ever saw in your life.[6]

I'm sorry that you find yourself with a cold, fever & cough; but as for your economic trifles, don't keep any longer the heart of a Riba-Coa whore[7] & everything will go marvellously. How dare you exercise the attribute of omnipotence that only belongs to the eternal father? Do what you can, without troubling your physique unnecessarily: and laugh at the rest of this buzarona[8] life.

3 Letter not found.
4 Magellan was suffering from a venereal infection; see Magellan to Sanches, 9 October 1778.
5 i. e. cadeau.
6 The Parcae, or three Furies of Greek mythology, were often associated with childbirth, hence Magellan's image of Nelly's pudenda as their grotto.
7 Colloquially understood to be penny-pinchers.
8 deceitful. See Magellan to Sanches, 23 June 1775, note 5.

78.10.30 I have already with me the book by Lucas[9] that says something reasonable about cold baths & their use &c. However I could not yet get the thesis from Edinburgh.[10] The man is now in England & they say only a few miles from London: but I haven't been able to see him, or to speak to anyone who has seen him. He is a young man & there is little more he can say than repeat what others have said. I have here several theses from Edinburgh to send to you, & you will share them with Dr. Begue du Presle.

Dr. Price is the one who wrote on Civil Liberty,[11] showing the political danger of breaking with our Americas. Nobody believed him, and now all see the effects that he predicted. That is why I call him the good Cassandra.[12] He is a dissenting minister, a good friend of Dr. Priestley & also of mine.

Next Monday Saldanha, who was governor of Macao, departs for Lisbon, together with his brother and nephew who came to fetch him.[13] He has recovered from his past illness. Our Minister[14] is sending his son (the oldest, 4 years of age, who had stayed with him in London) to his mother at the same time, that is, on the same ship on which these noblemen, the Saldanhas, go, accompanied by a maid & one servant. In this way he will stay alone without family, other than the two servants & the secretary. Truly it is a sad life for a man like him: and though it gives me great pleasure to see him employed here, I wish he would also go home to be good to his dependents &c. and not live abroad, spending his wealth and his life serving those ingrates who don't acknowledge what is done for them.

Of politics I don't know anything and don't want to: let them break their heads: I don't care whether the Portuguese, or the Spaniards, or the English, or any others are cringing with fear or not.

Vale.[15]

---

9 From Magellan's letter to Sanches, 5 January 1779, we learn that he is here referring to the three-part work by C. Lucas, *An Essay on Waters* (London, 1756), not to the extract from this work that was later published as *The Theory and Uses of Baths* (Dublin, 1772).

10 Samuel Byam Athill, *Exercitatio therapeutica, ehibens observationes quasdam de usu aquæ frigidæ externo* (Edinburgh, 1778).

11 Richard Price, *Observations on the Nature of Civil Liberty, the Principles of Government, and the Justice and Policy of the War with America* (London, 1776).

12 See Magellan to Sanches, 9 October 1778.

13 See Magellan to Sanches, 8 September 1778.

14 Luiz Pinto de Sousa Coutinho (1735-1804), Portuguese Ambassador in London.

15 Farewell.

To Ribeiro Sanches[1]

Mon cher Docteur, & très cher Respectable Ami,

Je vais vous annoncer un remede nouveau & très simple, que Mr. *Mudge*, membre de le Societé Royale de *Londres*, chirurgien à *Plymouth*, vient de publier. Ce remede guerit infailliblement la toux catarrale en très peu de tems, particulierement lorsqu'elle n'est pas ancienne. C'est d'après un grand nombre d'observations très constatées, que l'auteur parle si positivement de l'efficace de son remede, d'autant plus estimable qu'on en peut faire usage dans tout pays du monde, presque sans frais, & sans l'assistance de medecin, ou d'apothicaire. Ce remede est d'ailleurs si innocent en lui-méme, qu'il ne peut point nuire, méme en l'appliquant avec peu de discretion. Ces qualités doivent le rendre, on ne peut plus, recommendable à tous ceux, qui, comme vous, ont si fortement à coeur le bien & les interets de l'humanité. Je suis sur de vous faire un vrai plaisir en vous le communiquant; et j'en juge d'après celui, que j'ai senti moi-meme, lorsque j'en eus la premiere notice. Notre maniere de penser & de sentir, lorqu'il s'agit du bien public, est parfaitement analogue: & c'est peut-étre la source unique de l'amitié intime, dont vous m'honnorez depuis si long tems.

---

1 Letter not found. The text is from a printed pamphlet headed "Extrait d'une Lettre de Mr. *Magellan*, Membre de Societé Royale de *Londres*, à un de ses amis de *Paris* [A.R.S.], datée le 4 *Novembre*, 1778" that is filed with Magellan's letter to Johann III Bernoulli of 23 November 1781 (Öffentliche Bibliothek der Universität Basel, MS L I a 709, f. 85). The pamphlet was published in late November or early December 1778; see Magellan to Velasco, 24 November 1778. The same text was published, but without date or initials of the addressee in the heading and with more regular use of accents, in *Observations sur la physique, sur l'histoire naturelle et sur les arts*, 14 (1779), 225-7, in the issue for September 1779, under the heading, "Extrait d'une Lettre de M. Magellan, Membre de la Société Royale de Londres, à un de ses Amis de Paris." This was in turn republished in *L'esprit des journaux françois et étrangers*, 2 (1780), 343-7, in the issue for February 1780, but without the illustration and with a slightly modified description of the apparatus to compensate for this. A slightly different version of the text was appended as pp. 255-8 to the final item in Magellan's *Collection de différents traités sur des instrumens d'astronomie, physique, &c.* (Paris/London, 1780), his account of the collection of instruments ordered by the Spanish Court. This was headed "Extrait d'une Lettre de Mr. Magellan, Membre de la Société Royale de Londres, à un de ses Amis de Paris (Mr. A.R. Sanches, Docteur en Médécine, & Membre de l'Académie Impériale des Sciences de Petersbourg), datée le 4 *Novembre* 1778". A footnote explained: "Cet Extrait contient une notice assez utile au Public, pour trouver ici une place, quoiqu'elle soit d'un autre genre fort divers du Traité précédent". For the most part, differences between this and the other two texts are matters of grammar and punctuation only. More significant differences are noted below.

78.11.04 Mr. *Mudge* considere avec la plus grande raison, que la toux catarrale n'est, que la suite d'une vraie inflamation, du moins partiale, de la membrane qui tapisse les organes de la respiration: &, pour la guérir, il applique le topique le plus simple, le plus innocent, & le plus sur; c'est à dire, la vapeur de l'eau mediocrement chaude.[2] Pour mieux reussir dans cette application, avec le plus grand avantage, il a inventé un instrument, qu'il appelle *inhaler* en *Anglois*, & que je crois pouvoir nommer *respirateur*, à cause de son usage. En voici la description, avec les changemens, que j'y ai faits, pour le rendre plus commode dans le practique, en lui conservant dans le même tems tous les avantages, dont cet instrument est susceptible.

Le figure ci jointe represente le *respirateur*, dans l'etat, oû il faut en faire usage. *a* est un vaisseau cylindrique; qu'on peut faire, si l'on veut,

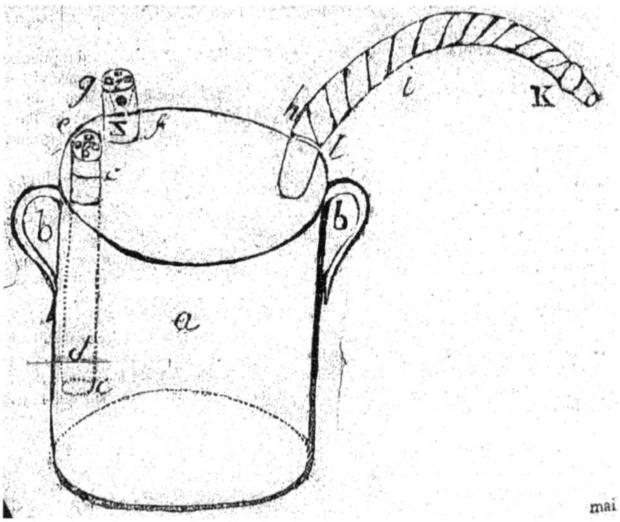

d'or ou d'argent: mais qui, etant d'étain, ou même de fer-blanc, est également bon pour son objet. Il est tout soudé à l'entour,[3] & doit contenir

2 The following is inserted in the alternative version at this point: "N'est-il pas bien surprénant, qu'un remede si directement indiqué par la nature même de la maladie, ait échappé si long-tems l'attention des Médécins! Tous les médicaments qu'on avale, ne peuvent aller aux poumons, que par la voye de la circulation. Au lieu que la vapeur qu'on respire, va tout-droit à l'endroit, où l'on a besoin de son influence: & assurément il n'y a pas, dans toute la *Matière Médicale*, aucun autre emollient, ou anodyn, si effectif & si innocent, comme l'eau modérément chaude."

3 The alternative version runs as follows from this point: "pour que l'eau, qu'on y met, n'en puisse point sortir que par l'embouchure *hl*. Autrement elle se répanderoit aisément dans le lit. C'est assez que ce vaisseau puisse contenir…".

environ une pinte: sa forme est à peuprès comme celle d'un pot à faire du Thé, ou comm'un petit arrosoir de jardin, garni d'une ou deux anses *bb*. On y trouve un tuyau *ecdc*, qui est soudé au plan superieur, descend audedans, jusqu'à la distance d'un demi pouce du fond; comm'on le voit marqué par des points, jusqu'à *d*. Ce tuyau a un petit couvercle *ce*, avec des trous pour laisser entrer l'air qu'on veut humer. Il y a à coté une autre embouchure *fg*, garnie d'un couvercle pareil avec des trous: mais elle est faite, au dedans, en entonnoir: desorte qu'en y mettant une petite boule z de liege,[4] elle y fait l'office de soupape, laissant èchaper l'air du dedans ou dehors; mais empechant qu'il n'y entre[5]. Les diametres, de l'embouchure *k*, du tuyau *cd*, & du trou z, doivent etre assez grandes, pour ne pas rendre difficile, le passage de la respiration. C'est assez de leur donner environ quatre[6] dixiemes de pouce *Anglois*. 78.11.04

Enfin *hik*. est un tuyau flexible de cuir,[7] qui renferme un fil de metal, couvert de soye, en forme spirale, pour le rendre tout à fait flexible; & garni d'une embouchure d'ivoire *k*, qu'on applique à la bouche, lors qu'on en fait usage[8].

Maniere d'employer le Respirateur.

Lors que la personne afligée de la toux catharrale, ou du mal de gorge (car il paroit, que ce topique doit étre aussi, bien avantageux dans ce dernier cas) và se coucher; on mettra de l'eau chaude dans le *respirateur*, par l'embouchure *hl*, après en avoir oté le tuyau de cuir *hik*. On ne le remplira pas tout à fait, mais seulement jusqu'aux deux tiers, ou environ: on l'envelopera dans une serviette, & on le mettra au lit du malade près de son aisselle: il attendra jusqu'à ce que l'eau soit un peu moins chaude, pour qu'il puisse en humer la vapeur, sans se brûler. -- Cet instrument agit de la manière suivante. L'air qui entre par le tuyau *cd*, passe àtravers l'eau modérément chaude, s'y charge de la vapeur aqueuse: & entre ensuite, par le tuyau *hik*, dans les poumons du malade; qui peut jetter l'expiration[9] par

4 The following is inserted in the alternative version at this point: "très bien arrondie dans le tour,".

5 The alternative version adds "point".

6 The alternative version adds "ou cinq".

7 The alternative version runs as follows: "doublé avec de la soye vernissée ou cirée. Ce tuyau renferme...".

8 The alternative version adds: "& qui a la même figure *applatie* & *angulaire*, comme celle de la bouche humaine".

9 The alternative version adds "s'il le veut".

78.11.04 le meme tuyau; parcequ'alors l'air sortant par la soupape *fg*, avec quelque partie de la vapeur chaude, & se repandant entre les draps, servira, au bout de quelque tems, à exciter le malade à un transpiration salutaire. On doit continuer cette operation pendant 20 minutes, ou une demi-heure. Si la toux est récente, on ne manque pas de se trouver soulagé, & tout à fait gueri, le jour suivant. Mais si la toux est ancienne, alors il faudra continuer ce remede pendant quelques nuites de suite.[10]

Quand on voudra faire usage du *respirateur*, l'Auteur conseille de prendre quelqu'opiate, ¾ d'heure avant de se coucher, comm'un remede concomitant pour obtenir la guérison de la toux. En conséquence, il prescrit environ 3 dragmes (ou 3 petites cueillerées, comme celles qu'on use en prennant le Thé) de l'*elixir Paregoricum* pour les adultes; une pour les enfans moins de 5 ans: & deux pour ceux entre cet age & les dix ans.

Le titre du livre, dont j'ai extrait cette relation, est: A radical & expeditious cure for a recent catarrhous cough, by *J. Mudge*, F.R.S. &c. *London*, 1778, in 8vo. Le lecteur y trouvera un grand nombre des discussions theoriques fort ingenieuses, & des observations très utiles dans la pratique. Entre les dernieres je ne puis omettre celle du bon effet, qu'il a vû dans les crachemens de sang, tendans à la fievre hetique, en faisant prendre au malade une demi-dragme de nitre dans un verre d'eau, deux ou trois fois par jour: & dans la toux seche & fatigante, des pillules faites de *gomme ammoniaque* avec quelques goutes de *Laudanum*, prises avant aller coucher. Je souhaite que ces remedes ne soient pas oubliés, par ceux de la Faculté, entre les autres qu'on connoit propres à ces maladies; parceque j'ai plus de confiance aux remedes authorisés par l'observation bien constatée, qu'à tous les autres, qui n'ont en leur faveur que l'authorité & la theorie de ceux qui les ordonnent.

P.S. Mr. *Mudge* parle d'une experience qu'on peut faire avec le *respirateur*; mais que je n'ai pas grande envie de repeter. Il dit que pour se convaincre, que la toux catarrale provient d'avoir respiré un air froid & humide; il n'y a qu'à faire usage du *respirateur* avec de l'eau froide. Car on ne manquera pas d'exciter, par ce moien, cette espece de toux.

10 The alternative version adds: "J'ai vû un grand nombre de guérissons, avec ce remede: & je l'ai essayé moi-même, avec le plus grand succès."

To Matthew Boulton[1]

78.11.06

Birmingham Central Library, Boulton and Watt collection, MBP 244, 60

London 6 Nov^{ber}. –78

Dear Sir

I hope you'll excuse me for troubling you with this Letter, on account of M^{r}. Jarry not being seen by me; tho' you noticed by yours of the 16 October,[2] that he was coming to London, and from hence to return to France. I received also a Letter directed to M^{r}. Jarry at my Lodgings, on the 30 October, which I thought to be from you, and I do keep it till he comes. but I begin to apprehend some accident has happn'd to him: and think proper to give you this notice; assuring you that nothing will be wanting on my part, to direct him to my acquaintances at Paris, where I have no doubt but he will manage the whole affair to mutual satisfaction.

Pray my Resp^{s}. to M^{r}. & M^{rs}. Watt and ever believe me

Dear Sir
your most obed^{t}. h^{ble}. serv^{t}.
Magellan

P.S.

I hope you have not forgoten the Picture of your manufacture for M^{de}. de'Herouville:[3] and the pieces of soft iron in y^{e}. shape of straps I bespoke for my rasors.

N.B. the Letter you sent me for Count d'Herouville,[4] will be much better to go and be presented by M^{r}. Jarry along with one from me, to the same Nobleman; than to send it alone by the post. for this reason I keep it, in waiting for M^{r}. Jarry to carry the same.

I have not yet seen the Commissioners from Prussia.[5] God knows, if these warlike affairs have not hindred, or retarded at least, their coming to England.

To
M^{r}. Math^{w}. Boulton
at Redruth
in Cornwall

---

1 Annotated "Madgellan London November 8th 1778".

2 M. Boulton to Magellan, 16 October 1778.

3 See Magellan to M. Boulton, 16 October 1778.

4 Presumably the letter Boulton had first intended to send to Magellan *via* Jary; see M. Boulton to Magellan, 16 October 1778.

5 See M. Boulton to Magellan, 16 October 1778.

78.11.10 *To Comte Joseph Velasco del Pico*
Archief Van Arenberg, Enghien/Edingen, Belgium

Londres 10 Nov^{bre}. 78

Monsieur

Je viens d'expedier dans une petite caisse, adressée à M^{r}. Hennessy, vice Consul Anglois à Ostende, pour La faire remettre à S.A.S^{me}.[1] ou plutot à vous meme, les petits articles que vous m'avez marqués dans votre Lettre du 2 du Courant.[2] Savoir, un Canif à 4 lames de la meilleure espece & une paire de Lunettes pour l'age de 40 ans environ. J'y ai ajouté un autre encrier de cristal, pour Remplacer celui qui fut cassé: & vous y trouverez deux petits paquets, que Je vous prie de faire passer à leur adresse.

Les petits deboursements que J'ai faits ne valent pas la peine d'en parler pour Le present: & seront remis à une autre occasion.

Je vous prie de me dire à quel age mourut Monseig^{r}. Le Duc? & s'il n'avoit pas jamais voulû etre inoculé? ou S'il croioit d'avoir deja eu La petite verole dans son enfance?[3] on me fait si souvent ces questions que je voudrois savoir quoi Repondre. Et quant à moi-meme je ne pris m'en rapeller cet acident, sans un regret acablant. en verité la mort du feu le Roy de France,[4] & tant d'autres Princes par La fatalité de la petite verole, ne pouvoit pas manquer de se faire Souvenir de ce grand preservatif, l'inoculation.[5]

Je vous prie de me repondre au plutôt sur ces articles. Dites moi aussi, oû fut deposé Son Cadavre.

Les Nouvelles que vous me donnez des Princesses qui se trouvent en Santé avec Le Reste de cette Illustre Famille, me font Le plus grand plaisir d'après les idees si ameres que J'ai toujours presentes dans mon ame de Leur Situation. La pauvre Duchesse douariere, que Je la vois souvent dans mon imagination, assisse & pensive dans Le malheur d'une si grande perte! ...! mon Dieu...! quel est le miserable sort des humains...! le plaisir est peu durable! je n'avois pas pensé sitôt seroit changé, celui ou Je les laissai à

---

1 Duke Louis-Engelbert d'Arenberg.

2 Letter not found.

3 Two forms of smallpox were recognized, presently known as two strains of the *variola* virus: *variola minor* and *variola major*. The first usually carried a reduced risk of death compared with the latter.

4 Louis XV died of smallpox in 1774, at the age of 64.

5 Inoculation grew in popularity in Europe during the 18th century. While in Britain it was common from the 1720s, in France there was considerable resistance to the practice.

Enghien le 30 Juin...! Mais je ne finirois jamais si j'avois à vous ecrire tous ce qui me passe dans l'imagination à cet egard. 78.11.10

J'ai l'honneur d'etre pour La vie

Monsieur

Magellan

A Monsieur
Monsr. Velasco, Secretaire de
S.A.S. Monseigr. Le Duc d'Arenberg
à Bruxelles

To Ribeiro Sanches[1] 78.11.12

Österreichische Nationalbibliotek, Vienna
A.N. Ribeiro Sanches Korrespondenz, codex 12714, ff. 233-234v

12 Novbro. – 78

Meo Amo. e Sr. e mto. do Coram.

Oxalá q em vez da milagroza escapadella, q V.M. teve nos varaes daquella [s]ege, donde apenas teve menos do q a contuzão, pudesse V.M. dar-me ainda noticias d'hũ Cavallinho, como o q a minha Nely me deo, ou pelo menos de hũ bom esquentamento: e entre nós, deve V.M. confessar q a sege, nem os seos varaes, não são tão agradaveis, como os cavalos. Eu estou tão bom, & são destes, q ainda não ha meia hora fui visitar uma comadre, q me convidou com uma franga q ainda não fez os 14 de Ide.: e q o faz tão bem, como as melhores Mestras do officio. mas falemos d'outra coiza por lhe não fazer crecer a agoa na boca.

a metade da carta de V.M. pa. o seo sobrinho, foi entregue por minha mão; pois V.M. não tem direito a privar-me do gosto de dar esta consolação a este moribundo. Elle não me dezagradou tanto ontem, como nas outras duas vezes. admirou-se q V.M. não saiba o seo nome, q he Salomão (ou Anto., entre christãos) Nunes, Framengo de alcunha. elle mesmo não sabe

1 Annotated "Respa a 30. 9bre" [Answered on 30 November]. Letter not found.

78.11.12 o nome de sua Mai, q morreo do parto do filho depois deste. elle naceo em Port^{al}.: elle veio p^{a}. Bordeos de 2 ou 3 annos: & lá foi q elle vio a V.M. eu creio q V.M. o não tornará a ver; pois apenas sahirá de caza por seo pé, senão p^{a}. a cova. lá lhe deixei a metade da carta, q lhe pertencia, de V.M., e parece q ficou mui consolado com ella – taes são os fins da vida ...!

Domingo passado jantei com Sua Exc^{a}., e o Bezerra me deo copia da Lista do Auto da Fé, celebrado na grande sala publica da Inquizição de Lisb^{a}. em 11 de 8^{bro}. pass^{o}. som^{te}. houve 10 criminosos, nenhũ judeo, mas todos christãos. os crimes são ler livros prohibidos & hereticos, negar a exist^{a}. de D^{s}., o mist^{o}. da Trind^{e}., Incarnação, divind^{e}. de Chr^{o}. differença de religião; ter Chr^{o}. por bom homem, mas não por D^{s}.: negar o Purgat^{o}., Ceo, Inferno. q'a simples fornicação não he pecado; q o celibato he contra a população e prejudicial ao Estado; q os suffragios & Missas são inuteis: q bastava a ley Nat^{al}. & ser homem de bem &^{a}. &^{a}. hũ delles tem, entre outros crimes deste genero, o de chamar pelo nome de S^{ra}. do Carmo & de Sta. Rita a huas certas raparigas q elle comunicava, ou fornicava. Verd^{a}.m^{te}. he hũa das listas mais risiveis q se tem visto. todos os castigos são de hir estar recluso, ou n'um convento, ou outro, como Rilhafoles, q he dos P.P. daMis[são das] Necessid^{es}., dos Capuchinhos &^{a}. por 3 annos, e degredo p^{a} Lamego, Vizeo &^{a}. confiscação de todos os bens &^{a}. som^{te}. o ultimo, q era estud^{te}. do 5° anno, na faculd^{de}. de canones na Univ^{de}. de Coimbra, q pretendeo converter outro p^{a} o seo sistema, & q dezafiou mesmo nos carceres, a Deos de converter a agoa em Sangue, p^{a} acreditar o christianismo &^{a}. som^{te}. este, q he filho de gente distinta de Valença, foi condenado a prizão arbitraria, açoites, carocha, mordaça, & 5 annos p^{a} galés. Ora veja V.M. lá como vai a nossa abençoada terra! dizem me q o Sonsinho (o D^{r}. Alvares) se faz tonto nes[sas] conversaçoens, e pretende não intender o q se fala. estou certo q não lhe hão de faltar palpitaçoens.

O govern^{or}. de Macáo ja lá partio p^{a}. Lisboa, com seo Irmão & Sobrinho, q o vierão buscar. recobrou boa saude, & lá se hirá encovar em alguã qu[in]ta, gozando das boas uvas ferraes, & meloens de Inverno, & tudo o mais q não falta em Portugal, q he o q lá se pode gozar sem risco.

Sua Exc^{a}. (o nosso Enviado) mandou o menino mais velho, q era o unico q ficava com elle, p^{a} sua Mãi: e Receio q elle tãobem não esteja aqui m^{to}. tempo. O nosso Enviado q está na Haya, chegará aqui brevem^{te}. p^{a}. hir p^{a}. Russia: e espera-se aqui hũ filho do Calhariz (sobrinho do D. João q V.M. ahi conheceo, & q ahi morreo em triste estado) p^{a}. hir Min^{o}. p^{a}. a Haya. o Abbé Ericeira vai p^{a}. Roma: pois o Almada lá cazou com a sua Cozinheira ou criada q o servia: & p^{a}. Turim vai hũ sobrinho, senão me engano desse S^{r}. q está em Pa[ris] Ora q Leve o Demo as novid^{es}. Pol^{as} ...!

Hum amigo q vai partir 4ª. frª. proxima pª. Paris, levará a V.M. os livros q quer, ecuido q tãobem a these sobre os banhos frios. alem de[stes] mandarei hũ tratadinho de Mr. Mudge sobre a tosse, q me parece excellte. a cura he infalivel qdo. ella he de pouco tempo. consiste em respirar o enfermo o ar q passa por dentro da agoa quente, sem outra mistura. he hũ vapor humido e emoliente, q faz imediatamte. cessar a irritação ou inflamação dos bofes. a repetida experiencia de hũ grande numero de casos decisivos, mostra a excellencia deste simplicissº. remedio. Eu venho agora de [ca]za do meo fer blanchier, onde fui (depois de sahir da Socieddє. Rl. e de hir dar o meo pino [n]a franguinha de 13 anos) encõmendar-lhe o instrumto. pª esta operação, q Mr. Mudge chama inhaler, e q mandarei a V.M. pelo mesmo amigo. a dizer-lhe a verdde. este será da minha invenção; pois o q vi hoje, q he da invenção do Author, não me agradou. alem disso o preço de meio guineo he exorbitante, pois he simplicissmo., e ficaria bem pago com a metade. aqui tem V.M. o meo, emqto. não vê o q Mudge descreve no seo Livro. O vazo a he de lata, e tem o tubo zx q communica com o fundo. 78.11.12

n he um piqueno tubo, onde está hua valva, ou soupape:
& ut he hũ tubo de coiro, com o bocal u de marfim. enche-se o vazo a de agoa quente até z: e qdo. o doente está na cama, poem [es]te vazo junto do peito, ou sovaco do braço. mete o bocal u na boca, & toma por elle a respiração. o ar entra por z; dece até x; atravessa a agoa quente, onde se carrega do vapor, e entra nos bofes. quando o doente expira o ar, este poderá ser lançado fóra, tirando da boca o tal bocal u: ou de outra sorte pode lançalo outra vez pelo mesmo tubo ut; pois neste cazo a expiração sahirá pela valva q tapa o buraco do piqueno tubo n. basta fazer esta operação por 20-minutos, ou por meia hora, em hũa ou duas noites consecutivas: & a tosse desaparecerá. Cuido q o amigo le Dr. le Begue du Presle fará serviço ao publico em traduzir este Livrinho de Mr. Mudge. V.M. lho emprestará até q eu lhe mande outro &ª. Mr. Mudge aconcelha a tomar a tomar 3 drachmas, ou 3 colheres de chá, cheias de elixir Paregoricum antes do doente se hir deitar. Depois de 3 quartos de hora, vai-se á cama. toma-se o instrumento cheio de agoa quente até os 2 terços pouco mais ou menos; embrolhado n'um pano: & faz-se a operação por 20 minutos, ou por meia hora de tempo.

ora a Deos q me vou deitar, tomando primº. uma das minhas pirolas do mercurius precipitatus per se, pª. me assegurar melhor da extirpação

78.11.12 do meo veneno, do qual não tenho a minima aprehensão, nem a minima molestia. o certo he antes quero destes cavalinhos, do q dos varáes da sege q V.M. lá provou. N.B. o uzo do precipitatus per se está hoje admitido, & absolutam[te]. aprovado com infinita vantagem ao sublimado: onde o acido produz sempre algũa desordem, mais ou menos a q[m]. o uza. em Lugar q o precipitado per se, he innocentissimo. não se toma senão desde meio grão até hũ grão: & não tem outro Regime senão o comer algũa coiza menos do costumado. eu jantei, a semana precedente á p[assa]da, lautamente, & bebi sem cautela: senti nessa noite algũ abalo de febre &[a]. mas logo c[om] a dieta do dia seg[te]., me achei são, & sem a minima molestia.

Vale & ama
tuum
Magellan

a Monsieur
Mons[r]. Sanchez D[or]. en Medecine
dans Le Cimiterre de S[t]. Jean, Rue de la Verrerie
à Paris

12 November –78

My Friend and very Dear Sir

Hopefully, instead of the miraculous survival that you had on the beams of that coach, where you got less than a bruise,[2] you could give me news of a horse ride like the one my Nelly gave me, or at least of a good dose of gonorrhoea: and between ourselves, you must confess that the coach, or its beams, are not as pleasant as the horses. I am so well, & healthy from these, that not yet half an hour ago I visited a woman friend, who offered me a little girl who is not yet 14 years old and does it as well as the best masters of the job. But let us talk about other matters in order not to make your mouth water.

Half of the letter to your nephew was delivered by me;[3] since you don't have the right to deprive me of the pleasure of giving this consolation to this moribund person. He did not displease me yesterday as on the other two occasions. He was surprised that you don't know his name, which is Salomão (or Antonio, among Christians) Nunes, nicknamed Framengo. He himself doesn't know the name of his mother, who died in giving birth to the son after this one. He was born in Portugal, came to Bordeaux at 2 or 3 years of age, & saw you there. I believe you

2 Sanches had evidently reported being involved in an accident.
3 Letter not found. See Magellan to Sanches, 8 Sepember and 9 October 1778.

won't see him again, because he won't go out of his home except to the grave. 78.11.12
I left with him the half of the letter from you that belonged to him, and he seemed very comforted by it – such is the end of life ...!

Last Sunday I dined with His Excellency,[4] and Bezerra[5] gave me a copy of the list of the auto-da-fé[6] celebrated in a big public room of the Inquisition on the 11th of October last. There were only 10 criminals, not one Jew but all Christians. The crimes were reading prohibited & heretical books, denying God's existence, the mystery of Trinity, Incarnation, Christ's divinity; difference of religion; taking Christ for a good man but not God; denying purgatory, heaven, hell; asserting that simple fornication is not a sin, that celibacy is against population and harmful to the State, that suffrage & masses are useless, that natural law & being a good man is enough &c. &c. One of them has, among other crimes of the kind, that of calling by the name of Our Lady of Carmo, & of St. Rita, certain girls with whom he used to get along, or fornicate with. It is truly one of the most ridiculous lists ever seen. All punishments involve entering one convent or another such as Rilhafoles[7] that belongs to the Fathers of the Mission of Necessidades,[8] to the Capuchins &c. for 3 years, and exile to Lamego,[9] Vizeo[10] &c., confiscation of all possessions &c. Only the last one, who was a 5th-year student in the faculty of canon law of the University of Coimbra, who wanted to convert others to his system, & who even in prison defied God to convert water into blood in order for him to believe in Christianity &c. — only this one, who is the son of distinguished people of Valença,[11] was condemned to arbitrary imprisonment, whip, cap, repression of speech, & 5 years in the galleys. You see how our beloved country is going! I was told that the Sly One (Dr. Alvares)[12] plays the fool in these conversations and pretends not to understand what is being said. I'm sure he won't lack palpitations.

The governor of Macao[13] left for Lisbon with his brother & nephew, who came to fetch him: he recovered his good health, & there he will bury himself in some farm, enjoying the good iron-coloured grapes & winter melons, & all the rest that can't speak in Portugal — that is, that can be enjoyed without risk.

---

4 Luiz Pinto de Sousa Coutinho (1735-1804), Portuguese Ambassador in London.
5 João Paulo Bezerra de Seixas (1756-1817); see Magellan to Sanches, 1 September 1778, note 13.
6 Either the process or the execution of the sentence of the Inquisition, often in public ceremony.
7 The Convent of St. Francis of Paula of the Oratorians near Lisbon.
8 i.e. the Oratorians.
9 A town in northern Portugal.
10 An old town in central Portugal.
11 A small town in the north of Portugal, bordering Spain.
12 Sanches' friend Dr J. Alvares, whom Magellan did not like; see Magellan to Sanches, 13 March 1778 and earlier letters.
13 Diogo Fernandes Salema e Saldanha, Governor of Macao in 1767-1770 and a second time in 1771-1777; see Magellan to Sanches, 8 September and 30 October 1778.

78.11.12 His Excellency (our Envoy) sent his eldest son, the only one who was staying with him, to his mother: and I'm afraid he too won't stay here much longer. Our Envoy in The Hague[14] will arrive here soon to go to Russia: and a son of Calhariz (the nephew of D. João whom you met there & who died there in a sad condition) is waiting here to go as minister to The Hague.[15] Abbé Ericeira[16] is going to Rome, since Almada[17] finally married his cook or maid who served him: & to Turin goes, if I am not wrong, a nephew of that gentleman who is in Paris. To the devil with political news ...!

A friend who is departing next Wednesday for Paris will take to you the books you want, and I suppose also the thesis about cold baths.[18] Besides these I am sending a little treatise by Mr. Mudge about the cough[19] that seems excellent to me. The cure is infallible when the illness is recent. It consists in the patient breathing air that passes through hot water, without other mixture. It is a humid and emollient vapour that immediately makes the irritation or inflammation of the lungs cease. Repeated experience in a great number of conclusive cases shows the excellence of this extremely simple remedy. I've just arrived from seeing my fer blanchier,[20] where I went (after leaving the Royal Society and going to give my f... to the young 13 year old girl) to order the instrument for this operation, that Mr. Mudge calls an inhaler, and I'll send you this through the same friend. To tell the truth, this one will be of my invention since the one I saw today, which is of the invention of the author, didn't please me. Besides, the price of half a guinea is excessive, because it is extremely simple and would be well covered by half that amount. Here you have mine, while you don't see what Mr. Mudge describes

in his book. Vase a is made of tin plate, and has a tube zx that communicates with the bottom. n is a little tube where there is a valve, or soupape,: & ut is a leather tube with a mouth of ivory. Vase a is filled with hot water up to z, and the patient when in bed puts this vase near the chest

---

14 Francisco José Horta Machado (1746-1817).

15 Augusto de Sousa Holstein (1741-1781) was Portugal's Envoy Extraordinary at The Hague, 1779-80. He was the son of Manuel de Sousa, Earl of Calhariz (1703-1759), who died in the dungeon of the Tower of Bugio, in the estuary of the Tagus River in Lisbon, after being incarcerated by the Marquis de Pombal. The "D. João" to whom Magellan refers is presumably Calhariz's younger brother, João de Sousa (b. 1719).

16 Henrique de Menezes, Count of Ericeira and Marquis of Louriçal (1727-1787), was Portuguese Envoy and Minister Plenipotentiary in Turin and then Rome.

17 Francisco de Almada Mendonça was a cousin of Pombal and Portuguese Minister in Rome.

18 Samuel Byam Athill, *Exercitatio therapeutica, ehibens observationes quasdam de usu aquæ frigidæ externo* (Edinburgh, 1778).

19 J. Mudge, *A Radical and Expeditious Cure for a Recent Catarrhous Cough* (London, 1778).

20 tinsmith.

or the armpit. He puts the mouth of the tube u into his mouth & breathes through it. The air enters through z, gets down to x, passes through the hot water where it takes up the vapour, and goes into the lungs. When the patient breathes the air out, this can be expelled by removing the said mouth of the tube u: from the mouth; or, in another way, by blowing it again through the same tube ut; since in this case the exhaled air will come out through the valve that closes the little tube n. It is enough to do this operation for 20 minutes or half an hour on one or two consecutive nights, & the cough will vanish. I suppose that our friend Dr. le Begue du Presle will do a service to the public if he translates this little book of Mr. Mudge. You may lend it to him until I send you another &c. Mr. Mudge advises taking 3 drachms, or 3 teaspoonfuls of elixir Paregoricum before the patient goes to bed. After 3 quarters of an hour he goes to bed. He takes the instrument up to two thirds full of hot water more or less, wrapped in a cloth, & does the operation for 20 minutes or half an hour. 78.11.12

Now good bye since I'm going to bed, taking first one of my pills of mercurius precipitatus per se[21] so as to be better sure of extirpating my poison, from which I have not the least apprehension or the least disturbance.[22] Surely I prefer these little horse-rides than the beams of the coach that you have tried. N.B. the use of the precipitate per se is today considered & absolutely approved as infinitely better than the sublimate[23] where the acid always produces some disorder more or less in those who use it. Instead the precipitate per se is totally innocent. Only from half a grain to one grain is to be taken & nothing else is prescribed except to eat a little less than usual. I dined plentifully the week before last & drank without caution. That night I felt some trouble from fever &c. but then by dieting the next day I found myself healthy, & without the least disturbance.

Vale & ama

tuum[24]

Magellan

21 Red oxide of mercury.

22 Magellan has been suffering from a venereal infection; see Magellan to Sanches, 9 October 1778.

23 Mercuric chloride.

24 Farewell and love | your | Magellan.

78.11.13 *To Anne-Robert-Jacques Turgot*
Biblioteca do Museu Nacional de Arte Antiga, Lisbon
Fundo Francisco Sousa Viterbo, Cx nº 1

Londres ce 13 Nov^{bre}. –78

Mon Cher Monsieur

Le jour de la poste prochaine partira d'ici pour Paris Monsr. Jarry,[1] Proprietaire des Mines du Nort de la Bretagne, dont j'ai attendu le depart, pour Lui confier les articles que Vous m'avez ordonnés, tant les thermometres que le Velour de tricot. Car la voye des Pays bas, est aussi longue & aussi dispendieuse, comme vous même avez essuyé dans les articles que Je vous envoyai d'ici en Juin dernier, & qui furent je crois deux mois & demi, ou 3 mois avant d'arriver chez vous. en effet je croiois qu'il partiroit plutôt, autrement aurois-je puis peut etre La meme resolution; quoiqu'à Regret. le demon de la guerre[2] derange toute sorte de correspondence directe: & il n'est pas permis à Calais de laisser passer dans le paquebot de la poste, que seulem^{t}. des caisses petites à titre de bagage de quelque passager. Je pense donc que cette occasion de M^{r}. Jarry est la plus à propos que J'avois en mon pouvoir. Il a passé 8 mois en Angleterre pour examiner la nouvelle Construction des machines à feu de Messrs. Boulton & Watt; qui est la plus avantageuse: l'epargne etant presqu'environ de 75 par cent, dans La Consomation du feu, ce qui fait un objet très Considerable, meme dans Les mines à Charbon. M^{r}. Jarry a fait un arrangement très avantageux pour La france. Si le privilege exclusif que le Roy Lui a accordé deja à Messrs. Boulton & Watt peut admettre la clause, de que les experiences de la comparaison entre la construction ancienne & la moderne, soient faites en Bretagne dans ses propres mines, au lieu d'etre faites à Paris.[3] Car aiant là la machine antienne pour faire La comparaison, & M^{r}. Jarry aiant acheté une de ces machines nouvelles, les Resultats seront decisifs, & la Consequence sera que La France deviendra en possession d'une mechanique, qui non seulem^{t}. est avantageuse pour elever des eaux des mines ou des Rivieres à peu de frais (car la depense de ces machines n'est pas plus que celle des

1 See Magellan to M. Boulton, 23 March 1778.
2 The American War of Independence.
3 In April 1778, the French Government issued a provisional *arrêt* in favour of the Boulton & Watt company, granting them an exclusive privilege to erect their new steam engines in France. Before the privilege was confirmed, however, one of their engines was required to be subjected to a trial, in France, in which it out-performed a traditional Newcomen-style engine.

artiennes) mais Messrs. Boulton & Watt l'ont deja arrangée en sorte qu'on peut appliquer Ce grand agent aux grandes manufactures, quoique alors la forme doit etre differente, que lors qu'on l'applique à lever les eaux &[a]. Je me flate, que vous, Monsieur, voudrez bien proteger Cet objet, qui est d'une tres grande consequence. Je donnerai à M[r]. Jarry une Lettre pour vous, qu'il vous presentera:[4] & c'est de vive voix, que vous pouvez vous informer bien mieux qu'il ne m'est pas possible de le faire par ecrit. 78.11.13

Je vous Suis infiniment obligé par la nouvelle de M[r]. Abildgourd[5] sur la transmutation du Shorle Transparent en Turmaline. elle est très curieuse, & je Souhaite qu'on la verifie Chez Vous.[6] J'en parlerai ici à mes amis.

Je vous Suis bien redevable pour l'explication de la Peridote. assurem[t]. Je ne pouvois croire qu'une pierre telle qu'une emeraude verte, melée de jaunatre, comme la decrit M[r]. Bomar dans son Dictionnaire,[7] pût servir aux micrometres de M[r]. l'Abbé Rochon: ainsi j'ai crus que c'etoit un nom nouvellem[t]. forgé pour signifier le Brasilian peble, qui est un beau cristal bien diaphane, avec un peu de jaune très delayé, & que Je connois d'aileurs avoir une double refraction très marquée. Apropòs on a Lû hier à la Soc[e]. R[le]. un Memoire du P. Boscowich,[8] oû il soutient etre le vrai

---

4 See Magellan to Turgot, 16 November 1778.

5 Søren Abildgaard (1718-1791), Norwegian-born Danish naturalist.

6 In 1772 Romé de l'Isle argued that tourmaline was nothing else than transparent schorl, and it has since come to be generally accepted that the two are the same thing. At the time, however, considerable uncertainty remained; thus in Valmont de Bomare's *Minéralogie* (1774), vol. 1, p. 300, schorl is said to be "une production pierreuse dont la nature est encore peu connue". In Magellan's edition of Cronstedt's *Essay towards a System of Mineralogy* (1788), p. 225, in a table of the comparative hardness of various stones, schorl and tourmaline are listed immediately after one another, with the same degree of hardness; but they are assigned different specific gravities. Magellan makes no reference in this work to Abildgaard's supposed transmutation of schorl into tourmaline. Even the name was uncertain: in Magellan's edition of Cronstedt's book, it is spelled variously as "shirl" (which Magellan says is the name adopted by the English), "schoerl" and "schorl".

7 Jacques-Christophe Valmont de Bomare, *Dictionnaire raisonné universel d'histoire naturelle* ... (Paris, 1764 and subsequent editions, one in 1776), article "Péridot": "Les Joailliers François donnent ce nom à une pierre précieuse d'un vert un peu jaunâtre, imitant la feuille de poireau".

8 The Royal Society's *Journal Book Copy* records (vol. XXIX, pp. 271ff.) that at the Society's meeting on 12 November 1778, "a translation of a French paper intitled Memoire sur le Micrometre objectif a prisme de verre simple par M. l'Abbe Boscovich was read" by Anthony Shepherd. The account of the paper notes that Boscovich "complains much of unfair usage from the Abbe Rochon who claims the invention of an Instrument the very same with this except that instead of common Glass the Abbe Rochon would use Rock Chrystal". Boscovich's paper was not published in the *Philosophical Transactions*, though an earlier paper by him was, "Account of a New Micrometer and Megameter",

78.11.13 inventeur de ce micrometre, & pas moins de deux fois, acuse M^{r}. Fontana de l'avoir divulgé, ou communiqué son invention à M^{r}. l'Abbé Rochon &^{a}.[9] M^{r}. Fontana etoit present à Cette Sceance, & je ne manquai de faire une risée avec Lui Sur Cette pretention du Ex Jesuite brouillon. M^{r}. Fontana & Fabroni se portent très bien, & me prierent de vous presenter Leurs très h^{bles}. Resp^{s}.

dans La poste prochaine, savoir, au depart de M^{r}. Jarry, Je vous ferai le detail de ce qu'il vous apporte: & pour Le present J'ai l'[honeur] d'etre avec mon vray attachem^{t}. de Cœur & pour la vie

Mon cher Monsr.

Votre très h^{le} & ob^{t}. Serv^{r}.

Magellan

a Monsieur
Monsr. Turgot, ancien Controleur gen^{al}.
Rue de l'Université
Fauxbourg S^{t}. Germain
à <u>Paris</u>

Philosophical Transactions, 67 (1777), 789-98 (this having been read at a meeting of the Society on 19 June 1777).

9 An account of Rochon's micrometer appeared in *Histoire de l'Académie Royale des Sciences, avec les mémoires*, 1777 (publ. 1780), Histoire pp. 64-8. The account ended thus: "Ces instrumens ont été connus de l'Académie, dans les premiers mois de 1777: l'Auteur en a lû la description dans l'Assemblée publique du 9 Avril de la même année. Ainsi quoiqu'il n'en ait point encore imprimé ni la construction ni les usages, on doit les regarder comme connus du Public depuis cette époque." This statement was a consequence of a protest lodged by Rochon when Boscovich read, at a meeting of the Académie on 16 May 1777, the paper he sent to the Royal Society a month later. In response to Rochon's request, the Académie established a committee to investigate the competing priority claims. This reported in Rochon's favour on 23 August 1777; see the *procès-verbaux* of the meetings of the Académie, vol. 96, meeting of 23 August 1777. The priority dispute is discussed by Danielle Fauque, "Alexis-Marie Rochon (1741-1817), savant astronome et opticien", *Revue d'histoire des sciences*, 38 (1985), 3-36, esp. pp. 31-2, and by August Ziggelaar in *The Philosophy of Science of Ruder Boskovic*, ed I. Macan (Zagreb, 1987), pp. 149-52.

To the Comte d'Hérouville[1] 78.11.16

Archives Nationales, Paris, MS Marine G110, ff. 184-5

Mon cher Comte & très respectable ami

Je suis on ne peut plus mortifié & inquiet pour l'etat de votre Santé, n'aiant point reçu reponse aux deux lettres que J'ai eu l'honneur de Vous ecrire àprès mon retour en Angleterre.[2] Peut etre avez vous eté en campagne, peut etre ont elles disparu de la poste sans arriver à vos mains; car on ne peut pas trop compter sur l'exactitude de ce Bureau dans les circonstances facheuses de cette malheureuse guerre.[3] J'espere cependant qu'à present Je sortirai de mes apprehensions à Votre egard, moïenant le porteur de cette Lettre, qui est M[r]. Jarry, Proprietaire des Mines du Nort en Bretagne;[4] il s'est chargé de vous remettre cette lettre, avec deux autres sous une envelope de M[r]. Boulton, & deux peintures de sa nouvelle manufacture à peindre,[5] que M[r]. Boulton vous prie de presenter à M[de]. & Mlle. d'Herouville, comme un petit tribut pour la protection que Vous donnez au beaux Arts, & pour celle que Vous avez donné à son objet d'obtenir le privilege exclusif pour ses nouvelles machines à feu, dans le Royaume de France &[a]. Vous verrez, Monsr., par les lettres de Monsr. Boulton que l'arrangement fait entre eux pour eriger une de ces nouvelles machines en Bretagne, oû il y a d'autres deja, construit[es][6] selon l'ancienne methode, qui peuvent servir de Comparaison, est entierem[t]. conditional, & dependant de Votre consentement. Car sans celà tout sera nul. D'ailleurs M[r]. Boulton sera toujours pret à eriger une autre machine aux Moeres,[7] ou quelqu'autre part que vous choisirez, avec les memes conditions & avantages qu'on av[oit] proposé auparavant. Je vous prie donc, Monsr., de reflechir sur ces circonstances: & Je me flate que vous approuverez avec votre consentement, & favoriserez avec votre protection & mediation la requete que M[r]. Jarry vous remettra, pour etre presentée au Conseil du

1 The letter is annotated: "Rep.", i. e. "Repondue". Letter not found.
2 Presumably Magellan is referring to his letters to d'Hérouville of 17 July and 1 September 1778. Magellan was in Paris in June of that year but left for London before the end of the month; see Magellan to B. Franklin, 20 June 1778.
3 The American War of Independence, which France had joined on the American side.
4 See Magellan to M. Boulton, 23 March 1778.
5 A product of Boulton's Soho Manufactory, Birmingham.
6 Obscured by binding.
7 Comte d'Hérouville's property at Les Moeres, near Dunkirk.

78.11.16 Roy, àfin d'obtenir que les experiences de comparaison soient faites en Bretagne, oû le meme M^r^. Jarry erigera pour ses mines la nouvelle machine, en cas que le privilege puisse etre obtenu: car sans celà M^r^. Boulton ne laissera pas passer aucune de ses machines en France, ce qui est une perte très considerable; non seulem^t^. pour M^r^. Jarry, qui est parfaitement convaincû du grande avantage & superiorité de ces machines nouvelles, & qui est egalement convaincu de l'impossibilité de faire construire leurs parties essentieles ailleurs qu'en Angleterre: Mais ce qui est de la plus grande consequence pour l'Etat, ces machines peuvent etre appliquées à un nombre considerable de manufactures (quoique dans une forme differente de celles pour ellever les eaux), & en effet M^r^. Jarry a bien vû qu'on les applique deja en Angleterre avec le plus grand succès aux grandes forges de fer, avec une epargne très considerable pour L'oeconomie, & produit de ces grandes manufactures.

J'ai pris la liberté de donner une lettre au meme M^r^. Jarry pour M^de^. d'Herouville,[8] pour qu'il lui presentat les Pinctures que J'avois promis de la manufacture à peindre du dit M^r^. Boulton, avec la petite brochure que Je viens de publier ici sur la mort du pauvre Rousseau,[9] à la fin de laquelle J'ai exprimé mes Sentimens & mes idées sur des affaires les plus communes de la vie humaine, que Je crois vous trouverez dignes de celui qui vous est devoué pour La Vie avec le plus de respect, d'attachement, e de consideration, aiant l'honneur d'etre

Monsieur Le Comte
Votre très h^ble^. & obligé serv^r^.
J.H. de Magellan

Londres ce 16 Nov^bre^. –78

8 Letter not found.

9 *Relation ou notice des derniers jours de Mons. Jean-Jacques Rousseau...* par Mons. le Begue du Presle... avec une addition relative su même sujet, par J.H. de Magellan... (London, 1778).

To Anne-Robert-Jacques Turgot 78.11.16a

Biblioteca do Museu Nacional de Arte Antiga, Lisbon
Fundo Francisco Sousa Viterbo, Cx nº 1

Londres ce 16 Nov^{bre}. –78

Monsieur

Le porteur de Celle-ci est M^{r}. Jarry, proprietaire des mines du Nort en Bretagne, dont Je vous ai parlé dans ma Lettre precedente.[1] Comme son objet est d'une utilité publique que vous avez toujours à Cœur, Je suis persuadé que vous ne Lui Refuserez La faveur & protection, qui dependra de vous pour Son affaire. Il s'est chargé de deux petites caisses, oû J'ai renfermé les articles, que vous m'avez ordonnés, & dont je vous ferai le detail par La poste suivante:[2] car a peine ai je le tems de finir Celeci avant qu'il parte.

J'ai l'honneur d'etre avec le plus grand Resp^{s}. & vrai attachem^{t}. de Coeur pour La vie

Monsieur
Votre très h^{ble}. & très ob^{t}. serv^{r}.
Magellan

a Monsieur
Monsr. Turgot, ancien Controleur gen^{al}.
Rue de l'Université
à Paris

1 See Magellan to Turgot, 13 November 1778.
2 See Magellan to Turgot, 20 November 1778.

78.11.20 *To Anne-Robert-Jacques Turgot*

Biblioteca do Museu Nacional de Arte Antiga, Lisbon
Fundo Francisco Sousa Viterbo, Cx nº 1

Londres 20 Nov^{bre}. -78

Mon Cher & respectable ami

Mercredi dernier partit d'ici pour Paris M^{r}. Jarry, qui a voulu bien se charger de deux petites caisses, qui vous Sont adresseés, entre son bagage: & me promit d'en avoir soin. Je lui ai donné une Lettre pour M^{r}. Caffieri, Directeur de la poste à Calais,[1] pour les depecher à la douane & donner la caution necessaire &^{a}. Sur quoi il ne manquera pas de vous faire part, & Recevoir vos instructions, en cas qu'elles soient arretées &^{a}. Le même M^{r}. Jarry vous presentera une de mes Lettres,[2] sur Le même affaire, dont Je vous ai parlé dans Celle du 13 Courant:[3] & comme son objet est tout à fait interessant pour La France je ne doute gueres que vous aiez la bonté de le proteger, dans ce qui pourra dependre de vous, & de vos bons amis.

La Caisse nº. 1. contient le velour de Tricot que vous avez ordonné, qui couta 5$^{£}$.15^{s}.6^{d} , 91$^{\#}$ 5^{s}/ 2do. Les bas de velvet silk hose, selon l'echantillon de grandeur que vous m'avez envoyé. Ce sont les premiers; & l'ouvrier souhaite, soient les derniers, qu'on ait jamais faits, par l'embarras qu'il a essuyé à y reussir. ils couterent £1:11^{s}:6^{d} , 36$^{\#}$ 0^{s} 5´/ dans la caisse nº. 2. Sont empaquetés 1º. 12 thermometres depuis 0 ou quelque chose au dessous jusqu'à 216º. ils couterent chacun à 30 shillings, ce qui fait la somme de 18$^{£}$: 0^{s}:0^{d} / plus 4 thermometres depuis 32º. jusqu'à 80º., qui couterent à 28^{s}: & leur total fait La somme de 5$^{£}$.12^{s}.0^{d} / plus 3 thermometres depuis 45º. (ou pour mieux dire depuis 32: car l'artiste a protesté que si on ne lui permettoit de prendre le degré de la glace, il ne voudroit repondre aucunem^{t}. de la graduation) jusqu'au degré 75: qui au meme pris de 28^{s}, couterent £4:4^{s}:0^{d} / en fin un thermometre depuis le 0/20 jusqu'au 216/0me. degré; qui couta £1:10:0: les deux caisses, porteurs messages & lettres 12^{s}:6^{d} / payé au carosse pour le carriage de ces deux caisses jusqu'à Douvre 4^{s}: 6^{d} / La somme totale est £ 37:10^{s}:0^{d} Esterl., 927$^{\#}$3^{s} [...][4] que

1 See Magellan to Lavoisier, 10 September 1776, note 2. Letter not found.
2 See Magellan to Turgot, 16 November 1778.
3 See Magellan to Turgot, 13 November 1778.
4 Illegible symbols.

vous aurez la bonté de me faire paier, ou permettre de les Recevoir de Mr. Bostok,[5] qui est deja à Londres Car Je l'ai rencontré un de ces jours, & il me demanda de vos nouvelles. 78.11.20

La pendule astronomique de Mr. L'Abbé Sigorgne, avec l'echapement à paletes de Saphire, & la rue[6] des seconds en acier très durement trempé, est deja finie, & marche actuellement ici dans mon parloir. elle me paroit aussi bonne qu'on peut La Souhaiter. j'aurai mardi la caisse, que J'ai ordonnée toute simple: & elle partira ensemble avec une autre de la même espece, pour Mr. Le Roy de l'Academie des Sciences, par La premiere occasion. Mais c'est cette occasion qui me chagrine très fort. car la longueur de la voye par Flandres, est aussi grande & la depense si extraordinaire que je me trouve dans l'embarras, Sans Savoir quel parti prendre. J'ai ici tout prets plusieurs articles pour d'autres amis à Paris depuis assez du tems: & aiant fait application à Mon correspondent à Douvre, il m'assure qu'on ne laisse pas rien passer, que seulement des petites Caisses à titre de bagage de quelque passager &a. Le Broker m'avoit promis que dans peu de tems il y auroit un navire licencié pour aller à Calais, & quand J'ai voulu envoyer ces caisses à bord on a refusé de les prendre, parce que toute la cargaison avoit eté prise par le compte d'un seul marchand &a. J'attendrai encore jusqu'à la semaine prochaine: & aussitôt que les deux pendules, de Messrs. Sigorgne & le Roy, seront pretes; il faudra me resoudre, je crois, à envoyer le tout par Flandres. Peut etre Mr. Jarry vous parlera de la permission qu'il souhaite obtenir pour avoir un vaisseau qui aille à Nantes pour Lui apporter La nouvelle machine à feu qu'il laisse ici ordonnée. Je ne scais pas encore quand est ce qu'elle sera finie, ni quand il obtiendra Cette permission. peut-etre me pouvez vous donner quelqu'idée là dessùs. Mais il faut encore considerer que le vaisseau en question ne partira d'ici de la Tamise, mais d'un autre port bien eloigné de cette Capitale: & alors il faudra envoyer & assurer Ces Caisses depuis londres jusqu'au port en question. en fin Je tacherai de prendre les mesures qui me paroitront les plus propres pour Reussir de mon mieux: & c'est tout ce que Je puis faire.

Sachant que vous etes du même avis que moi, sur l'article de ne pas etre pour nous Seuls, J'ai pris la Liberté de mettre dans vos deux caisses Les articles suivans, pour d'autres amis. 1°. Une petite Caisse ou boete, avec quelques brochures &a. de medecine pour mon intime ami Le Dr. Sanchez, qui demeure dans le Cimiterre St. Jean, Rue de la Verrerie. / un petit

5 Thomas Bostock of Bostock & Co, Cook Street, London.
6 roue?

78.11.20 pacquet avec quelques craions &a. pour Mr. Le Marquiz de Gerardin,[7] qui demeure dans La rue Ste. Anne, près les nouvelles Catholiques. / & enfin un autre pour votre voisin, Mr. Le Presidt. de Saron. Ces paquets, ou articles, sont adressés dans la caisse no. 1. / dans celle no. 2. vous trouverez, au dessous des thermometres, un pacquet contenant deux lunetes de poche pour Mr. le Chevr. de Bory, qui demeure dans l'hotel d'Aligre, Rue d'orleans St. Honoré: & deux autres paquets contenant 4 de ces Lunettes, pour Mr. Bezout de l'Acade. Rle. des Sciences, qui demeure dans La Rue Christine.

Nous aurons le 30 de ce mois l'election du nouveau Presidt. de la Soce. Rle. (car le Chevr. Pringle resigne): & nous ne savons pas encore qui aura plus de voix. Je crois que sera Mr. Banks selon que Je vois plusieurs de cette opinion. Quant à moi, je ne manquerai pas de lui donner ma voix, à moins qu'il declare de ne pas vouloir accepter. Car Je le crois un des plus dignes dans les circonstances qui sont requises.[8]

J'ai l'honneur d'etre avec tout Respt. & vrai attachemt. pour La vie

Monsieur

Votre très hble & très obeisst. servr.

Magellan

P.S. Je m'obliois de vous dire, que vous trouverez dans une de ces caisses, 3 copies d'une brochure que Je viens de publier, relative à la mort du pauvre Jean Jacques Rousseau.[9] Savoir pour vous, Mr. Sigorgne, & Mr. Rochon.

a Monsieur
Monsr. Turgot, ancien Controleur general
Rue de l'Université, Fauxbourg St. germain
à Paris

7 René-Louis, Marquis de Gérardin (1735-1808), a friend of Jean Jacques Rousseau and owner of the Ermenonville estate.

8 Joseph Banks was indeed elected President of the Royal Society in place of Sir John Pringle.

9 Jean Jacques Rousseau died in Geneva on 2 July 1778. Magellan arranged the publication, in London, of Le Bègue du Presle's *Relation ou notice des derniers jours de Mons. Jean Jacques Rousseau* (London, 1778), to which he added, as an appendix, an account of his own visit to Rousseau shortly before he died.

To Comte Joseph Velasco del Pico 78.11.24

Archief Van Arenberg, Enghien/Edingen, Belgium

Londres 24 Nov^{bre}. 78

Monsieur

Que Je vous suis Redevable, mon cher ami Velasco, par la nouvelle que Vous me donnez[1] d'avoir travaillé au Memoire de feu Mgr. Le Duc d'Arenberg! J'etois très inquiet ne sachant point que quelqu'un recueilloit des memoires de tant de belles actions heroiques, & exemplaires pour La posterité. Vous m'avez tiré de mon Souci; Je vous en Remercie de Coeur; & je brule de voir par ecrit un Si grand nombre d'anecdotes interessantes; & qu'en fin on perpetue, plus que par Le marble & Le bronze, la memoire d'un si grand Prince.[2]

Je vous prie de parler, ou ecrire, à M^{r}. Hennessy, sur les deux petites Caisses que J'ai envoyées à Son Altesse, la premiere le 23 Octobre, par le Navire London Merch^{t}, Cap^{tn}. Butchart: & la Sec^{de}. le 11 du Courant, avec votre Canif &^{a}.[3] par Le meme vaisseau. Je lui ai fait avis par la poste (au meme M^{r}. Hennessy) en lui envoyant, à chaque fois, le Reçu du Capitaine &^{a}. vos Lunettes couterent, avec La caisse en Roussette verte & jointes d'argent, 9^{s}. shillings: & le Canife à 4 tranchans, couta 8^{s}. Esterl.

Les grands verres à boire, comme ceux pour leurs Altesses, les deux Duchesses coutent à raison de 4$^{£}$: 10^{s}: 0^{d} par douzaine, ce qui fait 7^{s}. 6^{d} par piece. mais en achetant au dessous de la douzaine, alors on paye d'avantage. Les verres plus petits, comme celui pour S.A.Mlle. La Princesse, coutent à 3$^{£}$: 12^{s}: 0^{d} La douzaine.

Je ne vous dis rien, mon cher ami, sur Les couteaux, parce que vous avez rayé Vous meme l'article de votre Lettre sur ce sujet, en sorte que Je n'en ai pû Lire rien.

Pour ce qui regarde Les echantillons du velour à tricot, il est impracticable d'en envoyer, qui rende aucune idée de l'ouvrage. la grande varieté, & ce qui plus est la qualité des pieces ne permet pas d'en couper sans les gater. L'Artiste, qui est beaucoup de ma Connoissance, est venû exprès chez moi ce matin pour me satisfaire Là dessùs. Voici-ce que je me propose de faire à present pour seconder La Curiosité de Son Altesse. dittez moi par La suivante Lettre pour combien d'amis Son Altesse, voudroit avoir de ce

1 Letter not found.

2 Velasco's memoir of the Duke appears not to have been published. For Magellan's comments on the text, see his letter to Velasco, 15 December 1778.

3 See Magellan to Velasco, 10 November 1778.

78.11.24 velour à tricot, outre soi-meme. alors je choisirai une veste dont la couleur soit approchante du deuil, ou demi-deuil de ce velour à demi coupé, dont il a une grande varieté; une autre pour Mons^gr^. le Duc d'Ursel,[4] & une 3me. ou 4me. en couleur gay, pour quelqu'autre ami qui ne soit en deuil. Je crois que chaque veste ne coute que 2 guinees & demie, ou tout au plus 3 guinées: mais elles sont charmantes: très chaudes pour l'hiver, & très durables. Cet Artiste a fait d'une varieté etonante (c'est à dire Les hommes qu'il employe, car il a eu fort grand nombre de machines, plus de 100 travaillant à la fois). un de ces jours J'en ai vû chez Lui une espece de tricot avec du grand poil de soye, qui s'approchoit de celui de l'agneau: un autre en Laine pour des habits d'hiver comme les pellisses les plus communes des Laponois &^a^.

On vient de publier ici un Remede le plus immanquable pour La toux catharrale. C'est de respirer la vapeur de l'eau commune mediocrem^t^. chaude. il est etonnant l'effet de la guerison, qui est infalible, avec un remede si simple. l'Auteur qui est M^r^. Mudge, frere du Celebre Horloger de ce nom, a donné un instrument, que J'appelle respirateur pour cet effet.[5] Je viens d'y faire quelques amelioration, qui le rendent plus commode: & pense faire faire un en argent pour La Famille de S. A. S^me^.; car on ne doit desormais se passer d'en avoir un de ces instruments chez soi. il est à peu près comm'un pot ordinaire à faire du thé. on Les fait aussi en etain & en fer blanc. J'en ai deja fait executer de ceux-ci, pour mes amis: & J'imprimerai cette semaine la description, & la methode pour en faire usage; dont Je vous envoyerai quelqu'exemplaire.[6]

Je vous prie de dire de ma part à M^r^. Le Coq Tresorier de Son Altesse, que Je ne manquerai de Repondre à La Lettre qu'il eut La bonté de m'ecrire par ordre de Son Altesse,[7] aussitot que mes deboursements seront moins insignificants qu'ils Le sont à present.

à Dieu mon cher Monsr.
Je suis tout à Vous
Magellan.

---

4 Wolfgang-Guillaume, 3rd Duc d'Ursel (1750-1804), husband of Princess Flore d'Arenberg (1752-1832), a sister of Louis Engelbert d'Arenberg.

5 See John Mudge, *Cure for a Recent Catarrhous Cough* (London, 1778).

6 Magellan described Mudge's device, with his own improvements, in an account that was published as a pamphlet in late November or early December 1778 and later, under the heading "Nouveau Remède de Mr. Mudge pour la Toux" in *Observations sur la Physique*, September 1779, pp. 225-7, and in Magellan's *Collection de différens traités sur des instrumens d'astronomie, de physique, etc* ... (London, 1780). See also Magellan to A. Sanches, 4 November 1778, and Magellan to Duke Louis d'Arenberg, 29 December 1778.

7 Letter not found.

P.S. Est ce que feu Mgr. Le Duc expressa le desir d'etre enseveli aux capucins d'Enghien? parce je crois avoir vu au couvent à Everlê des Mausolées des Ancetres de la Famille de Croy[8] & Arenberg. Vous m'excuserez, mon cher ami, tant de questions impertinentes peut etre sur ce sujet. mais Je vous assure, que Je sens le Coeur blessé depuis la morte de S. A. jamais aucun acident de la vie humaine m'a fait tant d'impression. peu de jours après Le tremblem^{t}. de Lisbonne,[9] je commençai à guerir de l'impression qu'il me causa, à l'exception de la perte & malheurs de mes amis, & de ceux que Je connoissois. mais pour celle-ci, je vois qu'elle m'accompagnera jusqu'au tombeau, qui peut etre ne git trop loin de moi. J'ai actuellem^{t}. un autre ami intime fort malade ici: & Je crains beaucoup que Je le perderai aussi. Voilà pour quoi on vit d'avantage...! on voit partir ceux qu'on aime, sans retour, ni remede...! 78.11.24

Monsieur

Votre très h^{ble}. serv^{t}.

Magellan

A Monsieur
Monsr. Velasco, Secretaire de
S. A.S. Monseig^{r}. Le Duc d'Arenberg
à Bruxelles

8 The House of Croÿ was an important noble family with ancient roots that held a seat in the Imperial Diet from 1486. It had branches in Belgium, France, Prussia, Hungary and Serbia and many of its members distinguished themselves in the clergy, diplomacy, and the courts and armies of France, Burgundy, Germany, Spain and the Low Countries.

9 The great Lisbon earthquake of 1755.

78.12.02 *From Casimiro Gomez de Ortega*[1]

Royal Society of London, Letters & Papers, VII, 94

Monsieur et tres illustre Confrere.

Je m'empresse de vous annoncer une découverte interessante qu'on vient de faire au Royaume de Sta Fé, dans l'Amerique Espagnolle. Les Montagnes des environs de la Capitale du même nom, sont peuplées de deux especes d'arbres qui fournissent du Quinquina. L'examen le plus scrupuleux que j'ai fait de leurs fleurs et de leurs fruits, ne me permet pas de douter qu'ils n'appartienent au Genre Cinchona, Linnaei.

J'en ai comparé aussi les autres parties avec la description et le dessein que feu Mr. de la Condamine dressa à Loxa[2] du Cinchona officinalis; et quoique je me sents porté à croire que les arbres n° A et B, du Royaume de Sta. Fé ne sont que des variétés du precieux specifique du Peroû; Je n'ose rien affirmer; nos sçavans confreres de la Société Royale de Londres, seront bien plus en état que moy de decider cette question interessante.

C'est dans ce but aussi bien que pour constater les bons effets produits par Le Quinquina de Sta Fé, qui à été administré constamment avec un heureux succès ici à Madrid, et dans son païs natal, que j'ai l'honneur de vous addresser des échantillons des Plantes seches avec leurs fleurs et leurs fruits et une Boëte remplie de l'ecorce en poudre, du Quinquina n° A. qui sera incessamment suivie d'une autre n° B.

Le Roy qui a eû la bonté de vouloir sçavoir mon sentiment sur ces objets, a crû d'apres mon rapport, ne pouvoir mieux faire que de consulter la Société Royale de Londres, sur une découverte qui peut dévenir tres interessante, soit en elle même, soit en egard a la plus grande facilité du transport de ce Quinquina par la Riviere de la Magdalene a Carthagêne, et puis par Mer en Europe.

Monsieur Galvez Secretaire d'Etat et Ministre très eclairé du Departement des Indes,[3] m'a donné en consequence ses ordres. Mon dévoir donc, Monsieur, d'accord avec le desir que j'ai de merîter l'honneur que la Société Royale eut la bonté de m'anticiper en m'associant à cet illustre corps,[4]

1 Magellan received this letter on 5 March 1779 and forwarded it to Banks for the Royal Society; see Magellan to J. Banks, 6 March 1779.

2 Loja, a town in Ecuador.

3 José de Gálvez y Gallardo, marquis of Sonora (1720-1787), Minister of the Indies, 1775-87.

4 Casimiro Gómez de Ortega (1740-1818), first director of the Royal Botanic Garden in Madrid, had been elected a Fellow of the Royal Society in 1777, following a visit to England.

me fait saisir avec empressement cette occasion de lui temoigner ma vive reconnaissance, aussi bien que de lui fournir un nouveau moyen de se rendre utile de plus en plus au Genre-Humain, qu'elle soulage par ses bienfaits et eclaire par ses Lumieres. 78.12.02

Je vous prie donc, de vouloir bien présenter a la Société Royale avec mes plus profonds respects la premiere Boëte et les echantillons des Plantes sèches,[5] et de me croire avec tout le sentiment de la consideration la plus distinguee,

Monsieur, et illustre Confrere,
Votre tres humble et tres-obeiss^{t}. Serviteur
Casimire Ortega

à Madrid ce 2 Decembre 1778.

à Monsieur de Magellan.[6]

5 A translation of Ortega's letter was read to the Royal Society at its meeting on 11 March 1779, following which the Society established a committee to assess the materials he had sent.

6 The file includes an English translation of Ortega's letter.

To James Watt and Matthew Boulton[1] 78.12.04

Birmingham Central Library, Boulton and Watt collection, MBP 244, 59

London 4 Dec^{ber}. –78

Dear Sir

as soon as I received the inclosed in your favour dated y^{e}. 28 last month, directed to M^{r}. Perier, I sent it to his Lodging at n°. 35 Suffolk Street.[2] He had been with me before: said that your privilege at Paris was passed

1 Annotated "Mr Magellan 4 Decr 1778".

2 The French engineer Jacques-Constantin Périer (1742-1818), who had recently arrived in England to order two Boulton & Watt steam engines for the Paris waterworks company, to be erected at Chaillot to raise water from the Seine to distribute to Paris.

78.12.04 in your favour, and that he knew that I was there the Last Summer for that purpose &c^a^. I did not answer any thing to y^e^. contrary, as it would be ridiculous to deny, what was known to many. Just Now I received by this post the inclosed of M^r^. Jarry,[3] which will give the good news I had confirmed in a Letter by the same post from Count d'Herouville.[4] the privilege is agreed to change the place of essay into y^e^. mines of M^r^. Jarry. when M^r^. Boulton comes to Town I'll show him my letter from the Count, who is very ready to do every thing in your favour &c^c^. he desires me to present M^r^. Boulton with his thanks for y^e^. pictures.[5] every one is there in raptures with them &c^a^. I can not write more at present, and only present my respects to M^rs^. Watt, and company: and ever am

Dear Sir
your most obed^t^. serv^t^.
Magellan.

N.B. I approve very much of your moderation in regard to M^r^. Perier,[6] and at all times to every one else.

To
Messrs. Watt and Boulton
at Redruth
in Cornwall

3 Joseph Jary (1739-?), director of a colliery near Nantes.

4 Letter not found.

5 See Magellan to d'Hérouville, 16 November 1778.

6 During a previous visit to England, Périer had tried to persuade the Birmingham iron master John Wilkinson to make him a Watt engine without his going through the Boulton & Watt company; see J.R. Harris, *Industrial Espionage and Technology Transfer: Britain and France in the Eighteenth Century* (Aldershot, 1998), p. 300.

To Ribeiro Sanches[1]

78.12.08

Österreichische Nationalbibliotek, Vienna
A. N. Ribeiro Sanches Korrespondenz, codex 12714, ff. 227-228v

8 Dec^{bro}. –78

Meo am°. e S^{r}. e m^{to}. do Cor^{am}.

Apenas tenho tempo antes q se feche o Correio p^{a}. dizer a V.M., q o melhor oculo p^{a}. o seo amigo grand maitre des eaux & forets será hũ de 12 guineos q grossit environ 60 foit.[2] il est seulem^{t}. de 14 pouces q^{d}. fermé & de 52 qd°. aberto. hũ como este mandei eu p^{a}. o Emp^{or}. por ordem do seo Embaixador nesta Corte, q me honra da sua amizade & conhece o meo prestimo. ha outros menores de 10 polegadas, à 8 guineos outros de 9 polegadas a 5 guin^{s}. : & em fim outros de 5 polegadas à 2 ½ guineos. destes ultimos tenho eu mandado m^{tos}. aos meos amigos. Aqui se fez uma nova edição de Euclin pello D^{r}. J. Hunter de York. elle cita n'uma nota a M^{r}. de Buffon, dizendo q o escascar as arvores antes de as cortar (na primavera, e ao depois cortalas no inverno) he o melhor p^{a}. q a madeira seja boa. mas o texto do Author parece da opinião contraria à pag. 512. he hũ grosso livro em 4°. de 650 pagi[nas.]

Porem eu fui hontem de proposito á Socied^{e}. das Artes, e o Secretario M^{r}. More, q he bem intendido nestas materias, me allegava q em Stadfordshire se uza o escascar os carvalhos no tempo da Seva, em Abril ou maio: q as arvores continuão a dar folhas & verdura, mas q morrem no oitono ou inverno, e q então as cortão. isto he tudo q^{to}. parece de mais certo. mas eu cuido ter lido ha m^{tos}. annos em algũ tomo das Memorias da Acad^{a}. R. das Sciencias de Paris, esta mesma practica, & experiencias q se fizerão p^{a}. demostrar a realid^{e}. desta practica & sua excellencia.

o amigo Begue du Presle lhe dará um extrato impresso sobre o respirador, q agora lhe mando, e q eu dirigi à V.M. pondo som^{te}. as initiaes A.R.S. visto não merecer o ter o seo nome em semilhante ninharia.

---

1 Annotated "Recue à 15 x^{bre}" [Received 15 December].
2 fois?

78.12.08 fui eleito membro da Acad^a. de Petersburgo, mas não tenho, nem posso ter pensão, excéto algũ raro acazo. não creio q o Barão Klingstedt fizesse mais do q Euler p^a. esta honra q Recebi de nome academico &^a. &^a.

mandarei as encomendas. apenas tenho tempo p^a. correr á posta

vale & ama

tuum &^a.

a Monsieur
Monsr. Sanchez D^or. en Medecine
Cimiterre St Jean, Rue de la Verrerie
à Paris

8 December –78

My Friend and very Dear Sir

I just have time before the post closes to tell you that the best lunette for your friend the Grand maitre des eaux & forets[3] will be one of 12 guineas that grossit environ 60 foit. Il est seulement de 14 pouces quand fermé & de 52 when open. One of this kind I sent to the Emperor by order of his Ambassador to this Court,[4] who honours me with his friendship & knows my service. There are others shorter than 10 inches at 8 guineas, others of 9 inches at 5 guineas & lastly others of 5 inches at 2 ½ guineas. Of these last I have sent many to my friends. A new edition of Evelyn was made here by D^r. J. Hunter of York.[5] He quotes M^r. de Buffon in a note, saying that peeling the trees before cutting them (in spring, and then cutting them in winter) is the best way to get a good wood. But the author's text seems of the opposite opinion on page 512. It is a large book in 4° of 650 pages.

However I went on purpose to the Society of Arts and the secretary, M^r. More,[6] who is well acquainted with these matters, assured me that in Stadfordshire[7] it

---

3 Probably Jacques-Donatien Le Ray de Chaumont (1725-1803), "Intendant honoraire de l'Hotel des Invalides, Grand Maitre aussi honoraire des Eaux et des Forêts de France" between 1776 and 1785 (see John Bigelow, *Retrospections of an Active Life* [New York, 1909], vol. 3, p. 440).

4 Ludovico Barbiano di Belgioioso (1728-1801), Austrian Ambassador in London.

5 John Evelyn, *Silva: or, a Discourse of Forest-Trees*, with notes by Alexander Hunter (York, 1776).

6 Samuel More.

7 Staffordshire?

is usual to peel oaks at stripping time, in April or May, so that trees continue to grow leaves & green shoots, but these die in autumn or winter and then they cut them. This all seems more correct. But I think I've read many years ago in some volume of the memoirs of the Royal Academy of Sciences of Paris, about this same practice & experiments that were done to demonstrate the correctness of this practice & its excellence.[8] 78.12.08

Our friend Begue du Presle will give you a printed extract about the respirator that I'm sending you now, which I addressed to you putting only the initials A. R. S. since it doesn't warrant having your name on such a trifle.[9]

I was elected a member of the Petersburg Academy, but I neither have a pension, nor can have one except in rare circumstances.[10] I don't believe Baron Klingstedt[11] did more than Euler[12] towards this honour that I received of being named an academician &c.

I'll send the orders. I barely have time to run to the post.

vale & ama
tuum &a.[13]

8 Magellan probably has in mind the treatise by Duhamel du Monceau, *De l'exploitation des bois...* (Paris, 1764), especially Chapter V, "De la saison où il convient de l'abattre les arbres" (pp. 315-402).

9 i.e. Magellan's printed pamphlet, "Extrait d'une Lettre de Mr. *Magellan*, Membre de Societé Royale de *Londres*, à un de ses amis de *Paris* [A.R.S.], datée le 4 *Novembre*, 1778"; see Magellan to Sanches, 4 November 1778, note 1.

10 Magellan was elected a foreign member of the St Petersburg Academy of Sciences on 28 September 1778 Old Style. See also J. A. Euler to Magellan, 27 October 1778.

11 Baron Timothy de Klingstedt (1710-1786). See Magellan to Sanches, 14 November 1777, note 8.

12 Johann Albrecht Euler?

13 Farewell & love | your &c.

78.12.09 *To Johann Albrecht Euler*[1]

St Petersburg Branch, Archives Department, Russian Academy of Sciences, f 1, op. 3, Nr. 64, ll. 136-137ob

Londres 9 Decbre. –78

Monsieur & cher digne ami

Je ne vous remplirai non plus cette lettre de griffonage, avec des remerciements de parole;[2] parce que Je me flate que vous savez juger de mes sentimens, & qu'en toute occasion, oû je puisse vous marquer de [fait][3] ma gratitude, vous me trouverez tel que vous me faites l'honneur de me juger à présent. La lettre ci-jointe pour Mr. le Directr. [p]our l'Acade. Imple.,[4] n'est peut etre ce qu'elle devoit etre, si quelqu'autre l'auroit ecrite; mais je m'aime qu'à passer pour ce que je suis, c'est à dire, simple, vrai, & tout uniement reconnoissant. D'aileurs je me remets à vous entierement sur celà, avec une pleine faculté d'y chan[g]er, & raier tout ce que vous jugerez à propos. L'autre lettre de Mr. Bala,[5] vous prouvera qu'il se trouve encore ici: &, comm'il m'assure dans cette qui l'accompagnoit[6] (car il est dans un vilage à 10 milles de Londres) sa santé se rafermit de plus en plus; ce qui est l'objet le plus interessant de son voyage dans ce pays.

Vous n'avez qu'à vous servir de l'occasion que vous trouverez convenable, pour m'envoyer le diplome, soit du Marchd. que vous nommez, ou de quelqu'autre: & pour ce qui regarde à la dorure reele des 200 roubles, qui pour un homme dans ma situation, en doit augmenter de beaucoup le merite, soiez bien tranquile là dessùs, c'est à dire que ce ne sera pas faute de mon silence qui en empéchera le succès.

---

1 Annotated: "reçu le 20 Decembre 1778".

2 Magellan had just learned that he had been elected a foreign member of the St Petersburg Academy of Sciences; see J. A. Euler to Magellan, 27 October 1778.

3 The letter has been disbound from a volume of manuscripts that were once tightly bound. Words at the ends of some lines are now illegible.

4 See Magellan to the President and Members of the St Petersburg Academy of Sciences, 9 December 1778.

5 Presumably the Collegial Adviser (*kollezhskii sovetnik*) Balla mentioned in *200-letie Kabineta Evo Imya Rumichestva 1704-1904* (St Petersburg, 1911), p. 381, as having on 16/27 January 1782 received a payment of 1,200 roubles.

6 Letter not found.

Je me suis informé assez particulierem[t]. des 3 Messieurs Sibthorp, de Maresque, & Morton:[7] & Je trouve que tous trois vivent. je connois les deux derniers, & n'a pas long tems que J'ai vû le D[r]. Morton. Quant à M[r]. Desmaresque, il alla s'ensevelir dans une espece de cure, ou Benefice, qu'il a loin de Londres,[8] & ne correspond presqu'avec personne, quelqu'un m'a dit qu'il avoit perdu la vue, du moins d'un oeil. C'est un vrai honet homme; aimant la solitude & les [...] & assurem[t]. la faute de vue doit lui etre fort dure. 78.12.09

Je vous ai envoyé, par la voye d'Hollande, une petite brochure que J'imprimai ici, sur les derniers jours du pauvre Jean Jacques Rousseau:[9] & J'envoyai par la voye du Secretaire de M[r]. Mouchin Pouchkin[10] l'annonce d'un remede pour la toux catarrale, qui est tout simple, & infailible.[11] Ces circonstances, annoncées par un honet homme, come je sçais que l'auteur l'est, m'excitera à en repandre la notice pour le bien de l'humanité. C'est seulement la vapeur de l'eau chaude, qu'on inspire, par un instrument, qu'il nomme inhaler en Anglois & que J'appelle respirateur. 20 ou 30 minutes de cette simple operation en se couchant le soir, ne manquent pas de reussir.

J'imprime actuellem[t]. la description des instrum[s]. circulaires à reflection,[12] dont je parle dans mon Traité des Octa[nts] à pag. 212[13] n[c]. 364. Je vous en envoyerai une copie. C'est que mon ami le Chev[r]. de Borda ne donna pas la description qu'il pensoit de publier, avant son depart dans l'Esquadre du Comte d'Estaing:[14] & dieu sçait, s'il en

---

7 Humphrey Sibthorp (1713-1797), Sherardian professor of botany at Oxford from 1747, Daniel Dumaresq (1712-1805), chaplain to the English factory in St Petersburg, 1747-62 and 1764-66, and Charles Morton, M.D. (1716-1799), Secretary of the Royal Society of London, 1760-74, were all corresponding members of the St Petersburg Academy of Sciences, having been elected in 1754, 1762 and 1764 respectively; see B.L. Modzalevskii, *Spisok chlenov Imperatorskoi Akademii Nauk, 1725-1907* (St Petersburg, 1908).

8 Dumaresq was rector at Yeovilton, near Ilchester in Somerset; see John H. Appleby, "Daniel Dumaresq, D.D., F.R.S. (1712-1805) as a Promoter of Anglo-Russian Science and Culture", *Notes and Records of the Royal Society of London*, 44 (1990), 25-50.

9 A.G. Le Begue du Presle, *Relation ou notice des derniers jours de Mons. Jean Jacques Rousseau...* Avec une addition relative au même sujet, par J.H. Magellan (London, 1778).

10 Alexei Semenovich Musin-Pushkin, Russian ambassador to England, 1766-7 and 1770-79.

11 John Mudge, *A Radical and Expeditious Cure for a Recent Catarrhous Cough...* (London, 1778). See also Magellan to Sanches, 4 November 1778.

12 Magellan, *Description des nouveaux instrumens circulaires à reflection, pour observer avec plus de précision des distances angulaires sur mer* (London, 1779).

13 The correct page reference is p. 112.

14 Borda served as an officer in the French squadron commanded by d'Estaing that sailed from Brest in April 1778 to provide naval support for the American revolution.

78.12.09 reviendra du tout.[15] Ces instruments sont les plus commodes pour les observations sur mer. on m'en demande souvent pour differentes parts de l'Europe (car les Anglois ne les ont encore adopté, peut etre en obseque du sextant de Hadley, qui est fort au dessus des avantages de ceux dont je parle): & je n'ai plus de patience à en ecrire à la main les particularités.

notre collegue, le charlatan Wilson, va vous envoyer probablement le grand Memoire[16] qu'il a fait sur les experiences électriques pour contredire & refuter les conducteurs pointûs, dans le meme tems que tout ce qu'il y raporte de faits sont en leur faveur. Je gagerai qu'il aura l'impudence de vous prier de presenter le sot ouvrage à Sa M. Imp^{le}., comme Je sçais qu'il l'a deja fait à une autre Acad^{mie}. de l'Europe, oû l'on s'est mocqué de lui: car un ami intime affirme qu'on n'en fera rien. tout son objet est de decrier que les conducteurs pointûs ne valent rien, mais toutes ses experiences montrent qu'ils attirent mieux que les boules ou conducteurs émoussés la matiere electrique. lors qu'on voit un si grand appareil pour prouver ce que tout le monde sçait aujourd'hui, on ne peut pas se tenir sans rire. s'il auroit montré que le feu electrique etant attiré par un bon conducteur pointû, ne le suivoit point, & aloit faire du mal à quelque batiment; alors, en montrant que ses conducteurs à boule ou émousses, n'avoient pas le meme defaut, il auroit soutenu sa cause. mais il ne fait, ni peut faire rien de cela!

Le conducteur qui etoit dans la maison de Purflet, n'a pas empeché, que le crampon de metal qui etoit sur le parapet, à une certaine distance, & qui assurement n'avoit rien de pointû, ne fut point frapé. Mais heureusement il y avoit une pointe metalique du coin d'une goutiere de plomb, je crois qu'à <u>6</u> pouces de distance de ce crampon; ainsi la foudre y sauta tout de suite,

---

15 In his *Description des Octants et Sextants Anglois...* (London, 1775), p. 112ff., Magellan described the double sextant he had devised, but first gave credit in the following words to Borda for having invented it simultaneously: "Avant d'entrer en matiere je dois rendre justice à M. le Chevalier de Borda...qui sans avoir aucune notion de mes idées sur ce sujet, a reconnu aussi les mêmes proprietés & avantages d'un instrument à réflection construit sur ces principes, & en a fait l'application à son instrument circulaire, dont il fera peut-être bientôt part au Public. Je fus fort agréablement surpris lorsque ce Savant me communiqua, dans une conversation, ses vues sur ce sujet, & il ne le fut pas moins à la vue de la feuille de ce Traité, où j'avois déja donné les mêmes principes, & qui se trouvoit alors sous presse." When he published his account, four years later, of his circular reflecting instruments, Magellan indicated (p. 3) that he had not included this in his earlier treatise because he expected that Borda would be publishing an account of them. The new work was prompted, he said, by Borda's onerous duties with the French Navy having prevented him from doing this.

16 B. Wilson, *An Account of Experiments made at the Pantheon, on the Nature and Use of Conductors* (London, 1778).

& se dechargea, sans plus de domage que jetter en bas quelques briques. 78.12.09
En consequence de celà la commitée de la Société R[le]. jugea qu'on devoit faire une communication metallique avec tous les crampons, & mettre plus de pointes metalliques &[a].[17] Cependant comme cette doctrine des pointes metalliques appartient au grand Franklin,[18] qui est Ambassadeur des Americains à Paris, & que les armées Britaniques n'ont pû encore vaincre & defaire: du moins les sages de la Cour angloise feront charmés, que la doctrine du charlatan Wilson soit mise en triomphe aux depens de la vraie philosophie & du bien de l'humanité. telle est, fut, & sera sempre la Philosophie courtisante.

mes h[bles]. resp[s]. à M[r]. votre digne Pere que je revere avec tous qui le connoissent, & que Je serois enchanté de voir en lui rendant visite à Petersburg, ce qui tout seul suffiroit bien à m'engager au voyage, mais mes affaires pour le present ne m'en donnent aucune esperance. mes amitiés au D[r]. Guthrie,[19] que J'estime beaucoup.

P.S. M[r]. Banks fut elû Presid[t]. de la Soc[e]. R[le]. car le Ch[r]. Pringle resigne.[20]

Vale.
tuum
Magellan[21]

A Monsieur
Monsr. Jean Albert Euler, Membre
& Secretaire de l'Acad[e]. Imp[le]. des Sciences
à S[t]. Petersbourg

17 After one of the buildings at the British Navy's powder magazine at Purfleet was struck by lightning despite being furnished with lightning conductors, the Royal Society at the request of the Board of Ordrance established a committee to investigate. Benjamin Wilson, who was a member of the committee, filed a dissenting report. For the committee's report, see *Philosophical Transactions*, 68 (1778), 236-8. See also R. W. Home, "Points or Knobs: Lightning Rods and the Basis of Decision Making in Late Eighteenth Century British Science", in P. Heering, O. Hochadel and D. Rhees, eds, *Playing with Fire: Histories of the Lightning Rod* (Philadelphia, 2009), pp. 97-120.

18 Benjamin Franklin.

19 Dr Matthew Guthrie.

20 Joseph Banks was elected President of the Royal Society at the Society's anniversary meeting on 30 November 1778, following Sir John Pringle's resignation, allegedly after incurring the King's displeasure by not rejecting the report of the committee set up by the Society to investigate the respective merits of pointed *vs.* rounded lightning conductors, which found in favour of pointed conductors.

21 Farewell | your | Magellan.

78.12.09a *To the President and Members, St Petersburg Academy of Sciences*[1]

St Petersburg Branch, Archives Department, Russian Academy of Sciences, f. 1, op. 3, Nr. 63, ll. 130-131^{ob}

À Son Excell^{e}. Monsr. Sergius Domachnew,[2] gentil-homme de la Chambre de S.M. Imp^{le}. de Toutes les Russies, Chev^{r}. de l'Ordre de Vasa, & Directeur de l'Acad^{e}. Imp^{le}. des Sciences de S^{t}. Petersburg
&
À Messieurs les Membres de Cette Illustre Academie

Messieurs,

L'honneur extraordinaire que Vous venez de me faire, en m'aggregeant au nombre des Membres externes de Votre Illustre Acad^{e}.,[3] montre la grande generosité de vos dispositions à encourager ceux, qui se sont dévoüés au service des sciences, que Vous cultivez avec tant d'eclat. C'est encourager les talents, que d'en supposer à quelqu'un: Mais n'aiant, moi-même, d'autre [ti]tre[4] à Vos bontés, que mon zéle pour les sciences, je sens qu'à cet egard ma reconnoissance augmente; puisque vous daignez considerer ma bonne volonté, comm'un mérite reel. Pendant que par vos travaux, vous étendez les effets de la philosophie, en éclairant les hommes, j'employerai tous mes efforts à seconder vos vuës, du moins en recuissant quelques fruits pour moi-meme, à fin de me rendre en quelque façon digne de l'honneur que Vous me faites. Puisse cette nouvelle liaison, dont vous venez de m'honorer, servir à me faciliter les occasions de Vous montrer, Messieurs, toute ma gratitude par mes services, qui malgré leur peu de valeur, seront toujours executés avec un coeur reconnoissant, & avec un sincere attachement pour le reste de mes jours.

J'ai l'honneur d'étre avec toute Consideration, & respect,

Messieurs,
votre très h^{ble}. & très obligé serviteur
Jean Hyacinthe de Magellan

à Londres ce 9 Dec^{bre}. –78.

---

1 Annotated: "reçu le 20 Decembre 1778 et présenté à l'Académie le 21 du même mois".
2 i. e. Sergei Domashnev.
3 Magellan was elected a foreign member of the St Petersburg Academy of Sciences on 28 September 1778 Old Style. See J. A. Euler to Magellan, 27 October 1778.
4 Word partly obscured by binding.

To Joseph Banks[1] 78.12.12

British Library Add. 33977, f. 112[2]

Sir

The article of Dr. Ortega's Letter,[3] says that he has already sent a parcell or collection of dryed plants for you: which when arrived, I will take care of forwarding accordingly. He says also to have sent me, for the Rl. Socy., some especimens (dried) of ye. quinquina[4] from the Province of Santa Fé in America meridional,[5] by the way of Lisbon, together with some powder and a Letter, in order to be tryed by our Physicians: and a Copy of his Treatise on ye. waters of Trillo.[6] He adds that the botanists sent to Lima, are arrived safe:[7] and that there is now another botanist[8] named by the King of Espain (at ye. request of Dr. Ortega) to go to ye. Province of Santa Fé, to make the Flora of that Country; from whence, as well from other parts of South America he expects seeds &a. which he will communicate to us in England. That the ArchBishop of Valencia, in Espain, has succeeded with the upland rice I have sent to Spain: where he intends to propagate it at large &a. and desires the Rl. Socy. be informed of this happy success, as he terms it, of naturalising in Europe this usefull especies of rice. the Dor. begs the favour of some seeds of the

1 Marked "Mr Magellhan Jan. –79".

2 Published in *Scientific Correspondence of Sir Joseph Banks, 1765-1820* (London, 2007), I, 219-20.

3 It would appear from what follows that Magellan is referring to a different letter from the one Ortega wrote to him on 2 December 1778. Casimiro Gómez de Ortega (1740-1818) (FRS 1777), was first director of the Royal Botanic Garden in Madrid. The arrival of the package he had sent to Magellan is discussed in Magellan to Banks, 6 March 1779.

4 i.e. quinine.

5 i.e. Santa Fe de Bogotá, in present-day Colombia.

6 Gómez de Ortega, *Tratado de las aguas termales de Trillo...* (Madrid, 1778).

7 The party, consisting of the Frenchman Joseph Dombey (1742-1794) and the Spaniards Hipólito Ruiz López (1754-1816) and José Antonio Pavón y Jiménez (1754-1840), arrived in Lima in April 1778; this became their base for nearly four years while they botanized extensively in the surrounding region. For an account of their travails, see Arthur R. Steele, *Flowers for the King: The Expedition of Ruiz and Pavón and the Flora of Peru* (Durham, N.C., 1964).

8 Probably Sebastián José López Ruiz (1741-1832), one of the discoverers of the new source of quinine, who in 1778 at Gómez Ortega's instigation was placed in charge of all quinine gathering in New Granada.

78.12.12 self-moving plant (the hedisarum from Daca,[9] now well propagated in our gardens): and I beg you the favour of some, if possible, to send him by the post.

I will write the next post to y[e]. Prince Gallitzin desiring him to send me, or you, a copy of his Memoir on Conductors:[10] and I am sure, that if he has one, he will not fail of sending it immediately.

I have a very good account of M[r]. Achard experiments, on pointed Conductors in manuscript, which he did me the pleasure of sending to me from Berlin. this I[n]genious Philosopher, who has already made so many usefull discoveries, found, upon appearent experiments, that a round thin plate of metal put at the end of a conductor, did exhaust better than a sharp point. but on my trying it her[e] with D[r]. Ingenhouz, we could not find such a fact to be true. in consequence of this I wrote to him, desiring to repeat carefully his first experiments, and to send me his own method in case they should turn out as before. This he did, and, as a true philosopher, recants his former opinion with great candour. If you please to have these experiments read to the R[l]. Society, I'll send, or present them very willingly, provided that I may have them back again in case the Commitee of Papers do not chuse to print them.[11] This Custom of the R[l]. Society in appropriating to her Archives the accounts, which are not printed in the Phil[al]. Transactions, nor giving back, at least to the authors, a copy of them, nor allowing, as some pretend to say, the inspection of the same, has [det]erred[12] some persons from sending papers to [...][13] Society. As you, Sir, have exp[ressed][14] a desire of being acquainted by the members with whatever they find less agreable or advantageous to the Society, emboldens me to speak on this topic, in the present occasion of M[r]. Achard's Memoir:

9 Dacca, Bengal. The plant in question is the remarkable *Hedysarum gyrans* (now *Desmodium gyrans*); see J. Loudon, *Encyclopaedia of Plants* (London, 1829), p. 631.

10 "Lettre de M. le Prince de Gallitzin sur la forme des conducteurs électriques", *Mémoires de l'Académie Impériale et Royale des Sciences et Belles-Lettres de Bruxelles*, 3 (1780), 2-10. Gallitzin's letter is dated 6 July 1778 and was read to the Academy on 9 November of that year.

11 No account of Achard's experiments has been found in the Royal Society's *Journal Book Copy*, and neither were they published in the *Philosophical Transactions*.

12 Conjectural reading; paper damaged.

13 Paper damaged.

14 Conjectural reading; paper damaged.

and will be ready at any time to perform, what you may judge proper in the Case. 78.12.12

I ever am
Dear Sir
you most obed^{t}. obli^{d}. Servt
Magellan

this 12 Dec^{ber}. –79[15]

To Joseph Banks Esqr
Soho Square

15 Despite what Magellan writes here, the correct date for this letter appears to be December 1778, since it clearly precedes his letter to Banks of 3 March 1779 (q.v.). The earlier date is consistent with Banks's annotation, and also with the comment about the party of botanists arriving safely in Lima.

To Comte Joseph Velasco del Pico 78.12.15

Archief Van Arenberg, Enghien/Edingen, Belgium

Londres 15 Dec^{bre}. 78

Monsieur

Votre precis historique,[1] mon Cher Velasco, des promotions & charges de feu notre maitre & de Ses ancêtres, est aussi concis, & aussi net qu'il peut L'etre; mais vous n'avez pas autre assez, je crois, dans ce qu'on appelle caractere, & qui fait distinguer le phantome de la graduation civile (dont il y a un si grand nombre dans la premiere classe, comme dans La classe la plus inferieure) de ce que fait le grand homme, & le vrai heros. Je ne doute pas que vous aurez des occasions de recueillir de ces faits de la grandeur d'ame, d'humanité, generosité, & de toutes les autres vertus sociables, de feu Mgr. Le Duc; & que du moins vous les laisserez entre Les manuscripts de La Biblioteque de Son Altesse, pour encourager sa posterité à l'imiter.[2] Car c'est le seul moien qui nous reste après une perte Si grande, pour en tirer quelqu'avantage.

1 Not found.
2 Velasco's memoir of the Duke appears not to have been published.

78.12.15 J'ai ordonné les deux douzaines des grands verres, selon votre Lettre derniere.[3] il n'en a pas de faits actuelemmt.: & on n'en pourra avoir qu'après Noel.

Je puis envoyer tout de Suite le velour tricoté pour les deux vestes qui ne sont pas de deuil, fort beaux & que je suis sur de plaire: mais pour le tricot à grand poil, pour faire un habit, Je trouve une objection que Le meme Artiste me fiz. la voici: le tricot dont Je parle, a les poils enchainés comme ceci à peu près

or il est evident que dans un habit, ou aura le desagrement d'etre achroché souvent à des pointes, ou même aux gallons des habits d'autrui. Mais pour une veste, alors celà ne fait rien. J'avois deja choisi une piece de cette derniere façon, pour son Altesse (non pour un habit mais pour une veste) toute en Soie noire: & un'autre en Soie noire & blanche. Mais à present j'attendrai Sa Resolution. Vous pouvez Lui dire que s'il veut, ou lui fera ce qu'il faut en soye noire, mais coupée, comm'il en a ici chez le meme artiste en d'autres couleurs. Repondez moi, s.v.p., au plutôt là dessus. Pour ce qui regarde mon Respirateur Je l'envoyerai à S. A. en fer blanc.[4] Celui en argent a toujours la valeur du metal, & j'en disposerai un tems ou l'autre.

L'imprimé ci-joint vous donnera une idée de cet instrument.[5]

Voici le complimt. le plus laconique & significant que Je connois.

Vale & ama

tuum Magellan[6]

A Monsieur
Monsr. Velasco, Secretaire de
S.A.S. Monseigr. Le Duc d'Arenberg
à Bruxelles

---

3 Letter not found.

4 See Magellan to Velasco, 24 November 1778.

5 Magellan no doubt included with his letter a copy of the pamphlet he had just published describing John Mudge's *respirateur* and his own improvements to it; see Magellan to Sanches, 4 November 1778, and Magellan to Velasco, 24 November 1778.

6 Farewell & love | your Magellan.

To Joseph Banks[1] 78.12.28

British Library Add. 33977, f. 86[2]

Dear Sir

here I send the article of the Letter I received from my friend D^or^. le Begue du Presle of Paris,[3] relating to y^e^. Botanical Collection of y^e^. late M^r^. Aublet, in order to avoid any sort of mistake &^a^. the article runs thus

"On va vendre le Cabinet de feu notre ami Aublet: marquez moi si quelqu'un de votre pays en voudroit. Il y a au moins quatre mille plantes étrangeres dont beaucoup peu connues: & surtout les plantes d'Amerique, de l'Isle de France, de l'Isle de Bourbon. J'estimerois celà environ cinquante Louis. Tachez de me rendre reponse là dessùs avant Le 15 janvier, tems oû on a remis la Vente."

By this you see, dear Sir, that every specimen comes to about three pence only: but, what I think more valuable, is that as far as I recollect, the late M^r^. Aublet told me there were many and many plants never published, in this collection: that publication he did make of those of Guyane[4] not comprehending any of Bourbon, Isle de France &^a^. perhaps in this I may be mistaken but it will easily appear by looking into the said printed work of the same. Please to send me your answer, if any, on this subject till to morrow evening[5]

I am
Dear Sir
your most Obed^t^. h^ble^. serv^t^.
Magellan

this 28 Dec^ber^. –78

P.S.

I have received many congratulations from my friends abroad, on account of your election of Presid^t^. of y^e^. R^l^. Soc^y^.[6] and particularly from Messrs.

1 Marked "M^r^. Magellhen Decr. 28 -78".

2 Published in *Scientific Correspondence of Sir Joseph Banks, 1765-1820* (London, 2007), I, 191-2.

3 The physician Achille Guillaume Le Bègue de Presle (ca.1735-1807). Letter not found.

4 Aublet, *Histoire des plantes de la Guiane Françoise...* (London/Paris, 1775).

5 Aublet had died on 6 May 1778. Banks subsequently bought his extremely valuable collection, which included many drawings and manuscript descriptions as well as actual plant specimens; cf. Magellan to Banks, 29 January 1779.

6 Banks was elected President of the Royal Society at its anniversary meeting on 30 November 1778.

78.12.28 Mann,[7] Needham,[8] Baron de Poederlé[9] &c^{a}. this last came to London together with y^{e}. Late Duke of Arenberg,[10] and I companied them to your house in Burlington-Street, a little after your Circumnavigation with Cap^{tn}. Cook. You made a present of various curiosities to the Duke, which he deposited in the Cabinet of Prince Charles[11] at Bruxelles. the following are the words of the first, viz. of M^{r}. Mann

"Notwithstanding my sincere regard & respect for Sir Jno. Pringle, whose civilities to me in London I shall never forget,[12] I must own that Mr Banks's election to that Illustrious Chair, gave me the most sensible pleasure & satisfaction. I steem him & love him from my heart; & in return, he honours me wih a particular friendship, which I will do my best to cultivate. His Caracter & great Celebrity will do honour to the Society."

Joseph Banks Esqr.
Soho Square

7 The Abbé Theodore Augustus Mann (1735-1809). Mann's letter not found.

8 John Turberville Needham (1713-1781), Director of the Imperial Academy of Sciences in Brussels. Needham's letter not found.

9 The agricultural writer Eugene Joseph Charles Gilain Hubert d'Olmen, baron de Poederlé (1742-1813), best known for his *Manuel de l'arboriste et du forestier belgiques* (Brussels, 1772, and subsequent editions). Poederlé's letter not found.

10 Charles-Marie-Raymond, 5th Duke of Arenberg, died on 17 August 1778.

11 i. e. Prince Charles of Lorraine.

12 Mann visited London in November 1777 to study newly developed means of preserving buildings from fire. Sir John Pringle (1707-1782) was President of the Royal Society from 1772 to 1778.

To the Duke Louis Engelbert d'Arenberg 78.12.29

Archief Van Arenberg, Enghien/Edingen, Belgium

Monseigneur

L'honneur que V. A. S. s'a daigné de me faire en date du 18 du Courant,[1] n'est parvenu à mes mains que Samedi dernier, ensemble avec La lettre de M^r^. Velasco du 21 suivant.[2] Je puis assurer V. A. S. que rien n'est plus precieux, pour un bon serviteur, que de se voir honoré avec les ordres de celui, auquel, outre le devoir, il se trouve sincerement Lié d'ame & de Coeur: & que me trouvant moi-meme dans Le cas, Je sçais bien connoitre tout Le prix de la faveur dont V. A. S. vient de m'honorer.

Je suis assez bien informé que le prix d'environ 70 ou 80 guinées est celui d'une voiture, que les Anglois appellent <u>chaise de poste</u>, & meme <u>chaise</u> (ou <u>charet</u>) pour La <u>ville</u>: car tout depend de la qualité des materiaux. mais Je me trouve un peu embarassé sur la circonstance des 4 places: car cette espece de voitures ont des glaces par devant, & quoiqu'en effet on y met quelques fois deux places pour deux personnes, les anglois entendent toujours des <u>coches</u> lorsqu'on leur parle des voitures à 4 places. J'ai ajouté avec le craion deux figures dans la page suivante, que son secretaire expliquera à V. A. S. La premiere est celle des <u>chaises</u>: & la seconde est celle des coches anglois. Lors que V. A. me fera donner en detail les circonstances de la voiture qu'Elle souhaite, Je crois que Je ne manquerai pas d'y Reussir. mais pour ce qui regarde mon Souvenir de celles que J'ai vu ou meme entré à Enghien & à Everlé,[3] V. A. S. peut bien me croire sur parole d'honneur que je n'en conserve aucune idée assez distincte pour me conduire.

Aprés demain partira par un vaisseau la petite Caise,[4] contenant les articles du velour tricoté pour la veste en forme de poil entortille &^{a}. comme V. A. S. vient de l'ordonner,[5] ensemble avec les deux autres, & le Respirateur d'argent.[6]

---

1 Letter not found.

2 Letter not found. Joseph Velasco del Pico was the Duke's secretary.

3 Two of the Duke's residences.

4 caisse.

5 See Magellan to Velasco, 15 December 1778.

6 Magellan is referring to an inhaler invented by John Mudge (1721-1793), described in Mudge's *Cure for a Recent Catarrhous Cough* (London, 1778). Magellan described the device, with his own improvements, in a letter to Ribeiro Sanches that was published as a pamphlet in late November or early December 1778 and later, under the heading "Nouveau Remède de Mr. Mudge pour la Toux" in *Observations sur la Physique*, September 1779, pp. 225-7, and in Magellan's *Collection de différens traités sur des instrumens d'astronomie, de physique, etc* ... (London, 1780). See Magellan to Sanches, 4 November 1778, and also Magellan to Velasco, 24 November 1778.

78.12.29 J'ai tant d'exemples de l'utilité de cet instrument dans une famile[7] que Je me flate du succès chez V.A.S.

Je compte envoier cette Caise par Bruges dans l'idee qu'elle arrivera avec moins de retard à Bruxelles que les precedentes.

J'ai L'honneur d'etre pour La vie

de V.A.S.

Serv^{r}. le plus h^{ble}. & plus obligé

J.H. de Magellan

Londres 29 Dec^{bre}. –78

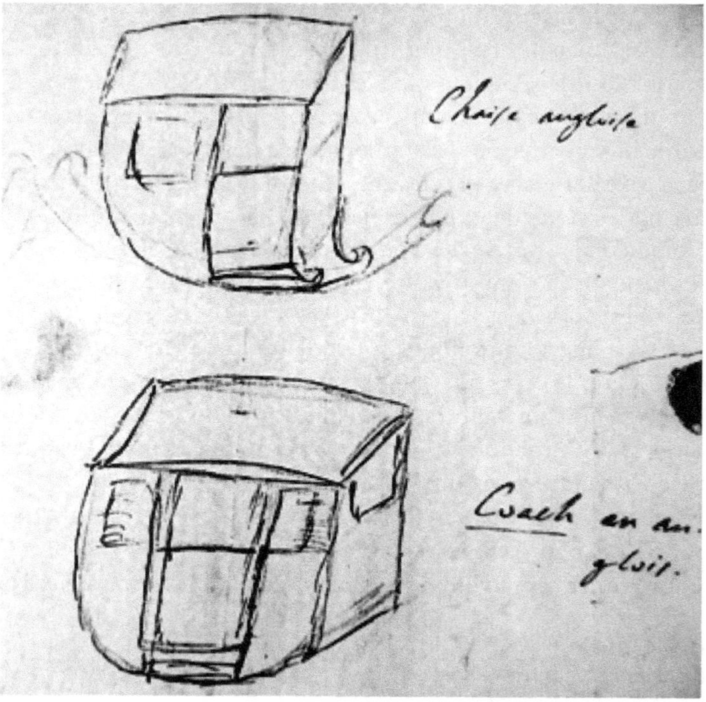

tant le coach, que la chaise[8] ont 4 roues chacun, & sont suspendus sur des resorts. Je voudrois savoir si S. A. S. ne voudroit plutot avoir de la nouvelle espece de roues dont les cercles sont entiers, ou en deux morceaux, pliés par art. elles sont plus fortes & dans le meme tems plus legeres.

---

7 famille.

8 The diagrams are labelled respectively "Chaise angloise" and "Coach en anglois".

To Giovanni Fabbroni[1] 79.01.00

American Philosophical Society Library, Philadelphia, Fabbroni papers, B/F113 no. 1

To M^{r}. Fabroni
at M^{r}. Bernardi's Italian ware shop
in ye Haymarket

M^{r}. Fabroni est prié de la part de Son Exc^{e}. le Chev^{r}. Pinto,[2] d'aller diner chez lui mercredi prochain, (20 Janvier), & M^{r}. Magellan s'est chargé de le lui faire savoir. il souhaite que M^{r}. Fabroni lui dise, s'il peut ou non y aller ce jour là.

1 The letter is undated but, given that it was written in January, must have been written in January 1779 during Fabbroni's visit to England from July 1778 to November 1779; in confirmation of this, between 1773 and 1790, the only 20 January that was a Wednesday occurred in 1779.

2 Luiz Pinto de Sousa Coutinho, Portuguese ambassador in London.

To Anne-Robert-Jacques Turgot 79.01.05

Biblioteca do Museu Nacional de Arte Antiga, Lisbon
Fundo Francisco Sousa Viterbo, Cx n° 1

Londres ce 5 Janvier –79

Monsieur

Je viens d'envoyer par La Voye d'Ostende deux Caisses avec differens articles pour mes amis de Paris, & Je n'ai pû m'empecher d'y mettre un que Je crois pouvoir etre de votre agrement. C'est une petite caisse qui vous est adressée & qui contient les articles suivans. 6 morceaux de brasilian pebles, dont un est plus grand que deux noix. Je trouve que les lapidaires memes donnent ce nom à des cristaux fort differents: ainsi vous verrez qu'il y a de plus, & moins foncés. Je n'ai pas voulu vous envoyer d'une grande depense: on m'offroit un beau morceau, comme la poignée, pour 2 guinées: mais Je n'ai pas voulû l'avoir. Ces 6 morceaux couterent 14^{s}. 6^{d}. Je vous envoye le nouveau respirateur, instrument pour guerir de la toux catarrale, en inspirant la

79.01.05 vapeur de l'eau chaude comme l'on trouvera dans la direction imprimée,[1] qui l'accompagne, il couta 7ˢ.6ᵈ. de plus Je vous envoye du tricot de soie noire pour une veste d'hiver, qui a du poil tortillé & fait une apparence assez singuliere, celui ci couta £ 2:5ˢ.0ᵈ / Ces depenses Respectives de la caisse, embalage, assurance &ª. montent à 6ˢ.6ᵈ. total £ 3:13ˢ.6ᵈ. Je suis invité à diner demain avec Mʳ. Bostock,[2] qui me payera cet argent pour votre Compte, de meme qu'il le fit pour le precedent des Thermometres.

J'ai deja la pendule astronomique de Mʳ. L'Abbé Sigorgne, à echapement de Saphire, chez moi. elle me semble aller à merveille, & Je me flate qu'il en Sera Content. il s'agira d'avoir quelqu'opportunité de la Lui envoyer. du moins il faudra nous prevaloir de la voye d'Ostende.

Mʳ. Le Dʳ. Ingenhouz, & l'autre ami qui fait la Collection des ouvrages du Dʳ. Franklin, & qui en deja imprimée une bonne partie,[3] me demandent des nouvelles de ce que Je vous ai demandé: Je souhaite savoir s'il est possible d'en avoir les deux articles en question. le premier est la lettre ou brochure de Mʳ. Gatti à l'occasion de la petite verole, que Mᵈᵉ. la Marquize, ou Duchesse, ou Comtesse Bonfleurs eut une Seconde fois après avoir eté inoculée.[4] le 2ᵈ. est l'edition que le Dʳ. du Bourg a fait des pieces politiques &ª. de ce grand homme de nos jours.[5]

un mot de Reponse sur les deux objets, obligera infiniment

Monsieur

Votre très hᵇˡᵉ. & obligé servʳ.

Magellan

a Monsieur
Monsʳ. Turgot, ancien Controleur general
Rue de l'Université
à <u>Paris</u>

---

1 J. Mudge, *Cure for a Recent Catarrhous Cough* (London, 1778).

2 See Magellan to Turgot, 20 November 1778, note 5.

3 Benjamin Vaughan was preparing a collection of Benjamin Franklin's papers that was published later in 1779 as *Political, Miscellaneous and Philosophical Pieces … written by Benj. Franklin …* (London, 1779).

4 Angelo Gatti, *Nouveaux éclaicissemens sur l'inoculation de la petite vérole* (n.p., n.d)? Marie Françoise Catherine de Beauvau-Craon, Marquise de Boufflers (1711-1787) was inoculated against smallpox by Gatti in 1763, and afterwards assured by him that she was now safe from infection. Two and a half years later, however, she was struck down by the disease.

5 Magellan is presumably referring to *Oeuvres de M. Franklin*, ed. Barbeu Dubourg (Paris, 1773), the second volume of which includes some of Benjamin Franklin's essays on political economy, as well as a much larger number of his essays on scientific subjects. Dubourg did not produce any other edition of any of Franklin's papers.

To Ribeiro Sanches 79.01.05a

Österreichische Nationalbibliothek, Vienna
A.N. Ribeiro Sanches Korrespondenz, codex 12714, ff. 237-238v

Londres 5 Jan°. – 79

Meo Am°. e S^{r}. e m^{to}. do Cor^{am}.

Ainda q escrevi a V.M. em 8 do passado com a Resposta á encomenda do seo Am°. Grande Maitre des Eaux & Forets, deixei outros artigos p^{e}. esta posta, visto não ter então tempo p^{a}. o fazer. Alem disto os frios q começão a ser asperos nesta terra, me fazem lembrar de V.M. q talvez os tenha igoaes em Paris, e por consegt^{e}. se achará atacado, como de ordinario, com elles. [E]u tãobem cá apanhei hum ou dois resfriamentos, mas brevem^{te}. me dezembaracei delles; q^{m}. arr-ta e fo-de, com todo o mal pode. Vejo pelo contexto da sua, q V.M. tem suas aprehensoens de q eu cá ficasse com algũm raspado do galico no Corpo. Posso porem assegurar a V.M., q nada disso tenho: estou são & galhardo como hũ pero malapio de ribacôa. Hoje ninguem tem medo a esta galantaria escolastica. e posso assegurar V.M. q se agora tivesse a escolha, antes o quizera ter tido, como tive, do q não o ter. achome mais lesto, melhor apetite: bom sono, boa digestão, menos gordo, & mais à mon aise. tomára eu q V.M. tivesse outro tanto; pois creio q se havia de achar melhor: e não zombo do q digo. o precipitatus per se não he o q V.M. imagina, o pó negro do mercurio triturado. verd^{a}.m^{te}. he a cal metallica do mercu[rio]. a operação de a fazer hé mui longa, pelo menos de 5 ou 6 mezes. tomão-se vazos de vidro, q sejão bem chatos como V.M. aqui vê. poem-se com o mercurio dentro (até o meio do bojo, p^{a}. ter grande superficie exposta á ação do ar) em banho de areia, mas destapados.

o mercurio q se eleva com o calor, torna a cahir dentro do mesmo vazo porq se esfria antes de chegar á boca do cólo. e por este modo se forma p[ou]co a pouco a cal mercurial q tem huã cor vermelha como o vermilhão &^{a}. eu comprei desta cal em Paris, na logea de M^{r}. Cadet,

79.01.05a cuido q à hum Louiz por cada onça, pª. as experiencias de meo amº. o D^{r}. Priestley. Por[em] nós temo-lo aqui em Londres m^{to}. mais barato. nos Laboratorios, q sempre tem fogo, custa pouco o metter destas garrafas nos cantos do banho de areia &ª.

O Sobrinho de V. M. já não sofre nada. elle morreo alguns dias antes do Natal, & talvez nos fins de Nov^{bro}.: pois falando nelle ao D^{r}. Sequeira, elle me disse era já partido. recõmendeilhe de saber o conteudo no seo testam^{to}., alguns dias depois. disse-me q deixara tudo á sua filha, q tinha viva na Jamaica: & som^{te}. coiza de 11(?) Livras Esterlinas á Prima q elle tem em Bordeos, e com a qual elle vinha cazar-se. mas não me soube dizer quem era esta Prima póde ser q seja a sobrinha de V.M. Eu procurei assegurar-me desta particularid^{e}., e fui deproposito aos Doctor's Commons, onde se [de]positão todos os testamentos. paguei pela busca, mas não se acha lá tal testamento. provavelm^{te}. os testamenteiros tem hũ certo tempo concedido pª. fazer este deposito: eu faço tenção de lá tornar daqui algum tempo pª. averiguar este ponto.

não me admira q os 6 tolinhos q fugirão de Portugal, se achem embaraçados por estas terras estranhas. Certam^{te}. elles devião ter prudencia: goardar para si o conceito das redicularias religiozas: & não se metter a publicar ideias livres, em terra de escravos & de bigots. Porem tal he a pequenhez do espirito humano. Cuidamos fazer bem em dizer o q pensamos: e em fim pagamos pelas nossas loucuras.

Lá mandei a V.M. hũ quartilho do oleum ricini em hũa Caixa q já partio pª. Ostende. o amigo Begue du Presle, terá cuidº. de lhe mandar a V.M. o q não será talvez tão cedo; pois este caminho das Remessas por Ostende, he m^{to}. longo. Tãobem lhe mandei o Med^{al}. Commentaries nº. 18. 19. 20. & o seo Index. mai o Monro's prelectiones. os 3 outros vol. de Lucas on waters. e cuido q Sheridan the Revolution of Sweden q hé hũ bom livro, pª. V.M. se divertir vendo a corrução de hũ Estado Livre. q prova como a raça humana hé peor, q a dos macácos seos semilhantes na Africa & America &ª.

[...]o epigraphe eripuit calo fulmen &ª. hé excellente. eu amo o sujeito porq tem excellentes virtudes sociaes & grande capacid^{e}. tanto no civil q no Philosophico.

o Practical Remarks on West India diseases, ainda não se imprimio nem se sabe qdº. se imprimira.

não creio no xarope pª. o gallico, [de]sse sujeito Francez. o caminho da estrada real (do mercurio) he o mais [ce]rto & mais seguro.

eu me acho ás mil maravilhas com elle, e espero q não sera a ultima vez q gozarei do seo beneficio. dezejára q V.M. fizesse outro tanto, e q goze tanta saude & Felicid[e]., bons annos & bons reis &[a]. &[a]. como lhe dezejo. 79.01.05a

vale.

a Monsieur
Monsr. Sanchez Doc[r]. en Medecine
Cimitere S[t]. Jean, Rue de la Verrerie
au Marais
à Paris

London 5 January – 79

My Friend and very Dear Sir

Although I wrote to you on the 8th of last month with the reply to the order of your friend the Grand Maitre des Eaux & Forets,[1] I left other matters until this post, since I had no time to deal with them then. Besides, the cold weather, which is starting to be harsh in this country, reminds me that you may be having the same in Paris, and then you are perhaps being affected, as usual, by it. I also have caught one or two colds, but soon got rid of them; he who b-lches[2] and f-cks, withstands every evil. I see by the context of your letter[3] that you have apprehensions that I could still have some remnant of pox in my body.[4] I can assure you that I have nothing of that: I'm fit and sound as a bell. Today no-one fears this scholastic gallantry and I can assure you that if I had a choice now, I would prefer to have it, as I did, than not to have it. I feel more active, better appetite, good sleep, good digestion, less fat, & more à mon aise. I wish you had as much, since I believe you would feel better; and I'm not mocking about this. The precipitatus per se[5] is not what you imagine, the black powder of ground mercury. It is truly the metallic calx[6] of mercury. The operation of making it is very lengthy, at least 5 or 6 months. Glass vases, very flat as you see here, are taken. They are put with the mercury inside (up to half of the bowl, to have a large surface exposed to the action of air) in a sand bath, but uncovered.

1 See Magellan to Sanches, 8 December 1778.
2 belches; see Magellan to Sanches, 7 July 1775, note 19.
3 Letter not found.
4 See Magellan to Sanches, 9 October 1778.
5 See Magellan to Sanches, 12 November 1778.
6 Calx – any powder obtained by strongly heating a substance in air. Almost always an oxide.

The mercury, which rises with the heat, falls again into the same vase because it gets cold before arriving at the mouth of the neck. And in this way the mercurial calx, that has a red colour like cinnabar &c., is formed little by little. I bought this calx in Paris, at Mr. Cadet's shop,[7] I suppose at one Louis per ounce, for the experiments of my friend Dr. Priestley. However we have it here in London much cheaper. At laboratories that have a fire, it is no problem to put these bottles in the corners of the sand baths &c.

Your nephew suffers no more.[8] He died a few days before Christmas, & perhaps at the end of November, since speaking of him to Dr. Sequeira, he told me that he had gone. I urged him to find out the contents of his will. Some days ago, he told me that he left everything to his daughter, who lives in Jamaica: & only some 11(?) Pounds Sterling to his (female) cousin who is in Bordeaux, and whom he was to marry. But he couldn't tell me who this cousin was, maybe she is your niece. I tried to confirm this particular matter, and went on purpose to Doctor's Commons,[9] where all wills are deposited. I paid for the search, but there is no such will there. Probably the will's executors have a certain time allowed to deposit this. I intend to go there after a while to investigate this point.

I'm not surprised that the 6 foolish fellows who escaped from Portugal feel embarrassed in this strange country. Certainly they should be prudent, keep to themselves the concept of religious trifles & not publish ideas freely in a country of slaves & bigots.[10] Such, however, is the pettiness of the human spirit. We think we're doing well by saying what we think, and at the end we pay for our madness.

I sent you a pint of oleum ricini[11] in a box that has already left for Ostende. Our friend Begue du Presle will send it to you but not very soon since this way of shipping via Ostende is very lengthy. I also sent Medical Commentaries nos.

---

7 Either Louis Claude Cadet de Gassicourt (1731-1799) or his brother Antoine Alexis François Cadet-de-Vaux (1743-1828), both of whom were pharmacists in Paris.

8 See Magellan to Sanches, 12 November 1778.

9 The buildings occupied by the College of Doctors of Civil Law in London, where matters of ecclesiatical law, prosecutions for heresy, divorce suits, licences for marriage, testamentary affairs etc. were dealt with.

10 Neither the "foolish fellows" nor the concept of religious trifles that they apparently espoused have been identified.

11 Castor-oil.

18. 19. 20. & the index to them,[12] plus Monro's prelectiones,[13] the 3 other vols of Lucas on *Waters*[14] and I presume Sheridan the Revolution of Sweden[15] which is a good book for you to get fun from by seeing the corruption of a free state, which proves how the human race is worse than the monkeys who are their fellow creatures in Africa & America &c. 79.01.05a

The epigraph eripuit cælo fulmen[16] &c. is excellent. I love the fellow because he has excellent social virtues & great capacity both civil and philosophical.

The Practical Remarks on West India diseases[17] is not yet printed nor do we know when it will be.

I don't believe in that Gallic fellow's syrup for the pox.[18] The royal road (of mercury) is the most certain & sure way.

I feel very well with it, & hope it won't be the last time I'll enjoy its benefit. I wish you did the same, and would enjoy as much health & happiness, good years & good epiphany days[19] &c. &c. as I wish you.

vale[20]

---

12 *Medical and Philosophical Commentaries* [by A.D.] (Edinburgh, 1773 onwards).

13 D. Monro, *Prælectiones medicæ ex Cronii instituto annis 1774 et 1775: et oratio anniversaria ex Harveii instituto, die Octobris 18 ... 1775, habitæ in Theatro Collegii Regis Medicorum Londinensium* (London, 1776).

14 C. Lucas, *An Essay on Waters* (3 vols, London, 1756). Magellan had sent another work by Lucas, his *The Theory and Uses of Baths* (Dublin, 1772), in an earlier package; see Magellan to Sanches, 9 October 1778.

15 Charles Francis Sheridan, *History of the Late Revolution in Sweden* (Edinburgh, 1776).

16 The opening phrase of the description of Benjamin Franklin attributed to the French finance minister Turgot, "Eripuit coelo fulmen sceptrumque tirannis" [He seized lightning from the heavens and the sceptre from tyrants], that was illustrated by Fragonard in a famous allegorical print first published in 1778.

17 Despite what Magellan says, this work appears to have been published several years earlier; *Practical Remarks on West India Diseases ...* (London, 1776).

18 Magellan may be referring to the "syrup" developed by François Chaussier (1746-1828) as a treatment for syphilis, the active ingredient in which seems to have been sodium hyposulphite.

19 Epiphany day is 6 January, the day after Magellan wrote this letter.

20 Farewell.

79.01.13 *To Matthew Boulton and James Watt*[1]

Birmingham Central Library, Boulton and Watt collection, Box 22/B3/11

London 13 Jan^{y} –79

Dear Sir

I have sent yesterday for y^{e}. Letter and inclosed proposals, or account, you mentioned in yours dated y^{e}. 9th. instant:[2] and I forwarded them by the post to Count de Herouville writing him at the same time, in a separate Letter,[3] on the same purpose. the whole appears to me extremely well conducted: and can not think, but the Count will approve of the same. It is very reasonable, and even necessary, to reduce to a sum of money the premium for each machine you dispose of abroad: it being near to impossibility, to estimate the value of fuel in foreign countries. I hope you'll do the same hence forth, and particularly with those spoken of for y^{e}. Prussian mines. By this time you must have seen the Baron de Waiz and his brother, to whom I gave a Letter of recommendation for you.[4] They are to see the effects of your Machines: and very probably there will be a positive order to come to terms.

As to the 25 Actions you have taken, or intend to take, in M^{r}. Perier's undertaking,[5] they may turn out well with time, and good Luck: but I should rather be contented with half the value in ready money: because I have no great opinion of great undertakings under despotic governments, unless they may be performed in a very short time.

Pray remember M^{r}. Boulton the soft iron straps I desired him long while ago to make for my use.[6] I wished to have at Least a dozen of them.

---

1 Letter not found; MS is a clerk's transcription.

2 Letter not found.

3 See Magellan to the Comte d'Hérouville, 13 January 1779.

4 Letter not found. Friedrich Sigismund von Waitz, Baron von Eschen (1745-1808) was a mining entrepreneur who started by mining salt in Mecklenburg, then silver and later lignite.

5 The French engineer J. C. Périer had recently travelled to England to order two Boulton & Watt steam engines for the Paris waterworks company.

6 See Magellan to M. Boulton, 6 November 1778.

You may depend at any time on my readiness to serve you in any occasion: and heartly am 79.01.13

Dear Sir
Your most obed^{t}. h^{ble}. ser^{t}.
Magellan

To
Messrs. Boulton & Watt
at their Fire Engine Manufacture
near Soho
at Birmingham

To the Comte d'Hérouville 79.01.13a

Archives Nationales, Paris, MS Marine, G110, ff. 197-8

Londres ce 13 Janv^{r}. –79

Mon cher Monsieur

Il n'est pas aisé de vous decrire le plaisir que votre Lettre de la fin de Nov^{bre}. dernier[1] m'a fait, depuis un etat d'apprehension pour votre santé & pour votre bien etre. Ce n'est que l'indisposition de Mad^{e}. d'Herouville, qui en a pû diminuer une grande partie: & J'ai à vous prier La grace de me faire savoir, lorsque Mad^{e}. se trouvera retablie en la santé comme Je le souhaite avec tout le reste du monde, qui a la bonheur de la connoitre. Je dis autant à l'egard de Mlles. d'Herouville & de Mondion,[2] qui n'auront rien de plus à desirer, si la mesure de leur bonheur se fait par celle de mes voeux.

Rien de plus genereux & de plus obligeant que ce que vous avez [fait][3] pour Messrs. Boulton, Watt & Jarry.[4] Je viens de recevoir d'eux c'est à dire des deux premiers, une lettre, avec la relation de ce qu'ils sont près d'arranger

1 Letter not found; but see Magellan to J. Watt & M. Boulton, 4 December 1778.
2 Comte d'Hérouville's daughter and niece; see Magellan to d'Hérouville, 17 July 1778.
3 Word obscured by binding.
4 Joseph Jary; see Magellan to J. Watt & M. Boulton, 4 December 1778.

79.01.13a avec M[r]. Perier, qui est venû ici, tout exprès pour avoir deux des Nouvelles machines:[5] & ils ne voudroient point [ag]ir[6] sans votre approbation là dessùs. Je leur ai dit qu'assurem[t]. vous ne pouviez qu'approuver, ce qui tendroit au bien commun de la France & dans le meme tems aux interêts des particuliers: car il faut toujours unir le bien particulier avec le bien commun, pour que celui-ci soit durable.

J'envoye cette meme poste la lettre & relation que ces Messieurs, Boulton & Watt, vous adressent sur cette affaire;[7] parcequ'ils ne veulent pas agir sans vous consulter, & soumettre tout à vos Lumières, qui avez si puissament protegé leur entreprise, & que sans vous ne reussirt probablem[t]. du tout.

Je ne puis bien concevoir le projet de faire entrer l'eau dans les Couretes, reservoirs, & tuyaux des administrateurs. Si c'etoit chez eux & dedans de leurs maisons, celà ne seroit un bien public aussi decidé, comme le projet de fournir toute la Ville avec de l'eau &[a]. J'ai bien chargé M[r]. Boulton qui vint en Ville dernierement, de penser à etablir quelque chose de manufacture avec ses machines: & je ne doute pas qu'il fera tout ce qui ne dependra que de lui.

Je n'ai pas aucune nouvelle certaine de M[r]. le Comte de Chastenet, ni de l'Esquadre, oû il se trouve.[8] Les gazetes ne font que mentir, affirmer un jour, & se dedire dans le suivant. Ce sont les fruits de cette malheureuse guerre,[9] que tout le monde voudroit voir finie, à l'exception de ceux qui s'enrichissent aux depens du reste des citoyens. tel est le sort de la pauvre race humaine.....!

Il est probable que les mots de la chanson de Rousseau ne lui appartiennent aucunement. Je crois que ce sont les mêmes mots de la traduction Françoise de Shakespear.[10]

---

5 The French engineer J. C. Périer had recently travelled to England to order two Boulton & Watt steam engines for the Paris waterworks company.

6 Word partly obscured by binding.

7 See Magellan to M. Boulton and J. Watt, 13 January 1779.

8 Comte d'Hérouville's son-in-law, Antoine Hyacinthe Anne, Comte de Chastenet de Puységur (1752-1807), was a naval officer and was currently at sea with the French fleet.

9 The American War of Independence.

10 "Un papillon badin caressoit une rose: Chanson mise en musique par M. Rousseau..." (*Mercure de France*, June 1737, in Jean-Jacques Rousseau, *Oeuvres complètes*, ed. B. Gagnebin and M. Raymond [Paris, 1959-1995], vol. 2, p. 1163)? Others have also expressed doubts about Rousseau's authorship of the song, though without attributing the words to Shakespeare; see the editors' note in the *Oeuvres complètes*, II, 1906, citing J. Peyrot, "Le premier air gravé de Rousseau", *Zeitschrift der internationellen Musikgesellschaft*, September 1913, 365-66. No Shakespearean source has been identified.

J'ai l'honneur d'etre avec le plus de respect & vrai attachement de coeur pour le reste de mes jours, 79.01.13a

Monsieur Le Comte,
Votre très h[ble]. & très obligé serv[r].
Magellan

A Monsieur
Monsr. Le Comte de Herouville
Lieutenant gen[al]. &[a].
n°. 6. rue du Fauxbourg S[t]. Honoré
à Paris
France

From Jan Henrik van Swinden[1] 79.01.15

Leiden University Library, MS B.P.L. 755, no. 311

J'ai receu Monsieur il y a quelque tems la Brochure que vous avez publiée au sujet de Rousseau:[2] tout ce qui concerne cet homme celebre est interessant aux yeux de ceux qui savent apprécier le merite. Je me flatte que quelque jour nous jouirons des oeuvres posthumes de cet ecrivain si celebre, si malheureux et si singulier. Agreez s.v.p. mes remercimens du present que vous avez bien voulu me faire.

La maniere d'agir de Wilson dont vous me parlez me paroit etonnante.[3] Je ne doute pas que les conducteurs en pointes ne soyent les meilleurs. toutes les Exp. toutes les bonnes observations deposent en leur faveur: c'est en un mot un point qui me semble decidé. Je suis mortifié d'apprendre qu'un homme d'une reputation aussi justement meritée que Wilson se soit abaissé à ne pas mettre dans ses experiences toute la sincerité et la bonne

1 Original not found; text from file copy retained by van Swinden and annotated "A M. Magellan à Londres". Published in David Willemse, "Suite d'un voyage aux Pays-Bas: João Jacinto de Magalhães (1722-1790) et ses rapports avec Jean Henri van Swinden (1746-1823)", *Arquivos do Centro Cultural Português*, 7 (1974), 225-278.

2 *Relation ou notice des derniers jours de Mons. Jean Jacques Rousseau; circonstances de sa mort; et quels sont les ouvrages posthumes, qu'on peut attendre de lui,* par Mons. Le Begue de Presle, avec addition relative au même sujet par J.H. de Magellan (London, 1778). See Magellan to Van Swinden, 27 October 1778.

3 See Magellan to Van Swinden, 27 October 1778.

79.01.15 foi requises. si cette methode reprenoit faveur, c'en seroit bientot fait de la Physique comme de bien d'autre choses.

Oserois-je vous prier Monsieur de vouloir me faire parvenir un Eudiometre, semblable à celui que vous m'avez fait parvenir au commencement de l'année dernière par le canal du Pr. de Gallitsin.[4] Celui que vous m'avez envoyé m'a beaucoup satisfait, comme je vous l'ai marqué dans ma lettre du 22 de Mai dernier, qui vous sera j'espere parvenue. Mais malheureusement cet instrument s'est cassé. N'y auroit-il pas moyen de faire la boulle ou l'on enchasse les deux petites bouteilles, d'un verre en peu plus épais? Cela defendroit beaucoup l'instrument contre les chocs ou autres accidens qui peuvent lui arriver. Voudriez-vous me faire le plaisir d'y ajouter 8 bouteilles à fonds ronds avec leurs tubes, représentés fig. 9 du 2ᵈ. Volume de Priestley,[5] et en outre l'appareil représenté 3ᵉ. Vol. fig. 2 et 3. Je ne doute pas qu'il ne se trouve par rotterdam des occasions d'expedier ce petit envoy parmi d'autres marchandises: de cette façon cela diminueroit beaucoup les ports. ayez s.v.p. Monsieur [la bonté][6] d'addresser le paquet à M. Van Swinden, Avocat à La Haye.[7] Je vous ferai rembourser les frais par qui vous voudrez, à la Haye ou à Amsterdam, ou peut etre que M. Dentan (s'il retourne par la Haye) voudra bien se charger de vous rembourser: en ce cas mon frere auroit soin de le satisfaire à son retour. Excusez Monsieur si je vous donne de la peine: mais je me fonde sur l'offre que vous avez bien voulu me faire, de me faire parvenir ce dont je pourrois avoir besoin. Je serois bien aise de mon coté de pouvoir vous etre de quelqu'utilité dans nos quartiers: vous n'auriez qu'à parler, et je me ferois un devoir et un plaisir de vous satisfaire d'abord.

S'il y a quelque chose de nouveau en instrumens sur les Airs ou les Electrometres, vous m'obligeriez de m'en envoyer s'il est possible une liste qui contient les prix pour les différens Articles.

Il n'y a absolument rien de nouveau ici en fait de Physique.

J'ai &ᶜ.

15 Janv. 79.

P.S. Je vois par l'introduction à son 3ᵉ. Volume que M. Priestley a un grand nombre de bouteilles fig. 2 et 3, de différentes grandeurs: je n'en

4 See Magellan to Van Swinden, 5 May 1778.

5 i.e. Joseph Priestley, *Experiments and Observations on Different Kinds of Air* (3 vols, London, 1774-7).

6 Van Swinden omitted these words in his file copy of the letter but presumably intended something along these lines.

7 Van Swinden's brother.

desirerois s.v.p. que de médiocres. Surtout puisque je n'en sais pas les prix. 79.01.15
Si M. Priestley avoit eu la bonté de les marquer on pourroit donner des commissions plus precises, sans risque d'en donner de trop fortes.

To Comte Joseph Velasco del Pico 79.01.15a

Archief Van Arenberg, Enghien/Edingen, Belgium

Londres 15 Janvier 79

J'ai deja acheté le Nocturnal, mon Cher Velasco, pour Son Altesse S^{me}. chez l'inventeur, & Je l'envoyerai en semble avec les beaux verres à boire[1] qu'on finira dans la quinzaine. en attendent je vous prie de m'expliquer ce qu'on entend par plume perpetuelle. Car par example celle avec la quelle J'ecris cette Lettre, qui est de Tunenough, ou cuivre blanc de la Chine[2] est appellée ainsi par quelques uns; tandis que d'autres appelent de ce nom une espece de petit encrier en forme d'etui de curedent, oû il y a de l'encre, & aiant à un de ses bouts une plume soit metallique, soit d'oiseau, l'encre de soi meme tombe à mesure qu'on le secoüe en peu &^{a}. Je vous prie donc de me dire au plutôt ce qu'on entend par ce mot de plume perpetuelle, & Je l'envoyerai au meme tems, que Le nocturnal.[3]

Il y a environ 8 jours que J'ai envoyé à S.A.S.[4] une petite Caisse par Brugges, dans l'idée qu'elle arrivera plutot à Bruxelles, que les precedentes. Je la Recommendai aux soins de M^{r}. Stanfield, & j'espere que celà reussira mieux. Cette caisse contient les articles suivans. 1°. du tricot de velour demi coupé, en blanc & noir pour une veste. prix £2.12.6 c'est à dire deux guinées & demie. une autre piece à poil tortillé, tout noir, prix £2.5:0 / une de couleur de velour tricoté sans etre coupé £1.14.0 une autre de tricot

1 See Magellan to Velasco, 15 December 1778.

2 Probably tutenag (toutenague in French), a whitish alloy of copper, zinc and nickel wih a little iron, silver or arsenic.

3 The pen and the "nocturnal" writing device were presumably requested by either Duke Louis d'Arenberg in his letter to Magellan of 18 December 1778 or Velasco in his letter of 21 December, neither of which has been found but to both of which Magellan refers in his letter to Duke Louis of 29 December. For descriptions of both devices, see Magellan to Velasco, 2 March 1779.

4 Duke Louis d'Arenberg.

79.01.15a de velour raié tout coupé £2.12.6. une troisieme du meme en couleur, demicoupé. £.2.12.6: & enfin une 4me. à long poil, qui couta £2.5.0. On trouvera dans la meme caisse le respirateur d'argent, avec les armes de Son Altesse:[5] & un autre en fer blanc japané.[6] Ces instruments font ici des merveilles dans les toux catharrales, qui ne sont pas anciennes. il y a des directions imprimées en françois,[7] pour en faire l'usage à propos.

Monsr. Le Comte de Belgioioso, Amb[or]. de l'Empire & ancien ami de S.A.S. en a eu trois, l'Ambass[r]. de Venise, le general Paoli[8] &[a] &[a]. tous en ont eu; & beaucoup du monde en a trouvé les effets aussi certains, & prompts qu'on en a eté etonné.

Lorsque J'envoyerai les verres à boire que Vous m'avez ordonnés, je crois pour M[r]. Le Duc de Duras[9] &[a]. J'en ferai une liste de toutes ces depenses & l'envoyerai à M[r]. Le Coq,[10] selon qu'il me l'a demandé il y a plus d'un ou deux mois par ordre de Son A. S[me].

Le tems presse pour fermer: ainsi Je finis en vous assurant que Je suis à jamais

Votre très h[ble]. vrai ami
Magellan

---

5 See Magellan to Velasco, 24 November 1778.
6 i.e. varnished in such a way as to acquire the appearance of Japanese lacquer-ware.
7 Presumably the pamphlet that Magellan had published a month earlier; see Magellan to Velasco, 24 November 1778.
8 Filippo Antonio Pasquale Paoli (1725-1807), Corsican statesman who lived in London between 1770 and 1790.
9 The French aristocrat Emmanuel-Félicité de Durfort, duc de Duras (1715-1789).
10 Duke Louis' treasurer; see Magellan to Velasco, 24 November 1778.

79.01.19 *To the Comte d'Hérouville*

Archives Nationales, Paris, MS Marine, G110, ff. 199-200

Londres 19 Janv[r]. –79

Mon cher Monsr.

La faveur gracieuse que vous avez fait à M[r]. Boulton & Compagnie, pour obtenir leur privilege exclusif en France, fait que ces Messrs. ne voudroient pas prendre aucune resolution sur les propositions de M[r]. Perier, sans savoir vos volontés, & recevoir vos avis.[1] À ce qu'il paroit de quelques indices, M[r]. Perier ne manquera pas de faire tout son possible pour invalider leur privilege: M[r]. Boulton qui est revenû à Londres, & a passé hier chez moi, croit que ce M[r]. Perier n'a pas d'autres idées, que de saper cette manufacture: car il a deja ordonné un cylindre tel, qu'on l'employe dans les nouvelles machines, disant dans le meme tems que ne pouvant faire les arrangemens necessaires avec Messrs. Boulton & Watt, sans reponse de ses associés de Paris, il vouloit eriger une machine de la construction commune, qui dans la suite, il pourroit tourner dans la construction nouvelle &[a]. Monsr. Boulton me pria donc instament de vous ecrire cette poste, pour savoir 1[o]. quel est votre avis à l'egard des propositions de M[r]. Perier? & 2do. s'il n'est pas possible de prendre quelques moiens de bien surs, pour que le privilege que vous lui avez obtenu soit rafermi à ne pouvoir pas etre ébranlé par les maquinations de quelques enemis? 79.01.19

La machine pour M[r]. Jarry de Nantes sera achevée en Mars prochain; & on pourra bien l'etablir dans le printems.

Je m'oubliois de vous dire que le calcul fait de la depense en charbon, pour elever la quantité d'eau demandée par M[r]. Perier, monte à près de cinq mil livres Esterlines par an, si la machine est de la construction ancienne. mais en la faisant de la construction moderne de M[r]. Boulton, elle leur epargnera près des trois quarts de cette depense. Le calcul ordinaire de ces machines de M[rs]. Watt & Boulton, est, qu'environ <u>88</u> livres pesantes du meilleur charbon de terre, ou <u>112</u>[lb]. du charbon ordinaire, elevent 500,000 pieds cubics d'eau à la hauteur d'un pied, ou 250,000 à deux pieds: 125,000, à la hauteur de 4 pieds: & ainsi de suite.

Quels sont les <u>reservoirs</u> dont vous me parliez dans votre lettre pour les gouverneurs de Paris? est ce les reservoirs particuliers de leurs maisons? ou ceux des fontaines publiques de Paris?

Je vous ai ecrit sur ce meme sujet il y a <u>8</u> jours[2] & Je vous ai envoyé (par la poste separement) une grande Lettre de M[r]. Boulton & Watt, sur les memes affaires. mais dans ce tems ci, on n'est jamais sur, que les lettres sont delivrées fidelem[t].

1 See Magellan to d'Hérouville, 13 January 1779.

2 Magellan mis-reports when he wrote; the previous communications to which he refers were dated 13 January 1779.

79.01.19 Je vous prie de mes très h^{bles}. Resp^{s}. à M^{de}. La Comtesse: donnez m[oi][3] des nouvelles de l'etat de sa sante & des Demoiselles. J'ai l'honneur d'etre pour La Vie

Monsieur Le Comte
Votre très h^{ble}. & très obligé serv^{r}.
Magellan

A Monsieur
Monsr. Le Comte d'Herouville
Lieutenant gen^{l}. des Armées du Roy &^{a}.
rue du Fauxbourg S^{t}. Honoré n°. 6
à Paris

---

3 Obscured by binding.

79.01.21 *To James Watt*[1]

Birmingham Central Library, Boulton and Watt collection, Box 22/B3/12

London 21 Jan^{y} –79

Dear Sir

I think that in about a fortnight, I'll have a good opportunity of sending to Paris your drawings, with a good security. as to y^{e}. post, it is at this time doubtfull, wether any parcell above the sise of a com̄on letter, arrives there. this I say, having Lost already some. perhaps the persons charged with revising them, are not Carefull of sending the same, tho' nothing be contained against the views of the Ministry. I have seen but once M^{r}. Boulton, and wrote again to Count d'Herouville,[2] from whom I expect an answer, unless my letter, as well as yours, have been gone astray.

I ever am, Dear Sir
Y^{r}. m. obed^{t}. h^{le}. serv^{t}.
Magellan

---

1 Letter not found; MS is a clerk's transcription. Annotated "M^{r}. Magellan Jany 21st 1779".
2 Magellan to d'Hérouville, 19 January 1779?

To Anne-Robert-Jacques Turgot 79.01.26

Biblioteca do Museu Nacional de Arte Antiga, Lisbon
Fundo Francisco Sousa Viterbo, Cx nº 1

Londres 26 Janvier –79

Monsieur

J'envoyai par L'occasion de M^r^. L'epinay[1] La pendule astronomique, que vous m'aviez ordonnée pour M^r^. l'Abbé Sigorne,[2] avec ordre à Calais de l'envoyer à l'adresse, que vous m'aviez indiquée, de M^r^. le Comte d'Argivillier,[3] ecrivant un mot d'avis à M^r^. L'Abbé Rochon. Cependant je viens de Recevoir avis de Calais par cette poste, que les Messrs. de la Douane n'ont pas voulû consentir à que cette caisse fut leisée[4] sous cette adresse: & M^r^. Caffieri prit la Resolution de la faire passer à l'adresse de M^r^. Necker,[5] faisant avis de celà à M^r^. L'Abbé Rochon. J'espere que vous ou M^r^. l'Abbé Sigorgne aurez le moien d'arranger Cette affaire, & que La pendule arrivera à sa destination. Je crois que M^r^. L'Abbé Sigorgne aura une excell^e^. pendule. les pendules de mon dernier plan coutent 5½ ^guin^. Celle-ci a les palettes de saphire, que je paiai 1½ guinées. differentes additions pour arreter les pivots des roues, & pour faciliter l'arrangem^t^. de la pendule ^s^10.^d^6. Caisse & embalage ^s^7.^d^6. porteur & carriage jusqu'à Douvre ^s^5.^d^4. total £10: 1^s^: 10^d^, que je prendrai, si vous voulez Chez M^r^. Bostock[6] pour votre compte.

Je trouvai aujourd'hui, chez moi en revenant de diner à l'Hotel, le nom du D^r^. Ingenhouz, qui peut etre est venû pour Savoir Si quelqu'un de vos amis philosophes avoit trouvé la Brochure du D^r^. Gatti, dont il a besoin pour l'ouvrage oû il est occupé sur la petite verole.[7] Je prends La Liberté aussi de vous souvenir des ouvrages (ou Collection d'elles) du D^r^. Franklin.[8]

---

1 Not identified.
2 Sigorgne; i.e. the Abbé Pierre Sigorgne (1719-1809).
3 Charles Claude Flahaut de la Billarderie, Comte d'Angivillier (1730-1810), Directeur général des Bâtiments, Arts, Jardins et Manufactures de France.
4 laissée?
5 Jacques Necker (1732-1804), Genevan economist who occupied high positions in the French administration.
6 See Magellan to Turgot, 20 November 1778, note 5.
7 See Magellan to Turgot, 5 January 1779. No work on smallpox by Ingenhousz is known.
8 See Magellan to Turgot, 5 January 1779.

79.01.26 Il y a 12 jours que j'ai mis à travailler ma machine pour faire les cristaux de Mr. Achard.[9] elle travaille très bien: & si je ne reussis pas à la cristalisation, J'ecrirai exprès à l'inventeur, pour lui demander des eclaircissemens nouveaux là dessùs. Je n'y ai employé que simplemt. la terre d'alum, que Mr. Fabroni m'a preparé Lui-même. l'operation que je fis cet eté avant d'aller à Paris, ou j'employai de l'argile au lieu de la terre d'alum, ne dura plus de 10 jours: Cependant J'ai pû voir au bout de ce tems des commencemens de la formation pierreuse, comme des pelicules solides &a., tandis que moi je n'ai pû voir Ce matin aucun principe, ou meme Soupçon de cristallisation. Je ne sçais pas si le tems froid, est egalement bon pour Cette operation, comme le tems chaud. Je vous Serai bien obligé par la communication du succès de vos machines de cette espece à la muete. l'objet est interessant: & on n'en peut pas trop multiplier les essays.

J'ai l'honneur d'etre avec tout Le Respt. & vrai attachement

Monsieur

Votre très hbl. & obt. Servr.

Magellan

a Monsieur
Monsr. Turgot, ancien Controleur genl.
Rue de l'Université Fauxbourg St. Germain
à Paris

9 Achard, "Copie d'une lettre ... contenant la découverte qu'il a faite sur la formation des cristaux & des pierres précieuses", *Observations sur la physique, sur l'histoire et sur les arts*, 11 (1778), 12-14. More formal reports were published as the final section (pp. 347-350) of Achard's paper, "Chemische Untersuchung verschiedener Edelgesteine", *Neue philosophische Abhandlungen der baierischen Akademie der Wissenschaften*, 1 (1778), 219-350, and as an appendix to Achard's *Bestimmung der Bestandtheile einiger Edelgesteine* (Berlin, 1779), pp. 122-8. The process Achard described involved a slow crystallization over many weeks.

To Nathaniel Pigott 79.01.29

Royal Astronomical Society, London, MSS Pigott, 43

the 17th. Comet discovd. by our friend Messier, was seen ye. 19th. instant near the head of ye. Swan. it grows brighter, and may be seen without Tellescope.

19th. Jan. 79. manè.[1] 6h. 26'. 54''. ye. Right asc. was 289°. 6'. 45''. North decl. 29°. 32'. 41''.

23 do. at 6h. 25'. 48'' manè. Right asc. 283°. 44'. 40. North decl. 31°. 16'. 8''.

24. at 6h. 23'. 33''. manè. Right asc. 282°. 22'. 37''. North decl. 31°. 36'. 44".

vale & ama
tuum
Magellan[2]

this 29 Jany –79

Nathanael Pigott Esqr.
Frampton House, near Cowbridge
in Glamorganshire

1 In the morning.
2 Farewell & love | your | Magellan.

To Joseph Banks[1] 79.01.29a

Royal Society of London, Letters & Papers, VII, 84

Dear Sir

The following articles I received just now by ye. post from Paris. the first relates to a new Comet, of which I beg you may please to send a copy made by ye. Secrety. to Greenwich to the Astronomer Rl., and to other astral. Gentlemen of ye. Socy., that no observation be lost, if possible. it is

1 Published in *Scientific Correspondence of Sir Joseph Banks, 1765-1820* (London, 2007), I, 193-4.

79.01.29a in a Letter from M^{r}. Messier, fellow of y^{e}. R. S. and runs thus. the letter is dated y^{e}. 24 Instant.[2]

"Le 19 Janvier –79 vers les 5 du matin Je decouvris une nouvelle comete, près la téte du Cygne: & c'est le 17me. que J'ai decouverte. Son asc. droite etoit de 289°.6'.46''. & sa declinais. boreale de 29°.32'.41''. à 6^{h}.26'.54''. de ce jour là au matin

Le 23^{j}. matin à 6^{h}.25'.48''. son asc. droite etoit 283°.44'.40''. & sa declinaison 31°.16'.8" boreale

Le 24 jour du mois au matin à 6^{h}.23'.33''. son asc. dr. 282°.22'.37''. & sa declinais. bor. 31°.36'.44''.

Cette Comete augmente en Lumiere, & on la voioit dejà sans lunette.["]

the other article relates to y^{e}. Plants, or bota^{l}. collection of y^{e}. late M^{r}. Aublet. it runs thus in a Letter from my worthy friend D^{or}. Begue du Presle,[3] dated[4]

"Je vous prie de presenter de ma part à Monsr. Banks que je ne puis pas certifier ni le nombre exactement, ni si tou[tes] les plantes ont eté cueillies dans leur pais natal. mais comme je ne voudrois pas que M^{r}. Banks se trouve mal de cet arrangement, je vous prie de lui faire la proposition suivante. J'envoyerai tout l'herbier à Mr. Banks: & après qu'il l'aura examiné, il le payera le prix qu'il jugera à propos & convenable."

you see by this, Sir, that my Friend intends to put every thing in your hands: and to abide by your own judgement.[5] Please a word of answer to,

Dear Sir
your most obed^{t}. h^{bl}. serv^{t}.
Magellan

this 29 Jan^{y}. –79

To Joseph Banks Esqr.
Soho Sq^{re}.

---

2 Letter not found. The section of Magellan's letter relating to the new comet was read at the next meeting of the Royal Society, on 4 February 1779 (Royal Society, *Journal Book Copy*, XXIX, 341).

3 Letter not found.

4 Magellan evidently intended to check the date and write it in but, in the event, he neglected to do so.

5 Cf. 78.12.28. Banks bought Aublet's extremely valuable collection, paying ten louis more than the fifty initially requested by Le Bègue de Presle (Henrey, *British Botanical and Horticultural Literature before 1800* [London, 1975], vol. 2, p. 173).

To Anne-Robert-Jacques Turgot 79.02.19

Biblioteca do Museu Nacional de Arte Antiga, Lisbon
Fundo Francisco Sousa Viterbo, Cx nº 1

Londres 19 Fev^r. – 79

Monsieur

Il y a près de 8 jours que J'ai Reçu l'honneur de votre lettre du 3 du courant, avec 2 pour Messrs. Bostock & Elmsley,[1] que que[2] leur envoyai tout de Suite. Mais je n'ai pas vû, ni sçû, qui fut la personne qui m'apporta votre Lettre. ainsi, si c'est par La meme personne, que vous m'avez envoyé les 13 thermometres, il ne m'est pas possible de les faire chercher chez Lui. pour ce qui regarde la valeur de ces instrumens Je tacherai d'arranger celà avec M^r. Nairne lorsque Je les Receverai: & je me flate qu'il ne se prevalera point d'une meprise pour refuser d'en accepter du moins une partie.

J'ai eté en consequence de votre Lettre avec M^r. Bostock:[3] & le tout Sera fait avec Lui à votre satisfaction. La pendule de M^r. Sigorgne n'etoit pas comprise dans l'argent qu'il me paya pour votre Compte; car je ne prends jamais aucun argent des correspondens de mes amis, sans avoir un ordre par ecrit pour celà; & vous savez que celle-ci (votre lettre du 3 du courant) est la premiere dans laquelle vous me parlez du paiement de la pendule en question.

Le meme M^r. Bostock me montra l'article de votre Lettre derniere qui regardoit cet objet, & etoit parfaitem^t. d'acord avec la mienne. Mais il etoit etonné que vous ne Lui disiez pas un Seul mot sur le sujet de la Lettre qu'il vous ecrivit le 5 Janvier dernier. peut etre, lui dis-je, que cette Lettre fut perdue à la poste: car depuis Cette malhereuse guerre à la sourdine, J'ai perdu plusieurs lettres moi-meme. Je Lui promis donc de vous parler de celà dans Cette Lettre-ci, pour que vous l'assurez dans votre premiere si vous avez, ou non, reçu Sa ditte Lettre du 5 Janvier dernier.

A propòs de lettres egarées voici le malheur qui vient d'arriver à M^r. Bailly, que vous connoissez sans doute à Paris. Je lui envoyai d'ici le 30 Nov^bre. un paquet de feuilles avec les titres de quelques manuscripts Arabiques, sous l'envelope d'un de vos Ministres, & Je lui ecrivis par la poste une petite Lettre, avec l'avis de cet envoy.[4] Apparement la sagesse

---

1 Letters not found.
2 je?
3 See Magellan to Turgot, 20 November 1778, note 5.
4 Letter not found.

79.02.19 de notre Excellent Ministre à donné des ordres exprès à la poste d'arreter tout ce qui porte l'adresse de vos gens en place: & je ne puis faire que rire, malgré la perte de Ces manuscrits, de la fatigue des decifrateurs, qui à cette heure auront rongé les ongles pour trouver quelque sens d'affaires politiques dans les titres arabiques de l'astronomie des anciens Arabes. Voilà comme un petit individu, tel que moi-meme, trouve à se divertir quelques fois, & avec raison, aux depens des Messieurs les grand hommes en place …! Si vous voyez Mr. Bailly, dites lui, Je vous prie, ou faitez[5] lui dire par l'Abbé Rochon, que Je vais ecrire de nouveau à Oxford pour Reparer, s'il est possible, la perte des titres Arabiques ci dessùs. N.B. à cause de cet acident, Je ne mettrai plus dans votre adresse le titre de votre place ou charge ancienne.

Je vous envoyerai de la Canne trouëe;[6] à moins que Mr. Parker, qui est l'unique qui en fait faire pour Ses machines,[7] ne me badine là dessùs, en disant que les artistes françois, si habiles comme vous avez la bonté de me l'assurer, à faire des tuyaux excellents pour les thermometres, n'auront pas aucune difficulté à faire aussi les battons troués dont vous parlez. Mais Je ne connois pas, ni ai-je entendu jamais parler de votre Mr. le Duc, ou Comte de Schouwalof. oû est ce qu'il se trouve? qui sont ses amis dans ce pais-ci? est-il françois, ou allemand, ou Russien?[8] pouvez vous m'indiquer quelque moien de le decouvrir dans cette Babilone de Londres?

Je vois par votre dernier article de la verge etalonée, en pieds & pouces anglois, que quelqu'une de vos lettres pour moi, s'est perdue, egarée dans la poste. Car assuremt. vous ne m'en avez jamais parlé de celà, en aucune lettre que J'aie Reçue. Je m'en informerai du prix, & Je vous en ecrirai avec precision une autre fois. J'ai l'honneur d'etre

Monsieur

Votre très hble & obligé servr

Magellan

a Monsieur
Monsr. Turgot
Rue de l'Université Fauxbourg St. Germain
à Paris

5 faites?

6 trouée?

7 William Parker (fl. 1784-1817) of Fleet Street, London.

8 Probably the Russian Count Andrei Petrovich Shuvalov (1743-1789), who spent a number of years in western Europe at this period.

To Anne-Robert-Jacques Turgot 79.02.23

Biblioteca do Museu Nacional de Arte Antiga, Lisbon
Fundo Francisco Sousa Viterbo, Cx nº 1

à Londres ce 23 Fev^r^. – 79

Monsieur

En Recevant l'honneur de Votre Lettre, datée le 18 du mois Courant,[1] je fus, on ne peut plus, etonné de votre decision = que ma pendule destinée à M^r^. L'Abbé Sigorgne, etoit entierement Sur le même principe, que celle de Franklin … & qu'une telle construction devroit couter beaucoup moins, que Celle-ci ne coutoit. = Assurement, Monsieur, en vous envoyant cette pendule astronomique, J'ai commis une meprise beaucoup plus forte, que Celle des Thermometres. Ainsi, comme cette pendule est encore à moi (car je n'en ai pas Reçu la valeur); la grace unique que J'ai à vous demander, C'est de la faire mettre, telle qu'elle est arrivée, cassée, ou non, dans La meme caisse; & de l'envoyer chez M^r^. Le Begue du Presle, qui demeure dans la Rue de S. Jacques, presque vis à vis celle des Maturins. Il y est assez connû, etant Doct^r^. en Medecine & Censeur Royal: et Je lui ecris, par cette poste, pour me garder Cette Caisse, pour mon compte, & jusqu'à nouvel ordre. Pour ce qui regarde la depense du transport depuis Douvre jusqu'à Paris; aussitot que J'en sçaurai le montant, je ne manquerai pas de le satisfaire.

Il faut que ce soit aussi, Monsieur, par meprise, que je vous aie annoncé (comme vous Le dites) la piece de tricot noir, Le respirateur, & les brazilian pebles; comm'etant empaquetés ensemble avec La pendule ci dessùs. Car assurement ces articles etoient deja partis, separement dans une autre Caisse par la voye de Flandres, bien avant La caisse de la pendule, qui partit par Calais.

Mais j'ai donné deja des ordres, par cette meme poste,[2] pour arreter Ces articles pour mon Compte; puisque vous ne vous Souciez pas de les avoir.

J'attendrai avec impatience les thermometres, que Je vous avois envoyés par meprise: & que vous me renvoyez en arriere. Aussitot que Je les Recevrai, je passerai chez Monsr. Bostock, pour Le rembourser, pour votre Compte, de leur valeur, & de celle de l'article precedent.

1 Letter not found.
2 Letter not found.

79.02.23 Pour ce qui regarde la commission de la verge etalonnée selon le pied anglois, & les battons de verre à trous capillaires, Soiez persuadé, Monsieur, que Je ne manquerai pas de prendre le moien le plus infaillible, pour ne plus commettre de Ces meprises.

J'ai l'honneur d'etre, avec la plus grande Consideration, & avec un profond Respect

Monsieur,

Votre très h^ble^. très obeiss^t^. & très obligé Serviteur

J. H. de Magellan

a Monsieur
Monsr. Turgot
Rue de l'Université, Fauxbourg S^t^. Germain
à Paris

79.02.26 *To Ribeiro Sanches*[1]

Österreichische Nationalbibliothek, Vienna
A.N. Ribeiro Sanches Korrespondenz, codex 12714, ff. 239-240v

Londres 26 Fev^ro.^ – 79

Meo Am^o^. e S^r^. e m^to^. do Coração

As de V.M. de 14 & 18 do passado me dérão o maior gosto e verd^o^. contentamento d'alma. Continûe V.M. a pegar de vez em quando na pena, escreva os seos pensam^tos^. aos verdad^os^. amigos, ainda q gaste m^tos^. dias com cada carta: e esteja convencido q sem este comercio reciproco de trocar as suas sensaçoens e as suas ideias, com os q V.M. está certo q lhe querem bem e q o estimão, necessariam^te^. vem hũ homem à ficar só, e a sentir o dezagradavel pezo da sua existencia.

o pó negro q Boerhaave achou pela grande agitação do azougue, não era propriam^te^. a sua calx, mas hũa separação de particulas minutissimas do azougue, recheadas do seo phlogiston. o Nosso amigo Priestley, q está agora em Londres, e me pergunta ás vezes por V.M., está imprimindo hũ Vol. de experiencias philos^as^. onde trata desta experiencia com m^ta^. curio-

1 Annotated "recue a 5 Mars" [received 5 March] and "R^a^. a 18 Mars" [Answered 18 March].

zid^e. e novid^e. tome-se hũ vidro bem tapado, com mercurio e agoa dentro. sacuda-se violentamente: e dentro depouco tempo toda a agoa toma a cor negra: deixa-se pouzar, separa-se o sedim^to.: & tanto q se seca, appare[ce] o mercurio em globulos minutissimos da sua côr natural: porem o a[r] q se achava dentro do vidro, fica phlogisticado. tanto q esta obra sahir do prelo, eu lhe mandarei hũm exemplar p^a. se divertir. 79.02.26

Eu tão bem não estou convencido, pelas experiencias de Fontana, q [os][2] nervos não sejão m^tas. vezes a verd^a. sedes morbi, quando se achão fóra [da]quelle tom, ou estado a q chamamos da saude. pois o veneno da vibor[a] pode mui bem não ter acção alguma, quando toca n'um nervo, [sem][3] q por isso se possa concluir, q o nervo não hé susceptivel de infermidade. o modo com q influe no corpo animal o ven^o. da vibora, ainda [não][4] he conhecido: pois ninguem afirmará, q seja por corrozão, nem por fermen[ta]ção, nem por incisão mecanica; visto q nem o acido mais forte, nem o f[er]mento mais acrescente, nem o escalpello, produzem effeitos semilhantes: por isso bem pode a ação desse veneno carecer de analogia, ou affinid^e p^a. se fazer sentir o seo estrago nos nervos: sem q dahi se siga q são inatacaveis por outros agentes &^a.

Pelo q toca ás opinioens do D^r. La Cour, V.M., e eu, temos ambos completa ideia do homem. he bom vivente, sabe fazer-se estimado, e o q mais hé sabe tirar lucro da ignorancia dos outros. Por isso se ri de tudo, e vai conservando o seo caracter, e o seo Estado: mas não posso dizer q me pareça tão feliz como outros o serião na mesma pêle delle. tem m^tas. apprehensoens, ás vezes ridiculas, varios whims, & como lhe falta aquella sensibilid^e. n'alma, p^a. ser feliz com o q faz a felicid^e. alheia, nunca os seos gostos e contentam^tos. podem ser completos: e por força deve encontrar mil occazioens q lhe são dezagradaveis.

ainda não vi o D^or. Seq^a. depois q recebi a ult^a. de V.M.: e por isso não sei dizer nada do Testament^o. Brandon de Seo Sobrinho. como este tinha herd^a. legitima na Jamaica, he mais q provavel, q não deixou mais nada do q esse pouco ou nada q V.M. sabe, p^a. a sua futura Espoza.

mandarei a onça do mercurio precipit. per se rouge pela prim^a. occazião. Mas sem verd^a. Cauza, não quizera q V.M. uzasse delle. lembre-se do epitaphio d'um Inglez, cuido q em Bristol, q diz assim = Here lyes N. for y^e. desiring of being beter = Eu cuido q a dose deste precipitado ou p^a. melhor dizer, cal verm^a. do Mercurio, não he mais de meio grão &

2 Conjectural reading.
3 Conjectural reading.
4 Conjectural reading.

79.02.26 raras vezes hũ grão inteiro. deve-ser bem levigado in porphirio com algũ Lambedor, ou outra materia indifferente

O Exmo. S^r^. Pinto da mesma sorte bem de saude, e estimado de todos como merece.

Eu estou occupadissimo, fazendo empaquetar os Instrum^tos^. p^a^. Lisboa. cuido q a semana q vem partirá a 2da. colleção. os destinados p^a^. Espanha, partirão depois dos ultimos de Lisboa, pois lhe são posteriores na ordem do tempo em q forão encomendados.

faça dizer à M^r^. Messier, q aqui anunciei o seo novo cometa, e executei q^to^. pude oq elle dezejava.

à M^r^. Le Begue du Presle q tenha a bond^e^. de mandar pela posta, com a adresse qu'elle sabe os Livros seg^tes^., brochés q são p^a^. o comũ amigo M^r^. Delaval. à saber. Recherches Historiques, & geogra^es^. sur le Nouveau Monde. par Jean Benedict Scheter. Paris 1778. Les Synonimes latins par M^r^. Gardin Dumenil, Profess^r^. de Rhetorique. Paris. 1777. & en fin le cahier du journal de Rozier du mois de Mars. 1778. que ces livres soient brochés & non pas reliés: & par differents jours de la poste de Londres: a M^r^. Robert Bigg. M &^a^.

vale & ama
tuum
Magellan

a Monsieur
Monsr. Sanchez, Doct^r^. en Medecine
Cemitere de S^t^. Jean, Rue de la Verrerie
à Paris

London 26 February –79

My Friend and very Dear Sir

Your letters of the 14th & 18th of last month[5] gave me the greatest pleasure and true satisfaction of mind. Keep holding the pen now and then, write your thoughts to your true friends, even if you take many days with each letter, and be convinced that without this reciprocal commerce of exchanging your feelings and your ideas with those whom you know wish you good and esteem you, a man will necessarily find himself alone and feeling the unpleasant weight of his existence.

The black powder that Boerhaave found on strong agitation of quicksilver, was not properly its calx, but a separation of very minute particles of the quicksilver,

5 Letters not found.

full of its phlogiston. Our friend Priestley, who is now in London, and asks me about you sometimes, is now publishing a vol. of philosophical experiments where he treats this experiment with much curiosity and novelty.[6] Take a glass vessel well closed, with mercury and water inside. Shake it violently and soon all the water assumes the black colour; let it settle, separate the sediment &, as soon as it dries, the mercury appears in extremely minute globules of its natural colour, but the air that was inside the glass becomes phlogisticated. As soon as this work is published, I will send you a copy to please you. 79.02.26

I too am not convinced, by the experiments of Fontana,[7] that nerves are not often the true sedes morbi,[8] when they are out of that mode or state that we call healthy, since the viper's poison may well not have any effect when it touches a nerve, without one's being allowed to conclude from this that the nerve is not susceptible of infirmity. The means by which viper's poison influences the animal body is not yet known, since no-one will maintain that it is corrosion, or fermentation, or due to mechanical incision; because not even the strongest acid, or the most effective ferment, or the scalpel, produces similar effects; thus the action of this poison can well be without analogy or affinity in bringing about its damage in nerves: that does not mean that they are not susceptible of attack by other agents &c.

Concerning the opinions of Dr. La Cour, you and I have a complete idea of the man. He lives a good life and knows how to make himself esteemed, or rather he knows how to take advantage of the ignorance of others. That is why he laughs at everything, and keeps up his character, and his state; but I can't say that he looks as happy as others would be under the same skin. He has many apprehensions, sometimes ridiculous, several whims, & as he lacks that sensibility of spirit to be happy with what brings about other people's happiness, his satisfaction and enjoyment can never be complete and he must necessarily have a thousand unpleasant occasions.

I haven't seen Dr. Sequeira since receiving your last letter, and so I can't say anything about the Brandon will[9] of your nephew. As he had a legitimate heir in Jamaica, it's more than probable that he hasn't left anything else than that little or nothing that you know about to his future wife.[10]

---

6 The first volume of Joseph Priestley's *Experiments and Observations relating to Various Branches of Natural Philosophy* (London, 1779).

7 Felice Fontana (1730-1805), court physicist in Florence, reported his experiments on snake poisons in *Ricerche fisiche sopra il veleno della vipera* (Lucca, 1767). He extended this research while visiting Paris in 1778-9 and later in London, where he investigated the effects of curare; see his "Memoria sopra il veleno Americano detto Ticunas", *Philosophical Transactions*, 80 (1780), 163-220. Fontana argued that such poisons acted on the voluntary muscles rather than on the nerves or heart.

8 Seat of disease.

9 Unknown term.

10 See Magellan to Sanches, 5 January 1779.

79.02.26 I'll send the ounce of the red mercury precipitate per se at the first opportunity. But I hope you won't use it without true cause. Remember the epitaph of an Englishman, I suppose in Bristol, that says this = Here lies N. for y^{e}. desiring of being better = I suppose that the dose of the precipitate or, better said, red calx of mercury, is no more than half a grain & sometimes one whole grain. It must be well levigated in porphyry[11] with some kind of powder, or any other inert matter

His Excellency M^{r}. Pinto[12] is also in good health, and esteemed by all as he deserves.

I'm extremely busy, doing the packing of the Instruments for Lisbon. I suppose that next week the 2nd collection will leave. Those addressed to Spain will leave after the last to Lisbon, since they were ordered later.[13]

Please tell M^{r}. Messier, that here I announced his new comet,[14] and have done as much as I could of what he wished.

To M^{r}. Le Begue du Presle, may he be so kind as to send me through the post, with the address he knows, the following books, brochés are meant, for our common friend M^{r}. Delaval[15] namely Recherches Historiques & geogr^{es}. sur le Nouveau Monde, par Jean Benedict Scheter. Paris 1778[16] Les Synonimes latins par M^{r}. Gardin Dumenil, Prof. de Rhetorique. Paris. 1777[17] & en fin le cahier du journal de Rozier du mois de Mars, 1778. Que ces livres soient brochés & non pas reliés: & par differents jours de la poste de Londres: a M^{r}. Robert Bigg.[18] M &^{a}.[19]

vale & ama

tuum[20]

Magellan

---

11 A slab of porphyry used for grinding (levigating) drugs and the like to a fine smooth powder.

12 The Portuguese Ambassador in London.

13 Magellan was supplying the governments of Portugal and Spain with instruments for re-surveying the boundary between the Portuguese and Spanish territories in South America.

14 Magellan communicated Messier's discovery of a new comet to Joseph Banks (see Magellan to Banks, 29 January 1779) and Banks reported the discovery at the next meeting of the Royal Society, on 4 February 1779 (Royal Society, *Journal Book Copy*, XXIX, 341).

15 Edward Hussey Delaval (1724-1814).

16 Jean Benoit Sherer, *Recherches historiques et géographiques sur le Nouveau-Monde* (Paris, 1777).

17 Jean-Baptiste Gardin du Menil, *Synonymes latins et leurs différentes significations, avec des exemples tirés des meilleurs auteurs* (Paris, 1777).

18 Bigg worked at the General Post Office in London; see Magellan to Sanches, 18 February 1777, note 10.

19 Magellan seems to be saying that the works requested should be sent separately, presumably to reduce the postage charges, addressed to Robert Bigg but marked "M" to indicate that they were intended for him.

20 Farewell & love | your.

To Joseph Priestley[1]

Dear Sir,

Prince Gallitzin, the Russian ambassador to the States of Holland, in a letter dated the 17th instant informs me of an extraordinary cure of a putrid fever by the internal application of fixed air, according to the method of Dr. Hulme,[2] both in draughts and in clysters; and I have now before me this case written by Dr. Janssens, an able physician of Operhout, near Breda, in the Dutch Brabant. The patient was a married woman, of thirty-two years old, whom he was called to attend on the ninth day by the assistant physician, to be consulted, in an alarming circumstance, which the patient was already in, almost all covered with exanthems, of a red and livid colour, shewing the greatest tendency to the last stage of general putrefaction. All her limbs were in a state of slow convulsion, and particularly with cold sweatings. The bark and all other means pointed out by medical art had been properly applied, but without any success. Dr. Janssens availed himself of the hints he received in a conversation with Prince Gallitzin on the subject, and ordered that the decoction of the bark, till then ineffectual with this patient, should be administred mixed with the salt of tartar, and vitriolic acid, both in draughts and clysters, relying on the effect of the *fixed air*, which was to be disengaged within the body of the patient. The success fully answered his expectations, for in three days time all bad symptoms were over, and a perfect recovery was the consequence of this new treatment. Dr. Janssens in this letter to Prince Gallitzin, says, that although he considered himself obliged to employ the quinquina (by the apprehension of a general putrefaction or gangrena, which he feared in so alarming a case) nevertheless he believed, that the *fixed air* had greatly contributed to this cure.

N. B. Prince Gallitzin in his said letter, submits to farther consideration, whether the antiseptic virtues of the bark might not depend chiefly upon

1 Letter not found. The text given here was published as Appendix VII to the first volume of Priestley's *Experiments and Observations relating to the Various Branches of Natural Philosophy* (London, 1779), pp. 472-3, under the heading: "*A Letter from* Mr. Magellan, F. R. S. *to* Dr. Priestley, *on the Efficacy of* fixed Air *for dissolving the Stone, and in putrid Fevers, tried in* Holland."

2 Nathaniel Hulme, *A Safe and Easy Remedy proposed for the Relief of the Stone and Gravel, the Scurvy, Gout, &c....* (London, 1778).

79.02.27 the large quantity of *fixed air*, it contains, as he had found by the analysis of this substance.

I apprehend the above information will be of some satisfaction to yourself, and to every one who like yourself, has at the heart, whatever is good to mankind,

I am with the greatest regard,
And truest affection,
Dear Sir,
Your most obed. humb. servt.
J. H. Magellan.

London, 27 February –79.

79.03.02 *To the Duke Louis-Engelbert d'Arenberg*
Archief Van Arenberg, Enghien/Edingen, Belgium

Monseigneur

L'Ecuier de V.A.S. m'a remis Sa Lettre, oû Elle me fit l'honneur de vouloir bien employer mes Soins à Son égard.[1] assurement il n'en a pas eu besoin; & à ce qu'on m'a assuré (car en effet je n'entends rien de chevaux) il a fait des achats, qui feront honneur à Son intelligence. Il me dit dimanche soir que sans doute il partiroit ce matin pour Douvre.[2]

Je lui ai donné un petit pacquet pour M^r. Velasco,[3] contenant le porte feuille de nuit, & quelques plumes, qu'on a voulû appeller eternelles, ou sempiternes, comme S'il en avoit aucune Chose dans Ce Globe qui meritasse un nom pareil. Le même Ecuiyer de V.A.S. vouloit bien se charger de porter la caisse avec les verres à boire que V.A.S. a ordonné, je crois pour M^r. Le Comte de Duras.[4] Mais il n'etoit pas practicable de la porter à cheval, sans un risque evident de les faire rompre en mil morceaux:

1 Letter not found.
2 Dover.
3 See Magellan to Velasco, 2 March 1779.
4 Emmanuel-Félicité de Durfort, duc de Duras (1715-1789) was a French politician, diplomat, peer, and freemason. He was made a marshal of France in 1775. See Magellan to Velasco, 15 January 1779.

ainsi je vais faire partir cette petite caisse par Le premier vaisseau pour Ostende, avec recommendation de l'envoyer par eau jusqu'à Bruxelles. 79.03.02

Par la poste suivante J'envoyerai au Tresorier de V.A.S. la liste, qu'il m'a demandée par Son ordre, de mes petits deboursemens; car j'ai attendû que ce dernier article fut prêt, à fin de ne pas multiplier le trouble de ces petites remises, subdivisées en plusieurs fois.

V.A.S. m'ordonna, l'été dernier, deux de mes pendules,[5] je crois pour Messrs. Les Contes de Spangen[6] & de Duras. voici comm'elles ont eté executées. l'une est une pendule astronomique qui va 8 jours; a la correction pour la chaleur & le froids & je l'ai fait essayer deja par un astronome. elle est fort reduite en sa figure, du moins aussi petite que celle que J'envoyai à Feu Monseig[r] Le Duc[7] pour regler sa montre: même je crois qu'elle est plus petite de quelques pouces. L'autre est une pendule qui marche en silence: frape les heures, les repete á volonté jusqu'à la demiheure: & a un reveil, ce qui est une circonstance que je trouve très commode pour se rapeler des rendez-vous, ou l'heure precise pour faire quelque chose &[a]. Je les fais mettre dans des caisses simples, mais nettes & propres de Mahogany: & toutes les deux partiront aussitot que V.A.S. le voudra.

J'ai l'honneur d'etre avec le Respect le plus profond, & attachement le plus sincere & pour La Vie

Monseigneur

de V.A.S.

serviteur le plus hum[ble]. & le plus obligé

Jean Hyacinthe de Magellan

P.S. Je prends la liberté d'envoyer 6 plumes de metal blanc de la Chine, montées en ivoire, qui ecrivent fort bien, & très long tems, sans avoir besoin d'etre coupées. tout au plus on peut les aiguiser avec une Lime. V.A.S. conçoit bien que mon bout a eté que leurs Altesses Mesdames & Mlle. en puissent faire usage; sans oublier M[de] la Duchesse d'Ursel.[8]

à Londres 2 Mars –79

---

5 Magellan's improved pendulum clock was described in *Observations sur la physique, sur l'histoire naturelle et sur les arts*, 17 (1781), 283-6; see Magellan to J.-A. Mongez, April 1781.

6 Probably Count Charles François Pierre von Spangen von Uyternesse (1715-1792).

7 Charles Marie Raymond d'Arenberg, Duke Louis' father.

8 Marie Flore Françoise, princess and duchess d'Arenberg (1752-1832), Louis Engelbert's sister. She married Wolfgang Guillaume Joseph Léonard Vital, duc d' Ursel.

79.03.02a *To Comte Joseph Velasco del Pico*
Archief Van Arenberg, Enghien/Edingen, Belgium

Londres 2 Mars 79

Monsr. & Mon Cher ami.

Voici le contenû dans le packet, que je delivrai dimanche soir au Ecuyer de S. A. S^me^.[1] Premierem^t^. le nocturnal, ce qui est un petit Livre de tabletes, avec une ouverture, par laquelle on ecrit dans l'obscurité: car la main est guidée par la Raignure, ou ouverture, qui se trouve sur chaque face du couvercle. quand on a fini d'ecrire une regle de chaque coté: on pousse le bouton, & le resort fait que les tabletes s'arretent montrant un autre espace blanc à ecrire.

La plume à fontaine, est comm' l'etuy d'un crayon en argent. on ouvre le dessùs a: on tire le liege, ou bouchon qu'on y trouve: & on le remplit d'encre assez liquide. on ferme le bouchon: on y met le dessùs a: on devisse la partie inferieure b: & on y verra une plume coupée de metal. alors en donnant une petite secousse avec La main qui tient la plume, sur la main gauche, l'encre sortira par le petit trou qui est au dedans, & on pourra ecrire, sans avoir besoin d'autre encrier, tandis qu'il y aura de l'encre au dedans. Car à chaque fois qu'on epuise, en ecrivant, l'encre qui est dans la plume, en lui donnant une autre secousse (comme ci dessus) l'encre descendra du dedans au dehors.

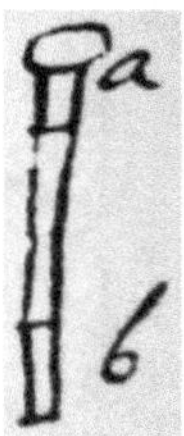

Vous y trouverez 5 autres bouts de metal en forme de plume; pour suppleer celui qui y est, lorsqu'il sera usé; ce qui ne Sera pas si tot. Car la plume avec laquelle je vous ecris celle-ci, me dure il y a plus de 3 mois, sans avoir d'autre besoin que de l'aiguiser avec une Lime lorsqu'elle s'use en devenant trop grosse pour changer les bouts de plume metallique, otez le colet o, avec le bout de la plume c; mettez-y un autre bout de plume c, & remettez-le avec son colet o, dans la même place.

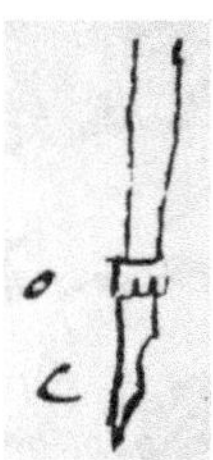

1 See Magellan to Velasco, 15 January 1779.

Ce metal blanc de la chine ecrit excellentment bien: mais il devient usé en moins de tems que l'acier: & Je vous envoye une Lime, pour les éguiser ces bouts de plume en metal blanc, lorsqu'ils ecriront trop gros. 79.03.02a

Je vous envoye 6 des plumes d'acier: qui en effet durent beaucoup plus que celles de metal blanc: mais l'encre noir qui a du vitriol, les corrode bientôt: de façon qu'en moins de 5 ou 6 semaines, les pointes d'acier deviennent mangées & détruites: au contraire que les plumes de ce metal blanc de la Chine, ne sont pas attaquées par le corrosif du vitriol, qui se trouve dans l'ancre noir, quoiqu'on les laisse plongées dans l'encrier.

Les 6 plumes de metal blanc en manche d'Ivoire sont pour Mesdames les Duchesses, & Princesses, sans oublier la M^{de}. La Duchesse d'Ursel, comme Je le dis dans la Lettre à Monseign^{r},[2] ni La Princesse Leopoldine non plus. Il en restera (par ce calcul) une pour Monseig^{r}. Le Duc, & une autre pour vous, si celà vous conviendroit mieux que la plume de fontaine

Vale & ama
tuum[3]
Magellan

P.S. Est-elle arrivée la Caisse que J'envoyai par la voye de Bruges à Son A.S^{me}.? Contenant les tricots de Soye &^{a}.[4] Je souhaite savoir 1°. si ces tricots ont plû à Ses Altesses, & aux personnes qui les on eu? 2°. si cette caisse arriva plus vite à Bruxelles, que celles envoyées par La voye d'Ostende? Car à vous dire le vrai le retard depuis Ostende jusqu'à Bruxelles par la voye de M^{r}. Hennessy, est terriblem^{t}. enuyeux. Ditez moi votre avis là dessùs.

Pour Monsr. Velasco,
Secretaire de S.A.S^{me}. Le Duc d'Arenberg.

---

2 See Magellan to Duke Louis d'Arenberg, 2 March 1779.
3 Farewell & love | your.
4 See Magellan to Velasco, 15 January 1779.

79.03.06 *To Joseph Banks*[1]

British Library Add. 33977, f. 90[2]

Dear Sir

The inclosed is a Letter I received yesterday from our common friend D^{or} Ortega,[3] intended to be presented to the R^{l}. Society. Along with it I received the especimens of the new Species of Quinquina, lately found in great quantities over the mountains of y^{e}. Province of Santa Fé:[4] and a tin box containing 44 ounces of y^{e}. pouder of this bark.

This box is solder'd round with tin, tho' not perfectly so: but as to the dried Specimens, I keep them unfolded, and still tyed as I received them, for fear of hurting the flowers.

You'll observe by the Letter of D^{r}. Ortega, that it is by the King's desire, that this present is sent to the R^{l}. Society: and that his Exc^{y}. M^{r}. Galvez,[5] Minister and Secretary of State for the Spanish Indias, gave His Majesty's orders on this Subject to D^{r}. Ortega, with two views, the 1^{rst}. to ascertain this tree, whether it is, or not, a species of y^{e}. cinchona Linnæi: and the Second to ascertain the good effects of it's application, as a febrifuge.

I intend therefore, to move, with your Leave, in the next meeting of the Soc^{y}.; that a Commitee be appointed for this purpose, composed of Members conversant both in Botanicks, and in the practice of Physic, in order to consider and examine these specimens according to the two Views of his Catholic Majesty, and to deliver their opinions to the Society, a copy of which by leave of y^{e}. R^{l}. Society I will transmit to D^{r}. Ortega; either thro' the hands of the Spanish Ambass^{or}., or by another way.

---

1 Marked: "M^{r}. Magellan March 7 –79 send him".

2 Published in *Scientific Correspondence of Sir Joseph Banks, 1765-1820* (London, 2007), I, 195-6, with the note that we have treated as a separate letter and numbered 76.00.00 presented as an additional postscript.

3 See Ortega to Magellan, 2 December 1778. Casimiro Gómez de Ortega (1740-1818), first director of the Royal Botanic Garden in Madrid, had been elected a Fellow of the Royal Society in 1777, following a visit to England.

4 i.e. Santa Fe de Bogotá, in present-day Colombia. The identification and medicinal value of the supposed new species of cinchona was a matter of great controversy among Spanish botanists for many years: see Arthur R. Steele, *Flowers for the King* (1964), chap. 11.

5 José de Gálvez (1729-1786), Minister of the Indies from 1775.

I should think that Dor. Pitcairne,[6] Dr. Fothergill,[7] Dr. Watson,[8] 79.03.06
Dor. Solander,[9] and Mr. Hudson (the Author of ye Flora Anglica)[10] ought to be named for such a Commitee: and that it will be better to keep the above samples closed, without unfolding, nor opening them, till the said Commitee meets; for fear, that the dried flowers, if handled by unskillfull hands, must be considerably hurted.

But this, and the whole of the matter, I must leave to your own judgement, and consideration:[11] and remain, waiting your further orders, with the greatest regard,

Dear Sir
your most obedt. hble. Servt.
Magellan

this 6 March –79
nc. 12 Nevils Court Fetter Lane.

N.B. I did intend to wait on you the last evening, but it was rather Late, when I was at liberty to go: and perhaps, I may not find you at home this morning.

6 William Pitcairn, M.D. (1711-1791) (FRS 1770).

7 John Fothergill, M.D. (1712-1780) (FRS 1763).

8 William Watson (1715-1787) (FRS 1741).

9 Daniel Carl Solander (1736-1782) (FRS 1764).

10 William Hudson (1730?-1793) (FRS 1761), author of *Flora Anglica, exhibens plantas per regnum Angliae sponte crescentes, distributas secundum systema sexuale...* (London, 1762; enlarged ed. 1778).

11 At its meeting on 11 March 1779, the Royal Society, noting "the desire of the King of Spain that the Royal Society of London may be consulted on this important subject", established a committee comprising the Fellows here suggested by Magellan, "to investigate the botanical distinctions of these trees, and the Medical Virtues of their barks" (Royal Society, *Journal Book Copy*, XXIX, 375-6).

79.03.09 *To Johann III Bernoulli*
Öffentliche Bibliothek der Universität Basel, MS L I a 709, f. 74

Londres 9 Mars –79

Pour M^{r}. Bernoulli

Monsieur & très cher ami

Je vous dois une Reponse à celle dont vous m'avez honoré le 24 octobre dernier.[1] J'ai attendu la Reponse de M^{r}. Hornsby sur la suspension de son Secteur; apparem^{t}. il vous l'a envoyée en droiture, car je n'en ai pas reçu.

Vous saurez sans doute que J'ai payé 81$^{£}$.2^{s}.0^{d}. Esterl. sur votre Lettre de change en date du 23 oct^{bre}. dernier, qui me fut presentée par Messrs. Batter & Zormlin.[2] Le reste pour 80 guinées fut depensé en gages, ou reconnoissances qu'on demande dans les Bureaux par les Clercks & en autres depenses relatives au même objet.

Vous n'avez pas raté à la Soc^{e}. R^{le}. comme vous le croyez, dans votre election. C'est que le Conseil de la Soc^{e}. observant Le grand nombre de candidats qui se presentoient pour l'association Etrangere de la Societé, prirent la resolution de fermer la porte à tous, sans exception.[3] ainsi il paroit que desormais il n'y aura que des membres honoraires, qu'on choisira de son Chef, & sans qu'il y ait aucune concurence de Candidats.

Les nouvelles qui Vous me donnez de la Soc^{e}. Literaire de Varsovie sont interessantes:[4] & Je crois que ce pais là a grand besoin de Societés pareiles pour sortir de la barbarie Escolastique & Jesuitique, qui y a fait la loi en

---

1 Letter not found.

2 Battier, Zornlin, and Co., merchants at no. 10 Devonshire Square, London.

3 In November 1777, the Council of the Royal Society, noting that "the Number of Fellows on the Foreign list is become immoderately great", resolved "that no certificates in favour of foreign Candidates be received until the Council shall have had time to resolve upon such measures to be taken for the future". This ban remained in force for almost ten years, until March 1787, by which time the number of foreign members had dwindled to 81. A new statute was introduced at that stage that limited the total number of foreign members to 100 (later reduced to 50) and also subjected nominations to much closer scrutiny than had formerly been the case. These restrictions did not, however, apply to foreigners such as Magellan who were resident in Britain, they being eligible to be elected on the so-called Home List.

4 In his *Nouvelles littéraires de divers pays*, 4th cahier (Berlin, 1778), pp. 34-5, Bernoulli announced the recent formation in Warsaw of a "Société pour l'avancement de la Physique, des Mathematiques & de l'Histoire naturelle" which at the time of writing was still a private organization but which was expected shortly to receive government backing.

fait de sciences & connoissances utiles. Mais quand celà arrivera, ce n'est pas encore bien certain. 79.03.09

Vous aurez sçû de la derniere comete decouverte par M^r^. Messier à Paris le 19 Janvier dernier. Sans doute il vous l'aura annoncée Lui même, comm'il le fit ici à la Soc^e^. R^le^. par ma voye.[5]

on va envoyer 6 Astronomes (par La Cour d'Espagne) pour faire des observations dans l'Amerique, & 5 Astronomes seront envoyés par La Cour de Portugal pour Le meme endroit.[6] On croit que c'est pour determiner la ligne de Division entre ces deux Puissances, c'est à dire entre leurs etats dans l'Amerique. On m'a ordonné pour cet effet autant de collections de bons instrumens; & Je vous assure que ces commissions m'ont donné bien du travail, pour avoir des instrumens bien faits, & dans peu de tems. Il y aura deux Quarts de Cercle de deux pieds de Raion, & 9 autres de 12 pouces de raion; faits & finis avec la plus grande attention. J'en ai deja envoyé 3 excellens, à deux rangs de divisions, c'est à dire de 90° & de 96 dans le meme arc. Cette division est la Seule dont on fait usage actuellement, du moins ceux qui aiment l'exactitude. Car le n°. 96 est capable de la bisecti[on] & je puis repondre de celle-ci avec tout escrupule, dans ces instruments, les aiant examinés avec tout Soin. Vous savez que Maskelyne, & Ludlam ont publié la table de la valeur de chaque division des 96 & leurs subdivisions,[7] qui fut calculée par le D^r^. Bli[ss][8] mais M^r^. De la Lande l'a oublié dans son Astr^e^.[9] parcequ'il ne seroit pas le grand usage que les bons observateurs en font aujourd'hui.

---

5 Magellan reported Messier's discovery in a letter to the President of the Royal Society, Joseph Banks; see Magellan to Banks, 29 January 1779. The section of the letter dealing with the comet was read at the next meeting of the Society, on 4 February 1779 (Royal Society, *Journal Book Copy*, vol. 29, p. 341). The report was not published in the *Philosophical Transactions*.

6 Bernoulli published a brief announcement of the forthcoming expedition in his *Nouvelles littéraires de divers pays*, 4th cahier (Berlin, 1778), p. 61, including the news that Magellan had been commissioned to supply a large number of the requisite instruments. Magellan's detailed account of the preparations for the undertaking was passed by Bernoulli to J. E. Bode and published by him in German translation in*Astronomisches Jahrbuch oder Ephemeriden für das Jahr 1782* (Berlin, 1779), pt 2, pp. 148-9. Magellan described the instruments supplied in his *Collection de différens traités sur des instrumens d'astronomie, physique, &c.* (London, 1780).

7 N. Maskelyne, *Tables for computing the Apparent Places of the Fixt Stars, and reducing Observations of the Planets* (London, 1774), Table LVI (p. 48). Ludlam's name is not mentioned in the work.

8 Part of name obscured by binding; completed from printed version. Nathaniel Bliss (1700-1764) was Maskelyne's predecessor as Britain's Astronomer Royal.

9 J. Lalande, *Astronomie* (2nd ed., Paris, 1771-81).

79.03.09 Il y a 11 Instrumen[s]. circulaires à Reflection, dont je viens de publier la description, & dont je vous envoyerai un exemplaire. J'en ai parlé à la page 112 de mon Traité sur les octants. / plus 11 barometres faits par Nairne de ceux de mon invention, qui reunissent tous les avantages de ceux de M[r]. de Luc, sans en avoir les inconveniens. / 11 theodoletes, avec mon addition pour reconnoitre avec la plus grande exactitude la variation de l'aiguille. / 6 pendules à seconds, 3 à demi-seconds, & deux (de mon invention) / qui pourront battre des demi-seconds, ou des seconds entiers, comm'on voudra, en y adaptant une verge plus longue. / 11 Lunettes achromat[es]. de 3½ pieds. / 5 plus petites de 18 pouces / 11 montres à seconds / bussoles pou la mer, & pour la terre. / thermometres / barres magnetiques / Etuis complets d'ingenieur &[a]. &[a]. &[a].

Il ne m'a pas eté possible d'expedier encore que les 3 premieres collections des instrum[s]. ci dessùs. Envoyez je vous prie cet autre feuillet à M[r]. de Castillon.[10]

Vale & ama
tuum
Magellan[11]

---

10 Magellan's note to Castillon not found.

11 Farewell & love | your | Magellan.

79.03.11 *To Ribeiro Sanches*[1]

Österreichische Nationalbibliothek, Vienna
A.N. Ribeiro Sanches Korrespondenz, codex 12714, ff. 241-242v

Londres ce 11 Mars – 79

Mon cher Monsieur, & respectable ami

Celle-ci vous sera remise par M[r]. Balla, gentil-homme Russien,[2] qui sans autre titre, meriteroit votre attention & bonne disposition en sa

---

1 Annotated "Recue a 22 Mars" [Received 22 March]; "R[a]. a 26 Avril" [Answered 26 April]; and "1) 1. part. Phil. Tr. 1778 | 2 Gibbons Defense | 3 Clare upon des absces | 4. Meth. mercur. exhibendo [= Method of applying mercury?] | le tout pour moi".

2 See Magellan to J.-A. Euler, 9 December 1778, note 5.

faveur: mais il est en outre ami particulier de ceux qui nous honnorent de Leur amitié, comme sont Le Prince Gallitzin de la Haye, Le Prince Orloff & tous les autres Seig[res]. de distinction de ce paiz là. Son Caractere personel & toutes les bonnes qualités qui peuvent recommender quelqu'un sont en sa faveur: & vous y ajouterez je crois avec plaisir, la circonstance de l'amitié dont il m'honore. J'ai donc souhaité qu'il puisse vous voir, & me donner de Vos nouvelles, comme temoin occulaire, tandis que je n'ai pas ce plaisir là, ce que J'espere pourra etre L'eté prochain, à moins de quelqu'embaras insurmontable. Et attendant portez vous aussi bien que je vous Le souhaite, jouissant de la belle saison qui commence de bonne heure; car en effet nous n'avons pas eu d'hiver cette année dans ce Pais-ci. 79.03.11

J'ai l'honneur d'etre
Mon cher ami
Votre très h[ble]. & obligé serv[r].
Magellan.

N.B. Je viens d'envoyer à M[r]. le Begue du Presle la 1e part du vol. 88 des Trans. Phil. qui vient de paroitre, avec la defense de Gibbon contre les ataques des Fanatiques,[3] & une brochure de M[r]. Clare sur les abscès, & la nouvelle methode d'appliquer Le mercure.[4] le tout pour vous, & dont M[r]. le Begue aura soin.

a Monsieur
Monsr. Sanchez, Doct[r]. en Medecine.
Cimitere St Jean Rue de la Verrerie
au Marais
à Paris

3 Edward Gibbon, *A Vindication of Some Passages in the Fifteenth and Sixteenth Chapters of the History of the Decline and Fall of the Roman Empire* (London, 1779). The work consists mainly of a reply to an attack by Henry Edwards Davis on the two chapters named of Gibbon's famous *History*.

4 Peter Clare, *An Essay on the Cure of Abscesses by Caustic, and on the Treatment of Wounds and Ulcers: also a New Method of Curing the Lues venerea* (London, 1779)?

79.03.23 ## *To Jan Henrik van Swinden*[1]

Leiden University Library, MS B.P.L. 755, no. 343

Londres 23 Mars – 79

Je vous ai envoyé, Mon Cher Monsieur, il y a plus de 20 jours, les verres que vous m'aviez demandés, par Votre Lettre du 15 Janvier.[2] Mais il faut vous avouer, qu'il ne m'a pas eté possible de vous ecrire sur celà jusqu'à ce moment. Ces verres sont 1 Eudiometer, qui coute 15ˢ. Cet Instrumᵗ. pourroit bien etre fait plus epais: mais les ouvriers ou pour mieux dire les maitres de la Verrerie n'en veuillent point pour l'ordinaire. la raison en est, que tout verre fort epais a plus de danger à crever de soi même; à moins de prendre un tems infini & des soins particuliers à le refroidir regulierement. moi-meme j'ai eu souvent l'experience: & je suis revenû de cette idée à mes propres depens.

plus 8 phioles à tuyau recourbé = 16^{s}. plus une suite des verres fig^{re}. 3 & fig^{re} seconde du 3me vol. de Priestley. une couta 18^{s}. & l'autre $7^{s}.6^{d}$. la caisse, couta 1^{s}. somme totale £2.17.6. que vous ferez paier à votre loisir. Cette petite Caisse fut envoyée au dedans d'une autre pour M^{r}. Le Prince Gallitzin à la Haye, qui sans doute vous La fera remettre à M^{r}. votre frere, ou Parent du Meme nom, Avocat dans La même place.

Le D^{r}. Priestley publiera incessament un Volume nouveau de ses experiences de Physique, oû il y a de belles nouveautés.[3] Mr. Atwood publie aussi $actuellem^{t}$. c'est à dire en peu de semaines, car l'ouvrage est presqu'imprimé, un Excellent traité sur le mouvement rectiline & circulaire.[4] Il a un talent extraordinaire pour rendre palpables, par l'experience, les plus hautes theories. tout le monde qui a vû ses experiences a resté etonné. Il a ouvert une Souscription pour imprimer un Cours Complet de Physique experimentelle en 3. Vol. 4°. prix 3½ guinées.[5] Il ne demande point des avances d'argent pour la souscription, mais $seulem^{t}$. des noms: & Je vais

1 Published in David Willemse, "Suite d'un voyage aux Pays-Bas: João Jacinto de Magalhães (1722-1790) et ses rapports avec Jean Henri van Swinden (1746-1823)", *Arquivos do Centro Cultural Português*, 7 (1974), 225-278.

2 Van Swinden to Magellan, 15 January 1779.

3 i.e. the first volume of Priestley's *Experiments and Observations relating to Various Branches of Natural Philosophy* (London, 1779), which was often treated as a continuation of the three volumes of *Experiments and Observations on Different Kinds of Air* (London, 1774-7).

4 G. Atwood, *A Treatise on the Rectilinear Motion and Rotation of Bodies* (Cambridge, 1784).

5 The projected work appears never to have been published, only a single-volume abstract (London, 1784).

lui donner le Votre, entre plusieurs de mes autres amis: si vous ensuite ne voudrez pas l'ouvrage, je m'en charge d'en disposer pour mon compte. 79.03.23

Je vous prie de m'envoyer, l'ouvrage dont vous parlez sur les experiences de M^{r}. Was (Je crois c'est le nom) sur l'expansion des corps dans Le pyrometre &^{a}.[6] C'est dans votre ouvrage des Thermometres, oû vous en parlez dans une note.[7] Un de mes amis a emprunté ce Livre: ainsi je ne puis Rectifier le nom de celui que vous citez.

Je crois vous avoir demandé, du moins Je vous en prie de nouveau, la Communication de Vos decouvertes & inventions sur le magnetisme. Car Votre Memoire qui fut coronné en France, ne m'est jamais parvenû a la main: peut etre n'est il pas encore imprimé, & dieu scait quand?[8]

Le D^{r}. Priestley ne pouvoit pas donner les prix de ses verres. plusieurs il les arrangeoit de ses mains: d'autres lui furent envoyés d'ici sans detail. d'aileurs la taxe qu'on vient d'ajouter aux verreries d'Angleterre, fait qu'ils sont devenus plus chers qu'auparavant.

Vous vous trompez fort sur le compte du caractere de Wilson.[9] il a eté toujours un mauvais sujet, faux, charlatan, enteté & très rusé. ainsi sa Reputation n'est qu'au loin, oû il ne'st pas connû personnellem^{t}., ou chez les gens de Cour, vis à vis desquelles il rampe, &^{a}.

Je demanderai à M^{r}. Dentan, s'il pense retourner par Hollande: & alors il ne manquera pas de me rembourser de cette bagatelle, que Je viens de vous envoyer.

A propòs de Wilson demandez à Mr. le Prince Gallitzin le Recit, que Je lui ai fait, de la jolie experience electrique de M^{r}. Long, qui n'est pas savant, mais qui est bon electricien de pratique.[10] Cette experience demolit de comble en fond la charlatanerie de Wilson, sur l'avantange qu'il veut etablir des conducteurs emoussés, contre les conducteurs pointûs. elle est charmante & très à propos: Vous ne manquerez d'en etre enchanté. J'en ai

---

6 *Specimen physicum inaug., sistens nova quaedam experimenta de expansione corporum solidorum ipsi flammae actioni exhibitorum* (Franeker, 1775), by Henricus de Wal (1754-1782).

7 J.H. Van Swinden, *Dissertation sur la comparaison des thermomètres* (Amsterdam, 1778), pp. 3-4.

8 Van Swinden's memoir was destined to appear in the *Mémoires présentés ...par divers savans* series of the Académie Royale des Sciences, but had not yet appeared; see Van Swinden to Magellan, 22 May 1778.

9 See Van Swinden to Magellan, 15 January 1779.

10 "Mr. Long of *Little Compton Street, Soho*, an experienced Practitioner, who has made Medical Electricity great part of his Study and Employment for many years"; see the anonymously published *New Thoughts on Medical Electricity, or An Attempt to Discover the Real Uses of Electricity in Medicine in Two Letters to a Friend* (London, 1782), p. 8.

79.03.23 fait un Recit assez clair, je crois, dans une de mes Lettres à Mr. le Prince:[11] & il ne manquera pas de Vous le communiquer.

Vale & ama
tuum Magellan[12]

Mr. Le Prince Gallitzin me parle d'une cure qu'on a fait chez vous avec l'air fixe. Mr. Dentan doit me faire voir le Recit: & probablemt. J'aurai le tems de l'inserer dans l'appendix de l'ouvrage du Dr. Priestley qui est actuelmt. sous presse.[13]

A Monsieur
Monsr. J.H. Van Swinden,
Professr. de Philosophie dans l'Université de
Freneker
Hollande

---

11 Letter not found.

12 Farewell & love | your | Magellan.

13 See Magellan to Joseph Priestley, 27 February 1779.

79.04.09 *To Alessandro Volta*

Istituto Lombardo, Milan, Cartellario Voltiano, P 2[1]

Londres 9 Avril –79

Monsieur

Quoique Je n'ai pas l'honneur de recevoir de Vos nouvelles depuis fort long tems; je ne me suis point oublié de la petite partie qui manquoit dans l'appareil que Je vous envoyai en Octobre dernier,[2] faute de l'ouvrier

---

1 Published in *Epistolario di Alessandro Volta*, vol. 1, pp. 338-40.

2 See Magellan to Volta, 27 October 1778. The earlier shipment had chiefly comprised apparatus for experiments with gases. A further indication of what was involved comes from the 1790 catalogue of the physics cabinet of the University of Pavia, where the following items that are listed are said to have been acquired from London during Volta's first years at the University: (1) "[Apparato pneumatico-chimico] di Mahogany per le arie acide, ed altre troppo miscibili all'acqua, che debbon perciò confinarsi nel mercurio"; (2) "Una copiosa provvista di campane, boccie, matracetti, tubi, e sifoni smerigliati per le sperienze sulle arie, con dippiù tre Eudiometri ad aria nitrosa di Magellan"; (3) "Doppio apparato di Parker p. impregnar l'acqua di aria fissa".

qui ne l'avoit finie dans ce tems là. J'ai profité donc de l'occasion d'une Caisse d'instrum^s. que Je viens d'envoyer à M^r. le Chev^r. Landriani à Milan, qui ne manquera pas de vous faire tenir la petite parcelle qui vous appartient. J'y ai mis dessùs votre adresse: & elle contient les articles suivans. 79.04.09

6 supports de metal pour soutenir les cylindres ordinaires sur la planchete de la Couvette, ou batte, oû l'on fait les experiences sur differentes especes d'air. 2do. & deux autres plus etroites pour les Cylindres minces. 3. huit ronds en mahogany pour l'effet que le Duc de Chaulnes decrit à la page 337 en parlant de la figure 11, dans le Vol. 3me. de la Traduction Francoise de Paris, de l'ouvrage du D^r. Priestley sur differentes especes d'air.[3] Ces ronds en bois avec la base couverte de drap, font le meme effet & sont du moins aussi commodes que ceux de tôle verni couvert de Cuir gras que le Duc y decrit. Vous trouverez dans la meme parcelle, deux de mes brochures, que je vous prie d'accepter, comm'un foible témoignage de mon egard & estime pour votre merit: et le prospectus du cours complet de Physique experimentale, que Mr. Atwood, Membre de la Soc^e. R^le. publiera en 4 Vol. in 4°. Ce sera l'ouvrage le plus complet, & le plus instructif dans ce genre. l'auteur est un excellent Mathematicien, & possede un talent extraordinaire pour rendre sensibles en practique experimentale les phenomenes les plus abstraits. Il se trouve actuellem^t. à Londres (car sa Residence est dans le College de la Trinité de Cambridge, oû il donne des cours de physique experimentale avec le plus grand aplause) & j'amenai hier matin chez Lui mes deux Respectables amis le D^r. Priestley et le D^r. Price avec M^r. Dentan[4] pour voir sa methode de demontrer mechaniquement la plupart des theoremes sur la vitesse, force, aceleration, & uniformité du mouvement rectiligne: & tous trois furent on ne peut plus enchantés. La simplicité de la machine qu'il inventa pour ces objets, et la justesse des experiences, sont etonantes. C'est une colomne verticale d'environ 7 ou 7½ pieds de hauteur, avec deux poids bien equilibrés, oû le frotement est presque aneanti, & la resistance d'inertie mise en compte. il y a une pendule à seconds: & l'on y voit par exemple que les graves prendent à raison de 16 pieds pour descendre dans une seconde, par un mouvement aceleré: qu'au bout de ce tems ils parcourent des espaces egaux avec le mouvement uniforme: que toute la dispute sur les forces vives, & les forces mortes, ne dependoit que

3 Joseph Priestley, *Expériences et observations sur différentes espèces d'air* (3 vols, Paris, 1777).

4 Probably Pierre Gédéon Dentan or Dentand (1746-1780); see J.H. Van Swinden to Magellan, 7 November 1777, note 2.

79.04.09 d'y faire entrer, ou non, la quantité du tems requis &^{a}. &^{a}. Mais ce M^{r}. publie actuellem^{t}. un Traité sur le mouvement rectiligne, & rotatoire, oû l'on voira cette matiere developée dans le plus grand jour possible, ensemble avec la description de son instrument nouveau.[5] J'ai crû vous faire plaisir de mettre votre nom dans la Liste des souscripteurs de son grand cours de Physique ci dessùs: mais celà ne vous impose point aucune obligation de le prendre, car l'auteur ne demande point d'argent, mais uniquement les noms de ceux qui voudront l'avoir, pour juger du nombre qu'il faudra imprimer d'exemplaires.[6]

Enfin vous trouverez dans la même caisse les trois listes des articles que Je vous ai envoyes d'ici, avec leurs prix respectifs vis à vis de chaqu'article; & signées par les Artistes, & commissaire que J'employai, avec leurs quitances. la liste (a) monte à £15.18^{s}.2^{d}. La liste (c) monte à £17.7^{s}.6^{d}: & la liste (b) monte à £2.3^{s}.2^{d}.[7] La somme totale en est £35.8^{s}.10^{d}: en y rabatant les livres 32.7^{s}.2^{d}. que J'ai reçû en octobre dernier de M^{r}. Songa, vous restez à devoir £3.1^{s}.8^{d}. pour les quelles je passai au meme M^{r}. Songa avant hier une autre petite lettre sur vous, dont il me paia la meme valeur: & par consequent tout notre Compte est arrangé & satisfait.

La machine pour les cristaux de M^{r}. Achard, que Je fis faire ensemble avec la votre, pour mon usage manquoit trois petits trous dans le bord exterieur du grand couvercle de metal: apparem^{t}. la votre en manquoit aussi. faitez les y faire avec un foret, pour passer la ficelle qui l'attache aux 3 anneaux de la petite table. J'ai deja achevé le procedé de cette experience sans reussir à la formation des cristaux. M^{r}. le Prince Gallitzin a manqué aussi avec sa machine. Il a ecrit à M^{r}. Achard, qui lui repondit disant que le succés depend essentiellem^{t}. de ce que le fond de terre cuite, oû l'on fait la filtration, doit etre fort serré en sorte qu'il n'y passe l'eau que très lentement. Je crois que chaque goute doit pendre une demi-heure à filtrer. Il a envoyé sa propre machine à M^{r}. le dit Prince Gallitzin. Celui ci va reprendre le meme procedé de nouveau; et J'espere qu'il reussira. Tachez donc d'emploier un fond tel dans votre machine, pour bien reussir.

---

5 Atwood's treatise was in fact not published for several years; George Atwood, *A Treatise on the Rectilinear Motion and Rotation of Bodies; with a Description of Original Experiments relative to the Subject* (Cambridge, 1784).

6 Atwood's proposed four-volume work seems never to have been published, only a single-volume abstract, *An Analysis of a Course of Lectures on the Principles of Natural Philosophy, read in the University of Cambridge* (London, 1784).

7 Lists not found.

Le volume que le D^r. Priestley imprime actuelem^t. de ses experiences Philosophiques sortira dans peu de tems.[8] Je suis faché qu'il n'etoit pas encore fini d'etre imprimé, pour vous l'envoyer dans cette parcelle. 79.04.09

Vale & ama
tuum
Magellan[9]

A Monsieur
Monsr. Alexandre Volta,
Professeur de Philosophie dans
L'Université de Pavie
par Milan
Italie

8 Joseph Priestley, *Experiments and Observations relating to Various Branches of Natural Philosophy* ..., vol. 1 (London, 1779).

9 Farewell & love | your | Magellan.

To Anne-Robert-Jacques Turgot 79.04.13

Biblioteca do Museu Nacional de Arte Antiga, Lisbon
Fundo Francisco Sousa Viterbo, Cx nº 1

Londres 13 Avril –79

Monsieur

Je viens de Recevoir, par La voye de M^r. Bostock, l'honneur de votre Lettre du 19 Mars,[1] qui m'a convaincû de mon tort dans celle de 23 Fevrier.[2] Il est bien malhereux pour Le genre humain, que les effets produits par les objets externes soient moiens[3] proportionés à leurs causes, qu'à la disposition respective des organes qui les reçoivent: & que nos jugemens soient si fortement dependens de nos Sensations, faute de cette Lumiere intuitive, dont il ne nous est point permis de jouir, dans l'etat actuel de la vie, pour discerner mieux entre l'apparent & le vrai. Je vous prie donc instament, Monsieur, de vouloir bien condamner à l'oubli, tout le contenû

1 Letter not found.

2 See Magellan to Turgot, 23 February 1779.

3 moins?

79.04.13 dans ma dite Lettre: & d'etre très persuadé que Je ne connois d'autre ambition plus flateuse, que celle de me rendre serviable, à tous ceux qui, comme vous, ont du vrai merite, & qui m'honnorent de leur estime & de leur bienveillance; pourvû qu'on me permette de reclamer en homme livre, Ce bien, que je ne dois à Personne; c'est-à-dire, la honnette satisfaction d'avoir agi avec honneur & sans reproche.

Mon ami Le D^{r}. Le Begue du Presle doit m'envoyer dans peu de tems, une caisse avec differens articles de Paris, par la voye de Flandres: ainsi Je vous prie de lui faire remettre les Thermometres, dont vous n'avez pas besoin: car j'avois deja arrangé cet affaire avec M^{r}. Nairne, bien avant Le mois de Fevrier. J'ai presque toujours des Comptes ouvertes avec Lui, & meme un peu considerables: à peine y a-t-il une semaine que Je lui paiai un bill d'environt Cent livres Esterl.: & avant la fin de ce mois j'aurai un autre pareil; car il a presque fini, pour mon Compte, un excellent Quadrant Astronomique d'environ 70 guinées, outre d'autres articles assez Couteux, qu'il me fournit. Ainsi on conçoit bien qu'il ne devroit me refuser de reprendre les thermometres en question, pourvû qu'ils n'aient pas soufert, ceux-ci etant des articles qu'il debite en sa boutique.

Dans la premiere occasion Je vous envoyerai les canes à trous capillaires de verre, que vous Souhaitez & qui sont dejà ordonnées. Pour ce qui regarde la verge etalonnée, je souhaite de savoir si vous vous-contentés d'avoir seulement quelque nombre de pouces anglois, puis[4] dans une Ligne sur la planche de metal, mais tel qui puisse etre divisé par la bissection, comme par exemple 16. 64. 128 &^{a}. Car si vous avez à Paris quelqu'artiste assez exact pour Cette operation, alors cet etalon ne coutera que peu d'argent: au lieu qu'un bon etalon divisé en pouces, dixiemes, & jusqu'à 500mes de pouce, ne coute moins de 5 à 6 guineés, & meme plus s'il excede la longueur de 3 pieds & demi.

Le D^{r}. Inghenhouz m'a chargé de vous presenter les Remercimens pour la brochure & information que vous avez eu La bonté de lui donner, sur l'affaire de l'inoculation de la Duchesse de Boufleurs.[5]

La Construction de votre Respirateur est fort ingenieuse: mais les vües de l'inventeur M^{r}. Mudge etoient de pouvoir repandre la vapeur de l'eau chaude, qui s'eleve au dedans du vaisseau, entre les draps du malade. Je crois bien, que celle qui sorte de la bouche est suffissante: & qu'il ne faudroit plus qu'ouvrir les levres à chaque expiration: Ce qui epargneroit la depense de la boule de liege, qui en effet est difficile à faire: & je ne connois qu'un seul artiste qui les fait ici avec perfection.

4 pris?
5 Boufflers. See Magellan to Turgot, 5 January 1779.

L'ouvrage de Mylord Mahon sur l'electricité est deja dans la presse:[6] & le nouveau vol. du Dr. Priestley sur des experiences de Physique paroitra dans la quinzaine à ce que Je crois.[7] il y a d'excellents articles nouveaux dans Les deux ouvrages & je ne manquerai pas de vous les envoyer au plutôt. Mr. Atwood publiera aussi en peu de semaines son ouvrage sur le mouvement rectiligne & rotatoire, oû il y a des experiences très curieuses & savantes:[8] car il est au meme tems un très bon mathematicien theorique: si vous voulez, je vous l'envoyerai avec les deux autres. J'ai l'honneur d'etre avec respect & vrai attachemt. 79.04.13

Monsieur

Votre t. h. & obt. servr.

Magellan

a Monsieur
Monsr. Turgot ancien Controlr. genal.
Rue de l'Université, Fauxbourg St. Germain
à Paris

---

6 Charles Stanhope, Lord Mahon, *Principles of Electricity* (London, 1779). See also Magellan to Turgot, 17 July 1778.

7 Joseph Priestley, *Experiments and Observations relating to Various Branches of Natural Philosophy...*, vol. 1 (London, 1779).

8 George Atwood's *Treatise on the Rectilinear Motion and Rotation of Bodies* was not published until 1784.

To Ribeiro Sanches[1] 79.04.13a

Österreichische Nationalbibliothek, Vienna
A. N. Ribeiro Sanches Korrespondenz, codex 12714, ff. 243-244v

Londres 13 Abril – 79

Meo Amo. e Sr. e mto. do Coração

Vejo pela de V.M. de 18 do passado, q V.M. não recebeo hũa parcella de livros onde hião os 2 ou 3, de q lhe falei, sobre os banhos de agoa fria, pois

---

1 Annotated "Recue le 18 Avril" [Received 18 April] and "Ra. a 26 Abril" [Answered 26 April].

79.04.13a com estes hia o primº. tomo de M^{r}. Lucas, de cuja obra V.M. agora recebe os 3 ultimos volumes. neste primº. tomo he q elle trata dos banhos de agoa fria. eu lá escrevi à M^{r}. le Begue; e elle pela Reposta q me deo ultimam^{te}., disse q havia de procurar os tais Livros na mão do Abb^{e}. Rochon &ª. onde eu os mandei, ou na de M^{r}. Turgot, q talvez recebeo a dª. caixinha &ª. he necessº. ter paciencia em semilhantes piquenas remessas: pois as difficuld^{es}. de fazelas em direitura em tempo de guerra não são faceis de se vencer sem pachorra.

Aqui dentro achará V.M. hũa oitava do mercurio Calcinado, ou precipitado per se. eu o comprei ja levigado pª. evitar algũa imposição do seo boticario. agora não tem mais do q passalo bem na pedra com alguma conserva, ou lambedor pª. lhe dar a consistencia de massa, e formar as pirolas. este precipitado per se, antes de ser levigado, he de hũa cor tão vermelha, como o mesmo vermilhão, e se V.M. quizer eu lhe mandarei outra oitava antes de ser levigado. eu compreio na melhor botica de Londres, onde não se pode recear a minima vilhacada ou impostura, q he o Apothicaries Hall. Como alli ha contin[ua]m^{te}. fornalhas com fogo de areia, q vão dia & noite, por isso tem nellas v[a]zos espalmados com mercurio pª. o calcinar nesta forma. disserão-me q pª. obter hũ arratel deste precipitado per se, he necessº. nove mezes de fogo continuo; tempo em q hũ bello rapaz, ou rapariga, po[de] sahir do ventre de Sua Mãi, pª. a seo tempo necessitar tão bem deste mercurio calcinado. N.B. o preço de cada onça aqui, hé de 16 shillings, q^{do}. certam^{te}. em Paris não carregão menos de 24 ou 30 Livras de França. Na carta seg^{te}. mandarei a V.M. a ou[tra] oitava; pois qdº. hoje passei pela Apothecaries' Hall, não tinha na aljibeira a carta de V.M.; e agora vejo q queria ter duas o[i]tavas, em lugar de hũa só.

Hontem jantei com Sua Excª. o S^{r}. Pinto de Balsemão, q sempre gosta de noticias de V.M. e rio m^{to}. com a histª. da moça q antes queria ter as garr[...]cas[2] dentro, do q o vê-las fóra, q^{do}. o seo mangalház lh[as] mostrava. V.M. tem m^{ta}. rezão no q diz do Arlequim de Voltaire: pelo q toca a Rousseau, elle tinha a infelicid^{e}. de ter sempre o coração em carne viva; pois tudo sentia com irritação doloroza, como q^{do}. temos hũa ferida antes de encarnar péle, q o minimo toque nos poem em convulsão dorîda. Coitado do nosso amigo le Begue! diz-me na sua Carta q tem estado mui doente, e q hia pª. o Campo à restabelecer-se nestas festas da Pascoa. He sojeito digno de m^{ta}. estimação, e de toda a felicid^{e}. q aqui se pode gozar neste formigueiro de macácos humanos ...!

Sequeira passa bem, e tem ocupação no seo emprego da Faculd^{e}. &.ª.

---

2 Partly illegible.

a semana passada tivemos aqui hũ cazo raro nos annais de Cupido. hũ moço de 25 annos, q se vio frustado na propozição de receber por Espoza a Maitresse do Secret^{o}. d'Estado (q ao q me dizem não era bella, e tinha já tido varios filhos [do] nobre amante com q^{m}. vivia) vendo-a passar no Coche p^{a}. a Comedia, tomou duas pistolas na aljibeira, com intenção de se matar a si mesmo aos seos pés q^{do}. ella sahisse do Theatro. Porem vendo q vinha accompanhada por hũ militar, mui galante & moço, descarregou a prim^{a}. pistola sobre a dulcinea, q cahio logo por terra e espirou immediatam^{te}.; porem descarregando a 2da. pistola ao seo proprio ouvido, a bala não entrou no casco, e elle não pode acabar de quebrar a cabeça com a culatra antes q o segurassem, & levassem p^{a}. a cadeia: elle será enforcado na semana q vem. porem ella foi até [su]a morte a mais feliz, como disse o outro general, ouvindo a do grande Turenne &^{a}. 79.04.13a

ainda aqui está o S^{r}. D. Augusto de Souza Holstein, filho do Infeliz Calhariz, & sobrinho do Bailio q V.M. ahi conheceo. elle vai Enviado p^{a}. a Haya: & o q está na Haya vai p^{a}. Petersburgo. O S^{r}. D. Augusto he sujeito de m^{to}. bom genio, proposito, & agrados. Tãobem com elle veio hũ fidalgo Jozé de Mello Brayner, Portuguez mui instruido, e Capitão de Mar & guerra. elle foi o q tomou aquella Náo Espagnola senão me engano, q^{do}. lá nos tomárão a Ilha de Sta. Catherina no Brazil &^{a}.

Saud^{es}. a M^{a}. Joanna & Familia Adoptiva. diga-me como vai lá petite Marie com a sua gravura: & q^{tos}. suit-hearts lhe andão já pelo rasto atraz della?

vale & iterum vale.

a Monsieur
Monsr. Sanchez, Doct^{r}. en Medecine
Cimetiere St Jean Rue de la Verrerie
à Paris

London 13 April –79

My Dear Friend

I see by your letter of the 18th of last month[3] that you didn't receive a parcel of books with the 2 or 3 that I mentioned, about the cold water baths,[4] since with these was the first volume of M^{r}. Lucas, of whose work you have now received the 3 last volumes.[5] It is in this first volume that he treats the cold water baths. I wrote

3 Letter not found.

4 See Magellan to Sanches, 9 October 1778.

5 Magellan appears to have first sent Sanches a one-volume work by Charles Lucas, possibly his *The Theory and Uses of Baths* (Dublin, 1772), and then, in a later package,

79.04.13a to Mr. le Begue;[6] and by the answer he gave me lately,[7] he said he would look for the books in the hands of Abbé Rochon[8] &c. to whom I sent them, or in Mr. Turgot's, who perhaps received the said box &c. It's necessary to be patient with such little parcels because the difficulties of sending them successfully in a period of war are not easy to overcome without phlegm.

Inside this you will find one drachm of the calcined mercury, or precipitate per se. I bought it already levigated[9] in order to avoid some imposition of your apothecary. Now you just have to mix it on the stone with some liquid or syrup to give it the consistency of paste, and form the pills. This precipitate per se before being levigated shows a colour as red as cinnabar, and if you wish I will send you another drachm not levigated. I bought it in the best apothecary's shop in London, where one need not fear the least villainy or fraud, namely the Apothecaries' Hall.[10] While there are in that place continuously operating sand furnaces[11] with a fire that goes day & night, there are also, in these, flat vases in this shape[12] containing mercury in order to calcine it. I was told that to obtain one pound of this precipitate per se, nine months of continuous fire is necessary, the time it takes for a nice boy or girl to leave the mother's womb, in order to be also in need, given time, of this calcined mercury. N.B. the price of each ounce here is 16 shillings, whereas certainly in Paris they won't charge less than 24 or 30 French livres. I will send the other ounce in the next letter, since when I went today to the Apothecaries Hall, I didn't have your letter in my pocket and now I see that you wanted 2 drachms rather than one.

Yesterday I had dinner with His Excellency Mr. Pinto de Balsemão,[13] who always likes to hear news of you and laughed very much at the story of the girl who preferred to have the [...] inside to seeing it outside, when her boyfriend showed it to her.

You are absolutely right in what you say of Voltaire's Arlequin; in what concerns Rousseau, he was unfortunate to have his heart always sore because he felt everything with painful irritation, just as, when we are hurt, before the skin

---

Lucas's *An Essay on Waters* (3 vols, London, 1756). See Magellan to Sanches, 9 October 1778 and 5 January 1779.

6 Letter not found.

7 Letter not found.

8 The astronomer Alexis-Marie Rochon (1741-1817).

9 i.e. reduced to powder.

10 A large and handsome building in London in which medicines of all kinds were prepared and sold at a cheap rate. The building was finished in 1670.

11 An arrangement in which a shallow metal vessel containing sand is placed on top of a furnace and phials, retorts, etc. are placed on this to provide even, long-term heating. A pencil drawing of the arrangement used ca. 1810 at the Apothecaries' Hall is reproduced in T.D. Whittet, *Clerks, Bedels and Chemical Operators of the Society of Apothecaries* (London, ca.1980), p. 27.

12 See Magellan to Sanches, 5 January 1779.

13 The Portuguese Ambassador in London.

grows the least contact puts us into a painful convulsion. Poor friend le Begue! He tells me in his letter that he has been very ill, and that he was going to the country during the Easter holidays to recover. He is a fellow worthy of great esteem, and of all the happiness that can be enjoyed here in this crowd of human monkeys...! 79.04.13a

Sequeira is well and finds occupation in his job at the faculty &c.

Last week we had here a case rare in Cupid's annals. A young man 25 years old, who was frustrated by the proposal to take as his wife the Secretary of State's Maitresse (who as I am told was not pretty and already had several children from the noble lover with whom she lived) seeing her passing by in a coach to the comedy, took two pistols from his pocket with the intention of killing himself at her feet when she came out of the theatre. However, noticing that she was accompanied by a very handsome & young army officer, he fired the first pistol at the Dulcinea, who fell at once to the ground and expired immediately; but on firing the second pistol at his own ear, the bullet didn't enter the skull and he could not finish breaking his head with the butt before being arrested & taken to jail. He will be hanged next week. But she was even in death most happy, as the other general said on hearing of that of the great Turenne[14] &c.

Here still is D. Augusto de Souza Holstein, son of the unhappy Calhariz[15] & nephew of the bailiff you met there. He is going as envoy to The Hague, & the one who is at The Hague[16] is going to Petersburg. D. Augusto is a very good natured, well intentioned & pleasant fellow. With him there also came a Portuguese gentleman, Jozé de Mello Brayner, very knowledgeable, a captain in the Navy. He was the one who took that Spanish ship, if I'm not mistaken, when they took from us the island of S^t^. Catherine in Brazil &c.[17]

Regards to M^a^. Joanna & the adoptive family. Tell me how does la petite Marie go with her engraving: & how many suit-hearts[18] are after her?

vale & iterum vale[19]

---

14 Henri, Count of Turenne (1611-1675), French general and hero of the Thirty Years' War, at whose death the Imperial general Montecuccoli, who had opposed Turenne in several famous battles, is said to have written: "Il est mort aujourd'hui un homme qui faisoit honneur à l'homme".

15 In the *Gazeta de Lisboa* of 9 October 1778, mention is made of the nomination of Augusto de Sousa as Envoy to The Hague. His father, Manuel de Sousa, Earl of Calhariz (1703-1759), died in the dungeon of the Tower of Bugio, in the estuary of the Tagus River in Lisbon, after being incarcerated by the Marquis de Pombal.

16 Francisco José Horta Machado (1746-1817).

17 In 1776, the Spanish Navy attacked the Portuguese colony of Sacramento and the island of Santa Catarina in southern Brazil. José de Mello Breyner (1745-1791) in command of the battleship *Prazeres* played a decisive role when, together with Arthur Phillip (1738-1814) commanding the battleship *Pilar*, he attacked a much stronger Spanish battleship, the *San Agustin*, leading her to surrender. Phillip was later the founding Governor of the new British colony in New South Wales.

18 sweethearts?

19 Farewell and again farewell.

79.04.29 *To James Watt*[1]

Birmingham Central Library, Boulton and Watt collection, Box 22/B3/13

London 29 April –79

Dear Sir

I take the Liberty to Recommend to your favour and acquaintance the two Noblemen and the young Gentleman, bearers of this Letter, who are my country-men,[2] and go to see the curious manufactures of Birmingham, chiefly those of Mr. Boulton; I know you'll exert your kind disposition to oblidge me, in affording the information and assistance, they may be in want of towards accomplishing their object; without forgetting the excellent performance of your fire-engine, and particularly that built for furnishing the Birmingham Chanel with water in dry seasons.[3]

I'll only add that you certainly shall not have now any dislike, as in the Last instance, with the noble behaviour and distinguished caracters of these foreigners.

I heartly am

Dear Sir
Your most obedt. obliged servt
J. H. de Magellan

P.S. I was told yesterday that Mr. Boulton is in London; but I have not yet seen him.

1 Letter not found; MS is a clerk's transcription. Annotated "Mr Magellan april 29th 1779 recommending 3 Portugueze Noblemen".

2 Not identified.

3 The Smethwick engine was used to pump water for the Birmingham Canal Navigation Company's canal at Smethwick, near the Soho Foundry. This steam engine was brought into service in May 1779, so it was ready to be shown at the time of the letter.

From Jan Henrik van Swinden[1]

79.05.02

Leiden University Library, MS B.P.L. 755, no. 359

2 Mai 1779

J'aurois repondu plus tot Mon cher Monsieur à votre lettre du 23 Mars,[2] si je n'avois attendu que je pusse vous donner des nouvelles de l'arrivée de la caisse que vous avez eu la bonté de m'envoyer. J'ai receu hier des lettres de mon frere qui m'a marqué qu'elle venoit d'arriver à bon port. Je vous suis obligé Monsieur des peines que vous vous etes données pour cet object. J'aurai soin de vous rembourser les frais de £2.17.6: pour cet effet j'ai prié mon frere, de demander à M. Dentan, qui est actuellement à la Haye, si avant son depart, il vous avoit paié pour moi, auquel cas mon frere lui remboursera des debours: si non je chercherai quelqu'autre voye, ou me servirai de celle qu'il vous plaira de m'indiquer.

J'attends avec beaucoup d'impatience le nouveau volume du Dr Priestley,[3] que je serai bien curieux de voir. Le Pr. Gallitzin doit avoir envoyè il y a plus d'un mois, à M. Dentan la relation des Experiences que mon digne ami et Collegue M. Coopmans[4] a faites ici avec l'Air fixe: j'en avois ecrit aussi au meme Dentan, mais ma lettre s'est trouvée dans la Male du Capt Stone qui a été pris par les francois: le mal par rapport à cette lettre n'est pas bien grand, d'autant plus qu'elle n'auroit pas trouvé M. Dentan à Londres.

Ce que vous me marquez des talens de M. Atwood me feroit bien desirer de voir ses ouvrages. Je vous suis obligé de ce que vous voulez bien me mettre au nombre des souscripteurs pour son Cours de Physique: je serois bien charmé d'en etre: mais l'extreme cherté du Livre, qui nous reviendra bien au dela de la[5] de 40 florins de Hollande me force à y renoncer, quoiqu'à regret: ainsi Monsieur, si vous n'avez pas deja fait usage de mon nom, vous

---

1 Original not found; MS is a file copy retained by van Swinden and annotated "A M. Magellan. Londres". Published in David Willemse, "Suite d'un voyage aux Pays-Bas: João Jacinto de Magalhães (1722-1790) et ses rapports avec Jean Henri van Swinden (1746-1823)", *Arquivos do Centro Cultural Português*, 7 (1974), 225-278.

2 Magellan to Van Swinden, 23 March 1779.

3 Joseph Priestley, *Experiments and Observations relating to Various Branches of Natural Philosophy*, vol. 1 (London, 1779).

4 Gadso Coopmans (1746-1810) was at this period professor of medicine at the University of Franeker. He reported some remarkable cures effected by the application of fixed air, to which Priestley referred in his book: see Van Swinden to Magellan, 26 November 1779, note 5.

5 *sic.*

79.05.02 m'obligerez de n'en point faire: j'en suis bien faché, car d'apres ce que vous me dites de l'Auteur, je ne doute pas que ce Cours ne soit excellent.

L'ouvrage de M. de Wal sur la dilatation des metaux est très à votre service:[6] j'en demanderai un exemplaire à l'auteur, c'est un en 4° de 70 pp: mais comment vous l'envoyer? ayez la bonté de m'indiquer une addresse: vous y verrez combien les Exper. faites avec les Pyrometres ordinaires, où la verge qu'on eprouve repose par un bout contre un point fixe, et meut par l'autre l'indice du Pyrometre sont peu propres à faire connoitre la dilatation. Ces Experiences m'ont fait conjecturer la cause de quelques irregularités dans les Thermometres: et, elles me donnent la clef des enormes différences qui se trouvent entre celles du Chev Shuckburgh et du Col. Roy[7] sur les dilations qu'éprouve la Colonne du Barometre par une chaleur de 32 – 212 degrés.[8] Je me suis proposé depuis bien des années un plan d'Exper^{s}. sur la dilatation des fluides mais je ne sais quand j'aurai le tems de l'executer: cependant il y a encore bien à faire sur ce sujet. Peut etre quelqu'autre l'entreprendra-t-il avant moi.

Vous me demandez Monsieur quelles sont mes decouvertes sur le Magnetisme: il me sera difficile de repondre sur ce sujet en detail: je vais, cependant tacher d'y satisfaire plus ou moins, en vous indiquant les ouvrages où elles se trouvent, et leur contenu.

1°. Un Essai de Theorie mathem. sur l'aimant: 4° 1772, en latin,[9] il a été présenté à la Societé Royale en 1773 et doit se trouver dans sa bibliotheque: Les expériences roulent sur les loix de la communication; sur quelques nouveaux phenomènes de ce que j'ai nommé points d'indifférence et points culminant: et une formule mathematique pour calculer la position du centre magnetique, et des explications mathematiques du tout, qui je crois est neuf.

---

6 Henricus de Wal, *Specimen physicum inaug., sistens nova quaedam experimenta de expansione corporum solidorum ipsi flammae actioni exhibitorum* (Franeker, 1775).

7 George Shuckburgh, "Observations made in Savoy to ascertain the Height of Mountains by Means of the Barometer", *Philosophical Transactions*, 67 (1777), 513-597; William Roy, "Experiments and Observations made in Britian, in order to obtain a Rule for measuring Heights with the Barometer", *ibid.*, pp. 653-770.

8 Marginal note: "Car selon le 1er elle est de 10/637 ou 157/10,000 et selon l'autre de 5117/10000 de pouce sur une colonne de 30 pou, ou de [171?]/10,000: différence très considerable, et uniquement due à la différence des methodes comme je pourrois le prouver en détail."

9 J.H. Van Swinden, *Tentamina theoriae mathematicae de phaenomenis magneticis* (Leiden, 1772).

2°. Mon grand Memoire, couronné à Paris, sur les aiguilles aimantées.[10] 79.05.02
L'impression de ce Mem. est entierement achevée: il contient 600 pp. in 4°: j'en ai actuellement un exemplaire sous les yeux, mais il n'est pas encore publié, je ne sais d'ou vient. ce memoire contient 2 parties: la 1re roule sur la fabrique des aiguilles, l'autre sur leurs variations.

Dans la 1re, en partant de ce principe qu'une ligne droite qui auroit ses deux extrémités aimantées s'arreteroit dans la direction du meridien magnetique, j'ai etabli une Theorie mathématique entierement nouvelle sur les directions des aiguilles: j'ai fait voir quelles sont celles quelles acquierent selon qu'elles ont deux ou plusieurs poles, selon que ces poles sont situés, selon la position du centre de mouvement, la figure des aiguilles, &c: croiriez vous qu'une aiguille qui n'a que deux poles, peut etre placée de façon a indiquer le sud? Rien cependant de plus vrai, je l'ai prouvé par Theorie et par Expériences. Je suis parvenu a une formule infiniment simple pour la situation du centre magnetique: j'ai fait voir par 85 Exp. et meme d'avantage que les forces des aiguilles sont dans un changement continuel: et j'ai déduit du tout, de quelle façon il falloit construire et suspendre des aiguilles pour en avoir de bonnes. Mrs Dentan et de Luc[11] ont vu mes appareils. Ma pièce contient 3 cent experiences, toutes nouvelles, et tout y est calculé. Voila bien en gros ce que j'ai fait: je ne saurois dessiner de plus long d'extrait sans figures, et sans entrer dans quelques détails. bientot j'espere que l'ouvrage sera publié.

Dans la 2e partie je traite des variations diurnes de l'aiguille. Vous savez que l'aiguille se meut le matin vers l'O, le soir vers l'E. Le journal de M. Canton, imprimé dans le vol LI des transactions[12] est ce qu'on avoit de plus complet. J'ai recueilli tout ce qu'on avoit sur ce sujet, et je l'ai discuté: d'après plus de 40000 Observations faites par moi meme j'ai determiné les Loix de ces variations et les différentes sortes de mouvemens de l'aiguille: j'ai fait faire des Observations ailleurs, comme à la Haye par mon frere, et j'ai trouvé qu'il y a des endroits où cette variation n'a pas lieu: en un mot

10 J.H. Van Swinden, "Recherches sur les aiguilles aimantées, et sur leurs variations régulières", *Mémoires de mathématique et de physique, présentées à l'Académie Royale des Sciences [de Paris] par divers savans...*, 8 (1780), 1-576. Van Swinden's paper shared the prize awarded by the Académie in 1777.

11 Jean André de Luc (1727-1817) and Pierre Gédéon Dentan (1746-1780), who made measurements together while climbing in the Alps. De Luc subsequently settled in England, and Dentan in the Netherlands.

12 J. Canton, "An Attempt to Account for the Regular Diurnal Variation of the Horizontal Magnetic Needle, and also for Its Irregular Variation at the Time of an Aurora Borealis", *Philosophical Transactions*, 51 (1759-60), 398-445.

79.05.02 le travail que j'ai fait est immense. Je continue toutes mes observations et mes recherches: j'ai quelque fois pensé à les présenter en abregé à la Societé Royale: mais je ne sais si cela seroit gouté.

3°. Dissertation sur l'Analogie de l'Electricité et du Magnetisme: couronnée à Munich en 1778. J'ai considéré cette analogie sous deux points de vue: 1° par rapport aux Loix que les deux sortes de forces suivent; et j'ai fait voir par un grand nombre d'expériences, que ces Loix sont très différentes: 2° par rapport à l'influence que l'electricité peut exercer sur l'attraction, la direction, l'inclinaison, et les forces de l'aimant, et j'ai trouvé qu'elle est ou nulle, ou très petite, ou irréguliere, excepté peut etre sur le dernier article. On imprime cette dissertation, qui aura 200 pp. in 4°: elle paroitra incessamment dans la 2e partie du 1er vol des Nouveaux Mem. de L'Académie de Munich.[13]

4° Dissertation sur un phenomene magnetique paradoxe, imprimée dans la 1ere partie du 1er Volume des Nouveaux Mem. de Munich, qui a été publié au mois de novembre dernier.[14] Le Phenomene est que tous les ecrivains disent que l'aimant attire plus fortement le fer, qu'un autre aimant. J'ai fait voir par Exp. 1° que cela n'est pas toujours vrai: 2° quels sont les cas ou cela a lieu: 3° par quelle raison, et comment cela peut etre: ce qui m'a mené a quelques nouvelles idées sur les Phenom^{s}. de l'attraction magnétique.

Voila M. une legere indication de ce que j'ai fait. j'avois envoye en 1772 à M. Maty une dissertation sur quelques derangemens de l'aiguille par l'AB,[15] avec priere qu'il la présentat à la Societé Royale:[16] je ne sais s'il l'a fait, ou si ce Memoire est resté parmi ses papiers. J'ai encore un très grand nombre de choses dans mes portefeuilles, et je pourrois donner bien des dissertations entierement neuves sur l'aimant, si j'en avois le tems.

Je suis, &c.

Quand vous m'ecrirez addressés s.v.p. vos lettres ainsi

à M. Van Swinden, Avocat, p^{r} remettre au Prof. Van Swinden à La Haye.

---

13 Van Swinden's prizewinning dissertation was in fact printed, in Latin, in the second volume of the Bavarian Academy's proceedings: "Dissertatio de analogia electricitatis et magnetismi", *Neue philosophische Abhandlungen der baierischen Akademie der Wissenschaften*, 2 (1780), 1-404. A French translation was published as the first volume of his *Recueil de mémoires sur l'analogie de l'électricité et du magnétisme* (La Haye, 1784).

14 J.H. Van Swinden, "De paradoxo phaenomeno magnetico, magnetem fortius ferrum purum, quam alium magnetem attrahere", *Neue philosophische Abhandlungen der baierischen Akademie der Wissenschaften*, 1 (1778), 353-388.

15 Aurore Boréale.

16 Van Swinden's memoir was not published in the *Philosophical Transactions*, and is not in the Royal Society's archives.